Machining and CNC Technology

2024 Release

Michael Fitzpatrick
Everett Community College Community and Corporate Education

Keith Smith
Shoreline Community College

Mc Graw Hill

MACHINING AND CNC TECHNOLOGY, 2024 RELEASE

Published by McGraw Hill LLC, 1325 Avenue of the Americas, New York, NY 10019.

This book is printed on acid-free paper.

1 2 3 4 5 6 7 8 9 LKV 29 28 27 26 25 24

ISBN 978-1-265-47493-5 (bound)
MHID 1-265-47493-1 (bound)
ISBN 978-1-265-19223-5 (loose-leaf)
MHID 1-265-19223-5 (loose-leaf)

Executive Portfolio Manager: *Beth Bettcher*
Marketing Manager: *Lisa Granger*
Lead/Senior Content Project Managers: *Melissa M. Leick, Laura Payne*
Senior Manufacturing Project Manager: *Laura Fuller*
Senior Content Licensing Specialist: *Gina Oberbroeckling*
Cover Image: ©*Mike Fitzpatrick*
Compositor: *MPS Limited*

All credits appearing on page or at the end of the book are considered to be an extension of the copyright page.

Library of Congress Cataloging-in-Publication Data

Names: Fitzpatrick, Michael, 1945- author. | Smith, Keith (Machinist), author.
Title: Machining and CNC technology / Michael Fitzpatrick, Everett Community College, Community and Corporate Education, Keith Smith, Shoreline Community College.
Other titles: Machining and computer numerical control technology
Description: 2024 release. | New York, NY : McGraw Hill LLC, [2024] | Includes index.
Identifiers: LCCN 2023018009 (print) | LCCN 2023018010 (ebook) | ISBN 9781265474935 (bound edition ; acid-free paper) | ISBN 1265474931 (bound edition ; acid-free paper) | ISBN 9781265192235 (loose-leaf edition ; acid-free paper) | ISBN 1265192235 (loose-leaf edition ; acid-free paper) | ISBN 9781265194703 (ebook)
Subjects: LCSH: Machining–Handbooks, manuals, etc. | Machine-tools–Numerical control–Handbooks, manuals, etc.
Classification: LCC TJ1165 .F566 2024 (print) | LCC TJ1165 (ebook) | DDC 621.9/023–dc23/eng/20230506
LC record available at https://lccn.loc.gov/2023018009
LC ebook record available at https://lccn.loc.gov/2023018010

mheducation.com/highered

While there were countless others along the way, these people made all the difference in my career and life. Without them I question whether this book would have been.

To Linda, my wife,
for never complaining about the time taken from us to do this, for believing, giving, and forgiving.

To Jan Carlson
for demonstrating with acts, what a caring professional should be, and especially for the encouraging space to grow.

Gene Haas
and the Gene Haas Foundation
for the total support of machinist education worldwide. You make me proud to be an American machinist. Thanks for the HTEC Network, Gene.

Contents

About the Authors

Courtesy Mike Fitzpatrick

Mike Fitzpatrick—Journey Machinist, Tool and Die Maker As if it was yesterday, I remember carrying my new toolbox down the aisle at Kenworth Trucks of Seattle. Scotty, the crusty drill press operator, stepped away from his machine and planted himself right in front of me. Without a welcome, he raised his bushy eyebrows, poked two fingers into my chest, and said, "You see all these men here?" He waited. At 18, I recall only nodding, unable to speak. He went on: "Each one of us will show you everything we know if you pay attention. We'll give you lifetimes of experience, but know this, lad, it comes with an obligation. Someday you'll pass it on."

Having served a machinist apprenticeship at Kenworth Trucks and a tool and die apprenticeship at Boeing, I have 14,000 hours of formal trade education in two very different industries and at the workbench of many craft-masters. I began teaching machining part-time, then moved to full-time many years ago. I can call myself a journeyman! And a veteran teacher. I am fulfilling Scotty's challenge–learn the trade well then at mastery pass it forward! I am so very honored to be your teacher in print alongside Keith Smith.

Courtesy Keith Smith

Keith Smith—Instructor of Manufacturing/Shoreline Community College The manufacture of machined parts is in a state of constant evolution. The challenge is to recognize changes as they happen and embrace them. Embracing change is also what I most appreciate about this book that Michael Fitzpatrick created. This book has been laid out to cover the basics that beginning machinists need to know while at the same time looking forward to the advancements taking place in manufacturing. I feel blessed that I've had the opportunity to work with Mike and be a part of this book. I am also grateful that Mike had enough confidence to allow me to take the lead in creating this new revision of *Machining and CNC Technology*. Our sincere hope is that you find this book helpful to your journey in this challenging and interesting career.

Preface

Programmed machine tools now represent most machine operation and, of greater impact to you, of new jobs. *Entry-level people usually start in the shop as CNC operators.* Flexible and friendly, the machines and programming systems are so quick and easy to learn that they are now practical even for one-of-a-kind work such as mold making and die work, as well as production. *Schools integrate and teach CNC as an entry-level subject–starting from the first lesson on the first day.*

This book was specifically written to serve this type of modern student. To do so, subjects have been grouped into four large career partitions:

Part 1 Introduction to Manufacturing

Manufacturing is a world of its own. Chapters 1 through 7 are designed to open the door. They provide the background needed to fit into the shop, to understand the rules, to read and interpret the drawings, to be comfortable with extreme accuracy, and especially to be safe.

Part 2 Introduction to Machining

Chapters 8 through 15 teach how to cut metal the right way. These lessons assume that you will eventually perform them on CNC equipment but will probably practice first on manually operated machines because they are a simple, safe place to learn setups and operations.

Part 3 Introduction to Computer Numerical Control Machining

Now we get to the text core: how to apply Parts 1 and 2 to setting up, programming, and running CNC machine tools. In Chapters 16 through 23, we will learn how to professionally manage a CNC world. Because they move at lightning speed with lots of power behind them, safety must be integrated into everything we study.

Part 4 Advanced and Advancing Technology

Chapters 24 through 28 set the tone for your career after graduation. The best is yet to come, so let's get started!

So, many thanks to those who are using our book to start your manufacturing careers. It's an honor to be your instructors. Here's what we can pass on about our trade.

The 2024 Release has been revised to include a number of new features:

- **New chapters on Mastercam and SolidWorks with student downloads of test software**
- **Multiaxis setups and operations**
- **An index to help you map this book to any national machining standard** Recognizing the growing trend toward standardized instruction, this textbook is now skill mapped and indexed to help fit it to any national machining standard. A **glossary of terms** can also be found in the Online Learning Center.
- **More on 3-D metal printing** Still an emerging technology, will it fall within the machinist job title? Where will it be in 10 years?
- **Interactive chapter feature** There have been many advances in mobile technology since the last edition, and this new textbook incorporates many of them. Students can now use their mobile devices to view interesting websites, videos, and articles when they scan the special codes throughout the book. Think of these **Xcursions** as a virtual field trip of material that will enhance student understanding of the text.

Acknowledgments

Our deepest gratitude goes to these major contributors, without whom this book would not have been possible:

CNC Software, Inc. Mastercam *Mark Summers, President; Dan Newby, Training Director*

Thank you, Mark, for believing in education, and Dan, for your editing and guidance, and many thanks to your entire team for improving our trade and supporting education worldwide.

CGTech *(Irvine, California) Scott Michalek, CGTech Sales Engineer (recently retired); Bryon Jacobs, Marketing Manager*

Special thanks to Scott, Bryon, and everyone at CGTech who provided assistance putting together the chapter on program verification.

Renishaw, Inc. *Rafael Herrera, Marketing Specialist*

Your time, effort, and energy helped illustrate the impact that probe technology has on the machining industry.

3DX Industries, Inc. *(Ferndale, Washington) Roger Janssen President & CEO, CFO*

Thank you for your support and help with updating the additive manufacturing unit of this book.

Milwaukee Area Technical College *(Milwaukee, Wisconsin) Dale Howser and Patrick Yunke, Lead Instructors*

Offering a nationally recognized, 2-year tool-and-diemaking diploma. MATC graduates learn die and mold making and qualify for Wisconsin's apprenticeship certificate. Thus, they often serve full apprenticeships in the highly paid tooling area of manufacturing.

Dale Howser Sr.: Apart from 28 years of journeyman toolmaking experience, with 15 years of teaching these subjects, Dale holds degrees in tool and die making from Milwaukee Area Technical College and in Vocational Education from Stout University. He also develops and works on educational materials for the Precision Metalforming Association and Wisconsin's Apprenticeship programs.

Patrick Yunke: A graduate of Wisconsin's Madison Area Technical College die-making program and Stout University for Vocational Education, Patrick brings many years of experience in all aspects of precision die, metal, and plastic mold making to MATC, where he has taught for 15 years. He has also been a consultant to industry for manufacturing and custom educational programs.

Many thanks to you both for your expertise and for supplying great photos from your beautifully organized shop.

NTMA–National Tooling and Machining Association *Dick Walker, President*

Many thanks for being at the root of this new book in the beginning, for investing time and energy in it, and for the 45 drawings donated from your training materials.

NTMA Training Centers, California *Max Hughes, Dean of Instruction*

Thanks, Max, for the assistance with the CNC portion of this book.

Haas Automation, Inc. *(Oxnard, California) Scott Rathburn, Marketing Manager, Sr. Editor CNC Machining*

Thank you, Scott, for supporting education, educators, programs and especially aiding and encouraging students.

Boeing Commercial Airplane Co. *Apprentice Instructor*

Thank you for giving me the best education possible when I began my career, and to Tim for your ongoing support of quality apprenticeship, for help in planning and executing this book, and for being a lifelong friend.

Brown and Sharpe Corp. *(Rhode Island) Metrology Equipment*

For your commitment to education in metrology in technical schools and colleges.

Kennametal Inc. *Kennametal University, dedicated to finding better methods and to educating wherever machining is taught or applied.*

Thanks for the data, advanced tooling photos, text, and charts.

Iscar Metals *Bill Christensen, Advanced Tooling Photos and Text*

Advancing knowledge through research and education; thank you, Bill, for the HSM article.

Sandvik Coromant

Thank you for photos from *Modern Metal Cutting.*

SME–Society of Manufacturing Engineers
Westech Machine Tool and Productivity Exposition

In addition, I would like to thank the following instructors who reviewed this textbook and provided suggestions for improvement:

Bob Dixon *Walters State Community College*

Bret Holmes *Ogden Weber Applied Tech Center*

Mike Wyckoff *Grand Rapids Community College*

Mike Cannon *Pensacola State College*

Dean Duplessis *Northern Maine Community College*

Mark Dodge *Nashua Community College*

Arthur Martz Thomas Wertman *Kaplan Career Institute*

Jim Kitchen *Pensacola State College*

Samuel Obi *San Jose State University*

Andrew Zwanch *Johnson College*

Patrick Mason *Morehead State University*

Michael Mattson *Clackamas Community College*

Dexter Hulse *University of Cincinnati-Batavia*

Richard Buhnerkemper *Gateway Technical–Elkhorn*

Thanks to the following individuals for recommending companies for the "Made Right Here" opening segments in this book:

Gerald Berry *Wayne County Community College District*

Herbert Skinner *Southeast Technical Center*

Ismail Fidan *Tennessee Tech University*

Erica Matthew *Florida Career College*

Teje Sult *Jackson State University*

Machining and CNC Technology, 2024 Release, provides the most up-to-date approach to Machine Tool technology available, with totally integrated coverage of manual and computer numerical control–based equipment.

Every other chapter opens with a **Made Right Here** profile. This profile contains a photo and a brief text describing a company or product manufactured today. It is meant to inspire students and show them all of the opportunities available to them in the industry.

In July 1996, my father, Mike Paulson, my brother, D.J., and I started a small machine shop called Straightline Precision Industries, Inc. We started with a small manual mill and lathe in our garage with the intention of doing general machining and repairs. In 1999, we purchased our first CNC mill, which allowed us to go after more complex and more lucrative work. We have embraced technology from day 1, constantly reinvesting in the best and newest equipment we could afford, always keeping our eye on future trends in manufacturing. In 2005, we launched Straitline Components, a line of high-end aftermarket mountain bike components, specifically to meet the evolving needs of downhill and extreme free ride athletes. To meet the demands of this added business, we have had to add a lot of new technology, such as robotics and multitasking mill-turn machines, as well as move into more advanced CAD/CAM software. This all means that the job has become very complex and at the same time very rewarding. It is a great feeling to design a product, take it to production, and then finally use it, and see other people enjoy your efforts!

Courtesy Dennis Paulson and Straightline Precision Industries, Inc.

—In memory of Mike Paulson 1950–2011
Dennis Paulson

Courtesy of Dennis Paulson

ARMCHAIR FIELD TRIPS will be found throughout the book. Scan the QR code and you will be taken to the **Haas Tip of the Day** YouTube website. Search for the suggested video to view additional materials relevant to the topic being discussed at that point of the chapter.

ARMCHAIR FIELD TRIP

Watch this Haas Tip of the Day video about thread taps.

In this video you will learn how to select the right tap for the job! Go to: **https://www.youtube.com/haasautomation** and search for this video called: **Tapping Essentials-Every Machinist Needs to Watch This Haas Automation Tip of the Day**

The **Terms Toolbox Challenge** allows students to review the vocabulary terms using their mobile devices. Students can scan the code to access the flashcard-like exercise.

TERMS TOOLBOX

Chips Metal particles of waste removed from the workpiece by machining.

Natural fibers Cotton or wool cloth, which tends to resist hot chip damage and melting, thus protecting the wearer from burns.

Synthetic fibers Plastic cloth such as nylon and polyester, which tends to melt when hot chips touch it.

Z87 or Z87.1 The mark found on safety glasses approved for shop work, which means they will protect your eyes from the front and side in a dangerous environment.

TRADE TIP

Besides using letters to track revisions, in your shop, drawing revisions can also be tracked by number or by the date on which they were updated.

KEYPOINT

When receiving a job, always verify that

A. The part number matches the W/O.

B. The drawing revision level and work order revision level agree.

SHOPTALK

Replacement Parts Some shops produce replacement parts for older products—*not the latest version!* In that case, detective work must be done. The changes made over time must be backtracked from the current Rev level back to the ones ordered. Each change must be investigated to see if it is compatible with the latest parts. Will they interchange, or does the planner need to modify them to fit the older product? These can be incredibly complex research issues that must be done before releasing the job to the shop. The good news is that beginning machinists never have to do them! After the planner solves the issues, he or she writes a work order covering the steps needed to make the older version of the part, guiding you to make the right parts.

MOTIVATIONAL CHAPTER FEATURES such as *Key Points, Trade Tips,* and *Shoptalk* are included to show students the practical side of the subject.

All topics in the textbook have been **indexed for programs needing to offer standard skills certificates**. The complete index can be found on the instructor side of the online learning center.

Figure 1-7 Hair does catch on moving machinery. Keep it out of harm's way.
Photography by Prographics/McGraw Hill

TOOL & DIE STEEL		
A 2 Annealed Air Hardening 5% Chrome	D 2 Annealed Air Hardening High Carbon/Chrome	O 6 Annealed Oil Hardening Graph-Mo®
A 6 Annealed Air Hardening Low Temperature	H 13 Annealed Air Hardening Hot Work	S 5 Annealed Oil Hardening Shock Resisting
A 10 Annealed Air Hardening Graph-Air®	O 1 Annealed Oil Hardening	S 7 Annealed Air Hardening Shock Resisting

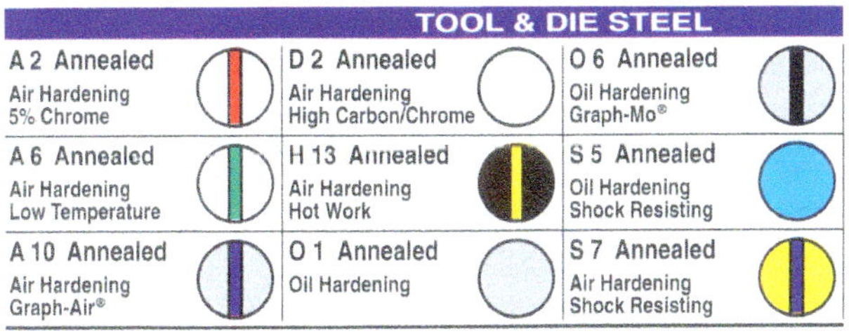

Figure 1-13 Correctly stored and color-code-identified raw materials.
Source: Pacific Machinery & Tool Steel Co.

Photography by Prographics/McGraw Hill

TERMS TOOLBOX

Chips Metal particles of waste removed from the workpiece by machining.

Natural fibers Cotton or wool cloth, which tends to resist hot chip damage and melting, thus protecting the wearer from burns.

Synthetic fibers Plastic cloth such as nylon and polyester, which tends to melt when hot chips touch it.

Z87 or Z87.1 The mark found on safety glasses approved for shop work, which means they will protect your eyes from the front and side in a dangerous environment.

Full-color **photos and illustrations** make concepts easier for students to understand and apply to the information presented.

THE ONLINE LEARNING CENTER contains a wealth of resources for instructors, including an Instructor's Manual with teaching tips and handouts and enhanced PowerPoint slides with videos.

PART 1

Introduction to Manufacturing

This book is about getting a job in a machine shop. But it's also about keeping that job and advancing career responsibility and pay. For career success, you need to know what is expected of you from the very first day. Like any workplace, there are tasks, procedures, and rules to be followed. Some are formal skills or rules; some are informal and generally accepted by your fellow workers.

Part 1 is designed to impart basic manufacturing knowledge and skills and to clarify trade expectations to help you in the following ways:

Chapter 1 Know how to look, act, and be professional

Chapter 2 Understand, use, and speak the precision numbers of the trade

Chapter 3 Read early lab drawings and be ready for more engineering drawing instruction

Chapter 4 Identify and use geometric dimensioning and tolerancing (GDT) found on drawings

Chapter 5 Work with the materials, documents, and processes before and after the machining occurs

Chapter 6 Use five basic measuring tools, the right way always with the best repeatability

Chapter 7 Use specialized measuring tools when the five basic tools aren't the right choice

CHAPTER 1

Professionalism in Manufacturing

Units

1-1 Dressing for Career Success

- Select the correct protective equipment for a machine shop
- Select the correct clothing

1-2 Handling Materials

- Safely lift heavy objects correctly and explain why bending your knees is the right thing to do, but the last resort for heavy lifting
- Safely move metals within the shop
- Store metals and machine accessories

1-3 Handling Shop Supplies

- List the six possible dangers of shop chemicals
- Ask for and read SDS sheets when in doubt about handling new chemicals
- Handle and store lubricants, solvents, and coolants
- Know how to correctly dispose of waste

1-4 Maintaining Equipment and the Work Environment

- Lubricate complex machinery
- Safely remove metal chips
- Understand what a lean green shop means
- Review fire prevention and safety

INTRODUCTION

This book represents a manufacturing world where computer-aided design/computer-aided manufacturing (CAD/CAM) and computer numerical control (CNC) have changed everything. Planned by a group of industry leaders and instructors, we used today's job market and tomorrow's career needs as our guide. To make room for the new subjects needed for career success, every effort was made to eliminate old technologies and skills no longer relevant to mainstream employment. Our goal was to equip you, the beginning machinist, with the competencies to get and keep that vital first job.

But we also knew that students often breeze past the usual opening chapters to get to the "real training." So why start with a chapter on professionalism and safety?

Because a critical part, perhaps the most critical part, of your training has nothing to do with measuring, reading prints, and running programs on machines, yet getting it right will have everything to do with your success! It might be called work ethic, team spirit, or job readiness. It's often called attitude on an employee evaluation or a grade report. No matter what you call it, it adds up to how you walk the walk of a skilled craftsperson. A large part of the separation between ordinary and skilled workers is a professional attitude.

Although a whole lot more could be said on the subject, these units are enough to get started fitting into a machine shop environment and starting the lifelong process of being a skilled professional. Taken to heart, the message of this chapter will make a real difference in your career.

Unit 1-1 Dressing for Career Success

Introduction: In Fig. 1-1, which person would you want making precision parts for your new car or outboard motor? In truth, they might all be good machinists but–well, you get the picture. The concern here isn't styles of clothing or grooming; it's about being right for a precision shop environment.

Figure 1-1 Which machinist looks right for the job—and more important, for a career?
Lake Washington Technical College, Kirkland, WA/McGraw Hill

TERMS TOOLBOX

Chips Metal particles of waste removed from the workpiece by machining.

Natural fibers Cotton or wool cloth, which tends to resist hot chip damage and melting, thus protecting the wearer from burns.

Synthetic fibers Plastic cloth such as nylon and polyester, which tends to melt when hot chips touch it.

Z87 or Z87.1 The mark found on safety glasses approved for shop work, which means they will protect your eyes from the front and side in a dangerous environment.

1.1.1 Getting Ready for the Work Environment—Eye Protection Always

Figure 1-2 shows several types of eye protection. Most shops supply one or more varieties. The best choice is the one you find comfortable and tend to leave on 100 percent of the time! Many prefer full-vision, wraparound lenses because they are all clear material so you don't see a frame.

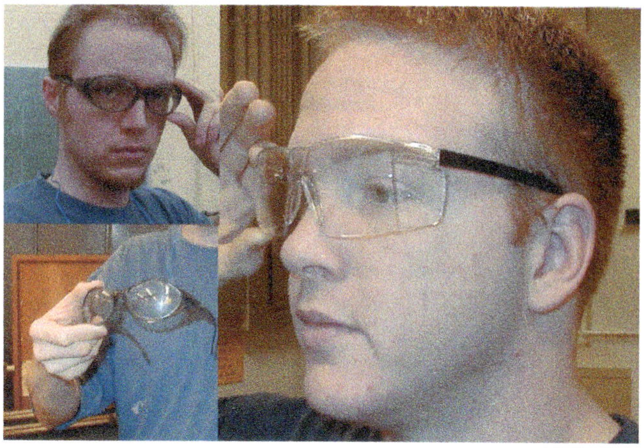

Figure 1-2 The best choice for eye protection is the one you find most comfortable and tend to leave on.
Lake Washington Technical College, Kirkland, WA/McGraw Hill

KEYPOINT

Safety glasses will bear the mark **Z87** or **Z87.1** on the earpiece if they have passed strict testing and are acceptable for shop work.

Clear or Yellow Lens Color

Either clear or yellow lens color is acceptable. Yellow lenses offset the blue of common fluorescent lighting, and many feel that the correction boosts their ability to read precision tools. Never select dark glasses unless you must work near electric welding flashes because they dull your ability to see details.

Prescription Eyewear

Most wraparound safety glasses can be worn over prescription glasses. But the law requires prescription glass lenses to be made from tempered glass or high-impact plastic, so it is acceptable to wear them alone as long as side shields are added. It's a fact that many eye injuries occur from the side rather than straight on, so shatterproof front lenses aren't enough protection by themselves.

Extreme Danger Areas

When performing tasks such as disk grinding with lots of flying debris, protect your eyes by adding a full-face shield over safety glasses.

Just Do It!

Be a trendsetter—wear safety glasses even when others don't. Modern CNC equipment usually has containment shielding so the operator feels safe from flying metal particles, but don't forget you must occasionally walk through the shop and pass by other unprotected machines. Make safety glasses a habit by putting them on when entering the shop. Here's an attitude check: You've got it right when you feel strange without your safety glasses. No kidding, I've gone home still wearing them! If those provided aren't comfortable, then find an industrial safety supplier and buy a pair just right for you—that's the kind of pro we're talking about.

1.1.2 Hearing Protection

Machine shops can be noisy places. Some operations are loud enough to cause hearing loss over time. Prevent permanent damage right now while your hearing is good. Get in the habit of wearing ear protection (Fig. 1-3) where noise gets above moderate—see the chart. The two common types of hearing protection are expandable foam inserts that fit every ear shape and the muff type that fit over your ears.

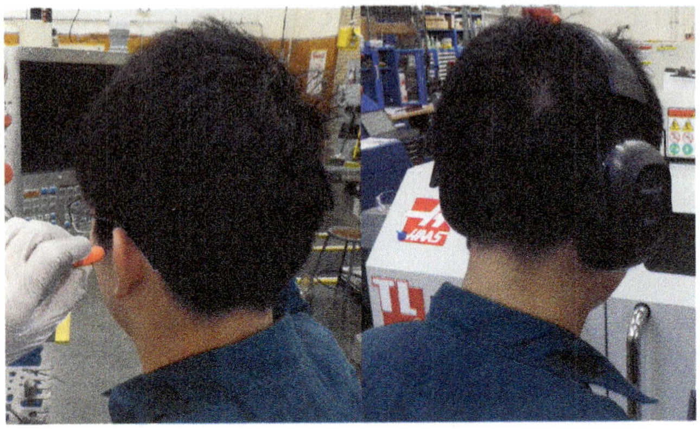

Figure 1-3 Machinists protect their hearing when shop noise is loud or high pitched.

Keith Smith

TRADE TIP

Hearing Is Your Primary Control Protecting your hearing is more than personal. Just as when driving a car, the machinist almost always hears a problem developing before seeing it, especially on fast CNC equipment.

Typical Range of Common Sounds

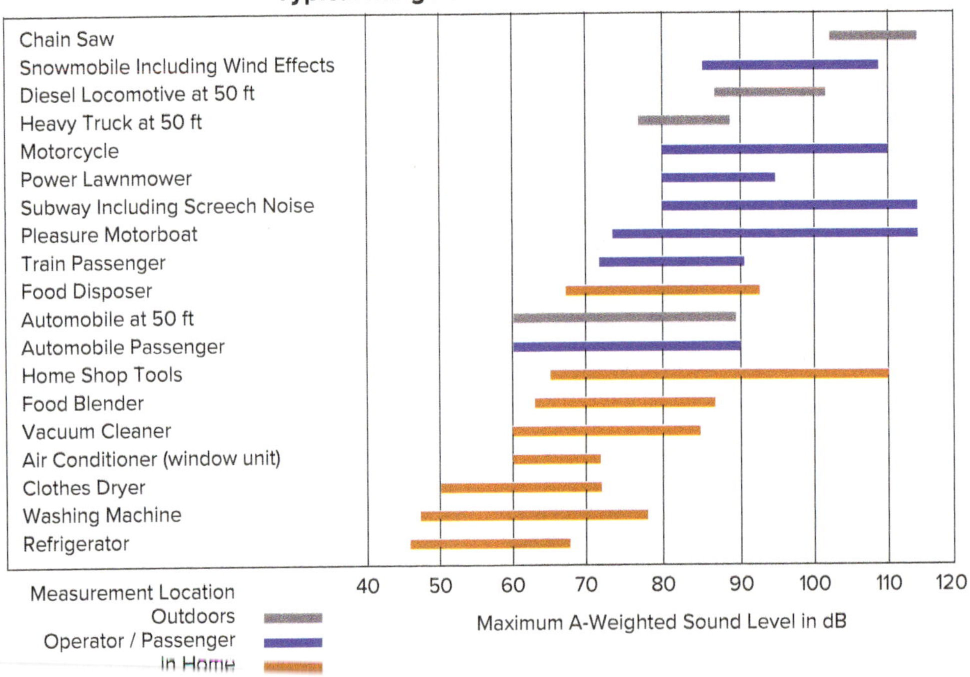

Source: Protective Noise Levels, EPA, 1979, pg. 5.

ARMCHAIR FIELD TRIP

Be sure and watch this Haas Tip of the Day video about hearing safety.

You will learn how to protect your hearing in loud environments like machine shops. Go to: **https://www.youtube.com/haasautomation** and search for this video called: **Are You Protecting Your Hearing?—Haas Automation Tip of the Day**

Earplugs remove the loud spikes but allow controlled hearing. For extreme situations, earmuffs are supplied by your employer. Either way, don't give away your most immediate control sense by not protecting it. And by the way, to control your machine it is important to not use personal music earphones–sorry, but they're not acceptable when running a machine.

1.1.3 Shop Clothing

The main danger of loose clothing is that it can be caught by moving machinery–but you already knew that. Although Fig. 1-4 was set up, really being caught in a machine is no fun–think about it! Wear nothing loose–sleeves, necklaces, untucked shirts–nothing! Well-fit **natural fibers** such as cotton or wool, without pockets or tie cords hanging out, are essential.

Figure 1-4 Seriously, loose clothing does get caught in moving machinery!
Photography by Prographics/McGraw Hill

Why Natural Fibers?

The shavings made when cutting metal are called **chips**. They're hot, as much as 1,000°. While today's CNC machines won't operate with the safety guard open, many older machines will. Flying hot chips must be dealt with. As we study machine operation we'll see several ways to control them, but one action the pro takes is the kind of fabrics worn on the job.

When hot chips contact **synthetic fibers** such as polyester, rayon, or nylon, they stick, then melt through (Fig. 1-5). So your shirt or pants are ruined and the hot metal is held against your skin. Ouch! Not only that, but that makes it hard to concentrate on the task at hand.

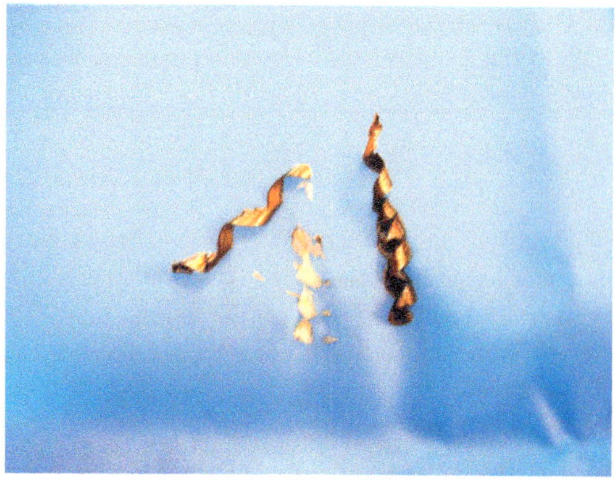

Figure 1-5 Two hot chips melted onto this nylon windbreaker while the center one went completely through to possibly burn the person within!
Courtesy Mike Fitzpatrick

An apron or shop coat can be a good choice. But be aware that not all are designed for machining. Some are made for lab work, where machinery won't grab loose ties or pockets. Find one with internal pockets and no loose belts tied in front. Long sleeves are not smart for obvious reasons. A professional approach is to keep a short-sleeved work shirt in your locker.

How About Shoes?

Your work shoes have three safety aspects. I'll bet the third one will surprise you. Work shoes provide the following:

- Protection for your feet from falling objects
- Nonslip soles designed for a shop where chips, coolants, and oils are often on the floor
- Protection from fatigue

Athletic shoes are a poor choice. Even though comfortable, they aren't designed to stand up to a shop environment.

Steel Toes

Shoes or boots with steel toe caps are better than shoes without and may be required on the job. No matter how careful you are, there's always the inevitable falling heavy object. Don't fall for the old tale that someone knows someone who had his or her toes severed by the steel insert collapsing when something really heavy fell on it. Think about it: If the thing was that heavy, it would have done the same damage with or without the steel protection!

SHOPTALK

Quality Could Be Linked to Good Work Shoes or Boots As a machinist, you're going to be on your feet all day, usually on concrete. Guess when most folks make the most mistakes. That's right, at the end of the day when they're tired. Good-quality work shoes offset some of the problems and help keep your mind sharp.

No Accessories

Jewelry catches on moving machinery (Fig. 1-6). Here's another aspect you may not know: Jewelry also conducts electricity and heat. In addition to the safety aspect, jewelry should be left in your locker.

Figure 1-6 Jewelry is dangerous. It catches on machines and chips and conducts heat and electricity.
Photography by Prographics/McGraw Hill

Hair Up

Think about this: Would you reach up and pull a handful of hair out of your scalp? Painful but that would be just a few hairs! So now, imagine a machine wrapping up most of your hair along with a patch of scalp. No kidding! I've seen it happen twice with my own eyes (Fig. 1-7). It gets caught up by static electricity produced during machining and is blown by the air currents swirling around moving tools. A hair band, a bandana, or a hat is right to wear when your hair is long enough to be caught.

That's it. With these dress-for-success guidelines, you're ready to step into the shop and, if you see the bigger picture, into a career.

Figure 1-7 Hair does catch on moving machinery. Keep it out of harm's way.
Photography by Prographics/McGraw Hill

UNIT 1-1 REVIEW

Replay the Key Points

- Eye protection is best if it's comfortable.
- Safety glasses will bear the mark Z87 or Z87.1 on the earpiece if they have passed strict testing and are acceptable for shop work.
- Ear protection is necessary in many work areas.
- Well-fitting coats or aprons made of natural fibers are best.
- Footwear must be designed for shop use and comfort.
- Athletic shoes, although comfortable at first, do not stand up to the shop environment.
- Any loose item is a danger, including jewelry and hair.

Respond

1. Are yellow-tinted safety glasses acceptable in the shop? Are brown- or green-tinted glasses acceptable?
2. Describe the two reasons to avoid synthetic fibers and wear natural fiber clothing when running machines.
3. What are three aspects of footwear with regard to safety?
4. Why is hearing such a vital issue for a machinist?
5. For what two reasons does long hair become tangled in moving machines?

Unit 1-2 Handling Materials

Introduction: Working in a machine shop requires the use of many materials. Some are technical chemicals with specific precautions and earth issues. Some are consumables meant to be used up but not wasted. Others are heavy and expensive. This unit lays out the ways to handle them all—like a pro. Doing so is an excellent way to demonstrate a well-tuned attitude.

TERMS TOOLBOX

Choker strap A nylon strap that cinches tightly around heavy objects with a strong eye loop on each end to be attached to the crane.

Heat lot number The original quality control number for a specific metal bar.

Intervertebral disk (disk) The flexible cushion between spinal vertebrae that can be damaged by the wrong lifting techniques.

Jib crane A heavy lifting device that may be fastened to a wall or column to swing in an arc or on portable rollers and often called a cherry picker.

Traceability The ability to link a given bar of metal from "birth" all the way to its final shape and individual serial number.

1.2.1 Lifting Heavy Materials

Always Use a Machine If You Can!

After all, we are *machinists*. The very last thing we use to get a job done is muscle power (Fig. 1-8). In most facilities, you will find one or more of these devices:

- Overhead cranes moving on rails
- Rolling lift tables
- Floor jacks for flat pallets of workpieces and boxes
- Forklift trucks
- Portable **jib cranes**, often called "cherry pickers"
- Jib cranes fastened to a column to pivot around a circular area

a. b. c.

Figure 1-8 Smart machinists use equipment, not muscle, to lift heavy objects.

Photography by Prographics at Milwaukee Area Technical College/McGraw Hill

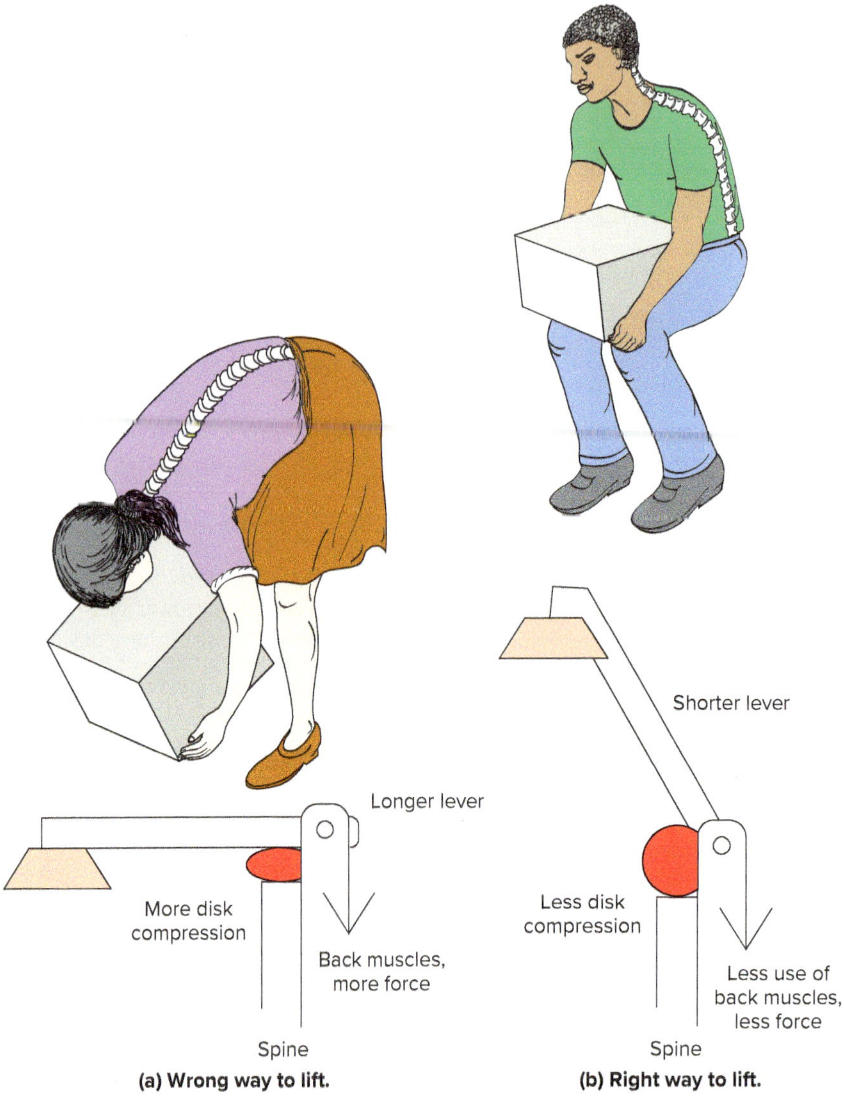

Figure 1-9 Lifting the wrong way and the right way. The worker on the left is setting up a painful future! Note that the spinal disk is under a lot more pressure because this person is bending the back, not the knees.

Use Your Legs (and Your Head), Not Your Back

However, there comes a time when lifting by muscle power is the only way to get the job done. At these times, the wise machinist asks for help. Two lifters are better than one. In North America, it's estimated that two of five adults suffer from back pain that could have been prevented.

Here's the point: It doesn't happen later in life; it starts right now–today! Unlike the common view, common back injuries don't actually happen catastrophically–during one bad lift! They are the product of lifting wrong over a long time. Everyone has heard about the straight back, bent legs method. It's best, but why?

To understand, let's look at your spine as a piece of machinery: A human lifter is a crane of sorts. In Fig. 1-9, the person on the left has turned the back into a long lever with the pivot point right at the lower back. *This wrong action causes pressure in the range of hundreds of pounds per square inch right on the lower back.* In contrast, the person on the right, bent at the knees, is focusing the pressure on the leg muscles, not the back.

Disks in Your Back

Now focus on that pressure point, the lower back. Your spine is well engineered with cushions between the vertebrae, called **intervertebral disks** (or just **disks**). They provide flexibility and they absorb shock. They also keep the vertebrae apart so nerves are not pinched as you move. Disks are critical and can be damaged (Fig. 1-10). When they are damaged, it can really hurt and even immobilize you!

The disks are tough, fluid-filled sacs with several layers containing the fluid center. Lifting wrong overpressurizes them, resulting in a blowout. Here's the part most people do not know: When the disks first stretch and break, it's an inner layer or two that goes, and usually the injured person doesn't even know. The layers do not often break all at once! But, with continued bad technique, the last one finally ruptures, resulting in enough pain that the person cannot stand up straight!

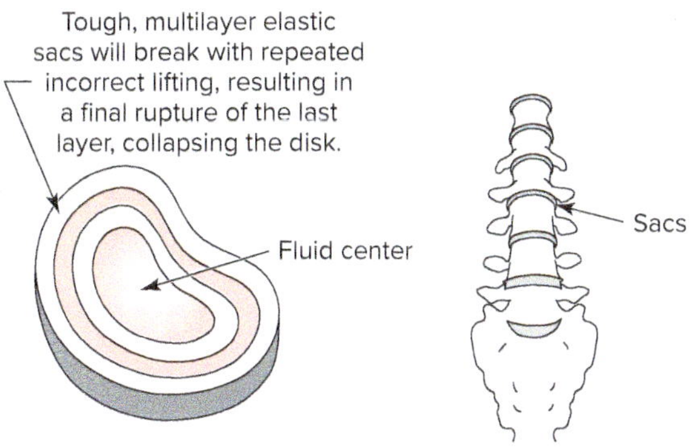

Figure 1-10 A spinal disk is a tough, multilayered shock absorber. But lifting incorrectly can cause progressive damage, leading to a disaster in the future.

KEYPOINT

Lifting wrong is not only the start of a painful future; it also creates a poor professional image.

Carrying Materials

Another aspect of handling heavy objects is carrying long bars of metal. Be cautious of the forward end–it's a ram! If the bar is too heavy to safely hand carry (beyond 40 pounds) or if it's over 8 feet long, there are two acceptable methods:

1. **Crane Carry–One Person**

   In the unbalanced carry, hold the bar with a nylon **choker strap** near the leading edge and pull the remaining bar along on the floor using the overhead crane. One person can perform this carry by being at the forward end. A choker strap is a strong nylon flat strap with loop ends that slip closed with pressure. Your instructor or shop lead will demonstrate this aspect of lifting (Fig. 1-11).

2. **Two Person**

   For the balanced carry, choke the bar near the middle and carry it with both ends off the floor. This requires a second person to walk at the leading edge, to prevent hitting people or objects (Fig. 1-12).

WARNING: Crane Overswing

A suspended heavy bar is a mass in motion that can't "stop on a dime." It will swing forward when you stop pushing or driving the crane forward. Anticipate a short swing forward upon stopping.

Figure 1-11 Use a crane and choker to haul heavy bars, while an assistant helps control the metal bar!
Photography by Prographics at Milwaukee Area Technical College/McGraw Hill

Figure 1-12 Two people hand carry long bars by preventing the leading edge from harming people and objects.
Photography by Prographics at Milwaukee Area Technical College/McGraw Hill

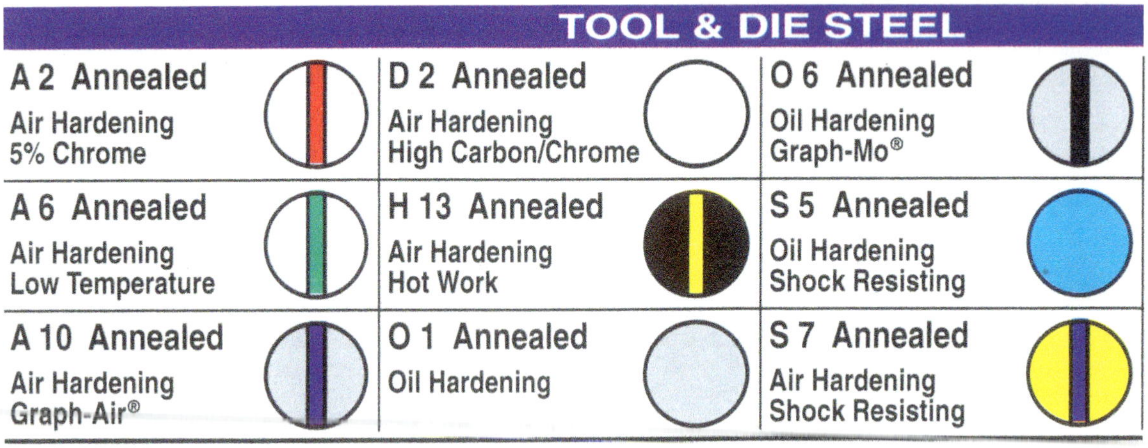

Figure 1-13 Correctly stored and color-code-identified raw materials.
Source: Pacific Machinery & Tool Steel Co.

Heavy lifting is a skill requiring knowledge of rigging (the lifting devices that contact the work), lift machines, and the physics of lifting. Due to the specialized skills and knowledge required, many shops employ a lifting crew. But this subject is also considered part of the machinist's duty. If you are the slightest bit unsure about how to lift or move an object, get help.

KEYPOINT

Never reach under or, worse, walk under a heavy object during a lift.

1.2.2 Storing Metals

Our school has received many donations of expensive metal from local shops because careless workers cut the color-coded identification off the bar end or lost the identification tag. Once gone, the bar is useless in most cases, especially when the shop makes people-moving products. In situations where lives could be at stake, once the **heat lot number** (the original manufacturing quality control number) is lost, the metal is not only worthless; it's also illegal to use!

When the design specs must be followed exactly, as in making airplane parts, for example, a continuous history must be traceable from the foundry to final part number. This is known as **traceability**. A specific part will have a specific number assigned to it, called the serial number (S/N). The manufacturer must be able to provide documents tracing the S/N to the heat lot of the metal's original manufacturer. Material can be tested in an independent lab and then recertified, but that's more costly than just buying more metal!

One of several good methods of storing metals is a rack. Note that the bars cannot fall out of the safety storage rack and that short bars are not kept here. "Shorts" are kept on smaller shelves where they won't fall through. Note that they are sorted by type and alloy such that the color-coded identification is facing out (Fig. 1-13). Later, we'll take a look at material identifications and alloys.

KEYPOINT

These are not small details. You must have exactly the right material for the design, and it often must be kept in storage for a specific job.

UNIT 1-2 REVIEW

Replay the Key Points

- Always use a lifting device for your first choice.
- If physical lifting is absolutely necessary, get help—all should bend their knees.
- Carry long bars safely by protecting the leading edge.
- Store long bars in a safety rack and short pieces away from the rack on shelves designed for this purpose.
- Mind the color coding and/or stamps—*never cut them off the bar.*

Respond

1. Name at least three shop lifting aids other than human muscle?
2. A choker strap can be used to carry long material through the shop as long as the forward end of the bar is high in the air. Is this statement true or false? If it's false, what could make it true?

3. Why is material identification important for many types of work?
4. Describe a disk in your spine.
5. Where must short bars be stored?

Unit 1-3 Handling Shop Supplies

Introduction: Today's professional must know the rules when handling materials. Beyond the cost of waste and a ruined planet, the fines for ignoring the rules can cost dollars and company image. Machinists must use chemicals and consume materials with an awareness of the issues concerning each. That means

- CONSERVE, REUSE, RECYCLE, and DISPOSE RESPONSIBLY.

TERMS TOOLBOX

Hydrostatic bearings Found on modern CNC machines, these require an exact oil viscosity to separate two sliding components apart as they move.

Safety Data Sheet (SDS) The product of the federal "Right to Know" act that ensures that workers must be provided the information needed to handle materials safely–found on Safety Data Sheets.

Spindle oil Lubricant designed to prevent rolling friction.

Viscosity The rated property of lubricants to resist thinning out and flow rate.

Way oil A lubricant used to control sliding friction in machines (the part that slides is called the machine's ways).

1.3.1 Your Right (and Obligation) to Know about Shop Chemicals

Promanagement means knowing hazards a material might have and taking the right actions. Those issues and the correct use of each chemical are found in documents called **Safety Data Sheet (SDS)** (Fig. 1-14). Mandated by federal law, manufacturers of chemicals must supply these sheets, and employers must have them on file for all to read.

In general, machine shop chemicals are not overly dangerous. Still, some will ignite, or skin can react, and some produce fumes. Many involve precaution.

The six precautions are

- Fire hazard
- Chemical stripping of skin oils–direct contact
- Fume toxicity and oxygen exclusion (displaces breathing oxygen)
- Eye irritation
- Allergic reactions
- Contamination of other fluids or chemicals, with possible side reactions

For example, we sometimes use acetone to clean and remove dyes from parts. Using the data sheet (SDS) in Fig. 1-14, we find the following:

- Acetone must be stored in a fire-resistant container.
- It should not be in contact with skin or eyes.
- You must avoid breathing excessive or concentrated acetone fumes.

So, using acetone requires a fireproof container. Wear chemical-resistant gloves and work in a place where air circulates–never a small room with closed doors and windows.

KEYPOINT

Although shop chemicals are generally not extremely dangerous, each has its own set of precautions and procedures found in its own SDS. Each also has a correct disposal procedure that must be followed. Knowing is your right and obligation.

1.3.2 The Five Kinds of Shop Chemicals

1. Lubricants
2. Coolants
3. Solvents/coatings
4. Cleaning products
5. Gasses (compressed and liquefied)

You might be called on to use compressed gasses for welding and brazing. We also use liquefied nitrogen for deep chilling when shrinking metal parts for special assemblies. But these tasks are beyond beginners, so we do not study gas safety here. However, be aware there are some highly specialized skills needed to use them. That leaves the other four categories.

1.3.3 Using Lubricants the Right Way

Other than fire hazard, today's machine lubricants are safe and *benign* (not toxic)

SAFETY DATA SHEET

1. IDENTIFICATION

Product identifier	ABC Cleaner
Other means of identification	
Product code	XXXXX
Recommended use	Cleaner
Recommended restrictions	None known.
Manufacturer/Importer/Supplier/Distributor information	
Company name	ABC Company
Address	123 Main Street Anywhere, USA 12345
Telephone	
General Information	123-45-6789
Technical Assistance	123-45-6789
Customer Service	123-45-6789
24-Hour Emergency (CHEMTREC)	123-45-6789 (US) 123-45-6789 (International)

2. HAZARD(S) IDENTIFICATION

Physical hazards	Flammable aerosols	Category 1
	Gases under pressure	Compressed gas
Health hazards	Acute toxicity, oral	Category 3
	Skin corrosion/irritation	Category 2
	Serious eye damage/eye irritation	Category 2A
	Reproductive toxicity (the unborn child)	Category 2
	Specific target organ toxicity, single exposure (oral)	Category 1 (eyes, central nervous system)
	Specific target organ toxicity, single exposure	Category 3 narcotic effects
	Specific target organ toxicity, repeated exposure	Category 2 (liver, lung, kidney, brain)
	Aspiration hazard	Category 1
Environmental hazards	Not classified.	
OSHA defined hazards	Not classified.	

Label elements

Signal word Danger

Hazard statement Extremely flammable aerosol. Contains gas under pressure; may explode if heated. Toxic if swallowed. May be fatal if swallowed and enters airways. Causes skin irritation. Causes serious eye irritation. Causes damage to organs (eyes, central nervous system) by ingestion. May cause damage to organs (liver, kidney, lung, brain) through prolonged or repeated exposure. May cause drowsiness or dizziness. Suspected of damaging the unborn child.

05079 Version #: 03 Revision date: 09-14-2016 Issue date: 08-11-2014 SDS US 1 / 12

Figure 1-14 An SDS sheet. When in doubt, read these bulletins. It's your right to know.

TRADE TIP

You Should Know! Fast, modern CNC machines are fussy about getting the exact oil for the duty. An expensive machine can be damaged by adding the wrong lube nearly as fast as not adding any at all. Of highest priority: Oil thickness and resistance to flow, called **viscosity**, are absolutely critical—both from the standpoint of the precision action of the machine and to prevent wear.

Special Oils Must Be Identified and Kept Clean

Several application-specific lubricants are necessary, especially where there is a variety of CNC machines. Both sliding and rotating equipment must have lubrication, but faster machines equipped with high-tech **hydrostatic bearings** depend on an exact thickness of oil between moving components. Components actually "float" on a predictable thickness of oil much like an air hockey puck. Introducing the wrong oil causes a double problem: It can create inaccurate machine movements, but even worse, a thinner film than that required can let metal touch and rub with sure damage to follow!

KEYPOINT

Keep oilcans labeled and stored in a marked, fire-protected area with their tops closed to avoid contamination (Fig. 1-15).

Figure 1-15 Specialized lubricants must be stored in fireproof surroundings and kept free from contamination.
Lake Washington Technical College, Kirkland, WA/McGraw Hill

Sliding Lubricants and Rolling Action Lubes

Oils are used in two very different ways: to prevent either sliding friction or rolling friction. The sliding type oil is called **way oil** because the machine parts it lubricates are also called "ways" (discussed in lathe and mill training coming up). Way oil is thicker than spindle oil. Even thicker yet, grease is occasionally used to prevent sliding friction in some older machines, but it's far less common today. However, if the machine needs grease, then using way oil would be a big mistake!

For bearings that rotate, **spindle oil** is the correct lubricant. There are more kinds of spindle oils than ways due to the wide differences in the kinds of spindle bearings. This is the area where you must be doubly sure to have exactly the right oil. Some bearings are made of ceramic composites, whereas others are ceramic/steel, and still others are steel alone. Some spindles are refrigerated to maintain perfect accuracy, while others aren't. Spindle accuracy and tool life depend on lubrication.

KEYPOINT

Never leave open oilcans in the work area or even in the storage area. They become contaminated with chips and dirt. They could even catch on fire from heated flying chips.

How Do You Know Which Lube to Use?

Lubrication instructions are usually found on a metal plate fastened to the machine, right at the input point. If no plate or other tag is visible, look it up in the owner's manual (Fig. 1-16). Or just ask.

Figure 1-16 Be absolutely certain to add the right oil, especially on CNC equipment!
Lake Washington Technical College, Kirkland, WA/McGraw Hill

SHOPTALK

Pros Do It! It's often the little things that identify the pro. Here are a few examples. When putting oil in a machine, clean the funnel and spout and do not pour oil while chips are flying. When finished with the fill, wipe the oilcan top before storing it so that it won't collect dirt. Last, put the can in the cabinet with the label turned out for the next user.

1.3.4 Using Cutting Fluids (Coolants)

Later in Chapter 8 you will learn the technology of coolants, also called cutting fluids. Here we'll look at them as shop chemicals. Coolants often make machining safer and faster. Most of today's coolants are water-based fluids that are either flooded or sprayed over the cut area. When used correctly, tools last longer, productivity increases, and horsepower is reduced along with excess heat. Today most shops mix coolants such that one dilution can be used throughout the shop, but others prefer to custom mix for a specific application. For example, a thicker mixture is needed for milling, but a thinner one works best for grinding. The important issues for coolants are as follows:

- If a vapor or spray is produced in the machine operation, it must not be inhaled. Containment barriers and a breathing mask are two good solutions.
- The undiluted syrup used to make coolant is ultracostly. A barrel of the unmixed syrup can cost a week's wages for an experienced machinist!!
- Coolants must be maintained by checking density and adding fresh water because the heat from machining evaporates the water and thickens the dilution ratio.
- Coolants become contaminated with lube oils from the equipment and dirt from machining, called "tramp oil." Tramp oil must be skimmed or filtered out.
- Coolants can be revitalized by chemical treatment and filtration, but at some point coolants must be disposed of. When the time comes, disposal must be done in the right, legal way!

Coolants come in three forms: synthetic, organic, and special threading compounds. *Synthetic coolant* is the most common, especially for CNC work. Synthetics are nonclinging, meaning that they run off the workpiece easily; thus, parts come away clean from the machine, and the shop stays cleaner, too. Synthetics last longer in the machine compared to older organic oils. *Organic coolants* can rot due to bacterial action, over time. Finally, even though their main ingredient is water, synthetic coolants prevent rust on machines and on steel or iron parts similar to antifreeze mixed with water in your car's engine block.

TRADE TIP

Knowing Coolant Ratios While there is a general shop ratio usually around 40 to 1 for most products, different applications may require an exact dilution ratio. For example, a given coolant might work best at 25/1 for milling but 45/1 for grinding. Too much syrup adds nearly no benefit but adds needless cost. But more important, too little degrades the performance of the mix, and it even rusts the machines! Cutting different metals on the same machine can even require differing ratios: Steel can be successfully cut with thinner coolant mixes than titanium, for example.

Highly engineered synthetics (Fig. 1-17) are designed to be both nonallergenic to people and nonreactive to metals.

Figure 1-17 Synthetic coolants are highly designed shop chemicals that are mixed with water to make machining more efficient and safer.

Lake Washington Technical College, Kirkland, WA/McGraw Hill

Cutting oils and cutting compounds (Fig. 1-18) are lubricants that tend to cling to the work, called the wetting property. Different from synthetics, these oils and compounds stay on the tool or work by design. Each has its use. They are either brushed or poured on during the operation. A few of these products are designed for thread machining and are exceptionally effective, but they're also expensive! They pay for themselves in results, but please don't waste them. If not used sparingly, they can contaminate the water-based coolants in the machine reservoir, which becomes tramp oil.

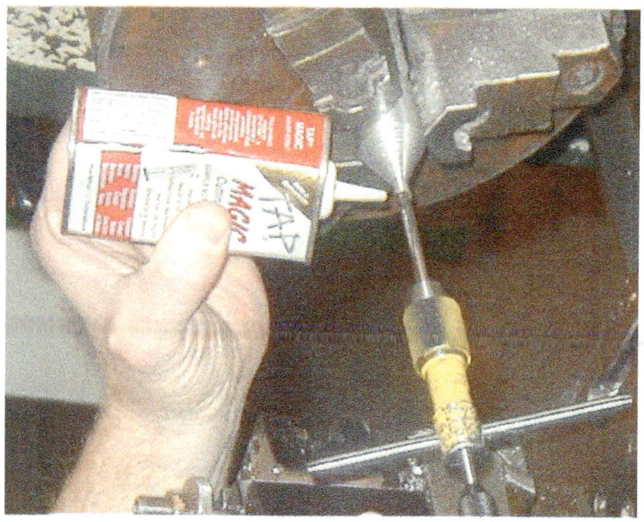

Figure 1-18 Thread-cutting compound makes a huge difference in thread quality and tool life, but it is expensive.

Lake Washington Technical College, Kirkland, WA/McGraw Hill

1.3.5 Recycling and Disposing of Waste

All responsible persons working with chemicals today must understand how to sort and dispose of them. That, too, can be found in the SDS sheet. Very few chemicals can be released as is back into the environment.

Most coolants, solvents, and oils must be turned over to a company equipped to either burn or chemically neutralize them based on the type. This is a very specialized operation that is regulated by state and federal agencies. Not surprisingly, it can be more expensive to dispose of chemicals than to buy them. Recycling within the shop whenever possible makes good sense. Your shop will undoubtedly sort the various kinds of scrap and have a well-defined waste program.

UNIT 1-3 REVIEW

Replay the Key Points

- Know the correct use and potential hazards of all shop chemicals.
- Store all chemicals and oils in designated containers and areas.
- Be absolutely certain that you're using the right lube for the application.
- Two kinds of coolants are used in the shop: Synthetic coolants are syrups mixed with water, and cutting oils and compounds are wetting lubricants for cutting metal.
- Always ask for supervision or read the SDS sheet if in doubt about how to handle a chemical.

Respond

1. We find information about shop chemicals on SDS sheets. Is this statement true or false? If false, what will make it true?
2. Synthetic coolants are mixed at dilution ratios of water to syrup from
   A. 20 to 1 up to 50 to 1
   B. 10 to 1 up to 20 to 1
   C. 1 to 20 up to 1 to 50
   D. 1 to 10 up to 1 to 20
3. Machine lubrication falls into two general types (other than grease used on old machines). Name them.
4. Without looking back, name as many precautions as you can recall with regard to shop chemicals (there are six).
5. The sliding parts of machines that must have a specific kind of lubrication are called.

Unit 1-4 Maintaining Equipment and the Work Environment

Introduction: Our next professionalism subject is not unlike taking care of your car. We all want to believe a clean car gets better mileage but more to the pro point, a messy car says a lot about the driver! Organization equals efficiency. Need proof? When working on your car, how much time do you spend looking for that wrench you just used a few minutes ago?

While the mileage issue might be false, long life and precision results on machine tools are true if they are maintained with care. Operating a new CNC machine means management has trusted you with the value of a dozen or more high-dollar automobiles and, more, with the profit that can be made with it if it keeps running.

TERMS TOOLBOX

Axis lock (override) One of the options a machinist might have to prevent any possibility of a machine moving while cleaning chips from the bin.

Chip breaking The action of breaking chips into easier-to-handle short segments.

Flow glass A glass window in an older machine that indicates the pump is working as oil flows over the window.

Lean manufacturing The study of organizing and managing an efficient work environment.

One-shot lubricator A manual lubrication pump found on smaller machine tools such as school machines.

Sight glass Functions like a dipstick by allowing the fluid level to be seen and compared to a line.

The following duties fall on the machinist. Done well they keep your machine and shop humming. We'll look at three categories:

1. Keeping chips under control
2. Doing under-the-hood checks: lubrication, coolants, and air pressure
3. Maintaining your work area

1.4.1 Removing and Handling Chips

Chips are the ever-present challenge for machinists. They come off the work in two varieties: short and long. Short broken pieces are the ones we try to produce because they're less dangerous and easier to clean up. The long ones can be nasty since they can snag and cut hands. But whatever the chip shape, here is a list of possible hazards to control (Fig. 1-19):

- Chips are *hot* right after they are made—they may be up to 1,000° Fahrenheit in standard machining and even higher when special ceramic cutters are used.
- Chips are *sharp*—as sharp as a razor at times—sharper actually since the edge may be only a few microns wide due to the way they are sheared off the workpiece!
- Chips *fly like bullets*—as much as 150 miles per hour on manually operated machines but up to 250 or more for CNC high-speed machining!
- Chips are *strong*. The long string type can catch and drag the unwary machinist into machines—this is the greatest danger. They can also cut deeply if drawn across your skin.
- Chips are *slippery* on bare concrete floors.

Figure 1-19 Long, stringy chips are strong *razor wires* that catch clothing and cut skin. All chips, even the safer "C" shapes on the right, can be dangerously hot.
Lake Washington Technical College, Kirkland, WA/McGraw Hill

Thinking of resigning? Hold on, with some preparation and prevention, these nasty critters can be tamed and controlled. There are lots of ways that we'll learn as we study drilling, milling, and turning.

Two Different Ways to Clean Up Chips

For now, let's talk about getting them out of the area. There are two ways we remove chips from machinery, depending on what the machine is doing:

- When the machine is operating and making chips
- When it's stopped; for example, when the chip bin is full at shift change or at the end of the job

For safety, each requires a very different action on your part.

Machine Not Running

Chips pile up fast (Fig. 1-20). Many CNC machines feature some form of automatic removal system, while others require the machinist or helper to intervene by blowing, raking, or brushing them away. We dispose of chips as we go but sometimes they get ahead of us and even the automatic chip conveyors get backed up. Whatever the reason, it's time to stop machining and get them out of the way. That means cleaning the chip bin, which almost always puts some part of your body at risk if the machine were to start.

Figure 1-20 Chips can pile up quickly and must be removed from the machine.
Courtesy Mike Fitzpatrick

Lock It Up or Lock It Out

Before reaching in to clear chips out, be absolutely sure the machine is locked or blocked from accidental startup, in case you bump it on or someone else might start it by accident (Fig. 1-21). Many CNC machines feature an **axis lock (override)** function. Other machines such as manual lathes require throwing the main power switch. But that might be a disaster on an older computer-driven machine because of the loss of data and/or setup positions.

Figure 1-21 Lock it out or lock it up—never perform maintenance on a machine that can be started accidentally.
Photography by Prographics at Milwaukee Area Technical College/McGraw Hill

KEYPOINT

Never reach into a machine to remove chips unless it is locked up or out. If you must go to the back of the machine, tell someone you are there so it can't be accidentally started.

If neither locking out nor turning off is possible, then do the following:

A. Turn CNC axis overrides to zero whereby no movement will occur.

B. Place a sign over the panel: "*Operator Cleaning Chip Bin.*"

C. Tell fellow machinists you are cleaning in the back or out of sight.

TRADE TIP

Chip Breaking There are two ways of breaking stringy chips into safe little packets. In the first, using heavier machining action forms thicker, stiffer chips that cannot take the bending action of being removed from the parent metal. They then break into small "C" shapes. The other answer is using a cutting tool with a chip breaking dam (Fig. 1-22). The obstruction redirects the flowing chip such that it snaps off. We'll see more on both techniques in tool geometry later.

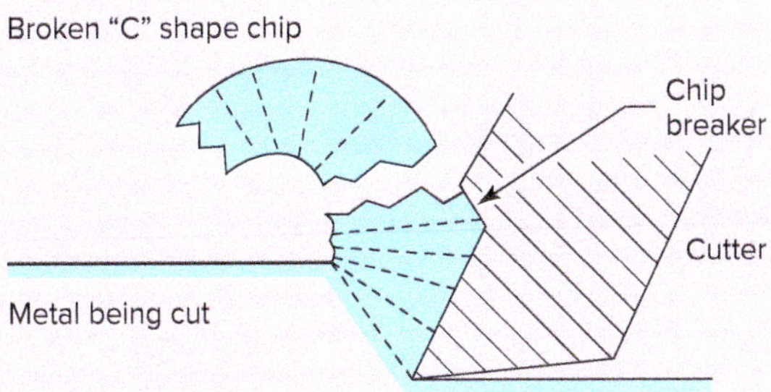

Figure 1-22 A chip breaker bends the flowing chip so quickly that it snaps off into small "C" shapes rather than a long string.

Now with the machine safely locked up or locked out, with no possible machine movement, put on gloves and rake the chips out safely. If you are assisting other machinists (you aren't the machine operator), be absolutely sure that the person who is the operator knows you are working in the machine chip bin.

Clearing Chips from Operating Machines

This is a very different procedure from the standpoint of safety. It's often necessary to clear away chips from machines while they're being made. This is a machinist's prime duty. Here are the details:

- If your machine is moving *do not wear gloves.* Lacking sensitivity, gloves put your hands at increased risk of being caught in the machine.
- Never reach in the machine with your hand or a gripping tool such as a pliers or vise grip. Can you guess why? Because you can't let go fast enough if the chips or machine catch the gripping tool.

Here are four safe ways to remove chips from moving machines:

- A sturdy brush
- Coolant stream
- Compressed air
- A chip rake or chip hook

The brush is a safe, self-explanatory tool used on drills and mills but not lathes and CNC operations since your hand is too close to the action. Although they sometimes get "eaten" by the machine, you can't hold on tightly enough to a brush to get hurt. A brush is the first choice of beginning machinists in training where chip volume is low.

SHOPTALK

Recycling Metal Chips Is Good to Mother Earth Most modern shops recycle as part of an environmental effort and because of cost savings, too. Larger chunks of metal are kept separated from the chips since each is handled differently at the recycling center and has different values. Correct sorting ensures value. The wing spar in Fig. 1-23 started out weighing over 500 pounds but when finished weighed only 8 pounds—96 percent of the weight became chips. Recycled aluminum chips require *80 percent less energy* to remelt and pour back into useful aluminum again compared to refining raw material from the earth.

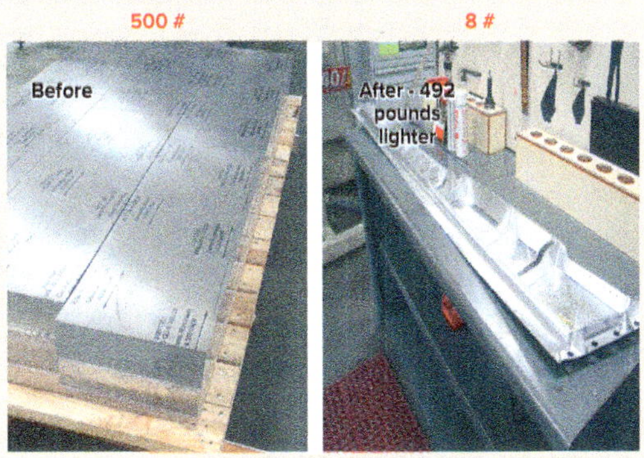

Figure 1-23 Sorting chips from scrap metal, then recycling each improves profitability plus it saves 80 percent of the energy required to mine and then refine raw metal.

Courtesy of Aerospace Manufacturing Technologies (AMT)/Arlington, WA

Using a Chip Hook or Chip Rake

The longer chip hook pulls long, stringy chips out of the machine, while a rake removes broken chips. Both tools are made with handles that can't grab your hand should the business end accidentally get caught in the machine. The handle should not turn back into a closed loop. Figure 1-24 shows the right shape. Notice, too, that the hook end is not a full 90° either. The open hook tends not to catch on the machine and also to release the chips easily.

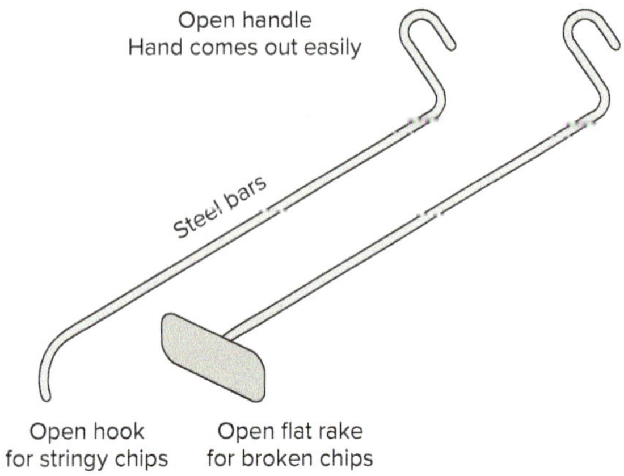

Figure 1-24 Chip hooks and chip rakes feature open handles designed to not get caught in moving machinery.

Compressed Air

Nearly every shop uses compressed air to power hand tools, to actuate machine actions such as pneumatic clamps, and also for cleanup work. Air is a safe and efficient tool used for chip removal in modern machining. But it can be misused against both people and machines. Use these two pro guidelines to stay out of trouble (Fig. 1-25).

Figure 1-25 Compressed air is a good chip removal and cleanup tool when correctly used.
Lake Washington Technical College, Kirkland, WA/McGraw Hill

Guide 1. Personal Danger—Never blow off chips from your skin or clothing or point the air at fellow workers!

You can embed chips in skin or even inject air under the skin. It is possible to introduce air into veins and arteries, where serious consequences can occur. It should be obvious: Never point the airstream, with or without ejected chips, at someone or at other machines. At the *very least*, damage to computers, delicate measuring tools, and your reputation can occur. When using air on a manually operated machine it's best to place a guard wall behind or beside it to protect others.

Guide 2. Machine Safety—Never point compressed air at seals on the machine.

All machine tools are equipped with seals designed to keep dirt and metal chips away from bearings and precision flat sliding surfaces. However, these tough seals are not able to stop pressurized air. Air can force chips under the seals, defeating their ability to stop further invasion.

KEYPOINT

A carelessly air-injected chip in a machine seal lifts the seal, thus defeating its usefulness. The embedded chips can also scratch precision surfaces.

Used correctly, compressed air works well to keep chips out of the immediate machining area, especially when the machine is moving. An airstream enables chip ejection without holding any solid object between you and the machine. It can't get caught and pull you into the machine. However, air cleaning and chip ejection are controversial. Many shops have an air policy.

KEYPOINT

Ask First

Although it's used commonly, be aware that due to the potential for machine damage some shops have a strict rule against using air as a chip cleanup tool.

TRADE TIP

Coolant Air Substitute Where air is banned for chip control, try a directed coolant stream with less velocity but more mass. It can do the job every bit as well with no danger to seals. Plus the hand-aimed stream can improve tool life since the operator can concentrate coolant right when and where it's needed during heavy machine cuts. This tip works especially well on CNC machines with strong coolant systems and full containment guards to catch the splashing. A tee-fitting put in the coolant delivery hose is connected to any flexible hose and garden nozzle. This little accessory works wonders—try it!

1.4.2 Maintaining the Work Area

As boring as a list might be, it can't be avoided here if you are to know what the other machinists will expect of you when you enter the craft. Here are a few unwritten rules on how to keep your shop lean and mean:

- As much as possible, keep measuring tools away from the machine and in their case.
- When laying measuring tools or wrenches down, never lay them on sliding machine parts such as lathe ways. Do put a shop towel under them on your workbench.
- Keep clutter such as rags, metal scraps, and especially old paperwork picked up. Keep personal items to a minimum, too.
- Keep chips swept and picked up.
- If you see something that needs maintaining, fix it or report it, but don't ignore it.

Checking Machine Fluid Levels

Oils are consumed and coolant levels fall due to splashing, evaporation, and small amounts leaving on the work itself, called *clinging*. Nearly all newer CNC machines feature sensors and low-fluid alarms, but not all are so equipped and no manually operated machines have this protection. To determine fluid levels in these machines, there might be a dipstick or a **sight glass**, such as the one in Fig. 1-26. This is simply a little glass window into the side of the reservoir. Add the lubricant or coolant to the indicated full line.

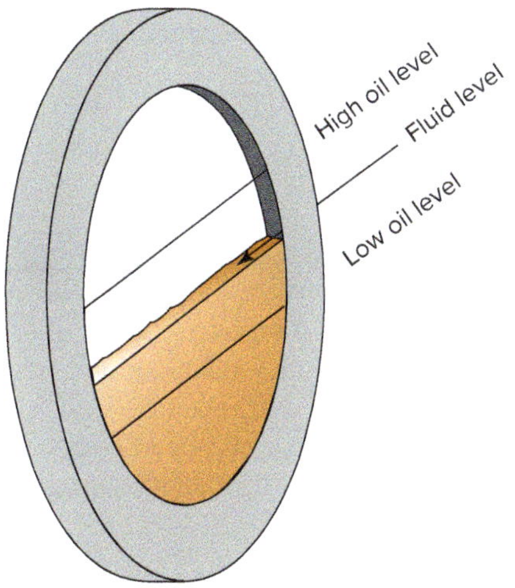

Figure 1-26 Sight glasses are small windows in the machine's oil sump that show the fluid level.

TRADE TIP

After wiping the oil and dirt from exposed, precision sliding surfaces, remember to replace with a wipe or spray of clean oil of the correct type. While these surfaces will be automatically lubricated by the next motion of the machine, however, the oil is absent unless you put it there for the first movement.

Lubrication Flow Glasses

When maintaining older machine tools, you must know the difference between a sight glass for checking fluid levels and a **flow glass** for coolants and lubrication. A *sight glass* has a line on the clear window; its purpose is to see critical fluid levels. It's like a dipstick, telling you how much oil or coolant to add.

No line on the glass or around the window's outside rim means it is a *flow glass*. It's a foolproof way of seeing that the fluid pump is working. A flow glass shows when oil is moving. When the pump is working, the oil flows down over this glass.

KEYPOINT

Caution, machines may have more than one sump with different oils in each.

On many smaller machines like those used in a typical machining course, there is probably a central, manual lubricating oil pump for sliding surfaces. Its handle must be lifted or pumped each time you step up to the machine and once each hour of operation. This is sometimes referred to as a **one-shot lubricator** (Fig. 1-27).

Figure 1-27 A one-shot central lubricator. Use at least once per half day or on the schedule set by your instructor.
Lake Washington Technical College, Kirkland, WA/McGraw Hill

Adjustments to Equipment

Naturally, there are many routine adjustments that must be made on machining equipment—to walk the walk, don't walk away from them. Make adjustments when they need to be done and on a schedule, too. Safety guards, for example, or eye shields, and doors that close for containment, chuck guards, limit switches, and belts—they're everywhere in a machine shop, and they are part of your job.

Tool rests on bench grinders (Fig. 1-28) are a perfect example. As the grinding wheel is used, the gap widens. To minimize the pinch point created between the grinding wheel and rest, keep adjusting it to a minimum of 1/16 in. or less. Zero clearance is acceptable because items cannot be pulled into the gap, but it could result in the tool rest being eventually worn away.

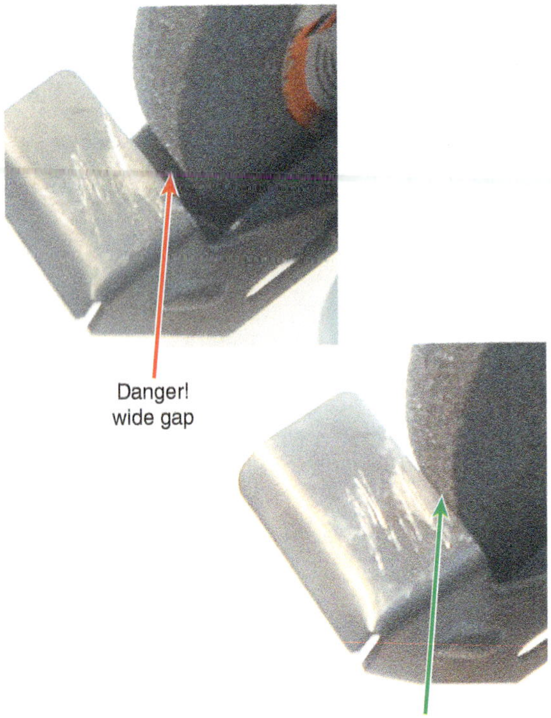

Figure 1-28 Due to grinding wheel wear, this tool rest gap has become dangerously wide and must be adjusted closer to the wheel to avoid pulling in small parts and fingers!
Courtesy Mike Fitzpatrick

1.4.3 Managing the Workplace—Lean and Green

Along with the machines and tools of the trade, the work area is also a tool that requires attention and proper care. A recognized trait of professional machinists is the pride they show in the way they organize and maintain their work area. But the subject involves more than pride–if managed correctly, tasks go faster and have more reliable results.

There are several names for the science of efficient work areas and environments, but the most common is **lean manufacturing**. The heart of lean manufacturing is finding the most efficient, logical way to organize and manage a working space. Using the lean concepts we count the steps and minutes required to complete a task, then do whatever it takes to reduce them. After observing the task, or a process and the people doing it, we ask what tools and supplies are needed and whether they can be placed closer to the task. Once provided, a system is created to keep them there consistently.

So you can see that lean focuses on common sense applied to shop procedures. Another aspect of lean lies in reducing unused inventory and clutter–old tools, leftover materials, that item "you might need sometime"–called the 5S process. While there's a lot more to the science, it starts with people having exactly what they need to do a job right where they need it and eliminating clutter. It also means returning tools to their place throughout the shift–not just at quitting time. Lean people constantly control the space in which they work. It's a habit.

A picture says it all (see Fig. 1-29). Your company may sponsor workshops where teams work together to get their area up to this level of efficiency. But there's nothing wrong with you becoming a committee of one! Start with your toolbox and the immediate area in which you work. Here are a few more ways to keep the workspace humming.

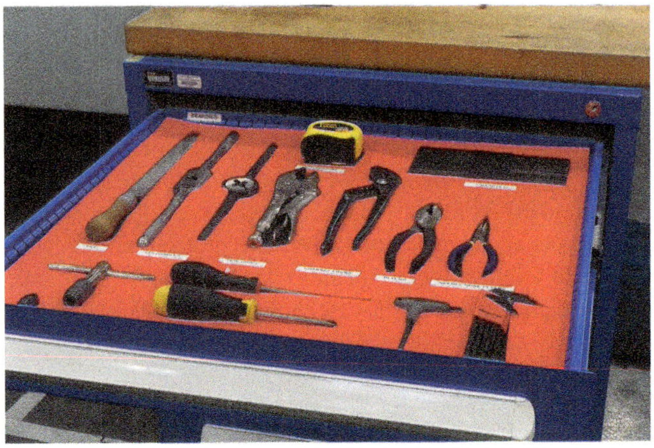

Figure 1-29 A lean shadow drawer.
Keith Smith

Safely Storing Machine Accessories

Machine accessories are precise, heavy, and expensive. Use lifting equipment to move them to avoid back injury and to minimize the chance of dropping them. Store them on shelves made for the load they represent and set them back from the edge (Fig. 1-30). If the stored item is a machine accessory or finished metal part prone to rusting and if it will not be used for some time, a light coat of protective oil is in order before storage.

The 5S Process to Become Lean

The obvious goal of becoming a lean machine shop (or company) is to be more efficient, thus profitable, by identifying and eliminating several categories of waste:

- People time
- Unused inventory
- Process time
- Inefficient processes
- Overproduction

Figure 1-30 Machine accessories must be stored and transported correctly. Make sure they are on shelves made for heavy storage.
Courtesy Tyee Aircraft, Everett, WA

But in a much greater sense, it's a way to change the essential culture of how you go about manufacturing. Shops wishing to become lean follow a set of steps called the *5S process:*

Sort (Inventory everything)

To be certain of upcoming decisions to surplus or not, a team surveys tools, parts, inventory, and fixtures; then determines which are useful and which must be eliminated. They then surplus the unneeded, unused, and clutter! This is not an easy step, but it is necessary!

Straighten (Organize)

Everything–tools, supplies, and instruction–gets a specific location such as a shadow board or labeled shelf. Figure 1-29 is as close to the point of use as possible. This phase is also about creating a culture of putting it back continuously–not just at the end of the day or week.

Sweep (Make it look lean) (see Fig. 1-30)

This is as much a visual kickoff to the newly improved shop as it is a lean-making tool in itself. Clean, paint, label where things go–make the shop shine! Completing the sweep, add to the lean culture that everyone keeps it that way! Taken as a whole, the sweep phase is a celebration of the new you.

Standardize (Make it permanent and ingrained)

This is where you solidify all the changes–create third-order change, not just paint and labels. Lean becomes the shop norm–planning and scheduling upkeep and preventive maintenance, for example, assigning responsibility, anything that makes your decision to be lean a permanent change with 100 percent buy-in from all workers.

Sustain (Set up schedules and audits–reevaluate)

Now lean is a way of doing your job. Schedule and post audits about how it's going. The objective is to be on and stay on track and to get even leaner as you learn. This is the long-term investment: Investigate and improve systems and culture–encourage each other, and celebrate success by recognizing improvements in quality, costs, and customer satisfaction!

ISO 14001 to Become Environmentally Responsible—Green

Reacting to several earth-related issues facing us all, the International Standards Organization (ISO) has developed standards that help organizations take a proactive approach to managing environmental issues. ISO 14000 sets the standard and 14001 guides implementation.

The ISO 14000 family of environmental management standards can be implemented in any type of organization in either public or private sectors–from companies to administrations to public utilities. ISO is helping to meet the challenge of climate change with standards for

- Greenhouse gas accounting
- Verification and emissions trading
- Carbon footprint measurement of products

ISO develops informative documents to facilitate the fusion of business and environmental goals by encouraging the inclusion of environmental aspects in product design. ISO offers a wide-ranging portfolio of standards for sampling and testing methods to deal with specific environmental challenges (see Fig. 1-31). It has developed some 570 international standards for monitoring:

- Quality of air, water
- Soil, as well as noise, radiation
- Control of the transport of dangerous goods
- The technical basis for environmental regulations

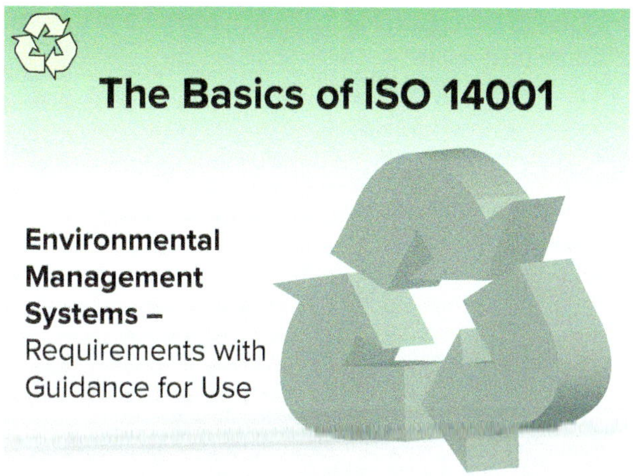

Figure 1-31 The basics of ISO.
Source: International Organization for Standardization, ISO Central Secretariat.

ISO Environmental Standards 14000 and 14001 for business, government, and society as a whole make a positive contribution to the world we live in–and are accepted in 160 countries. They ensure vital features such as quality, ecology, safety, economy, reliability, compatibility, interoperability, conformity, efficiency, and effectiveness. They facilitate trade, spread knowledge, and share technological advances and good management practice.

ISO develops only those standards that are required by the market. This work is carried out by experts on loan from the industrial, technical, and business sectors who have asked for the standards and have subsequently put them to use. These experts may be joined by others with relevant knowledge, such as representatives of government agencies, testing laboratories, consumer associations, and academia, and by nongovernmental or other stakeholder organizations that have a specific interest in the issues addressed in the standards.

1.4.4 Fire Prevention and Safety

You will be given specific training in your school lab and on the job about fire prevention. Here are a few rules for safely controlling a shop fire:

1. Know where the exits are located in your shop.
2. Never attempt to fight a fire without knowing how you can escape. Fight a fire with your exit clear.
3. Know which kind of extinguisher fights which kind of fire:

    A = Common burning objects–wood, paper–anything that leaves an **A**sh.

    B = Volatile liquids–lubricants, gasoline–anything that goes **B**ooom!

    C = Electrical fires–anything with an energized **C**ircuit.

    Remember: Never attempt to fight an electrical fire with an extinguisher not designed for the purpose. Turn the power off before attempting to fight an electrical fire.

    Most shops employ A-B-C extinguishers, but read the label to be certain (Fig. 1-32)–you can get it wrong with unwanted results!
4. Be certain you CAN extinguish the fire–schools may have specific rules on this.
5. Above all else–be certain that all persons have been informed and evacuated–your first concern before extinguishing the fire!
6. Be certain that someone calls the EMS fire department immediately–don't wait until you KNOW you can't control the fire. This is a critical step–do it simultaneously with the effort to control the fire. (Author: "I speak from experience; the fire department doesn't mind being told that the emergency is under control–while racing to the fire, they like to hear a radio call to 'lower your code to an inspection.' But they do mind finding they could have controlled the fire if called in time!")
7. Review the PASS process for fighting a fire:

    Pull the pin.

    Aim at the base source of the flame.

    Squeeze the trigger.

    Sweep over the whole source to exclude oxygen from reaching any area.

Class	Fire type	Examples
A	Common Combustibles	Wood, paper, cloth, etc.
B	Flammable liquids and gasses	Gasoline, propane, and solvents
C	Live electrical equipment	Computers, fax machines (see note!)
D	Combustible metals	Magnesium, lithium, titanium
K	Cooking media	Cooking oils and fats

Figure 1-32 Notice the flammable metals category—titanium and especially magnesium and the new symbols.

UNIT 1-4 REVIEW

Replay the Key Points

- Depending on whether the machine is working or stopped, safe chip removal actions are very different. Identify or describe them.
- Air is a safe chip removal and cleanup tool, but some shops ban its use in this capacity.
- Lubrication is technical and often the machine operator's responsibility. Using the wrong oil can damage technical equipment almost as soon as using none at all!
- Savvy machinists keep their eyes open for conditions that need attention, and they fix problems themselves.

Respond

1. Why is air not always permitted as a cleanup tool?
2. Gloves are acceptable when removing long, stringy chips. Is this statement true or false? If false, what makes it true?
3. Why are flow glasses not found on modern CNC machines?

4. Name the four safe methods of removing chips from moving machinery.
5. Complete this sentence: Shop-made chip hooks and rakes are made such that.
6. List and describe the 5S process.
7. What is the ISO standard that aids shops in becoming environmentally responsible?

CHAPTER 1 REVIEW

Unit 1-1

No part of Chapter 1 has been filler material. Demonstrating a professional attitude in the way you dress for the shop environment is one of the best tools you can develop to show your instructor and employer early on that you are on your way to becoming a top machinist.

Unit 1-2

Back injuries are a serious concern for any machinist. Once you damage your back, it can become a lifelong disability. Pay close attention to the guidelines in this unit for handling and moving shop materials safely.

Unit 1-3

I once took my CNC class on a tour of a local shop. As we walked in I saw an apprentice I had previously trained, putting away a 5-gallon oilcan. Not realizing we were watching, the apprentice carefully wiped the top, then rotated it so the label could be read. Only then did the apprentice close the door to the fireproof cabinet. I was truly proud but also noticed her supervisor nod toward me with approval!

It's details like that–taking care of shop supplies, knowing their value, and using them responsibly along with having a good respect for machine accessories and tools–that give the beginner the walk of a skilled worker. They will be noticed or, more important, when they are ignored, they get noticed even faster.

Unit 1-4

Maintaining a well-organized work environment is an important step in keeping the workplace safe. It is much easier to spot potential problems in a clean and organized space than a dirty and cluttered one.

Keeping your machines in good repair with all safety guards in place is an important way to create a safe working environment. Equipment that does not work properly can lead to disastrous accidents.

QUESTIONS AND PROBLEMS

1. Describe a well-dressed machinist. Use at least five dos and two don'ts. (Unit 1-1)
2. True or false? It's okay to listen to music with a personal player as long as you leave one earphone out to be able to hear your machine. If it is false, why? (Unit 1-1)
3. List the best to worst ways to lift heavy objects. (Unit 1-2)
4. Name two (of three) professional precautions for storing metal bars. (Unit 1-2)
5. Explain why we use the bent knees lifting technique in 10 words or less. (Unit 1-2)
6. What would you expect to find on an SDS sheet? (Unit 1-3)
7. Lubricants fall into two general groups, to prevent and friction. (Unit 1-3)
8. True or false? It's okay to substitute lubricants as long as the new one has a higher viscosity than the required oil so it prevents friction better than the original. If it is false, why? (Unit 1-3)
9. Why do we not wear gloves around moving machinery? (Unit 1-4)
10. True or false? Pressurized air is permitted as a cleanup and chip removal tool in all modern shops. If it is false, why? (Unit 1-4)

CRITICAL THINKING

11. A bar of spring steel has been misplaced, but you have found another that you are sure is the same material. Can it be substituted for the job? Is there a way to maintain the traceability? (Unit 1-2)

12. True or false? Dark glasses impair vision, and they are never worn in the machine shop. If it is false, why? (Unit 1-1)

CNC QUESTIONS

13. Your CNC mill has stopped after completing a part cycle and a red light flashes while your operator's panel tells you that the way lubricant is low. Name three or more professional steps needed to get the machine up and running again. (Unit 1-2)

14. Why do you suppose the machine completed the cycle in Question 13 before it shut down due to low supply of a vital fluid? (Unit 1-2)

15. Why must a specific lubricant be used in most modern CNC bearings? (Unit 1-3)

CHAPTER 2
Math Skills Self-Review

Units

2-1 Understanding Precision

- Say decimal inches using the language of the shop
- Use measurements at one thousandth of an inch or smaller
- Convert between metric and imperial units (review)
- Appraise your needs for further review

2-2 Using Feature Tolerances

- Do a quick review

INTRODUCTION

Machining means math—lots of it. It doesn't matter whether you're a top toolmaker, a programmer, or a beginner, nearly all your actions in the shop will be based on numbers, and most must be calculated in some way. There's no escaping it. Every step from the design to final inspection requires math.

Today, calculators are as common in the machinist's toolbox as micrometers. This need for number skills is only going to increase as we commit further to technology. Are you ready, or would a little review be in order? Chapter 2 helps you answer that question and offers some hints about how to approach shop math. In addition, Chapter 2 is provided as a self-appraisal warmup. If after trying the problems your score is below expectation, then a refresher math course or some self-study is indicated. Here are the goals of Chapter 2 to get those gray cells humming.

Unit 2-1 Understanding Precision

Introduction: What does precision mean? The answer varies depending on the science or profession in which the word is used. In machining, it means cutting and measuring part features to within a few thousandths of an inch or decimal parts of a millimeter. Miss the target by too much or too little and the work is scrap! There is always a target and a limit as to how much variation is allowed–the allowable variation is called **tolerance** (see Unit 2-4).

Working constantly with small decimal numbers, machinists have developed their own way of pronouncing them. Besides sounding in-the-know, learning this trade lingo helps eliminate misunderstandings and, more importantly for the student, it greatly speeds up mastery of measuring tools, print reading, and machine operations.

TERMS TOOLBOX

Imperial The units based on Imperial England–the ruler that includes feet and inches.

SI–International System of Units The system of measurements based on metric values.

Tenth (of a thousandth) 0.0001 in.–to a machinist one tenth of our basic spoken unit, the thousandth of an inch; thus, it's a tenth in shop lingo.

2.1.1 "Talking Precision"—Thousandths of an Inch

When working in inches we machinists refer to the inch as our working unit, but we speak as though it is the thousandth of an inch, and that's different from the regular nonprecision world. How do you pronounce *0.01 in.*?

KEYPOINT

In the shop, we pronounce inch decimals as though they were all extended to the third column, and 0.01 becomes *0.010 in.*

If you said either "one hundredth inch" or "point zero one inch," you were correct for general math, or when using metric tools and dimensions "point zero one" is correct. But when machining to inch dimensions, you will say "ten thousandths of an inch"! Or it might be shortened to ten thousandths. Or even "ten thou" in slang.

TRADE TIP

When a decimal inch number isn't extended to the third column, add the missing zeros to make it into thousandths of an inch.

TRY IT

A. 0.13	C. 1.31	E. 0.25
B. 0.013	D. 0.2	F. 0.303

ANSWERS

A. One hundred thirty thousandths of an inch—or one hundred thirty thousandths
B. Thirteen thousandths
C. One inch, three hundred ten thousandths
D. Two hundred thousandths
E. Two hundred fifty thousandths
F. Three hundred three thousandths

How Small Is a Thousandth of an Inch?

It's difficult to visualize 1 inch divided into one thousand parts (0.001 in.); however, here are some everyday examples:

A human hair is from 0.002 to 0.005 in. thick.

The paper this book is printed on will be from 0.003 to 0.004 in. thick.

A dime is about 0.04 in. (forty thousandths), and a ball-point pen is around 0.310 in. in diameter—three hundred ten thousandths of an inch or shop-shortened to three hundred ten thousandths.

TRY IT

Pronounce the horizontal and vertical dimension on the part print in Fig. 2-1 and the hole diameter.

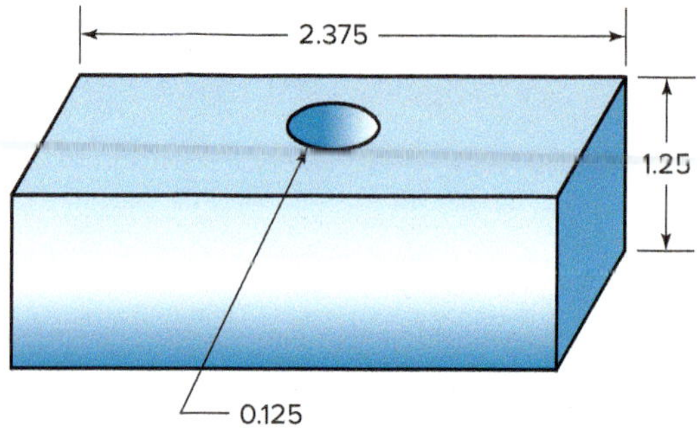

Figure 2-1 Say these dimensions to the thousandth of an inch.

ANSWERS

Two inches, three hundred seventy-five thousandths

One inch, two hundred fifty thousandths

One hundred twenty-five thousandths

2.1.2 Smaller Yet—Tenths of a Thousandth

Thousandths of an inch aren't precise enough for some designs, so we further divide them into 10 smaller parts. As a machinist, you will pronounce these as **"tenths."** Again, quite different from the outside world, not tenths of an inch (0.1) but tenths of our spoken base unit, the thousandth.

For example, *0.0001 in.* is one tenth of a thousandth.

A human hair might measure 0.0037 in. thick, larger than three thousandths by seven tenths. That's pronounced three and seven tenth thousandths or shop-shortened to three and seven tenths. As odd as that sounds to an outsider, any person in the trade would understand exactly. Say to an inspector or engineer, "The diameter is five tenths too small," and he or she would know that 0.0005 in. more must be taken out of the hole to be the right size (Fig. 2-2).

TRY IT

0.1259

ANSWER

One hundred twenty-five thousandths *and nine tenths*

Figure 2-2 This machining inspector needs math skills to read the drawing and then program this computer coordinate-measuring machine (CMM).
Courtesy Contour Aerospace/Everett, WA

TRADE TIP

First pronounce the thousandths; then add on the tenths.

TRY IT

Practice pronouncing these with another student.

A. 0.0067
B. 1.5678
C. 0.9999
D. 0.0878
E. 0.0087

ANSWERS

A. Six thousandths and seven tenths or six and seven tenths thousandths (Either expression would be okay even though it isn't correct outside the machining field.)

B. One inch, five hundred sixty-seven and eight tenths thousandths

C. Nine hundred ninety-nine and nine tenths thousandths, or nine hundred ninety-nine and nine tenths

D. Eighty-seven thousandths and eight tenths

E. Eight and seven tenths thousandths

2.1.3 Getting the Most Accuracy Using Your Calculator

When the calculations are to be made into solid metal, there simply can be no answering less than 100 percent correct. In the real world, anything less results in scrap and lost profit! So before moving on to the review problems, here are a few suggestions on how to get the best results when using a calculator in applied math.

Rounding

Never round any number until the final result.

Then, *always round to the nearest tenth (0.0001 inch).*

Use the whole decimal number during calculations.

Keep all numbers and results on the calculator using the memory keys.

Round up if the trailing digits are 5 or greater. Round down if they are less than 5. Example: rounding to the nearest tenth (of a thousandth of an inch):

1.3466*52 in.* = 1.3467 (Rounded up)

1.3466*48 in.* = 1.3466 (Rounded down)

Take Advantage of "The Mystery Digits"

The display isn't the entire amount the calculator is using. In some calculations where decimals extend very far from the point, the processor will be using a number that extends two or three places beyond the display, depending on the calculator sophistication.

If the screen number is written on paper to be used later, the unseen decimals are lost. However, by using the calculator's memory to record intermediate results, the entire number is stored, including the unseen digits. Usually we do not need this exceptional accuracy, but it can make a difference when doing several multiplications of the number in a series. While the difference might be small, keep in mind we deal in small. Missing this hint might result in a loss of one or two tenths–just enough to cause a problem.

Calculator Rounding

Your calculator may have a decimal fix function (rounding). Use it to round results to the job tolerance of thousandths or tenths. But doesn't that diminish the accuracy? *No.* Unlike pencil and paper rounding, nothing is lost in the calculator. Only the display is rounded. The internal number continues to the full capacity of the processor.

Little Hints

Use Memory to Record Intermediate Results By all means, write them on paper, too, but *do not depend on written numbers;* they create a chance to introduce errors.

Use Fresh Data When given the choice between two numbers to solve a problem, always choose the number that is closest to the original information on the print. Do not use calculated results if they can be avoided. They may be rounded, transposed, or wrong. Next is a hint to avoid that.

Talk to Yourself During calculator entry avoid accidental transposition and wrong key errors by saying them aloud slowly as you touch the keys. While doing so, look at the display. No kidding, this works when writing numbers on paper, too. It's okay to talk to yourself in shop math!

Simplifying Fractions To convert a fraction into a decimal number divide the numerator (top) by the denominator (bottom); that's the rule. But modern calculators often feature a direct fraction key. It may be an A/B on the keypad. There may be several fractional keys involving A/B and C as variables. These keys allow dual inputs in a single problem; both decimal and fractional numbers can be used at the same time. If an older print is dimensioned with fractions, using this key can speed up and simplify the math. Find it and teach yourself to use it. It will be very helpful in the upcoming problems.

KEYPOINT

Use your calculator for rounding numbers with no loss of accuracy, for simplifying fractions, and for storing intermediate numbers carried out to the capability of the calculator.

2.1.4 Conversion Review: Metric and Imperial Units

Occasionally, it becomes necessary to convert between **imperial** (foot-inch) and metric **(SI)** units, especially when making products for the world market. As we enter the world competition, expect more work to be dimensioned in SI units. Here's a brief refresher on conversion. Most calculators feature direct conversion functions–read the manual or experiment with these examples:

Converting to Metric from Imperial Units

Multiply	by	To obtain
inches	25.4	millimeters

Examples:

How many millimeters are in 2.125 in.?

2.125 × 25.4 = 53.975 mm

Convert 0.0935 in. to millimeters

0.0935 × 25.4 = 2.3749 mm

Converting from Metric to Imperial Units

Multiply	by	To obtain
millimeters	0.03937	inches in thousandths

Examples:

How many thousandths are in 23 mm?

23 × 0.03937 = 0.9055 in.

Convert 245 mm to inches

245 × 0.03937 = 9.6457 in.

TRY IT

A. On a separate sheet of paper or on keys, using Fig. 2-3, convert the imperial dimensions to millimeters.

B. From Fig. 2-4, convert the metric print to imperial units.

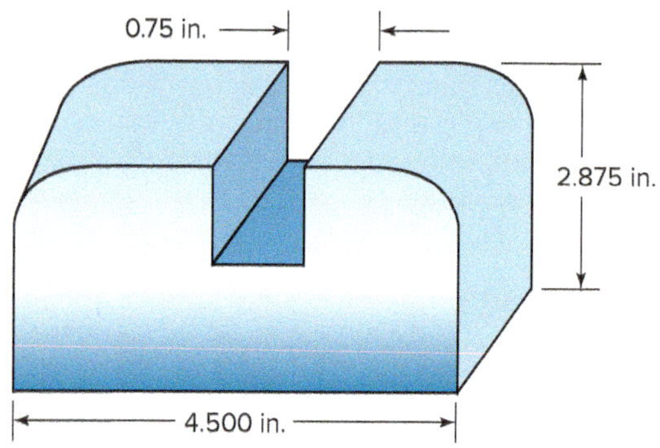

Figure 2-3 Convert these dimensions to metric units.

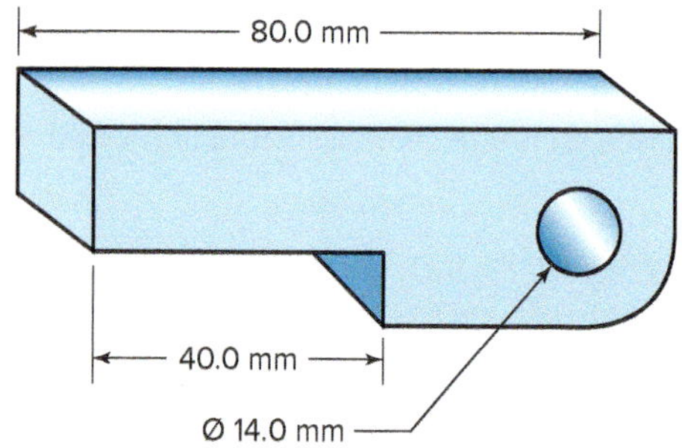

Figure 2-4 Convert this metric print to imperial (inch) units.

ANSWERS

A. Converting to metric units:

Width 4.500 in. = 114.3 mm
(4.5 × 25.4 = 114.3 mm)
Height 2.875 in. = 73.025 mm
Slot 0.75 in. = 19.05 mm

B. A metric block 80 × 40 × 14 mm is 3.1496 by 1.5748 by 0.5512 in.

UNIT 2-1 REVIEW

Replay the Key Points

- Machinists pronounce decimals as though the thousandth of an inch was their base.
- Using a calculator to its fullest requires study of its various functions.

Respond

Move on to Unit 2-2 to solve the first set of self-evaluation problems.

Unit 2-2 Using Feature Tolerances

Introduction: The essence of machining a feature on a part (e.g., drilling a hole or tiring a diameter) requires achieving a quality target. The hole must be the right size and in the right location on the part. It might also have a specification for roundness or straightness–those targets are called the **nominal dimensions** for the feature.

But we cannot realistically achieve perfection, so a range of acceptability is added to the nominal in the form of a **tolerance**–the amount of allowable variation. The best quality is right at the specified size (nominal size), but the feature is okay within the range.

Look at Fig. 2-5 to see examples as we discuss several kinds of tolerances.

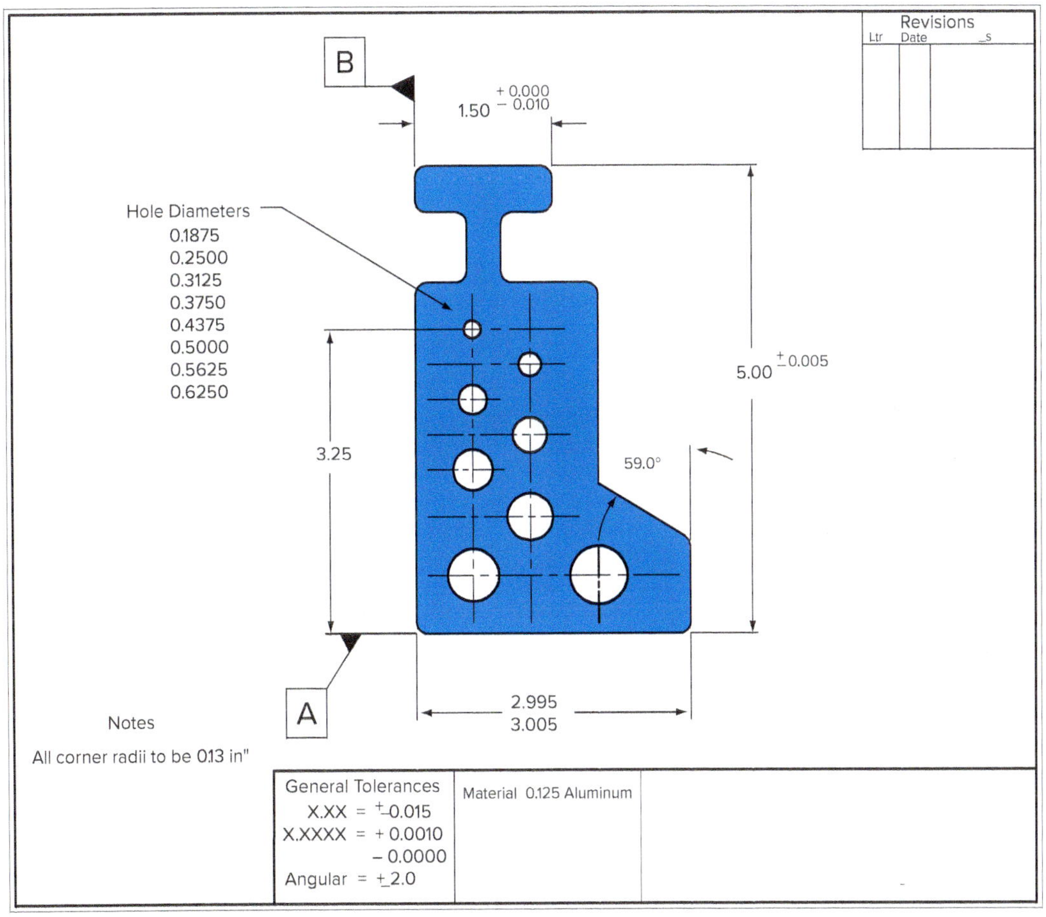

Figure 2-5 Different types of dimensioning and tolerancing.

TERMS TOOLBOX

Nominal dimensions The desired target value.

Tolerance The amount of acceptable variation from nominal.

Bilateral Tolerance that extends both plus and minus values from nominal only.

Unilateral Tolerance that extends one direction plus or minus from nominal.

Limits Tolerance expressed as a maximum and minimum value.

2.2.1 Bilateral Tolerances

Bilateral tolerances are easy to understand but not always easy to accomplish. That's how we earn our pay! For example, on the engineering drawing you read that a hole diameter specification is to be 0.6250 diameter; then, in the table on drawing, you see the tolerance for a four-place number is "±0.0010."

Question: So, what size range would be acceptable?

Answer: 1.624 in. on the low and 1.626 in. on the high.

Expressing it as a plus/minus range is called a **bilateral tolerance**, meaning its acceptable range extends both directions from nominal.

2.2.2 Unilateral Tolerance

Expressing the tolerance in one direction from nominal is a **unilateral tolerance**. Look at the 1.50-in. width dimension in Fig. 2-5.

$$01.50\frac{+.000}{-.010}$$

And now what is the range? Answer 1.500 to 1.490 in.

You cannot machine the width larger than 1.50 in., but ten thousandths smaller is okay.

2.2.3 Limits

Sometimes the designer will express the tolerance as **limits** of size; for example, the bottom width:

2.995 in.

2.305 in.

That means there is no specific quality target–any result within the limited range is okay.

Final point: No matter how it's expressed, your task will be to machine and measure the controlled feature well within the tolerance!

CHAPTER 2 REVIEW

Introduction

Now, with a review of the previous problems, see how much improvement has been achieved. Remember, ask for help with unfamiliar terms or symbols.

Instructions

A. This is a self-test–not a warmup.

B. Do not look at each answer–be certain you are correct on all problems.

C. Correct your own work and conclude readiness for lab experiences.

QUESTIONS AND PROBLEMS

Write these decimals in words. (Units 2-1 and 2-2)

1. 0.809 in.

2. 0.056 in.

3. 2.345 in.

4. 6.09 in.

5. 0.12 in.

6. 0.0089 in.

7. 0.0324 in.

8. 3.0506 in.

9. 0.5427 in.

10. 5.3387 in.

Convert imperial numbers to metric. (Unit 2-1)

11. 4 in. = _____mm

12. 2.5 in. = _____mm

13. 4.75 in. = _____mm

14. 20.0 in. = _____mm

Convert metric numbers to imperial inches–to the nearest tenth of a thousandth.

15. 25 mm = ______in.

16. 120.5 mm = ______in.

17. 358 mm = ______in.

18. 225 mm = ______in.

19. 4.75 mm = ______in.

20. 2.5 mm = ______in.

CRITICAL THINKING

21. A print calls for one rectangular part made from sheet steel. It is to be machined to $3\frac{3}{4}$ by $5\frac{3}{8}$ in. when finished. (See Fig. 2-6.) Operation 30 requests 0.150 in. extra machining metal added to every edge of the rough sawed material. The saw kerf consumes 0.060 in. for each cut. How much material will be consumed from the sheet for the one rectangle? (Units 2-1 and 2-2)

22. A customer asks for 50 tool hooks. This job is made from $\frac{3}{8}$-in. round steel rod. Each hook requires a length of $4\frac{3}{8}$-in. material plus 0.06 lathe, parting tool, kerf where it is cut away from the bar. How much material in inches must you bring to your lathe to make all 50 parts–rounded to the nearest inch? Answer in feet and inches, rounded to the nearest inch. (Unit 2-2)

23. The print requires a custom bored hole of 0.875-in. diameter. You need to select a drill bit that is $\frac{1}{32}$ in. smaller than the finished bore size, for machining excess (metal left to machine after drilling). What size drill will be used to predrill this operation? (Unit 2-2)

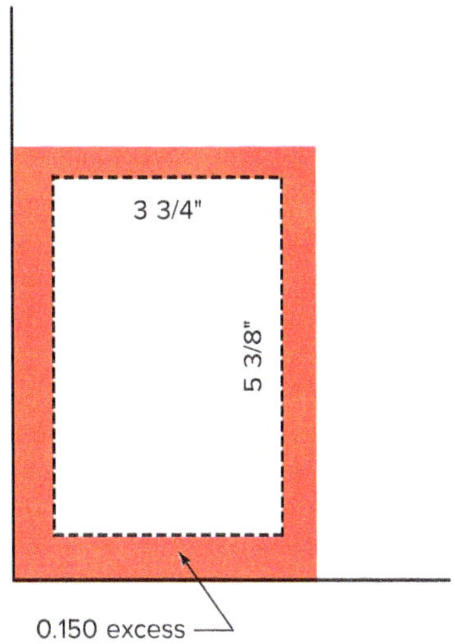

Figure 2-6 Problem 21.

24. Using the decimal chart in *Appendix I*, what is the next nearest whole-millimeter metric drill diameter that is *smaller* than the inch drill of Question 23?

CNC QUESTIONS

25. In writing a drill routine for a CNC mill, you need to complete the **S** command word to tell the controller to revolve the given spindle **S**peed. The numbers following the **S** prefix will tell the controller the RPM speed. Calculate the RPM; then fill in the command statement **S**_____.

For this hardened steel workpiece, the recommended surface speed is 80 ft per minute. The drill diameter is $\frac{7}{16}$ in. The formula to find the correct drill RPM for drilling a steel workpiece is

$$\text{RPM} = \frac{\text{surface speed} \times 12}{\pi(3.1416) \times \text{drill diameter}}$$

$$\text{RPM} = \frac{\text{surface speed} \times 12}{\pi \times \text{drill diameter}}$$

$$\pi = (3.1416)$$

(Units 2-1 and 2-2)

Review: Pi (π) is the ratio comparison of a circle's diameter compared to its circumference equal to 3.1415926 in. Any circle is just over three times bigger around than it is across.

CHAPTER 3
Reading Technical Drawings

Units

3-1 Orthographic Projection

- Identify the six orthographic views
- Visualize objects using third-angle projections
- Solve simple visualization problems typical of entry-level training
- Identify auxiliary, detail, and cross section views

3-2 The Alphabet of Lines

- Visualize object, hidden, and phantom lines on drawings
- Recognize leader, break, center, extension, and dimension lines on drawings
- Define the correct weight and form of a given line

3-3 Putting It All Together—Challenge Problems

- Select the correct objects given limited views
- Visualize three-dimensional objects given orthographic views

INTRODUCTION

Technical drawings and their companion documents, work orders (WOs), are the means by which machinists receive instructions. These orders tie the industry together, linking management, customers, engineers, planners, programmers, and quality assurance people. These two very different documents work together to create a foolproof way to deliver the goods. That is, they're foolproof if all involved understand how to read them!

Drawings are the master engineering document. In pictures, numbers, symbols, and words, they show exactly how the parts are to be shaped and/or assembled, but they don't tell the machinist how to make them. The work order serves that purpose. The main body of a work order is a carefully planned sequence of operations. The work order includes step-by-step instructions on how to follow the drawing and produce the part. Learning to use WOs is so important that we'll explore them twice in later chapters. Chapter 3 is about reading the drawings, to equip machining students with just enough baseline information to get started with lab assignments.

ONLINE ELECTRONIC SOLID MODEL DOCUMENTS OR PAPER

Paperless Trend

Today nearly all designs are created with computer software called CAD (computer-aided design) (Chapter 1, Chapter 24, Solid Modeling). That's the way most of the illustrations in this book were drawn.

Drawings printed on paper can be useful in the shop. They probably won't ever be completely eliminated. But more and more, machinists are receiving their drawings and work orders in digital form, at their CNC workstations.

Many CNC controls have evolved from single-purpose program managers to multitask computers. Machinists can access programs, drawings, and work orders right at their CNC stations. They can communicate with programmers and with other support people who bring cutters or lubricants, for example, or with other machinists in other locations. Using these stations, the machinist becomes a data manager and can input job tracking, information, quality assurance data, and program requests and can return suggested edits to the programmer and perform many other tasks—all from the CNC control. The shop lead person can instantly see the number of parts completed, the number that were not acceptable (hopefully, none), and generally do a much better job of managing the shop.

This paperless trend not only saves storing thousands of hard-copy drawings and the people required to manage them, but it also makes it possible to control and instantly distribute design updates to everyone working on the same job, worldwide. It also saves forests.

SHOPTALK

Technical drawings are also called several different names: prints, drawings, engineering drawings, and blueprints. But why "blue" when they are black and white? Read on.

Unit 3-1 Orthographic Projection

Introduction: In Unit 3-1, we'll look at the system used by the designer to lay out the various views you see on paper—**orthographic projection**. Projecting means showing something on a flat surface, a computer screen or paper, for example. The Latin root *ortho* means "90-degree relationships," and *graphic* refers to images. But an orthographic drawing is more than images; it's also about how they are arranged on the paper.

When assigned a job with a set of instructions, the machinist must be able to get to work as soon as possible. But it's critical that no metal is removed until the true nature of the part is fully understood. That means visualizing the part shown on the drawing.

TERMS TOOLBOX

Auxiliary view A view taken as though an extra sheet of glass were added to the box that shows details not seen from any other perspective.

Cutting plane (line) The thickest line on a drawing indicating where the part has been sliced open to reveal interior details.

Detail view A view that is magnified larger than the general drawing to show fine details not clearly seen otherwise.

First-angle projection The second-most popular method of projecting orthographic drawings–one that does not change the image but rearranges its location on the paper.

Fold lines Imaginary lines where a paper drawing could be folded to re-create a model of the glass box; fold lines simulate hinges between the glass surfaces.

Orthographic projection Views taken at various 90° perspectives.

Projection lines Lines that are not on the drawing but when added connect details in one view to the next; when drawn a projection line runs 90° to the fold lines.

Section view A view taken internally to show details not clearly seen from the outside of a part.

Third-angle projection The more common system of arranging the views in North America; the imaginary glass box in which the part is envisioned is the third of a possible four.

***Review the key terms in the Terms Toolbox Challenge!** Just scan the code in every Chapter Review, or go to www.mhhe.com/fitzpatrick4e.

3.1.1 Print-Reading Skills Needed

No matter in what form the drawing comes to you, on a computer screen or on paper, the skill of understanding and applying it to your work remains the same. Print reading divides into two categories:

1. *Visualization* is the ability to see the image and compile an idea of what the designer saw when creating the drawing. It is the ability to look at a flat, two-dimensional image on screen or paper and then create a three-dimensional (3-D) object in your mind. This skill is where we'll concentrate our efforts in Chapter 3.
2. *Interpretation* is the second and far more complex set of print-reading skills. It comprises what might be called the "legal" aspects of taking a design from drawing to reality. Interpreting a drawing requires an understanding of the rules, symbols, nomenclature, and procedures of manufacturing.

3.1.2 The Six Standard Orthographic Views

There are six possible standard views, but it's rare that more than three or four will be needed to convey the image for any one part (Fig. 3-1). They can be envisioned as projections of the part out to one of six possible sides of an imaginary glass box surrounding it: *top, front, bottom, rear, right,* and *left.*

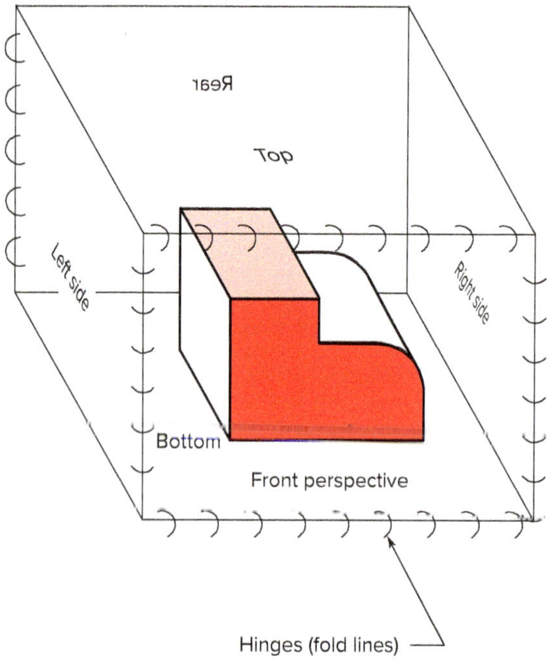

Figure 3-1 The *imaginary* glass box surrounding the object.

The imaginary glass box provides a model of how each view relates to the overall drawing. It can be envisioned as the surfaces upon which the views were projected, from the object out to the glass–then folded out to become a drawing. Although drawings aren't actually made in this way, Fig. 3-2 illustrates the best way to begin learning visualization. Each view on the paper is as though the box had hinges and each surface was unfolded.

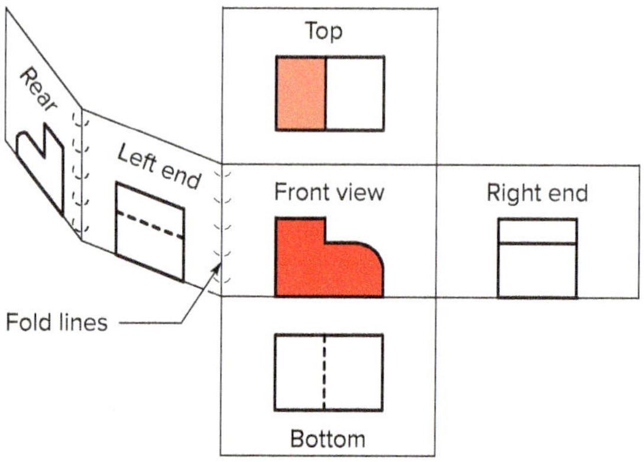

Figure 3-2 Each drawing view is rendered as though the viewer is directly facing the various surfaces of the glass box.

All views that show the height of the object might be called *elevation* views: front elevation, right-side elevation, back elevation, and so on. But they are usually just called front, right-side, rear, left-side (or left/right-end) views. The term *elevation* is more correctly used in architectural (building) drawings.

Third-Angle Projection

While there are other systems of projecting objects to flat surfaces, the most common in North America is **third-angle projection**. In this system, print readers envision each view as though they were standing outside the box and the view they see (top, front, and so on) is projected to the glass surface *between the object and viewer*. In other words, each view is as though you were walking around the box and looking in from each side.

The symbol used to notify the reader that the drawing was produced with third-angle projection is a cone as seen from the front and side views as if it were in the box (Fig. 3-3). Where there is a chance that there might be more than one method used, such as for an international customer, you will see this symbol in the drawing title area to indicate that the drawing is a *third-angle* projection. With third-angle projection, the top front edge of the box is considered the prime edge (or hinge) out of which all other views unfold.

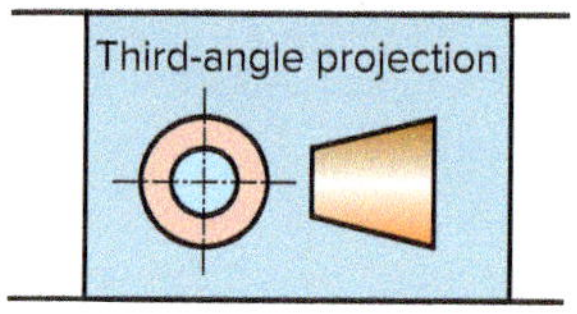

Figure 3-3 The symbol for a *third-angle* orthographic projection drawing is a cone from the top and the right side.

KEYPOINT

Third-angle projected images are created as though the projection surface (glass) is between the viewer and the object. These are the drawings you will use in this book and probably in your lab training, too.

First-Angle Projection

The other projection method used works equally well. Here the views are arranged differently as though the image were projected out of the object to the far glass surface–the surface away from the viewer on the other side of the box. It does not change the appearance of the six standard views, but it changes where they are placed on the drawing. Drawings produced with **first-angle projection** rearrange the image placement on the paper.

SHOPTALK

Why are drawings called *blueprints*? The answer is found back down the technical timeline, before drawings were files, when they were hand-drawn pencil copies on semitransparent paper called vellum (tracing paper). Copies of that precious master were made for the shop.

There were no photocopy machines then, either. Copies were made one at a time on chemically treated paper that turned dark blue with exposure to strong light. To make the duplicate, the vellum was laid over the treated paper, and a strong light was shown on the pair. The treated paper below turned blue if not shadowed by the pencil lines on the master above. The pencil shadows became white lines, but the exposed paper turned blue: truly a blue print. Those old drawings were hard to read if the design was very complex, plus they tended to fade over a long time. The process has passed, but the term lives beyond the technology. See Fig. 3-4.

Figure 3-4 A page out of history, a blueprint.

To see the difference in first- and third-angle drawings, we examine the four possible quadrants created by intersecting a pair of glass planes, as seen in Fig. 3-5. This is the universe of four boxes. Much of the industrialized world uses the third box; fewer use the first.

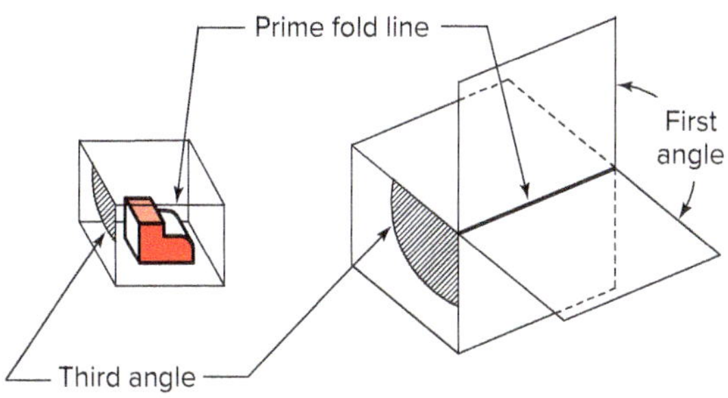

Figure 3-5 The third-angle projection box comes from the third quadrant of a possible four.

The Glass Box Universe

If two imaginary glass planes are intersected, they create four possible glass box areas. Their mutual intersection line becomes the prime edge or hinge to each. In third-angle drawings, the object is placed in the third possible box of the four. First-angle drawings are then taken from the first possible box. Due to common usage, from this point forward in this book we will study only third-angle drawings. We'll leave first-angle drawings for your formal print-reading course. Neither system is better. Once understood, they are used with equal ease.

> **KEYPOINT**
>
> The third-angle box uses the top front hinge as its prime or parent hinge. All other views radiate out from this front view and the prime fold line (studied next).

Fold and Projection Lines

To speed up visualization of the object, mentally place imaginary **fold lines** between each view. They represent where the hinges would have been between the glass surfaces if the box were real. As long as the views are correctly arranged on the paper, it is possible to cut, then fold the print along these lines to make a paper model of the box, complete with the projected views.

The designer used the concept of fold and **projection lines** when generating the image on paper or computer screen.

You can use them to visualize the object. For example, notice in Fig. 3-6, in both front and end views, the bottom of the object is touching the bottom of the box, and the top is the same height in adjoining elevation views. The bottom of the box is a projection line. Details are found directly across the fold line from each other. They project from view to view. Again in Fig. 3-6, to ensure that the step was the same height in both views, a temporary projection line was extended from the front view to the right side.

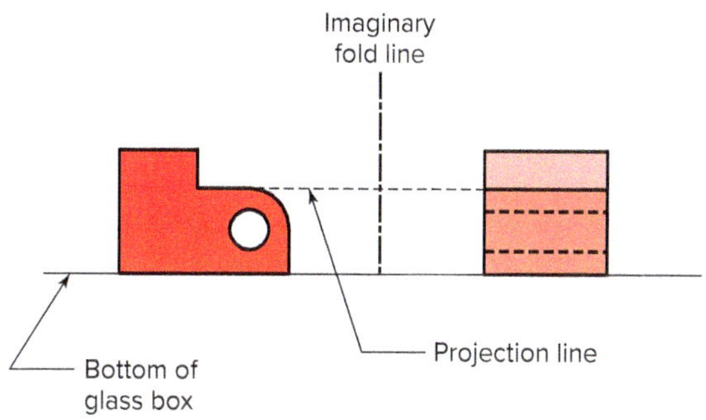

Figure 3-6 Details project to the same height above the bottom of the glass box in both views.

Ruler Check

To see how this works, check out the hole in Fig. 3-6 with a straightedge or ruler. Extend the dotted lines representing the hole, on the right, over to the front view.

> **TRADE TIP**
>
> **Ruler Projection** Use this projection line trick to test for the relationship between confusing lines or details from one view to the next.

3.1.3 Auxiliary, Section, and Detail Views

There are several other views that can be added to clarify details not shown well in any of the six standard perspectives. We will examine the three most common: **auxiliary views, section views, and detail views**.

Auxiliary Views

Sometimes, the six-sided glass box doesn't provide a view perspective that reveals the true size or shape of the object, or some detail on the object in the right perspective. For example, in Fig. 3-7, in every standard view the hole in the angular surface appears as an ellipse or it is hidden from view. To show this hole in true perspective, an extra (*auxiliary*) sheet of glass is added to provide a straight-on view of the detail–in other words, to look directly into the hole where it appears as a circle. Notice that the hinge/fold line is on the top of the box where the angular surface is shown as a straight line. This is the only view from which the new auxiliary view can be hinged out as a true surface.

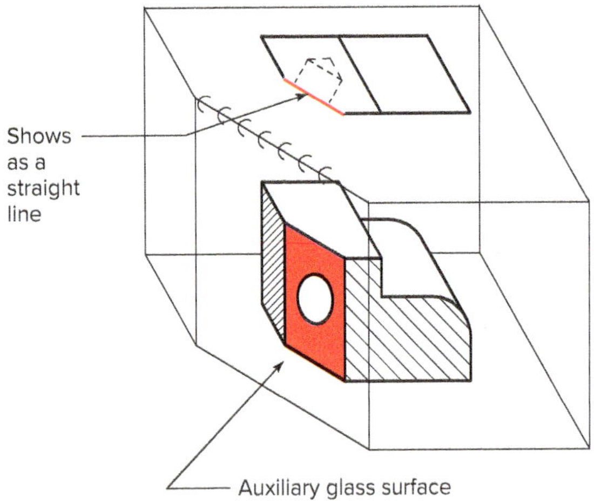

Figure 3-7 An extra (auxiliary) sheet of glass is added to show the hole in true size and shape.

Removed or Rotated for Convenience

Due to that auxiliary projection rule, we often find auxiliary views projecting or folding out to odd positions on a standard orthographic drawing, as seen in Fig. 3-8. To solve this problem, they are usually transferred to a better location on the drawing, and a note is *usually* provided next to the view to let the reader know this has been done–that the view is not found directly across the imaginary fold line but put elsewhere to make the drawing more compact. The view will be found in a more logical place on the paper but not in the place it would have been projected (Fig. 3-9).

KEYPOINT

When the auxiliary view is no longer in its correct ortho position, the projection line ruler trick will not work. Remember, auxiliary views are drawn to show some detail(s) of the object not shown clearly in the standard view perspective.

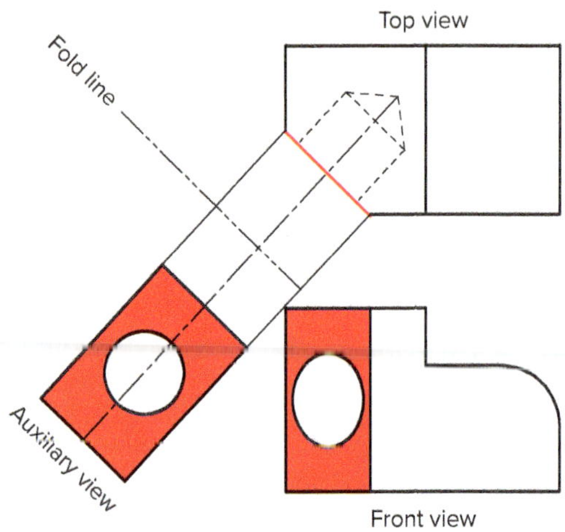

Figure 3-8 This auxiliary view is in its correct position, across the fold line.

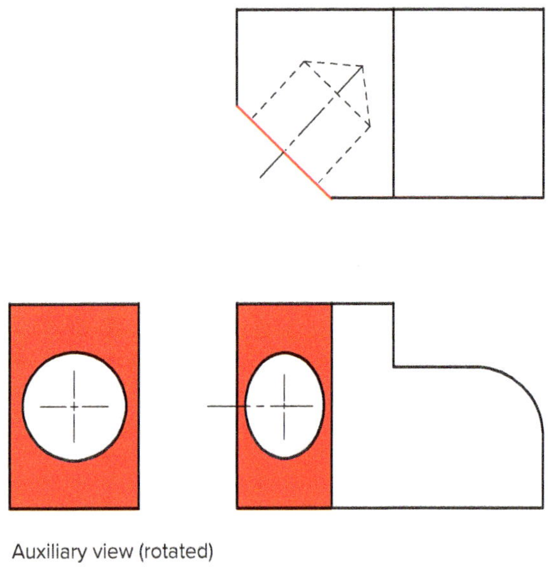

Figure 3-9 The auxiliary view has been moved to a better location on the paper but no longer projects back to the top view.

Section Views

Another secondary view arises when the designer needs to show some internal feature not shown clearly from the outside of the object. There are several types of section views, but they all share the concept that they are drawn as though material has been sliced away from the object and you are seeing the inside of the part. **The section view is depicted as it would appear if the material between the viewer and the remaining object had been removed.**

Here are two common section views, the *full section* and the *broken out section*. Each clarifies inside details of this machined-brass hose fitting (Fig. 3-10).

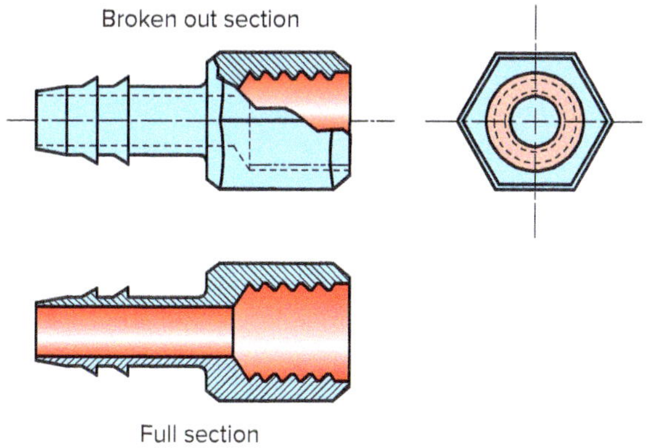

Figure 3-10 Two different kinds of section views show internal detail.

Two rules are used as guidelines for creating section views:

1. **Cutting Plane Rule**

   The theoretical surface along which the material has been sliced open is called the **cutting plane**. In Fig. 3-11, note that the extra heavy, dashed line with arrowheads at each end is the cutting plane line. It represents the edge of the slice. The section view created by the cutting plane will be seen as though the viewer were looking in the direction of the arrowheads. The material that will be removed in the section view is behind the arrowheads, between the viewer and the remaining object. If at all possible, the section view is placed in its correct orthographic position on the paper. But as with auxiliary views, it may be necessary to locate it elsewhere on the drawing (Fig. 3-11). The person making the drawing may or may not choose to draw the cutting plane line, as was the case in Fig. 3-10. Also, there may or may not be arrowheads, depending on the simplicity of the section view.

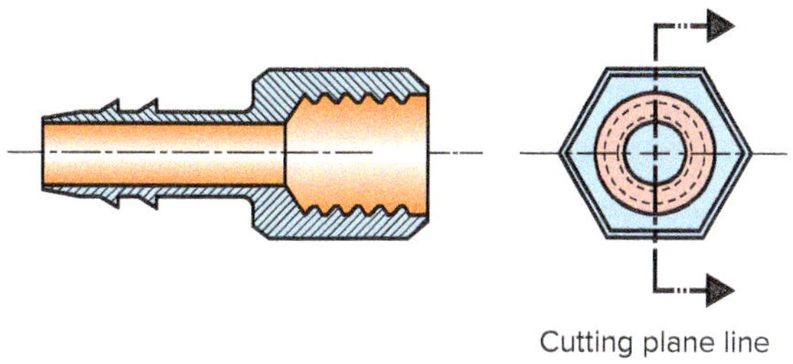

Figure 3-11 The cutting plane line shows where the slice has been taken. It's wider than all other lines on the drawing.

2. **Sectioning Rule**

   After the section view is drawn, light lines will probably be added to clarify where the internal cut has been made. Section lines are not required but are added only when the designer feels they clarify a view or detail. Sectioning lines are also called *crosshatch* lines, as seen in Figs. 3-10 and 3-11.

Detail Views

Detail views are magnifications of small features not seen clearly at the scale of the entire drawing. They might be labeled as to how much they have been scaled up (blown up), and a circle might be shown to indicate from where the view has been taken—again, these are options the person drawing may or may not use. For example, "3X" in Fig. 3-12 means the view is three times the normal drawing size.

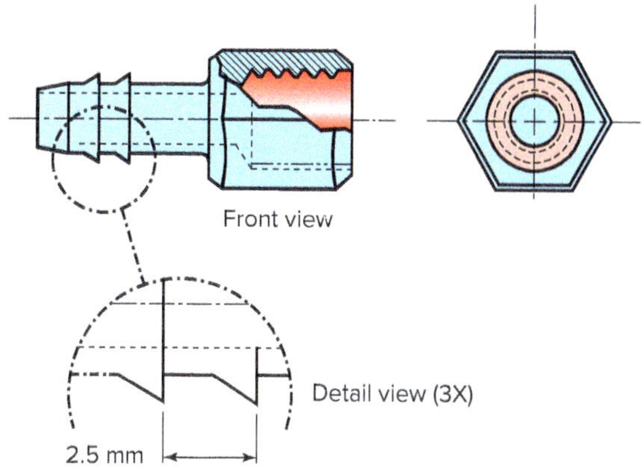

Figure 3-12 A typical detail view—three times the size of the actual drawing.

UNIT 3-1 REVIEW

Replay the Key Points

- There are six standard orthographic views: front, top, right and left end, bottom, and rear.
- All the standard views are pictures at 90° to each other.
- The *front* view should depict the object best. Study it first when starting a new job. It is the parent view out of which all other views originate.
- Auxiliary, detail, and section views clarify a detail not clearly shown in the standard views.
- Section views show internal detail obscured by material. These views show the details as though the intervening material has been removed.

Respond

1. In your own words describe an orthographic projection.
2. To interpret a drawing means to collect all the views together in your head, then create a 3-D image in your mind. Is this statement true or false? If it is false, what will make it true?
3. In North America, do we use first- or third-angle projections most commonly?
4. Although they aren't on real drawings, name two imaginary lines that help in understanding orthographic projections.

5. Match the letter to the number(s):
   - **A.** Section view
   - **B.** Auxiliary view
   - **C.** Detail view
   1. A view that expands some small feature that can't be seen clearly otherwise.
   2. A view that is projected to an imaginary glass surface that is not one of the six standard views.
   3. A view that shows some internal aspect of the object.
   4. A view that has some material removed from the object.
   5. A view that is at a larger scale than the main drawing (magnified).

SHOPTALK

Why Aren't Drawings Provided with a Pictorial View in Color? After all, the extra pictorial view would speed up the visualization process and end a lot of guesswork.

The first answer is that the objects used here were simple. Pictorial renderings become difficult to draw for complex shapes.

However, when using CAD programs such as our chosen example SolidWorks, drawing the illustration is as easy as drawing any view—one click of the mouse! When drawing in SolidWorks, color is a prime discrimination tool between different kinds of lines. Since color is already a part of all CAD drawings, online versions will be in color. It's only the reproduced paper copies that do not have color due to printing costs.

Unit 3-2 The Alphabet of Lines

Introduction: A second family of visualization clues helps build the 3-D image by making various lines trigger different images. We will look at seven basic types. We've seen the eighth and ninth as cutting plane and section lines in Unit 3-1. Once you've become familiar with these different lines, it will be much easier to see the image.

TERMS TOOLBOX

Break lines Depict material that has been removed from the drawing *but not from the real part* to simplify or compact the drawing only.

Dimension line A thin, continuous line that indicates size and shape; usually terminates in arrowheads.

Extension line A thin, continuous line that leads the eye off the object to provide a dimension location.

Hidden line A lightweight, dashed line that depicts surfaces and edges not directly seen in that perspective.

Leader line A thin, continuous line that conveys special information or directs a note to a specific location.

Object line A solid line of heavy weight that depicts surfaces and edges that can be seen in that perspective.

Phantom line A dashed line of light weight that shows an auxiliary position or a related part not actually on this drawing.

Symmetrical The same on both sides of a central axis.

3.2.1 Form and Weight of Lines

The clue as to what a line conveys is its form. For example, whether the line is composed of dashes or solid lines or combinations of both changes its meaning. Weight is the relative width compared to other lines on the drawing. Weights range from light section lines to the extra-heavy cutting plane lines (the widest line on any drawing).

1. **Object Lines**

   When visualizing, the foremost lines are **object lines**. These solid, heavy lines represent edges and surfaces that are visible. Thus, these lines are also called visible lines. Other than cutting plane lines, they are the heaviest line on the drawing.

2. **Hidden Lines**

   These dashed lines depict details unseen to the eye, as though an X-ray has revealed further information inside or on the opposite side of the object. **Hidden lines** are lightweight lines and are composed of equally spaced dashes. Hidden lines represent real surfaces, edges, and features obscured by other material. They are lighter in weight than object lines but thicker than section lines.

3. **Phantom Lines**

   Phantom lines are repeating double dashes and slightly longer lines. They are used in two different ways. They represent alternative positions of an object. For example, they might show the remaining part of the hinge in the full open position. Second, phantom lines depict an object that relates to the object being drawn but is actually not in the drawing. As an example, see the second part of the hinge in Fig. 3-13.

4. **Centerlines**

   Centerlines do not depict material; they show the central axis of an object or feature. They are medium-weight lines with a single dash and a longer line. Often, but not always, the object may be **symmetrical** (the same) on both sides of the line. If this is so, there may be a note or symbol stating this. There are several centerlines shown in the common usage in Figs. 3-13 and 3-14.

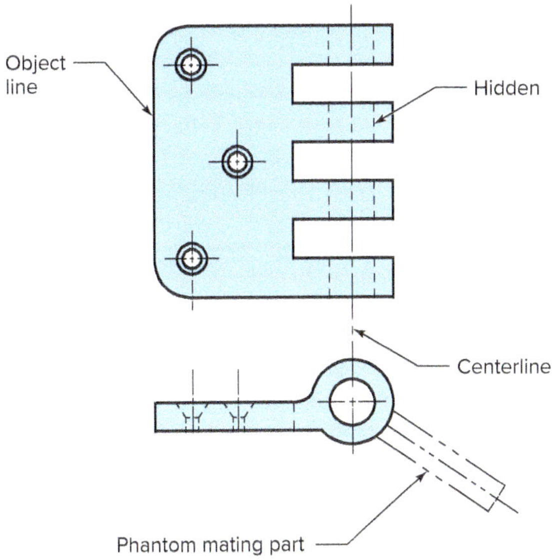

Figure 3-13 Phantom lines are depicting alternative positions or related parts that aren't on this drawing.

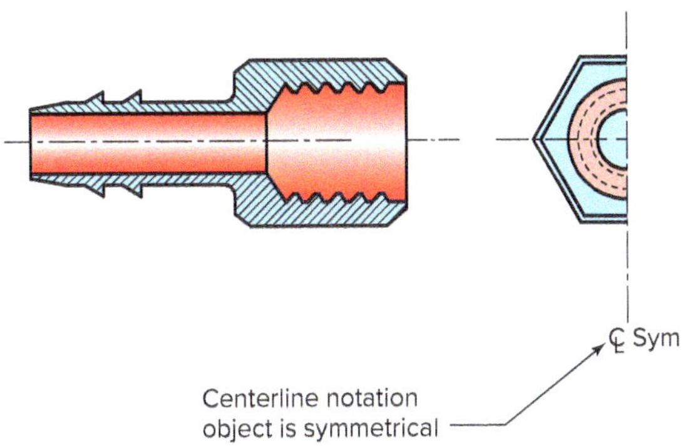

Figure 3-14 This fitting is the same on both sides of the centerline axis.

5. Dimension Lines

Dimension lines and the next two types of lines are of the same weight and form. They do not depict material but are used to clarify the meaning of dimensions and notes. They are lightweight indicators of a distance or size on the object. They almost always terminate in arrowheads. They may be broken by a dimension, but their form is continuous. They often are used in conjunction with extension lines, as seen in Fig. 3-15.

6. Extension Lines

Extension lines are detailing lines that extend a feature out to where a dimension line can be used to show the size without interfering with the view. Because they don't depict material, extension lines do not connect to the object by a small gap. The gap helps to avoid leading your eye off the object. They are the same weight and form as dimension lines, continuous and light

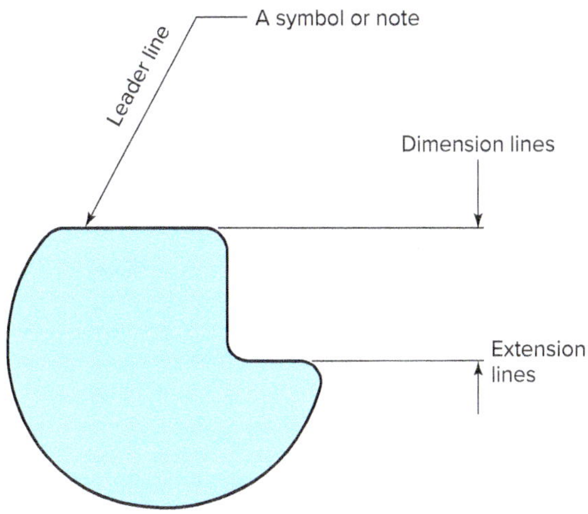

Figure 3-15 Leader, extension, and dimension lines all have the same form and weight.

7. Leader Lines

Leader lines are utility lines and are similar to both dimension and extension lines in that they are not part of the object but clarify details about the object. They point to a particular zone or detail with an arrowhead at the end. They point out where a symbol or note applies to the part. Or, when the note is large they may include a letter or number keyed to a note found elsewhere. Leader lines might be curved but more commonly they are slanted, straight lines. The curve or slant is to avoid confusion with object lines.

8. Break Lines

Break lines show where material has been removed to simplify the drawing. For example, a flagpole drawing needn't be 50 feet long if the pole is the same from top to bottom, or as seen in Fig. 3-16, this roller is the same throughout its length and can be broken to save space. There are several types of break lines seen on drawings, but they are all easy to recognize.

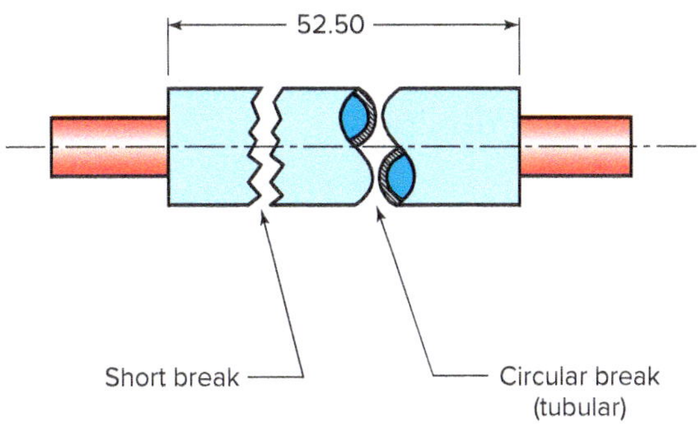

Figure 3-16 Two kinds of break lines show removed material to save space.

UNIT 3-2 REVIEW

Respond

1. Why does the designer place arrowheads on a cutting plane line?
2. True or false? A phantom line shows material that is hidden behind other features.
3. Why do designers use break lines?
4. A centerline depicts the ______ of a feature.
5. What is another name for object lines?

Unit 3-3 Putting It All Together—Challenge Problems

Introduction: Now, let's investigate combining the clues to begin the visualization process. After a few hints on how to assemble views in your head, we'll tackle a few puzzles. Remember, your goal is to gain enough understanding to use drawings in training.

TERMS TOOLBOX

Nicknaming The mental trade trick that helps put a visual handle on a drawn object.

View-to-view (V-T-V) The process of mentally assembling an orthographic drawing.

3.3.1 Quick Tips and Tricks

It's critical that you have a clear picture of the object before putting the cutter to the metal. Costly mistakes are made doing otherwise! But time is limited; the job needs to move forward. Here's a set of mental tools to see the image ASAP:

1. **Study the Front View First**

   It will depict the object best. But don't stay there too long; just get the rough outer shape at first, a rectangle, "L" shape, or oval, for example. Don't try to see every detail.

2. **Choose One Major/Obvious Front View Feature**

   This is a single feature that shows clearly in the front view: a hole, a thread, or a large surface, for example. Then relate it to other views using the ruler projection method and the alphabet of lines. This is called the **V-T-V (view-to-view)** process.

3. **Repeat the V-T-V Process**

   Continue using features you *do understand,* V-T-V, several times more, choosing finer details each time. This solidifies general shape and gets the image ready for the next phase.

4. **Now, Choose Details That You Don't See Clearly in the Front View**

   Start to investigate lines, and features V-T-V that you didn't see clearly. Use the projection line trick between adjacent views until they begin to clarify.

5. **Mentally Sum Up a More Detailed Overall Image of the Object**

   At this point, you should clearly see a refined definition of the overall shape. For example, a cylindrical shape with a tab on the top, a cubical shape with two big holes, an elbow with a slot through one end, and so on. If this does not work, start over with the V-T-V process.

6. **Nickname the Object If You Can**

   As soon as possible try **nicknaming** the object with a handle of some everyday thing. This does wonders to complete the imagery and it really helps in talking about the part with others in the shop. For example, I have worked on parts we all called hockey sticks, frying pans, pork chops, paper hats, and Tootsie Pops! Can you see them even without a print (Fig. 3-17)?

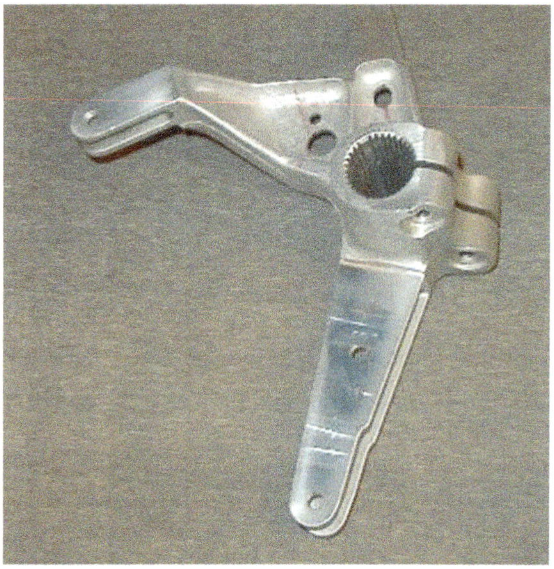

Figure 3-17 Can you guess what this part was nicknamed in our shop?
Courtesy Mike Fitzpatrick

   Using this communication tool, I might say, "We need to drill the hole in the handle of the fry pan." Instantly you know approximately where the hole is to be drilled and where to look on the drawing for the dimensions! It works!

7. **Recognize Common Drawing Symbols**

   Because they save time and have exact meaning, symbols are common on drawings. Go to Appendix V to begin your vocabulary.

3.3.2 Up and Running Production

There are times in industry when the product is complex but there isn't time to visualize it. It's time to cut metal. This is an undesirable situation we all must face at some point. There are three ways to help get over the queasies of making a part you truly haven't visualized.

Sample Parts

For parts made previously but scrapped, the shop may keep an old one around that cannot be passed on to the customer. Called a sample part template (SPT), it might be painted a bright color, usually red, and it may be deliberately cut or stamped with the word *scrap* to guarantee that this one is never returned to the production line.

Follow the Work Order and Ask Your Supervisor

During times when you must "shoot without seeing the target," rely on the step-by-step instructions in the work order. Ask others who may have made these parts previously when you are in doubt. Primarily, your shop lead's job is to be there to assist in such tough situations, too.

KEYPOINT

If you don't have a complete visualization of the work but need to get going on it, then don't guess; ask!

Digital Images

There may be a pictorial image of either the CAD-drawn solid model part or the CNC toolpath, or both, as shown in Fig. 3-18. The toolpath is a likely source of information since it is almost always checked by the programmer before releasing the new program to the shop. These files can put a turbo-boost to the visualization process. By the way, we called the parts shown in Fig. 3-17 boomerangs.

UNIT 3-3 REVIEW

Replay the Key Points

- Never guess at the shape to be machined: always ask shop leads.
- Use the view-to-view process to speed up the visualization process.
- Nicknaming parts is a healthy way to increase visualization and communication.

Respond

1. Why would you study the front view first when presented a new drawing?
2. How do the work order and drawings work together to produce a finished product? What are their purposes in the manufacturing process?
3. When performing the view-to-view process of visualizing a drawing, what three clue sets are used? (Hint: They are all lines.)
4. Other than visualizing the part using orthographic projection, name some other ways to "see" the part before machining it. In other words, what are some other sources of information as to how the three-dimensional part looks in reality?
5. Four kinds of special views can be added to the drawing to clarify details not shown well on the standard six orthographic views. What are they, and what information do they add to the visualization?

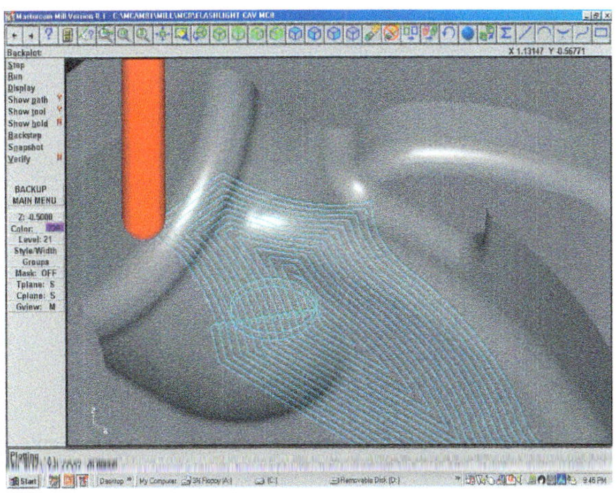

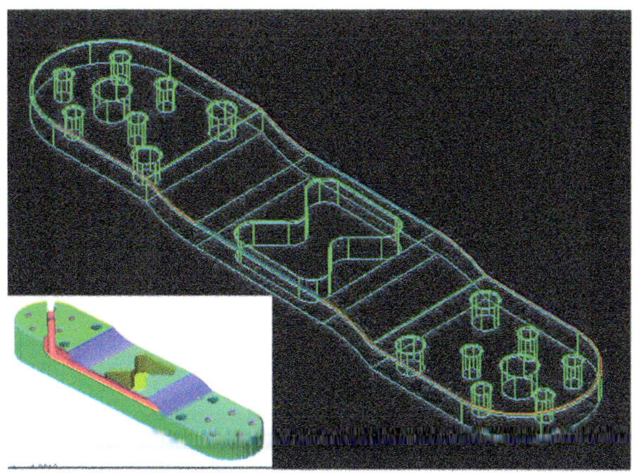

Figure 3-18 Screen images of the program toolpath can often help visualize objects.
Courtesy Mastercam (toolpath) and Metacut Utility Software (solid model)

CHAPTER 3 REVIEW

Unit 3-1

Being a machinist also means you'll be an information manager. Working hand in hand with work orders and programs, prints are *the* prime directive in our trade. They show us *what* but not *how*. While there are other ways to visualize a part shape using sketches, models, and high-tech screen images, projecting views outward to flat planes, from several sides of the object, is the undisputed best method not only to convey part shape but to record specifications and dimensions. The views fall into the following categories:

- Six standard views taken from mutual 90-degree perspectives
- Details blown up beyond the drawing scale to magnify small features
- Auxiliary views taken from nonorthographic perspectives of the part section views
- Internal views that reveal details not seen clearly from outside the object

Unit 3-2

Visualizing an object is a matter of adapting your mind to using a universal set of clues—or as Unit 3-3 puts it, putting it all together, to reassemble the flat views into a 3-D concept. One of the strongest sets of clues is the alphabet of lines. Using relative weight and form they show

- Visible lines and surfaces—directly seen from that perspective
- Hidden lines and surfaces—obscured by other parts of the object
- Phantom lines and surfaces—represented but not actually on this object
- Dimension and centerline details—to add clarity to specifications

Unit 3-3

This chapter was provided not only to help you understand the first few lab assignment sheets but also to demonstrate the need for further training in print reading. Time and again, industry studies show that frustrated trainees say that the greatest blockage to career success is a lack of print-reading skills. Be sure to take a formal course in this skill. But far beyond that upcoming course, the learning process for print reading continues as lifelong self-education. Thanks to rapidly changing technology, the long-term practicum will never end!

QUESTIONS AND PROBLEMS

1. Print-reading falls into two general categories. They are
   - **A.** Visualization and tolerancing
   - **B.** Interpretation and dimensioning
   - **C.** Visualization and interpretation
   - **D.** Projection and orthographic (Unit 3-1)
2. Name the orthographic view that should convey the most information about the general shape of the object. (Unit 3-1)
3. True or false? In North America we use the first-angle system, but a few countries use the third. If false, what makes it true? (Unit 3-1)
4. When rotating between views, what two imaginary lines, not found on the drawn object, help relate details? (Unit 3-1)
5. A centerline has what form and weight? (Unit 3-2)
6. List at least four lines used on orthographic drawings that are continuous (unbroken) and their relative weights. (Unit 3-2)
7. A view must be taken internally—with material removed to see it clearly. What kind of view is that? (Unit 3-1)
8. Referring to Question 7, what line is used to show exactly where the internal view is taken on the object? What form and weight are used for this line? (Units 3-1 and 3-2)
9. Referring to Question 8, what do arrowheads show when attached to the line that shows where the section view has been taken? (Two facts) (Unit 3-2)
10. On a separate piece of paper, sketch the section view as it would appear in the indicated space on the drawing in Fig. 3-19. (*Hint:* See Problem 11 for more information.) (Unit 3-3)
11. The arrowheads have been omitted from the cutting plane line in Fig. 3-19. If you were to add them, would they correctly point to the left of the page or to the right? Explain. (Units 3-2 and 3-3)
12. Given the front view in Fig. 3-20, identify which object or objects are depicted as pictorial views. Note that object A, the center hole, extends through the part. (Unit 3-3)
13. On a separate piece of paper, assuming Fig. 3-20 is object A, sketch the correct top view, and assuming Fig. 3-20 is object C, sketch the correct top view. (Unit 3-3)

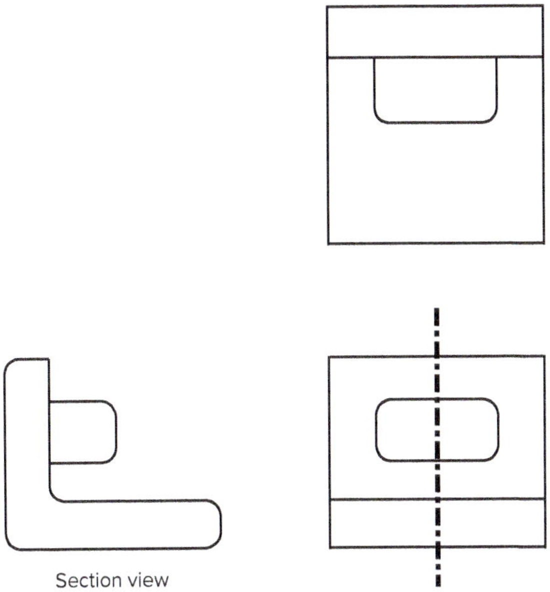

Figure 3-19 Sketch the section view on a piece of paper.

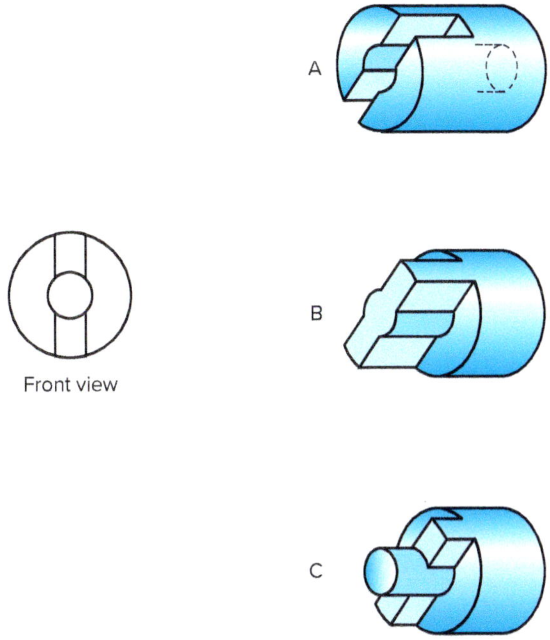

Figure 3-20 Which object or objects are shown with this front view, A, B, or C?

CRITICAL THINKING

14. Why must a rear view always be hinged to either the right- or left-side view? See Unit 3-1, Figs. 3-2 and 3-3. (Unit 3-1)

15. In your own words, describe why a designer might choose to use an auxiliary view. (Unit 3-2)

CNC QUESTIONS

16. A work order and print have been given to you to make a rush order of parts. The object is complex, and the supervisor tells you the job is "hot." "Load the program and make the setup; then start machining the parts right away," you are instructed. You are uncomfortable with your situation because there's no time to visualize the object. What can you do to get a visualization of the object ASAP?

17. **A.** Which object in Fig. 3-21 is depicted in Fig. 3-22?

B. Are the two orthographic views in Fig. 3-21 complete without the pictorial view? Explain. (Unit 3-3)

Figure 3-21 Using Fig. 3-22, identify this object.

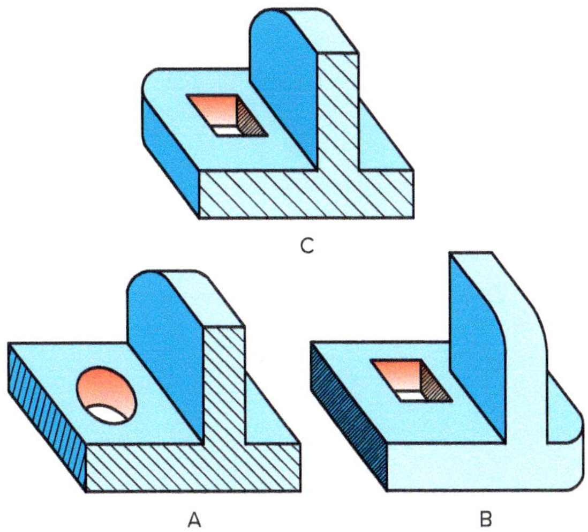

Figure 3-22 Which object is depicted in Fig. 3-21?

18. On a separate piece of paper, sketch three views of Fig. 3-23 in their correct orthographic relationship. No size is necessary; fit your drawing to your paper. Note the view that is to be the front perspective in your drawing is indicated by an arrow.

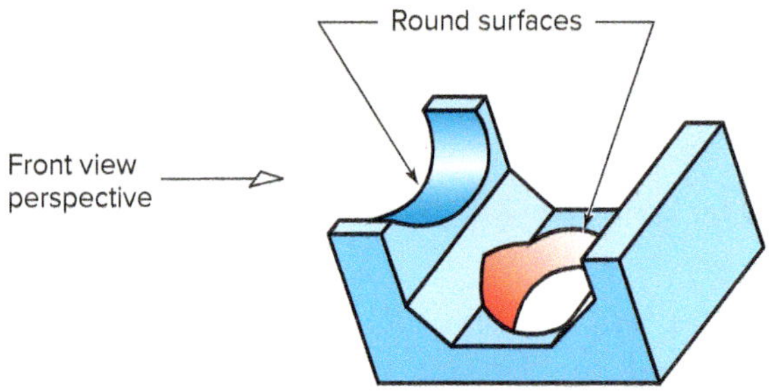

Problem 18—Saddle block

Figure 3-23 Using this pictorial view, draw three orthographic views of this saddle block.

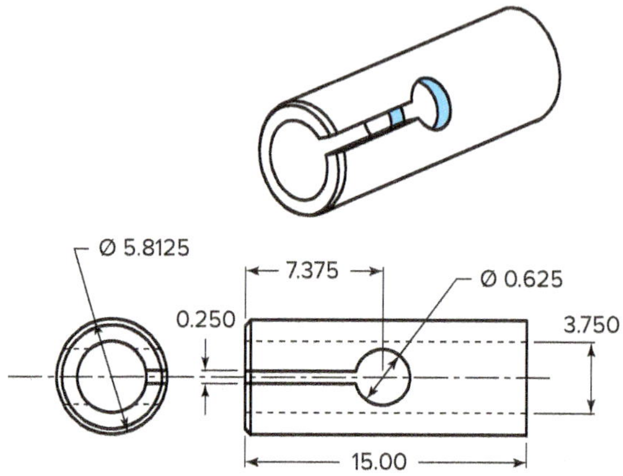

Problem 20—Slotted bushing

Figure 3-24 What's wrong in the orthographic views?

19. Regarding the answer to Problem 18 (Fig. 3-27), in your opinion, did the engineer select the best front perspective view of the saddle block to use for the front view (Fig. 3-23)? Explain.

20. Given these views of the slotted bushing (Fig. 3-24), find the error in the orthographic drawing (not the dimensions). Explain.

CHAPTER 4

Introduction to Geometrics

Units

4-1 Geometric Dimensioning and Tolerancing—What and Why?

- Express the terms *feature* and *characteristic* in your own words
- Define the five groups of geometric tolerances
- Describe why GDT is better by comparing it to older methods

4-2 Recognizing Datums in Designs and Using Them in the Shop

- Find datums on drawings
- Use datum reference the right way in your setups and programs
- Explain why datums define the right way to start a job and the right way to measure it

4-3 Geometric Control Characteristics

- List and describe 14 geometric characteristics by symbol and purpose
- Describe element and axis control and be able to recognize the difference on drawings

Professor Emeritus Don Day and his daughter, Wendy Patterson (who holds a master's degree in mathematics), formed Tec-Ease, Inc. in 1996. Their goal was and continues to be teaching the manufacturing industry geometric dimensioning and tolerancing (GD&T) through classroom instruction, self-paced learning texts, and computer-based training. A highlight of Day's career was being selected as a member of the ASME Y 14.5 Committee on Dimensioning and Tolerancing in 1995. Don recognizes that GD&T is enabling the new technologies required to design, produce, and inspect parts. The need to apply and understand GD&T is greater today than ever before.

Photograph courtesy of Don Day (2015 Passed)

INTRODUCTION

At the leading edge of the computer revolution, as the jet age gained altitude and programmed machines operated on punched tape, it became obvious that the dimensioning and tolerancing for engineering drawings needed upgrading. Previous methods no longer provided the flexibility or the feature control required to serve the developing products or of the foreseeable future.

Military leaders, engineers, and manufacturers came together to solve that need. They could see that manufacturing would continue to evolve. So, they set out to create a dimensioning and tolerancing system to serve technology of the time and had the ability to adapt as new needs arose. Their system eventually became known as geometric dimensioning and tolerancing (GDT).

GDT was to be a living system with users meeting regularly to oversee updates. As visionary as they were, they could not have foreseen today's exploding technical age. Yet the system they compiled still serves well.

In the United States, the American Society of Mechanical Engineers (ASME) Y14 Committee assumes the duty of writing and publishing revisions. Similar groups do the same in every industrialized nation. They then regularly bring their ideas to the International Standards Organization (ISO).

As in Chapter 3, our goal in this chapter is to support tech school lab assignments and to familiarize students with enough information to get started in training. It will also demonstrate the different way you must proceed when the drawing uses geometric dimensioning and tolerancing.

Our third goal is to recognize the geometric symbols and what they mean.

Unit 4-1 GDT—What and Why?

Introduction: This lesson is designed to provide proof of why we use GDT today. The GDT system is designed to better serve *function*. **Function** is the way machined features work, assemble, and operate. In order to understand this improved focus, we'll examine a few baseline definitions and look briefly at the meaning behind the symbols used in the system.

TERMS TOOLBOX

Characteristic One of 14 model objectives (for example, roundness, straightness, or location). See Fig. 14-2.

Feature Any single aspect of a part that can be dimensioned and toleranced (for example, a thread, a slot, and a hole).

Form Controls the outside surface shape of regular objects (one exception–an axis, which can be controlled with straightness).

Function The way things assemble and work in application.

Location Controls the position of a feature's axis with respect to a datum; the controls are position, symmetry, and concentricity.

Orientation Controls the angle of a feature's axis or surface with respect to a datum; angularity, parallelism, and perpendicularity are the controls of orientation.

Profile Controls the outside surface shape of objects where the shape cannot be defined with form controls.

Runout Controls surface wobble on objects that rotate around a datum axis.

KEYPOINT

GDT was invented to better serve *function*.

4.1.1 Symbolic Language

Let's compare the two different systems: GDT and standard dimensioning and tolerancing. In Fig. 4-1, the location and tolerance for a hole are shown using examples of both. A second specification controls the roundness of the hole to 0.5 mm (the circle in the second box). It may seem that the GDT callouts only restate in symbols that which was always there. **Not so**! Each and every geometric symbol was created to solve an unmet need or to expand some capability not found in standard dimensioning and tolerancing.

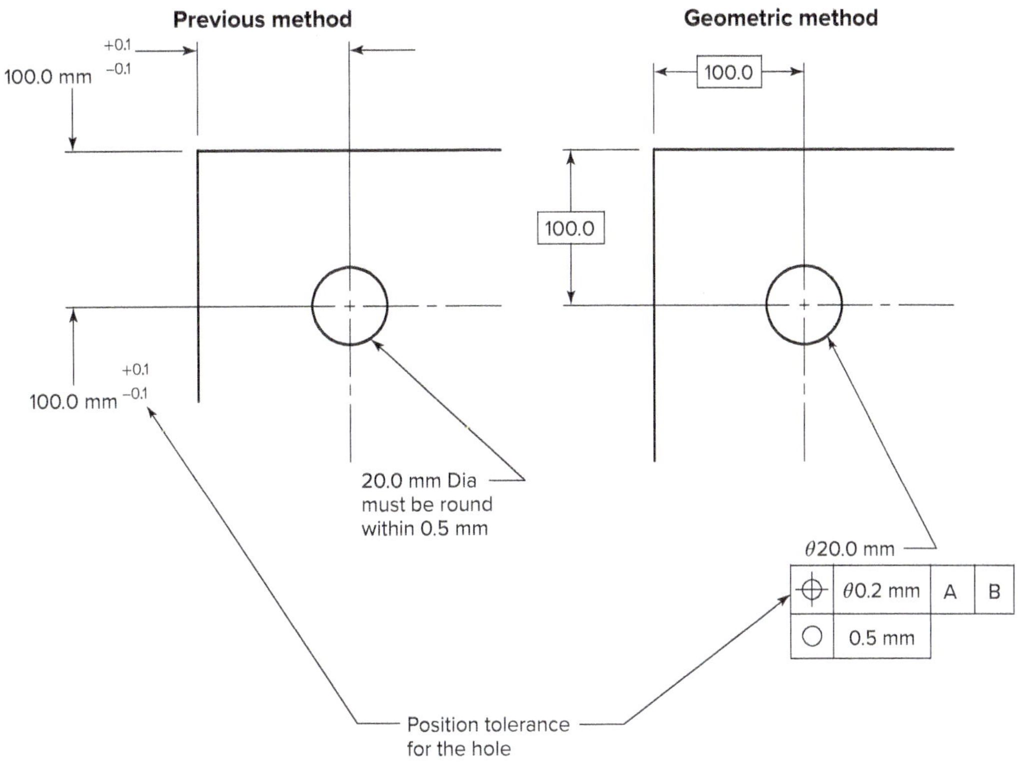

Figure 4-1 Geometric compared to standard dimensioning methods.

We will study what many of those symbols mean in Unit 4-3, and the full set can be seen in Appendix V. But we've now come to the first advantage to GDT: It's a symbolic language that eliminates words on drawings. Words can be ambiguous, but with a bit of study, everyone knows the exact meaning of a symbol.

Of greatest import, symbols are clearly understood worldwide. Nations can share drawings using geometric controls regardless of the language spoken.

However, symbols clear up drawing information only when the print reader studies their meaning. For example, the boxes around the 100-mm dimensions in Fig. 4-1 have a definite meaning and purpose—we'll discover it later.

> **KEYPOINT**
>
> Every box, circle, or symbol on a print using GDT conveys a meaning that the machinist must understand.

4.1.2 Basic Machinist Principles?

Without background geometric concepts seem to only be focused on dimensioning and tolerancing of prints. They seem like a new way to convey the same old information, and the underlying information itself is the same. One might think that geometric or not, the machinist's actions remain the same—this is definitely **not so**! The concepts behind the symbols affect much that you do in setting up machinery, planning cut sequences, writing programs, especially in measuring results, and making holding tools. These are all based on geometric principles.

Two Important Definitions—Characteristics and Features

Take a moment to scan Fig. 4-2. The 14 symbols in the right-hand column are the keys to the system. We'll return to this chart again. You might want to put a marker here.

Characteristics Are the Target

Each symbol in Fig. 4-2 represents a desired condition or relationship (the target) for one feature of the workpiece, the geometric characteristic. What each means is a main goal of this chapter. Once the characteristic is defined, then a tolerance will be placed on it.

> **KEYPOINT**
>
> A geometric dimension and tolerance do this:
>
> 1. **Define the Target Characteristic**
>   A desired *characteristic* is called out. For example, the hole in Fig. 4-1 is to be round.
> 2. **Set Allowable Variation—The Tolerance**
>   This is the allowable variation from the perfect target (the tolerance). In this example, it defines how far the hole may depart from perfect roundness 0.5 mm, found in the lower frame.

	Type of tolerance	Characteristic	Symbol
For individual features	Form	Straightness	—
		Flatness	⏥
		Circularity (roundness)	○
		Cylindricity	⌭
For individual or related features	Profile	Profile of a line	⌒
		Profile of a surface	⌓
For related features	Orientation	Angularity	∠
		Perpendicularity	⊥
		Parallelism	//
	Location	Position	⌖
		Concentricity	◎
		Symmetry	⌯
	Runout	Circular runout	↗*
		Total runout	⌰*

*Arrowheads may be filled or not filled.

Figure 4-2 The five groups of geometric controls.
Source: American Society of Mechanical Engineers.

SHOPTALK

Advanced Manufacturing Working skills in geometrics are especially useful when using computer-based measuring equipment, computer-assisted drawing, or computer-assisted programming or when applying statistical process control to your job. These topics will be studied in Chapters 25, 27, and 28—don't miss them.

Features Are Anything That Can Be Dimensioned and Toleranced

A characteristic applies to a single **feature** on the workpiece (a slot or a thread or a hole, for example). The characteristic then is the target for that individual feature.

4.1.3 The Big-Picture Definition

Let's put this all together: GDT is a system of 14 controls that define a desired condition (characteristic) for some feature on the part. They also show how much variation is permissible. All 14 characteristics solve some functional need that was unmet before GDT was introduced.

Five Characteristic Groups

GDT controls fall into five families, sorted by purpose. Returning to Fig. 4-2, notice that the symbols are grouped as

Form Controls the outside surface shape of regular features—round, straight, flat, and cylindrical shapes. (There is one exception to this where straightness can be applied to a feature's centerline.)

Profile Controls the outside surface shape of contoured features—that is, items that aren't straight, flat, or round. Profile is very similar to form, except for the nature of the outside shape.

Orientation Controls the relationship of one feature with respect to a datum (for example, the angle of a surface or the perpendicularity [squareness] to a datum).

Location Controls the position of a feature's center with respect to a datum.

Runout Controls the wobble on the outside of a spinning object.

Geometric Advantages

Geometric Principles Underlie All That We Do

The prints you use may or may not use GDT, but the geometric design is by far the more thought-out version. Even more to the point, *the challenge of the work itself is, and always has been, geometric.* That's the truth on which the Y14 Committee based its work.

KEYPOINT

Manufacturing is geometric whether or not the design print is GDT.

Factoids—Why GDT?

- More *control and flexibility* for the designer and the shop compared to older methods.
- All the *full, natural tolerance* possible within the function of the object, yet make the best quality product. (GDT doesn't lead to large tolerances. What GDT does is remove the need for excessively tight tolerances "just to be sure.")
- An instant understanding of the *functional priorities* of the features to be machined.

Understanding design priorities tells the users where to begin machining. Functional priorities are the order of importance of features on the part. Understanding these priorities is important to the machinist because they indicate the following:

How to hold the part for machining
Which cuts to take first
How to measure the results

KEYPOINT

There will be a difference in setups and measuring methods when the drawing uses geometric dimensioning and tolerancing, compared to working from a drawing using the older dimensioning and tolerancing system.

As a machinist, you do not need to understand how the object being made actually works, but you do need to know what part features are the most critical–that is, which have the highest priority. From a geometric design, you will quickly identify the critical functions and features. Then you will base setups and the sequence of operations on those features and functions. We'll see examples of this in Unit 4-2. Those priorities are noted in the datum frame–coming next. Datum A is higher than B, then C, and so on.

The GDT system better serves the way we make objects and the way they work. One final point: GDT is a true system. As you learn more, you will begin to see many relationships and interconnections between controls and principles. Looking for the connections between concepts and controls is called a systems approach to learning. We'll explore it a bit in upcoming units.

UNIT 4-1 REVIEW

Replay the Key Points

- Every box, circle, or symbol on the GDT print conveys a meaning that the machinist must understand. Not to know is to work with a handicap!
- The geometric system was invented to solve technical function problems arising in manufacturing.
- Every aspect of the system is built around function.
- GDT enables flexible control of the entire manufacturing process.
- GDT is a living standard. It is overseen by an international group that ensures that it evolves with the times and the technology. The machinist must stay up to date with revisions as they occur.
- Because it's symbolic, GDT transcends language barriers.

Respond

1. What group oversees GDT in the United States?
2. What is the basis of the GDT system? What does it serve better than the previous system?
3. When making a machine setup using a GDT drawing, what do the functional priorities show the machinist to do? How do you start a GDT job? List three items.
4. List and briefly describe the five characteristic control groups.
5. What is the one exception to the form control of straightness?

Unit 4-2 Recognizing Datums in Designs and Using Them in the Shop

Introduction: The use of datums is one of the most important differences between a GDT design and one that is not. Nine of the 14 characteristics originate from a datum reference. For example, the location of a hole must be positioned "from" something. That reference is always a datum in GDT. In Fig. 4-3, two datums are called out: datums A and B. They are symbolized by the placement of boxes around a capital letter.

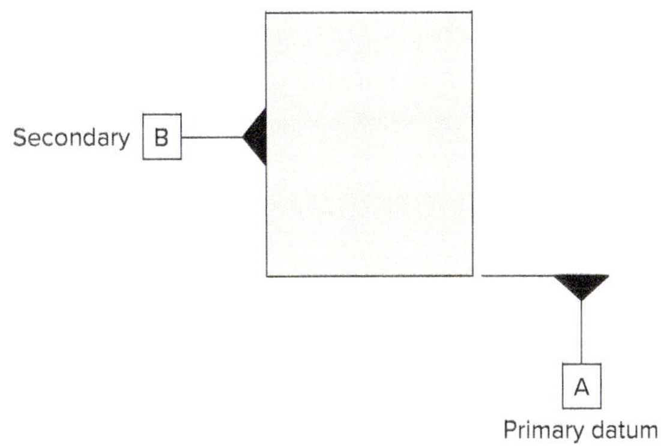

Figure 4-3 Datums are indicated as capital letters in a square box, connected to the object.

TERMS TOOLBOX

Datum A theoretically perfect surface or axis used for reference. The datum is established by features on the part–it is not on the part. It simulates the assembly world outside the part.

Datum feature Any regular part feature that is used to establish a datum on a drawing.

Datumization (author's term) Accounting for the irregularities of a real object by placing it against a theoretically perfect reference surface for measuring or machining. Or an averaging of the irregularities on the feature to derive an axis datum. (Figure 4-18 uses a precision ground pin in a hole to datumize its location.)

Feature Any detail on the print that is dimensioned and toleranced.

Formal datum A feature assigned a capital letter datum on the print.

Informal datum Any feature used for reference for another feature. Does not have an assigned letter on the print.

KEYPOINT

In GDT the referencing of a feature "from something" is always based on a datum.

4.2.1 Datums Perform Three Vital Tasks

1. **Establish an Exact Basis for the Geometric Control**
2. **Eliminate Ambiguities (Uncertain Facts and Guesswork)**

   Without datums, craftspeople find themselves guessing at setups and measurements, as will be clearly demonstrated.
3. **Set Priorities**

   A geometric print helps decide how to start the job. By the datum priorities, it shows what machine operations should be first, how to hold the work, and what features are critical. The job is easily organized. It's not always possible to go right to the first priority, but even if intermediate cuts must be made, the target is to establish Datum A (if not already established) as soon as possible, then B, C, and so on.

Make sure you understand all three of these concepts when we complete Unit 4-2; they are the main points. These concepts will appear every time you machine and measure a product.

4.2.2 Datums—The Exact Basis, the Starting Point

Without a **datum**, many measurements (and machine operations, too) are not exact. To illustrate, answer the question in Fig. 4-4: Exactly how **TALL** is the object with its exaggerated irregularities? It's difficult to tell. Where do you place the measuring tool? Measuring in different places produces different answers. The problem is that the measurement has no formal basis, and the question of tallness cannot be answered. **Thickness can be measured but not tallness.** However, if this part is to be assembled with another, the functional question is how tall, not how thick it is. The problem lies in the way the drawing is dimensioned. It simply shows that a pair of surfaces must be 40 mm apart but does not clarify how to treat the measurement.

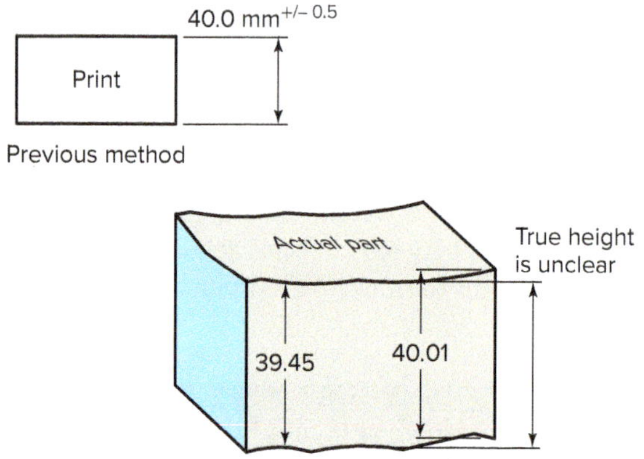

Figure 4-4 How tall is this machined block?

A datum solves the problem. In Fig. 4-5, by laying the part on a precision flat table, called a layout or inspection table, the answer becomes clear. The *functional* height is the tallest point above the datum—**above the table on which the part rests**.

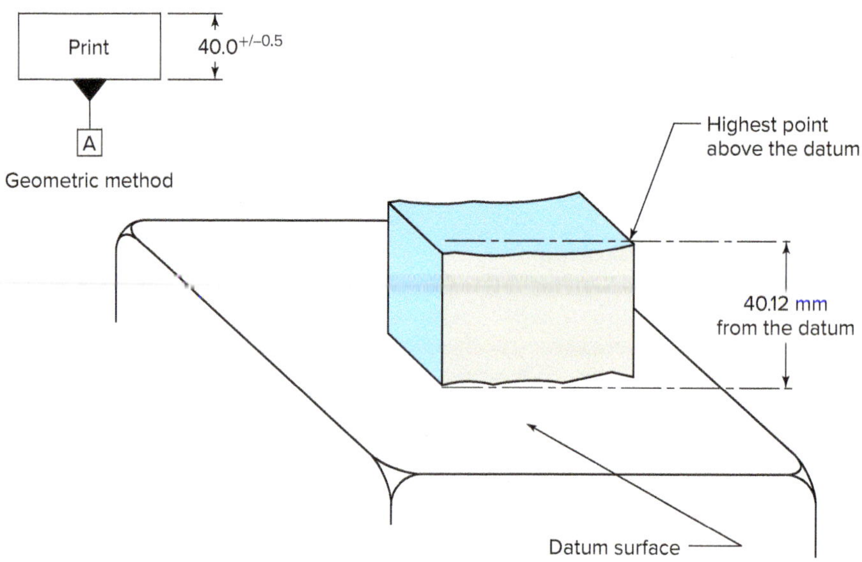

Figure 4-5 The flat layout table is the datum. It answers the question.

So, the designer has assigned a datum to the bottom of the block as shown by the A in the box. It is the reference surface from which the functional height *must* be determined. To measure otherwise is wrong. **It must be measured up off the datum**.

The datum priority for the part of Fig. 4-3 is that Datum A holds the highest functional priority (order of importance to the function of the part) while Datum B is secondary. The alphabetical order shows their ranking in terms of importance to the part.

A datum might be indicated by an extension of the surface, or connected to the surface itself, as with Datum B. A box drawn around a capital letter is called a *datum identifier*. That symbol signals to the machinist that the feature to which it is connected is significant and that the datum will be used for reference for other features on the part.

KEYPOINT

The datum is not the part but rather is represented by the flat table upon which the part rested. The datum, which is considered perfect, accounted for irregularities on the real part. It simulates the assembly world outside the part.

The datum is needed not only to measure the part but is also used as a foundation to machine it as well.

The Example Datum:

1. Is NOT on the part, but rather it is the perfect surface touching it.
2. Accounts for irregularities on the part surface.
3. Simulates the functional assembly world around the part.
4. Must be used for setups and measuring.

Datums on Prints

Note that the crowfoot, triangle in Fig. 4-5, connects to the outside of the part surface—that symbolizes the datum is touching the outside of the feature (Fig. 4-6).

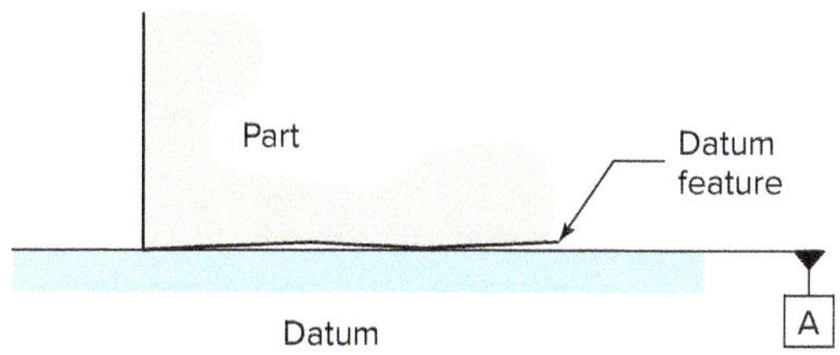

Figure 4-6 Datum A is established by the feature's surface, called datum feature A.

Two Kinds of Datums—Surfaces and Axis Centers

Datums Usually Contact Outside Surfaces

Any regular feature of the object can be assigned a datum. The designer chooses the order of importance based on the function of the object or the assembly procedure.

Surface Datums

When a datum is connected to a part surface, the feature it contacts is called a **datum feature**. So when establishing datum contact during machining or measuring, the datum feature is held against a gauge-quality surface such as a precision table, a ground vise jaw, or a milling machine table.

SHOPTALK

About Targets You can read more about datum targets in the standard document ASME Y14.5 M.

When the entire work surface isn't needed to establish the datum, specific points on the surface can be assigned that are called *datum targets*. The drawing will specify where the specific points of contact are to be located.

Center Axis Datums

For functional reasons, it's sometimes necessary to use the center of an object as a datum. An example is an automobile axle. Its most critical function is that all features are centered around the axis—so the axle's centerline is Datum A. Axis datums can also be flat planes at the center of a slot or a square tab, for example.

Axis Datums on Prints

Figure 4-7 illustrates one way an axis datum might be specified on a print: If the identifier touches a dimension then the datum is the center axis.

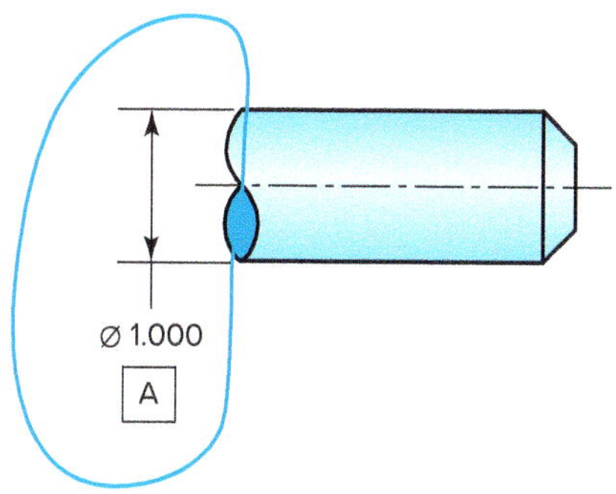

Figure 4-7 Datum identifier attached to the dimension.

> **KEYPOINT**
>
> If *any* GDT callout connects to the feature's dimension, then it applies to its center axis, not its surface.

> **KEYPOINT**
>
> A well-written job plan or program sequences operations such that functional priorities are achieved in order, based on datum priorities.

Axis Datums Are Mathematical

Working with the center of a feature requires a bit of detective work because the datum feature does not exist as an entity. To define an axis the entire surface that establishes it, along with its irregularities, must be summed up. For example, in Fig. 4-8, a hole's center has been found by inserting a precision round pin. The largest pin that will just slip into the hole determines the largest circular space available—the functional hole. Then the theoretically perfect pin's center becomes a reliable datum.

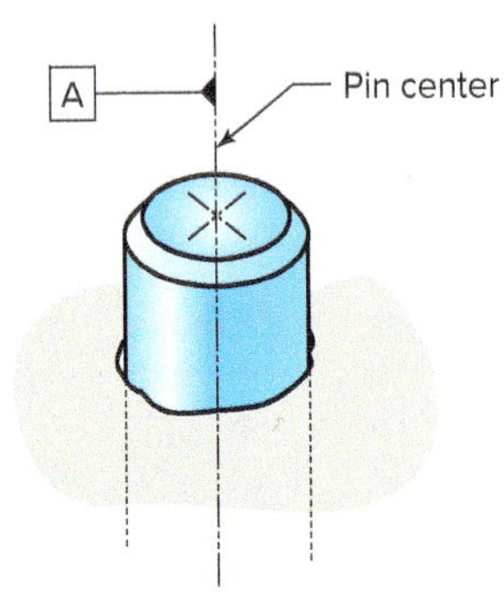

Figure 4-8 The round test pin establishes the centerline datum.

4.2.3 Your Shop Actions—Machining and Measuring

Datums Define Priorities for the Job

While there can be many datums on the print, there is one master, called the *primary* datum. It is the one that should be used to start the job as much as possible. There are exceptions, but the goal is to make machine cuts that establish that reliable datum feature surface as soon as possible if it doesn't already exist. If the primary datum is already there, then establishing Datum B using Datum A as the reference becomes the first machining objective.

Sometimes, it's necessary to make one or more intermediate cuts that lead to establishing the datum feature. As explained earlier, the machinist plans early operations around the datum priorities in their order on the drawing. The lowest letter of the alphabet found on the print (usually A) is the primary datum for that design. The designer created the design outward from it. Certain letters are not used to designate datums, O and I, for example, due to possible confusion with numbers or other letters.

Measuring a Perpendicular Edge—First Example

If datum priorities are ignored, several errors can occur, causing quality problems with the product and extra costs, too. Consider measuring a square corner, defined in Fig. 4-9. The print specifies a 2 × 4 rectangle with the left edge at 90° to Datum A.

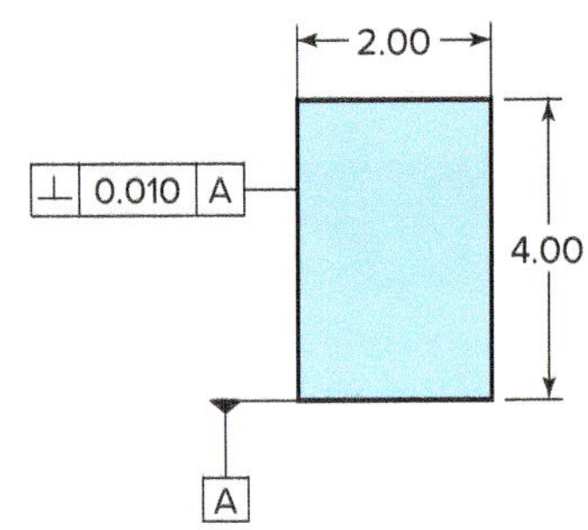

Figure 4-9 This edge must be perpendicular (90°) within 0.010 in. with respect to Datum A.

We'll discuss the tolerance meaning in just a bit; for now note which edge is datum feature A—the bottom edge in the drawing. The information next to the part is enclosed in a special GDT symbol called a feature control frame. The information found in the frame states that

> This surface shall be perpendicular, within 0.010 in., with respect to Datum A

(Fig. 4-10). Once the datum reference is established, it must be proved that the left edge is within perpendicularity tolerance. The left edge can vary in any way as long as it does not violate the 0.010-in.-wide tolerance zone, which is perfectly square to Datum A.

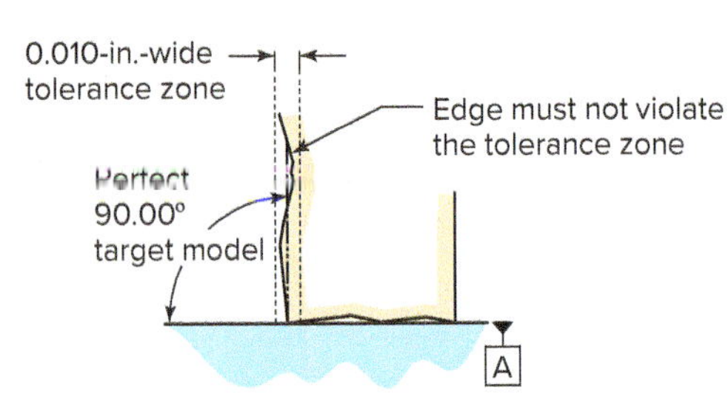

Figure 4-10 The controlled edge must not violate a 0.010-in.-wide control zone built around a perfect 90° model extending from Datum A.

Figure 4-11 shows the right way to measure it. A precision square is used. Notice that in both methods, feature A is against a flat datum surface, the square's base or the flat table, and either becomes Datum A. Only then is the variation correctly checked along the controlled edge. As long as no gap is found to exceed 0.010 in., the edge is within the perpendicularity tolerance.

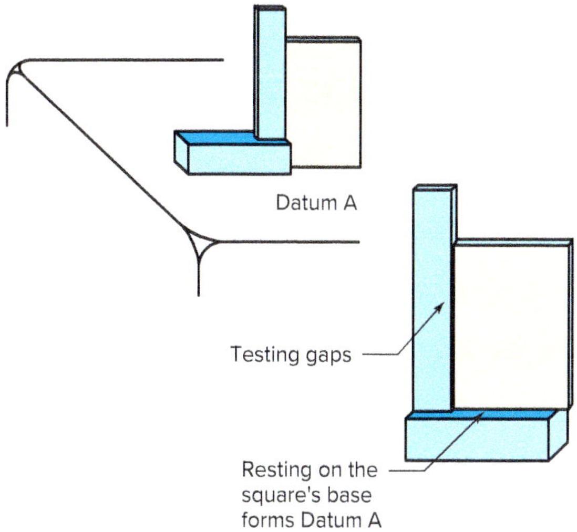

Figure 4-11 Two correct ways of establishing Datum A, then measuring the controlled edge.

Now, using the right method, suppose that the error found is beyond the allowance—that is, the corner fails the square test when a 0.012-in. gap is detected (Fig. 4-12). But suppose also that by not heeding the datum priority, the bad part is incorrectly measured in this way (Fig. 4-13).

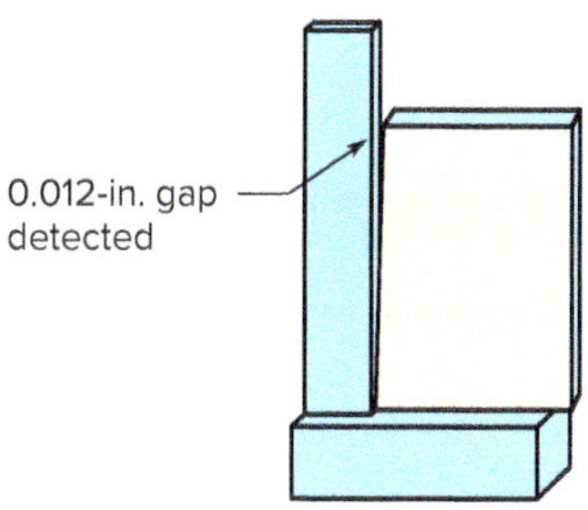

Figure 4-12 This part fails the test by showing a 0.012-in. gap.

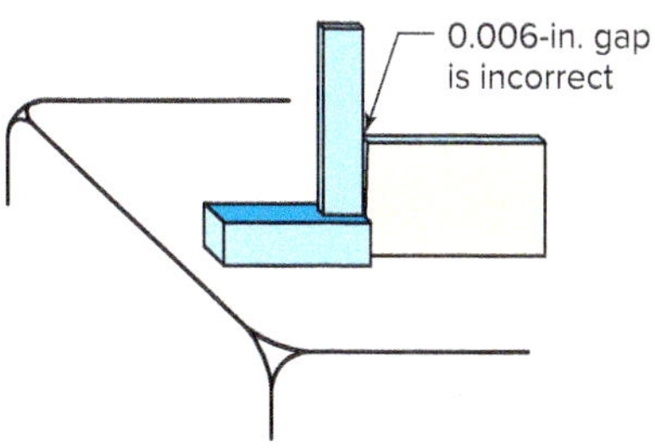

Figure 4-13 The wrong way to make the measurement—a bad part might be accepted!

In Fig. 4-13, the machinist has incorrectly placed the wrong feature against the datum, then tested for a gap on the wrong edge. Clearly under the GDT system, this is the wrong test! Since the incorrectly tested edge is shorter, it shows less error.

Guessing Without a Datum

From this example, can you see that without a geometric datum basis for the control, there could actually be two answers to the square corner? Recall that datums eliminate ambiguities and clarify functional priorities. Before GDT, the dimensioning and tolerancing methods defined only that two edges were to be square to each other, not which edge was to be square to which.

Machining the Square Corner

There is a right way to machine this example, too. It must be started by establishing datum feature A first (Fig. 4-14). Then the rest of the procedure follows, referenced from it. Track datum feature A as it progresses through the next drawings. In the first operation, datum feature A is machined. In Figs. 4-15 and 4-16, the second setup is made such that Datum A (the precision vise jaw set up parallel to the milling machine) touches feature A. Both 90-degree cuts are made relative to that datum.

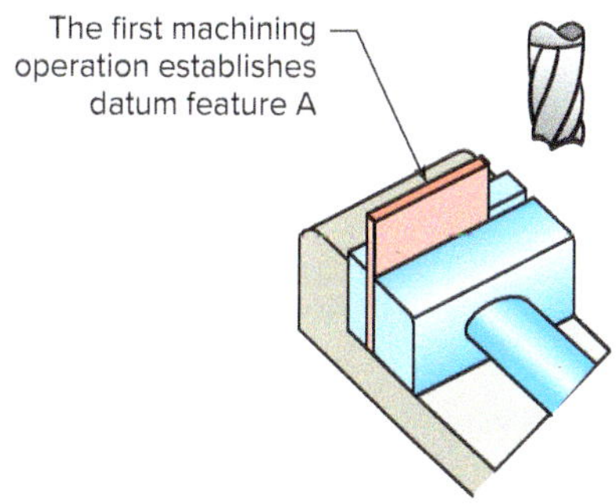

Figure 4-14 The first correct cut establishes datum feature A.

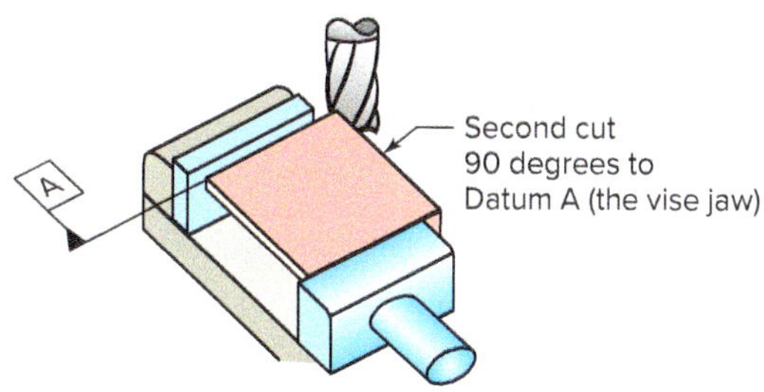

Figure 4-15 The second cut establishes the 90° relationship to the vise jaw Datum A.

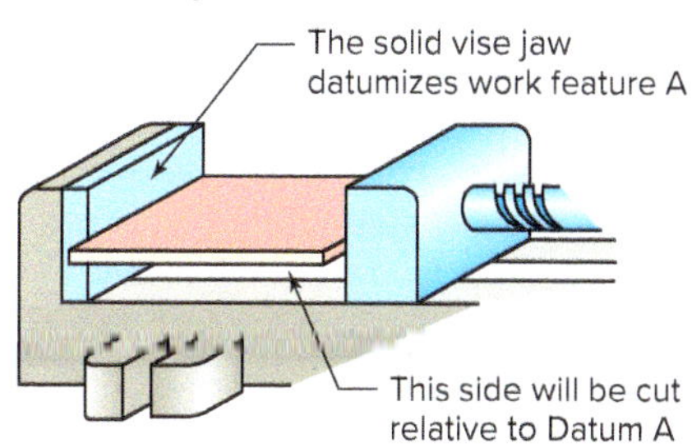

Figure 4-16 The third cut is parallel to the controlled edge—square to Datum A.

KEYPOINT

The precision (nonsliding) vise jaw becomes Datum A for cuts 2 and 3.

The vise jaw, which is considered perfect, averages out the part irregularities. Here in *Machining and CNC Technology*, we'll call the contacting of the real part surface against a theoretically perfect surface **datumization** (that's my definition).

TRADE TIP

Vise Jaw Datums For this setup to work right, the vise must first be aligned to the machine's main axis, and it must be smooth and flat.

Now, in Fig. 4-17, a second control has been added–the edge opposite Datum A must be parallel. So the part is set up vertically, and the floor of the vise becomes Datum A. At this time, it would also be machined to the 4.00-in. dimension. Cuts 3 and 4 could be performed in different orders, but all these operations depend on a well-maintained, accurate mill vise touching datum feature A.

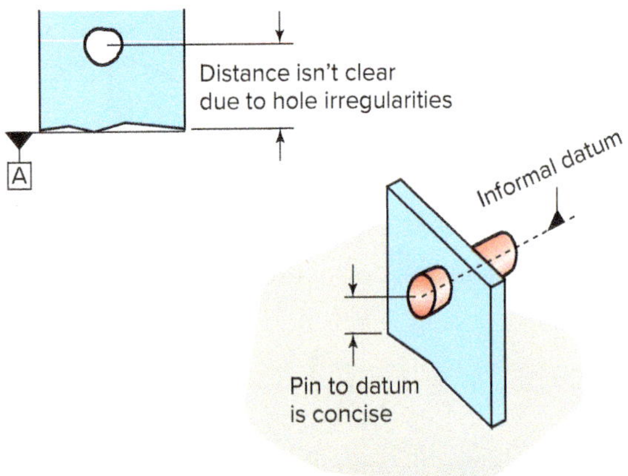

Figure 4-17 How far is this drilled hole from Datum A? The pin solves the problem.

While this squaring up of edges is a simple example, it represents much more complex machining plans for larger objects. They must be sequenced with datum reference. Planning the operations by the datum priorities becomes even more critical when we use CNC machining, because doing otherwise introduces needless variation. Each and every dimension and/or control on the print that refers to Datum A must be set up so that the holding tools contact feature A to establish a reference. It follows similarly for datums B and then C, if they exist.

4.2.4 Formal and Informal Datums

The datums shown thus far have been assigned a capital letter by the designer. They are **formal datums** on the print. But there are times in the shop when you'll see a need for the concept without formal assignment. Figure 4-17 poses a question to aid in seeing this concept. How far above the datum is the hole?

This hole has feature irregularities, as all do. It's unclear where the hole center might lie. To answer accurately, we must datumize the irregularities to find the functional round space available by inserting the largest round test pin that just fits. Considered perfect, the pin's surface establishes the **informal datum** at its center axis. An exact measurement of the functional distance above the datum to the hole is now possible.

Sometimes an informal datum is expressed on a drawing as shown in Fig. 4-18. The drawing makes it clear that an assembly condition exists that is dependent on the spacing between the two holes. Also shown is the fact that the left hole is a datum to the right.

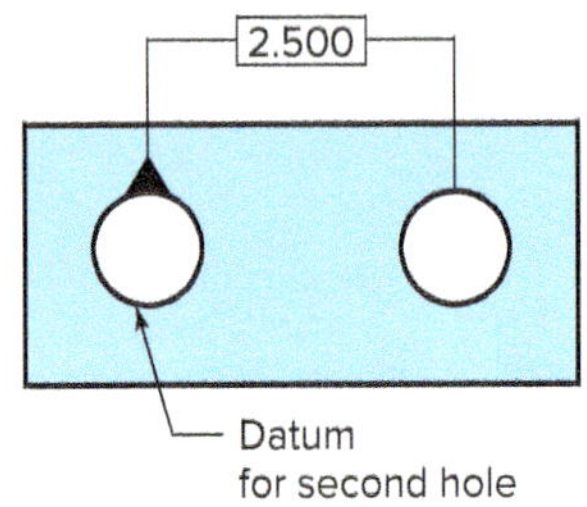

Figure 4-18 An informal datum.

Datums on Drawings Before 1994

You will encounter older geometric drawings and texts that show datums slightly different than the way I've been illustrating them. Shown in Fig. 4-19, the datum will still be in a box but it will be between two dashes: **-A-** or **-B-**, for example. The leader line to the datum feature will not terminate in a crowfoot triangle. Why? Because a change was made to make all drawings similar, worldwide–the living system at work.

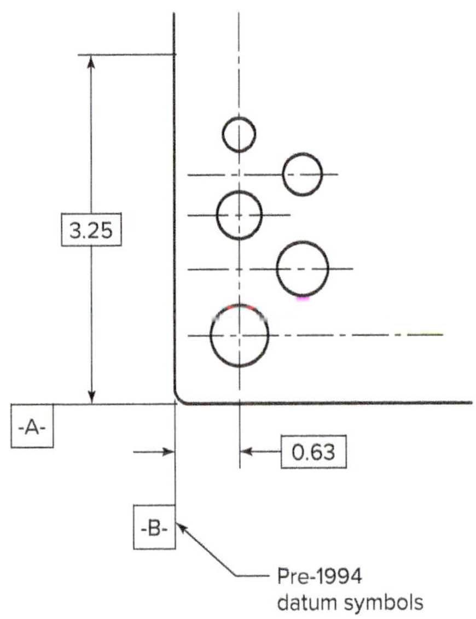

Figure 4-19 Datums on older drawings.

UNIT 4-2 REVIEW

Replay the Key Points

- A datum can be assigned to any regular feature on a part: a surface, a hole, a centerline, or even a set of *designated points* (not covered in the text).
- A datum is not on the part, but it is established by the part.
- A datum is considered perfect for the purpose of controlling individual features.
- A datum simulates the assembly world outside the feature.
- A datum accounts for and averages part irregularities.
- A datum eliminates ambiguities and sets functional priorities.

Respond

1. On a drawing, a rectangular box around a capital letter is called a(n) _______.
2. What single word sums up the reason or purpose behind the GDT system?
3. Explain why starting a geometric machining job from a GDT drawing is easier than from one that does not use GDT.
4. List the five types of tolerance groups for feature control symbols.
5. Many of the 14 feature controls require a starting point basis. What is it?
6. Describe a datum that is assigned to a feature surface on a drawing.
7. Besides surfaces, datums can be assigned to _______.

Unit 4-3 Geometric Control Characteristics

Introduction: In controlling a feature's surface or its axis, all 14 geometric characteristics perform the same two tasks:

1. Define the perfect model of the desired characteristic
2. Denote the allowable variation from that perfect model—the tolerance

And, a new concept:

3. Imply how the characteristic must be measured

   The characteristic doesn't tell the machinist which measuring tool or method to use, but it does specify datum reference (or not). That is the first factor—must you set up one or more datums to correctly measure the feature? There are other unique factors to verify that each control has been satisfied.

As we study individual control symbols, keep in mind that all 14 follow the three-stage process of defining the perfect target, then the tolerance. Then by knowing the nature of the control, you understand the method of proving that it's right—how to measure it.

TERMS TOOLBOX

Basic Dimension The box around a dimension means it's the perfect target—no tolerance.

Bonus Additional tolerance possible based on function and actual measured feature size.

Contour Controls that deal with irregular-outside-shaped features. Contour may or may not need a datum reference, depending on function.

Elements Surface lines running in the direction of the control.

Entity Either a surface element or a feature center that must be shown to lie within the tolerance zone to be within compliance.

FIM (full indicator movement) The measured difference in high and low points on the surface of an object, rotated about its axis through 360°.

Form Four GDT controls that define the outside shape of regularly shaped objects. Form has no datum reference.

Location Three controls that define the tolerance for the center of a feature with respect to a datum.

Orientation Three controls that define the angular relationship of a surface or feature center, with respect to a datum.

Perfect model The goal of the desired characteristic.

Runout Two controls that deal with surface wobble on rotating objects (axis datum).

4.3.1 Surface or Axis Control

The measurement method is determined by whether it applies to the feature's surface or its axis. We need to discuss both types.

KEYPOINT

In GDT, the control applies to either one or the other—to the feature's outside surface or to its center axis.

Elements on the Outside

If the control deals with a feature's surface, such as roundness in Fig. 4-20, then the control entity is called an *element.* **Elements** lie on the outside surface of a feature. Elements are lines that could be drawn on the surface with a pencil. They run in the direction of the control. Figure 4-20 illustrates two sets of elements–the roundness and straightness on a cylindrical pin. For the pin to be within the roundness tolerance, every circular element tested must be able to fit within a perfect pair of concentric circles–the tolerance zone. The distance between the two tolerance zone circles is the tolerance. When the control is cylindricity, then straightness elements are added, and the whole surface must lie within the tolerance.

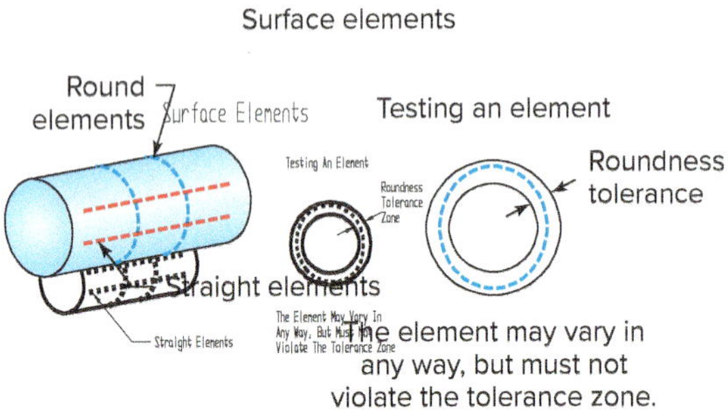

Figure 4-20 Elements are lines on outside surfaces. There are two kinds here—round and straight elements.

If the control is straightness, the elements run in a straight line, with the direction indicated by the control symbol line. We'll see examples shortly. Recall when measuring of the square edge in Unit 4-2, the machinist was testing straightness elements running perpendicular to the datum.

There is an infinite number of elements on any given surface that could be tested to prove the feature is within tolerance. The number chosen to measure is a matter of experience. For example, suppose the pin is 1 in. long. Then testing at two or three circular elements would probably be okay. But suppose that it was 3 in. long and perhaps had been made on a worn lathe. In such a case, testing more elements would be wise.

KEYPOINT

Elements

A. Elements are the outside surface control entities.

B. For the feature to be within tolerance, elements must be shown to lie within the tolerance zone.

C. There is an infinite number of elements on any surface.

Control of Center Axes Is Mathematical

When the control deals with the axis of a feature, rather than a surface, the axis exists only after some inspection of the surface that creates it. For example, in controlling the location of a drilled hole (Fig. 4-21) its centerline must fall within the tolerance zone (0.05 mm). But there is no centerline until the sum of the hole's roundness and straightness (cylindricity) is accounted for.

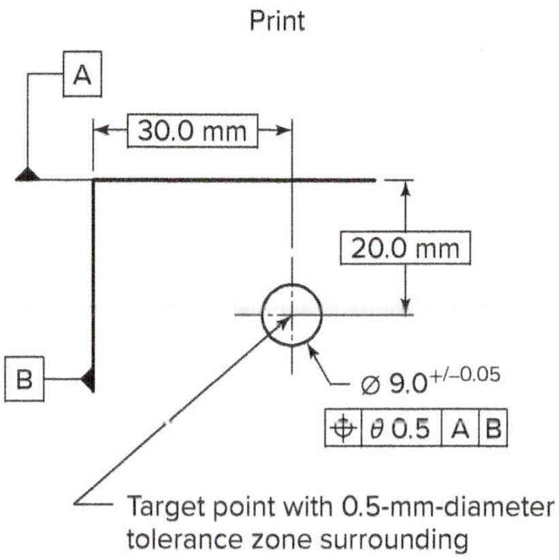

Figure 4-21 The feature control frame specifies that the position of this hole's centerline must be within a round tolerance zone of 0.5 mm with respect to datums A and B.

In Fig. 4-21, the three-step process is defined within the *feature control frame.* We will further study the characteristic of position later; for now it's used to illustrate the use of the feature axis as a control entity. This drawing states that

> The position of this hole's centerline shall fall within a round 0.5-mm tolerance zone located with respect to datums A and B.

KEYPOINT

We know the tolerance zone shape in Fig. 4-21 is round because the 0.5 mm is preceded by the Greek letter theta, meaning "diameter" in the GDT system and throughout mathematics: a circle with a line through it.

Depending on the CAD software, the theta diameter symbol might be smaller than the tolerance numbers—or it might be of the same font size. In either case, if it precedes the tolerance, the tolerance zone is circular or cylindrical.

Figure 4-21 follows the three-step process:

1. For a *perfect model* the target for the hole's axis location is a point that's 20 by 30 mm from datums A and B (from the datums, not the part edges).
2. A *tolerance zone* of 0.5-mm diameter is defined around the target. The hole's centerline must fall inside the round tolerance zone.

3. It *implies* how it is to be *measured.* This control applies to an axis positioned relative to two datums. Therefore, any test of its location must use the datum concept and determine where the axis lies with respect to datums. This aspect is never shown in the control callout, but it must be known by the user.

How do you know the control applies to an axis? The control frame is attached to the hole's size dimension, the control applies to the axis. We've seen this same rule used for indicating datums that are axes.

Functional Test

Proving the hole axis location with the largest ground pin that will just slip into it with no movement shows where the largest round space lies. That's the function if, for example, a bolt is to be put into the hole. The ground pin occupies the largest cylindrical space available. We call that kind of measurement a *functional test.* It is functional because it does two things: It detects the biggest bolt that could be placed in the hole and it simulates the axis location of that bolt hole. The test pin's centerline then becomes the control element.

There are only the two possible control entities: outside elements or center axes. You must determine which by the way the feature control frame is applied. If the frame connects directly to the center of the feature or to its dimension, as on the two drawings on the left in Fig. 4-22, then the **entity** is an axis.

The Special Case—Straightness of an Axis

Figure 4-22 illustrates the special case for straightness whereby it is applied to the center of the pin. In this application, the axis must lie within a cylinder of 0.005-in. diameter. On the right-hand drawing, the frame connects to the feature's surface; thus, it is controlling straightness elements.

Basic Dimensions Define Perfect Targets

Another geometric symbol is the box around the two dimensions in Fig. 4-21. They indicate that the dimension defines the **perfect model**. When boxed in this fashion, they are called *basic* dimensions. Basic dimensions always originate from datums and they specify perfect targets (true location in this example). They show where the bull's-eye is located. They have no tolerance themselves, but tolerance will be built around their target location.

That is very different from standard dimensioning where the distance to the hole's center would have a tolerance. Functionally, in the nongeometric method, the target point can move. It has tolerance, *but not in the geometric system.* Here the bull's-eye remains stationery while the tolerance is built around it.

The shape of a geometric tolerance zone built around the target can vary, but the tolerance is always the distance across the zone. In Fig. 4-21 it's a circle 0.5 mm across. Okay, it's time to see how the three-step process applies to all of the controls.

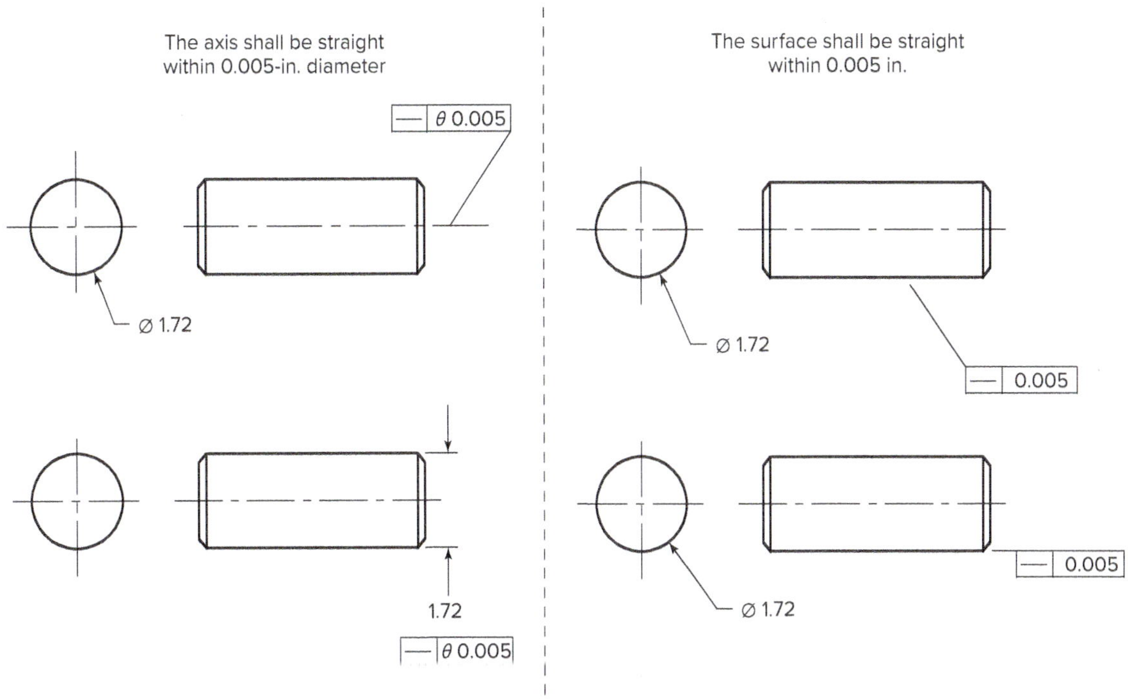

Figure 4-22 Different ways of specifying either the axis or a surface element—control entities.

KEYPOINT

Summary

To correctly *measure* any geometric control, you must determine

Does the control require datum reference or not?

Does it apply to surface elements or to an axis?

4.3.2 The Fourteen Geometric Characteristics

As we go through these controls, our objective is familiarity. Still, you might see deeper into the system to discover interrelationships between controls. Many contain further controls buried within as subsets. Measuring a feature for compliance to the stated control will also verify a second and even a third characteristic. For example, testing cylindricity (to be a perfect cylinder) to be within 0.005-in. tolerance also tests for roundness and straightness within 0.005 in. When one control is a subset of another, it's called *embedded.*

You might also see ways controls could be combined as building blocks. Two or more can be put together to form something new. For example, how might we combine controls to define a perfect cone (see Fig. 4-23)? Actually, there are several ways that it could be accomplished, some without a datum reference and some with. The composite control will need some way of defining the angle of the cone and its roundness. Keep this challenge in mind as you'll be asked to build the cone control in the final questions. To do so, you'll need to select controls from the 14 choices that follow.

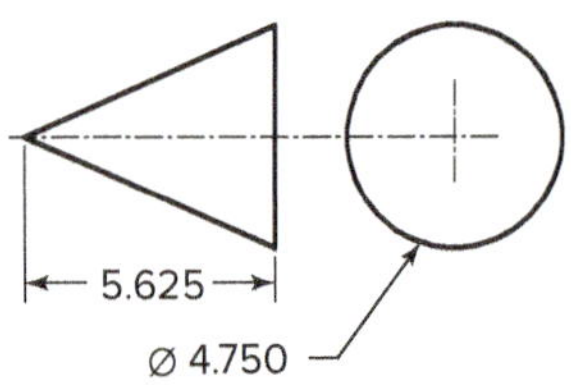

Figure 4-23 In GDT, controls can be put together to define functions such as this cone.

Group 1—Controls of Form

All four controls in the **form** group apply to the outside surface of a feature; thus, they prove *elements* are within tolerance (except straightness, which can be applied to an axis):

- **Straightness** —
- **Flatness** ⏥
- **Roundness** ○
- **Cylindricity** ⌭

Straightness in Two Dimension

In Fig. 4-24, a straightness control has been called out for the top of the part. To pass inspection, it must be proven that each element tested lies within a tolerance 0.003 in. wide. Each element can vary in any way, as long as it does not go outside the 0.003-in. sandwich.

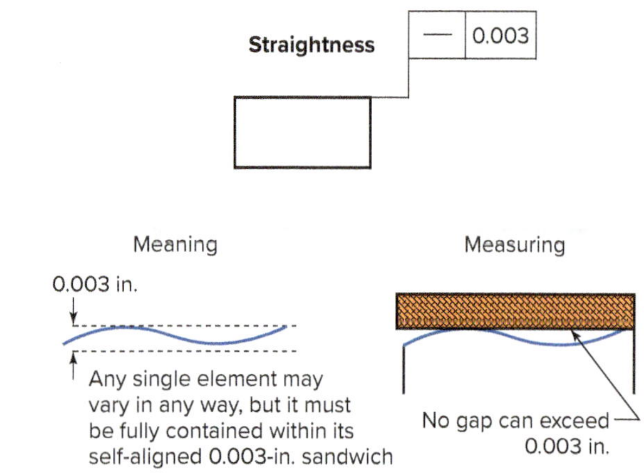

Figure 4-24 Straightness per print, and its meaning.

The most convenient way to test straightness is to place a gage quality straight template against the surface, with the edge running in the direction of the control element (left to right in the drawing). A precision ruler or straightedge could be used for the test. Then using a 0.003-in.-thick gage (a thin metal feeler gage), check to see that no gap is greater than the drawing tolerance between the surface and the straightedge. See Fig. 4-25.

Figure 4-25 Testing a lathe bed for straightness.
Courtesy Mike Fitzpatrick

Single-Element Test–Measuring Factor 3. Straightness elements are tested one at a time, which is known as a *single-element test.* Another way to word this is that a straightness test is two dimensional. It's important to understand this concept, for it's a third factor that dictates how a given control is to be inspected. Several controls are also single-element controls.

When the control is single-element, each element checked must pass its own test. This means each must be shown to fit between two lines 0.010 in. apart in the example. None of the tests are connected. Figure 4-26 illustrates the concept. Both of these objects are straight in one direction only. A precision straightedge could be used to inspect them. Straightness does not control the surface except one element at a time, and that element has a defined direction.

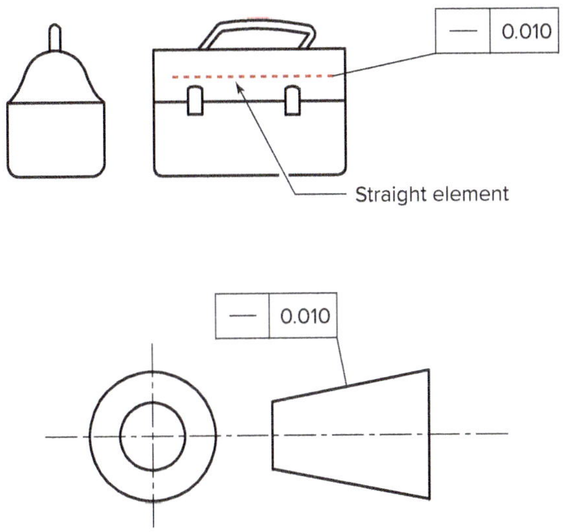

Figure 4-26 A control of straightness applied to the top contour on a lunchbox and to a cone.

To test the lunchbox, look for gaps exceeding the 0.010-in. tolerance compared to a straightedge. How many elements are tested is a matter of how reliable the feature might be. Because the box top is a formed dome, front to back, many elements should be checked, while the cone might be tested in three or four locations.

> **KEYPOINT**
>
> Straightness is a single-element control. It is two-dimensional.

Exception—Straightness Applied to Axes. As mentioned, straightness can be applied to the center axis of a feature. Figure 4-27 is a good example of how much meaning can lie behind GDT symbols. Not the surface but the pin's axis has been toleranced to be straight within 0.003 in. We know the control applies to the centerline because the feature control frame connects to the dimension. Second, we now know that the tolerance zone is not two-dimensional (2-D) because of the theta symbol just before the 0.003 in. This tells the reader that the tolerance zone is cylindrical.

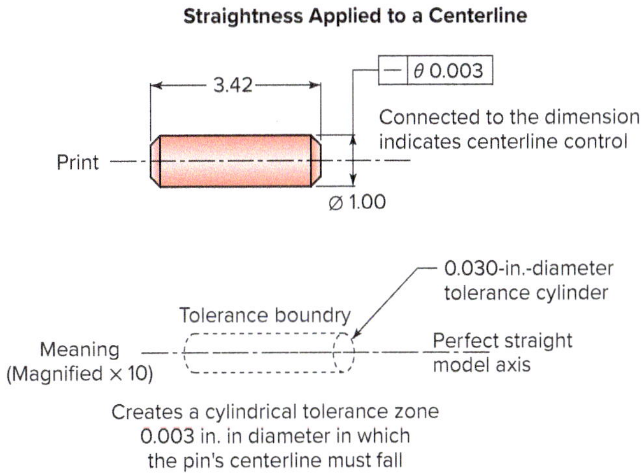

Figure 4-27 Here straightness has been applied to the center axis of a feature, not to the surface.

The axis of the pin can vary in any way, as long as it doesn't violate that tolerance zone cylinder.

Flatness in Three Dimension ▱

Flatness is the three-dimensional extension of straightness (Fig. 4-28). Flatness is an *all-element* control, meaning that the entire surface must be tested at one time. When the control requires an all-element test, individual elements exist, but they are connected into a surface. Here, the lunchbox bottom has been given a flatness tolerance of 0.2 mm. The feature control frame reads "this surface shall be flat within a tolerance of 0.2 mm."

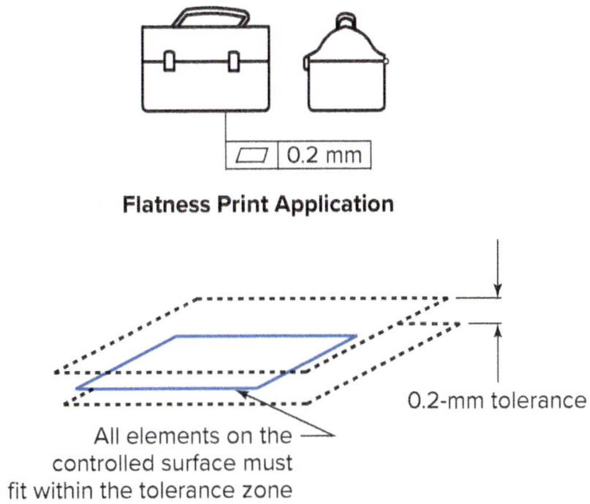

Figure 4-28 Flatness as seen on the drawing and its interpretation.

Testing Flatness. Theoretically, to measure flatness, a perfect, flat template is placed against the surface and gaps are measured similar to checking straightness. However, that is impossible.

To test this flat surface with a straightedge template first using right-to-left elements, then turning it 90 degrees and testing front-back elements **will not work**—that's a test of bidirectional straightness. Think this through: Every single line element both left to right and front to back could pass a 0.2-mm straightness test, yet their total variation would not fit into the 0.2-mm flat, three-dimensional (3-D) sandwich.

So, how does one test for flatness? There is one practical shop method that uses a dial test indicator (see Chapter 6) to sweep the surface and look for the highest high point compared to the lowest low point. However, while this test works okay, it is truly a test of parallelism—not flatness. It shows that the entire surface is parallel to the table datum. Go to Fig. 4-35 to see this test. While it tests parallelism, flatness is embedded within. Therefore, if the object passes the parallelism test within 0.2 mm, it must be flat within 0.2 mm as well. It might be tilted slightly yet be perfectly flat, but nevertheless, it passes the test.

KEYPOINT

Flatness is the 3-D version of straightness.

Keep this 2-D/3-D relationship in mind as we explore further controls. Understanding the difference aids in choosing the right way to measure a given control. Here's the next example: Roundness is a two-dimensional control, while cylindricity extends roundness to a three-dimensional control.

Roundness 2-D Control ○

Continuing our GDT systems approach, roundness can be envisioned as straightness rolled into a perfect circle. The tolerance zone then becomes a pair of concentric circles, whose distance apart forms the tolerance zone for each single element.

In Fig. 4-29, roundness is shown the way it is specified. The second illustration is how it is evaluated. Each circular element must be round within 0.005 in.—that is, each element must be able to fit into a pair of perfect circles that have 0.005 in. between them.

Roundness

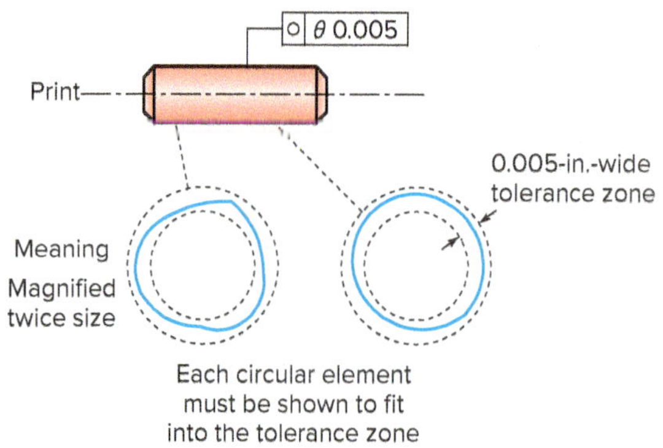

Figure 4-29 A control of *roundness* is a single-element control.

Each element could be a different diameter but cannot exceed the overall diameter tolerance. Again, each is its own roundness test; for example, as with a tapered roller bearing, each element must be round within its own 0.005-in. tolerance. In other words, on the cone the roundness tolerance zone conforms to each element.

TRADE TIP

Don't Fall into the Round Trap! Roundness can be tricky to evaluate at times. It is common to evaluate it using a distance-measuring tool called a micrometer (Chapter 6). However, the problem is that micrometers measure distance, not roundness. The shape shown in Fig. 4-30 is rarely produced if the machine is set up correctly. But it does happen every so often! Called a trichoid, it is not round with three lobes and rounded corners. Even though the trichoid is not round, when tested with a standard micrometer, every point measures the same distance across!

A Trichoid

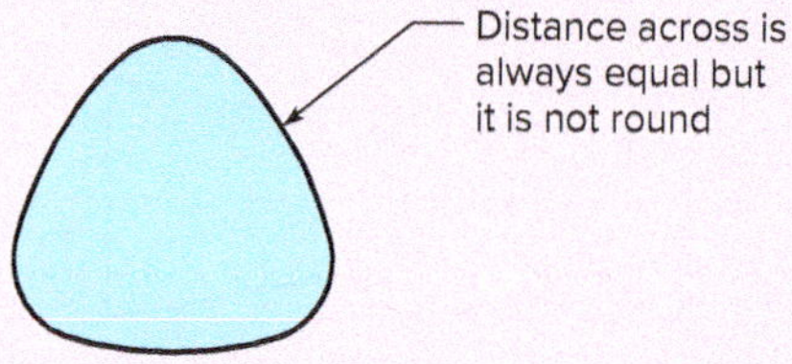

Figure 4-30 A trichoid measures the same distance across, everywhere, yet it is not round.

To detect the trichoid, a second type of micrometer is used that supports the object in a three-point contact, called a tri-anvil micrometer. We will study micrometers and other testing methods later, but for now, be aware there is much more to learn about geometric evaluation.

Form Controls Use No Datum. Have you noticed that none of the form controls were referenced from a datum? That's because they don't need one. Form control needs no outside reference. Think of the control zone as a custom-fitting sandwich that applies to each element or surface "as is." If the element can be shown to fit into its own custom sandwich zone, then that element passes the test. In other words, an object is simply straight without any outside reference.

Cylindricity 3-D Control ⌭

Cylindricity is a three-dimensional, all-element control extending roundness into a cylindrical surface. The tolerance zone then is a pair of perfect floating cylinders into which all circular elements on the controlled feature must lie (Fig. 4-31). In other words, it must be proven that the entire cylinder can fit into two theoretical tolerance cylinders that are the specified distance apart (0.015 in. in the example).

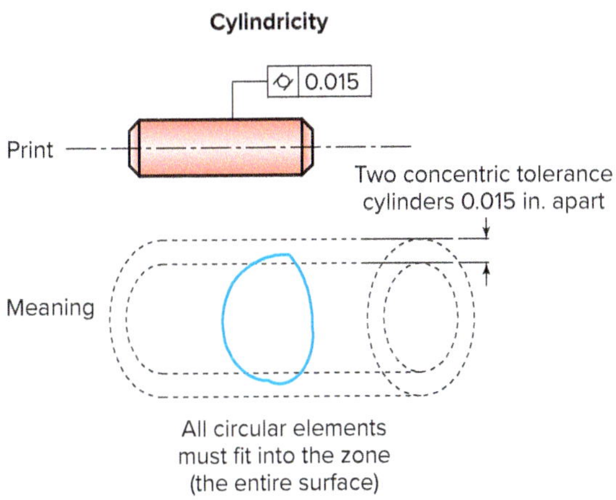

Figure 4-31 A form control of cylindricity.

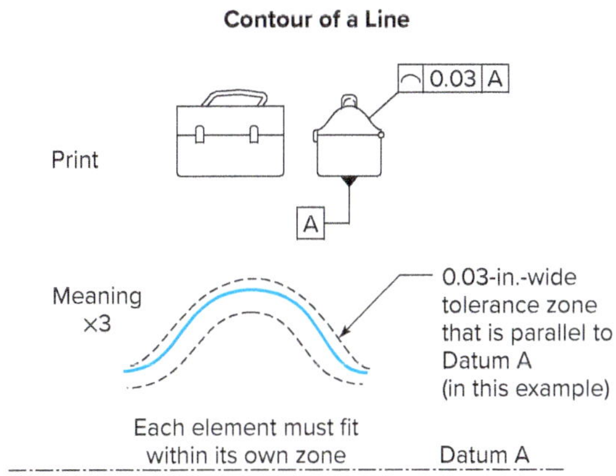

Figure 4-32 Contour of a line is a 2-D control (single element). This contour is oriented to the assembly world with reference to Datum A.

Group 2—Controls of Contour

The **contour** controls were created to do two things: primarily to define and limit the surface of odd-shaped objects and to orient (position) the shape relative to datums. For example, they are used in making a die to form a rain gutter or the top of the lunchbox we've been looking at.

Both are similar to form controls of straightness and flatness. If we were to take a straightness control and bend it into some special shape, then it would become the contour of a line (2-D). Then if we were to take flatness and bend it in one direction only, as with a sheet of paper, without wrinkling, its special shape would become contour of a surface (3-D).

- **Contour of a Line** ⌒
- **Contour of a Surface** ⌓

There are two differences between contour and form: For contour, the perfect model must be defined. This can be by standard dimensioning or by a master template. The master contour can also be defined with a table of *X* by *Y* data points, usually in a database, or the shape can be defined with equations.

The other difference from form is that these two controls may or may not require a datum, depending on function. While the form controls never refer to a datum, the contour may need a reference to orient the shape to the assembly world. For example, in the lunchbox in Fig. 4-32, since the top dome shape must be pointed straight up, the bottom must be assigned a datum, and the contour referenced from it.

KEYPOINT

Controls of contour may or may not require datum reference, depending on function.

Contour of a Line 2-D Control ⌒

Line contour is similar to roundness in that a single element is tested against a template cut to the right shape for evaluation.

> Figure 4-32 states:
>
> This surface shape shall be contoured within 0.03 in. with respect to Datum A.

You can see that the function is that the overall dome should be parallel to the bottom of the box or it would appear skewed. Therefore, this control refers to a datum. If the contour was all that mattered, similar to roundness, then there would be no datum reference shown—stamping out a rain gutter, for example. What difference would it make if the die was tilted, as long as the finished gutter was the right shape?

Contour of a Surface 3-D Control ⌓

Following the 2-D/3-D relationship, line contour is extended out with a straightness element to form a surface control. All elements must now fit into a shaped sandwich as shown in Fig. 4-33. The elements in question are the form shape plus straightness, at 90° to them. Think of the control zone as a flatness control—but bent on one axis only. Within current ASME standards, there will always be a straightness element on the surface in one direction. The GDT system is not used to define surfaces that curve in two directions, called compound curves, other than spheres (next).

KEYPOINT

Since the top of the object in Fig. 4-33 is being formed only, there is no datum to control the orientation of the contour in this example. The shape is all that matters.

Contour of a Surface

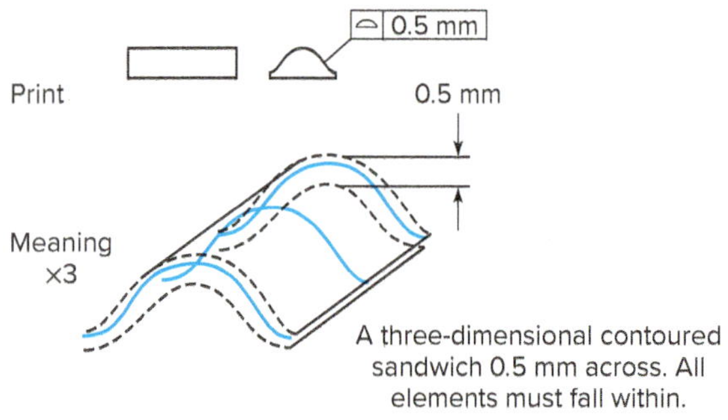

Figure 4-33 Contour of a surface. This control has no datum reference. It is its own reference in this application.

Contour Footnote—Spheres

There is one exception to 3-D shapes defined by the geometric system: the sphere. A sphere is curved in two directions and has no embedded straightness element (ASME Y14.5 M 1994 Standards).

Group 3—Controls of Orientation

Now we look at the three controls that establish the *relationship* of a feature with respect to a datum. They have two new properties. Each **orientation** control can be either a single-element 2-D control or a 3-D all-element control, depending on the function of the object. Second, controls of orientation can apply either to a surface or to an axis, depending on function.

- **Parallelism** //
- **Perpendicularity** ⊥
- **Angularity** ∠

These three controls are nearly the same. Each defines a perfect angular relationship with respect to a datum. The difference concerns

The angle at which the model lies relative to the datum.

Whether the control applies to a surface or to an axis.

Whether it is a single-element or an all-element control.

KEYPOINT

From this point on, including orientation, all controls refer to a datum.

Evaluating Orientation

The controls of orientation are much easier to evaluate than those of form due to the addition of the reference datum. Typically, a dial test indicator (a precision distance measuring tool) is used on the layout table. The table becomes the datum.

In Fig. 4-34, the control of parallelism is applied to the top of the lunchbox. In this case, it is a single-element test. The parallelism does not apply to the entire surface but to each right-left element tested. You might envision that test as straightness that refers to a datum. In fact straightness is embedded in the parallel test.

Control of Parallelism

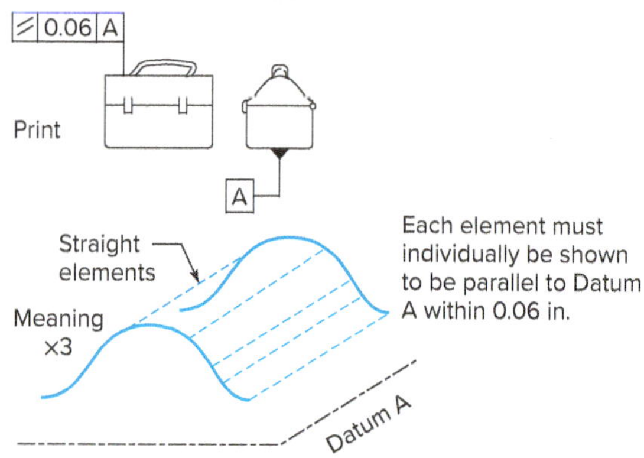

Figure 4-34 A control of parallelism for the top of the lunchbox. In this example, it's a single-element control.

Parallelism //

The perfect target lies at 0° with respect to the datum. The controlled feature can be a line, surface, or axis depending on how it's specified on the design. For a line example, in Fig. 4-35 each element must be parallel to the datum but not to the entire surface. This is a single-element control. The feature control frame states that

Each element must be parallel within 0.060 in. of Datum A.

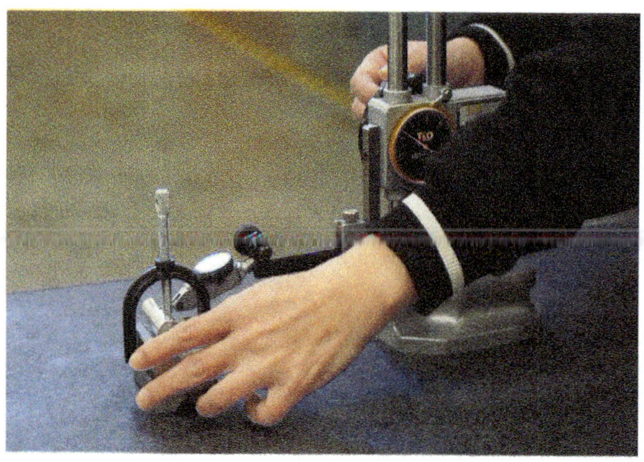

Figure 4-35 Using a dial test indicator to evaluate the parallelism of a surface.
Keith Smith

Critical Thinking Question 1

Okay, let's test your understanding of the system. Parallelism can be an all-element control as with the object shown in Fig. 4-35. That means the entire surface must be parallel to the datum table. Therefore, two controls are embedded within. Can you name them? This is easy, but the next question isn't. The answer is found just before the Unit 4-3 review.

Perpendicularity ⊥

Manufacturing is an orthographic, *X-Y-Z* world. Ninety-degree relationships are common. Perpendicularity is a convenience control for the often occurring, special case where the orientation is to be 90° with respect to the datum. While we could use angularity to do the same job, the GDT system includes special cases such as this.

In Fig. 4-36, the perpendicularity has been applied to the entire end of the lunchbox, therefore it's an all-element control. The whole surface must fit into the 0.15-mm vertical tolerance zone sandwich. Similar to the parallelism of a surface, this perpendicularity control embeds flatness. Flatness then contains straightness as a further subset.

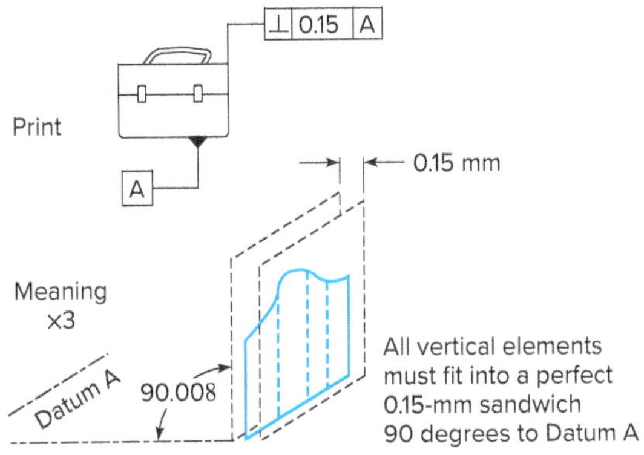

Figure 4-36 A control of perpendicularity.

Critical Thinking Question 2

Talk this over with a fellow student. In solving it, you will have gained insight into GDT. Can embedded controls be toleranced differently than the parent control? For example, in Fig. 4-36, with the perpendicularity tolerance of 0.15 mm on the lunchbox end, could its end flatness then be toleranced to 0.2 mm or to 0.1 mm–which or both? After making your decision, find the answer. Part of the solution is coming up in a Key Point, with the full answer found just before the Unit 4-3 review.

Angularity ∠

Notice in Fig. 4-37 that the 60° angle is placed in a basic frame. This signifies that the angle is the perfect model. Other than the basic frame, the concept is no different from perpendicularity or parallelism, where the target angle was understood to be a perfect 90° or 0°.

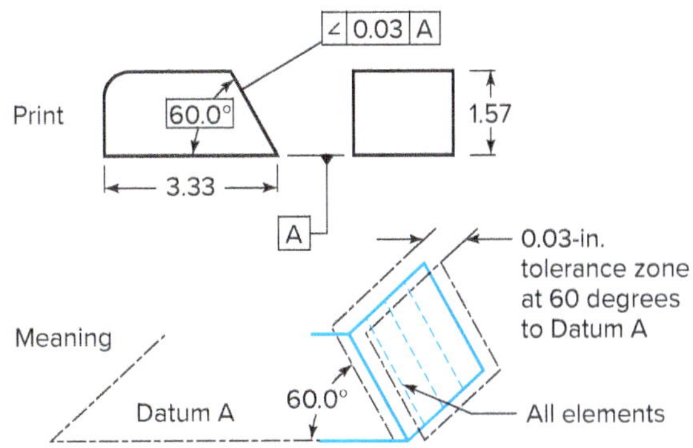

Figure 4-37 A control of angularity.

For perpendicularity, it's understood that the 90° has no tolerance; it is the target. So, too, the 60°, also has no tolerance. The tolerance is built around it.

Measuring a Geometric Angle. One simple way to measure a geometric angle is shown in Fig. 4-38 using a precision protractor (an angle measurement tool discussed in Chapter 7). The protractor is *locked* at the exact 60° angle. The user then checks several elements for gaps that exceed the tolerance when compared to the protractor blade. Feature A must rest on the datum that is the basis for the measurement. Another way to envision this test is that it's similar to testing perpendicularity except the *square* in this case is at 60°.

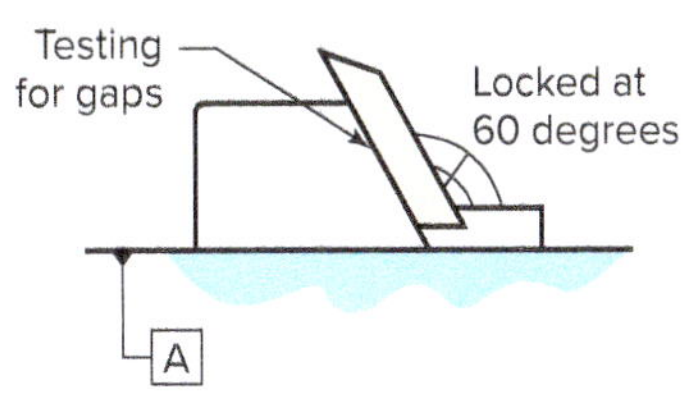

Figure 4-38 Measuring a geometrically defined angle by testing for gaps that exceed the tolerance.

> **KEYPOINT**
>
> **Conclusion**
>
> Embedded (subset) controls may be given a smaller tolerance than that of their parent control, but they cannot be given a greater tolerance since they cannot violate the boundary of the parent that contains them.

Group 4 Controls of Location

The three **location** controls always position the axis of a feature with respect to a datum.

- **Position** ⌖
- **Concentricity** ◎
- **Symmetry** ⌯

KEYPOINT

All three location controls always refer to datums and always control the *center axis of a feature*, never a surface element.

Position ⌖

Following the three-step process, (1) this control defines the perfect model position with respect to one or more datums. (2) Then a tolerance zone is built around that perfect target. (3) Then understanding the shape of the tolerance zone, the way to measure it is understood. The tolerance zone will take on different shapes, depending on the feature to be controlled. We will limit our discussion to the position of a machined hole, which produces a round tolerance zone. (It's actually a cylinder if extended through the part, but we'll restrict this demo to 2-D for now.)

In Fig. 4-39, first, using 30.0-mm and 20.0-mm basic dimensions, the model position is defined. Next, in the control frame, a 0.5-mm round tolerance zone is specified where the hole's centerline must lie. To be in compliance, the hole's axis must fall within the circular tolerance zone built around the target location.

The feature control frame of Fig. 4-39 states that

> The centerline position (of this 9.0-mm hole) shall fall within a round tolerance of 0.5 mm with respect to datums A and B.

In Fig. 4-39, the hole has been drilled slightly off target to the upper left, but its center falls within the 0.5-mm target circle. It is acceptable.

KEYPOINT

The 9.0-mm hole's center in Fig. 4-39 can vary in any direction as long as it doesn't exceed a 0.25-mm radius from the perfect target point.

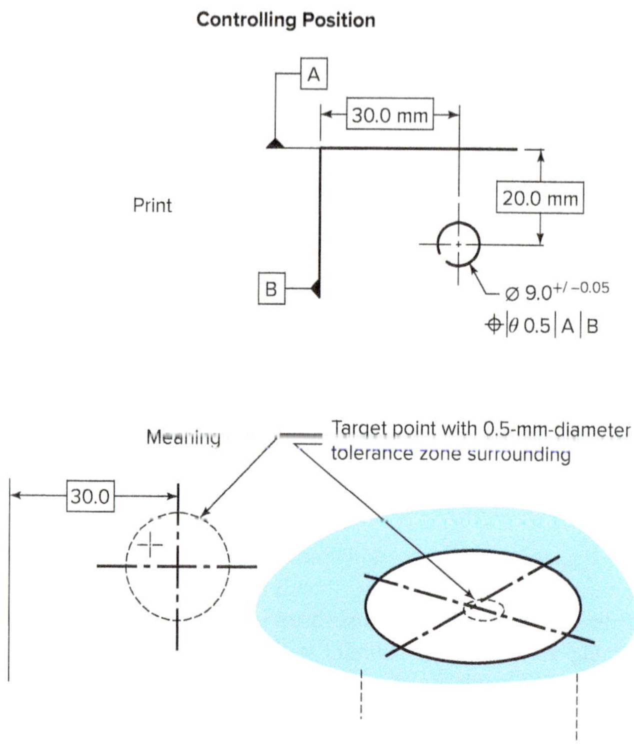

Figure 4-39 A control of position.

Concentricity ◎

Concentricity controls the center of two objects axis to axis. It's similar to position in that the center of a feature is being controlled relative to a datum. The difference is that the datum is an axis, too. Concentricity specifies that the center of one feature must lie within a tolerance zone built around the axis of another feature (the datum). It's the axis-to-axis control shown in Fig. 4-40. For something to be concentric means it will share centers with something else; recall that in GDT, the "something else" is always a datum.

KEYPOINT

Concentricity is correctly applied to the location of features that spin.

TRADE TIP

Give Yourself a Bonus (Tolerance That Is)!

Depending on function, the tolerance for a given feature isn't always the "bottom line." **Bonus** tolerances can be earned as long as you know the rules. When the Ⓜ symbol is present within the control frame, there is a possibility of additional tolerance (Fig. 4-40). The idea is that tolerances must always be based on the tightest possible case for assembly. For example, suppose you were drilling a 1.00-in. hole with a plus/minus diameter tolerance of 0.020 in.

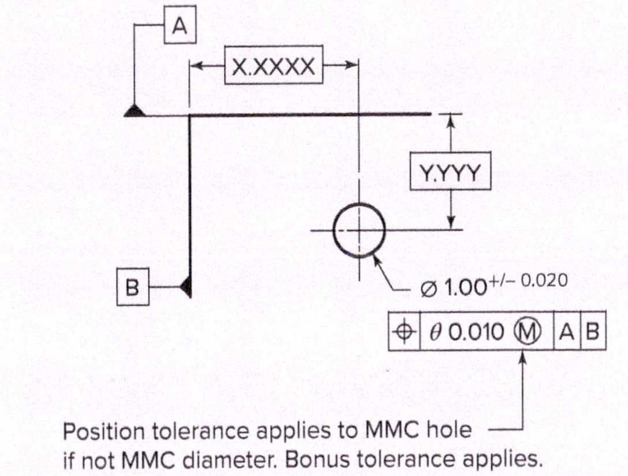

Figure 4-40 This feature control frame includes the MMC symbol.

The tightest assembly case, within the size range, would be a hole at 0.980 in. That's the maximum material condition (size) and is abbreviated as MMC. It is symbolized as an M within a circle, as seen in Fig. 4-40. Then as the feature departs from the tightest condition (a bigger hole), the target location size can increase in many instances, called bonus tolerance.

The feature control frame of Fig. 4-40 states, "The centerline position of this hole shall fall within a 0.010 in. diameter tolerance, when the feature is at MMC size, with respect to datums A and B."

KEYPOINT

MMC is the size where the feature has the most metal on it within the tolerance—or in other words, the tightest assembly. The 0.010 position tolerance applies to a hole of 0.980-in. diameter.

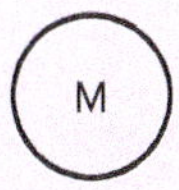

What if the hole wasn't 0.980 in.? Any diameter greater than the MMC could have more location tolerance due to the larger assembly space it creates. That hole could move around a bit more, and whatever was to go in it would still fit. Thus, based on function, if the designer can permit the bonus, the MMC symbol will be added to the control frame.

You'll need to study this more in your formal geometrics training, but in a condensed form, here's the four-step rules if the Ⓜ symbol is shown:

- Determine actual machined size by measuring it.
- Determine MMC size. By applying the feature's tolerance to its nominal size.
- Find the difference in the two numbers, which is the bonus.
- Add that bonus to the stated tolerance in the frame.

Let's try it. Suppose the hole in Fig. 4-40 is machined to 1.003 in. Compute the position tolerance for this hole accounting for bonus. Try it before looking.

The bonus tolerance answer is this: 1.003 is 0.023 in. greater than the MMC size of 0.980. Your bonus is 0.023 in. added to the location tolerance. The answer is 0.033 in. for that hole. Doing this is possible only if the Ⓜ is present, but there may be a bonus just waiting to be claimed by the informed machinist. This is an example of the extra control that the geometric system imparts to manufacturing.

In Fig. 4-41, the tolerance zone is specified to be a 0.5-mm cylinder. The diameter of the cylinder is the limit for the object's centerline.

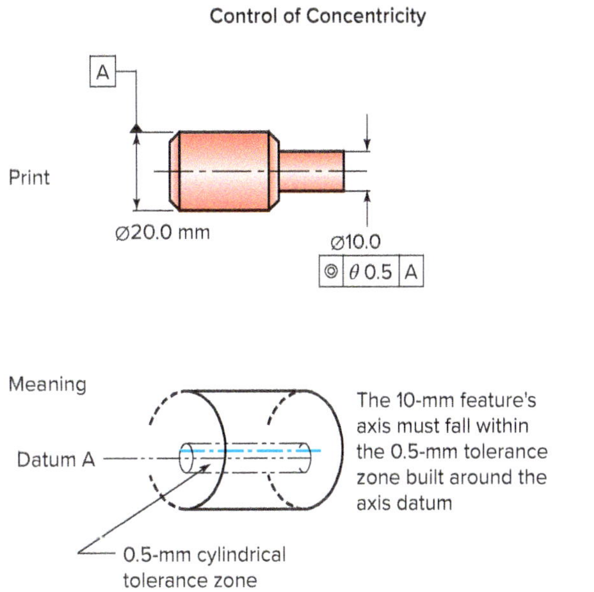

Figure 4-41 *Concentricity* positions a feature centerline with respect to another feature centerline datum.

Never a Bonus for Concentricity. Because concentricity is used to control functions such as vibration or the center of mass on spinning objects, it cannot be given a bonus tolerance (Trade Tip). The function doesn't permit it. This is true for the next control as well; no bonus tolerances apply to symmetry.

Symmetry ⌯

Symmetry infers equal distribution of a feature about a central target (the same on both sides). Similar to position, symmetry requires that the controlled feature's centerline must lie within a tolerance zone built around a perfect target (Fig. 4-42). The difference is that the target is usually a centerline, too. Symmetry is a convenience control that locates objects that are not spinning. In truth, position can be used for the same purpose and has been in the past.

Control of Symmetry

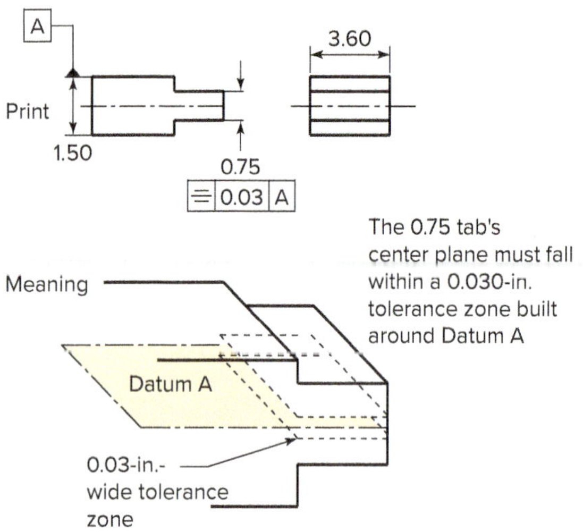

Figure 4-42 Symmetry controls a feature that is equally distributed around a datum.

Group 5—Controls of Runout

- **Single-Element Runout**
- **Total Runout**

Runout is designed to control the outside surface wobble of rotating objects with respect to a central axis datum. Different from concentricity, runout always applies to surface elements. Looking a little deeper, you might see form controls embedded in these controls, too. In fact, runout can be described as a kind of form control with respect to a datum. Also similar to controls of form, the 2-D/3-D relationship is again involved.

SHOPTALK

Changing Standards One geometric system advantage is it is overseen by a committee. As improvements are found, they are added. For example, between 1984 and 1994, symmetry was not being used as a symbol here in North America since it is virtually the same control as geometric position—except no bonus tolerance is possible.

Symmetry is truly a position control—it specifies where a feature's axis must lie with respect to a datum. However, there are times when symmetry is closer to the way we see the function; to think of "centering" rather than locating, as the objective. So, it was reintroduced in 1994.

Single-Element Runout 2-D Control

Two-dimensional runout, sometimes called circular runout, can be applied to round or tapered surfaces, as shown in Fig. 4-43, or to a flat end surface. Each element is an individual test. In Fig. 4-43, to be within tolerance no element may exceed 0.01 in. Full indicator movement (FIM) is explained next.

Control of Runout (single element)

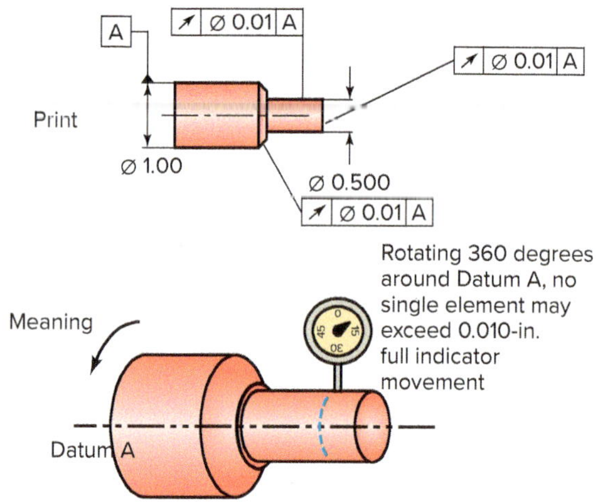

Figure 4-43 Runout as applied to three surfaces of the object.

Measuring Runout. To measure runout, a dial indicator is placed against the surface in question, and the part is spun 360° about its datum axis. The indicator shows movement in thousandths of an inch or decimal parts of a millimeter. As the object spins, the highest point and the lowest point are noted.

The difference between the high and low points is called the **FIM (full indicator movement)** or sometimes it's called the *total indicator reading (TIR)*. The FIM for one circle is the runout of that single element. To comply to a runout tolerance, the FIM for each element tested cannot exceed the tolerance (Fig. 4-44).

Figure 4-44 Testing runout on a lathe, using a dial test indicator to find the FIM.
Lake Washington Technical College, Kirkland, WA/McGraw Hill

KEYPOINT

Each element requires an individual test and none may exceed the tolerance.

Total Runout—Control of a Surface ⌰

This is the all-element, 3-D version. It controls an entire rotating surface with respect to a datum axis. It is tested the same way as single-element runout, but this time the indicator must be moved over the entire surface to test all elements. The highest point is found and compared to the lowest point, and the difference is the total FIM for the surface. In Fig. 4-45, one might envision this control as cylindricity with respect to a datum.

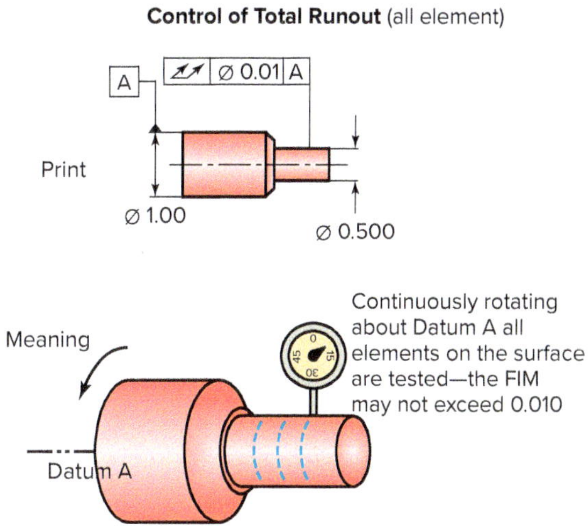

Figure 4-45 Total runout—an all-element, surface control is symbolized by the double arrows.

SHOPTALK

The Fully Informed Machinist It would be a good idea to remember the name of the document that controls GDT—*ASME Y14.5 M* (Fig. 4-46). You will see it cited as a source reference on many technical drawings.

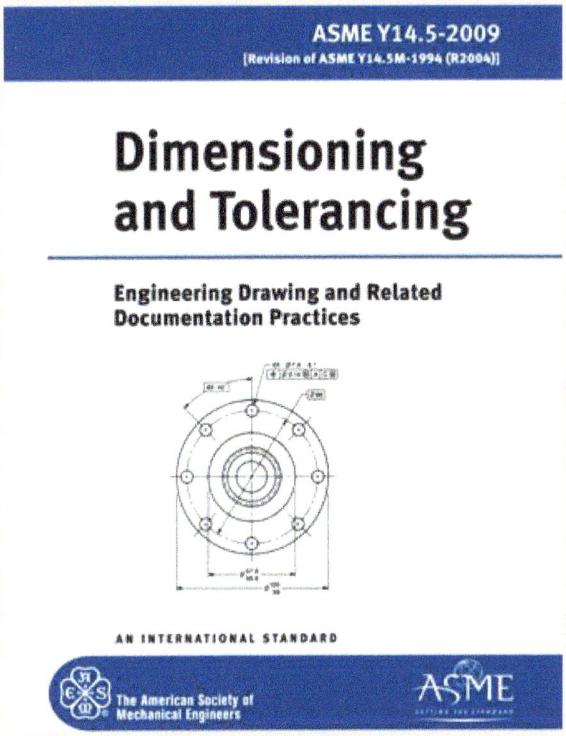

Figure 4-46 *ASME Y14.5*, the living document that controls geometric application in the United States.

Comparing Concentricity to Runout

At first it seems that concentricity and runout are the same controls, but while they both deal with rotational features and a central axis, they are different. Here are a couple of examples. Consider a camshaft lobe (Fig. 4-47). Its profile must rotate around an exact center–it must be concentric. But, to work correctly, the lobe has a designed amount of runout. In the second example, the hex head is concentric to the bolt, yet it also has a lot of runout. From these, you should see that

> *Concentricity* controls the *axis* of a feature with respect to the axis of a second datum feature where both have the same center. *Runout* controls the surface of a feature, with respect to an axis datum.

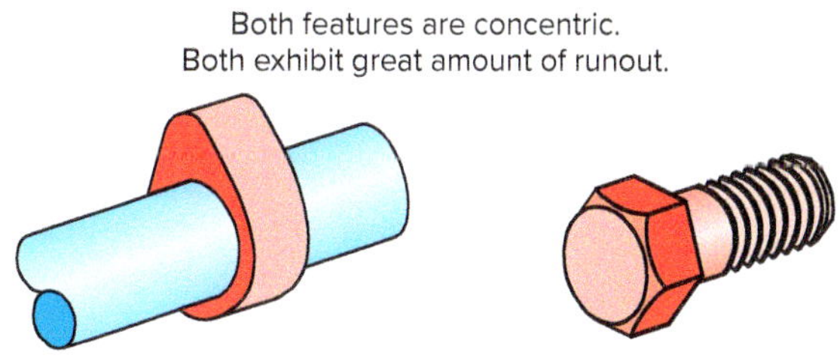

Figure 4-47 Compare runout to concentricity.

Critical Thinking Answers

Answer 1

Straightness and flatness are embedded within the test of surface parallelism.

Answer 2

The flatness of the lunchbox in Fig. 4-34 could be toleranced tighter at 0.1 mm, but to tolerance it to 0.2 mm is meaningless since it must already be flat within 0.15 mm to meet the perpendicularity requirement.

UNIT 4-3 REVIEW

Replay the Key Points

- A geometric control applies to either the feature's surface elements or its center axis.
- Elements are the outside surface control entity.
- For the feature to be within tolerance, each element tested must be shown to lie within the tolerance zone.
- There is an infinite number of elements on any surface.
- Each element runs in the direction of the control.
- Elements could be drawn as pencil lines on the surface of the feature.
- There are interrelationships within the GDT system, and many individual controls can be combined as building blocks, to create control definitions.

Respond

1. List the five groups of controls.
2. What is the reason for using a GDT system compared to the standard dimensioning and tolerancing methods, in 10 words or less?
3. Repeat Question 2, but respond using only one word.
4. A datum is always a reliable surface on the part. Is this statement true or false?
5. A drawing has datums A, B, and C assigned to surfaces. If it isn't already established on the part, which datum feature would probably be machined first? Why?

CHAPTER 4 REVIEW

Ongoing Study

Chapter 4 was written to get going in the lab with a basic GDT background. A second goal was to demonstrate that a lot of geometric knowledge remains unrevealed, but it would be too much information at this point in your training. After studying Chapter 4, as machining challenges come up in the shop, think about function and how the principles we've studied apply. You'll soon be ready for more formal study.

Unit 4-1

Relatively recent rapid advances in technology showed that old ways fell far short of meeting the functional needs of modern manufacturing. The new controls were carefully crafted by engineers, scientists, and manufacturing leaders not only to express drawing dimensions and tolerances but also to

Set functional priorities using datum references

Specifically serve function

Eliminate ambiguities

Unit 4-2

Datum reference is at the heart of the GDT system and is a major improvement over older methods. Datums tie the part to the real world by referencing its dimensions to theoretically perfect surfaces and lines outside the object. The design is built outward from a complete set of these references, starting with the most important function. The highest priority feature is assigned to the highest priority datum, and that makes all the difference in how you proceed as a machinist, inspector, or programmer. Using nongeometric prints, machinists find themselves guessing at how to get started. But datum reference clearly defines the functional targets and where to begin cutting. It doesn't show how to cut them. That aspect remains within the realm of skills. But using datum references, a geometric design clearly shows the functional goals.

Unit 4-3

The GDT system puts more responsibility on the machinist because so much information is implied by the symbols. Not knowing the rules of the game creates needlessly high costs in manufacturing. However, once the system is understood, the savvy craftsperson has control that just wasn't available to earlier machinists using previous dimensioning and tolerancing methods.

QUESTIONS AND PROBLEMS

Answering the following questions about the GDT system will help prepare for the final test.

1. What characteristic does this symbol represent? (Unit 4-1) ⌭
2. What two controls are embedded within the control of Question 1? (Unit 4-3)
3. What does this symbol represent? (Unit 4-1) ⌖

4. Describe the control of Question 3. (Unit 4-3)
5. What is a datum? (Unit 4-2)
6. True or false? A datum cannot be an axis since an axis is a mathematical entity. If it's false, what makes it true? (Unit 4-2)
7. Name the document that standardizes GDT in the United States. (Unit 4-3)
8. When a geometric control applies to a feature's surface, the control entity is a(n) _____. (Unit 4-3)
9. On a sheet of paper, sketch the GDT symbols for
   A. Profile of line
   B. The all-element (3-D) counterpart to straightness
   C. The three controls of orientation—in any order
   D. The three controls that define the angle of a feature's axis with respect to a datum
   E. The location control(s) that *can have* a bonus tolerance
   F. The controls that are *never* referenced from a datum
   G. The four controls of form (Unit 4-3)
10. Explain how a GDT drawing shows the machinist how to organize the first few cuts on the part. (Unit 4-2)

CRITICAL THINKING

These are tough—talk them over with a fellow student or two. Question 11 leads to 12, then to 13 with increasing difficulty. In Fig. 4-48, the larger diameter datum feature A is perfectly centered in the lathe chuck and being rotated about its axis. The smaller diameter is being tested with a dial indicator. The FIM is found to be 0.0035 in. over the *entire surface.*

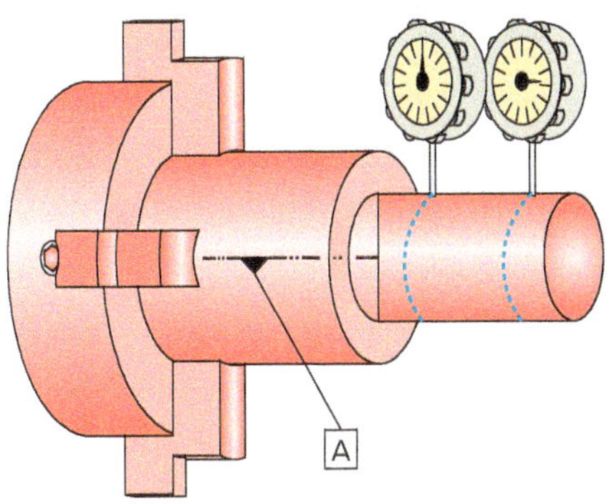

Figure 4-48 What is this test, and what other tests are embedded within it?

11. Is this a test of concentricity or runout?
12. Which controls are embedded in this test?
13. What *range* of possibilities are there for the concentricity of the tested feature? In other words, how little and how much might its center be off from the axis of Datum A? Warning—this is a genuine brainteaser!
14. The 0.750-in. hole in Fig. 4-49 has a size tolerance of ±0.005 in. It has a position tolerance of 0.010 in. The actual drilled hole is measured at 0.7467 in. What is its position tolerance after considering the drilled size?

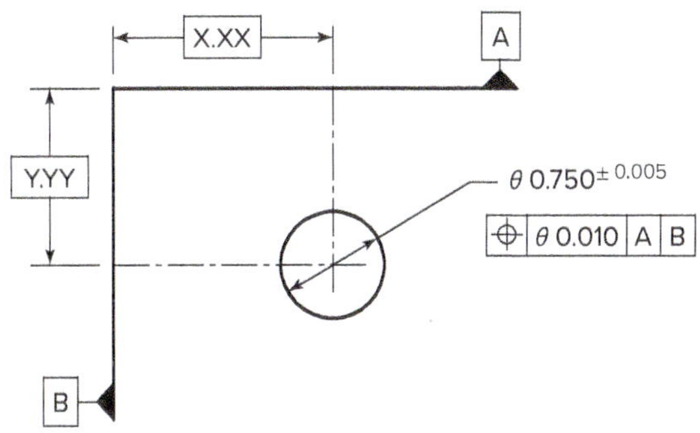

Figure 4-49 What is the position tolerance if this hole is drilled to a 0.7467-in. diameter?

CNC QUESTIONS

15. Lathes machine round objects. On a CNC lathe *X* axis dimensions denote diameter positions of the cutting tool, while the *Z* axis dimensions denote length. If you command the tool to go to *X*3.500, it will be ready to cut a diameter of 3.500 in. You are planning out a program for a CNC lathe to machine an axle with many critical diameters. In terms of function, where would you expect the primary datum to be?
16. Referring to Question 15, at what value would the *X* axis be when it was at that primary datum?

CHAPTER 5 Before and After Machining

Units

5-1 Using Work Orders—Learning Job Planning

- Use work orders, part numbers, job planning, and revisions
- Read and use a typical job sequence
- Understand part and detail numbers on prints

5-2 Selecting the Right Material for the Job

- Identify five basic metals

5-3 Sawing Material

5-4 Part Finishing

- Perform part burring, edge breaking, and chamfering

5-5 Performing Secondary Offload Operations and Assembly

- Production
- Threads and making them by hand
- Hand-reaming
- Press fit assembly

5-6 Part Marking and Identification

INTRODUCTION

Chapter 5 has two overall goals: to learn to read work orders and to learn to perform a collection of tasks found on them that aren't performed on the high production CNC machines. They get material ready for machining or finish it afterward.

Some are machining, since they shape the workpiece, but they are performed by hand at a workbench or on small machines that support the big CNCs. As such, they might be called offload, secondary, or benchwork. Perhaps they are less challenging than CNC work, but not less important, for without them the job wouldn't be complete. Secondary and offload work is often assigned to the newly hired employee.

Chapter 5 will seem long, but in actual fact it's a collection of basic skills that you must have. They were presented in several "benchwork" chapters in older textbooks but our planners streamlined them as much as possible. The knowledge and skill in Chapter 5 isn't glam, but it does make you a well-educated craft candidate! Don't take it lightly!

Unit 5-1 Using Work Orders—Learning Job Planning

Introduction: Nearly all shops big and small use some form of work order. They complement the drawing with step-by-step directions guiding the job to completion. Work orders determine how the job will proceed through the shop. They support consistent quality and safety. There's much vital information on them beyond the job planning sequence, although for the machinist, that's their central purpose. The dated work order lists

Part Number The specific product to be made, abbreviated *P/N*

Batch Quantity Abbreviated *Qty*

Revision Level Showing the design level, abbreviated *Rev*

Sequence of Operations The logical step-by-step instructions, abbreviated *Opp*

The Material Type The form and size of the raw metal provided

The Standards to which the product is to be built

Instructions for Quality Assurance

Finishing and Packaging and other handling instructions

Special Requirements for tooling or other items needed

TERMS TOOLBOX

Detail/dash number Individual identification of a part.

Job planning A set of logical steps found on a work order, chosen to guide a job safely and efficiently through the shop maze.

Part number Drawing and detail numbers combined to identify a single part.

Revision level A letter showing how many times the drawing has been redrawn.

Work order A master document initiating a job in the shop.

KEYPOINT

In addition to symbols, there are a lot of abbreviations used on manufacturing documents. They save time and space both on paper and in the database for the instructions.

5.1.1 Job Sequence Planning

The heart of a work order is a set of carefully planned, specific instructions that guide the job from start to finish. They outline each step and operation in the required sequence. For example, operation 45 tells you to drill a 0.375-in. hole that's located 3.000 in. from Datum A, its reference. To drill these parts the planner has scheduled them at a particular machine called a Bridgeport mill.

Operation	Instructions	Station
Opp 45	Drill 0.375 dia @ 3.0 ref Dat A × 4.375 ref Dat B	Bridgeport
Opp 50	Drill and tap 5/8-11 threads	Bench

Finding the Best Sequence

Each new design requires a puzzle-like solution to find the best way to go about making it. Often, that challenge can be solved several different ways. For example, the job just discussed could also be drilled at the drill press, perhaps back at operation 20. Each path solution creates a completed product, yet one or two paths will be faster, of higher quality, and safer. That best plan requires a full knowledge of all the machining processes available in the shop and those of other shops with which your company may offload the work when yours cannot do it.

Because so much depends on good planning, work orders are written only by highly experienced and well-paid people. As a new employee your goal will be to follow the work order exactly without deviating unless given direct orders by your supervision. One way to learn job planning is to consider, Why was this particular path chosen?

Here in Chapter 5 we will investigate how to follow work orders. Then, later in this text after the study of lathes, mills, grinders, and drill presses, in Chapter 15, we'll look at the logic of creating a job plan. Then we'll study CNC program planning in Chapter 21. To get the ball rolling, at the end you'll be given a chance to solve a few planning puzzles yourself.

Followed—In Sequence

The work order sometimes overrides the design print. Often we must deviate from the drawing by following special instructions on the work order. For example, you may leave extra material on dimensions so the parts can be sent out to be heat-treated, then returned to the shop for final machining.

Opp 55	Mill 3.00 to 3.060 (grind excess)	Bridgeport

Here at operation 55, they are being rough machined, to be ground to size later. Missing this step and making the parts exactly like the print, 3.000 in. instead of 3.060 in., would be a huge mistake! Sometime later, they will be moved to the precision grinding machine to finish. This is so critical that in some industries such as passenger aircraft, each operation in the plan must be certified complete by the machinist or an auditing inspector before moving the job forward. Records must be kept to prove the certification of the steps.

KEY POINT

All journey machinists are expected to keep their "paperwork" straight. It seems complicated at first but not after using it for a while.

5.1.2 What You Must Know About Job Planning

Poor **job planning** leads to expensive or even dangerous traps in machining. It is possible to "machine yourself into a corner." That is, the poorly planned part is unfinished, yet there is no practical way to hold it or to machine the remaining operations. Or due to bad planning, it could be finished but only at great cost or risk to the machine and its operator.

To simulate a real job, here in Chapter 5 we'll track the work order instructions for a batch of 12 parts. We'll make a dozen drill gages (QTY-12) using Work Order Number S5-U1-1 (Fig. 5-1) and the drawing in Fig. 5-2. Bookmark them; we'll be refer to them several times. That's typical of the way jobs proceed in the shop, going back to the work order and printing it often.

Part Numbers and Drawing Numbers

Often, a technical drawing shows a single detail part. However, it's also common for drawings to show several details–individual parts but on the same drawing number. A **detail** is a single unique part–sometimes called a **dash number**. Thus, the drill gage is FY101-1.

KEYPOINT

Remember, different dash numbers can be found under the same drawing number.

Figure 5-3 depicts three dash numbers. To make any one of them, you would need the right drawing and the dimensions of the dash number called out on the **work order**.

203B-605-3 (Piston)

203B-605-5 (Rod)

203B-605-7 (Pin)

Work Order S5-U1-1	McGraw Machining Company	Date: Jan 6, 20xx
Drawing FYM-101	Detail-1	Revision: New
Quantity 12	Customer: Student	Issued By: Fitz
Sequence	Operation	Complete
10	Saw blanks 0.125 aluminum 0.1 excess—all sides	
20	Square up—3.0 3 5.0 vertical mill	
30	Burr blanks	
40	Inspect operation 20	
50	Layout holes per print	
70	Locate, drill, and ream holes—drill press	
80	Burr holes	
90	Layout gage details	
100	Saw details with excess 0.1	
110	Profile mill shape	
120	Part finish—sharp edges and clean out cutter radii 1 place	
130	File 0.12 corner radii	
140	Part mark all hole sizes per print	
150	Final inspection	
160	Annodize—black	
170	Count wrap and package	
180	Ship to customer	
190		
200		

Figure 5-1 Typical work order plan.

Part Number Breakdown

There are industry-to-industry differences; however, **part numbers** generally break down this way: In our example, the overall product is a 203B. For example, let's say it's a chainsaw Model 203, and it's been engineered to level B. Then this drawing is a detail sheet of the 605 section of the 203B saw–engine parts. Finally, there are three details on this drawing: the 203B-605-3 and the -5 and -7 (called the "dash 5 and dash 7" in shop lingo). The more complex the assembly on which the parts are to go, the longer will be the part number. This part number has three data fields and one dash number; others might have five or six depicting the complexity of the work they control.

KEYPOINT

When the drawing number (203B-605) is combined with the detail number (-3), it forms the part number: 203B-605-3.

Sometimes, there is little difference between one detail and the next. For example, a dash 5 part and the dash 6 might be mirror images of each other–the same but right- and left-hand versions. Or the -5 is 1.500 in. long while the -7 is 1.550 in. long; these are small things that matter a lot.

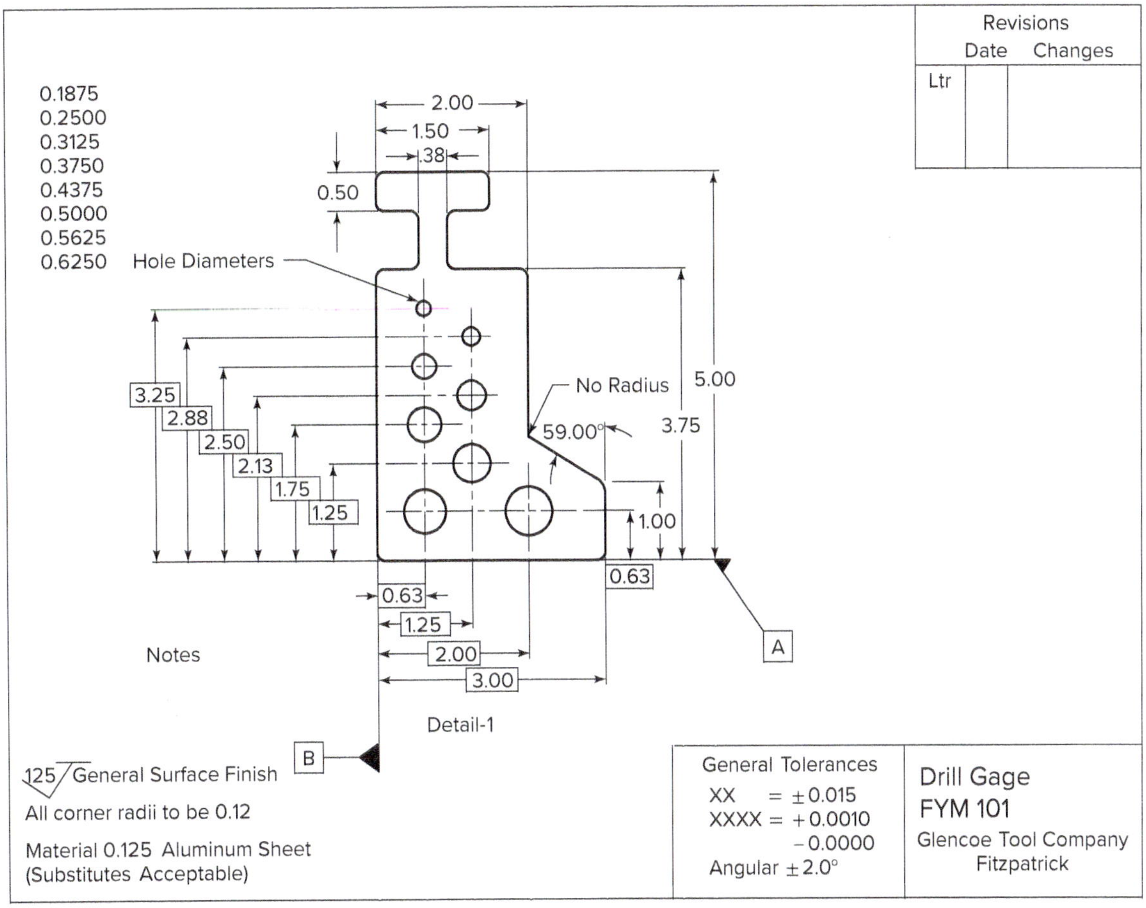

Figure 5-2 Drill gage drawing—sheet 1 of 1.

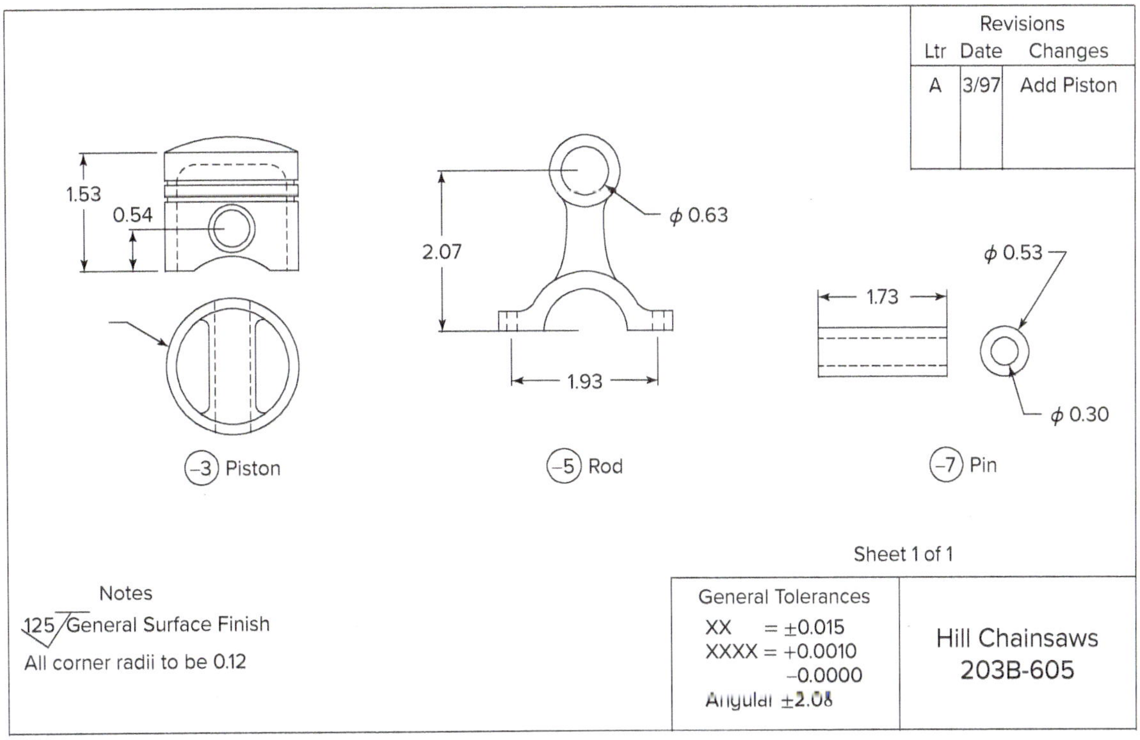

Figure 5-3 Drawing number 203B-605.

Multiple Sheets for One Drawing

Drawing number 203B605 might also have more than one sheet. Be on the watch for the note

Sheet 1 of XX

While it's correctly called "a" print or "a" drawing, it's not necessarily a single sheet of paper. This drawing might have several pages:

203B-605 Sheet 1 of 3

203B-605 Sheet 2 of 3

203B-605 Sht.3 of 3 (typical abbreviation)

If the drawing is a single sheet it might still say Sheet 1 of 1 to eliminate guessing.

TRADE TIP

General Notes on Multiple Sheet Drawings *Caution!* Don't get caught without the needed information! The important, general notes for multisheet drawings are found on Sheet 1 or in a document that goes with the drawing. **Never work a job without having or knowing Sheet 1 even when the assigned detail is on another sheet.**

Revisions—The Right Version of the Part Number

The final fact to be coordinated by the machinist is the *version* of the part to be made called the Rev level. In addition to ensuring that the drawing number and sheet number match the work order, you need to verify that the drawing you've been given is the right *version* of the part. That, too, will be recorded on the W/O (abbreviation for work order).

As original designs are improved and mistakes are corrected, drawings are updated to the latest version. These improvements are rigorously tracked and recorded in what is called the **revision level** system. Each time the drawing is changed, it's given a new *REV* letter. Drawings start as *NEW* releases, then on the first design/drawing change become *Rev A*. The second redraw is *Rev B*, and so on. Fig. 5-3 is at Rev level A—it's been upgraded once.

Each progressive revision letter signifies that the drawing has been redrawn and any number of changes could have been incorporated into the design print at that time. Look in the upper right corner of Figs. 5-2 and 5-3, the sample prints. You must know the level of the drawing.

TRADE TIP

Besides using letters to track revisions, in your shop, drawing revisions can also be tracked by number or by the date on which they were updated.

Check the Revision Box

Look in the upper-right corner of the drill gage drawing (Fig. 5-2). There have been no revisions; this is a NEW release drawing. It may be a very old design, but it has never been revised. The second print 203B-605 is at Rev A. It has been updated once, probably when the saw was improved. Brief notes in the Rev box give clues as to what those changes were. Normally you needn't worry about what the changes were; they are history as far as you are concerned. They have been incorporated into the print. The important thing to ensure is that the next key points are followed.

KEYPOINT

When receiving a job, always verify that

A. The part number matches the W/O.

B. The drawing revision level and work order revision level agree.

SHOPTALK

Replacement Parts Some shops produce replacement parts for older products—*not the latest version!* In that case, detective work must be done. The changes made over time must be backtracked from the current Rev level back to the ones ordered. Each change must be investigated to see if it is compatible with the latest parts. Will they interchange, or does the planner need to modify them to fit the older product? These can be incredibly complex research issues that must be done before releasing the job to the shop. The good news is that beginning machinists never have to do them! After the planner solves the issues, he or she writes a work order covering the steps needed to make the older version of the part, guiding you to make the right parts.

UNIT 5-1 REVIEW

Replay the Key Points

- Learning the information presented here and in the next units may not change tech school experience greatly, but it will make you very capable on the job.
- Even though the design details are found on all of the drawing sheets, it would be wrong to proceed without the first drawing sheet, which shows the notes and general tolerances.
- You could have the right drawing number and the right dash number and still make the wrong part. How? The wrong revision level!

- Often there are intermediate steps required before machining the final shape. They are found in the job planning within the work order.

Respond

Are you ready to use a work order? Answer the following to find out:

1. Name at least two reasons why sequence number 40 might be on work order S5-U1-1 (Fig. 5-1).
2. Why would it be a big mistake to drill all the holes to the sizes shown on drawing FYM 101 (Fig. 5-2)? Which W/O sequence covers this?
3. According to W/O S5-U1-1, what revision level drawing is to be used?
4. What does the work order instruction mean at sequence 120?
5. To what size must the 0.125-in.-thick blanks be sawed at sequence 10?

Unit 5-2 Selecting the Right Material for the Job

Introduction: The first aspect of getting a job ready for the machines is to choose the right metal, in the right condition and form. There are literally hundreds of different *alloys* (metal combinations), but this unit focuses on the five most common. As a new machinist you should be able to roughly identify them by color, relative weight, code marks, and other characteristics that we'll investigate.

It's not enough to get the right alloy; machinists must also see that the form and condition of the metal are right for the job, too. It starts with knowing the necessary terms.

TERMS TOOLBOX

Alloy Noun: The specific metal formula within a group. An exact combination of a base metal, secondary metals, and lesser amounts of minerals. Verb: To put one metal within another.

Alloy steels A family of steels used for production parts. Various alloy steels range from plain mild steel through highly engineered high-carbon steels.

Aluminum A lightweight, white-silver metal that does not adhere to a magnet. Has a very high machinability rating.

Brass/bronze Two relatively heavy metal groups with a base of copper. Colors run from yellow-gold to reddish. Not all alloys adhere to magnets.

Cast iron A dull gray heavy group of metals. They do adhere to magnets.

Castings Molten metal is poured into a mold. Castings are weaker than forgings, but they can have many internal details.

Composite A custom-made material composed of plastic resins, metals, and fibers. Composites provide strength-to-weight ratios not possible in standard metals.

Extrusion A preform made by forcing heat-softened metal out through dies. Aluminum, copper, and brass are often supplied as extrusions.

Forging A preform made by hammering and forcing metal into progressively more detailed shapes. Stronger than castings but less complex in shape. May be accomplished on heated or cold metal. Costly, but very strong.

Grain The strength bias characteristic created by the way the metal was originally formed. Different metal forms (bars or extrusions, for example) have different amounts of grain structure.

Mild steel/low-carbon steel A general utility steel with less than 0.3 percent carbon. Simple heat treatment to change its physical characteristics is not possible. Work hardening is not common or problematic.

Stainless steel A metal grouping based on iron with added chromium and nickel. Some alloys will adhere to a magnet and others will not. Ranging in machinability from moderate to very low, corrosion resistant steel (CRES) will work harden.

Tool steels A family of steels that are more complex in formula and more specific in application than alloy steels and more expensive. They are generally used to make cutting tools, dies, and molds that must endure extreme conditions.

Work hardening A characteristic of some alloys that change hardness due to incorrect machining actions. Once work hardening has occurred, it is difficult or impossible to machine the object further until heat-treated back to a softer state.

SHOPTALK

Alloying and Alchemy For hundreds of years, creating a new alloy has been a trial-and-error process with a bit of magic thrown in. At the dawn of the science, *alchemists* tried to make gold by melting and putting combinations of heavy and shiny metals together—but no luck. But from those early experiments, they discovered that many of their creations were greater than their parts, that they exhibit characteristics beyond those of their base elements! But today, given deep computing power and newly discovered math models of those mysterious properties, *metallurgists* (new metal magicians) can model the alloy before ever melting and combining ingredients. Using scientific methods, do you suppose they might conjure up gold? Yes? No? Why?

5.2.1 Material Identification and Characteristics

Work Order S5-U2-1 Op 10 (Opp 10)

For your first trip into the world of metallurgy, we'll look at

Aluminum, Cast Iron, Brass, Steel, and Stainless Steel

These materials can be differentiated by color, weight, magnetic attraction, spark color and pattern (when ground), and their factory label. On completion of the reading, you will be asked to go into the shop and find examples of these metals, then answer questions about them.

Machinability

One important characteristic of various metals is machinability. This is a rough guide to the comparative ease or difficulty with which a particular alloy can be cut. It's based on the hardness and toughness of the metal. A high rating indicates an easy metal to cut like aluminum, while a low rating means extra caution is needed. The low-rated metals, such as hardened steel or stainless steel, require slower cutting speeds, extra sharp tools, and coolants. The higher-rated metals, such as brass, can be cut much faster with far less concern for heat, cutting tool wear, and breakdown.

Metal hardness is the key. (Hardness is studied in Chapter 14.) As hardness increases, the machinability rating decreases. It becomes more difficult to machine. Cutters wear quickly; plus size and finish are more difficult to control. Some alloys exceed the limits of practical machinability due to their hardness. They cannot be cut with standard cutting tools. These must be machined by grinding or electrical spark erosion (all studied in this book). Or they must first be softened through heat treatment, then machined.

Metallurgy is a big subject, but for now a background in the five groups is all that's required to get started. Each group has very different machinability. It makes all the difference in the speed with which the work may be cut without destroying tools and/or the work.

Groupings and Alloys

Metal is classified with two headings–the *general family* and the *specific alloy*. The family name is based on a single, major ingredient. For example, in our five, steel is based on iron, while brass is based on copper. But, furthermore, there are many **alloys** within a group determined by their exact composition. Different alloying ingredients customize the metal for different purposes. So, starting with the base metal, each added ingredient provides some improvement for its use on a product.

KEYPOINT

In most cases, the exact alloy called out on the work order and print must be used for that job. Your job during material prep is to make it so—get and saw the right stuff.

For example, starting with iron, the mineral carbon is added for hardness and chromium and molybdenum (both metallic elements) are added for strength and toughness. With this unique composition of ingredients, it becomes *chrome-moly* steel, common in performance bicycle frames.

5.2.2 Start-Up Metal Facts

The field has a name: *metallurgy with college study needed to become a metallurgist.* Machinists need to know about metals, too. Collecting all the facts will be a career-long activity. For this lesson, associate only the highlights with the metal. Most of all, remember the facts that affect your machine setups like the metal's machinability, the cutting tool's life expectancy when cutting it, and whether or not coolant is needed.

Physical Forms—Extrusions, Forgings, and Castings

We'll start by looking at three forms in which a given job might come to your machine other than in simple sawed blocks, sheets, or bars:

Castings Forgings Extrusions

All three save machining time because they are roughly preshaped toward the finished product. Our five metals vary as to the forms in which they may be supplied.

Forgings (Fig. 5-4) have been forcefully hammered from a solid block (called a billet) into a nearly finished shape. The forging hammers are massive machines that deform the metal by hitting it with great force, smashing it down in stages into precision dies. Forging might be accomplished on hot or cold metal. Forgings are exceptionally strong due to the working of the metal. Due to the initial tooling required to produce the dies and the individual handling of each part when forged, they are costly to produce, yet they save much time in machining.

Figure 5-4 These parts will become drive joint sets, so they must be the strongest physical form of the metal, a forging.
Photography by Prographics at Milwaukee Area Technical College/McGraw Hill

KEYPOINT

When the duty to which the part will be put requires the strongest version of the metal, then a forging is usually the answer.

Extrusions are long, complex shapes that are preshaped by forcing the metal through extrusion dies, much like toothpaste from a tube. To make an extrusion, the metal is heated to soften it to a plastic state, whereby it can be extruded through the die (Fig. 5-5). Not all alloys can be extruded practically. Aluminum, brass, and copper, all relatively soft metals, are the most common metals to be extruded.

Figure 5-5 An extrusion is forced out through shaping dies into long lengths.
Courtesy Universal Aerospace/Arlington, WA

An example of an extrusion might be the frame around a screen door. First, at the extrusion mill the metal is extruded in a continuous length and then cut into 20-ft sections. The extrusion is further cut to the job length in the shop, then machined to final shape. Extrusions are also costly to create due to tooling costs to make the dies, but they are super-economical for production since they require far less handling than forgings.

Castings are the result of molten metal being poured or injected into molds or dies. Castings are made by two different processes, depending on the accuracy required: Sand castings are less accurate; die castings are closer to the final shape and smoother.

Sand Casting Everyday castings are made by pouring the liquid metal into a hole in compacted sand. Beforehand, the cavity is created by tightly packing lightly moistened sand around a model of the casting. This is called a *pattern.* After packing, the pattern is carefully removed to leave behind a roughly accurate hole in the sand (Fig. 5-6).

After the liquid metal cools, the sand is broken away, leaving the final shape. The surface of a sand casting is as rough as the sand from which it was molded. Typical examples might be the engine block in your car or an iron frying pan.

Die Casting When more precise castings are required, the liquid metal is poured into a precision die mold. These are the more costly version of a casting, since the dies are complex. But die cast metal is often accurate enough that many features require no further machining. Typical examples might be the metal frame in your car's radio or the carburetor body.

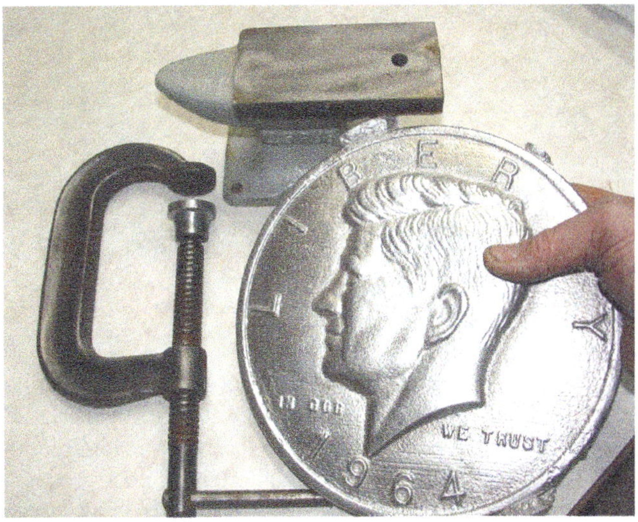

Figure 5-6 Castings are made by pouring molten steel (clamp), iron (small anvil), or aluminum into a sand mold, which results in a rough surface as seen on the large coin.
Courtesy Mike Fitzpatrick

Both casting types look similar to forgings, but they are not as strong. Since they are molded and not hammered into shape, castings can be made with complex internal features not possible with forgings, but they exhibit a large, crude crystal structure, so they haven't the strength of a forging or extrusion.

KEYPOINT

Castings are less strong than forgings, but they may have intricate internal details possible only through casting.

5.2.3 The Five Basic Metals

Aluminum

Aluminum (abbreviated AL) is a comparatively lightweight metal that is a joy to machine. It has the highest machinability rating possible. It can be machined and sawed at very high removal rates with very little tool wear. (*Caution:* There are exceptions to that statement–called the super-high-silica alloys.) Of our five metals, aluminum shows the whitest hue of the silver-colored metals we will examine (Fig. 5-7).

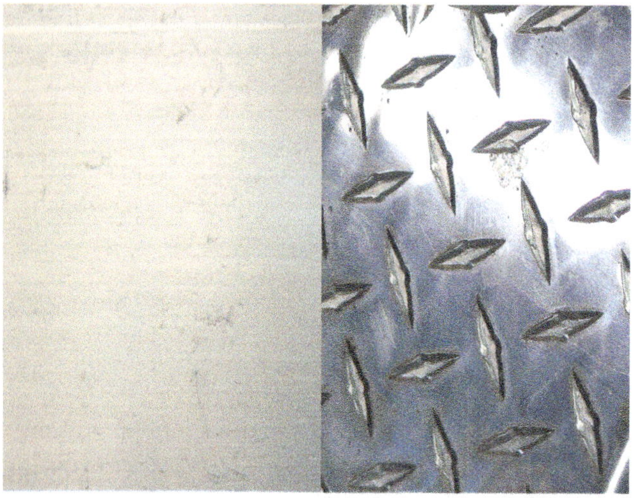

Figure 5-7 Aluminum is a light metal with the whitest hue of the metals we will look at.
Courtesy Mike Fitzpatrick

Some aluminum alloys can be heat-treated to exhibit differing physical characteristics (toughness and hardness), but none ever become hard enough to change their machinability. Aluminum is used extensively on aircraft and on other products where weight to strength is a factor. It is supplied in all forms: long bars, blocks, thin rolled sheets, castings, forgings, and extrusions. Cutting tools have a very long life when cutting most aluminum alloys–that is, they do not become dull for a long time.

SHOPTALK

Aluminum Rocks (Literally)! Pure aluminum is so soft that it's nearly worthless in mechanical applications. The mineral silica is added to give it the needed strength. Moderate amounts are added to most alloys; thus, the alloy is easy on tools. But, the newer superalloys eat cutting tools as quickly as hard steels! Why? Silica is the stuff in rocks, sand, and window glass! It's harder than the cutting tools! Today, cutting tools with diamond particles grown on their cutting surfaces seem to be the best solution to resisting the abrasion of the silica.

Aluminum machines best with a cutting fluid on the cutter, but it can be machined dry. The fluid, or coolant, reduces heat, aids with chip removal, improves finishes, and lengthens tool life. Softer alloys of aluminum have a tendency to jam and clog cutters, known as "loading the cutter." When this happens, the necessary space between cutter teeth is jammed with unwanted aluminum. All cutting comes to a stop, and the cutter can even break if the situation is not corrected. Soft metal such as this is referred to as a *gummy* alloy, and cutting fluids are the main solution to the problem. See Fig. 5-8.

Figure 5-8 Machining aluminum can result in long strings if not deliberately broken.
Lake Washington Technical College, Kirkland, WA/McGraw Hill

Aluminum Factoids

Aluminum will not adhere to a magnet.

Aluminum will not spark when ground. Do not grind it because it "clogs" the grind wheel.

There is a family of extremely high-silica aluminum alloys that wear on cutting tools faster than most steels. These superalloys are brittle but very wear resistant. They are used in automotive applications–piston cylinders, for example.

Hot aluminum chips do not change color during machining (many other metals will).

Aluminum's *malleability* (ability to be squished, stretched, or otherwise deformed without breaking) tends to cause long, stringy chips on the lathe. This is dangerous and should be prevented.

Brass and Other Copper-Based Metals

Brass is an alloy of copper with added tin. Adding zinc and various other ingredients creates a bronze. Bronze is tougher, harder, and stronger than brass but due to cost is less common in school labs. However, you might see brass in training and perhaps the reddish-colored parent metal copper. Copper itself is very soft; thus, little is machined from it. Due to its great *malleability*, copper is extruded, rolled, and drawn (pulled) into pipes and electrical wire. Copper without other alloying ingredients is gummy when machined.

Brass The alloying ingredients of brass (Fig. 5-9) turn the soft red copper into yellow or red-yellow brass and make it tougher, harder, and durable under mechanical forces and friction. Copper is a heavy metal; thus, brass and bronze are relatively heavy metals, too, about twice the weight per volume of aluminum. There are many varieties of both brass and bronze differing in hardness and toughness. Brass and bronze can be cast and forged and are also commonly supplied as bars and plates. Brass is also extruded but not as much as aluminum.

Figure 5-9 Typical brass products. Note the color range.
Courtesy Mike Fitzpatrick

Many alloys of both brass and bronze have good antifriction bearing qualities and are used where metal-to-metal contact friction is encountered. They are also used for decorative items since they can be polished to high luster (shine).

Most brasses are considered free machining (a term used for good machinability) just below aluminum. Coolants are used only to remove heat and flush out chips. Coolant does not improve the machine finish or tool life in most alloys other than the hard bronzes. Since brass can be machined without coolant it is called a "dry" metal.

KEYPOINT

Brass can be machined without coolant, but it's needed for the harder bronzes.

Brass and Bronze Factoids

Brass/bronze will not hold to a magnet.

Brass/bronze will not spark during grinding; however, it is considered poor technique to grind it. The soft metal tends to clog the pores in the wheel and render it useless until it is dressed to remove the contaminating metal. This is true for aluminum, too—do not grind it on standard grinding wheels.

Due to the high copper content, both alloys are used where electricity is conducted as in contact springs.

Common yellow brass is the right stuff to make your small machinist hammerhead.

Weird: Adding just 2 percent beryllium and 2 percent other metals (cobalt and/or vanadium and/or nickel) to plain copper yields beryllium copper, which can then be hardened enough to machine soft steel even though it's made of 96 percent soft copper!

With some brass or bronze alloys, the chips turn fascinating rainbow colors during machining while other alloys remain unchanged.

TRADE TIP

Work Hardening Work hardening is a machining problem *caused* by the machinist. It is the direct result of some wrong cutting action; cutter rubbing, dull cutting tools, and over-speeding are the main factors. Rubbing happens when a tool remains in contact with the work while it spins, yet no chip is made. We call that *dwelling* the cutter. Dwelling can be useful in a program for improving finish, but if too much dwell is applied on some alloys, the work becomes impossible to machine further. Additionally, the tool is dulled and made useless as well.

Cutting speeds that are too fast can also cause work hardening. These errors can cause surface hardness in some alloys and not in others. You must know which are prone to hardening and which are not. For example, common brasses will not work harden; however, many alloys of bronze will harden almost immediately during a dwell. For comparison, all aluminum alloys will not work harden.

Cast Iron

Cast iron (Fig. 5-10) is a dull gray metal used extensively to make heavy parts. Cast iron, abbreviated to CI, is easy to machine with a relatively high machinability rating; however, a few sensitive alloys can work harden beyond practical machinability. Also, due to the casting process, CI can form a very hard, thin outer crust. Another unique challenge for CI: surprise pockets of unwanted trapped sand left over from the mold can cause catastrophic tool failure. These highly undesirable pockets always break the cutting tool. While CI is nearly always supplied in the form of castings, it is sometimes brought to the machine in cast blocks.

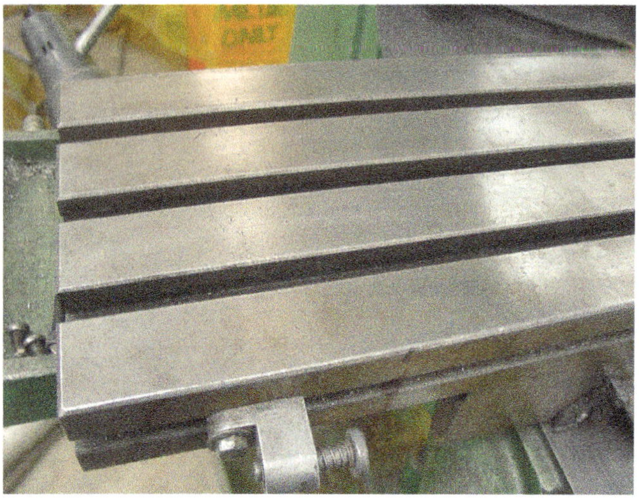

Figure 5-10 A typical cast iron object, this machine table appears dull gray even after machining and grinding.
Courtesy Mike Fitzpatrick

CI is considered a dry metal. It requires no coolant during machining except to wash out the chips and the ugly free carbon dust produced and to cool the cutting tools.

Tools exhibit a medium-long life span when machining CI. The major reason for tool wear is due to work hardening and to the surface crust. If you do not have to cut through the crust, tools last longer.

TRADE TIP

Cast Iron Safety Precaution When machined, CI produces a fine black carbon dust (Fig. 5-11). Wear protection from breathing the dust. Cleanup after machining it is important, too, both for yourself and to keep the dust from clogging machines.

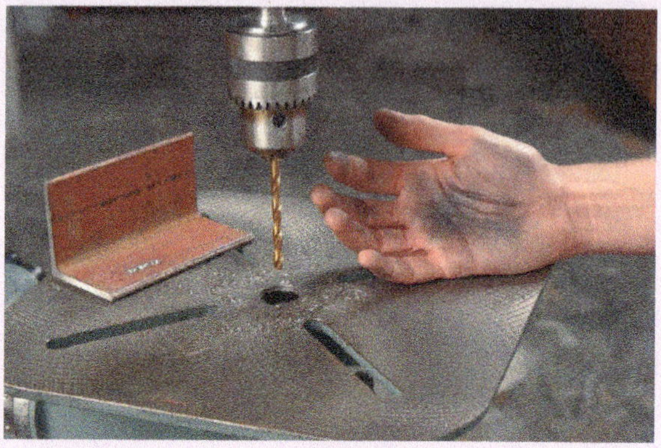

Figure 5-11 Cast iron chips always break into small packets but they also create black carbon dust, which is a health issue.
Photography by Prographics/McGraw Hill

Cast Iron Factoids

CI chips rarely change color, but some alloys turn dull brown.

Cast iron will hold on a magnet very well.

When CI is ground, the sparks are a very dull red and compact (do not form starlike bursts). See Fig. 5-12.

Cast iron rusts very fast when exposed to moisture.

Cast iron is overloaded with free carbon, which must be removed by smelting (burned off by oxygen injection into the molten metal) before the iron can be turned into steel.

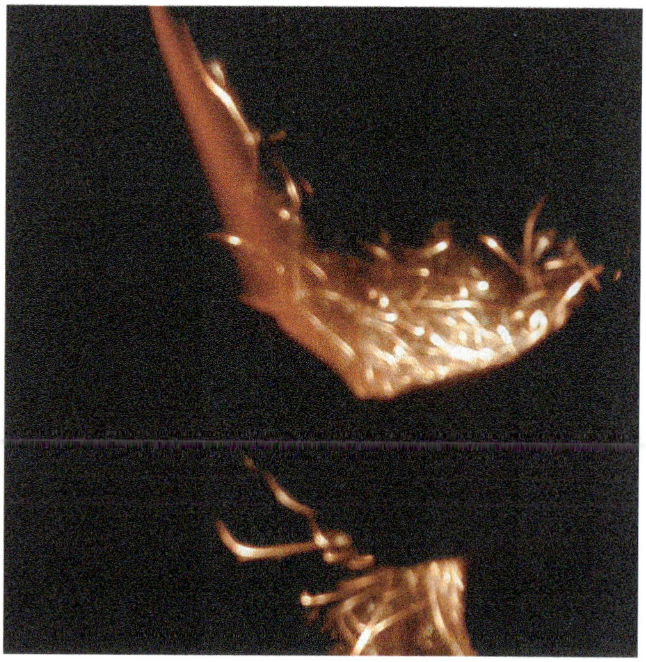

Figure 5-12 When ground, cast iron creates a very dull red spark due to the excess carbon.
Courtesy Mike Fitzpatrick

Due to the excess carbon, coolants become quickly contaminated when machining CI.

When machined, CI produces small, broken chips along with carbon dust.

The carbon dust gets into your skin and is difficult to wash out.

Steel Alloys

When iron is refined, the excess carbon is first completely removed, then very controlled amounts are reintroduced to produce steel. There are many alloys of steel, subdivided into two broad categories by application:

1. **Tool Steels**

   Not often used on consumer parts, tool steels are more expensive and generally the more technically complex. Tool steels are used to make cutters that machine other metals, to make dies and punches to shape or cut, or for any application demanding extreme hardness, friction resistance, durability, and strength. Due to their low machinability rating, tool steels can be very challenging to machine.

2. **Alloy Steels**

   Alloy steels are commonly used for consumer parts (for example, automobile axles, fenders, and fence posts). However, some alloy steels are very highly designed for a given purpose and are nearly as challenging to machine as tool steel.

Both types are supplied as castings and forgings, but bars, billets, and sheets are more common but never as extrusions.

Steel Factoids

A little carbon goes a long way in steel! Similar to silica in aluminum, the mineral carbon in steel is the most important ingredient after the base iron. Increasing the amount of carbon makes the difference in steel hardness, thus its machinability. Below 0.3 percent (3/10 percent) the steel is called a *mild steel*. Above the 0.3 percent level, the steel can be hardened to some degree, depending on the amount of carbon in the alloy. Carbon content rarely rises above 2.5 percent in most steels, with 4 percent the upper limit.

Mild steel is an alloy steel that's also called **low-carbon steel**. Mild steel has a single soft state. In Chapter 14 you will learn of a special process called carburizing, whereby the physical state of mild steel can be changed by adding carbon.

Steel machines best with a coolant.

All steels will hold to a magnet.

Steel sparks when ground. Mild steel sparks with bright white/orange stars, but as the alloy's carbon content rises, the sparks become globular and progressively redder (Fig. 5-13).

Mild steel rarely work hardens, nor can it be heat-treated to change its hardness. Other alloy steels do harden depending on carbon content.

Free cutting steel (FCS) is a steel alloy with a uniquely high machinability rating. It's made by adding lead and/or sulfur to the molten steel during the original pour in the *smelter* (a factory making raw metal from base elements). While the lead nearly disappears in the alloy, it modifies the steel structure to magically machine easily. FCS exhibits nearly all the characteristics of mild steel in application, yet it can be cut at nearly twice the speed.

Steel's machinability ranges from a bit more difficult than cast iron for mild steel to the very difficult, more technical and harder states of both tool and alloy steel, which must be cut with extreme care by slowing the cutter, adding flood coolants, and paying strict attention to cutter sharpness.

> **KEYPOINT**
>
> Remember, mild steel has less than 0.3 percent carbon present in its makeup.

Mild steel is not prone to work hardening, although it can occur. Usually, when it does happen, the hard spot can be machined away using a new sharp tool and reduced cutting speeds. Even so, it can burn up tools.

Steel chips commonly turn yellow-brown, then blue when machined as frictional heat rises. The iron combines with room oxygen to form various oxides when heated above 400° Fahrenheit. In Fig. 5-14, the chips cut at a lower temperature are on the left, and those cut at the highest temperature are on the right.

Two Steel-Forming Methods: Hot and Colds

Bars and sheets are rolled into shape in the mill, in one of two different ways: hot rolled steel (HRS) and cold rolled steel (CRS). The cold rolled steel is the stronger and finer finished material due to the cold working (stretching of the structure in the metal). The elevated temperature required to soften the steel during hot rolling steel (Fig. 5-15) forms a black scale of carbon on the outside. The shape of HRS is not as well formed compared to CRS; thus, it requires more final machining.

While it's more expensive initially, cold rolled steel is smooth and shiny on the outside from the rolling action and often requires less machining to bring to size. Due to its clean appearance and better shape, CRS can often be used as is.

High carbon　　Mild steel

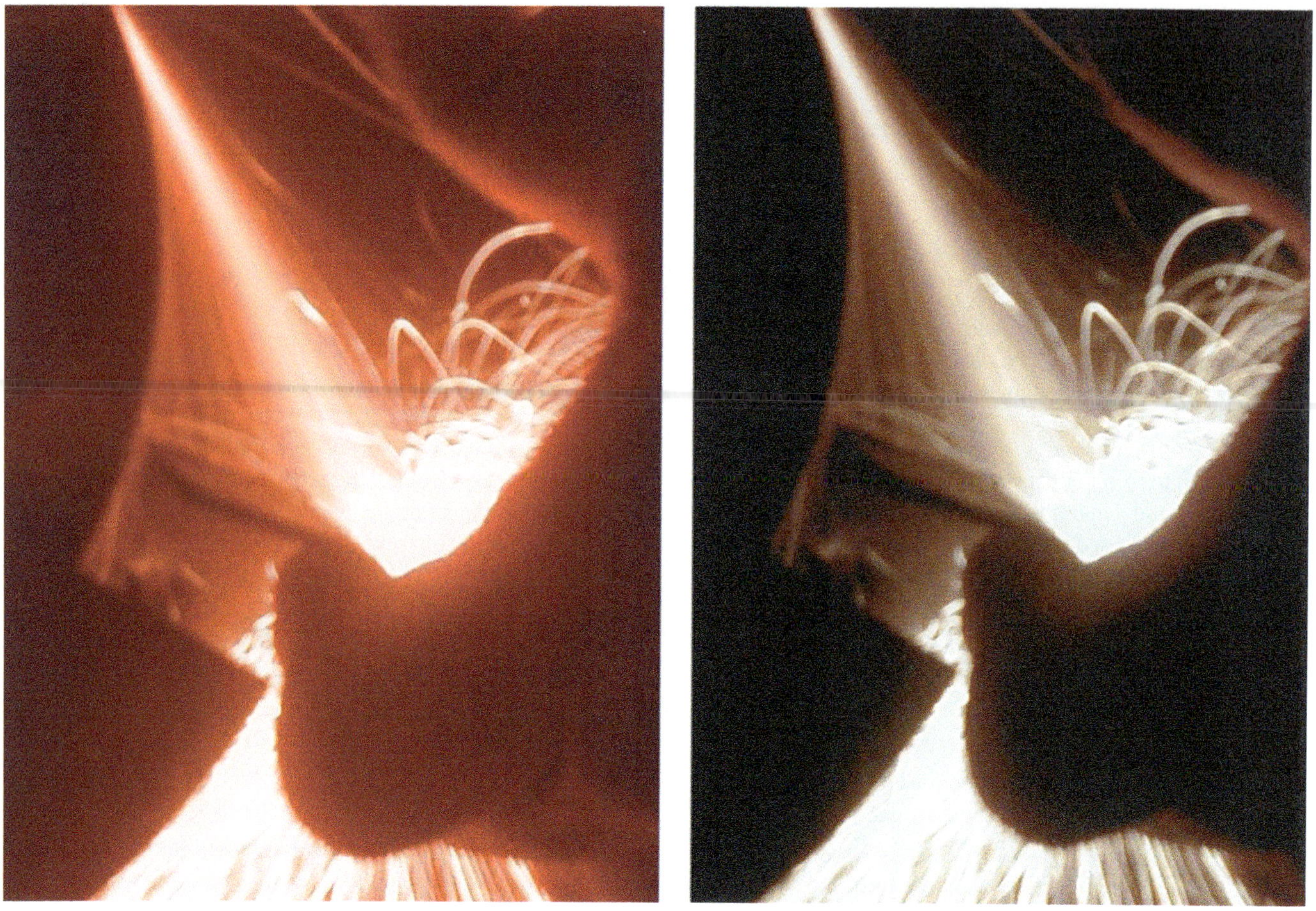

Figure 5-13 Steel sparks range from a dull red for the higher carbon to bright orange stars for the lowest carbon steels.
Courtesy Mike Fitzpatrick

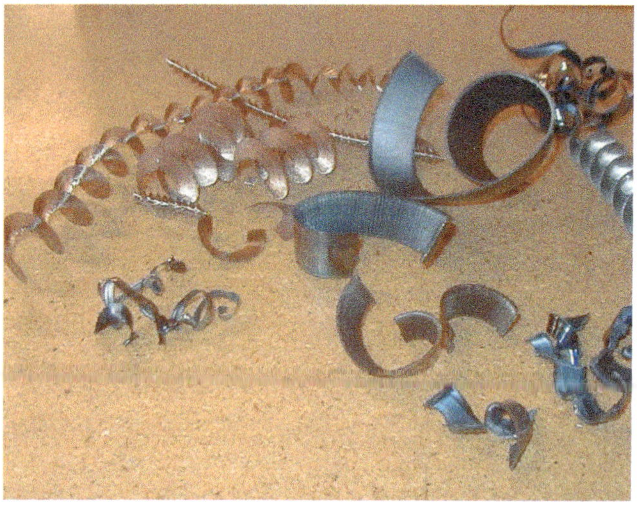

Figure 5-14 Steel chips change color with increasing machining heat.
Lake Washington Technical College, Kirkland, WA/McGraw Hill

Stainless Steels

Stainless steels are also iron/steel alloys but with higher nickel and chromium contents than tool or alloy steels. **Stainless steel** can be cast and forged, but it's far more commonly supplied in bars and sheets. Stainless steel is also called corrosion resistant steel, abbreviated CRES.

Within the CRES family, there are several subgroupings for different applications. The groupings could be characterized as soft but very resistant to corrosion (called the 300 series), the harder types used in jet engines and surgical instruments (the 400 series, which can be heat-treated as is), and the third series called *PH* (precipitation hardening), meaning that other elements can be added to the metal to make its surface harder.

(a) Hot rolled steel

(b) Cold rolled steel

Figure 5-15a-b Compare the appearance of two steel forms: cold rolled steel (CRS; shiny surface, right side) with the dark hot rolled steel (HRS) on the left.
a,b: Courtesy Mike Fitzpatrick

Stainless steel looks slightly different from regular steel in color. It is difficult to describe in words, but it is easy to see the difference between regular steel and stainless (Fig. 5-16). I perceive stainless to be a bit tanner in hue–or perhaps just brighter. You'll get the chance to view samples of both coming up.

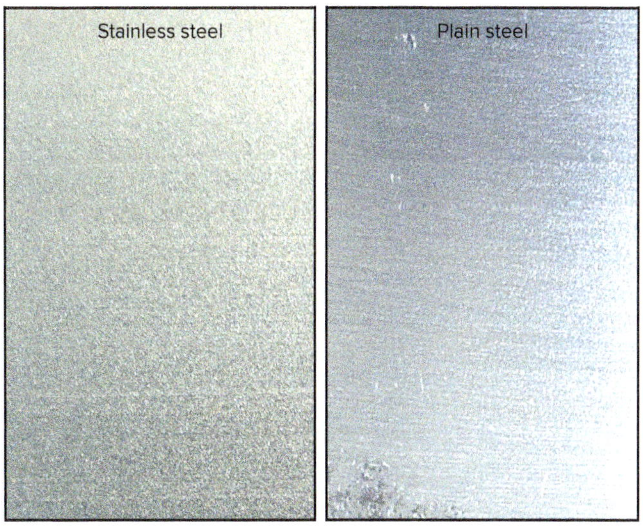

Figure 5-16 Compare the color of stainless steel to plain steel.
Courtesy Mike Fitzpatrick

> **KEYPOINT**
>
> Rust is a good indicator that the metal *isn't stainless!*

Cutting tools don't wear as well when machining stainless. Various stainless steels exhibit machinability ratings from moderately difficult to downright nasty. Absolute control of cutting speed is essential. Maintaining a steady stream of cutting fluid and monitoring cutter sharpness are also critical. Even a hint of dullness and most alloys will work harden. When machining stainless, examine the cutters often and at the slightest hint of dullness, change them. Any slight change in sound or vibration or the finish on the work surface is a sure sign that disaster is just around the corner when machining stainless.

Stainless Steel Factoids

Magnetic attraction is an issue with the various stainless alloys. Some do; some don't.

Common stainless alloys (300 series) will not hold on a magnet at all. The added nickel and chromium break up the metal's crystal structure to the point where there is no net attraction to a magnet. In theory, each crystal grain would adhere to a magnet but combined, their microscopic nature cancels this property. Some with a higher iron ratio will hold weakly (400 series). Still others hold nearly as well as regular steel (PH series).

Some stainless will spark similar to CI, with dull red sparks when ground. Other alloys will barely spark at all.

Stainless chips generally do not change color; however, there are a few special exceptions.

> **KEYPOINT**
>
> A heavy, bright silver metal with no sign of rust that does not hold to the magnet is most likely stainless, but titanium is another possibility.

SHOPTALK

Look It Up or Ask Before machining any new material, find out the machinability and hardness either from a journey-level machinist or by looking it up in a machinist's reference book. There are lots of surprises within the subject of metallurgy. For example, there are two metals that actually catch on fire, given the right conditions. They are magnesium and titanium. Solid blocks of these metals are impossible to ignite, but when their chips are surrounded with room oxygen and heated as during machining, they will burn! When machined, some metals create carcinogenic particles that must not be breathed or exposed to bare skin!

Composite Materials

Composites are manufacturing materials made of combinations (Fig. 5-17) of a honeycomb core, an outer skin, and a resin to hold it all together. The cores can be metal or plastic fiber, and the skin can be plastic or metal. These highly engineered materials are extremely strong compared to their weight. Because the cutting of these materials requires special cutters and processes, you probably won't see them in the training lab, but they are used in industry and gaining popularity.

Figure 5-17 Composite materials are combinations of fibers, resins, and metals. They have an amazing strength-to-weight ratio.
Courtesy Mike Fitzpatrick

In machining composites (Fig. 5-18), you must wear breathing protection from fumes or dust. Another precaution is fire prevention, as many composites are part metal and part plastic. Special cutters must be used that look more like a meat slicer than regular metal cutters. Holding composites during machining is a challenge as they are extremely strong for their weight along a given axis, but they can be crushed if not protected.

Figure 5-18 A composite being machined.
Courtesy Mike Fitzpatrick

5.2.4 Physical Characteristics of Metals

Beyond selecting the right alloy for a job, there are more work order requirements that must be certified–the grain direction and the heat-treat condition, how hard it is. We'll leave the study of hardness for later.

KEYPOINT

The hardness of the workpiece may be changed as it progresses through the work order sequence.

Grain Structure and Direction

When a metal is first cast after smelting, it becomes a solid ingot or billet. At that time, it is a block of connected crystals–not unlike plain ice. But when the block is worked into a consumer shape, bars, castings, extrusions, sheets, and forgings, for example, the working stretches the crystals into *stringers*. They create a **grain** direction in the metal. Each form exhibits varying amounts of grain structure. The grain creates a strength bias similar to a wooden board.

When flexed parallel to the grain direction, the metal breaks more easily while it resists breaking when bent across the grain. Depending on the function of the finished product, you can see that the grain direction in the raw material might be very important to identify before sawing raw stock for machining. When this is a critical factor (it often is), the need for grain identification and tracking will be noted in the work order.

Figure 5-19 is a computer simulation of how grain structure occurs in metal. On the left, a highly magnified section of cast metal shows large grains with no bias, but as the computer elongates the billet, simulating rolling into bars or sheets or hammering it by forging, the grains form the stringers—stretched-out grains. It's easy to see that as the bar becomes longer so, too, do the stringers.

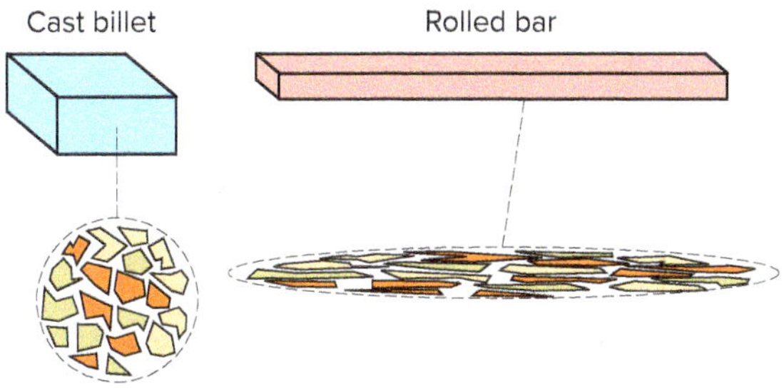

Figure 5-19 Grain structure is caused by elongating crystal grains, as illustrated in this computer-enhanced drawing.

Determining Grain Direction

There are three ways to know which way the grain bias lies within a piece of raw material:

1. **Parallel to Long Bars**

   In long bars, the grain stringers are elongated with length because that is the direction in which they were stretched during rolling.

2. **Parallel to Printed Identification**

   Lettering and numbers on sheets parallel the grain. When the metal is run through the final roll, the alloy and other information are printed on its surface. That's the grain direction because it's the direction the metal traveled through the rolling mill.

3. **Grain Stamp**

   A stamp mark that looks like an arrow (Fig. 5-20), sometimes with the word "GRAIN."

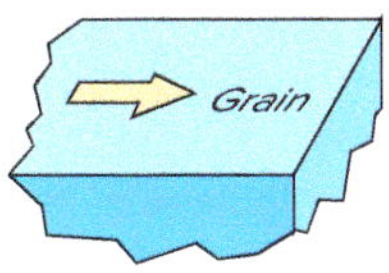

Figure 5-20 Grain structure is shown with a stamp similar to this.

Material Identification and Traceability

Often, not only the exact alloy must be used for a job, but it's heat-treat history and alloy certification must be linked all the way back to its original formation as raw metal. That certificate must stay with the job on certain people-moving applications and for aerospace and military work. That critical paper trail is known as *traceability*. In military and commercial aircraft manufacturing, for example, each flight critical part (parts affecting the safety of the aircraft) must show a serial number traceable all the way back to its birth by its *heat lot* number. The heat lot is the certificate number of the metal sample from which the individual part has been made. It is a pedigree that identifies the *crucible* (melting pot) from which it was born.

KEYPOINT

When traceability is required, without that thread of evidence, even though the metal is truly right for the application, parts made from nontraceable metal are not legal to put on items where people's lives are at risk!

Suppose the material order falls three parts short of completing the job? In that case, more certified material can be substituted into the batch, *but not casually by the machinist!* The extra material can be substituted officially when records prove it's the right stuff.

Color Coding

To quickly identify steel alloy bars, they are sometimes color coded on the end (Fig. 5-21). In a well-organized shop, they are placed in the rack with that ID toward one end.

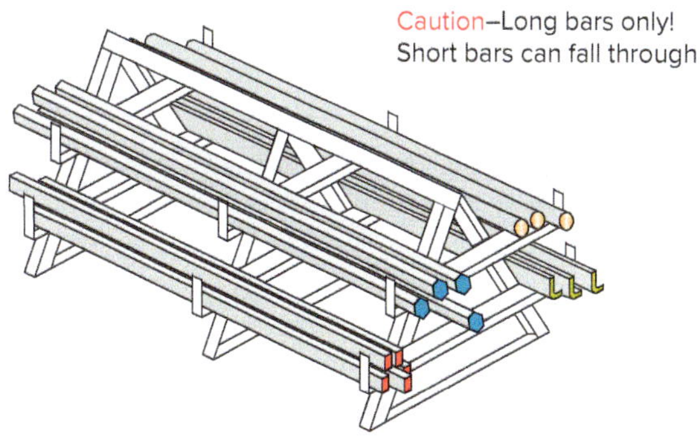

Figure 5-21 Color coding keeps raw metal identified until the last piece is used (with the coding on it).

KEYPOINT

When cutting sections from a color-coded bar, *never cut away the color coding unless you are taking the last piece of the metal.*

As discussed in Chapter 1, without the proper ID, metal becomes useless, even dangerous to keep around as it might accidentally be put to the wrong use!

UNIT 5-2 REVIEW

Replay the Key Points

- In many manufacturing situations, the exact alloy, heat-treat lot, grain direction, and history of the metal must be kept.
- The machinability of a metal is a relative rating that defines how fast the material can be cut and how long one can expect tools to last.
- There are a few metals that "can" be cut without coolant; however, it does help to remove heat and clear chips. Using coolant is rarely the wrong thing to do.
- Work hardening usually results from a machinist not heeding the correct cutting speed for a given metal or allowing a tool to dwell too long.
- Most forms of metal exhibit a grain bias that must be heeded when flexure of the product is a critical function.

Respond

1. A print calls for the work to be made from CI. Describe the material you are seeking. In what form will it likely be found?
2. The work order requires aluminum extrusion to be cut for the job. Describe what you will be looking for.
3. Name the forms of metal that a work order might require to be sawed into blocks or lengths for a job.
4. Since forgings are more expensive than castings, when would a job require one?
5. Name the best way to identify the right CRES for a job that, once completed, is to be sold to the military.

TERMS TOOLBOX

Abrasive sawing *Cutoff* or *chop* sawing accomplished with a resinoid fiber reinforced grinding blade.

Annealing Softening of a metal–annealing a band saw blade just next to the new weld area.

Cleanup cut Machining raw material to create a reliable machined surface by removing a minimum amount of material from the workpiece.

Excess A calculated amount of material left to finish dimensions; also called machining allowance or cleanup material.

Friction sawing Cutting metal by melting rather than cutting but accomplished on a band saw, not an abrasive saw.

Kerf The slot created when sawing. The kerf is wider than the blade due to the tooth set. Kerf must be calculated into excess when sawing material.

Pitch The number of teeth per inch in a saw blade.

Scroll sawing Sawing to a curved line possible only on a vertical saw.

Set (saw blade set) Bending or warping saw teeth out from the base blade. Blade set creates a wider kerf than the blade to prevent drag and allow scroll sawing.

Silicon carbide A hard abrasive used in resinoid cutoff saw blades.

Skip tooth Band saw blades with extra space between their teeth for sawing soft material.

Stripping A dangerous event where teeth are torn out of a blade, caused by selecting a blade pitch too course for the material.

Surface speed The recommended speed of the cutting action expressed in feet or meters per minute of band saw movement when sawing.

Unit 5-3 Sawing Material

Introduction: Okay, you've selected the right alloy, and the material is the form specified on the work order. The paperwork is in order, so now it's to be sawed into blocks, bar lengths, or other shapes for machining. There is a chance to scrap the job right here, before the machinists even get it! The work order usually notes the amount of extra metal that must be left on each blank to allow the machining to occur, but not always. If not, deciding how much bigger the raw blanks must be before machining can be simple or complex depending on the part design. In this unit we will explain how to calculate excess material, then discuss the sawing.

5.3.1 Planning Excess Material

Go back and note operations 10 and 100 on the sample work order in Fig. 5-1. At Opp 10, the raw material is sawed into blanks suitable for the upcoming machining. Then at Opp 100, an intermediate saw cut is taken to speed up the machining of the final profile. It's important to note that at each operation, the sawing includes enough *excess* to machine the final shape. **Excess** is a layer of extra metal to be machined away. It might also be called **cleanup cut** *material* or *machining allowance* material, which provides the amount needed to remove all rough material, thus leaving a 100 percent cleaned (finished) surface.

Besides providing metal to finish, excess material might be planned for holding purposes during machining. For example, an extra amount is needed to hold the part in a lathe chuck while it is being turned into a shape; then it is removed at the last cut—this excess is called *chucking allowance.*

The Right Amount of Excess?

The answer can be as simple as reading the rough-sawn size from the WO. Or in many job plans, the amount is noted, but the machinist must calculate the overall size including the excess. However, on a few, the machinist is left to make the decision of how much is right.

KEYPOINT

How much excess? Usually, from 0.100 in. up to a maximum of 0.250 in. for normal machining (with many exceptions). For precision grinding, from 0.005 to 0.060 in. excess.

Planning too much excess robs profit due to extra machining time to remove it; plus the cost of consuming too much metal. However, too little excess and the part will not *clean up*. That is, the surface will not be completely machined when it reaches the indicated size or shape. Saw marks will still remain. So, an error on the heavy side is better until your skills improve. But the goal is always to add a minimum amount.

Here are the factors to be considered:

1. Complexity of the part to be made—complex parts often require a bit more.
2. Previous experience with this part—proven production requires less excess.
3. Cost of the material itself—less costly material might have more excess.
4. How is the part to be machined? What machines are to be used? Are there fixtures to hold the material? What cutting tools are available?
5. Nature of the material—since aluminum machines so easily, more excess might be better (Fig. 5-22). However, leaving a large amount of excess on a stainless steel blank could be a real nightmare!

TRADE TIP

While learning, or whenever in doubt, leave 0.25-in. excess on all surfaces to be machined (or as directed by your instructor). This is too much by industrial standards, but it's right for trainees.

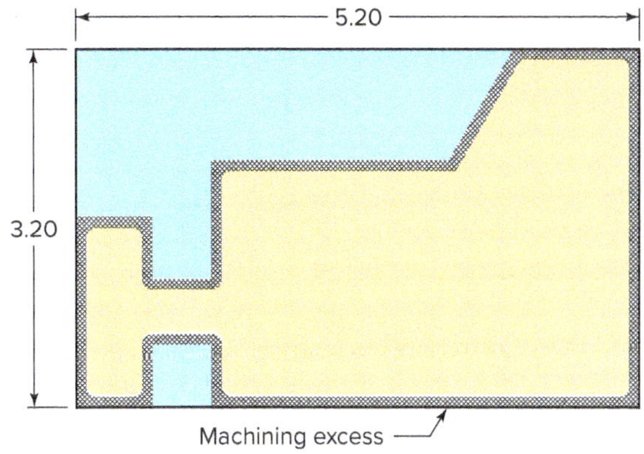

Figure 5-22 Excess material must be left on the rough blank.

Allowing for Saw Kerf Thickness

Here's another consideration: When sawing you must also allow for the thickness of the saw cut itself, called the **kerf**. If a second part was to be sawed from the sheet in Fig. 5-22, approximately 0.100 in. would be needed for the kerf as well as the machining excess on both parts. Each type of sawtooth pattern has its own kerf thickness.

5.3.2 Sawing Metal

In today's shop, there are 10 very different possibilities for cutting blanks for machining. The choice can make a big difference in job success. This is a perfect example of the challenge of job planning. At the end of this chapter, you will be quizzed as to which might be best, given a specific material situation. The first few in the list would be expected in a school shop so we'll study them here, while the remaining are CNC—high-tech methods found in industry.

Common Methods

1. **Hand Sawing—Hacksaw**

   While too slow for production work, hacksawing is a needed machinist skill.
2. **Vertical and Horizontal Power Band Saws**

   A continuous cutting action is caused by a looped saw blade running on wheels. A complete shop must have both types.
3. **Abrasive Sawing**

   Cutting metal with a thin, specially reinforced grinding wheel; this method is primarily dedicated to cutting long steel bars into smaller lengths.
4. **Oxygen-Acetylene Flame Cutting**

   Produces a very rough edge so this option isn't suitable for most job plans. It cuts by rapid oxidation, so only ferrous (iron-based) metals can be cut this way. It creates a heat-affected zone (HAZ), an undesirable band of modified metal next to the cut. Usually a last resort in machining, this method is used extensively in welding and fabrication work.

5. **Shearing**

   Shearing cuts with a slicing action and is used for relatively thin material such as sheet metal or for the drill gage blanks. Shears might be found in schools but are more likely in industry where shears cut up to $\frac{1}{2}$-in.-thick steel or more! This machine is banned for use by minors due to its extreme danger factor.

Industry Methods

1. **Plasma Cutting**

   A high-energy state, electrified gas cuts by injecting an electric current into a compressed airstream. The extreme heat allows a greater range of material to be cut compared to oxyacetylene; but it creates an unwanted band of modified metal next to the cut, which as mentioned is called a HAZ.

2. **High-Energy Laser Cutting**

   This method uses highly focused heat to remove a narrow band of material. It can cut many kinds of metals but is usually applied to those not easily cut by other means. Using CNC drives can cut intricate shapes. Leaves a HAZ. (See Chapter 17.)

3. **High-Pressure Waterjet Cutting**

   This method uses a highly focused stream of water to cut an amazing range of materials from glass to chocolate to baby diapers without getting them wet! To cut metals and other hard materials, abrasive grains are injected into the stream. Using CNC drives, it can cut complex shapes. It does not leave a HAZ but is relatively slow compared to sawing and shearing. (See Chapter 17.)

4. **Rigid Sawing**

   Similar to a very sturdy version of a table saw for wood, a rigid saw has a round disk cutter that spins while either the material or the saw blade is moved, depending on machine construction. It is used mostly to accurately cut very large plates to size and is found only in major industry.

5. **Reciprocal Sawing**

   This sturdy, power-driven saw is similar to a big hacksaw. Due to the blade cost and pinch point danger, these saws are fading from the industrial scene.

Hacksawing

Although unlikely as an industrial operation, hacksaws are basic to a machinist's toolbox. With all sawing including hacksaws, there is a right blade for the alloy and shape thickness.

Selecting Blade Pitch

Saw pitch is the number of teeth per inch (Fig. 5-23). In general, the coarser blade will cut faster. They are used for heavy cuts or for softer metal. Softer metal cuts better with bigger teeth but fewer per inch of blade. However, there is a limit for coarseness both for safety and for tool life.

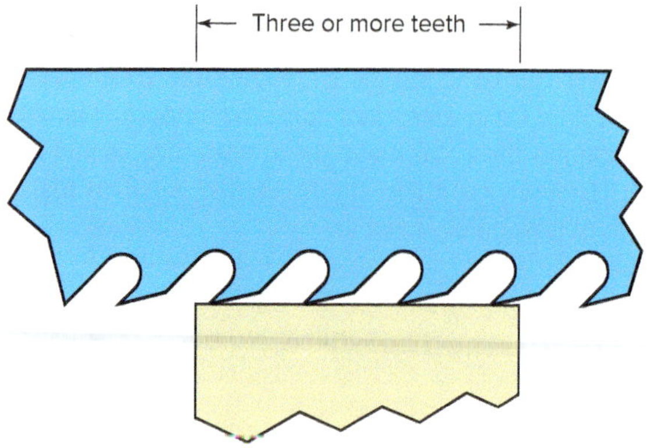

Figure 5-23 Select blade pitch such that three or more teeth are in contact with the work.

> **KEYPOINT**
>
> **Three-Tooth Pitch Rule**
>
> Choose the pitch such that a minimum of three teeth remain in contact with the work surface at all times.

Three teeth touching prevents thin metal from slipping between the teeth. The blade in Fig. 5-24 is too coarse–too few teeth per inch. When this happens, teeth will be broken away. Choosing a blade with this contact is critical for all kinds of saws, both hand or power. Teeth ripped out, as shown in Fig. 5-24, is called **stripping**. For hand sawing, stripping isn't dangerous but will damage the blade.

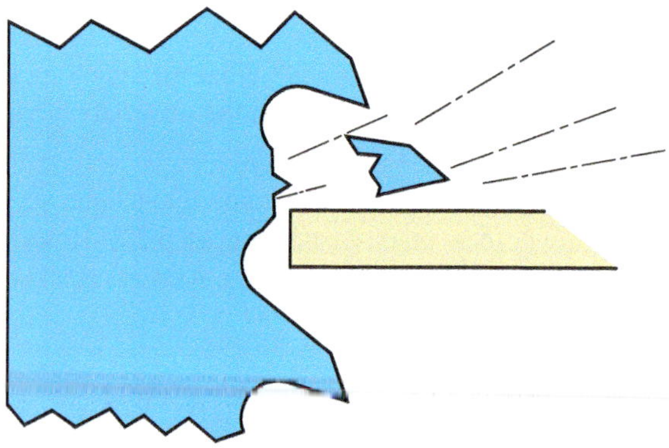

Figure 5-24 Too coarse—teeth can be stripped.

Blade Set Causes a Wider Kerf

Tooth set is a saw blade feature necessary to make the cut wider than the base blade. Without tooth set, the blade would drag in the cut and be unable to cut curves. To produce a kerf wider than the blade, the teeth are either bent to alternate sides, as shown (Fig. 5-25), or for very fine teeth not easily bent, the entire blade is bent in a wavy pattern.

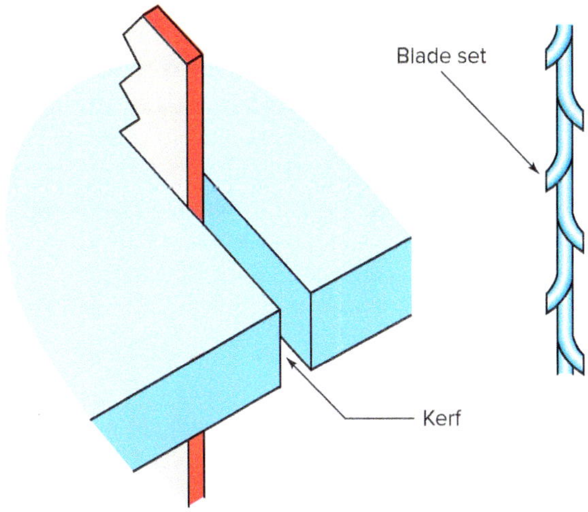

Figure 5-25 Saw blade set causes a kerf wider than the blade's thickness.

SHOPTALK

Stripping Hurts! When the teeth strip off power band saw blades, especially when sawing fast as with aluminum, they come out like machine gun bullets! No kidding, I have seen a student make this mistake, and within a fraction of a second, the student had a row of hot teeth embedded in the student's safety glasses! We hung those glasses on the saw as a reminder—*change the blade when sawing thin material or find another sawing process.*

Correct Use of the Hacksaw

There are five guidelines beyond pitch selection when using a hacksaw:

1. **Mount the Blade so the Teeth Face Away from the Handle**

   They cut on the forward stroke (Fig. 5-26). Often an arrow is painted on the blade indicating in which direction the teeth cut.

2. **Do Not Force the Cut**

   From experience you will learn there is a "sweet" pressure beyond which any greater push results in broken or dull blades.

TRADE TIP

New Saw Blades in an Old Kerf A problem is encountered when a worn blade is replaced with a new blade in the same cut both for hacksawing and for power sawing. When the old blade dulled, it lost its set. Its kerf became narrow. However, the new blade's set is wide again.

It's a disaster to begin sawing at the original place. The set on the new blade will be instantly flattened or the teeth will be stripped. To prevent this, start sawing some distance back, where the kerf was correct, then slowly saw back to the last position. When this reentry becomes time-consuming, it might be better to just start a new cut from the opposite side of the work to meet the first.

3. **Apply Pressure on the Forward Stroke**

   Depressurize the blade for the return stroke or lift it up instead of dragging it. This simple act will lengthen blade life.

4. **For Hard Materials, Slow Your Stroke Rate**

   Even when hand sawing, if the strokes are too fast, a blade quickly dulls. Saw from 50 to 70 strokes per minute—around one per second. *Adjust your speed according to the machinability of the alloy being cut.*

5. **Hacksaw Blades Can Shatter**

   Most hacksaw blades feature hard teeth with a soft back. They will not break all at once. However, some higher quality blades are hardened all the way across the blade (a better blade but brittle). These break instantly, and if you are pushing too hard, a good knuckle skinning follows! Don't lean into the cut under any circumstance but especially when using full hard blades.

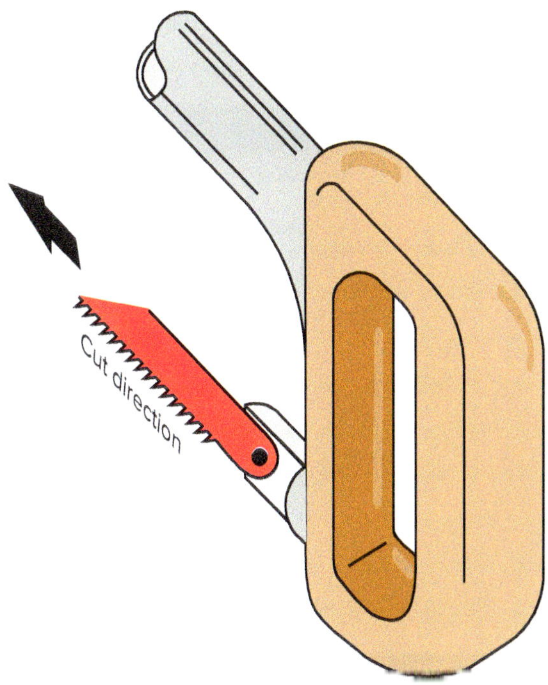

Figure 5-26 A hacksaw blade is correctly mounted when the teeth cut on the forward stroke.

TRADE TIP

Sawing Thin Material For material that's too thin, or where only coarse blades are available, rotate the material to cut along another dimension (side) or tilt the saw blade to make a longer cut; thus, more teeth come in contact (Fig. 5-27). For thin tubing, insert a wooden plug to keep the teeth from stripping. For thin sheet material that can't be tilted, you can sandwich clamp it between plywood or sacrifice metal plates of a softer metal before sawing.

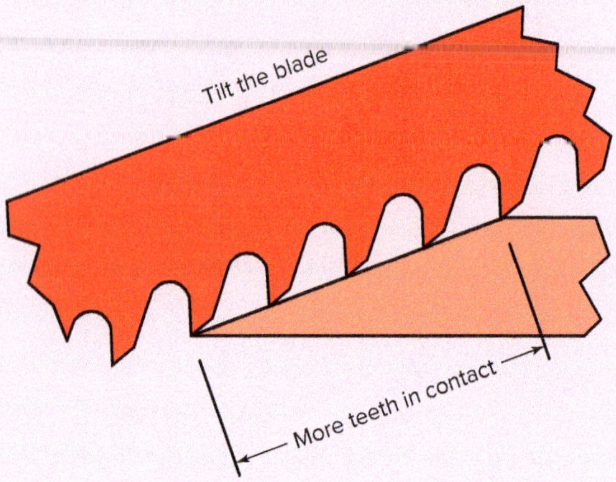

Figure 5-27 Tilt the blade to achieve three or more teeth in contact.

Power Sawing

Three types of saws are commonly found in training labs and industrial workshops:

Vertical Band Saws	Horizontal Band Saws	Abrasive Saws (Cutoff or Chop Saws)

Saw Blades Selection

To saw effectively using a vertical band saw, the machinist needs to understand saw blade selection. Blades need to be changed for the specific application.

Pitch is the main factor, with three-tooth contact even more important to avoid breaking teeth from the blade. Stripping of teeth on a vertical band saw set to a high blade speed for aluminum or brass can be a rapid-fire, dangerous event. Three or more teeth contacting the work is more important on vertical saws where the user applies the forward movement of the work into the blade. On horizontal saws equipped with automatic down feed, two-tooth contact will work since the saw can't surge forward to strip teeth.

Tooth Pattern and Set

Power saw blades differ from hacksaw blades in their cutting teeth patterns and in the types of **set** (Fig. 5-28). In general, **skip tooth**, *claw* or *hook*, and *buttress teeth* (Fig. 5-29) are chosen for cutting softer materials. The extra clearance is necessary to allow each tooth to temporarily contain the larger chip created due to the quicker cutting action. It's held in the hollows between teeth, until it exits the kerf to be ejected, before returning to the cut again.

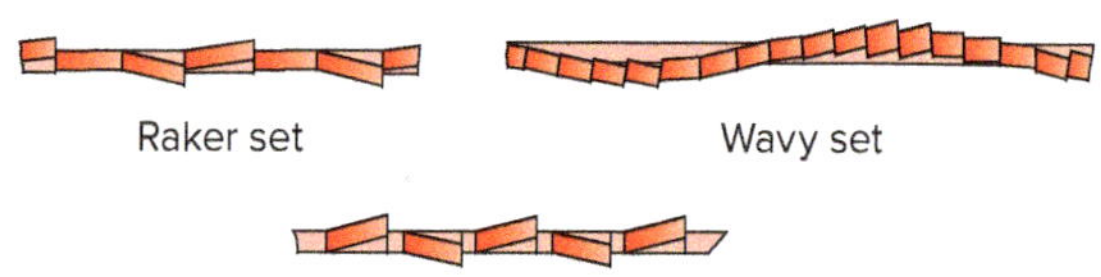

Figure 5-28 Blade set creates a kerf wider than the blade to prevent rubbing and allow curved cuts.

Blade Alloys

Band saw blades are supplied in three varieties, high-carbon steel, high-speed steel, and bi-metal. Each has its correct application.

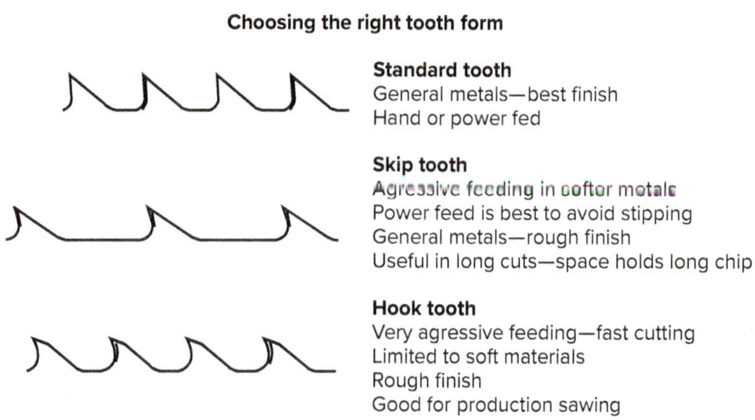

Figure 5-29 Choose the right tooth form based on the sawing method, the alloy hardness, and the size of cut. Choose coarser blades for longer cut surfaces or softer metals.

SHOPTALK

Chunks Recycle Better Compared to cutting away metal chips, sawing away chunks from the workpiece is an economical plan for another reason beyond fast removal: They command a better payback fee at the recycling plant. Clean, large blocks can be recycled at the smelter more easily because they are closer to pure uncontaminated metal; thus, they command a higher return yield to the shop. But to take advantage of this, the chunks must be stored separate from the chips.

Carbon-steel blades are the budget version. Their teeth are not as strong or as hard as those of the other two blades so they wear the fastest. But they have a great advantage in that they can be welded together. This is useful for several reasons.

Sawing Internal Features When cutting internal holes in the work (donut holes), a pilot hole is drilled near the cut line; then a blade is inserted through to be welded closed inside the hole. After sawing out the central *slug* (leftover unwanted metal–donut hole), the blade is cut open with shears, then welded back together for other work.

Using Bulk Blades Because carbon steel can be welded, it's economical to buy in 100-ft rolls, then make blades as needed. That also means blades that break during use can be rewelded or even short sections can be welded in place to repair stripped sections.

High-speed steel blades feature the hardest teeth; thus, they last the longest when sawing difficult metals such as stainless steel. They cannot be welded back together without special equipment and processes, therefore they are considered a specialty blade in most shops. These are more costly than the other two blades.

Bi-metal blades fall between high-speed steel and carbon steel, in both price and performance, although in most applications they stand up almost as well as the higher priced high-speed steel blade. They are constructed with very hard teeth but softer backing material. You can recognize them by a dark band of heat-treated metal on the tooth side of the blade. Due to the dual advantage of price and performance, bi-metal blades are the most popular in industry, while carbon-steel blades are more popular in training labs.

Vertical Band Sawing

Often, an efficient job plan includes metal removal by sawing, as in operation 100 on our work order. Even in the modern world of high-speed CNC machining, sawing can be the fastest metal removal tool in the shop when used correctly. Sawing takes away whole "chunks" of metal rather than making chips. When planning a job, always consider the saws as an operation. In addition to efficient metal removal, those large scrap pieces are far more valuable for recycling. See Fig. 5-30.

Major Saw Types and Nomenclature

Saws vary greatly as to their features and size. We'll concentrate on the two basic types most likely found in a tech school (Fig. 5-31).

Vertical Saw Safety Is a Critical Issue

Because the blade moves fast and the work is fed through by hand, you need to know about

Specific safe moves

Holding and pushing work through the saw

Calculating blade speed

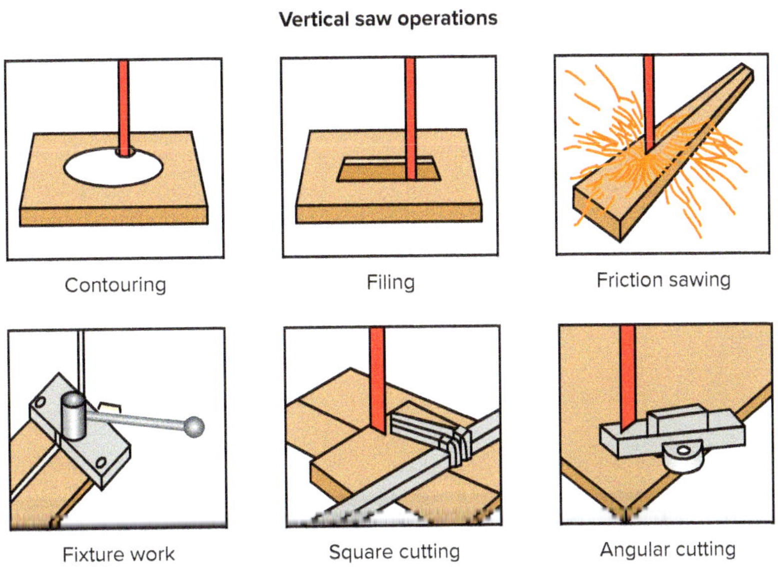

Figure 5-30 Standard and special band saw operations.

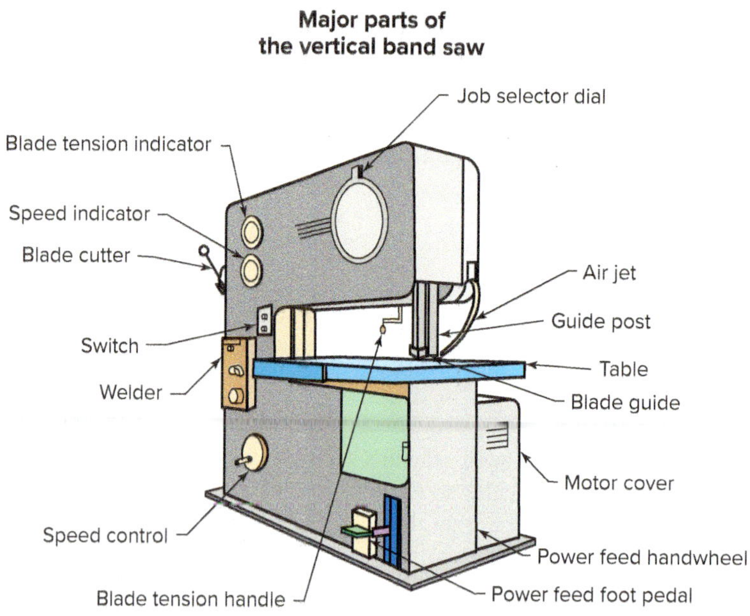

Figure 5-31 The basic parts of a vertical saw.

Even with correct guards a saw blade must necessarily be exposed over some part of its length and those teeth face directly at you. Coarse saw blades, moving fast, will immediately cut anything–including your hand. To be safe follow the guidelines listed next.

Specific Safety

1. **Adjust the Guard**

   Always have guards in place covering as much blade as possible.

2. **Brace Yourself**

   Brace your hip or knee against the saw base to avoid slipping forward if something moves unexpectedly (Fig. 5-32).

Figure 5-32 Always brace your body against the saw to avoid slipping forward.
Lake Washington Technical College, Kirkland, WA/McGraw Hill

3. **Hands Clear**

   Never put your hand directly in line with the saw blade. If pushing the work from directly in front of the blade is absolutely necessary, use a pushstick (wood block or metal) to take the place of your hand. Always use a pushstick or saw vise rather than your hand.

4. **No Gloves**

   Gloves catch on things faster than bare skin does.

5. **Correct Guides**

   Make sure the blade guides are as low as practical for the material and that they are adjusted right. This keeps the blade tracking right, keeping it from wobbling as it enters the work.

6. **Nonskid Flooring**

   If standing on a bare concrete floor, keep the small saw chips swept up, as they become slippery. (Note that bare floors in front of saws are banned under OSHA guidelines.) Better yet, make sure you are standing on nonskid surfacing (paint with added sand for traction). Even better yet, stand on a rubber skidproof mat if at all possible. That allows the tiny saw chips to fall through to the floor.

7. **Brace Your Hands**

   To avoid sudden, unexpected movements of your hands if the work should accidentally shift, drop your wrists down onto the table (Fig. 5-33). Do not hold work from the top, but rather put your hands on the saw's table; then push the work. This action ensures that if the work slips, your hands won't go with it!

8. **Danger Area to the Right**

   Do not allow others to stand to the right side of the saw table as that is where a broken blade will tend to eject.

Figure 5-33 Wrists or palms in contact with the saw table, and not in front of the blade, create safe stability. If something moves unexpectedly, your hands won't move.
Courtesy Mike Fitzpatrick

9. Stable Work Only

Finally, never saw a workpiece that might tend to tip forward as the blade pulls downward. Sawing disks from round stock is the prime example of this potential danger. When the blade grabs it, it tends to rotate forward out of control, which often breaks the blade and jerks your hand with it. There are two ways to avoid unstable work. The best idea is to take it to the horizontal saw where the bar is firmly held in a vise and no hand movement is needed. If the work must be sawed on a vertical saw, use a special saw vise to firmly clamp the unstable work. Firmly clamped in the vise so it cannot shift or tip forward.

Set the Saw at the Right Blade Speed

In setting up all machines, drills, lathes, mills, and saws, three things must be considered for choosing the right cutting speed:

Material type

Material hardness

Coolant availability

Cutting tool hardness is also a factor on most other setups, but for saws it's always a constant as the three blades are about the same hardness.

In general, the harder the work material, the slower the saw must cut to avoid burning or dulling the blade or work hardening the material. Vertical saws don't have coolant because there is no way to collect it to a reservoir, so your job will be to shift the saw's transmission to achieve the right blade speed based on the work material alone.

Correct blade speed can be obtained from a chart like the one shown in Fig. 5-34 as the job selector dial. To use a rotating selector such as this, place the material in the window then read the speed in the index. If there is no speed chart, refer to any comprehensive machinist reference book such as *Machinist's Ready Reference*©. It will list blade speed for sawing a given material in feet per minute for materials (generally under the topic of cutting speeds).

Figure 5-34 A metal plate attached to this vertical band saw shows the correct blade speed for various materials.
Photography by Prographics at Milwaukee Area Technical College/McGraw Hill

KEYPOINT

Band saw cutting speeds are specified in feet per minute abbreviated as F/M or FPM.

Surface Speed

For all machining where a cutting action takes place, there exists a magic number called *recommended* **surface speed**, sometimes also called *cutting speed.* This important number is needed to set up all equipment including saws.

We will discuss it again for drilling, then for turning on a lathe and for milling as well. For sawing, it's an easy number to envision. It's the velocity of the blade expressed in feet per minute (FPM), or it might also be in meters per minute (m/m). If your saw has no speed chart, remember these ballpark guides.

Material Being Sawed	Cutting Speed in FPM
Aluminum	300
Brass	200
Soft Steel	100

Although very low for industry, these numbers are easy to remember at the start. They are biased toward survival and safety. After some experience in sawing, they can be increased, especially when sawing aluminum. Notice that these numbers reflect the metal's machinability. Steel, a hard metal, must be sawed slowly, while soft aluminum is sawed the fastest.

KEYPOINT

For any machine cut where chips are made, there is an optimum surface speed in feet per minute.

If the selected surface speed is too slow, the cutting action is poor and production is low. Conversely, if it's too fast, the blade will dull quickly, then burn up if not removed from service, and worst of all, the material can be ruined if it work hardens.

Industrial surface speeds commonly range from a slow 30 feet per minute for sawing hard steel to 2,000 feet per minute for some alloys of aluminum.

Step by Step—Setting Up a Vertical Saw

Here's a checkoff guide:

- ☐ Enter the blade speed table with the material type/hardness—find the correct cutting speed.
- ☐ Shift the saw's belts and/or transmission to obtain that recommended speed.
- ☐ Check the blade pitch for minimum tooth contact.
- ☐ Position and check all guards and blade guides.
- ☐ Be ready with a pusher or saw vise to force the work through the blade. Never have your hands in direct line with the blade!

Lab Verification

With correct shop attire, and with permission, take your book to the vertical band saw in your training facility. Answer these questions:

1. Most saws have 10 or 12 blade speed selections. How many speeds does your saw have? What do you do when the chart recommends a blade speed between two of the speeds? (*Answer:* Choose the lower speed and observe the results. From experience, you will come to know if you can shift up to the next speed.)
2. Does your saw have a speed selection plate or job selector dial, similar to the one in Fig. 5-34?
3. If so, what range of speeds does it show and which materials are shown?
4. What speeds could you select to saw steel? (*Note:* They will probably fall between 60 and 200 FPM.)
5. Observe—does your saw have a blade welding attachment? (Training is coming up on this.)
6. If your saw doesn't have a surface speed chart (blade velocity), where might you find this information in your shop?

Saw Vises

A special vise made for vertical band sawing is a must (Fig. 5-35). It not only provides a safe way to push the work, but also the handlebars help steer through curved cuts. While keeping your hand out of the blade path, saw vises amplify the ability to direct the work using the extended handles.

Figure 5-35 A saw vise provides a safe way to control scroll curve cuts on the vertical band saw.

Photography by Prographics at Milwaukee Area Technical College/McGraw Hill

Saw Accessories

There are several vertical saw accessories used for safety and efficiency (Fig. 5-36). For long work, the outboard support is used. For complex or specialty cuts, the chain jaw vise is used. It can be power fed if the saw has the capability. By stepping on a foot pedal, the chain pulls the work forward through the cut. Other accessories include work guides and stops.

Scroll and Friction Sawing

These two sawing operations are less common but are within the beginner's skill level.

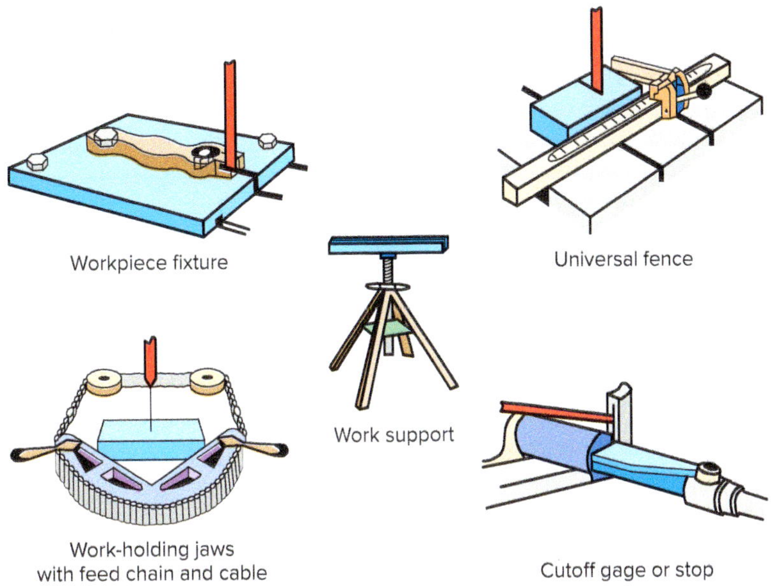

Figure 5-36 Saw accessories.

Scroll Sawing (Curved Cuts)

Scroll sawing is possible only on vertical saws. Cutting curves is a major advantage of a vertical band saws. The reason it can cut curves is that the kerf width allows the user to rotate the work relative to the blade within its own cut. Well-adjusted blade guides and sharp blades with a wide blade set are essential.

From Fig. 5-37, you can see that if a tighter curve (smaller radius) is needed, then a wide blade must be replaced with a narrower blade, to be able to turn within its kerf. When doing so, you must readjust the trust wheels on the saw guides such that the tooth set does not contact the side guides. If not, the blade set will be flattened on the wider blade.

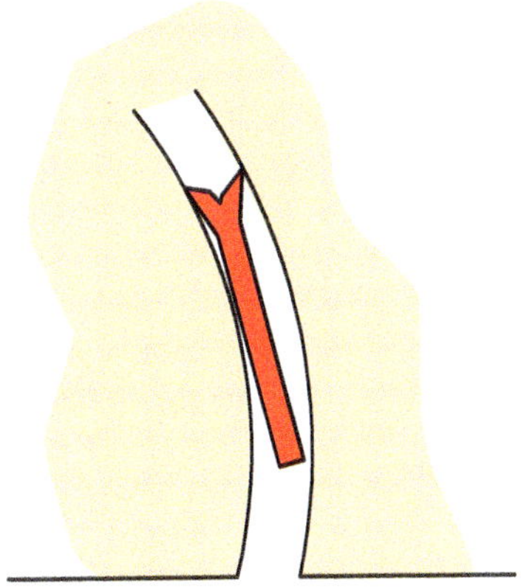

Figure 5-37 Kerf width makes scroll sawing possible.

KEYPOINT

When changing from a wide to a narrow blade, always readjust the thrust rollers. Otherwise, the narrow blade's set will be flattened if it touches the side rollers.

Adjusting blade guides When replacing one blade for another of the same thickness and width, no adjustment is needed. But, when a new blade size is mounted on the saw, the guides must be set to control it. This step is often overlooked by beginners, yet it will have everything to do with cut accuracy and blade life.

Notice in Fig. 5-38 that a power band saw features two sets of blade guides, one before the cut and one after. These guides stabilize the blade from wobbling and vibrating. In Fig. 5-39, a top view looking down on a guide set, notice that there are three rollers in a set. Two support the sides of the blade, the third is a thrust bearing. When you push the work against the saw, this bearing provides a rolling brace to keep the blade vertical.

Both upper and lower guide sets must be adjusted to fit the blade. Each set creates a three-point suspension: the sides and the back of the blade. Set the thrust roller with as much blade contained within the side rollers as possible *but not touching the tooth set.* The teeth must never touch the blade guides.

Some saws use solid guides rather than rollers for side support of the blade. These guides are usually made of brass/bronze and due to wear must be constantly maintained (Fig. 5-40). When cutting curves they are consumed by the twisting force and friction of scroll work.

Figure 5-38 Here the safety guard has been removed to expose the upper blade guides. There is a second set below the table as well.
Courtesy Mike Fitzpatrick

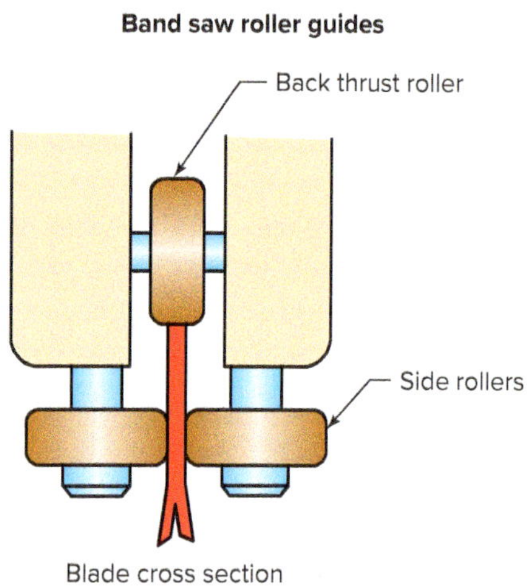

Figure 5-39 Adjust the three-point suspension to fit the blade width and thickness but never touching the tooth set.

Solid guides work okay for light sawing but aren't practical in industry. Solid guides are adjusted close to but not touching the blade. They are common on saws meant to cut wood, rather than metal, due to the reduced side loads on the blade.

KEYPOINT

Bringing the correctly adjusted upper blade guide down close to the work surface creates the best control over the saw path.

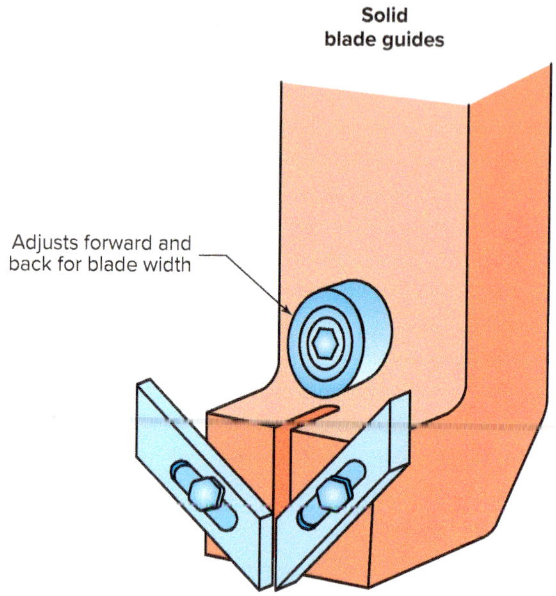

Figure 5-40 Solid guides are found on some saws.

Friction Sawing

This sawing technique requires a powerful saw capable of *surface speeds of 2,500 feet per minute or greater.* **Friction sawing** might not be considered a beginner skill in your shop due to the list of possible dangers and the problems it causes for both the saw and the workpiece. However, in sawing certain problem materials it works wonders when other methods won't do the job. For example, you can saw through another band saw blade using friction sawing. Ask for a demonstration and always check with your instructor before attempting friction sawing.

Friction sawing does not cut metal; it melts the work ahead of the blade. Why doesn't the blade melt, too? Because any portion is in contact with the work for a fraction of a second; then it rotates through the entire saw loop to cool before once again contacting the cut. The workpiece cut area remains hot. Friction sawing has several advantages and four disadvantages.

Advantages

- You will be amazed at how fast you can saw once the metal starts to melt.
- Material that's too hard or too thin to be sawed by standard methods can be cut.
- Thin material that would strip teeth is easily friction sawed.
- A worn blade with no sharp teeth works best; thus, old, completely dull blades work best and extend their life span.

Disadvantages and safety precautions

- Friction cutting leaves a ragged burr.
- Heat may work harden a thin zone around the cut (the heat-affected zone).
- Fire! See the upcoming key point.
- Freight train blade crashes.

The blade is traveling at such high speeds that should it break, it begins to fold upon itself as rear parts catch up to the crash. Thinner blades don't just stop when they break.

KEYPOINT

When friction sawing, make sure all blade and safety guides and guards are as close to the work as possible and that nobody stands near and especially to the right of the saw frame where an escaping blade might go.

Heat and Fire

The work will heat rapidly near the cut line during friction sawing. Plus the outgoing swarf (tiny chips being removed) is molten. Before friction sawing, inspect the swarf bin and the saw's interior for wood or plastic waste left behind from previous cuts (Fig. 5-41).

Figure 5-41 From above, nothing looks unusual while friction sawing this steel, but looking under the saw table, the stream of sparks and melted metal tells of the high heat in this process.
Courtesy Mike Fitzpatrick

Remove all flammable material from the catch bin and around the saw.

Supersonic tooth stripping Be aware, tooth stripping can occur, except now at bullet speeds! To avoid this problem, start your cut gently; then increase the pressure. Wear double eye protection—safety glasses and a full-face shield are good ideas.

Horizontal Power Sawing

This technique is used to cut large bars and heavy work. There are four major differences and advantages of a horizontal saw over a vertical:

1. **Self-Feeding**

   Built with a pivoting band saw frame on a horizontal bed, horizontal saws lower their frame and blade to automatically control the cut rate. This means they can be left unattended to saw large work and that the three-tooth rule doesn't apply as long as the drop rate is adjusted such that tooth stripping cannot occur.

2. **Any Length Workpiece**

   Between the saw guides, the saw blade is warped about 45° relative to the plane of the saw wheels. This means that the frame sits back from the cut and therefore the work does not need to pass through the frame, as it does on vertical saws. On a horizontal saw, any length bar can be cut.

3. **Vise Held Work**

   A long-range vise is part of the bed frame, which means no hands are needed to steady the work as with a vertical saw. The work remains stationary while the pivoting support arm and blade progress down into the work. This means that the machinist is safer without pushing the work through the saw by hand. These saws are used to cut off bars or blocks with little operator intervention. Straight and angular cutoff work is the only possibility on a horizontal saw (Fig. 5-42). Scroll and friction sawing are not done here.

4. **Coolant Systems**

   Horizontal saws feature a pressure coolant system. Their construction allows coolant recovery into a sum pan at the base. Coolant means longer blade life and higher sawing speeds.

Figure 5-42 A horizontal saw self-feeds down through the work at a controlled rate.
Keith Smith

With guides adjusted and sharp blades, horizontal saws can cut stock to length with repeatabilities of 0.010 in.! Start to learn the various parts of the horizontal saw from Fig. 5-43. The saw is shown with the safety shield open to expose the idler wheel at the front but not the drive wheel at the back.

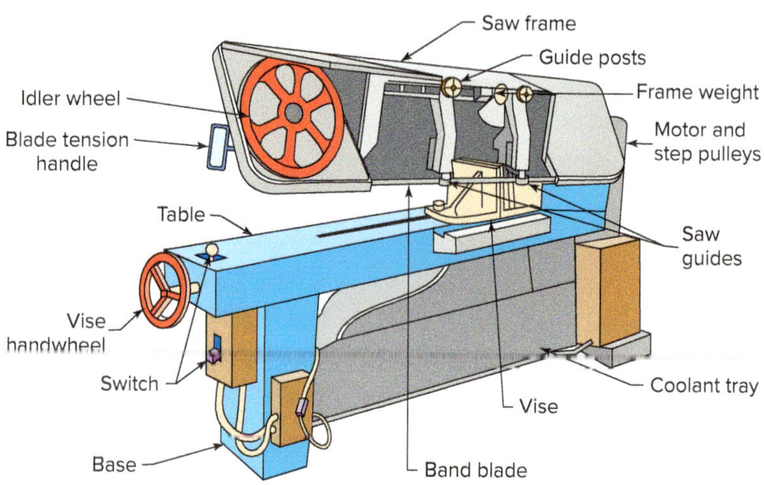

Figure 5-43 The main features of a horizontal saw. Note that the blade is warped 45° to the plane of the wheels to allow sawing long stock.

Special Features

Where large quantities of material are sawed, several extra production features are added to horizontal saws (Fig. 5-44). The newer versions are also CNC controlled for quick setup and operation. Industrial versions of the horizontal saw, called cutoff machines, include

- Automatic or CNC stops, once the saw completes the cut.
- Automatic or CNC bar feeds. When a long bar is to be cut into short blanks, the saw can be set to automatically advance, then saw, then advance again.

Figure 5-44 Fitted with a bar feed, a horizontal saw can advance and cut raw stock blanks.
Courtesy Contour Aerospace/Everett, WA

Horizontal Saw Safety

Horizontal saws are relatively safe compared to vertical types. But, this saw has its own particular precautions:

1. When lowering the moving blade to the work, go easy. New machinists often contact the work too fast, and then blade breakage and stripping can occur.
2. When setting the work in the vise to the correct cutoff mark, never have the saw blade moving. *It must be turned off until the work is clamped in the vise and your hands are out of the saw.* Incorrectly clamped work that suddenly shifts is a major reason that blades break on horizontal saws.
3. Do not try to catch the work with your hand as the part is cut off. Let it drop to the floor or set up a catching chute. Occasionally, at the moment of disconnection, the work will catch and pivot toward the blade. If you are holding it, your hand could be carried into the blade!

KEYPOINT

Blade guide adjustment is even more important on a horizontal saw because the guides warp the blade 45° out of the plane of the drive and idler wheel to allow the sawing of long stock.

Abrasive Sawing

Abrasive sawing uses a thin abrasive cutoff wheel (Fig. 5-45), somewhat like a grinding wheel. It too is used to cut pieces off metal bars. Cutting steel is its primary purpose, but other metals can also be cut with this process as long as they don't clog the cutoff wheel. Compared to band sawing, abrasive cutting is very fast.

When using mineral grains to remove material, rather than cutting tools, we call the process *abrasion* or *abrading.* It's also called *grinding* in many instances.

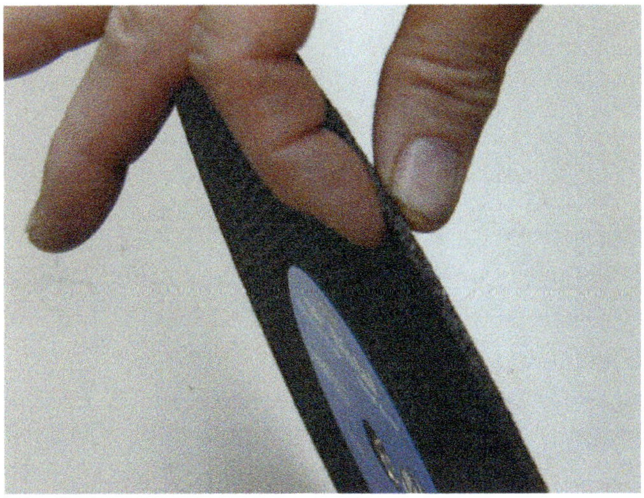

Figure 5-45 An abrasive cutoff blade is strong on its rim but easily broken from side loads.
Courtesy Mike Fitzpatrick

Abrasive Cutting Makes Chips, Too

The definition of abrasion is to scratch or rub away material. But, looking closely, we discover that abrasive processes remove metal by cutting with sharp-edged tools. There are chips being made, but they are nearly microscopic, and thousands are made per second. While a grinding wheel will abrade at slow speeds, it becomes efficient only if its speed is boosted up many times faster than that of cutting tools.

The cutting tools in this case are the sharp edges on each individual grain within the grinding wheel. Those grains are much harder than cutting tools and are able to withstand extreme heat. So when abrasive sawing steel, for example, we see a spray of sparks exiting the cutting wheel (Fig. 5-46). These happen because the tiny hot chips react with oxygen in the room and literally burn up.

Figure 5-46 An abrasive saw in action.
Courtesy Mike Fitzpatrick

Safety Precautions for Abrasive Sawing

1. **Always Inspect the Wheel Visually Before Using It**

   Lift the guard and rotate it to be sure it is okay to use. Do this every time you use the saw!

2. **Double Eye Protection and Dust Masks**

   Both hot flying debris and the possibility of exploding wheels make it important to wear a full-face shield over your safety glasses, and a dust mask is advisable if you are doing a lot of this kind of metal cutting (Fig. 5-47).

3. **Stand to the Side**

   Avoid the hot swarf and possible flying wheel fragments. Stand to the side, never in line with the wheel when it's cutting!

4. **Check Guards**

   Be certain the wheel scatter guard is in place and drops down! These wheels can explode when used incorrectly.

5. **Fire Prevention**

   Be certain that nothing behind the saw is flammable. If walls or other items are in the path, place a sheet metal spark catcher in the swarf path.

6. **Look Before Operating**

   Let others in the area know when you intend to use the "chop saw."

7. **Wear Ear Protection**

   Abrasive sawing is loud. If more than one quick cut is made, ear protection is a must.

Figure 5-47 It's wise to wear double eye protection when using the "chop saw."
Courtesy Mike Fitzpatrick

Reinforced, Flexible Saw Wheels

To keep the kerf narrow and concentrate the cutting action, abrasive saw wheels must be thin, but that makes them prone to shattering. To help them take the punishment, abrasive wheels are made differently from regular grinding wheels in three special ways:

- The binder (the glue) holding the abrasive grains together is a plastic resin, which after curing under low heat can flex with the changing loads of this kind of abuse. The resulting flexible grinding wheels are called *resinoid* wheels.
- In addition to the grains and binder, *cutoff wheels* contain fiber reinforcement similar to radial tires for your car.
- The special grains, called **silicon carbide**, are even harder than normal grinding wheels. This is too technical for now, but we'll explore the various abrasive materials when we come to grinding operations.

Portability Advantage

The compact size of abrasive saws is another advantage. They are small enough to be moved to the steel rack (Fig. 5-48). Instead of taking a heavy bar to the saw, the cutoff saw can be put on a rolling cart and taken to the steel rack. This increases safety and saves time. The bar is pulled out, the saw is cranked up to the bar's level, and it's cut off right in the rack!

Figure 5-48 Not a heavy machine, the abrasive saw can be taken to the steel rack rather than move heavy bars to the saw.
Photography by Prographics at Milwaukee Area Technical College/McGraw Hill

Power Cutoff Saws

Normal cutoff wheels are 12 or 14 in. in diameter. The industrial big brother is much larger in size, using wheels as big as 2 ft, and they feature power-driven coolant. They are CNC controlled or semiautomatic; thus, large material requiring lengthy cuts can be started and monitored by the machinist, who can do other tasks during the cut. It's unlikely you will encounter them in tech school.

Heat Factor

High-carbon steels that are prone to work hardening can and will form a HAZ when abrasion sawed. They can create a glass-hard surface at the cut line (the heat-affected zone). Additionally, a jagged burr forms at the exit point of the cut, and the parts are often too hot to handle with bare hands.

Not for Soft Metals

Aluminum and other gummy metals are not usually cut with manual abrasive saws. The problem is that the air spaces between the grains become clogged with removed material and they stop cutting. With no space between grains, there is no way for the edges to contact the work surface. They float over the work and only rub. This problem happens on all grinding wheels and is called *loading*. On small saws where no coolant is available, there are rub-on lubricants that do work, but most shops set a policy about cutting soft metal using abrasion. On the larger power saws, coolant solves the problem (Fig. 5-49). The dry version makes a real mess. Be sure to cut where the slurry of burned metal and spent abrasive won't contaminate other machines.

Figure 5-49 A grinding wheel loaded with soft aluminum cannot cut.
Courtesy Mike Fitzpatrick

5.3.3 Band Saw Blade Welding

Our final sawing skill is making a blade from rolls of bulk material. To save time, most shops buy prewelded blades ready to mount on the saw. However, the competent journey machinist must know how to weld a blade for four reasons:

1. Broken blades that are still sharp can be repaired by welding.
2. Blades cost far less when bought in 100-ft rolls.
3. Without a blade welder, it is impossible to saw the inside of a part. Here the blade is inserted through a drilled hole, then welded together once inside (for example, when sawing a circle shape out of the center of a plate).
4. Otherwise, sharp blades that have a few stripped teeth can be repaired by cutting out the bad part and inserting a new section.

Note the blade welding attachment shown in Fig. 5-50.

Figure 5-50 As the electric current passes through the blade joint, heat melts the joint together.
Courtesy Mike Fitzpatrick

Blade welders (Fig. 5-51) use electrical resistance to create the high heat needed. The trimmed blade ends are clamped into copper or brass jaws and then forced together. An electric current is passed between the jaws and blade joint. The joint heats, and as it melts, the jaws push the joint together (Fig. 5-52). When the electricity turns off automatically, the two halves are fused together.

Annealing the Weld

As the blade weld rapidly cools, the metal just next to the weld becomes extra brittle and would snap instantly during use. Before putting the newly welded blade into service, it must be softened by **annealing**. The blade welder has provision for opening the jaws wider than the weld, to hold a longer portion of the blade than the weld area. Reheating the general area more slowly, then cooling it even more slowly will soften the hard spots.

> **KEYPOINT**
>
> Annealing is a softening process that happens when metal is cooled slowly from an elevated temperature.

Final Shaping

After annealing, the bulging weld must be ground or filed thin and smooth to allow it to pass through the saw guides. When shaping the weld flat, make sure the joint is as thin as the original blade to allow passage through the saw guides. Many welding attachments feature an onboard grinder and a test slot to try the blade trim; however, a micrometer or caliper will also tell you if it is thin enough.

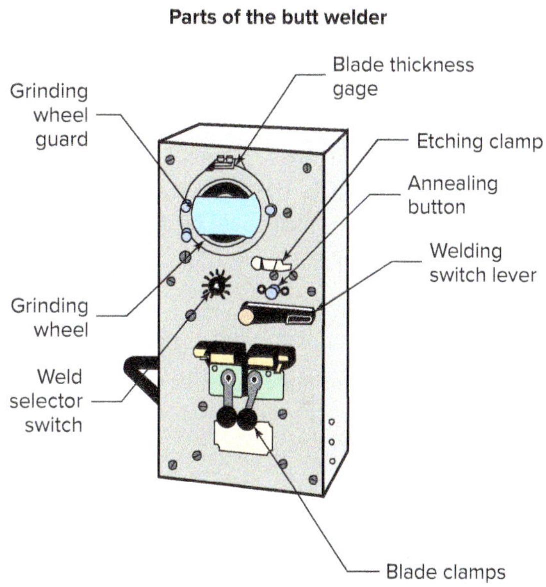

Figure 5-51 The features of a band saw blade welder.

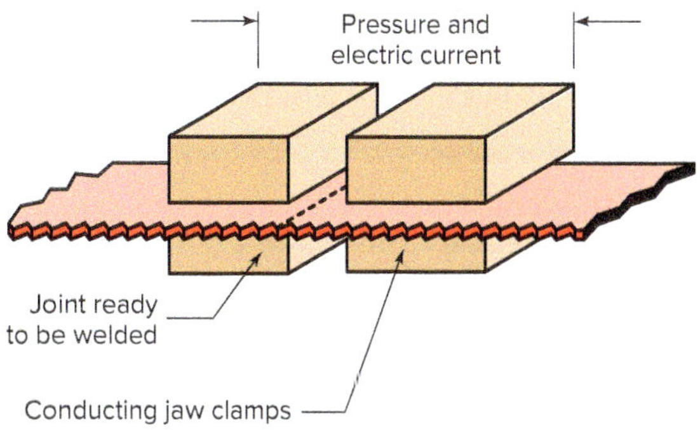

Figure 5-52 The conducting jaws force the melting joint together.

UNIT 5-3 REVIEW

Replay the Key Points

- To be safe a vise or pushstick must be used on vertical band saws, and your wrists should contact the table. Never push on the part alone.
- For all sawing operations, blade pitch should be chosen such that three teeth are in contact with the work at all times.
- Horizontal sawing is for cutting large metal parts.
- Friction sawing is performed on vertical saws capable of 2,000- to 3,000-FPM blade speeds. It is used when the material is too hard to saw otherwise.
- Abrasive sawing is used primarily for quickly cutting long steel bars to length.

Respond

The questions here are based on these material cutting methods: vertical band saws, hacksaws, horizontal saws, abrasive saws, and shears.

1. You have a sheet of $\frac{1}{8}$-in.-thick aluminum from which approximately 2.25-in. × 5.25-in. rectangles must be cut for the drill gage. Of the five methods listed, which could be used?
2. Why would you not choose the abrasive saw for Question 1?
3. You have a sheet of $\frac{1}{8}$-in.-thick aluminum from which 400 4-in. squares must be cut. Of the five methods listed here, which would be most practical?
4. The drill gage is to be made from $\frac{1}{4}$-in.-thick sheet steel. Would a horizontal saw be a good choice to cut the blanks?
5. Would the abrasive saw be a good choice for cutting the drill gage blanks? They are made from low-carbon steel.
6. You are going to first saw the blanks of Question 1, then the steel blanks of Question 4. What changes should be made to the saw?

Unit 5-4 Part Finishing

Introduction: Work Order Operations 30, 80, 120, 130. Although we try to avoid it, machining often leaves raw, sharp edges, burrs, and occasionally incomplete transitions between surfaces. To correct these imperfections, handwork and part finishing are needed. Because this kind of work adds to manufacturing cost, programmers make every effort to eliminate the need for secondary finishing work. But clearly there are times when it cannot be avoided.

Apart from the danger of cutting the hand of the next person to lift a raw part, sharp edges look and feel wrong. As a craft master, you will be able to put your hand to a machined part and know it needs finishing. Here we'll cover a general background in part finishing—enough to get those lab assignments finished and ready to grade. There's a lot more to the subject, but because it becomes shop specific, the rest is left for on-the-job experience. On completion of this unit, you should clearly understand how to handle machined products such that you improve their quality and finish while never marring or damaging them during handling.

TERMS TOOLBOX

Aluminum oxide The common gray abrasive used in general grinding wheels.

Burrs The unwanted jagged metal left after machining is completed; the tool used to remove burrs is also called a burr or a rotary file.

Draw filing Holding the file nearly sideways. Produces a finer finish and better control of shape.

File card The multifunction brush used to clean out loaded (pinned) files.

Lathe file A flat file with teeth on two sides.

Mill file A flat file with teeth on four sides.

Part-finishing wheels A soft fiber and abrasive wheel that is very effective at finishing part imperfections. Often called a 3-M wheel.

Vitrified The hard grinding wheel made by firing a clay binder and an abrasive until it forms a material similar to china or porcelain.

5.4.1 Putting the Final Touch on Work—Part Finishing

Work Order Operation 120, "File 0.12 inch Radii," and Operation 130, "Remove Cutter Radii 1 Place."

Three Kinds of Part Finishing

Part-finishing tasks fall into three categories:

1. *Surface finishing*
   - **A.** Removing burrs and sharp edges
   - **B.** Improving surface finish
   - **C.** Final shaping

2. *Part marking* numbers and other information
3. *Secondary machining operations*
   - **A.** Threading with taps and dies
   - **B.** Assembly press fits
   - **C.** Drilling, chamfering, reaming, and honing

These are mostly hand skills (Fig. 5-53); however, new technologies are providing some interesting methods of improving and marking parts. In this unit and Unit 5-5, we'll look at these new skills as well as the more traditional ones. Work imperfections can be corrected both with power equipment and by hand, using a variety of implements and tools.

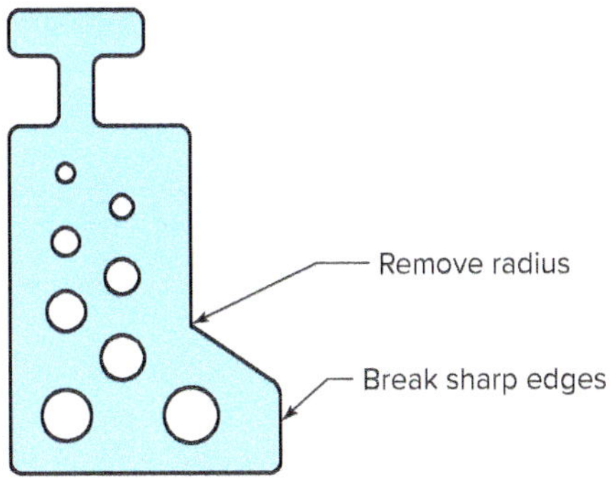

Figure 5-53 Edges and a corner requiring hand operations.

Hand Tools

We're talking good-old hand-eye coordination. It comes with practice with files, power tools called rotofiles and burrs, along with a variety of other finishing tools (Fig. 5-54).

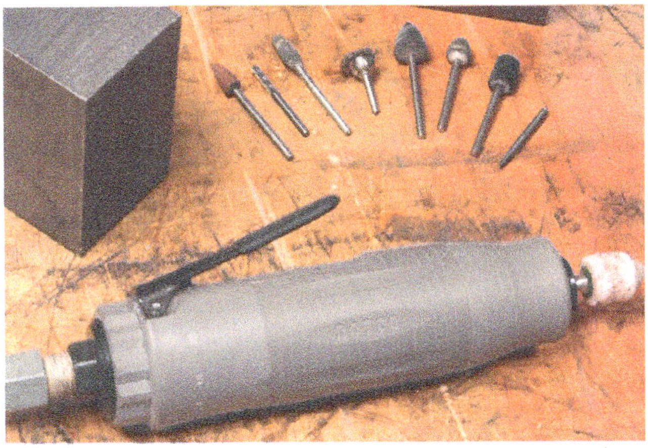

Figure 5-54 A selection of hand work burr tools and an air motor tool.
Photography by Prographics at Milwaukee Area Technical College/McGraw Hill

SHOPTALK

What Makes Burrs, and How Can You Prevent Them? Burrs are the ragged, last chip that should have been removed at the end of the cut but wasn't because the metal could flex and the cutter wasn't able to shear it off (Fig. 5-55). Burrs cost money, cut hands, scratch adjacent parts, destroy the accuracy of measurements, and delay production. Professional machinists produce the fewest possible burrs, and they remove them at the machine if time and the situation permits.

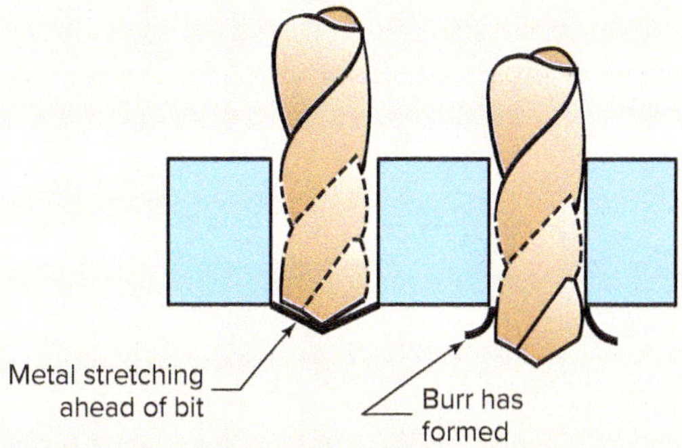

Figure 5-55 Left: The burr is forming ahead of the bit. Right: The drill has burst through and bent it to where it cannot be cut off.

Using a drilled hole as an example, notice that as the drill nears the bottom, a dome of uncut metal forms ahead of it. That's because the metal in front of the drill is able to deform, and it's stretching out. With no resistance to the cutting edge, it cannot be cut. On the right, the bit has burst through the dome, bending the still-attached burr back beyond where it can be cut away—a burr has formed.

A little machinist attention to solving this problem can save hours of secondary work! Using a sharp drill will tend to form far fewer burrs, and it will cut the shell away as the drill breaks through. Correct machining speeds and coolants, and correct force called "feed," will also reduce burring, but clearly tool sharpness is the number one answer.

Let's look at the various operations to finish the order of 12 drill gages. Operation 30 reads, "Remove all burrs" and also the general note on the drawing: "Break all sharp edges 0.005 in." That means to remove the sharp edge by creating a small chamfer or corner radius no more than 0.005 in.

At this stage, the blank rectangle has been machined to the finished 2 × 5 in. size and the holes drilled. We will start by *breaking* the outside edge that is to put a small rounded corner on the sharp edge by hand. An edge break is from 0.005 to 0.015 in. There are five choices and any combination might be used:

1. Hand filing
2. Edge-breaking tools
3. Hand power tool
4. Pedestal burr wheel
5. Slurry vibration ("vibraburring")
6. Tumbling with burring stones

Hand Files

Files come in many shapes and with different tooth patterns and coarsenesses. For this job, you would select a flat file with fine teeth and a pattern of single- or double-cut rows. Let's look at each characteristic.

File Shape

Little training is required for this aspect of files. Most of the shapes available are shown in Fig. 5-56. Many file shapes are supplied in two varieties: tapered and straight (Fig. 5-57).

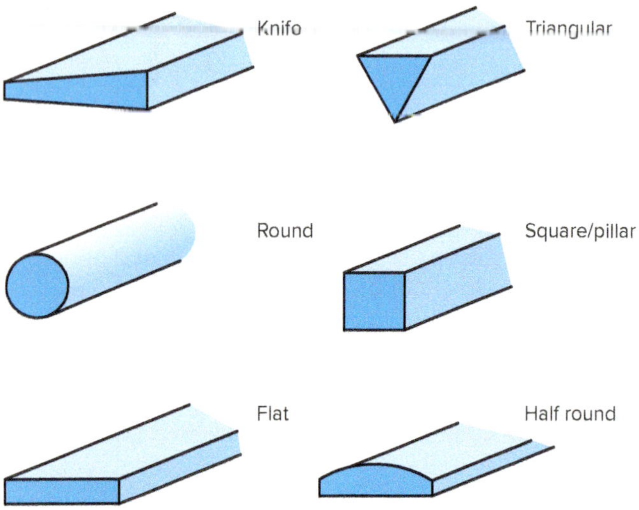

Figure 5-56 Common file shapes.

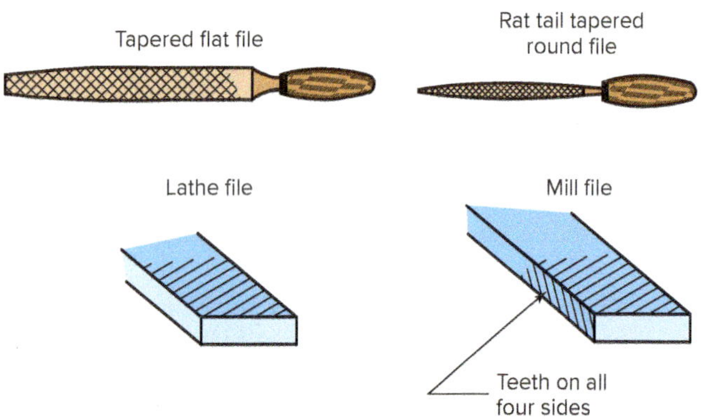

Figure 5-57 Tapered files and mill files compared to lathe.

Mill and Lathe Files

Files with teeth on all four edges that are able to cut down and sideways are called **mill files**. Those with teeth on the two larger surfaces are called lathe files. **Lathe files** can be used to file against an edge without metal removal from the edge itself (see Fig. 5-57).

Tooth Pattern

Machining files have four kinds of teeth, as shown in Fig. 5-58. Each has a purpose:

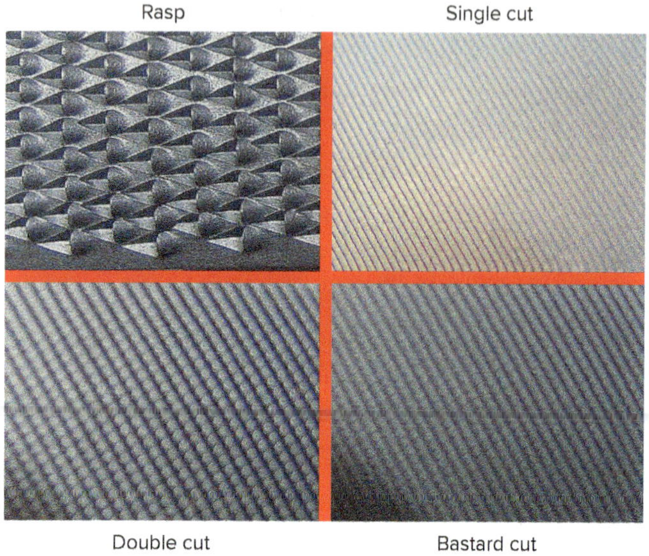

Figure 5-58 Four file tooth types. Each has a different purpose.
Courtesy Mike Fitzpatrick

Single-Cut Files: Finishing Work With one row of teeth, they are used where metal removal is small and finish should be good. They would be a good choice to break the edges of the drill gage blanks. Due to the small space between teeth, single-cut files tend to clog or "load up," also called "pinning," especially when filing soft metal such as aluminum. The loaded metal then scratches the finished product.

> **KEYPOINT**
>
> Single-cut files are the slower metal remover but produce the best finish.

Double-Cut Files: Semifinishing and Roughing This file will shape metal well but leaves a rougher surface behind. Double-cut files remove more material faster than single-cut files. They wouldn't be the best choice to break the edges.

Bastard Cut Files These have teeth set at an angle to each other but not in a double-cut pattern. Notice the discontinuous teeth in the pattern. Because of the way they lay to each other, this file cuts very fast. It is used for rough filing and heavy metal removal. It is a very poor choice to break the edges.

Rasp Cut Files Used for fast removal where surface finish is of no concern, this file works well on soft metals like aluminum, where tooth loading can be a problem with finer files. A rasp is made differently than a standard file in that individual teeth are formed by lifting metal up off the file body. They are sharp enough to cut flesh!

File Handles—Always Use Them!

File handles (Fig. 5-59) protect palms from a nasty stab from the sharp *tang* (the pointed piece that is forced into the handle). Never use a file without a handle. This is doubly true if filing on the lathe, where the machine action could catch and send the file back at high speed.

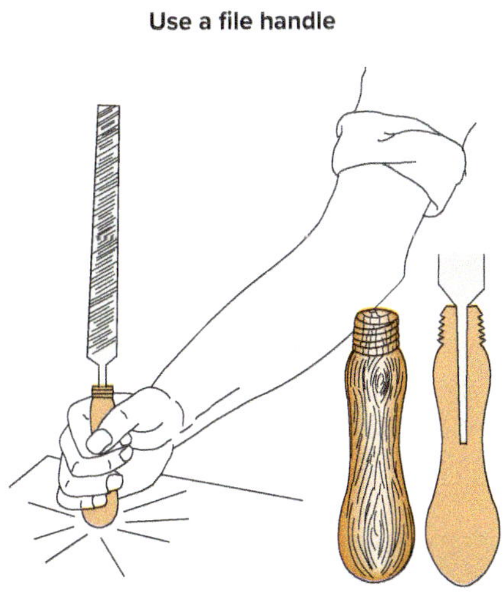

Figure 5-59 Always use a handle on files.

Filing the Drill Gages

To file the sharp edges of the drill gage requires a couple of skills:

1. Producing good straight edges without overfiling
2. Avoiding marking the part in some other way while holding it

There are two ways to remove metal using a file:

1. Reciprocal (back and forth) (Fig. 5-60)
2. **Draw filing** (angular shearing) (Fig. 5-61)

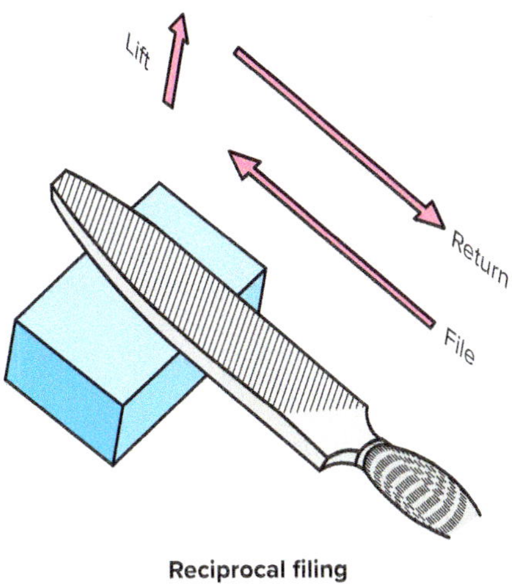

Figure 5-60 Straight filing makes removal faster but can make it difficult to control shape.

The draw file method might work better for the edges. To use the correct method, place your second hand on the end opposite the handle.

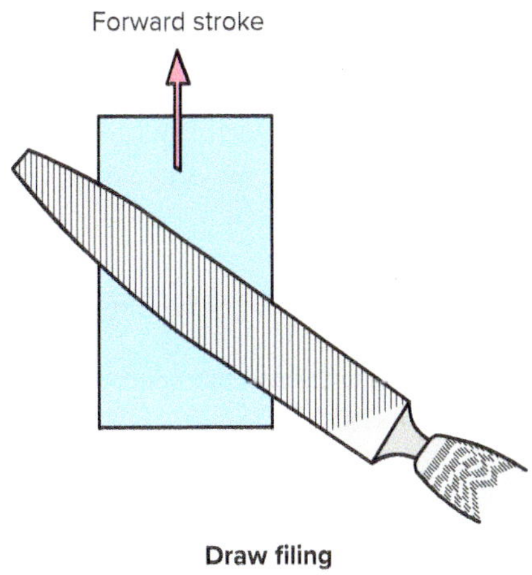

Figure 5-61 *Draw filing* controls the shape better and produces a better finish.

File Cleaning

Use a **file card** to remove particles of built-up metal in the file, a process called pinning. Brushing a light coat of cutting oil on the work (or plain chalk on the file) can often help, but there comes a time when you must stop because unwanted loaded metal is marring the surface as it is produced.

To clean a file, brace it against a bench and use the three parts of a file card/brush (Fig. 5-62). First, use the stiff wire "card" for loosening and removing big particles. Then use the bristle side to clean the entire surface. For difficult lumps of metal that remain, the file cards feature a foldout metal pick.

Figure 5-62 Cleaning with the wire and then bristle brush sides of a file card can remove built-up metal from the file to keep it cutting cleanly.
Lake Washington Technical College, Kirkland, WA/McGraw Hill

Do No Harm to the Work— Holding It for Finishing

Many employers complain of tech school graduates causing more work than they solve by marring and scratching parts while working on them.

KEYPOINT

This is very important: Always protect the work from marring. Don't cause mars by the way you handle or set parts down.

When holding parts in a visc, always use jaw protectors. These can be shop-made from a material softer than the work or they can be rubber or plastic protectors made specially for your vise. Here's another set of little things that do matter–the game rules of part handling:

Always set parts down gently.

Keep parts separated if possible by paper or in racks or boxes–don't allow them to bump together.

Never hammer or wrench against machined surfaces.

In machine setups, watch out for clamps and vise marks.

Never stack parts up where they could fall.

Using Rotoburrs

These are little files that cut by spinning at very high rates–from 10,000 to 20,000 RPM. The hand instruments that spin them, called roto-motors, die grinders, or air motors, are mostly air driven to keep them light yet powerful. Electric burr motors are also available for use where the work is not demanding (Fig. 5-63).

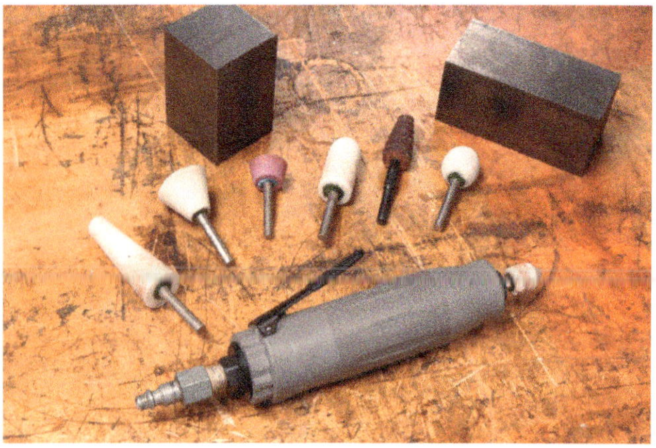

Figure 5-63 Sanding tools can be held in an air motor.
Photography by Prographics at Milwaukee Area Technical College/McGraw Hill

SHOPTALK

RPM Galore! Air motors are turbine driven and they spin anywhere from 10,000 to 20,000 RPM, depending on shop air pressure and turbine design. In addition to eye protection, be sure your hands are braced; the burr's cutting action can grab and quickly pull the tool across the work.

In addition to burrs, we also mount sanding tools in the spindle of a roto-motor (Fig. 5-64). The shaped sanding tools might be a good choice for the drill gage edges. There are disks, cones, and cylinders. With practice these tools can shape very well to a fine finish.

Figure 5-64 Breaking the edge with a sanding cone (beehive) might be a good choice.
Photography by Prographics at Milwaukee Area Technical College/McGraw Hill

Rotary Files (Sometimes Called "Burrs" or "Rotoburrs")

The files are made from carbide or high-speed steel. They are supplied in a wide range of shapes: cones, balls, cylinders, and many others. They are fast metal removers and can be used to form transitions and other part features. They are called rotofiles or rotoburrs (Fig. 5-65).

Figure 5-65 Rotary files (rotoburrs) are too aggressive to slightly break aluminum edges unless they are used with exceptional care.
Courtesy Mike Fitzpatrick

Figure 5-66 Mechanical burring and edge-finishing machine.
Courtesy Universal Aerospace/Arlington, WA

For breaking the drill gage, burrs are too aggressive. They would be good at removing large burrs from abrasive sawing or the rough edge of friction sawing, for example. Using them requires practice and a steady hand. They are used only for fast metal removal where further sanding or filing is expected. They are not appropriate for our drill gages.

Lubricating Air Tools

The turbine bearings require twice daily lubrication. The easy way is to put a few drops of air tool oil right into the air pressure inlet. In shops where they are used extensively, the air lines themselves are injected with the oil.

Power Machine Edge Burring and Breaking, Tumbling and Vibrating Burr Machines

There are several machine methods of removing burrs and breaking edges. Two are made specifically for the purpose, the part finisher and the part tumbler (Fig. 5-66). These finishing machines both use loose abrasive in a slurry (water, abrasive, and sometimes a cleaner). This type of burring/finishing is automatic and very good for large qualities of parts. The parts are either vibrated rapidly or rotary tumbled against an abrasive medium. The tiny collisions remove sharp edges and burrs. The challenge is to find an abrasive medium that will get at the problem areas and not get stuck within details of the parts or remove too much material. There are hundreds of abrasive shapes and sizes from which to select (Fig. 5-67). These artificial stones are manufactured from abrasive grains and a resin binder.

Figure 5-67 A variety of abrasive shapes for mechanical finishing of parts.
Courtesy Mike Fitzpatrick

The advantage of this type of burring and finishing is that it can finish hundreds of parts at one time while requiring no operator intervention. The disadvantage is that this process is slow, and the entire surface is given the tumbled-matte surface. Many vibrating machines are noisy, so operators must wear ear protection when working around them, or the machine must be placed in a soundproof room.

Common Shop Machines for Part Finishing: Pedestal Burr Wheels, Disk and Belt Sanders

The final method by which you could finish the edge of the drill gage uses one of three common machines in the shop.

Pedestal/Bench Grinders

Used to sharpen tools and perform many utility removal operations, they are the same machine other than the addition of a pedestal (Fig. 5-68) to mount the grinder to the floor. Since our parts are aluminum, doing them on a regular grinding wheel would be banned due to loading the wheel; plus it would do a very bad job of breaking the edges of the drill gage. However, there are special wheels made for part finishing that can be mounted on pedestal grinders (Fig. 5-69). We will look at these special wheels after some discussion about bench grinding and grinding wheels.

Figure 5-68 A pedestal grinder is used for tool sharpening but might be too rough for part finishing.
Lake Washington Technical College, Kirkland, WA/McGraw Hill

Grinding Wheel Construction

Grinding wheels are composites of three materials:

Abrasive grains

A *binder* to hold the grains together and provide strength to the wheel

Air spaces between the grains

Without the air spaces, the wheel wouldn't work. They create the necessary chip clearance, help coolants get to the cut (air in the case of a bench grinder), and provide temporary places for the removed chip to stay until thrown out by centrifugal force.

Vitrified and Resin Wheels

The two kinds of grinding wheel binders used in machining are rigid (**vitrified**) and flexible (resinoid discussed previously under abrasive sawing) wheels.

Figure 5-69 A variety of grinding and finishing wheels for a pedestal grinder. From left to right: Standard abrasive, high purity fine abrasive, brass wire wheel, steel wire wheel, and resinoid cutoff wheel.
Courtesy Mike Fitzpatrick

Resinoid—The Flexible Binding

These more shockproof wheels are pressed and heated using a plastic binder and abrasive grains. The abrasive is usually silicon carbide, but other abrasives are also used. These wheels can take more mechanical shock (for example, the wheels used on handheld right-angle grinders for cleaning up welds). But they can be broken, and they do fly apart when misused.

Slitting Wheels

On the pedestal grinder, when we need to cut a narrow slot in a part, a resinoid wheel would be used. An example might be cutting off a few coils of a spring. Or they (Fig. 5-70) might be used for any task where a regular grinding wheel would be too wide to do the job.

Figure 5-70 A thin and flexible slitting wheel cutting a slot in a steel part. (Safety guards removed for photo clarity only.)
Lake Washington Technical College, Kirkland, WA/McGraw Hill

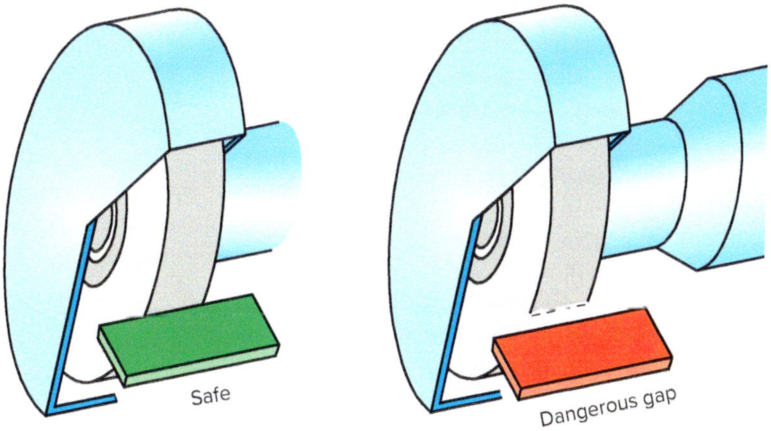

Figure 5-71 The tool rest must be within 0.100 in. or less from the wheel.

Vitrified—The Rigid Binding

The common, hard gray grinding wheels used for most bench and precision operations are held together with clay. This binding material is mixed with the abrasive grains, then fired similar to a ceramic cup or porcelain. The way the air space is created is unique–can you guess how it's done? Chapter 12 answers the question.

Aluminum Oxide—Everyday Abrasive

In addition to the clay, the abrasive in most pedestal grinding wheels is made from **aluminum oxide**. Higher-quality, pure aluminum oxide wheels are bright white. Not usually used for pedestal applications, the white wheels are used on precision grinding machines where long life and shape retention is needed and the expense is justified. Everyday aluminum oxide is gray.

Due to brittleness, vitrified wheels do break when misused. When struck or when work shifts rapidly under them, they can crack.

"Green Wheels on the Pedestal Grinder"

Special vitrified grinding wheels are made from silicon carbide rather than aluminum oxide. They are for grinding extra-hard carbide tools only. The binder in these wheels is colored green for easy recognition.

A Few Tips the Pros Use

1. As the wheel wears, the gap between the tool rest and the wheel increases. Keep this gap to less than 2 mm or $\frac{1}{16}$ in. Otherwise thin objects and the fingers holding them can be instantly sucked in! Loose clothing and hair go in through any gap, big or small.
2. Be certain there is a scatter shield on the upper portion of the wheel and that it is lowered close to the wheel. Its main purpose is to contain hot swarf, but in the rare incidence of a wheel explosion, it will also contain wheel fragments. (The upper guard is not shown in Fig. 5-71.)
3. Stand to one side when starting a grinding machine. If the wheel was damaged by the last user, it will blow up during acceleration.
4. Inspect mounted grinding wheels before starting the machine by rolling them while looking for cracks and chips.

Mounting a New Wheel

Before the ingoing wheel is put on the grinder, perform the following:

A. *The Visual Test*–Look them over for tiny cracks. However, cracks can be missed this way.

B. *The Ring Test*–Holding vitrified wheels on your finger or by a rod through the center hole, tap them lightly with a screwdriver handle or other plastic object at several locations. They will make a thud if they are cracked and will ring clear if unbroken. Think of a broken porcelain cup in the same test. It would sound similar. See Fig. 5-72.

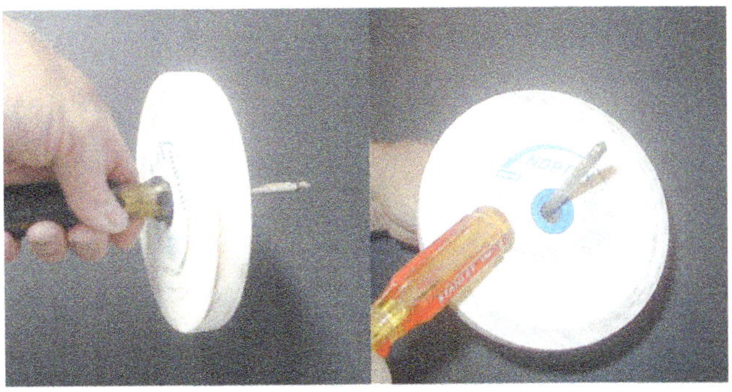

Figure 5-72 Perform a ring test of any vitrified wheel before mounting it on a machine.
Courtesy Mike Fitzpatrick

5. Perform an acceleration test on new wheels. On initial turn-on, after mounting a new wheel, touch the start button; then quickly press the stop button. Allow the wheel to coast for a second. Now press the start button again and let it accelerate a bit more before touching the stop. Do this three times, each time letting the wheel go a bit faster. Finally, start the wheel for 1 minute and just let it run with no load. This ensures that it is stable and safe. Of course, stand to the side when doing this.
6. When mounting any kind of wheel on any grinder, always use a paper gasket on both sides of the wheel. This ensures even pressure from the mounting flanges against the wheel and it helps to absorb shocks to the wheel. Do not overtighten the wheel nut.
7. Be cautious of the heat generated during grinding (Fig. 5-73). Handheld objects can heat rapidly when ground. Keep a water pot close by to cool the work.
8. When grinding, as much as possible, move your work over the entire surface of the wheel to avoid wearing a groove in the flat surface. And do not force it—you will find the right amount of minimum push that works best with practice.

Figure 5-73 For safety and performance, always use paper gaskets when mounting grinding wheels on spindles.
Photography by Prographics at Milwaukee Area Technical College/McGraw Hill

Dressing Grinding Wheels

Grinding wheels self-sharpen to some degree. That is, the binder will break as dull grains drag on the work. The extra pressure causes them to either crack or shear away completely, thus exposing new cutting edges. However, with prolonged use, grinding wheels become dull and their surface becomes irregularly shaped. These wheels require a touch-up dressing. There are three choices (Figs. 5-74 and 5-75):

Wheel dressing tools
Diamond dressers
Carborundum sticks and blocks (a hard material)

Figure 5-74 The three methods of dressing a wheel: a wheel dressing tool, a diamond dresser, and a carborundum stick.
Photography by Prographics at Milwaukee Area Technical College/McGraw Hill

Figure 5-75 The wheel dresser features a set of star-shaped cutters and guide lips to hold it straight on the tool rest.
Photography by Prographics at Milwaukee Area Technical College/McGraw Hill

Diamonds and carborundum are used more for technical precision grinding and will be studied later. For pedestal grinding, we use the wheel dresser.

Using a Wheel Dresser

Note in Figs. 5-75 and 5-76 that there are small guide tabs on the leading edge of the dresser. They are meant to hook over the front edge of the grinder's tool rest, to help move the dresser in a straight line from side to side. However, they work only as long as the tool rest is straight. If there is a groove ground in the rest from adjusting it too close to the wheel, then the tabs cannot be used as guides. To use the guide tabs, loosen and then pull the tool rest back enough to allow them down past the wheel, then retighten the rest. The dresser is now tipped up until the dressing wheels contact the spinning grinding wheel lightly. Moving it side to side while increasing the pressure will cause the dresser wheels to spin rapidly and vibrate. That action plucks out grains. Have your instructor demonstrate how to do it the first time.

Part-Finishing Wheels

A very good option for breaking the edges of the drill gage, this soft part-finishing wheel is used on a pedestal grinder (Fig. 5-77). **Part-finishing wheels** come in several stiffness degrees and abrasive sizes, but they all have the ability to conform to the edge and remove just the right amount of material. It takes only a minute to learn to use one.

> **TRADE TIP**
>
> **Make a Straight Guide** Most of the time, the rest is grooved on the front edge from incorrectly adjusting it too close to the wheel. Here's a tip to get a perfect dress every time: Make a slip-on guide, as shown in Fig. 5-76. After pulling the rest back, slide it over the tool rest, so it's just past the worn edge. The guide must be made to fit the particular grinder, but most feature a similar rest. Make one in training; then bring it to your first job. Your coworkers will be dazzled unless they, too, studied in *Machining and CNC Technology*.
>
> **Don't Dull the Fresh Grains** When dressing a grinding wheel, remove the minimum amount required to create a new true surface. Once the wheel is trued all the way across its face, do not go back over the surface lightly "just to be sure." That only dulls the grains you have just exposed.

Wire Wheels

The last pedestal grinding wheel we might use to finish the edges of the aluminum drill gages is a wire wheel (Fig. 5-78). Wire wheels do take off burrs on work and can clean away rust or unwanted coatings. Their wires vary in stiffness and composition. The stiff variety is used to remove steel burrs and rust. However, only the most flexible version could possibly be used to break the edges of the drill gages. Wire wheels would be a wrong choice for aluminum parts, but not a bad one for finishing some steel products. Wire wheels come with two special precautions:

1. **Safety**

   The wire bristles fly out like arrows! Be certain to wear adequate eye protection. Watch your eyes and the eyes of others working around you.

2. **Contamination**

   Steel wire wheels (the typical type) will cause surface contamination in other metals. Stainless steel that has been brushed with a common steel wire wheel will rust from the microscopic particles of steel embedded in the material. There are stainless wire wheels and brass ones made to avoid surface contamination, but they are not common in most shops.

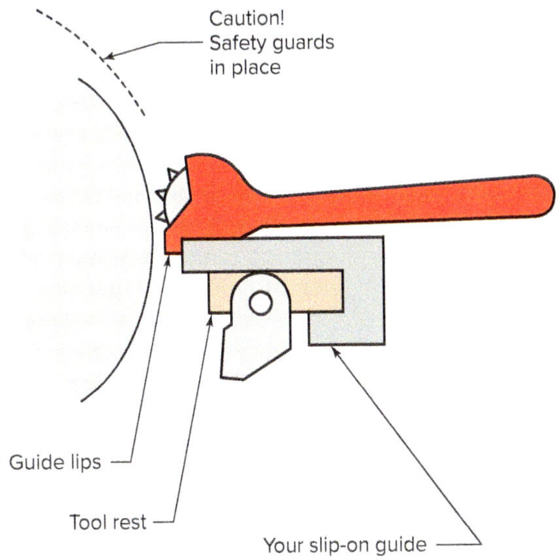

Figure 5-76 Dressing a grinding wheel using the guide lips.

Figure 5-77 A part-finishing wheel on a pedestal grinder.
Courtesy Universal Aerospace/Arlington, WA

Figure 5-78 Wire wheels remove burrs and sharp edges but would be too crude for the drill gages. Be cautious of flying wires and the tendency to pull objects into wire wheels!
Courtesy Mike Fitzpatrick

SHOPTALK

Robot Burring and Part Finishing The theme in this text is adapting to technology. It's the essential survival trait of the future-minded machinist. Here's the perfect example: A progressive French company added a robot to its part-finishing system to automate the tedious job of burring and polishing work. Working faster and more consistently, it paces the company's high-speed machining facility.

The point is, productivity goes up using about the same number of employees after the change as there were before, but their roles are drastically changed. The lower-skilled part finishers are replaced by programmers and support tooling people. Are they the same people with new skills? Or are they displaced by the changing technology? Either way—education is the key!

This robotic arm is breaking the sharp edges of machined parts.
Courtesy Mecafi and Staubli Robotics

Disk Sanders

The disk might be chosen to finish the edges of aluminum. But since it is an aggressive metal remover, a fine-grit disk would be used and still it might remove metal beyond the 0.005-in. target. The challenges would be not to overround the edges and to create an even amount of break for the entire edge. Both disk and belt sanding machines are very aggressive metal removers. Although they are simple, sanders have two serious safety hazards that are often overlooked:

1. **Keep Loose Clothing Away from the Disk**

   Both disk and belt sanders pull loose items into their pinch point at an amazing speed. Cloth instantly disappears! There's no time to react or pull back. If it's your shirt, you will be injured. See Fig. 5-79.

2. **Stay Out of the Disk Danger Zone**

   On disk sanders, the left side pulls the work down against the table—that's good. The right side lifts it up very fast, and it usually flips your knuckles over to the left side of the disk—that's bad! See Fig. 5-80.

KEYPOINT

Never sand on the right half of the disk!

Hand Burring Tools

There are some new and innovative burr- and edge-dressing tools in tool catalogs. Shown in Figs. 5-81 and 5-82 are a couple of examples. These tools are pulled along the aluminum, and both the top and bottom edge are given a chamfer at the same time.

Figure 5-79 A disk sander can pull loose sleeves into its pinch point faster than you can react!
Photography by Prographics at Milwaukee Area Technical College/McGraw Hill

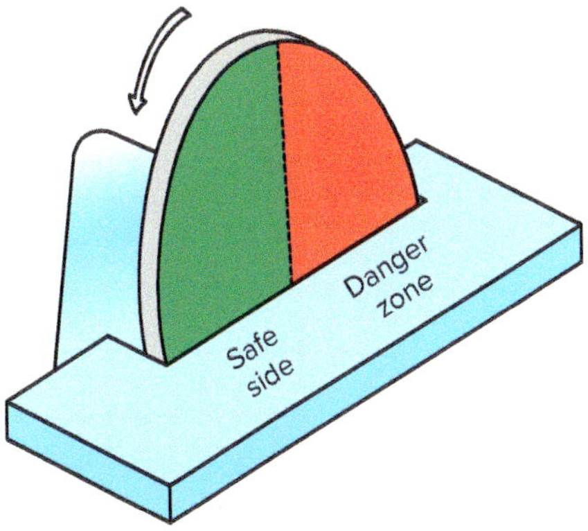

Figure 5-80 There is a good side and a bad side on a disk sander—know the difference!

Figure 5-81 With a little practice, you could do a good job of breaking the drill gage edges with this edge finisher.
Courtesy Mike Fitzpatrick

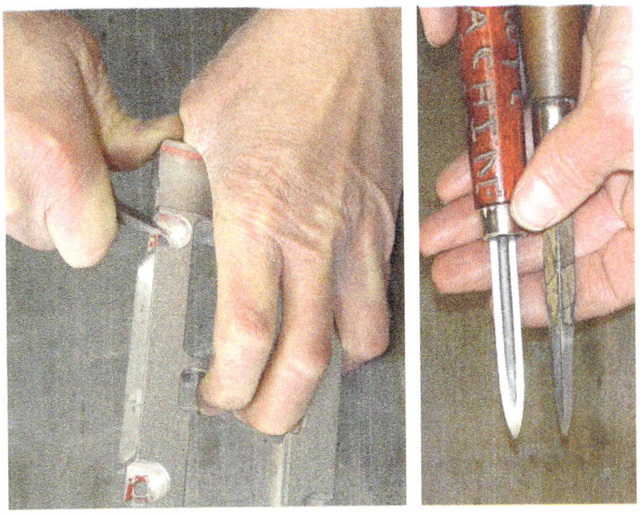

Figure 5-82 A hand burr knife made from a triangular file can be useful in removing sharp edges.
Lake Washington Technical College, Kirkland, WA/McGraw Hill

Getting a Feel for a Finished Edge—Hands On

A journey machinist is not likely to turn in a part for inspection or shipment without first testing for sharp edges. This test will tell whether or not a machined edge is finished.

Correctly dressed for the shop, find a piece of scrap steel and/or aluminum. For an experiment, lightly slide the unfinished edge across a piece of plastic such as a screwdriver handle. Using very little pressure, if it peels or slices a slight bit off the plastic or digs in and will not slide, the edge is too sharp to be a finished product.

Now select any method we've discussed, then break the edge just a bit, 0.005 to 0.010 in. Try it again, scrape the plastic and get the feel of a finished edge. Now touch it with your hand and compare it to the feel of the raw edge.

Finishing the Holes

So far, we have concentrated on the edges of the drill gage, but what about finishing the drilled holes? There are two methods beyond the vibrating and tumbling machines discussed: hand tools or a drill press with a countersink.

Countersink by Hand or Power

Countersinks are cone-shaped cutters used to machine small bevels at the entrance of holes–if the cone-shaped hole produced is deep enough, it provides a nest for a screw head. They are also very efficient tools for breaking the sharp edges of the holes. We'll look at them more closely in Chapter 9 on drilling skills.

A countersink mounted in a portable drill motor could do a fairly good job of breaking the hole edges (Fig. 5-83). A countersink mounted in a file handle also works for this task (Fig. 5-84).

Figure 5-83 A countersink mounted in a portable drill motor could break the hole edges.
Courtesy Mike Fitzpatrick

Figure 5-84 A handheld, rotary hole finisher could break the edges quickly. (We call these swizzle sticks.)
Courtesy Mike Fitzpatrick

Burr Knives and Other Hand Tools

Because of the way they are twirled around, these handy little tools are sometimes called swizzle sticks in shop lingo (see Fig. 5-84). They quickly break the sharp edge.

UNIT 5-4 REVIEW

Replay the Key Points

- Burrs can be minimized during machining by keeping tools sharp.
- Taking just a few minutes longer on setups and maintaining sharp tools can save hours of benchwork removing burrs.
- Air turbine rotary tools spin at 15,000 to 20,000 RPM—always protect your eyes when using them.
- When handling and finishing parts, do them no harm.
- There are many options for finishing work. Choose the one that delivers the quality required in the least amount of time.
- A true craftsperson would never turn a part in for inspection with sharp edges—unless the situation simply wouldn't allow doing otherwise. For example, the CNC machine cycles too fast to permit burring the work.

Respond

1. Burrs are caused by

A. Dull tools

B. Metal's ability to stretch ahead of the cutter

C. A cutting action that's too fast or hard

D. Poor or no coolant

E. All of the above

2. True or false? Files with teeth on all 4 sides are classified as mill files.

3. Rasps are primarily used to file.

4. You need to break the edge of 500 parts with many edges and features. What is the best method? Why?

5. Name the methods that might be used if the order in Question 4 was for 20 parts.

Unit 5-5 Performing Secondary Offload Operations and Assembly

Introduction: In this unit we'll look at several crossover operations that could be done by hand or on the machines. The reasons to perform them off the CNC machines are threefold:

1. To free up the high-production machines to do what only they can do
2. To break up bottlenecks and speed the work forward when it can't all be done on the CNC machines
3. To complete some operation not easily finished or even impossible to finish on a CNC

Offload Machining Avoided If Possible Today every effort is made to avoid moving the work to several stations if the planner can see ways to machine it with one CNC setup (*one-stop machining*). But there are times when it's more efficient to do the work elsewhere.

If the secondary task involves chip making it is called "offload" machining. In following this kind of planning, some shops have small, less costly offload machines that take work from bigger machines to remove burrs or finish operations when it speeds up workflow.

TERMS TOOLBOX

SECONDARY OPERATIONS

Arbor press A machine used for pressing and depressing assemblies.

Blind hole One that does not go all the way through a part.

Bottom tap Used to finish more threads in a blind hole.

Die A threading for making a bolt or stud–outside thread.

Honing An abrasive process that removes small amounts of metal to create a close tolerance hole of good finish.

Plug tap Cuts closer to the bottom of a blind hole than a starting tap but not as close as a bottom tap.

Press fit (P/F) Permanent assembly by forced insertion.

Shrink fitting Using heating and cooling to lighten the pressure of a press fit.

Secondary/offload operations Necessary tasks, usually not performed on the machines, needed to start or finish a job. Sometimes called offload work.

Tap A threading tool used to make internal threads–nuts.

THREADS

Class The fit of a thread–how tight or loose it is to be.

Clearance The deliberate space between bolt and nut to allow movement.

Coarse threads General-purpose threads.

Fine threads Sometimes called SAE threads. Fine threads have a shallower helix and thus create more translation force but less distance.

Form The shape of a particular thread. The most common form is the triangular thread for both ISO metric and Imperial Unified threads.

Helix angle The angle at which a thread's inclined plane works.

Nominal diameter The basic design size of the bolt. Nominal is the drawer size for the bolt or nut but not the real size.

Pitch The term used to express either the pitch distance or the threads per inch.

Pitch distance The repeating distance of a thread. On Imperial threads it's found by dividing pitch number into 1 in. For metric threads, it is specified in the callout.

Threads per inch Abbreviated TPI and often called pitch.

Translation The distance a nut moves on a bolt with one turn–the result of the wedge being driven by turning the nut on the bolt.

Truncation Clipping the top and bottom points off a 60 triangular degree thread.

SHOPTALK

TMI Sources At this stage, the body of information on threads and threading would be too much information; however, there are places to go when an in-depth reference is needed: *Machinery's Handbook® CD*, website, or hardbound book, or the Internet. Search for Standards + Screw Threads + Unified, or the American Society of Mechanical Engineers home site; then let your pointer find its way through the menus.

SHOPTALK

SAE Threads Fine threads in the United States are also referred to as SAE threads, because in the distant past there was a lack of standardization, so the Society of Automotive Engineers took on the task for use in its industry. We seldom use that term in machining because all threads were also organized by the American National Standards Institute (ANSI) some time ago, to become the Unified Screw Thread System. Then the entire jumble of thread types were collected together and standardized. That's where we get the UNF and UNC. So, UNF and SAE are the same thread—perhaps the first "double standard." Today, in an effort to streamline the entire gambit of manufacturing controls, the ANSI has become part of the Society of Mechanical Engineers (SME), the same group that deals with GDT standards.

5.5.1 Introduction to Threads

This is a first lesson in the technology of threads. There is enough to the subject that we'll discuss them twice more for now, staying with our theme of W/O secondary operations; this will provide what you need to know to make threads at the workbench.

Threads are inclined planes (wedges) wrapped around a cylinder. In the drawing (Fig. 5-85), which wedge could move the most weight with the least forward push? The answer is the one on the right–the shallow-angled wedge. By contrast, the one on the left could move the weight the greater distance (known as the **translation** distance), but it would also require more force to do so.

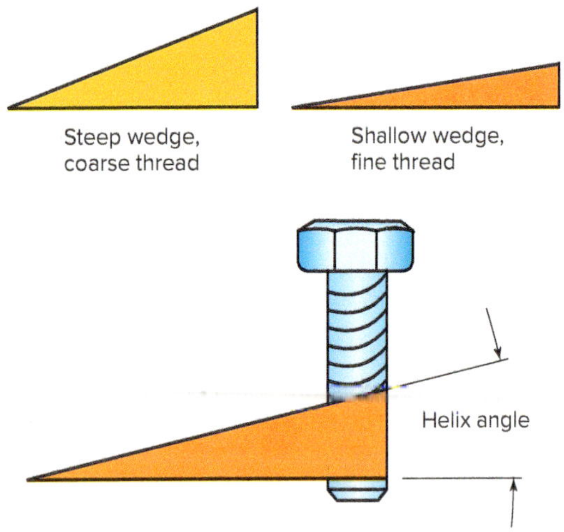

Figure 5-85 A thread is a wedge wrapped around a cylinder. A fine thread is a shallow wedge; a coarse thread is a steep one.

Threads have shallow and steep counterparts–the shallow version being a fine thread, and the steep version, coarse threads. The angle at which the thread lies with respect to the cylinder is called the **helix angle**, as shown in Fig. 5-85.

Standard Sizes

While there are many more kinds of threads, for every size of standard bolt (metric or Imperial) the American National Standards Institute (ANSI) and the ISO have designated a fine and coarse thread–for example, a half-inch fine and a half-inch coarse or the same for a 12-mm bolt (Fig. 5-86).

Figure 5-86 Coarse and fine bolts compared. The fine thread has a lower helix angle and smaller threads placed closer together.
Courtesy Mike Fitzpatrick

Fine threads, UNF (unified national fine), and ISO fine (metric threads) develop more closing force from a given amount of rotary torque, but they also require more turns to assemble compared to the coarse version. Designers tend to use fine threads for precision fastening and critical assemblies.

Coarse threads, UNC (unified national coarse), and ISO-C are quicker to assemble. Since there are fewer thread grooves per inch, each individual thread is bigger than a fine thread. Due to the larger thread size, coarse threads have more lateral resistance to thread breaking, called *stripping*. They are common for everyday mechanical assemblies. The majority of your car is held together with coarse bolts (or duct tape, depending on your budget).

Main Thread Designations—Nominal Diameter and Pitch

To select the right cutting tool as called out on the print and work order, and to cut a specific thread, there are two numbers beginners should understand (Fig. 5-87).

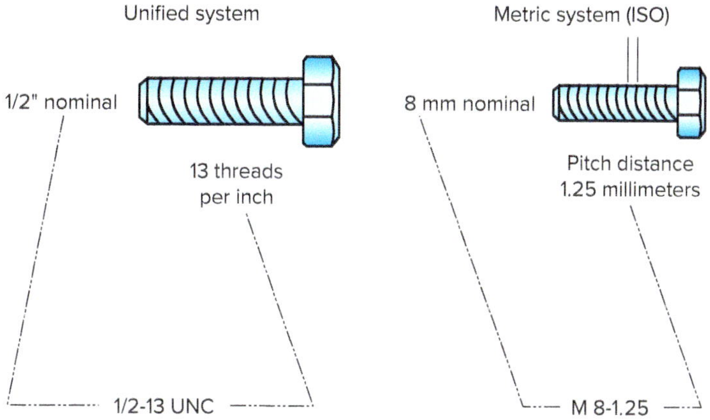

Figure 5-87 The thread terms a beginner should know.

Bolt Nominal Diameter

This first number is the theoretical outside diameter of the bolt. In reality, the outer diameter is just a bit less than nominal. That is called the *major diameter*. The tiny difference (from 0.003 to 0.010 in.) between major diameter and nominal provides mechanical **clearance** so the nut turns on the bolt. In Fig. 5-87, the UNC bolt is a nominal 0.5 in., but would measure a bit less.

Pitch and Pitch Distance

In the unified system, separated by a dash, the second designator tells the user if the thread is fine or coarse. For Imperial threads it is the *pitch number* abbreviated to **pitch**. It states how many **threads per inch** (TPI) are on the bolt–in this case, 13 TPI. This is an area where U.S. and ISO threads are slightly different.

The metric designator for the fine or coarse thread is a different number related to pitch. It's the **pitch distance**, the distance from thread crest to crest. It's often shortened to pitch as well, which might be confusing at first. Pitch distance is the measurement between corresponding points on a thread. In Fig. 5-87, the metric bolt has a pitch distance of 1.25 mm.

Critical Question

For an Imperial bolt then, how would you calculate the pitch distance for a 1/2-13 UNC thread? The answer is to divide 1 in. by 13 to get 0.0769-in. pitch distance.

KEYPOINT

Unified threads are specified as nominal diameter–pitch number. Metric threads are specified as nominal diameter–pitch distance.

Thread Forms

There are several shapes or **forms** for threads, with each having a specific mechanical function. We'll look at them more closely in Chapter 13. However, the vast majority are the 60° triangle used both on metric and Imperial threads. It is built on an isometric triangle–all sides and all 60° angles equal.

Truncation (To Cut Off)

If left as a theoretical triangle with sharp points, the bolt would have a serious weakness. Shown shortly in Fig. 5-89, it would tend to break at the bottom of the vee. So, modifications are made to the points to improve the strength–at both the top and bottom of the thread shape. Figure 5-88 shows the formulas used. It's not critical that you understand them at this time; just be aware of their existence for later study.

The sharp point of the triangle is removed top and bottom–called **truncation**. This is done both to the internal (nut) and external (screw) threads for two reasons.

SHOPTALK

Amazing Strength Interestingly, with their tops and bottoms gone, triangular threads retain almost all of their ability to wedge—that is, to not strip. A truncated thread retains 96 percent of the original lateral translation strength. Okay—maybe that's not "amazing," but it got you to look!

1. **Mechanical Strength**

   Look at the two threads in Fig. 5-89. Which would tend to break more readily and where? *Answer:* At the bottoms of the sharp triangles, while the clipped threads would be far less prone to breaking the shaft. Both the nut and bolt triangles must be truncated if they are to fit together.

2. **Less Jamming**

   Left on the thread, those pointed tops are prone to breaking off and jamming between the nut and bolt when heavily loaded by a wrench. But, with truncation, they resist breaking and jamming.

$D = 0.6495 \times P$ or $\frac{0.6495}{N}$

$C \& F = 0.125 \times P$ or $\frac{0.125}{N}$

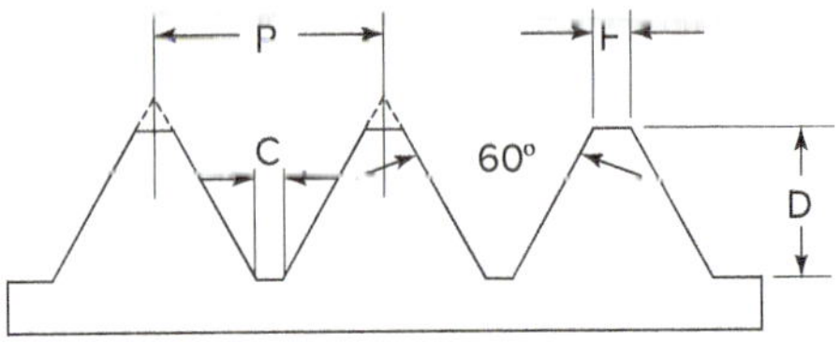

American National thread form

$D(MAX.) = 0.7035 \times P$ or $\frac{0.7035}{N}$

$(MIN.) = 0.6855 \times P$ or $\frac{0.6855}{N}$

$C \& F = 0.125 \times P$ or $\frac{0.125}{N}$

$R(MAX.) = 0.0633 \times P$ or $\frac{0.0633}{N}$

$(MIN.) = 0.054 \times P$ or $\frac{0.054}{N}$

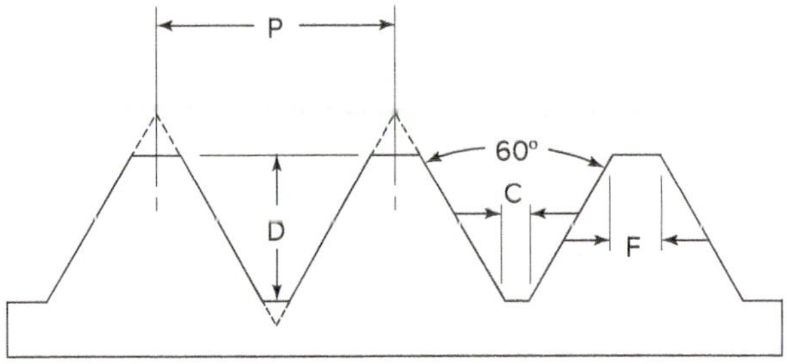

International metric thread form

Figure 5-88 Formulas for 60° thread forms.

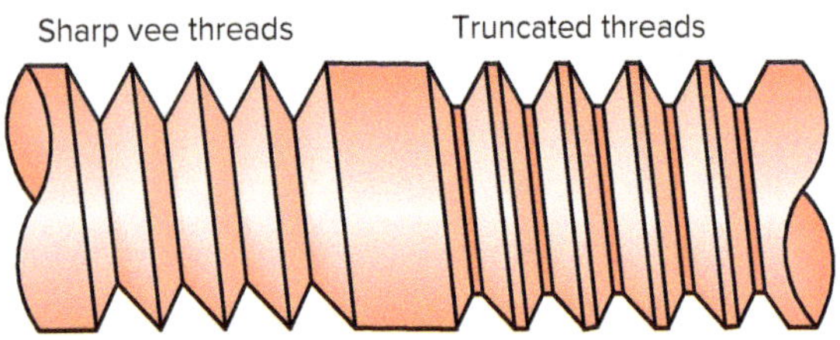

Figure 5-89 A sharp thread would create a serious weak point in the shaft, while truncated threads reduce this problem.

SHOPTALK

Question: I have an old print that designates a *1/2-13 UNC-3-A*. What does the *A* stand for? *Answer:* On some older prints, you might find an "A" meaning it's a bolt or a "B" meaning it's a nut. I don't know why they didn't use "B" for bolt, but it doesn't matter; the practice has been dropped today because all one needs to do is look at the drawing to see if it's a bolt or nut!

Thread Classes

Adding a third data field to the designator, the **class** shows the level of precision of the thread. This is only added to the designation when it's required by some special function. If not, the class is considered to be 2-everyday threads.

Class 1 designates loose threads where they must be constantly turned. A good example is a "C" clamp thread.

1/2-13-2

Class 2 is for general-purpose threads, such as the nuts and bolts you would buy at the hardware store.

1/2-13-3

Class 3 is a precision thread for critical assembly. It is the tighter tolerance of the three fits.

As the classes become tighter, the tolerance for acceptability becomes narrower. There are further refined classes beyond the three outlined here, but they are beyond this discussion.

KEYPOINT

Class 2 threading tools are what you normally find in any tool room.

Introduction to Pipe Threads

When the threads are to go on pipes either for mechanical connections as on construction scaffolding, for example, or for carrying liquids or gasses, the thread designator will include a letter P.

Pipe threads are similar to the fastener threads in that they also use the 60° triangle form. They are different in three ways. For now, if you understand these differences that should suffice.

Tapered Threads

There are two kinds: straight and tapered. The tapered version has its wedge wrapped around a gradual cone rather than a cylinder. The increasing size closes tighter as it is screwed together, to aid in sealing. The designator for tapered pipe threads is NPT (National Pipe Tapered).

Size Is Based on Hole in Pipe

A second difference compared to fasteners is how the nominal size is specified. With bolts and nuts, it's the outside diameter of the bolt, but not with pipe threads (Fig. 5-90). Their size is based on the inside diameter of the pipe. That means that a $\frac{1}{2}$-in. pipe thread is a great deal bigger than a $\frac{1}{2}$-in. bolt, as shown in the drawing.

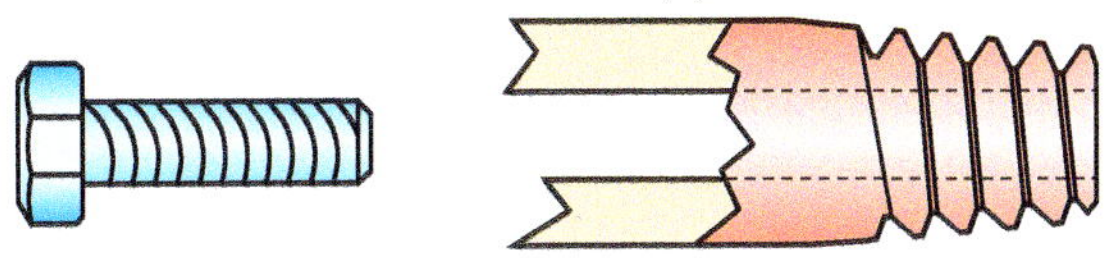

Figure 5-90 Pipe threads are designated by the pipe's carrying capacity.

Different Pitches

Pipe threads feature pitches different from those of everyday bolts and nuts. They cannot be interconnected, to avoid possible failures or misapplications. While there's a lot more to know about them, the chances are you will not be required to cut pipe threads during your early training, so we'll leave them until Chapter 13. But, when you see a designation of 1/2-14 NPT, you will know that it's a tapered pipe thread.

> **KEYPOINT**
>
> Pipe threads do not assemble with other kinds of threads for safety reasons.

Determining Thread Pitch

When we need to check a given bolt for its pitch, we use a small template gage called a *screw pitch gage* (Fig. 5-91). The leaves are spread out; then each is held against the bolt in question until an exact fit is found, as shown in Fig. 5-92.

For now, that's all the background needed to select and cut standard threads with a tap or a die. A tap makes an internal thread, a nut, and a **die** makes an external one. Now, here's how to cut the threads.

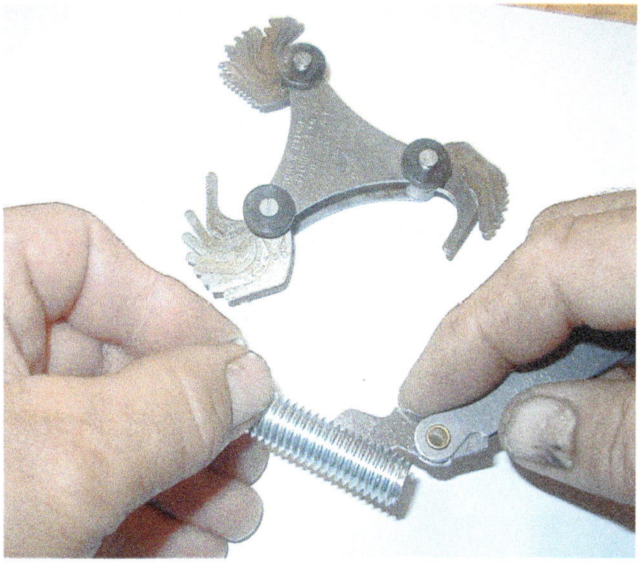

Figure 5-91 A screw pitch gage features leaves, each with a different pitch number or distance if it's metric.
Courtesy Mike Fitzpatrick

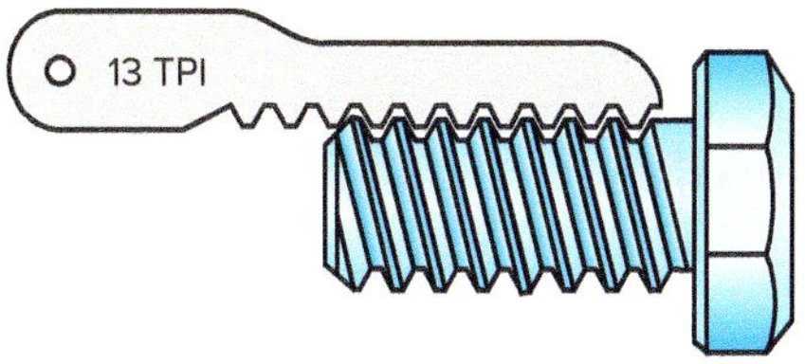

Figure 5-92 The pitch will become obvious when the correct leaf fits the thread.

Figure 5-93 A chasing die nut is designed to recut a damaged thread.
Courtesy Mike Fitzpatrick

5.5.2 Threading with a Die

Threading dies are cutting tools used to make or to restore outside threads (bolts and studs). Studs are bolts with no head and they have threads on both ends. Threading dies are supplied in two general forms:

Cutting dies To make a thread on a raw shaft

Chasing dies To restore a damaged thread (Fig. 5-93)

Chasing Dies

These tools are normally easy to use. They are started, then run down the thread just like a nut. In fact, their outside looks like an oversized nut, and they are turned onto the bolt using a standard wrench.

Skill Hints for Using a Chasing Die

Check the end of the shaft for damaged threads–if they are smashed, grind a small chamfer on the end before starting the chasing die. This helps the die start in the original threads.

Use cutting oil or thread compound and go slow even though they can be run on faster.

One side of the nut is designed to lead the cut–the side with the markings on it. Look for "start" on the die face.

Although it's possible in an emergency to cut a new thread on a raw shaft using a chasing die, it's the wrong thing to do. They aren't designed to cut metal chips, only to remove dents and smooth out damage on existing threads.

Threading Dies

The affordable kind you'll likely encounter in schools are called button dies for their compactness. They are designed to cut new threads on round features. Button dies (Fig. 5-94) differ from chasing dies in that they provide an adjustment for thread size. Because they are usually round, they must be mounted in a die handle, as shown in Fig. 5-95.

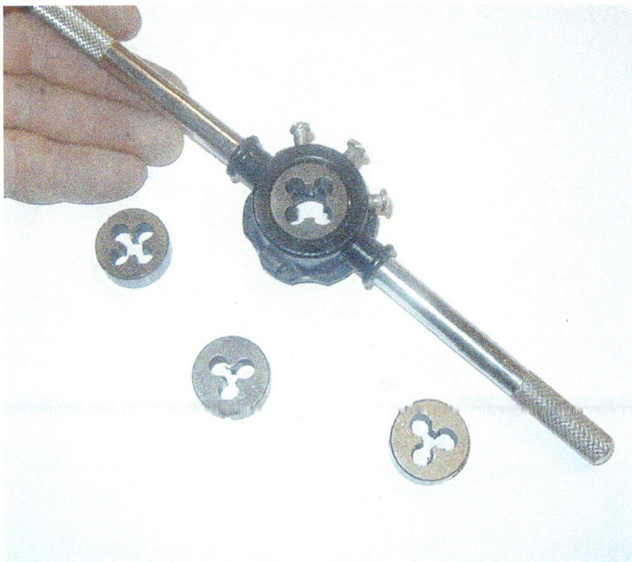

Figure 5-94 A button die in a handle is called a die stock.
Courtesy Mike Fitzpatrick

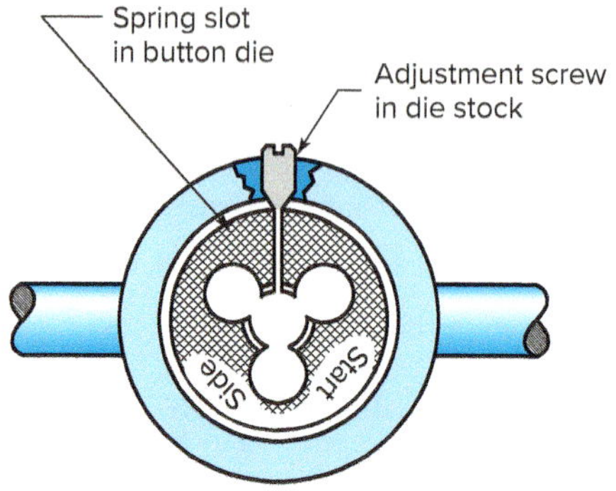

Figure 5-95 The adjusting screw opens the die slightly.

Adjustable

Quality button dies feature a relief slot through one side of their outer circle, as shown in Fig. 5-95. The spring slot and set screw allow adjustment to compensate for resharpening of a dull die. Each grind of the cutting edge makes it cut a slightly larger thread. Without the adjustment, the thread would become progressively larger until a nut would not screw on the bolt produced. The adjustable die is also used to create the different classes of threads. Finally, it can be used to set oversize for a first rough pass, then closed to the right diameter to make a final pass. This technique helps where the best finish and size are required on expensive work material.

In the drawing, notice the central adjustment screw wedges into a chamfer on the slot edges to force the die open. Holding screws on both sides (not shown), keep the die closed while the adjusting screw forces and holds it open against the spring pressure of the die.

Skill Hints for Threading Dies

Aligning the die to the shaft is important. Misaligned dies are the number one cause of failure when hand threading. To help, there may be an alignment guide as part of the die handle. It is composed of three movable pads and an adjustment ring that is used to ensure the die is going on straight. Use them; they help avoid threads that wander off. If using handles that lack the alignment guides, use a square in two directions to be certain the die is on perpendicular to the shaft (Fig. 5-96). When threading, constantly watch the die handle to ensure it is not tipping relative to the shaft.

Make sure the intended shaft has a starting chamfer on the end.

A cutting oil or compound is essential.

A threading die can be used as a chasing die.

Due to the lead-in angle, dies produce incomplete threads for the first two or three threads. To thread closer to the end of the shaft, finish the thread, then turn the die around, and thread with its reverse side.

To set the die to the right size, open then screw it onto a good-quality screw or model part of the right size, pitch, and class. Rotate the die slightly while turning the adjustment screw out and the clamping screws inward until the die drags a bit on the screw.

Figure 5-96 When threading with a die, watch the die handle to ensure that it is perpendicular to the shaft.
Courtesy Mike Fitzpatrick

KEYPOINT

When threading by hand, turn the die forward two or three turns; then reverse it to break the chip.

Other Kinds of Dies

Fixed Size Dies

You may encounter a button die lacking the size adjustment feature–a solid button die. They are tools purchased for their lower price only, since they can only be resharpened a few times before they must be thrown away due to progressively bigger threads. Solid dies of this type are usually supplied in Class 2 threads only and are considered throwaway tools.

Insert Jaw Dies

This is a very different type of threading tool (Fig. 5-97). Although more costly, they bring several features to the shop where a lot of hand threading is being performed. It's unlikely that you will encounter these dies in schools.

The inserts can be removed from the carrier jaw and easily resharpened on their sides using a standard tool grinder or surface grinder. This isn't so with button dies, which must be reground with shaped stones in the holes between cutting flutes. The jaw types can have their inserts replaced in just minutes where time is of the essence.

Taps for Internal Threads

Taps could be described as very hard bolts with cutting teeth. They cut threads inside a hole as they screw into it. Taps are supplied in several varieties based on the material to be cut, the method to be used, and the far end of the drilled hole to be threaded (Fig. 5-98). After determining the nominal size, pitch, and class of thread to be produced, in selecting a tap the questions are

- Does the hole go through or is it a blind hole?
- Must the threads go to the bottom of the blind hole or not?
- Can the chips go ahead of the tap or should they come back up the tap?

Depending on the answers, there are three shapes of taps (Fig. 5-99) from which to choose: starting, plug, and bottom tap.

1. **Starting Tap**

   Sometimes called a taper tap, this is the general-purpose tap that begins cutting best. Always use a starting tap first if you possibly can (even in a **blind hole**–one that isn't all the way through). Notice in the drawing that there is a tapered portion on the first two taps that cuts the initial threads. This starting portion of the tap leaves imperfect threads that are not cut to full diameter. The taper tap leaves the most imperfect threads.

2. **Bottom Tap**

   Used only to finish a blind hole where perfect threads are required to the bottom of a blind hole. A bottom tap is used after the start; then plug taps are turned in as far as they will go. It is not used to start a thread except in rare cases for extra shallow threads. These taps have no lead; thus, they cut poorly compared to the other two.

3. **Plug Tap or Second Tap**

   Halfway between the other two forms, this intermediate tap cuts better than a bottom tap, but it does not start the thread as well as a starting version.

Above the tapered portion, all taps produce a finished thread. There is no difference in the size of the thread produced by each tap except the lead part at the tip.

Figure 5-97 An insert jaw die has the advantage of easy resharpening and quick size adjustment.
Lake Washington Technical College, Kirkland, WA/McGraw Hill

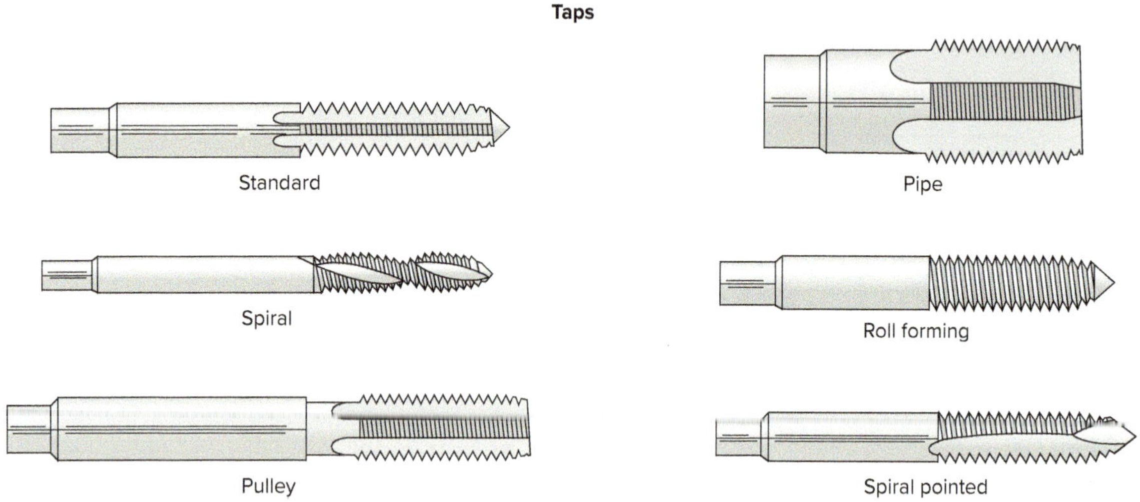

Figure 5-98 Different kinds of taps for different purposes.

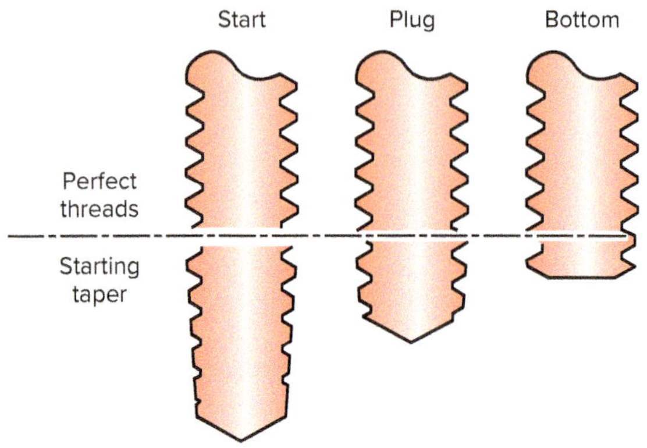

Figure 5-99 A starting tap, a plug tap, and a bottom tap. The difference lies below the line, in the length of the tapered portion.

Tap Flutes

Another difference in tap construction is the number of cutting edges with which they are supplied: two, three, and four *flute* (cutting edges) versions (Fig. 5-100). In general, taps with fewer flutes are used in softer material, while hard material is tapped using four-flute tools. For hand tapping, there is little performance difference in the three taps. However, when power tapping at a much faster RPM, the two- and three-flute taps tend to deliver chips out of the hole better while the four-flute tap has more cutting edges and a stronger center portion for threading harder materials but tends to choke on chips.

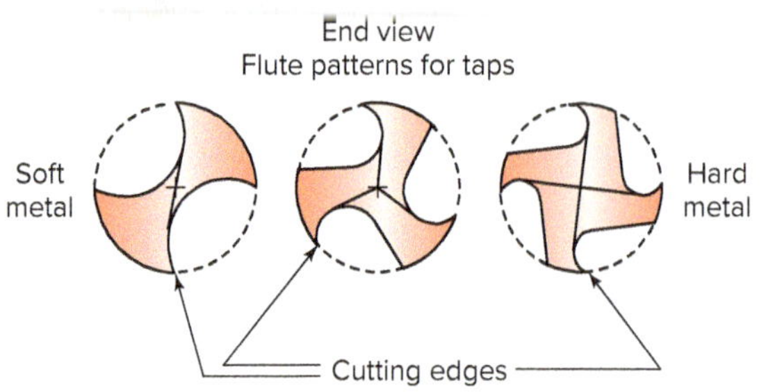

Figure 5-100 Compare two-, three-, and four-fluted taps—cross section.

Spiral or Gun Pointed Taps

The third difference in taps is the direction they drive chips as they cut the thread. By changing the angle of the cutting portion, the chip can either be sent ahead of the tap or back up the flutes, similar to a drill.

Notice in Fig. 5-98 that the spiral gun tap has an axial lead angle on its cutting edge. It sends the chip forward into the hole. This type works best in holes that are drilled through or are much deeper than the intended thread. The straight pointed tap lifts chips back out of the hole.

Taps break easily. Here's a group of hints on how to avoid the embarrassment of breaking yours.

Sharp Tools

This is critical. A dull tap breaks far easier than a sharp one. To recognize a dull tap, look at the cutting edge. If there is a tiny line of reflected light right at the cutting edge, caused by the rounding of what should be the clean intersection of two surfaces, it must not be used. A sharp edge will not reflect light. This visual inspection works for all cutting tools.

A sure symptom of dullness is if, during use, the tap makes any sound or drags even slightly more than it has been. When detecting even the slightest change, stop immediately and back it out! Any change in sound or feel usually signals that the next step is going to be a broken tap. Once broken in the hole, the tap is difficult to remove.

Removing Broken Taps

Once broken inside the work, the tap cannot be drilled out as it is as hard as the drill or even harder! There are two viable options if the tap has snapped off leaving nothing to grab for backing it out, which is always the case! A broken tap remover (Fig. 5-101) is a set of strong, shaped wires that fit into the holes between the flutes. You'll need an experienced person to demonstrate these little tools, but their success rate isn't wonderful. Most of the time, the broken tap is fragmented and tends to lock up inside the hole, in which case the part is thrown away or if it's costly enough, turn to the next option. Patience is the key to using tap removers.

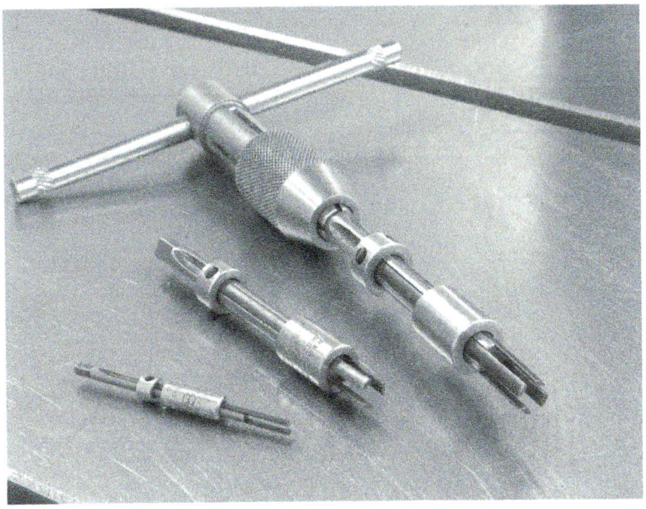

Figure 5-101 Tap remover wires and guides are supplied in sets, to fit a specific tap.
Photography by Prographics at Milwaukee Area Technical College/McGraw Hill

Tap Burners

These great little machines use electrical sparks to erode the broken tap out of the hole. They have a 100 percent success rate if available. We'll explore this very different metal removing process called electrical discharge machining in Chapter 26.

Tap Alignment

Next on the hint list for avoiding disaster is getting the tap pointing directly into the drilled hole. In other words, make sure the tap is parallel to the hole and that it remains so during the cutting (Fig. 5-102). This is the number one reason taps break during hand tapping. A small square is useful (Fig. 5-103) or a guide called a tap block can be used to start parallel to the hole. With some practice, you will be able to get taps reasonably aligned by looking at the tap handle. Here are two more accurate methods used to get a tap aligned to the hole.

Tap Blocks An alignment tool with a series of holes slightly bigger than the nominal size for several different taps. The block is clamped over the intended hole as a guide for starting the tap straight. After several threads are in, the tap handle is removed and the tap block is slid up and off the tap.

Drill Press Guidance Here a technique called a "dummy center" (Fig. 5-104) is being used. The machinist has aligned the drill press spindle over the hole to be tapped, then bolted the part in this position, drawing it back to guide the tap.

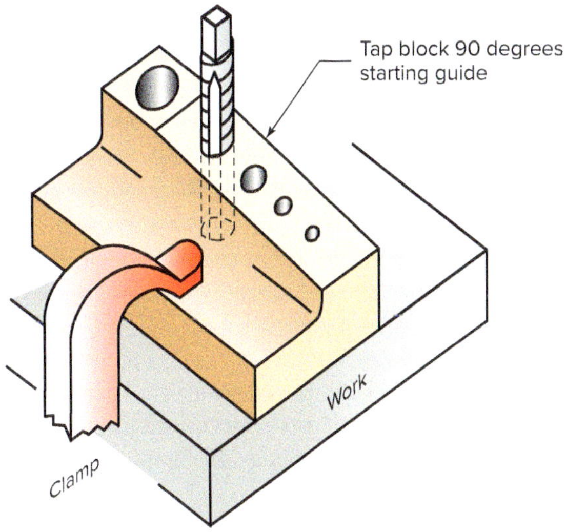

Figure 5-102 A tap block helps us to align taps and keep them parallel with the hole.

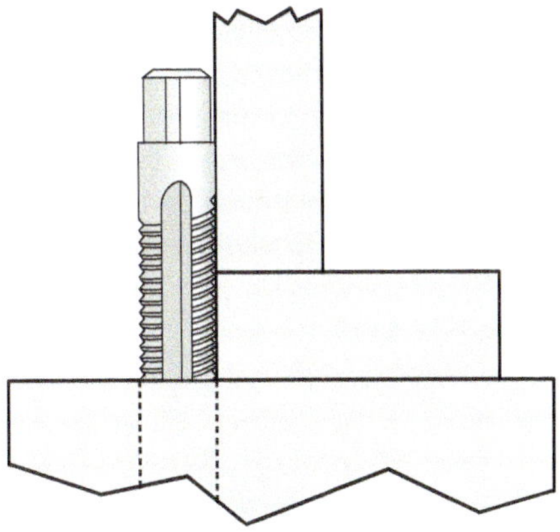

Figure 5-103 A precision square can be used to check the tap.

Figure 5-104 Use a drill press and center to keep the tap aligned to the hole.
Courtesy Mike Fitzpatrick

KEYPOINT

When using a dummy center on a drill press, no power is used. The press guides hand tapping.

Lubrication Is Essential

Brush a light coat of cutting oil or, even better, a special threading fluid on the tap and in the hole. And keep replenishing it as the cut progresses. Lubrication makes a big difference in cutting action and chip delivery. If the thread is deep, don't go all the way in. Reverse the tap and bring it out of the hole, then blow the chips out of the hole and off the tap, relube, and send it back.

The Right Moves—Tapping Actions

A major cause of broken taps is chip loading. In certain situations the chips can't come out of the hole. You must stop every two turns or less and break the chip by backing up one-half turn (Fig. 5-105). If the broken chips continue to clog the flutes, you will need to completely remove the tap after each succeeding two turns in, brush it off, and blow out the chips from the hole. Be careful in blowing chips out of a hole; hold a rag in the airstream to trap the oil and chips.

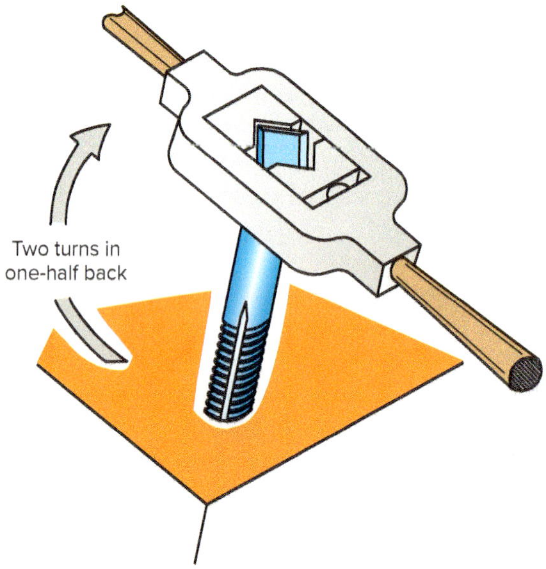

Figure 5-105 Breaking the chip is important when hand tapping.

Listen and Feel

If a tap squeaks while threading, it is from one of three causes:

A. Lack of lubrication–use cutting fluid.

B. It is dull and should not be used.

C. Some metals just squeak and nothing is wrong. Squeaking might occur in some brass/bronze and stainless and in harder types of steel.

Tap Handles

Taps must be mounted in handles to turn them by hand. The tee handle closes down on the tap similar to a drill chuck. It is compact and is used where the tap is small and you wish to limit the force to avoid breaking it, or where space is limited to turn the handle (Fig. 5-106). The standard tap handle is used to turn taps where more force is required (Fig. 5-107).

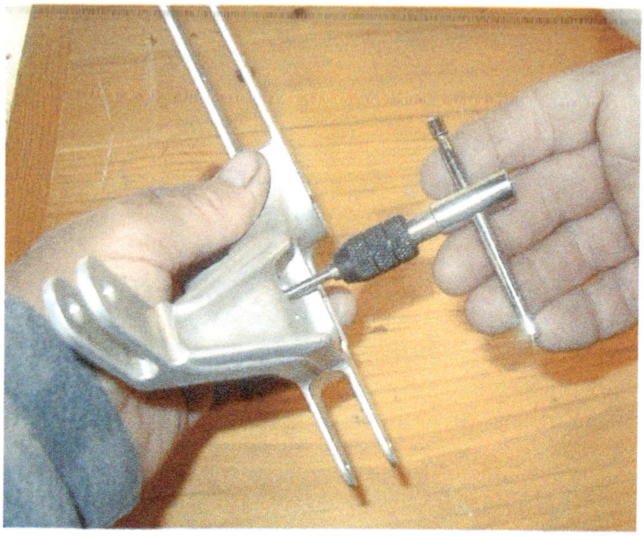

Figure 5-106 A tee tap handle is used for small taps.
Courtesy Mike Fitzpatrick

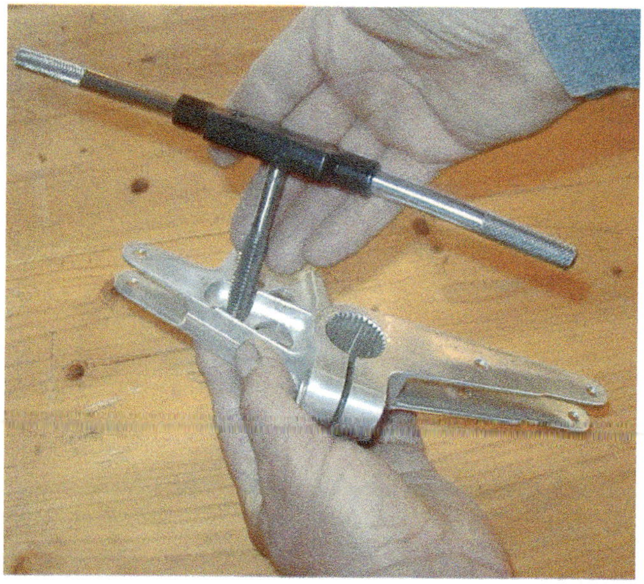

Figure 5-107 A tap handle is used for large and small taps.
Courtesy Mike Fitzpatrick

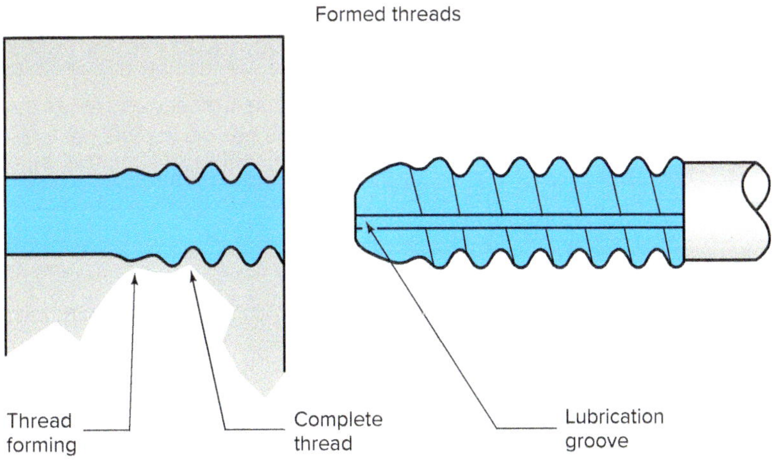

Figure 5-108 A roll tap displaces metal from the grooves to form the ridges.

Roll Form Tapping (Cold Formed Threads)

This kind of tapping isn't usually performed by hand, but it is worth mentioning here because it produces a stronger thread compared to cutting threads. It is performed with a tap somewhat similar to the others, except it has no cutting edges and removes no metal. See Fig. 5-98, right-hand column.

The tap deforms (ploughs) the top of the thread up by displacing metal from the grooves. The threads produced, due to cold working of the metal, are much stronger than cut threads. They are rounded in form, too, which adds strength and long life. The roll form tap in Fig. 5-108 has gone partly through the work to show the gradual thread formation.

The hole drilled before tapping is larger than the one for a cut thread. While getting the size right for regular tapping is critical, for roll threading, it's extra critical. Based on the 60° triangle, the shape of the threads formed is quite different (see Fig. 5-108). Roll form tapping is not an entry-level skill and usually is not taught in schools. We'll leave further discussion for on-the-job training.

5.5.3 Tap Drill Holes

For any tap, whether it's for cutting or forming threads, the hole size drilled before tapping must be the exact right size. There is a correct drill size for every tap–it depends on nominal thread diameter and fine or coarse threads.

KEYPOINT

There is a correct size drill for each tap size and very little tolerance for variation in that size.

Trying to tap a wrong-sized hole will either break the tap if the diameter is too small or make weak or nonexistent threads if it's too large. For information on tap drill selection, turn to the Appendix I decimal chart or go to Unit 9-4.

5.5.4 Finishing Holes—Work Order Operation 50

Two types of secondary operations that make drilled holes smoother, more accurate, and rounder are

Hand-Reaming Machine Honing

Reaming

Reaming finishes a predrilled hole. There are two types of reamers: hand and machine. The hand reamer is of interest here in Chapter 5. It removes very little metal but makes the hole round, straight, and smooth. Tolerances within 0.0005 in. (or closer) for diameter, straightness, and roundness can be met using a hand reamer (Fig. 5-109). Reaming done right can produce a 32-microinch finish or better.

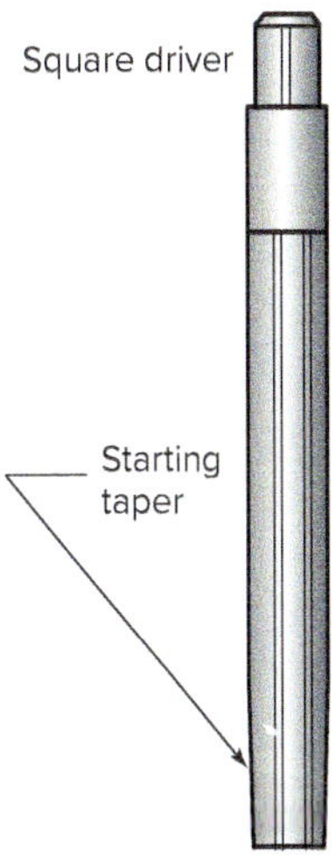

Figure 5-109 A hand reamer features a square top to be used in a tap handle and a lead-in taper to improve the starting qualities.

How much metal is removed during hand reaming depends largely on the size of the hole. From 0.003 in. up to around 0.010 in. max is normal. With all reaming, by hand or by machine, larger holes usually call for greater amounts to be left for reaming, but there is a maximum reaming excess, at around 0.015 in., even for machine reaming.

Hand reaming is often performed to resize a bushing after it is press fitted into a hole or to remove paint or other coatings during final assembly or to put the final touch on holes, as shown in Fig. 5-110. The hand reamer is mounted in a tap handle, then pushed in while being rotated.

Figure 5-110 Hand reaming might be used to finish the holes in the drill gages after drilling undersize.
Photography by Prographics at Milwaukee Area Technical College/McGraw Hill

Skill Hints for Using Hand Reamers

Always use cutting oil when reaming metals that are not easily machined (hard steel, for example). A specific tapping fluid designed for that metal works well also.

Keep constant in-pressure on the reamer. If the reamer progresses about one quarter of its diameter into the hole per revolution, you are pushing about right. Reamers need steady progress into the hole to get good results.

The lead-in diameter (see Fig. 5-109) will help ensure the reamer is parallel to the hole. Sight the tap handle from two directions to be sure before reaming very deep. Different from hand tapping, once the reamer starts right, it will tend to stay parallel and not wander.

Never turn it backward when in the hole—this will dull the cutting edges. When through, pull it out while turning it forward.

KEYPOINT

Reaming is a cutting process that removes from 0.003 in. diameter up to a max of around 0.010 in. for hand reamers and up to 0.015 in. for machine versions.

Honing

Honing is a secondary operation that finishes holes finer than reaming. It's an abrasive action used to improve diameter, roundness, and surface finish. A hole that's tapered or not straight can be corrected by honing, to bring it closer to a perfect cylinder. However, honing to achieve better straightness requires a lot of operator intervention; it's not simple to do. It's best if the hole is straight with no taper before starting to hone.

Although honing uses friction/abrasion rather than cutting, it's much slower than grinding, and it occurs at a very slow speed with no heat. In production machine shops, honing is performed on a machine specially designed for the task. In automotive machine shops, it's done using portable tools. Honing machines (Fig. 5-111) are common where extremely close hole tolerances and smooth finishes are a requirement, such as hydraulic cylinders, for example.

Figure 5-111 Looking at the front of a hone head. Observe the two pads and the adjustable stone.
Courtesy Mike Fitzpatrick

Honing is necessary when the metal must be hardened after drilling/reaming, then hardened beyond machinability. The carbon scale on the hole's surface, resulting from heating the metal, must be removed by honing.

Honing uses a three-point suspension process: two floating pads that rub and oppose the abrasive stone and an abrasive stone on the opposite side that removes a tiny amount of metal. While rotating the honing head in the hole with cutting oil, the stone is forced out against the hole. It slowly abrades material to a very fine finish of around 8 to 16 microinches or better. The machine floods the cutting head and works with cutting oil to lubricate and flush metal particles.

Honing removes far less metal than reaming: from 0.0002 in. up to a maximum of 0.003 in. with around 0.0015 in. being normal. Prior to honing, the hole would have been drilled and probably reamed. Honing is challenging in soft metal such as aluminum due to the loading of the stone, but it is possible. The constant challenge of operating a hone is to correct taper and bellmouthing, as shown in Fig. 5-112.

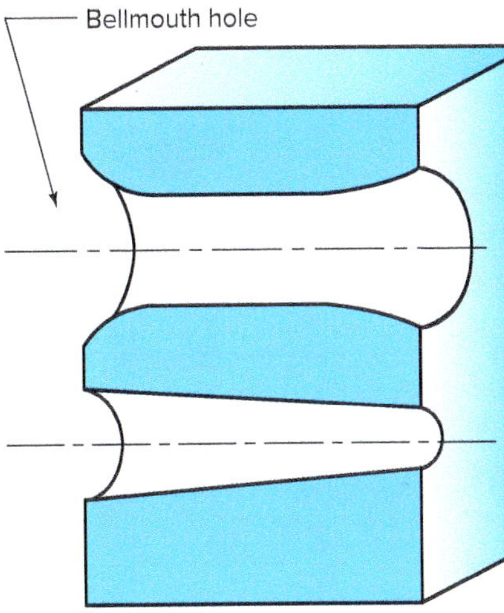

Figure 5-112 Preventing and correcting the bellmouth and taper are a constant challenge for a hone operator.

KEYPOINT

Honing is a slow, hole finishing, abrasive process used to remove from 0.0002 in. up to a max of 0.003 in. of material from holes.

ARMCHAIR FIELD TRIP

Watch this Haas Tip of the Day video about thread taps.

In this video you will learn how to select the right tap for the job! Go to: **https://www.youtube.com/haasautomation** and search for this video called: **Tapping Essentials-Every Machinist Needs to Watch This Haas Automation Tip of the Day**

5.5.5 Press Fit Assembly

Press fitting is a secondary operation used to permanently assemble one part into another. It uses friction to hold the part in place as a round bushing is pressed into a hole for example. While the object can be removed, press fitting isn't designed to be removed easily. Objects that are pressed in can be pressed out again, but it is considered a permanent assembly.

KEYPOINT

Amazingly, it is possible to force a pin or bushing that is slightly bigger than a hole into the hole! Why? Because metal stretches, but only a small amount.

A Typical Press Fit

For example, in Fig. 5-113 we'll consider pushing a $\frac{3}{4}$ (0.750) nominal outside diameter bushing (dash 2 part) into a reamed hole in the dash 1. Either the bushing or the hole must have the press fit allowance built into its size. Usually that's not more than a 0.001- to 0.003-in. difference in the two parts, depending on how tightly the press fit must be functionally.

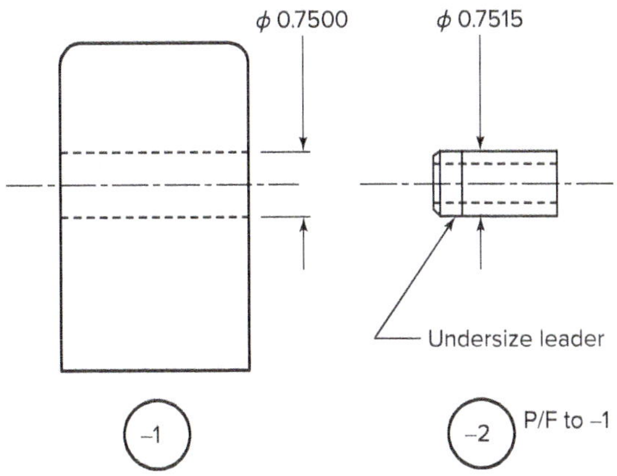

Figure 5-113 The press fit allowance (0.0015 in.) is built into the purchased bushing in this drawing. The mating hole then is produced by a standard 0.750-in. (3/4-in.) reamer.

Which Component Has the Allowance?

The **press fit (P/F)** allowance is the amount of interference in size between components. This calculated amount of interference is usually built into the bushing because bushings are mass-produced and designed to fit into a common size hole. For example, a 0.5-in. bushing might measure 0.5005 in., to be pressed into a 0.5000-in.-diameter hole. On the other hand, pins, due to their function, usually do not have the allowance built in and thus require the hole to be under the nominal by the interference amount.

KEYPOINT

P/F bushings have the press allowance built into their outside diameter. P/F pins have the press allowance built into the receiving hole diameter.

Either way, the bored or reamed hole must be made for the correct press fit application. Select either a standard size reamer or a specially made undersize reamer for the job. Read the work order and print to know which part has the allowance and always measure components before press fit assembly.

How Is the Press Fit Allowance Calculated?

There are three variables considered when computing a press fit allowance:

1. **Permanency of the Assembly, Called the Class of Press Fit**

   This is the tightness based on the duty to which the component will be put.

2. **Nominal Size of the Objects**

   As the nominal size of the assembly grows larger, so does the P/F allowance required. For example, a 1.000-in.-diameter pin might use a 0.001-in. press fit allowance, while a 2.000-in. pin requires a 0.003-in. press. The amount can be found in machining or engineering references.

3. **Metal Alloy**

   Although a less important variable, different metals have different holding characteristics. For example, a bronze bushing might require a bit more interference in a steel assembly than a steel bushing pressed into the same hole.

Rule of Thumb—Press Fit Amounts

A permanent press fit is considered from 0.0015 to 0.0025 interference for the first inch in diameter of the objects to be put together. Then lesser amounts of interference are added per additional inch of diameter as the size of the object increases.

Heating and Cooling Metal for Press Fits

When interference amounts become large, the outer part is heated and the inner part is cooled to allow the parts to go together more easily. Heating one component while freezing the other temporarily reduces the force required. Then as the components' temperatures equalize, the fit tightens and becomes more permanent. The heat/cold method is called **shrink fitting** and is considered a nearly nonremovable press fit if both items are the same alloy because it would be difficult to reheat or cool individual pieces once they are in intimate contact.

Many objects made for a press fit will feature a short, *lead-in diameter* (Fig. 5-114). It will be smaller than the target hole size for a short distance so the object starts in with no resistance. This assists in alignment of the bushing or pin.

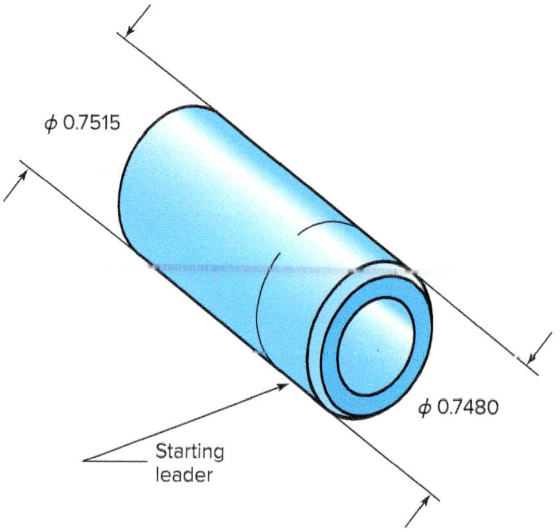

Figure 5-114 This bushing has been specially made with a built-in press fit allowance and a lead-in diameter.

Machines That Perform Press Fits

Arbor Presses

Whether hand or hydraulic, a press fit can be performed with vises and even hand-driven tools, but press fitting is usually accomplished with an **arbor press** (Fig. 5-115). Several types are found in production, repair, and automotive machine shops:

1. Powerful electric-hydraulic
2. Manually pumped hydraulic
3. Hand-operated rack gear presses (see Fig. 5-115)
4. Hand-operated screw presses

Skill Hints—Press Fit Assemblies

1. Always do a quick caliper check of both the hole and bushing before starting.
2. Make sure to place the lead end of the bushing or pin in the hole first!
3. A little lubricant such as grease or other high-pressure lube is recommended. Rub a thin coat on both the bushing and the hole. Check with your supervisor–some press fit assemblies cannot have a lubricant present.
4. Make sure the opening of the hole has a small chamfer, not a sharp edge that might scrape the bushing and possibly misalign it during assembly.

Figure 5-115 Arbor presses are used to force-assemble press fits.
Photography by Prographics at Milwaukee Area Technical College/McGraw Hill

5. Make sure there are no burrs, dirt, or chips before attempting assembly.
6. After starting the bushing in, stop and check alignment similar to a tap. Don't attempt to push it all the way in without checking alignment. They can and do start in crooked by scraping material off the wall of the hole. This is difficult if not impossible to repair.
7. After pressing, measure the inside of the bushing. It will probably be slightly smaller due to the metal squeeze from the press. Bushings that close down depend on a relatively thin bushing wall and may require a final hand ream.

UNIT 5-5 REVIEW

Replay the Key Points

- Standard threads feature a coarse and fine series both for metric and Imperial sizes.
- Both metric and Imperial standard threads feature the 60° triangular thread form.
- The thread designator tells the nominal size–the pitch or pitch distance–then the class (all separated with dashes).
- There is a slight difference between Imperial and metric designators in that the Imperial shows pitch number while the metric shows pitch distance.

Respond

1. Why do we truncate the standard thread form?
   A. To save material–less metal
   B. To make the bolt smaller than the nut
   C. To remove the fragile tips that would break
   D. To lighten the bolt and nut
   E. None of the above
2. Which of these threads would *not* use the 80° triangular form?
   A. 3/8-16 UNC
   B. M 12-1.25
   C. 1/8-27 NPT
3. Describe this thread: 3/8-16 UNC-3.
4. True or false? This thread has a tighter (smaller) manufacturing tolerance than the one in Question 3: 3/8-16 UNC-2.
5. Between these two threads, which has the smaller individual threads: 1/2-13 UNC or 1/2-20 UNF?

CHAPTER 5 REVIEW

Unit 5-1

Today's machinist is a manager as well as a craftsperson. There are many daily items that must be organized. And there are game rules–lots of them: to manage materials, to use and dispose of chemicals, and other systems for part numbers, dash numbers, and program revisions to track. In this unit, we've looked at the general system used in most factories and shops. But there are differences. Be sure to find out how your new employer processes:

Work orders
Part numbers
Revisions to the drawing and to related programs

Unit 5-2

One of the greatest obligations is to guarantee that the right metal is used to make the product. Metallurgy is a big subject, in fact it's an entire college degree. But we've started your knowledge bank using five selected metals:

Steel

Cast iron

Brass/bronze

Stainless steel

Aluminum

Plus we've looked at the various physical forms in which metals are supplied beyond blocks, bars, and sheets:

Castings

Forgings

Extrusions

But simply getting the right alloy in the right form isn't enough in some cases. The hardness and grain direction often must be included in the history of the material. In some situations such as transportation equipment, where lives are at stake, the metal's pedigree must be proved from the smelter to the individual serialized part number. If not, the parts might be right physically, but they are illegal to use! Paperwork is the reality of our business, but computers are making it easier.

Unit 5-3

Modern sawing equipment can shave hours from production. It's a skill well worth learning. When the saw is fine-tuned with the right blade pitch and cutting speed, and the blade guides are well adjusted, repeatability tolerances can be held within 0.020 in. or even less! Additionally, sawing removes large chunks of material from the raw stock, all at one time—an advantage that only saws have. All other machines make small chips from the removed material. That not only saves time, but it makes recycling dollars and sense, too.

Unit 5-4

Part finishing is no fun! We all agree on that. But it must be done before the products can be delivered to the customer. Above all else, while removing burrs and smoothing rough spots in the machining, do no damage to the parts. Be careful how you handle and stack them. Protect parts from dropping or banging together. Do not mar them in any way.

Unit 5-5

In previous machining textbooks, this subject was called benchwork. *The bench* was the place most of us began our apprenticeship. But today, the focus is more on supporting the CNC equipment to keep it producing work that only it can do. Work is only taken off the expensive machines when it removes a bottleneck in production. Still, the complete bag of machining skills requires the ability to perform:

Hand threading with taps and dies

Hand reaming

Press fit assembly

QUESTIONS AND PROBLEMS

Are you ready to read work orders and then perform secondary operations when they are specified?

1. List at least four vital kinds of information found on a work order and briefly describe each. (Unit 5-1)
2. True or false? Part numbers are comprised of a detail number followed by a drawing number but separated by a dash. If it's false, what makes it true? (Unit 5-1)

Questions 3 through 8 refer to the five basic metals: aluminum, cast iron, brass, steel, and stainless steel. (Unit 5-2)

3. Which has the highest machinability rating (most easily machined)?
4. Which is the dullest in color?
5. Which is the lightest in weight?
6. Which is often abbreviated CRES?
7. Beyond bars and sheets, name three preshape forms that metal might be supplied in.
8. Of the two forms CRS and HRS, which is the strongest and why?
9. Identify and describe the saw blade factor that creates a wider kerf than the basic thickness of the blade (the portion behind the teeth). Identify two advantages imparted by this blade feature. (Unit 5-3)
10. When power sawing on a vertical band saw, potentially dangerous tooth stripping occurs because the machinist failed to do what? (Unit 5-3)
11. The common gray grinding wheel used on bench and pedestal grinders has its abrasive grains bonded together with fired clay. Identify the bonding process and abrasive type. (Unit 5-4)
12. There are two methods of removing burrs and finishing edges on parts that require no hand-work by the machinist. Identify and briefly describe each. (Unit 5-4)
13. List the five categories of part marking. Of these, which method or methods are not harmful to surfaces? (Unit 5-5)
14. Explain the difference between the two numbers 20 and 1.5 in these two thread callouts: 1/2-20 and M12-1.5. (Unit 5-6)
15. In Question 14, what is the pitch distance for the $\frac{1}{2}$-in. thread? (Unit 5-6)

CRITICAL THINKING

Your supervisor tells you, "Make twenty 501B-3456-75 Rev J parts."

16. You have found a Rev K drawing of the correct number. Is it okay to use to make these parts?

17. In view of Question 16, what kind of parts are you most likely making?

18. There are three documents you'll need to do this job. What are they, and why these three?

CNC QUESTIONS

19. A program has been written to machine parts from stainless steel; however, you have been asked to make one prototype (experimental example) from aluminum to prove out the program.

A. What editing can you do to the RPM?

B. What changes could you make to the cut sequences?

20. Assuming the prototype part of Question 19 was inspected and found to be within tolerance for all dimensions, can you return to the original speeds and sequences and assume the machine will also make a good part from stainless?

CHAPTER 6
The Science and Skill of Measuring—Five Basic Tools

Units

6-1 **Task 1: Match the Measuring Tool to the Tolerance**
- List the five categories of measuring
- Find and use dimension tolerances on design prints
- Recognize and apply the various tolerances to measuring
- Recognize protracted and geometric angles on drawings

6-2 **Managing Accuracy**
- Define the resolution and repeatability of the five tools
- Recognize and control inaccuracy factors found in each tool

6-3 **The Five Basic Measuring Instruments**
- Read micrometers to a 0.0001-in. accuracy when measuring real objects
- Read machinist's rules to 0.015-in. repeatability
- Measure to a repeatability of 0.002 in. using a caliper (dial or electronic)
- Measure with a height gage (dial, vernier, or electronic)
- Use a dial test indicator to measure

Courtesy Dennis Paulson and Straightline Precision Industries, Inc.

In July 1996, my father, Mike Paulson, my brother, DJ, and I started a small machine shop called Straightline Precision Industries, Inc. We started with a small manual mill and lathe in our garage with the intention of doing general machining and repairs. In 1999, we purchased our first CNC mill, which allowed us to go after more complex and more lucrative work. We have embraced technology from day 1, constantly reinvesting in the best and newest equipment we could afford, always keeping our eye on future trends in manufacturing. In 2005, we launched Straitline Components, a line of high-end aftermarket mountain bike components, specifically to meet the evolving needs of downhill and extreme free ride athletes. To meet the demands of this added business, we have had to add a lot of new technology, such as robotics and multitasking mill-turn machines, as well as move into more advanced CAD/CAM software. This all means that the job has become very complex and at the same time very rewarding. It is a great feeling to design a product, take it to production, and then finally use it, and see other people enjoy your efforts!

—In memory of Mike Paulson 1950–2011
Dennis Paulson

Courtesy of Dennis Paulson

INTRODUCTION

Previously, delivering quality work meant a lot to the machinist's career. It was the pathway to advancement and to keeping happy customers. But in today's world of Internet commerce, rapid transoceanic shipping, and free trade, quality products, along with price, mean nothing less than survival! Machining quality means accuracy; it's a do-or-die issue.

We remain strong in manufacturing not by working cheaper but by being efficient and delivering quality. Today's machinist must be able to measure in five different ways using a variety of instruments, all of which you will learn in the next three chapters.

Our theme in these next three chapters is that getting it right requires more than just studying the measuring tools. The fine-tuning happens after you learn about their purpose and the name of their parts, and how to read them. The best accuracy comes about with six kinds of skills:

Knowing the Measurement Task Understanding tolerances and which tool matches the exactness of the task.

Learning to Identify Inaccuracy Factors Understanding what might spoil a measurement using a given tool by knowing how accurately a given measuring tool can repeat results.

Controlling the Inaccuracy Factors Learning how to reduce inaccuracy factors to a minimum and how to choose the instrument and process that meets the need and has the least inaccuracy factors.

Uncompromising Attitude Of the greatest import, mastery happens only when students develop an uncompromising attitude about accuracy from the start. You don't get that from a book.

Practice-Practice-Practice Using the tools and processes in a true research fashion, with the goal to always refine measured results.

The Scientist's Approach To develop the greatest ability in the least time, approach learning as a researcher might, studying the principles, then participating in experiments to "zero in" on accuracy. This chapter offers exercises that set that lifelong pattern of awareness and control.

Unit 6-1 Task 1: Match the Measuring Tool to the Tolerance

Introduction: The first step is to choose the right measuring tool and process based on the tolerance for variation. While the objective is to eliminate variation, it is inevitable that some will occur.

KEYPOINT

We can't eliminate variation, but we can control it. The idea here in this chapter and in Chapters 7 and 8 is to not introduce needless variation *due to choosing the wrong measuring process.*

TERMS TOOLBOX

Bilateral tolerance A variation range distributed equally around the nominal dimension.

Geometric angular tolerance A linear variation built around a perfect model that lies at an angle with respect to a datum.

Limits A tolerance method where the range is expressed as the maximum and minimum size, usually with no nominal shown, but it can be added.

Nominal The target dimension on a print, for a given feature.

Protracted angular tolerance An angle originating from a single point, creating a fan-shaped tolerance zone.

Tolerance The range of variation that the design will tolerate and still work correctly.

Unilateral tolerance A variation range extending in only one direction from the nominal.

Variation The difference in the produced result and the nominal.

6.1.1 Five Kinds of Measuring in Machining

There are five aspects to measuring in a machine shop, each with a tolerance for variation. Ability in all five will be needed to complete your first few assignments in the lab. For example, consider drilling a hole in a part. You will need to measure (Fig. 6-1) the

- *Size*—Diameter and/or depth of the hole.
- *Position*—Is the hole located correctly with respect to datums?
- *Form*—Shape limits. Form is controlled to some degree by size tolerance, a subset of size. However, the design could require a hole to be far closer to perfect roundness than its size tolerance range. Form is an embedded control within size.

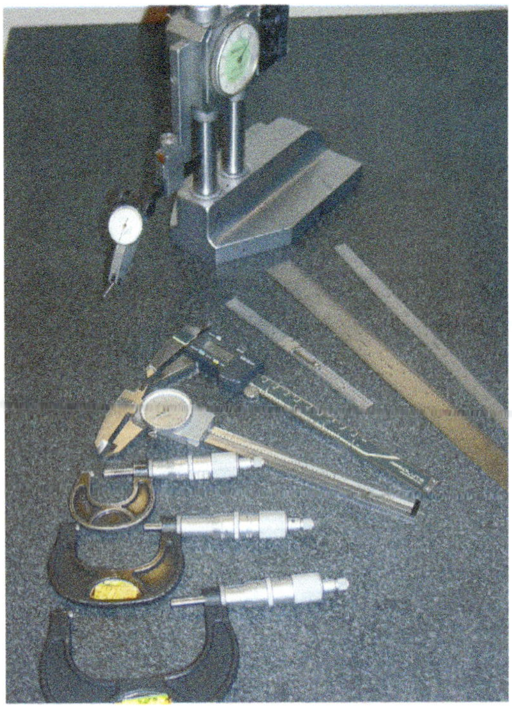

Figure 6-1 A source of pride: five basic measuring tools.
Lake Washington Technical College, Kirkland, WA/McGraw Hill

- *Surface Finish*–Machined objects are given a max roughness in microinches. This specifies how deep the microscopic marks left behind by the machining can be.
- *Orientation*–Is the hole perpendicular or parallel to a given reference datum? Orientation is also a subset of position. But it can be controlled separately.

6.1.2 Tolerances on Drawings

The first step is to find the tolerances that apply to your task and know how "tight" they are. Then choose the right measuring tool and process to deliver the needed results.

> **KEYPOINT**
>
> Although handy and easy to use, calipers and micrometers aren't always the best choice.

Quick Review of Tolerances

In Fig. 6-2, four different means of expressing linear tolerances are shown (straight-line distance).

1. **Bilateral Tolerance**

   The 5.00-in. dimension has a plus or minus 0.005-in. tolerance–a total range of 0.010-in. permissible **variation**.

2. **Unilateral Tolerance**

   The 1.50 nominal dimension is expressed as a minus 0.010-in. variation.

3. **Tolerance Limits**

   On the bottom we see a limit tolerance. It actually has a nominal dimension, but it is embedded in the middle of the range. If the nominal target is critical, then it can be inserted between the **limits**.

4. **General Tolerances**

   Used when the object is very complex with lots of dimensions. The tolerances are compiled in a table near the bottom edge of the drawing.

Can you find a dimension that has a tolerance of ±0.001 in. on the print?

6.1.3 Angular Dimensioning and Tolerancing

We will study how to measure angles in Unit 7-4, but while we're examining tolerances, here's how they are shown on a drawing. There are two very different methods of noting angles on prints, and these have different kinds of tolerancing: **protracted angles** and **geometric**. Each is measured differently but using the same tool.

Protracted Angles

Figure 6-3 expresses the angular tolerance to be plus or minus 2°. Expressed this way, variation is measured with an angle-measuring tool called a protractor. Protracted angles start from a single point and radiate outward. They yield a fan-shaped tolerance zone that totals 4°, as seen in the illustration. To measure a protracted angle, you will not lock the protractor in position but rather move it to best fit to the feature, then read the result in parts of degrees.

Geometric Angles

The second and more functional way of dimensioning and tolerancing angles in terms of assembly is the geometric method (Fig. 6-4). A geometric angular tolerance range is a pair of parallel lines built around the perfect definition of the angle. In this case, it is 59°. This yields a tolerance zone that is a parallel sandwich, not a fan.

Elements on the machined surface must lie within the sandwich tolerance zone lines. It is measured by placing an exact angular gage against the surface relative to a datum and checking for gaps exceeding the tolerance. While this gage will be the protractor again, its use is different. **This time, it's locked in position similar to a square; then gaps are checked between the blade and the object,** always with datum reference as we learned in Chapter 4.

The geometric definition of angles works far better for the assembly of two parts where a maximum space between the two is of concern. When gap is the function, using a protracted, fan-shaped tolerance makes control difficult, while the geometric version is quite simple.

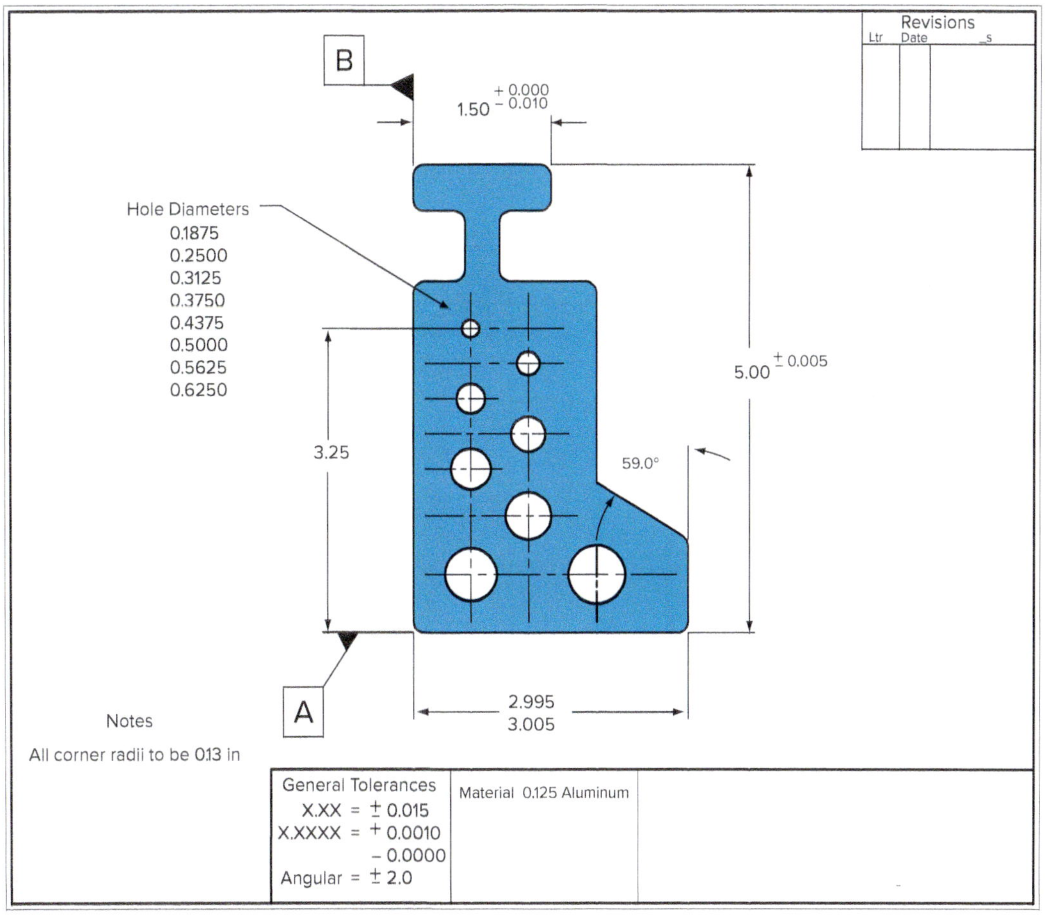

Figure 6-2 Different types of dimensioning and tolerancing.

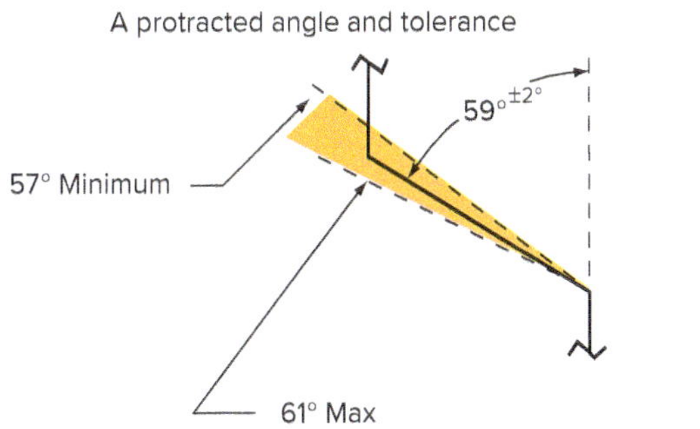

Figure 6-3 Protracted angular tolerance yields a fan-shaped zone.

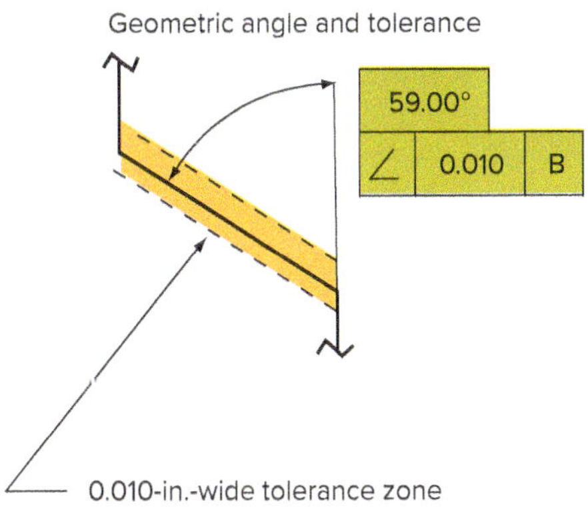

Figure 6-4 A geometric tolerance for angularity is linear.

UNIT 6-1 REVIEW

Replay the Key Points

- All dimensions must have an associated tolerance.
- The target for the dimension is called the nominal.
- Linear tolerances divide into four types:
  - Bilateral, equal distribution around the nominal
  - Unilateral, variation permitted in one direction from nominal
  - Limits, the range of variation shown with no nominal
  - General, tolerances not shown on the print but tabulated in a box
- Angular tolerances are either geometric or protracted.
  - Protracted tolerances are stated in degrees. Protracted tolerances produce fan-shaped tolerance zones—measure with angular tools called protractors.
  - Geometric tolerances are stated as sandwich-like zones around a perfect model of the angle.

- Datums expressed on drawings are starting points for measuring and they must be heeded for correct results.
- Any feature can be used as a datum and must then be accounted for when measuring other features relating to it.

Respond

1. Variation stems from what obvious sources? (Name three.)
2. What are tolerances?
3. Nominal is the range of tolerance permitted. Is this statement true or false?
4. Linear tolerances can be bilateral, unilateral, and (two more).
5. When there is no tolerance associated with a dimension, it will be found.

Unit 6-2 Managing Accuracy

Introduction: Resolution and repeatability define accuracy in our trade. It is important to know the difference and to understand how they affect results with any given measuring tool.

TERMS TOOLBOX

Calibration Testing the tool against proven standards to adjust out or remove all error and to certify that it agrees with those standards.

Coefficient of expansion The amount of linear growth with each degree of temperature change per unit of length.

Feel The amount of pressure that must be applied to a measuring tool to be coordinated to the calibration.

Parallax Not viewing a measuring tool from the correct perspective, which can cause errors in reading the tool.

Repeatability Delivering the same results time after time.

Resolution The way a tool is made; its smallest graduations or movement.

Standard (gage) A master comparison tool that is certified to be a given dimension. A standard is used to test and correct error in shop measurement tools and it can be proven to be theoretically perfect within certain tolerances, depending on the level of certification.

6.2.1 Resolution

Resolution is the finest graduation on the tool or the finest possible controlled movement of the instrument or machine. For example, one micrometer may be graduated with 0.001-in. divisions, while a second micrometer is graduated to 0.0001 in. The second has 10 times finer resolution.

KEYPOINT

Resolution is the result of the way the tool or machine is made.

Often, a skilled user can measure finer than the resolution of the tool. For example, when using a 0.020-in. graduated ruler or the 0.001-in. graduated micrometer, experience will show that it's possible to "read between the lines." While that improves accuracy to some degree, these are skilled estimates. In terms of repeatability, it's far better to find a measuring tool or process with a resolution finer than the job tolerance. We machinists call that the *decimal point rule*—if possible, use a measuring tool or process whose resolution is 10 times finer than the job tolerance.

6.2.2 Repeatability

Repeatability is a combination of many factors. It is the overall trusted accuracy time after time. But most important, tools need users; the repeatability that finally counts is when it's in your hands. **Personal repeatability** is the objective of this unit.

There is no absolute answer as to the best repeatability. However, buying quality measuring tools, taking good care of them, understanding their inaccuracy factors (next), and practice combined with habits gained over time are the vital ingredients. When describing the tools in this chapter and Chapters 7 and 8, I will provide a target repeatability for each. Use it as a goal. It might be difficult at first, but it's easy to achieve with practice. They are my estimates only—a journey machinist and toolmaker!

6.2.3 Ten Inaccuracy Factors

In all five kinds of measuring, there are spoilers that can degrade results. Each measuring tool has its combination that determines repeatability. As an example, suppose a job tolerance of plus/minus 0.001 in. An electronic caliper reads to the nearest half thousandth (0.0005 in.) on the screen. But can you actually trust the measurements to this accuracy, time after time?

The answer is no. Why? It's mostly because of the tool's built-in inaccuracies. Calipers must be closed by hand pressure; that along with alignment of the tool to the surfaces to be measured are the two big ones. This tool has feel and alignment challenges. It isn't the best choice for this tolerance.

The Ten Spoilers

This list is arranged in order of common occurrences for all measuring tools but every process and tool has its own special combination. Note, tool quality is not listed as it is assumed professional tools are being used.

1. *Pressure* or **feel**
2. **Alignment** to the object
3. **Dirt and burrs**
4. **Calibration**, testing the tool against proven standards to adjust out or remove error to certify that it agrees with those standards
5. **Parallax**, the error introduced by not sighting the tool from a correct perspective
6. **Tool wear**
7. **Heat**
8. **Damage** to the tool–bent, sprung, and so forth
9. The **measuring environment**–a place that is organized, clean, with good lighting, and a comfortable safe place in which to take the measurement
10. **Bias for a result**–expecting a result

The trick is to recognize which factors are greatest for a given tool and situation, then control each. Here's how.

Improving Pressure and Feel

Sensitivity is the challenge to many measuring tools and processes, even digital electronic instruments. Tools that are calibrated often require a feel. That is, they must be used with the same touch or pressure in measuring as that to which they were calibrated. The machinist must have a sense of that pressure. All craftspeople soon develop different sensitivities and preferences. Thus, it's critical to either calibrate a particular tool yourself or test and "get the feel" of it if you did not calibrate it.

When selecting a measuring tool from a tool room, always test its feel against a standard. First measure an object that is an exact, known size–called a **standard** or sometimes a *gage.* This will check the calibration of the tool while showing its feel at the same time.

Some feel-type tools require no standard for testing such as the 0- to 1-in. micrometer or the 6-in. caliper. Both tools are closed to their zero position upon themselves, then read to test their reading accuracy and feel. Tools of this sort are called *self-zeroing.*

Gaining the All-Important Sensitivity

Controlled experiments and practice are the keys. A good experiment to develop sensitivity is to measure an object that has a known size (but one that is unknown to you at the time to avoid biasing the results). Upcoming activities offer opportunities to practice this. Another method is to compare your results to those of a more skilled person(s). Third, measure any object using a given tool and process; then find a better process–one that has fewer inaccuracy factors, and make your own comparison.

Improving Tool Alignment

The second common inaccuracy factor is measurement along the axis intended. For example, when measuring the thickness of any object, if the measuring tool isn't aligned (perpendicular to the sides), it could be measuring an incorrect distance, as shown in Fig. 6-5.

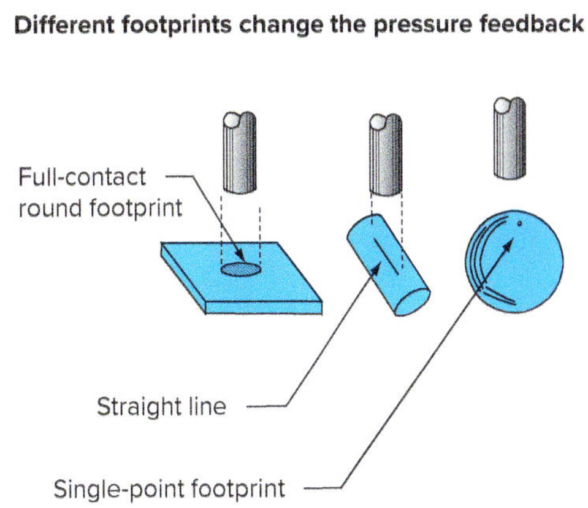

Figure 6-5 As part shape changes, pressure feedback challenges vary.

Control for this factor is exactly the same as for pressure. In fact, you will be correcting both factors at the same time–measure and compare, measure and compare.

Differently shaped objects present different challenges, too. For example, when measuring the thickness of the three objects illustrated using a micrometer, the contact area changes, thus affecting your perception of the amount of pressure being applied.

Eliminate Dirt and Burrs

Although listed as third, for inexperienced machinists this might be first. The smallest burr can be from 0.003 to 0.005 in. Dust and atmospheric grime can account for another 0.001 error–this factor matters! The measuring faces of a tool become contaminated simply sitting on a workbench. They should be cleaned often using lint-free towels or clean paper.

KEYPOINT

Keeping tools in their case when not in use is one simple solution to preventing a buildup of grime.

Setting the Tool to Zero and Calibration

Zeroing index error (shortened to zeroing) and calibration are two different acts. Both terms must be understood by the machinist.

Zeroing Index Error

Many but not all tools have a method of adjusting out error in the readings. Time, heavy use, wear, daily temperature changes, and shocks such as dropping the tool or allowing it to quickly heat up from contact with a hot metal part can all change the point where the tool reads zero (or its lowest value). Micrometers and calipers are such tools. When closed upon a standard (or upon themselves), they should read zero at their index line (training coming up).

Calibrating Measuring Tools

Calibration certifies that the tool reads perfectly all along its entire range of measurement. That certification is a legal document with traceability all the way back to national standards. A certified tool bears a sticker that shows the date of calibration, the lab that performed it, and when it must be recertified. Using the example of the micrometer, to calibrate it means to first zero the index error, then compare its readings all along its range to various size standards.

To ensure constant and total calibration throughout the company, many shops require master tools be tested on a schedule. Whether personal tools are certified is a matter of shop management philosophy. Some shops do not allow personal tools in critical machining areas so that they can control the total calibration of their facility!

KEYPOINT

The mic in Fig. 6-6 was taken directly out of our lab's tool room—showing nearly 0.003-in. index error. The calibration of this tool is very suspect at this point. It might be fine, but after zeroing the index error, it should be tested against a few different size standards, 0.5000 and 1.0000 in. for example, to prove its calibration.

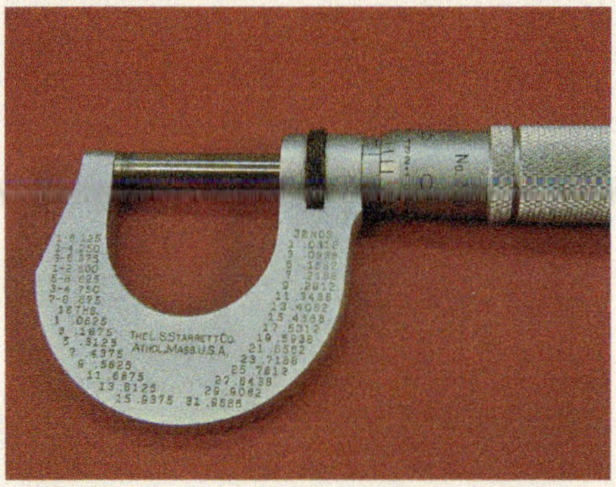

Figure 6-6 Be on the watch for a micrometer out of calibration.
Courtesy Mike Fitzpatrick

TRADE TIP

Detecting calibration error has two stages:

A. Check against a standard.
B. If wrong, stop and zero the error out right then—then test it against the standard again.

The purpose is not only to zero error but to refresh the feel/pressure at zero as well. While it's possible to add or subtract incorrect index errors until you can *get around to fixing the problem,* it's unprofessional to do so unless there is absolutely no time to do otherwise.

Controlling Parallax

This is a visual problem that's easy to eliminate in some instances and difficult to avoid in others. It's caused from not sighting the tool straight on when measuring and reading it (Fig. 6-7).

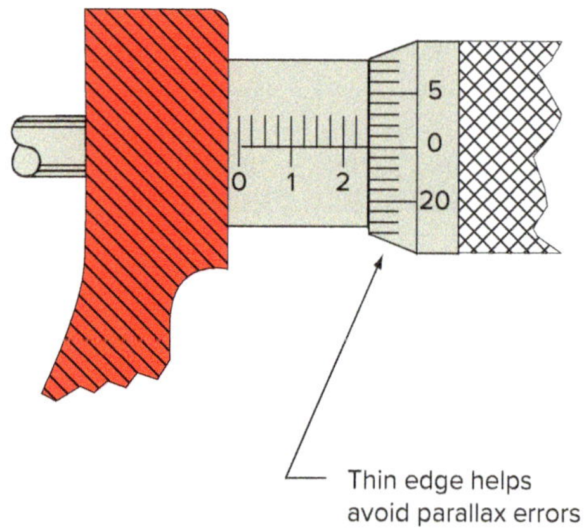

Figure 6-7 This micrometer's edge is thinned to avoid parallax error.

Parallax either affects the numeric reading or changes the alignment of the tool to an object, such as a ruler, to a part edge. Every now and then, we find ourselves in measuring situations where the viewing angle cannot be straight on. There are four possible control solutions:

1. Reach in "blind," get the measurement, and lock the tool in position. Pull it out to where it can be read without parallax. (The bad news is that locking often changes the reading.)
2. Read it in place and account for parallax.
3. Find another tool, process, or position so you can read at the correct viewing angle.
4. Anticipate the problem from the start. Avoid making setups that create these impossible-to-measure scenarios. This skill is usually learned by trial and error, over time.

TRADE TIP

Electronic Measuring Tools A major advantage of electronic, digital tools is that they eliminate parallax factors with clearly read digital results.

Accounting for Tool Wear

The problem here is undetected wear, especially when it is uneven. An example would be a micrometer used to measure round, machined pins, from say 0.250 to 0.450 in., repeatedly for a year. That area of the tool is worn in one area while the numbers out of this range have nearly no wear.

While there is a backlash adjuster to compensate for wear in most micrometers, for what is it set? If the worn region is corrected, the unworn sections will be too tight for use. Uneven wear usually means using the tool to a less important role within your toolbox. Many machinists keep two or more micrometers for this reason—one for daily use where tolerances aren't tight and one for extra precision tasks.

Another example of tool wear can be the faces of micrometers where they constantly rub against the work. They become dished or rounded. The most common example is rounded corners on rulers whereby the end zero becomes unclear. For rulers, I'll bet you already know the fix for this problem: Don't start measuring from the ruler's end but rather start from the 1-in. line.

Controlling Heat

Machining causes heat, which expands metals. If a part is significantly above room temperature, the measured size will be larger than when it returns to normal.

A predictable growth factor can be used, called the **coefficient of expansion**. Found in an engineering table, it is given in expansion units per degree, per inch of length. Every material has a different coefficient of expansion. Figure 6-8 shows two metal bars, one aluminum and one common steel. Notice they both start at 15 in. long at room temperature, but as their temperature rises, they grow at different rates. The aluminum expands faster!

Depending on tolerances, metal type, and overall size, expansion might be discounted as a problem since in the example we see only 0.033-in. growth in 250°F of a 15-in. object! However, where print tolerances are less than 0.002 in. and the machined object heats over 40° to 50°F beyond room temperature, you must do something about expansion and measuring. The control can take two different paths.

Solution 1: Compensate for Heat

Measure the object and its temperature; then mathematically compensate for the heat. Use the table of expansion values found in *Machinery's Handbook*® by Industrial Press. Search the CD or look under *Coefficient of expansion* for various materials within the printed index.

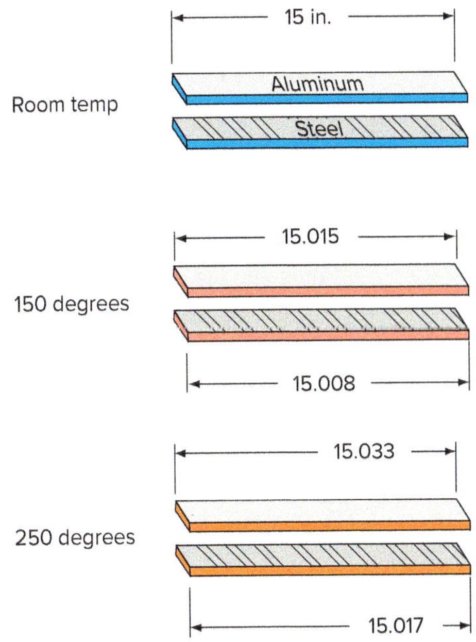

Figure 6-8 As metal heats, it expands, but various alloys expand at different rates.

For example, in Fig. 6-8 (from *Machinery's Handbook*®), aluminum has a coefficient of expansion of 0.00001244 per inch unit per degree. A bar 10 in. long will lengthen 0.0124 in. when heated 100°F, while a steel bar of the same length will expand 0.0063 in. over the same temperature change.

Multiply the change in temperature times the length, times the coefficient of expansion. The problem with this correction method is that an accurate thermometer must be present.

$$\text{Expansion} = \text{Coefficient of expansion} \times \text{Temperature rise} \times \text{Length}$$

ARMCHAIR FIELD TRIP

Learn more about using Machinery's Handbook as a resource to find out all kinds of information machinist need to know by watching this Haas Tip of the Day video, Go to: **https://www.youtube.com/haasautomation** and search for this video called: **Essential Reading for Machinists Haas Automation Tip of the Day**

Solution 2: Cool the Part or Reduce Heat in the Machining Process

Reduce the creation of heat in the work by adding coolants, changing tool geometry, and adjusting machine speeds and force. This is the more plausible solution within your control.

Heat in measuring tools themselves also has the potential to be a problem, but control is not difficult. Don't hold them in your hand for a long time and keep them out of the sunshine or away from other sources of heat. Heat in the machined object is usually a much greater problem.

Detecting Damaged Tools

Similar to wear, this inaccuracy factor is often hidden. For example, if the frame of a micrometer is closed using too much force, although stout, it can bend and the faces can become out of parallel (Fig. 6-9). The sprung tool will appear to be calibrated—that is, it will read zero when closed all the way, but, as seen in Fig. 6-10, it would be far from accurate on real measurements. With tools of this kind, check for damage by closing calipers and micrometers; then look for light gaps in the faces. There are many other kinds of damage, too numerous to list.

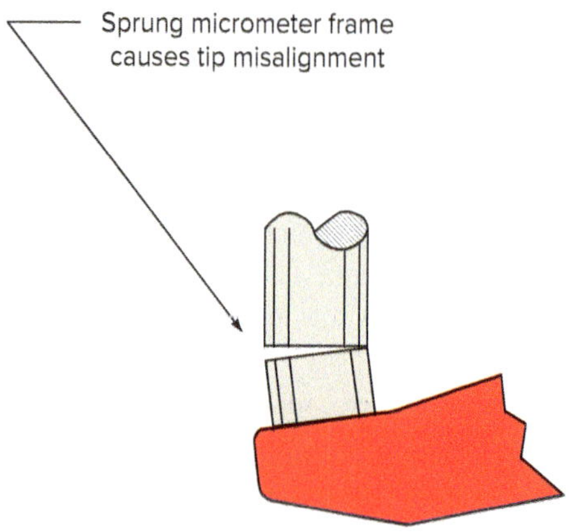

Figure 6-9 The frame of this micrometer is sprung, causing tip misalignment.

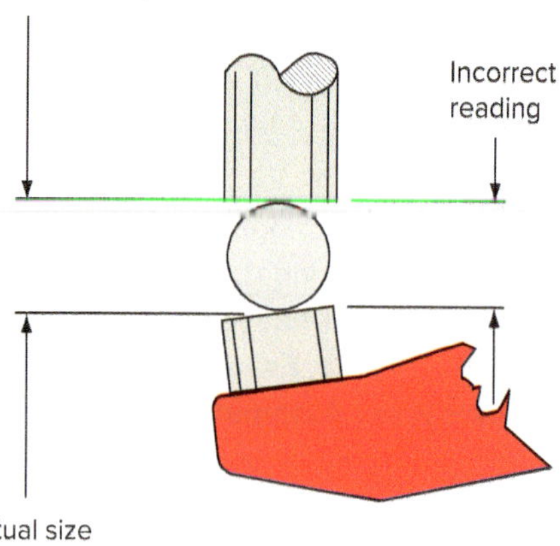

Figure 6-10 Results from the micrometer of Fig. 6-9 could not be trusted.

Take Charge of the Measuring Environment

Accuracy is improved when performed in the right setting or perhaps better put, is degraded in the wrong setting. That means

Comfortable body positions

Clean parts with no burrs

Exceptionally good lighting; for example, if shadows cause distortion, bring in a portable light or, better, keep a work light as part of your tools

A good magnifier is useful no matter how good your vision might be. Clean up chips, burrs, and dirt before measuring and keep measuring tools safe and clean. Develop an organized, patient approach to measuring.

Bias—The Built-In Error

This factor is mental but real. If the machinist starts out to prove the part is right, rather than to measure it, errors happen. This isn't a matter of integrity; it's human nature to "make the reading" rather than "get the reading." In other words, without conscious thought, your hand will either tighten the micrometer too much or too little, if that's what it takes to make the part the correct size.

The main control of bias is to be aware that it does exist, then check your actions to avoid over- or undermeasuring. But one can also look away as the final adjustment is made, then look back.

6.2.4 Options for Control and Accuracy

As you can see, other than damaged tools, machinists must develop control of all the various factors. Once detected, the damaged tool must be repaired or discarded. Feel, calibration, and alignment are the major challenges to accuracy, but practice solves them. Once calibrated, a tool should remain so for a long period with normal use. The main control is simple awareness of the factors.

In today's integrated shops, the ultimate instrument is the CMM, the computer coordinate-measuring machine (see Chapter 28). It's at the top of the evolutionary process. The good news is that CMMs eliminate all 10 errors studied, including heat in the work. If the computer knows the work temperature, it can automatically compensate readings. The bad news is they have their own list causing an uncertainty in the sub-ten-thousandth-in. range.

ARMCHAIR FIELD TRIP

Check out this Haas Tip of the Day video about measuring tools.

Learn about tools and terms that every machinist should be familiar with. Go to: **https://www.youtube.com/haasautomation** and search for this video called: **Measuring Tool Basics: Day 1, Start Off Haas Automation Tip of the Day**

SHOPTALK

An Experiment to Prove Bias A supervisor from a local shop set up an experiment to prove that bias is real. After his top machinists had gone home for the day, the supervisor secretly made three parts that were very familiar to them since they made dozens each week. They knew they should be 1.000 plus or minus 0.001-in. tolerance on a key dimension. But these three parts with a key dimension just 0.0001 in. too big (1.0011 in.).

The next morning, the supervisor gave them first to my tech school students, who had no idea of the required size, and most measured them at or near the real diameter. Back in the shop, using the same micrometer, only one or two machinists found the trap while the rest pronounced them good but right at the upper limit of 1.001 in.!

UNIT 6-2 REVIEW

Replay the Key Points

- Resolution of a measuring tool is the smallest readable graduation.
- Repeatability is your ability to deliver consistency.
- Every tool has its own list of inaccuracy factors.
- There are 10 possible factors that affect measuring:
  - Pressure-feel
  - Alignment to the object
  - Dirt and burrs
  - Calibration
  - Parallax
  - Tool wear
  - Heat
  - Damage to the tool (bent, sprung, and so forth)
  - Control of the measuring environment
  - Bias

Respond

1. Calibration error occurs when viewing a measuring tool from the wrong angle or perspective. Is this statement true or false?
2. Name two ways a conscientious machinist can control biasing results.
3. Name the two ways to deal with machining heat.
4. Describe three possible methods to help reduce pressure and feel-type errors, to zero in on accuracy. (There are four possible answers.)
5. What can you do to the measuring environment to improve accuracy?

Unit 6-3 The Five Basic Measuring Instruments

Introduction: These are the baseline universal tools. Each has its own factor list. Start now to teach yourself the correct uses; after learning about the tool, dig into experiments with it.

Buying Your Own Tools There is a real advantage in knowing the feel of your own tool. I will point out the necessary features and three levels of priority for tool purchase:

Priority 1—Buy as soon as you can. A P-1 tool is fundamental to your kit.

Priority 2—Buy after employment, perhaps within 6 months. Some companies have purchase plans set up with local suppliers for this.

Priority 3—Buy only after you are sure of the direction your career is going. Priority 3 tools are often supplied by the shop.

Also be aware some shops prefer to provide measuring tools; thus, they can control their quality and calibration.

TERMS TOOLBOX

Colinear When two lines coincide (line up). The way a vernier scale is read.

Cosine error False readings caused by tilting the travel indicator's axis away from the direction to be measured—causes amplified results.

Discrimination In measuring, the ability to perceive a difference in two or more graduations on a given tool.

Nominal zeroing The ability of electronic tools to set any point as zero—thus measuring a large quantity of parts by reading the variation only.

Progressive error The error created by a test indicator's probe swinging in an arc yet reading in linear thousandths of an inch.

6.3.1 Basic Tool 1—Rulers

You may not have considered rulers to be precise instruments—think again! With practice, it's not only possible to reliably measure to 0.005 in. using a ruler (even closer in some cases);

it's also expected of the journey machinist! Some tips for using a ruler are shown in Figs. 6-11, 6-12, and 6-13.

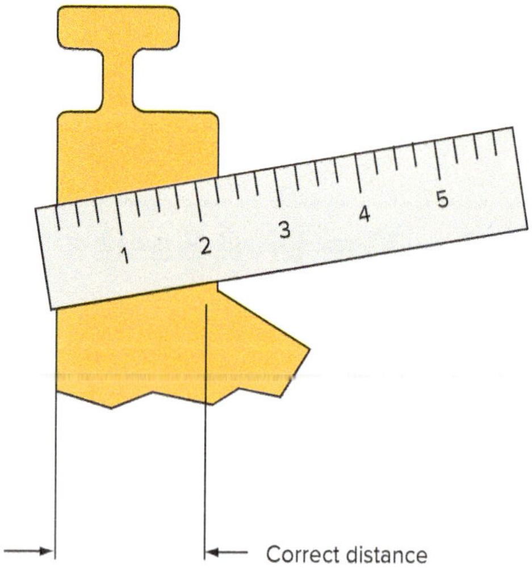

Figure 6-11 Alignment to the distance in question is a must.

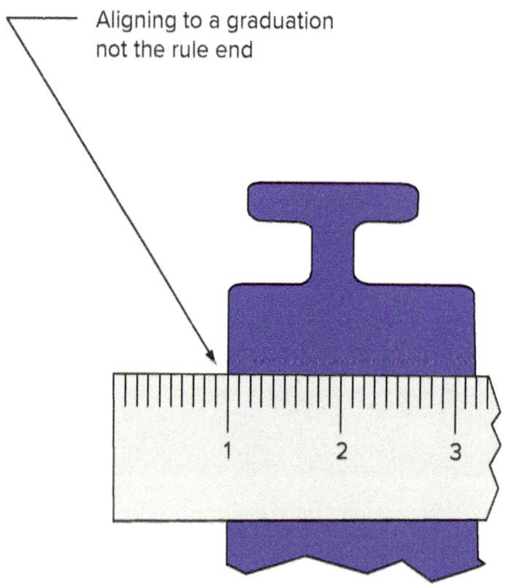

Figure 6-12 "Split" the lines on the rule.

Rules are used as a fail-safe, backup check of other process or tool. For example, after measuring a diameter with a micrometer, do a spot check with the rule. This double check avoids the "decimal place" errors such as 0.100 in. big or small.

KEYPOINT

Due to a difficult location or odd shape, once in a while rulers are the only tool to do the job.

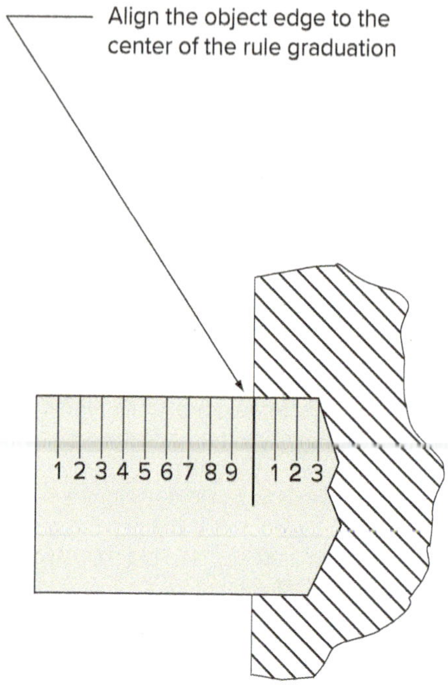

Figure 6-13 Using the center of the graduation line makes measuring more accurate.

Tech Specs for All Rulers

Rule measurement is common for tolerances of 0.015 to 0.030 in.

Target repeatability = 0.005 in. or 0.1 mm

This is the expected result of careful measuring with a rule after some practice and experience. You'll need an excellent rule, good lighting, practice, and perhaps a magnifier to achieve it.

Rule Inaccuracies

Be on the watch for

A. Alignment to the measurement
B. Parallax
C. Burrs and rounded part edges
D. Poor lighting
E. Damaged rules—especially the corners

Three Types of Machinist Rule

Type 1. Flex Rules

These are 6, 12, or 18 in. long (150, 300, or 450 mm). Flex rules are thin, universal tools (Fig. 6-14). The 6- or 12-in. rule is a priority 1 purchase. Longer versions are a priority 2.

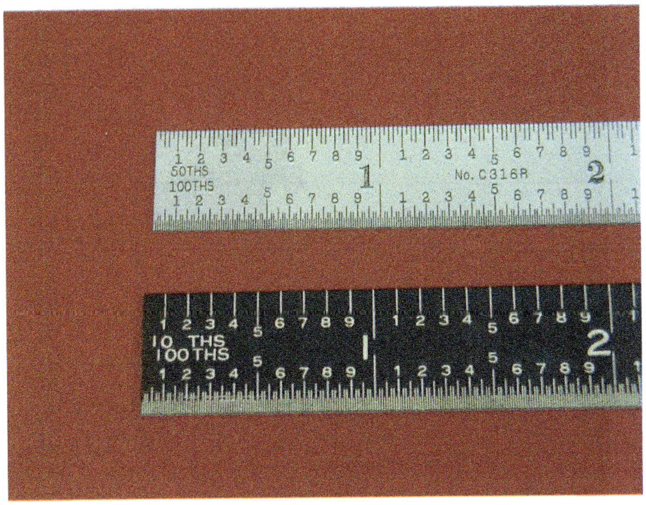

Figure 6-14 A pair of machinists' flex rules. Choose satin chrome or black finish, both excellent reading rules.
Courtesy Mike Fitzpatrick

Type 2. Spring and Hook Rules

Sometimes called flat rules, these are about twice as thick as and wider than a flex rule. These sturdy little rulers feature the same graduations as flex rules. Their advantage beyond rigid strength and long life is that some feature an attachment called a *hook.* It's useful in reaching through or across an object to measure its thickness. The upper rule in Fig. 6-15 has a drill point gage attachment used for resharpening bits.

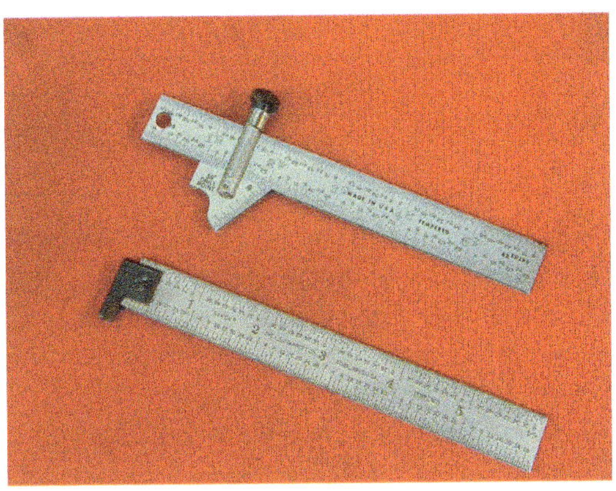

Figure 6-15 A spring or flat rule is useful when used with a hook attachment.
Courtesy Mike Fitzpatrick

Type 3. Ground or Rigid Rules

Because they are used in several trades, several levels of accuracy and quality can be found for this rule. Only the best are used in machining. Made of high-quality spring steel in 12- and 18-in. lengths, they are a priority 1 for purchase. The 24- and 36-in. lengths are priority 3 tools.

TRADE TIP

Quality Features for Rules *Surface finish* (see Fig. 6-14) is important. A dull (matte) finish is the best for eliminating reflected glare. Avoid the less-expensive shiny metal rules as they are difficult to read.

The *graduations* on quality rules feature etched or engraved, not stenciled or painted, lines. You can test this with your fingernail. The rule you select should have a variety of graduation patterns, one per edge. All rules should have one edge graduated with 0.100-in. marks. The opposite edge should be either 100th-in. (0.010-in.) graduations or 50th-in. (0.020-in.) graduations.

Rigid rules are often combined with a set of angle-measuring tools called a combination set. See Fig. 6-16. We will discuss the combination set later in Chapter 7. This set could be a priority 1 or 2 purchase, depending on the nature of local industry—check with your instructor.

Figure 6-16 An 18-in. ground rule with combination set. Left to right, a square, a protractor, and the centering head.
Courtesy Mike Fitzpatrick

Be aware that "bargain" rules are available stamped from plain sheet metal. These tools are okay for home shopwork but not for machining. The ground rule intended for machinist use features precision engraved divisions. Satin finish is a must. A high-quality ground rule is used as a template for testing straightness, too.

6.3.2 Basic Tool 2—Calipers

Calipers are the most versatile measuring tool in the machinist's kit. Modern calipers feature multiple measuring abilities using a dial face or an electronic digital readout. They are quick and easy to learn but require a bit of practice to master. They are priority 1. Buy an electronic digital type if your budget allows, or a dial type.

Calipers Are Not a Panacea

Since they are so quick and easy to use, there is a tendency among beginning machinists to overuse calipers. They have several accuracy factors affecting repeatability:

- Feel and alignment are the most prominent.
- Calibration can be a problem.
- Damaged or sprung jaws are not uncommon.

Four Functions

Modern calipers provide four different measuring abilities (Fig. 6-17):

1. *Inside measurement*
2. *Outside measurement*
3. *Depth measurements*–from a surface to an opposing surface
4. *Step measurement*–from a surface to a surface, with both facing the same direction

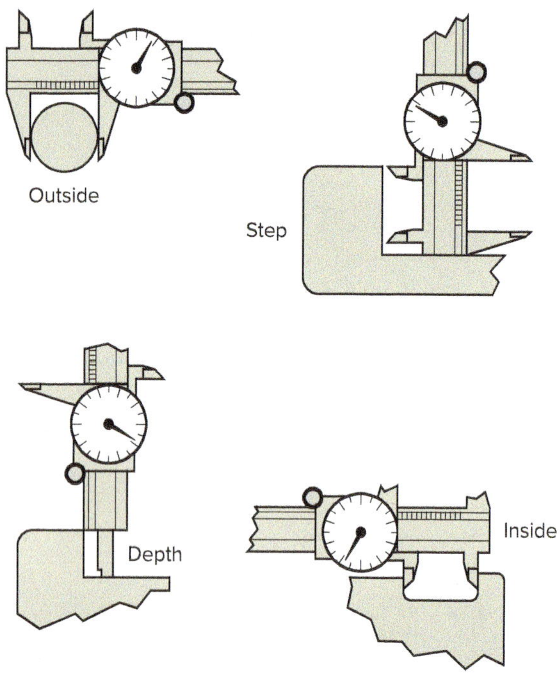

Figure 6-17 The four functions of a caliper: outside, inside, depth, and step measurement.

Note that step measuring is recommended only as an alternative when some other method isn't available. Step measuring is usually not the best process due to the estimated position of one or the other jaw. It also requires knowing the *offset amount* of the step, then adding it to the reading on the caliper (see Fig. 6-17).

SHOPTALK

Electronic calipers (Fig. 6-18) have three distinct advantages. The first is their ability to switch between metric and Imperial dimensions.

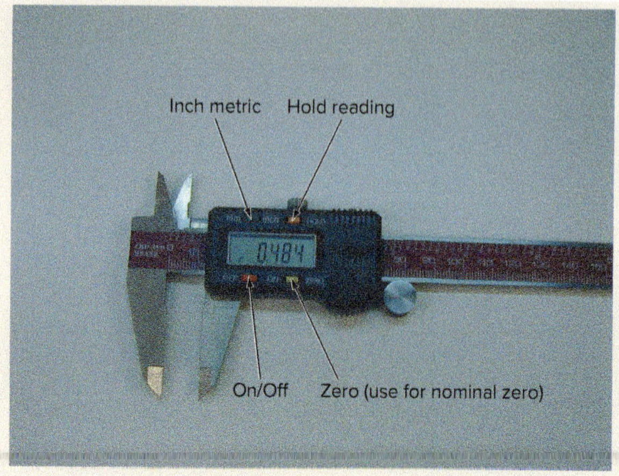

Figure 6-18 An electronic caliper has several data functions.
Courtesy Mike Fitzpatrick

The second is called **nominal zeroing**. Electronic calipers can be zeroed anywhere along their range. This allows you to set the nominal size for a large group of parts as the zero. Thus, set at zero, each new measurement reports the variation with no mental work to see if it's within tolerance.

The final advantage is *data output*. Electronic calipers can send their measuring data to a host computer concerned with graphing the size variation of the samples. This enables statistical process control by directly entering results to the database. Electronic tools either plug into a data cord or transmit via Wi-Fi.

Tech-Specs for Calipers

Correct Application

Both electronic and dial types are used for tolerances above 0.001 in., even though the electronic readout features a 0.0005-in. resolution. Both are used in the same way and have the same accuracy challenges, except parallax is not a problem with electronic readouts. The major challenge to accuracy with all calipers is their feel and alignment factors.

Target Repeatability

Due to different contact footprints and feel factors, various functions on the caliper produce slightly different repeatabilities.

Outside Measurement

Dial types measure from 0.001 in. to 0.002 in., while electronic types (slightly better only because of the ease of reading the scale) measure from 0.001 in. to 0.0015 in.

Inside Jaws and Depth Probe

Both types measure from 0.001 in. to 0.003 in.

Step Measurement

Due to the visual estimation of one jaw to the distance, in most cases, these measure 0.003 in. to 0.010 in.

Older Vernier Calipers

You might encounter a third variety of caliper mostly displaced by technology: the vernier caliper (Fig. 6-19). A vernier caliper has no readout or dial.

Figure 6-19 Vernier calipers, an earlier measuring tool, are useful in tight spots, due to their compactness.
Keith Smith

If you require vernier training, turn to height gages in this chapter, then return here. Vernier calipers are sturdy and less costly for 24- and 36-in. types. They are compact tools, meaning they can fit into tight places.

Caliper Familiarization

Take a look at the calipers found in your training lab. It would be beneficial to locate the following features.

On dial types (Fig. 6-20) look for

A. Rotating bezel ring
B. Bezel lock screw
C. Slide lock screw
D. Depth probe
E. Index line
F. Fine movement thumb wheel

On electronic types, you'll see

A. On/off button
B. Slide lock
C. Depth probe
D. Metric/imperial button
E. Data port (optional)
F. Zero reset button

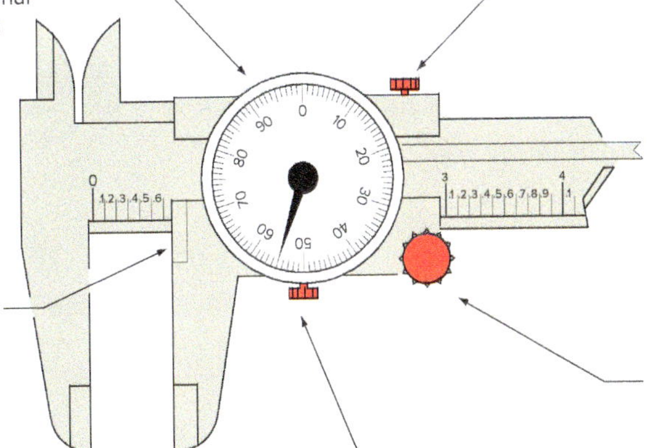

Figure 6-20 The parts of a dial caliper.

Zeroing Dial or Digital Calipers

After cleaning the jaws, use the fine movement wheel to close them to the pressure you feel is right. Then be sure the readout or dial reads zero (Fig. 6-21).

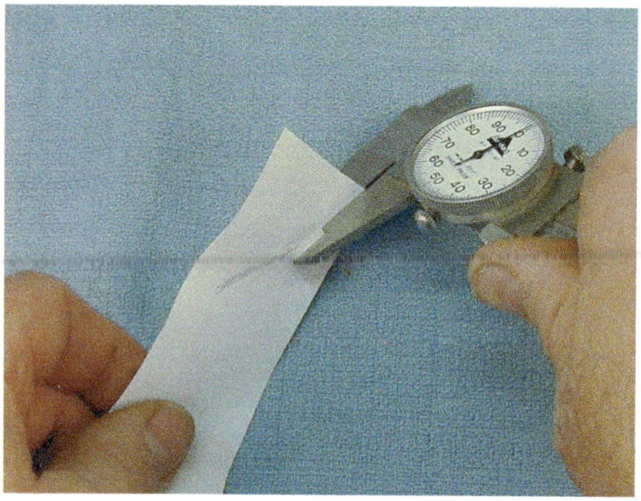

Figure 6-21 To clean caliper jaws, close them on paper, then slide it out.
Courtesy Mike Fitzpatrick

Dial Caliper Ranges

Calipers using mechanical dials (Fig. 6-22) are supplied in one of two possible dial ranges: 0.100 in. per dial revolution or 0.200 in. per dial rotation. The 0.100-in. dial is slightly better due to the expanded numbers on the face. This is called the **discrimination** of the tool. The improved discrimination of the 0.100-in. dial aids accuracy.

Figure 6-22 Dial face calipers are supplied in two ranges.
Courtesy Mike Fitzpatrick

Reading the Dial Caliper

It's a snap to learn to interpret the numbers on these and even easier to read electronic types.

TRY IT

Read the caliper using the magnified dial face in Fig. 6-23.

On the caliper of Fig. 6-23, the index line is between 0.6 and 0.7 on the horizontal scale. The dial is at 0.055 (the index has not passed a full inch yet). The total is 0.655 in. Now take up your caliper and set it at 0.655 in.

On dial types, first the full inches and 0.100-in. increments are noted on the horizontal rule, at the index line. To that, the dial or vernier amount is added. Now read the three calipers shown in Figs. 6-24, 6-25, and 6-26.

ANSWERS

Figure 6-24 = 1.238 in.

Figure 6-25 = 3.192 in.

Figure 6-26 = 9.815 in.

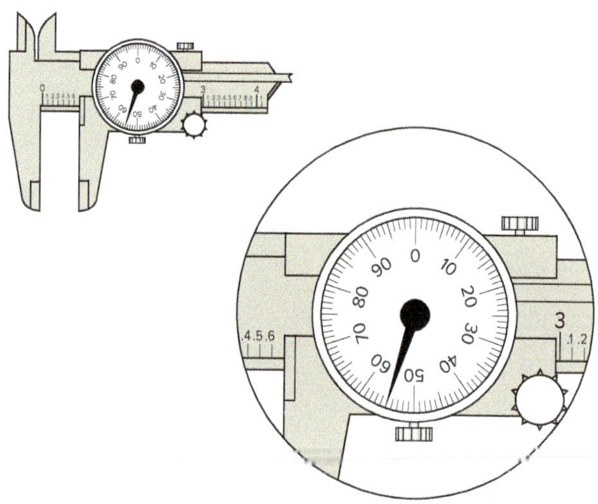

Figure 6-23 Read this dial caliper position.

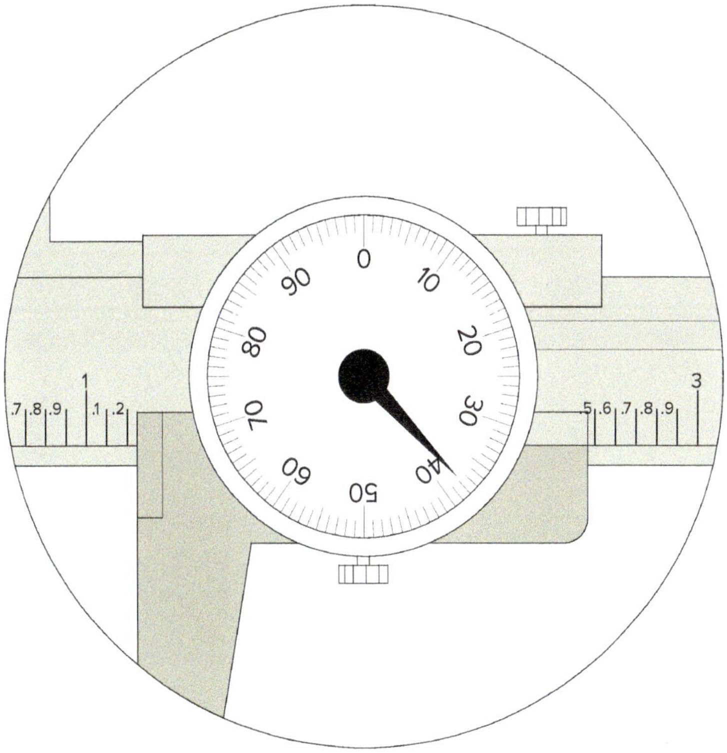

Figure 6-24 Read this dial caliper position.

Figure 6-25 Read this caliper setting.
Courtesy Mike Fitzpatrick

Figure 6-26 Read this caliper.
Courtesy Mike Fitzpatrick

KEYPOINT

Note that all three functions—outside, inside, and depth probe—are at the same reading at the same time. Modern calipers that feature this three-way action are known as *offset jaw* calipers.

6.3.3 Basic Tool 3—Outside Micrometers

Micrometers are such basic trade tools that they are used as the logo of the craft. Sometimes they are formally called micrometer calipers, but more often they're referred to as *mics* (sounds like "Mikes") in trade lingo. The smallest, 0- to 1-in. mic, and the next size, the 1- to 2-in. mic, are a priority 1 tool purchase, with the 3-in. mic being a priority 1 or 2, depending on your budget.

SHOPTALK

Absolute Accuracy and Range of Certainty As you follow the experiments provided here, working with other students and comparing results, you will see some variation among yourselves as to the size of any single feature. That's normal, and it's caused by many factors. Although the greatest contributor to the disagreement is the human factor, every measurement process has some variation in the process itself. Even the best computer CMMs are not absolute. But the drive is always to find a better process that has less variation. As understanding of metrology improves along with skills, your experimental group should produce a smaller and smaller variation between you. This is an excellent indicator of improvement. Make narrowing the gap between the group a game. Test yourselves every so often to see how close any three of you can get to perfect agreement. Try it; it works to improve your skills. Then, try challenging any other three students.

Micrometers are divided into three general types: outside, inside, and depth. Each is a separate function of the dial caliper, but, due to improved resolution and fewer inaccuracies, they are from 10 to 20 times more accurate than calipers. Figure 6-27 shows basic mechanical micrometers; Fig. 6-28 shows a direct reading electronic version. Inside and depth mics will be studied in Chapter 7.

Specialized Micrometers

There are many specialized kinds of micrometers that we will not study in this unit. For example, Fig. 6-29 shows an electronic thread mic. In industry, special micrometers are used to measure deep grooves, or with very deep frames to measure sheet metal thickness. After learning the basics of micrometers here, adapting to the specialized mics will be easy when encountered on the job.

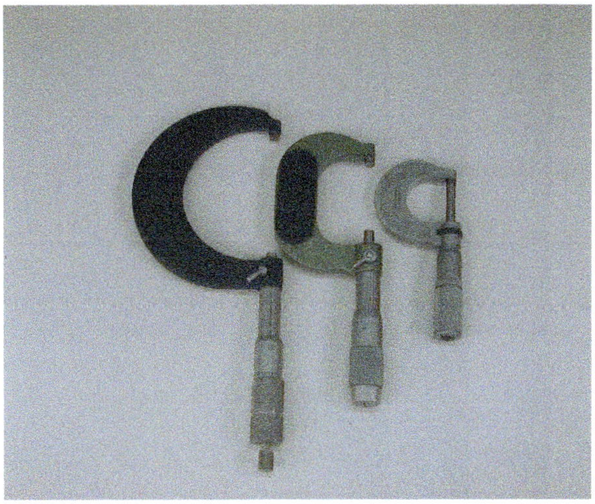

Figure 6-27 Micrometers from 0 to 3 in. are normally owned by the machinist.
Courtesy Mike Fitzpatrick

Figure 6-28 Both direct reading and electronic micrometers are easy to use and calibrate.
Lake Washington Technical College, Kirkland, WA/McGraw Hill

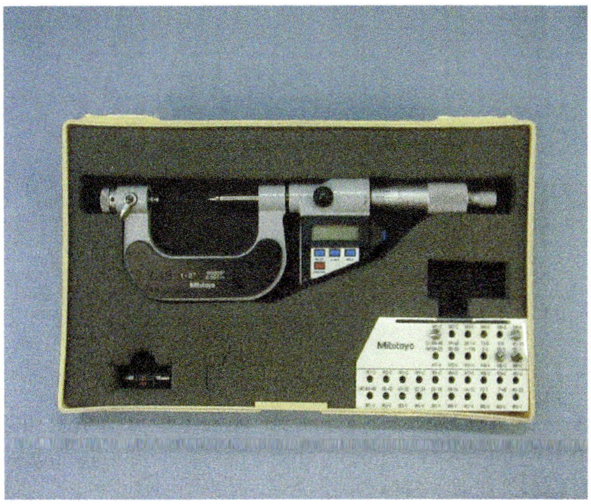

Figure 6-29 A thread micrometer set.
Courtesy Mike Fitzpatrick

We will also not study reading electronic mics (see Fig. 6-28) as there is virtually nothing required beyond what you will learn using mechanical types. Getting the right feel and alignment are the real skill challenges and they remain the same for all micrometers, no matter their construction.

Tech-Specs for Outside Micrometers

Correct Application

Outside micrometers are used to measure thickness and diameters. Tolerances below 0.002 in. require micrometers. They are also used to measure and calibrate other measuring tools such as small hole gages or adjustable parallels (see Chapter 7).

Target Repeatability

Inch Micrometers	
Vernier Scale	0.0001 to 0.0002 in.
Without Vernier	0.001 in. to ≈ 0.0005 in. by estimating
Metric Micrometers	
Vernier Scale	0.001 to 0.002 mm
Without Vernier	0.01 mm to ≈ 0.005 mm by estimating

Control of Inaccuracies

The five factors for outside micrometers are feel, calibration, dirt, burrs, and alignment. Wear and damage can be problems, too. Parallax can also affect results. Use the joggle/null method to arrive at correct alignment. Here's how.

After cleaning the measuring faces as shown in Fig. 6-30, hold the micrometer as shown in Fig. 6-31 with one hand; then move it in smaller and smaller sweeps to obtain the lowest possible reading–the movements are described as joggles. The micrometer will stop changing at lowest possible value, called a *null.*

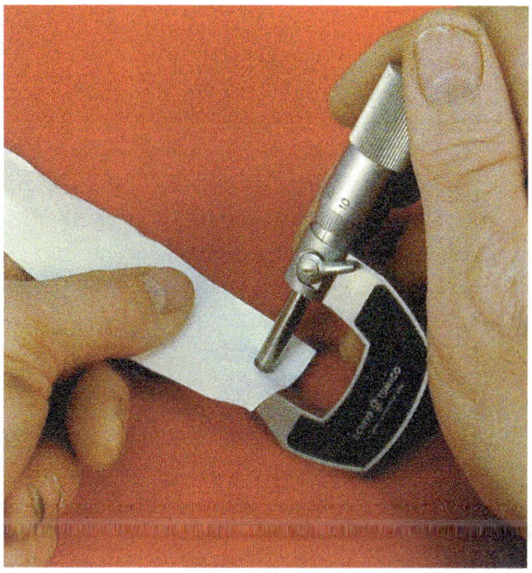

Figure 6-30 Always clean the tips before making a critical measurement.
Courtesy Mike Fitzpatrick

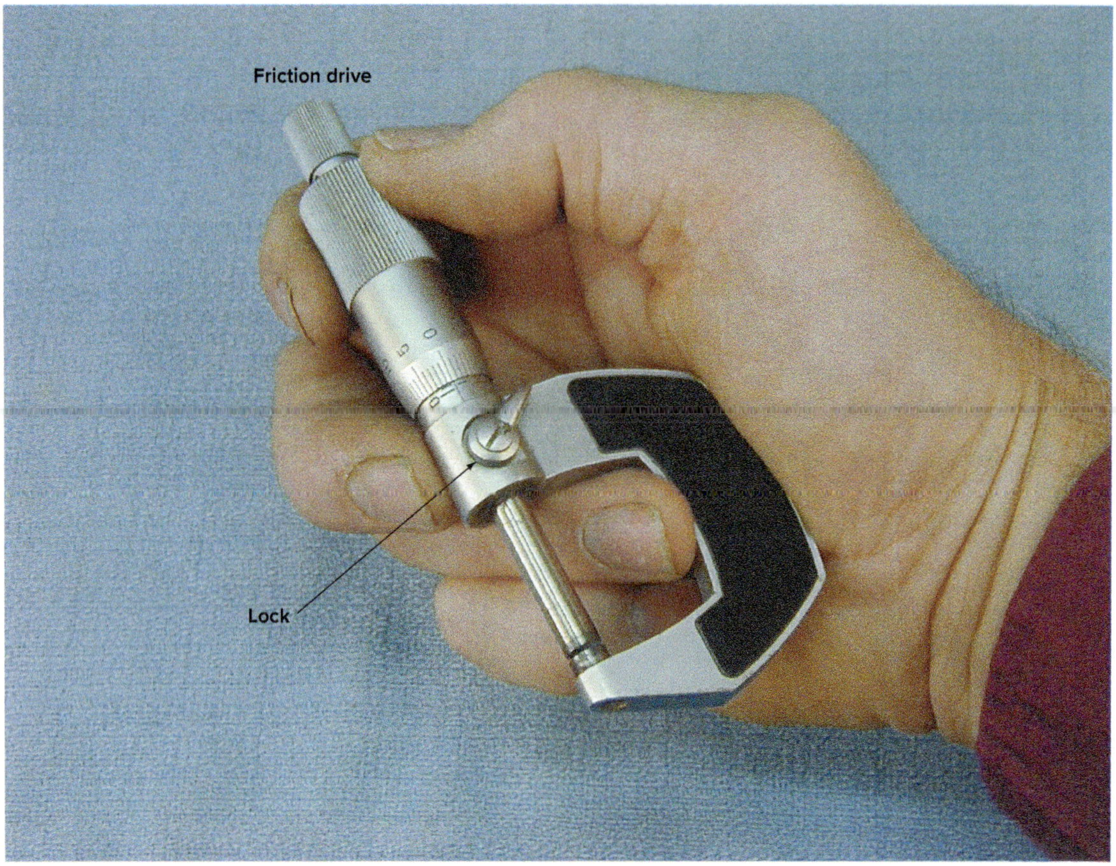

Figure 6-31 Holding it correctly, all functions are within reach of one hand.
Courtesy Mike Fitzpatrick

To avoid overtightening, use the friction drive on the micrometer for developing your own personal feel. Parallax and tool heat can be factors but to lesser degrees than alignment and feel.

Reading a Micrometer

Please obtain a micrometer and check its features as we proceed.

Hand position is important to allow full control of all functions with the least disturbance to measurements and to achieve alignment. The position is different for small and large outside micrometers. For small micrometers up to about 4 in., it is held and aligned with one hand, as in Figs. 6-31 and 6-32. This allows joggling the mic while reaching the friction/pressure control drive and spindle lock.

The friction/pressure drive is used to regulate the maximum amount of force brought to bear on the object being measured. Friction drives are supplied in two forms: a ratchet at the end of the micrometer or a friction drive on the thimble. The friction type is a better solution because it is easy to reach. (See Fig. 6-34.)

Friction or ratchet drives reduce variation where several machinists must share the same tool. They are also useful for learning about feel.

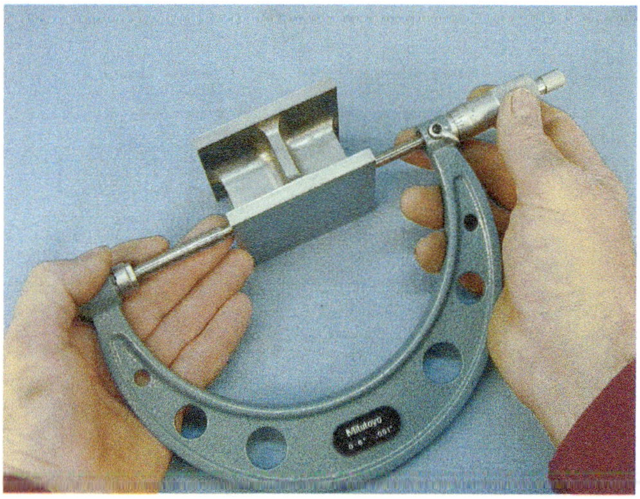

Figure 6-32 This is the correct method of holding a larger micrometer.
Courtesy Mike Fitzpatrick

Precision Features for Micrometers

When purchasing a micrometer, check for the following features (Fig. 6-33). Please examine the student micrometer or your own as we proceed.

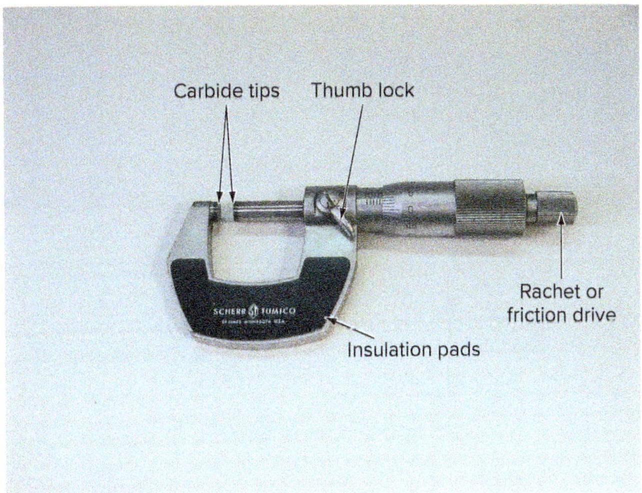

Figure 6-33 Standard features on a micrometer.
Courtesy Mike Fitzpatrick

1. Carbide wear tips
2. Tenths scale: visible as 10 numbered lines running parallel to the index line on the sleeve
3. Hand heat insulation–both sides of the frame are insulated with a pad.
4. Wear and backlash compensator–In a clean environment, unscrew the thimble until the internal parts are visible. This is not part of normal calibration and is adjusted only over a long time. Caution! Be careful to not allow dirt into the precision threads when reclosing the micrometer!
5. Protective case

Developing Micrometer Skills

First, we'll learn to read the micrometer, then do some practice measuring. Remember the best repeatability results from recognizing and controlling the inaccuracy factors.

TRADE TIP

Caution! Never close your micrometer to zero when putting it away. If left in the toolbox drawer as it is now, for any length of time, the micrometer in Fig. 6-34 will begin to change shape. That tends to spring the frame over time.

Figure 6-34 Never close a 1-in. micrometer when storing it! Damage can result.
Courtesy Mike Fitzpatrick

Reading an Imperial (Inch) Micrometer

The micrometer shown in Fig. 6-35 reads 0.475 in. Notice the line parallel to the sleeve axis. This is the index line. The index line is where the measurement is to be read. Identify the index line on your micrometer; then set it to the same reading. When doing so, notice that the gap opens between tips, by 0.025 in. for each full revolution, and that the rotating thimble is divided into 25 parts. Each line on the thimble then represents 0.001 in.

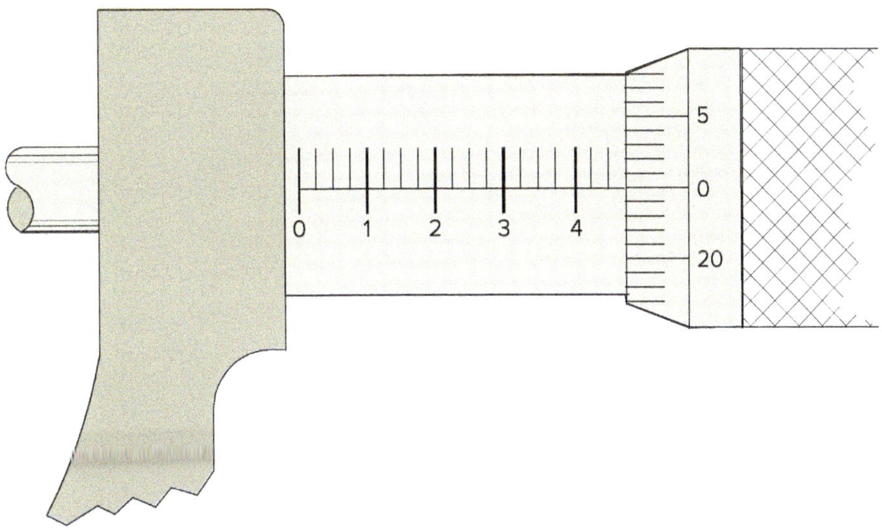

Figure 6-35 Read this micrometer.

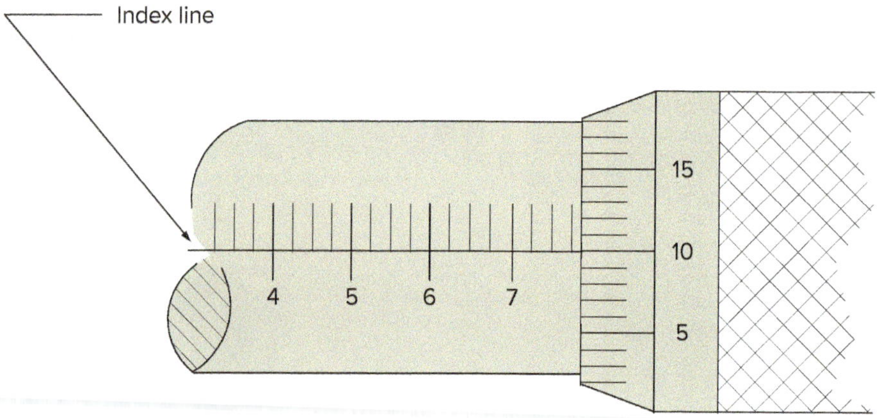

Figure 6-36 Read this measurement.

Now read the micrometer in Fig. 6-36. This illustration shows 0.785 thousandth. First, the 0.700-in. graduation is uncovered along the index line. Next, three 0.025-in. revolutions are visible, 0.775 in. Finally, the thimble is rotated an additional 0.010 in.

Set the student micrometer for the following numbers; then look at Figs. 6-37 and 6-38 to see examples.

A. 0.555

B. 0.840 (Notice that the number 15 is aligned with the index–to create the forty thousandths.)

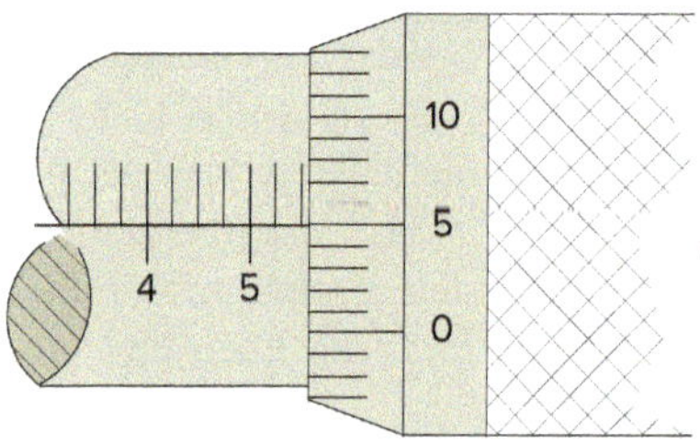

Figure 6-37 Reading, 0.555 in.

Reading the Vernier Tenths Scale

To increase resolution to 0.0001 in. we use a vernier micrometer–10 times finer. Obtain a vernier micrometer as shown in Fig. 6-39. It has 10 numbered lines running parallel to the index line on the sleeve. They subdivide each thousandth into 10 parts, 0.0001 in.

Tilt the micrometer up and observe the lines running parallel to the index line along the sleeve, starting at 0 and ending on 9. These are the vernier lines. (Some scales start and end on a zero.) To read it, find the vernier line that best matches a line on the micrometer's thimble. The number at the end of that vernier line indicates how many tenths (of a thousandth) the thimble has been rotated beyond a particular thousandth.

Read the micrometer in Fig. 6-40. Note that the drawing is as though the vernier scale is laid flat. Notice that only the vernier line numbered 3 exactly matches a line on the thimble. Set your micrometer to this position.

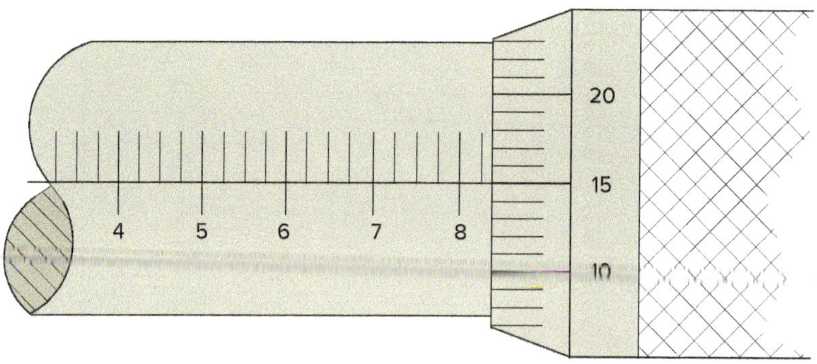

Figure 6-38 Reading, 0.840 in.

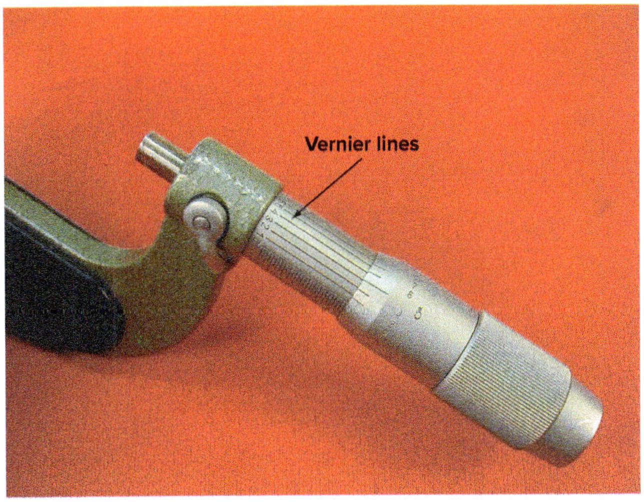

Figure 6-39 The vernier scale is wrapped around the sleeve.

Courtesy Mike Fitzpatrick

Set your micrometer to the following sizes. Remember, first set the thimble to the correct thousandth and estimate the tenths–then tilt it up to see the vernier scale. Now turn the thimble just a bit more until the correct numbered vernier line on the sleeve coincides with a line on the thimble.

1. 0.4477 in.? See Fig. 6-41 for the first answer.
2. 0.3689 in.? Have your instructor check this one and the next.
3. 0.3333 in.?

A Pattern for Improvement

Now it's time to use the micrometer in real measurement situations. Remember, the goal is to obtain consistent results that fall within the expected repeatability of the tool. That takes practice.

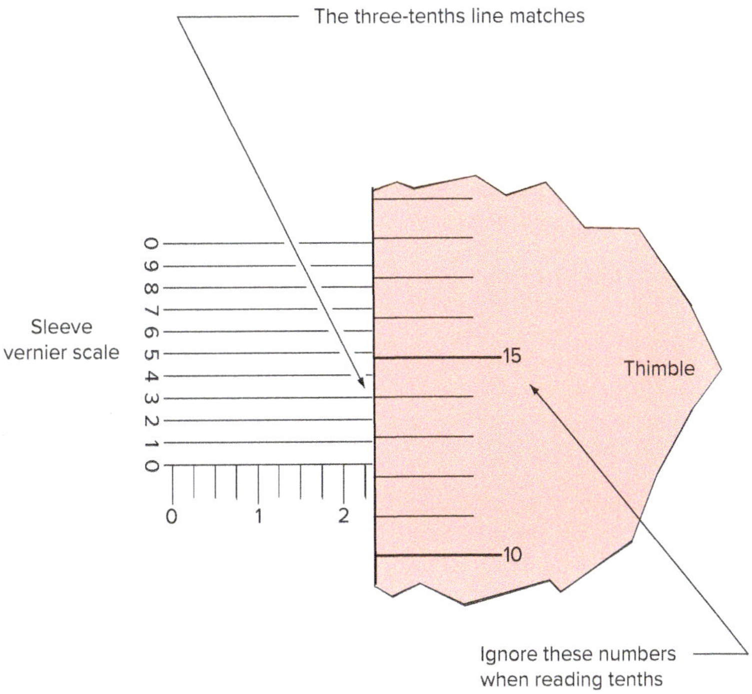

Figure 6-40 The micrometer at 0.2373 in.

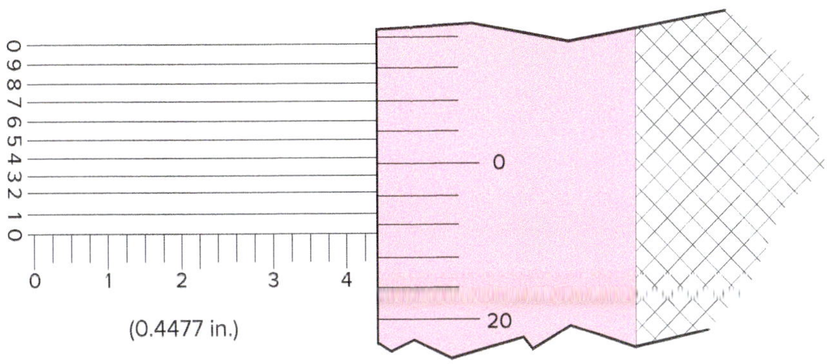

Figure 6-41 The vernier scale will be on the side of the sleeve.

Improving ability with all measuring tools is a five-stage process:

1. *Measure* one or more objects (unknown size at the time, to avoid bias).
2. *Compare* your results to the best answer available of true size.

   The best are gage blocks (see Chapter 7) or standards or prepared items with answers.

   The next best is to have a more skilled person test the measurement.

   The third method is to test the same measurement with a better tool or process.

   A last is to compare your results to those of two others of the same skill level, but you can see the potential problem with this.
3. *Determine tendency*. Did you tend to make a consistent error? Too much by, say, 0.003 in., or too small repeatedly.
4. *Correct* the problem (less pressure, more, better alignment, and so on). Have your instructor show you ways of aligning the tool to the measurement.
5. *Try again.*

Your instructor may have an activity planned with objects and an answer sheet. If not, make one up and test each other. Make up an accuracy challenge between yourself and fellow students.

Zeroing Out Index Error Before Measuring

Before measuring with a micrometer it's important to test the feel and the zero index. First, clean the two measuring tips with lint-free wiper. Next, close the mic to zero or upon a standard that's the lowest value of the micrometer. For example, to check a 3- to 4-in. mic, use a 3.000-in. standard.

If the mic reads zero at a pressure you feel is okay, proceed. If not, it's time to realign the mic. Realigning the zero index requires moving the thimble relative to the spindle. The task requires loosening but not disconnecting the spindle from the thimble. The numbered thimble is then adjusted to read zero when clean tips are closed at the right pressure. After adjustment, the thimble is relocked in place.

SHOPTALK

Caution! A reminder that zeroing may not be allowed at your level of training or even for beginners in the industry where specialists perform the task. Ask about the policy.

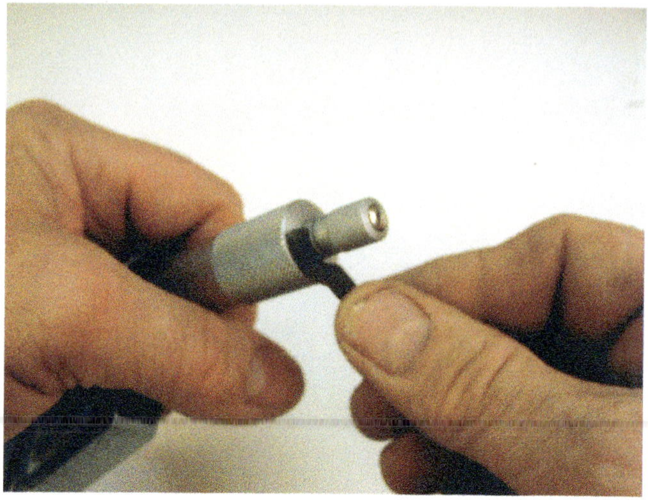

Figure 6-42 Calibrating against a standard.
Courtesy Mike Fitzpatrick

Always double-check calibration after locking the mic (Fig. 6-42). The force can shift the new zero!

If allowed, practice first on a 1-in. micrometer, then on a larger mic using a standard. There are two methods of loosening the thimble from the shaft:

Straight shaft—set screw

Tapered shaft—self-holding

The set screw type requires no instruction. If it is a self-holding taper, then a special spanner wrench is required (Fig. 6-43). After loosening it one-quarter turn, tap the thimble with a screwdriver handle or some other firm, non-metal material; this breaks the taper loose. Ask your instructor or a machinist to demonstrate this action.

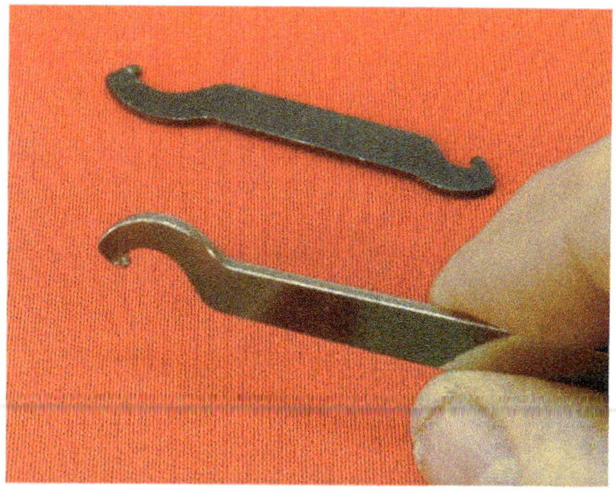

Figure 6-43 Calibrating requires a special spanner wrench to unlock and relock the thimble.
Courtesy Mike Fitzpatrick

SHOPTALK

Vernier Metric Micrometers Metric micrometers are also supplied with vernier divisors. They read to 0.001 mm, which is about twice as fine compared to an imperial micrometer's resolution using the vernier scale.

Reading a Metric Micrometer

Metric measuring is important in a world economy. As with all metric implements, these measuring tools are easy to understand. The index line is graduated in full millimeters above and half millimeters below.

KEYPOINT

Each full rotation of the thimble opens 0.5 mm.

The thimble is divided into 50 parts, thus each part equals 0.01 mm. To read the micrometer, add the uncovered full and half millimeters on the index line to the hundredths on the thimble.

Figure 6-44 reads 4.79 mm–4 full millimeters plus one 0.5 graduation plus 0.29 from the thimble. Now read the metric micrometers shown in Figs. 6-45 through 6-48.

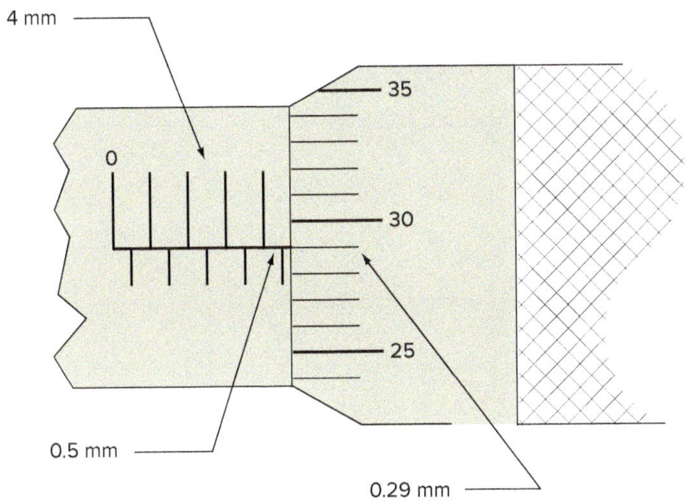

Figure 6-44 Metric micrometer at 4.79 mm.

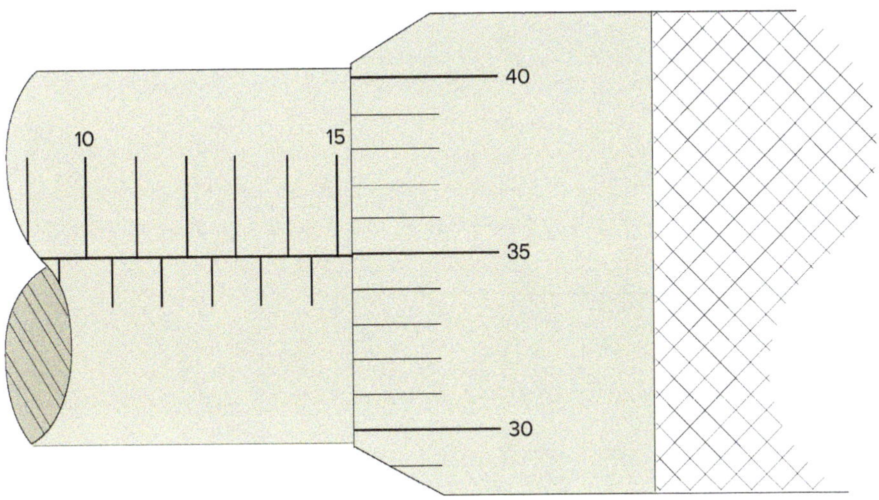

Figure 6-45 Reading, 15.35 mm.

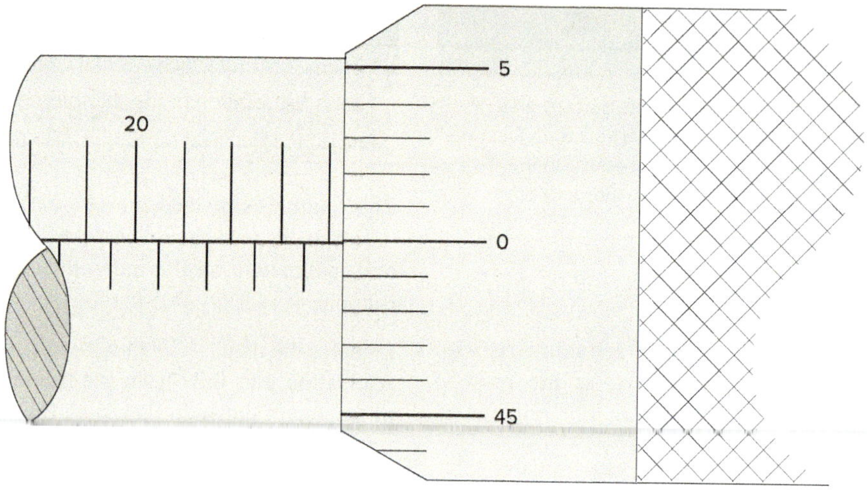

Figure 6-46 Reading, 24.50 mm.

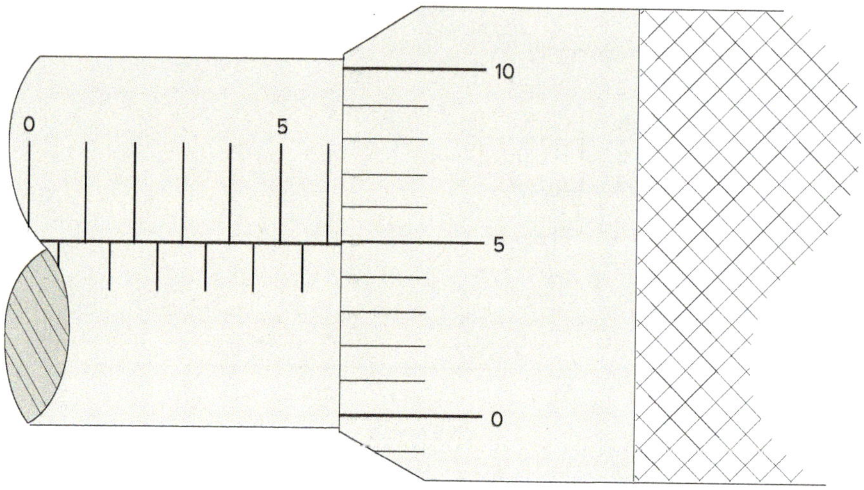

Figure 6-47 Metric mic reading, 6.05 mm.

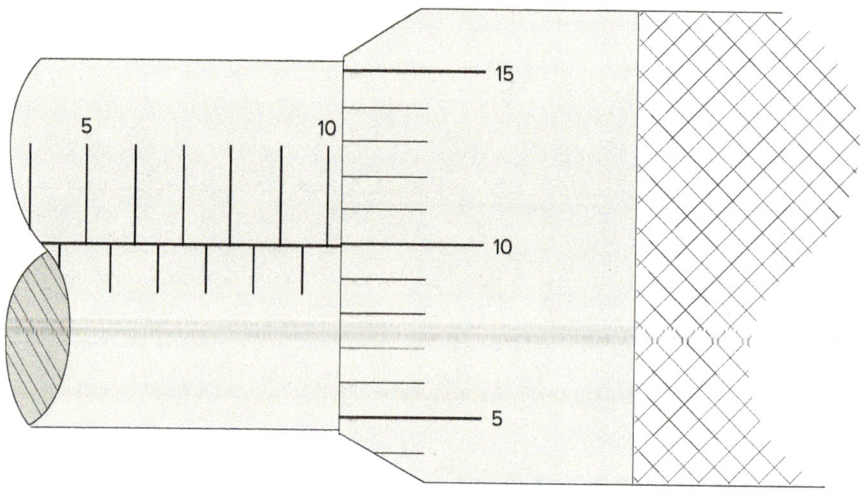

Figure 6-48 This metric micrometer reads 10.10 mm.

6.3.4 Basic Tool 4—Height Gages

Height gages are dual-purpose tools used for measuring and layout work. Layout work is a premachining activity taught in Chapter 8. Here we concentrate on measuring applications. Height gages sit on and measure up from a flat datum surface such as a granite table or a milling machine's table.

Height gages are a priority 3 for ownership. Due to their high cost, they are usually provided by the shop.

Scribe and Indicator Attachments

When measuring, height gages are used in two ways:

- With a hard scribe as a touch probe. In Fig. 6-49 the machinist is measuring the height of a part above a vise bed. To do so, they have adjusted the dial with the scribe reading zero on the vise bed before loading and machining the part. The machine table creates the flat datum, but the vise bed is calibrated to be zero height above the datum. The part height is read directly on the height gage.
- Using an indicator (Fig. 6-50) as a more sensitive touch probe for improved accuracy–this superior process eliminates feel problems. In Fig. 6-51, the indicator is zeroed on the workpiece of Fig. 6-49. We'll study indicators next.

Reading a Vernier Height Gage

It's possible to find three kinds of height gages in industry: vernier, dial, and electronic. The high initial cost of vernier height gages means they will be used for some time; plus they stand up to daily machine use. Since applying and reading electronic height gages require little training beyond the skills of the dial or vernier type, we will concentrate our efforts on vernier types. The real technique is not in reading but in using them.

Figure 6-49 A scribe can be used as a measuring probe in the height gage.
Lake Washington Technical College, Kirkland, WA/McGraw Hill

Figure 6-50 An indicator is mounted on this height gage.
Lake Washington Technical College, Kirkland, WA/McGraw Hill

Figure 6-51 The indicator could be used to measure the height of the part.
Lake Washington Technical College, Kirkland, WA/McGraw Hill

In simplest terms, a height gage is a precision rule standing upright on a sturdy base (Fig. 6-52). We could attach a sliding straightedge as a probe and read its intersection with the vertical rule. However, using this simplified height gage, the best repeatability would only be slightly better than the rule graduations.

The Vernier Concept

Vernier scales are math-based devices that divide graduations on a rule or scale into finer parts than can be seen or estimated with the eye. They improve resolution several times over. Once the vernier concept is mastered, transferring to a new tool becomes far easier. We've seen them on micrometers; reading a vernier height gage is a good place to learn the principle.

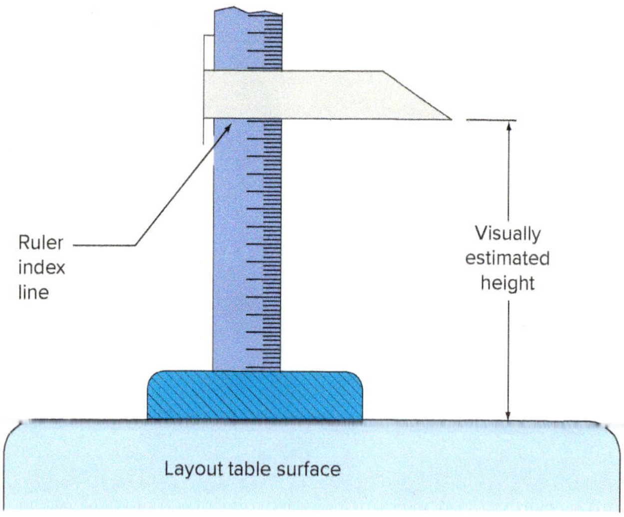

Figure 6-52 A height gage is an upright ruler with a pointer.

The Vernier Divisions Are Slightly Smaller

When used to measure distances, vernier divisions allow us to create a 0.001-in.-resolution tool without actually dividing the rule into 1,000 parts. The secret lies in a concept called colinearization, meaning that it is possible to tell when two lines coincide (line up)—better than we can read graduations. You saw this on the micrometer vernier scale.

A linear vernier scale is divided such that each graduation is 0.001 in. shorter than the rule graduations (Fig. 6-53). Each time a further vernier line coincides with a rule line, the vernier has moved 0.001 in. So, if the vernier is colinear with the rule at the fifth line, it has moved five thousandths beyond zero.

First, read the tool roughly (see Fig. 6-53). The index line points slightly beyond the 0.200-in. line on the rule. Estimating the answer helps avoid big errors. Next, the true index position is determined by counting the vernier lines until the colinear pair is found. This vernier then reads 0.205 in. This reading could be on a height gage or on a vernier caliper.

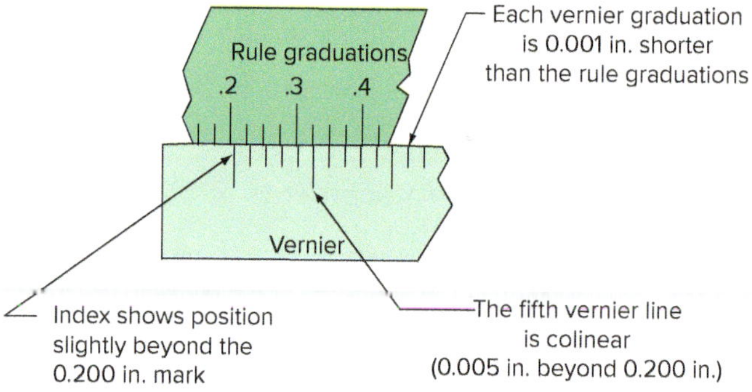

Figure 6-53 Detail view of vernier at 0.205 in.

KEYPOINT

On this linear scale, the vernier graduations are exactly 0.001 in. smaller than the rule graduations. But they could be divided into many other ratios on other tools.

There are two common Imperial height gage vernier sets:

The rule is graduated with 0.025-in. spaces.

The rule is graduated with 0.050-in. spaces.

It takes a moment to determine which is which by looking at the rule and vernier scale on your height gage. Try it yourself. Read the height gage in Fig. 6-54. Answer and instructions follow.

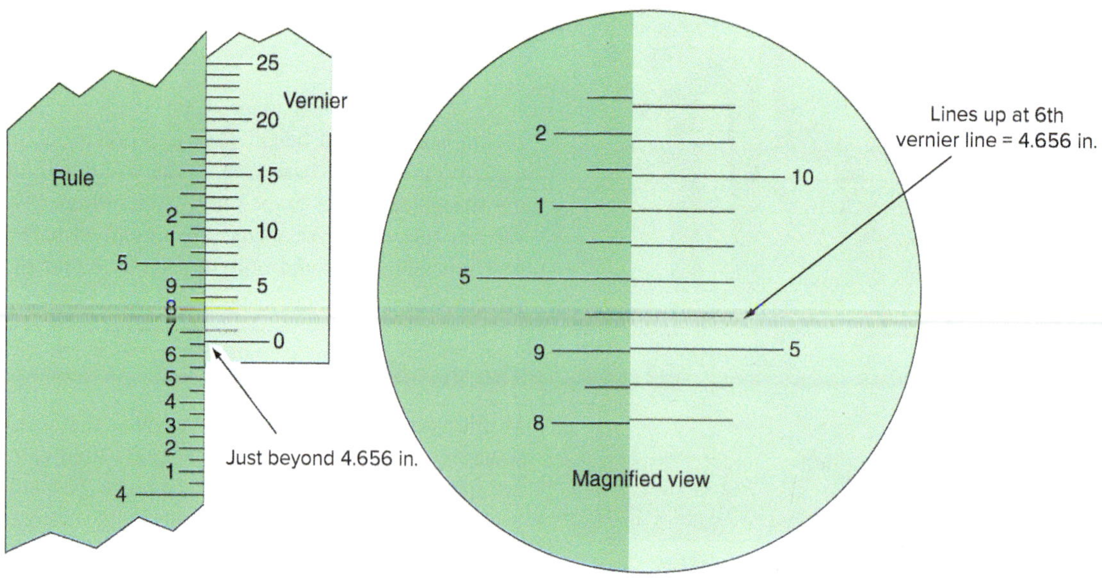

Figure 6-54 Vernier scale reading, 4.656 in.

The rule of Fig. 6-54 is graduated with 0.050-in. lines. First, add up full inches and hundred thousandths plus the graduations passed. In the illustration, the index line shows 4 in. plus 6 full hundred thousandths lines plus one fifty thousandths line = 4.650 in. (a large total).

To that add the vernier distance 0.006 thousandths = *4.656 in.*

Precision Features Height Gages

Compare the photographs in Figs. 6-51 and 6-55 to your height gage, if available. There will be differences depending on the kind: dial, vernier, or electronic.

1. *Fine movement screw.* It consists of a slide with a lock and a travel screw attached to the vernier slide (see Fig. 6-55).
2. *Calibration lock and adjustment screw* (vernier types). Many but not all height gages have some form of setting the height gage to zero. On vernier height gages, this is accomplished by moving the actual rule relative to the upright body. This is where electronic height gages are clearly better since they can zero anywhere on their scale by pushing a button! Some height gages have a small range for calibration adjustment while others allow a long range of calibration positions.
3. *Slide lock screw.* This locks the slide and attachments into position once they are set (all types).
4. Most important, *the index line.* The index points to the exact position of the tool on the rule. To avoid errors, always estimate the position using the index line on the rule, before reading a vernier (Fig. 6-56).

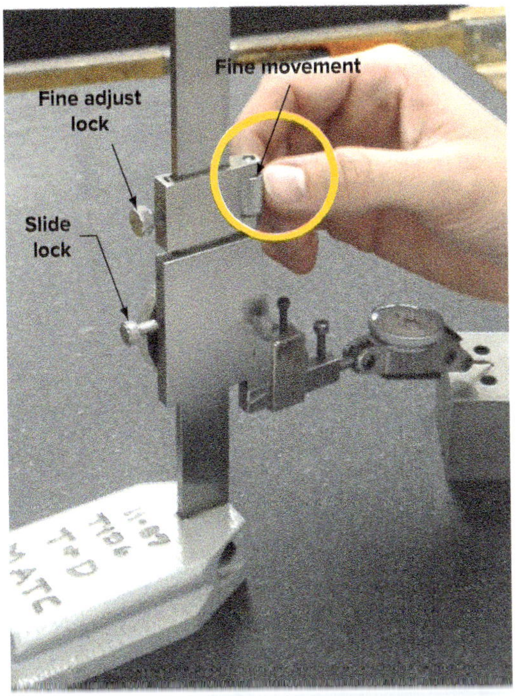

Figure 6-55 The fine movement screw helps with final adjustment of the slide.
Photography by Prographics at Milwaukee Area Technical College/McGraw Hill

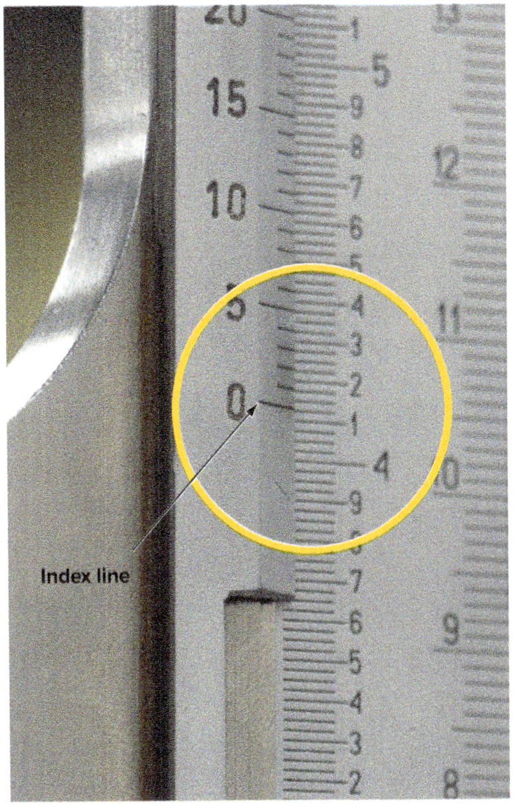

Figure 6-56 Always estimate first by reading the index line's position on the rule.
Photography by Prographics at Milwaukee Area Technical College/McGraw Hill

TRADE TIP

Paper Feelers for Improving the Feel Inaccuracy
When using a scribe as a probe, the problem is always knowing when it has just touched the surface in question—it's not always easy to tell. We often press down too hard, thus losing accuracy.

To help control this factor, use a feeler gage or a piece of paper of known thickness as a sensitivity gage between the part and scribe (Fig. 6-57). Pull it in and out of the touch zone while slowly lowering the scribe toward the surface just until the paper drags. That's the light touch! But then, after locking the slide screw and reading the height gage, don't forget to deduct the thickness of the feeler from the total.

Figure 6-57 A paper feeler improves sensitivity, but its thickness must be accounted for in the measurement.
Lake Washington Technical College, Kirkland, WA/McGraw Hill

Zero Calibrating Height Gages with a Scribe

Often, the height gage needs to be zeroed when the scribe touches the datum upon which it sits (Fig. 6-58). Zeroing is the same for all types, no matter how they are constructed. After cleaning the table and height gage base, lower the scribe until it touches the table. Once the scribe is locked at the table surface, check the reading–is it calibrated? If not, reclean and try one more time. Still off zero? Then either unlock the vertical rule from the height gage and reset it to read zero against the vernier slide or turn the dial bezel to read zero.

Figure 6-58 Calibrating a gooseneck scribe at zero against the datum table.
Left: Photography by Prographics at Milwaukee Area Technical College/McGraw Hill. Right: Lake Washington Technical College, Kirkland, WA/McGraw Hill

Layout Tables

We'll study them more in Chapter 7, but a word on the correct use of the layout table would be good here. Height gages are commonly used on special granite or older cast iron tables, also called surface plates (also called a layout table, "the rock," or inspection plate). The granite table is an ultra-precision implement right at the heart of precision in a shop (Fig. 6-59). Layout tables form the master *X-Y* datum for the shop.

Figure 6-59 The granite table is the prime datum and the basis for precision in a shop.
Lake Washington Technical College, Kirkland, WA/McGraw Hill

While layout tables are made from several kinds of granite, black is preferred because it is the most stable over a long period. In recent years, composite granite made from granite chips and epoxy resins has gained popularity, too. Since granite is harder than the tools slid upon it, its surface should last indefinitely as long as it's treated right! There are some serious expectations for correct use.

1. **Avoid Contamination**

   Clean your work before touching the table. Never set dirty parts upon the table. Avoid leaning your hands or arms on the surface, too. Oil in your skin will make tools drag when moved. Wash your hands first to remove oils. When using the table always wipe the surface before and during use. Dust settles out of the shop atmosphere all the time. A final wipe with a clean, oil-free hand is the best. You will feel any speck of dirt or chip.

2. **They Chip and Break!!!**

   Never hammer, punch, or otherwise hit any object on the layout table. Move your work to the bench for that type of operation. Note that some shops permit light taps of automatic or even pilot punches that will be deepened later using a center punch but away from the table. Some shops will not even permit that risk! Don't guess; ask first!

3. **Precision Work Only**

   Use the table exclusively for layout, measuring, and machine accessory setup. Do not store or lay unrelated items on the table. Many will feature a cover that is placed over the table when not in use (Fig. 6-60).

4. **Care for the Table**

   Follow the shop recommendations for cleaning using residue-free solvents such as acetone (read the MSDS) and lint-free shop towels—not paper towels. Occasionally, a scrub with cleansing powder and a little water or a specialized cleaner, followed with an acetone wipe and a final dry wipe, is necessary. Each shop will have its own preference. Then, cover the table when it is cleaned but not in use and set nothing on the cover.

Figure 6-60 When not in use, cover the granite table to protect and keep it clean.
Lake Washington Technical College, Kirkland, WA/McGraw Hill

A well-maintained layout table is a delight to use. Precision tools slide as though they were on magic carpets. In fact, be sure to hold onto precision tools; more than one has been given a slight push to see it go all the way across, then crash to the floor if not held back!

6.3.5 Basic Tool 5—Indicators

Indicators are super universal measuring tools. As such, one or more, along with one or two indicator holders, are a priority 1 purchase. They are used throughout the shop in three different ways:

1. **Axis Control**

   To monitor machine movements where exact motion is required. Example: on the *Z* axis of a manual lathe.

2. **Setup Alignment**

   To set up machine accessories or align accessories such as vises true to the machine or to align a workpiece in a chuck or vise.

3. **Measure**

   Improved measuring at the inspection table and on the machines.

It is this measuring aspect we will study. The other uses will be covered in the lathes and mill chapters. The indicator will be mounted in a height gage.

Two Common Types

Dial test indicators (often abbreviated to DTIs) are supplied in two basic forms: probe and travel types (Fig. 6-61). The differences in these two types are discussed later. Both are precise and delicate and have a particular application. Both measure in thousandths or ten thousandths of an inch or decimal parts of a millimeter depending on construction. The indicators graduated with 0.0001-in. divisions are referred to as *tenth indicators* in shop lingo.

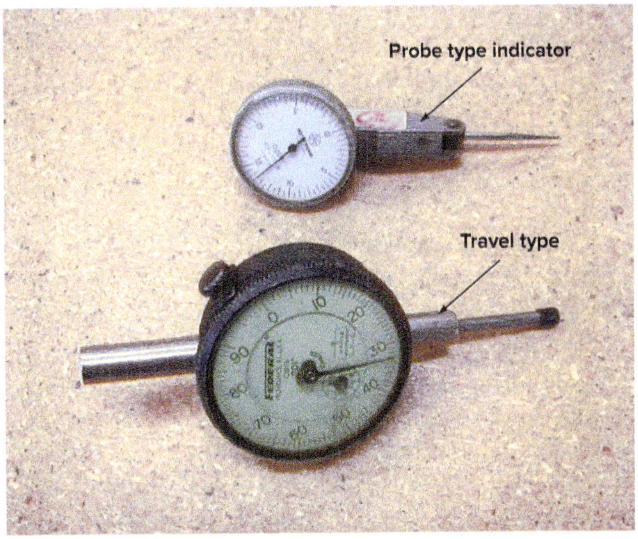

Figure 6-61 Two kinds of indicators—travel on the bottom and test or probe type on the top.
Lake Washington Technical College, Kirkland, WA/McGraw Hill

Travel indicators are sturdier, and they measure greater ranges of movement. Probe indicators are smaller and versatile, thus they can be used in a greater number of applications. We will be using the probe type for all height gage measurements

Here at last is a measuring instrument free of human feel problems. They are, however, subject to parallax errors and wear. That's the good news. But the bad news is that both types of indicators have their own unique factor that we will study in detail coming up. Once understood, it can be controlled.

Tech-Specs for Indicators

Correct Application

Job tolerances below 0.001 in. require indicators. The tenth indicator (0.0001 in.) is one of the most accurate instruments in a machinist's toolkit.

Target Repeatability

Resolution varies with indicators supplied in 0.001-in., 0.0005-in., or 0.0001-in. graduations. Repeatability should be close to the resolution due to the lack of inaccuracy factors.

> **SHOPTALK**
>
> **Buying Tools** Here's a great hint on how to get started buying your toolkit: Find a retired toolmaker or machinist. They are often glad to see their life's collection go to an up-and-coming craft candidate. Usually the box you will get is filled not only with expensive measuring tools but with bits and pieces collected over many years. I bought one full toolbox from a retired journey machinist when I started, and a second donated tools to me upon learning how serious I was about the trade! This philosophy fits the trade saying of "get your skills from us for free but then give them back to the next generation."

Indicator Holders

All indicators require mounting accessories that either clamp or magnetically hold the indicator in the setup or measuring situation. For the measuring discussion, the indicator is held in a height gage using an attachment bar supplied with the indicator, the same size as the scribe (Fig. 6-62).

For working in the machine area, one or two types are shown: the spindle clamp for milling machines (Fig. 6-63) and the magnetic mount used on all machines (Fig. 6-64). Both are priority 1 tools.

6.3.6 Three Methods of Measuring Distances

This is a perfect example of choices that must be made to match the job tolerance to a measuring process. Although performed with the same three implements–height gage, indicator, and table–the three methods discussed next differ greatly in total distance that can be measured, expected repeatability, and time required to perform the task.

- Dial Measuring
- Comparison Measuring
- Height Gage Scale Measuring

Each has its correct use and resolution.

Figure 6-62 The indicator bar replaces the scribe.
Lake Washington Technical College, Kirkland, WA/McGraw Hill

Figure 6-63 A spindle mount indicator on a vertical mill.
Lake Washington Technical College, Kirkland, WA/McGraw Hill

Dial Measuring

This method directly detects the distance in question on the dial face. Therefore, the distance to be measured must fit within the indicator's range. In Fig. 6-65, parallelism is being measured. The indicator is mounted in the height gage, then moved over the entire top surface. The total difference in the highest and lowest point is observed in thousandths of an inch.

Comparison Measuring

The second method's advantage is near perfect accuracy. In Fig. 6-66, the indicator is used as a transfer tool to compare a part to a master standard of the expected nominal result.

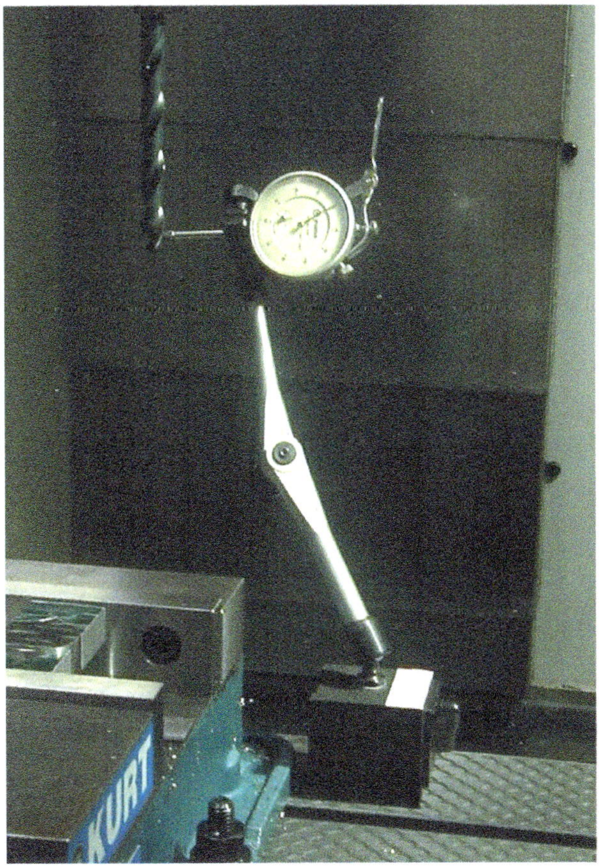

Figure 6-64 A magnetic base is useful for mounting indicators on machines.
Lake Washington Technical College, Kirkland, WA/McGraw Hill

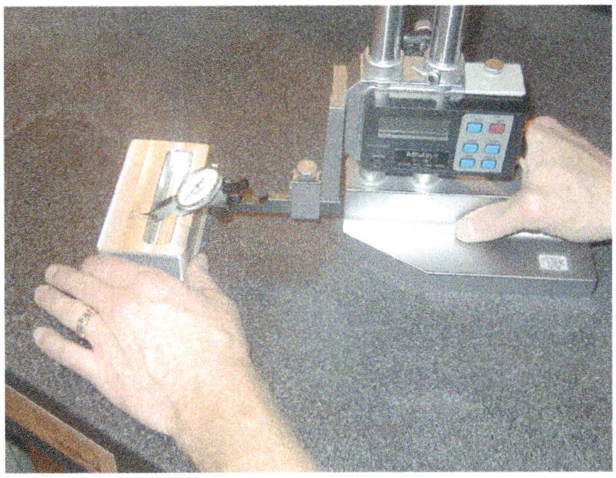

Figure 6-65 Reading parallelism FIM.
Lake Washington Technical College, Kirkland, WA/McGraw Hill

The nominal dimension for the drill gage (2.5000 in.) is set up in precision gage blocks (hard steel blocks of the exact nominal size). Then the indicator is set to read zero touching the master. Next it is moved over the part, and the difference shows on the indicator.

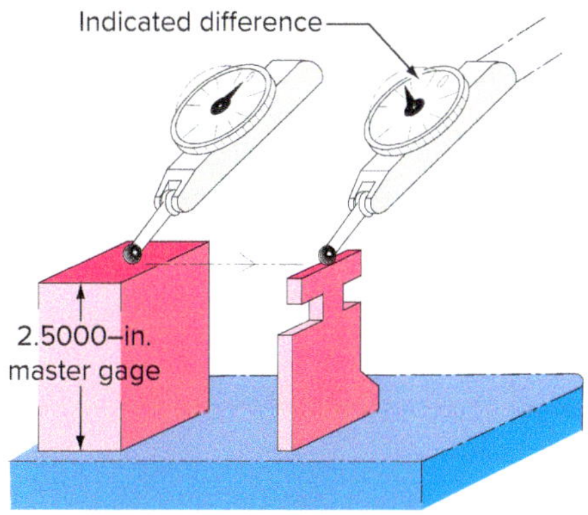

Figure 6-66 Comparison measuring requires a master gage close to the size of the object.

The difference on the dial face is either added to or subtracted from the standard amount. Free of alignment, feel, and calibration problems, this method is accurate to 0.0001 in. as long as the indicator is so graduated and the standard is clean and accurate.

KEYPOINT

Other than CMMs, this is the most accurate measuring method you will learn in this book.

Height Gage Scale Measuring

Here the indicator is used solely as a sensitivity probe. Accuracy will be only as good as the resolution of the height gage since we will be reading distances from its scale. In Fig. 6-67 we are measuring the height of this transmission case from the lower shelf to the top. First the height gage is lowered until the indicator reads zero on its dial face at about half of its total travel. That sets how much pressure is brought to bear on the surface. Now the height gage is read at the first position location. Or it is reset to zero if it is electronic.

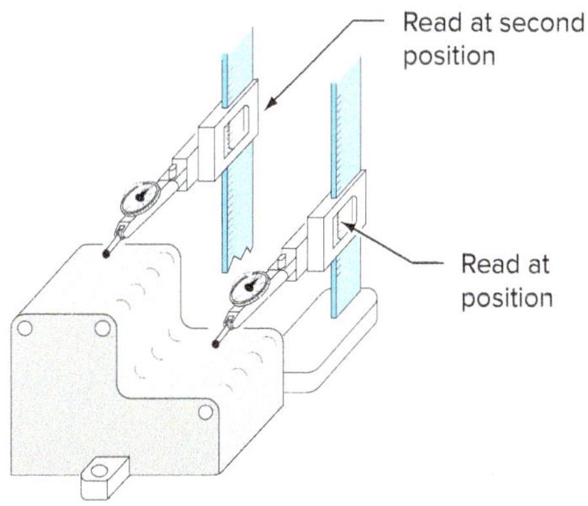

Figure 6-67 Measuring using the height gage as the instrument and the indicator as a pressure probe.

Gently pulling the probe off the surface to not disturb its position on the height gage, the slide is lifted above the second surface, then slowly lowered until reaching zero on the dial face again (same pressure). The position is now read directly from the height gage's readout if it's electronic. Or if it's mechanical, the first position is subtracted from the present to yield the measured height.

6.3.7 Selecting the Gage Standard

The best master for measuring (Fig. 6-68) is chosen from precision gage blocks (studied in detail in Chapter 7). It should be as close as possible to the expected part size.

Figure 6-68 A set of precision gage blocks can be used for comparison measuring.
Lake Washington Technical College, Kirkland, WA/McGraw Hill

6.3.8 Comparing Indicators

Let's look more closely at the two indicator types: test probe and travel indicators. Both do the same job but in a different way.

Travel or Plunger Type Indicators

Sometimes called long-range indicators, these DTIs are used for alignment and axis control; they can be used for measuring, too. The plunger moves in a straight line, while the dial rotates to indicate total distance traveled. They are supplied in many ranges of travel from 0.250 in. up to specialized versions detecting several inches. The 1-in. travel indicator is the most popular. Many but not all DTIs of this type feature one full revolution of the dial equal to 0.100-in. or 4-mm movement of the plunger.

Each single rotation of the main dial is called one *clock*. To track the total number of clocks, a second dial hand is included. Read the label near the center of the face to discover the resolution and what one clock represents. Read the indicator shown in Fig. 6-69. This travel indicator reads 0.445 in. (Four clocks plus 0.045 in. on the dial face.)

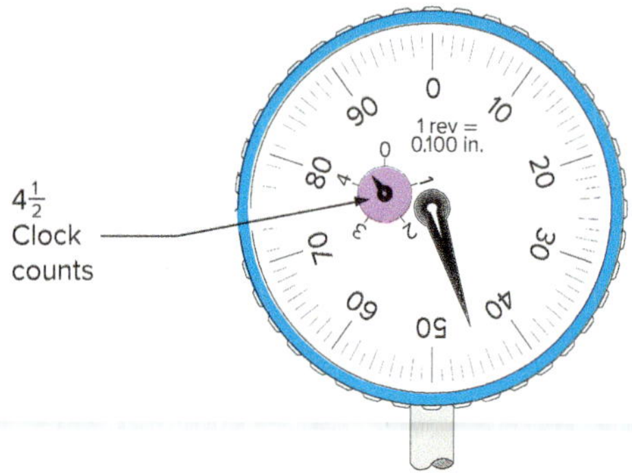

Figure 6-69 A travel indicator with a clock counter.

Control of Cosine Error

Travel indicators are subject to a particular kind of operator-induced problem called **cosine error** (Fig. 6-70). It is caused by not setting the plunger axis perpendicular to the direction to be measured. In the drawing, notice the distance A-B compared to the distance detected by the indicator B-C. The tilt of the probe axis *amplifies the result* to a larger false FIM (full indicator movement). Preventing cosine error is easy; if the probe axis is eyeball close to square to the measurement, results will be well within shop accuracy.

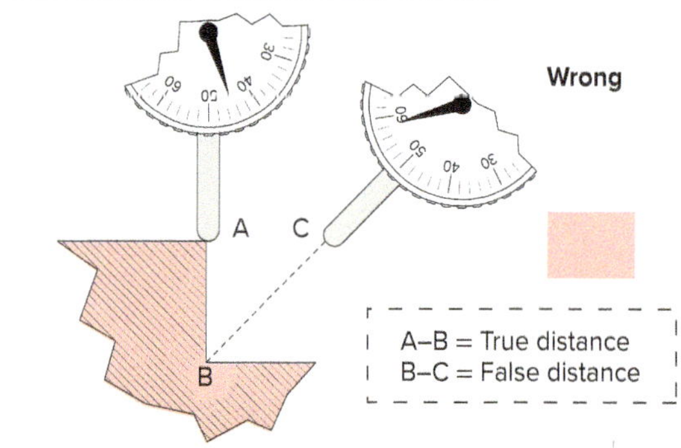

Figure 6-70 Cosine error is caused by tilting the indicator.

Test Indicators

Due to their construction, these indicators are limited in total range to around 0.030 in. or 1 mm. Beyond that, they lose accuracy.

Body Position versus Probe Position

In using a test indicator, there are two setup positions: the body and probe angles. The body position is noncritical. It's adjusted to keep the indicator out of the way of the setup and to position it such that it can be read (Fig. 6-71).

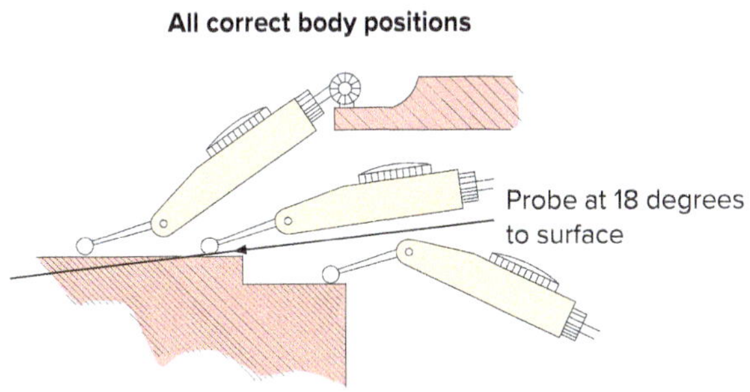

Figure 6-71 Indicator body position has no effect on measurements.

KEYPOINT

The test indicator's body can be set at any convenient angle with respect to the work being tested with no effect on accuracy. It's the probe angle that matters.

Probe Position Is Critical

Setting the probe angle relative to the surface to be measured is very important. It should be a specific angle, usually 18°. However, from 15° to 20° will produce shop accurate results. This 18° angle is true for most (but not all) brands of test indicators–known as its base angle.

If not correctly positioned, a subtle inaccuracy is produced. The resultant reading will either be amplified or reduced relative to the true distance as the probe departs from the correct base probe angle. Look at Fig. 6-72 where the probe is measuring from the upper surface to the lower and see if you can reason out why the probe angle must be right. The secret is that as the probe moves in an arc, not in a straight line, the gearing inside translates that into thousandths of an inch on the dial face. Truthfully, there is no arc that equals the true distance to be measured; however, the movement of these indicators is internally corrected to read within shop requirements when the probe is set approximately at 18° relative to the surface being tested.

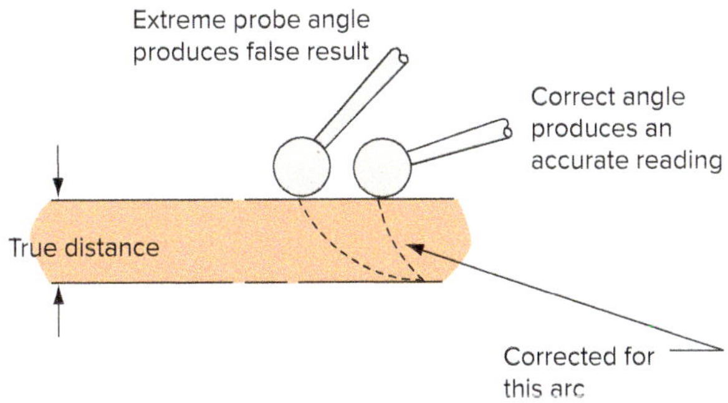

Figure 6-72 Progressive error caused by missing the indicator's base angle.

However, the extreme probe angle on the left produces a much longer arc as it swings from the top to the lower surface. It shows a false distance bigger than the truth. Conversely, if the probe angle is set much less than 18°, closer to parallel to the surfaces, the result will be reduced. The effect is known as **progressive error**–the farther from the center of the base angle the probe swings, the worse the error becomes.

Setting the Base Angle

While any change in probe angle will affect results, the good news is that the 15° to 20° range is acceptable or even a bit more when tolerances aren't very tight. Now, can you see why these probe type indicators must have a limited travel range? Otherwise, the longer swing would produce too much progressive error. Keep in mind that the body angle in Fig. 6-73 has no effect on the result–only the probe angle matters.

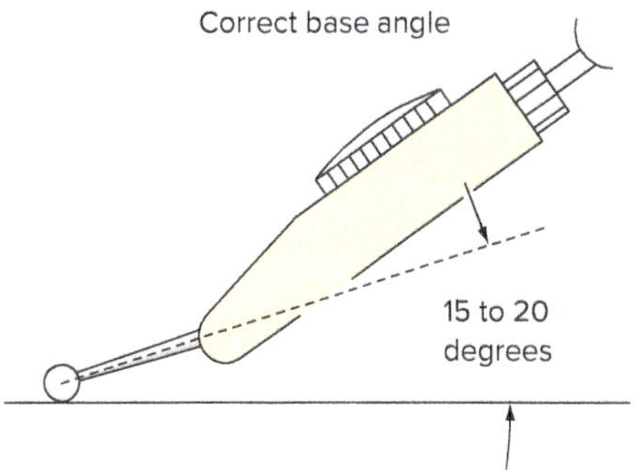

Figure 6-73 The correct base angle for accuracy.

TRADE TIP

Know Your Indicator's Base Angle Not all test indicators feature the 18° base angle. There are some compensated for 0°. How do you know? Three possible ways: New indicators come with a certificate that shows the base angle. Some have their base angle engraved on the body. Third, a few manufacturers machine the body with a template as shown. This is not meant to dictate what position the body must be, but it does provide a visual reminder of the size of 18°. To find if yours does, check the body with a protractor! See Fig. 6-74.

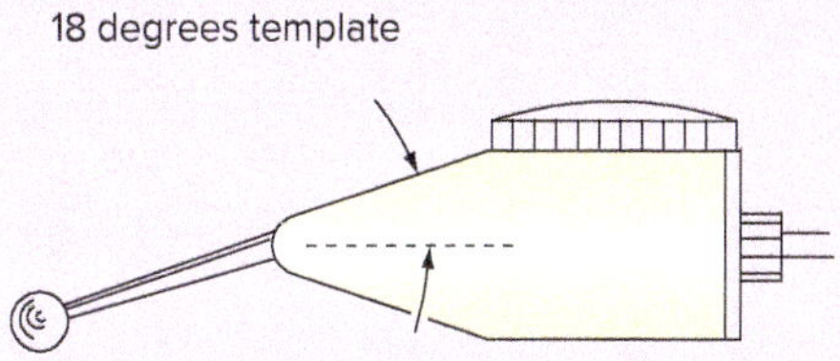

Figure 6-74 Some indicators feature their base angle as part of the case shape.

6.3.9 Smart Height Gages

Combining a microprocessor with a height gage, smart height gages perform several tasks that previously required calculators and multimovements of the probe (Figs. 6-75 and 6-76). For example, with two touches of a probe, smart height gages can automatically calculate slot or hole centerlines! They can indicate with color–in or out of tolerance, save locations, and compare heights–and output distances between features without operator calculation.

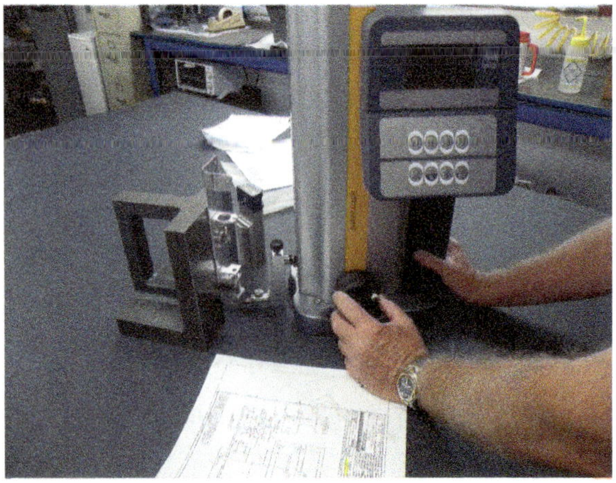

Figure 6-75 Multistep measuring routines are simplified with a smart height gage.
Courtesy Universal Aerospace/Arlington, WA

Figure 6-76 Preprogrammed routines are accessed with function buttons.
Courtesy of Universal Aerospace, Arlington, WA

UNIT 6-3 REVIEW

Replay the Key Points

- There are two types: plunge types, also called travel indicators, and test types.
- Both types have a possible error due to misaligning the indicator to the measurement to be taken.
- To avoid progressive error, the test probe must be set at 15° to 20° to the direction to be measured.
- To avoid cosine error, the plunge type must have the plunger probe directly in line with the measurement to be taken.

Respond

1. How fine might a skilled machinist read a precision rule?
2. At what minimum (tightest) tolerance should the machinist forsake the calipers for outside micrometers?
3. Why are calipers not used as very accurate tools–what challenges their accuracy?
4. If a travel indicator is tilted relative to the direction to be measured, the result will be an amplified reading. Is this statement true or false? If false, what will make it true?
5. The error described in Question 4 is a progressive error. Is this statement true or false? If false, what will make it true?

CHAPTER 6 REVIEW

Unit 6-1

One of the most basic mistakes made is to begin a job without knowing the exactness of the upcoming operations. Without that knowledge, the machinist might choose the wrong process both for shaping the part and for measuring the dimensions. Always look at the print to determine tolerances before beginning the job.

Unit 6-2

Although we've discussed many ways to control results and improve personal repeatability, the process has only just begun. Staying accurate is a lifelong process–no different from playing a musical instrument. It takes regular practice and awareness.

Unit 6-3

I taught machining for engineers for 1 year in New Zealand. On my first day in class, the veteran lab assistant handed me a micrometer and asked me to "check it out for feel." After I took it and measured a gage block, the lab assistant said, "I can see you are going to work out just fine by the way you held that micrometer." No kidding, little things like correct hand position show you to be a well-trained pro.

QUESTIONS AND PROBLEMS

1. Of the four kinds of tolerancing found on drawings–*bilateral, unilateral, general,* and *limits*–

- **A.** Which is stated as the minimum and maximum size?
- **B.** Which usually has no stated nominal size?
- **C.** Which is determined from a table on Sheet 1 of the drawing?
- **D.** Which is determined by counting the number of places to the right of the decimal point?
- **E.** Which extends one direction from nominal? (Unit 6-1)

2. A *protracted* angle is dimensioned and toleranced in what units? (Unit 6-1)

3. True or false? Developing the best possible resolution for any given measuring tool should be a personal goal. If it's false, what will make it true? (Unit 6-2)

4. Define parallax error in two words. (Unit 6-2)

5. List the 10 factors that might plague a given measuring tool. (Unit 6-2)

6. What advantages are there to electronic calipers compared to dial? (Units 6-2 and 6-3)

7. Complete this: Cosine error occurs when a(n) type dial test indicator is relative to the measurement. (Unit 6-3)

8. Of the three measuring techniques involving a test indicator and height gage,

- **A.** Which involves gage blocks?
- **B.** Which requires a datum table?
- **C.** Which does *NOT* use the DTI as a measuring tool but rather a pressure sensing probe? (Unit 6-3)

9. The friction drive on a micrometer helps avoid. (Unit 6-3)

CRITICAL THINKING PROBLEMS

10. You have two micrometers in your toolbox, one with a friction drive and one without.

- **A.** Which would be the most accurate for you alone?
- **B.** Which would be the most accurate when you and another machinist use them? Why? (Unit 6-3)

11. Why would the repeatability of a dial caliper with graduations of 0.001 in. be similar to that of an electronic, digital caliper that reads to 0.0005-in. increments? (Units 6-2 and 6-3)

12. You wish to dimension and tolerance the diameter of a hole in which a precision pin must fit. The design calls for a slip fit between the pin and the hole not closer than 0.001-in. clearance between the two. In other words, under no circumstance can the pin be any closer to the hole size than 0.001 in. so it will always slide freely within the hole. The pins are commercially supplied at 0.875 in. with a guaranteed unilateral tolerance of plus 0.000 in. and minus 0.0005 in. (plus nothing, minus a half thousandth). You are to allow three thousandths total variation on the hole size. Using a bilateral tolerance, what should the hole's nominal size be and the tolerance? (Unit 6-1)

13. Explain how progressive error might be used to your advantage. This was not directly covered in the reading–discuss it with other students. (*Hint:* Review the uses of an indicator in the shop.) (Unit 6-3)

14. Describe the principle on which all vernier scales are based (not how to read them but how they work). (Unit 6-3)

15. If a micrometer was heated above the temperature of the object being measured, would it measure larger or smaller than the true distance? (Unit 6-2)

16. A. You are to machine a very accurate hole. What five controls might the engineer require for that hole? (Unit 6-2)

B. If the drawing used GDT, which of the five would require a datum reference? (Unit 6-2)

17. Describe a geometric angle tolerance zone. (Unit 6-1)

18. Describe one method of measuring a geometric angle using a protractor; use three facts. (Unit 6-2)

19. The graduation lines on the sleeve of a metric micrometer represent how much distance each? (Unit 6-2)

CNC QUESTION

20. You are running a high-speed turning center, making 500 parts. Due to your sharp editing of the program, the part cycle time (one part completed) has been cut to less than 3 minutes! You are trying to burr the parts as they come off and stack them carefully in egg crates, so there's not a lot of time to measure a critical diameter to a tolerance of five thousandths inch. What is the fastest measuring method you can think of that will reproduce within that tolerance?

Design element (gear): ©Design Pics/PunchStock

CHAPTER 7 Single-Purpose Measuring Tools, Gages, and Surface Roughness

Units

7-1 Measuring with Inside and Depth Micrometers

- Assemble and read an inside micrometer to a repeatability of 0.001 in.
- Set up and read a depth micrometer to a repeatability of 0.001 in.

7-2 Setup, Use, and Care of Precision Gage Blocks

- How to select blocks
- How to use gage blocks

7-3 Gage Measuring

- Measure inside diameters using four types of small-hole gages
- Set up and measure using adjustable parallels
- Measure radii using radius gages

7-4 Measuring Angles

- Use decimal degree and degree–minute–second angles
- Use protractors to measure angles
- Set up sine bars

7-5 Measuring Surface Roughness

- Interpret surface roughness symbols on prints
- Estimate and gage roughness using a comparison plate

Final Review Quiz—Choosing the Right Measuring Tool

- Choosing the right tool and process to inspect a sample part
- The review of Chapter 7 simulates the real task in the shop—choosing the right measuring tool for the job—it's a summary of both Chapters 6 and 7 and a brainteaser

INTRODUCTION

The measuring tools and processes of Chapter 7 are more specialized. They fill in gaps in the ability to perform close tolerance measuring in special situations. We'll also study measuring angles and surface roughness. On completion of Chapter 7, you'll have a fairly well-rounded set of entry-level skills.

Similar to other technologies, measuring is a never-ending lesson that's evolving with advancing technology. New instruments become available regularly. The best way to stay informed is through tool catalogs and flyers sent to your shop or home. Go to the Internet and search for precision + measuring tools; then sign up for their mailing lists.

Unit 7-1 Measuring with Inside and Depth Micrometers

Introduction: These two micrometers measure linear distances using contact pressure applied by touch. In addition to touch variables, each has its own list of inaccuracy factors. Each requires practice to use it correctly.

TERMS TOOLBOX

Base size The smallest size possible to measure for inside or depth micrometers.

Drag The relative tightness or looseness of a measuring tool, set by the wear adjustment ring within the head.

Jacking The undesirable act of lifting the base of a depth micrometer up off the surface upon which it rests.

Joggling Author's term for the act of lightly moving a measuring tool until the null point is detected.

Null–null point A maximum (or minimum) value detected on a measuring tool when it is moved into alignment.

7.1.1 Inside Micrometers

Inside micrometers (Fig. 7-1) measure hole diameters within a range from around $1\frac{1}{2}$ in. up to 6 in. for the standard set, or for expanded sets to 12 in. There are much larger sets owned by the shop, not the machinist, that measure up to several feet. A standard set would be a priority 2 or perhaps 3, for your purchase plan.

Tech Specs for Inside Micrometers

Correct Application

These measuring instruments are applicable where diameter tolerances are closer than 0.002 in. but not closer than around 0.0005 in. depending on the situation. Closer inspection of hole diameters requires bore gages (discussed next) and other hole-measuring technologies beyond the scope of this section.

Inaccuracies

In addition to the usual factors of parallax, wear, heat, and burrs, there are two new challenges to control for accuracy. Please obtain an inside micrometer now as we look at the features and factors (Fig. 7-2).

Figure 7-1 The complete inside micrometer set includes the $\frac{1}{2}$-in. range head, extension rods, and the $\frac{1}{2}$-in. spacer.

Prographics at Milwaukee Area Technical College/McGraw Hill

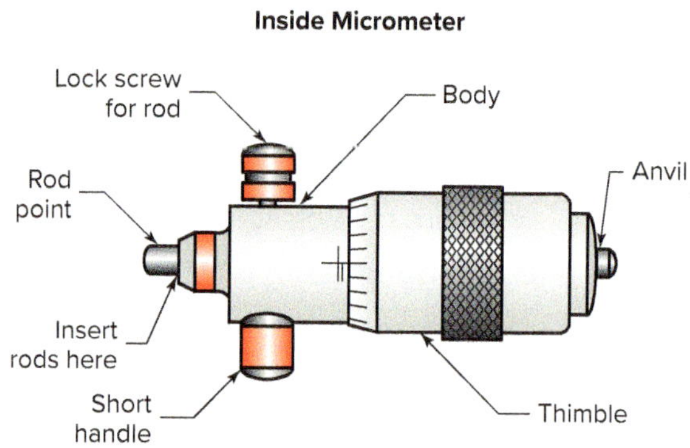

Figure 7-2 Inside micrometer components.

Double Calibration

Note that the set is composed of a measuring head and several precision rods. On some sets, not only must the micrometer head be calibrated to remove zero error, but each interchangeable rod must be adjusted, too. Adjusting out error in these tools is time-consuming.

Inside micrometers can be checked for zero index error with outside micrometers, but introducing a tool with its own inaccuracies as a checking gage degrades results. The best solution is a precision ring gage (Fig. 7-3). A ring gage is a hardened steel, ground donut made to an exact inside diameter. Several must be owned by the shop if it is to calibrate inside mics and other inside measuring tools as well. Often, shops will purchase the major sizes such as 3.0-in., 4.0-in., 5.0-in. diameter, and so on.

Figure 7-3 Inside micrometers and dial bore gages (studied next) are best calibrated against precision ring gages.
Lake Washington Technical College, Kirkland, WA/McGraw Hill

Feel Challenge

Differing from other kinds of micrometer measurement, the inside mic is not adjusted when the micrometer head is in the hole. That's because your fingers may not fit in. So, finding the **null** or **null point** when measuring (the largest micrometer size that will just pass through the hole) is a matter of trial and error. With the micrometer set to a trial size, insert it to see if it fits. Then take it out, adjust up or down, and try again until it slips through at the right touch for which it was calibrated. It's a rocking motion feeling for the right drag as it goes through.

The alignment method of lightly moving a measuring tool is called **joggling** in this book; you've already used it for other measuring tools in Chapter 6. It's a light wiggling side to side and back and forth, rocking it through the hole. If it falls through easily, remove it and turn the head out one or two thousandths and try again.

> **KEYPOINT**
>
> To joggle the inside mic, hold the measuring head end against the hole wall, then joggle the other end through.

Set with More Drag

The inside mic's head has no lock. Instead, it is set much tighter than all other micrometers. That's so it won't move when joggled and removed.

Target Personal Repeatability

Your target should be 0.0005 in., depending on hole diameter (larger holes are easier to measure) and the machined roughness in the hole.

There is no vernier scale on an inside micrometer to subdivide thousandths of an inch. It would be pointless, since their discrimination is hampered by the extreme feel factors and the way they must be used inside a hole.

> **SHOPTALK**
>
> **The Rule of Ten** Conscientious machinists live by this rule (as much as possible): Any measuring tool or process chosen should be at least 10 times more accurate than the tolerance for the feature. Otherwise, needless accuracy is given away by introducing variation in measuring tool or process not in the machining.

Setting Up, Reading, and Using an Inside Micrometer

Setting Up

Make absolutely certain both the rod and mating head surfaces are clean before insertion; then tighten the retaining cap. Test it against a micrometer to get the feel factor started.

Half-Inch Range

Now, notice that the inside micrometer head is smaller than other micrometers. It is limited to a 0.5-in. (or 13-mm) range. That keeps it compact enough to be used in smaller diameter holes.

The basic micrometer with the shortest rod measures from around $1\frac{1}{2}$ in., depending on the makeup, up to $2\frac{1}{2}$ in. Each rod covers a range of 1 in. but in half-inch increments. The first half-inch of the range is achieved without the spacer. To set the micrometer up for the second half of the range, the half-inch spacer must be added. Identify the spacer now. It is also shown in Fig. 7-1. Examples of this setup are seen in Fig. 7-4.

> **KEYPOINT**
>
> The $\frac{1}{2}$-in. spacer in an inside micrometer set is a simple, small item that can be mistaken for a small bushing and misplaced easily. Without it, the set becomes useless for half of all its potential range.

Skill Tips—Reading Inside Micrometers

Here are four little hints to help obtain the best accuracy from an inside mic:

1. **Ruler Double-Check**

   After completion of the measurement, it's a good idea to double-check the results against a precision ruler.

Inside Micrometer Ranges

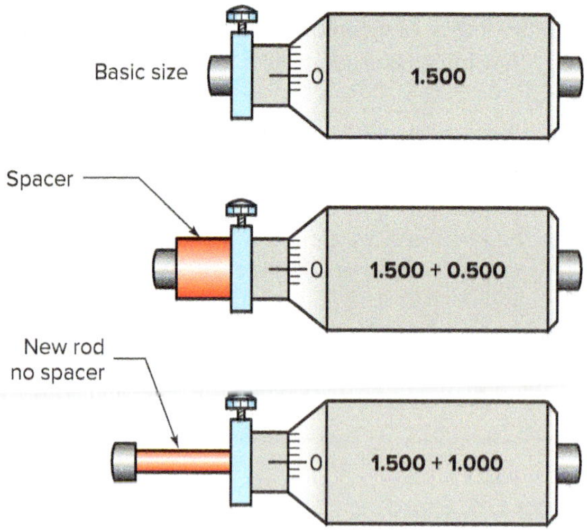

Figure 7-4 Three ranges by the way the inside mic is set up.

2. **Extension Attachment**

   The set may include an extension handle similar to the one in Fig. 7-5. It is used to get the micrometer into places where you would not be able to reach with your hand alone.

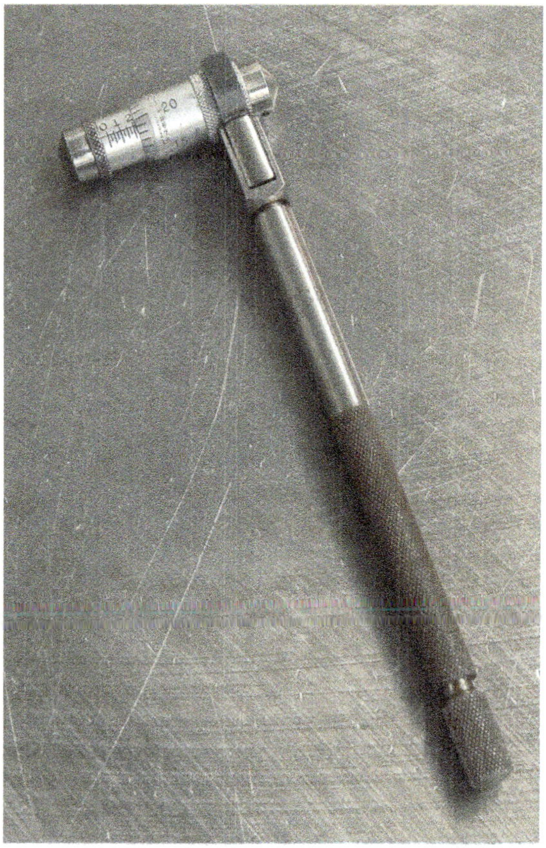

Figure 7-5 The extension handle helps reach into deep holes or tight places.
Prographics at Milwaukee Area Technical College/McGraw Hill

3. **Check with an Outside Micrometer**

   When taking an inside mic out of your toolbox always test the basic head against a calibrated outside micrometer unless a ring gage is readily available. It's also a good idea to double-check final results using an outside mic–double-check just to be sure!

4. **Test Several Locations**

   It's good practice to test the hole in several locations to verify cylindricity. It might not be round or cylindrical! Three tests minimum at each circular element are recommended. In other words, measure each element three times, then test other circular elements. Vary the contact point with each test.

Reading the Inside Micrometer

When reading an inside mic, you will add three or four factors together for the final measurement, depending on whether the spacer is used or not.

Base Size First

Determine the basic or **base size** of the inside micrometer–that is, determine the smallest measurement it can make with the first rod and no spacer. That size will often be engraved on the thimble or body of the tool. For example, in Fig. 7-6, if the thimble of this mic was at zero (0.000) it would be at 1.500 in. What does it read as set? *Answer* 1.502 in.–just slightly larger than its base size on the thimble.

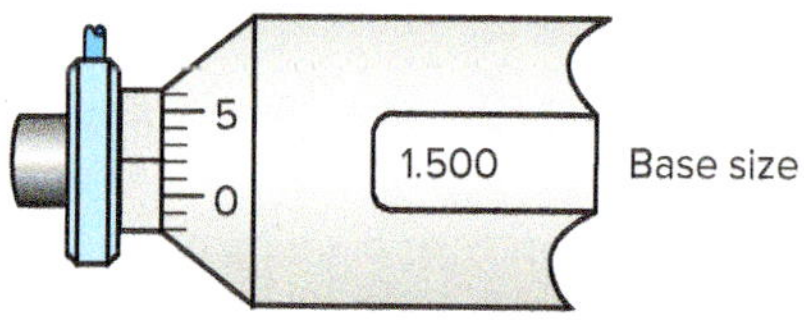

Figure 7-6 Read this inside micrometer.

From that we see that any given measurement will be the total of

Base head size + Head reading
+ Rod + Spacer (if it's used)

Sound difficult? Okay, maybe a little, but the double-checks with a micrometer or ruler will help offset the added complexities.

KEYPOINT

We'll assume a base size of 1.500 in. for all exercises here, but that's not always true with all inside mics! If it's not engraved on the tool, use your outside micrometer to find out for sure the smallest size with the base rod and no spacer.

TRY IT

Question? What would the reading be if the inside mic in Fig. 7-7 had the spacer? Now read Fig. 7-8.

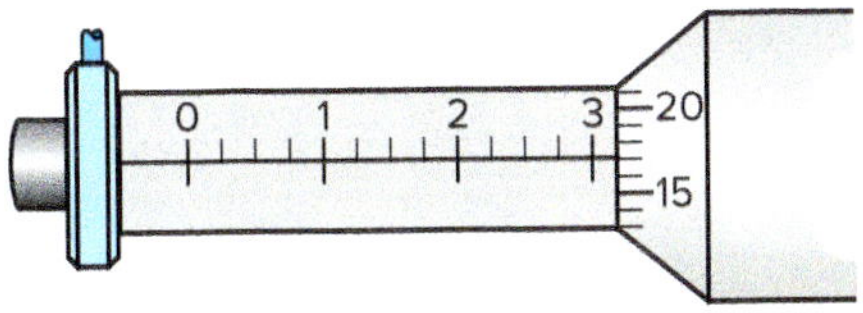

Figure 7-7 Read this inside micrometer.

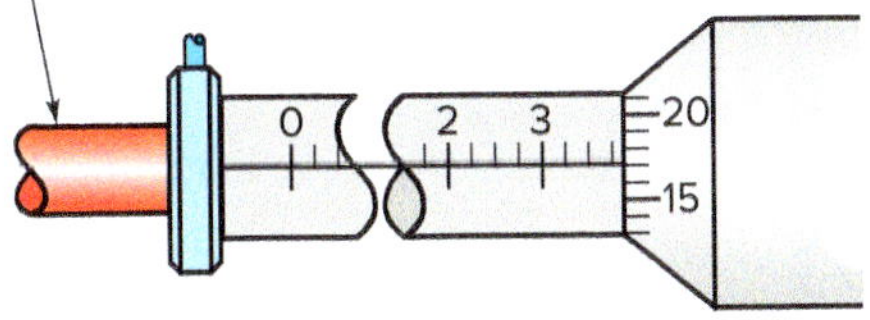

Figure 7-8 Read this micrometer with a longer extension rod.

ANSWERS

A. 1.5 + 0.317 = 1.817 in. (no spacer)

B. 1.5 + 0.500 + 0.317 = 2.317 in. (spacer adds 0.500)

7.1.2 Depth Micrometers

Like inside micrometers, depth micrometers are purchased in universal sets measuring from 1.5 in. up to 6 or 12 in. They are also a priority 2 or 3 in your purchase plan. Depth mics have their own inaccuracies that must be recognized and controlled, but they are more accurate than caliper depth measuring.

Tech Specs for Depth Micrometers

Correct Application

Depth micrometers (Fig. 7-9) are used when measuring linear distance between two surfaces. With practice, they deliver personal repeatabilities finer than 0.002 in. but not finer than 0.0002 or 0.0003 in. due to their feel factor and head construction. Reading finer than 0.001 in. requires visual estimation as there is no vernier scale (Fig. 7-10).

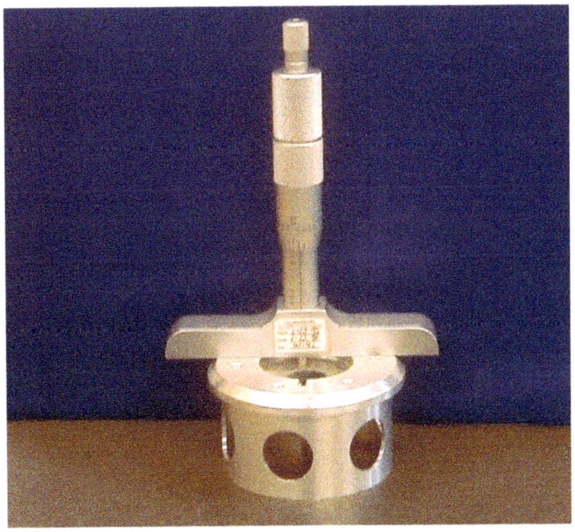

Figure 7-9 This depth micrometer is being used to measure this precision ratchet part in the shop.
Lake Washington Technical College, Kirkland, WA/McGraw Hill

SHOPTALK

Inside Mics Are Too Slow for CNC Production! For faster measurements with better repeatability, the dial bore gage is one of the best hole-measuring tools (see Fig. 7-10). Bore gages are set up to measure a specific diameter with a repeatability of around 0.0002 to 0.0001 in. Using a ring gage, they are set to read zero at the nominal size for a specific measurement. With no math and being fast and repeatable within tenths, these tools are absolutely necessary for CNC production where messing around with inside mics would be way too slow and inaccurate. More to come on bore gages in Unit 7-3.

Figure 7-10 A dial bore gage is far more repeatable and easier to use than an inside micrometer.
Lake Washington Technical College, Kirkland, WA/McGraw Hill

Target Personal Repeatability

This needs to be from 0.0003 to 0.0005 in., depending on the situation.

Control of Inaccuracies

The major factors for depth micrometers are calibration and feel.

Calibration

Notice that these too have calibrating rods similar to the inside mic. Using the basic rod size (0 to 1 in.) with the head set to 0, test it against a flat precision surface to check for zero error. Performing this test is a good way to get the feel of the zero pressure as well.

Calibrating the measuring rods requires measuring down from a stack of precision gage blocks against the granite plate (Fig. 7-11).

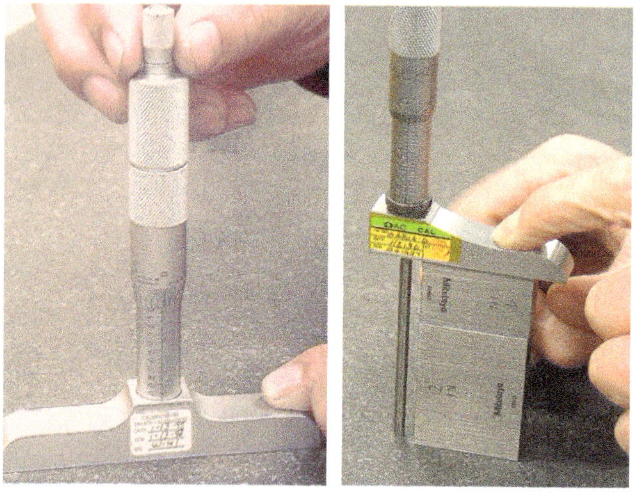

Figure 7-11 Calibrating the base rod (left), zero against the table. To check longer rods, use gage blocks.
Lake Washington Technical College, Kirkland, WA/McGraw Hill

Feel

The major bug with depth mics is the contact footprint of the tip of the small diameter probe rod. It's not easy to tell when it touches the surface being measured. One can accidentally lift the base up off the surface on which it rests–known as **jacking** the mic or lifting it (Fig. 7-12).

Outside, inside, and depth micrometers all feature a small drag adjustment ring inside the head that allows changing the clearance between the spindle threads and the threaded thimble. It's used to set a given amount of **drag** between the two (the relative tightness or looseness of a measuring tool). To see this adjustment ring, the thimble must be backed out beyond normal measuring position.

Reading a Depth Micrometer

It would be best if you had a depth micrometer set as we proceed. There's an oddity about these instruments.

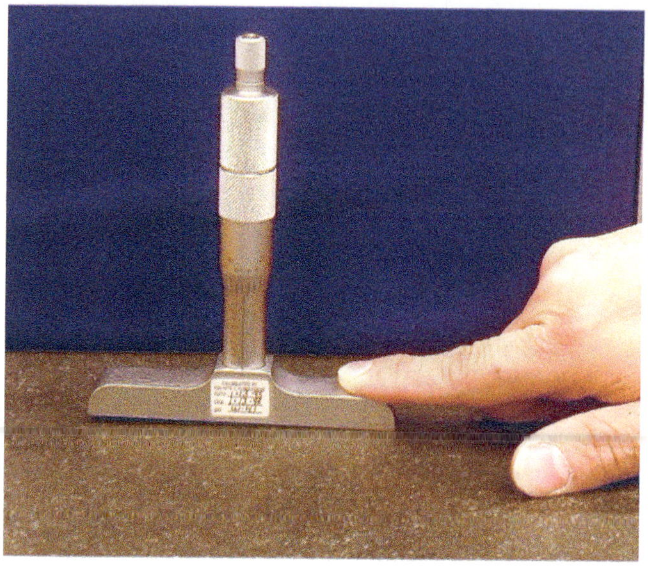

Figure 7-12 Lifting of the micrometer off its base can occur.
Lake Washington Technical College, Kirkland, WA/McGraw Hill

Backward Readings!

In Fig. 7-13, note that the micrometer head reads opposite previous instruments. It is at zero when the thimble is backed all the way out (unscrewed). *You must read it by seeing how much has not been covered up.* Look at Fig. 7-13. What is the reading? Now read the magnified depth mic setting in Fig. 7-14. It's tricky.

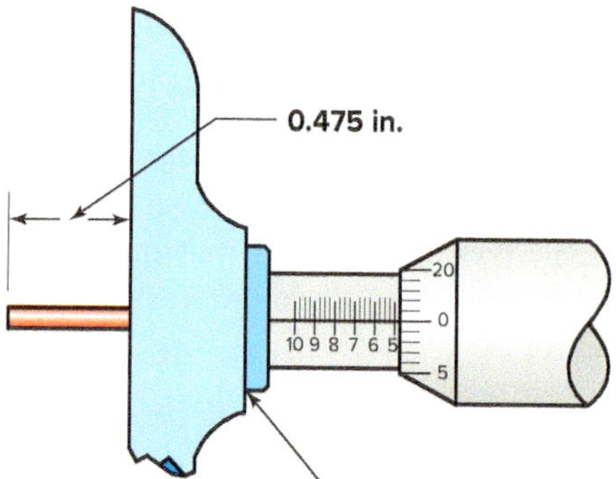

Figure 7-13 This depth micrometer reads 0.475 in. Can you see why?

Answer

It's 0.012 in. beyond the 0.050 line but not yet to the 0.075 graduation line, a total of 2.662 in.

Now set your depth micrometer to the same readings as in Figs. 7-13 and 7-14.

Skill Tips for Depth Mics

- Be certain the measuring tip is clean before using it.
- Use ultralight fingertip pressure on the thimble.

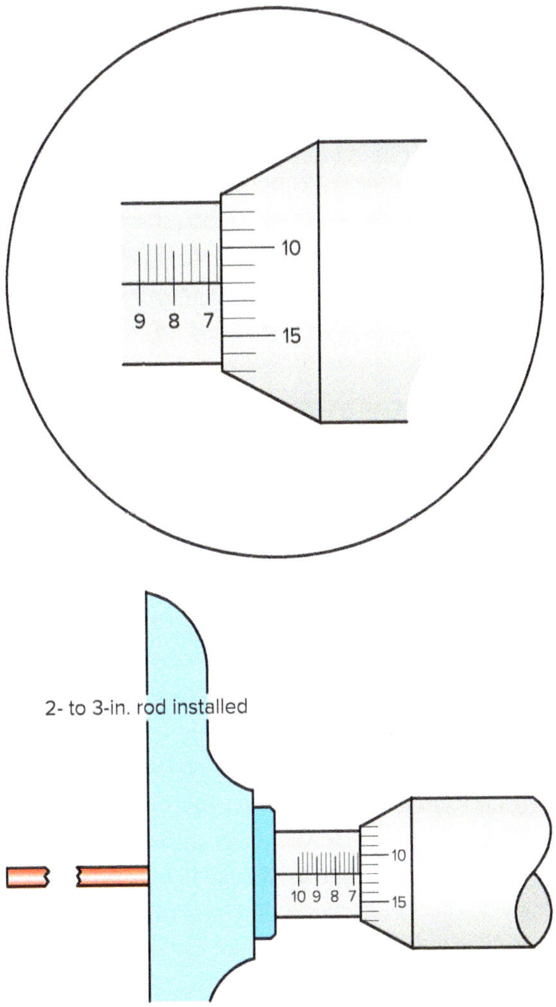

Figure 7-14 Read this depth micrometer.

- Apply hard pressure to the topside of the base to avoid jacking it above the surface.
- After contact is made, back up and recontact three or four times to be certain.
- If the width of the base is too narrow for the measurement, it's possible to bridge the gap with a ground parallel bar, as shown in Fig. 7-15. Remember to deduct the parallel thickness from the actual measurement!

Precision Features for Depth Micrometers

- *Different Base Widths* There is no best size; however, a narrow base can fit where a wide one wouldn't. (Note the half base in Fig. 7-11.)
- *Pressure Regulators* Depth mics may have a friction or ratchet pressure control device, which is good if more than one machinist is to share the tool (see Fig. 7-13).
- *Carbide Tips* Less important than outside mics since they do not rub with as much pressure, some depth mics feature wear-resistant carbide tips on the measuring rod faces.

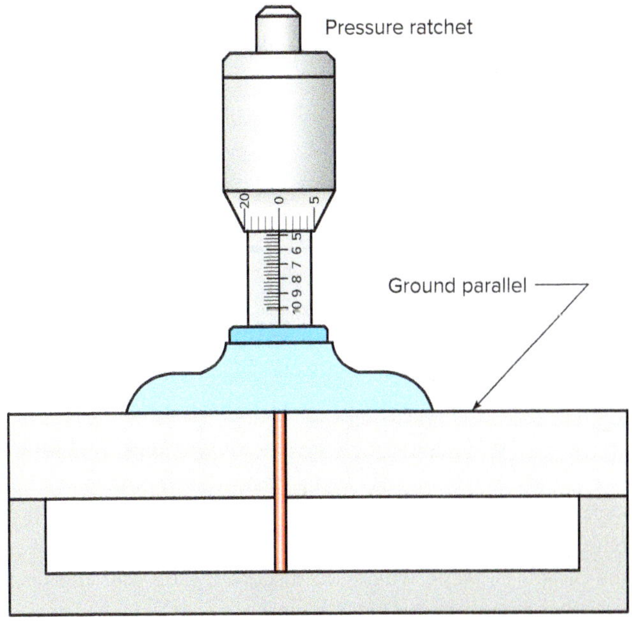

Figure 7-15 A ground parallel bar of known size can bridge larger gaps.

UNIT 7-1 REVIEW

Replay the Key Points

- Every inside mic has a basic smallest diameter size that must be known to the user.
- Inside micrometer heads feature a range of $\frac{1}{2}$ in. to keep them compact.
- The inside micrometer kit includes a small spacer to extend the range of each measuring rod.
- Depth mics are read by determining the graduation lines covered–just backward from all other micrometer heads.
- Both depth and inside mics do not have a vernier scale as their feel inaccuracy makes their repeatability too coarse to measure to tenths.

Respond

1. True or false? Depth micrometers measure backward–the largest reading on the head is with the thimble backed all the way out (unscrewed to the end of their range).
2. The range of inside micrometers is ___________ inches or millimeters. Explain why the head of inside micrometers is different than all others.
3. The final reading of an inside micrometer is the sum of three or four elements. What are they?
4. What is the greatest accuracy challenge common to both inside and depth micrometers?

5. When tolerances dip below around 0.0005 in., the inside micrometer might not be the tool to use, especially where many measurements might be needed of the same dimension. What instrument would be better?

Unit 7-2 Setup, Use, and Care of Precision Gage Blocks

Introduction: The word *gage* is used in many different ways within our trade and outside, too. The classic definition, when applied to shop measuring, means a tool that's used as a comparison template of the perfect shape and/or size. It is held against or next to the part in question and compared in one of several ways, which are discussed here. But the term has other meanings in our trade, too.

In this unit, we'll look at a variety of gages, each of which is normally owned by the machinist. The exceptions to ownership are the gage block sets, which, due to their high cost, are shop supplied. We'll start with them because they are at the root of precision in most shops along with the granite table.

TERMS TOOLBOX

Functional gage A gage that detects MMC condition. A functional gage simulates the worst-case assembly to the mating component.

Gage (gauge) In measuring, a gage is a template of the size or shape of an object. This definition is one of several as the word *gage* is used in other ways within measuring.

Jo blocks Trade term for precision gage blocks, stemming from the original manufacturer, the Johansson Tool Company.

Master gage Any calibrating tool not used for shop or day-to-day inspection work.

Wring Pronounced "ring," the act of placing two precision gage blocks together whereby they cling to each other. Wringing is proof of surface accuracy for gage blocks.

7.2.1 Precision Gage Blocks

Precision gage blocks are a set of extremely accurate hard steel or ceramic blocks, supplied in graduated sizes. They are used in several ways within the overall measuring system. Many shops keep two or more sets, one for each duty listed below. They are manufactured in several different tolerance grades depending on the application starting with the closest to perfection:

1. **Master Calibration**

   The parent reference standard found in calibration labs but not in manufacturing, the **master gage** is used to calibrate tools farther down the pyramid. It will be used daily within calibration labs or rarely within inspection areas in manufacturing but is not often used for direct contact with work at the machines.

2. **Inspection Shop**

   They are used in the inspection area of the shop but not at the machines.

3. **Shop Grade**

   Working gages are used for machine setups and in-work inspection.

KEYPOINT

No matter the tolerances to which they are made, even the shop-grade precision gage block could be considered a perfect tool within a machinist's frame of reference. The only way to be more accurate is through scientific means!

Target Repeatability

Correctly used, gage blocks (Fig. 7-16) can be combined to reproduce a comparison size model to the closest 0.0001 in. with an overall repeatable tolerance of plus 0.00004 to minus 0.00002 in.* per block–inspection-quality blocks. Most sets can be combined to create a stack of from around 0.010 in. up to 8 or 10 in. The goal is always to choose the fewest number of blocks to achieve the target size–that's a trick we'll explore here.

Figure 7-16 A set of precision gage blocks.
Lake Washington Technical College, Kirkland, WA/McGraw Hill

*That's not a misprint! These tools are made at subtenth tolerances–pronounced "plus forty to minus twenty millionths" in shop lingo.

Control of Inaccuracies

The whole point of gage blocks is that there are only three possible spoilers to repeatability–dirt, damage to the precision surface, and heat. All are easily overcome, but the damage aspect requires some training.

SHOPTALK

How Are Gage Blocks Made? Many parts of this process are guarded trade secrets, but even the known basic steps are complex and labor intensive. First, they are machined from a highly refined tool steel without impurities. The alloy is custom designed to exhibit a very low coefficient of heat expansion and shape stability. The steel is progressively hardened through a series of steps that leave it without the internal stresses of normal hardened tool steel—it won't warp over time as other heat-treated steel will. Also, it will not warp with temperature change.

Next, after machining to within a few thousandths of the final size, they are precision ground in stages to within tenths of the target. Then the blocks are abrasion lapped (rubbed with loose abrasive grains) against other flat master lap blocks. This time-consuming process occurs in stages with finer and finer abrasive. The final step is to remove microscopic imperfections, then inspect and grade them. They are sorted to become one of the four grades, then rust-proofed in a variety of ways and put in sets. The resulting gage blocks have pedigrees that are just one or two generations removed from international standard master-masters. Very few shops, worldwide, can produce gage blocks to these exacting standards and tolerances. That explains why seemingly simple little steel blocks cost so much!

KEYPOINT

Radical Care Required! Because they are simple little blocks, untrained beginners might miss how carefully they must be handled and used. No kidding, one single thumbprint left behind will ruin them within just minutes! The surface is so fine that skin acid will quickly corrode it.

7.2.2 Caring for Gage Blocks

Here are the game rules for the savvy user:

Wash Your Hands First. Especially if you've been working hard enough to perspire.

Do Not Touch the Gage Surface. Even with clean hands. To help, the storage box is made such that the gages can be tilted up, then grasped by the edge (Fig. 7-17).

Clean Them Immediately If You Do Touch the Surface. Nearly every time I demonstrate how to not touch the surface, I accidentally do, with the whole class watching! It happens. But the pro-fix is to keep clean white paper or tissue handy to wipe off the oils and acids right away.

Clean the Surface Every Time They Are Stacked Together. Using clean white printer or notebook paper lying on a flat surface, pull the block across it three or four times. Lens cleaning tissue is ideal for this task, but not paper towels–they are not pure enough.

Keep the Box Closed. Even normal atmospheric contaminants can ruin gage blocks! But the dust that settles on them must also be wiped away before stacking together.

Preserve the Surface. To prevent corrosion, an oil spray or preservative wipe is required for storage in the box if they won't be used for a while.

Never Lay the Stack on a Machine Surface. Once the stack is set up, lay it on a clean rag on the workbench.

Figure 7-17 The right way to lift precision blocks from their case.
Lake Washington Technical College, Kirkland, WA/McGraw Hill

As extreme as that all sounds, one sure way to identify yourself as a beginner in your new job is to mistreat the **Jo blocks**.

Correct Application

In machining, you will use gage blocks in four different capacities:

1. Calibration of other tools such as outside micrometers.
2. As a gage measuring tool. In Fig. 7-18, the machinist is using a stack of blocks as a limit test of the slot width. The test shown is called a **functional gage** test because it determines the true geometric width of the slot by accounting for irregularities. A functional gage tests a feature such as this slot (or a group of features) by simulating assembly to the mating component at its MMC size. In other words, it simulates the worst case for assembly. The largest gage block stack that will slip in is the true

functional width. It is determined from the maximum material condition of the mating part. The stack can be adjusted up in size by one tenth of a thousandth at a time.

3. As a comparison standard for indicator measurement, studied in Chapter 6.
4. To set up other tools or processes (see sine bars in this chapter).

Figure 7-18 Gage blocks are inserted in the slot as a functional test of the slot's machined size.
Lake Washington Technical College, Kirkland, WA/McGraw Hill

Gage Block Accessories

Gage blocks are supplied in imperial and metric sets. A shop might buy carbide end blocks to resist wear or clamp holders to keep them in a stack. Other accessories include special probes, jaws, and scribes that can be added to the blocks to use in special applications. These are items you will learn on the job as required. The important thing here is to know how to handle gage blocks and how to set up a stack.

SHOPTALK

Jo Blocks You'll hear machinists call gage blocks by their nickname; here's why: They were originally made by the Johansson Tool Company of Sweden. (**Gage *or* gauge**—either spelling is correct, but here in North America we tend to use gage.)

7.2.3 Skill Tip 1—Wringing Gage Blocks

The nature of gage blocks can be demonstrated. They are so flat and smooth that they will tightly hold together without any mechanical fastening. This is called **wring** (as a bell would sound). Ask your instructor to demonstrate this or see if you can wring the gage blocks yourself.

Back on earth, here's how we wring blocks (Fig. 7-19):

Put two clean blocks together at an angle to each other.

Push lightly and rotate into alignment.

At a certain point, they will adhere; this is a perfect test of their cleanliness and surface condition.

They will not come apart easily without rotation or sliding sideways, if correctly wrung! Several blocks may be wrung together to form an exact measuring stack.

SHOPTALK

Why Do Jo Blocks Wring? At one time it was thought that atmospheric pressure alone kept Jo blocks together. However, if we were to perform the mental experiment of putting them together in orbit where there is no air pressure, would they wring? Answer—yes. We now theorize that it's either due to a residual microscopic film of oil between surfaces or because they are approaching the flatness where the forces that hold all metals together can take over—that is, molecular bonding. However, ask yourself, if it was molecular bonding, then would not the two blocks become one indivisible material? The bond between molecules in the two blocks would be as strong as between molecules within one block. Okay—it's your turn: Why do they wring? Can you think of an experiment to prove the reason? Is it a combination?

7.2.4 Skill Tip 2—Making Up the Gage Block Stack Size

It's easy to set up a stack of gage blocks if you follow this example. There is a trick to getting the right blocks out in the correct order.

KEYPOINT

Always eliminate the smallest digit to the right of the decimal point first, then the next smallest, and so on.

Observe the gage block set in your lab. You will see three series of sizes: one with nine blocks having 0.0001-in. graduations (0.1001, 0.1002, 0.1003, and so on). There are two more with 0.001- and 0.010-in. increments, then the larger blocks in quarter, half, and full inch increments. Metric blocks are far simpler due to the nature of the ISO system.

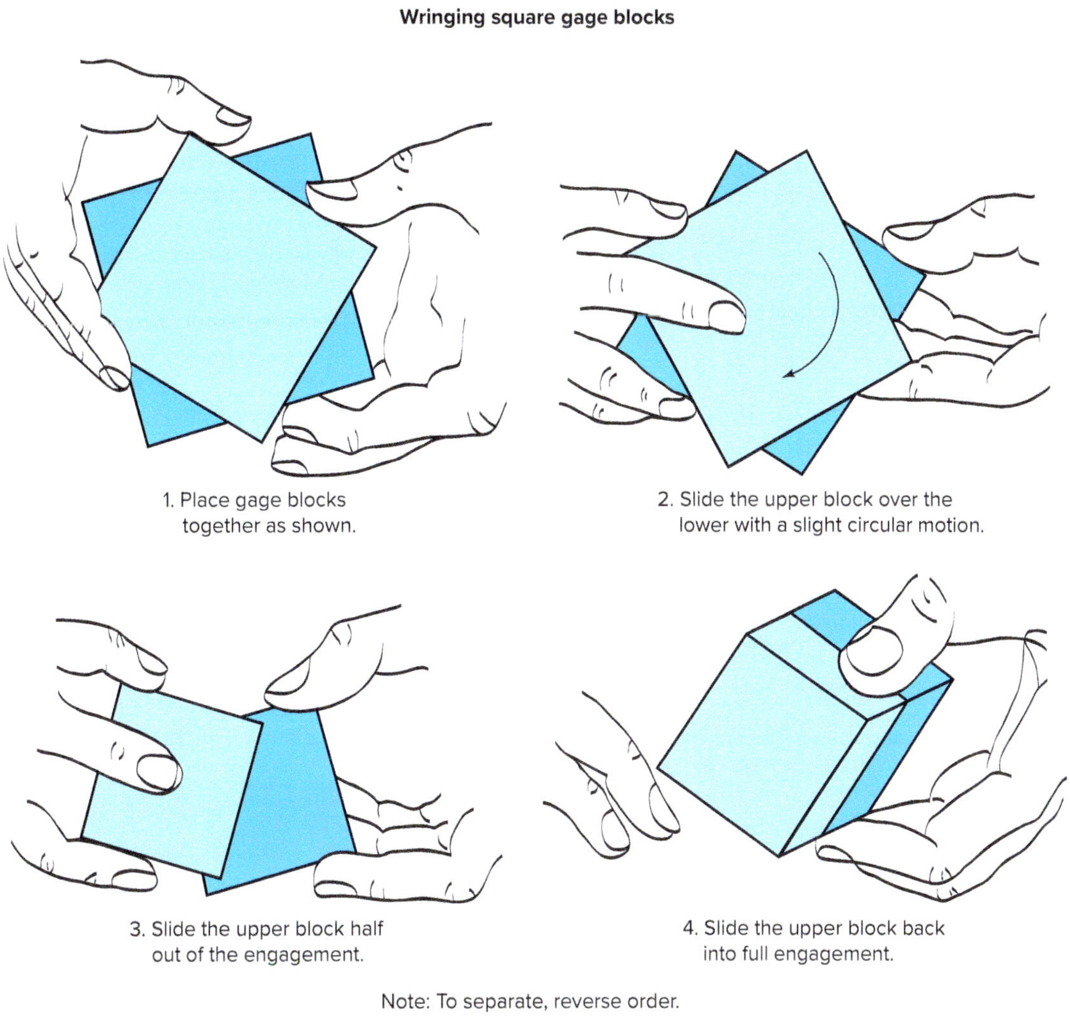

Figure 7-19 Procedure for properly stacking gage blocks together.

For this example (Fig. 7-20), you are to make up a stack of blocks that totals 3.7834 in. The first block to select is the 0.1004-in. block from the tenths series. Now subtract that from the total stack:

$$3.7834 - 0.1004 = 3.683 \text{ in. left to go.}$$

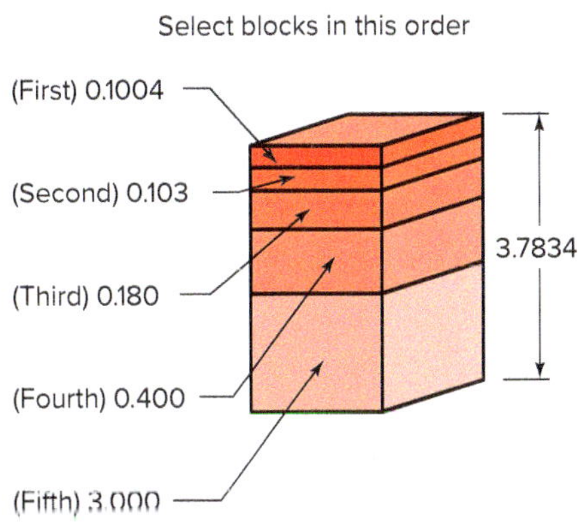

Figure 7-20 Setting up a stack at 3.7834 in., eliminate digits from the right (smallest value) as you select and wring blocks.

Now take out and clean the 0.103-in. block from the thousandth increment series and so on.

$$3.683 - 0.103 = 3.580 \text{ in. left}$$

Next select the 0.180-in. block from the ten thousandth increment series. Clean and wring it in; then subtract:

$$3.580 - 0.180 = 3.400 \text{ in. left}$$

Finally, remove the 0.400-in. and 3.0-in. blocks to complete the stack. The final total is 0.1004 + 0.103 + 0.180 + 0.400 + 3.00 = 3.7834 in.

UNIT 7-2 REVIEW

Replay the Key Points

- Precision gage blocks are considered perfect in machine facilities.
- Avoid making and immediately clean fingerprints off the surfaces of the blocks.

- They are the standard for measuring in a shop.
- To set up a stack, remove the smallest digit first.
- Wringing is the test of perfection.
- Always store them in their special box, and if it is to be for a long term (1 day or more) wipe them with a preservative.

Respond

1. What two major challenges affect inside and depth mic accuracy?
2. True or false? An inside micrometer is read differently than an outside micrometer. If it's true, then how is it read differently?
3. True or false? A depth mic is read differently than an outside micrometer. If it's true, then how is it read differently?
4. List the blocks that should be taken out of the box, in order, to make up a stack of gage blocks to equal 4.4089 in.
5. Complete: Gages are used to ___________ .

Unit 7-3 Gage Measuring

Introduction: In this unit, we'll look at a group of measuring tools that determine size and shape. We'll start with the hole gages, then look at several more tools that determine either size or shape elements that are affordable and might all be considered a priority 1 for your measuring kit.

TERMS TOOLBOX

Go, no-go gage A functional gage that tests the feature against upper and lower limits of its tolerance.

Planer gage A type of adjustable parallel–might also be called a machinist's precision gage in tool catalogs.

Radius gage A small, flat template featuring several arcs all of which are the same radius curve.

Ring gage A hardened steel donut with the inside diameter ground into a perfect circle of known size.

Snap gage The same tool as a telescope gage.

Telescope gage An expanding feeler used to determine inside diameter. After setting at best fit, it is removed and measured with a micrometer.

7.3.1 Small-Hole Gages

Ball Gages

These inexpensive tools are for measuring the diameter of moderately small holes–from around 0.7500 in. down to a 0.125-in. diameter. Found in sets of four (or more), each measures a range that overlaps with the next. Ball and telescope gages aren't measuring tools by themselves. Set to best fit, then removed, they transfer the size out of the hole to be measured with a standard micrometer.

Correct Application

Ball gages (Fig. 7-21) are used where tolerances are at or above 0.001 in. However, in a practiced hand they might deliver repeatabilities close to 0.0005 in. when no other method is available. Again, practice and awareness of their shortcomings are the keys to repeatability. Due to the double feel (one time while the gage is in the hole and once when measured with a micrometer), one can never be sure of the result with these (and with telescope gages, too!). As such, they must be considered utility, universal tools as opposed to dead accurate methods. Ball and telescope gages are okay for many measuring situations where a dial bore gage isn't set up. But they are slow to use.

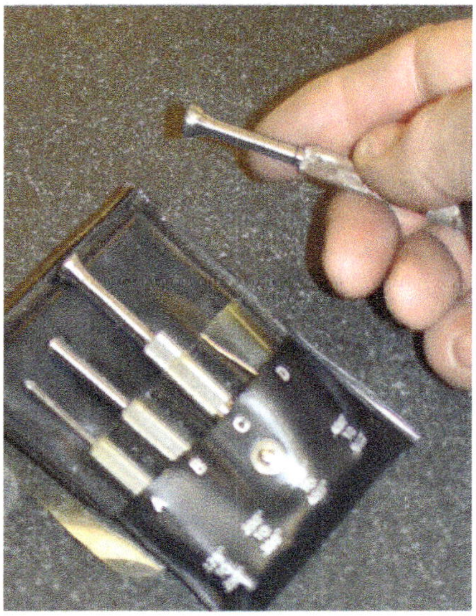

Figure 7-21 A set of ball gages has a range from 0.1 in. up to around 0.5 in.
Lake Washington Technical College, Kirkland, WA/McGraw Hill

Gaging the Hole

Ball gages are used as an adjustable, expanding feeler to determine the best-fit diameter. In the hole, while moving the gage in and out, the expansion nut is turned to expand the ball (Fig. 7-22). Once the ball drags in the hole, approximately the same as the micrometer that will be used to measure the gage, it is removed and measured.

Only practice will develop that sensitivity. Using them this way, ball gages are too slow for most modern CNC production.

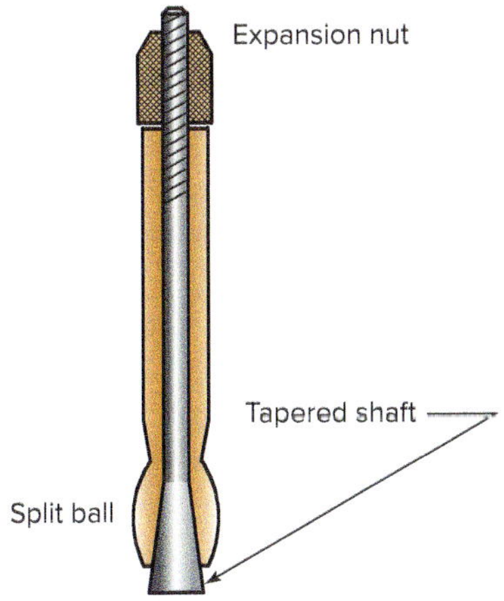

Figure 7-22 The ball gage in cross section.

KEYPOINT

Due to the way they are used, ball and telescope gages bring a double-feel challenge to the accuracy test. The feel within the hole must be the same as that applied by the micrometer to measure the ball. Both ball gages and telescope gages require experimentation and practice to get consistent results.

Telescope Gages

Telescope gages are another inexpensive, universal set for gaging holes bigger than those checked by the ball gages (Fig. 7-23). Starting at the smaller size of around 0.50-in. diameter, they cross over the inside micrometer's range at around 1.50 in.

Correct Application

Similar to ball gages, telescope gages are used in an adjustable feeler test of the hole size that's removed and measured with a micrometer. They can deliver accuracy within 0.0005 in., but if there is an inside micrometer that can do the job, it's usually the better choice. Telescope gages require more practice to use correctly than ball gages, due to their alignment/feel problem.

Different from ball gages, the control of the drag is preset with a telescope nut.

TRY IT

AN EXPERIMENT

Set up your own test comparing yourself with one or two other students to see how close you can come to consistency. You'll soon see what I mean. But, with good technique and patience, it's possible to reduce the variation between the test group. Here's how.

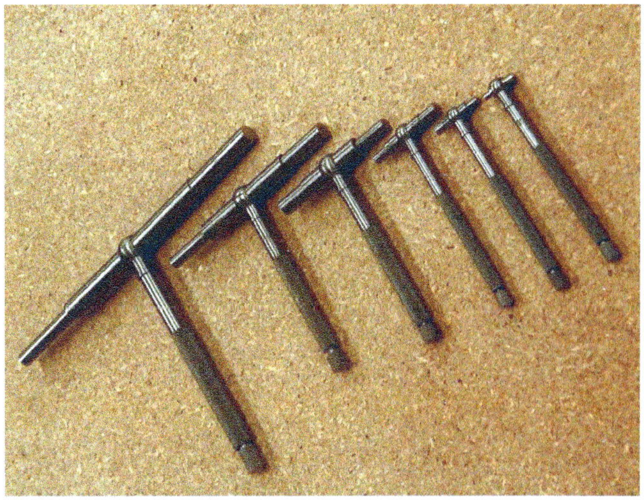

Figure 7-23 A full set of telescoping gages.
Lake Washington Technical College, Kirkland, WA/McGraw Hill

Using the Telescope Gage Right

These gages are inserted in the hole, then tilted at an angle, as shown in Fig. 7-24. The friction lock is released, which allows the telescope to expand larger than the hole's diameter. The friction lock nut is re-snugged "just the right amount"; then the tool is rocked through the axis of the hole–*one time only.* Not back. Once again, that right amount will be determined only with practice.

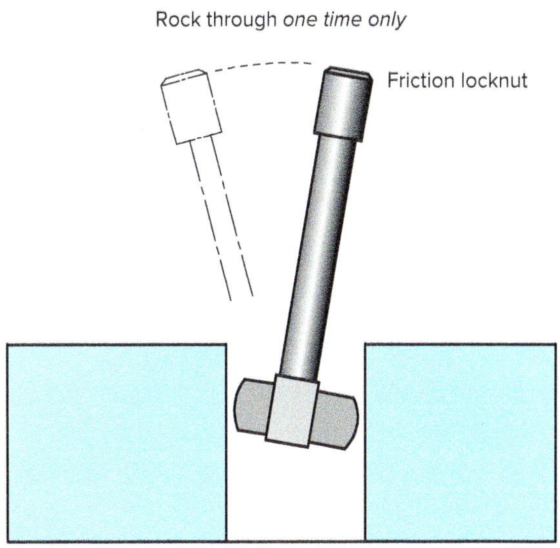

Figure 7-24 After expanding the telescope in the hole, snug the lock. Then rock it through, remove it, and measure it.

KEYPOINT

Do not rock the telescope gage back and forth—pass it through in one direction, one time only. The tightness of the friction lock determines the feel of the gage within the hole.

As the telescope rocks through, it is squeezed down to the hole size and does not change. At least that's the theory! The right amount to tighten the nut is light but firm–I can't describe it better than that.

TRADE TIP

Due to the uncertainty of telescope gages, always make five or more tests of the same hole. Record results and look for a consistent average. This is not just a beginner's move—we all do it if there's no other instrument available to do the measurement.

As shown in Fig. 7-25, the micrometer is closed slowly and lightly on the telescope gage. Note the correct hand position. The problem is, using the fine-threaded micrometer, the contact feel can actually push the telescope farther closed to create a false reading. The solution would seem to be to tighten the lock after removal from the hole, but that often disturbs the position of the telescope. The only real control is to go easy on the measurement!

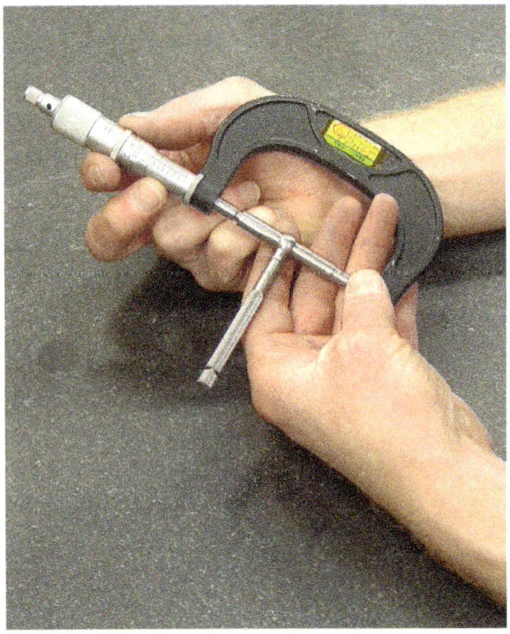

Figure 7-25 The correct hand position for measuring a telescope gage with a micrometer.
Lake Washington Technical College, Kirkland, WA/McGraw Hill

SHOPTALK

Snap Gages You might hear telescope gages called **snap gages** due to the way they spring open when the locknut is released. Some feature a single sprung telescope section while others use a double expansion system to encompass a wider range of possible sizes per individual gage.

7.3.2 Dial Bore Gages

We've come to the best everyday solution for fast, accurate hole measurement, but it is expensive compared to the previous two. Dial bore gages incorporate a three-point contact with the hole, which produces a better picture of roundness.

Not strictly following the definition of a gage, they report size on a dial (or electronic readout), as shown in Fig. 7-26.

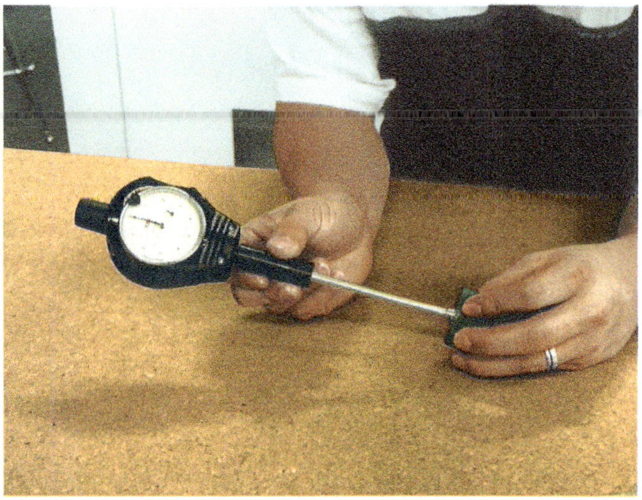

Figure 7-26 Dial bore gages for measuring hole diameters.
Lake Washington Technical College, Kirkland, WA/McGraw Hill

Each dial bore gage covers a range of diameters by installing different probe rods (Fig. 7-27). But each is still limited by the curvature of the hole and the head size; thus, shops must own several to cover a full working range of sizes.

Tech Specs for Dial Bore Gages

Correct Application

Different sized dial bore gages can measure small holes from around 0.100-in. diameter up to an unlimited size. They are suitable for tolerances finer than 0.0005 in. As mentioned in Chapter 6, they are one of the best tools for fast production where there is no time to use the telescope or ball gages and a reliable test is required. They're nearly foolproof to use with no inaccuracies other than the original setup to make them read zero at the nominal hole size. Dirt or burrs in the hole can destroy accuracy.

For CNC production, you may be supplied with a shop-made ring gage of the nominal size, to quickly check that the bore gage hasn't slipped from nominal.

Skill Tips for Use

Similar to a telescope gage, the head is placed in the hole at an angle, then rested on the contact points (see Fig. 7-27). The probe is then rocked though the diameter while you watch the dial. Holding the handle parallel to the hole's axis, in the side-to-side plane, it is rocked up and down. Doing so, the reading will swing from small to large, then small again. The high point null is the correct diameter. This is not difficult to do and requires little practice to get near-perfect results. It's a good idea to move the gage to several other locations within the hole, to test for taper.

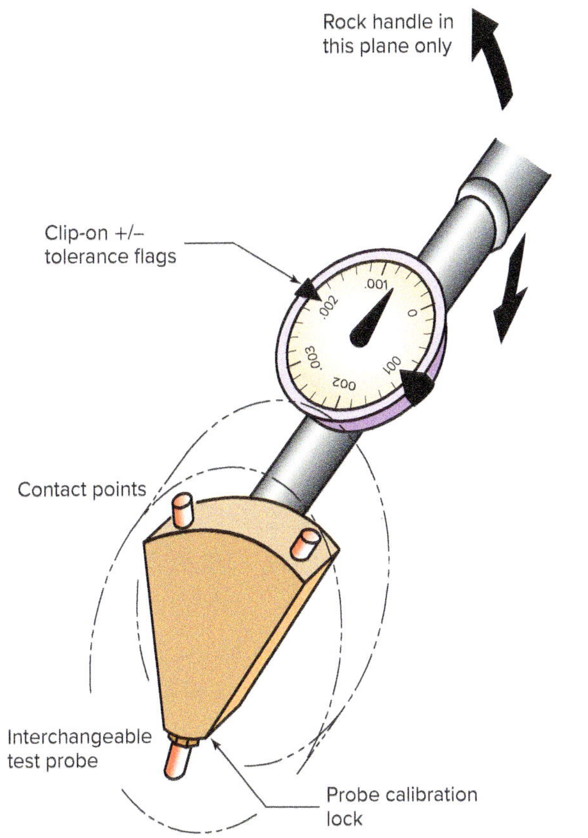

Figure 7-27 The right way to rock a dial bore gage.

Repeatability

Quality dial bore gages deliver from 0.0001- to 0.0002-in. repeatability when correctly set up, maintained, and used. Because the probe spring always pushes against the hole's surface, the challenge of feel is eliminated. One challenge to accuracy is the hole's roughness.

Setting Nominal Zero

Adjustment to their reading is made in one of two possible ways, depending on the gage construction: Either the probe is screwed in or out, then locked with a jam nut or the readout is adjusted to zero at the right size. (Both adjustments are common on many dial bore gages.)

No Math! Dial Bore Gages Are Plus/Minus Reading

Note in Fig. 7-27 that dial bore gages register variation in a plus and minus amount from zero, similar to probe indicators. This test shows the diameter at 0.0007 in. larger than nominal zero. Most are equipped with small clip-on tolerance flags that can be set at the upper and lower limits, as depicted in Fig. 7-27.

> **TRADE TIP**
>
> **Note the Size** In high-speed CNC work, where more than one dial bore gage might be required to inspect one part, it's a good idea to put a small piece of tape on the gage face or the handle, with the nominal size noted on it. Also, keep the **ring gage** handy and test the setup twice or more per shift; heat and other factors can cause them to slip as much as 0.00005 in. (fifty millionths). That's not much, but it could be a half to a third of your entire tolerance!

7.3.3 Ground Gage Pins

This gage can be used to check the size of very small holes. They are supplied in large sets similar to gage blocks, but not in series as with blocks. Pins simply start at the smallest diameter and increase in increments of 0.001, 0.0005, or 0.0001 in., depending on the set.

To gage a hole diameter, one starts testing with increasingly larger pins until the size that fits the best is found. Like gage blocks, they too are shop-owned tools due to extreme cost. Like gage blocks they must be treated with ultracare (Fig. 7-28).

Figure 7-28 A set of gage quality pins must be treated with great care.
Prographics at Milwaukee Area Technical College/McGraw Hill

> **TRADE TIP**
>
> While they aren't as accurate as manufactured gage pins, machinists make their own utility pins in sizes they need often for a given job. While in tech school, if the opportunity arises, make as many test pins as you can find time to do. Don't forget to engrave the size on them if you have the ability!

Inaccuracies for Gage Pins

Pins might be envisioned as round gage blocks, and they too are nearly free of challenges. When testing holes, they do have the inaccuracy of feel, but it's minor compared to that of ball and telescope gages. The possibility exists of a damaged pin, but that's rare as they are made of hardened tool steel.

Correct Application

Using best fit/feel, test pins can gage hole diameters of a few thousandths inch up to around 1.000 in. In addition to diameter, they are also a geometric functional test of cylindricity. With the MMC pin inserted, you have found the largest cylindrical assembly space available. Left within the hole, that MMC pin can also be measured for position and orientation to a datum (Fig. 7-29).

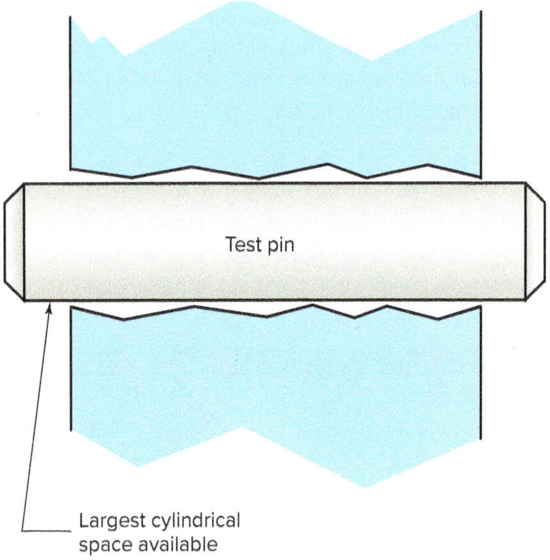

Figure 7-29 A gage pin tests for the available cylindrical space. This pin could now be measured for location and orientation.

7.3.4 Choosing the Right Tool

Comparing Hole-Measuring Methods

The real skill, and theme, of this entire chapter is deciding which tool to use: calipers, inside micrometers, ball or telescope and dial bore gages, or test pins. You'll base your choice on

What's available.

How tight the tolerance is.

How quickly the test must be made.

TRADE TIP

Pins as a Go, No-Go Test Often, two pins will be used: one at the lower tolerance diameter and one at the upper. If the "go" fits in and the "no-go" doesn't, the hole falls within the tolerance. When doing so in production, it's useful to tape them together in a pair, as shown in Fig. 7-30, then mark on the tape red and green or go and no-go.

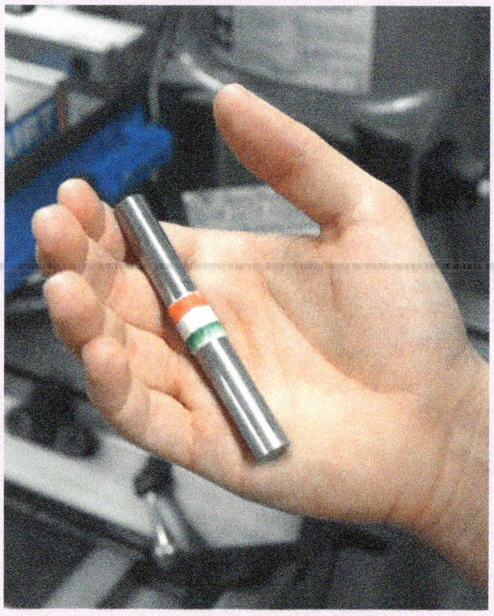

Figure 7-30 With the lower hole size marked green and the upper limit in red, joining two gage pins forms a go, no-go tester.
Prographics at Milwaukee Area Technical College/McGraw Hill

Here Are a Few Deciding Factors

- Ball and telescope gages have a double-feel challenge, but they are usually readily available.
- Bore gages are the best, but they must be set up to a given nominal before they can be used.
- Unless already set up, a bore gage would be a slow choice for measuring only one hole!
- Gage pins determine not only diameter but also the geometric cylinder.
- Inside micrometers are universal, but they are slow and cannot measure small holes.
- Calipers are fast but can't reach down into the hole; plus they have a low resolution compared to the other methods. Calipers make a good double-check to ball and telescope gages, though.
- There simply is no perfect answer! Every measuring task brings decisions—that's why we're called *skilled* machinists.

7.3.5 Adjustable Parallels

These are adjustable distance gages. They are useful tools for both measuring and setup work (Figs. 7-31 and 7-32). They can be used as a functional feeler gage for slots and grooves or can be preset as a **go, no-go gage** of thickness of an exact size (but are not as accurate as gage blocks).

Tech Spec Comparisons for Hole Measuring

Name	± Repeatability (in.)	Skill Level	Speed
Calipers	0.001 to 0.0015	Easy (overused by beginners)	Moderately fast
Micrometer	0.0001 to 0.0002	Practice required	Medium slow (math)
Ball gages	0.001	Practice	Slow (repeat test)
Telescope gage	0.001	Extra practice	Slow (must repeat)
Dial bore gage	0.0001	Easy, expensive	Fastest once set up
Ground pins	0.0001	Very easy, expensive	Moderate (try several)

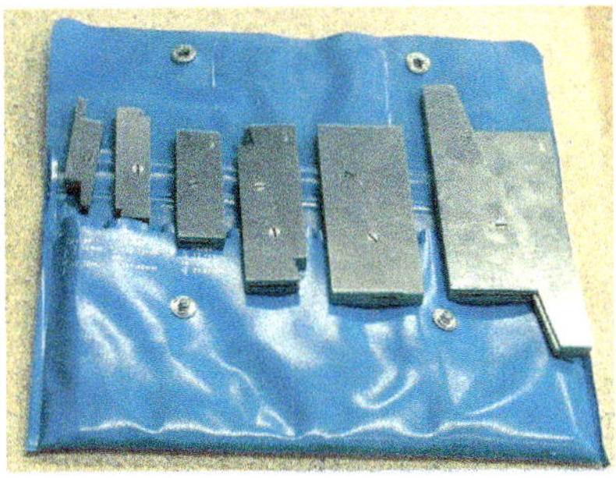

Figure 7-31 A set of adjustable parallels are very useful in your measuring tools kit.
Lake Washington Technical College, Kirkland, WA/McGraw Hill

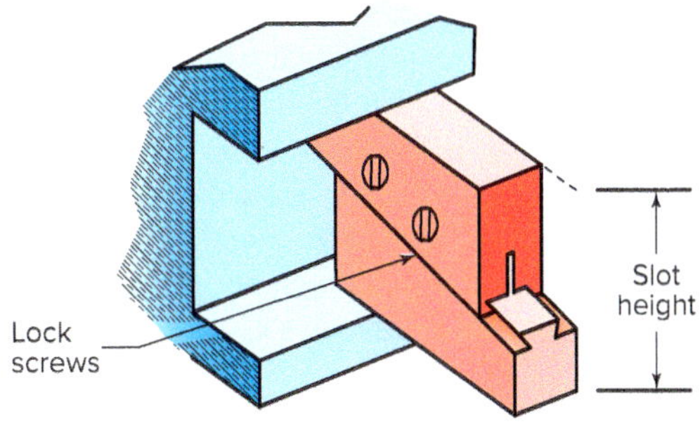

Figure 7-32 One use of an adjustable parallel is to measure the width with an outside micrometer or caliper.

TRADE TIP

Planer Gages Illustrated in Fig. 7-33 is a second type of adjustable parallel called a **planer gage** or *adjustable precision gage* in modern tool catalogs. They are a larger, more rugged version of the adjustable parallel, with potential uses all over the shop! Called "Machinist Precision Gage" in new tool catalogs these tools are useful and versatile.

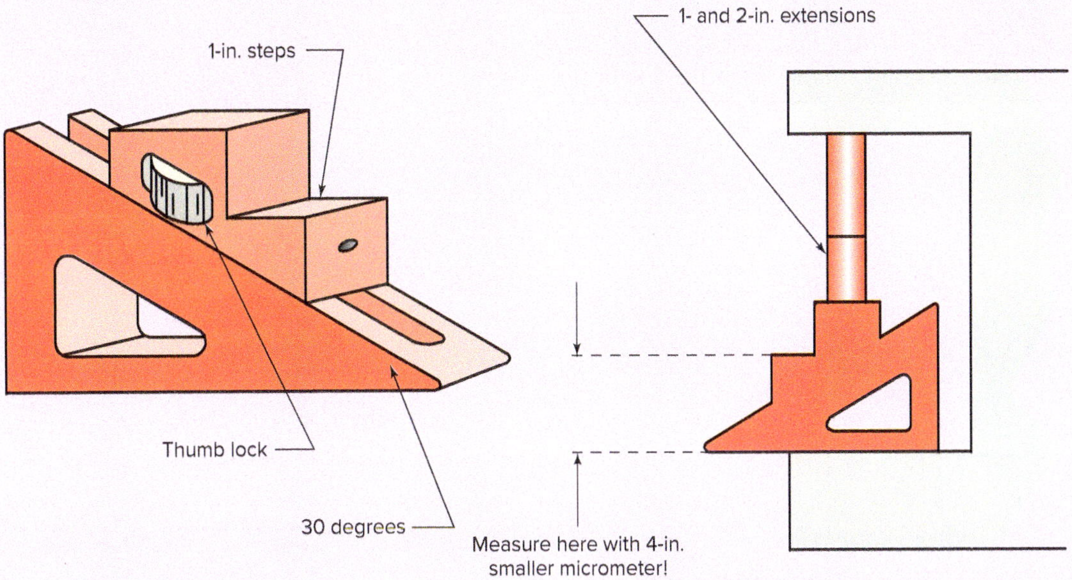

Figure 7-33 A planer gage has many shop applications.

Planer gages feature two screw-on extension rods that increase their range by 1- and 2-in. increments. Also, there are several steps of exactly 1 in. built into the tool. They are also a 30/60/90-degree triangle, which is useful for checking common angles. The carriage can be removed to use the triangle alone. Some even feature a bubble level set in the frame!

Invented originally to set tool heights above the table on an uncommon machine in modern industry—the horizontal planer—these tools are being forgotten. But this is one item that shouldn't be made obsolete by changing technology. They are excellent measuring and gaging tools. Here's just one example: As shown in Fig. 7-33, a slot width can be measured when the needed micrometer size is unavailable.

Adjustable parallels are a priority 2 purchase. As you become more skilled, you will find many uses of these tools not only in measuring and layout but also throughout the shop.

7.3.6 Radius Gages

These little flat templates are both a size and form evaluation tool (Fig. 7-34). For use, hold them against the round feature in question to find the best fit (Fig. 7-35). The clue will be the visual gaps. Similar to evaluating straightness, you are looking for the widest gap between the gage and surface in question. **Radius gages** are a priority 2 purchase. They are supplied in metric, fractional, or decimal inch sets.

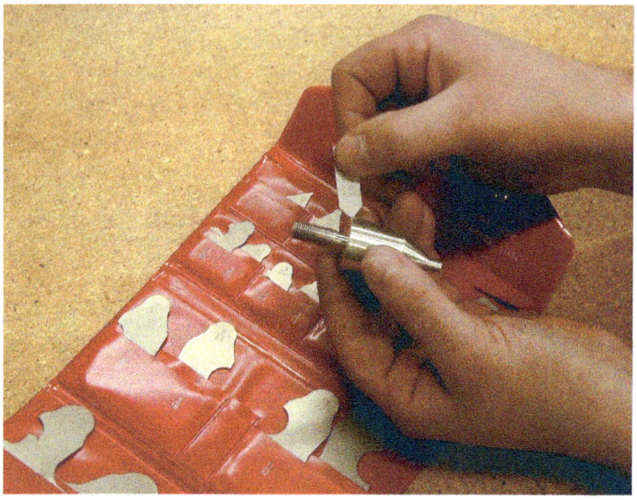

Figure 7-34 A full set of radius gages in a protective pouch.
Lake Washington Technical College, Kirkland, WA/McGraw Hill

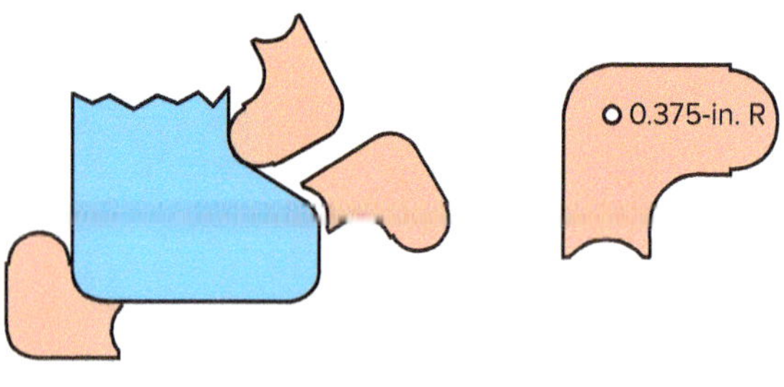

Figure 7-35 Using a radius gage to test the corners of a drill gage.

Reading the Gaps

It's not difficult to read the gaps to tell if the gage is too small or too big for the feature. In Fig. 7-36, three radius gages are shown testing the 0.750-in.-diameter round nose pin. One gap indicates the part's radius is bigger than that of the tool. Two gaps on either side tell the user that the gage is too big.

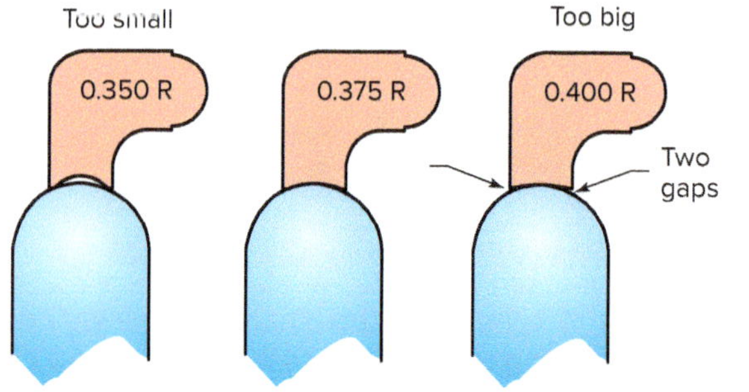

Figure 7-36 Gaging a radius using three progressive sizes.

TRADE TIP

Use Light or a White Background When evaluating a radiused feature, a light background or a light source is very useful. To do so, hold the part and gage up toward a light or place a piece of white paper in the line of sight beyond the object. This Trade Tip is useful in many ways throughout the shop, wherever visual estimating is done.

UNIT 7-3 REVIEW

Replay the Key Points

- Both ball and telescope gages have serious feel challenges in terms of accuracy.
- Dial bore gages are the most accurate tool of the hole gages we have studied.
- Planer gages are a useful form of adjustable parallel.
- Radius gages can be fit to the feature by looking for gaps.

Respond

1. Name the possible measuring tools to determine a diameter of 1.500 in.
2. Same as Question 1 but within a tolerance of 0.0003 in. in one part. Rank your choices.
3. What tool must be used to calibrate a dial bore gage?
4. True or false? A planer gage is a versatile version of the adjustable parallel.
5. Radius gages evaluate two related items, which are: ___________ and ___________ .

Unit 7-4 Measuring Angles

Introduction: Unit 7-4 is all about angles and how to measure them. Before we tackle the measuring aspect, we need to review how angles are expressed and toleranced.

TERMS TOOLBOX

Bracketing Determining a colinear pair of lines by looking for the symmetrical mismatch to either side.

Complementary angles Two angles that add to 90°.

Decimal degrees (DD) Angular expression in whole degrees and decimal parts.

Degrees, minutes, and seconds (DMS) Angular expression in whole degrees, minutes (1/60°), and seconds (1/3,600° [60 × 60]).

Hypotenuse The longest side in a right triangle–always opposite the right angle.

Protracted angular tolerance A tolerance zone that radiates outward from a vertex point.

Protractor An angle-measuring instrument.

Quadrant A 90° segment of a circle; four quadrants to one circle.

Sine bar An angular setting and measuring tool based on the sine ratio of right triangles.

Sine ratio The ratio comparison of the side opposite the angle in question to the hypotenuse side in a right triangle.

Supplementary angles Two angles that add to 180°

Vernier bevel protractor An angular-measuring tool with a resolution of 5 minutes of arc.

7.4.1 Angular Expressions and Tolerances

Protracted and Geometric Angular Tolerances

For a quick review, recall that protracted angular tolerances start at a point and radiate outward. They produce a fan-shaped zone of plus or minus so many degrees. A geometric angle is a sandwich-shaped tolerance zone around a perfect definition. A geometric angle's permissible variation is a width. Knowing this, we'll apply the protractor to measuring the angle differently. See Fig. 7-37.

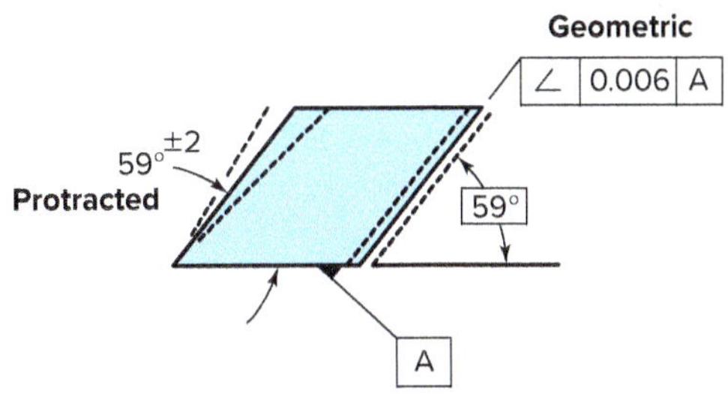

Figure 7-37 A comparison of a protracted and a geometric angle.

Decimal Degree (DD) Angles or Degrees, Minutes, and Seconds (DMS)

Using either expression, **DD** or **DMS**, a full circle is divided into 360°. Then degrees are further divided into finer units, by either extending the decimal or adding on minutes, then seconds.

DD	60.22833°
DMS	60°13′42″

When angles are expressed in DMS, degrees are divided into 60 minutes, and minutes are divided into 60 seconds.

Either expression might be seen on drawings. DD and DMS (or the symbols ° ′ ″) appear on your handheld calculator.

DMS

While less useful from a mathematical and resolution perspective, DMS angles are still found on older prints, and they are used on the vernier protractor coming up. Degrees are signified by a circle superscript. Minutes receive a prime mark (′), and seconds are shown with a double prime mark (″).

Example: 38°30′45″

It's pronounced "38 degrees, 30 minutes, and 45 seconds."

SHOPTALK

I Have a DRG Key on My Calculator—What's That?
In engineering and mathematics, there are two more ways the circle is divided into major parts other than degrees. They are more useful for engineering and

scientific calculations where extreme revolutions or long or fast cycles are part of the computation.

A full circle contains 2π radians or 400 gradients. While using gradients would simplify shop math, don't expect to see them showing up anytime soon. Changing tradition, redoing drawings, and buying new measuring tools would all need to be addressed. The DRG key (degrees, radians, and gradients) toggles from degrees to radians to gradients, then back to degrees again. Make sure it's in degree units when we come to sine bars in this unit.

Figure 7-38 A protractor for angular measurement comes in several forms.
Courtesy Mike Fitzpatrick

Decimal Degree Angles

Simpler to use in every way, whole degrees are followed by decimal parts. Given the choice, always use decimal angles. Conversion is a good idea, too–before starting a problem involving angles.

For example, DMS angle of 38°30′45″ becomes *38.512°*. This decimal angle is pronounced as with all math–"thirty-eight point five, one, two degrees" or "thirty-eight and five hundred twelve thousandths degrees"–either expression is okay.

TRADE TIP

Angle Conversion on Keys Many scientific calculators have a DMS/DD function that makes conversion easy. Manual conversion is not difficult, but this calculator function can save a lot of time. Each brand has its own way of accepting and outputting angular data. Read the instructions. This is a time-saving ability; learn to use it for shopwork.

7.4.2 Measuring Angles

There are only three low-tech choices for measuring angles. Moving toward technical solutions, we find the computer coordinate-measuring machine and optical comparitor–both covered later in this text.

Plain Protractors	Vernier Protractors	Sine Bars and Plates

Plain Protractors—1° Resolution

These utility angle-measuring tools have a resolution of 1°. They are all called a flat or plain **protractor**. They are used to measure and scribe angles where tolerances are modest (Fig. 7-38). One tool from this group should be a priority 1 for your kit.

Correct Application

With a good magnifier, it's possible to use plain protractors slightly finer than 1° but not closer than 30 minutes. Because the graduated arc is larger, reading between the graduations is a bit easier on the combination set protractor. In addition to measuring angles, you might use these protractors for layout work, hand tool grinding, and to set up a machine where angles are not closely toleranced.

Vernier Bevel Protractors—5-Minute Resolution or .08 Decimal Degrees

The **vernier bevel protractor** is shortened to the *vernier protractor* in this text. This angle-measuring tool is a universal, high-quality instrument that might be purchased as a priority 3. They are often shop supplied. The vernier version (Fig. 7-39) is common; therefore, we'll study reading it. However, they are also supplied in dial and digital versions, a very good improvement. Both digital and dial types are far easier to read because, as you will soon see, the vernier on these instruments is challenging!

Figure 7-39 A vernier (bevel) protractor and measuring blades in a kit.
Prographics at Milwaukee Area Technical College/McGraw Hill

While the word *bevel* is another name for a chamfer, these tools (see Fig. 7-39) are far more universal than just checking bevels, when measuring angular surfaces of machined parts. Figure 7-40 shows just a few of the ways these tools can be used with and without the 90° attachment. Note in Fig. 7-40 that two applications are referenced from a datum surface. Also note in Fig. 7-40 that the long and short blades have different angles on their tips 30°, 60°, and 45°. These can be used as templates.

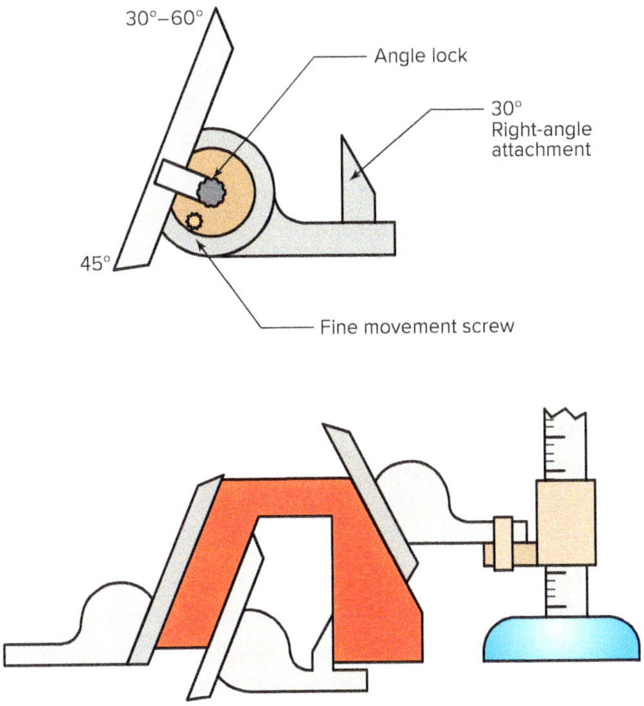

Figure 7-40 Three possibilities for measuring angles with a vernier protractor.

The vernier protractor is supplied in a kit in which the main blade, a short blade, a central measuring head with a built-in magnifier, and a right-angle blade are included.

Tech Specs for Vernier Protractors

Correct Application

Vernier protractors are the right tool for measuring or setting up angles where tolerances are finer than 1° but not finer than 5 minutes of arc. Besides measuring angles, vernier protractors are used for layout work and for machine setups where attachments or slides must be set at a precise angle to the vise or machine table.

Target Repeatability 5 Minutes

It is not possible to read finer than the tool resolution even when using a magnifier. A magnifying lens is usually supplied as part of the kit. If not, it's a good idea to use one when viewing the scale.

Control of Inaccuracies

The major inaccuracy factors for vernier protractors are

Parallax–a real problem using the curved magnifier lens. Be certain to view it straight on.

Visual estimation of the blade alignment to the work surface.

Reading the small vernier lines is difficult and it's possible to read the wrong vernier scale (training is coming).

Reading a Vernier Protractor

It would be best to obtain a vernier protractor as we proceed. Notice on yours or the one in Fig. 7-41, where it's shown with the blade removed, that the outer part of the measuring circle is divided into four 90° sections called **quadrants**. Two quadrants start from zero, then they are graduated up to 90°. The adjacent quadrant descends back to zero. *For any given measurement, the index line will be one of the four. It's important to note that in the quadrant in which the index lies dictates which of the two vernier scales is to be read.*

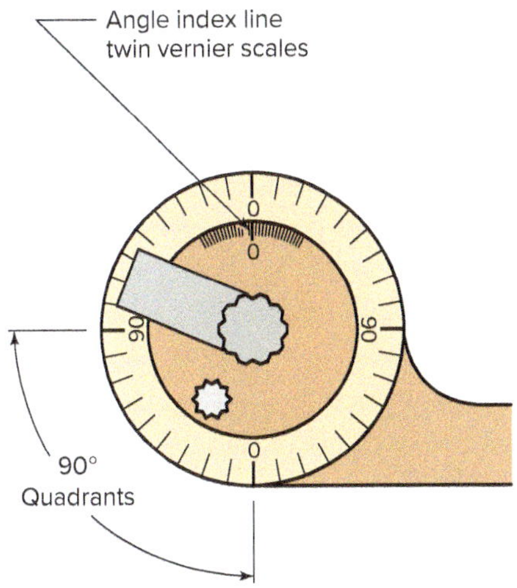

Figure 7-41 The blade has been removed to show the four 0 to 90° quadrants.

> **KEYPOINT**
>
> **Quadrant Rule** The direction in which the quadrant increases in value (clockwise or counterclockwise), where the index line points, will dictate which vernier scale is used. Use the corresponding vernier scale—clockwise or counterclockwise.

Twin Vernier Scales

Notice in the magnified view of Fig. 7-42 that the vernier scale on the inner circle has a central index line that points at the actual angle on the outer circle. Then there are two vernier graduation sets that extend opposite directions from the center. They are mirror image scales extending clockwise and counterclockwise. Use the quadrant rule to determine which vernier is used. Notice also that the vernier lines divide 1° into 12 parts $\frac{60}{12} = 5$ minutes of arc—that is the resolution of this tool. The vernier tells the user how many 5-minute divisions the index line has passed beyond a whole degree. Figure 7-43 illustrates a reading of "42 degrees and 25 minutes." It is 42 whole degrees and $\frac{25}{60}$ of the distance across to 43°.

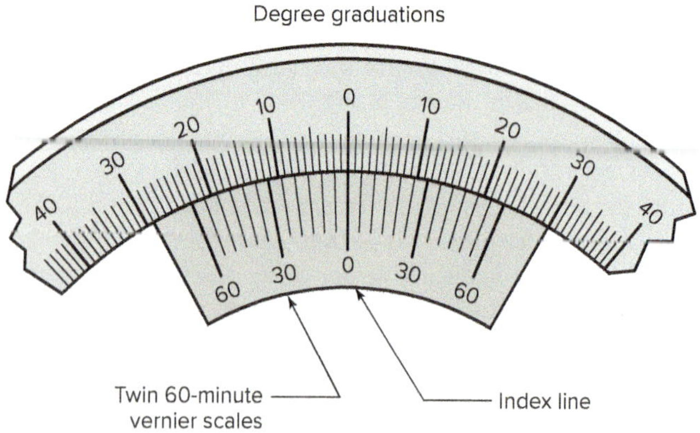

Figure 7-42 The vernier scale features an index line and two vernier scales. Use the "quadrant rule" to choose which vernier scale to use.

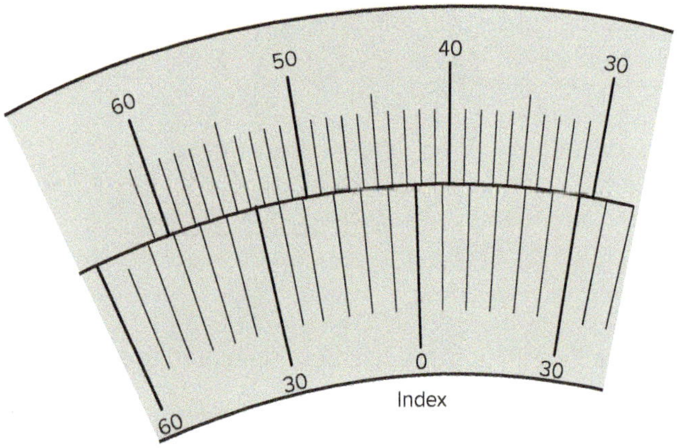

Figure 7-43 This vernier protractor reading is at 42°25′.

TRADE TIP

Zeroing in on the Colinear Pair To help find the lined-up pair (see Fig. 7-43) look for equally mismatched lines surrounding the colinear pair—the lines at 30 and 20 minutes. This trick, called **bracketing**, works for all vernier tools—the mismatch will be symmetrical in pairs around the centered lines.

Now read the protractor shown in Fig. 7-44 to be sure you have the concept, before moving on.

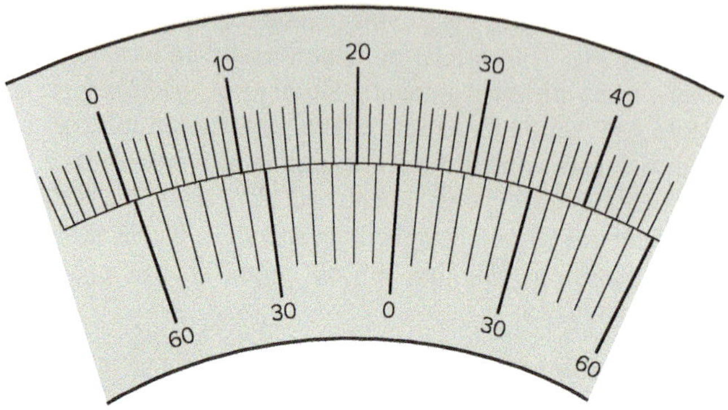

Figure 7-44 Read this vernier protractor.

The answer is 23°35′.

Measuring Protracted Angles—Best Fit to Surface

If the angle is protracted as in Fig. 7-45, then place the protractor base on the edge of the object and use the fine movement screw until the blade fits best on the edge to be measured. The angle should fall within the stated **protracted angle tolerance**. In this example, it's plus or minus 2°.

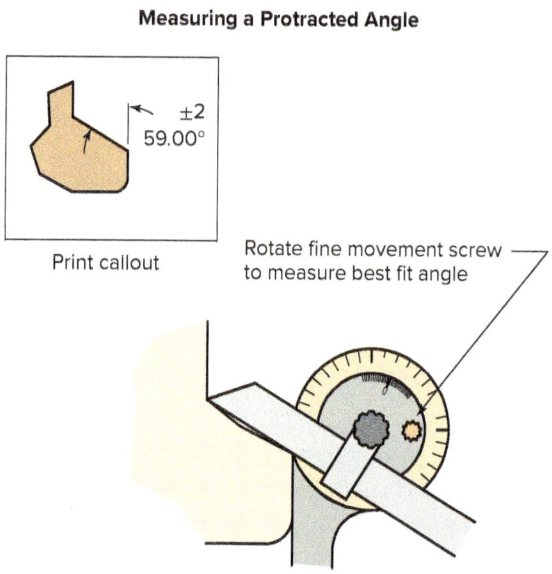

Figure 7-45 To measure a protracted angle, seek the best fit by rotating the fine movement screw.

KEYPOINT

When measuring protracted angles with a plus/minus tolerance in degrees, you are moving the blade until it fits the object best. Sometimes that's hard to decide if it's bowed.

Measuring Geometric Angles—Gage Surface

If the angle is geometric, then instead of measuring by moving the protractor you will be gauging the fit. First, set the protractor at the stated nominal (perfect model) angle–59°–and lock it in position.

KEYPOINT

You are testing the surface with the blade locked in position at the target angle, looking for gaps. The protractor's base forms a datum of the reference surface. There is no guesswork.

Now hold the object in the protractor or place the set on a datum (Fig. 7-46) reference; then using a feeler gage or precision wires, check that no gap exceeds the tolerance along the blade. The geometric definition of angles is far more useful when fitting two machined parts together where the functional objective is to limit the maximum permitted gap between.

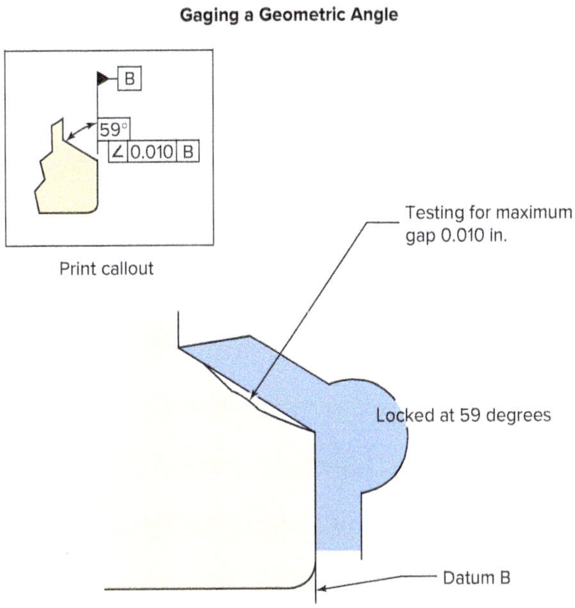

Figure 7-46 Gaging a geometric angle with a locked protractor.

TRADE TIP

Stay Out of the Complementary Trap! Avoid this predictable error when using any kind of protractor by being certain the angle measured is the target angle that you intend to measure. This problem arises from a geometry twist called **complementary angles** and **supplementary angles**. Angles are complementary when any two add to a total of 90°; they are supplementary when they total 180°. In some cases, it's easy to confuse the target angle with the complementary or supplementary angle on an object.

This drill gage would be a prime example (Fig. 7-47). The angle needed must be tested from the supplementary side. Awareness of the trap is the real key. Want proof that it's easy to fall into this trap? I set you up. Return to Fig. 7-46 and see that truly, the protractor isn't measuring the 59° angle, but it's gaging the supplement. That's okay as long as you've done the math: What the protractor should read in Fig. 7-46 is 180° − 59° = 121°.

Variation Is Reversed! Using the right-angle attachment can eliminate this problem in some cases, causing the reading to be direct. However, most of us just determine which angle we are actually measuring—the target angle, the supplement, or the complement. Then, when it is the complementary or supplementary angle that we have found, we know that the variation is reversed. In other words, if the supplement in Fig. 7-47 measures *big* at 123°, the resulting target is actually *small* at 57°.

The Final Confusion! But we aren't through yet. Notice in Fig. 7-48 that flat protractors have two scales running clockwise and counterclockwise, 0 to 180°. These are supplementary values for the blade position. Setting the arm index at 59°, as shown in the blown-up view, also sets it at 121°. When angular problems get complex, always use a flat protractor to check the vernier protractor to be sure. The flat protractor double-check can solve many of the headaches.

Remember the vernier protractor quadrant only goes up to 90°, then starts over.

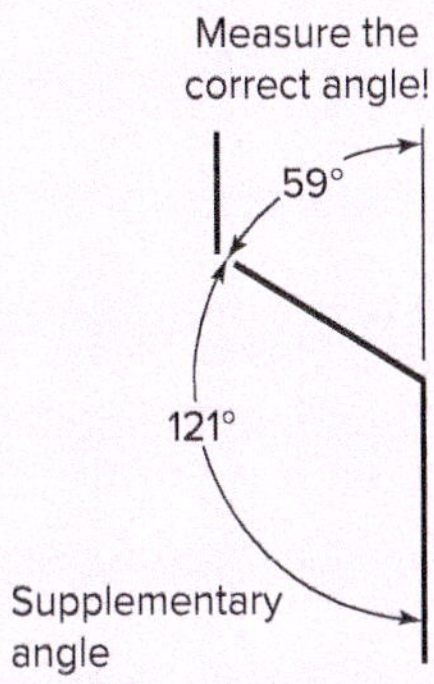

Figure 7-47 Be cautious when measuring that you do not confuse the target with the complementary or supplementary angle.

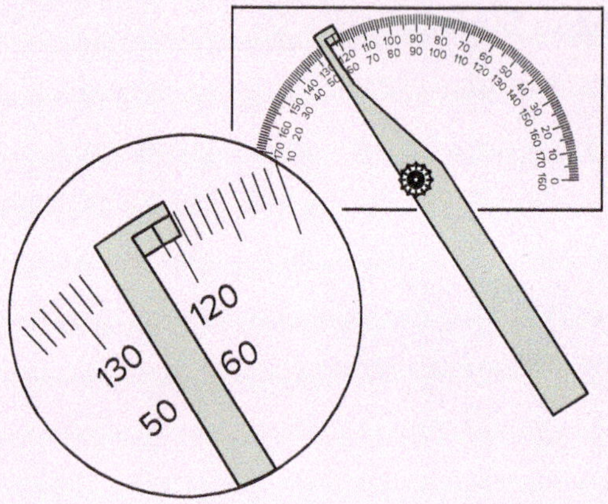

Figure 7-48 Use the double scales to verify the angle or its supplement.

Measuring and Setting Angles with Sine Bars—0.001° Resolution

When angular tolerances are tighter than 5 minutes, we turn to a very different tool–the **sine bar** or its wider hinged version, the sine plate. They are the most accurate angular tools in the shop.

Sine bars function as parallel bars that tilt relative to a datum surface (Fig. 7-49). Although ultraprecise, they are also very sturdy, so they can be used during machining setups–but they are more commonly used in measuring and making setups, where they would be removed once the angle is established.

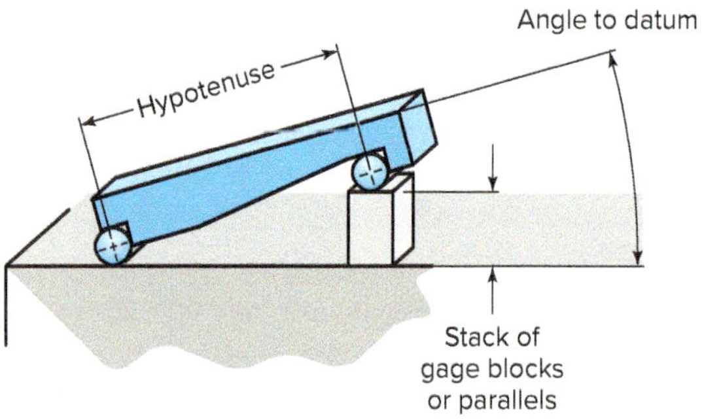

Figure 7-49 A sine bar is the ultimate angular measuring and gauging tool.

Make Your Own

Sine bars are often made by the trainee along with parallel bars. To make them it's necessary to cut and grind the pads that establish the distance between the circular rods to an exact number, usually 10 in. (but 5- and 1-in. sine bars are also used). The 10-in. bar is the more useful because of both its size and that it simplifies math. If purchasing your sine bar, it would be a priority 2.

Sine plates are wider tilting tables with their top surface hinged to a bottom plate that can be firmly held on the mill or grinder table. Sine plates usually feature threaded holes or magnets on their top surface to hold parts during light machining or angular grinding operations.

Both bars and plates are based on the trigonometry principle that any given angle will produce a specific ratio between the stack under the front of the bar, as shown in the drawing, and the distance between the round rods–called the **hypotenuse** (longest side of a right triangle). This ratio is called the **sine ratio** and abbreviated to *SIN* on your calculator (see Fig. 7-49).

KEYPOINT

The sine ratio is a comparison of the length of the side opposite a given angle to the length of the hypotenuse side of a right triangle.

Here are two examples. Please follow them on your calculator (Figs. 7-50 and 7-51).

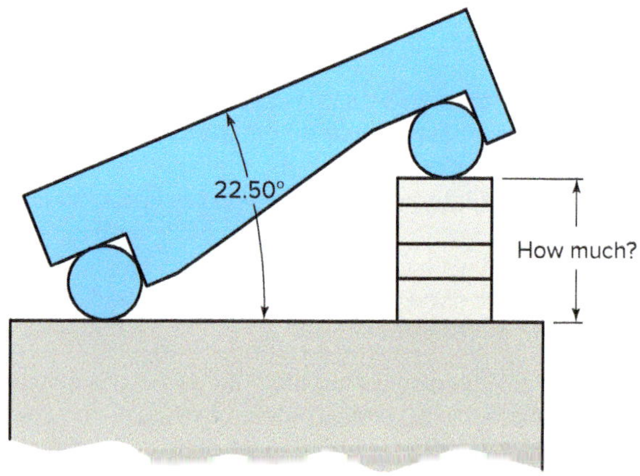

Figure 7-50 What stack height produces 22.5° on this 10-in. sine bar?

Setting Up an Angle

In this case, we know the angle and want to set up a perfect model–for example, to hold a part up at 22.5° relative to the floor of a milling machine vise. This setup will cut the top off a part at 22.5°. Once the part is tilted at the right angle, then clamped in the vise, the sine bar would probably be removed (see Fig. 7-50).

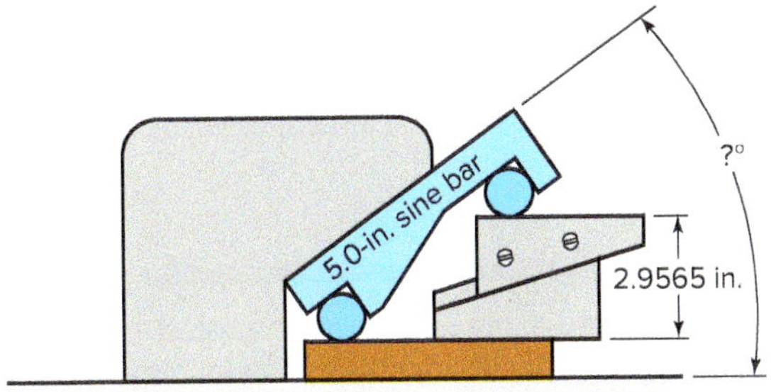

Figure 7-51 Measuring an angle with a sine bar.

We need to calculate the stack height for *a 10-in.* sine bar when it is elevated up to 22.5°. Be certain the calculator is working in degrees. Touch the DRG button until it does. Most calculators default to (start at) angles expressed in degrees when turned on. First we need to look up the SIN ratio.

Depending on your calculator's logic, enter either

22.5, then touch SIN

or

SIN, then touch 22.5 = **0.3826834**

There are two ways to look at that number:

Percentage of Hypotenuse It's a comparison of the opposite side (the stack) compared to the hypotenuse. In this case the opposite side of a 22.5° triangle is about 38 percent of the hypotenuse–38.26834 percent to be more exact. So the last step is to calculate

0.3826834 × 10 in. = 3.8268-in. stack height

Compared to One Because 0.3826834 decimal ratio can be envisioned as compared to one unit (0.3826834/1), the ratio we've just obtained would be the stack height if the sine bar was 1 in. long, but it's 10 times bigger, so multiply times 10.

0.3826834 × 10 in. = 3.8268-in. stack height

KEYPOINT

Either way you look at it, you arrive at this conclusion:

Stock height = Sine ratio × Bar length

Multiply the sine bar's hypotenuse times the sine ratio to obtain the stack height.

TRY IT

A. Calculate the stack height for a 10-in. sine bar to create an angle of 27.35°.

B. Calculate the stack height for a 5-in. sine bar to set up an angle of 7°.

C. Calculate the stack height to create an angle of 18°27′ for a 10-in. sine bar.

ANSWERS

Note that answers are rounded to nearest tenth of a thousandth inch.

A. A stack for a 10-in. sine bar must be *4.5942* in. to create an angle of 27.35°.

Sin 27.35 × 10 = 0.459424845 × 10

B. The stack height for a 5-in. sine bar set up to 7° is *0.6093* in.

Sin 7° × 5 = 0.121869343 × 5 = 0.6093 in.

C. The stack height for an angle of 18°27′ under the 10-in. sine bar will be *3.1648* in.

18°27′ = 18.45°

Sin 18.45° × 10 = 0.316476967 × 10

= 3.1648 in.

Measuring an Angle

This time we need to measure to the best accuracy possible the angle on the undercut block of Fig. 7-51. Here the adjustable parallel comes in handy along with a ground parallel bar under the sine bar. It's placed under the front rod, then opened until the bar fits the angle surface perfectly. Now locked, it's removed and measured with a micrometer and found to be 2.9565 in. tall. To proceed we reverse the process of the last examples. We have the height of the opposite side and need the angle it produces on a given sine bar. In this example, it's a 5-in. sine bar.

First, divide 2.9565 by 5 to obtain the sine ratio for the mystery angle.

$$\frac{2.9565}{5} = 0.5913$$

Then, depending on your calculator's logic, enter

0.5913, then touch 2nd + SIN 36.2493°

or

2nd + SIN 0.5913 = 36.2493°

TRY IT

Round your answer to the nearest thousandth of a decimal degree (three places, which is accurate enough for most but not all shopworks).

A. A 5-in. sine bar is inserted into an angle on a part and the stack height measured to be 2.5566 in. What is the angle?

B. A stack height of 0.476 in. under a 1.0-in. sine bar indicates an angle of 3 pica ___________ °.

C. An adjustable parallel is measured to be 0.7654 in. under a 10-in. sine bar. At what angle is the bar elevated?

ANSWERS

A. A stack of 2.5566 in. in a 5-in. sine bar indicates an angle of *30.752°*.

2.5566/5 = 0.51132 ratio;

2nd sin 0.551132 = 30.752°.

B. A stack height at 0.476 in. in a 1.0-in. sine bar indicates an angle of *28.424°*. 0.476 2nd Sin = 28.42447991 (Division by 1.0 was not needed; the sine bar is 1.0 in. long, so the stack height is the ratio.)

C. 0.7654 in. under a 10-in. sine bar calculates to *4.390°*.

KEYPOINT

Here's what we just learned about the calculator: Touching the ratio button (SIN) with an angle on-screen produces the corresponding ratio. Touching 2nd + ratio button (SIN) with a ratio on-screen produces the corresponding angle.

UNIT 7-4 REVIEW

Replay the Key Points

- Protractors are used to measure or gage angles depending on the drawing specification.
- Angles can be specified as DMS (degrees minutes, and seconds) or as decimals (DD).
- Flat and combination protractors are semiprecision tools.
- Vernier protractors have a resolution of 5 minutes of arc.
- Use the rule of quadrants to select the correct vernier scale.
- Watch out for complementary and supplementary angle confusion when measuring angles.
- Sine bars are the most accurate measuring tool for angles.

Respond

1. Name the two big differences in measuring a protracted angle and a geometric angle.
2. True or false? Because DMS are the units found on many drawings and tools within the shop, they are obviously the more useful angular expression. If it's false, why?
3. State the quadrant rule.
4. What inaccuracy factors could spoil vernier protractor accuracy?
5. Which statement is true?
   A. The SIN ratio is a comparison of the opposite side of a right triangle to its hypotenuse.
   B. The SIN ratio is a comparison of the hypotenuse of a right triangle to its opposite side.
6. Do supplementary or complementary angles add up to 90°?

Unit 7-5 Measuring Surface Roughness

Introduction: Now we turn to the fifth element, determining how rough or smooth the machining action has made a feature surface. When operating machine tools, you must control not only the size, shape, and position of features but often the finish product.

TERMS TOOLBOX

Cutoff The lateral span from which the profilometer extracts the average.

Lay The direction and nature of the surface roughness marks.

Microinch One millionth of an inch, the base unit for Imperial surface finish callouts.

Micron One millionth of a millimeter–the metric unit for surface finish callouts.

Profilometer An electronic device that determines surface finish roughness.

Scratch plate Used by scratching samples with fingernail, then comparing to machined feature. Also called a surface finish comparitor.

Surface roughness The smaller cyclic effect of the cutting action on machined surfaces.

Waviness The larger cyclic effect in machining caused by vibration.

7.5.1 Microinches (0.000001—one millionth of an inch)

Roughness is expressed in **microinches** symbolized as μin. There are other systems used elsewhere in the world, but they are similar enough that conversion is simple. Metric roughness is stated in **microns**, which are 0.001 of a millimeter.

SHOPTALK

μ is the Greek letter mu, for *micro*, meaning "one millionth."

The marks left behind from machining are the *result of working the metal.* If we look closely, they are like regular hills and valleys produced from the cutting, grinding, abrading, and eroding electrically and also from vibrations in the process. Machinists are challenged with controlling these irregularities to not rougher than a given microinch tolerance. Functional examples where surface finish control might be necessary are for

- Customer appearance.
- Smooth mechanical sliding action.
- Just the opposite, minimum roughness because bonding is going to occur to the surface (paint or glue).
- To create specific machined lines. This control specifies how the surface is made. If the tiny surface lines were parallel to a flexing action, cracks could start along them. This becomes very important in some high-stress applications.

KEYPOINT

One microinch is 0.000001 in. and is symbolized as 1 μin.

7.5.2 The Symbols Found on Prints

In Fig. 7-52 you see a closeup of a hole in the drill gage. The bracket symbol adjacent to it is the way the designer communicates a finish requirement.

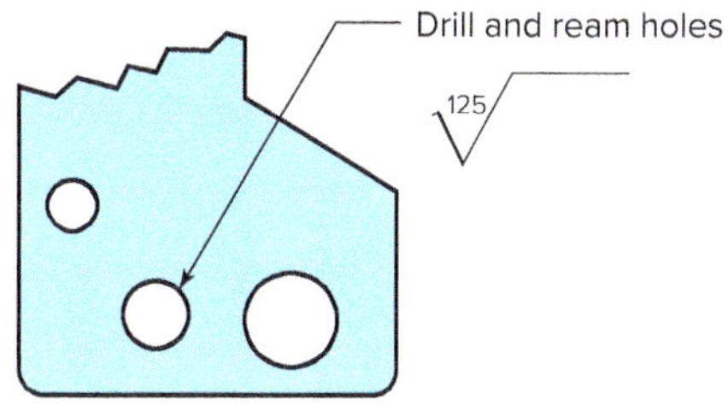

Figure 7-52 The bracket specifies a maximum of 125 microinches (125 μin.) roughness.

Maximum Roughness

In Fig. 7-52, the number in the bracket, *125*, is the upper microinch limit for the finish of that hole–the roughest allowed. Any smoother finish is acceptable unless otherwise stated in the control bracket. Specification of minimum roughness is rare, but it can be used where a rough surface is desired. If both upper and lower were required, there would be a second number below the 125.

On engineering drawings, these bracket symbols might be found relating to individual features as with this hole. Or, you might also find a general finish requirement in the title block. It would specify the requirement for all machined surfaces not given a specific finish on the print: "Unless otherwise stated all surfaces to be 125 μin." Often the μin. will be left off but is understood.

TRADE TIP

General Finish In the United States, if no finish is specified on the surface or in the general tolerances box, then a 125-μin. roughness is considered standard machine finish.

The drawing symbol used today for a finish requirement evolved from a simple check mark to indicate surfaces to be machined, to the one you see in Fig. 7-52. It has three data fields–the first and most commonly used is inside the bracket crook where the 125 appears; a second would be above the horizontal line, and the third within the bracket.

Let's see what the 125 actually means. In Fig. 7-53, you see a magnified cross section of a machine cut. Points 1, 2, 3, and 4 would be averaged as would P5, P6, and P7. The sum of the two averages must not exceed the 125-μin. limit. It could be smaller, however–smoother. The 125 μin. then is the averaged distance from a median line up to the hills and down to the valleys. It's calculated over a horizontal span called the **cutoff** distance.

KEYPOINT

The 125 μin. is not the distance from the peaks to the low points; it is the average of their deviation from a median line.

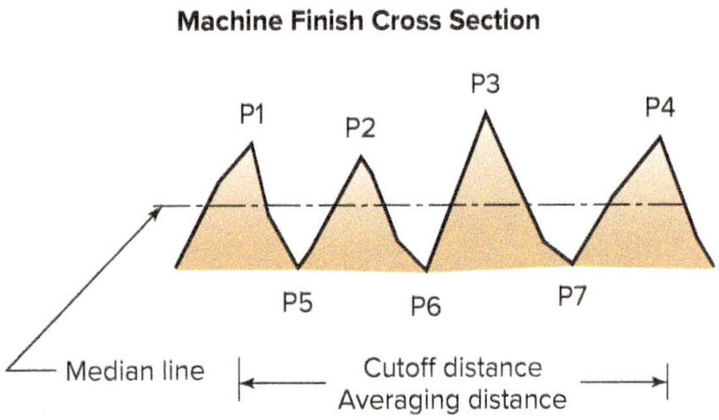

Figure 7-53 Roughness peaks and valleys are specified relative to the median line about halfway between.

Cutoff Distance

The cutoff distance is specified in the bracket (as shown in Fig. 7-57 later in the unit). It's stated at 0.010 in. wide. If this data field is blank, then a 0.030-in.-wide cutoff span is assumed.

Industrial Surface Evaluation

It only becomes important to use cutoff values when it's in the finish symbol on the drawing. Then it becomes necessary to turn to an electronic roughness computer, such as the one in Fig. 7-54. This portable electronic machine, called a **profilometer** (pronounced with a short *i*), is common in inspection departments. The profilometer can go to the machine to check results without moving the work to the inspection department.

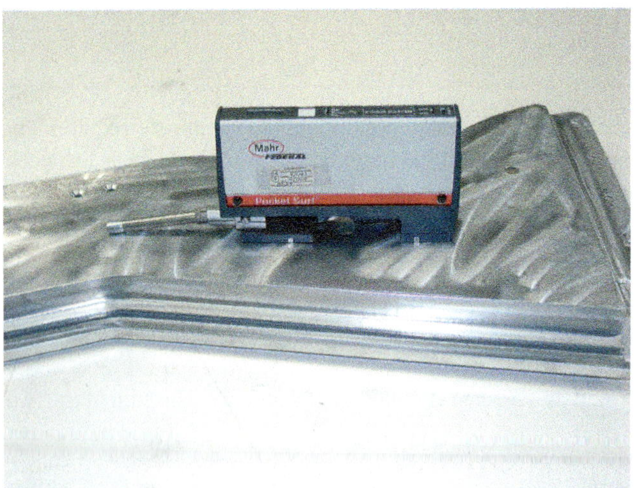

Figure 7-54 This portable profilometer is checking for a 64-in. (64 μin.) surface finish on this aluminum part.
Courtesy of Aerospace Manufacturing Technologies (AMT)/Arlington, WA

The contact stylus is a pointed diamond or similar material so it can fit between the peaks in order to get a clear picture of the surface. As its central processor collects data, it computes the roughness average over the specified cutoff distance, then reports on-screen, dial, or on paper.

7.5.3 Surface Roughness Comparison at Your Machine

Getting a Feel for the Surface (Literally)

Profilometers are necessary when surface finish specs are given with cutoff distances or they are finer than 32 μin. But for the everyday application at your machine, many machinists use the surface finish comparitor plate (Fig. 7-55), or sometimes called a **scratch plate** because of the way it's used.

Figure 7-55 A scratch plate helps estimate finishes.
Courtesy Mike Fitzpatrick

The plate features different sample patches of various kinds of machine finishes. Finding the closest representation of the finish you've made, the plate is scratched with a fingernail; then with the feel for the **surface roughness**, immediately scratch the surface in question! Go back and forth–it works amazingly well with a little experience. Be sure to scratch across the marks–not parallel to them. A scratch plate would be a priority 2 or 3 for purchase for your box but is handy to have. Looking at the roughness is of some help, too, but you can be fooled by how shiny the metal is due to surface conditions smaller than that of machining marks.

SHOPTALK

Improving Smoothness Versus Tool Life, Benchwork, Profit or Loss, and Quality The journey machinist's duty is to seek methods of producing better finishes by finding better cutters, keeping them sharp, calculating the right speed and feed, and maintaining coolant streams. However, this is not at the expense of extra time to make the parts unless the improvements save bench finishing or increase tool life. If, by slowing your feed rate, for example, the time to make a part goes from 9 to 12 minutes but saves a half hour of bench-work, or extends tool life by a measurable amount, then it's justified. But if the change only makes the surface substantially smoother than the specs, with no other gain, then it only costs profit without a significant gain in quality!

Lay Symbols

As Fig. 7-55 shows, there are different ways a finish might be produced that make swirls, straight lines, or crosses on the feature's surface—called the **lay** of the finish. The various lay marks will become familiar as you operate equipment. If it's necessary to specify how the finish lays on the feature, then a symbol will be placed inside the bracket. Similar to geometric symbols, each conveys meaning to the machinist. Figure 7-56 depicts a few of the more common ones, and Fig. 7-57 shows a closeup of the bracket with the multidirectional lay symbol in the correct location.

Symbol	Meaning	Illustrated
=	63-μin. finish required with machined marks parallel and running parallel to view. Finish method optional.	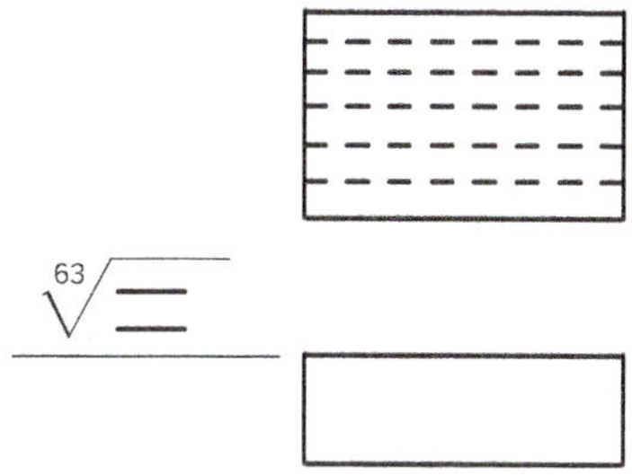
⊥	125-μin. finish with marks perpendicular to view. Surface must be machined.	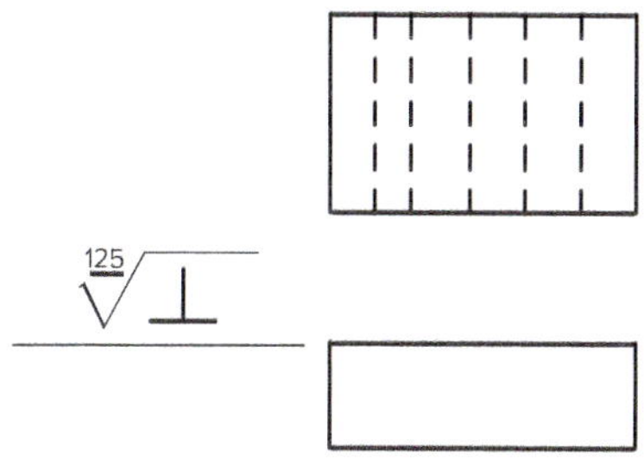
X	32-μin. finish with marks running angular to view.	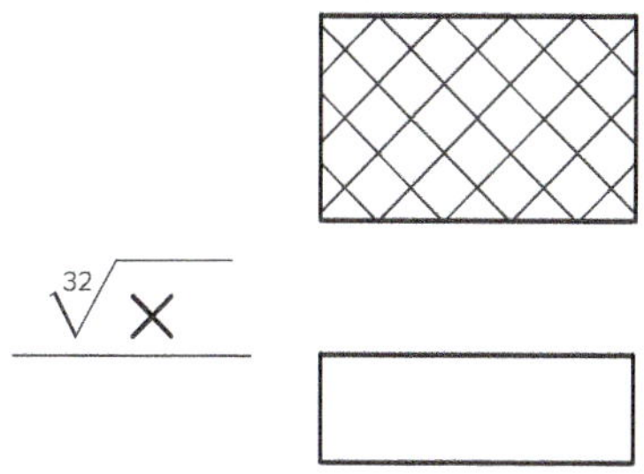
	No machining permitted on this surface.	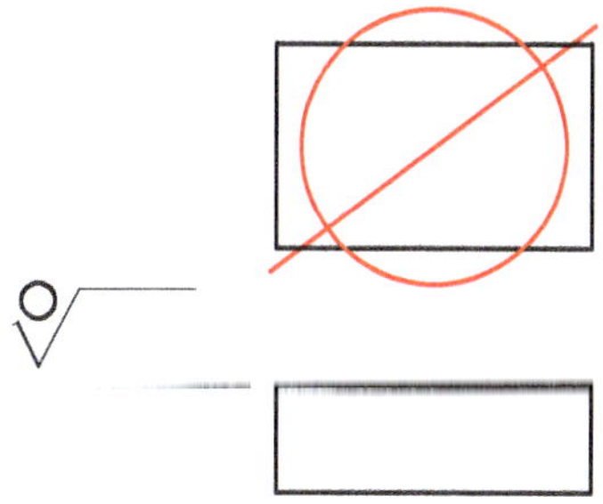

Figure 7-56 Typical lay symbols and their meanings.

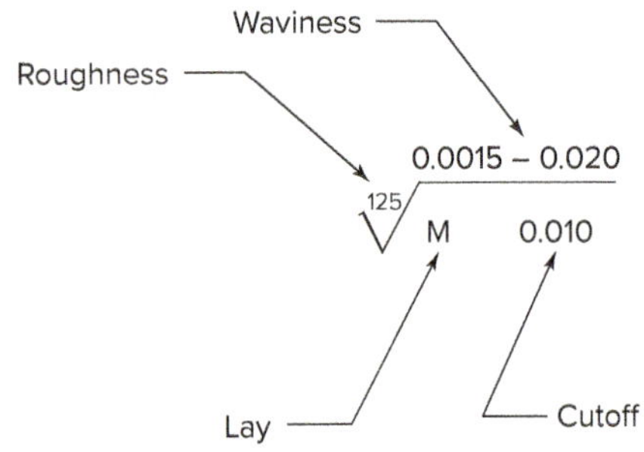

Figure 7-57 Waviness, roughness value, and cutoff distance can all be found in the bracket.

> **SHOPTALK**
>
> **More Information** The remaining facts are beyond our objective here, but there is a lot more about surface roughness to know. If you would like to know more, turn to *Machinery's Handbook*© under surface finishes or you might try searching the Net under finishes + surface + metal + machined.

It is important to keep in mind that the roughness can look quite different between, for example, the straight lines made by a lathe and the swirl lines made by a mill cutter. Yet both would measure the same microinch roughness.

Waviness

Sometimes, a cutting action will produce two different types of patterns on the work surface. The first type is caused by individual marks made by the cutting point or teeth, which we can think of as roughness. The second type of pattern is caused by vibration of the cutter, or by work vibration, or even the machine itself. When vibration patterns occur, a wave is formed. This is called **waviness**. When it is important to control waves on the surface, a waviness spec will be placed above the bracket. Again, this matters only for highly technical surface controls and cannot be measured except by electronic means.

> **KEYPOINT**
>
> If a lay is specified, it must be achieved as much as the other specs and tolerances on the drawing.

7.5.4 Standard Finish Applications

Under normal conditions, specified finishes run from around 125 μin. down to a modestly smooth 32 μin. or perhaps 16 μ in. Notice in the chart that the numbers are each one-half the previous finish. For example, a moderate 125, 63, 32, and a fine 16.

You might encounter a rough 250 μin. or even rougher 500 μin., depending on the nature of products made at your company. As you scan the examples notice that the various categories have gray areas where their functions overlap. Here are a few examples where these common finishes might be applicable.

> **SHOPTALK**
>
> **Don't Confuse Shiny with Smooth** Light reflection is one of the inaccuracies of a scratch plate. Often, a finish will reflect light when it is not as smooth as it might seem. Don't be fooled by shiny surfaces—they may not be smooth.
>
> Reflectivity is caused by closing all pores in the work surface. It's a microscopic function below the resolution of normal machining. Reflectivity is a gage of the surface between the machine peaks and valleys.

μin	Function and Description
125√	**General Machine Finish** Noncritical parts of an automobile engine or a lawnmower. Standard operations produce this finish such as drilling, turning, and milling. Fairly rough to the feel. Will leave a trail of fingernails if scratched.
63√	**General Machine Finish—Better Quality** Gasket surfaces or a smooth nonbearing shaft for a sliding control rod in disk player, for example. Made in the same way as the last but taking one extra cut. Will also remove a bit of fingernail when scratched.
32√	**Quality Finish** A shaft bearing or "mag" wheels before polishing. Generally requires special machining such as roughing and two finishing cuts. Or might require an abrasive (grinding) process to achieve. This finish costs more to produce. The last finish to remove fingernail. Any smoother will not.
16√	**Precision Finish** Hydraulic cylinder wall for a jack or a plastic mold surface. Must be planned carefully with roughing and finishing operations, then final abrasive work. This costly finish is only justified by special conditions.

µin	Function and Description
8 √	**High-Quality Precision Finish** Surgical instruments or a contact lens mold, for example. In addition to standard machining and grinding, these finishes often require secondary abrasive processes to reach this level of smoothness.
4/2 √	**Lapped and Polished** Farther along in your career, you might encounter a 4 µin. and a remote possibility of a 2 µin. (might be on your scratch plate, too). These are produced with very technical methods requiring exceptionally fine abrasives and a well-thought-out, three-level machining, grinding, and finishing plan.

UNIT 7-5 REVIEW

Replay the Key Points

- Unless otherwise specified, surface finish is specified as a maximum value; any smoother finish is acceptable. The 125 µin. is standard if not otherwise specified.
- Missing a surface callout can lead to scrap–these are tolerances.
- The cutoff distance is 0.030 in. unless otherwise specified.
- One of the main reasons for poor finish is caused by a combination of tool vibration, work vibration, and machine vibration! The main control is reduced cutter speed.
- The symbol for "surface must be established by machining" is a horizontal line in the bracket.
- The circle in the finish bracket indicates no machining is to occur.
- The numbers specify the average height of the marks above and below the median.

Respond

1. True or false? When no surface roughness is called out in the general tolerances, then all surfaces are to be 125 µin. If false, explain why.
2. Explain the two entries in the surface roughness symbol shown in Fig. 7-58.

32 √ X

Figure 7-58 Describe this roughness callout.

3. What data are missing that might be in Fig. 7-58?
4. Which is the smoothest finish–125 µin., 64 µin., or 32 µin.?
5. At your machine there are two instruments that might test surface finish. They are
   A. Micron meter
   B. Optical comparitor
   C. Surface finish comparitor
   D. Scratch plate
   E. Profilometer

CHAPTER 7 REVIEW

Unit 7-1

In this unit we strove for an initial repeatability of 0.001 in. even though, in the hands of a skilled user, micrometers can deliver around 0.0005-in. repeatability (or slightly better–depending on situation and user). The two major challenges to using them right are feel and zero-error calibration.

Unit 7-2

To an untrained observer it would be hard to see why a box of little steel blocks could cost a week's wage for a machinist. But you now know the reason. These little gems are made with the greatest precision of any measuring instrument in the shop. In order to make them, the machine shop must own some of the most sophisticated equipment on the planet. Not only that, but they also undergo many steps in their manufacture and each is performed by ultra-skilled people. Along with a couple of other gages and standards, gage blocks are the bottom line for precision in most machine shops. Even where CNC and CMC machines are used, the gage blocks still rule as the basis for accuracy.

Unit 7-3

Here we lumped together a series of small measuring tools that each perform a specialized measuring function. They all have some inaccuracy. One final thought about them (and all others we've studied) is to be on the lookout for technical advancements that improve what they do. They leave plenty of room for improvement!

Unit 7-4

There are *two kinds of angle callouts on prints:* the protracted emanating from a single point outward and the geometric lying within a parallel sandwich tolerance zone with respect to a datum. And although each requires a very different measuring setup, they are measured with the same tools used differently. There are also *two ways of expressing their value*: degrees-minutes-seconds (DMS) versus decimal degrees (DD). Then, as with all other measuring tools, the final skill was to know the various choices available to measure angles depending on tolerance.

Unit 7-5

While we use electronic profilometers to measure surface roughness, we also develop the ability to scratch a machined surface and come fairly close to its measured roughness. Similar to looking at a spinning cutter and knowing if it's right or wrong, working with surface finishes for a while will bring a sense of them with no scratch plate to use for comparison.

Final Review Quiz—Choosing the Right Measuring Tool

When faced with proving a part is within tolerance, nearly every dimension creates questions. Which measuring tool and process will prove beyond doubt that the feature is of the best quality? This final review is designed to simulate that daily game. The upcoming problems are designed to stimulate discussion about ways others might get the same job done! As you'll see in the answers, the basic tools we've learned thus far can be combined in many different ways.

Different from a classic test, the upcoming problems may have several answers. If your choice delivers the accuracy required, then you've got it. It's more than possible that you will come up with a method the planning committee and I didn't think of, not on the answer sheet! That's great if you do.

Your instructor or another skilled machinist may have differing opinions or expanded ideas from those in the answers, and that too is to be expected. Your instructor may also have real parts similar to those shown in Fig. 7-59.

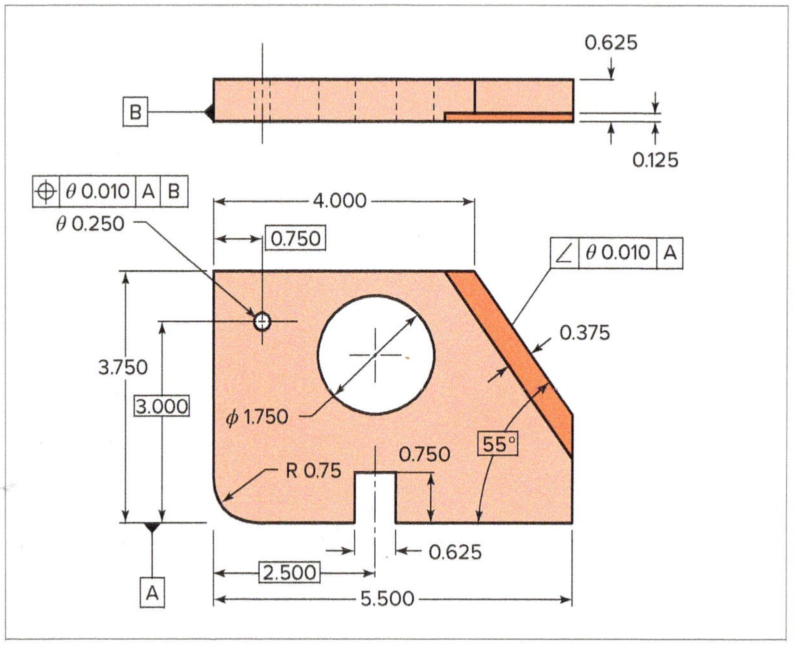

Figure 7-59 All process measuring test.

Guidelines for Choices

- First, before choosing the tool and process, know the tolerance(s) for the feature in question! The tolerance eliminates some tools and dictates which methods will deliver the accuracy required.
- Decide how quickly you must deliver results. Time added to a part cycle due to a poor measuring choice cannot be justified in industry. This is especially true on fast CNC equipment. For a one-time-only test, it's okay to use the slow method if it delivers. But when parts come off every few minutes, there's no time for it.
- Know your personal repeatability with each tool in the arsenal.
- *Never assume the tool you choose is in calibration!*
- Look for the spoilers and control them as you take the measurement.
- Always plan a backup, double-check process especially if it's an expensive or critical job.

> **SHOPTALK**
>
> The old masters in the shop gave two bits of advice about measuring: "Measure twice, then cut once!" and "It's a poor machinist who blames his tools." The meaning is if you get it wrong because your tools are out of calibration, it's still your fault!

Problems 1 through 10

Beat the Author by earning over 100 points total on the first 10 problems. Challenge another student or your entire class in this game.

Game Rules

- **Legal Tools** Use only measuring tools we've discussed in Chapters 6 and 7 except the smart height gage, but you may use them in any way possible that delivers the accuracy.
- You may also choose to use the granite table with right-angle plates to hold parts perpendicular to the table datum. Ground parallel bars are okay, in any way you choose. You may choose to lightly clamp the work on these utility items.
- **Base Points** The first bracket for each question sets the point value for finding a given number of solutions, the target for the question. *Finding fewer solutions than the target, no points are earned.*
- **Bonus Points** Each solution you find that is on the answer list, above the target number, earns the bonus value for that question. One bonus per answer above the target number.
- **Losing Points** If your number of solutions falls short of the target value, then deduct the bonus value multiplied times the number below target. (You'll notice that it's possible to score well below zero in this contest!)
- **Double Bonus** Discovering a reasonable answer that measures within the resolution of the question that isn't on the answer list adds *twice* the bonus to your score for each solution found.

If you identify the *best accuracy solution*, add one double bonus to your total. There may be more than one best accuracy solution—but you get only one double bonus for best accuracy.

QUESTIONS AND PROBLEMS

In addition to the objectives called upon in this chapter, you will be using the tools and processes of Chapter 6 as well.

1. For Question 1, you may use any item of known thickness to test for gaps. Describe or sketch *three* ways the 55° angle might be measured within the print tolerance. (Unit 7-4) (point value: 10; bonus value: 3)
2. Name *eight* methods that could measure the 0.375-in. dimension within ±0.005 in. (Units 7-1 and 6-3) (point value: 12; bonus value: 2)
3. Identify *three* methods to measure the 1.750-in.-diameter hole within 0.0005-in. tolerance. (Units 7-1, 7-3, and 6-3) (point value: 10; bonus value: 4)
4. Identify *two* methods to measure the 0.750-in. component of the 0.250-in. hole's position. Discrimination must be ±0.001 in. or better for your choices. (Units 7-2, 7-3, and 6-3) (point value: 8; bonus value: 5)
5. Find the *three fastest* ways to measure the 0.250-in. hole's diameter within ±0.005 in. in order of time consumed. Note: If your speed order matches answers add one double bonus! (Units 7-3 and 6-3) (point value: 6; bonus value: 3)
6. Identify *three* methods of measuring the width of the 0.625-in. slot within ±0.0005-in. discrimination. (Units 7-2 and 6-3) (point value: 12; bonus value: 4)
7. Find at least *five* ways to measure the 0.125-in. depth within ±0.002 in. (Units 7-1, 7-2, and 6-3) (point value: 10; bonus value: 3)
8. Using the tools and processes of Chapters 6 and 7, find *one* way to measure the 2.500-in. dimension to the 0.625-in. slot's centerline within a tolerance of ±0.0003 in. (Units 7-2 and 6-3) (point value: 15; bonus value: 10)
9. How many ways can you list to measure the 0.625-in. thickness of the part within ±0.010 in.? Target is *nine* solutions. (Units 7-1, 7-2, and 6-3) (point value: 12; bonus value: 3)

CRITICAL THINKING PROBLEMS

10. Is there any practical way to measure the size of the corner 0.75-in. radius other than radius gages? (point value: 5; no bonus value here)

    End of contest, add up all points including deductions and check your score against your opponents (and me); then answer the remaining questions.
11. What does the box around the 55° angle mean? (Unit 7-4)
12. Return to Question 2. What method would be the most practical and quickest? (Unit 7-1)
13. If the surface of the hole in Fig. 7-59 was given a finish callout of 32 μin, would it be possible to achieve using a standard drill? (Unit 7-5)
14. The position tolerance for the 0.250-in.-diameter hole applies to its (A) surface, (B) centerline, or (C) axis?
15. What is the distance of the 1.750-in. hole from Datum B? How do you know?

CNC QUESTIONS

16. A program is running well. All key dimensions are within tolerance and the surface finish is better than 125 μin. However, with each part cycle it seems to be getting a bit rougher. What might be the cause and fix?

17. CNC lathe is turning out gear shafts that have just one key dimension—they must be 4.500 in. long with a tolerance of ±0.003 in. The machine cycles (makes one part) every 1 minute and 45 seconds. It has an automatic bar feeder and a finished part catcher so you are free to measure and record data with no duties between taking out the sample part. According to your job plan you are to measure every tenth part, then enter the data into the SPC software on your controller to track real-time quality.

A. What facts must be considered to make your choice of measuring process?

B. Name at least three ways this dimension might be measured.

18. Explain what a 125-μin. finish means. Four facts minimum. (Unit 7-5)

19. A surface finish bracket has only the 63 μin. with no other information. What, then, is the cutoff distance? (Unit 7-5)

20. What is the classic definition of a measuring gage? (Unit 7-3)

Design element (gear): ©Design Pics/PunchStock

PART 2

Introduction to Machining

Now in the second part of *Machining and CNC Technology*, we focus on the skills needed to shape metal using a cutter or grinding wheel. Chapter 8 is not to be missed, because in it you'll learn the basis of all machining, the physics of making chips, and how to contain the forces encountered on various kinds of machines.

Today's CNC machines are so fast and powerful that we machinists can no longer estimate or guess how to write a program or set up a machine; everything must be right. The subject is called "Speeds and Feeds."

But the chances are your lab hasn't enough CNC machines for every student, so we'll present this material assuming you might be practicing basic setups, operations, and safety using manually operated machinery. Sometimes, however, there are big differences in using your hands to start a machine motion compared to the computer taking control. Those differences will be pointed out so that when you get to Part 3 on CNC machining, you'll also understand their capabilities.

But don't get the idea that manually operated equipment is gone in industry. Not even! To prepare for this book I estimate I've visited 50 machine shops, large and small. In every one, there were some manually operated machines. Sure, they were far from the heart of the shop, but they were used for secondary operations when it was not practical to tie up the CNCs to do a job. Manually operated machines are also used for simple production when the tooling has been built and the cost of moving the job to the CNC doesn't make dollar sense. Manually operated machines are especially useful for building tooling for the shop or for repair work.

In Part 2, you will learn

Chapter 8 To gain complete control of the machining process by studying the physics of cutting metal and how to vary the cutters doing the work; to control longer tool life, better accuracy, finer finishes, and more efficient production, always with safety as the overall goal.

Chapter 9 To drill holes to the specified size, at the exact location—a baseline requirement for machinists. While your early training will be on manually operated drill presses, the skills gained will apply to all CNC programs that call on drill bits to remove metal.

Chapter 10 That lathes spin the workpiece, then cut metal away using a moving cutter; they create more potential danger, and you need to completely understand setups and turning operations. Although this book assumes you'll be practicing early setups and operations using a manually driven lathe, the lessons are aimed at their safe, efficient performance on a CNC machine.

Chapter 11 That mills perform most metal shaping today. Using a computer to drive multiple axes of motion and to feed the work through any complex path, almost any shape can be milled.

Chapter 12 To turn to metal removal by abrasion (grinding it away) when extreme accuracy is required or when finish requirements or workpiece hardness exceed the ability of chip making—a skill that applies across the whole range of processes; every machine type has a counterpart that grinds.

Chapter 13 The range of threads to be machined and a bit about how to machine them. While this is our second investigation into this subject, we have only just begun, but you will now have the background needed when specialized threads are required.

Chapter 14 More complete control through an in-depth understanding of metal; alloy identification and makeup; and how to change steel and aluminum's hardness and how to test the results. Beyond Chapter 14, metallurgy is one of the most fascinating frontiers with the potential to affect your future.

Chapter 15 To accept and succeed at the challenge of challenges! Planning out a logical sequence of operations that lead to the perfect part, in the least amount of time, with the safest possible path is an incredibly complex task that can make the difference in profit and loss—and survival as a company!

CHAPTER 8
Cutting Tool Geometry

Units

8-1 Four Universal Cutting Tool Features

- Identify rake, clearance, and cutting angles
- Know the benefits of corner radius on tools

8-2 The Physics and Forces of Chip Formation

- Define the shear line in a chip
- Identify two sources of chip heat

8-3 Cutting Fluid Technology

- Know what kinds there are and how to apply them
- Define the advantages of using cutting fluids
- Know when cutting fluids are not the right choice in a setup

8-4 Taking Control—Adjusting Variables

- Control variables for learners
- Improving results

INTRODUCTION

We first take a close look at what happens when any tool cuts a chip away from a metal part, and why the tools that do it are shaped as they are. This is bedrock theory for all machining, even grinding, where the chips are microscopic, but chips just the same.

Surprisingly, with over 200 years of learning about the subject behind us, cutting tool technology is currently undergoing its greatest development other than during the period when tungsten-carbide became commercially affordable! Cutters are one of the few hardware issues evolving as fast as computers and the software that drives them. In fact, the two—cutting tools and CNC computers—are locked in a leapfrog growth, with each spurring the other to improve.

Each new generation of central processors crunches faster, enabling them to drive the machines at unheard-of velocities and RPMs. Then tools must be improved to take full advantage of the increased speed. Once the tools catch up, for a time, we cut with new cutter compositions, new coatings, or even gem crystals grown upon their surfaces, always aiming to improve performance. Tools are also shaped in radical new ways because computer simulations provide deeper insight into how a chip flows past the cutter at accelerated speeds. So, for a time, the tools can take all the abuse this generation of CNCs can dish out and the process goes on. But before we can discuss the basic operations and then the future, we need a starting point.

The first two units of this chapter explain how all cutting tools produce a chip (called subtractive manufacturing—removing material). Then we'll look at applying cutting fluids and end with how to control the whole process using the variables that can be adjusted by a skilled machinist to improve finish, accuracy, tool life, and safety. Make sure you fully understand the material in the presented sequence. These units are steps on a critical ladder, one leading to the next, with understanding and control at the top.

SOLVE THE SELF-HEATING CHIP MYSTERY!

There's an enigma to solve in Chapter 8: Metal chips become hot due to extreme friction during machining. Okay, that's no surprise. But the mystery is that once they're cut away and lying on the floor, they seem to get hotter. Here's proof: Due to that puzzling rising heat, steel chips begin to oxidize and turn rainbow colors right before your eyes, after the chip is fully formed. Why, then, does this occur when the friction is over? As we go forward, we'll pick up clues to solve this puzzle.

Unit 8-1 Four Universal Cutting Tool Features

Introduction: All cutting tools share four common shape features–rake angle, lead angle, corner radius, and clearance angle–each of which must be present to varying degrees for the cutter to work correctly. They are the basis of tool geometry.

The cutting bit shown in Fig. 8-1 is generic. It could be the tip of a drill bit as it rotates in the metal, moving to the left in the illustration. It could be the tip of a lathe bit as the work spins past to the right, or it could be the cutting edge of a milling machine cutter as it rotates and is fed sideways into the material. Representing all three, the figure shows two of the four features needed to cut metal.

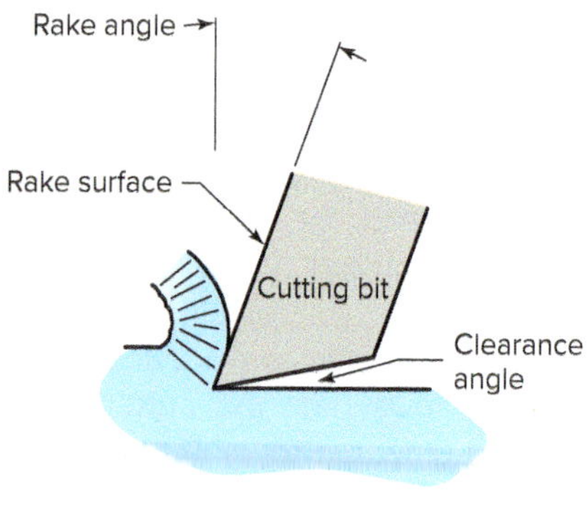

Figure 8-1 A generic cutting tool showing the rake and clearance angles.

For now, as a clear example, let's relate it to a familiar operation, drilling. Envision that this bit is spinning. We are looking at it from the side as it makes a spiral chip. While the sharp edge cuts, the action isn't a simple slice as one might assume. There are a few surprises here. As always, understanding the terms will be very useful in your career and to unravel the hot chip mystery.

TERMS TOOLBOX

Chatter Unwanted vibration in the tool, the work, or the machine. Chatter can destroy tools and ruin finishes.

Clearance angle A relief behind the edge to enable cutting contact.

Corner radius (nose radius) The rounding of the point or junction of the front and side of a tool bit's cutting edge.

Deformation The permanent bending of any metal, in this case, of the chip as it is raked.

Lead angle/cutting angle The angle of the bit compared to the angle of the cut. Lead spreads the chip width over a larger area.

Negative rake The tool that bends the chip the most.

Nose radius Term used for corner radius on lathe tools.

Positive rake The tool that bends the chip the least.

Rake angle/rake surface The surface upon which the chip forcefully slides.

8.1.1 Two Distinctly Different Actions: Cut and Rake

Two different actions occur together when a chip is made. First, the cut occurs where the sharp edge slices the chip away from the parent metal. Then immediately after the rake begins the now flowing chip is forced up, across the surface or face of the bit–the rake.

The Rake Surface

During this brief phase, the chip comes into forceful contact with the bit. As it slides up over the rake surface of the cutter, the chip is redirected, bent to flow in another direction.

There is a lot of friction at the contact point between chip and tool bit face, where the chip is raked. That pressurized rake zone is critical to the cutting action and can be in hundreds of pounds per square inch.

Beyond the rake zone, the bent chip is fully formed. It continues to slide up the extension of the surface and out of the drilled hole in our example or off the bit in some other way in other operations. But the true rake occurs from a few thousandths to no more than one eighth of an inch beyond the cutting edge. The distance depends on how hard the bit is being pushed through the metal, a process called the *feed rate.*

If the feed rate is light, the rake can be a minor action compared to the cut. Or it can be the major part of the chip removal. In the drilling example, you might be exerting 5 or 10 pounds of pressure downward on the drill press feed lever, producing a small chip. In that case, there is a very short rake.

The rake becomes ever more significant as feed rates climb. Pushing down with a force of 25 pounds, the chip pressure increases on the bit face, and the rake zone widens away from the cutting edge. The rake contributes more to the formation of the chip.

In CNC production, we're always looking for ways to cut more metal per minute. As we feed ever harder, the rake phase becomes more important. We experiment with varying rake angles, improving cutting fluid formulas, and using better antifriction materials on the bit to help the chip slide up the rake face, all to achieve faster cutting and longer tool life.

KEYPOINT

As feed rates increase, the rake becomes more significant relative to the cut, and we use that fact to our advantage.

Chip Deformation

The **rake surface** of the bit redirects the chip as it flows. It bends it away from the direction it would take if the rake face wasn't in its path, somewhat like a wedge. This permanent bending action is called chip **deformation**. Remember that; it's one of the clues that solve the hot chip mystery! But here's another odd fact and perhaps another clue: Of the total heat produced in making the chip, ***only 25 percent of the total heat comes from the friction between the bit and rake face during the deformation***.

KEYPOINT

The bit does not split the chip away from the work as when wood is split, but rather, right after the cut, the chip is deformed to another direction by friction against the rake face.

8.1.2 Rake Angles from Positive to Negative

Changing the rake angle relative to the cut direction can have a dramatic effect on the cutting action and on the chip. We'll return to the various angles involved later, but notice in Fig. 8-2 that a **positive rake** deforms the chip the least, while a **negative rake** bends it more. Changing rake angle is one of the control variables.

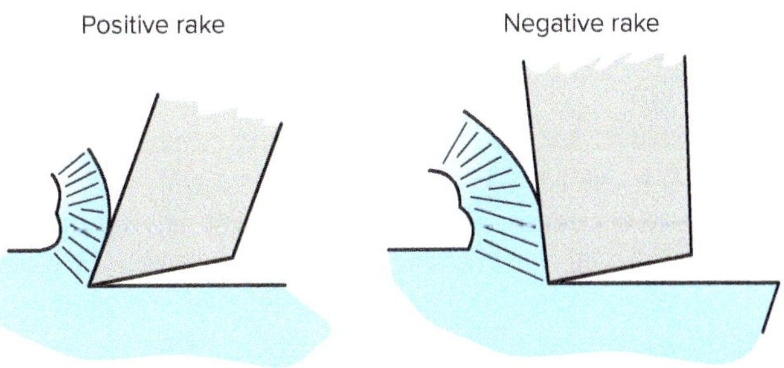

Figure 8-2 Positive and negative rake bits.

Investigate It!

Ask your tool room attendant for a relatively large drill bit of about a 0.5-in. diameter or bigger. Look for one that has been used for some time. Now starting at one of the two cutting edges, trace a pencil up the spiral groove. The groove is called the *flute*.

Similar to the microphotograph in Fig. 8-3, there will probably be a small shiny portion next to the cutting edge where the dark or dull finish has been worn shiny. That's evidence of the rake zone friction.

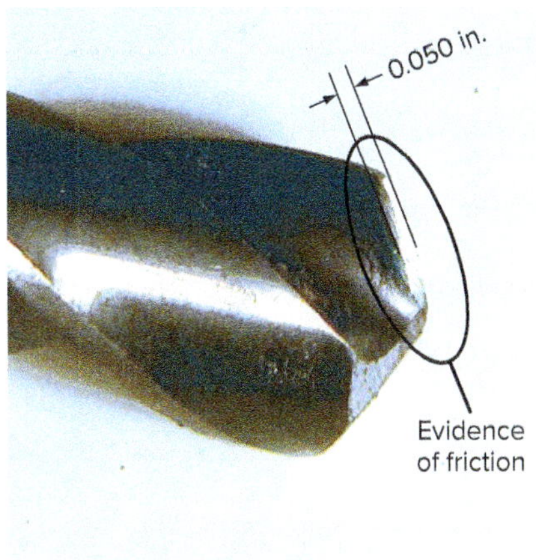

Figure 8-3 The small shiny spot is evidence of chip friction against this drill bit's rake face.
Courtesy Mike Fitzpatrick

Note that, beyond the rake around 0.050 in., the drill's surface returns to the normal color. Keep this drill close by; we'll look at it again.

Critical Ladder Check Point

Before moving on to clearance angles, be sure you can define the two different phases a chip undergoes during its formation. Can you describe them to another student? What happens to the chip during the rake phase?

8.1.3 Clearance (Relief) Angle

Return to Fig. 8-1 and note the **clearance angle**. It serves a very different purpose from rake–it enables the cutting edge to contact the work, to dig in, with no rubbing behind the edge. You might also hear it called a relief angle because that's what it does; it relieves the cutter behind the cutting edge, to prevent unwanted friction.

In Fig. 8-4, you see a correct bit compared with an incorrect bit with zero clearance. The bit on the right will not cut. Not only that; it will quickly heat up due to the rubbing between the clearance surface, sometimes called the heel of the bit, and the work.

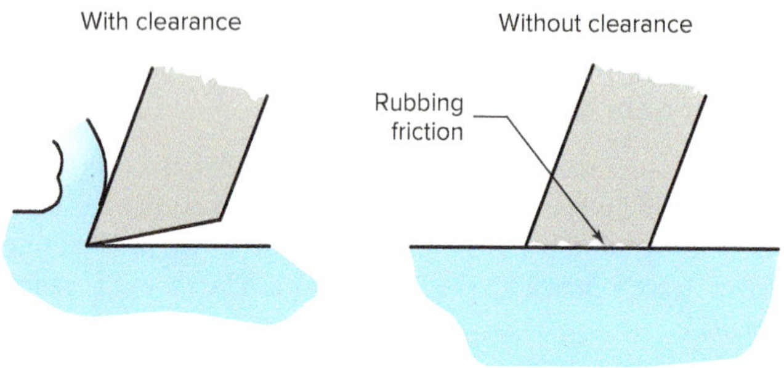

Figure 8-4 The bit on the left will cut while the right bit with *zero clearance* will only rub.

Just Enough and No More!

As long as a minimum amount of clearance is present, the bit will work. Adding more has no effect on the cutting action. But with too much clearance, the bit will cut only for a while. Excessive clearance makes the bit weak due to the thinning of the cutting edge, as shown in Fig. 8-5. It usually breaks before it dulls. The correct angle is usually around 3° to 7°.

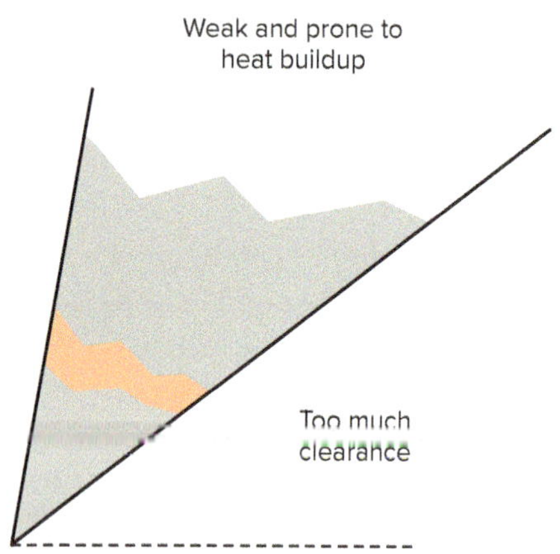

Figure 8-5 *Too much clearance* creates a weak bit that is prone to heat buildup and breakage.

> **SHOPTALK**
>
> When a bit with inadequate clearance rubs the work, it's sometimes called "heel drag."

> **KEYPOINT**
>
> Changing clearance has little effect on the actual chip formation. It is necessary to have only a minimum amount. We do not adjust clearance to achieve a result. It is not a control variable.

Excessive clearance not only weakens the bit; the thin edge also makes it a target for heat buildup as well. With less mass to conduct heat away, bits burn up quickly with excess clearance.

There is one positive but *very minor* aspect to adding more clearance to a bit; the triangular, hollow space created by the relief angle face enables coolant to collect, which is a good thing. However, when balanced against the potential of a bit breaking or burning up, coolant induction doesn't justify more clearance!

Identify the clearance angle on your example drill. In the shop, you'll need to identify the clearance angle on every cutting tool selected. Note that when resharpening a drill bit, you will be grinding the surface that makes the clearance, called the *heel* of the bit. (More on this in Chapter 10.)

Critical Ladder Key Points

- Machinists change rake angles to achieve different results but not the clearance angle.
- Clearance is a matter of "just enough" and no more.

Got the idea? If so, we step closer to solving the chip mystery–which, as you may have guessed, has nothing to do with clearance.

8.1.4 The Lead or Cutting Angle

The third universal tool angle is called either the **cutting angle** or the **lead angle** on many cutting tools. (It's often shop-shortened to "lead," which rhymes with feed.) The cutting angle is an important chip control feature on all bits.

Lead is the angle of the cutting edge compared to the cut direction. It imparts many benefits to the action. To see the cutting angle on the example drill bit we rotate it 90°. In Fig. 8-6, the cutting angle on both edges forms the point of the drill bit.

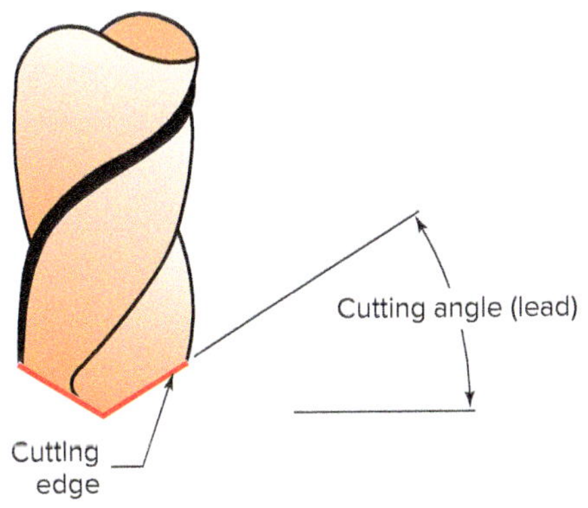

Figure 8-6 The *cutting angle* may also be called the *lead angle* on other cutting tools. The combined cutting angle of both flutes forms the *point angle* of this drill bit.

What Is Lead?

To understand the action a lead angle imparts, envision yourself whittling wood with a knife. If you hold the knife square to the direction you wish to carve, the chip (shaving, in this example) will be the same width as the work (Fig. 8-7). The knife enters the wood all at once and exits the same way, abruptly. That's zero lead causing abrupt entry and exit.

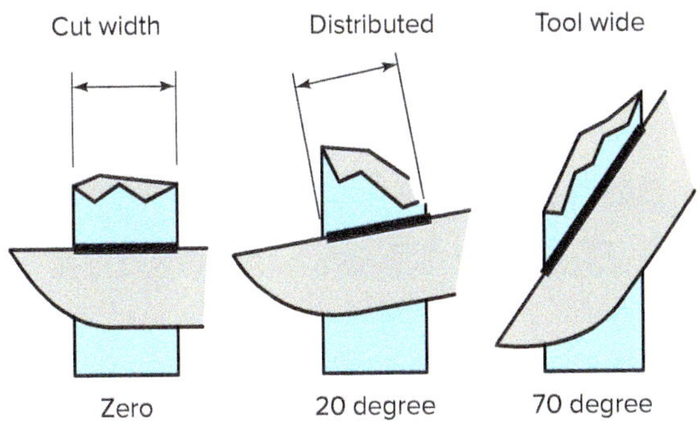

Figure 8-7 A comparison of three lead angles.

That action would be awkward and would not work well, so you would naturally rotate the knife to an angle of about 20° relative to the direction of the whittle–that's a lead angle. Here's why.

Easy Entry and Exit

With lead, the knife would start into the work at a single contact point and the contact width would grow wider gradually to the full chip width. It would then slice cleanly as it moves forward, then exit the work gradually, too.

Look at Fig. 8-7. More cutting edge contacts the work with lead than without lead. Lead angles on metal cutting tools vary from 0° up to a maximum of around 30°, as on a drill bit.

Additionally, when lead is used, the chip produced is wider. Because it spreads over a longer cutting edge, the tool also lasts longer in most cases. The chip is thinner and easier to shear away when lead is used.

KEYPOINT

Lead angle produces

A. Smoother entry and exit.

B. Longer tool life due to more cutting edge doing the work.

C. Better finish on the work.

We change lead as a control variable.

Too Much Lead Can Cause Chatter

There is an upper limit to the benefits of increasing the cutting angle lead. Imparting too much produces a chip that's too wide. Beyond a certain point, when too much cutting edge comes into contact with the work, an unwanted vibration starts, known as **chatter**. If left uncorrected, chatter can ruin the finish on the work and even break the tool.

That's one way lead is used as a control variable: to reduce chatter when it occurs and lessen the lead angle. This reduces the contact length between bit and work. Several other controls can also reduce chatter. We'll learn them under troubleshooting (Unit 8-4).

Too Little Lead

Returning to the drill bit example, assume there is no lead angle. The bit would be flat on the tip, as shown in Fig. 8-8. On that flat-pointed drill, the chip width is the radius of the bit. That drill would be a nightmare to get started drilling. Trying to take the entire chip width the instant it touches the work, it would grab, try to walk off center, and then snap back repeatedly before it would start drilling. It might even break the drill.

However, if we grind it with a point angle (a cutting angle on both edges), the drill bit would tend to start into the work more smoothly, at a tiny circle that gets wider as it goes down. The chip width increases to the full length of the angular cutting edge. So comparing this to the wood carving example, we see that the pointed bit exits out the far side of the hole the same way, with a tiny circle breaking through that gradually widens to the full diameter finish. That's because it now has a lead angle.

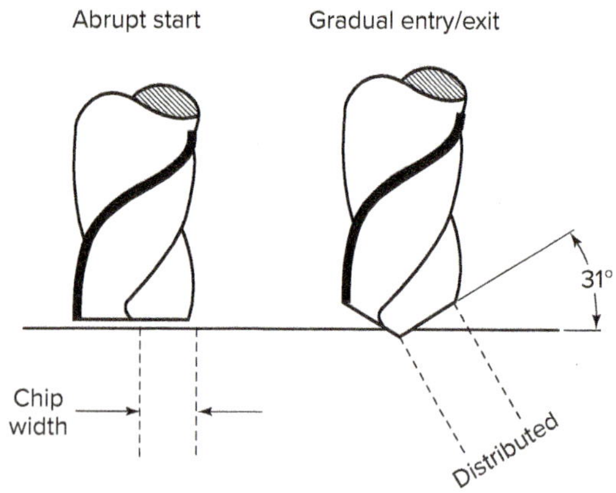

Figure 8-8 A drill ground without a point angle (lead) will be difficult to start into the work.

8.1.5 Adding and Changing Corner Radius to the Cutting Edge

Corner radius is the fourth basic feature of tool shape. In Fig. 8-9, a drill is shown with a **corner radius**, which is a rounding of the cutting edge between the intersection of its front edge and side. The lines on the right represent the standard grind, unmodified, sharp-cornered drill bit. Corner radius is far more common on lathe and mill cutters, but it imparts the same benefits on drills or any cutting tool. (See Trade Tip.)

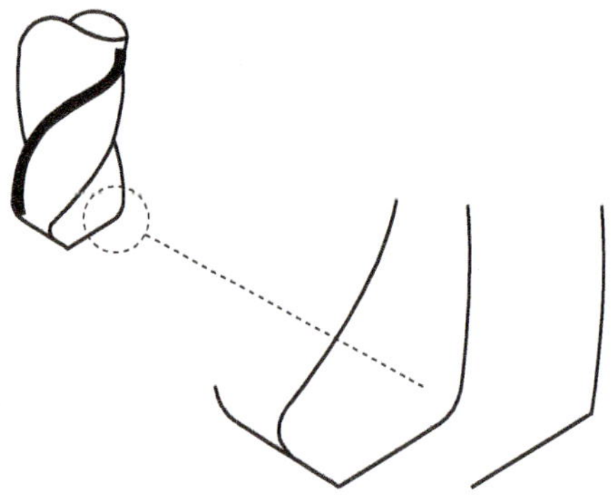

Figure 8-9 Corner radius is more common on other cutting tools but can be ground on a drill bit, too.

TRADE TIP

When no reamer is available have a journey machinist demonstrate grinding a corner radius on your drill bit. The main improvement is better finish in the hole, as well as improved size control. Corner radius also prevents edge breakdown when drilling harder materials; the radiused bit lasts longer.

There are three benefits:

1. **Smoother Cutting Action and Chip Distribution** Adding radius to a cutting edge has an effect similar to adding lead angle. It smoothes the cutting action and distributes the chip over a longer cutting edge. But corner radius adds two more positive benefits to the tool's geometry.
2. **and 3. Better Finish and Stronger Tools** With a corner radius, the tool is stronger and less prone to breaking at the point intersection. The finish a radiused tool produces is also smoother. Figure 8-10 shows two lathe tools machining metal. With radius, the tool leaves tiny wave-like marks in the finished surface compared to sawtooth marks for the sharp-cornered tool. When that corner radius is put on a lathe tool, as in Fig. 8-10, it's called **nose radius**.

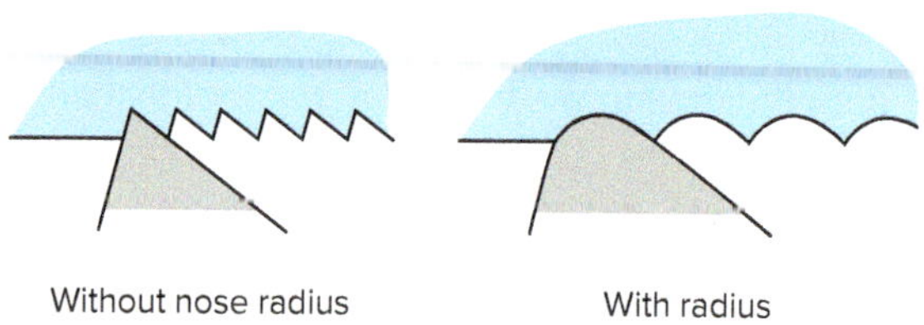

Figure 8-10 Finish is improved with corner radius.

Too Much Radius

A large corner radius can cause chatter in the same way as too much lead angle; there is too much bit in contact with the work. If chatter arises, reduce radius to shorten the contact edge of the bit with the work. There is a time when chatter is such a problem that a sharp-cornered tool is the answer.

KEYPOINT

Corner radius is one of the control variable features.

Adding corner radius brings three benefits:

- Stronger tools
- Smoother machining action
- Improved machine finish

UNIT 8-1 REVIEW

Replay the Key Points

- Rake angle is a control variable.
- As feed rates increase, the rake becomes more significant relative to the cut.
- During the rake phase, the flowing chip is deformed into another shape.
- Clearance is a matter of enough and no more.
- Clearance is not a control variable as long as the bit isn't rubbing on its heel.

- Lead is a control variable.
- Lead imparts gradual entry/exit in the cut, longer tool life, and better finish.

Respond

1. List and describe the three basic tool angles.
2. Of the three answers to Question 1, which are control variables?
3. True or false? The cutting angle is varied to achieve different results in machining.
4. Name the two phases of chip formation.
5. Of the two phases of Question 4, which becomes more significant as the feed force is increased?
6. What is the range of common rake zone widths?
7. What are two benefits of tool radius?

Unit 8-2 The Physics and Forces of Chip Formation

Introduction: In Unit 8-1, we didn't explore all that happens inside the chip. After we complete this study, hard, unbending metal might seem more like clay or hot plastic as it yields to the cutter.

Machinists are expected to operate their machines at maximum efficiency yet not break or destroy tools. To do that, choices must be made based on the upcoming knowledge. Cutting just a little too fast can burn up or break the bit, requiring time to resharpen or replace it, and, often even more costly, lost production time to get the setup back up and running again. But running too slow to preserve the cutter brings reduced productivity. It's a decision to be balanced based on experience and knowledge of the facts presented here.

TERMS TOOLBOX

Boundary zone The small space between interference point and cutting edge on a bit.

Built-up edge The result of chip weld holding on too long, then allowing a small part to escape under the bit to reweld to the surface of the work.

Chip load The amount of chip being taken as a combination of feed rate and depth of cut. See *Removal rate.*

Cratering The erosion of the rake surface of a bit due to welding.

Deformation The bending and piling up of the chip during the rake phase.

Interference point The concentrated center of contact pressure between chip and bit.

Negative rake The bit deforms the chip through an acute angle, causing more bend than positive rake bit.

Plastic flow The characteristic of a chip that causes it to deform without returning to its original shape.

Positive rake The bit deforms the chip through a lesser angle, causing less deformation to the chip.

Removal rate The combined result of depth or cut, feed rate, and cutting speed (RPM) rated in cubic inches per minute (or liters per minute). Equates directly to horsepower requirements for the cut.

Shear line The line inside the chip along which the deformation occurs.

Welding The instantaneous, heated adhesion of the chip to bit. Leads to cratering and built-up edge.

SHOPTALK

Removal rate = speed and feed = volume in cubic inches/minute There are two ways to machine more metal per minute: by cutting a thicker chip, known as the depth of cut, and by feeding the cutter faster through the work. Then by speeding up the rotational RPM, they combine to become the **removal rate** per minute, in cubic inches. Removal rate is the measure of efficiency. We'll use the term here, then calculate it later.

8.2.1 Force, Friction, and Heat

Choosing the right cutting speed and feed can prolong tool life in ways that might surprise you. To understand how, look closely at a chip being made. Notice that there is a pressure curve against the rake face. It starts out light near the cutting edge, then increases to a focus point farther back from the edge, and finally lightens up again until the chip no longer contacts with the bit (Fig. 8-11).

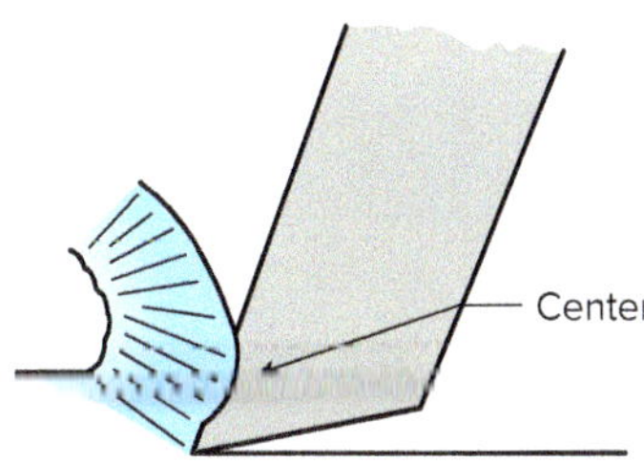

Figure 8-11 The pressure gradient on the rake face centers at the interference point.

Evidence of that gradient pressure can be demonstrated by looking at the rake face of a tool that's been used to cut steel chips for a long time with no coolant. When we do so, we see a worn spot that formed due to the extreme friction wearing away small particles of the bit. We'll look at ways to prevent **cratering** because it will eventually destroy the bit. But for now, we're using it to demonstrate how the pressure is distributed.

The center of the crater where the pressure was the greatest is a distance back from the cutting edge. That central focus point is called the **interference** (or *contact*) **point.**

> **KEYPOINT**
>
> Good cutting speed and feed rate choices push the interference point up the rake face, away from the cutting edge, such that the mechanical pressure, friction, and heat are distributed over a stronger part of the tool.

External Friction and Heat

Pressure at the interference point can be as high as 100,000 pounds per square inch between chip and bit. That's not a misprint; the total force is concentrated on microscopic contact points as the chip slides by. In a setup with no coolant or an extreme cutting speed, the resultant heat can actually cause **welding** of the chip to the bit. But each tiny weld must break as the chip pushes forward. This weld-break-weld cycle not only causes cratering but can also affect the finish.

That high-pressure contact causes "external friction and external heat." Some of the external heat is conducted away in the bit and in the coolant, if used, but most is conducted to the chip and the uncut workpiece as well. Everything heats up; work, bit, and chip can all become dangerously hot! Again, by knowing the forces and actions, a skilled machinist can control heat and even use it to good advantage as in high-speed machining (see Chapter 27).

So, here's a big clue to solving the "hot chip mystery." While the removed chip receives its share of the external heat, this clue doesn't quite explain why the chip continues to heat after removal. There is no more friction once it passes the rake surface.

> **KEYPOINT**
>
> A chip can reach as much as 1,000°F or more.

Surprisingly, only 25 percent of the total heat created in machining is attributable to external friction.

Uncontrolled, heat burns up the bit and possibly work hardens the material. It's also personally dangerous as the heated chips fly off the work. Metal expansion due to heat can also cause havoc with measuring, as we've already seen (Fig. 8-12).

Figure 8-12 Uncontrolled friction and its crime partner, heat, can make any cutting tool fail too soon.
Lake Washington Technical College, Kirkland, WA/McGraw Hill

Deformation Heat Is Internal

The final clue to solving the mystery lies inside the chip as it is deformed up and away from the work. Notice in Fig. 8-13 that as the chip presses against the rake face, friction causes it to stack up. The chip actually gets thicker than the depth of cut.

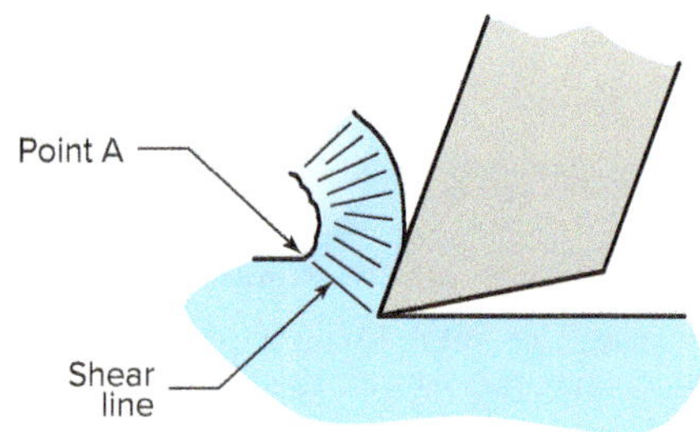

Figure 8-13 As the chip deforms it gets thicker than the depth of cut—along the shear line.

As the chip is raked, there is a continuous, forced "flow" of metal along a real line called the **shear line**. It occurs at the "shear angle," which is in the chip. Due to external friction against the bit, the chip is stacked thicker. As it does so, grains of metal are rubbed along the line that extends from point A, close to the bit tip in Fig. 8-13.

Mystery Solved!

The greater part of the total heat in the chip-making process is from internal friction. The heat originates inside the chip and conducts outward.

It is desirable to increase the shear angle, thus shortening the shear line, which in turn reduces internal friction and heat. The shear angle can be changed with three elements of control:

- Coolants to lubricate the external friction
- Harder bit faces that resist friction so that chips slide past
- Varying rake angles on the bit

Investigate It

Machinists and programmers must understand these chip-making principles, especially when pushing the limit with fast CNC equipment. Here are four experiments that bring tool geometry concepts into clear focus.

Experiment 1—Modeling Clay

Using plastic clay and a wooden bit, we can simulate nearly all of the concepts we've discussed–rake, clearance, deformation, and shear line. Figure 8-14 shows a bit being fed through the material. Note the deformation and chip thickness relative to depth of cut. If you perform this experiment, watch closely and you will see the shear line (emphasized in the photo).

- Try varying the rake angle and observing the chip thickness.
- Try cutting with no clearance.
- Try cutting with an unsanded, rough wood tool, then a smooth rake face and note the difference in the stacking. (Modern tooling is coated on its surface with exceptionally hard materials, and some are even micropolished on the rake face.)

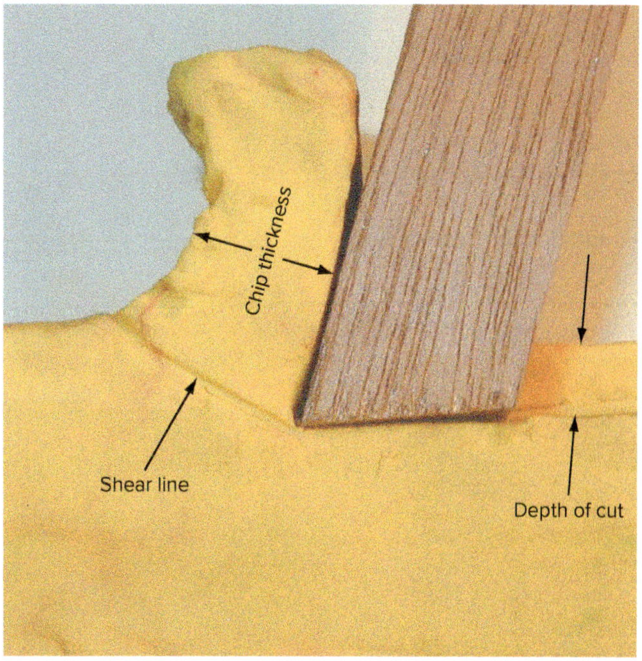

Figure 8-14 Chip deformation can be demonstrated with clay.
Courtesy Mike Fitzpatrick

Experiment 2—Look at Any Chip

To see firsthand evidence of this stacking effect, with safety glasses on, go to the shop and find any example chip made by any process and notice that it is smooth on one side and rough on the other. Do not take it from a moving machine. Is it obvious which side was toward the bit and which was away? With experience, you'll learn to read that difference. When cutting speed is too slow, the rough side shows signs of poor, "lumpy" **deformation** (Fig. 8-15).

Figure 8-15 The smooth and rough sides of typical chips indicate deformation.
Courtesy Mike Fitzpatrick

Experiment 3—Internal Heat

Hold a piece of thin sheet metal with the sharp edges removed to protect your hands, and bend it repeatedly. What happens at the bend? It heats up from the shifting metal grains. Eventually the metal's granular bond fatigues, and it begins to break along the bend line.

Experiment 4—Blue Steel Chips

Your instructor may have examples of this phenomena. Evidence of chip heating can be seen when machining steel (Fig. 8-16). When it gets above 375°F, it begins to form surface oxides from the air in the room. Different temperatures produce different colors. The lower chip in the photo runs from a near-straw color at 430°F to brown at 450°F. The upper example starts at deep brown at 500°F and ends with deep blue at 550°F. These chips were made by programming constantly increasing downward force on the drill while increasing the RPM.

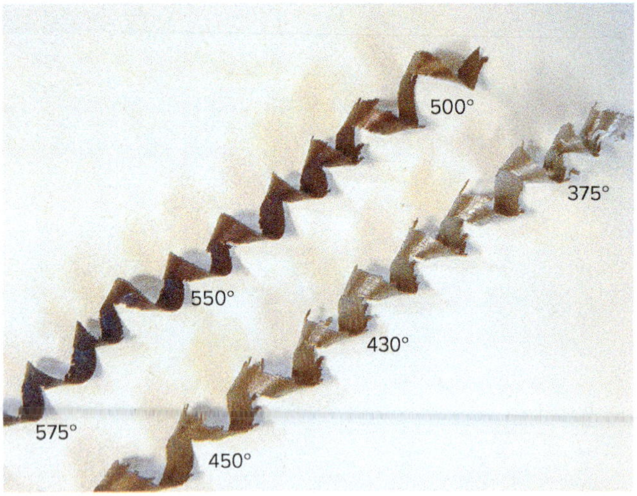

Figure 8-16 Steel chips change color due to oxidation. Various colors indicate different temperatures.
Courtesy Mike Fitzpatrick

But the odd fact is that the chip comes off the work a shiny steel color; then lying on the floor, it begins to turn light yellow, then light blue, then dark blue as it heats from the inside out.

8.2.2 Machinability

Resistance to deformation differentiates various metals as to the relative ease or difficulty with which they can be machined. For example, soft aluminum will easily deform, while steel resists from 3 to 10 times more compared to aluminum. The softer metal will heat less, both externally and internally, and requires less horsepower to machine. That can be envisioned in the clay experiments. Warm clay versus cold–how would they each deform when the model bit is passed through?

> **KEYPOINT**
>
> There are two sources of chip heat: external from the rubbing of the chip against tool and internal caused by the deforming or internal sliding of the chip against itself.

Now that you understand the heat source, let's see how to control them.

8.2.3 Changing Rake Angles

Figure 8-17 depicts exaggerations of three bits making a chip. Notice that the depth of cut remains the same, only the rake angle is changed, and that changes the thickness of the chip. As the rake becomes more negative, it's a thicker wedge; thus, the chip must deform more.

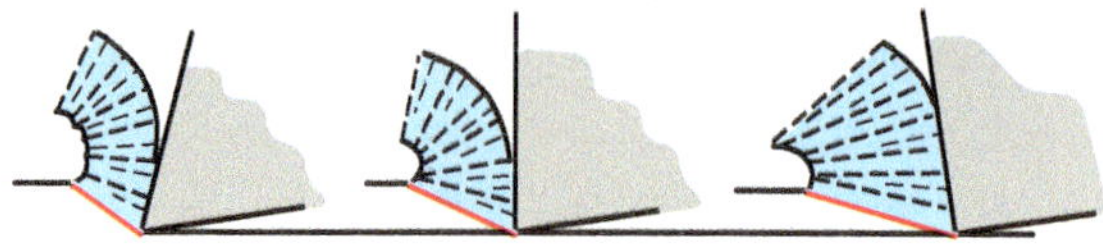

Figure 8-17 Changing the rake angle affects the shear line length, altering chip thickness, which produces heat and changes the force required to machine.

> **KEYPOINT**
>
> Keeping all other variables the same, a chip made by a positive rake tool is thinner, closer to the depth of cut compared to one made with a negative rake tool. **Positive rake** bits require less energy to produce the chip.

As the rake surface becomes more negative, more deformation occurs, which results in more heat both externally from more pressure and internally from a longer shear line. A **negative rake** tool also consumes more energy (machine horsepower).

Coolants (Cutting Fluids) Shorten the Shear Line

Coolants also lubricate. They reduce external friction, which makes the shear angle go up, resulting in a shorter deformation and far less overall heat. So coolants play a double role in the reduction of heat.

Harder Bits Shorten the Shear Line

If the bit material is harder, it offers less frictional resistance to the chip. In other words, the chip slips past easier. It has the same effect as adding a lubricant to the interference point.

Critical Question

Assuming no other variable changes, what differences would you expect to find in the machining if you changed to a more negative rake bit?

1. More external and internal friction
2. More force required to do the job
3. Hotter chips

So, from these traits, one might assume that a positive rake tool is the way to go–but it isn't always so. In fact, in production machining, negative rake tools are the more common. Why? The answer requires investigation.

SHOPTALK

Neutral Rake In theory, in Fig. 8-17, the center tool is a neutral rake. However, in actual shop practice, a neutral rake bit performs much the same as a negative rake.

Why Not Always Use Positive Rake Tools?

That's a natural question. It seems that all factors are improved when we do. Less heat is produced, less force is required, and the chips are cooler. The simple answer is we would if we could, but the bit with positive rake is the weakest due to its shape. If we wish to push a positive rake bit harder—that is, to machine more metal per minute—the tip can snap off due to its thin shape. Also, positive rake tools don't conduct away heat as well due to their lack of mass. They are more prone to burning as well. But strength and mass aren't the sole reasons for using a negative rake tool. The longer answer is coming up—pushing the boundary.

Each rake angle has its correct application. Your objective is to understand them both and when to use each. Unit 8-4 troubleshooting and some experience machining will both help gain a feel for rake angle decisions.

KEYPOINT

Increasing the removal rate is the prime reason to select a negative rake tool.

8.2.4 Preventing Cratering and Built-Up Edges

As removal rates increase toward maximum, two problems can arise, both caused by friction at the interference point: cratering and edge buildup. As mentioned previously, cratering (Figs. 8-18 and 8-19) can destroy bits, while edge buildup can ruin the finish on the workpiece. Both are traced back to the high pressure at the interference point.

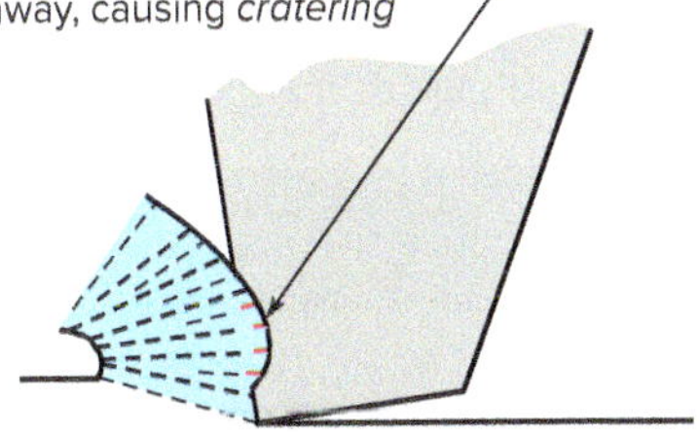

Figure 8-18 *Cratering* will eventually break the tip off the bit due to the weak point it creates.

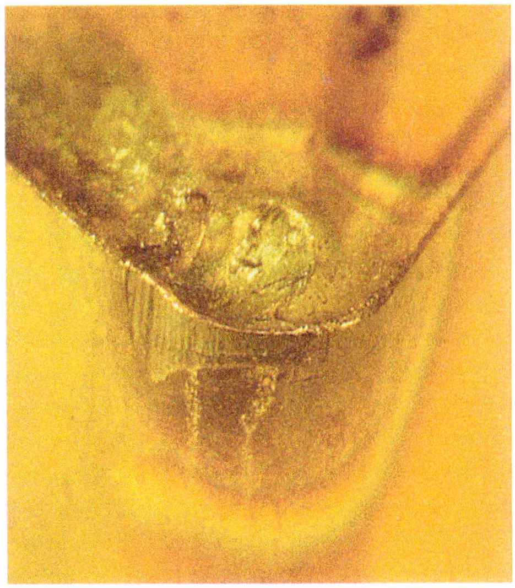

Figure 8-19 Cratering is seen on the tip of this carbide cutter after it has cut metal for some time.
From "Modern Metal Cutting" 1996, ©Sandvik Coromant

Edge Buildup

A second common problem associated with welding is called **built-up edge**. Here, the weld occurs, but this time holds on a little tighter. It doesn't instantly snap off. The flowing metal within the chip begins to back up. It disturbs the area upstream within the chip behind the jam until the pressure becomes great enough to shear away the stubborn weld. Envision a large rock in a river. The water upstream piles up behind it, then escapes around it when the pressure is great enough.

The problem in machining is that if the buildup is not corrected, some of the disturbance escapes *under* the bit to become an irregular and permanent part of the machine finish. Figure 8-20 shows this disturbance in the clay model.

To simulate the weld a nail was driven into the rake face of the wood bit (Fig. 8-21). Notice the disturbed area has extended ahead of the cutting edge, ready to break away and become part of the work surface. Red clay was added to emphasize the disturbed area. But if you try the same experiment with a clay model, you'll see it with no marker.

KEYPOINT

If your setup has edge buildup, the resulting finish appears smooth, then rough, then smooth, in a repeating pattern. Sometimes the roughness looks like small wires protruding from the surface.

The Solutions

The same control variables that prevent cratering also eliminate edge buildup.

Figure 8-20 A depiction of built-up edge escaping under the bit.
Courtesy Mike Fitzpatrick

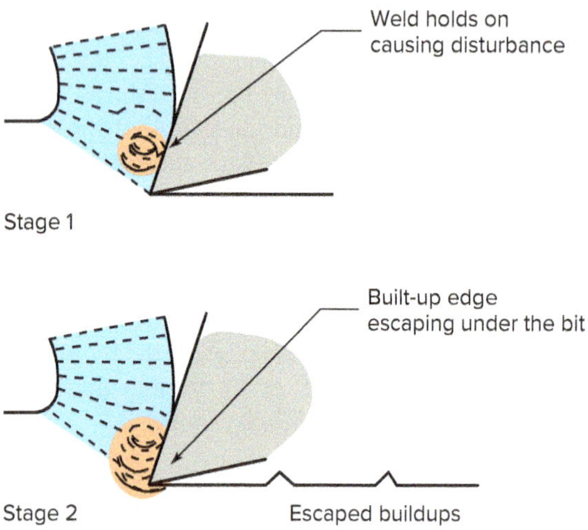

Figure 8-21 The weld holds on a little longer, the disturbed area backs up behind the weld, and small amounts escape under the bit.

Harder Bits or Electrocoated Rake Surfaces

The harder the rake face, the less the chip will drag upon it. That's why most negative rake tools are made of tungsten-carbide due to its extreme hardness.

Lubricants in the Cutting Fluid (Coolant)

Either cutting oil or coolant forms an antifriction barrier between the bit and chip. Reduced friction at the interference point shortens the shear line, thus lowering heat and pressure—the weld cannot form. However, there are several other advantages gained when coolants are introduced.

> **KEYPOINT**
>
> The list of coolant benefits is growing:
>
> A. Prevents cratering
> B. Reduces internal and external heat
> C. Makes safer, cooler chips
> D. Improves machine finishes
> E. Prolongs tool life
> F. Ends edge buildup
> G. Stabilizes measuring problems due to heat expansion. We'll see even more in Unit 8-3.

Coolants Cause Thinner Chips

With less rake friction, the shear angle is reduced and the chip produced is thinner. Internal heat goes down along with the power required to make the cut. The overall result is that the bit performs like a more positive rake tool without changing its shape. The reduced heat means better tool life and better machine finishes (Fig. 8-22). And safety!

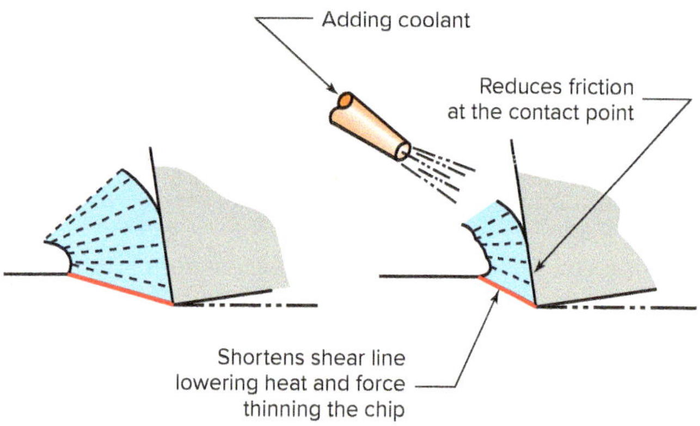

Figure 8-22 Coolant reduces interface friction, thus reducing deformation force. The result is a thinner chip, and a reduction in the heat and horsepower required.

8.2.5 Pushing the Boundary—Why We Use Negative Rake Bits

Why does tool life often improve when we machine more metal per minute or switch to a negative rake bit that causes more friction and heat? The partial answer is that by pushing the point of contact away from the edge, we distribute the force over a stronger part of the tool. But there's more.

> **SHOPTALK**
>
> **Reduce Heat—Save Tools** Experiments show* that a slight reduction in external heat of as little as 50°F can lengthen tool life as much as six times!
>
> *Cincinnati-Milacron Company, "Cool Chips."
>
> Source: Cincinnati-Milacron Company, "Cool Chips."

The deforming chip concentrates great pressure against the bit but not right at the cutting edge. A small area exists between the central force point and the cutting edge called the **boundary zone**. Return to Fig. 8-18. The boundary is illustrated as a small, flat spot between the crater and the cutting edge and, in fact, it can appear that way on real tools, too. The zone is subject to less force and friction than exists at the center of the crater farther up the bit.

The boundary zone, shown in Fig. 8-23, can be beneficially extended, farther back from the cutting edge, by pushing the bit harder (more feed rate) or by taking a deeper cut (more bite) or by changing to a negative rake tool. Often, when the boundary zone is extended away from the vulnerable cutting edge, the tool lasts longer because the force, heat, and friction have been shifted up to stronger parts of the bit. Can you see how this might prolong tool life?

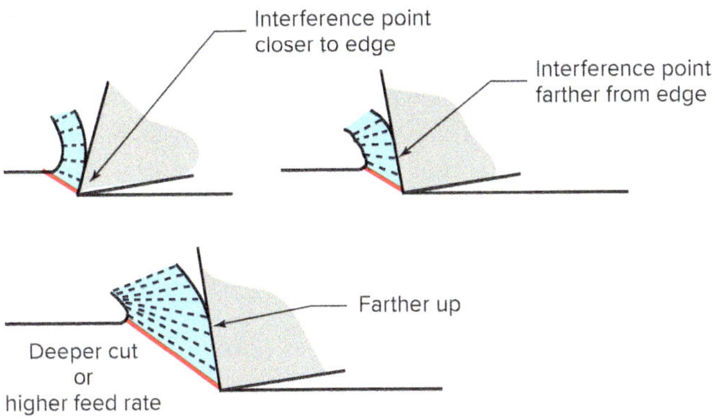

Figure 8-23 The *boundary zone* can be shifted away from the cutting edge by increasing the cutting force.

However, the trade-off concerns how much force the tool and the setup can withstand. There is an upper limit to how hard one can push the setup. Knowing the limit is a matter of practical experience and calculating the right feed rates.

Plastic Flow

Plastic flow also concerns how speeding up the cut affects the way the chip deforms. If the deformation rate is increased –that is, faster machining–then the internal heating softens the chip and it flows more easily.

In each machine chapter, there is a unit on selecting speeds and feeds. This is the final information you need to make the right choices about correct removal rates.

8.2.6 Chip Breaking

There are two ways of breaking stringy chips into safe little packets. In the first, using heavier machining action forms thicker, stiffer chips that cannot take the bending action of being removed from the parent metal. They then break into small C shapes. The other answer is using a cutting tool with a chip-breaking dam (Fig. 8-24). The obstruction redirects the flowing chip such that it snaps off. We'll see more on both techniques in tool geometry later.

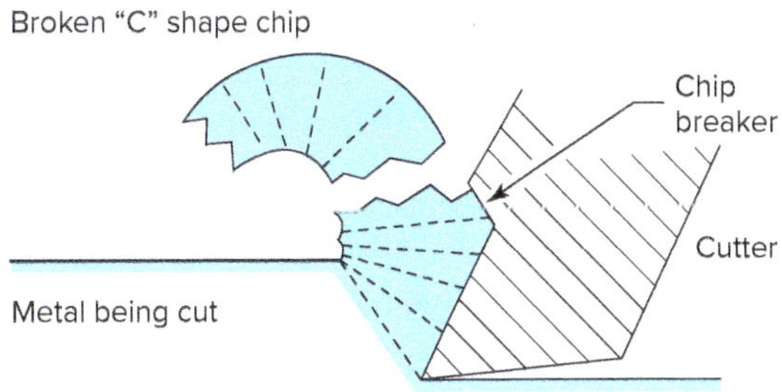

Figure 8-24 A chip breaker bends the flowing chip so quickly that it snaps off into small "C" shapes rather than a long string.

KEYPOINT

Tool life can often be skillfully extended in one of two ways:

A. Increasing cutting *force* shifts or redirects the boundary zone away from the weak cutting edge, which increases depth of cut or feed rate.

B. Increasing cutting *speed* causes the chip to flow more easily.

However, both techniques have practical upper limits where the tool itself is compromised from either force or heat. You will balance these factors in your selection of speeds and feeds and depth of cut.

TRADE TIP

Heavy Chip Load in Titanium and Stainless Steel
When cutting certain difficult metals, tools will dull quickly unless the cutting force is sufficient. Keeping the tool working hard is called a "heavy" **chip load.** Machining titanium, most stainless, and hard bronze are examples of metals that require a heavy chip load to prolong tool life.

When machining these metals, you must "bury" the cutter (a deeper cut depth) and keep it working hard (a heavy feed rate). Light cuts dull the cutter every time. A heavy chip load extends the boundary zone and relieves the cutting edge.

UNIT 8-2 REVIEW

Replay the Key Points

- Chips deform as they contact the rake surface of the bit.
  - Chips get hot from external friction–25 percent of total heat.
  - Chips get hot from deforming internal friction–75 percent.
  - Bits are worn away by forced contact with the chip–welding occurs, causes cratering.
  - Several factors can affect the shear angle with differing results in the machining process.
- Changing the rake angle affects friction.
  - Positive rake causes less friction and less deformation in the chip.
  - Positive rake is the weakest, thinnest bit and thus heats faster.
  - Negative rake tools can push harder.
  - Negative rake tools are usually made of harder material to better resist the additional friction.
- There are several other control factors that affect the shear line and chip formation:
  - Coolants
  - Bit hardness
  - Speed and feed changes
  - Cutting angles
  - Depth of cut

Respond

Determine readiness to move up by answering these questions. This is a critical point beyond which you should not pass until you feel you have a solid understanding of shear line, rake angles, and chip formation.

1. Why do negative rake tools often last longer in a setup?
2. In what two ways do coolants reduce heat in the machining process?
3. Identify and explain two ways to control built-up edge problems in machine finish.

Critical Thinking Questions

4. A tool keeps breaking before it's dull. Identify at least two ways to prevent this.
5. From where does most of the heat originate in machining a chip?

Unit 8-3 Cutting Fluid Technology

Introduction: Cutting fluids are commonly called coolants, but cooling isn't their only advantage in machining. Correctly applied to the right metal, focused right on the cutter, and delivered in a *consistent stream,* cutting fluids can work wonders. But adding them isn't always the right thing to do. Unit 8-3 will teach you when they are right to use, what they can do, or when they might be the wrong choice for your setup.

TERMS TOOLBOX

Carry out Cutting fluid that clings to the work when it's removed from the machine.

Rancid–Rancidity Bacteria reacting with organic compounds in the fluid–bad odor.

Semisynthetic fluids Modified or partial petroleum.

Synthetic fluids Non-petroleum based.

8.3.1 Advantages of Applying Using Cutting Fluids

Cutting fluids come in four varieties (discussed next), but all are used to achieve many or most of these advantages:

- Lubricate the interference between the chip and cutter's rake face, which reduces friction, deformation heat, and pressure leading to most of the remaining features except rust prevention
- Reduce total heat in the process–both external and internal
- Reduce horsepower
- Increase removal rates
- Help prevent welding and built-up edge
- Prevent loading (chips bunching up in the cutter flutes)
- Improve tool life
- Improve surface finish
- Safely and efficiently cool and flush away chips
- Prevent rust or oxidation on some metals
- Extend service life of the machine

8.3.2 Fluid Makeups and Applications

Today the selections are narrowing as the semisynthetic and full **synthetic fluids** become ever improved. Engineered for a wide range of applications, they have become the mainstay of the industry, especially for CNC milling and turning. Here are your choices and the reasons why you might choose to use them.

Petroleum Oils

Oils, often with added sulfur and/or other fats and minerals to increase lubrication, are used mostly in threading and honing operations–not for CNC work. This is the original fluid employed in machining.

Good properties: They cling and lubricate very well for slow cutting operations.

Bad properties: They tend to make a mess–stay on the work when removed from the machine–called **carry out**, and some can even catch fire if overheated. These oils are not very good at cooling due to their viscosity.

Tapping Fluids and Paste Compounds

Closely related to oils, a special group of fluids and compounds is used for making internal threads especially effective when cutting them by hand. Some are said to be universal for all metals, while many are formulated to work in a family of metals–one for steel, another for aluminum, for example.

Good properties: Highly effective at clinging and lubricating to avoid tap breakage.

Bad properties: Expensive and never used in great quantities, as in a sump.

Soluble Oils

These oils are used in some CNC work and in manual machines but mostly only as a cost savings. They are composed of organic petroleum base oil and an emulsifier (which breaks the oil down to stay mixed with water); when water is added to the oily syrup at around a 40:1 ratio, they form a white fluid we often call "milk."

Good properties: They do a fair job of lubricating and rust prevention, plus an excellent job of removing chip heat externally.

Bad properties: They can become **rancid** (due to bacterial action), and some machinists even react to them on their skin. They also tend to become contaminated faster than synthetics. In general they are being replaced in modern machining due to their shorter life in the machine and the superiority of synthetics.

Semisynthetic Fluids

Used in CNC work and manual machining, they begin as soluble oil but have added ingredients to help reduce unwanted properties. They, too, are mixed with water, at approximately a 50:1 ratio (varies with the application). These hybrids work better than oil, but they don't perform as well compared to synthetics. They are the "tweener," usually chosen over synthetics when extreme demand isn't required.

Good properties: They do an excellent job of both lubrication and cooling.

Bad properties: Not all unwanted bacterial and other issues are solved due to the organic components.

Full Synthetic Fluids

Engineered for the purpose, synthetic fluids start with inorganic and organic compounds (but no petroleum oils) along with a corrosion inhibitor plus proprietary ingredients formulated by each supplier. Most machine shops are adopting them today especially in large universal systems.

Good properties: Long life in machines. Can be reconstituted with additives and water (same for semisynthetics). Mixed at a high dilution ratio with water–the syrup goes a long way! They tend to not go rancid; plus they resist contamination. Synthetics bring the full range of advantages listed earlier.

Bad properties: Initial cost, but as with many such items they often pay for themselves in long life.

8.3.3 How Cutting Fluids Are Applied

There are six possible ways to apply cutting fluids:

1. By hand with a squirt bottle or brush for oils and compounds (manual machine application). Not a highly recommended method for CNC, but often used in tooling and offload work.
2. Flooded over the cut area through a pumped stream, with adjustable nozzles or a handheld hose. This (and several others that follow) requires a recycling collecting sump and filters to keep chips out of the pump.
3. By a high-pressure jet–usually an added machine accessory. These very high-pressure systems can force coolants into tight spots and get right where they are needed.
4. Through the tool–requires special tooling that can pump the stream through a collar into hollow tools on lathes and mills. Using extra coolant filters to keep passages open, this method also gets the coolant right where it's needed and is especially efficient for deep hole drilling where the chips are forced out of the hole (see Fig. 8-25).
5. Mist–a spray is driven by an air jet streaming over a nozzle to inject the coolant into a fine mist. This method can be used on machines with no coolant system, such as a tool room surface grinder, and also for deep or hard to reach places in CNC work. It is sometimes added to CNC programs when flooding or high pressure can't reach the action.
6. Adjustable and programmed CNC nozzles–*teach learn.* On some modern CNC machines, the operator aims the coolant at the best location using two axis control knobs and changes the aim as the program progresses, which saves reaching in. Then, on a few systems, the control remembers those movements and reproduces it on the next cycle–aiming automatically.

Figure 8-25 Coolant being forced through the tool to lubricate and eject chips.
Courtesy of Haas Automation, Oxnard, California

8.3.4 When NOT to Use Coolants

Coolants can be the wrong choice or are just not needed sometimes, for three reasons:

1. **Dry Metals**

   While machining a few metals that we often call "dry" materials (e.g., cast iron, brass, and even aluminum in some cases), coolants are not needed. One can successfully machine these metals without coolants, although you can still use them to flush and cool with no harm to the operation.
2. **Inconsistent Delivery**

   Thermal shock–hot-cold-hot–often destroys carbide cutters! If coolant isn't delivered in a steady flow, it's often better to not use it at all! In these situations, it's far better to adjust cutting speeds and feeds accordingly and cut dry.
3. **Plastic Flow**

   The high-tech reason to cut dry is, as cutters become ever more capable, many researchers and tooling companies are advocating not using coolants. By adjusting speeds and feeds upward in some metals (steel, for example) without applying cutting fluids, we reach a sweet spot whereby deformation heat reduces the hardness of the chips–they begin to flow and horsepower requirements flatten out as we continue to increase removal rates. Experiments with this method are also seeing better tool life in certain applications. Heat goes out with the chips, the rake surface does most of the work, and the cutting edge isn't challenged, for reasons you just learned.

Cutting Fluid Maintenance and Earthwise Disposal

As mixed fluids are used, the water evaporates from heat and time and the mixture becomes too thick. Other ingredients and properties can become depleted, too. The cutting fluid must be reconstituted with more water to return it to the correct dilution ratio. The liquid is tested with small handheld optical and pH meters (acid/alkaline ratio). Unwanted lube oil mixed in–called "tramp oil"–must also be removed by skimming. Suspended metal and other particles can be settled out or filtered. Last, depending on the product, a biocide might be required, along with a syrup reconstituting fluid.

Finally, as coolant wears out, loses properties, becomes rancid or too contaminated, it's time to change it! Disposal becomes an issue. Check the MSDS (Chapter 1, Sections 1.3.1 and 1.3.3) for correct disposal for a given product. It's rare that spent coolant can be put directly back into the environment (sewer or landfill), even when properly treated for disposal. More commonly it must be sent to a recycler that knows how to treat and dispose of it.

UNIT 8-3 REVIEW

Replay the Key Points

- Cutting fluids must be delivered in a consistent stream to not shock cutters.
- Cutting fluids aren't always needed.
- The more commonly used products in CNC work today are synthetics and semisynthetics.
- All coolants eventually wear out and must be disposed of properly.

Respond

1. Without looking back, name as many advantages of using synthetic coolants as you can recall.
2. What is the advantage of petroleum oils?
3. What is a major disadvantage of petroleum oils?
4. What are at least two advantages of full synthetic coolants?
5. Name three times cutting fluids should not be used.

Unit 8-4 Taking Control—Adjusting Variables

Introduction: This unit is about putting what we've just learned to work. It's about what happens when we change the tool geometry variables and the setup factors and why. At the end, we'll take a brief look at troubleshooting a few common problems.

Read through the following suggestions but don't try to retain it all. Then, when improving a setup or when things don't go as expected, return for suggestions. Experience on the machines will bring it into focus. Also ask your instructor and several journey machinists, but don't be surprised if each offers a different fix.

Nearly every problem or improvement can be dealt with using several fix combinations. For example, if a tool is chattering, you would probably slow the cutting speed and possibly change the lead angle on the tool. Another control that would decrease chattering is to increase the feed rate and/or change the rake angle of the cutter. The setup may need to be strengthened, too, since the vibrations may be emanating from the work-holding method, not the workpiece. The corner radius might need to be lessened. It's a case-by-case challenge.

You'll soon develop a sense of what's wrong and your own ways of solving the problem. Doing this requires more than applying knowledge; it involves intuition based on experience–and that is the essence of being skilled!

TERMS TOOLBOX

Carbide grade The relative hardness of the cutter running from very hard to tougher but softer at the opposite end of the scale.

Interrupted cut A noncontinuous cutting action causing shock to the cutter.

Loading Chips clog the cutter, destroying finishes and even breaking cutters.

Maximize To make a setup more efficient with the goal of more parts per time unit.

8.4.1 Summing Up the Control Variables

Here are the possibilities in brief.

Changing Tool Shape

There are three shape variables:

Corner Radius Cutting/Lead Angles Rake Angles

Change corner radius to

Prevent tool failure

Improve finish

Reduce chatter

Form an inside rounded corner on the workpiece

Increase radius when

Finish is poor

Tools are breaking or burning up frequently

Decrease radius when

Chatter is a problem

Rake Angle Changes

Both positive and negative rake tools can deliver a fine finish when correctly applied.

Select positive rake when

The job requires moderate removal rates or for softer work such as aluminum or softer brass.

Very slow speed machining is performed, such as threading on a manually operated lathe.

Chip ejection is a problem with negative rake tools. It is more commonly encountered when machining softer alloys of aluminum. The chips are welding and clogging the cutter, called **loading**.

Change to negative rake tools when maximum metal removal is desired.

Harder work materials must be machined.

The tool breaks or burns up with positive rake.

Changing Cutting/Lead Angles

Lead angles are changed for three reasons:

1. To smooth out the cutting action.
2. To increase the removal rate.
3. To form a shape as on the right in Fig. 8-24. The part is rotating while the lathe bit moves to the left through the work, leaving an angular surface and a radius.

Increase lead angle when

The zero lead tool (left side) is common for moderate metal removal setups.

The 20° bit would be used for a high-volume cut.

Decrease lead angle when

The 20° lead tool might chatter, at which time a lead of 10° might be selected.

Forming

Sometimes, even on CNC lathes, the situation forces use of a cutter such as the 20° tool in Fig. 8-26 to create the shape of a part. The shape of the bit produces the shape of the work. A form tool has the effect of a very long lead angle.

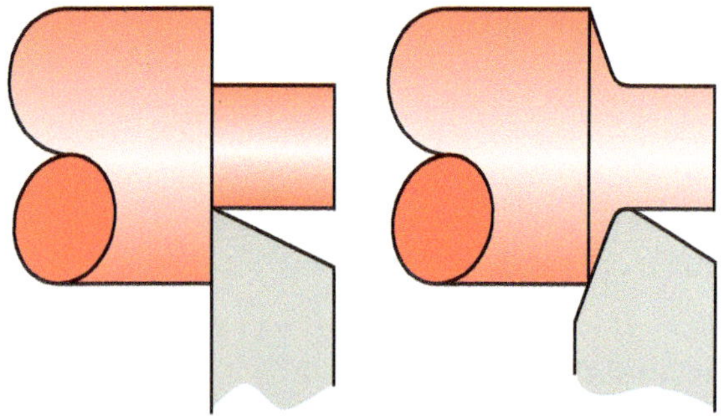

No lead
no radius

20-degree lead
0.050-in. nose radius

Figure 8-26 Looking down on lathe bits, the bit on the left has zero lead while the bit on the right has 20° and a nose radius.

With form tools, select a slow cutting speed and use lots of coolant or cutting oil. Also be sure the setup is very rigid. Due to the large amount of contact area between bit and work, chatter is the enemy when forming. The fixes are in this order:

Slow the RPM.

Improve setup rigidity.

Add cutting oil rather than coolant.

Sometimes, try fiddling with the feed rate (up or down case by case), but it's the last item to try.

Changing Tool Hardness

Tool hardness change usually accompanies a change to negative rake to increase production, but not always. Every so often, we need to change to a less hard, but tougher, cutting tool.

Hard Versus Tough Tools

Carbide-grade tools are supplied in a range of hardnesses that we'll discuss in detail later on. The range goes from very hard at one end of the scale to tough but softer at the other. High-speed steel tools are softer than the lowest grade of carbide, but they are tougher, too, and more able to withstand shocks and vibration.

Change to a Softer Tool When

The carbide tool is breaking or chipping on the edge, before it becomes dull from normal use.

HSS tools resist breaking on intermittent cuts, too—that is, an in-out-in, hammering action such as a lathe turning a non-round object or one with a hole in its side. The bit must enter, exit, and enter the work for each revolution. This kind of abrupt action is called an **interrupted cut**.

Change to Harder Tool When

The job requires a large removal rate.

The tool is burning up before becoming dull from normal use.

The work is too hard and dulls the tool quickly.

Changing Cutting Speeds and Feed Rates

These are two vital controls that can be varied with the twist of dial for a CNC operator and a few shifts of levers or belts on manual machines. They are almost always the first variable to be adjusted for improvement or troubleshooting. In the drill, lathe, and mill chapters of this book, we'll calculate the right speed for the job. Here, we discuss this control in a general way.

A quick reminder—cutting speed is the speed with which the chip is flying off the cutter. It is the velocity of the cut. In drilling and milling operations, the cutter's rim is set to turn at the specified feet-per-minute rate. With lathe work, it's the work's rim that must revolve at the specified speed. In either case, the objective is to calculate and adjust the RPM to achieve the cutting speed in feet per minute (F/M) or meters per minute (M/m).

KEYPOINT

Cutting speed is the speed in F/M or M/m of the metal passing the tool or the tool passing the work. Changing cutting speed means adjusting RPM.

The goal is to find the sweet spot where the best finish is accomplished but tools also last well. Not calculating the correct RPM based on recommended cutting speed is one of the most common errors we instructors observe with new machinists.

Safety

Machining too fast means that something is spinning very rapidly. Either the work or cutter might be exceeding safe limits. Lathe work is the more dangerous place to overspeed a machine. Here the heavy work and chucking device are rotating. Also, with increased speed, chips become hotter and fly faster.

Horsepower and chip ejection are also practical limits to adjusting speed upward too far. Often the chips can't get out of the cut area, and they load it regardless of whether a positive or negative rake is being used. The first fix is to bring coolant to the cut.

Decrease Cutting RPM When

Slightly below calculations; when setting up a machine for a first cut, it's smart to start out slightly below the calculated speed then work upward, observing the results to both the work and tool.

Tools burn up before dulling due to work hardness.

Chatter occurs. Reducing RPM is the most immediate fix for chatter.

The setup appears unsafe or unusual.

Increase RPM

The action doesn't shear chips away correctly. The chip is occurring in big chunks, not flowing layers.

Tools break before dulling.

Here's a mind experiment THAT YOU SHOULD NOT TRY! If a machine's power is turned off, but left to cut until it coasts to a stop, it will slow down while making a chip. Commonly, just before stopping completely, the bit will break due to poor chip deformation. This is more common with brittle carbide, but HSS will do the same in heavy cuts.

When examining a chip produced under this low-speed condition, you will see the back side is exceptionally rough due to the large chunks of deformed metal sliding, then sticking (Fig. 8-27).

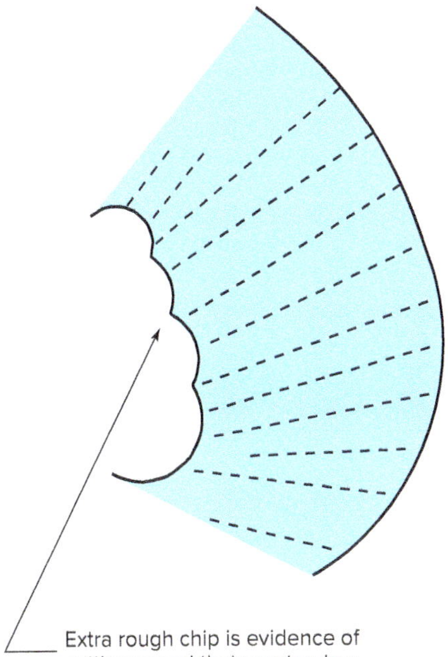

Figure 8-27 An extra rough chip may be a sign of poor deformation.

KEYPOINT

A cutting speed that is too slow results in tool breakage, poor finish, and slow production—inefficient removal rates.

Adjusting Feed Rates

Similar to cutting speed, feed rate changes have an effect on both tool life (Fig. 8-28) and work finish and accuracy. Calculating the rate of cut is important but less critical than cutting speed because there is normally a range of acceptable feed rates.

Figure 8-28 Early tool failure from chipping can be caused by several factors.
Lake Washington Technical College, Kirkland, WA/McGraw Hill

The upper limit is due to the strength of the setup and the roughness of the finish produced. Fast feeds are the objective in terms of efficiency. Slow feeds are usually not a problem except that it's low-volume machining. They are commonly used for a final cut where finish and accuracy are of prime concern. Here are two exceptions to that.

Increase Feed Rates When

Work hardening is occurring due to rubbing in some sensitive metals such as stainless, high-carbon tool steel, or certain bronzes.

Chatter is occurring–sometimes a faster feed rate can stabilize chatter because a larger chip is forced against the rake face of the cutter.

Plan Heavy, Then Fine Feeds

A slow feed leaves finer machine marks and a smoother finish. Also, finer feeds exert less force on the tool, thus improving accuracy. A good plan usually includes one or more rough, heavy feed passes; then a semifinish cut to stabilize the accuracy; then a final, fine-feed cut.

Decrease Feed Rates When

The work and machine are subject to questionable forces.

Machined finish is rougher than expectation.

Accuracy and repeatability are below the job tolerance.

Tools are breaking before dulling.

8.4.2 Troubleshooting Setups

Here are a few common problems and their associated causes. As you work toward more efficient operations keep in mind the first goal is safety, then accuracy. Speeding up, or **maximizing**, is also desirable but never at the risk of the first two: safety and quality.

Problem	Possible Cause
Poor finish	Dull tool Excessive feed rate and/or depth of cut No coolant Inadequate tool clearance Low cutting speed Wrong rake angle No tool radius No lead angle on tool
Built-up edge	No coolant or poor injection Dull tool Tool too soft
Chatter/vibration	Excessive RPM speed Large tool radius Large lead angle on tool Poor holding method for workpiece Workpiece too thin or weak for operation Feed and/or depth of cut too light Tool too deep; depth of cut too great
Inaccurate/inconsistent repeatability	Dull tool Poor or loose setup/work-holding method No coolant or cutting fluid Excessive feed rate Excessive depth of cut Inadequate tool clearance Tool flex—too much overhang/excessive force for tool size
Excessive work heat	High RPM Dull tool Rake too negative Poor cooling action
Tool breakage	Excessive feed or depth of cut (number 1 cause) Weak tool Tool is too hard Setup not rigid or not tightened well Positive rake not strong enough Slow cutting rate causes poor deformation
Tool cratering	No coolant Coolant doesn't reach cut area—chips pushing it away Tool too soft—use harder grade
Chip load-up	RPM too high No coolant Dull cutter Negative rake tool
Parts too hot	RPM too high Too much cut depth Excessive feed Dull tool No clearance on tool

Problem	Possible Cause
Tool burn-up	Use carbide (harder) tool Feed rate too high RPM too high Depth of cut too great No cutting fluid No tool clearance Spindle in reverse (backward, it happens!) Hitting (or creating) a work hardened zone
High-pitch squeal	Hard workpiece Excessive speed No clearance Excessive tool overhang—too long
Dangerous chips hot and fast	No coolant Excessive RPM No safety shield Not cleaning up chips as they buildup close to spindles and rotating work
Long, stringy chips	Feed or depth of cut too low Continuous feed No chip breaker on tool

TRADE TIP

Squeals Sometimes a very high-pitched squeal will occur with no evidence of chatter marks in the work. This pesky irritant doesn't affect tool life or work accuracy, but it can set your teeth on edge. It occurs when machining very hard steels, some stainless steels, and hard bronzes. More than irritating, prolonged exposure can result in hearing loss—wear external earmuffs for this kind of noise.

ARMCHAIR FIELD TRIP

Watch this Haas Tip of the Day video to learn more about how machinists determining correct feed rates. Go to: **https://www.youtube.com/haasautomation** and search for this video called: **YOUR FEEDRATE IS WRONG!—Haas Automation Tip of the Day**

UNIT 8-4 REVIEW

Replay the Key Points

- Surface speed is the velocity of the rim of the cutter or work (whichever is spinning).
- Achieving the right surface speed means calculating and adjusting RPM up or down.
- Choosing a surface speed too low results in poor finish, broken tools, and generally inefficient machining.
- Choosing a surface speed too high results in burned tools, work hardened material, extra-hot fast flying chips, and dangerous conditions.
- Feed rates are important, but there is a wide range that produce acceptable results.

Respond

Check your comprehension of troubleshooting.

1. Identify at least three ways that chatter might be stopped.
2. True or false? Feed rates are based on surface speed. If false, what makes it true?
3. Which control variables lower chip heat in a setup? Briefly explain each.
4. What three tool bit shape variables can be adjusted?

Critical Thinking Questions

5. When is it desirable to change to a softer tool bit?

CHAPTER 8 REVIEW

Unit 8-1

We've only touched on the basics of tool geometry. As skills build with machine experience, you'll learn about three more kinds of rake angles, for example. Look for compound rakes on lathe tool bits that create a bias for the tool's movement right or left. Look for milling cutters with various combinations of positive or negative radial (center outward) and axial rakes (parallel with the spindle) and for hooks in the rake faces. But they all start with the four basics: rake, clearance, lead, and corner radius. Every tool has these, even though in some cases they might be zero, as with corner radii. Be certain that you can identify them on any tool you put your hand to. That's the bottom line–you must know cutting tool geometry.

Unit 8-2

Part of the machinist's paycheck comes from doing the research. Today, cutting tools are evolving at a fantastic rate, driven by the need for speed. Better tooling is one of the frontiers that I suspect (and hope) we'll never completely cross. As a journey machinist you will be bombarded with samples of new tools, with articles provided by the makers of them, and by end users like yourself test driving them. There will be demos, brochures, and sales pitches to wade through. Although expensive, some new tools pay for themselves in days, yet some don't live up to their promise in your particular application. Part of the job includes doing smart research based on the knowledge.

It's easy for even a small shop to own hundreds of thousands of dollars in tooling! No exaggeration. In large companies, excess costs tied up in cutting tools are one of the most common challenges.

Unit 8-3

Cutting fluids can work wonders when used in the right situation in the right way. But keep your eye on dry cutting advances.

Unit 8-4

If ever there was a perfect test of logical thinking combined with guesswork, cutting tool troubleshooting has to be it. I can recall bets and near shouting matches between three or more machinists in our shop as to exactly what the problem might be and how to fix it. Each of us could and did make it work better, but the truth is the absolute answer doesn't exist. It's a science of variables. CNC machining makes it even more complex as we demand more parts per minute and per day. Learn this stuff well.

QUESTIONS AND PROBLEMS

1. Identify the tool geometry (shape) features that can be used as control variables. (Units 8-1 and 8-2)
2. Briefly describe the purpose of each answer to Question 1. (Unit 8-1)
3. Of these four, which is *not* a control variable: rake, lead, clearance, and shear? Explain. (Units 8-2 and 8-3)
4. Of the four features in Question 3, which is *not* on the bit? Explain. (Unit 8-2)
5. What happens to the chip if we change to a more negative rake angle? (Units 8-2 and 8-3)
6. In terms of machine requirements, what happens when we switch to a more negative rake tool? (Unit 8-2)
7. What is the most likely reason to have a corner radius above 0.125 in.? (Unit 8-1)
8. Name two corrections or changes to solve a built-up edge. (Units 8-2 and 8-3)
9. Is this statement true or false? The boundary zone is defined as the space between the point of interference and the cutting edge. (Unit 8-2)

CRITICAL THINKING

10. In a given setup, the bit is prematurely breaking along the cutting edge before it becomes dull from normal use. Why and how might it be corrected? Identify at least four solutions. (Unit 8-3)
11. A bit is burning up before becoming dull. Identify at least four causes and their solution. (Unit 8-3)
12. A bit has a gold coating on its surface. In terms of what we've just learned, why might that be?
13. The bit is changed from a noncoated to a coated version, as in Question 12. Identify at least two benefits assuming all other tool variables and setup factors remain the same.
14. What is the main reason to change to a carbide bit from an HSS?
15. The machine finish on a milled surface is too rough to pass inspection. Without looking at the Unit 8-3 troubleshooting chart, what changes might solve it?
16. In what order might you try your fixes for Question 15?
17. A job must be taken to a manual milling machine that's limited to eight spindle speed selections. After doing the math (to be learned in Chapter 12) you determine that

the calculated speed of 375 RPM falls between two speeds at 250 and 400 RPM. What should you do?

18. Using the same machine as in Question 17, the needed RPM calculates to 300. Now what must be done?

CNC QUESTIONS

19. A standard carbide bit is producing 15 parts on a CNC lathe before it dulls and must be changed to a new, sharp cutter. However, you suspect a different cutter with artificial diamond coating can make 40 parts per cutter at a nearly 50 percent decrease in cycle time per part.

What questions should you ask yourself before making the change?

20. If your career is aimed at shops with CNC equipment, why must you stay informed of advancements in cutting tools?

CHAPTER 9
Drilling and Operations and Machines

Units

9-1 Basic Drilling Tools

- Select the right drills for the job
- Measure, name, and describe an industrial drill bit
- Select a pilot drill for a reamer
- Select the right shank type and adapter

9-2 Drill Presses and Setups

- Identify the various features on standard drill presses
- Know the safe work habits and holding methods for vises, plates, fixtures, and clamps

9-3 Speeds and Feeds for Drilling

- Calculate or look up a starting RPM on any drilling or reaming operation
- Access recommended surface speed charts in reference books

9-4 Drilling per Print and Secondary Operations

- Set up and perform drill location by the layout-punch method to a 0.030-in. repeatability
- Set up and perform drill location by the center find method to a 0.010-in. repeatability
- Choose the right tap drill for a specific tap
- Cut threads using taps on the drill press
- Select and use a counterbore and countersink/chamfer tool
- Drilling setups for parallel and angle holes

9-5 Sharpening Drill Bits

- Safely and correctly sharpen a drill bit using a pedestal grinder
- Recognize a dull drill needing regrinding

The Viking Yacht Company, based in New Gretna, New Jersey, produces approximately 90 percent of every component that goes into the yachts it builds. The company has a massive machine shop equipped with CNC technology (three vertical CNC milling machines and one CNC Haas lathe). In the shop, workers produce more than 500 different parts from aluminum, copper, plastics, and stainless steel. Viking not only designs the boats in-house but also produces the molds for the boats in the 810,000-square-foot facility. The incorporation of CNC machines has revealed spectacular results due to the machines' accuracy. In this photo, the company's five-axis profiler carves an epoxy-covered foam plug. It will be used to make a mold for the enclosed bridge and open bridge for two boats.

Courtesy of Viking Yacht Company, New Gretna, NJ

INTRODUCTION

Learning to operate a drill press is less challenging than other kinds of machining, but don't mistake your experience there to be of lesser value. Beyond learning to use a manual drilling machine safely and efficiently, Chapter 9 is about an important family of operations where we spin a cutting tool, then plunge it straight into the work to create a precise hole. Machining exact diameters in the right location to a specified depth is one of the most fundamental ways metal is shaped. The odds are that within any CNC program calling up three tools or more, at least one will perform some form of drilling to complete the work shape.

No matter whether the holes are produced by high- or low-tech means, a competent machinist must be able to make a drill bit start and stay in the right location and must know how to drill deep holes while keeping the form straight and round. A third challenge is to avoid clogging the bit with chips and eventually breaking it. Last, drill setups must also include safety and long tool life. Today, the focus isn't on the minor cost of halting to replace or regrind an abused drill, due to poor setups; rather, it falls on the higher cost of lost production due to stopping for any reason!

Beyond basics, there's another reason to learn drilling. Today in CNC mill programs, drilling can be used for rough profile machining of extreme metals that otherwise would be difficult to cut. Due to the ability of CAM software to quickly create drill locations around the perimeter of the part, programmers are able to create cut sequences that remove initial excess, drilling it away rather than cutting around the outside. The advantage is that a hard but brittle carbide drill will withstand an end push much better than a carbide mill cutter can take a sideways thrust. Drilling can be an efficient CNC option compared to milling, and if done properly, the drilling cutter lasts longer than other kinds of cutting tools.

There are nine challenges to successful drilling:

- Choosing the right tool
- Drilling in the right location
- Surface finish per drawing specs
- Drilling to the correct depth
- Calculating speeds and feeds
- Making safe and strong setups
- Removing chips from deep holes
- Prolonging tool life
- Achieving size, including roundness and straightness

Virtually all the skills required to meet these challenges can be practiced on an everyday manual drill press like those found in your tech school lab. Once mastered, your ability will transfer to higher-level machine tools.

Unit 9-1 Basic Drilling Tools

Introduction: Drills and reamers are supplied as small as a human hair and as large as 6 in. in diameter or more in special cases. The depth of standard drilling operations is limited to around 5 to 10 times the drill's diameter; however, it's possible to drill far deeper using special techniques called pecking, or by switching to tools called spade drills and gun drills (all studied in this chapter).

We begin the conquest of hole making by looking at the basic tools: drills and reamers. And as usual, there are terms to learn. If you are to reach the Term-Master level in the vocab game, read on!

TERMS TOOLBOX

Blind hole A hole that does not go through—this leads to chip and coolant challenges, especially when tapping threads.

Dead center The inefficient center of a drill bit that does not cut.

Drift pin (drill drift) A small, flat wedge used to break the holding force of taper-shank tools.

Land Similar to margins, a small surface on a cutting tool.

Margin A narrow band where drills are full diameter.

Morse taper A self-holding machine taper of $\frac{5}{8}$ in. per foot.

Morse taper adapter Either a *sleeve* for upsizing or a *socket* for downsizing.

Reamer float (tool float for other kinds of cutting tools) The tendency of cutting tools to pick up and follow previously machined shapes, such as a reamer following an incorrectly located drill hole even though the press is in the correct location.

Shank The mounting end of the tool.

Tang The torque drive tab on taper shank drills and reamers.

Turned-down shank A drill larger than $\frac{1}{2}$-in. diameter, but featuring a straight shank. May be made from a damaged taper shank drill.

Twist drill A high-speed steel cutting tool made with spiral flutes.

SHOPTALK

HSS Tough When Hot High-speed steel retains its full strength up to around 450°F and most of its strength up to an amazing 650°F! When buying drill sets, never fall for high-carbon steel drills, because they will snap off and the edge breaks easily. Also you may see "titanium" drills, which is misleading, because they are coated with titanium nitride (gold color)—a good feature as long as it's over HSS steel.

9.1.1 Choosing the Right Drill and Reamer Alloy

The everyday **twist drill** (helical flutes) that you'll be using in lab assignments is commonly made from high-speed steel. HSS is also used for many other cutters because it has three great characteristics for this duty:

1. It can be resharpened using standard grinding wheels.
2. It flexes to withstand the unique forces of drilling.
3. It holds a sharp edge well, and strength, even when it's heated to high working temperatures.

When selecting a drill for a job here's what you need to look for:

- Diameter
- Tool-holding method—shank shape—straight for chucking and tapered for holders
- Drill geometry

Four Different Drill Size Systems

Fractional drills (for example, $\frac{3}{16}$-in. diameter) are common. But using fractional increments doesn't create enough size differences for manufacturing. Therefore, three additional size systems are in use, and they fit between the fractions.

Number series	Sizes 1 to 80 (big to small)
Letter series	Sizes A to Z (small to big)
Metric size drills	Sizes 0.5 mm and up

To organize the process, machinists use a chart such as that shown in Fig. 9-1, found on the wall of all shops, and a large one in Appendix I of this book, Drill charts are also found in the lid of machinists' toolboxes, and in reference books such as *Machinist's Ready Reference* by Prakken Publications (ISBN: 0-011168-90-7). Using a drill chart is a snap with just a bit of investigation. Divided by big red lines, this chart has four vertical columns with three data entries:

1. Drill fractional diameter or size designation
2. Decimal diameter
3. Tap size drill for a specific thread (e.g., 1/4-20 threads require a #7 drill)

For another example, note on the chart in Fig. 9-1 that near the $\frac{3}{8}$-in.-diameter drill (0.375 in.), a "U" size drill is found that measures 0.368 in. and a "V" drill at 0.377 in. That's not much difference in size, but it's enough to make a big difference on the final product.

Question?

What drill size is recommended for tapping 1/2-13 thread?

Answer:

27/64

Review

To be certain you have the idea, here are a few questions with answers found at the end of Chapter 9.

INCH / METRIC TAP DRILL SIZES & DECIMAL EQUIVALENTS

Small Drills

FRACTION OR DRILL SIZE	DECIMAL EQUIVALENT	TAP SIZE
NUMBER SIZE DRILLS 80	.0135	
79	.0145	
1/64	.0156	
78	.0160	
77	.0180	
76	.0200	
75	.0210	
74	.0225	
73	.0240	
72	.0250	
71	.0260	
70	.0280	
69	.0292	
68	.0310	
1/32	.0312	
67	.0320	
66	.0330	
65	.0350	
64	.0360	
63	.0370	
62	.0380	
61	.0390	
60	.0400	
59	.0410	
58	.0420	
57	.0430	
56	.0465	
3/64	.0469	0-80
55	.0520	
54	.0550	
53	.0595	1-64,72
1/16	.0625	
52	.0635	
51	.0670	
50	.0700	2-56,64
49	.0730	
48	.0760	
5/64	.0781	
47	.0785	3-48
46	.0810	
45	.0820	3-56
44	.0860	
43	.0890	4-40
42	.0935	4-48
3/32	.0938	
41	.0960	
40	.0980	
39	.0995	
38	.1015	5-40
37	.1040	5-44
36	.1065	6-32
7/64	.1094	
35	.1100	
34	.1110	
33	.1130	6-40
32	.1160	
31	.1200	
1/8	.1250	
30	.1285	
29	.1360	8-32,36
28	.1405	

Bigger

FRACTION OR DRILL SIZE	DECIMAL EQUIVALENT	TAP SIZE
9/64	.1406	
NUMBER SIZE DRILLS 27	.1440	
26	.1470	
25	.1495	10-24
24	.1520	
23	.1540	
5/32	.1562	
22	.1570	
21	.1590	10-32
20	.1610	
19	.1660	
18	.1695	
11/64	.1719	
17	.1730	
16	.1770	12-24
15	.1800	
14	.1820	12-28
13	.1850	
3/16	.1875	
12	.1890	
11	.1910	
10	.1935	
9	.1960	
8	.1990	
7	.2010	1/4-20
13/64	.2031	
6	.2040	
5	.2055	
4	.2090	
3	.2130	1/4-28
7/32	.2188	
2	.2210	
LETTER SIZE DRILLS 1	.2280	
A	.2340	
15/64	.2344	
B	.2380	
C	.2420	
D	.2460	
1/4 E	.2500	
F	.2570	5/16-18
G	.2610	
17/64	.2656	
H	.2660	
I	.2720	5/16-24
J	.2770	
K	.2810	
9/32	.2812	
L	.2900	
M	.2950	
19/64	.2969	
N	.3020	
5/16	.3125	3/8-16
O	.3160	
P	.3230	
21/64	.3281	
Q	.3320	3/8-24
R	.3390	
11/32	.3438	
S	.3480	
T	.3580	

Largest

FRACTION OR DRILL SIZE	DECIMAL EQUIVALENT	TAP SIZE
23/64	.3594	
LETTER SIZE DRILLS U	.3680	7/16-14
3/8	.3750	
V	.3770	
W	.3860	
25/64	.3906	7/16-20
X	.3970	
Y	.4040	
13/32	.4062	
Z	.4130	
27/64	.4219	1/2-13
7/16	.4375	
29/64	.4531	1/2-20
15/32	.4688	
31/64	.4844	9/16-12
1/2	.5000	
33/64	.5156	9/16-18
17/32	.5312	5/8-11
35/64	.5469	
9/16	.5625	
37/64	.5781	5/8-18
19/32	.5938	
39/64	.6094	
5/8	.6250	
41/64	.6406	
21/32	.6562	3/4-10
43/64	.6719	
11/16	.6875	3/4-16
45/64	.7031	
23/32	.7188	
47/64	.7344	
3/4	.7500	
49/64	.7656	7/8-9
25/32	.7812	
51/64	.7969	
13/16	.8125	7/8-14
53/64	.8281	
27/32	.8438	
55/64	.8594	
7/8	.8750	1-8
57/64	.8906	
29/32	.9062	
59/64	.9219	
15/16	.9375	1-14
61/64	.9531	
31/32	.9688	
63/64	.9844	1 1/8-7
1	1.0000	
1 3/64	1.0469	1 1/8-12
1 7/64	1.1094	1 1/4-7
1 1/8	1.1250	
1 11/64	1.1719	1 1/4-12
1 7/32	1.2188	1 3/8-6
1 1/4	1.2500	
1 19/64	1.2969	1 3/8-12
1 11/32	1.3438	1 1/2-6
1 3/8	1.3750	
1 27/64	1.4219	1 1/2-12
1 1/2	1.5000	

Met. Taps

METRIC TAP DRILL SIZES

METRIC TAP	TAP DRILL mm	DECIMAL EQUIV. Inches
M1.6 x 0.35	1.25	.0492
M1.8 x 0.35	1.45	.0571
M2 x 0.4	1.60	.0630
M2.2 x 0.45	1.75	.0689
M2.5 x 0.45	2.05	.0807
M3 x 0.5	2.50	.0984
M3.5 x 0.6	2.90	.1142
M4 x 0.7	3.30	.1299
M4.5 x 0.75	3.70	.1457
M5 x 0.8	4.20	.1654
M6 x 1	5.00	.1968
M7 x 1	6.00	.2362
M8 x 1.25	6.70	.2638
M8 x 1	7.00	.2756
M10 x 1.5	8.50	.3346
M10 x 1.25	8.70	.3425
M12 x 1.75	10.20	.4016
M12 x 1.25	10.80	.4252
M14 x 2	12.00	.4724
M14 x 1.5	12.50	.4921
M16 x 2	14.00	.5512
M16 x 1.5	14.50	.5709
M18 x 2.5	15.50	.6102
M18 x 1.5	16.50	.6496
M20 x 2.5	17.50	.6890
M20 x 1.5	18.50	.7283
M22 x 2.5	19.50	.7677
M22 x 1.5	20.50	.8071
M24 x 3	21.00	.8268
M24 x 2	22.00	.8661
M27 x 3	24.00	.9449
M27 x 2	25.00	.9843
M30 x 3.5	26.50	1.0433
M30 x 2	28.00	1.1024
M33 x 3.5	29.50	1.1614
M33 x 2	31.00	1.2205
M36 x 4	32.00	1.2598
M36 x 3	33.00	1.2992
M39 x 4	35.00	1.3780
M39 x 3	36.00	1.4173

PIPE THREAD SIZES (NPSC)

THREAD	DRILL
1/8-27	11/32
1/4-18	7/16
3/8-18	37/64
1/2-14	23/32
3/4-14	59/64
1-11 1/2	1 5/32
1 1/4-11 1/2	1 1/2
1 1/2-11 1/2	1 3/4
2-11 1/2	2 7/32
2 1/2-8	2 21/32
3-8	3 1/4
3 1/2-8	3 3/4
4-8	4 1/4

Figure 9-1 A standard drill decimal equivalent and tap drill chart. Often found in plastic "pocket charts" and on most shop walls. You will use this chart every day in this trade. Larger chart, see Appendix I.

Courtesy L. S. Starrett Co.

1. What are the two drill sizes surrounding the common fractional size of $\frac{3}{16}$ in.?

2. Looking at the drill chart, what is odd about the number size series? Why might that be?

3. What is the smallest number size drill and its decimal inch size on the chart?
4. What full millimeter drills are closest to $\frac{3}{8}$ in.? (Next metric size larger and smaller?)
5. What is the letter size drill closest to $\frac{1}{2}$ in.? What size is it?

9.1.2 Selecting Drills and Reamers for Size

Okay, from the chart you know the size you need. Now it's critical to know how to check the tool to be certain it's actually that size. There are three ways of determining the diameter of a drill bit or reamer:

- Size markings
- Measuring the drill
- Drill size gage

The drill size gage shown in Fig. 9-2 is similar to the one we've been using as an example in other units—a series of ascending hole sizes, used as a best-fit test of drill diameters. But, as we've seen, there can be such little difference between some of the smaller drills, it must be considered a quick, rough sorting guide only. Like the drills, the size gages are supplied in each of the four series: metric, inch fractions, number, and letter sizes.

Figure 9-2 A drill size gage can quickly sort drills, but it must be taken as a rough measurement only.
Photography by Prographics at Milwaukee Area Technical College/McGraw Hill

Diameter Markings

When they are made, drills and reamers (a hole-finishing cutter that follows a drill) are size marked near the **shank**, their holding end (Fig. 9-3). If the marks are readable, trust that the drill is the size shown, *but never trust a reamer's size* due to resharpening (see discussion on *nonstandard sizes* next). However, on drills that are gripped by chucks, the size is often obliterated because the bit spun during use and the identification was wiped out. The final word on any tool's diameter is to measure it before use.

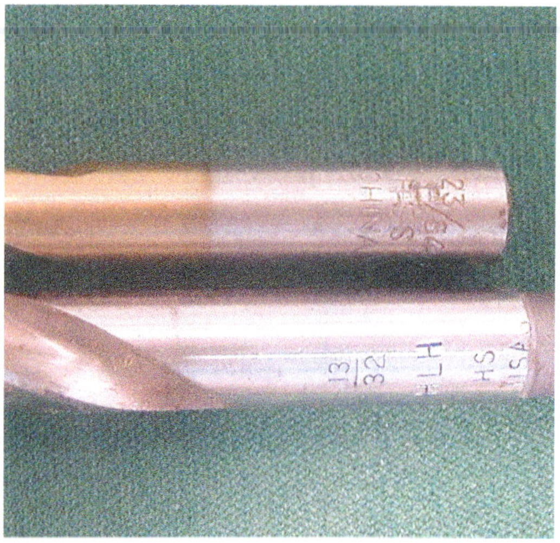

Figure 9-3 Drill sizes stamped or engraved on the drill's shank may be obliterated.
Courtesy Mike Fitzpatrick

> **KEYPOINT**
>
> Make it a habit to measure a cutting tool before use. Like using the uncalibrated micrometer, if the cutting tool isn't the right size and you use it to make parts, the fault for the scrap or rework falls on the machinist, not the person handing the bad tool to them.

Measuring on the Margins

In Fig. 9-4 (looking at the tip of the drill) and in Fig. 9-5, notice the narrow band along the length of the drill on each side, the **margin**. Always measure the size of the bit on these margins, near the cutting edge, as this is the only full-diameter portion. The narrow margins provide side clearance in the hole. Behind them, the bit is relieved to reduce rubbing and allow coolant to penetrate down to the cut from above.

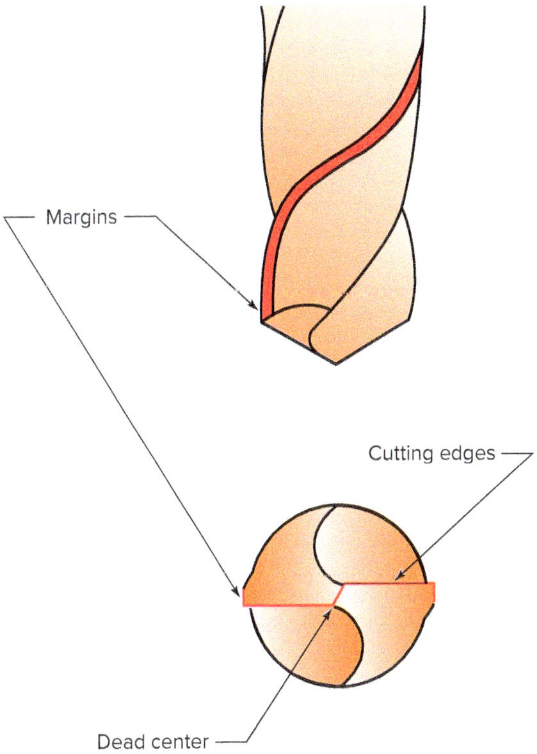

Figure 9-4 Always measure on the margins of the drill bit.

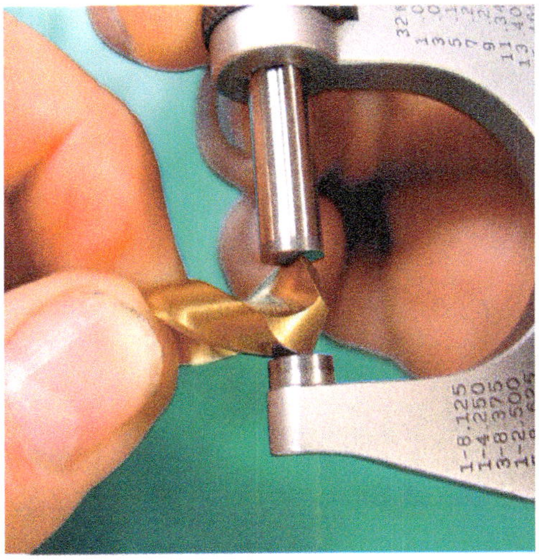

Figure 9-5 The micrometer is placed on the margins.
Courtesy Mike Fitzpatrick

Standard and Nonstandard Diameters

While both drills and reamers are supplied in each of the four size series, reamers can be found in the shop in any diameter. There are two reasons that these odd-sized reamers lurk about the shop just waiting to trap the unsuspecting machinist.

Special Undersized Reamers

Recall the press fit discussion of Chapter 5; to press a 0.500-in. bushing into a plate, one would ream the hole to 0.4985 in. That reamer is made by reducing a 0.500-in. standard size reamer by grinding the side cutting edges. When a reamer (or any cutting tool diameter) is modified, the tool grinder is *supposed* to remark the new size, and they *should* be kept in a special drawer away from standard sizes.

Resharpening Makes Them Smaller

When a drill bit is sharpened, only the clearance face is ground; thus, it stays nearly the same diameter for its entire service life. However, the side teeth of reamers eventually wear. They, as well as the end cutting edges, must be reground. At that time, the objective is to grind away a minimum amount of cutter diameter to restore the sharp edge. But no matter how little is ground away to sharpen it, the reamer ends up a nonstandard size.

Oversize Reamers

Or perhaps, a special-diameter reamer might be needed just slightly bigger than 0.500 in. to allow a precision slip fit to the hole. It is possible to buy a specially made reamer from 0.0001 in. up to 0.005 in. larger than a standard size.

KEYPOINT

Size markings may not be the true diameter with reamers. Always measure!

9.1.3 Selecting the Holding Method—Chuck or Taper

Straight- and Taper-Shank Tools

These are the two ways drills and reamers are held in drill presses. Each imparts some advantage.

Straight-shank drills (Fig. 9-6) must be held in a chuck that squeezes them between jaws. Under a heavy load, they can spin in the chuck. In contrast, taper-shank tools are mounted by tapping them into a receiving socket—they are more solidly held both for centering and for torque drive.

KEYPOINT

Small drills up to $\frac{1}{2}$ in. are usually straight shank, while sizes above are normally taper shank.

There are exceptions to the size rule, where drills larger than $\frac{1}{2}$ in. feature straight shanks, called **turned-down shank** drills. They are used where the drill press hasn't the capacity to hold taper shanks, or they are made from damaged taper shank drills to continue using them. Using them isn't a recommended practice as they strain chucks beyond designed limits.

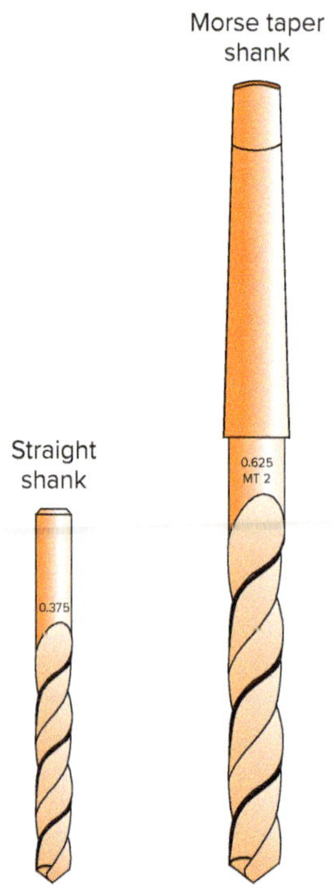

Figure 9-6 Straight-shank and taper-shank drills.

Chucking Straight-Shank Tools

Straight-shank drills and reamers require a chuck to hold them in the machine (Fig. 9-7). It may be either the key type often called a "Jacobs" chuck, due to a well-known manufacturer, or the more accurate and expensive keyless chuck. Both feature three jaws that close simultaneously and remain on the center of the chuck as they do—called a universal jaw chuck.

Figure 9-7 Two varieties of drill chucks—with and without chuck keys. If using a key type chuck, never leave the key in the chuck!
Lake Washington Technical College, Kirkland, WA/McGraw Hill

Chuck Limitations

While chucks are quick and easy to use, the grip force on the drill is limited with both types, key type or keyless. The keyless chuck has some advantage over the key type, in that as the drill torque increases, it tends to tighten its grip on the tool. However, that can lead to overtightening if the bit hangs up in the hole.

But both chuck varieties can allow the drill to slip. Even though the jaws are made harder than the drill shank, with time, this action degrades the centering accuracy of chucks. Their jaws can be reground, but most often they are replaced.

TRADE TIP

Tightening Drill Chucks When tightening either type of drill chuck, lightly tighten the tool, then loosen it just a bit, two or three times, while joggling the tool, before the final squeeze. This helps eliminate pinching the bit, off-center, between two of the three jaws (Fig. 9-8).

Figure 9-8 Avoid pinching the bit between two of the three jaws.
Courtesy Mike Fitzpatrick

Safety Tip

Never leave a chuck wrench in a chuck. If the press is accidentally turned on, the wrench suddenly becomes an unguided missile!

Holding Taper-Shank Tools

Taper-shank tools have three big advantages over straight-shank tools:

1. **Positive Drive**

   They are designed to carry greater torque loads without slipping. The flat **tang** (Fig. 9-9) transfers torque from the spindle to the tool. The inner end of the tapered hole features a slot where the drive tang fits.

2. **Accurate Centering**

   Tapers stay more accurately on the center of the spindle, as long as neither the spindle nor the taper is damaged.

3. **Self-Holding Solid Design**

   If cleaned, then taped lightly into a clean spindle, taper shanks stay put until driven out again. After wiping both taper and hole, insert, then tap them with a soft hammer. To remove them, use the **drill drift pin** (or just drift pin) shown in Fig. 9-9.

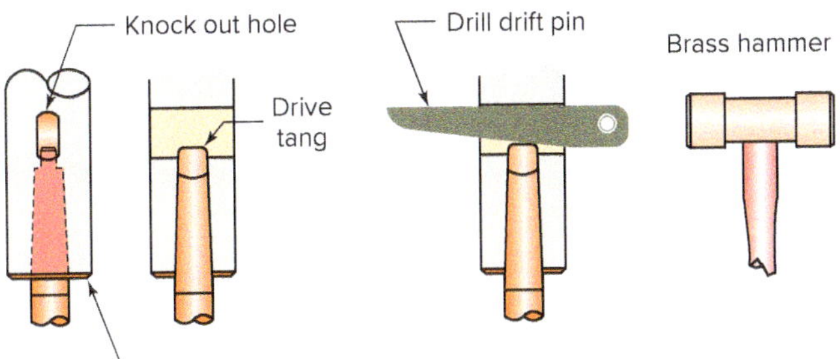

Figure 9-9 Clean all tapers before assembly. Then wedge out with a drill drift pin.

Morse Taper Sleeve Adapters

Made popular by the Morse Tool Company, tapers for drill tools bear their name–the **Morse taper**. The cone shape of the taper shank is a universal, worldwide standard. (There are several other tapers used to hold tools in machines but only one for drilling tools.) The standard is $\frac{5}{8}$*-in. taper per foot.*

But there are six sizes of Morse taper: MT1 (smallest, nearly 0.5 in.) up to MT6, the largest (Fig. 9-10). They are actually different segments of the same $\frac{5}{8}$-in.-per-foot cone. When mounting taper tools, a **Morse taper adapter** might be needed to increase or decrease the size of the tool to match the size of the drill press spindle.

Taper *sleeves* might be needed to increase the tool taper up to the spindle taper (Fig. 9-10). Taper *sockets* are used to reduce a taper down, as shown in Fig. 9-11. Both sleeves and sockets are marked as to their smaller and larger size, for example, MT 1-3, meaning that it converts a number 1MT to a number 3.

Figure 9-10 A variety of Morse taper adapters.
Photography by Prographics at Milwaukee Area Technical College/McGraw Hill

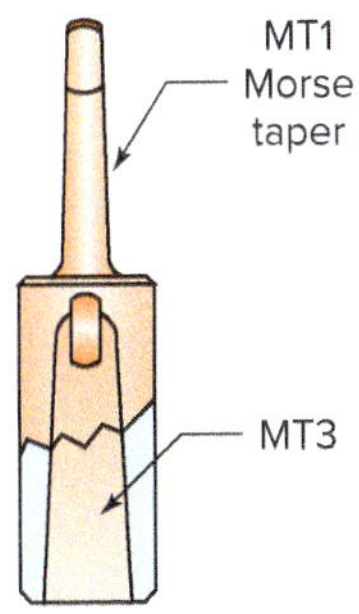

Figure 9-11 A Morse taper socket increases taper size.

Assemble Taper Tools Right—*Clean!!!*

When assembling taper-shank tools or adapters, it's critical that both parts are perfectly clean. A single chip of even the softest metal caught between the taper and receiving tapered hole will dent the shank of the tool, which then destroys the centering and self-holding of the tool (until it's repaired).

The taper in the machine spindle is deliberately made with harder steel than the tools it accepts so it will tend to survive such a careless act. But they can be damaged, too. A fast way to be seen as an amateur is to slam them together without taking the time to clean both components!

9.1.4 Drill Geometry

Varying the Point Angle

As we've discussed previously, the three basic tool angles are seen on any drill bit. The point angle is created by two lead or cutting angles on both edges relative to the axis of the drill. They combine to make the point angle.

> **KEYPOINT**
>
> A point angle of 118° is standard—for soft to moderate steel, aluminum, brass, and cast iron.

Most metal is drilled well with an included angle of 118° (Fig. 9-12). To check this angle we use a *drill point gage,* which tests that each cutting edge is 59° relative to the drill centerline—half the point angle.

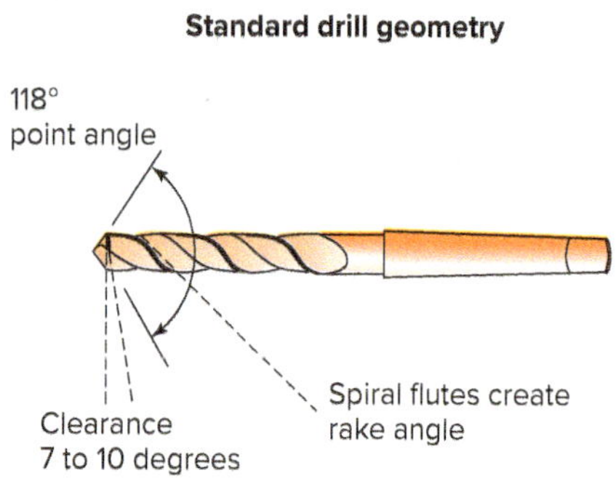

Figure 9-12 Standard drill point geometry shows the three primary angles.

When drilling materials such as hard steel, bronze, or stainless, we sometimes flatten the point angle to an included angle of 135° or more, not because it cuts better but because it makes the drill's outer corner stronger and therefore less prone to chipping and to heat buildup. The corner radius between the cutting edge and margin also helps the drill withstand drilling hard metals (Fig. 9-13).

Because they have less point angle, these flatter drills do not start well without a pilot hole. They tend to wander off location as they touch work material. To prevent *wandering,* the hole can be started with a standard drill point of the same diameter, but it is fed down only enough that the flattened drill will begin drilling fully within the hole.

For drilling plastics we sometimes use a more pointed drill than the standard 118° angle. But all soft or thin materials tend to grab and tear as the bit bursts through the opposite far side of the work. That leaves a jagged edge on the hole's rim. To prevent this problem, there is a different modification that can be ground on the point for plastics or thin sheet metal, called the *pilot point* or *spur-pointed drill.*

Pilot Point Drills

The pilot point stabilizes the drill from wandering as the V-shaped pilot starts before the cutting spurs cut the rim of the hole, thus producing a good finish on the outer edge. These drills are sometimes called sheet metal points. They are also used for wood and plastic materials of any thickness (Fig. 9-14).

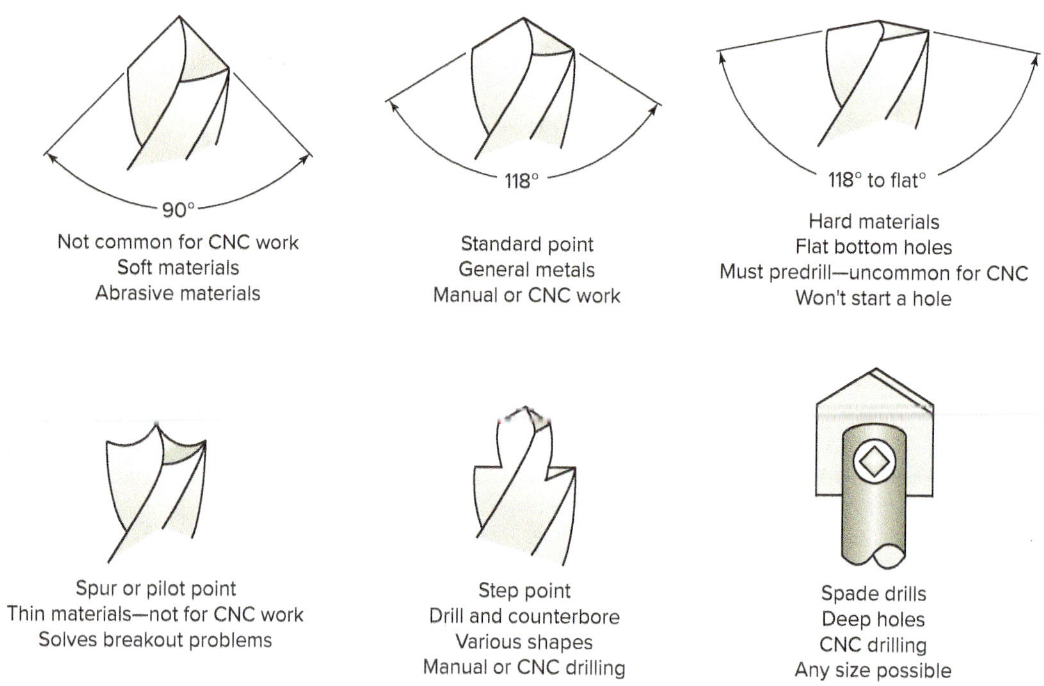

Figure 9-13 Modifications to standard points for various materials—some for CNC work, some not.

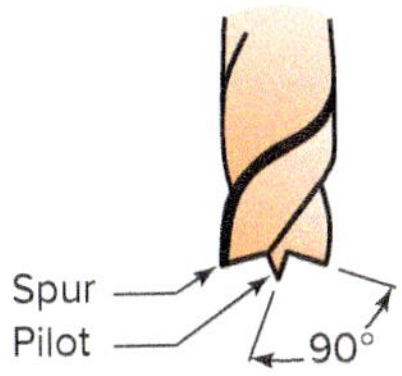

Figure 9-14 A spur or pilot point drill is ground for thin cutting materials, plastics, and wood.

Although drills can be purchased preground with this configuration, the spur point can be ground onto any twist drill with a large enough diameter (above $\frac{3}{16}$ in. or so). To create the tiny V into the cutting edge on both sides of the point, the grinding wheel must be dressed with a 90° sharp corner, as shown in Fig. 9-15. The challenge is to tilt, elevate, and rotate the drill to create the necessary side clearance on the point and the end clearance on the cutting edge spurs. Correctly held against the grinding wheel, the front and side faces of the grinding wheel produce all the necessary angles at the same time. Have your instructor or an experienced journey machinist show you how.

Figure 9-15 Grinding a spur point requires a well-dressed wheel. Note—safety guards are removed for clarity only!
Courtesy Mike Fitzpatrick

Adding a Negative Rake Angle to Prevent Self-Feeding

Due to the helix angle and digging action of the rake face, drills sometimes self-feed into soft or thin materials. They literally screw themselves into the work as a wood screw might. In certain materials such as soft brass, self-feeding can occur even when no pilot hole has been drilled. This unwanted action can break the drill and pull work out of a setup. It's happened to me! An unmodified drill literally screwed its way through a brass door striker plate! There was no hole; the drill went right through but snapped off as it broke through the opposite side of the work. It fed down so fast that it pulled the quill feed lever out of my hand!

To avoid this happening to you, grind a tiny neutral or even negative rake **land** on the rake face of the drill at the cutting edge, as shown in Fig. 9-16. (Land means narrow, flat spot similar to a margin.) Land no wider than 0.015 in. is needed to retard the self-feed tendency. Grinding more adds nothing to the action. However, the excessively large negative rake land is far more difficult to grind away, to restore the bit back to a normal rake when it's no longer needed on the drill.

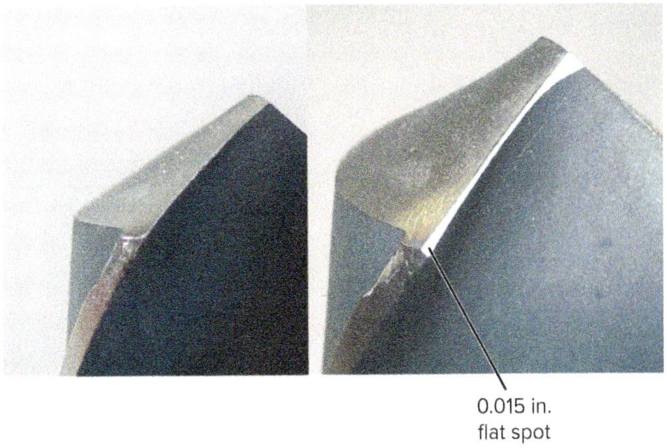

Figure 9-16 Compare an unmodified drill to one with added negative rake to stop self-feeding.
Courtesy Mike Fitzpatrick

TRADE TIP

Waiting to Change Tools We want to use tools to their maximum, to machine for the longest time before halting to replace or sharpen. However, overestimating and using any cutting tool beyond the time when it should be resharpened can cause all sorts of extra work and problems. The damaged drill in Fig. 9-17 is the perfect example. It will require a lot of work to restore it to usefulness.

A nearly sharp cutting tool can be reground in just a few seconds, whereas a very dull or damaged one takes a long time to reestablish the cutting edge, or it may even be beyond repair. But even worse, waiting too long can ruin the work, too!

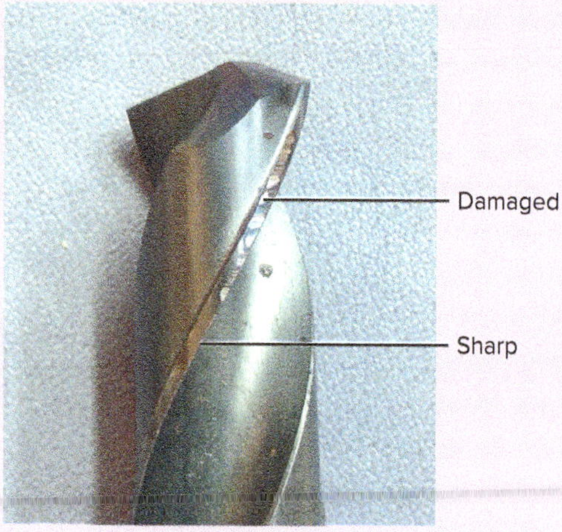

Figure 9-17 Severe damage such as this, to both cutting edge and margins, requires much time to repair.
Courtesy Mike Fitzpatrick

Sharpness—Cutting Edge and/or Margin Wear

Drill damage is determined with your eye before mounting the tool in the machine. Look for obvious nicks and chips in the cutting edge or rounded edges, especially at the outer corner (see Fig. 9-17).

However, when the cutting edge has worn normally from use, it requires a closer look. The dull edge, which may be as small as a few thousandths of an inch, will look like a bright line right at the junction of the cutting edge with the rake surface. To detect it, rotate the drill until the room light reflects off its worn radius. Cutting edge dullness is easily resharpened. Margin wear is not so easy to fix. The entire length of damage must be cut away using an abrasive saw, then a new point reground from there.

Clearance

When selecting a drill, verify that it has a correct clearance angle of around 6° to 10°. In industry it's unlikely that you will encounter a poorly ground bit–however, in training, experience has shown that clearance is often missed by beginners. Verifying the clearance angle (Fig. 9-18) and regrinding if needed can prevent a burned drill or work hardened material.

The Dead Center

The flutes of a drill bit are held together by a solid core of metal in its center, called the *web*. In Fig. 9-19, looking at the tip of the bit, the web creates the **dead center**, so called because it has no rake angle, clearance, or cutting edge. Without modification, the dead center does not cut; it simply scrapes metal away. There are two choices as to how to proceed:

1. *Leave the dead center as is.* While it's inefficient, it is acceptable point geometry in many cases, for example, when drilling aluminum, mild steel, or brass. Unmodified, the drill requires more pressure to force through the cut, but it will drill successfully. Predrilling with a smaller pilot drill relieves the pressure.
2. *Thin the web with some extra grinding.* There are two ways to do that. The simple method shown in Fig. 9-19 makes the web thinner but doesn't create very efficient cutting edges. It's quick and easy to grind onto the point but less efficient. It is created by nicking the point on both sides using the corner of a grinding wheel.

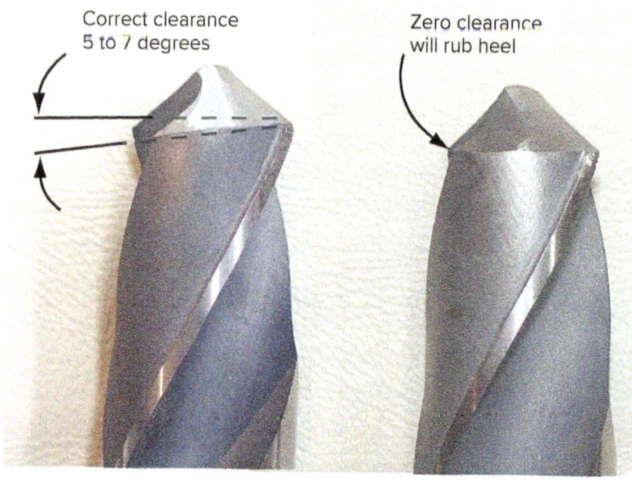

Figure 9-18 Compare two drills—one with correct clearance (left) and one ground incorrectly without clearance.
Courtesy Mike Fitzpatrick

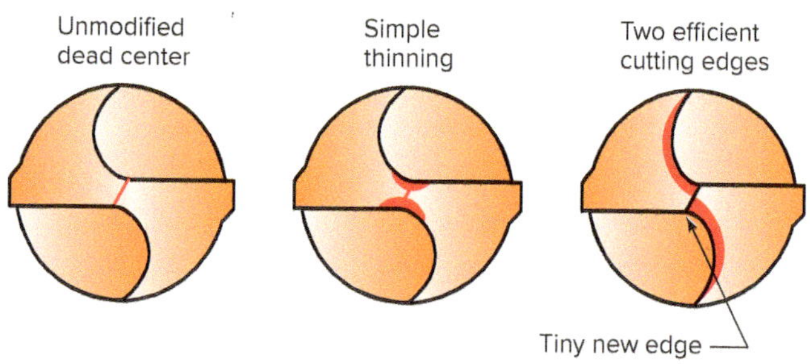

Figure 9-19 Thinning the dead center helps the drill cut with less down-feed pressure and heat.

The drill point on the right in Fig. 9-19 has been ground with two new cutting edges. This second method requires a sharp corner on the grinding wheel, a steady hand, and some practice, but it creates cutting edges at an angle to the original edges. Have your instructor demonstrate this skill after practicing grinding drill bits in Unit 9-5.

TRADE TIP

When hand-feeding a drill through work that has not been pilot drilled, it's obvious when the dead center is thinned and when it is not. Relieving a dead center can reduce the pushing force by as much as 50 percent in some cases.

9.1.5 Reamers

Machine Reamers

These reamers are designed to be turned by a drill press or many other machines (Figs. 9-20 and 9-21). They are supplied in both straight- and taper-shank versions. Machine reamers have multiple flutes of either straight or spiral design. The spiral fluted reamer tends to not chatter as readily as the straight flute. Straight flutes work okay, but RPM must be kept to around 25 percent of the possible speed to prevent chattering. The spiral design not only tends to dampen chatter but also delivers chips out of the hole better and makes a finer finish due to the lead angle of the side-cutting edges in the hole.

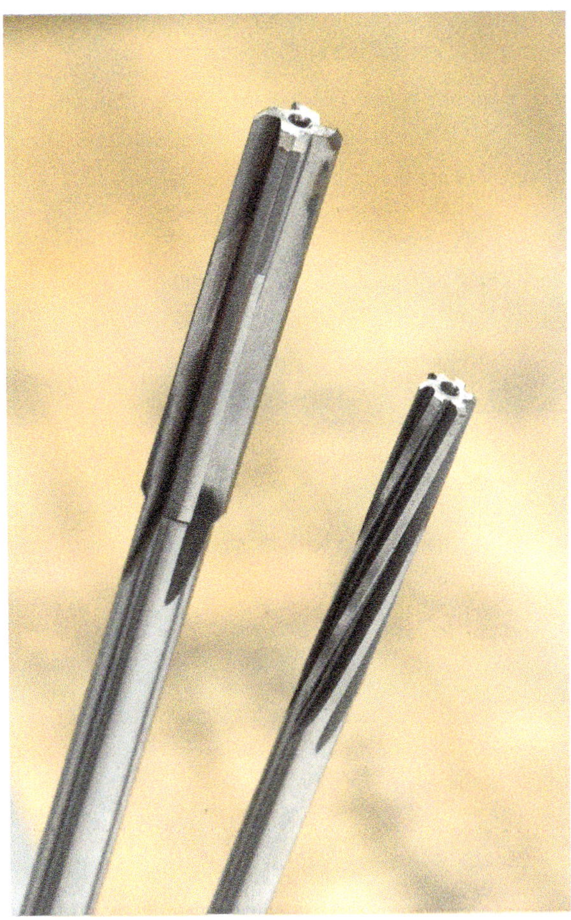

Figure 9-20 Spiral and straight fluted reamers.
Photography by Prographics at Milwaukee Area Technical College/McGraw Hill

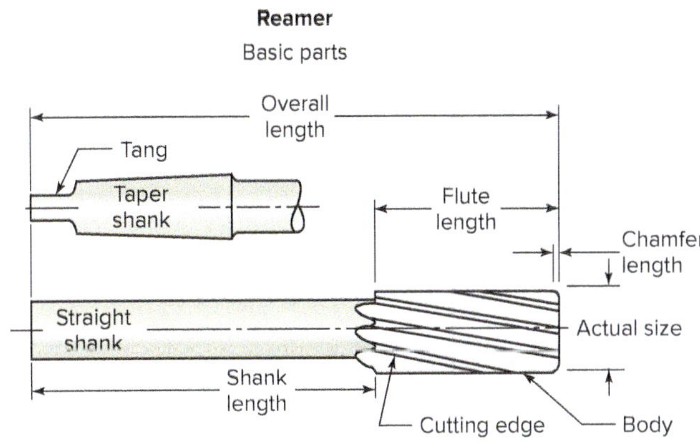

Figure 9-21 The main parts of a reamer.

Specialty Reamers

There are several other choices of reamers beyond the standard machine reamer (Fig. 9-22):

Shell reamers are common in production where they have the cost advantage of interchangeable cutting heads on one shank. These compact tools are very easy to resharpen as well.

Expansion reamers can be adjusted to any size within a small range. Expansion reamers are supplied in both hand or machine versions.

Taper reamers are made to finish tapered holes (e.g., the internal taper in a Morse taper sleeve). When using tapered reamers, use caution because pushing them just a little too deep can cause grabbing, whereby the reamer often breaks.

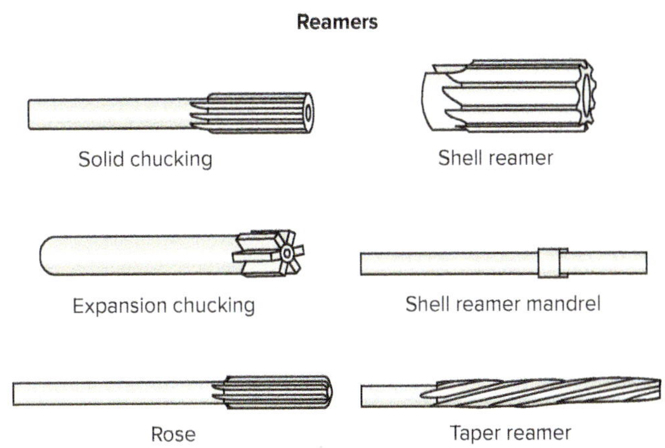

Figure 9-22 There are several other choices for reamers beyond the standard version.

Reverse helix reamers are used in place of standard reamers, which are made with a right-hand helix. Although the right-hand helix helps to bring chips up and out of the hole, occasionally a reamer with the standard helix will grab and tend to self-feed into a workpiece similar to a drill. This happens in soft material such as brass or lead. When self-feeding on a power feed machine, the reamer will tend to pull a taper shank out of the spindle. The correction is the left-hand spiral reamer, which cancels self-feeding but requires more force to push through the hole. They push chips forward into the drilled hole rather than removing them, so they work best when the hole goes all the way through the work.

If the *pilot hole is too large*, there isn't enough metal remaining for the reamer to cut an efficient chip. It may chatter, and the drill marks may not be completely removed as well.

Accurate location of the pilot hole is also important. The reamer will tend to align to the drilled hole and follow it even if the spindle is misaligned to the drilled hole. The reamer will flex as it rotates and follow the drilled hole a bit like a plumber's snake. We refer to this as **reamer float**.

TRADE TIP

When Reaming First, use liberal amounts of coolant or cutting oil for machines without coolant.

Second, push the reamer through the hole at a moderately high rate. (This is known as *crowding* the reamer in the trade.) In-feed should equal approximately one-fourth the reamer's diameter per revolution. Often, instead of slowing the RPM, increasing the feed rate will produce a larger chip load and stop the chatter. When practicing on the manual drill press, experiment with various combinations to see this effect for yourself.

Finally, there's *backing the reamer out of the hole*. Normally it's okay to pull the spinning reamer out of the hole before turning off the spindle on drill presses or CNC machines. However, in softer metals such as aluminum or brass, that action can remove more material on the way out—making an oversize hole. To solve this problem, on manual presses, turn off the spindle at the bottom of the stroke, wait until it nearly stops, then quickly withdraw the reamer as it coasts to a stop. But don't let it stop turning completely, as that makes straight marks along the hole's wall, resulting from contact with the reamer's cutting edges. If it does stop, rotate the spindle forward a half turn or so by hand as you withdraw the reamer. *Caution:* Do not rotate the reamer backward while in the hole, as that can destroy the cutting edge. If a CNC reaming operation has this double-cut problem, edit the program to slow the RPM below 150 before backing it out.

SHOPTALK

Blind Holes When a hole doesn't go all the way through the work, it's called a **blind hole** because you cannot see through it.

9.1.6 Drilling the Reamer Pilot Hole

Reamers require a pilot hole slightly smaller than the intended finished size. The size and location of the drilled pilot hole, sometimes called the *reamer drill*, is important for accurate, smooth results.

If the *pilot hole is too small*, the reamer will clog on the chips. While reamers are not as efficient at delivering chips up out of the hole as are drills, to help, always use a cutting fluid. Lots of it. Oil or flood coolant are essential for good finish, tool life, heat reduction, and chip ejection.

Selecting Drills for Reamers

As the reamer size increases, so can the excess left for it to cut, because the flutes become larger and can deliver more chip volume up out of the hole. There is some discretion here, based on material being reamed and the depth of the hole–the sizes are recommended. The chart is a rough guide to reamer drill excess to be left for reaming.

For an example, consider what fraction size bit would work well as a pilot for a $\frac{5}{8}$-in. reamer. Solve this from the drill chart or calculate the drill size.

Answer: $\frac{5}{8}$ falls within the $\frac{1}{4}$- to $\frac{3}{4}$-in. range–subtract $\frac{1}{32}$ in. on the drill chart to derive a $\frac{19}{32}$-in. pilot drill.

KEYPOINT

This is a rough guide. A slight difference in pilot drill size, larger or smaller, will work well, too. If the material is hard, choose less excess than suggested in the chart; if soft, slightly more machining excess is okay.

TRADE TIP

Control by Sound During lab practice, make an effort to learn the normal sounds of sharp tools drilling, reaming, and tapping. Then, if the sound changes, that's a strong signal that the tool might be dulling or loading up with chips. Stop immediately if there is any doubt.

Reamer Excess Chart

Inch Sizes

Reamer Diameter Range	Excess for Reaming (Deduct from Finished Diameter)
$0-\frac{1}{4}$	$\frac{1}{64}$
$\frac{1}{4}-\frac{3}{4}$	$\frac{1}{64}-\frac{1}{32}$
$\frac{3}{4}-1$	$\frac{1}{32}-\frac{1}{16}$
1 and up	Can be greater than $\frac{1}{16}$ in. but serves no real purpose; $\frac{1}{16}$ in. is adequate.
Metric Sizes	
0–6	0.5 mm
6–12	0.5–0.75 mm
12–24	0.75–1.0 mm
	1.0–1.5 mm
24 and up	Can be greater but serves no purpose.

UNIT 9-1 REVIEW

Replay the Key Points

- The five factors that determine the right drill for the job are *size, shank type, point angle, sharpness,* and *dead center.*
- Drills and reamers can be specified either as *metric, inch fractions, letters,* or *numbers.*
- Always measure a reamer before use.
- Measure the drill or reamer diameter on the margin.
- The pilot hole needed for a reamer is a calculated amount smaller than the reamer.

Respond

1. Grab the drill gage drawing, which can be downloaded from the Online Learning Center or supplied by your instructor. Choose reamer drills for each hole size listed on the print up to $\frac{1}{2}$ *in.* Use the guide here in Unit 9-1 to find the correct excess. Choose the greater excess (smaller hole) of the two possible.
2. Using the drill chart, select the metric reamer drill to ream a 12-mm-diameter hole.
3. Of the holes in the drill gage, which will probably be taper-shank tools?
4. What is the standard drill point angle?
5. To drill hard bronze, what point angle might be needed to prevent edge breakdown, 108° or 135°?
6. Why choose the point angle of Question 5?
7. What problem arises when we choose to modify the drill point geometry as in Question 5?
8. Describe the three possible options a machinist has in dealing with a drill's dead center.

Unit 9-2 Drill Presses and Setups

Introduction: Before learning setups, let's take a brief look at the manual machines commonly found in labs and industry. There are four general types with many variations within each. They are classified by whether they have power feed and what type of drilling head they feature:

Standard drill presses

Power feed drill presses

Radial arm drill presses

Production drilling machines–multispindle/multihead

CNC milling machines (often called machining centers) are rapidly replacing many of the production drills because they can also cut and shape the metal as well as drill it, all in one efficient setup—called a one-stop setup. So, we'll limit our investigation to the first three. However, the knowledge gained here will transfer to the production machines should they be encountered on the job.

TERMS TOOLBOX

Bushing (slip bushing) A hard steel guide to ensure drill and reamer location.

Fixture (jig) A work-holding device that accurately secures odd shapes and shortens part turnaround time.

Peck drilling The in-out-in action used to break long chips into short segments and to clear chips during deep drilling.

Quill A cylinder that contains the spindle bearings and spindle. The quill slides in and out to create the *Z* axis of most machines.

Radial arm drill press A drill press for large work wherein the head is moved over the intended hole using the radial movement of an arm.

Sensitive drill press A lightweight, high-speed drill press for small drills.

9.2.1 Standard Drill Press

At least one is found in every shop. These simple drilling machines are designed for light-duty drilling and reaming work up to around $\frac{1}{2}$-in. diameter. They subdivide into two general types.

Sensitive or Bench Top Presses

Because they break easily, drills smaller than 3 mm or $\frac{1}{8}$ in. require very high RPM for efficient cutting and a light in-feed touch by the operator. The **sensitive drill press** meets that need because it can achieve spindle speeds as high as 12,000 RPM or more. To increase operator feel, the sensitive press has a lighter **quill** and less powerful return spring compared to its big brother presses.

What Is a Quill?

Quills are used on many machines where the tool must spin and slide in and out along an axis. In a CNC program, that axis is always identified as *Z*. On a drill press, the quill is the precision cylinder that contains the spindle bearings and rotating spindle (Fig. 9-23). The quill slides down by hand lever pressure and returns by spring action.

KEYPOINT

When a spinning cutter slides in and/or out, it's always named the *Z* axis if it's the only quill on the machine.

Figure 9-23 A sensitive press is best for small holes below $\frac{1}{16}$ in. (0.5 mm) as it can achieve 20,000 RPM and has a light quill return spring.
Courtesy Morgan Branch CNC/Arlington, WA

The operator can sense or feel how the drill is progressing. The sensitive press uses only straight-shank drills.

Standard Floor-Mounted Presses

Figure 9-24 shows the larger, standard drill press, the everyday workhorse of all shops. Its major difference from the sensitive press is its greater size and power. These presses feature a lower RPM range from around 500 up to 3,000 or so, depending on make and motor, but always more horsepower compared to their little brother bench models.

Most of these presses feature a spindle taper with an MT2 or MT3. Taper-shank drills can be mounted directly in the spindle taper, or chucks can be inserted in the spindle taper. The chuck is normally left in the spindle since that makes exchanging tools faster. Similar to bench drill presses, most standard presses cannot reverse their spindle rotation.

Both bench and floor-mount presses feature the following:

A. Variable speed for differing size drills. The speed selection may be by step pulley, which offers a limited number of RPM selections (Fig. 9-25). Or RPM may be changed with a variable cone pulley transmission, offering infinite speed adjustment (Figs. 9-26 and 9-27).

Figure 9-24 Standard drill press.
Courtesy Mike Fitzpatrick

Figure 9-25 Step pulleys provide a limited number of spindle speeds.
Courtesy Mike Fitzpatrick

Figure 9-26 A variable speed dial.
Courtesy Mike Fitzpatrick

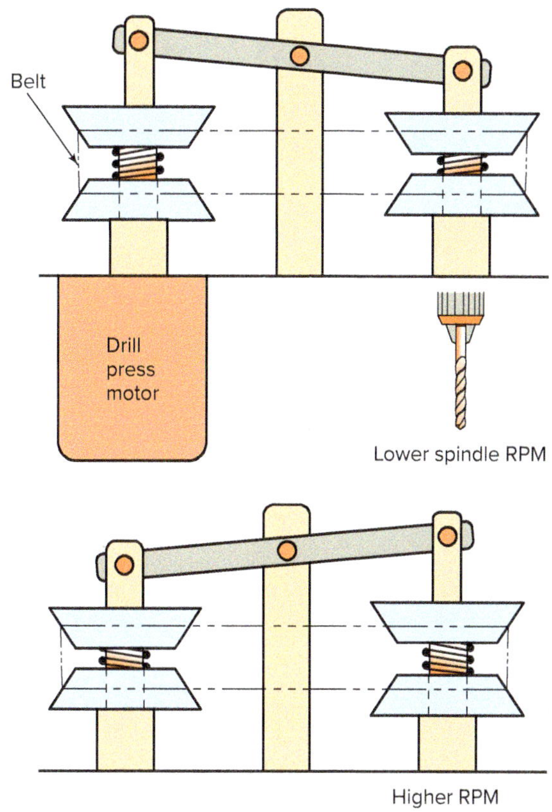

Figure 9-27 The principal of a variable cone drive.

Variable Speed Drives This is a good time to understand variable speed cone transmissions as they are used on other small machines in the shop, too. Figure 9-27 is a simplified illustration of a cone transmission. Rotating the speed selection dial (see Fig. 9-26) causes one pair of facing cones to move apart, thus creating a smaller circumference for belt contact, while the other two cones move together at the same time, creating a larger circumference. This action changes the drive-to-driven ratio.

TRADE TIP

Changing RPM the Right Way If the spindle speed is selected by variable drive, ***only change speeds when the spindle is running.*** Incorrectly cranking the speed dial with the spindle turned off can dislodge the belt from the pulleys—similar to shifting a 10-speed bike while it's not moving.

B. Vertically adjustable table. The worktable can slide or be cranked up and down to accommodate large or small work.

C. Horizontally rotating table. The worktable can swing around on the column to accommodate different shaped work or to completely clear the table to mount extremely tall work on the base plate (Fig. 9-28).

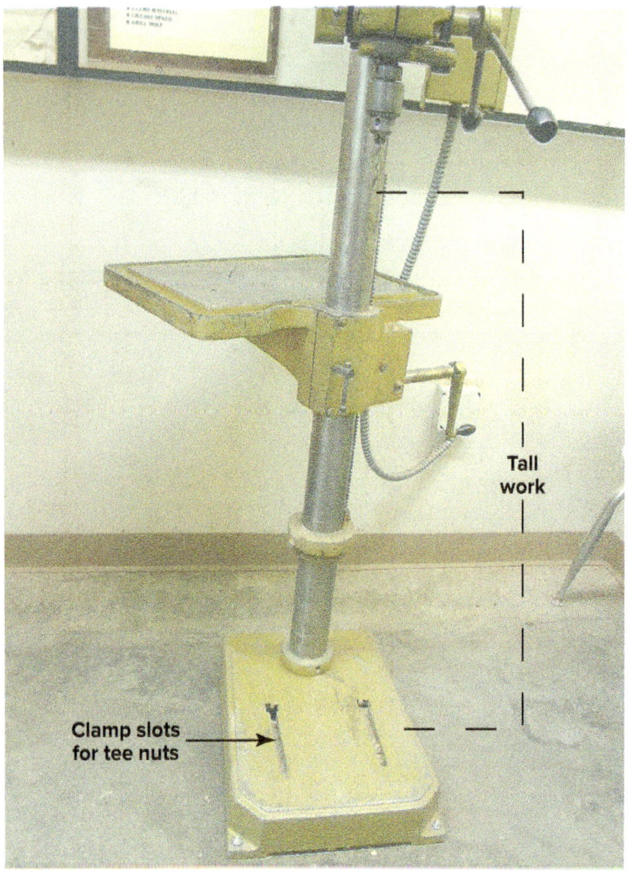

Figure 9-28 A drill press table swung out of the way to accommodate a part too tall to mount on the table.
Courtesy Mike Fitzpatrick

D. Quill depth control (also called a quill stop) to limit and control the hole depth. Most quill stops (Fig. 9-29) are graduated with a vertical rule and division lines. One full turn often equals an increment of $\frac{1}{16}$-in. change, or a 2-mm change, for example. Take the time to investigate what those graduations equal on your lab's press.

E. Some standard presses feature a tilting table.

Figure 9-29 The quill stop will allow the spindle down another 20 mm. One full turn of the stop equals 1.0 mm.
(both): Courtesy Mike Fitzpatrick

9.2.2 Power Feed Drill Press

Evolving up in size and strength, this heavier machine features a power feed and a larger quill to accommodate bigger drills (Fig. 9-30). These presses are less common in schools but are well suited to industry, where they are put to work drilling holes from 12 mm up to 30 mm or greater ($\frac{1}{2}$ in. up).

Figure 9-30 A *power feed* drill press or sometimes called a gear head drill press.
Lake Washington Technical College, Kirkland, WA/McGraw Hill

Due to the extra force they must withstand, many of these presses do not use belt drives. Instead a series of gears is used to shift speeds. For that reason, you might also hear them called "gear head presses."

In addition to the features of the standard press, these presses feature

A. Varying power feed rate selections–The quill can be pushed down by hand, or the operator can pull the hand lever sideways to engage the power feed. This feature not only prevents operator fatigue, but it also imparts consistency to drilling operations.

B. Micrometer quill stop–Halts both hand and power down feeding.

C. Coolant systems and reservoir–An option but very common on these presses.

D. Larger spindle, usually MT3.

E. Reversing spindle to allow power tapping. Some feature automatic stops to reverse the motor at a preset depth.

F. Gear head to create slower speeds but greater rotational torque (force) for tapping and reaming.

9.2.3 Radial Arm Drill Press (Radial Drills)

This machine is generally the most powerful of the three (but there are crossover exceptions). In industry, there are **radial arm drill presses** large enough to drive an automobile on the worktable. There are also *radials* with two complete drilling heads at each end of the worktable to enable reaching extremely large work.

Radial Arm Presses Take the Head to the Work

The presses discussed thus far require the operator to push or tap the work into vertical alignment with the drill, then clamp it down. However, as work gets very big or bulky, that's difficult or even impossible to do. Using the radial arm capability on these large presses, after swinging the arm out of the way, the operator can move the work directly over the table with an overhead crane, to set it down. The crane is removed, and the work clamped firmly in place on the table (Fig. 9-31). The operator then positions the arm and drill head back over the work, by moving in or out along the rail and in a circular arc about the column. Once the drill is in position over the work, an electric lock (or a manual one on smaller radials) is actuated to secure the drill head in place.

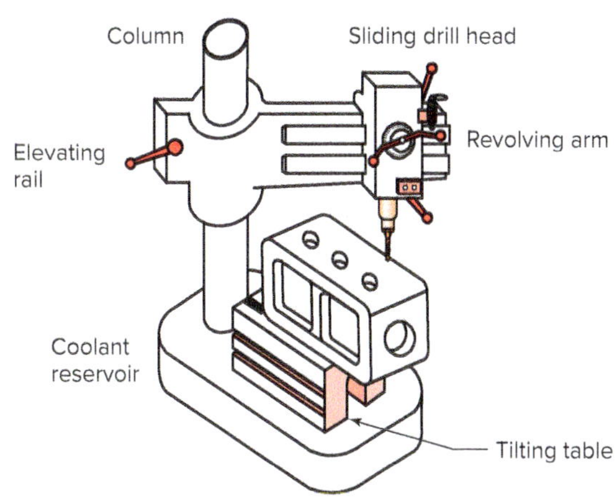

Figure 9-31 A radial arm press can position over the worktable or swing to allow the crane over the table.

In addition to the features of the power press, radial drill presses (Fig. 9-32) have these features:

A. The entire radial head moves up and down on the column, by hand crank or by power. The table does not move up or down. Some tables can tilt.
B. The motor reverses by clutch action.
C. A true clutch is included that allows gently starting the spindle rotation and reversing by slipping the clutch. This is a great feature for power tapping and other operations requiring operator control such as spotfacing (coming up).
D. Extra large worktable is included with tee-slot-type hold-down capability.

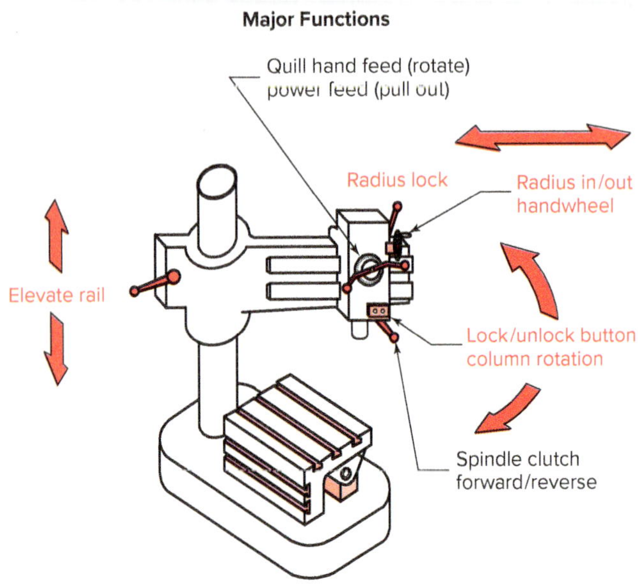

Figure 9-32 The main parts and features of a radial arm drill press.

9.2.4 Drill Press Safety Always

Because drill presses appear to be simple, and harmless, many machinists both new and experienced, let their attitude slip when running them. We instructors see the following three accidents occur far too often–don't let them happen to you!

Unclamped Parts Spinning out of Control

It's tempting to put a workpiece on the table and drill it without clamping, or even dumber, to try to hold it by hand! Don't do it! The problem is that as the drill breaks through the far side of the unclamped work, it grabs the workpiece, pulling it up the drill bit much like a screw in wood, then it spins and goes out of control on the bit. Depending on the bit's RPM, the wildly rotating part can hit your hand several times before you can react. But the accident isn't over yet!

As the unbalanced object spins out of control on the bit, the bit often breaks; then the part flies out of the drill press. It's a "fire-drill" easily avoided by taking a few moments to secure the work in a vise or to the table with clamps before drilling.

If due to incorrect clamping the work comes loose during drilling, your actions are to immediately turn off the press and step back. There are no heros in machine shops, only smart people. Don't try to grab the wildly spinning work with your hand! See Fig. 9-33.

Figure 9-33 Warning! This accident happens when work is not clamped!
Courtesy Mike Fitzpatrick

KEYPOINT

Pay close attention to instruction on how to clamp work the right way to completely avoid this accident!

Hair or Clothing Caught in the Spindle

The spindle and bit can grab and instantly wrap up and pull any loose item that gets too close. This happens for three reasons. One is that the friction of drilling causes a static electric buildup in the tool. Items like hair or clothing are attracted to the tool. Once contact is made, they wrap up lightning fast!

The second is that there are air currents created by the spinning tool. Like a fan, air is slung out and away. If it spins out, more air must rush in elsewhere to replace it, bringing along the hair or cloth.

This accident commonly occurs when a machinist with long hair leans over to see how the drill is doing.

KEYPOINT

Wear a hat or hair band and appropriate clothing for the shop environment. No long sleeves or jewelry while operating any machine, especially a drill press!

Chips Flying Out (and Rarely a Broken Tool)

The third way to get caught by the drill press is the long, stringy chips they produce (Fig. 9-34). If not broken by operator action, they fling out in an expanding radius as they grow in length. Protect yourself from them by wearing safety glasses, no long or loose clothing, no fabric gloves that will be caught by the chips, and clear away anything on the table that might get caught up by them. If, as in the upcoming Shoptalk, long chips become a real problem, use an in-and-out motion on the feed lever, called *chip breaking* or **peck drilling**.

Figure 9-34 Unbroken chips can be dangerous.
Courtesy Mike Fitzpatrick

Drilling to a Controlled Depth

Clearing and Breaking Chips During Drilling

Peck Drilling This is a good exercise to practice for getting to know drill operations and limits before writing programs. As drills progress deeper into the hole, at first, small up-down pecks safely break the chip, but at a given depth (learned from experience), you must begin chip-clearing pecks completely out of the hole, then return. These moves are repeated with drilling cycles in CNC work. Get the feel for the need to peck drill and practice it on the manual drill press. The only way to go much deeper than six to eight times the drill's diameter using a standard twist drill is to perform peck drilling.

KEYPOINT

As a rough guide, switch to peck drilling when hole depth is five times the hole diameter.

Small pecks or even no pecks at all work well for hole depths up to around five times the drill's diameter, depending on alloy and coolant supply. A $\frac{1}{8}$-in. drill, for example, will not need to clear chips until drilling down to around $\frac{5}{8}$ in. or so, depending on the kind of metal being drilled. But drilling deeper requires switching to big pecks.

Coolants help lubricate the chips and prevent clogging. But as the depth progresses, chips become packed more tightly in the drill's flutes and prevent the coolant from getting down to the cut area. Again, to avoid burning the bit, big pecks momentarily allow coolant to flow down to the contact area to cool the drill and prepare for the next peck.

SHOPTALK

Peck Drilling by Hand or Program CNC machines that drill feature two different peck drilling cycles that either break the chip by repeatedly halting and pulling back a few thousandths, then commencing forward again, called *small pecks or a chip breaking cycle.* Or, for very deep drilling, they pull the drill all the way out of the hole with each peck, called a *chip-clearing cycle or big pecks.* After the upstroke, the control remembers the last depth to which the drill had progressed and quickly moves it forward into the hole, near to that depth before drilling again. Big pecks also allow coolant to flow down to the cut depth while the drill is out of the hole.

9.2.5 Drill Press Setups

Now let's look at four common methods of holding the work in a drill press: vises, clamps, plates, and fixtures. Each has its own purpose.

Drill Press Vises

These specialized vises are different than standard milling vises in two ways (Fig. 9-35). The most obvious is the fast open and close feature for quick size adjustment and turnaround time (changing from one part to another).

The second feature that sets drill vises apart is the shelf in the jaws that enables suspending and holding work up off the floor of the vise to drill entirely through the work without the drill touching the vise bed.

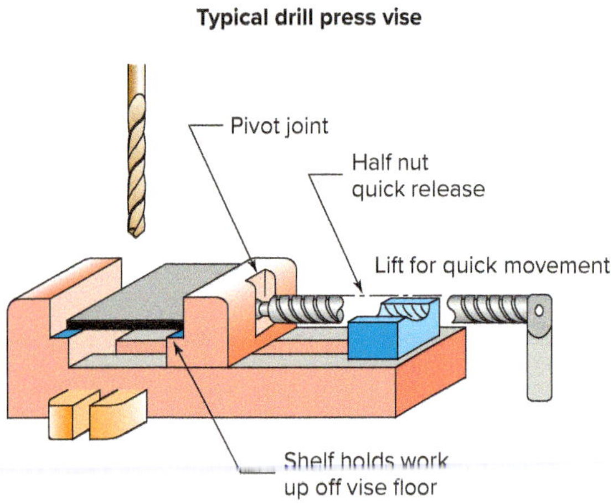

Figure 9-35 A drill press vise has two unique features.

> **KEYPOINT**
>
> **Clamp the Vise to the Drill Press!** Once the work is secured in the vise, you aren't done with the setup. Next, clamp the vise to the table in one of the ways described next.

> **TRADE TIP**
>
> **Drill Press Vises Cannot Hold Work for Milling!** Never use drill press vises for operations that produce a side thrust such as milling. They are designed for vertical pressure only.

Clamping Parts

There are a variety of clamping tools for drill press work. Some are used on other machines, and a few are made expressly for drill press work since they would not withstand forces other than vertical drilling.

Table Clamps

The most common clamp used throughout the shop is a simple bar with a slot through which a bolt is placed (Fig. 9-36). It is supported at the back with a heel block. There are a variety of modifications made to these clamps, which you'll see when studying milling. There is a right and a wrong way to use table clamps.

Test Your Powers of Observation

In Fig. 9-37, there are four things wrong with the clamp on the right. What are they? But there is something wrong with the clamp setup on the left, too. What is it? Answers are found just before the unit review.

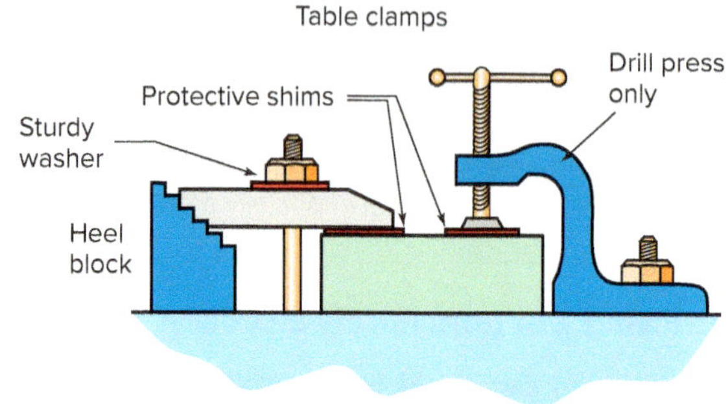

Figure 9-36 Two kinds of table clamps.

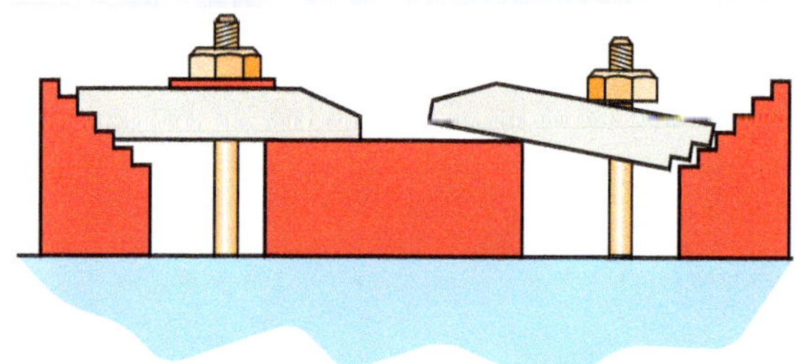

Figure 9-37 What is wrong with both clamps?

> **KEYPOINT**
>
> To concentrate the pressure on the work, not the heel block, place the bolt close to the work, not close to the heel block. Also, set the heel block such that the clamp is level or tilted slightly down toward the work, which prevents pushing the work out from it.

Other table clamps are used exclusively on drill presses as well. A typical quick clamp example is shown in Fig. 9-38. Common "C" clamps are also used on drill presses at times.

Prevent Drilling the Vise or Table!

When locating and clamping work directly on the drill table, or any machine, be certain that the setup provides cutting tool limits that will not allow it to contact the table surface. There are two solutions on drill presses:

- *Use the clearance hole* (Fig. 9-39). Place the workpiece with the intended hole location directly over the clearance hole in the machine table. Before clamping, first test the location with the drill bit to the clearance hole. Rotate the table if needed, such that the drill passes through the hole.
- *Elevate the work* up off the table by clamping over parallel bars (see Fig. 9-39) a block of metal or even a sheet of plywood to provide drill clearance to the table. In this case, use the quill stop on the press to prevent the drill from reaching the table as it breaks through the work.

Figure 9-38 A quick-action table clamp for a drill press.
Courtesy Mike Fitzpatrick

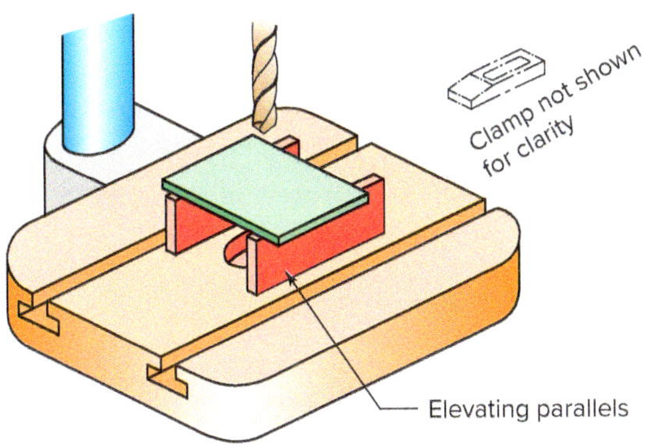

Figure 9-39 Two ways to protect the drill press table.

Work-Holding Plates Create Vertical Datums

When holes or other cuts must be parallel to a datum, or when work is difficult to clamp in a vise or directly to the drill press table (Fig. 9-40), one solution is the shop-grade, right-angle plate. Recall that these utility plates are different from the layout version that are not to be used during machining. They are heavy-duty, work-holding plates with holes for bolts or tee slots for clamping work. The clamps we've discussed could be added to Fig. 9-40 to firmly hold the part on the plate.

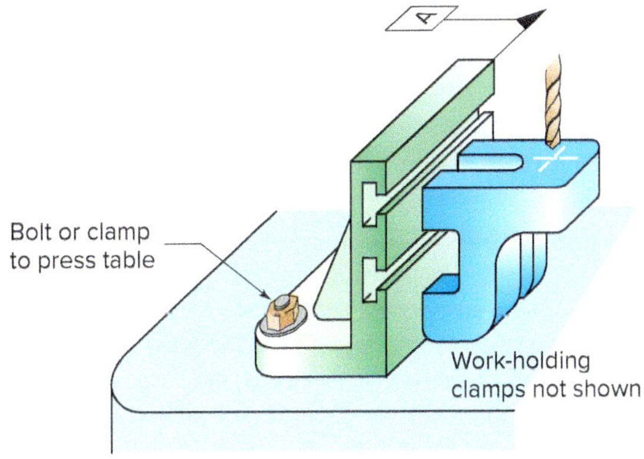

Figure 9-40 Drilling parallel to a datum or surface.

Drill Jigs

In manufacturing, the word ***jig*** means a holding tool that secures parts that are otherwise difficult to hold for machining, welding, or assembly. Jigs ensure accurate location of the workpiece on the machine and accuracy of the operations to be performed. They also make it faster to exchange one part for the next during production. In Fig. 9-41, the top of the jig slips off two pins, to exchange parts. This jig locates the face of the part on Datum A, similarly to the angle plate setup of Fig. 9-40. By using the left-right position of the part in the jig, Datum B is obtained by slipping the central slot in the part over a locator tab in the jig. When in position, a clamp is screwed against the workpiece. Once in position, the hardened slip bushing ensures that the drill will not wander.

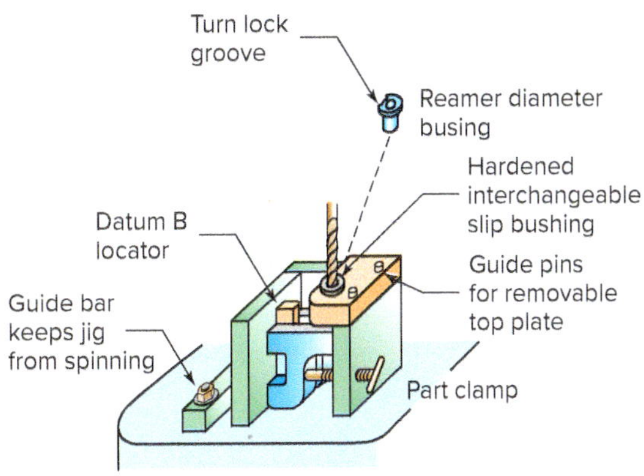

Figure 9-41 A drill jig holds the work and guides the drill.

Drill jigs take on two forms:

- *Custom-made* tools are specifically built for a single part or perhaps to hold a family of very similar parts. Figure 9-41 is a custom made jig consisting of four steel plates fastened together.
- *Universal kits* enable the machinist to make up a jig using components. These are sometimes called box jigs, as they

start with several sizes of square boxes. A fourth plate on the near side of Fig. 9-41 would simulate a box jig, drilling through the work.

SHOPTALK

Jigs and Fixtures Tradespeople abbreviate "drill jig" to "DJ," but the word **fixture** is often interchanged with jig (Fig. 9-42). What's the difference? Not much. They both perform the same task—hold the work firmly for machining. But when they are used on a mill or lathe, custom holding tools are called fixtures. While on a drill press, they are called jigs.

Figure 9-42 After drilling, the removable guide bushing will be rotated to unlock if from the jig, then exchanged for the reamer bushing.
Lake Washington Technical College, Kirkland, WA/McGraw Hill

Like everything else in our trade, CNC has changed the need for jigs and fixtures. Since we often can place the part in a single machine to drill and shape it, fewer jigs or fixtures are needed when programmed machine tools are used.

Interchangeable Guide Bushings

Where drilling and reaming must be at a precise location, and the work will be done on a manually operated press, drill jig guide bushings are used in Fig. 9-42—called **slip bushings**. First, the drill size bushing is inserted in the fixture with a one-quarter twist that locks it in position. Then the hole is drilled. Next, the drill bushing is exchanged for one that fits the reamer. Each bushing guides each tool to an exact location.

Observation Answers

The right-hand clamp is tilted down; it will push work out or slip off. There is no washer; the nut is touching on the corner only. There is no shim, so it will mar the workpiece. Finally, the bolt is too far from the workpiece.

The left-hand clamp has no protective shim; it might mar the workpiece!

UNIT 9-2 REVIEW

Replay the Key Points

- Adjust RPM on variable speed transmissions only when the spindle is turning.
- When setting up any machine, be certain the cutting tool will not contact the machine table.
- Drill presses can be dangerous even though they look simple and safe.

Respond

1. What is the normal upper limit of diameter of straight-shank drills? For the standard drill press? Metric limit?
2. Which drill presses are equipped with a reversing motor to accommodate power tapping?
3. What is different about a drill press vise compared to a standard vise?
4. True or false? For safety, you must either clamp your part to the table or angle plate or put it in a vise. The vise must also be clamped down. If this statement is false, what will make it true?
5. Name the type of drill press shown in Fig. 9-43.

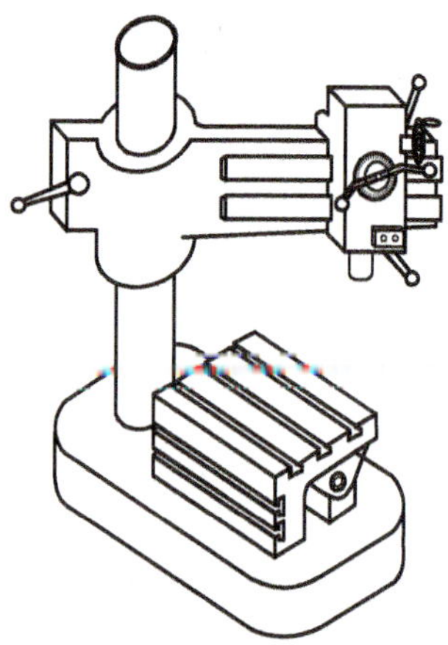

Figure 9-43 Identify this industrial drill press.

6. What four methods exist for holding work on a press?
7. Which method in Question 6 runs the risk of drilling the table?
8. How do you prevent the problem of Question 7?
9. Name a method of preventing long, stringy chips on a drill press.
10. Describe the four methods of work holding for a drill press.

Unit 9-3 Speeds and Feeds for Drilling

Introduction: Setting the correct RPM and feed rate is necessary for good finish, tool life, and size accuracy on all machines. All the methods we'll discuss determine the best *initial* values. After testing them on the machine, they are almost always adjusted up or down to suit the particular situation. For example, does the setup tend to chatter, how is the particular tool performing, is coolant available, and so on? The skill of fine-tuning a setup improves with experience. It's nearly intuitive at times. It's also critical to recognize safety limits for both speed and feed rates and not to exceed them.

When practicing at the manual drill press observe closely the result of changing RPM (cutting speed) and feed rates: down pressure. That vital knowledge will help ensure success in your CNC programs!

TERMS TOOLBOX

Empirical data Data collected by experimentation rather than calculation. All speed and feed charts are empirical.

Feed rate Cutter advancement through the workpiece, expressed in inches of decimal parts of a millimeter per minute.

Surface speed The speed in feet per minute or meters per minute of the drill rim.

9.3.1 Drill RPM

As discussed earlier in Chapter 9 there is an optimum surface speed when drilling a given material and that translates into the RPM of the various tools in your setup.

For example, from Appendix II in this book, a surface speed of 100 feet per minute (F/M, or 30 meters per minute) for drilling steel is recommended and 175 F/M when drilling brass. *But that is not the RPM! It is the rim speed of the object that is spinning in feet per minute.*

KEYPOINT

A. There is a "best" **surface speed** for any material being drilled.
B. Different diameter drills must be set to turn at different revolutions per minute—such that their rim is at the correct surface speed.
C. Smaller drills require *faster* RPMs, while larger drills must go *slower* to achieve the same surface speed at their rim: big drill = small RPM and little drill = big RPM.

Five Methods for Accessing RPM

There are five ways to determine the right RPM for a drill or any other spinning tool:

1. *RPM chart* based on material being machined and tool diameter.
2. *Formula* using surface speed and tool diameter as the variables.
3. *Insight* based on experience.
4. *Dedicated speed and feed, slide-rule calculator,* or smartphone app (Sanvik.com).
5. *CAM,* built-in RPM calculation within the software.

SHOPTALK

Estimating RPM We'll study RPM charts and formulas in this chapter, but while using them, you'll also be developing your built-in sense of the right speed.

Drill Speed Chart

This is the simple way to obtain tool RPM. You'll find them on the wall in most shops, on plastic pocket charts, metal plates on the machine itself, and in reference books. The chart is entered with two cross-referenced variables: *drill diameter* and *type of material.* See Appendix III.

KEYPOINT

Drill charts assume the tool is made from HSS (typical of most shop quality drills). But, if using a carbide-tipped drill, the RPM can be increased by a factor of three or more.

There's not much training required to learn to use an RPM chart. The fastest way is to try a few practice problems.

Check Your Understanding

Use Appendix II, the drill speed chart. Answers are found at the end of this chapter.

1. How fast should a $\frac{3}{8}$-in. drill be turning when drilling mild steel?
2. What is the correct RPM for a $\frac{1}{4}$-in.-diameter drill in aluminum?
3. What is the RPM for a 1.0-in. drill in brass?

Critical Thinking

4. RPM changes as drill diameter changes. Look at Appendix II under aluminum and compare RPM for a $\frac{1}{8}$-in. drill to a $\frac{3}{16}$-in. bit (diameter difference = $\frac{1}{16}$ in.). Now compare the RPM for a $1\frac{1}{8}$-in. to $1\frac{3}{16}$-in.-diameter bit (also $\frac{1}{16}$-in. difference). Explain why the RPM difference is so great between the two smaller drills but the change is very little between the two larger drills.
5. Looking at the chart, what can you say about the machinability of carbon steel, compared to annealed stainless?
6. Not covered in the reading: What should you logically do if the RPM you find isn't available on the machine you are operating? (This happens on manual machinery such as step pulley drill presses, where there are only eight speeds available, but never on CNC, where spindles are computer controlled.)

Calculating RPM

Drill charts are specific, they don't cover diameters that lie between two common drills, and they cover only a range of sizes, so any drill larger or smaller than the chart must be calculated. On the other hand, formula results cover any diameter, large or small. Also when using the short formula coming up, the calculations can often be performed in your head, thus there is no need for a chart. Formulas use two variables:

- Diameter of the drill (*DIA*)
- Recommended surface speed (for the metal being drilled; *SS*)

Another advantage of a formula is that it can be programmed into your calculator with stops for each variable (surface speed and tool diameter). Since the calculator is always by your side, it's a ready method, but you must remember recommended surface speeds, or have an SS chart handy or store them in memory.

Surface Speed

To use either formula, long or short, one must start with a recommended surface speed chart. Appendix III provides the student biased numbers. It also features rounded numbers that are easily recalled. This should aid in developing the journey-level feel for RPM.

RPM Formulae—Long and Short Version

The short is derived from the long. Both produce acceptable results with only 5 percent difference in RPM. That's well within acceptability, since initial RPM is a ballpark (best estimate) process anyway.

When working in the inch system, surface speed is entered in feet per minute while diameters are given in inches.

ARMCHAIR FIELD TRIP

Don't miss this Haas Tip of the Day video about cutting speed calculations! You'll learn how to determine correct speed and feeds. Go to: **https://www.youtube.com/haasautomation** and search for this video called: **How To Calculate Speeds and Feeds—Haas Automation Tip of the Day**

Long RPM Formula (Diameters in decimal inches, PI = 3.14159)

$$\text{RPM} = \frac{\text{Surface speed} \times 12}{\pi(\text{PI}) \times \text{Diameter}}$$

Example calculation: A $\frac{3}{8}$-in. drill cutting mild steel. Turn to Appendix III to obtain the surface speed.

$$\text{RPM} = \frac{100 \text{ F/M} \times 12}{3.14159 \times 0.375 \text{ DIA}} = 1018.6 \text{ RPM}$$

Now compare this result to the Drill RPM Chart (from the Online Learning Center or instructor handout) at 1,066, just slightly faster. The long formula is the version you would program into your calculator and the one found in CAM software.

The short version shown next is useful because it can often be performed in your head. It's derived by shortening PI to 3.0, not 3.14159. That reduces the formula to

Short Formula (Diameters in decimal inches)

$$\text{RPM} = \frac{\text{Surface speed} \times 4}{\text{Diameter of rotating object}}$$

Now, resolve for the $\frac{3}{8}$-in. drill example using this formula. Try it in your head first.

It's 400 divided by a number slightly less than 0.4–the result will be close to but slightly larger than 1,000.

$$\text{RPM} = \frac{100 \times 4}{0.375}$$
$$= 1{,}066 \text{ RPM}$$

Metric Short Formula

If the spinning tool diameter (or lathe workpiece) is dimensioned in millimeters and the feed rate in meters per minute, substitute 300 for the 4 in the preceding formula.

Metric Short Formula (Diameters in mm and SS in meters/minute)

$$\textbf{RPM} = \frac{\textbf{Surface speed} \times \textbf{300}}{\textbf{Diameter of rotating object in mm}}$$

9.3.2 Reamer RPM

In theory, the reamer RPM should be the same as an equal-sized drill. Since both tools are HSS and the diameter difference is very little, the same surface speeds and resulting RPM should be nearly the same. However, in real practice this is usually not so.

Reaming brings the additional challenge of chatter due to the thin chip they produce and the reamer's multiple flutes. To stop reamer chatter, we usually slow them to less than half the drilling RPM; sometimes even lower. The goal is to find the nearest RPM to the correct speed that does not chatter. If the reamer does vibrate, the hole will be oversize and of poor finish.

SHOPTALK

Recommended Surface Speed Is Empirical Data
While RPM charts are reliable data, you will find variations between charts. That's because surface speed (SS) is the underlying data used to calculate RPM. Surface speed is derived by observation of experimental machining. The objective is to raise speed until tool failure occurs and track tool life up to that limit. The chart author then backs up until his or her particular objective RPM is met. So results can vary.

Depending on who is creating the chart, the data will represent a particular bias. Some push closer to failure, trading shorter tool life for higher production to promote their tools. Others favor long tool life but slower speeds, as those found in machinist's reference materials. Instructors often place safety higher on the list and thus even slower speeds. The small surface speed chart in Appendix III of this book is biased toward beginner safety. The RPM obtained there will be low for production work but a good platform for start-up.

9.3.3 Drill Feed Rates

Drill in-feed, the **feed rate** at which the drill advances into the work, is expressed in **thousandths of an inch per revolution** or in *decimal parts of a millimeter per revolution.* When pulling the feed lever by hand, in-feed on a standard manual press is a matter of feel, sound, and observing the chip–all quickly learned with practice. Feeds on power presses and on CNC machines must be selected before drilling.

KEYPOINT

Drill feeds range from a light 0.0005 in. (0.1 mm) per revolution to a heavy 0.040 in. (1.0 mm) per revolution, or even higher on large radial arm machines.

Selecting the feed rate is a three-way balance between tool life, cycle time, and machine finish. Here's a list of the factors to consider:

1. **Depth of Hole–Ability to Deliver Chips out of Hole**

   As the hole gets deeper, the removal of chips becomes a problem. Often, you must slow the feed or begin peck drilling at a given depth.

2. **Material Type**

   Some materials produce chips that tend to feed up and out of the drill hole easily: cast iron, free-cutting steel, hard aluminum, and some brasses are good drilling materials. On the other hand, softer aluminum and the tougher stainless steels are difficult to drill because the chips tend to build up in the flutes of the drill.

3. **Coolant Availability**

   Flood coolant not only reduces friction during the rake phase, but it also lubricates the chip's passing up and out of the flutes. If a flood stream isn't available, typical of most small drill presses, then a cup of cutting oil brushed on the tool is an adequate second choice, or a spray bottle with synthetic coolant can be used.

4. **Size of Drill**

   As shown in the sample chart coming up, smaller drills require lighter feeds to avoid breakage.

5. **Strength of Setup**

6. **Expected Finish and Accuracy of Hole**

   Pushing harder causes rougher, less accurate holes.

Drill Feed Rate Chart

Drill Size	Feed Rate (Range) in IPR
$\frac{1}{16}$ to $\frac{3}{16}$	0.001 to 0.004
$\frac{3}{16}$ to $\frac{3}{8}$	0.002 to 0.006
$\frac{3}{8}$ to $\frac{1}{2}$	0.003 to 0.008
$\frac{1}{2}$ to $\frac{3}{4}$	0.004 to 0.010
$\frac{3}{4}$ to $1\frac{1}{2}$	0.005 to 0.015

Student note: Remember, these are moderate rates.

9.3.4 Reamer Feed Rates

Reamers require faster inward feed than drills. Scan the drill fee rate chart above; each value can be doubled or tripled if reaming. For example:

Reamer Feed Rate Chart

Reamers Dia.	IPR Feed Rate
$\frac{1}{16}$ - $\frac{3"}{16}$	= 0.003 to 0.012
$\frac{3}{16}$ - $\frac{3"}{8}$	= 0.006 to 0.018
$\frac{3}{8}$ - $\frac{1"}{2}$	= 0.009 to 0.024
$\frac{1}{2}$ - $\frac{3"}{4}$	= 0.012 to 0.030
$\frac{3}{4}$ - $1\frac{1"}{2}$	= 0.015 to 0.045

So when you get to the power drill press or the Bridgeport mill (see Chapter 12), you CAN select reamer feed rates as shown, but it's far more common to drive them through with the quill lever by hand on the drill press or mill. Just push the reamer through faster; experience will show you how fast. The sound will tell if you are doing it right–chatter or not. But when programming CNC, follow the rule of thumb–multiply drill feeds by 2 to 3 times for reamer feeds.

UNIT 9-3 REVIEW

Replay the Key Points

- There is a "best" surface speed for any material being drilled.
- Different diameter drills must be set to turn at different revolutions per minute–such that the rim is at the correct surface speed.
- Smaller drills require *faster* RPMs while larger drills must go *slower* to achieve the same surface speed at their rim.
- Drill RPM and feed rate charts assume an HSS tool. If the tool is carbide, rates may be increased.
- Reamer RPM is often adjusted to half the drill RPM to stop chatter.

Respond Calculating Speeds and Feeds

Unless directed otherwise, find surface speeds in Appendix III.

1. Use the long formula to solve correct RPM for a 2.0-in.-diameter drill in brass.
2. Use the short version (no calculator): a $\frac{1}{2}$-in. drill in annealed stainless steel.

Critical Thinking Problems

3. A line in a program includes the code *S400*, meaning the programmed RPM is to be 400 (spindle at 400). It's written to drill $\frac{1}{2}$-in. hard stainless, using a $\frac{7}{8}$-in. drill. Is it correctly written? *Assume a cutting speed of 70 FPM for this material.* Use the short formula to analyze this in your head before calculating it.
4. Use both formulas to compute RPM. Then compare percentage difference between the faster and slower result: Letter "A" drill in aluminum?
5. Calculate the RPM for a 12-mm drill cutting steel at 30 m per minute surface speed.

Extended Reference

6. Using the short formula, and *Machinery's Handbook* or CD to find recommended surface speed, compute the RPM for a Letter J Drill, cutting red brass. *Hints:* The main ingredient in brass is? (Unit 5-2). You need speed and feed tables for drilling to access the correct surface speed.
7. Using *Machinery's Handbook* data and the long formula, compute the correct RPM for a 2.0-in. drill cutting AISI alloy steel, Type 4140 (a chrome-moly steel discussed in Chapter 15). This material is in its softest annealed state.
8. Still drilling the material in Question 7, you must now switch to a 36-mm diameter drill. It should be turning at ____ RPM. Which formula should be used?
9. A 0.625-in. access hole must be drilled and reamed in the side of a cast iron transmission case. On a separate piece of paper, fill in the setup facts listed next. Use only data found in

this book's Appendix III and the short RPM formula.

Reamer drill diameter

Reamer drill RPM

Drill feed rate range

Reamer RPM

10. Using *Machinery's Handbook* and the long formula, determine speed and feed for a 1.0-in. drill, cutting 1010 alloy, plain carbon steel in its annealed (softest) state.

Recommended surface speed

Drill RPM

Drill feed rate range

Unit 9-4 Drilling per Print and Secondary Operations

Introduction: Now we investigate producing the hole in the right position on the work, within the size tolerance (Fig. 9-44). We'll explore a couple of standard drill press methods, where we move or tap the part into alignment with the spindle. One produces a repeatable location accuracy of 0.030 in. and the other around 0.010 in. But that's not good enough for many jobs.

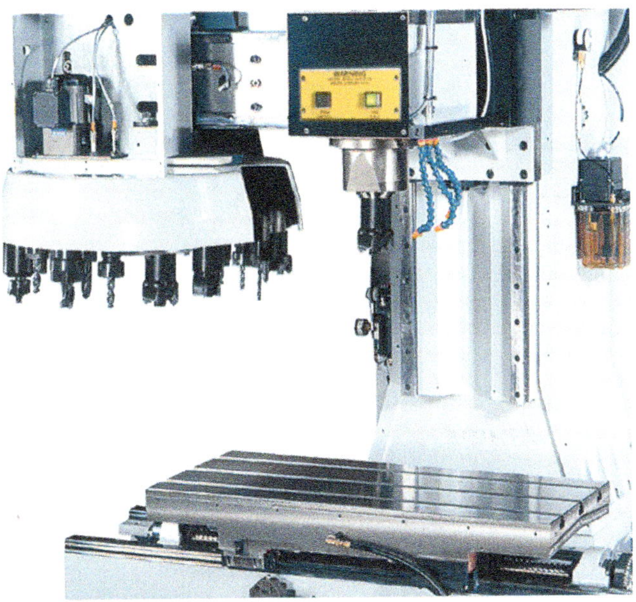

Figure 9-44 A CNC machining center can position, center drill, drill, and ream to tolerances below one thousandth of an inch.

Courtesy Bridgeport Machines, Inc./Bridgeport, CT

But once the workpiece is positioned correctly under the spindle, there still remains the challenge of getting the drill to bite in and drill at that location.

TERMS TOOLBOX

Center finder An adjustable pointer used to pinpoint the exact center of a rotating spindle.

Core drill A multiflute cross between a drill and reamer that must follow a pilot hole. Used mostly to drill out rough cores in castings.

Counterbore A round opening surrounding a hole, usually to receive a bolt head. See also *spotface*.

Countersink A cone-shaped opening at the entrance of a hole, or the tool that produces this shape.

Engagement (thread) The amount of contact between a bolt and nut, compared to the theoretical 100 percent; 75 percent engagement is common.

Gun drills Specialized tools for drilling extra deep holes.

Spade drill A flat cutting tool with two cutting edges, used on more rigid machines than drill presses.

Spotface A smooth surface around a hole, created for a washer or nut.

Step drills Drill press tools that cut multiple features and diameters at one time. May create angular transitions between diameters.

Wiggler A set of utility holding devices that can be used as an indicator holder, an edge finder, and a center finder.

9.4.1 Two Drill Press Location Methods

The first method is fast but not very accurate. The second improves accuracy but takes more time to perform. The best accuracy for either method starts with good, sharp layout lines with high contrast between the line and dyed metal. Since you'll be using your eyes to determine alignment in both cases, good lighting is also very important.

Layout/Punch Method—Repeatability Approximately 0.030 in.

Step 1 Lay Out the Hole Locations

Always reference part datums as a basis for the layout lines (if they are shown on the drawing; Fig. 9-45*).*

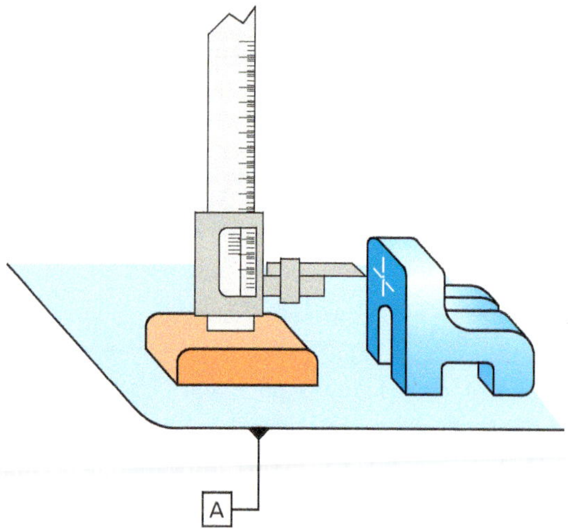

Figure 9-45 The layout must be clear and accurate.

Step 2 Pilot Punch (Prick Punch)

Use a very light tap of the pilot punch while holding it perpendicular to the work surface (Fig. 9-46). After checking the accuracy of the mark, a second slightly harder tap is then performed.

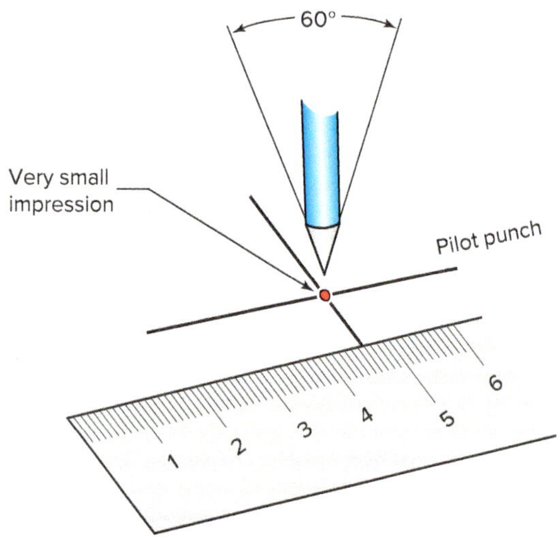

Figure 9-46 A pilot (prick) punch creates a small, 60° indentation at the layout location.

Step 3 Center Punch

Follow the pilot punch with a broader, deeper center punch mark (Fig. 9-47). Again, be certain the punch is held vertically. Remember, don't do any of this hammering at the layout table!

Step 4 Visually Locate and Spot Drill to Test Location

Before pilot drilling, test the work location to be certain the drill is lined up over the layout lines. With the work lightly clamped in place, a center drill is mounted in the drill chuck and spun slowly by hand as the part is tapped into alignment with a plastic hammer. Tighten the clamps a bit more when the work appears directly beneath the spindle.

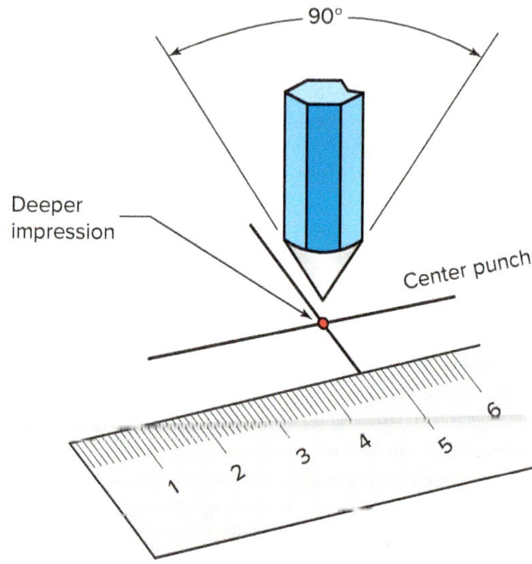

Figure 9-47 Following the pilot, the 90° center punch drives the mark deeper and wider.

Then spot the location. Turn on the spindle and very lightly touch the work surface with the center drill's pilot. This is not a drilling operation–do not remove metal yet. Stop and read the shiny round spot created. If it's right on target, a bright ring will be created around the center punch mark. Now tighten the work clamps one last time and center drill to the depth shown (later) in Fig. 9-49.

KEYPOINT

The objective of a spot drilling is to remove no metal, just layout dye—at the exact place where the center will start when fed down.

Center drills are best for this operation as they are short, rigid tools. They feature a precision ground body and a very small pilot on their tip (Fig. 9-48). Center drills are supplied in four sizes, #1 smallest to #4 largest. On the drill press we generally use a #2 unless the pilot is bigger than the intended drill diameter.

TRADE TIP

A center drill's RPM should be computed based on the tiny pilot. As such, it will require a high RPM to ensure good location accuracy. However, the correct RPM will often exceed the drill press's range of speeds. When it does, select the highest possible RPM and push down very gently on the quill feed lever. Center drills are harder than drill bits, and the tiny pilot breaks easily.

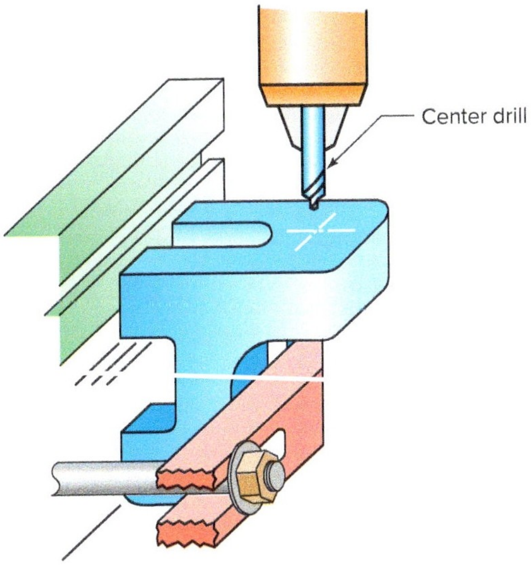

Figure 9-48 To *spot* the location, touch the work surface with the spinning center drill; then read the marks.

Step 5 Pilot Drill to Correct Depth

Change to the pilot drill, then follow the center drill dimple with the pilot. Drill deeply enough (halfway or all the way through the work) such that the next drill will tend to follow it (Fig. 9-49).

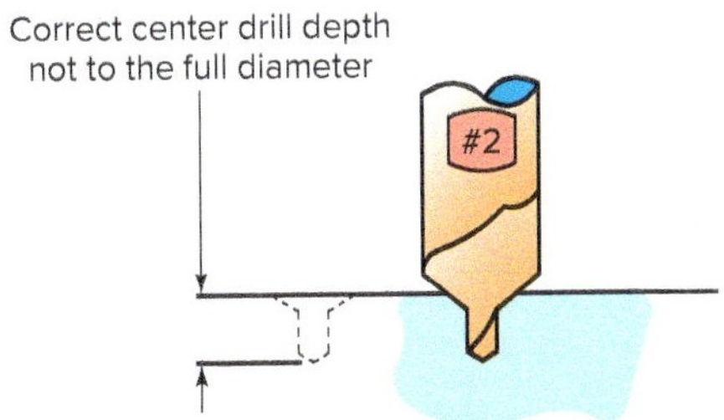

Figure 9-49 A center drill does not drill below this distance.

TRADE TIP

Pulling a Mislocated Drill Back on Location A started drill that has wandered off location can be pulled back toward the correct location with this trade trick. But it only works if you detect the problem before drilling downward to the drill's full diameter (Fig. 9-50).

With a hammer or chisel create deep marks in the angular cone surface that is the started hole. The marks are made on the side to which you wish the drill to return. They catch on the cutting edge and draw the drill toward themselves. They can bring the drill location back on target.

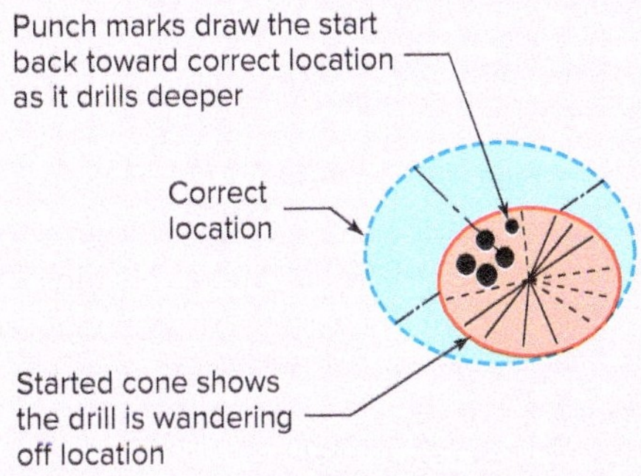

Figure 9-50 It's possible to pull a mislocated drill back on location, if the start isn't yet at full diameter.

Step 6 Drill to Size

Check the sharpness and geometry of the drill before chucking it. Set the right RPM on the press; then test the location by gently touching the hole rim with the spinning drill bit.

Drill Press Method 2—Layout/Center-Finder Pickup—Repeatability Approximately 0.010 in. or Better

This method uses an alignment tool, the **center finder**, to refine location (Fig. 9-51). A center finder is a two-piece pointer whose tip can be gently tapped or nudged onto a wobble-free, perfect center on its rotating base. Theoretically, a precision ground, solid metal pointer could be used, but in practice, every chuck and spindle has wobble. The center finder eliminates any hole misalignment from spindle or chuck wobble.

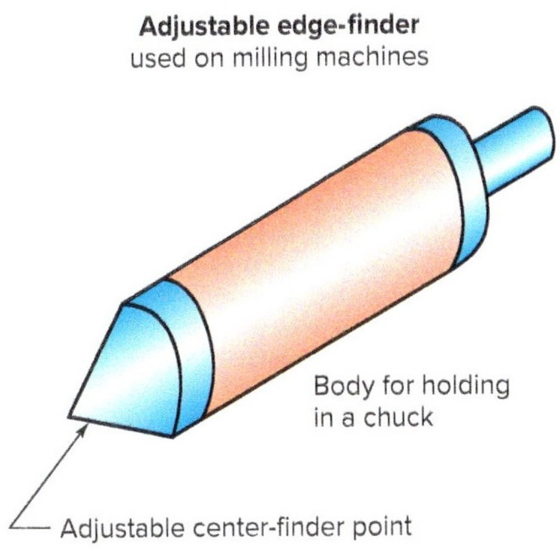

Figure 9-51 A center finder is used to improve spindle location over layout lines.

Some steps and photos are omitted because they duplicate steps in method 1.

KEYPOINT

To use the center finder correctly, no punch marks are made in the layout.

Step 1 **Layout** Cross-hair scribe the location

Step 2 **Rough Position the Work** Add the holding method and tighten lightly only.

Step 3 **Center-Find the Location** With the center finder held lightly in the drill chuck, set the drill press at a moderate RPM from 400 to 500 maximum. If the RPM is set too fast, the center finder can be damaged by centrifugal force. The adjustable point is held on the body by a spring. It can be stretched beyond usefulness if the point is flung out from excessive RPM.

Step 4 **Center the Point** Lightly nudge the wobbling tip until it settles on the exact axis of the spinning spindle. Use a pencil, rule, or other small but hard instrument for this purpose–not your hand!

Step 5 **Align the Work Under the Rotating Center Finder** With the spindle rotating, bring the quill down until the point is approximately $\frac{1}{16}$ in. from the layout surface. With light taps to the workpiece with a dead blow mallet or plastic hammer, move the part until the point is directly over the intersection of both layout lines (Fig. 9-52). Be sure to position your eye in direct line with the *X* direction line, then the *Y* to avoid parallax error.

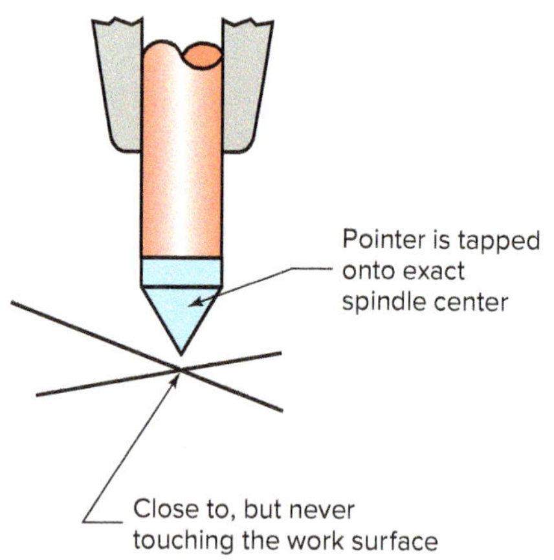

Figure 9-52 Using a center finder is easy.

KEYPOINT

The center finder never touches the work. *It does not leave a mark!*

Step 6 **Clamp the Work in Position** Use the center finder again to see if the clamping has disturbed location.

Step 7 **Spot Drill to Test Work Alignment**

TRADE TIP

Using a Center Finder and the Colinear Concept
Here's an optical trick for improving location accuracy using the center finder. The secret lies in that you can see alignment of two lines better than estimating the position of a point relative to a line.

The two lines that will be made colinear are the layout line and its *reflection* in the shiny cone of the center finder (Fig. 9-53). To avoid parallax error, place your eye directly in line with the layout line; then tap the work until its reflection lines up. Then repeat with the other layout line. Go back to check if the first has been disturbed.

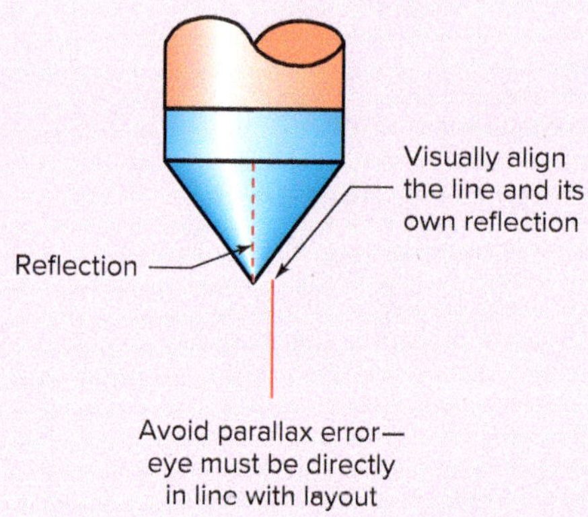

Figure 9-53 Another way to use a center finder.

There is a second, older tool that performs the pointing task of the center finder, called a **wiggler** (Fig. 9-54). The wiggler is a universal tool that also holds indicators in spindles and performs edge finding (see Chapter 12) as well. It cannot be used for the optical trade trick. If your shop has this tool ask for a demonstration.

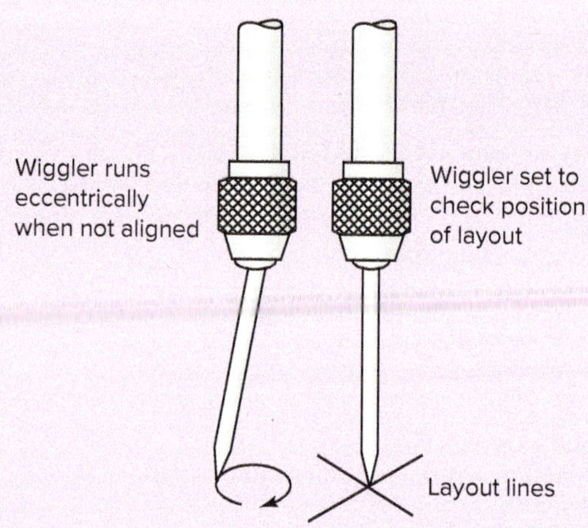

Figure 9-54 A wiggler is another spindle locating tool.

9.4.2 Drilling and Tapping Threaded Holes on the Drill Press

Selecting the Correct Tap Drill

Recall from Chapter 5 that there is a designated drill for a specific tap. If not predrilled to the correct size, the tap will break if the hole is too small while a hole too large will produce weak or nonexistent threads. When you read a print callout such as

Drill and Tap $\frac{1}{2}$-13 UNC

or

$\frac{1}{2}$-20 UNF

or

M8-1.25 ISO

it's specifying not only a certain tap but also a specific drill size. That size selection is found on the drill chart or in standard machinist's references.

> **KEYPOINT**
>
> Even two different pitches of the same size thread require different tap drill diameters. Every thread designation has its own tap drill size.

Why the Specific Tap Drill Size?

You may see a note on some tap drill charts that the recommended drill produces a 75 percent **engagement** thread. That's the standard fit. Here's what that means. The drill size is calculated such that a bolt threaded into the correctly drilled and tapped hole will touch on 75 percent of their thread faces; 75 percent of what?

> **SHOPTALK**
>
> **Wrong Tap Drill** A common error made by trainees is to choose the drill size equal to the thread size (e.g., to drill a $\frac{1}{2}$-in.-diameter hole for a $\frac{1}{2}$-13 tap). If you do this, the tap drops through because there is no metal left for the threads.

A theoretical thread, with pointed triangle tops and zero mechanical clearance between the bolt and nut, would produce a 100 percent engagement. But, as we've discussed before, the thread triangle is truncated, clipped at the tiny points; plus there is a built-in clearance between the two. Using the recommended chart diameter forms the truncations on the inner diameter of the thread, as shown in Fig. 9-55.

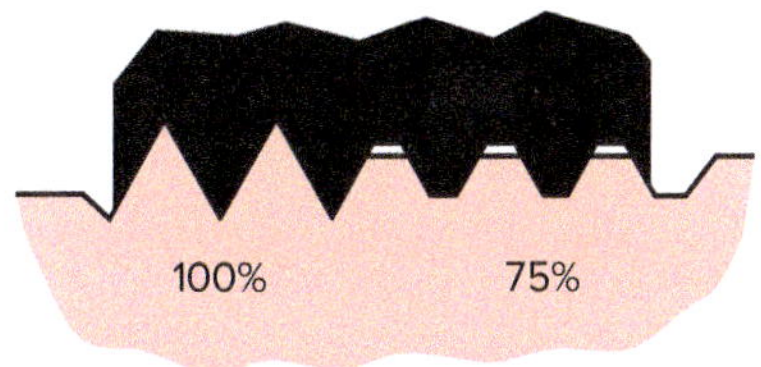

Figure 9-55 The correctly chosen tap drill produces internal threads with a 75 percent engagement.

Drill Press Thread Tapping Methods

There are three common methods used on drill presses for tapping threads:

1. Hand turning
2. Power turning
3. Tapping heads

Hand Turning the Tap Using the Drill Press and a Tap Guide

This method is used for single or a limited number of tapped holes (Fig. 9-56). To begin, a pointed center is mounted in the chuck. It will be used as a guide to keep the tap perpendicular to the work. **The spindle motor is not used in this method.** With one hand on the quill lever to maintain light down pressure against the tap, turn the tap handle with your free hand. Be sure to use a tapping compound or cutting oil.

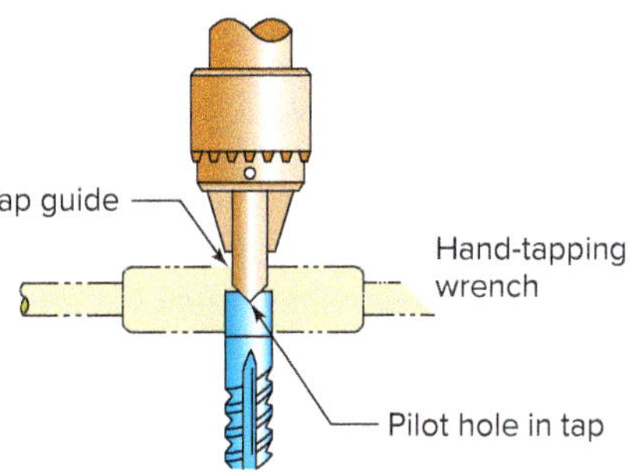

Figure 9-56 Hand-turning the tap in the drill press is useful for one or two parts.

Turning the Tap by Power

This operation is accomplished on a gear head or radial arm drill press, with a reversing spindle and lower speed ranges (Fig. 9-57). Standard drill presses cannot be used this way.

Safety Precautions

Here the press screws the tap in and out while you keep light pressure on the feed lever. When power tapping on a drill press, be sure you do not accidentally engage the power feed when the tap is in the hole. The power feed will advance the quill down at a different rate than the tap must lead into the work. Pressure will instantly build up, and the tap will break.

Figure 9-57 Tricky to do, this skilled machinist has the tap mounted in a chuck, within a reversing gear head press.
Courtesy Mike Fitzpatrick

Another factor to control is that the spindle of most gear drive presses will coast some distance with the power turned off. When nearing the bottom of the threads, be sure to allow for the extra distance the tap will advance while the spindle decelerates and reverses. Power tapping requires practice and a demonstration by your instructor! The chances of breaking the tap are near 100 percent, at least for trainees.

Tapping Head Attachments

Tapping heads are good solutions for driving taps on any drill press with or without reversing spindle motors (Fig. 9-58). When many parts must be tapped, this attachment prevents tap breakage and operator fatigue. It turns any drill press into an efficient production tapping machine.

A tapping head adds four or five features to a standard drill press:

- Geared-down RPM–safer for tapping.
- Quick reversing mechanism–as the tap reaches its lower stroke limit, the operator holds back on the quill lever arm, and the head instantly shifts to reverse without reversing the motor on the press.
- Efficient collet-type holder for taps (not a chuck)–often the collet is rubber isolated such that it absorbs much of the shock that occurs during tapping, to avoid breakage while rotating and keeping the tap centered in the spindle.
- A "floating" tap holder mechanism–a major advantage of tapping heads, this feature permits the tap to move up and down in the spindle but not to move sideways. This results in some forgiveness to overrun of the spindle motor and operator lack of skill.

Figure 9-58 A tapping head adds several advantages when used in a standard or gear head drill press.
Photography by Prographics at Milwaukee Area Technical College/McGraw Hill

- Adjustable torque friction that allows the tap to slip rotationally in the holder if it jams during the cut. (Not found on all tapping heads.)

TRADE TIP

Don't Forget the Custom-Designed Tapping Fluids
These work wonders for cutting threads. Tapping fluids reduce friction far more than standard cutting fluids. They also improve thread finish and reduce tap breakage. There are general fluids for most alloys and specific compounds for aluminum, titanium, and other metals. Although costly, they pay for themselves in tool life and consistent results.

9.4.3 Counterboring and Spotfacing Operations

Two specialty operations are performed on the drill presses using cutting tools designed for this purpose (Fig. 9-59). These operations follow a drilled hole and produce a flat, smooth, circular surface around the hole. They are different terms for the same operation. The depth of the cut is their only difference. A **counterbore** is a deep cut, while a **spotface** is a shallow cut. The cutting tools are correctly called either a counterbore or spotface.

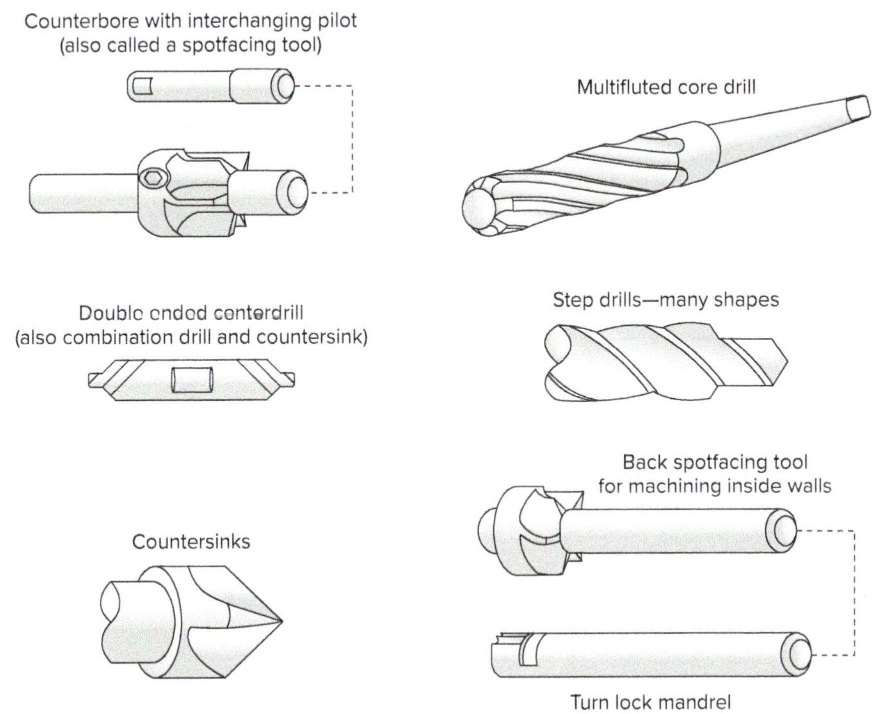

Figure 9-59 Secondary drill press tools you'll need to recognize, some used for CNC operations.

In Fig. 9-60, the computer symbol for counterboring/spotfacing is a wide "U," and the depth is the down arrow symbol.

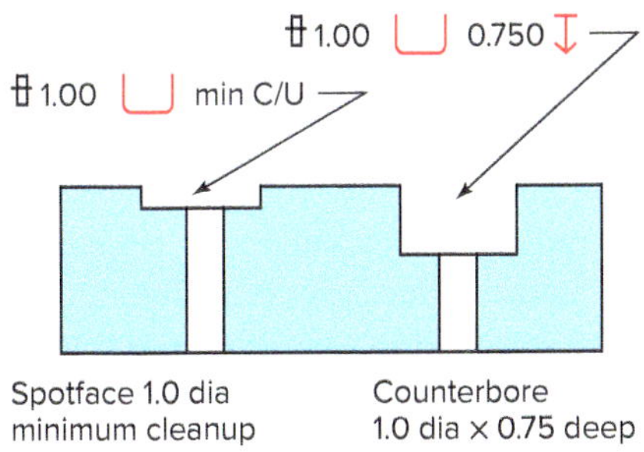

Figure 9-60 Compare the CAD symbol (top) for spotface/counterbore to the previous word callout below.

Spotfacing a shallow smoothed circle for a washer or bolt head is usually required around a bolt hole in a rough casting. Spotfacing depth is commonly called out as in Fig. 9-60 as *minimum cleanup* or sometimes abbreviated to C/U.

Counterboring is machined to a controlled depth. A typical use of a counterbored hole is to receive the head of a bolt below the surface of the part. When counterboring, you must set a quill stop to achieve depth accuracy.

There are three versions of counterbores. Some are single-piece cutters with a solid pilot, but the more popular version features interchangeable pilots (Fig. 9-61). By exchanging the pilot, these counterboring tools can follow a range of hole sizes. The third tool that counterbores is a custom-made step drill (coming up next). They are mainly used to drill and counterbore standard bolt head sizes all in one operation.

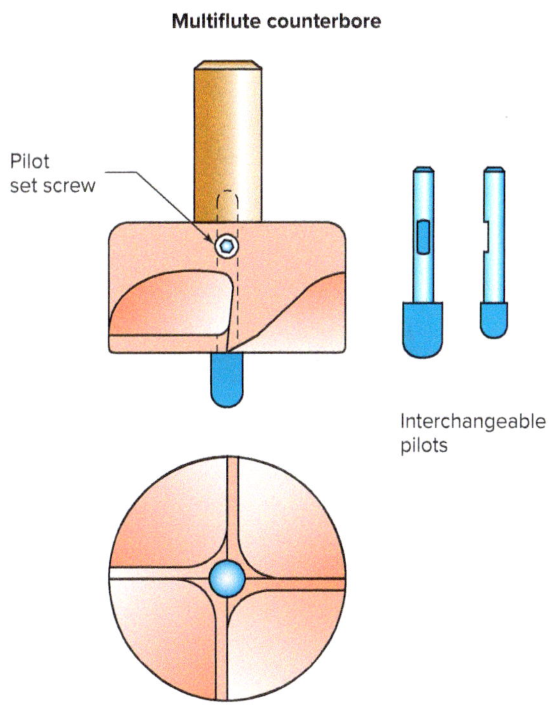

Figure 9-61 Counterbores are also used for spotfacing.

All three feature a central pilot that ensures that the cutter stays centered in the hole. The pilot also ensures that the 0° lead on the cutting edges doesn't wander or grab as the cutting edge touches the work surface.

TRADE TIP

Counterboring Cutting fluid performs an additional function here; It lubricates the pilot, which can jam from heat and chips. Because their cutting edges are flat (zero lead), they tend to chatter. Similar to reaming, reduce spindle RPM to half the calculated value or even slower to prevent counterbore chatter.

9.4.4 Countersinking

Countersinking is a drill press operation that produces a cone-shaped opening around a drilled hole, for several reasons:

- To remove sharp edges and burrs (recall the term *breaking the edge*)
- To provide a starting slope for tapping threads
- To finish the raw edge of a hole for functional requirements or for appearance.
- To provide a conical nest for a flat head screw (as shown shortly in Fig. 9-63)

Countersinking is similar to spotfacing. The tool must be turned at a lower rate than normal cutting speed (50 percent or less) to avoid chatter. The closed-face countersink (Fig. 9-62) is the least prone to chatter and tends to produce the better finish, but since it has only one cutting edge, it doesn't last as long as the others before regrinding is necessary.

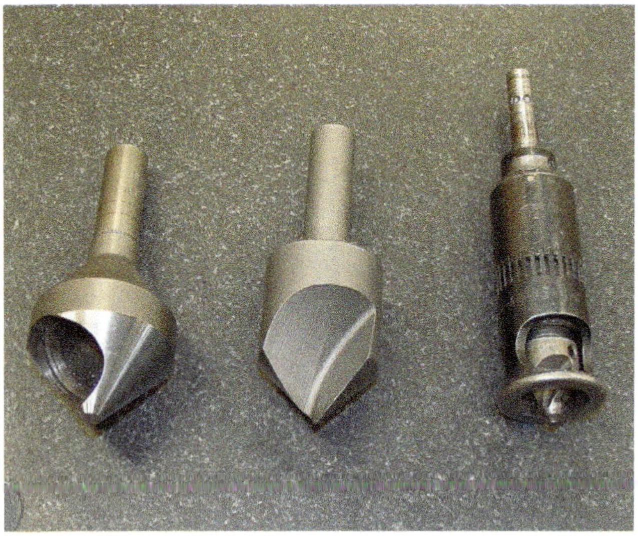

Figure 9-62 Countersinks: closed face (left), open faced, and micrometer stop (right).
Lake Washington Technical College, Kirkland, WA/McGraw Hill

Eliminating the Confusion—Countersinking Versus Chamfering

You might see either term used on a print (Fig. 9-63). While they are related, they are different terms. A chamfer is a small angular cut at the intersection of two surfaces, and it can be around a hole entrance.

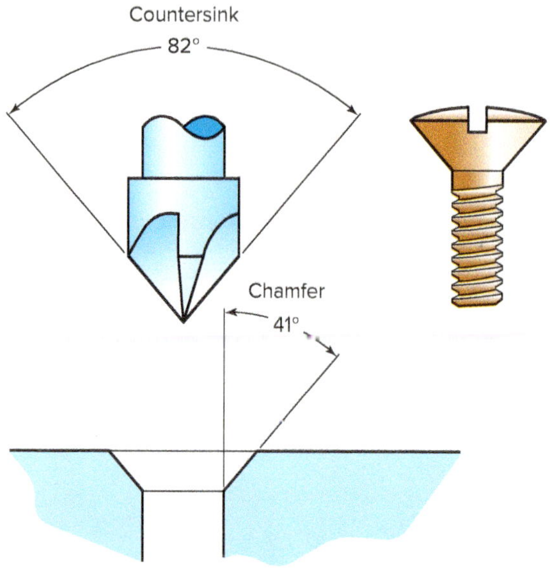

Figure 9-63 An 82° countersink produces a 41° chamfer.

KEYPOINT

A chamfer is specified as the angle of the chamfer as compared to a vertical axis line.

A countersink is called out as to the included angle of the cone.

The two common included angles for countersinks are 82° and 90°, although others are used in industry. Flathead and oval-head screws (see Fig. 9-63) are made with 82° shoulders.

KEYPOINT

An 82° countersink produces a 41° chamfer around a hole.

Micrometer Stop Countersinks

When countersink depth must be controlled within a tighter tolerance than is possible using the quill stop, on a drill press, we use the micrometer stop countersink (Fig. 9-64). It allows the countersink to extend down through the contact skirt to a preset depth. This tool is extra useful where material thickness varies or is rough, yet the countersink depth must be held to a close tolerance. Countersinks can also be used away from the drill press, in hand drill applications.

Figure 9-64 A micrometer stop countersink accurately controls depth.
Courtesy Mike Fitzpatrick

9.4.5 Specialty Setups

You'll soon come to appreciate that making many setups is similar to constructing a Lego® block masterpiece. You must work with basic components to assemble something custom fit to the part shape and the task at hand. I believe setups are the fun stuff of machining. Here are a few suggestions that work on a drill press.

Drilling Through Round Work

The challenge is to get the drill to start on the center of the round surface. First, a pair of layout lines are needed. One from the end of the object, 0.750 in. in the example (Fig. 9-65).

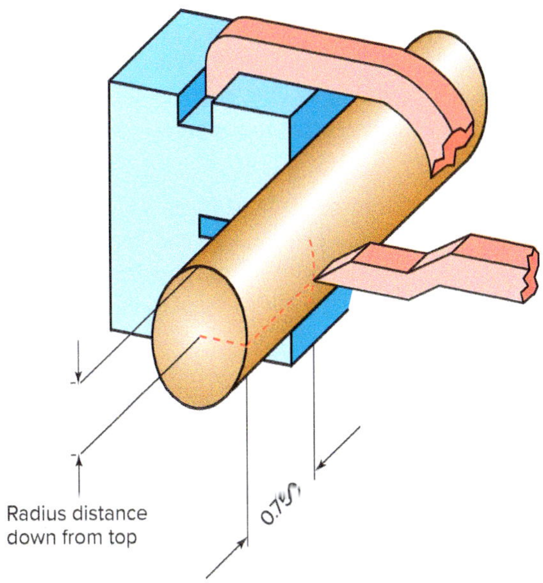

Figure 9-65 Laying out a centerline starts the process.

Next, a centerline is laid out with the work clamped in the vee block but with it lying on its side. Touch the upper surface of the round work with the scribe. Now, with the work clamped in the vee block, turn it upright and set it in the drill press vise. The marked intersection will be vertical and ready for center-finder location.

Note that a second vee block is supporting the work on both sides of the hole. The pair of vee blocks can be held in a drill press vise (Fig. 9-66) or clamped directly to the table.

Figure 9-66 Supporting the work in two vee blocks for drilling.
Courtesy Mike Fitzpatrick

Drilling Holes at an Angle to a Work Surface or Datum

Here are two ways to drill at an angle. The simple method is to rotate the drill press table as shown in Fig. 9-67, if the drill press is so equipped. If not, the work is tilted and held at the correct angle, in a tilting vise.

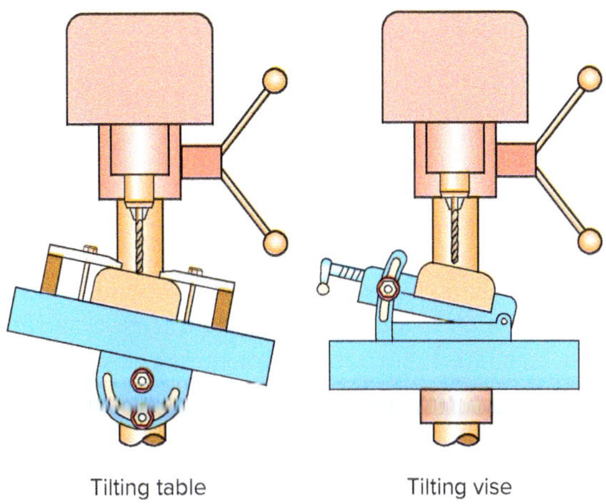

Figure 9-67 Two angle drilling setups.

The challenge in either setup is to start a drill location accurately on a tilted surface (Fig. 9-68). The bit tends to walk downhill before it bites in. To prevent this, start with a number 1 center drill set at a high RPM. Lightly touch the surface with the drill, then back it away immediately. Repeat this action with a series of light pecks, each just a few thousandths deeper than the last. Once the entire pilot of the drill has entered the work, you can increase the force.

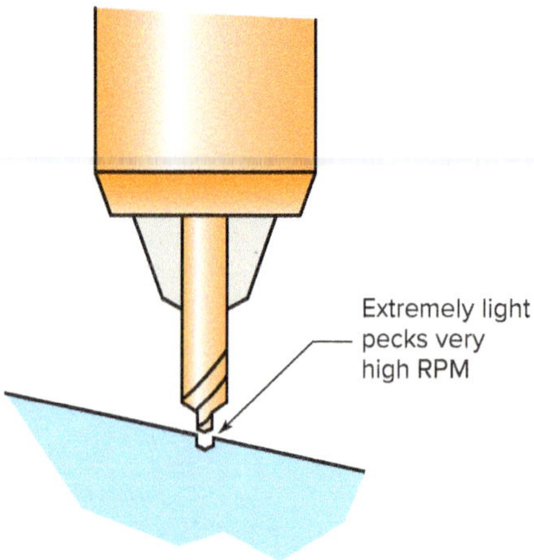

Figure 9-68 Starting a drill on location on an angular surface, without wandering.

If the surface is slanted beyond 15° relative to the drill's axis, nothing can prevent wandering, especially if the work material is hard. The illustrated surface is at 12°, and you can see the problem—without a pilot hole, the drill is going to slip to the right. Depending on the kind of material, hard or soft, even that much slant might not be possible to drill accurately on a press unless a drill guide bushing can be set up to prevent wandering.

In these cases, we move the job to a vertical milling machine with a more rigid spindle. There, a circular landing spot can be created that's perpendicular to the intended hole's axis—called prespotting the hole. The cutter is a rigid, flat-bottom, two-flute tool called an end mill. It's similar to a drill with 0° point angle (Fig. 9-69). We'll study these cutters in detail in Chapter 12.

9.4.6 Specialty Drills

There are four types of drills beyond the common twist drill that you might encounter in your first year of employment but not likely in tech school: the spade drill, the gun drill, the step drill, and the core drill. Here is a brief description of each.

Spade drills are used to drill large holes above 2 in. in diameter or for odd sizes. A spade drill can remove large quantities of metal at a high rate. It has a removable cutting edge that can be carbide or HSS. These drills are not common on standard drill presses due to need for rigidity, but they are used in production CNC turning, on larger radial arm presses, and occasionally on manual lathes. Spade drills can drill deeper than twist drills because of their chip-removing ability.

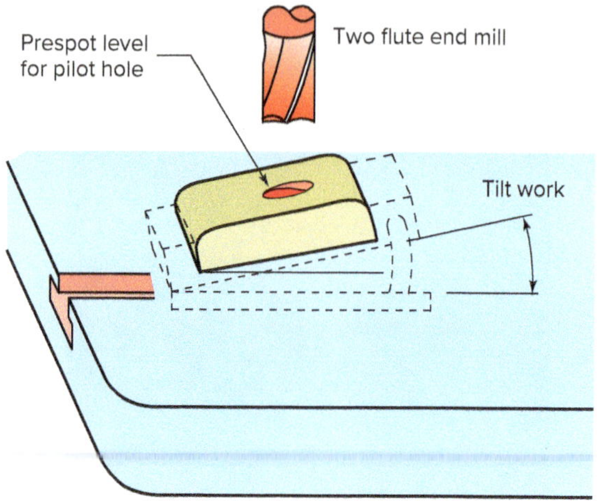

Figure 9-69 Sometimes it's necessary to create prespots to allow the drill to bite on location.

Very precise changes in hole diameter are possible by custom grinding the inserted cutter (Fig. 9-70).

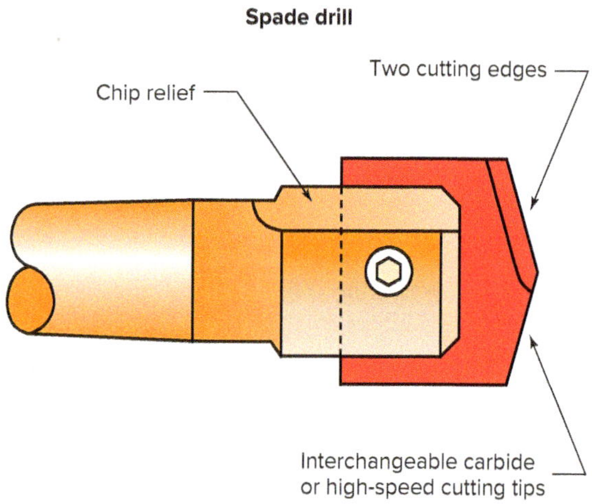

Figure 9-70 An insert cutter, spade drill.

Gun drills are specialized tools and processes designed specifically for drilling extra deep holes. When drilling deeply, standard twist drills tend to clog and wander into a curved axis—and they eventually break in the hole.

A gun drill corrects these problems by using a three-point suspension head that supports itself as it cuts (Fig. 9-71). A gun drill has a single carbide cutting edge and two support pads spaced at 120° from the cutter. They also feature a straight flute that better sends chips out using pressurized cutting oil or fluid injected down to the tip, to force the chips out. The fluid is pumped through a gallery (oil hole) that runs through the tool's body. It is injected right to the cut.

Gun drilling can be accomplished on regular lathes using special oil injection pumps and attachments. However, if a lot of gun drilling is needed in the shop it is usually assigned to specialized machines having the ability to pump the high-volume cutting fluid and retrieve it into the sump again (Fig. 9-72).

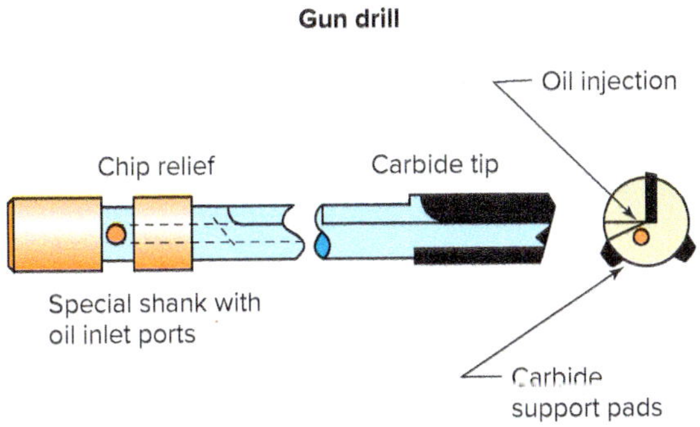

Figure 9-71 A gun drill can cut extremely deep holes.

Figure 9-72 A gun drilling machine in industry.
Courtesy Smiths Aerospace-Actuation Systems/Yakima, WA

Core drills are a combination of a drill and a reamer (Fig. 9-73). The main purpose is to machine holes in castings that already have a rough cast hole–called a *cored hole.* A core drill features a very strong web and has several flutes (three, four, six, or more for larger tools). It will not start a hole without a core or a pilot hole in the metal. These drills tend to produce a fine finish, often nearly as good as a reamer.

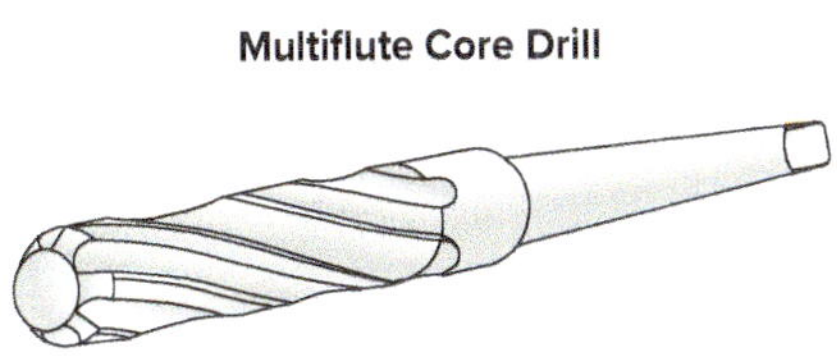

Figure 9-73 A core drill is designed to follow rough holes in castings. It resembles a reamer but with deeper flutes.

Step drills are designed to accomplish two tasks at one time (Fig. 9-74). They can drill and chamfer, drill and counterbore, or any combination of pilot, drill, counterbore, and chamfer, depending on their ground shape.

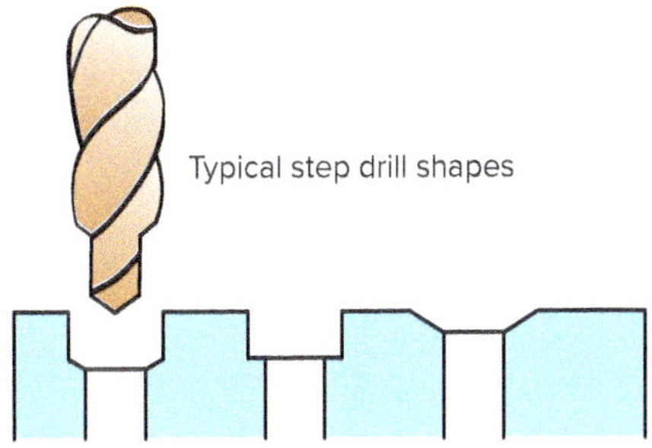

Figure 9-74 Various step drills.

UNIT 9-4 REVIEW

Replay the Key Points

- The objective of spot drilling is to remove no metal, just layout dye, at the exact place where the center will start when fed down.
- No center punch marks are made when picking up hole locations from layout using a center finder. The center finder neither touches nor marks the workpiece surface.
- Two different pitches of the same size thread require different tap drill diameters. Every thread designation has its own tap drill size.
- A chamfer is specified as the angle of the chamfer as compared to a vertical axis line.
- A countersink is called out as to the included angle of the cone.
- An 82° countersink produces a 41° chamfer around a hole.

Respond 9-4

1. Using Appendix I, what is the correct tap drill for a $\frac{5}{16}$-24 UNF thread.
2. True or false? Two common methods of locating a part exactly on the axis of a drill press spindle are the center finder (or wiggler) and the layout punch method. If it is false, what will make it true?
3. Name a third, more accurate method of locating a drill on a standard press.
4. Draw a sketch or describe the method used to "pull" a drill back toward the right location if found that it has wandered off position. At what point in the drilling can this be accomplished?
5. True or false? To mount a Morse taper-shank reamer in a drill spindle requires a shell adapter. If it is false, what will make it true?
6. Name two methods of tapping by power in a drill press capable of reverse rotation.

7. Draw a sketch or describe the colinear method of aligning a drill bit to a layout line. What tool and processes would you need?
8. Name the operations in the sequence needed to *ream a hole* in undrilled material (the punch mark method).
9. True or false? Machining a chamfer around a hole is called flatspacing. If it is false, what will make it true?
10. What is the difference between a *spotface* and a *counterbore?*
11. Identify two drilling methods for cutting extra deep holes.
12. Of the above two methods, which requires a special machine or attachment, and why?
13. Identify the tools shown from left to right in Fig. 9-75.

Figure 9-75 Identify the three drill press tools from left to right.
Lake Washington Technical College, Kirkland, WA/McGraw Hill

Unit 9-5 Sharpening Drill Bits

Introduction: While there are easy-to-use manual sharpening jigs and automatic machines available to sharpen drills (Fig. 9-76), we will concentrate on hand-sharpening drill bits. This is an ability no machinist should be without. You'll use it on the job and in the home workshop, too. More than useful, hand grinding a drill is a fast way to increase tool geometry knowledge.

But, there's an even greater reason for Unit 9-5. The techniques and hand coordination skills required to grind a drill bit will apply to touching up or hand-grinding many other tools in the shop. We try to avoid hand-ground tools in production machining, but occasionally, they become the only short-term solution to keep the machine moving. Handmade cutting tools are more common for the tool and die industry and, especially, the mold maker. Even today in an era of CNC tool-grinding machines, every so often the only way to produce some wild-shaped tool is to do it by good old hand skills.

Figure 9-76 A drill sharpening machine in industry.
Lake Washington Technical College, Kirkland, WA/McGraw Hill

Before sharpening a bit, it's important to recognize if it's time to grind it or not. Grinding reduces its shop life span. While nonproductive downtime can be created by the machinist stopping too soon to sharpen or replace a dull tool, waiting too long, pushing the tool beyond normal wear usually results in disaster both for the workpiece and the cutting tool!

TERMS TOOLBOX

Dead center (drill) The noncutting portion of a drill point—caused by the central web.

Overtempering Reducing the hardness of any HSS tool bit by heating it during grinding.

Scatter shield The metal grinding wheel guard above the grinding wheel, designed to contain a broken wheel if it explodes.

9.5.1 Recognizing a Dull Bit

There are two ways to detect a dull bit.

Inspection Before Mounting

Check the cutting edge especially at the outer corners and along the margins near the cutting edge. Look for chips in the cutting edge and rounded corners at the margins. But not all dull drills are so obvious. So, for the normally worn bit, hold it up to catch the light. The cutting edge will appear as a bright line. Compare a sharp and a dull drill—you'll see the difference (Fig. 9-77).

Figure 9-77 This drill cannot be easily reground. Notice the margin damage.
Courtesy Mike Fitzpatrick

Observing Signs While Drilling

There are five signals that indicate the bit is failing:

1. Excessive noise—squeal, vibration, and chatter.
2. Chips jamming in the flutes. This can also be caused by deep drilling, lack of coolant, and/or wrong speeds or feeds.
3. Yellow, brown, or blue chips emerging from the hole when drilling steel. Discolored chips can also be caused by excessive speed or feed rates, but whatever the cause, it will soon lead to a dull drill.
4. The drilled hole measures too large.
5. Excessive heat—steam if using coolant or smoke if using cutting oil.

See Fig. 9-77.

9.5.2 Regrinding the Bit

There are two features to be controlled while sharpening a normally worn standard twist drill:

1. **Point Angle (lead angle)**

   Both cutting edges are of equal length and angle.
2. **Clearance Angle**

   The clearance angle is 6° to 10°.

Holding the Bit for Sharpening

As shown in Fig. 9-78, hold the bit with the cutting edge *horizontal* (parallel to the floor) with the cutting edge facing up.

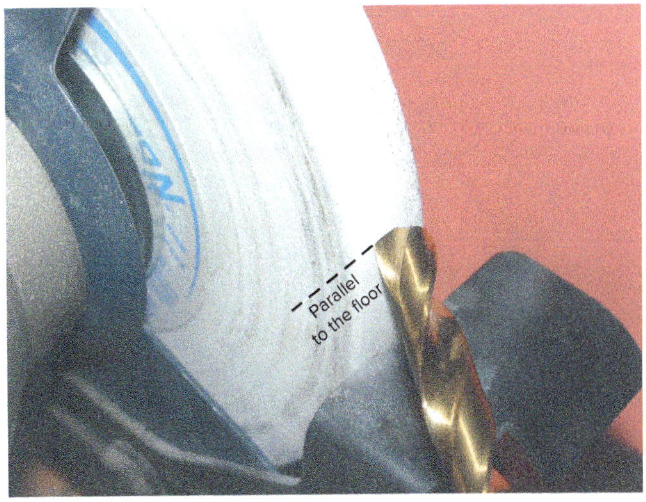

Figure 9-78 Hold the cutting edge level to the floor during the entire grind motion. Guard removed for clarity.
Courtesy Mike Fitzpatrick

KEYPOINT

The cutting edge will remain horizontal the entire time it's in contact with the grinding wheel.

Your right hand supports the front of the bit and your left is on the shank. Note, hand positions can be reversed by preference.

Safety Check List

- ☐ Put on safety glasses.
- ☐ Make sure the tool post is adjusted close to the wheel ($\frac{1}{16}$-in. to $\frac{1}{32}$-in. space).
- ☐ Check that both the eye guard and **scatter shield** (upper guard) are in place.
- ☐ Make a visual inspection of the wheel for cracks or broken out chunks. Also inspect to see that the grinding wheel surface is straight and not clogged by soft metal.
- ☐ Make sure the water pot is ready—heat will be generated.

TRADE TIP

Hand Position Practice Here's a safe "dry run" to develop the hand position and movements needed to sharpen a drill. Use a large bit around 0.5 in. or bigger that is already correctly ground. *Make sure the grinding wheel is not spinning.*

Place the cutting edge at position 1, at the bottom of the grinding stroke, against the wheel as shown in Fig. 9-79. Now, while pressing the bit lightly against the wheel, raise the bit tip to position 2, as in Fig. 9-80, and in so doing carry the wheel up with the movement. As you do so, keep the cutting edge horizontal—do not twist it! Now, move the bit (and wheel) back to position 1. Repeat this motion, always keeping the cutting edge horizontal.

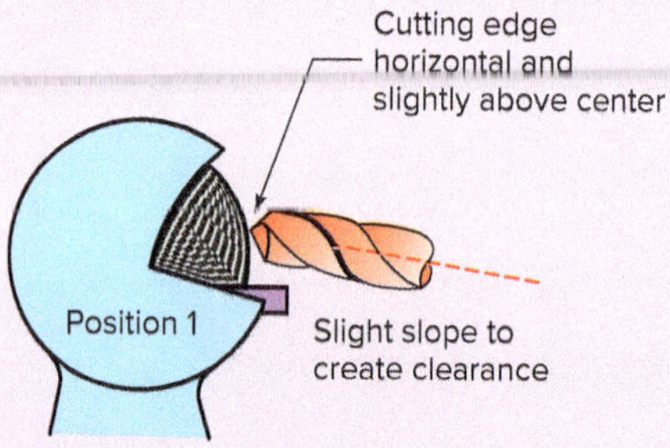

Figure 9-79 The lower limit of the motion, position 1 is forming the cutting edge.

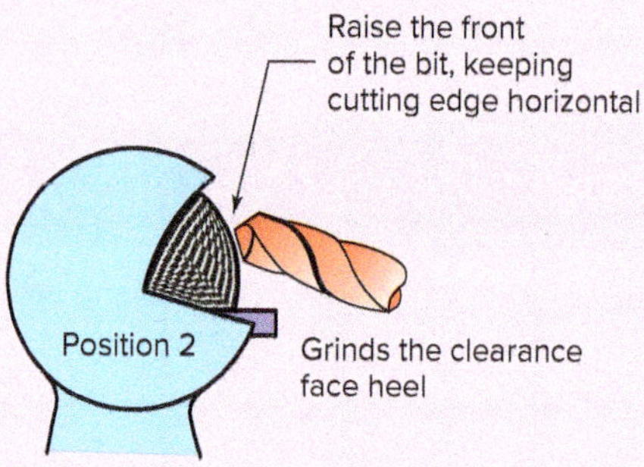

Figure 9-80 Position 2 creates the necessary clearance angle.

Safety Note—The eye guard is not illustrated in these figures for clarity only; be sure to use this important guard in actual grinding practice.

Repeating this up-and-down motion is the movement needed to grind the bit.

Why is the horizontal factor for the cutting edge so important? (Turn to Fig. 9-86 later in the unit to see why.)

Now with the grinder running, place the bit up at position 2 close to but not touching the wheel. Move it forward and with a light touch until very light sparks show up. Touching on the heel of the bit at position 2 ensures that you won't accidentally damage the cutting edge.

Then immediately lower the bit from position 2 down to position 1, keeping the cutting edge horizontal; then back up to position 2 again. Repeat the action several times; then stop and do the same on the other cutting edge. Do not force it; let it just touch the wheel at first to get the feel of the action.

KEYPOINT

The goal is to grind equal length cutting edges and equal angles relative to the drill's axis.

Control by Reading the Sparks

The spark pattern can be used as a visual control of the cutting edge angles you are producing (Fig. 9-81). Sparks will be equally distributed across the entire edge if the angle on the bit is being reproduced. To change the cutting angle, concentrate the sparks on one edge or the other until the surface is fully reshaped.

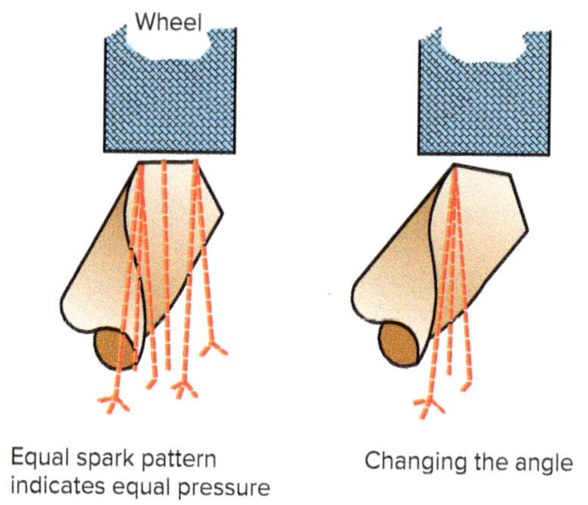

Figure 9-81 Read the spark pattern to control the cutting angle.

To control the clearance angle on the downstroke, you must stop at position 1. Do not go lower or the cutting edge will be rounded off and the bit will not cut. The spark pattern can be used to gage when the drill is lowered to position 1 (Fig. 9-82). Pivoting the bit below position 1 rounds the cutting edge clearance and the bit will not cut (a common error made by beginners). See also Fig. 9-83.

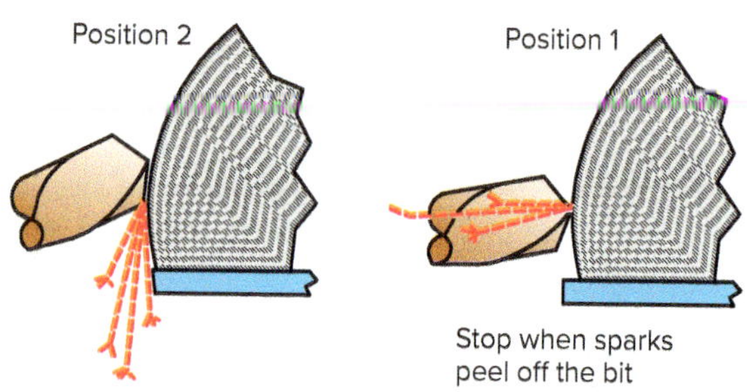

Figure 9-82 Read the sparks to create a perfect cutting edge!

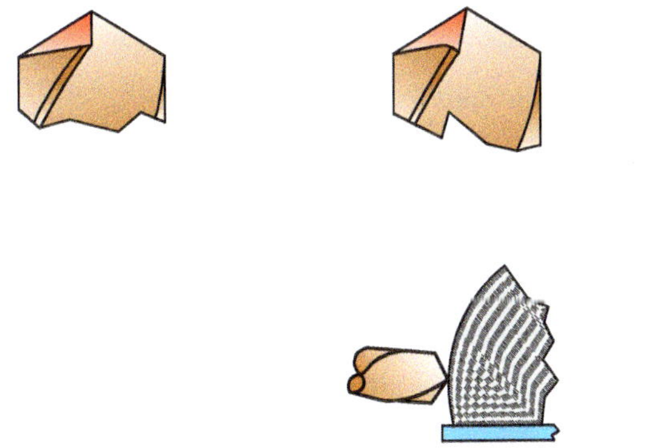

Figure 9-83 Lowering below position 1 creates a drill that will not cut!

At the instant the cutting edge is at position 1, sparks will peel off the upper side of the bit across the rake surface as a chip would. Before reaching position 1, the sparks will go past the cutting edge—straight down.

KEYPOINT

When the sparks peel off the top rake surface, you are at position 1—do not go any lower.

TRADE TIP

A Visual Template of 59° If you have difficulty obtaining the 59° cutting angle, here is a hint that can help. Using a protractor, mark a line at 59° on the tool rest of the grinder. Sight down this guideline as you hold the bit. The correct angle will soon become natural to reproduce (Fig. 9-84).

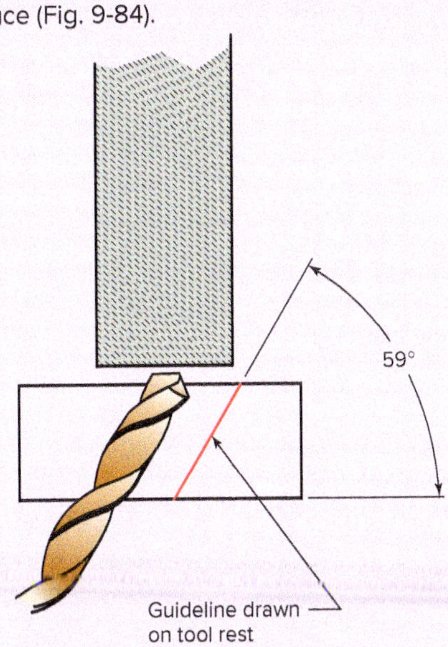

Figure 9-84 A 59° guideline helps maintain the right angle.

9.5.3 Checking and Correcting the Shape

During the grind, use a drill point gage (Fig. 9-85) to check the cutting angle and centering of the point.

Figure 9-85 Two typical drill point problems.

Wrong Angle

Two problems can arise. Different angles on each cutting edge cause early failure due to only one edge contacting the work and thus doing all the cutting. Equal angles but off-center cause an enlarged hole (Fig. 9-85). The longer edge forms the radius of the hole.

TRADE TIP

The enlarged hole made by an off-center point is sometimes deliberately used by machinists to produce a slightly larger hole than the drill. This "trade trick" will work only when there is no pilot hole. Without a pilot hole, the off-center drill pivots around the dead center, which produces a hole not quite equal to twice the longest cutting edge.

Keep the Bit Cool

Keep dipping the bit in the water pot while grinding. Don't wait for it to get too hot to hold. Never let it discolor at all. If you do heat it up, the hardness can be reduced and will become dull more quickly. This is called **overtempering** the metal.

Checking the Dead Center

If you have held the drill correctly during grinding, with the cutting edge horizontal, the web will be at its shortest length. The angle will be approximately 115° relative to the cutting edge (Fig. 9-86). However, if you have rotated or tilted the cutting edge, the **dead center** appears as the bit on the right. The bit on the right can have its center thinned, but the better fix is to regrind it.

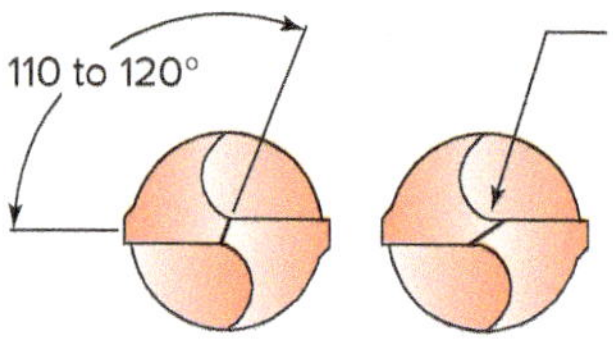

Figure 9-86 The dead center looks like this when the drill isn't held horizontal while grinding.

Long Dead Center on Short Drills

You may produce a dead center that looks like the one shown in Fig. 9-87. The angle is right relative to the cutting edges, but the length is too long. You did nothing wrong. For added strength, the twist drill's web is made progressively thicker closer to the shank. On bits that have been reground many times and have lost over half their total length, the dead center becomes extra wide. You must thin this dead center for all drilling circumstances (Fig. 9-87).

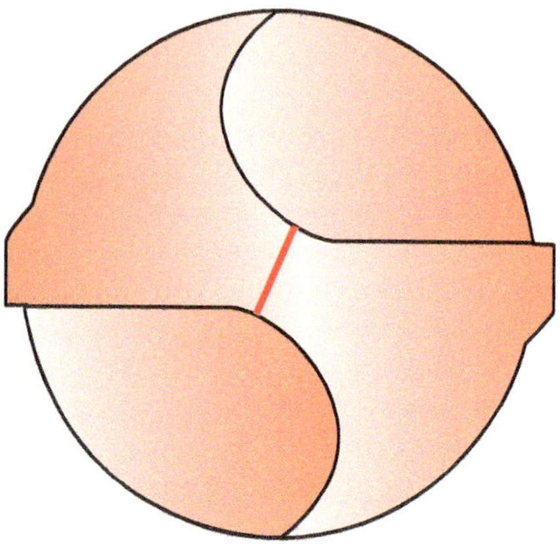

Figure 9-87 A long dead center appears on a drill that's shortened from many resharpenings.

UNIT 9-5 REVIEW

Replay the Key Points

- During drill bit sharpening, the cutting edge will remain horizontal the entire time it's in contact with the grinding wheel.
- The sharpening goal is to grind equal length cutting edges and equal angles relative to the drill's axis.
- When grinding a drill bit, when the sparks peel off the top rake surface, you are at position 1—do not go any lower.

Respond

1. Explain why you must hold the cutting edge of a drill bit horizontal during grinding.
2. When sharpening a drill bit, which face or surface is contacting the grinding wheel?
3. To produce a standard drill point, at what angle are you grinding the cutting edge, relative to the drill's axis?
4. How do you know when you are reproducing the angle on the bit or changing it during grinding?
5. On the downstroke toward position 1, how do you know to stop lowering the bit?

CHAPTER 9 REVIEW

Unit 9-1

Drill and reamer combination is the best method to produce a hole that's round, smooth, and of consistent accuracy. There are two ways a drill or reamer is held in a drill press, either straight or taper shank, and we saw that there are several modifications that can be made to the drill point shape to achieve a special goal.

Unit 9-2

Standard drill presses divide out by their sizes and drilling heads. Starting with the light-duty, sensitive drill press used for extremely small holes, we looked at presses in advancing horsepower and size all the way up to the giant radial arm press big enough to drive a car onto the table. Perhaps the greatest difference between presses is their ability to rotate their spindle forward and backward. Without that ability, a tapping head is required.

Unit 9-3

There are several ways a machinist might determine drill and reamer RPM: charts, calculating formally or in your head, from built-in speed calculators in CAM software, and from good old intuition based on experience. That last category will come in handy not only when setting up manual machines but when trying out a brand-new program. A good operator must be able to see the tool's spin and know instantly whether or not it's close to right.

Unit 9-4

By now, if you've had any lab experience on the machines, you've seen that setups are a lot like working with Lego® blocks. Every new part shape and cut sequence brings some new puzzle to solve as to how to hold the object safely and reliably using a few components such as vises, clamps, angle plates, parallel blocks, and such. We touched on only a few of the myriad possibilities. Along with planning the cut sequence, great setups are the most complex and challenging part of our trade!

Unit 9-5

Grinding a drill bit that looks right and works well takes practice. Since you undoubtedly have a machine or jig that also does the job, was it worth the effort to learn to grind them by hand? Yes. Even in the age of computer machining, there are many situations where a fully competent journey machinist must be able to modify or quickly sharpen a tool by hand. Not just drills; we sometimes find ourselves whipping up a special lathe or mill cutter, too.

QUESTIONS AND PROBLEMS

Test your total recall about drilling. The answers follow the review.

1. True or false? The best material for an industrial twist drill is high-carbon steel. If it is false, what will make it true? (Unit 9-1)
2. You see the following callout on a technical print: *Drill and Tap $\frac{5}{16}$-18 UNC*

   What drill bit must you select for this hole? (Unit 9-1)
3. Now you have checked out the supposedly correct drill bit for tap drilling the $\frac{5}{16}$-18 UNC thread of Question 2. Name the five remaining factors you should examine on the bit before mounting it in a press. (Unit 9-1)
4. A Morse taper shank has a taper ratio of ____ inch per foot. (Unit 9-1)
5. Explain why machine shops use tapered-shank drills for holes above ____ in., even though these tools are more costly. (Unit 9-1)
6. Identify the drill bit features A through G in Fig. 9-88. (Units 9-1 and 9-5; and from previous discovery in Chapter 5)

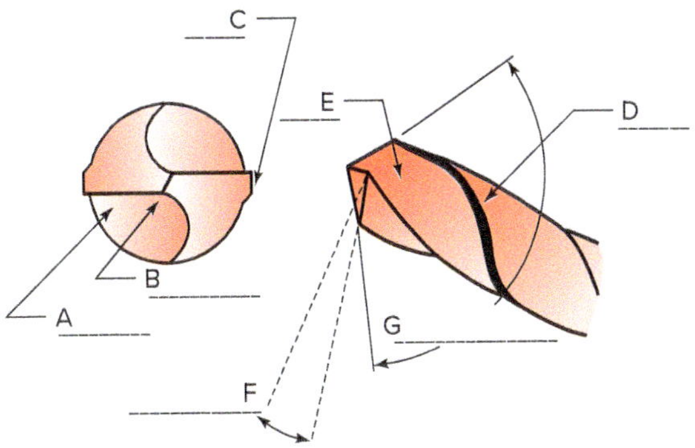

Figure 9-88 Identify the main features of a standard drill bit.

7. In Fig. 9-89, the same part is set up to drill in two different vises.

   A. Which vise is designed to be a drill press vise in Fig. 9-89?

   B. What features set drill press vises apart from the other type?

   C. For what safety reason is it important to know the difference in the two vises shown?

   D. What's wrong with the setup using the upper vise? (Unit 9-2)

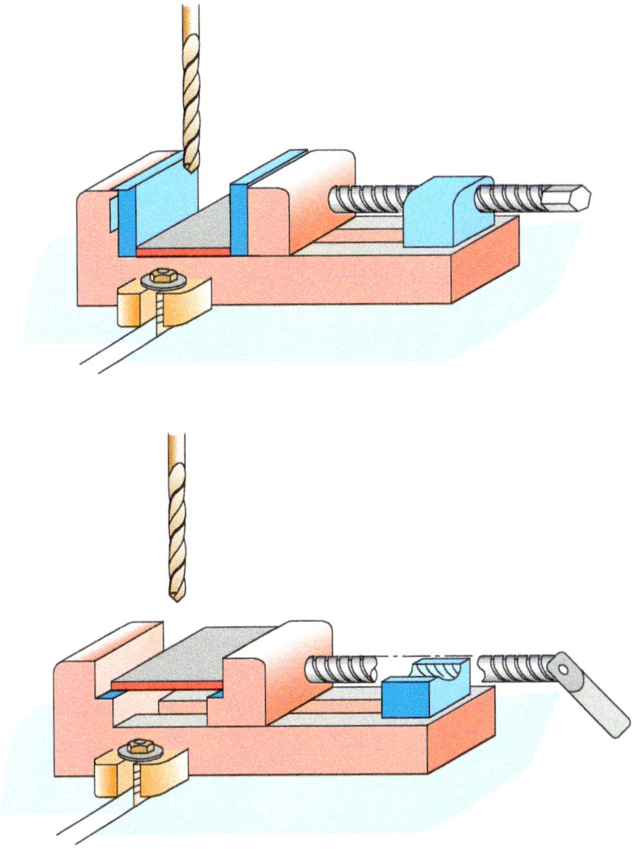

Figure 9-89 Identify the drill press vise by its features.

8. What two variables are used to enter a drill speed chart? (Unit 9-3)
9. Use the short formula and the recommended cutting speeds here in Appendix III, to calculate the correct

RPM for a 0.4375-in. ($\frac{7}{16}$-in.) drill bit cutting cast iron. (Unit 9-3)

10. Use the long formula to solve Problem 9. Will the answer be higher or lower, and by how much?
Use Appendix III to access surface speeds. (Unit 9-3)

11. Name the two different methods for drilling a hole to a drawing location on a standard drill press. List each method's expected repeatability. Describe each briefly. (Unit 9-4)

12. A center finder can be used in two ways to pick up a location: as an accurate spindle *pointer* or as an *optical colinear* locator. Which method is more accurate? What is the repeatability of each? (Unit 9-4)

13. There are two methods of *power tapping internal threads* on some drill presses but only one on a press that has no spindle reverse.

A. Identify the tapping method that drives the tap in, then out on presses without reverse?

B. Which presses have reversing spindles? (Unit 9-4)

14. True or false? A counterboring tool requires a pilot to keep it from chattering and to keep it concentric to the hole. A spotfacing tool needs no pilot since it cuts less deeply than the counterbore. If this statement is false, what will make it true? (Unit 9-4)

15. A standard flat head screw has an included angle of _____ °. What countersink must you select to produce a cone-shaped location for that screw head:
(A) 41°, (B) 82°, or (C) 59°? (Unit 9-1)

16. When drilling soft brass or sheet metal, how can you modify the bit for safety? (Unit 9-5)

17. After regrinding a standard twist drill, the dead center is inefficient.

A. What modification can be made to the drill's point geometry to improve its cutting action?

B. What two improvements does this modification add to the use of the drill?

C. Under what three conditions would this modification not be needed? (Unit 9-5)

CRITICAL THINKING

18. In Fig. 9-90, something is wrong with one or more of the bits. Name the problem and tell what the result will be if it is used to drill. Describe what needs to be done to correct each bit.

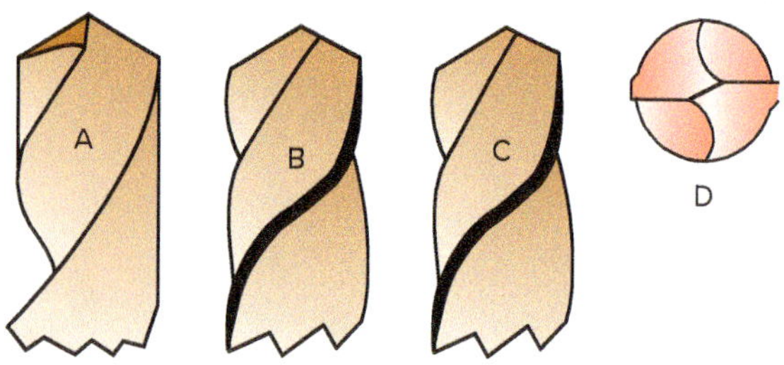

Figure 9-90 Identify the problem with each drill and the result.

19. A new drill bit is checked out from the crib and it has a golden-colored plating on the cutting portion but not the shank. What might that added feature be for? See Fig. 9-91. The solution was not covered in the reading, but you have the facts to solve it.

Figure 9-91 Why would this coating be on the drill bit?
Courtesy Mike Fitzpatrick

CNC QUESTIONS

20. On CNC machines that perform drilling operations there are two kinds of cycles that aid in drilling deep holes. Name and describe them.

21. Solving this may require discussion with fellow students—but it is an important difference in CNC machines and manuals for tapping operations. Why might a CNC machine not need a tapping head to cut internal threads using power feed?

CHAPTER 10
Turning Operations

Units

10-1 Basic Lathe Operations

- Identify the name of the operation needed to machine a given feature based on a part drawing
- List and describe the 11 entry-level lathe tasks
- Recognize basic setup components for an operation

10-2 How the Lathe Works

- Know the parts and functions of the manual lathe
- List a few vital bolt-on accessories

10-3 Work-Holding Methods

- Select the right chuck or other holding device for the job and for safety
- Develop a working ability in selection factors for work holding

10-4 Turning Tool Basics

- Identify and choose the right cutting tool for the job
- Mount that tool in the right tool-holding accessory

10-5 Lathe Safety

- Identify the potential for danger on lathes
- Plan, practice, and refine your emergency actions when operating a lathe

10-6 Lathe Setups That Work Right—Troubleshooting

- Set up and perform a typical job on a manual lathe
- Identify the actions you must take to turn a part correctly and safely

10-7 Single-Point Threading

- Make your setup to thread
- Monitor your operator actions during threading
- Measure the finished thread

10-8 Measuring Threads

- Access and define pitch diameter
- Use thread wires
- Use thread gages

INTRODUCTION

Some time ago, manually operated lathes were considered the heart of the shop and the primary machine for two reasons:

Products of the time were mostly round, for example, axles, gears, and shafts.

In theory, lathes were at the base of the machine pyramid. Theoretically one of those ancient lathes could make another working lathe.

But that's changed. Two factors put turning operations behind milling. First, turning spins the workpiece, not always a good idea compared to milling, where it's only the cutter that spins. Also today's CNC mill can create round surfaces of bearing quality in roundness and smoothness. But the CNC mill can also machine a variety of shapes not possible on a standard CNC lathe. Still, there are times when turning is the efficient (or only) way to get the job done. Knowing when it's right or not can make a big difference in quality, safety, and profit. But stay tuned, Chapter 17 shows us how lathes with milling capability are making new kinds of parts with mill-turning (Fig. 10-1).

KEYPOINT

As you learn turning operations keep in mind that it's a mass spinning. Double-check everything before starting that spindle!

There's another reason to learn turning: Machine evolution isn't over! Today's computing power has given us fifth-generation machines that multitask. The hard line between lathe and mill is fading. Today, we find CNC lathes with milling heads and feed axes. A second universal lathe employs bolt-on cutting tools attached to the lathe tool turret. Hydraulically powered, they rotate milling cutters, drills, and other spinning cutters and move them to cut the work held in the lathe chuck—called live tooling lathes. On the other side of the fuzzy line, new multitask mills equipped with powerful auxiliary rotary axes can spin the work when it's the better way to cut the feature.

These multiplex machines open up new operations for job planning. Operations can now be written that don't simply mill or turn; they combine them. Rotating the work as a lathe would, while contacting it with a spinning mill cutter leads to new ways to shape metal unseen previously—called *generating* the shape. So don't miss Chapter 17 where we explore these new one-stop pieces of equipment.

Here's an example in Fig. 10-1. A hex head is machined onto a spinning round shaft. Using a cutter with the exact tooth spacing needed, the CNC control ensures that the cutter's RPM is coordinated to the work speed such that the teeth pass in a straight line creating a six-sided object. In just seconds the hexagon is machined. As these machines become accepted in industry, and grow even more capable, new ideas will arise and you'll be there to invent them!

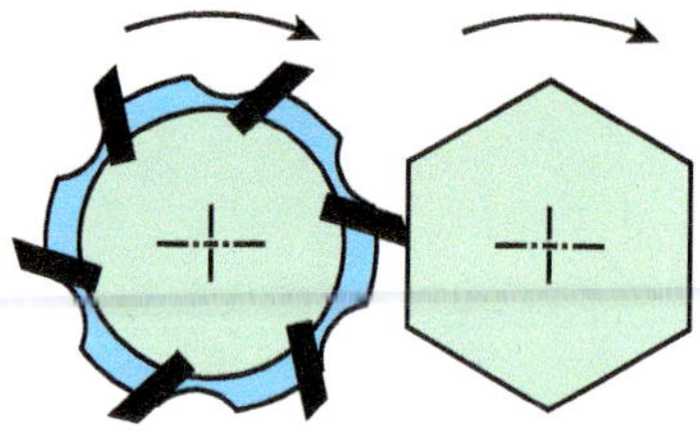

Figure 10-1 On the frontier, simultaneously turning and milling.

But before you can write or manage programs that use these fantastic possibilities, you've got to clearly understand the basics of turning. Chapter 10 is about spinning the work when it's the right process and how to do it efficiently and safely.

Unit 10-1 Basic Lathe Operations

Introduction: Unit 10-1 introduces the basic ways lathes shape metal. It defines the name and nature of the operation, a typical cutting tool required, and in some cases, a bit of basic information about the setup. The purpose is familiarization before lab training; this unit introduces what a lathe can do. In the review problems, you'll be given a part print and asked to identify the various lathe operations needed to complete it and in what logical sequence they might be machined.

Manual Versus Programmed Lathes This unit assumes you'll be practicing operations and setups on a manually operated machine before trying them on a CNC. But there are some differences in how these two kinds of lathes perform them. When it's important to know the difference, it will be pointed out. These operations are not everything that can be accomplished on a lathe. But by mastering them, you become ready for your tech school, then industrial, experience.

KEYPOINT

Keep in mind as you scan, the goal is an understanding of what can be done, not how to do it! That comes later.

Accuracy Comparisons Some operations are more accurate than others when performed on a manually operated lathe. On CNC lathes, the accuracy and repeatability are far better. The numbers are my relative estimates, provided for comparison of the various operations at the beginner level—they are not *a standard.* The first time, you might find them challenging but a good goal. In most cases, a skilled machinist would be able to produce results finer than those described here. The age and condition of the lathe on which you will be practicing are also factors.

TERMS TOOLBOX

Boring A single-point tool operation (boring bar) used to machine a cylindrical hole, but can be used for other internal features.

Chamfer An angular surface usually specified in degrees with respect to the work centerline, on a corner of the workpiece.

Drive dog The clamp-on accessory used to transfer spindle rotation to a workpiece.

Facing A lathe operation that cuts the end of the work to be flat and perpendicular to its axis.

Included angle Specifying a conical surface from one side to the other–twice the half angle. Sometimes called the whole angle in lathe work.

Knurling A rough pattern deliberately produced for grip or to increase size.

Parting A relatively narrow groove that severs the work from the stock remaining. Accomplished with a special tool designed for this purpose.

Taper A slightly changing diameter specified in change per foot or degrees. Might also be specified in half-angle degrees or whole-angle degrees.

10.1.1 Operation 1—Straight Turning

During straight turning, the diameter of the work is reduced to a precise size and probably also to a length dimension, creating a cylinder. It's accomplished by positioning the cutting tool inward along the *X* axis (Fig. 10-2) to a given diameter, then moving the tool along the *Z* axis (left-right motion as you face the lathe) to cut the metal away.

During a turning operation, the work rotates at the calculated RPM, while the cutting tool is forced laterally at a specific *inches per revolution* feed rate past the work (e.g., 0.006 IPR). For every revolution the tool advances 0.006 in. That's how lathe feed rates are expressed. On both manual and CNC lathes, the *Z* axis moves to your left/right. The *X* axis moves inward and out. On a manual lathe, the *X* and *Z* tool motion is provided either by a hand crank or by a power feed. Exact position is achieved by hand dial with micrometer graduations or by digital readouts.

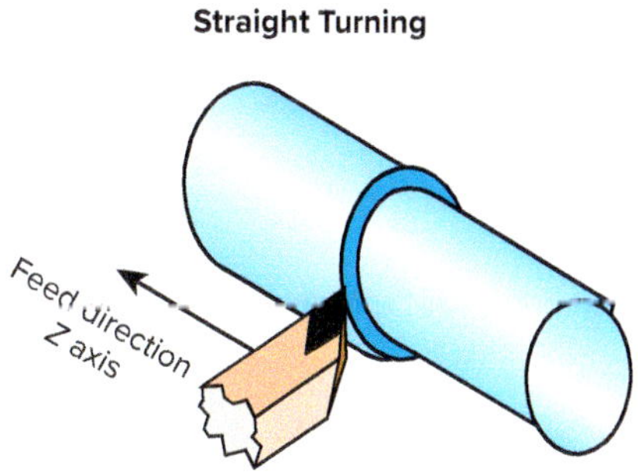

Figure 10-2 Straight turning machines a cylinder.

KEYPOINT

Lathe feeds are normally specified in inches per revolution (e.g., 0.006 IPR). Lathe axes are *Z*, left/right, and *X*, in/out, on the diameter of the work (Fig. 10-3). The *Z* axis of a lathe (or any machine) is parallel to the axis of the main spindle.

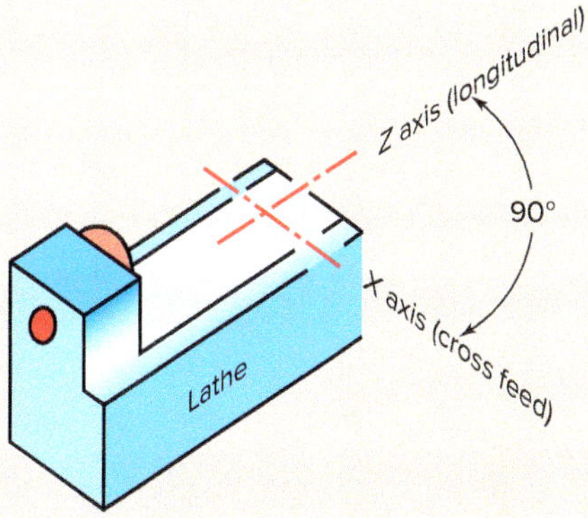

Figure 10-3 The *X* and *Z* axes on a lathe.

Turning Accuracy

On most manual lathes, there are differences in the repeatability between the two axes. The *X* axis is cranked into position with an accurate micrometer dial graduated in thousandths of an inch. Expect a repeating accuracy of around 0.001 to 0.0015 in., depending on the lathe's quality and condition. That means a repeatability on the work diameter at 0.002 to 0.003 in. plus or minus. In contrast the accuracy of a CNC lathe for all axes is finer than 0.001 in.

Because the *Z* axis is not moved by a precision lead screw as the *X* is, but by a simple spur gear pulling the tool on a rack gear, the *Z* axis repeatability is not as accurate unless the lathe is equipped with electronic digital positioning. Using the *Z* axis hand dial, position repeatability varies from 0.005 in. on the better models up to around 0.030 in. on everyday lathes.

SHOPTALK

When discussing manual lathe motion, machinists might refer to *Z* axis motion as longitudinal (or long axis), while *X* is often commonly called cross feed.

Z Axis Stops

Lacking the digital positioner, there are two ways to improve tool *Z* position and part length results for the *Z* axis. A precision micrometer stop attachment (Fig. 10-4) is bolted to the lathe bed. Its probe stops *Z* axis movement and can be adjusted in 0.001- or 0.0005-in. increments, depending on its construction. Alternately a dial indicator mounted on a magnetic base can also be used to control fine movements of the *Z* axis. Your instructor will show you how.

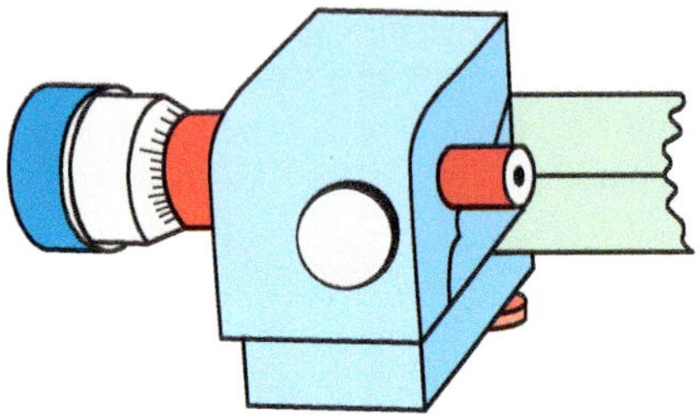

Figure 10-4 Some lathes feature a micrometer stop to assist *Z* axis depth stop control.

KEYPOINT

On manually operated lathes, there are differences in the mechanical method by which each axis is located, creating different repeatabilities. This problem doesn't exist on CNC lathes.

Length-to-Diameter Ratio Determines Setup

During straight turning, work holding is an important issue for both safety and accuracy. We'll devote Unit 10-4 to the subject. For now, the way the work is held for straight turning will vary, depending on the length of the raw stock compared to its diameter. Shorter parts are held on the chuck end only (Fig. 10-5).

Anything over an approximate 5:1 ratio (length to diameter) usually requires extra support at the outboard end. Parts beyond 10 to 12 compared to 1 will also require support at both ends and in the middle, too. The end support shown in Fig. 10-5 is called a tailstock. It ensures that the medium-length shaft doesn't bow away from the tool, bend, or vibrate out of control. Heavier chip removal rates determine which holding chuck will be used and how much material must be reserved for gripping the work. This grip issue is a huge part of lathe planning.

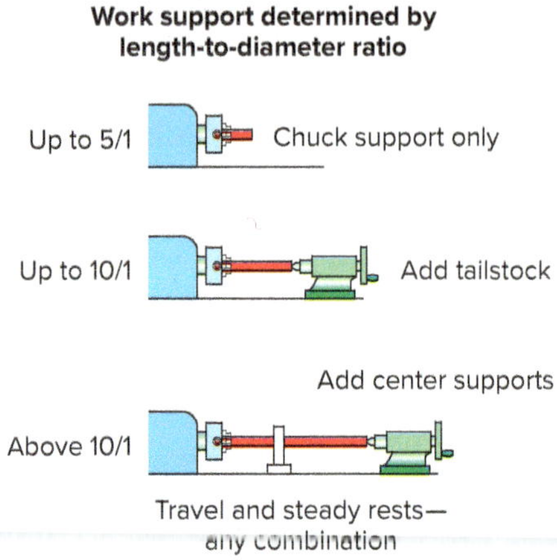

Figure 10-5 The amount of support needed depends on work length.

SHOPTALK

How Fine Can You Go? In contrast to manual lathes, all CNC lathes feature a resolution well below 0.001 in. (usually 0.0001 to 0.0002 in.), with repeatabilities near 0.0005 in. or better. A few superlathes have resolutions of 0.00005 in.—they work at 50 millionths in. resolution! No kidding.

KEYPOINT

Length-to-diameter ratios beyond 5:1 require tailstock support. Ratios beyond 10:1 require mid-part support. These are rules of thumb; there are many variables. Early setups of long stock must be checked by a journey machinist or instructor before taking a cut with them.

Notice that in Fig. 10-6, a turning tool features the three basic tool angles: lead (cutting angle), rake, and clearance plus a nose radius to improve finish.

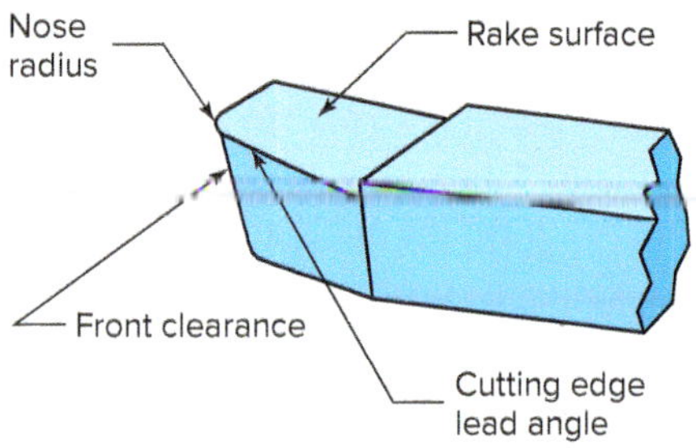

Figure 10-6 A tuning tool features the four universal geometry features.

10.1.2 Operation 2—Facing

Here the tool cuts across the end (face) of a part, moving on the *X* axis. The objective is to make it flat, perpendicular to the *Z* axis and to a given length. Or, as shown in Fig. 10-7, to machine a step in the work to a given dimension from a starting reference–usually, the outer face of the work. During **facing**, the *X* axis moves, while the *Z* axis is locked in position. On manual lathes, the *X* axis can be hand cranked inward (or outward) toward (or away from) the center of the work. It can also be power fed.

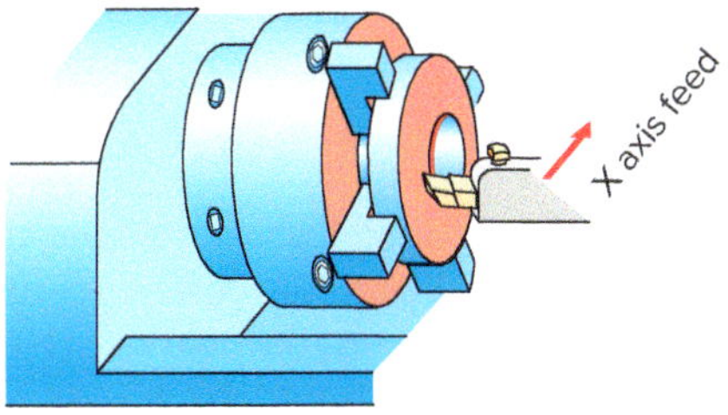

Figure 10-7 A facing cut is taken by moving the *X* axis.

On manual lathes, the repeatability of facing work to a specified length dimension can be from 0.002 to 0.030 in., depending on how the lathe is equipped for *Z* axis cutting tool positioning. The dial indicator attachment previously discussed can be used; however, a digital readout positioner is the best solution. But these options are not always found on manual lathes.

Facing to a Step Dimension

In Fig. 10-8, we need to cut a face step 0.125 in. from the work face. The first thing to do is to coordinate the cutting tool to the position readout (digital or indicator). Either the tool is just touched to the outer end of the work or a light face cut is taken. Then the position is set to read zero. The hand dial numbers would be set to read zero as well, as a double-check of the tool's position.

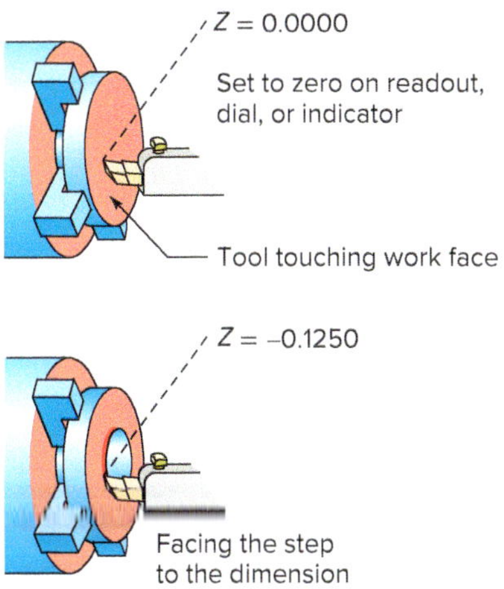

Figure 10-8 Prepping for a face cut, the machinist touches the work, then sets the axis to zero.

Second, the tool is moved to the first cut position at *Z*0.080 in., for example. This cut, called a *roughing pass,* takes the majority of the metal to be removed. Next a semifinish pass at *Z*0.100 is taken. This cut might also be called a *spring pass* because it neutralizes tool or work deflection from the rough pass. Finally, the tool is positioned at *Z*0.125, and the finish pass is taken.

Digital Positioning

An attachment added to the lathe, positioners report tool location relative to any given zero point preset by the operator, usually to a 0.0005-in. resolution (or finer).

KEYPOINT

Learning to use digital tool positioners and tool to work coordination is excellent preparation for CNC training.

RPM—The Need for Speed

One big challenge to facing operations is that the effective rim diameter is constantly changing as the tool moves in or out along the *X* axis. Thus, a change in RPM is needed to maintain consistent cutting speed. Some manual lathes feature a variable speed drive; their RPM can be increased or decreased during the cut. However, most feature gear-driven spindles to accommodate heavy work. They must be shifted to a single RPM and they cannot be shifted on-the-fly (while the spindle is turning). So the cutting speed for facing is chosen as a compromise. It will either be too slow or too fast during some part of the face cut.

TRADE TIP

CNC lathe controls feature a constant surface speed command that, when invoked, continuously changes RPM as the facing cut progresses.

10.1.3 Operation 3—Drilling and Reaming

On manual lathes, drilling operations are accomplished using the tailstock with its sliding quill. It's always hand-fed by crank. The tailstock holds cutting tools on the center of the work and drives them forward by a sliding quill. The quill features a Morse taper hole (the same as on drill presses). Differing from a drill press, the cutting tools do not rotate; it's the work that spins here (Fig. 10-9). Drills and reamers are mounted either in chucks or directly into the Morse taper.

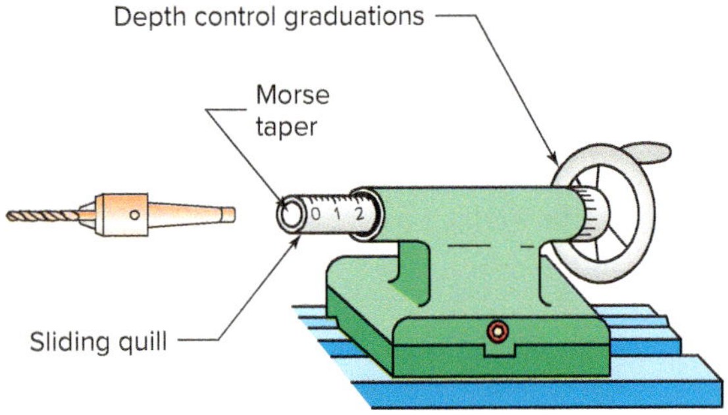

Figure 10-9 The lathe tailstock holds and drives the drill tools.

Hole Diameter Accuracy

There are four factors affecting diameter accuracy when drilling on any lathe. Tool sharpness and size are obvious. The other factors are the rate at which the drill is fed inward and heat/coolant issues. Diameter repeatabilities of around 0.005 in. for drilling and 0.0005 in. for reaming are typical. Coolants are a must in deep holes, but they tend to spray off the rim of the chuck. As shown in Fig. 10-10, when drilling parts in a chuck, coolant must be contained in a chuck guard not only to keep coolant from flying out but to safely contain chips, too.

Figure 10-10 Without a chuck guard the coolant would be flung out everywhere!
Courtesy Mike Fitzpatrick

Hole Depth Accuracy

Hole depth is an accuracy challenge, it will repeat from 0.005 to 0.030 in. (0.1 to 0.5 mm), depending on tailstock dial graduations. Better lathes feature graduations as fine as 0.005 in. (see Fig. 10-9).

KEYPOINT

Always Check Tailstock Centering Before Drilling!
For taper turning operations (coming up later) the tailstock can be offset sideways—off the horizontal center of the spindle. However, to not break drills and reamers and to create accurate hole sizes, the tailstock must be set back on-center for drilling.

Starting the Hole On-Center

Even though the work is spinning about the spindle center and the tailstock is centered, a drill can and does wander at initial contact. To prevent this, a face cut should be taken to clean up the work flat and perpendicular, as long as there is material to allow it. Then a small pilot or a center drill should be used to spot the exact center, the same as for drill press location.

Safety Note—Chuck RPM Limits

When the workpiece is held in a standard lathe chuck, sometimes RPM cannot be shifted to the correct speed for a small drill, even though the lathe may be able to achieve it. Most shops limit RPM for chuck work. In my shop, it's 1,000 RPM, and that assumes the work is well balanced! Due to centrifugal forces and vibration, cast iron chucks can explode at their rim! Poorly held parts can and do fly out of the lathe. The worst accident I have seen in my entire career was caused by a large part flying out of a lathe.

KEYPOINT

Find out about RPM policy. Though I'm repeating myself, always have your setups checked!

There are chucks made for both manual and CNC lathes that safely withstand greater RPM limits, as we'll see examples in Unit 10-4. However, on a CNC lathe operating under constant surface speed mode, a dangerous condition can develop when a facing cut is going toward the center of the work, or when using a very small drill. The control will attempt to speed up the spindle beyond safe limits. To counter that potential, a red-line code word can be added that limits RPM to a specific maximum.

10.1.4 Operation 4—Turning Angles and Tapered Surfaces

Conical work (changing diameter as a cone, not a cylinder) may be called tapered, chamfered, or when the conical surface connects two other surfaces, might be called a *transition* (Fig. 10-11).

When turning tapers, there are two related characteristics to verify within tolerance:

The *diameter* of the surface at the large and small ends

The *angle* specified in degrees or sometimes in taper amount per foot

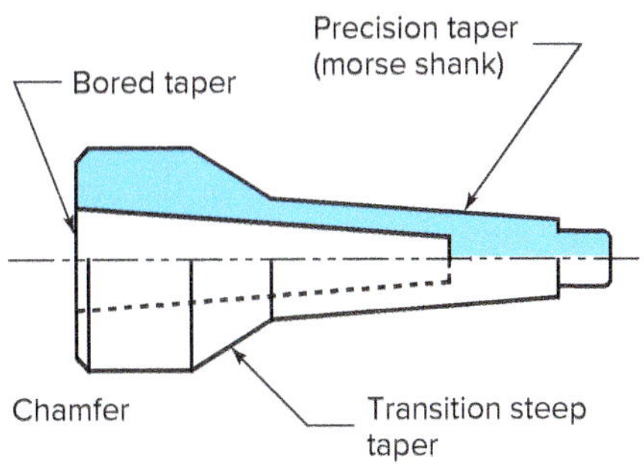

Figure 10-11 This taper shank adapter exhibits four angular surfaces.

Angular Feature Expressions

To cut tapers one must understand two terms found on engineering drawings:

Chamfers *(steep transitions) are specified in degrees relative to the work centerline;* for example, 45° chamfer or a 45° transition.

Shallow **tapers** *are usually specified with diameter change per length or in degrees;* for example, Morse tapers are $\frac{5}{8}$ in. per foot of shaft length, or it can be expressed as an angle at 1.49° taper.

Be Certain It's Either the Included or Half Angle

When reading a taper callout on a print, determine which angle is being shown, the half or full angle (Fig. 10-12).

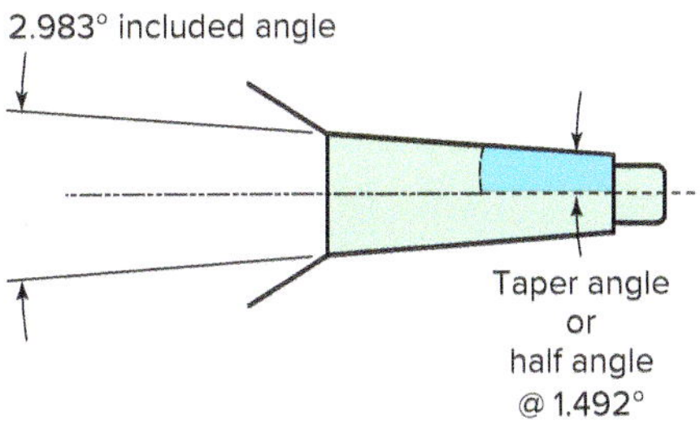

Figure 10-12 Tapers are noted in either the *half* or *included* (whole) *angle.*

KEYPOINT

The *half* angle is measured from one side of the work with reference to the centerline. In contrast, the **included angle** compares one side of the part to the other side—the overall taper angle of the cone.

Four Angular Setup Methods

On CNC lathes there is no difference in the way any taper cut is set up and made, since the computer can move the X axis relative to the Z axis at any specified feed ratio. However, on manual lathes, there are four different setups, depending on how steep and long the taper is to be.

Method 1—Swiveling the Compound Axis

Manual lathes feature an auxiliary axis that can be rotated to any desired angle with respect to the X-Z framework. Hand-cranked only, the compound provides a simple way of producing a taper of moderate accuracy. Note the perspective for these drawings is from above, looking down on the lathe (Fig. 10-13).

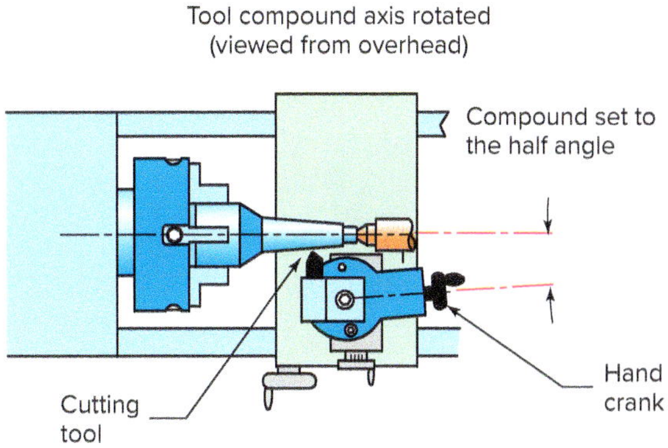

Figure 10-13 The *tool compound rest* allows machining steep angles by rotating it to the correct half angle.

Must use complementary or supplementary angles When swiveling the compound a common mistake is made by beginners. Most compound bases are graduated in 90° quadrants with a few supplied in 180° segments.

The compound axis on most manual lathes is at 0° when it is parallel to the X axis (Fig. 10-14). However, engineering drawings usually specify the part angle off the centerline Z axis.

On most lathes, you will be setting the compound to the complementary angle of that to be machined. However, on a rare few, 0° on the compound occurs with it parallel to the Z axis, and you may be setting the compound to the supplementary angle. Confused? See the Trade Tip.

TRADE TIP

To avoid confusion during compound setup, set the axis parallel to the work axis. Then count the degrees moved as it is swung around. Ignore the graduations, and count the degrees moved. Then double-check the position with a protractor.

Using the compound to cut angular surfaces can be as accurate as $\frac{1}{2}$° and 0.002 in. on the diameter. But, since both the X axis micrometer dial and compound dial are involved, repeatability must be assumed not better than 0.005 in.

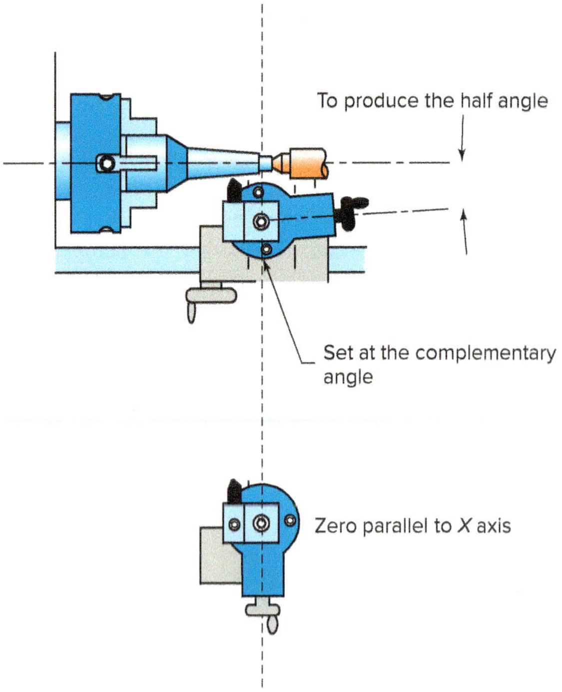

Figure 10-14 Avoid complementary angle problems associated with the compound axis.

Method 2—Forming the Angle

This short, steep-tapered surface is machined with a tool ground for the purpose. We'll see other versions of the form tool coming up (Fig. 10-15). Forming a cone or radius is accurate and fast in production situations. Form tools are even used for tricky CNC lathe work, where the control could duplicate the shape but the form tool does a better job.

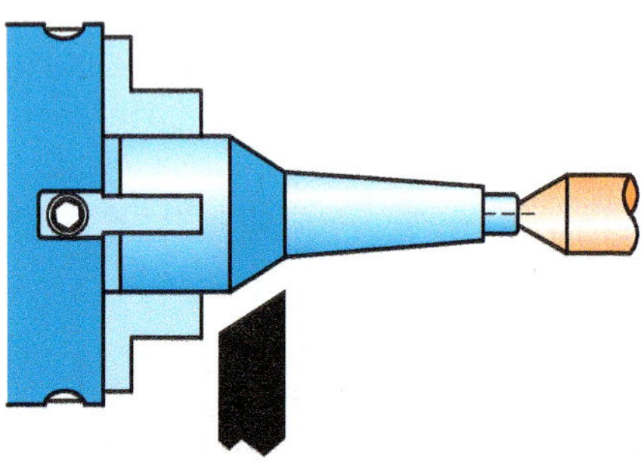

Figure 10-15 Steep angles can be formed. The tool bit is ground and set up to the shape of the angle required.

Accuracy For both diameter accuracy and repeatability, this method is the best of the four, as long as the form tool is set to the right angle. The tool is cranked inward along the *X* axis to the desired size, and therefore repeatability is as good as the lathe's *X* axis.

Method 3—The Taper Attachment

Once it's set up, turning with a taper attachment is similar to straight turning. The lathe is engaged to move along the *Z* axis in the usual way, by power feed or handcranking. While it moves laterally, the angled taper attachment moves the tool in or out along the *X* axis at the ratio set on the attachment (Fig. 10-16).

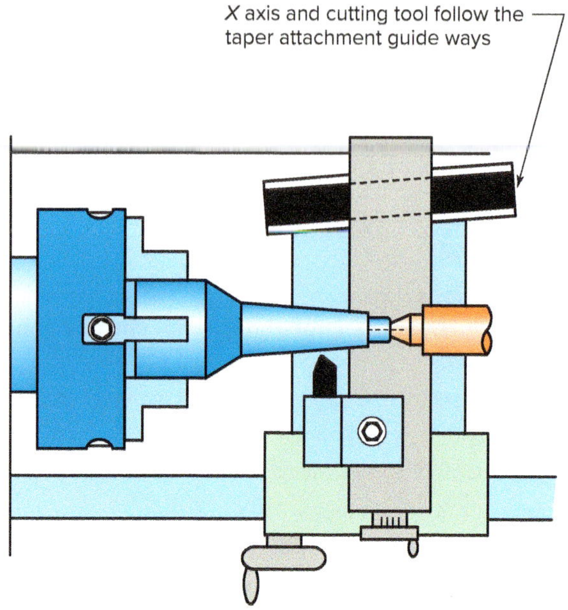

Figure 10-16 The tool is moved in two axes at the same time, producing the taper. Not all lathes have a taper attachment.

Accuracy The taper attachment can be adjusted for any taper angle between 0° and around 25°. The angular resolution is around $\frac{1°}{2}$ using the attachment's graduations alone but can be improved greatly with sine bars, dial indicators, and vernier protractor tricks you'll learn later.

Although diameter is determined by *X* axis position, repeatability isn't as good as the lathe's *X* repeatability because the taper attachment includes a mechanical clearance to allow it to be operated freely—it has backlash. Expect around 0.005-in. repeatability, but that can vary depending on the quality and maintenance of the attachment. The taper angle will repeat perfectly once the attachment is adjusted to the correct angle and locked in position.

> **KEYPOINT**
>
> Remember, this is a visual task inventory. It's unimportant that you fully understand the attachment right now, just that it is one of the options for producing a taper on a long workpiece.

Method 4—Offsetting the Tailstock

This method makes very shallow taper angles. You would choose this method because of the ease with which the part can be removed and quickly replaced in the setup with no disruption to accuracy or because the work is very long. Notice there is no chuck, but rather the part is held between pointed centers that we'll discuss more in Unit 10-3 (Fig. 10-17).

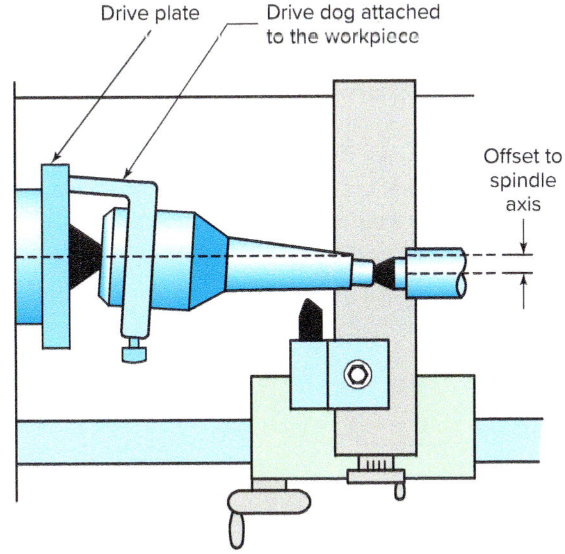

Figure 10-17 Looking down, the tailstock is moved off-center to make a shallow taper.

KEYPOINT

In this method, these centers are not parallel to the *Z* axis by *half* the taper amount.

Using the offset method of producing a taper, the tool action is the same as for straight turning except now the work is *not parallel* to the *Z* axis of the lathe. The work is spun by a clamp-on driver called a **drive dog** because it has a tail. Its tail is inserted into a slot in the drive plate.

10.1.5 Operation 5—Boring

Boring is an accurate method of machining an inside diameter or another internal feature on the work. Bored holes are usually straight (cylindrical), but they can also be machined with a taper using the compound or the taper attachment (Fig. 10-18). This lathe operation requires a cutting tool called a boring bar.

Reasons to Bore a Hole

The advantages of boring over reaming are twofold. One is that any size can be produced with a repeatability of 0.0005 in., assuming the lathe's *X* axis is accurate. Second, a boring tool will machine to the exact centered location without wandering as a drill/reamer combination might.

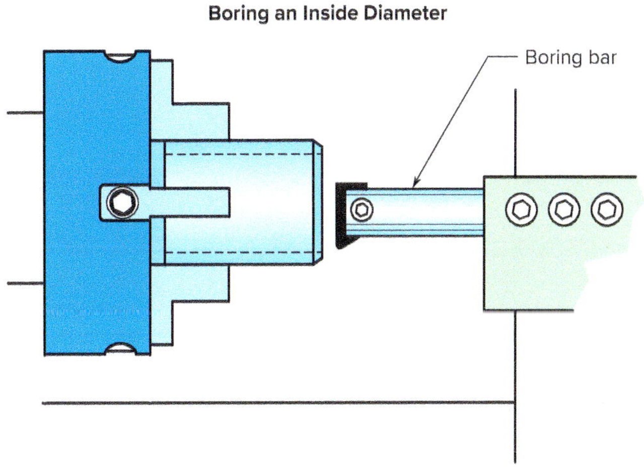

Figure 10-18 This lathe is set up to bore a hole with a boring bar.

Two or three passes (light sequential cuts) of a boring bar will correct a misaligned drilled hole. During boring, the tool is either hand- or power fed, usually inward along the *Z* axis. Reversing the *Z* feed, outward boring is also possible and less risky as the cutter is powering away from the bottom of the hole.

There are three challenges to boring operations:

1. **Boring Bar Deflection**

   This causes inaccurate sizes and chatter due to the long, thin boring bar.

2. **Ejecting the Chips out of the Hole**

   Coolants help, but chips tend to wind around the boring bar and jam at the bottom of the hole. Operators must always watch and listen for chip buildup when boring. The jammed chips rub and ruin the newly machined surface and/or break the cutter.

3. **Blind Operation**

   The operator cannot see the cut as it proceeds. The depth of the bore must be controlled without being able to see the tool.

In addition to producing internal cylinders and cones, other operations are performed with a form-shaped boring tool. Features such as corner fillets, round or square internal grooves, and threads are made with a formed boring bar (Fig. 10-19).

SHOPTALK

We seldom use a form-ground tool on a CNC lathe for two reasons. First, it is a custom-ground tool, which we try to avoid. Second, the controller can produce most shapes by driving standard tools in the shaped tool path.

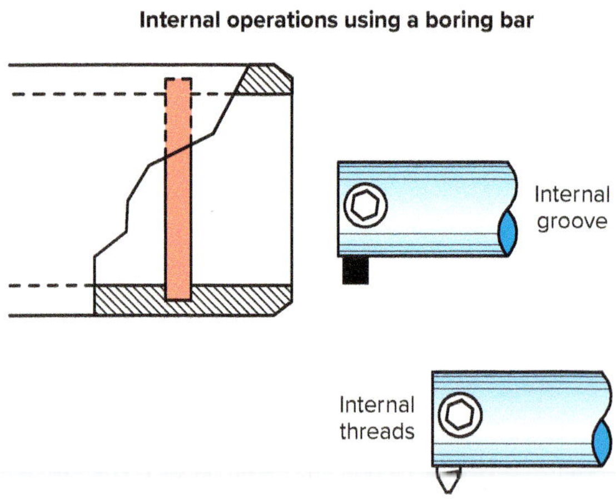

Figure 10-19 Internal grooves and threads are produced with a boring bar but using a shaped bit.

10.1.6 Operation 6—Forming

Forming is used more on manual than CNC lathes to create special shapes that are impossible to produce using tool axis motion. The tool bit is ground to reproduce the desired shape (Fig. 10-20).

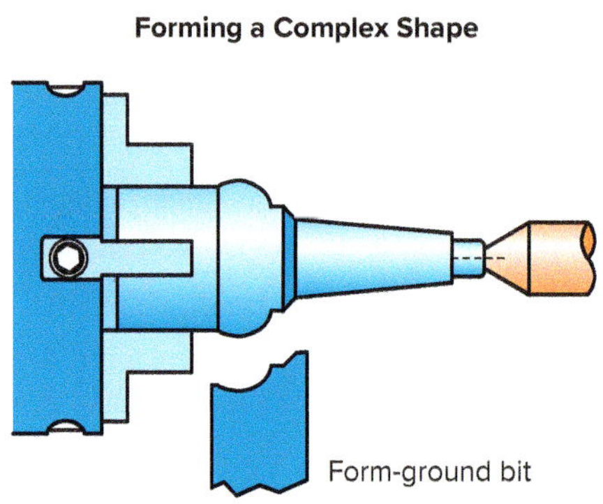

Figure 10-20 A forming tool making a complex, shaped cut.

> **KEYPOINT**
>
> When forced against the rotating workpiece, the form tool produces a reverse image of itself. Cutting fluid and slow RPM are vital for success in forming, and due to the large tool contact area, chatter is always a potential problem. Accuracy is difficult to determine for forming due to the complex shapes, but within an overall 0.0015 in. might be expected.

10.1.7 Operation 7—Threading

Cutting threads can be accomplished in three ways on the manual lathe: taps mounted in the tailstock, dies in regular handles, and using a forming tool to cut the helical groove between threads, called *single-point threading*. Single-point threading (*single pointing* in shop lingo) is a vital lesson in using the manual lathe and, in your first year machining course, is likely to be the most complex operation to set up and to perform. It's time-consuming and prone to errors due to the many details and actions the machinist must handle.

Still, it's a vital competency to have because any custom or standard outside or inside thread can be produced using single-point methods. There are times when single pointing the thread is the right choice; perhaps one thread is needed when a die or tap isn't available, or the CNC lathe is doing other things. This is the full journey machinist's fallback method!

The setup uses a cutting tool ground to the correct thread form with large clearance and rake angles. Since the spindle will be rotating slowly, these extra angles help to improve the finish (Fig. 10-21). As with all threading operations, cutting fluid is an absolute must. As we've just seen, single-point threading can also be performed on internal threads using a boring bar and a form-ground cutting tool.

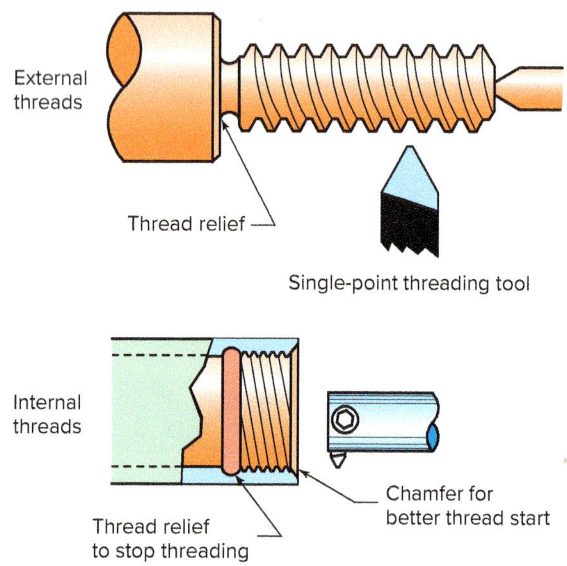

Figure 10-21 A threading tool is a type of form tool.

Single-point threading can repeat at around 0.001 in. if the lathe is accurate and well maintained. But, since it requires the use of two hand dials, the X and the compound axes, care must be taken to control axis backlash. We've got a bang-up trade trick coming up in Unit 10-7 to make this aspect a snap.

Tapping and Die Threading

These are two less complex options for producing threads at the lathe.

Taps in the Tailstock

To produce internal threads, regular taps can be mounted in a drill chuck in the tailstock. The spindle is placed in neutral, then rotated by hand to cut threads where tap breakage might be an issue. For sturdy taps of $\frac{1}{2}$ in. or greater, the spindle can be placed at a low RPM, then started forward under power. When using power, it's wise to slip the clutch and never remove your hand from the start lever. If the lathe can be reversed, the tap can be removed under power, too. If not, the spindle is placed in neutral and rotated backward by hand to reverse the tap. Alternatively it can be unchucked, then unscrewed with a regular tap handle.

KEYPOINT

The tailstock handwheel is not used for this kind of tapping on the manual lathe. Advancing the tap into the hole (and out) is done by hand pressure on the entire tailstock. Left unlocked on the ways, it's slid forward by hand pressure to engage the tap with the work (Fig 10-22).

Tapping from the Tailstock

Figure 10-22 Tapping can be done by pushing the loose tailstock forward and back.

As the tap nears full depth, the spindle is stopped and then reversed. At the same time, the machinist pulls the tailstock outward. It's a coordinated action requiring a demonstration before you try it yourself.

Handheld Dies (and Taps)

It is possible to thread with a handheld tap or die, mounted in a handle, pushed against the work as the chuck is rotated slowly by hand—*not by power!* If it is power driven, there's a chance to pinch hands between the handle and lathe bed or saddle. Also, it's prone to alignment error of the cutting tool to the work axis. However, if your instructor approves of this method, ask for a demonstration. It can be used to quickly make utility threads, although they are not customer quality—we call this a "quick and dirty" method. See the Trade Tip.

Production Taps and Dies

In contrast, there are times when large batches of parts are required with high-quality threads and the CNC machine isn't available. Then, on any lathe, manual, automatic screw cutting, or even the CNC, we turn to the collapsing tap or automatic die head. These are designed to cut the threads in one pass, on the inward stroke. Then using a triggering action, withdraw their cutting edges so they can be pulled back quickly rather than unscrewing them from the thread, which tends to dull the cutting edges in some materials. They are more commonly used on automatic lathes made expressly for threading, but if mounted in the tool post or tool turret and centered to the spindle, both vertically and in the *X* axis, *they can cut threads faster than any other method* with repeatabilities in the subthousandth range.

Another advantage to these kinds of threading tools is that they use inserted HSS or carbide cutters that are easily exchanged to a different thread. The die inserts can be removed from the head and quickly resharpened. This threading setup would be chosen when hundreds of parts are to be threaded and tolerances are tight. Dies and taps of this kind pay for themselves in large batch production (Fig. 10-23).

Figure 10-23 Trip dies and taps speed up production threading but would not be common in school shops.
Courtesy Landis Threading Systems

TRADE TIP

Hand Tapping Improvement in the Lathe. Recall the centering tip we learned when hand tapping on the drill press? To ensure tap alignment place a lathe center in the tailstock; then push the tap forward using the center hole in the tap's shank. The tap is rotated using the tap handle. See Fig. 10-22.

10.1.8 Operation 8—Parting (Cutting Off)

The **parting** operation is primarily used to cut the finished product away from the raw material remaining in the chuck, or occasionally to produce an extra deep groove in the cylindrical side of the work. Parting tools are special thin blades that produce a narrow, deep groove (Fig. 10-24, top view).

The parting tool is either hand or power fed in along the *X* axis. Parting tools are supplied in either high-speed tool steel or as brazed carbide tips and carbide insert types.

Accurate work length control is the objective. Repeatabilities vary, depending on *Z* axis positioning of the tool from 0.030 in., using handwheel graduations to 0.002 or 0.003 in. if digital readouts or long-range DTIs are used. Besides positioning the cutting tool, then locking it in place, another challenge to length control is the perfect alignment of the tool to the *X* axis and its tendency to wander to one side as it dulls. Coolants, well-calculated RPMs, and moderate feed rates are essential in a parting setup.

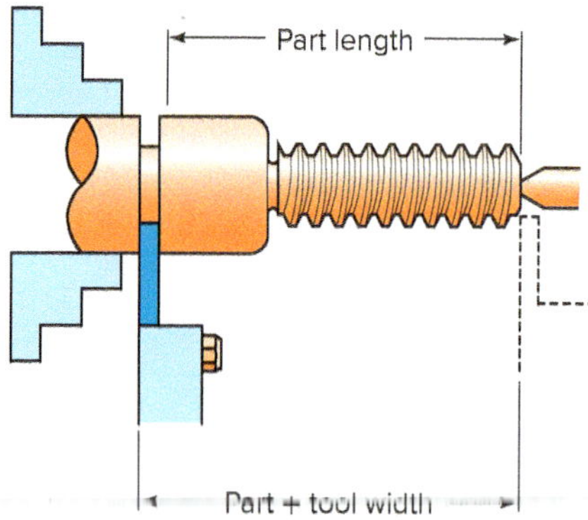

Figure 10-24 During parting, chip removal is a challenge.

KEYPOINT

Using the chuck face or the lathe's cross slide as a reference, the parting tool must be perfectly parallel to the *X* axis to maintain accurate cutoff lengths. Keep parting tool overhang to a minimum!

Parting Operation Challenges

In addition to part length, there are several factors making this a tough operation to perform consistently and safely. Chip removal, chatter, and tool digging and/or breakage are a concern.

Part Length

To achieve the required length, the tool's left edge is touched to the work end (dashed lines in Fig. 10-24), then after withdrawing it from the workpiece, the *Z* axis is moved the distance required (tool width plus part dimension). Most parting tools are made to an exact width to help with the math. But be aware that as the HSS version is reground, the effective width may change.

Chips Wider Than the Groove

Due to the flowing deformation of the metal, the chips grow slightly wider than the groove made by the tool—they rub in the groove and often coil up into a tight mass. Coolants are a must both to alleviate heat and to lubricate the chip so it can be ejected from the groove. (See the Trade Tip, Fig. 10-25.) Ideally, coolant should be induced from above and below the cut to get it to where it's needed. However, it's uncommon for manual lathes to feature dual spigots, so it must be flooded generously into the groove. A chuck guard is then necessary.

Chatter and Tool Shattering

It's likely you'll be using HSS parting tools initially when learning parting. While the carbide insert or carbide-brazed tip parting tool outperforms the HSS by a large factor, the tougher HSS tool forgives most of the digging and chattering problems associated with parting. Most—but not all!

KEYPOINT

Caution! Parting tools do shatter when the cut is deep and the tool is extended far from the holder. They dig in, pull down, then snap into dangerous projectile shards! Be sure to set the tool with just enough overhang to do the job and use conservative feeds.

TRADE TIP

Chip Folding Tool To help the wide chip made by parting come out of the groove, it's possible to grind a small V on the top rake surface of the tool. The chip flows over this surface creating a disturbance line in the chip that tends to fold it; thus, it does not hang in the groove so tightly. Many commercial carbide parting tools have this feature built into their top surface geometry. The groove is put in using the corner of a well-dressed grinding wheel—have your instructor demonstrate this modification (Fig. 10-25).

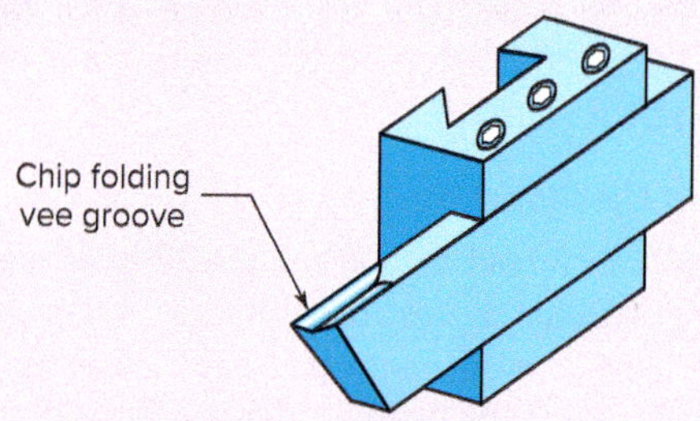

Figure 10-25 A top groove in the parting tool helps to crease and fold the chip.

Constant Surface Speed Needed

Similar to facing, the need for increasing RPM becomes critical as the tool approaches center. Due to the tight deformation problems of the chip piling up within the groove, surface speed is a must-do challenge. This requires a variable drive spindle or CNC control in constant surface speed mode to do it correctly Sometimes, as the tool creates an ever smaller diameter, the workpiece will break prematurely, leaving an unwanted protrusion as shown shortly in Fig. 10-26. However, RPMs closer to the right speed without exceeding the chuck limit will help reduce this problem.

Dropping the Part

Refer back to Fig. 10-24. There is a potential problem—can you see it? As the final cut is made, the part is going to fall down. But the support center is still in the right end. This can lead to real pinching problems for the loose part and chuck, unless the lathe is stopped instantly. There are a couple solutions but . . .

KEYPOINT

Safety None of the solutions involve reaching in with your hand to catch the part as it is finally severed from the stock. (The key words here are *severed* and *hand!*)

Stop and Withdraw the Center Slightly

Do this before recommencing to fully disconnect the object. Then when it does disconnect, let the part fall into a catch basket or fall to the chip bin if damage to the part isn't a critical issue.

TRADE TIP

Grinding a Slight Lead Angle on the Parting Tool If many parts are to be cut off a bar, then a small, unwanted point can be left behind on each finished part after the parts disconnect (Fig. 10-26). It must then be removed later, adding non-value-adding time to the product. To solve this problem, grind a face lead angle around 10° maximum on the parting tool, with the more pointed corner toward the work (see Fig. 10-26). The pointed tool tends to create a weak spot in the bump that allows it to break away or leaves it on the raw stock, not the work. The cutting tool then proceeds forward to cut the bump off the raw stock.

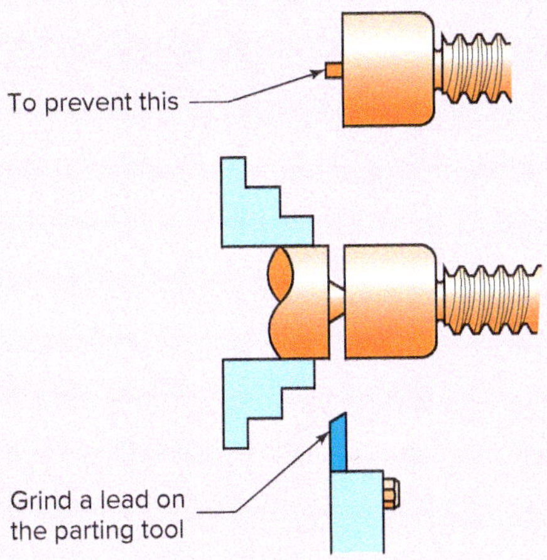

Figure 10-26 Modifying the lead, the raw stock retains the small protrusion rather than the workpiece.

If the Part Is Nearly Through with the Support Center in Place

Stop the spindle, then *hand saw* the object off the stock or bend it off. This is the safer solution, especially for beginners.

10.1.9 Operation 9—Grooving

There are two types of grooves made on the lathe. The first is similar to a parting tool cut, a groove into the outer round surface of a part, while the second is on the face of the work. Tools used for outside diameter grooving are similar in tip shape but shorter than parting tools. Face-grooving tools must be slightly modified with extra side clearance to fit within the outside curvature of the groove the tool makes (Figs. 10-27 and 10-28).

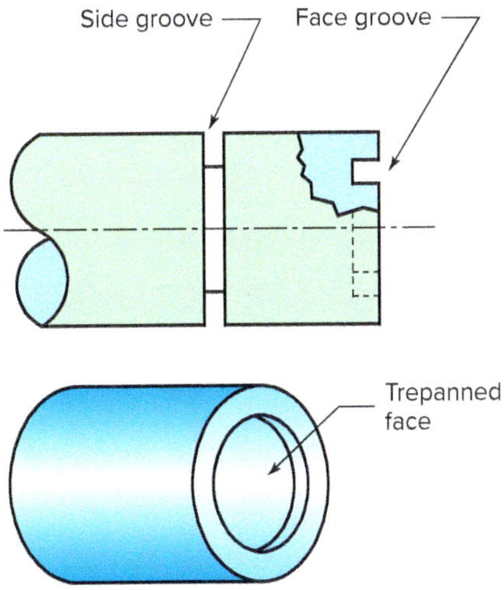

Figure 10-27 Face grooves require a modified tool bit shown in Fig. 10-28.

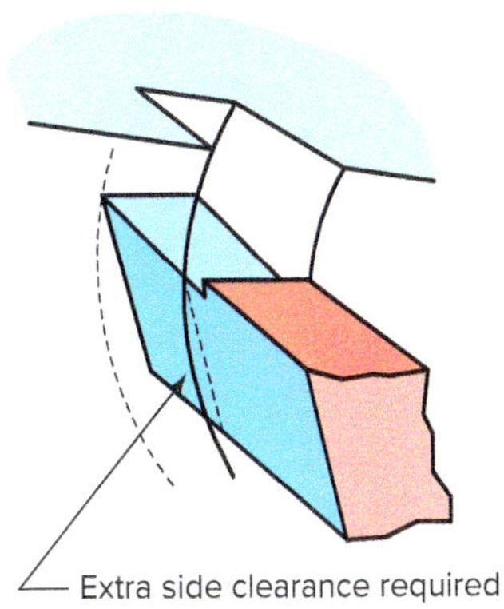

Figure 10-28 A face-grooving tool requires more side clearance on one side to allow for the shape of the groove produced.

Expect a repeatability of 0.0005 in. for the width of the formed groove, which is often a critical feature. Diameter repeatability depends on the positioning method of the lathe and is from 0.002 to 0.0005 in.

A face-grooving tool can also be used to remove material from the face of a part to a flat-rimmed dish or pan. This is often called *trepanning* and is probably a reference to even older wood-turning lathes.

SHOPTALK

Turning, Not "Lathing" Even though it's right to be drilling on a drill press and milling on a milling machine, when working on the lathe, we express it in trade lingo as turning. For some odd reason that I can't explain, we don't do "lathing."

10.1.10 Operation 11—Knurling

Knurling (rhymes with *sterling* silver) produces a deliberate rough pattern on a workpiece surface. Knurling does not remove metal but rather displaces it. With the work rotating slowly, a knurling tool creates small, sharp dents in the work. The displaced metal is deformed (shoved) up next to the dent, a little bit at a time, into the rotating forming dies.

You have seen this diamond or straight-line pattern on handles of tools or on thumb screws (Fig. 10-29). Grip and appearance are the objectives for knurling, so size is not the usual objective. Expect size repeatabilities of no better than 0.010 in.

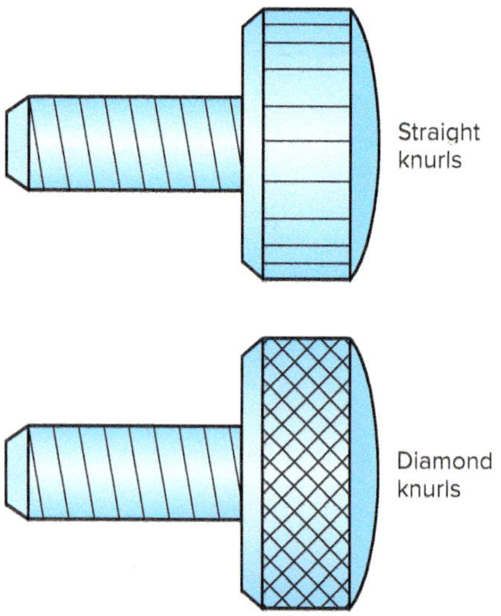

Figure 10-29 Two types of knurls on a shaft, straight and diamond.

Knurling is called out for two purposes: first, to improve hand grip or bonding qualities; second, to make a shaft bigger. Because the displaced metal is shoved up, knurling actually increases the effective size of a shaft. Suppose a worn shaft no longer holds a press-fitted bearing in place. It can be knurled, then turned back to the correct diameter to hold the bearing again. It won't be as good as the original since there is far less metal to form the surface, but it does work! (See Fig. 10-30.)

Figure 10-30 Knurling requires a rotary tool composed of two dies that displace but do not remove metal.
Lake Washington Technical College, Kirkland, WA/McGraw Hill

10.1.11 Operation 12—Hand Filing and Hand Sanding

We try to avoid handwork on machines as much as possible, but there comes a time when just a bit can make the difference in the following:

Minute size reductions. Filing or sanding can take smaller bites than a cutting tool, for example, sanding to remove a tenth or two to exactly bring a diameter into tolerance.

Surface finish and appearance. Sanding can create finishes not possible with cutting tools. It is useful for final details such as burr removal with a file, corner radii, or chamfers.

When filing and sanding, machinists achieve close to 0.0002-in. accuracy (repeatability is not appropriate here). But we always seek other solutions, since hand operations are not only slow and, on the low-tech side, they also bring your hand and arms dangerously close to the chuck.

Safety Precaution

Because you will be placing your hand close to a rotating spindle, there are specific safety precautions for both filing and sanding. Figures 10-31 and 10-32 show the right way to file and sand. Hand and arm positioning are critical as is your braced stance in front of the lathe! As a critical thinking exercise, you'll be asked to reason why and how in the review. Study the illustration closely.

Figure 10-31 This machinist is using the safe, skilled hand position for filing, along with a chuck guard and no loose sleeves or jewelry.
Lake Washington Technical College, Kirkland, WA/McGraw Hill

Figure 10-32 When sanding work, always hold the media as shown. Why? The answer can be found in the Unit 10-1 quiz.
Lake Washington Technical College, Kirkland, WA/McGraw Hill

UNIT 10-1 REVIEW

Replay the Key Points

- Lathe feeds are normally specified in inches per revolution (e.g., 0.006 IPR).
- Lathe axes are *Z*, left/right, and *X*, in/out, on the diameter of the work.
- The *Z* axis of a lathe (or any machine) is parallel to the axis of the main spindle.
- Because there are so many variables as to how to hold work, it's critical that your setups be checked by your instructor before starting them.
- For accurate drilling operations, and to not break drills and reamers, the tailstock must be set on-center.
- Know shop RPM policy and always have setups checked!
- The *half angle* (sometimes called the taper angle) refers to one side of the work compared to the centerline, while the *included angle* refers to side to side—the overall taper of the object.
- The compound axis is at 0° when it is parallel to the *X* axis on most machines.

Respond

Do you recognize the basic lathe tasks? Complete the following questions to find out.

1. On a piece of paper, name the operations needed to turn this axle from raw stock (Fig. 10-33).

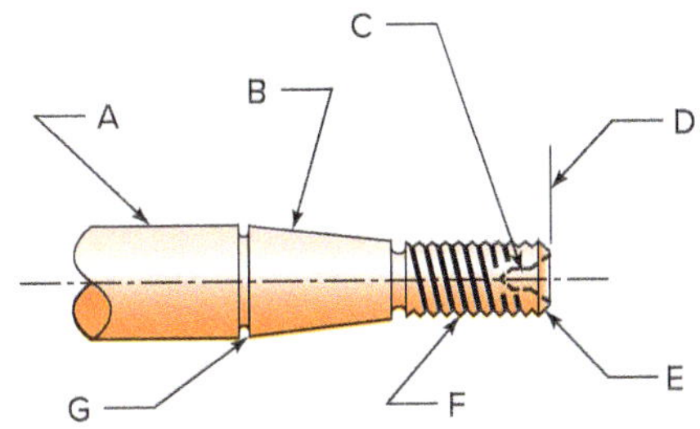

Figure 10-33 Name the lathe operations required to turn this axle.

2. Grooving the side and face of a part can be accomplished using a parting tool. Is this statement true or false? If it is false, what will make it true?
3. Without referring back to the text, list the basic operations performed on a manual lathe. Describe each briefly
4. Identify the operations required to make 25 of the fixture locator pin in Fig. 10-34. The pins will be machined from a 2.25-in.-diameter bar of

aluminum. In this single-view drawing, the circles with the cross lines indicate diameters.

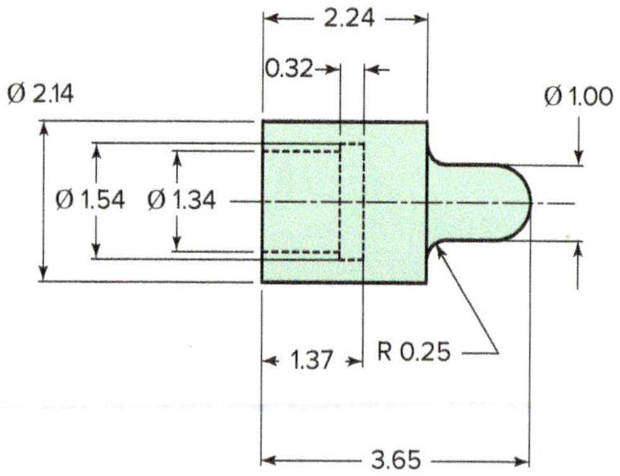

Figure 10-34 Identify the operations needed to make this pin. The bar is long, and many will be made.

5. Name the operations required to turn the camera tripod tip shown in Fig. 10-35.

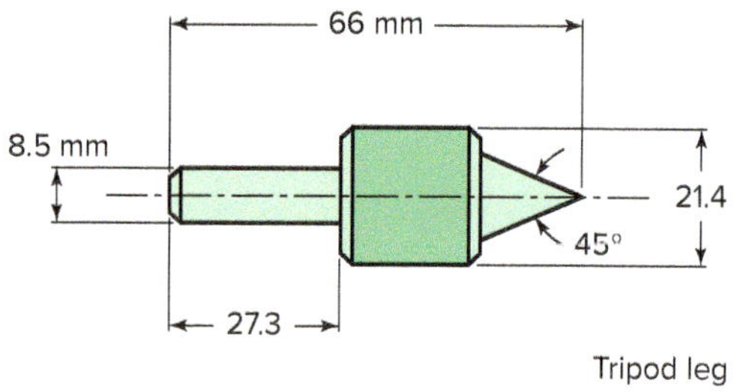

Figure 10-35 Name the operations needed to machine this tripod tip.

6. List the potential safety hazards presented in turning on a lathe not found on other machines.

Critical Thinking for Safety

7. Return to Fig. 10-31 and list the reasons for the hand and body positioning. Note, safety is a matter of common sense in most cases. This was not covered in the reading.
8. Look again at Fig. 10-32 and identify the reason for two hands on the sandpaper instead of pinching it into a loop with only one hand.

Unit 10-2 How the Lathe Works

Introduction: Now we look at the major components of a screw-cutting lathe. You might hear them called an *engine lathe* because they are the earliest machines to be powered with their own drive motors rather than by an overhead drive shaft powering the entire shop.

TERMS TOOLBOX

Apron The front, vertical part of the carriage where most of the operator controls are located.

Bed ways The precision spine of the lathe on which tool motion slides.

Carriage The assembled apron and saddle make up the carriage.

Chasing dial The indicator of when to engage the half nut lever to ensure that each new pass will fall in the same thread groove as the last.

Dovetail The guides upon which the cross slide and compound slide.

Gibs Small wedge-shaped shims that are used to compensate for wear and minimize unwanted motion in tool slide.

Half nut (lever) The device that closes the clamshell half nut upon the lead screw to start pulling the carriage for threading.

Saddle (lathe) The major component that rides upon the bed ways.

Tailstock The far end support and drilling accessory.

Tool compound (lathe) The pivoting axis of the lathe used to machine angular surfaces.

SHOPTALK

Ways Matter When installing a new lathe, the first challenge is to level the ways using extremely accurate levels or laser alignment transits. If you were buying a used lathe, the first item you would inspect is the condition of the ways. Damaged ways can be repaired, but sometimes, if the damage is extensive, it's the lathe's death sentence!

Building a Lathe One Component at a Time

10.2.1 Bed Ways and Frame

The **bed ways**, often shortened to *ways*, are the precision heart of the lathe. They are made of fine-grain cast iron or of alloy steel. Their surface is hardened and ground flat and parallel. The ways are attached to the frame, which must also be flat and parallel to the shop floor. Frames vary in their construction from single castings to welded components and assembled combinations of both. On some lesser-quality lathes, the frame and the ways are combined into a single casting, but that makes replacement of damaged ways impossible.

All sliding-axis components are supported by the ways and frame. The tailstock clamps to them, too. If a turned shaft is to be perfectly cylindrical, the lathe must have ways in top condition.

Lathe Size—Longest Supported Bar

The length of the ways determines the longest bar that can be supported for machining (Fig. 10-36).

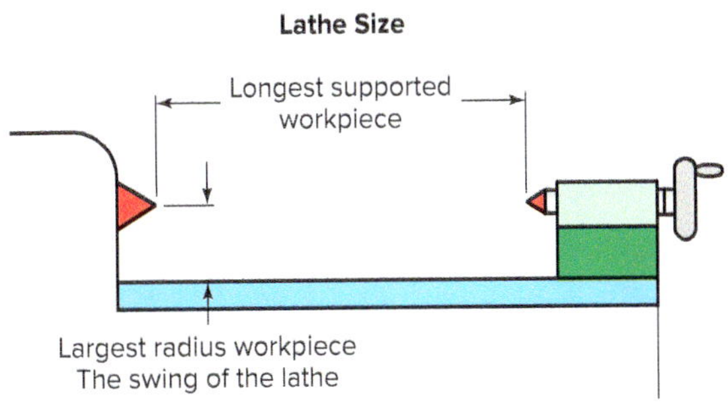

Figure 10-36 Lathe size is determined by longest supported part and largest radius to be rotated without hitting the bed ways.

There are four pro-care tasks that must be given to bed ways:

Keep the ways clean and oiled. After use, wipe the chips, coolant, and dirt off the ways with a shop cloth, then rewipe on a thin coat of machine lubricating oil. The oil provides initial lubrication for the next movement after cleaning. A reminder from Chapter 5, while compressed air is used in production to clean chips, **it is unwise to use air to clean the ways;** rather, use a brush or a rag. Chips can be forced into the dirt seals and wipers of the **saddle** (the device that slides on the ways). These seals are not designed nor are they able to hold out compressed air.

Never lay machine accessories, work material, or hand tools on the ways.

When adding accessories to the lathe, protect them from any falling object. In keeping this policy, you will find a special wood cradle, protecting the ways when changing chucks (Fig. 10-37).

Figure 10-37 Do it right—use a wood cradle when changing chucks.
Prographics at Milwaukee Area Technical College/McGraw Hill

Always check and use the lubricating system on your lathe. Lathes either feature an automatic system or manual oiling pump(s). Determine which and make sure the reservoir is above the add mark with the right type of oil for sliding friction ways (Fig. 10-38). Then pump the manual plunger system on a schedule.

Figure 10-38 If the lathe does not have automatic slide lubricators, then it will feature a manual pump like this.
Prographics at Milwaukee Area Technical College/McGraw Hill

10.2.2 The Headstock

Adding on to the lathe frame, the next major component, the headstock, contains the necessary gearing, shifters, and spindle bearings (Fig. 10-39). Highly engineered, spindle bearings support and rotate the spindle, chuck, and work.

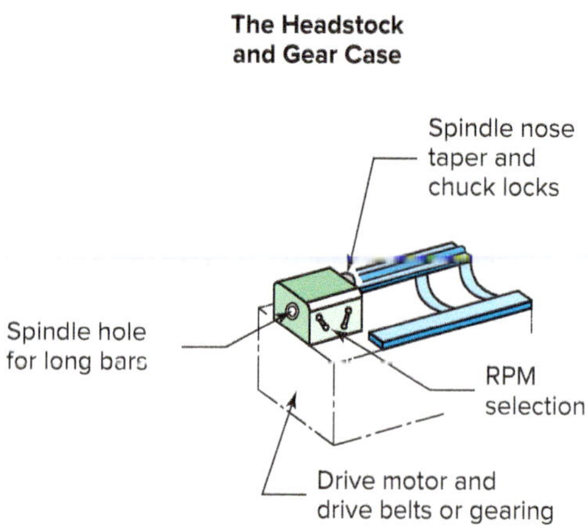

Figure 10-39 The headstock supports the work, drives the spindle, and contains the shifters for various RPMs.

Figure 10-40 Opening the outer cover, user-serviceable components are seen, but many more are beyond this within the inner headstock case.

Prographics at Milwaukee Area Technical College/McGraw Hill

Lubrication Is Critical To keep these special bearings alive and well, there will be a headstock lubricating procedure for your lathe separate from the ways (Fig. 10-40). Most likely, an oil level sight glass or dipstick will be present. There are several types of spindle oil, so be certain to read the placard or operator's manual for that machine. Keep the oil above the add mark but do not overfill, as that can sometimes blow out seals.

KEYPOINT

There are two lubrication systems on lathes: one for the ways using a viscous (thick) way lube and one for the spindle and gearbox—called spindle oil.

10.2.3 The Spindle and Spindle Taper

The external part of the spindle is the *spindle taper*, also called the *spindle nose*. It features a taper on which precision chucks or other work-holding attachments such as faceplates are firmly mounted. The nose will feature a set of locks to pull and lock the chuck or other holding device on the taper during turning (Fig. 10-41).

Figure 10-41 The spindle nose taper and chuck locks work together to securely hold chucks and other accessories on the lathe.

Prographics at Milwaukee Area Technical College/McGraw Hill

KEYPOINT

It's beyond critical that you never attach any spindle accessory (chucks, faceplates, and adapters) without first cleaning both spindle taper and the mating surface carefully (Fig. 10-42). Dented spindles destroy the accuracy of a lathe!

Figure 10-42 *Critical!* Always clean the tapers on the spindle nose and mating surface on the chuck before mounting and tightening.
Prographics at Milwaukee Area Technical College/McGraw Hill

Bar Feeding Through the Spindle

The spindle continues from the operator side, through the headstock bearings, to the far side. It's a hollow tube so that long bars can slide completely through. The diameter of that tube is another way we rate the size of lathe–the largest bar that can be slid through the spindle. Lathe spindle holes commonly range from 1.0 in. up to around 3.0 in., in training facilities and much larger in industry.

Making a part from a long length of metal is called a *bar operation.* A single part is machined, then cut off the bar. Then with the lathe stopped, the bar is unchucked, slid out, rechucked, and the next part is made. While you must stop to advance the bar on manual lathes, on CNC lathes (turning centers), bar feeding can be automatic and performed while the lathe spins. When a bar is within the spindle, several precautions must be taken.

Safety Note About Long Bars

Long bars running entirely through the headstock can present a serious safety problem. When holding and machining these bars be absolutely sure that the far end (out of the headstock into the shop) is guarded and supported. If this step is not performed

The Immediate Danger! The spinning bar could hurt other machinists as they walk by your lathe. Place a bar support and some form of barrier around the headstock, or better yet, select a bar length that does not protrude at all (Fig. 10-43)!

Figure 10-43 *Danger!* A long, unsupported bar is a death-causing accident ready to happen! See Fig. 10-44.
Prographics at Milwaukee Area Technical College/McGraw Hill

The Whip Far more dangerous, when the lathe is started rotating with a thinner unsupported bar such as in Fig. 10-44 protruding from the headstock, it will first vibrate, then wobble, then quickly bend to a 90° angle, which violently whips the bent rod in a very large circle.

Safety Key Point The force of this out-of-balance condition can actually move lathes right off their foundation. The danger to yourself and others is tremendous, from the bar tip swinging in a large circle with its tip moving at close to the speed of sound!

Figure 10-44 This is a real example. The instructor started the spindle and here's what happened!
Prographics at Milwaukee Area Technical College/McGraw Hill

Lathe Swing—Largest Radius Part

The third lathe size rating is the *swing*. The swing of a lathe is the largest part radius that can be rotated over the bed ways. It's the distance from center of the headstock spindle to the ways. Common swing radii are from 8 to 22 in. in schools, while swing capacities of several feet can be found in industry (Fig. 10-45).

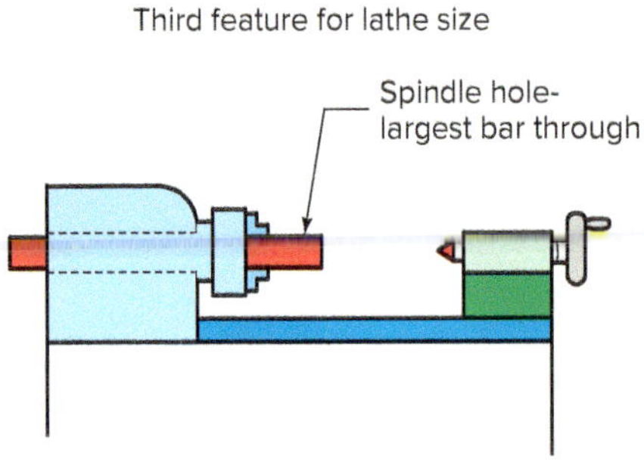

Figure 10-45 Size is also determined by the biggest diameter bar that can be passed through the spindle hole.

> **KEYPOINT**
>
> Lathe size is determined by swing radius, longest supported bar, and largest diameter bar through the spindle. Some feature a short piece of the ways removed nearest the headstock (or removable) to increase the swing—called a gap bed lathe.

10.2.4 Lathe Carriage

This lathe component is made up of two parts joined together, the horizontal saddle and the vertical apron (Fig. 10-46). The entire mechanism adds the left/right *Z* axis motion. The **carriage** slides upon the bed ways to provide the *Z* axis of the lathe. It is moved either by hand or by power for straight turning. It also supports the *X* axis, then the compound, and ultimately the cutting tool. When *Z* axis motion must be stopped, the saddle features a positive lock.

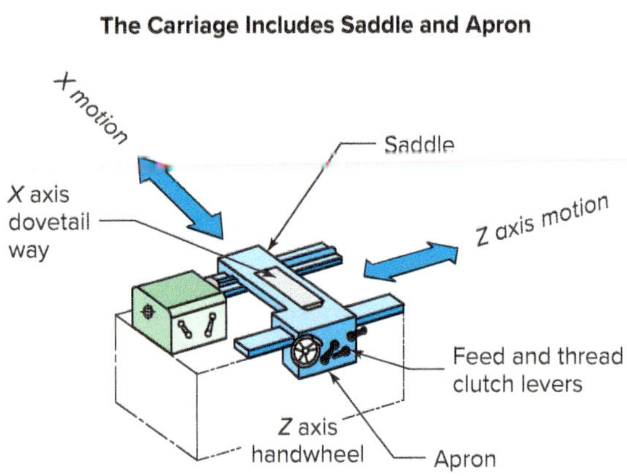

Figure 10-46 The *saddle* attached to the *apron* forms the *carriage* to provide Z axis tool movement.

The Apron

The vertical, front of the carriage is called the **apron**. The controls used during turning to start and stop the action are located here (Fig. 10-47). We'll look at them individually in a moment.

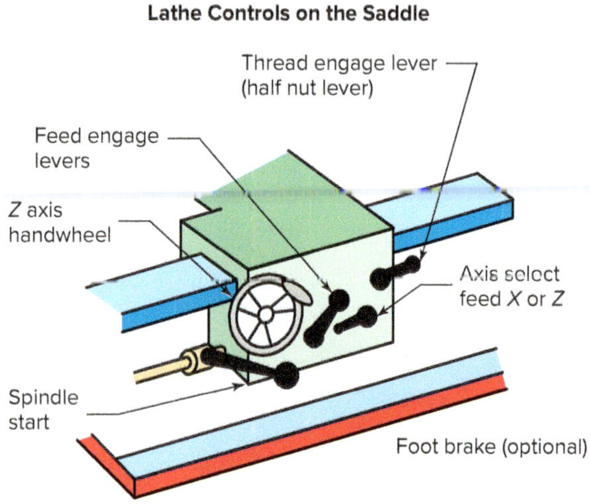

Figure 10-47 The *apron* contains most of the operator controls used during turning.

Cross Slide

Now assemble the cross slide over the carriage dovetail to create the *X* axis, perpendicular to the *Z* axis (Fig. 10-48). It can be moved by hand crank or by power feed for facing or parting and features a micrometer dial for accurate positioning. The *X* position can be locked on most lathes, to prevent tool slippage during turning, for example.

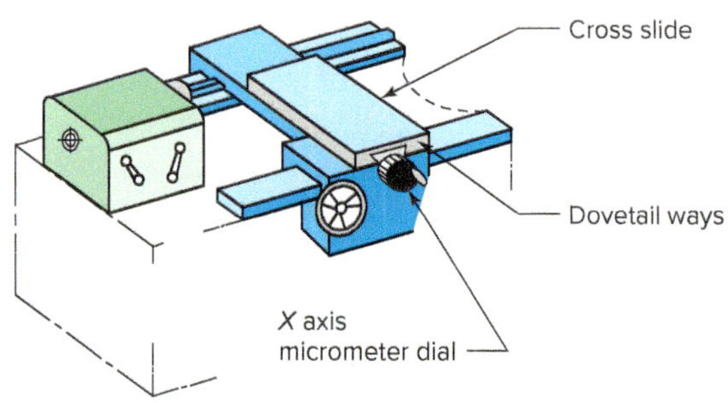

Figure 10-48 The *cross slide* moves the tool in and out along the X axis. It is held to the saddle by a dovetail way.

Diameter Graduations

The lathe's *X* axis is graduated for diameter values. For example, after taking a cut, then dialing the *X* axis forward by another 0.010 in., and taking a second cut, the measured part will be 0.010 in. smaller in diameter. Similarly, CNC lathes are usually programmed with *X* coordinate values representing diameters.

This useful feature is known as a diameter graduate lathe or sometimes as a direct reading lathe. So, when dialing the X axis in to a smaller diameter, by 0.010 in., for example, what is the actual tool movement? The tool goes forward half the amount (0.005 in. in this example), which reduces the radius 0.005 in. But that translates to a 0.010-in.-diameter change. See Fig. 10-49.

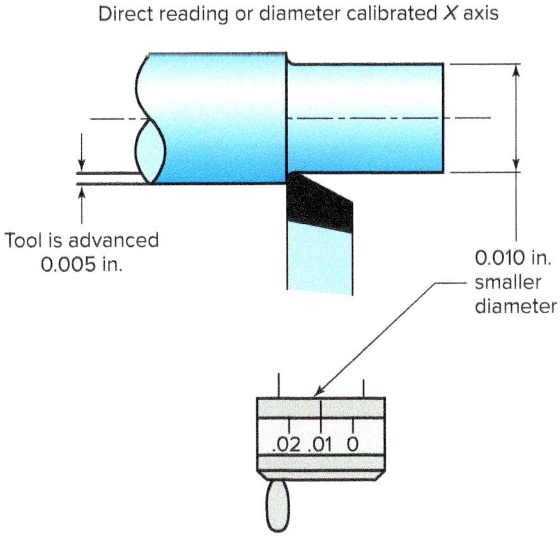

Figure 10-49 If the cross slide and tool are advanced 0.005 in., the diameter becomes 0.010 in. smaller. The amount dialed equals the amount of *diameter* change.

SHOPTALK

Radius Lathes Rarely, one might encounter a very old, indirect reading lathe that moves the tool in or out at the dialed amount. That bit of questionable engineering changes work diameter at twice the dialed amount! These confusing machines were known as *radius calibrated,* and they've mostly been scrapped or converted to diameter calibrated dials. I've only run one in my entire career, and it was a real nightmare to use!

10.2.5 Motion and Positioning Adjustments and Backlash

Lathes are made such that both backlash and unwanted axis movement can be adjusted to a minimum. What is "backlash"? See the Trade Tip coming up.

Gibs for Slide Adjustment

To minimize unwanted side or up and down movement of the X axis cross slide, which would destroy accuracy and the finish on the workpiece, small adjustable wedges are used in the **dovetail** part of the slide (Fig. 10-50). These wedges are called **gibs**. The gib screw requires periodic adjustment to compensate for wear. You'll see gibs on many other machines that feature sliding actions.

Figure 10-50 Adjustable gibs minimize unwanted movement of the cross slide and the tool compound too.
Prographics at Milwaukee Area Technical College/McGraw Hill

TRADE TIP

Understanding Backlash! Also called "mechanical play" or mechanical clearance, machinists must understand and control backlash. It affects the accuracy of all machines in the shop where a tool, machine table, or slide is positioned or moved during machining. Uncontrolled, backlash not only spoils accuracy, it can cause an accident!

For example, backlash is present in the cross slide of a lathe because it's moved by feed screw, turned by the X axis handwheel. Turning the handwheel moves the axis drive nut (and cutting tool) fastened to the cross slide. The axis is either pushed or pulled, depending on the direction the handwheel is rotated.

There must be a necessary small space between the nut and screw or the system will bind. That minimum required clearance is responsible for the backlash. Over time, the nut wears and the backlash increases. With regular lubrication, machinists minimize this wear. On most lathes it can be adjusted to a minimum. But not all backlash can be eliminated with this nut and screw system. Later in Chapter 17, we'll see another kind of axis positioning and drive called a ball screw, used on CNC machines. They eliminate backlash problems, but they are too expensive for manual machines. So, how does backlash affect your actions when running the manual lathe or mill? (See Fig. 10-51.)

The answer is that for a short time, the tool does not move even though the reversed crank and micrometer dial are indicating that it should be. That brief stall in movement, when the crank direction is reversed, is the backlash. It also means that for a short time the axis has no positive control should the cutting action grab the tool—it can jump forward or backward! That uncontrolled jump can lead to a broken cutter or worse.

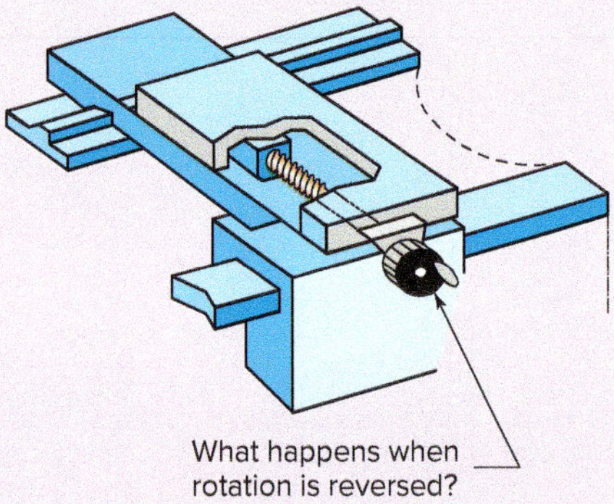

Figure 10-51 After cranking inward, what happens to tool movement when it's reversed (cranked outward)?

It is caused by the clearance between screw and nut—reversing. This might be as little as 0.002 in. or as great as 0.030 in. or more, depending on wear and maintenance. A minimum amount of backlash is necessary but is problematic on machines using standard nut/screw systems.

Compensating When referring to a micrometer dial position on a machine, you must always know in which direction the handle is turning with respect to the graduations and backlash. Your instructor or a journey-level machinist will demonstrate how to deal with backlash on your lab's machines. Remember—you must know how to compensate for backlash in all setups where a tool is positioned. Digital positioners eliminate mispositioning the cutting tool, but they do not overcome potential slippage when the tool is pulled across the backlash gap.

With your safety glasses on and with permission, take a moment to view the gibs on both the cross slide and the upper compound axis of a lathe in your lab.

Split Nuts Adjust for Backlash

There are several ways to maintain a minimum amount of backlash in an axis screw. The split nut is the more common method used on engine lathe cross slides (Fig. 10-52). By adjusting the gap between the two nut halves, a minimum amount of play is maintained, minimizing backlash in both directions.

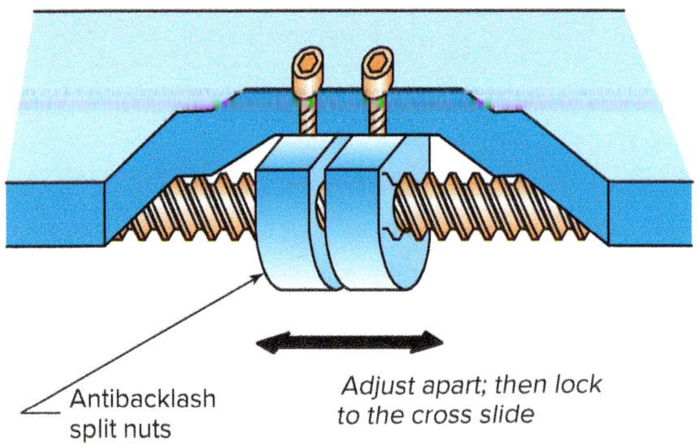

Figure 10-52 Adjusting split nuts apart maintains backlash at a minimum.

10.2.6 Compound Axis

The compound axis adds versatility to the lathe. Mounted on top of the cross slide, it has no power feed. It can be rotated to any angle with respect to the X-Z axes, then locked in position. The **tool compound** is used mostly for turning angular surfaces. The compound also features a slide lock for the times when movement is not needed, and it has a micrometer dial on the hand crank. The compound slide is fitted to its base using dovetails and gibs to control unwanted side movement (Fig. 10-53).

Tool Compound Axis

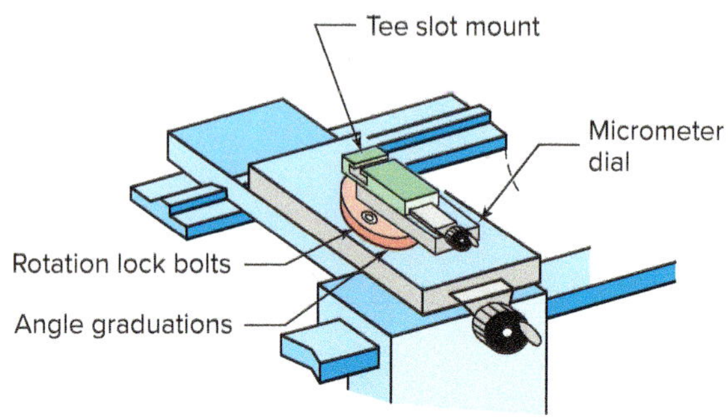

Figure 10-53 The tool compound is rotated by loosening the nuts.

The compound is subject to backlash error in the positioning screw, too. If diameter accuracy is important, then lock the compound axis at the supported position shown in Fig. 10-54, not hanging out over its dovetail.

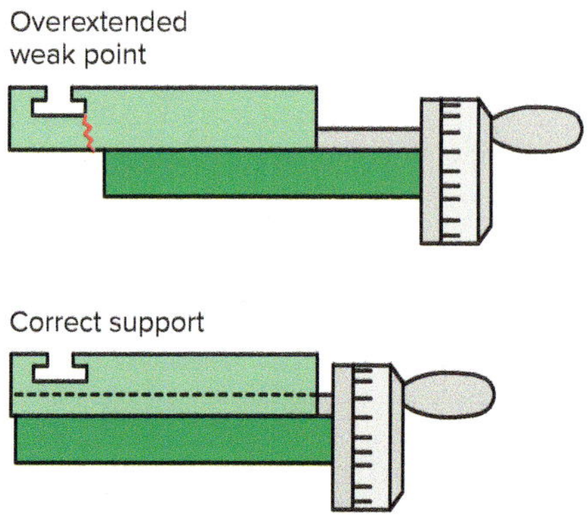

Figure 10-54 *Caution!* Overextending the compound can break it at the tee slot—the weakest point.

TRADE TIP

Lathe Compounds *Keep it supported for general work.* In Fig. 10-54, you see a common error. It is possible to crank the compound slide out quite far from its supporting dovetail. Occasionally, extending it this far is necessary for special or delicate work. However, avoid this position for everyday work. The slide is too weak when extended this far! It will deflect down, breaking tools, ruining work, and occasionally breaking off the compound itself!

KEYPOINT

When not using the compound axis, always keep it drawn in to the minimum extension and locked as shown in Fig. 10-54.

Angle Graduations on the Compound Most lathes (but not all) mark the angular graduations such that the compound is at *zero degrees* when it's parallel with the *X* axis. However, when the compound is swung to machine a steep angle, recall that the drawing usually specifies the needed angle with respect to the *Z* axis centerline (Fig. 10-55).

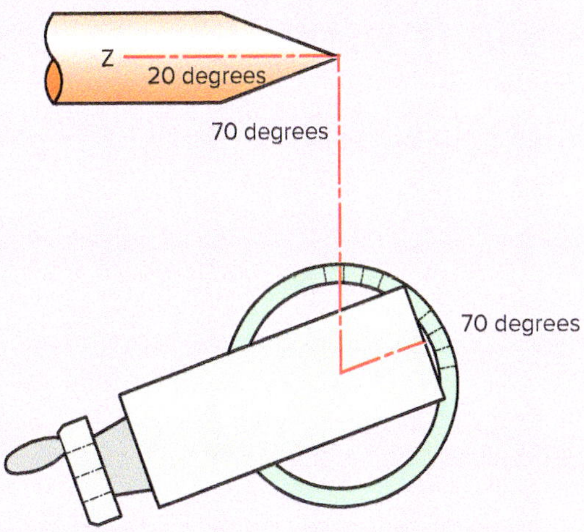

Figure 10-55 To machine the 20° half angle (reference to the *Z* axis), the compound is rotated to 70° from the *X* axis.

KEYPOINT

Set and lock the compound to the complementary angle.

10.2.7 The Gearbox

The area directly under the headstock is the gear case selectors. You need to understand the "quick change" shifting gears (Fig. 10-56). These gears provide control over four major setup selections:

Feed rate amount in inches per revolution

The feed rate direction–*X*, left/right, or *Z*, in/out

The RPM

Selection of feeding action or threading

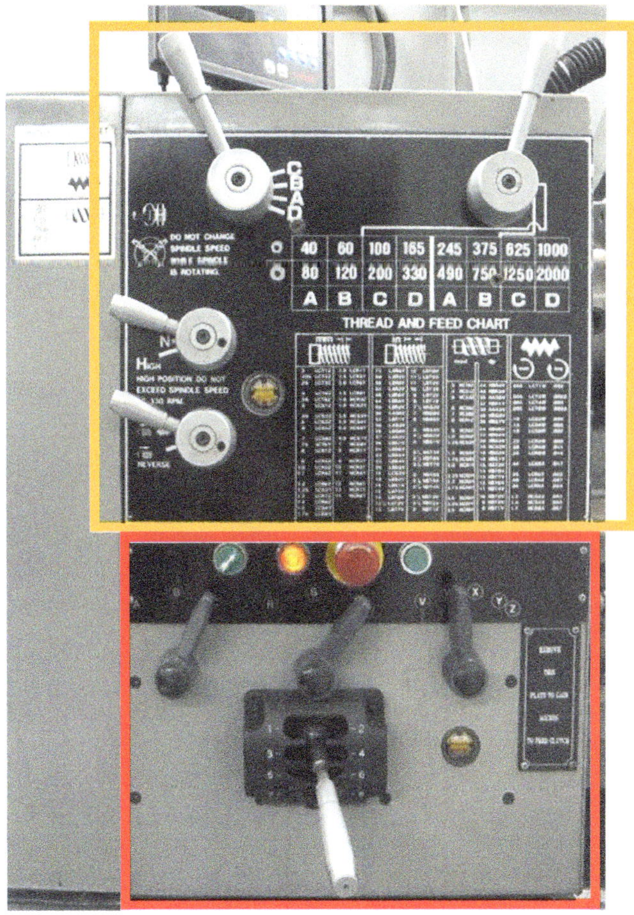

Figure 10-56 The quick change gear case contains most of the setup shifters and gear selections.
Prographics at Milwaukee Area Technical College/McGraw Hill

TRADE TIP

Lathe Gear Changing Shift like a "pro." Never attempt to shift these gears while the lathe is running. A single ruined cluster gear (the main selection gear) can cost as much as a month's wages or more for a machinist!

Second, when changing gears, you may need to rotate the spindle by hand just a bit to "drift" the gears into mesh. Some lathes (other machines, too) feature a jog button for this purpose. When you tap it briefly, the spindle motor jumps forward to gently bump gears into mesh.

The lathe will feature a built-in chart (Fig. 10-57) on the headstock to show which gear positions result in which feed rate or thread pitch.

10.2.8 Tailstock

The **tailstock** is an integral part of the lathe although it's a bolt-on component. It's not an accessory.

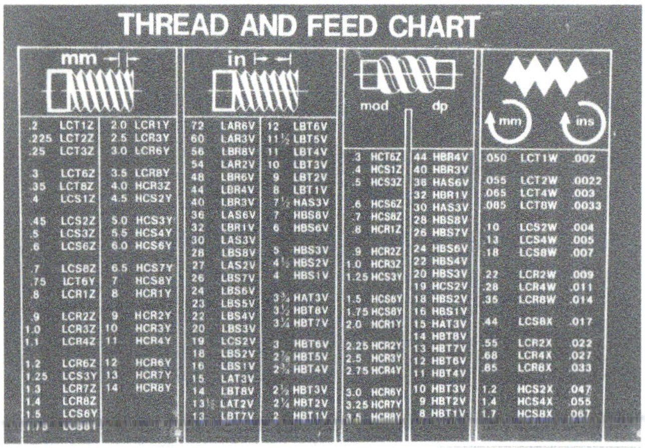

Figure 10-57 Most screw cutting lathes feature a feed rate and thread lead gear selection chart similar to this.
Prographics at Milwaukee Area Technical College/McGraw Hill

In addition to the quill in/out function (the drilling axis) the tailstock features two more movements. Both are for setup position only and must be locked into place during turning.

Z position The tailstock slides on the bed ways. This is used only for rough positioning. It is then locked for drilling or for supporting work. When sliding the tailstock, always wipe chips and dirt away from the ways to prevent scoring the bed ways. Recall in Unit 10-1, there is one operation where the tailstock is left unlocked on the ways–when tapping a threaded hole.

Quill feed Note in Fig. 10-58 that the quill is graduated for drill depth control. However, these graduations are ruler accurate only. The handwheel also has graduations of varying resolutions. A 0.005-in. resolution is common on better machines.

Lateral offset Mentioned in Unit 10-1, the tailstock quill frame can be moved sideways on its base such that it is no longer in line with the headstock, for tapering operations. It is shown set to the back in Fig. 10-58.

Tailstock Functions

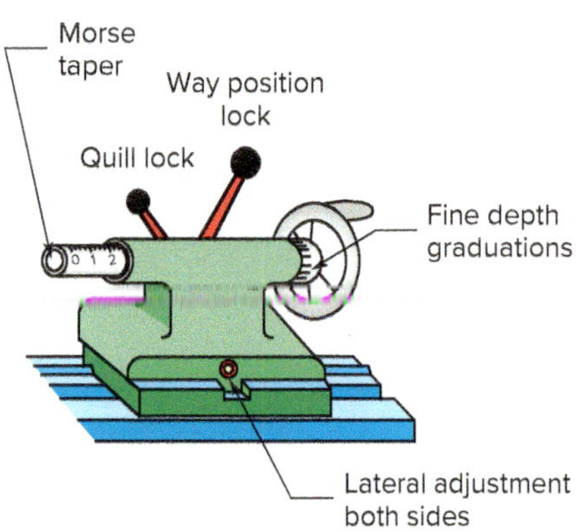

Figure 10-58 The tailstock has three possible movements: the *setup positioning* parallel to the *Z* axis, the *lateral positioning* off the centerline for taper operations, and the *quill slide positioning* parallel to the *Z* axis for drilling.

TRADE TIP

Tailstock Off-Center If you move the tailstock off-center for taper turning, then return it to alignment before leaving the lathe (or leave a note). Should the next operator not detect the misaligned condition, the surprise taper produced might result in a scrap part. When finishing a job, do not leave the machine with the tailstock off-center unless you leave a note that it is off (Fig. 10-59). When setting up the lathe, always check its centering.

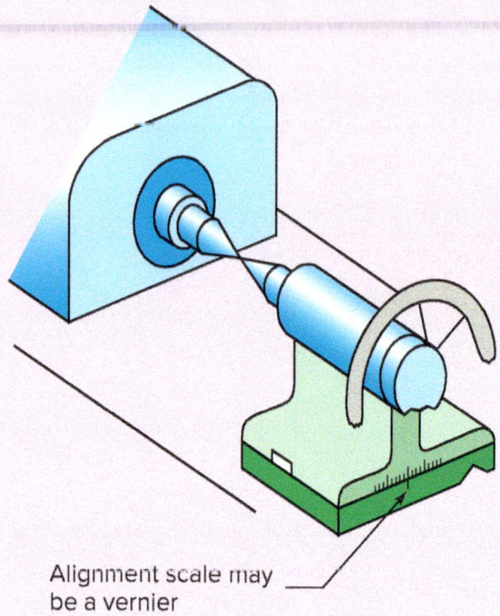

Figure 10-59 Recentering the tailstock, the off-set graduations below the crank are also a good rough guide.

In addition to the tailstock position scale, a quick eyeball check is made by sliding it all the way forward to compare its center point to one placed in the headstock.

An indicator is required for perfect alignment (Fig. 10-60). Precisely centering the tailstock requires the use of a dial indicator. The indicator holder is mounted on or in the chuck and spun around the tailstock quill.

Recentering the Tailstock

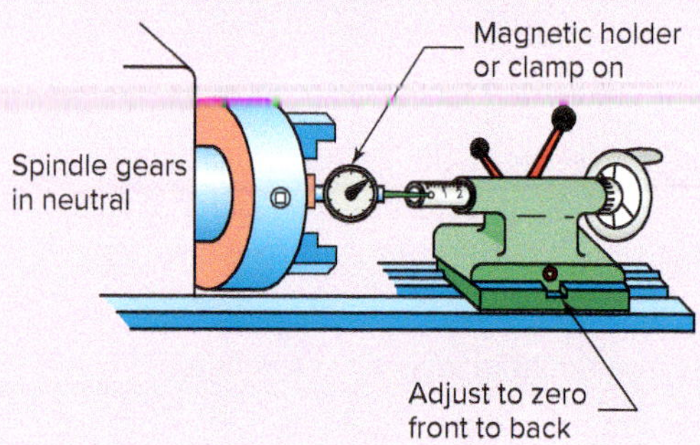

Figure 10-60 The best way to align the tailstock to the exact spindle center.

10.2.9 Work Support Accessories

Long work cannot be safely supported by the chuck and tailstock alone. To prevent the central portion from springing away from the cutter, or from riding up on it, one or more center supports are needed. There are two types, and they can be used individually or together.

Follow Rests

Because it's bolted to the cross slide or carriage, the follow rest (also sometimes called a travel rest), moves along with the tool in the Z axis direction. The follow rest is a two-point support opposing the work's tendency to spring up and away from the cutter. The follow rest can be bolted ahead of the cut to ride on the uncut metal or behind the cut to ride on the cut just made. Figure 10-61 shows the rest from the tailstock side.

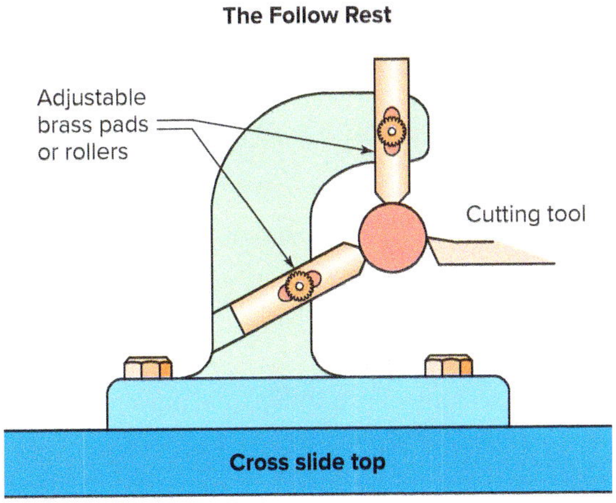

Figure 10-61 The *travel rest* is a two-point support that moves with the saddle.

Better follow rests feature roller bearings, but they also can use consumable brass rub pads that contact the work. If using the brass pad version, a heavy pressure grease lubricant is used to aid in reducing friction and to prevent smearing of the brass onto the work. Watch out for heat and too much pressure against the work with these type of supports (Fig. 10-62).

Placing the Follow Rest Ahead of the Cut

If you wish the cut to be concentric to an existing diameter, then the follow rest must be set to contact the work ahead of the cutter, on the original surface. In order to do that, the original diameter contacting the support must be smooth and round. Then, due to a changing diameter, the supports must be readjusted after each pass of the tool.

Placing the Follow Rest Behind the Cutter

This rest position is used when the original material is rough, not round, or deliberately not on-center, as in turning a cam roller. For this setup, it's necessary to create a smooth place to contact the rest pads to the work surface (Fig. 10-63).

Figure 10-62 On smooth stock with no runout, the travel rest leads the cut to machine the work concentric to the original surface.
Lake Washington Technical College, Kirkland, WA/McGraw Hill

Figure 10-63 To support and cut long stock such as this hexagonal bar, the travel rest is set up to follow the cut.
Lake Washington Technical College, Kirkland, WA/McGraw Hill

> **KEYPOINT**
>
> A follow rest can be placed ahead of the cutter or behind, depending on the work condition and effect desired. Two follow rests can be used, one on either side of the cutter.

Steady Rests

These three-point supports clamp to any position along the bed ways. *They do not move with the tool* (Fig. 10-64). One (or more) are used to support long work between ends.

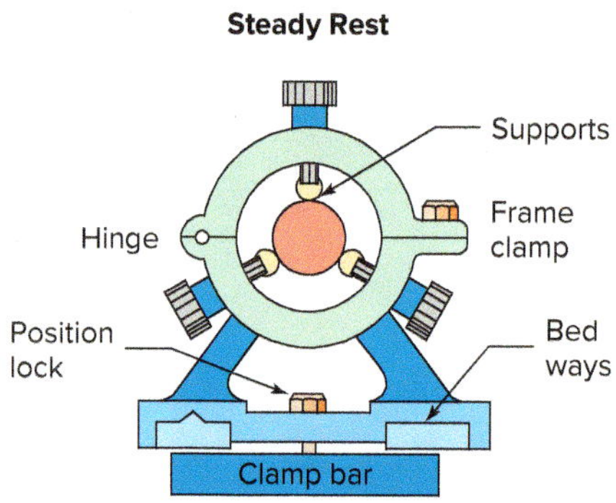

Figure 10-64 A steady rest clamps to the bed ways and does not move during turning operations.

To use a steady rest, the work must be round and smooth at the contact point. Where the raw stock isn't round or smooth, it's necessary to prepare it first by chucking it near the intended support location, then with a tailstock supporting the outboard end, cut a small section to any diameter that cleans up. It's then pulled out, and the rest supports are brought into light contact. Do not overpressurize them; they should be brought just in contact, then stop.

Controls Runout of the Work

As long as the work is round and smooth, runout of the workpiece before contacting the rollers is not an issue. As the rollers or pads are brought together, they will remove the wobble. (See the Trade Tip.)

KEYPOINT

Sometimes the only way to get the job done is to set up both steady and follow rests working together for very long or slender parts. Use caution as either work support can force the work off-center if not set up correctly.

TRADE TIP

Adjusting a Steady Rest to Remove Runout While Keeping the Work On-Center It's possible to force small-diameter work off-center when setting up a steady rest. First, with the lathe not moving, turn each adjustment pad inward until it's about 0.050 in. from the work; then snug but do not lock them. Next, start the lathe spindle at a slow RPM. Now begin turning each pad $\frac{1}{4}$ turn inward—all three pads, once around. Then again until the wobbling work just touches each roller or pad. Now, turn each toward the work $\frac{1}{8}$ of a turn—all three once around. Keep doing this with finer adjustments until the roller or pad touches the work 100 percent of the time. Stop and lock the steady rest. It's fully contacting the work and the work is on-center. Take a test cut and measure each end to be certain the work has not been deflected by pad pressure.

10.2.10 Axis Drive Systems

The final lathe components are the power feed and threading systems. Due to a difference in backlash control and precision, there are two different axis motion systems on a lathe: one for moving the cutting tool in the X and Z axes for general machining and one for driving the threading tool along the Z axis only.

KEYPOINT

When setting up a lathe, you must select either the machining *feed* or the threading *lead* system, depending on the operation to be performed.

SHOPTALK

Comparing CNC and Manual Axis Drives One of the greatest differences between CNC and manual lathes is that the CNC has no gearing between the feed axes and the spindle. On the CNC, each is driven by a computer-controlled motor of its own, while the manual lathe features one electric motor that powers both the spindle and tool axes.

Feed System

For general machining, the spindle is geared to a shaft that runs just below the ways—called the feed shaft. Take a moment to go to the lab and see it along the front frame extending out to the tailstock end. It's a round shaft with a square groove in it called a key way. In mechanical devices this kind of shaft is sometimes called a *jackshaft.*

The feed shaft rotates at the rate selected with the quick change gears, at a specific ratio with respect to the spindle. It passes through the apron, where a takeoff gear slides upon it and is driven through a key in the shaft key way—but this happens inside the apron out of sight (Fig. 10-65).

Feed Direction

Manual lathes feature a selector that determines the direction of feed. The X axis can move either toward center or away, and the Z toward the headstock or away. The direction selector is usually found on the gearbox near the feed shaft output, but on some lathes it might also be located on the apron.

Manual Drive Gearing

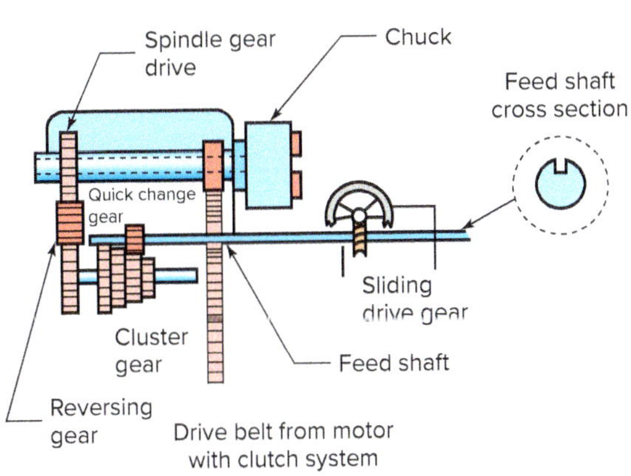

Figure 10-65 A simple sketch of the feed system on a lathe.

> **KEYPOINT**
>
> The feed direction also has a neutral position, in which no feed action can occur. Most lathes must be in that neutral position to shift to thread cutting.

Lead Engagement System for Threading

To make precision threads the cutter is pulled along the carriage using a screw action. This threading screw system would wear out quickly if used for daily machining, therefore it's reserved for threading. It's a long screw, called the *lead screw,* running next to the feed shaft, seen in Fig. 10-66.

Figure 10-66 For threading only, we use the *lead screw, half nut engagement lever,* and *chasing dial.*
Lake Washington Technical College, Kirkland, WA/McGraw Hill

> **KEYPOINT**
>
> When setting up for threading, the *feed direction must be set to neutral, and the lead screw must be engaged.*

Once the lathe's lead screw is selected for threading, a new engagement lever is used to start the tool moving, the **half nut** *engage lever* (see Fig. 10-66), located on the right side of the lathe apron. On most lathes, the half nut lever is locked out until the feed is set to neutral.

Chasing Dial

During threading, several passes of the cutting tool are usually necessary. The **chasing dial** tells the machinist when it's time to start the threading tool moving, such that the cutting tool enters the previously cut thread with each new pass. The chasing dial is a gear-driven indicator, as shown in Fig. 10-67. It indicates when the cutting tool will start in the same thread groove as the previous pass. The chasing dial is located near the half nut lever. We'll investigate its use in Unit 10-7.

Thread-Chasing Dial Mechanism

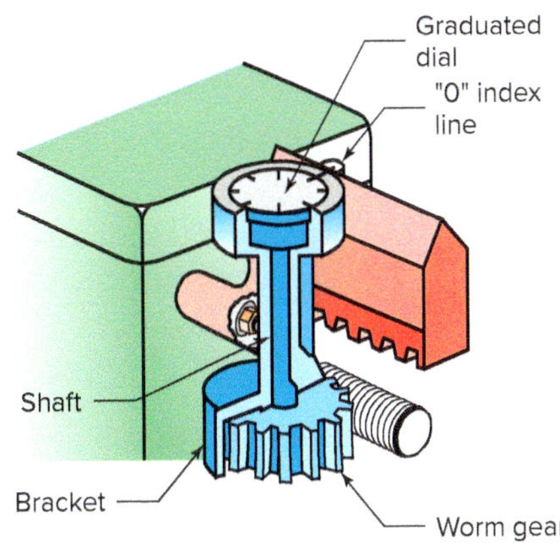

Figure 10-67 The chasing dial indicates when it's time to start threading.

UNIT 10-2 REVIEW

Replay the Key Points

- For power drive of an axis, you must select either feeds or threads—two different systems.
- The direction of feeding or threading is selected on the headstock gearbox. This causes either a right-hand spiral or left-handed threads, or when feeding in/out or right/left selection.
- When threading, a separate drive shaft and engagement lever are used. The engagement lever is not a clutch. Called the half nut lever, it is either engaged or not, but it will not slip as a clutch would.
- To engage the half nut lever at the right time, the lathe features a chasing dial that indicates when to drop the lever in gear.

Respond

1. On a sheet of paper, identify the indicated items in Fig. 10-68. Then with permission from your instructor, with the correct safety eyewear, identify the same items on a lathe in your lab. Quiz a fellow student as to each item's purpose. Do not interfere with machinists operating the machine.

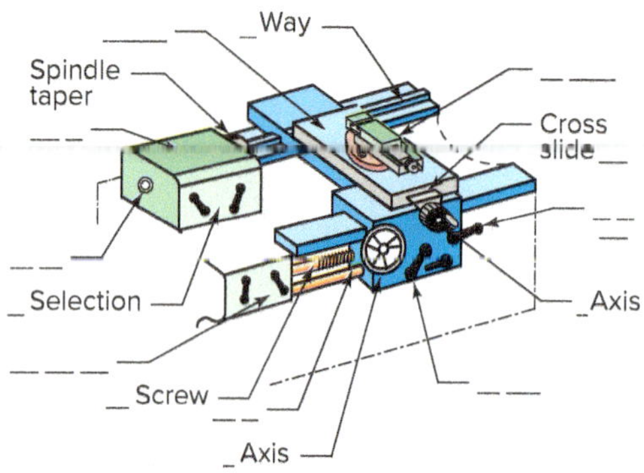

Figure 10-68 Identify the following lathe parts.

2. With a partner, identify and describe the function of these parts on a manually operated lathe:
   Bed ways
   Saddle–tool carriage
   Apron
   Tool compound
   Quick change gearing
   Z axis handwheel
   X motion
   Tailstock lock
   Tailstock quill
   Axis direction selector
   Dovetail and gibs (cross slide or compound)
   Chasing dial for threading
3. The compound is an adjustable axis that has a micrometer dial. Is this statement true or false? If it is false, what will make it true?
4. What danger is there in extending the tool compound out to the end of the stroke when machining? When is overextension of the compound necessary and justified?
5. Of the two different work rests available, which has three-point suspension?
6. Describe a "follow rest" and its use in machining. Be complete.
7. Name the three possible movements of a tailstock and a brief description of each.
8. What is the name for the coasting process that helps to engage gears while shifting a lathe?
9. Name the moving objects used during turning operations, both rotating and sliding, that could be a safety danger (i.e., things that could catch, cut, or pinch the operator but not control levers, handwheels, or dials).
10. Head stock spindles are hollow on lathes to save weight and reduce the horsepower required to drive them. Is this statement true or false? If it is false, what will make it true?
11. Name two reasons for wiping oil on the bed ways after cleaning them.
12. Match the axis with the sliding device.

A. *X* axis	Compound
B. *Z* axis	Saddle
	Cross slide

13. Screw cutting lathes feature two complete systems for moving the cutting tool. These are the ____________ and ____________ systems.
14. The splitting dial is used to ensure the tool runs down the same thread groove each time. Is this statement true or false? If it is false, what will make it true?
15. Gibs are small wedges used to minimize unwanted movement in the saddle, cross slide, and compound slides of a lathe. Is this statement true or false? If it is false, what will make it true?
16. The axis direction selector is found on the lathe ____________.
17. What is the name of the operation shown in Fig. 10-69?

Figure 10-69 Identify this lathe operation used on both manual and CNC machines.
Lake Washington Technical College, Kirkland, WA/McGraw Hill

Unit 10-3 Work-Holding Methods

Introduction: There are six different methods of holding material for turning. The choice depends on several factors. Your objective at this stage is to know the difference in their abilities. Which is safer for the job and holds best?

Work holding is a big issue with turning. Keep in mind that the work-holding device you select will keep the spinning mass under control. If the job is for a CNC lathe, your choice is even more critical as the speeds and forces multiply. Here in Unit 10-3, we explore the options for work holding on manual lathes. Most, but not all, apply to CNC as well.

TERMS TOOLBOX

Cam locks The most common method of holding chucks on a lathe, they use rotating cams in the lathe spindle that grip pins protruding from the chuck.

Collets Sometimes called spring collets due to the way they are made, these precision hardened and ground griping tools hold the work by being pulled into a tapered hole.

Dead centers A support center that has no bearings for rotation. They rotate with the work when used in the headstock but rub when used in the tailstock.

Fixtures Custom-made tools for holding special shaped work not possible to reliably hold by any other method. Sometimes called jigs.

Grip The comparative amount of bite a holding method is able to impart to the work.

Independent chucks Chucks that feature jaws able to move individually.

Live centers Support centers used only in the tailstock; they have bearings to eliminate friction.

Scroll plate The helical face thread that moves all three jaws simultaneously in a three-jaw chuck.

Soft jaws A custom-machined three-jaw insert used where concentricity or part shape is an issue.

10.3.1 Decision Factors

In planning your work-holding setup, there are five aspects to consider:

1. **Work/Operation Requirements**

   Above all else, you must know the operations that are to be performed. That includes the amount of raw stock to be removed and how. Find the facts on the drawing and work order, and compare the raw stock to be turned to the finished shape and size. For example, where the choice might be different: (A) It's a rough HRS (hot rolled steel) bar that will be completely machined away, or (B) it's a brass forging and must not be marred by the holding method since some of the original surface remains on the finished part. Stock has or doesn't have excess for gripping inside the chuck. Is it a long bar job–through the headstock? What are the job tolerances?

2. **Grip**

   Directly related to number 1, how much holding pressure is needed without marring or crushing the work, or how strong or delicate is the object to be held?

3. **Turnaround Time**

   Is this a one-off part or are you setting up to make several parts? How quickly should the setup be rechucked? This becomes a greater issue with CNC lathe work.

4. **Runout (discussed next)**

   This is an accuracy issue. What are the runout tolerances for the job? What is the typical surface wobble method?

5. **Special Characteristics**

   Each method offers special advantages or a definite type of work for which it is best suited.

We'll look at each of these categories as we examine the various holding methods.

10.3.2 Runout

From Chapter 4, runout is the geometric characteristic of surface wobble on rotating objects. It is always present in lathe work but unwanted. Each work-holding method exhibits various amounts of runout in the way they work. Your objective is to match the holding method to the tolerance.

Measuring Runout

In Fig. 10-70, an indicator is touched to the work surface while the object is slowly rotated 360° (usually by hand–placing the spindle in neutral). The indicator is adjusted to zero at the lowest or highest point in the rotation. Next, rotate the part and find the greatest difference in values, which is the FIM or full indicator movement (also called TIR or total indicator reading). Both ends of a shaft should be tested and possibly the middle also. It is possible for one end to show very little TIR while the other end might show a large runout.

KEYPOINT

Workpiece runout can be caused by irregularities of the work surface, but the point is, it can also be introduced into the setup by the holding method itself. It's usually a combination of both.

Testing Runout

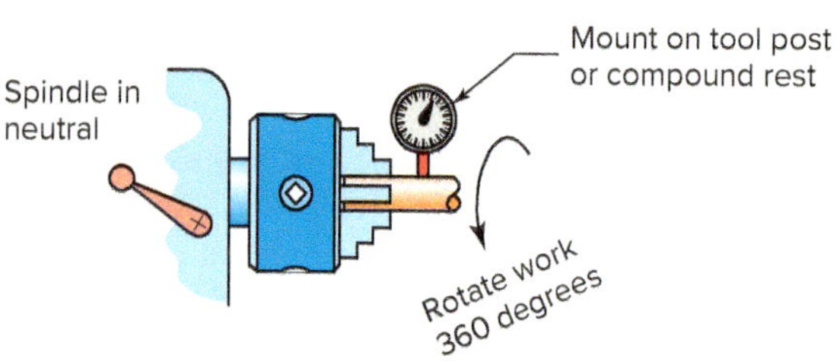

Figure 10-70 The total movement detected is the runout—a geometric term for surface "wobble" on a rotating object.

With some turning jobs, you need not give much consideration to initial runout–for example, when holding a rough metal bar to turn its entire length to finished diameters. Then when complete, it is parted off the remaining bar gripped in the chuck. Any wobble in the initial work surface is removed with the machining–the accuracy of the chuck isn't an issue.

However, now suppose you wish to machine a new feature on a previously machined axle. Functionally, the new diameter must be concentric to the original shaft's centerline and surface. Here runout control of the holding method is of the highest concern.

10.3.3 The Work-Holding Methods

Here are the seven choices for your setup.

Three-Jaw Universal Chucks

These chucks are simple to use and have an excellent grip on work where heavy cuts are to be made. Their advantage is that the jaws close simultaneously, similar to a drill chuck. As such, they feature moderately fast turnaround times.

As shown in Fig. 10-71, turning the chuck wrench gear rotates the **scroll plate** inside the chuck body, a flat disk with a helical groove. Moving jaws simultaneously gives rise to the term *universal jaws,* as opposed to chucks whose jaws open or close independently. In Fig. 10-71, one jaw has been removed to show the scroll plate beneath.

KEYPOINT

To replace the jaw in Fig. 10-71, all three jaws must are removed, then inserted onto the scroll in their numbered sequence. Otherwise, the chuck will not close on-center. Have your instructor demonstrate this task.

The jaws can grip on the outside or the inside of objects (Fig. 10-72). However, when objects become relatively large, the jaws of many chucks (both three- and four-jaw chucks coming up next) can be reversed such that they nest-grip, as shown in Fig. 10-73.

Three-Jaw Universal Chuck

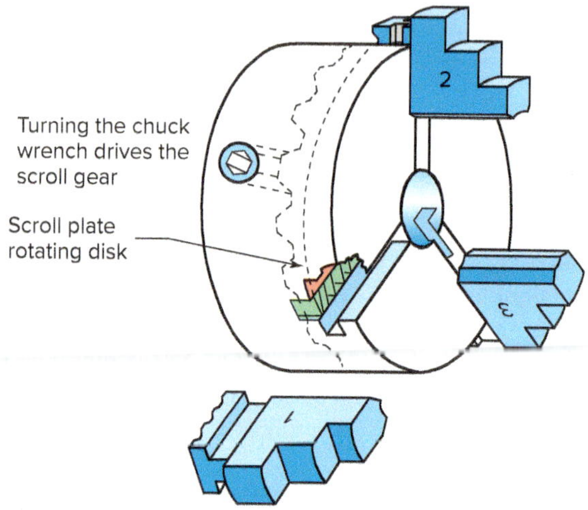

Figure 10-71 A universal *three-jaw chuck.*

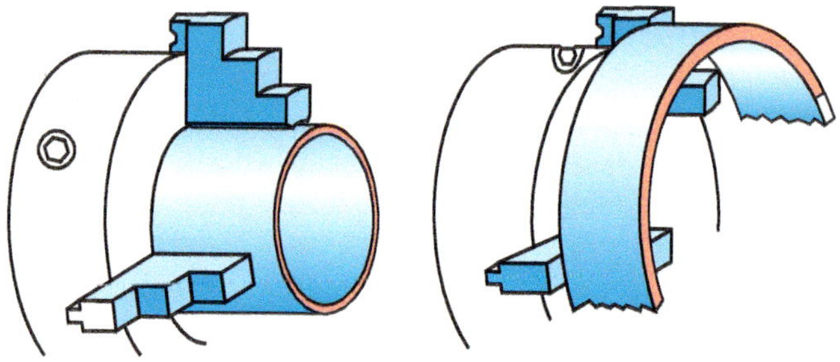

Figure 10-72 Three-jaw chucks feature *inside* and *outside* grip jaws.

Reversed Jaws for Extra Large Work

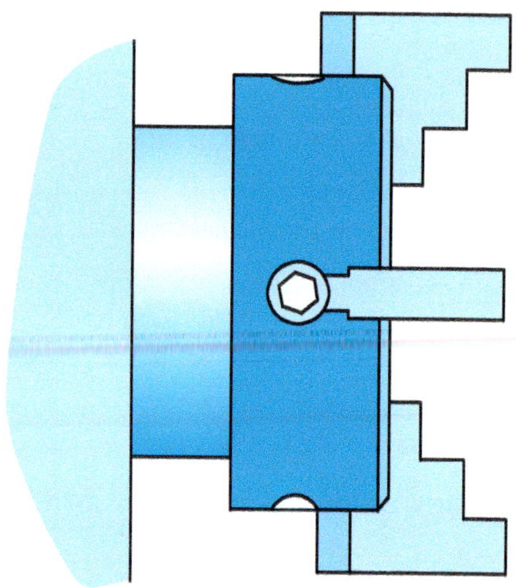

Figure 10-73 Here the chuck jaws have been reversed to nest grip a large diameter object.

Reversing Chuck Jaws

There are two ways chucks might be manufactured for reversing jaws:

Two-piece jaws (more expensive but more useful)

Total reversing jaws (remove from chuck carrier slot, turn around, and reinsert)

Two-Piece Jaws

To reverse these industrial quality jaws, two bolts are removed (Fig. 10-74). The jaw is tapped lightly with a soft hammer to remove it from its carrier, which remains in the chuck. Next, both the jaw and the carrier are cleaned of dirt and chips. Then the jaw is rebolted in the reverse position.

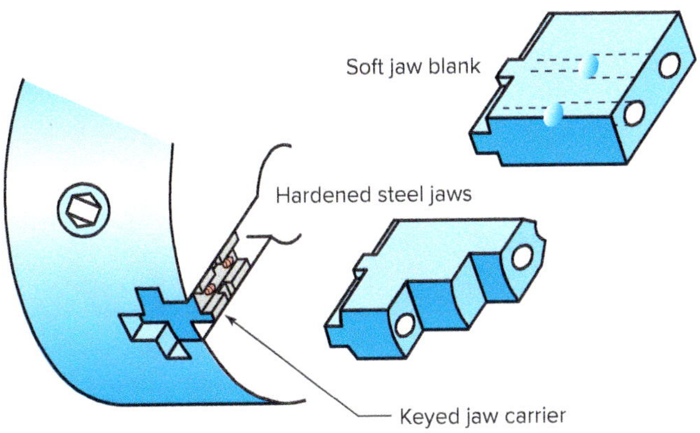

Figure 10-74 Two-piece jaws are reversed by removing two bolts.

Custom soft aluminum jaws are an option that bolts in place of the hardened steel jaw. We will look at the uses of soft jaws later.

KEYPOINT

When reversing chuck jaws, clean the mating key way and faces beyond all doubt that chips and dirt remain! Otherwise, the dirt left behind the chuck jaw adds extra runout to the setup.

Solid Jaw—Total Reversal

Here the jaws cannot be taken apart from the carrier. To reverse these jaws, back them all the way out of the chuck. Reverse, and then reinsert them in sequence while turning the chuck wrench.

TRADE TIP

Jaw Reversing Sequence After cleaning, make sure to replace the jaws in the correct order in the chuck. Notice that both the jaws and the slots in the chuck are numbered 1, 2, and 3. Insert and recatch jaw 1 on the scroll plate in slot 1. Then rotate the chuck and start 2 in 2, then 3 in 3 last.

Why Choose a Three-Jaw Chuck?

A high-quality, new three-jaw chuck might produce as little as 0.001 in. runout when gripping a perfectly round object. However, as they are used for heavy machining, runout degrades due to overtightening, crashes, and normal wear of the scroll plate and spindle mount taper.

Runout Not the best choice to control runout. Expect from 0.003 to 0.005 in., depending on the history of the chuck. Of all five methods, three-jaw chucks are the least capable of producing zero runout.

Grip Excellent on solid round work. They grip a large range of part sizes both outside and inside. Three-jaw chucks with steel jaws can mar the work.

Turnaround Medium fast.

Special Characteristics Besides its universal nature, the three-jaw's main advantage is the ability to create custom-shaped soft jaws to fit odd-shaped work not easily held in other chucks.

TRADE TIP

Flying Wrench Rule When tightening or removing work from a chuck, *never leave the wrench in the chuck!!*

On the very day I wrote this unit, a student in the next lab forgot the wrench in the chuck, then started the lathe. It hurled past the student and three other terrified students on its way to a plate glass window into the hall, where a group of amazed high school visitors stood watching. The student machinist was speechless for several hours. A bunch of visitors decided right then they would not enter the trade! Fortunately there were no injuries—this time.

To avoid this problem *never* leave a chuck wrench in the chuck for any reason. **DO NOT** *take your hand off the wrench when it's in the chuck. Remove that wrench—always!!!* (See Fig. 10-75.)

Figure 10-75 Leave it, and you'll launch it!

Four-Jaw Chucks

Four-jaw chucks feature individually operated jaws. Each jaw is cranked in or out independent of the others, hence their second name, **independent jaw chucks**. As shown in Fig. 10-76, each jaw need not remain concentric to the chuck; therefore odd-shaped work can be gripped. To reverse these jaws, each is backed out, then reversed and reinserted one at a time in any order since each is independent of the others.

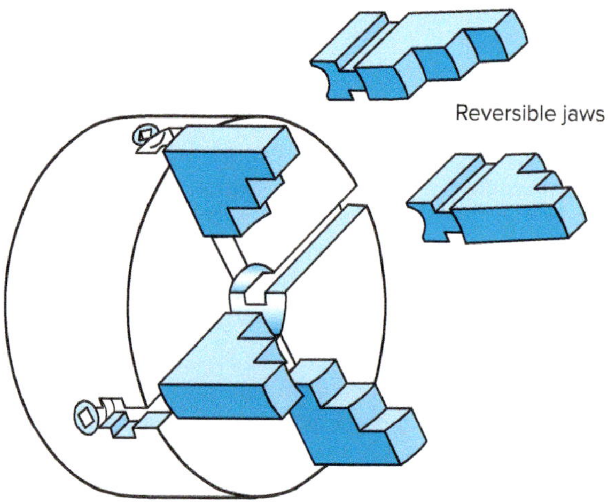

Figure 10-76 Each jaw is moved independently; thus, 100 percent of all runout may be adjusted out of the setup.

Why Choose a Four-Jaw Chuck?

Runout There is no inherent runout in the chuck–it depends on how skillfully the machinist tightens them on the work. Using an indicator, the machinist can adjust a perfectly round object such that all the runout is gone–perfectly centered. This is called "indicating the work" or "truing the work."

KEYPOINT

Because of its adjustable nature, the four-jaw chuck is the only method that can eliminate 100 percent of the runout when gripping a perfectly round object.

Grip A very strong grip can be had on round and square objects.

Turnaround Although there are a couple of trade tips that help speed up changing from one part to another, the four-jaw chuck is one of the slowest to turn around because each jaw must be tightened, usually while watching an indicator to center the work. Since they are such a slow part exchange method, they're avoided in production and CNC setups.

Special Characteristics Four-jaw chucks are used where runout is of the highest concern for the job. They are also good at holding square and odd-shaped work, something a three-jaw chuck cannot do.

TRADE TIP

Indicating Work in a Four-Jaw Chuck Here are several trade secrets that save time and frustration when indicating a job in a four-jaw. The indication procedure has three phases: (1) rough centering, (2) indicator centering, and (3) tightening. By following this sequence the entire procedure will take only minutes.

Rough Quick Centering Use the concentric lines scribed on the chuck face to coordinate centering the work. Using this trade trick, the object is placed close to center initially. Crank an edge of each jaw to the same line (Fig. 10-77), a little bigger than the workpiece. Then turn the wrench for each one-half turn inward for each jaw. Then one-quarter turn, then another quarter until the jaws grip the work loosely. The jaws all contact the work, approximately on-center. This should yield concentricity within about 0.050 in. or so, which is usually good enough to begin using the dial indicator, next.

Indicator Centering With the spindle in neutral, place the indicator probe on the workpiece and rotate the object until the lowest point is under the probe. (The lowest point is farthest away from the indicator.) Zero the dial, then very slightly loosen the jaw closest to the low point. Rotate the object 180° and tighten the opposite jaw an equal amount. Keep doing this until the runout is close to being within tolerance for the setup. But you are not done—at this point the grip is not tight enough for machining.

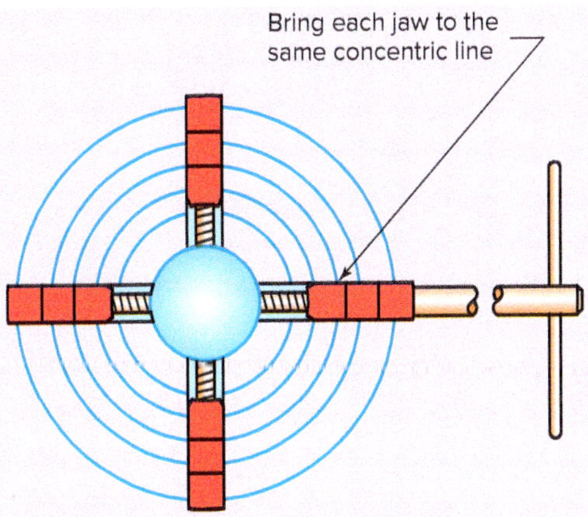

Figure 10-77 Trade tip—Use the centering guidelines on the chuck face.

Tightening the Chuck On-Center At a point learned from experience, you will switch from loosening the low jaw to increasing the grip. Only tighten the high side (Fig. 10-78). This adds holding pressure while bringing the object right on-center.

Figure 10-78 Centering the workpiece occurs in stages:

1. Rough-in (eyeball to lines on chuck)
2. Tightening and loosening opposite jaw
3. Final tightening of all jaws to gain grip

Lake Washington Technical College, Kirkland, WA/McGraw Hill

Make a Final Check Be sure you tighten all four jaws. Remember, two opposing jaws may be perfectly centered but not fully tightened. Test all jaws. The right amount of jaw pressure is a complex issue learned from experience. Factors are

1. How much metal removal is planned?
2. How solid is the work being chucked? Could it be crushed as with a thin-walled tube?
3. Are you chucking on finished surfaces or on excess material? Tight jaws will cause marks in the metal.
4. How long is the work, and is the end supported at the tailstock? Do you expect vibration? This can loosen a chuck.

Adding Leverage Precaution Lever tightening by placing an extension on the chuck wrench is not a recommended practice. However, in high-removal situations where the work absolutely must not slip in the chuck, we all do it sometimes. This is a risky procedure as it can mar the work and distort the chuck itself and even bend the wrench. Your instructor or supervisor may hold a different opinion on using a chuck wrench extension bar. Ask before using one.

Indicating Work That Is Not Round Occasionally you'll need to center work that is not round (Fig. 10-79). After roughly centering the work, set the indicator to zero, then rotate the spindle 360° and observe. If the work is not round, it will show multiple high and low points. Compare this to the runout of a round object off-center, which would produce a single high and low point. During indication of this kind of object, balance the highs and the lows on the average.

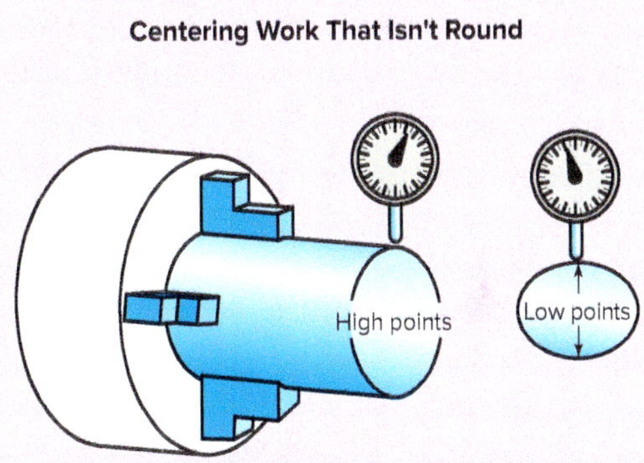

Figure 10-79 Centering an object that is not round is a matter of averaging opposing dial readings.

Collet Chucks

These precision chucks grip by pulling the tapered collet into a tapered hole, squeezing closed on the work. To allow it to close, then spring open again the collet is split along its axis in three or four locations around the rim, thus their other name, *spring collets* (Fig. 10-80). **Collets** are a precision device made of hardened and ground tool steel.

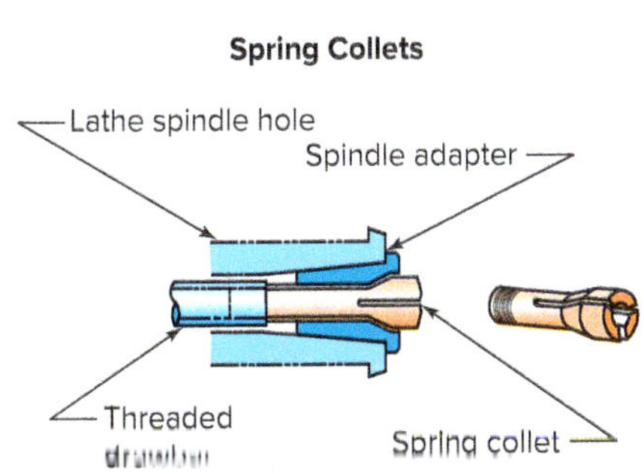

Figure 10-80 A collet grips the work when it's sprung closed by being pulled into a precision tapered hole.

Why Choose a Collet for Your Setup?

Collets are the best gripper of small-diameter rods, where light cuts are expected. On manual lathes, they are a poor choice if heavy turning is to occur–they are best suited to light machining. However, they are the most common gripping method on CNC lathes where hydraulic or pneumatic force is used to pull them closed; thus, they grip better than the manually closed version.

Grip Medium to light **grip** on manual lathes depending on the pulling method for closing the collet. However, on CNC lathes, they grip about as good as any method we'll discuss.

Runout Quality collets in good condition feature the least built-in runout of all the methods presented. They are guaranteed to have less than 0.001 in. of runout when new, with some higher, quality versions offering as little as maximum 0.0003-in. factory runout.

Turnaround Very fast. In fact, on most production CNC lathes, the spindle needn't be stopped to open the collet chuck if the lathe is equipped with a bar feeder. While still rotating, the collet opens enough that the bar can slide forward a preset distance; then the collet closes again. But even with manual collets, they are the fastest turnaround holding method.

Special Characteristics They can be purchased in regular shapes such as squares and hexagons or as blanks, whereby the shop machines the shape needed. Each collet has a small range of sizes that may be gripped without distortion to the collet or marring the work. The usual range is $\frac{1}{32}$ in. or 1 mm for each collet. This means that the shop must have 24 collets to grip a total range of sizes from $\frac{1}{8}$ in. up to 1.0 in. diameter.

KEYPOINT

Never squeeze the collet beyond its intended range—the fingers of the collet can break! The closer the part size is to the collet, the better the grip will be. It's best to use rubber collets when part size is more than 0.015 in. from collet size.

Soft and Custom Shapes

Collets that haven't been hardened can be purchased, then custom machined to a special work shape. Hardened and ground collets can also be purchased with several different grip shapes other than round (Fig. 10-81)–for example, to make bolts from preshaped hex bar stock.

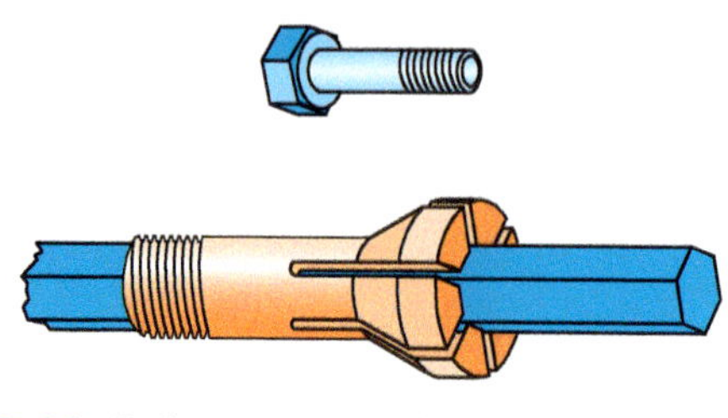

Figure 10-81 Collets are supplied in regular shapes, as shown, and in blank form ready for custom machining.

Setting Up Collets

There are two kinds of collet chuck accessories for the manual lathe: front closing and rear closing (Fig. 10-82). The front-closing version is a self-contained bolt-on accessory, while the rear-closing collet requires extra components mounted on the lathe to pull the collet closed. Front-closing sets are more expensive, but they are faster to use since one needn't walk around the lathe to operate them.

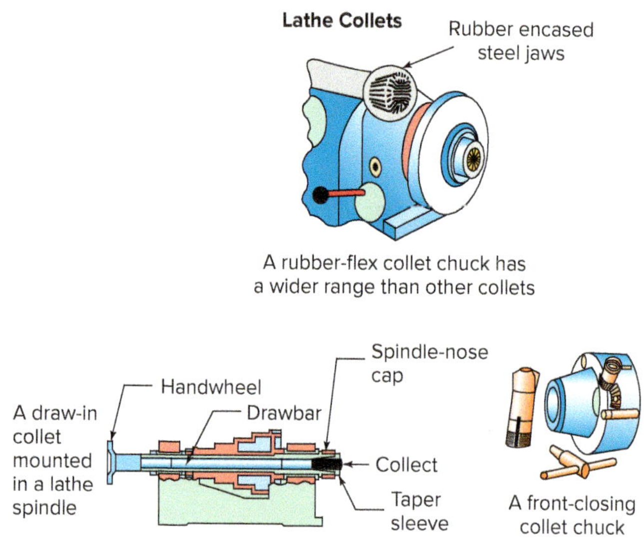

Figure 10-82 Above: front-closing rubber collets; right side: front-closing steel; and left side: rear-closing collets pulled in from the back.

There are two kinds of front-closing collets–spring and rubber isolated.

KEYPOINT

Rubber collets are different from spring collets. Their jaws float in flexible rubber such that each collet has a larger grip range than spring collets and can better accommodate small part irregularities (Fig. 10-82).

The rear-closing setup requires four components (Fig. 10-82):

1. **Spindle Collet Adapter**

   A solid metal sleeve that's lightly pressed into the spindle similar to a drill sleeve. The adapter converts the spindle taper to the correct internal angle matching the collet.

2. **Drawbar**

   A hollow round shaft that fits through the spindle. It pulls the collet into the taper adapter. The drawbar screws into the collet.

3. **The Collet**

4. **Collet Closer**

   This will be either a lever attached to the back of the headstock or a handwheel device. Both pull the drawbar into the headstock tightening the attached collet, They then remain in the pull position until they are reversed to release the work.

Threaded and Expanding Mandrels

Sometimes the work cannot be gripped in a chuck or collet. But, if it has a hole drilled through it, it can be held on one of two kinds of *mandrels*. A mandrel is a round bar that's inserted into a hole in the workpiece. At times, this may be the only solution. Here's an example–suppose you must reduce the outside diameter of 20, $\frac{1}{2}$-in.-diameter washers. How would you hold them in any of the methods above? The answer: You can't.

Threaded Mandrels

The solution in this case is the threaded mandrel (Fig. 10-83). It's not unlike a $\frac{1}{2}$-in.-precision shoulder bolt. The washers are placed on the mandrel, then the nut is tightened upon the stack. The grip force is the friction between washers, so the nut must be tightened very well.

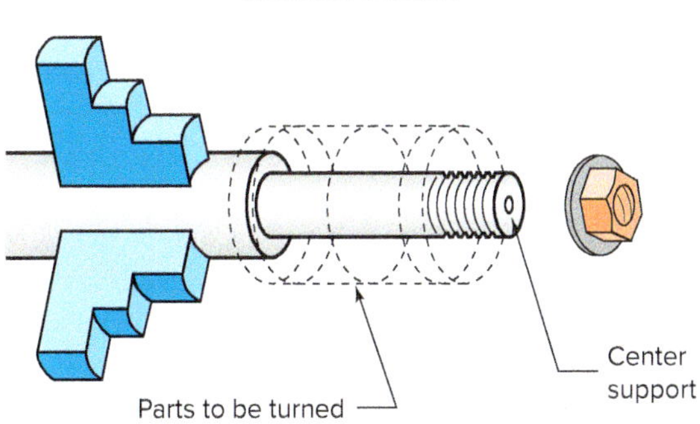

Figure 10-83 Threaded mandrels are used to hold small parts with holes drilled through.

These kinds of mandrels are often machined on the spot, as needed.

Tailstock Support

When making or using a mandrel to machine work, the 5:1 ratio of length to diameter dictates whether to use a tailstock support. However, it's not a bad idea to use a live center in the nut end, unless it interferes with very small work.

> **TRADE TIP**
>
> When one small item requires mandrel turning, an everyday bolt can be used, if concentricity isn't a big issue. If its head is gripped firmly in the three-jaw chuck or a collet, then nuts placed on both sides of the object will allow light turning. For support and safety, you may need to center drill the bolt, then insert a live center.

Why Choose a Threaded Mandrel?

Choose one of these when you have small work with a central hole, when no other method will do the job:

Grip Restricted to light duty, small work.

Runout Depends on the situation. Can be near zero if the mandrel is a precision shaft or when it's machined on the spot. Overall runout is a function of the mandrel's condition, how it's held in the lathe, and how well it was made.

Turnaround Time Medium slow, but considering it's probably the only way to hold the work, this isn't often an issue.

Special Characteristics Suitable for work that can be held with side pressure from the nut, as in the stack of washers. Could be used for a single washer in the example using two nuts.

Precision Expanding Mandrels

These are ground and hardened precision holding devices. They work like an internal collet. The split expanding part is inserted in the hole in the work, then the tapered shaft is driven into the expanding sleeve with a soft-face hammer until it fits tightly within. The work-holding force is friction against the hole's wall.

Once it is tightened inside the work, the mandrel and work are usually placed between centers on the lathe. But the mandrel shank can also be held in a chuck or collet.

> **KEYPOINT**
>
> The 20-washer job would be impossible with this mandrel due to variation in their hole sizes. Only the smallest inside diameter (ID) washers would be held. The rest could spin freely.

Machining a bushing (Fig. 10-84) is a good example of an expanding mandrel job. Suppose its outside diameter (OD) must be reduced by 0.030 in. and that new OD must be concentric to the ID within 0.001 in. The expanding mandrel is the right choice.

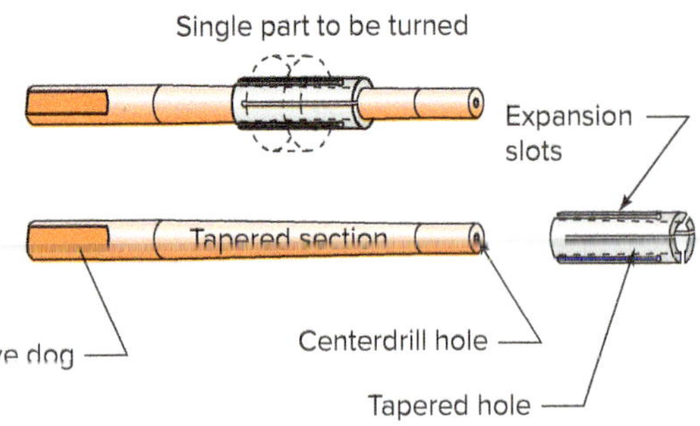

Figure 10-84 Expanding mandrels work like internal collets.

Grip Restricted to extra light duty, small work.

Runout 0.0005 plus or minus depending on condition and method it's held in the lathe.

Turnaround Time Very slow since it must be removed from the lathe to load and unload. But considering it's the only way to hold the work, speedy turnaround isn't often an issue here either.

Special Characteristics Suitable for work with a precision hole and where concentricity is a prime requirement. Will fit a modest range of hole sizes. Must have several sizes to cover holes from $\frac{1}{8}$ in. up to approximately $1\frac{1}{2}$ in.

Lathe Fixtures

In production, we might need to hold and turn castings, welded parts, and forgings that are an unusual shape–often an unbalanced object. Figure 10-85 is the perfect example.

Lathe **fixtures** are custom made expressly for a single part or family of similar parts. They feature a set of clamping devices to ensure safe and accurate part holding and some means of nesting (locating) the part accurately with respect to the center axis of the lathe. Fixtures are used in industry when part shapes are impossible or unsafe to hold by any other means.

Fixtures Are Potentially Dangerous

Two challenges are created: out-of-balance rotation and swiftly passing corners, clamps, and edges. Those spinning protrusions are scary to watch go whizzing by! Often the fixture features counterweights and guards, but they don't cancel all the dynamic problems or danger. Compounding the balance problems, as the object is machined, its center of mass changes. You will require further training in using a fixture and will probably not use one in school (Fig. 10-86).

Figure 10-85 A custom fixture is being used to hold this odd-shaped part in a production lathe.
Courtesy Tyee Aircraft, Everett, WA

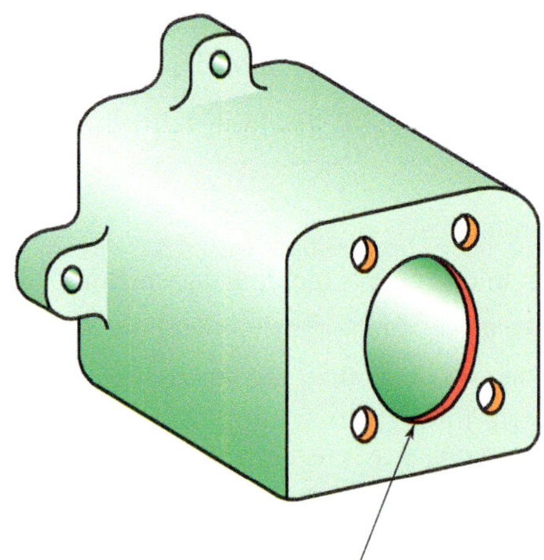

Figure 10-86 Boring the holes could be accomplished in a lathe fixture, but it's safer on a CNC mill.

Fixture Grip Very good as long as the machinist clamps carefully! Correctly made to withstand the forces, fixtures can be used where heavy machining is to occur.

Runout Since a large, odd-shaped object is mounted in a fixture, runout is difficult to quantify. It will vary with each fixture.

Turnaround Again, since every fixture is different, some are quick to use, while most are very slow for part removal and replacement. All locating surfaces must be cleaned, then checked to be certain the part is nested on them correctly and that all clamps are tight.

Faceplates

Faceplates are a kind of universal fixture. They are large, flat disks, mounted in place of the chuck, upon which we clamp and bolt the workpiece. They feature slots or tee slots for bolts and clamps, in a radial pattern (Fig. 10-87). Work is clamped directly to the plate using high-strength setup bolts, nuts, and clamps similar to the drill press or milling machine—with one critical exception.

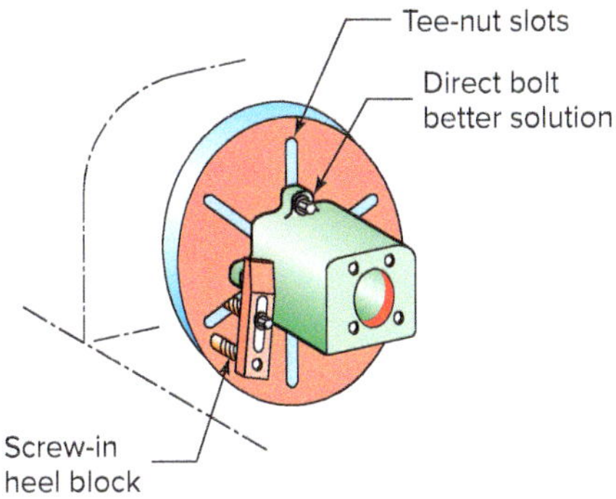

Figure 10-87 The transmission bolted to a faceplate.

> **KEYPOINT**
>
> When using clamps to hold work to a faceplate, use only screw-in heel blocks, never use loose heels. Even well clamped down, during the heat, impact, and force of machining plus centrifugal action, separate heel blocks can be flung out, thus letting go of the work. (See Chapter 12, Unit 12-4 for clamp training.)

Using the transmission case example, bolts could be put through the mounting ears if their holes are already drilled. That way, the part is secured well to the plate. If not, safe planning might include drilling them first. However, as shown, clamps are a necessary evil at times with faceplate setups.

Last Resort Choice

Faceplates are only chosen to hold objects that are impossible to chuck otherwise and cannot be machined elsewhere. Even more than fixtures, which usually feature work nests, pins, and custom-made clamps to retain parts, faceplate setups are dangerous due to clamping and holding issues. I do not recommend that they be attempted until you have much experience with turning. *Always have someone double-check your faceplate setup!* See the Trade Tip.

> **TRADE TIP**
>
> **Safety** If machining isn't required to the center of a job, then a solid wood pusher block can be forced against work by tailstock force, using a live center. This adds an extra protection from work coming loose.
>
> *Wood block support for setups, too.* The wood block support can be used when centering work on the faceplate or in the four-jaw chuck—before the clamps or jaws are tightened. The forced block holds the work in place but allows light tapping to move the workpiece closer to center.

Why Choose a Faceplate Setup?

Grip Medium well if all clamps are well tightened. Faceplates can be used for heavy machining by an experienced machinist, but that's not a recommended practice.

Runout Runout is a function of the care with which the work has been indicated or located on the plate. There are two runout issues with faceplates. *Axial runout* concerns having the work accurately centered around the spindle axis. Accurately centering the work is usually accomplished by roughly clamping it to the faceplate, then tapping it toward center while testing the object with a DTI. For *radial runout* the face of the plate itself must not wobble such that one of the work datums will be perpendicular to the lathe's axis. Sometimes, with use, the machinist may need to take a light face cut off the faceplate surface to return radial runout to zero. Never cut the faceplate without asking!

Turnaround As slow as it gets. Don't even think of hurrying a faceplate setup!

Special Features Used to hold odd-shaped work but must have one flat surface to place against the faceplate. They are a last resort work-holding method.

Holding Work Between Precision Centers

This final holding method isn't common in industry today, but it is used on precision cylindrical grinding machines. Even though we try to avoid this old holding method, there are times when the work has center support holes already drilled and it's the right choice to get the workpiece quickly aligned to the lathe. In Fig. 10-88, the work is drilled on both ends, with center holes. The holes are made with a center drill, which is what they were originally created to do. The drive plate is similar but smaller than a faceplate, usually with three slots for the drive dog.

Turning Between Centers

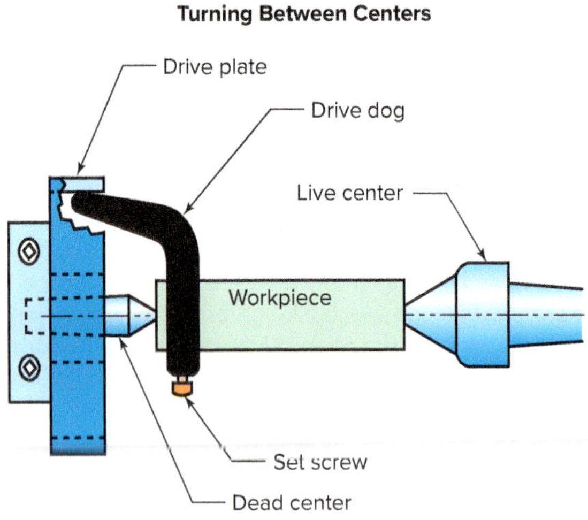

Figure 10-88 Four components for holding work between centers. Note, center drill holes must be present.

Lathe Centers

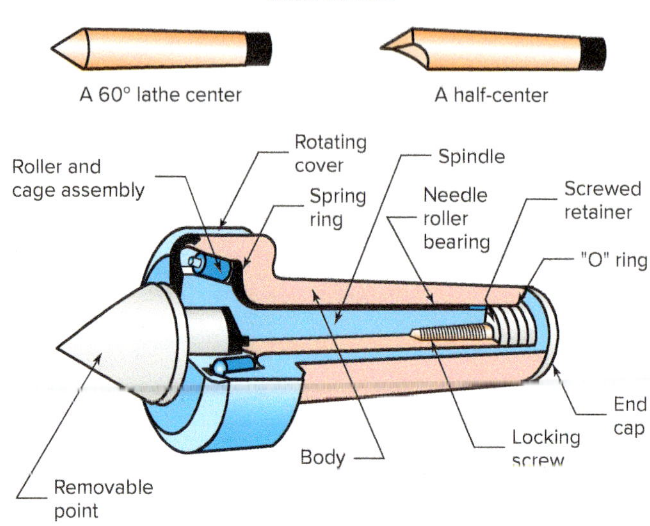

Figure 10-89 Two types of lathe centers *dead* or solid and the popular *live center*.

TRADE TIP

Faceplate and Fixture Safety Precautions

Use RPM Limits Never operate a faceplate above the limit set by your instructor or supervisor.

Use Tailstock Pressure If at all possible, always push the tailstock center against the work to force it against the plate.

Add Extra Stop Blocks In addition to the clamps, place blocks of steel with holes drilled through, directly bolted to the faceplate, alongside the work. They prevent the work from shifting sideways on the plate during heavy or interrupted cuts.

Stay out of the Danger Zone! Never stand in line with the work's rim, even if it's spinning without the cutter touching it! Be extra careful about loose clothing, too.

The center drilling can be accomplished on the lathe with the work held in a three-jaw chuck or collet, or it can be done on the drill press with the work held in a vee block.

Next, remove the chuck and place one pointed center in the spindle hole and one in the tailstock (see Fig. 10-88). Four accessories are added to the setup to positively center and rotate the part:

A *lathe dog* is bolted directly on the work. The "tail of the dog" is then placed in one of the slots in the drive plate.

A *drive plate* is mounted on the spindle nose taper.

A **dead** *(solid metal) 60°* **center** is placed in the headstock end. This center rotates along with the work (Fig. 10-89, left side).

A **live center** (roller bearing) is placed in the tailstock (see Fig. 10-89). The live center point rotates with the work while its shank remains stationary in the tailstock.

Grip Shown in Fig. 10-88, the work is rotated by a clamp on the drive dog. It transfers spindle torque to the workpiece via a set screw. That's not a very efficient grip. Between center work is reserved for light cuts only.

Runout Very low–less than 0.001 in. as long as the drilled center holes are round and the centers are in good shape.

Turnaround Once the initial setup is made, this is a very fast way to replace parts in the setup. Working between centers is especially useful when the part must be removed and replaced back on the exact center again.

Special Characteristics Used where a student part might not be finished during your training shift, so it can be removed and replaced exactly back on-center next time. Turning between centers is a good way to bring old parts that were originally turned between centers, back onto the lathe axis. Also used where other means of holding the work are not possible.

TRADE TIP

About Centers When the turning tool must come close to the work's center as in facing or threading a small object, the relieved half center is used, as shown in Fig. 10-89, but only at these special times. Due to the smaller support surface, it is the wrong choice for general turning. Also, since it has no bearings, it requires constant lubrication and attention as the work heats from machining.

The center drill, sometimes called a combined countersink and drill, was originally invented to prepare work for this holding method. To use it correctly, drill into the work about three-fourths of the distance up the cone portion (Fig. 10-90). This leaves a cone-shaped receptacle for the lathe center.

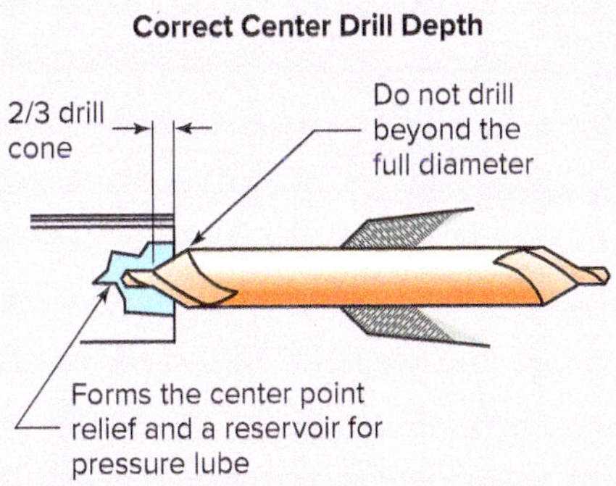

Figure 10-90 Use about three-fourths of the cone portion but do not drill beyond the cone.

While live centers in the tailstock are far more useful, sometimes a dead center is used when runout is a prime concern or because a live center is unavailable. Live centers are moving objects themselves and can add small amounts of runout.

When setting up a dead center in the tailstock, friction becomes a problem between it and the work. Grease must be used to minimize heat and wear. The pilot part of the center drill not only relieves the hole for the center point, but it also forms a reservoir where special high-pressure grease is placed before putting the part into the setup.

10.3.4 Mounting Chucks and Other Accessories on the Lathe

Chucks that screw onto the lathe spindle have been phased out of industry due to their inability to turn the spindle backward without the dangerous risk of unscrewing from the lathe. They will not be considered here, but you might encounter them in home or school shops–if you do, get extra instruction!

Cam Lock Chucks

The **cam lock** *spindle* is safe, strong, and universally accepted for mounting chucks and faceplates on a lathe, both manual and CNC. The lock system uses hardened steel pins extending from the back of the chuck or faceplate (Fig. 10-91) that are firmly gripped by rotating cams in the spindle nose of the lathe (Fig. 10-92). Correctly locked, the chuck holds equally well for forward and reverse spindle rotation (Fig. 10-93).

Figure 10-91 Cam pins firmly mount chucks and faceplates, allowing safe forward or reverse spindle rotation.
Lake Washington Technical College, Kirkland, WA/McGraw Hill

Figure 10-92 *Repeated Warning!* Clean cam holes, pins, and tapers before assembly, because one tiny chip left behind will dent them due to the great pressure between them!
Lake Washington Technical College, Kirkland, WA/McGraw Hill

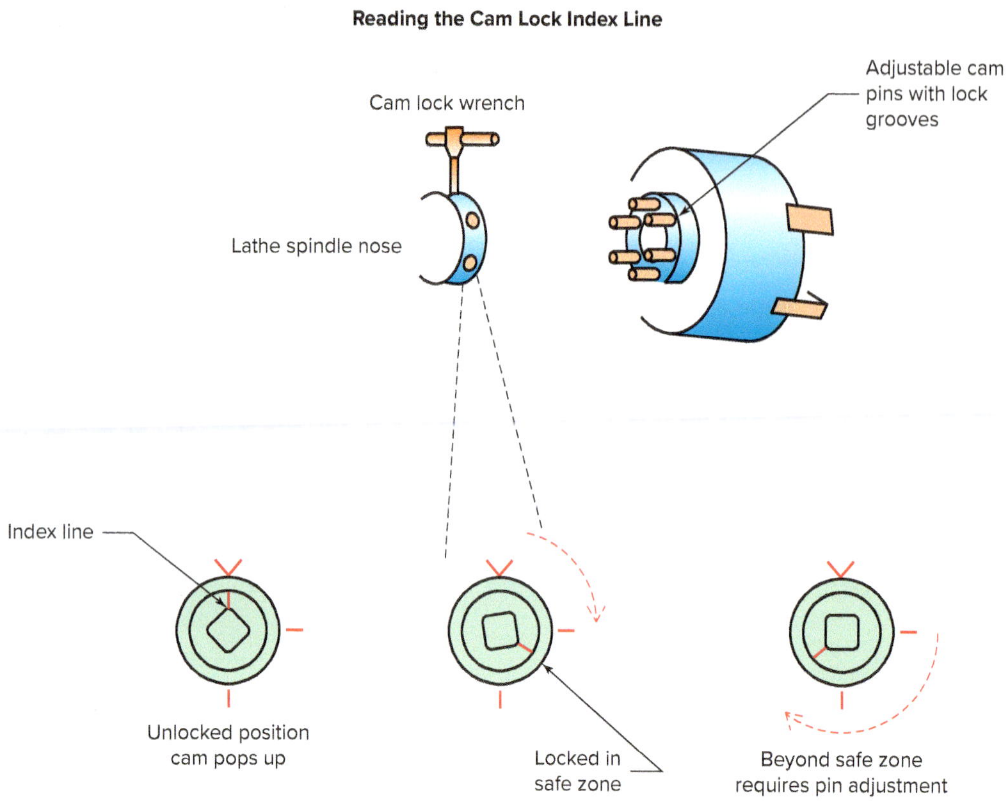

Figure 10-93 *Trade Tip*—The index line shows the lock status.

Removing a Cam Lock Chuck

Caution! Before proceeding be sure the machine is locked out–the main power is off, not just turned off.

1. Place a wood cradle under the chuck for protection of your hand should it slip and the bed ways should the chuck fall.
2. If an overhead shop lifting system is available, screw an eyebolt into the chuck–ask for instruction on this. If overhead lifting isn't available, ask a fellow machinist to stand by to assist. As the chuck is removed, do not extend your back out over the lathe bed–do not get into the wrong lifting position!
3. Using a special square or hexagonal wrench, turn all locks counterclockwise to the unlock position, as shown in Fig. 10-93. The socket will pop up when in the unlock position.
4. Now for safety, turn one cam back toward the hold position. But not tightly locked. This restrains the chuck from falling free as you pop it off the spindle taper.
4. To break the holding force of the spindle taper, using a soft hammer, strike the chuck at the rim with a moderate blow. The chuck will pop off but not fall away due to the single cam holding it.
5. With the chuck supported from above or with two machinists holding it, one in front and one behind the lathe, unlock the final cam, lift the chuck away, and store it in its rack or shelf.

KEYPOINT

During this entire operation, keep your hands out from under the heavy chuck.

Assembly of a Cam Lock Chuck (Pay Close Attention to Steps 2, 5, 6, and 7)

Again, be certain the machine is locked off or the main power is turned off.

1. Place a wood cradle under the spindle.
2. Clean all mating surfaces on the cams, chuck, and spindle first with a shop towel and then with a final wipe of your clean hand.
3. Turn all cams to the unlock position.
4. Lift the chuck into place using a lifting device or a helper as a second choice.
5. Lightly turn each cam clockwise until it just starts to snug up. Do this in an across-and-back pattern to equalize the initial seating of the chuck; work softly and not all on one side at first.
6. Retighten in this same pattern. Always turn the cam wrench in a clockwise direction as viewed from above. This can be repeated twice more with a bit more force each time. Apply 40 or 50 pounds of force on the wrench for the final tighten.

TRADE TIP

On the final tighten, read the lock index lines per Fig. 10-93. With normal use or especially after a serious crash, the cam pins can become out of adjustment for a safe lockup. If when locked, the index lies within the safe zone, it's okay to proceed. If the index goes beyond the safe zone when locked, then remove the chuck and adjust the cam pins as needed by removing the set screw, then turning them in one turn. Note that the cam groove will always be toward the inner circle.

7. Do a final test "around the circle" this time to ensure all cams are tight.
8. Remove the lifting aids, wrench, and the wood cradle.

10.3.5 Soft Jaw Work Holding

There are two different versions of soft jaws for two different purposes:

1. Protectors on regular jaws to prevent marring finished work (Fig. 10-94)
2. Specially machined aluminum jaws to hold difficult shapes and/or to better align the work on-center (Fig. 10-95)

TRADE TIP

Equal Thickness When using protectors be certain they are all three of equal thickness when used with universal three-jaw chucks. If not, the work will be held off-center! If using them with a four-jaw chuck, thickness isn't as critical as you will be centering the work with a DTI anyway (Fig. 10-94).

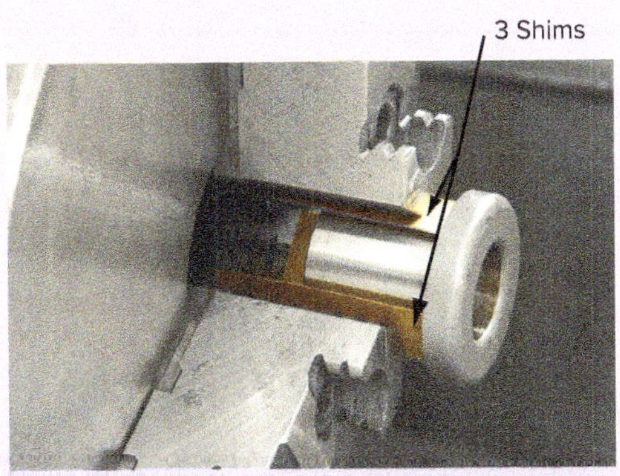

Figure 10-94 Soft sheet metal protects a finished part in a chuck.
Lake Washington Technical College, Kirkland, WA/McGraw Hill

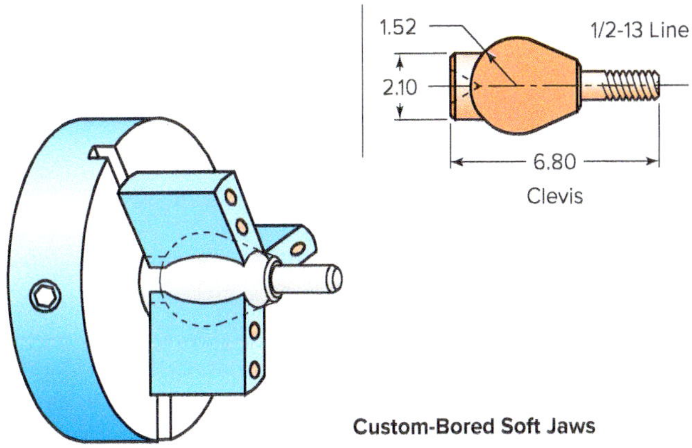

Figure 10-95 This clevis would be difficult to hold without the custom soft jaws.

Protectors

Usually cut from sheet aluminum, but any soft metal or even plastic would work, these little sheet metal pads are held with your hand or with tape until the chuck pressure keeps them in place. Since the chuck pressure distorts them, they are consumed in the setup. We usually shear several dozen at one time.

Machined Soft Jaws

Used often in real manufacturing but seldom in school, custom-machined **soft jaws** are chosen when runout is a prime issue or when the work can't be gripped by any other means (see Fig. 10-95). For example, this clevis (a round end on a rod) must be held to cut the threads. Soft jaws could be used. They are often created for a special job on a CNC lathe.

Soft jaws blank are usually aluminum, or occasionally mild steel for extra-demanding production of many parts. The backside is premachined such that it will key and bolt in place of the regular hardened jaws. Once mounted on the chuck, they are custom machined to fit a given workpiece. Soft jaws are used on three-jaw chucks. Bored in place, on the chuck, they bring the work perfectly onto center every time.

Exact Centering on a Three-Jaw Chuck

Cutting soft jaws correctly, on the same three-jaw chuck in which they will be used to hold the workpiece, ensures they then close on the work very close to center each and every time. But only as long as they aren't removed from the chuck and the chuck is not removed from the lathe. Using them without disturbance, runout of the work can be as repeatable as 0.0005 in. or better.

Soft jaws are often removed from the chuck to be used later, to hold the same job or similarly shaped objects. But, once they are removed then remounted at a later time, the centering must not be trusted—perhaps degrading to an estimated 0.005 in. plus or minus. That's due to refastening pressure and general alignment of the jaws to the chuck. So, when reusing them, a light skim cut of the grip surfaces restores their near-perfect centering capability.

Ensuring Chuck Grip Bias

To ensure perfect centering with each opening and closing of the custom-machined jaws, it's necessary that the chuck be forced tight against some object when the custom shaping of the jaws is being done. This brings the chuck jaws to the right pressure and alignment bias with each closing. If the chuck isn't stabilized with opening or closing force, the machining will not be correct, and it will be outright dangerous, as the jaws will move as the cutter hits them.

If the work is to be gripped from the outside, then the chuck must be closed downward toward center, as shown in Fig. 10-96. A hex nut or a round plug can do the job. See the Trade Tip.

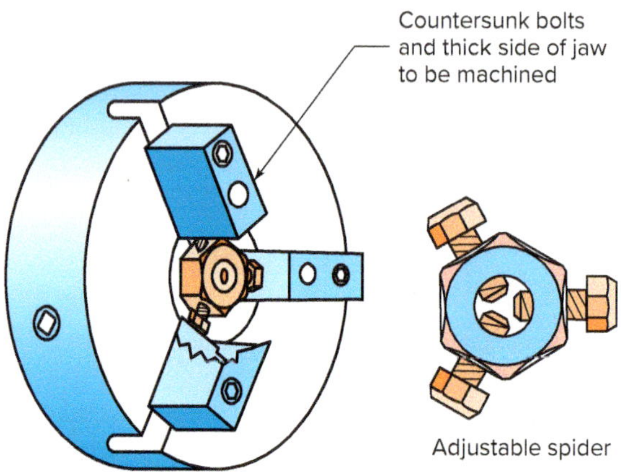

Figure 10-96 A spider helps pressurize the jaws correctly during their custom machining.

> **TRADE TIP**
>
> **Making Your Own Spider** A handy self-made tool for boring soft jaws with inward bias is the adjustable three-point spreader called a spider. Drill and tap three holes in a large hex nut. Then add the right length bolt to fill the space. Sometimes, when the entire jaw must be machined, the spacer can be placed against the jaw carriers farther into the chuck.

When the soft jaws will be used to grip the work from its inside—with an outward chucking force—a solid metal ring or other retaining device must be placed around the outer surface of the jaws to stabilize and bias the grip during custom machining. Check with your instructor or shop supervisor as to what is available.

UNIT 10-3 REVIEW

Replay the Key Points

- Workpiece runout can be introduced to the setup by the holding method itself.
- Solid reversing jaws can be reinserted in the wrong order, resulting in chuck closure at some point other than center.
- A collet grip is round at the specified size and distorted slightly when squeezed smaller. Never squeeze the collet closed beyond its range.
- In using cam lock chuck mounts, do not use a chuck if the index line goes beyond the safe lock position.
- Each and every time, clean the spindle taper and chuck before mating them. This is one of the most irritating brain lapses that beginners can make. *Just one trapped chip ruins expensive equipment!*

Respond

Choose the right chuck and support accessories for the following jobs: Note that there is often more than one correct answer. Although parts may be similar in shape, other conditions such as batch size, raw material, and tolerances dictate choices. Check each answer as you go. Compare yours to those of a fellow student.

1. Ten expensive A arm, titanium steering castings designed for an undersea rover must be turned round on the pointed end. They resemble a bent, 12-in.-tall letter A.
2. Fifteen 2.5-in.-diameter shafts must be turned to 0.750-in. diameter, then threaded with 1.0 of threads. They are precut to 8.250 in. long from a CRS bar—no finishing of the 2.5-in. diameter will be required. Runout requirement is 0.005 in. between the shaft and new threads. See Fig. 10-97.

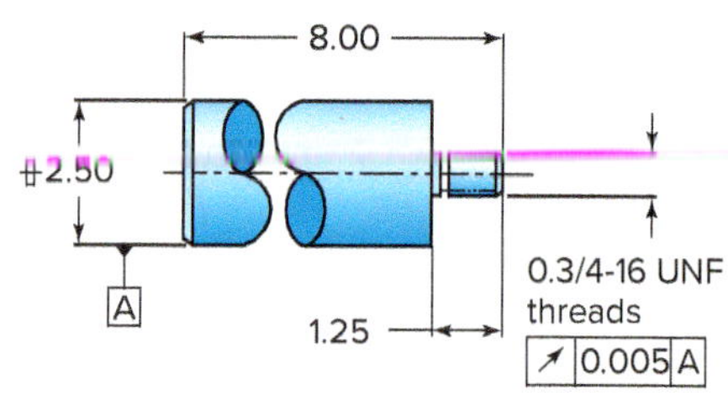

Figure 10-97 Which holding method would be best for this CRS cut bar job?

3. Five hundred finished 15-mm-diameter twist drills must be turned down to 0.350 in. on the soft part of the shank (to make turned-down shank drills), with a runout requirement of 0.001 in. to the original diameter. They are 100 mm long and require about 12-mm length of turned-down surface.

4. A photocopy machine needs its main roller just cleaned up to remove scratches, then repolished to the original 8-μin. finish. (*Hint:* Runout is tightly toleranced.) See Fig. 10-98.

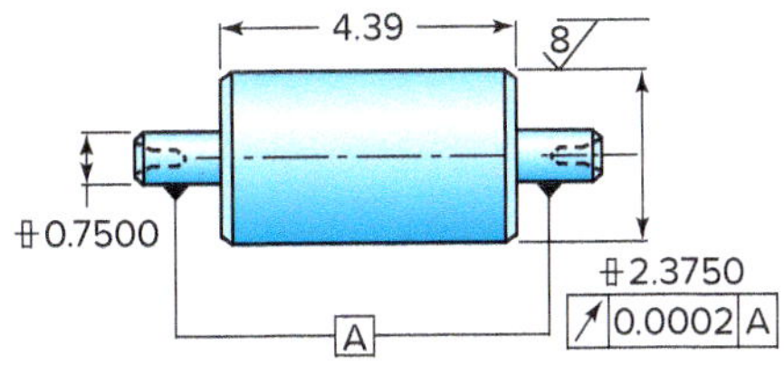

Figure 10-98 Turn one paper roll for minimum cleanup of scratches only; then repolish to a fine finish.

5. You are assigned to turn 300 aluminum window screen holder pins per work order and print. (*Hint:* They must be turned very fast for good cutting action.) See Fig. 10-99.

   Now, answer the following questions from what you know about chucks.

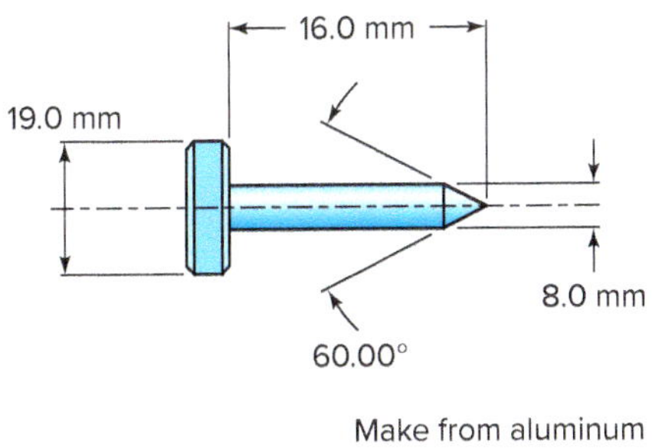

Figure 10-99 Turn the 19-mm aluminum rod to 8.0 mm, chamfer; then part off.

6. True or false? Due to manufacturing tolerances, industrial-quality four-jaw chucks will have a certified maximum runout when new; however, with use and abuse, runout can degrade due to stretching of the scroll plate. If it is false, what will make it true?

7. What are the special characteristics of a three-jaw chuck?

8. What are the advantages of a collet chuck? What are the disadvantages?

9. What type of work is best suited to three-jaw chucks?

10. Name two reasons for choosing a between-centers setup.

11. Why would you use a faceplate setup?

12. True or false? The three-jaw chuck is the most common chuck because it grips well and is quick to use, but it produces a moderate runout. If it is false, what will make it true?

Unit 10-4 Turning Tool Basics

Introduction: An important part of setting up a lathe is choosing the cutting tool and the tool holder. Often, depending on the tools on hand, several might work okay, but one or two might be best for the task.

TERMS TOOLBOX

Cemented carbide Tungsten carbide grains cemented together with a cobalt binder and a term sometimes used to describe a carbide tip is brazed on a steel lathe shank.

Chip breaker–chip dam A groove in a lathe cutting tool (breaker) or metal protrusion (dam) used to snap off the advancing chip into safe "C" shapes.

Cobalt A tough metallic element added to HSS bits for extra toughness or used as the "glue" in a sintered tungsten carbide.

Indexable insert An insert carbide (or other cutting material) tip that features more than one cutting edge per insert. When the insert is indexed to a new position, it doesn't lose the tool's position on the cutter holder.

Inscribed circle (IC) The size of a carbide insert, the largest circle that can be drawn on the insert. The tool is working efficiently when it is cutting a depth of two-thirds the IC.

Side rake (also compound rake) A lathe tool featuring a rake angle that has a bias to cut left or right.

Sintering Powered metal is pressed into a mold, then fired to form a solid composite with characteristics different from any cast metal.

10.4.1 Choosing and Setting Up the Right Tool for the Job

The factors to consider when choosing the right cutting tool are

- **Tool Shape**

  Will it be able to cut the surfaces and shapes required?
    Often more than one cutting tool will be needed to complete the part shape. If so, then the kind of tool post (cutting tool holder) becomes an issue, too.

  Positive or negative rake
    Depends on the alloy to be cut and the amount of metal to be removed.

  Nose radius and lead angle

  Right- or left-hand bias
    Turning, threading, and facing tools are usually ground or selected such that they cut in one direction.

- **Tool Construction**

  Standard technology
    Solid HSS

    Carbide insert tooling

  High-tech cutters for high production or for cutting extremely difficult alloys
    Polycrystalline cutting edges

    Artificial and natural gemstones embedded into the tool's cutting edge

  Ceramic or ceramic/metallic composite cutting tools
    Aiming at harder cutters, some are made of nonmetal materials. While they cut very well, they are also brittle.

  Brazed-tip carbide (mostly as form tools)
    A steel shank with a shaped piece of tungsten-carbide brazed on to form the cutting edge, they are being displaced due to carbide insert technology and CNC abilities to drive standard tools to create shapes rather than form them.

While all types of cutting tools can be found in industry, the standard technology options are the more practical choices for training. We'll concentrate our explorations there.

10.4.2 Tool Shapes

Other than the drilling, reaming, and tapping from the tailstock, the basic operations can be accomplished with the following five utility shapes. There are many variations within each including HSS and carbide insert tooth varieties. We'll look at both, but for now let's discuss their shape.

Turning Tools

Although there are neutral bias tools, cutting to the operators left or right, they usually do not produce the best finishes. They are used for utility setups where several different kinds of cuts must be made by the same tool or where tool changing is slow or difficult. For best results, we generally choose a tool with a cut bias. Depending on how the lead and rake angles are ground, turning tools are biased to feed to the operator's left, or they can face inward toward the work center or to the right and face outward on the work. There are a nearly infinite number of combinations of rake, lead, nose radius, and tool composition. In fact, like two artists choosing brushes, often one will work best for one machinist, while a different combination is chosen and used with good results by another (Fig. 10-100).

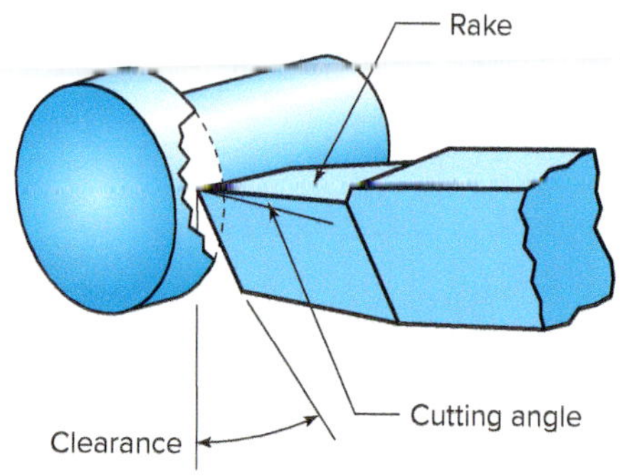

Figure 10-100 A turning tool is used for other kinds of operations, too.

Parting and Grooving Tools

Similar in shape and function, the difference is the depth to which these cutting tools must drive into the work (Fig. 10-101).

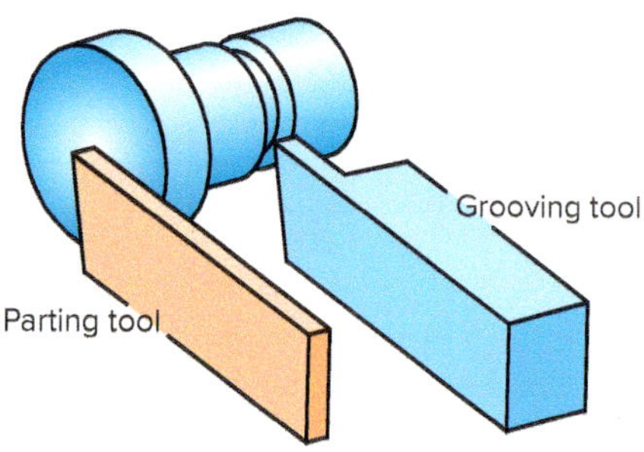

Figure 10-101 A parting tool and a grooving tool are very similar.

Form Tools

Form tools (Fig. 10-102) are custom made mostly from solid HSS, or for longer tool life, with steel shanks and brazed-on carbide tips. Today, however, CNC lathes are reducing the need for custom-shaped form tools. Programming allows complex movements of off-the-shelf, standard tools to reproduce custom shapes. Additionally, using standard cutting tools eliminates the challenges of slow cutting, chatter, and tearing of the surface associated with forming.

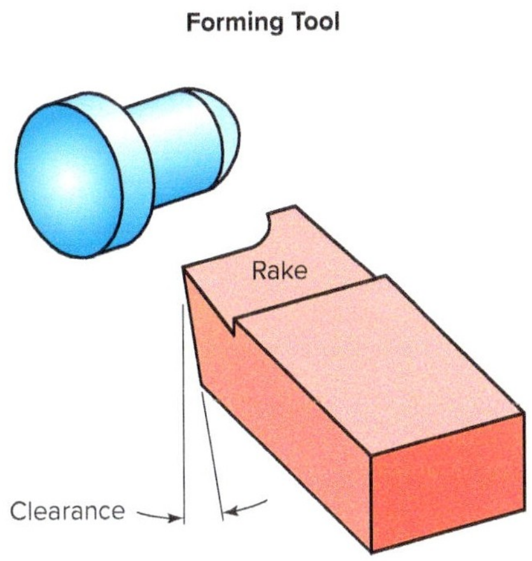

Figure 10-102 A forming tool will make an outside corner radius.

Threading Tools

Threading tools (Fig. 10-103) are purchased in preformed versions or as insert-type carbide tools. But it's likely that your first threading lesson will include grinding your own using a blank HSS bit. It must be ground with the included angle of the thread (60 °for Unified threads) and small, flat section on the tip to duplicate the thread form.

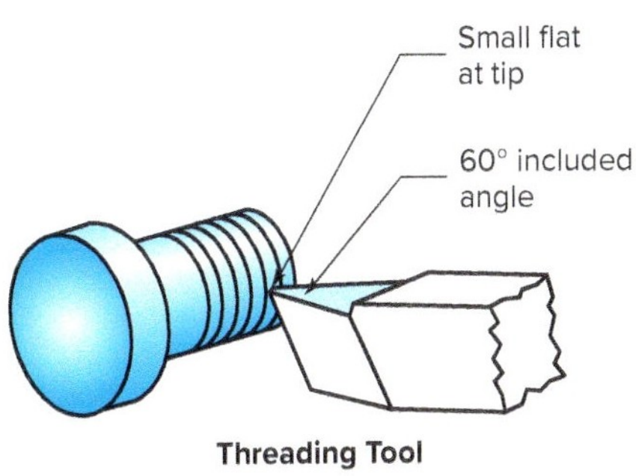

Figure 10-103 A threading tool is a type of form tool.

Boring Bars

Boring bars are found in machine shops in three varieties (brazed carbide, shop made, and insert tooth carbide) and come in a great many sizes. Many are used on the milling machines as well as the lathes, both manual and CNC. Many of the concepts we'll discuss here also apply to boring on the mill, where it's the tool that spins, not the work.

Preshaped Solid Boring Bars

One popular version for small holes is the "Swiss" type (Fig. 10-104) that's mounted in a tool holder block for lathe work. They produce excellent results when boring diameters below 1.0 in.

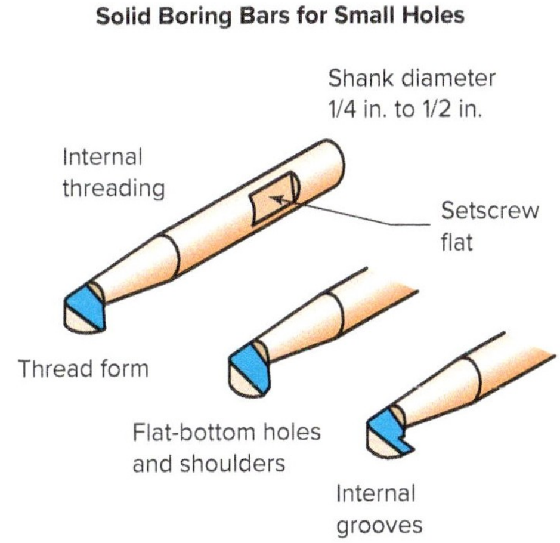

Figure 10-104 Boring bars for small diameter holes are usually the preshaped "Swiss" type.

Universal Boring Bars

These shop-made bars are common for tooling and production work. Using set screws to hold the cutter in a vee groove, they accept standard HSS tips, ground to any shape by the user (Fig. 10-105). Or small a carbide cutter can be used in the slot, as well. Machining the bar tip at an angle can be useful for reaching the hole bottom before the bar face rubs the workpiece.

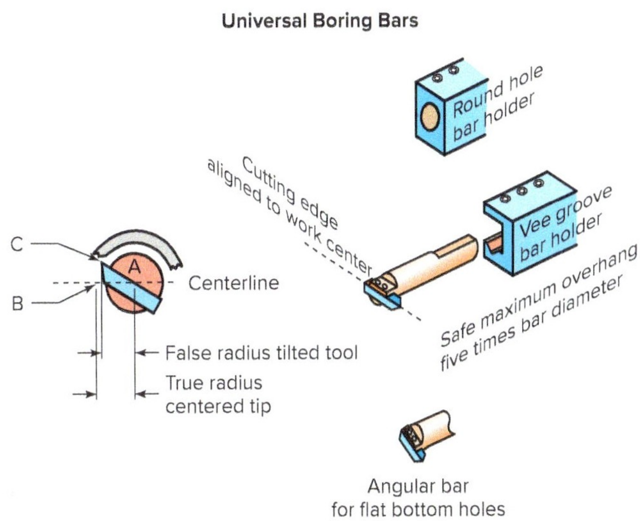

Figure 10-105 Shop-made boring bars are used on manual lathes.

KEYPOINT

To ensure correct cutting action, when making your own boring bar be certain that the upper edge of the tool bit slot is near the center of the bar's diameter.

Setting Up a Boring Bar

Four factors must be considered when setting a boring bar.

Factor 1—Clearances

There are four clearances to be checked for a successful boring bar setup:

1. **Bit Length**

   First, as shown in Fig. 10-105, the replacement bit might be too long. It might rub the backside of the hole on the first cut. The bit cannot protrude out the back of the bar beyond the radius of the drilled hole.

2. **Chip Clearance Between Bar and Hole**

   While the goal is to use the largest bar diameter that will fit into the hole, it cannot be so big that chips cannot eject. If they bunch up (a common problem during boring), then coolants cannot reach the cut. Heat builds rapidly with tool failure to follow.

3. **Cutting Edge Clearance**

   Next, as shown in Fig. 10-106, more cutting tool clearance than normal must be ground such that the cutter's lower edge does not rub on the hole's curvature.

4. **Tool Holder Clearance**

   As Fig. 10-107 shows, after setting the depth control, be sure the tool holder does not touch the face of the work when the cut is at full depth. That space is necessary to allow coolant in and chips out.

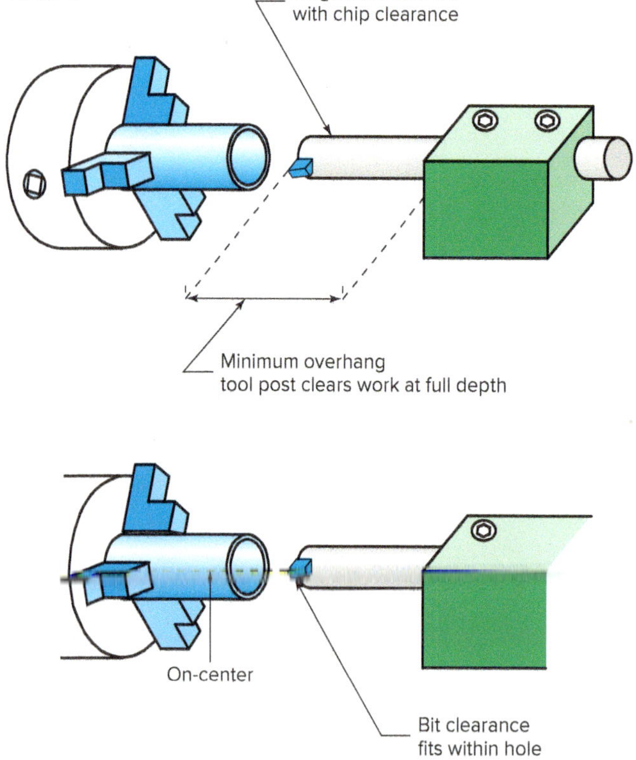

Figure 10-106 Factors for a correct boring bar setup.

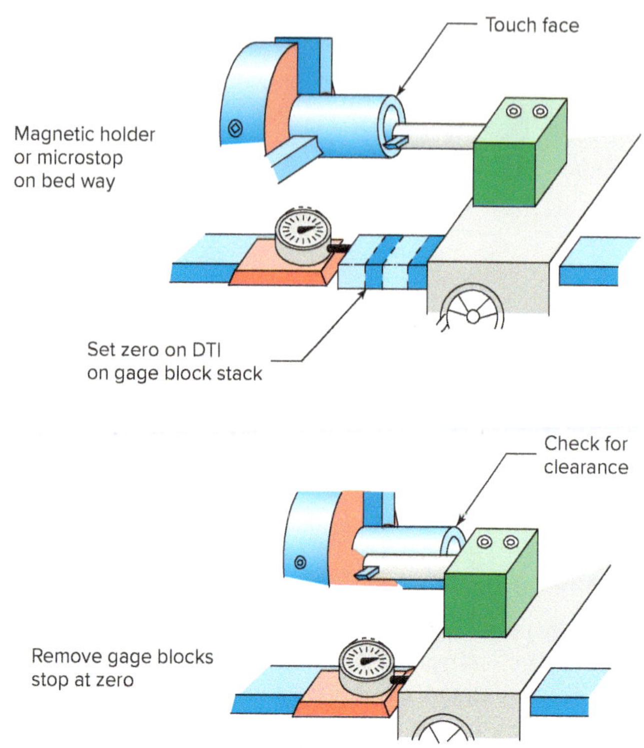

Figure 10-107 Setting depth control and checking tool holder clearance.

Factor 2—Overhang and Depth Control

When choosing a boring bar, remember the guideline of 5:1 overhang ratio, length to diameter (Fig. 10-106). Long, small-diameter boring bars tend to dig and chatter. Many boring bar holders allow the bar to slide in and out–use that function to set a minimum overhang.

Since boring is a blind operation, you cannot see when to stop cutting inward. You must then build in some form of positive stop to ensure depth control of the bored hole (see Fig. 10-107). If the lathe isn't equipped with a digital positioner, here is a way to set it up. With the bit's cutting edge touching the work face, set a zero against a gage block stack equal to the bored hole's depth. This can be on an indicator, as shown, or use a micrometer stop if the lathe has one. On many lathes, the microstop features a trip that halts *Z* axis movement as it contacts the stop.

A set of gage blocks can be used to coordinate set the depth (Fig. 10-107). After zeroing the dial, remove the stack and the saddle will come to the preset zero at the correct depth. Note that the full depth clearance must be checked too, as shown.

> **KEYPOINT**
>
> A 5:1 ratio of bar diameter to overhang is an entry-level safety guideline—but it is possible to exceed it only after a bit of experience with boring operations.

Factor 3—Vertical Centering

This tool setup parameter is important to ensure size control when boring. The cutting edge must be on the vertical center of the workpiece. Centering of the tool can be accomplished in one of three ways. The best solution is using the quick change vertical adjustment screw. The next choice uses shims under the tool in the holder, as with an open-sided tool post. The least desirable is by rotating the bar in the round hole mount, because that changes the effective rake angle as well.

KEYPOINT

If the bit isn't on-center, a given size change on the *X* axis micrometer dial won't yield the same amount of hole diameter change.

Study the end view inset in Fig. 10-105 to see why. Distance A to B is the true radius distance from work center to the inside of the hole, parallel to the *X* axis of a correctly centered tool. If the bit is set correctly, cutting edge movement will be along that horizontal line—so when making a diameter change on the *X* dial, the same amount of diameter change occurs in the hole.

Distance A to C represents the effective, smaller radius of the incorrectly tilted tool. A given change in the *X* axis would produce a false result. The farther the cutting edge is misplaced from the horizontal *X* axis, the greater will be this shrinking effect. Remember this concept. It's important when aligning boring bars on milling machines, too. The cutting edge must be at the work centerline to achieve diameter accuracy and correct form tool shape as well.

Carbide Insert Boring Bars

The third kind of boring bar is mostly used for larger holes and heavier cuts. However, they are also supplied for holes as small as $\frac{3}{8}$ in. (Fig. 10-108). We'll study the insert geometry and makeup in Unit 10-6. For now, their main advantage is the quick changeover to new sharp inserts without disturbing the setup position of the bar itself. Every cutter we discuss here can be purchased as a carbide insert version.

10.4.3 Direction of Tool Cut

The next setup choice is the *hand* of the tool (the direction it's designed to cut—to the left or to the right). Many turning, facing, and threading tools are given a compound rake angle—with a bias toward cutting on one edge (Fig. 10-109). This geometry modification is called **side rake**, or sometimes it's referred to as **compound rake** (see **Shoptalk**). The lead angle is also ground onto the tool to cut toward either the left or the right.

Figure 10-108 Carbide insert tooth boring bars with triangular inserts such as this are common in industry.
Lake Washington Technical College, Kirkland, WA/McGraw Hill

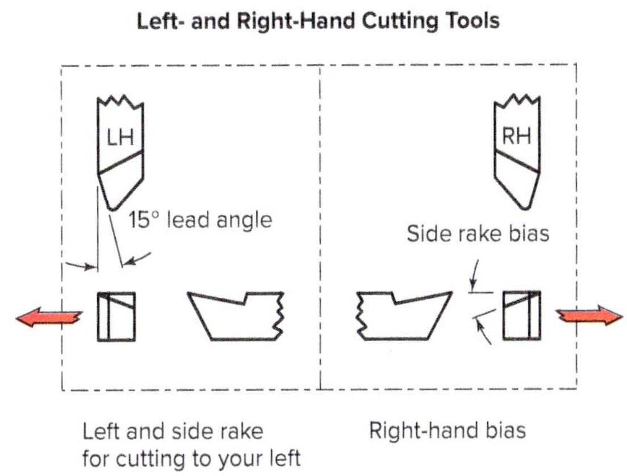

Figure 10-109 Lead and rake angles combine to create cutting tool hand bias.

SHOPTALK

Whose Right and Left Hand? Originally, the earliest lathes had their tools located on the far side of the spindle away from the operator. It was then that cutting tools were named right or left, since that's the direction they moved, to the operators' right or left.

Later, to aid in lathe setup, manual lathe manufacturers placed the tool in front of the spindle—closer to the operators where they could reach it easily. The manual lathes in your lab are probably made that way. On them, the left-hand tool actually cuts to the machinist's right and the right-hand tool to the operator's left—backward from their name.

Still evolving, many of today's modern CNC lathes with a single tool turret have moved the cutter once again, to the back of the *X* axis with the cutting tools pointing at the operator (Fig. 10-110). So on CNC lathes with the tool behind the spindle, left goes left and right goes right.

Figure 10-110 With the shield open, note the tool turret located behind the chuck on this CNC lathe.
Courtesy Okuma America Corporation

Confused? To keep it straight, always view the lathe cutting tool with it pointing toward you (see Fig. 10-109); the side bias hand will make sense in that perspective.

10.4.4 Cutting Tool Material

The third choice is the material from which the cutter is made. While there are several high-tech choices such as embedded diamond crystal tips or ceramic-metal composites, their cost is justified only in special application production. For now, we'll limit our investigation to the standard choices found in most tech schools and the average shop: HSS and carbide cutting tools.

High-Speed Steel Tools

For this lesson, the HSS tools also include two slightly harder but similar metals called cast alloy and **cobalt** HSS tools. As a beginner, the difference isn't significant. They do perform slightly better than plain HSS when cutting harder work, but their cost puts them beyond normal school budgets.

The similarities between cast alloy, cobalt, and HSS tools are significant—all of these tools are meant to be ground in your shop, using a standard gray aluminum-oxide grinding wheel. They are usually supplied blank or partially pre-formed, with the machinist grinding the final shape and regrinding them when dull or when a different cutter geometry is required.

Why Choose an HSS Tool?

They are the lowest initial cost cutting tool. While they dull quicker in service, compared to carbide, HSS tools offer an extremely long overall shop life because they can be reground many times, to any practical shape. The most important feature is that they are the toughest cutting tool. Toughness does not mean a cutting tool that can machine the most metal. It does mean it can withstand vibration, chatter, physical and thermal shocks (hit and miss coolants), and flexing without breaking (Fig. 10-111).

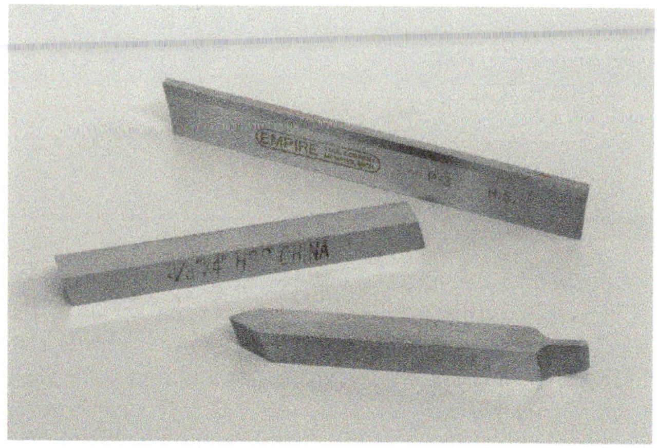

Figure 10-111 A blank HSS bit, a ground bit, and a new parting tool.
Prographics at Milwaukee Area Technical College/McGraw Hill

KEYPOINT

Choose HSS tools over carbide when an *interrupted cut* must be taken. These hammering situations (a non-continuous surface where the cutter must enter and come out of the cut with each revolution of the work-piece) break harder tools.

On the downside, HSS tools withstand the least heat from machining, therefore the cutting speed is the lowest of the two everyday choices. The cobalt and cast alloy tools perform a bit better at higher temperatures compared to plain HSS in this respect—but not by greater than 15 percent.

Carbide Insert Tooling

This is the tool of choice for industrial production. These tools are a combination of a precision, hardened alloy steel shank with a replaceable carbide tip—called an *insert.* Most inserts feature more than one cutting edge (Fig. 10-112). When the tip dulls it can be rotated or reversed to another edge. Inserts that have this feature are called **indexable inserts**.

Figure 10-112 A variety of insert tools used for turning.
Courtesy Morgan Branch CNC/Arlington, WA

Insert tooling is more costly for initial purchase compared to HSS, since the tool shanks must be purchased as well as a supply of inserts (Fig. 10-113). But, for the following reasons, they are very economical over a long period of use.

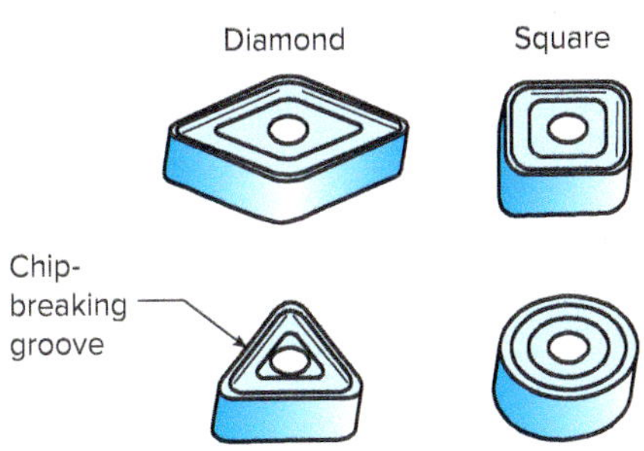

Figure 10-113 Indexable inserts are found in many shapes and sizes. These are the four most common.

Why Choose Insert Tooling?

Insert tools have several big advantages over HSS:

Quick Replacement The inserts feature guaranteed repeatability, usually to a tolerance of less than 0.001 in. The dulled cutting edge can be quickly rotated to a new edge or the entire insert can be replaced with very little size change in the setup. That means a minimum of down (nonproductive) time when the insert is indexed or replaced.

High Production Carbide's hardness allows cutting speeds at least three times faster than standard HSS and higher in some metals.

Multiple Choices Another big advantage is that many different combinations of hardness/toughness ratios, harder coated inserts, rake angles, and corner radii are available in the same shape insert. For example, if the insert is chipping before it becomes dull, then try a tougher insert with slightly less hardness. That change takes less than a minute to do. Or suppose the tool is chattering. Then change to a smaller nose radius insert (this will change work size). These changes are quick since they require changing only the insert, not the entire cutting tool.

SHOPTALK

Carbide—It's Not a Metal as You Know It! When we surround the heavy metal tungsten with carbon, then sustain it at high temperature without oxygen, it forms tungsten-carbide (commonly called carbide). The first firing creates a material that cannot then be formed into cutting tools in the standard ways because it melts at around 6,000°F (approximately three times the temperature of steel).

So, to make carbide cutting tools, the coarse carbide crystals are broken down into a uniform, fine powder using ball mills to crush them. It's then mixed well with a second metal powder made of cobalt. Next the mixture is pressed into a mold. The soft inserts have no strength until they are placed in a high-temperature furnace. During the elevated firing, the tough cobalt melts and glues together the hard carbide grains. The process called **sintering** forms a composite that is both hard and tough—more or less like concrete where the rocks are held together with cement (Fig. 10-114). In fact, another name for carbide is **cemented carbide**.

Hardness/Toughness Ratios By changing the ratio of carbide to cobalt in the insert, the hardness versus the toughness is altered. Higher cobalt content yields a tougher carbide tool but one that is less hard. More carbide relative to cobalt creates a harder tool but it's also more brittle and prone to chipping.

Figure 10-114 Carbide is a powdered metallurgical product with fine grains of hard carbide cemented together by other tougher metals.
Source: From Modern Metal Cutting, 1996, ©Sandvik Coromant

Tool Geometry Control

Insert tooling adds flexibility to your setup. Within a single insert tool holder, you can quickly change the nose radius, chip breaker, hardness, and even the rake angle in some cases by simply replacing the insert in the tool shank. Today, nearly all lathe tooling needs can be met with quick change insert tools with the single exception of custom-shaped forming bits.

10.4.5 Understanding Carbide Insert Tooling Numbering

When selecting an insert tool for your setup you must know industry standard numbering. The system includes data fields that identify the size, hardness, and shape of the insert. A similar system exists for selecting the tool holder.

When making a manual lathe setup, you have some latitude as to the cutting tool size and shape chosen, but when making a CNC lathe setup, most program setup sheets call out exactly which tool holder and insert are to be used. They will probably be called out using the following number system or they will be in picture form–often it's both. At that time, you'll need access to reference books such as *Machinery's Handbook* or the *Machinist's Ready Reference*.

Insert Size

This decision is about how big the insert shape should be. It's based on the tool post capacity, the availability of sizes in the shop, and the shape and size of the work. The amount of metal to be removed is a factor, too. Bigger tooling is better for strength, but remember that the larger tool is more costly. Make sure your setup will fully utilize it.

KEYPOINT

Choosing a large insert for small cut depths is not economical. But choosing a small insert that is overused is not smart either.

The Inscribed Circle

For comparative reference, the largest circle that can be drawn on the inset face is used for the size called the **inscribed circle (IC)**. Ideally, choose an insert such that it will be cutting at about two-thirds of its full IC cutting depth for roughing cuts (Fig. 10-115).

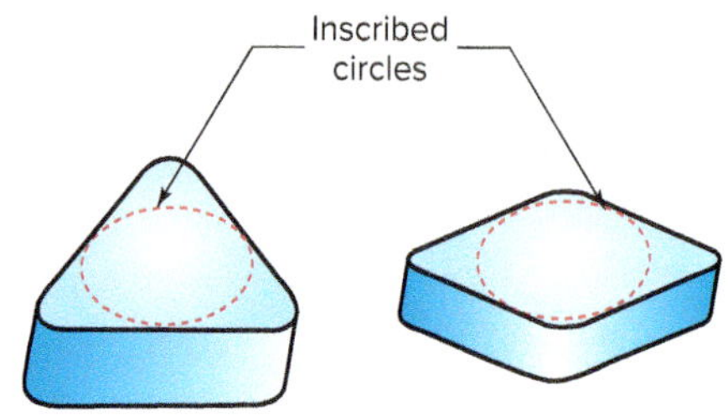

Figure 10-115 The inscribed circle describes insert size.

Insert Shape

Cutting tool shape is a more complex issue than size. Keep in mind a CNC document will probably call out exactly what insert shape must be used because there are issues you'll learn in programming as to how the cutter must approach and cut the material.

KEYPOINT

For CNC work, setting up the wrong shape (not what the program thinks it will be) can make scrap parts and even cause a crash!

The insert shape and the tool holder shape are linked. In the following three charts (Figs. 10-116, 10-117, and 10-118), note how the holder is chosen. It's not critical to understand it all at this time. You should be aware of the system and where to find it for future reference.

Toolholder Identification System

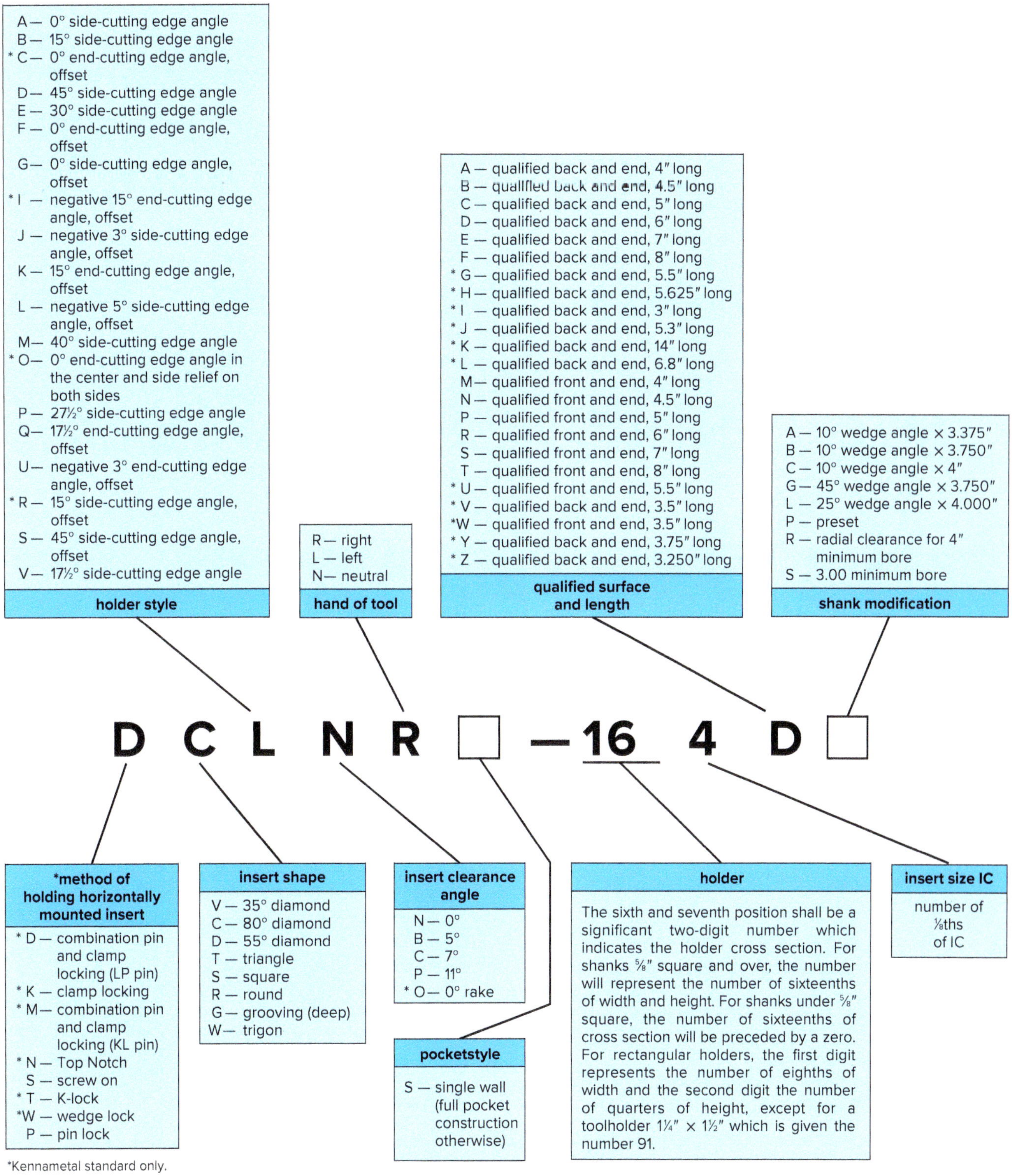

Figure 10-116 Scan for understanding—don't try to memorize this selection chart.

Metric Toolholder Identification System

M C L N R □ - 25 25 M 12 Q

holder style

A — straight shank with 0° side-cutting edge angle
B — straight shank with 15° side-cutting edge angle
C — straight or offset shank with 0° end-cutting edge angle
D — straight shank with 45° side-cutting edge angle
E — straight shank with 30° side-cutting edge angle
F — offset shank with 0° end-cutting edge angle
G — offset shank with 0° side-cutting edge angle
J — offset shank with negative 3° side-cutting edge angle
K — offset shank with 15° end-cutting edge angle
L — offset shank with negative 5° side-cutting edge angle
M — straight shank with 40° side-cutting edge angle
N — straight shank with 27° side-cutting edge angle
O — offset shank with 30° negative end-cutting edge angle
R — offset shank with 15° side-cutting edge angle
S — offset shank with 45° side cutting edge angle

rake

N — negative
P — positive

hand of tool

R — right
L — left

holder length

H — 100mm
J — 110mm
K — 125mm
L — 140mm
M — 150mm
N — 160mm
P — 170mm
Q — 180mm
R — 200mm
S — 250mm

method of holding insert

(horizontally mounted)
K —clamp locking
M—with locking pin (insert with hole in center) and clamp
N —Top Notch clamp
S —screw down
W—wedge locking

insert shape

C —30° diamond
D —55° diamond
K —55° parallelogram (Top Notch)
R —round
S —square
T —triangle
V —35° diamond
W—trigon

holder height

A significant two-digit number which indicates the holder height in millimeters.

holder width

A significant two-digit number which indicates the holder width in millimeters.

qualified

pocket style

S—single wall, (full pocket construction otherwise)

insert cutting edge length in millimeters

IC	¼" 6,35	3/5" 9,32	½" 12,70	⅝" 15,88	¾" 19,05	1" 25,40
triangle	11	16	22	27	33	44
square round		09	12	15	19	25
55° diamond			15	19		
80° diamond			12		19	25
35° diamond		16				

D —insert IC
L—length on insert cutting edge

Figure 10-117 Standard lathe tool selection identification chart.

Turning Insert Identification System

symbol	insert shape	shape	nose angle (degree)
S		square	90
T		triangular	60
C		rhombic (diamond)	80
D			55
E			75
F			50
M			86
V			35
W		hexagonal (trigon)	80
H		hexagonal	120
O		octagonal	135
P		pentagonal	108
L		rectangular	90
A		parallelogram-shaped	85
B			82
K			55
R		round	—
1 shape			

tolerances

IC: theoretical diameter of the insert inscribed circle
T : thickness
B : see figure below

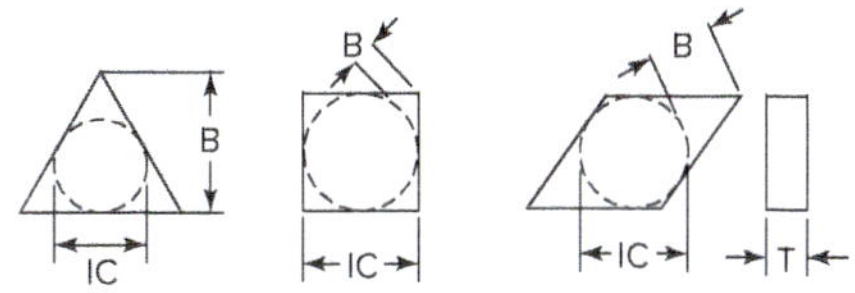

tolerance class	tolerance on "IC"		tolerance on "B"		tolerance on "T"	
	in	mm	in	mm	in	mm
C	±.0010	±0,025	±.0005	±0,013	±.001	±0,025
H	±.0005	±0,013	±.0005	±0,013	±.001	±0,025
E	±.0010	±0,025	±.0010	±0,025	±.001	±0,025
G	±.0010	±0,025	±.0010	±0,025	±.005	±0,13
M	see tables at right		see tables at right		±.005	±0,13
U	see tables at right		see tables at right		±.005	±0,13
3 tolerance						

example

inch	C	N	M	G
(metric)	C	N	M	G
position	1	2	3	4

relief angle 2
N—0°
A—3°
B—5°
C—7°
P—11°
D—15°
E—20°
F—25°
G—30°

4 insert type					alternate symbols	
symbol	hole	shape of hole	chipbreaker	shape of insert's section	ordinary system	IC less than ¼"*
N	without		without		N	E
R			single sided		R	
F			double sided		F	
A	with	cylindrical hole	without		A	D
M			single sided		M	
G			double sided		G	
W		partly cylindrical hole, 40–60° countersink	without		A	
T			single sided		M	
Q		partly cylindrical hole, 40–60° double countersink	without		A	
U			double sided		G	
B		partly cylindrical hole, 70–90° countersink	without		A	
H			single sided		M	
C		partly cylindrical hole, 70–90° double countersink	without		A	
J			double sided		G	
X	special				X	X

*inch system only.

5 size

inch	IC		metric cutting edge length						
	in	mm	C	D	R	S	T	V	W
1.2 (5)	5/32	3,97	—	04	03	03	06	—	—
1.5 (8)	3/16	4,76	04	05	04	04	08	08	—
1.8 (7)	7/32	5,56	05	06	05	05	09	09	03
—	0.236	6,00	—	—	06	—	—	—	—
2	1/4	6,35	06	07	06	06	11	11	04
2.5	5/16	7,94	08	09	07	06	13	13	05
—	0.315	8,00	—	—	08	—	—	—	—
3	3/8	9,52	09	11	09	09	16	16	06
—	0.394	10,00	—	—	10	—	—	—	—
3.5	7/16	11,11	11	13	11	11	19	19	07
—	0.472	12,00	—	—	12	—	—	—	—
4	1/2	12,70	12	15	12	12	22	22	08
4.5	9/16	14,29	14	17	14	14	24	24	09
5	5/8	15,88	16	19	15	15	27	27	10
—	0.630	16,00	—	—	16	—	—	—	—
5.5	11/16	17,46	17	21	17	17	30	30	11
6	3/4	19,05	19	23	19	19	33	33	13
—	0.787	20,00	—	—	20	—	—	—	—
7	7/8	22,22	22	27	22	22	38	38	15
—	0.984	25,00	—	—	25	—	—	—	—
8	1	25,40	25	31	25	25	44	44	17
10	1 1/4	31,75	32	38	31	31	54	54	21
—	1.260	32,00	—	—	32	—	—	—	—

NOTE: inch sizes in parentheses for "alternate sizes" D or E (under ¼ inch IC).

Figure 10-118 Standard numbering system for carbide inserts.

IC		± tolerance on "IC"							
		class "M" tolerance						class "U" tolerance	
		shapes S, T, C, R & W		shape "D"		shape "V"		shapes "S, T & C"	
inch	metric	inch	mm	inch	mm	inch	mm	inch	mm
5/32	3,97			—	—	—	—	—	—
3/16	4,76			—	—	—	—		
7/32	5,56								
1/4	6,35	.002	0,05	.002	0,05	.002	0,05	.003	0,06
5/16	7,94								
3/8	9,52								
7/16	11,11								
1/2	12,70	.003	0,06	.003	0,06	.003	0,06	.005	0,13
9/16	14,29								
5/8	15,88								
11/16	17,46	.004	0,10	.004	0,10	.004	0,10	.007	0,18
3/4	19,05								
7/8	22,22	.005	0,13	—	—	—	—		
1	25,40			—	—	—	—	.010	0,25
1 1/4	22,22	.006	0,15	—	—	—	—		

IC		± tolerance on "B"							
		class "M" tolerance						class "U" tolerance	
		shapes S, T, C, R & W		shape "D"		shape "V"		shapes "S, T & C"	
inch	metric	inch	mm	inch	mm	inch	mm	inch	mm
5/32	3,97			—	—	—	—	—	—
3/16	4,76			—	—	—	—		
7/32	5,56	.003	0,06			—	—		
1/4	6,35			.004	0,11	—	—	.005	0,13
5/16	7,94					—	—		
3/8	9,52					.007	0,18		
7/16	11,11					—	—	—	—
1/2	12,70	.005	0,13	.006	0,15	.010	0,25	.008	0,20
9/16	14,29					—	—	—	—
5/8	15,88					—	—		
11/16	17,46	.006	0,15	.007	0,18	—	—	.011	0,27
3/4	19,05					—	—		
7/8	22,22			—	—	—	—		
1	25,40	.007	0,18	—	—	—	—	.015	0,38
1 1/4	22,22	.008	0,20	—	—	—	—		

—	4	3	2	□	□	□
	12	04	08	□	□	□
	5	6	7	8	9	10

6 thickness			
thickness		symbol	
in	mm	inch	metric
1/32	0,79	0.5 (1)	—
1/16	1,59	1 (2)	01
5/64	1,98	1.2	T1
3/32	2,38	1.5 (3)	02
1/8	3,18	2	03
5/32	3,97	2.5	T3
3/16	4,76	3	04
7/32	5,56	3.5	05
1/4	6,35	4	06
5/16	7,94	5	07
3/2	9,52	6	09
7/13	11,11	7	11
1/2	12,70	8	12

NOTE: inch sizes in parentheses for "alternate sizes" D or E (under 1/4 inch IC).

hand of insert (optional) 8

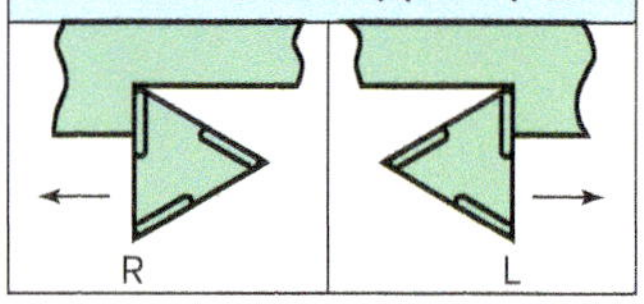

7 corner radius			
corner radius		symbol	
in	mm	inch	metric
.004	0,1	0	01
.008	0,2	0.5	02
1/64	0,4	1	04
1/32	0,3	2	08
3/64	1,2	3	12
1/16	1,3	4	16
5/64	2,0	5	20
3/32	2,1	6	24
7/64	2,3	7	28
1/8	3,2	8	32
round insert (inch)		—	00
round insert (metric)		—	M0

9 & 10 cutting edge condition or chip control features (optional)

- T — negative land
- K — light feed chip control, double sided Kenloc insert
- M — heavy feed chip control, deep floor Kenloc
- N — narrow land Kentrol insert with chip control on one side
- W — heavy-duty chip control, wide land Kenloc insert one side
- J — polished to 4-microinch AA (rake face only)
- UF— ultra-fine finishing

See Technical Section for additional conditions and chip control features.

Figure 10-119 Standard numbering systems for carbide inserts.

Similar to the holder, the insert itself has a numbering system. Figure 10-119 is an example of a standard number chart.

TRADE TIP

Chip Breaking Breaking the pesky long chips into safer "C" shapes improves your setup and workday as well. Insert tools usually feature a **chip breaker** groove or a protruding bump that bends the chip violently enough as it flows past that it breaks into small, safe, easy-to-handle packets. However, when you grind your own bit it is possible to add a chip breaker groove.

In place of grinding a breaker into the tool, machinists often set up a small piece of metal known as a **chip dam** or chip trap that also bends the chip so tightly that it breaks off (Fig. 10-120). It is simply clamped in with the tool, then adjusted to work best yet release the chip after breaking it.

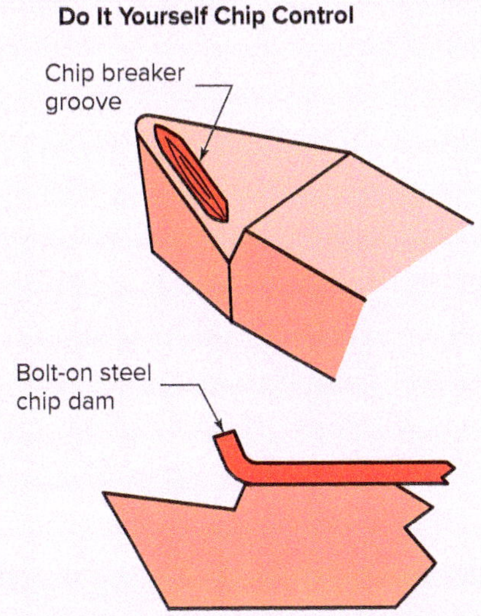

Figure 10-120 A chip breaker, or chip dam, aids in creating safer, easier to handle "C" chips.

10.4.6 Manual Lathe Tool Posts

A tool post holds the cutting tool on the lathe compound. There are four types. Each offers some advantage.

Industrial Varieties

These include

- Rocker tool post
- Solid tool post
- Quick change tool post
- Indexing turret tool post

Not often used in industry, the rocker tool posts are the most universal but the least strong.

Tool Post Strength

With each tool post, I'll provide a relative depth of cut and feed rate. The descriptions are meant to provide comparisons between the various kinds, not to set a standard. Each tool post is different in terms of sturdiness, and that's a prime factor in choosing the right one for your setup. If the work requires light cuts, then all of the following can be chosen. But as the cut forces increase, some posts should not be used.

Rocker Tool Posts and Tool Holders

This is the weakest of the four presented but is also the lowest cost and the most universal. The rocking action allows easy vertical and right and left positioning. Rocker posts hold standard HSS cutting tools. The cutting tool is clamped in a tool shank by tightening a square-headed set screw (Fig. 10-121). Then the pair are clamped into the post. Due to their tendency to move and slip, rocking posts are not often used in industry especially where heavy cuts are made, but when the standard tools don't do the job, they can be useful in some situations.

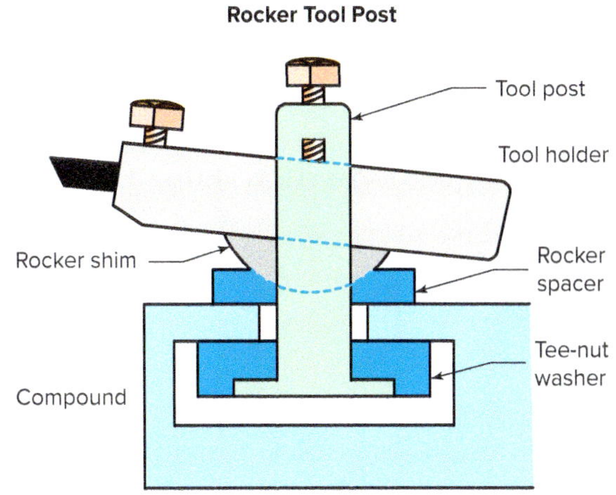

Figure 10-121 Six rocker post components: tee-nut washer, tool post, rocker disk spacer, rocker, clamp bolt, and tool holder.

In order to use a rocker tool post, the shop must have a selection of the tool holders (Fig. 10-122) made for the post. Note the brazed carbide cutter on the left-hand holder. These holders have a "hand" similar to tool bits. This allows the tool more "reach" in tight machining situations and provides better cutting angles in some setups. They are either right, left, or neutral (pointing straight ahead). Slow to exchange one for another, they are not suitable for jobs requiring several tools to complete the shape.

Figure 10-122 Left-, neutral-, and right-hand rocker post tool holders.
Prographics at Milwaukee Area Technical College/McGraw Hill

Why Use a Rocker Tool Post?

The main reason is that nothing else is available. However, rocker tool posts are able to rotate through 360°, and the tool can be pivoted up or down on the rocker. This makes them simple to set up and very universal. Carefully used, a rocker tool post can meet most of your training needs.

Solid Tool Posts

Also called open-sided tool posts, they are the opposite of rockers: heavy-duty brutes. Solid tool posts can hold HSS bits, or the tool holders for rockers, or carbide insert tools, as shown in Fig. 10-123, but only as long as the tool's cutting edge can be shimmed to the vertical center of the work. They are used for roughing work where extra heavy metal removal rates are needed.

Solid or Open-Sided Tool Post

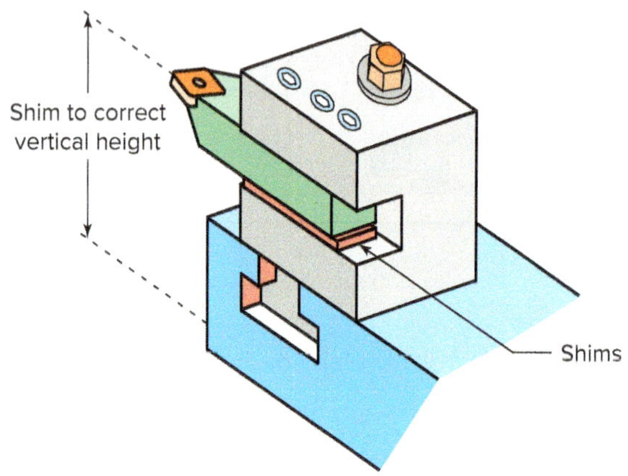

Figure 10-123 A solid tool post (open-sided) is the strongest and simplest of the four types presented.

> **KEYPOINT**
>
> For safety, limit cuts to a maximum depth of 0.045 in. and 0.010 to 0.015 IPR when using a rocker tool post—or to the limits set by your instructor.

But because the tool shank is held down with set screws, they are not meant for changing tools. Set up these tool posts when you expect to use a single tool throughout the turning of the product.

While they are easily rotated to point the tool in any direction, vertical adjustment is possible only by adding shims. As always, be mindful of the tool's overhang from the post–keep it to the minimum.

Quick Change Tool Posts

Also called cam lock tool posts, their purpose is to allow several different tools to be set up, then quickly exchanged–one at a time. Each tool is mounted on its own tool block. When the tool block with its cutting tool is removed, then returned to the post again, the cutting tool tip returns to its original position relative to the X and Z axes, within a repeatability of perhaps 0.0005 in. or less.

> **KEYPOINT**
>
> You would select this type of tool post for heavy roughing or when only one tool is needed. The depth of cut can be 0.100 in. or slightly more and feed rates as much as 0.025 IPR.

The quick change tool post features a dovetail method of removing and replacing the tool blocks in the same position every time. The vertical height of the tool is set with a locator pin and adjusting screw. Note in Fig. 10-124 that a drill or reamer can be held in the quick change tool post rather than the tailstock, but you are then faced with positioning it on the exact X axis center of the work.

> **SHOPTALK**
>
> **Standard Load Tools** In both CNC and manual lathe work, we often leave the most common turning tools mounted in the turret or in a set of tool blocks all the time. Then the required custom tools are added in other blocks (or turret stations) for a particular setup.

Why Choose a Quick Change Tool Post?

Set up a quick change tool post where several cutting tools are required to finish parts. They are especially useful when running batches of parts with multiple operations. After initial setup of the cutting tool in its own tool block, exchange time tool-to-tool is a matter of just seconds.

> **KEYPOINT**
>
> Quick change posts are appropriate for medium heavy cuts. Nearly as strong as open-sided posts, cut depths to 0.085 in. and feeds to 0.020 IPR are acceptable.

Quick change tool post sets (Fig. 10-125) are supplied in different sizes to accommodate various lathe swings.

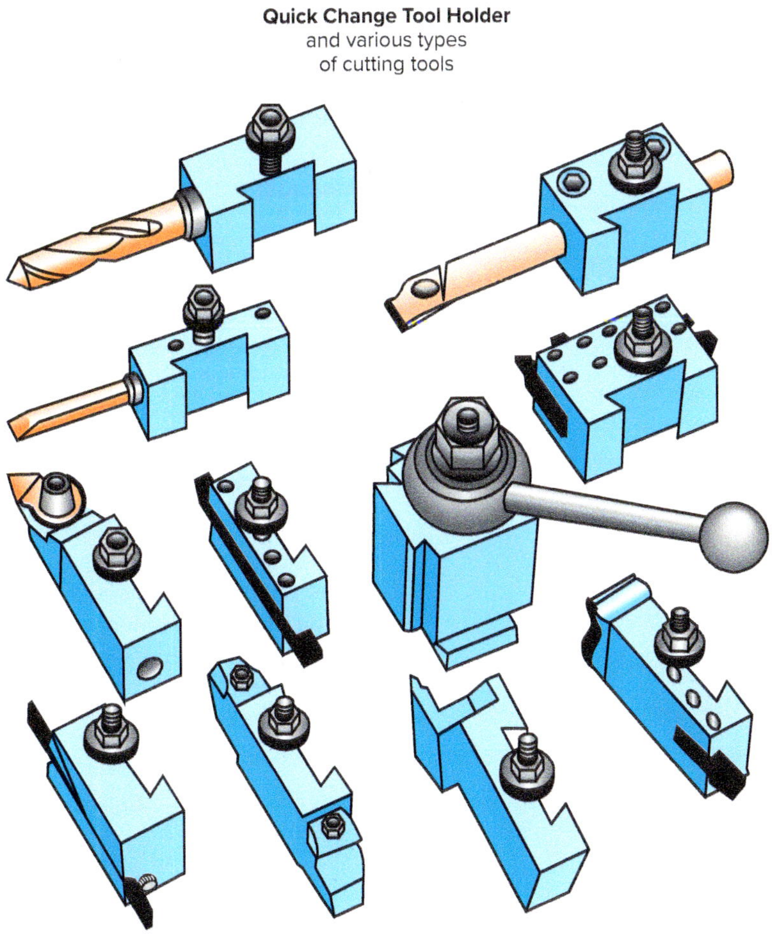

Figure 10-124 Quick change tool posts with preset blocks reduce tool changing time, plus they repeat position accurately when replaced.

Figure 10-125 A quick change tool post in action.
Courtesy Universal Aerospace/Arlington, WA

KEYPOINT

Safety

Minimize Tool Overhang As any cutting tool is extended out away from its mount, it becomes a lever trying to pull down, which digs into the work and may break the insert or cutter or even bend the work up over the cutter.

Stop the Lathe to Change Tools! With quick change tools, it's tempting to exchange one block for another without stopping the spindle, especially when racing the clock in production work—but don't!

Indexing Turret Tool Post

This accessory is another way of setting up multiple cutting tools, but here, they stay on the tool post and lathe. A screw, cam, or lock pin action allows the operator to unlock the turret, rotate it to several discrete positions relative to the work (usually 8 or 16 positions), then securely lock it in position.

Why Choose a Turret Tool Post?

These holders are the fastest for production work if the job doesn't require more than four cutting tools. When indexed to the next tool, they repeat position within 0.0005 in. or better.

Although the turret will hold the tool shank for which they were designed, close to vertical center, most will require shimming. Thus, they are slow to set up initially. It's also necessary to ensure that the other cutting tools not being used do not interfere with the work. For example, in Fig. 10-126, the carbide parting tool (upper right) might touch the tailstock center. If it does, then it must be moved to another position or the turning tool on the left could be slid farther out. Most CNC turning centers (production lathes) present the same challenge, except their tool turrets hold many more tools.

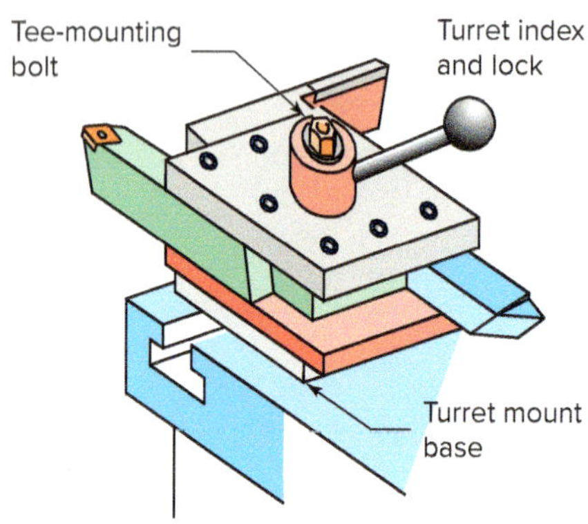

Figure 10-126 A square turret offers multiple positioning for several permanently mounted tools.

KEYPOINT

Tool interference is one of the many items that must be cautiously and slowly checked when the CNC program is tried for the first time.

Why Choose a Turret Tool Holder?

This tool post is designed for runs of many parts where two to four cutting tools are needed to turn a part. The turret post is the fastest for changing tools during turning operations and the second most solid holding method behind solid tool posts.

KEYPOINT

Cuts of around 0.090-in. depth and 0.025 IPM are possible.

The entire turret is preloaded with cutting tools and quickly bolted to the lathe for a superfast setup. For example, a right-hand turning tool, a left-hand tool, a parting tool, and a chamfer forming tool could be left on the turret; then custom tools could be added as needed. Turrets are costly but rugged, with a long life in the shop.

Indexing Quick Change Tool Posts

Combining the best features of both production tool posts, the quick change indexing tool post shown in Fig. 10-127 offers removable tool blocks, but it can hold four blocks on the turret at any time–one per side. Then the turret can be indexed to 24 positions–every 15°. This adds great flexibility to the setup.

Figure 10-127 An indexing quick change turret features 15° discreet positioning and quick change blocks.
Aloris Tool Technology Co., Inc./Clifton, NJ

Tool Post Application

Type	Relative Removal	Best Suited for
Rocker	0.045 in. at 0.010 to 0.015 IPR	Light work
Solid or open-sided	0.100 in. at 0.025 to 0.035	Roughing and simple work
Quick change	0.085 in. at 0.020 to 0.025	Complex batch requiring multiple tools
Indexing turret	0.090 in. at 0.025 to 0.030	Rough/finish, medium number of tools
Indexing quick change	0.090 in. at 0.025 to 0.030	Very complex setups for batches

10.4.7 Positioning Turning Tools on Vertical Center

When mounting a tool in any tool post, it's important to set the cutting edge at or near the vertical center of the work. As shown in Fig. 10-128, it will cut okay if positioned slightly below, but the low cutting tool can dangerously encourage thin work to bend and then climb up over the bit. Also, low tools tend to distort form shapes.

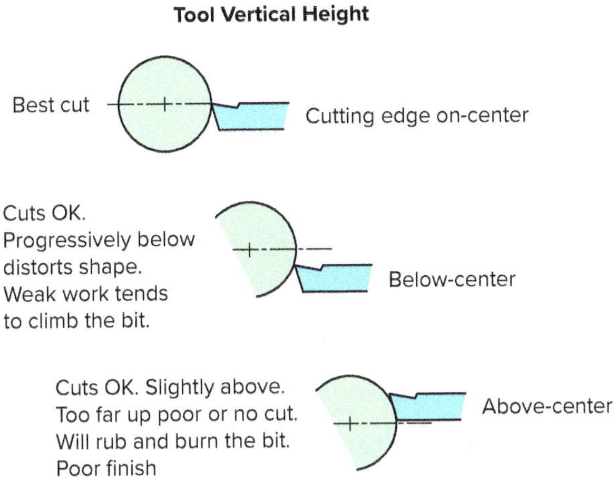

Figure 10-128 The vertical height of the cutting edge must be set.

> **KEYPOINT**
>
> If the tool is very far above center, the clearance face will rub and the cutting edge will not contact the work at all (see the Trade Tip).

> **TRADE TIP**
>
> There is disagreement among masters with regard to tool centering. Some recommend that a lathe cutting tool should be set just a bit above-center because the cutting forces will pull it downward during heavy cuts; thus, it ends up on-center. I do not agree, as it is a guessing game as to how far above to place it. I place it on-center.

Three Methods of Tool Centering

Here are the ways we ensure the tool's cutting edge is on-center.

1. **Dedicated Height Gage (Best Method)**

   The surest method is a simple, shop-made bent rod pointer. The rod is welded or screwed into a heavy steel base. The rod end is sharpened as a divider point, with a flat surface (Fig. 10-129). Set upon the top surface of the cross slide or bed ways (whichever is more convenient on that lathe), it makes tool alignment a snap.

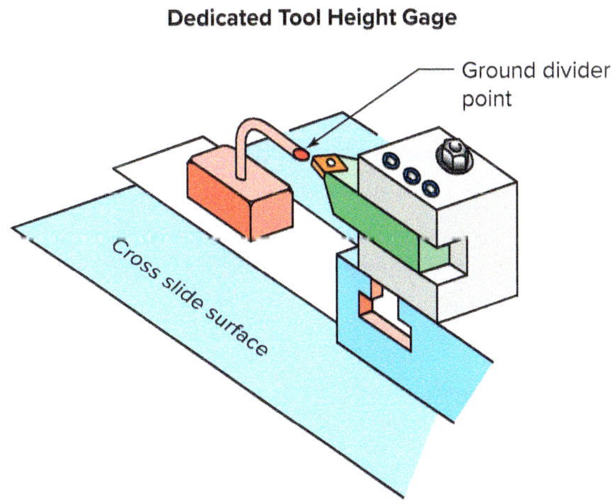

Figure 10-129 A shop-made height gage points out the spindle centerline.

2. **Tailstock Center**

   Sliding the tool and tailstock (or headstock) center together can also point out vertical centering (Fig. 10-130).

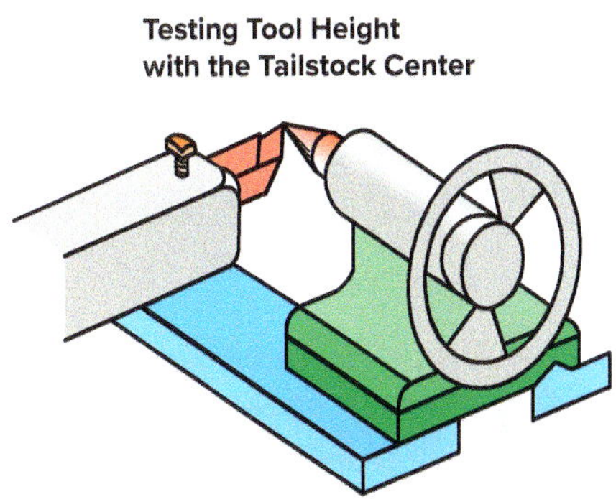

Figure 10-130 The cutting tool can be compared to the height of the tailstock/headstock center point.

> **TRADE TIP**
>
> A surface gage used for layout work or a small height gage, set to the correct vertical distance, can also be used to align cutting tools on the lathe if you know the vertical height from a reliable flat surface.

3. **Captured Flat Object (Quick and Dirty Method)**

   This method is a trade trick recommended only if you take it easy. Here, the cutting tool is cranked forward against the work surface to trap and hold any flat object (a wood popsicle stick or if you are extra careful, your pocket ruler) *lightly* against any centered round surface, the workpiece, the chuck side surface, or tailstock quill. Never use enough pressure to mar objects! The slant of the trapped object indicates vertical position (Fig. 10-131). Not all instructors recommend this method, especially when trapping a pocket rule, for it is inexact, potentially damaging to the ruler and the cutting tool's sharp cutting edge; plus it can chip brittle carbide tools if too much pressure is brought to bear. I do use it, as it's a quick way to check rough tool height.

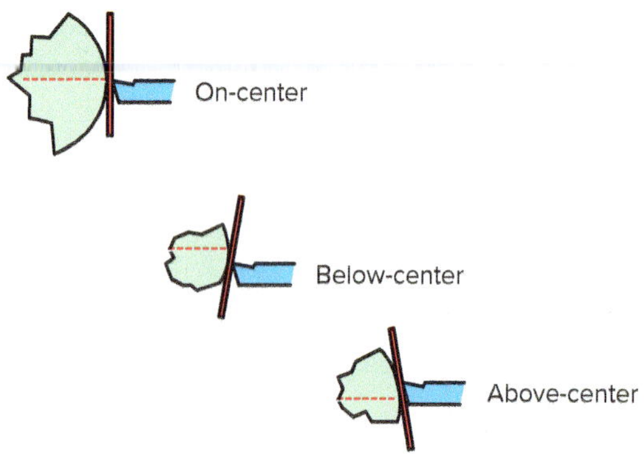

Figure 10-131 Capturing a flat object between the bit and round workpiece roughly shows tool height. Caution, this can chip carbide bits.

UNIT 10-4 REVIEW

Replay the Key Points

- Choose HSS tools over carbide when an *interrupted cut* must be taken that breaks harder tools (a non-continuous surface where the cutter must enter and come out of the cut with each revolution). They also work better where flex or chatter breaks harder cutting tools.
- Choosing a large insert for small cut depths is not economical. But choosing a small insert that is overused is not economical, either.
- For safety, limit cuts to around 0.050- to 0.075-in. depth and 0.008 to 0.012 IPM when using a rocker tool post—or to the limits set by your instructor.
- Never try to exchange a quick change tool while the spindle is turning or rotate tool turrets. *Stop the lathe to change tools!*
- Always have someone check your setup as you learn.

Respond

The CNC setup sheet notes that Tool #5 must be a KTMNR-164 tool holder. The following five questions refer to that holder. For outside research, you might go to *Machinery's Handbook* rather than the charts here.

1. What is the size of the insert?
2. What is the "hand" of this holder?
3. What size is the shank?
4. What is the shape of the insert used on this tool?
5. What is the recommended maximum working depth of the insert for this tool?
6. A cemented carbide tool is chosen when carbide is a must, but some custom shape is needed that is not available in insert tooling. Is this statement true or false? If it is false, what will make it true?
7. A rocker tool post is chosen because of its ability to swivel and pivot in any direction and take deep or heavy cuts. Is this statement true or false? If it is false, what will make it true?
8. Identify the tool post in Fig. 10-132.
9. Name the three tool posts most often used on standard lathes in industry (this excludes rocker tool posts) and give a brief description of each.
10. You have an order of 250 spring pins. They require a large amount of straight turning, facing, and a chamfer on the corner. Which tool post would be best for this production run?
11. Modifying the job in Question 10, you now need to knurl the work, make a deep groove on one end, and also perform an internal bore to a depth of 4 in. as well as the other operations required. Which tool post would be best?
12. On the very first cut for roughing away the excess metal on the pins in Questions 10 and 11, the HSS tool gets very hot, and the finish is very poor. What is wrong, and what can you do?

Note that there are many factors here. Take your time responding and consider all the possibilities.

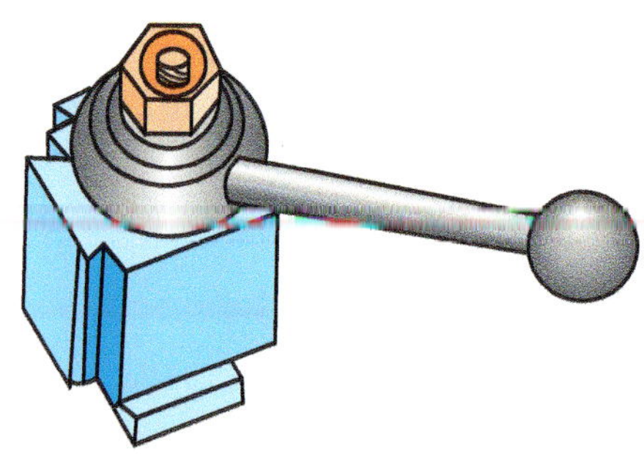

Figure 10-132 Identify this production tool post.

Unit 10-5 Lathe Safety

Introduction: Coming up in Units 10-6 and 10-7, we'll look at setups. From the start, it's beyond critical that you plan and execute every setup with safety as the prime directive. In this unit we'll consider some specific safety tips that pertain directly to lathe operation.

KEYPOINT

Unit 10-5 is short but is packed with *key points!* As a beginner, always have your instructor or shop lead check your setup.

This unit's solutions are by no means comprehensive; they are the common problems we instructors observe beginners experiencing. There are two overshadowing traits you must develop:

1. Safety first and always–it's easy to get too busy to remember safety!
2. If it looks dangerous, it most likely is. Discipline yourself to take the time to do it right.

From all the instructors who helped plan this book, here is our condensed best advice on how to stay in full control of the potential danger of turning.

10.5.1 Best Practice Safety on Lathes

Know Your Equipment

Here are two different aspects to equipment skills to develop:

1. Review the various dials and levers before early machining exercises and always ask yourself, "What will happen when I pull this lever or turn this dial?" Think out your actions before performing them, especially where the bake and stop levers are located.
2. Have an emergency crash plan ready and practice it! It's common for a beginner or even an experienced machinist to do the wrong thing during a developing accident. When something goes wrong–a part slips in a chuck or a tool breaks, for example–you must have quick reflexes to stop the machine. That takes practice. With a delayed or, even worse, a wrong action, accidents are amplified beyond their initial danger *by the operator* (Fig. 10-133)!

Figure 10-133 Never start the lathe without a review of how to stop it quickly and safely—the spindle lever or the foot brake.
Lake Washington Technical College, Kirkland, WA/McGraw Hill

10.5.2 Recognize Three Classes of Danger

Class 1 You hear or see a sign that something might not be right. Stop, investigate, and fix it before it does degrade. Examples include changing sounds, degrading finishes, smoke, steam, or blue chips when they weren't previously seen. In this case, you might let the cut finish before stopping.

Class 2 No doubt about it, you hear or see something coming loose or breaking. The danger is real, and you must act quickly to prevent it from degrading into a class 3 crash. Stop now, investigate, and fix the problem.

Class 3 This is the catastrophe. It's a fast-developing, all-out crash–there is nothing to do but turn your back to the machine, then get away as fast as you can. Once it's under control and everybody is safe, confer with your instructor or supervisor to be certain this never happens again.

The point is, there's no time to think out your actions when a class 2 or 3 accident occurs. Like a Kung-Fu master, be mentally prepared to act. This goes handinhand with your crash plan. Prepare now for the eventuality when there is nothing to do but duck and run! We'll review the three classes again in CNC machining, for there the crashes are far more spectacular!

KEYPOINT

No matter how valuable the machine, the machinist is more valuable.

Build in Safety with Every Setup

Setups that are not well planned are a common cause of accidents and ruined work. Choose work-holding and cutter tooling that's rigid with the absolute minimum overhang. Test cutting tools for chuck interference by rotating the spindle in neutral before engaging power turning.

In Fig. 10-134 how many things can you find that are wrong with setup B compared to setup A? Not only are setup B's tool and holder set too far out from the turret, but the turret itself is mounted out beyond the lip of the compound. Adding to the overhang, the bit is out too far from the holder. This is a weak, long lever arm just made for trouble.

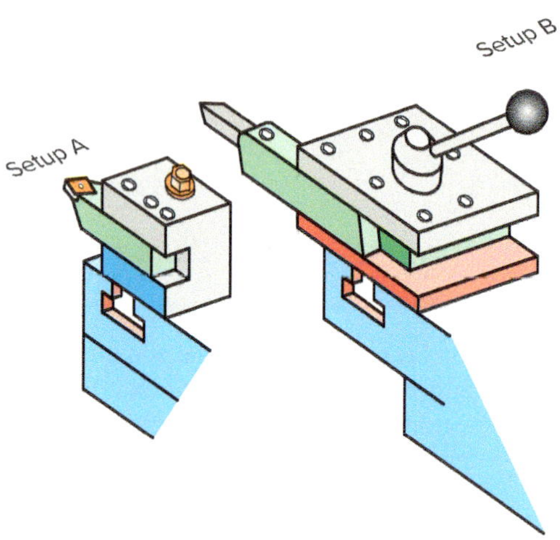

Figure 10-134 The tool on the right extends too far. It will flex down and vibrate, perhaps digging into the work.

What's wrong with the setup shown in Fig. 10-135?

Figure 10-135 What is wrong with this setup?
Prographics at Milwaukee Area Technical College/McGraw Hill

I'm sure you saw the unsupported bar now correctly held up with the tailstock. But did you see the compound was too far out (Fig. 10-136)?

Figure 10-136 Now it's safe to turn.
Prographics at Milwaukee Area Technical College/McGraw Hill

10.5.3 Make Sure Chuck Guards Are in Place

Don't forget to add the chuck guard to your setup. The chuck guard does three things (Fig. 10-137):

1. It keeps coolants inside the lathe rather than letting them fling off the chuck into the shop.
2. It prevents chips from flying out to hit you or others nearby.
3. It helps contain the workpiece should it accidentally be flung out.

Figure 10-137 This lathe has a chuck guard keeping the chips and coolant from flying out.
Prographics at Milwaukee Area Technical College/McGraw Hill

Guards are even more critical in CNC turning, where the forces and speeds are accelerated. Most CNC turning centers feature a full containment safety shield that surrounds the entire machining area. If that guard isn't in place, the lathe will not start except by a major adjustment to the control's software (called a parameter change). Defeating a safety feature–really, I don't think so!

10.5.4 Control Chips

Long, stringy chips can cause multiple problems. They cut and grab, and they also catch and wrap around chucks with an amazing speed causing a tangled mess, called a *rat's nest* (Fig. 10-138). There are two steps you can take to prevent chip problems.

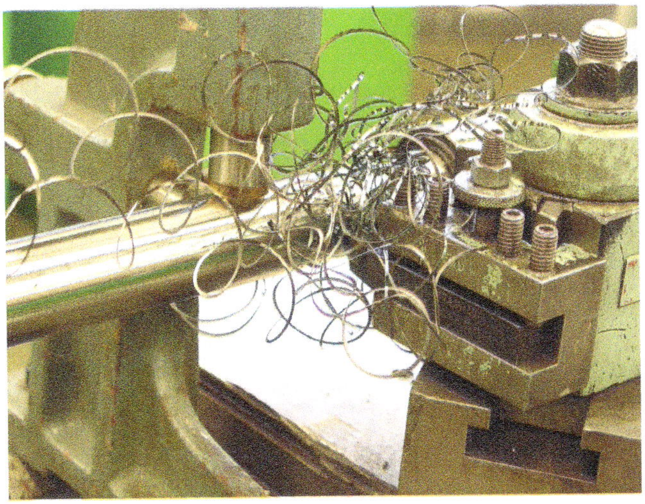

Figure 10-138 *Danger!* This setup is producing a sharp, hot, and very strong wire.
Lake Washington Technical College, Kirkland, WA/McGraw Hill

Break Chips into "C" Shapes

Do this by using better feed rates and tool geometry and by adding a chip breaker on top of the tool (Fig. 10-139). Ask your instructor to show you how to break your chip if it starts to dangerously curl up and out of the setup. If it's a roughing cut, you can briefly stop the feed movement, then start it again. Don't stop it long enough to let the tool dwell and rub, but the quick stall causes a weak place in the chip that breaks it. There are also times when without disengaging the feed, the handwheel can be moved forcefully forward, called *crowding the wheel.* Then, when it is let go, that also causes a weak place in the chip.

Keep Long Chips Cleared Away

Sometimes it isn't possible to break chips, especially on light cuts. When this happens use the chip hook shown in Fig. 10-140 to keep stringy chips pulled away from the chuck and the work. Also keep piles of chips swept away from the machine. They can cause feet to slip on floors, which then leads to further injury if you fall toward the moving lathe.

Figure 10-139 Broken "C" shape chips indicate correct tool geometry as well as good cutting speed and feed rate.
Courtesy Kennametal, Inc.

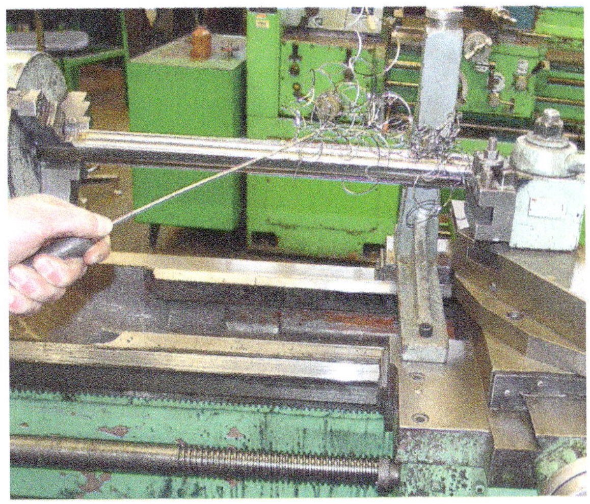

Figure 10-140 Use a chip hook to safely pull long chips out and away from the lathe.
Lake Washington Technical College, Kirkland, WA/McGraw Hill

10.5.5 Calculate Speeds and Start Moderately, Then Work Up to Maximum Rates

As a beginner, do not guess at the correct speeds. Calculate them and have someone check them with your first few setups.

10.5.6 In Case You Missed It: Safety Reminders

Don't Stand in the Danger Zone When Operating the Machine

Remember, when filing use your left hand on the file handle and during all turning operations, stand to the right of the area where chips will be ejected (Fig. 10-141).

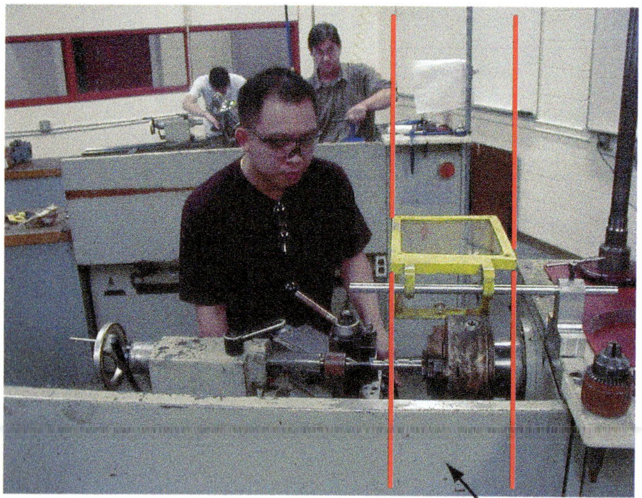

Figure 10-141 Stand to one side of the danger zone, away from the rim of the chuck and workpiece where the chips could fly out of the setup.
Lake Washington Technical College, Kirkland, WA/McGraw Hill

Mind That Chuck Wrench!

Remember, the forgotten chuck wrench left in the chuck is transformed into a missile when the spindle is turned on. Never take your hand off the wrench when it's in the chuck.

Dress for the Job

Also a reminder, no jewelry or loose clothing. Control your hair with a hat. Wear eye protection even if the machine isn't moving and protect your hearing with plugs or muffs if noise is uncomfortably high.

UNIT 10-5 REVIEW

Replay the Key Points

- Know your equipment.
- Consider your actions before trying them.
- Have an emergency crash plan ready and practice it!
- Build in safety with every setup.
- Instantly recognize three classes of danger.
- Make sure chuck guards are in place.
- Build in chip control. Use chip breakers. Keep dangerous stringy chips cleared away.
- Always turn off and lock out the power to the lathe or open main breakers for chuck or other accessory changes, such that the lathe cannot be accidentally started. This is supercritical for CNC setup work!
- Calculate speeds and feeds–start out moderately.
- Don't stand in the danger zone when operating the machine.
- Remember that chuck wrench!
- Dress for the job.

Respond

1. Name the four ways a pro controls chips on a lathe.
2. What operator action can you take when a class 1 event occurs?
3. What is the main characteristic of a class 3 crash?
4. What's a good rule to prevent flying chuck wrenches?

Critical Thinking

5. What would you say is the most common cause of accidents on a lathe?

Unit 10-6 Lathe Setups That Work Right—Troubleshooting

Introduction: After a discussion about calculating RPMs for lathe work, we'll look at setups and troubleshooting. Then, after completing an instructor demonstration plus any planned checkout activities, you'll be ready to set up and operate a lathe.

TERMS TOOLBOX

Coordinating the axis Setting the micrometer dial or digital readout to represent the true diameter of the work.

Capability of the process (Cp) The normal amount of variation in any process.

Statistical process control (SPC) A method of detecting normal and trend variation in any process.

10.6.1 Calculating Lathe RPM

There is an additional consideration for correct RPM on lathes beyond good cutting action: *maximum safe limits!* Recall that general limits for chucks are usually around 1,000 RPM or less on manual lathes. However, please check with your instructor for your shop limit.

KEYPOINT

Your shop will have its upper limit RPM for certain holding methods.

Computing RPMs for turning is similar to the process for drills. To compute the RPM for a lathe setup

1. Find the suggested surface speed based on the tool alloy and metal to be cut. To find a recommended surface speed, use the student chart in Appendix III, a commercial Recommended Surface Speed chart, or *Machinery's Handbook*.
2. Now calculate the RPM such that the work outer rim will run at the recommended surface speed if it is safe. Using the short formula

$$\text{RPM} = \frac{4 \times \text{Surface speed}}{\text{Diameter (of rotating object)}}$$

For example, a 3.0-in.-diameter mild steel part is turned using an HSS turning tool. In the Appendix III chart, we find the recommended surface speed to be 100 F/M.

In the formula

$$\frac{4 \times 100}{3} = 133 \text{ RPM}$$

133 is low—it is possible that after an initial trial, you can probably increase it. Calculating the RPM is an initial step critical to achieve good finishes, accuracy, and safety.

10.6.2 Planning Your Setup

Even a simple job such as the one in this exercise has many steps. Take a few minutes and give it a try, see how you would plan and execute this job. This exercise would be more beneficial if worked with a partner of equal skill. Not all of the steps have been covered in the reading but you have the knowledge to work them out. You will need to extend and read the Trade Tips along the way, too. We'll be following this W/O:

Work Order U11-6-142

Make Qty 15—Steel Shear Pins, Per Dwg 11-142 (Fig. 10-142)

Rough Stock: Cold Rolled Steel 1.375-in. Diameter

Stock has been pre-sawed into 9-in. lengths with excess for grip. The ends are rough. Machine this order using HSS tools only

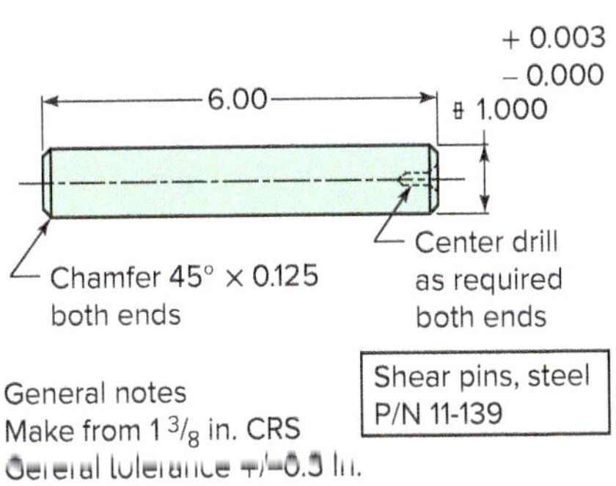

Figure 10-142 Plan and produce 15 of these shafts.

Instructions

Use a sheet of paper to plan every step in the setup and in the machining operations. The goals are safety, quality, and efficiency (in that order). Compare your finished setup and operation guide to the one provided. In making your setup choices, be aware of the print tolerances for various features of the part.

Here is a checklist of the things to be considered:

- ☐ Holding method
- ☐ Cutting tools
- ☐ Tool post type
- ☐ Work support
- ☐ Operation sequences

My Suggested Setup

This is one solution. There are other combinations that would produce good results.

1. **Holding Method—Three-Jaw Chuck**

   There's lots of metal to remove, so runout is not important, but good grip is. In making 15 parts, a reasonable turnaround time is needed. The parts have excess length to hold inside the chuck.

2. **Cutting Tools**

   Right-hand turning tool—cuts toward the chuck

   Facing tool (left-hand turning), cuts inward toward center

   Parting tool

   Chamfer form tool to remove sharp edges per drawing

   Center drill in chuck in tailstock—Note: Be sure the part can have a center hole!

3. **Tool Post**

   Multiple tool setup. Either a cam lock or a turret is right for this job.

4. **Work Support**

   Tailstock, live center. You will have about 7 in. of material protruding from the chuck. Thus, it's beyond the 5:1 ratio for length to diameter of unsupported objects.

10.6.3 Suggested Operation Sequences

Step 1 End Prep

The part protrudes out of the chuck a short distance—1 in. or less. At 500 RPM, face the end—clean up a minimum amount for a true face using a right-hand (RH) facing tool. At 1,000 RPM, center drill per Fig. 10-143 (or your shop's upper chuck RPM limit).

Step 2 Rechuck for Turning

There are 7 in. protruding from the three-jaw chuck. Set up tailstock live center support.

Figure 10-143 Ready for the center drilling of the bar.
Lake Washington Technical College, Kirkland, WA/McGraw Hill

Step 3 Reshift the Lathe for Turning
Set RPM to 400 (turning steel with an HSS tool). Set feed rate 0.015 in./rev.–a roughing rate for the OD (outside diameter).

Step 4 Turning
First coordinate the micrometer dial and digital readout (DRO) if the lathe has one. Just touch the tool bit to the OD of the work; then set the dial or digital readout to 1.3750 (0.375 on micrometer dials), then lightly lock the dial. Next take a short Z axis test cut of about $\frac{1}{4}$ in. long and from 0.030 in. to 0.050 in. deep (the rough pass amount).

TRADE TIP

Grouping Operations It is sometimes more efficient to perform the same operation to all 15 parts before moving on to the next step (Fig. 10-144). In this case, all could be cut to length, chucked, then faced on one end and center drilled. Another example would be in the actual turning of the diameter. Each part might be rough turned to within 0.030 in. of finished size, then set aside to cool while rough turning the remaining parts.

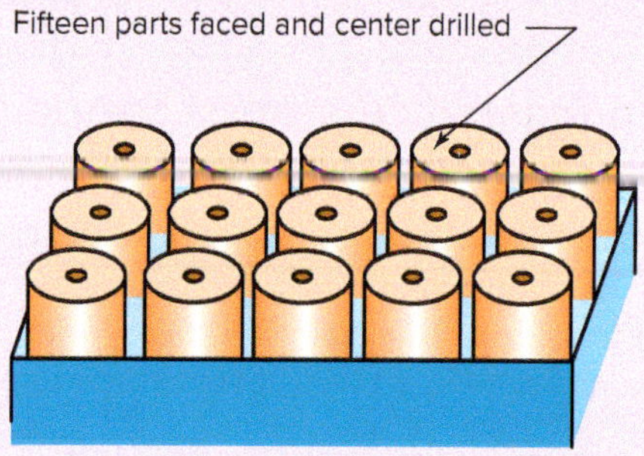

Figure 10-144 To achieve efficiency, all 15 parts could be faced and center drilled without changing the setup.

Stop the lathe and measure the result with a micrometer. Now refine the X axis position on the readout or dial to represent the amount remaining on the shaft until final size. This is an important step often overlooked by beginners. It saves time in getting a final result as quickly as possible. This act is known as **coordinating the axis** (Fig. 10-145).

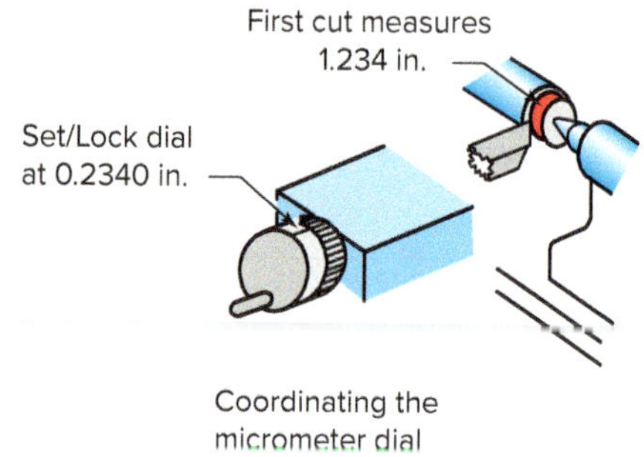

Figure 10-145 Coordinating the micrometer dial to represent the turned results is a time saver.

Step 5 Chamfering
To chamfer the part on the headstock end, make a groove with the parting tool. First, touch the parting tool to the right edge; then move it to the left a distance equal to the blade thickness plus the part length (Fig. 10-146).

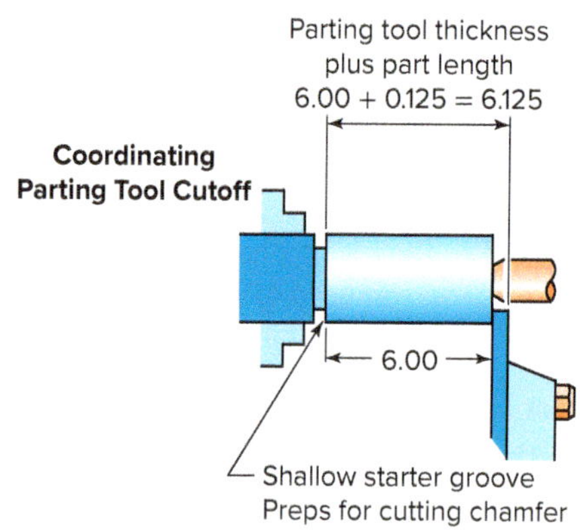

Figure 10-146 Touch the parting tool to the right end of the work; then move it to the left a distance equal to the part length plus the tool thickness. Make a groove for the chamfer tool.

TRADE TIP

Most parting tools are manufactured to an exact thickness of 0.100, 0.125, or 0.1875 in.

Step 6 Parting
After chamfering, remount the parting tool and position it at the groove. Lock the saddle and shift the X axis to feed at 0.005 per revolution–inward. Use coolant for parting to lubricate the chip. This is so even with metals that require no coolant. Lubrication helps the chip come out of the deep groove.

KEYPOINT

Safety Note—The final cutoff

The part will fall down when the parting tool severs it away from the stock remaining in the chuck.

KEYPOINT

Remember, it's unwise to catch the part with your hand! A part-catching tray is often placed under the part.

Be ready! The instant the part is severed from the parent metal still in the chuck you might need to stop the lathe. Have your foot or hand on the stop lever!

10.6.4 Operator Control

Problem Solving When Things Don't Go per Plan!

You need the ability to reason out how to improve a setup or fix a problem. On the previous job the 1.00-in. diameter's tolerance was plus/minus 0.005 in. If, after running 10 parts, you find the diameter varies by, say, plus or minus 0.001 in., then your setup is in control and producing consistent results. In the process, you have a normal variation of 0.002 in., spread evenly around an average size, say, 1.002 in. Parts range from 1.001 to 1.003 in.

The challenge now is to get the process variation centered around the target size, in this case, 1.000 in. To do so requires a final adjustment of the micrometer dial on the machine. Then monitor successive parts to see that the sizes remain centered around the target value.

You'll experience a natural variation from heat, mechanical play in the machine, your own actions in controlling the backlash, and other factors. These factors add up to an overall limiting repeatability of the setup called the **capability of the process (Cp)**.

Your responsibility as a journey machinist is to recognize how much variation is built into the setup and machine and how much is the result of factors that require control. If, for example, you observe a trend of parts growing bigger in diameter by 0.0002 in. with each new part, what would you suspect is the cause?

Tool wear would be the number one suspect, but could there be other factors? How about the tool slipping in the holder? The micrometer dial could be slipping too. So, your job is similar to that of a detective—to reason out the clues and decide the most likely cause.

Statistical Process Control

Beyond safety, the machinist's main duty is to find and control variation, to make parts as close to nominal as possible—that's the foundation of quality work. To do so, you must understand the nature of variation in the setup and cutting process. It divides into three types: common cause (normal), trend, and special cause. Common causes are the amount created by the equipment and process that cannot be eliminated altogether without changing the process (setup or even the machine). Special variation is that created by one-time-only events that might lead to a scrap part, but there is little that can be done to the process to avoid it. For example, every day at 12:45 p.m., a train goes by the factory, causing the machine to vibrate. One needs to detect the cause and avoid it, but you would not change the setup. That leaves trends as the real enemy to quality—something is changing or has changed away from normal variation.

A very good tool for finding and tracking all three kinds of variation is called **statistical process control (SPC)**. It is a math/graphic tool that works especially well on production runs in machining, but it can be applied to any process that has a quality outcome, machining or otherwise. We'll investigate SPC in detail in Chapter 27.

UNIT 10-6 REVIEW

Replay the Key Points

- Your shop will have its own upper limits on certain holding methods—RPM.
- Operations are frequently grouped into batches. The decision to run parts that way depends on turnaround time for the holding method.
- Coordinating the micrometer dial saves lots of extra measuring and testing.
- When a part is severed from the raw stock, do not try to catch it with your hands!

Respond

Taper in a turned shaft is frequently a problem. Here are some critical thinking challenges. The detective work and fixes are typical of the daily challenges we face. Anybody can run a machine when everything is going okay. It's the skilled master who can keep making good parts when things go awry!

1. *Taper in First Part After First Roughing Pass.* Turning a length of $\frac{3}{4}$-in.-diameter CRS stock down to 0.625 in., after the first rough cut you've found a problem. Using a micrometer on both ends and the middle, plus testing it with a straightedge, you find a straight taper (not barrel shaped) with

the tailstock end being the largest size. The difference end to end is 0.018 in. (Fig. 10-147).

A. List what the five possible causes might be.

B. Describe how you might test to eliminate each possibility.

C. Assuming the cause is as described in the answers, find the specific solution. For a quick-fix method, see the Trade Tip in the answers.

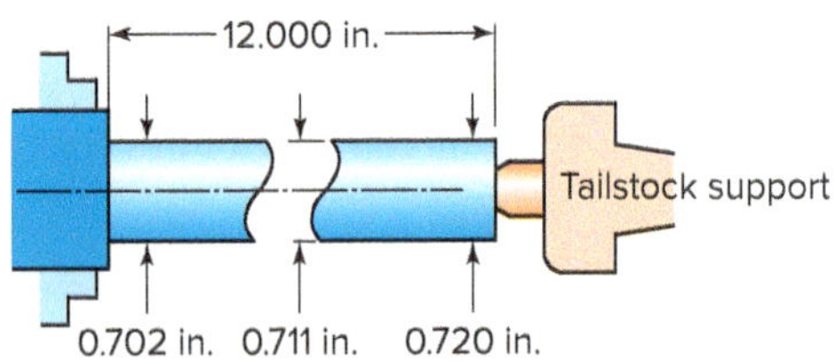

Figure 10-147 This part was produced on the first rough cut. What could the problem be?

2. (Same job as Problem 1) *After Five Parts with No Taper, the Sixth Shows Tapering.* Similar to the last problem, the right end is bigger than the left. List the possible causes.

3. *Poor Surface Finish on Rough and Semifinish Passes.* Solve this problem before taking the final pass.

Unit 10-7 Single-Point Threading

Introduction: Continuing setups, we look at single-point thread cutting on a manual lathe.

You will not be able to read this material, then go to the lathe and make single-point thread. Scan it now for understanding. Then attend the demonstration by your instructor. Finally, take the book with you to use as a reference when setting up and machining the first thread. Pay special attention to the Trade Tips for threading; they work!

TERMS TOOLBOX

Center gage The small arrow-shaped tool used as a setup template and thread checking guide.

Chasing dial The indicator of when to engage the half nut lever to make each pass in the same thread groove as the last.

Half nut The split nut engagement device for pulling the carriage along at the right pitch ratio to the spindle. The half nut lever engages the threading action.

Plunge threading Single-point threading where the compound is not rotated. The thread is cut on both edges of the bit.

Relief (thread relief) A groove formed in the shaft in which the thread bit stops cutting.

Shoulder thread A thread that allows the object to be screwed in up to the very end of the thread.

10.7.1 The Setup for Threading

For this exercise, we'll produce a single thread on this part (Fig. 10-148).

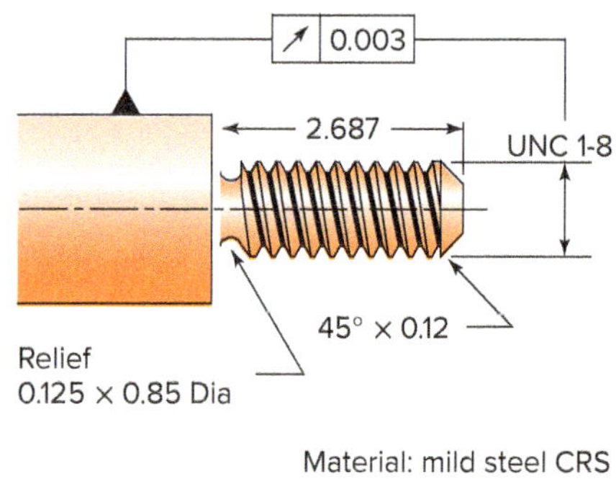

Figure 10-148 You are required to set up and thread one part.

Task 1 Holding Method

Runout of the thread to the existing diameter is tight (0.003 in.) so either hold the finished shaft in a four-jaw chuck, a collet, or between centers or machine the thread diameter from a larger shaft.

Task 2 Part Prep

First, be certain the shaft is at the major diameter for the thread. Next, machine a lead-in chamfer and a **thread relief** (if the design will permit it). Your actions will be different as the tool reaches the end of the thread, depending on the way it terminates (Fig. 10-149).

The thread on the top of Fig. 10-149 is called a **shoulder thread**, meaning that in use it can be screwed completely to the flat end–the shoulder. Doing so requires a thread relief. The threading action is simplified with a relief because it provides a place to stop the tool's Z axis travel, without ruining the thread. I strongly recommend that your first few threads be relief types if you have the choice.

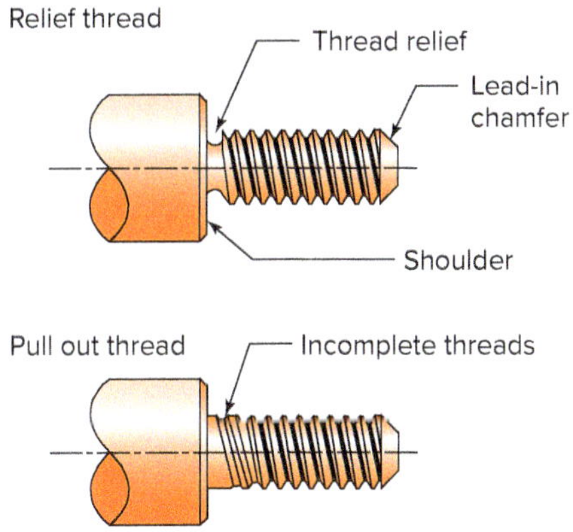

Figure 10-149 Your actions change depending on how the threads terminate.

However, it's sometimes necessary to thread without a relief, as shown on the bottom object in Fig. 10-149. The unrelieved, *pull out* thread requires a coordinated movement of both hands! As the bit reaches a mark on the part or a position on a dial indicator, you must quickly crank the *X* axis back so the tool clears the thread while simultaneously lifting the half nut lever to stop the carriage. That's a white knuckle time!

If you miss this simultaneous move, the thread is ruined. The pull out thread then has a short section where incomplete, tapered threads are produced as you crank the tool back, while threading to a relief allows simply stopping the carriage when the bit reaches the relief. Then crank the tool back.

KEYPOINT

If given the choice, always thread to a relief.

Task 3 Grinding the Thread Bit

For this exercise we will use the Unified thread form that you already know, and assume you aren't using preshaped threading tools (even though they save a lot of time). The thread form features a 60 ° included angle template and a small flat at the tip. You will need to custom grind the bit to produce the UNC 1-8 thread form.

First, form grind the bit to a 60 °included angle using the center gage.

The top rake is ground as a compounded angle–both to the side and back to improve the cutting action of one edge only. The tip flat is a different width for each thread pitch. Larger threads mean a larger flat. It will be one-eighth of the thread pitch distance (thread-to-thread distance).

TRADE TIP

When grinding your own bit, the point need not be in any specific relationship to the shank and is often slanted to one side to allow threading to a shoulder. In other words, it isn't required that the point is on the center of the bit or that it points straight off the bit (Fig. 10-150).

Checking Bit Shape with a Center Gage

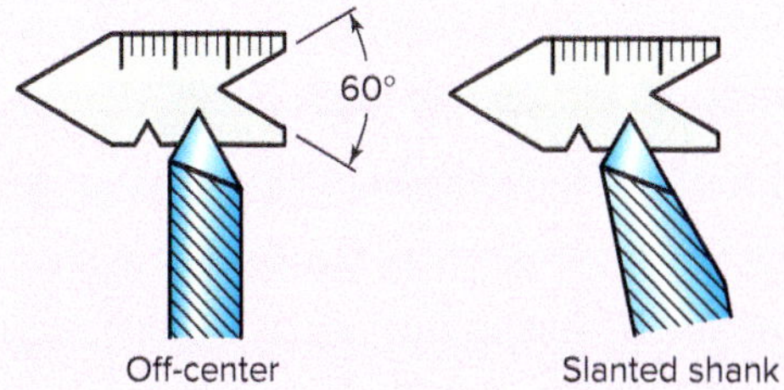

Figure 10-150 Grind the bit to 60°. The relationship of the point can vary to the shank as shown.

For our example, Fig. 10-151, 8 threads per inch (TPI), the pitch distance would be 0.125 in. Dividing that by 8 yields the flat width of 0.0156 in.

1-8 UNC thread

Pitch distance = 0.125 in. (thread-to-thread)

Flat width = 0.125/8 = 0.0156 or rounded to 0.016 in. flat

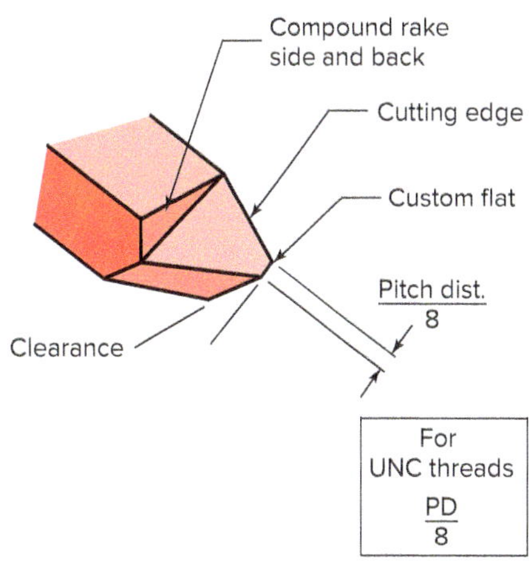

Figure 10-151 Notice the combined side and back rake makes this a right-hand bit—cutting toward the chuck using the left edge only.

This is so for ISO metric 60° thread forms, too. So a metric thread example would be M8-1.25 (Pitch distance = 1.25 mm already supplied)

$$\frac{1.25}{8} = 0.16 \text{ mm tip flat}$$

KEYPOINT

The small flat is found by dividing *pitch distance by 8.*

Use a Center Gage to Check the 60° Angle

Center gages are small, flat, arrow-shaped templates that have three purposes while threading. (Watch for two more uses coming up.) Shown in Fig. 10-150, it is being used as a template for checking and grinding the included angle on the bit.

TRADE TIP

Hold the center gage on the tool bit such that a light source is behind the pair. Any shape irregularities will show very well.

Task 4 Setting the Compound

Single-point threading requires several passes of increasing depth, to complete the total thread depth. To do so, we swivel the compound to an angle slightly less than half the included angle of the thread. For a Unified thread of 60°, turn the compound from 29° to 29.5°–in the counterclockwise direction (Fig. 10-152).

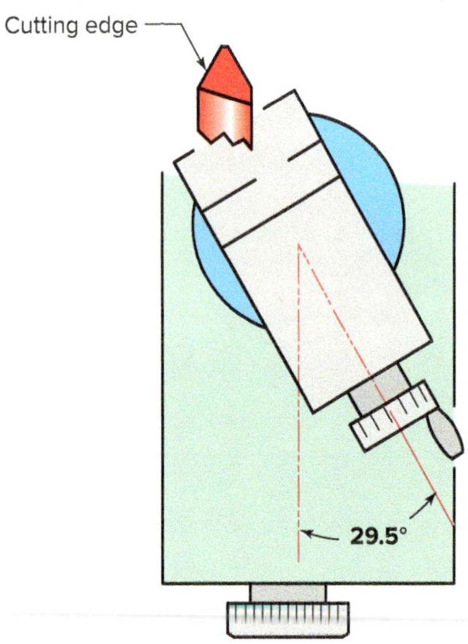

Figure 10-152 For a UNC thread, the compound is set at 29.5°, counterclockwise.

TRADE TIP

Why 29° to 29.5°? To aid in machine finishing at these low speeds, the tool must be as efficient as it can be. Therefore, a large, positive, compounded rake to the side and back is used.

Then to take advantage of that tool geometry, only one edge of the bit cuts best. The compound is turned such that each new pass advances the cutting edge into the work. In other words, the good edge does the cutting. If not, both edges would cut the thread resulting in lower-quality finish. (This is called **plunge threading**.)

But why not 30° instead of 29.5°?

Using a 29.5° angle rather than exactly half the thread included angle ensures that the back side of the thread is smoothed after each new pass. A very small amount of metal is removed by the back side of the bit to ensure that small steps aren't produced.

However, when threading on the CNC lathe where the controller starts and stops tool motion, the cutting speeds are restored to full efficiency. Then the single-edge cutting action described in Figs. 10-152 and 10-153 is not required. In CNC threading, there are advantages and disadvantages to both single-edge cutting and plunge threading.

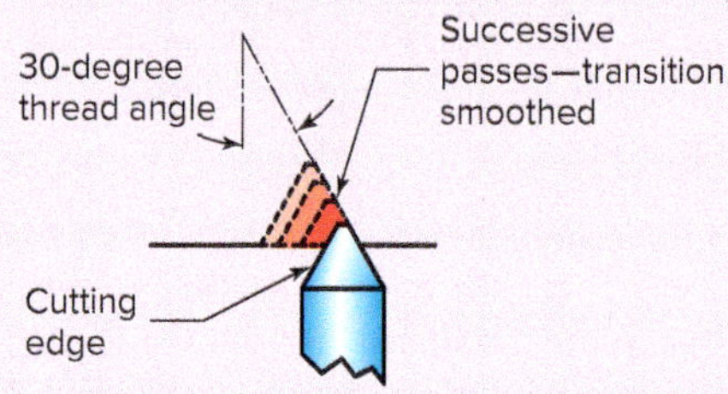

Figure 10-153 Each pass is taken by the best edge of the tool. The back edge smoothes the transitions of each pass.

Task 5 Positioning the Bit

Then after aligning the compound, there are two items to check when mounting the tool in the tool post:

Vertical center height–as usual

Bit angle relative to the axis of the work–checked using the thread center gage

Setting the Cutting Angle Using the center gage a second time to align the tool to the Z axis of the lathe, there is a right and wrong way of doing this task. For best results, *hold the gage firmly against the bit,* then move the X axis inward close to the work. Now, compare the gap between the gage edge and the work or chuck side (Fig. 10-154). Using the gage this way, the alignment is easy to see. Do not hold the gage against the work then put the tool in the template groove. That action makes it difficult to detect any misalignment, which would only show between the gage vee and the bit.

Tool Alignment with a Center Gage

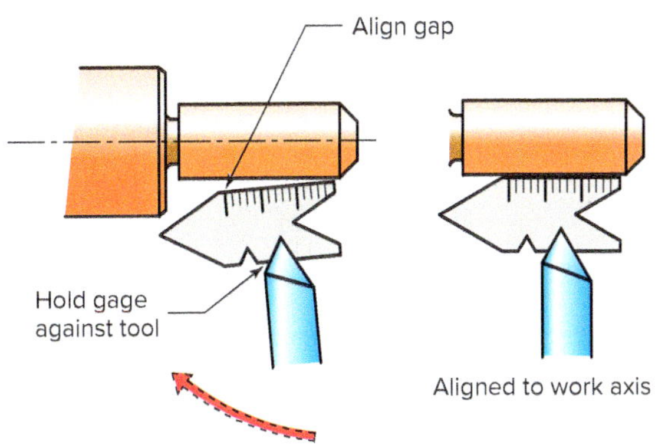

Figure 10-154 Holding the center gage firmly on the bit, move them both to the workpiece and compare.

Task 6 Coordinate the Micrometer Dials—*X* Axis and Compound

With the bit correctly positioned in the lathe and the compound swiveled and locked down, dial it forward to remove backlash in both the *X* and the compound axes. Check to see that the compound is not overextended at that position, then continue the *X* forward and touch it lightly to the workpiece. With the tool just touching the workpiece, zero both micrometer dials, compound and *X* (Fig. 10-155).

Coordinating Micrometer Dials

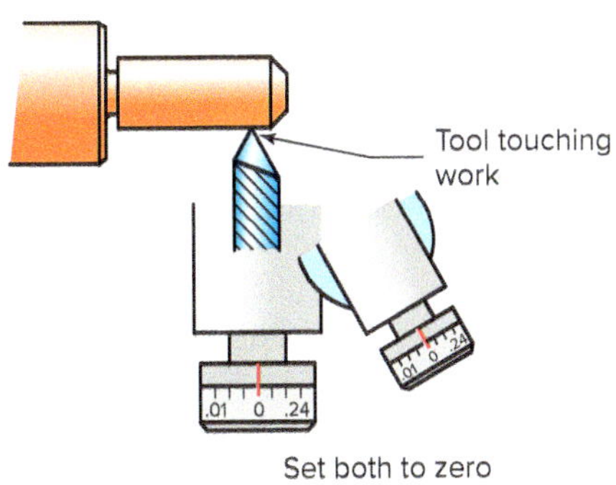

Figure 10-155 After removing backlash, lightly crank the bit to the work; then set both micrometer dials to zero.

Note, it is common to use the dials as you have now coordinated them to make the thread; however, there is a much more skilled version of coordination coming up in a great Trade Tip. Don't miss it.

Task 7 Gear Selections for Threading

Next, set the gearing on the lathe. There are five items to check off:

Select thread drive Shift such that the lead screw is engaged, not the feed system.

Right-hand spiral The tool will be moving toward the chuck when the half nut lever is engaged.

Thread pitch levers (Two or three shift levers) Set for a pitch of 8 for this example (maybe four shifters for dual metric/imperial lathes). Refer to the chart on the lathe front.

RPM Set for a low RPM (300 plus or minus). Although too low for efficient cutting, it is not possible to react fast enough at higher spindle speeds during single-point threading. Threading is difficult to control if the RPM is very high.

Engage the **chasing dial** This is done by loosening a screw and pivoting it into engagement with the lead screw.

10.7.2 Making the Thread—Your Actions

Okay—the lathe is ready to go. There are three actions you must take to be in control of threading. While I'll give a brief explanation, you'll need a demonstration before the solo:

Marking and checking the first pass

Completing the finished thread passes

Measuring the thread

First Pass—Positioning the Tool and Axes

Since you'll be working with the compound axis, which has no digital readout, even if the *X* does, use digital *X* as a backup reading only. Concentrate on the micrometer dial readings for threading.

With the *X* drawn away from the work 0.050 to 0.150 in., the saddle is moved to the right, off the work. Then the *X* axis is recranked back to the original zero position.

Next, the compound axis is cranked forward beyond zero, just enough to perform a light test cut on the work. From 0.003 to 0.010 in. are reasonable amounts. It is a good idea to snug the axis lock on the compound when making a threading pass—not locked, just snug.

KEYPOINT

On the first pass, you aren't cutting the thread but, rather, making a scratch that can be tested to see if the gear selections yield the pitch distance required.

After the first pass, test the thread for correct pitch. Use one of the methods shown in Fig. 10-156 or a caliper set to the correct pitch distance. This will indicate that the gearing is correct for the intended thread. Note that center gages have two or four scales, one of which will usually be divisible by the thread pitch. In this example, 8 threads per inch will divide into 24 every three divisions, but the 32-division scale could be used, too; 8 is prime to either scale.

Two Methods to Check Screw Pitch Distance

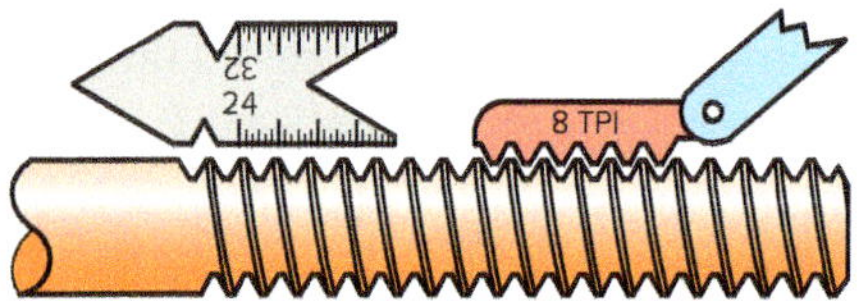

Figure 10-156 The witness marks are compared to the standard lines on the center gage or to a screw pitch gage.

Half Nut Lever Engagement

Before cutting that first pass, I strongly recommend practicing the following actions a few times, with the tool drawn away from the work, to get the feel of single point threading.

Your left hand's first action is on the *Z* axis, saddle hand-wheel rim, lightly. This is to hold back on the wheel and keep the backlash to the right against the lead screw. This is the way to get good repeatability of tool position on the lead screw. Once the thread is cutting, you'll be moving that hand position.

Your right hand is on the **half nut** lever. When the chasing dial gets close to lining up, drop the lever down lightly before the index lines completely coincide. This will ensure that you do not miss engagement, which must be made smartly on time. Doing so, you will feel the half nut touch, pause, then drop down into engagement. (See the chasing dial rules–next.)

KEYPOINT

When learning to thread (Fig. 10-157), it's wise to keep your right hand on the half nut lever throughout the pass! Then, if a problem occurs, you don't need to fumble for the correct lever.

Figure 10-157 Left view: Threading begins: left hand on the Z handwheel, lightly holding back, right on the half nut lever. Right view: With the half nut engaged, after the saddle moves, switch left hand to the X cross feed dial.

Lake Washington Technical College, Kirkland, WA/McGraw Hill

Left hand second action is as the thread begins to cut, transfer your left hand to the *X* axis dial ready to crank it back.

KEYPOINT

Remember to stop the *Z* axis motion, pull up on the half nut lever when the bit reaches the relief. Then crank back on the *X* axis.

Caution: When pulling up on the half nut lever before the thread relief, with the tool within a cut thread, you will produce an unwanted groove around the shaft, ruining the thread. So follow the next procedure.

Crash Procedure

To stop threading somewhere in the middle in an emergency, first, using your left hand, crank away on the *X* axis handwheel; then quickly pull up on the half nut lever with your right hand. This disengages the action without making any unwanted groove in the threads.

Threading Passes

To position for a new pass, with the *X* axis still cranked back, return the *Z* axis to the start position, to the right end. Dial the *X* back to the zero; then the compound in a reasonable amount of around 0.020 in. depending on the size of the thread, ready for the next pass. Watch the chasing dial for engagement timing and keep repeating these actions. For the best cutting action, paint thread oil on with a brush. Flood coolant isn't recommended as it obscures your vision of the cutting tool, which must be seen to stop exactly at the relief. Cutting compounds or oils do the trick.

10.7.3 When to Engage the Chasing Dial

With some thread pitches, the half nut can be engaged in the wrong position, which cuts the wrong groove and ruins the thread. A standard lathe features an eight-position chasing dial to prevent this from happening (Fig. 10-158).

The rules for its use are based on the number of threads per inch you wish to produce. The outer index line will coincide with one of the eight lines on the dial to indicate correct timing to start the thread. Look back at Fig. 10-67 for an example of the chasing dial.

Which line(s) you can use is based on pitch. If the thread has

Even number threads–for example, 16, 18, or 20 TPI

Then, drop the lever on any of the eight lines.

Odd number threads–for example, 13 or 27

Then drop the lever on the numbered lines only.

Rules for engaging the split nut for thread cutting

Threads per inch to be cut	When to engage split nut		Reading on dial
Even number of threads	Engage at any graduation on the dial	1 1 1/2 2 2 1/2 3 3 1/2 4 4 1/2	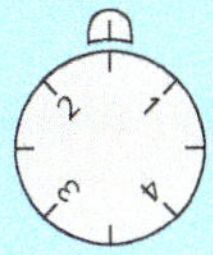
Odd number of threads	Engage at any main division	1 2 3 4	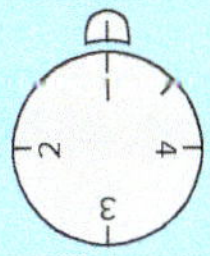
Fractional number of threads	1/2 threads, e.g., 11 1/2 engage at every other main division 1 & 3, or 2 & 4 other fractional threads engage at same division every time		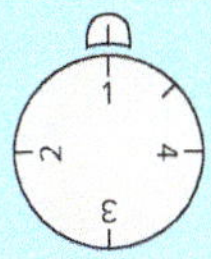

Figure 10-158 The eight possible engagement positions for a chasing dial.

Fractional threads—for example, $27\frac{1}{2}$ TPI

Then, drop the lever on any two opposing numbered lines—for example, 1 and 3, or 2 and 4 (see the Trade Tip).

TRADE TIP

When you must drop the half nut lever on selected lines, it's a good idea to use a highlighter to clearly mark the target line(s). Or, if there's any doubt, just use the same line every time; that guarantees correct alignment. (I use line 1 when doing this.)

Metric Threads

When cutting metric threads on older lathes (before the mid-1980s), it's necessary to open the gear case and add or change drive gearing between the spindle and the lead screw. However, many modern manual lathes are made to easily cut both imperial and metric threads by use of an extra quick change gear selection on the headstock. That complicates the chasing dial and rules. To solve the chasing dial issue, some feature two interchangeable chasing dials, one for imperial pitches and one for metric. However, other dual-system lathes are equipped with chasing dials with more than eight positions. For these, find and follow the manufacturer's instructions or use my Trade Tip and drop the lever on the same dial position for every pass.

TRADE TIP

Coordinate the Thread Depth on the *X* Axis—Save Calculating and Extra Measuring This tip simplifies cutting the thread to the correct depth. The classic method of dialing forward along the compound axis until the full depth is reached means moving the tool into the work along the long (calculated) side of a triangle, to create the full depth—or alternatively cutting and measuring until the depth is reached. But, adding these two simple steps after coordinating the *X* and compound dials to zero makes thread depth a snap!

Step 1 From the coordination position (bit tip touching—both dials at zero) *back away on the compound axis about three times the thread depth (Fig. 10-159).*

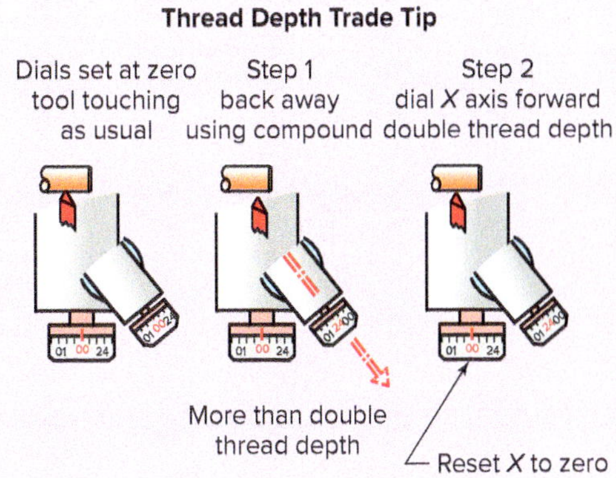

Figure 10-159 Use this trade tip to simplify threading.

Step 2 Now, with the bit off the work, *crank the X axis forward a distance equal to the thread depth.*

That distance is easy to find in *Machinery's Handbook* or to compute, depending on the series and root shape. For example, the depth of a $\frac{1}{2}$ – 13 UN thread = 0.54127 × Pitch distance = 0.0416 in.

Keep in mind that on diameter-calibrated lathes, that means you will actually dial in twice the thread depth to move the bit forward the real depth. For this example, that means cranking 0.0832 in. The tool will physically advance the required 0.0416 in. *Now reset X to zero.*

With this coordination correctly performed, you will still move the *X* in and out to its zero position for each pass, and will still be in-feeding along the compound axis. But you will stop when both dials reach zero. Presto, the thread will be at the right depth!

10.7.4 Setting Up to Cut Internal Threads

Cutting an internal thread is a bit more complex compared to the external setup (Fig. 10-160). You cannot see the tool as it approaches the end of the cut. That means a DTI or other means of showing when the half nut lever must be disengaged, at the exact *Z* axis position each time. Digital readouts can be used but require you to look away from the action. There are two choices:

Option 1 Progressing Inward To cut an internal thread with the cutting tool going to your left, the compound is rotated 29.5 ° in a clockwise direction (opposite normal rotation). Operator actions are similar to cutting external threads with two exceptions. After the thread pass is completed, you must crank the *X* axis inward to clear the tool of the work for the next pass—not out! (See the thread stop Trade Tip.) Second, to take successive passes, the compound axis is cranked outward, not in, until the full depth is achieved.

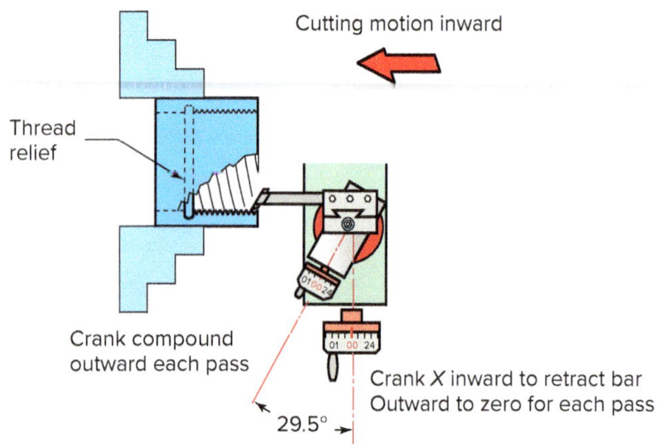

Figure 10-160 Boring an internal thread option 1—cutting inward.

Option 2 Threading Outward Here are two versions of a safer setup whereby each pass is started with the boring bar all the way in, at the bottom of the far end of the hole, then the cut feeds outward—to your right, away from the bottom of the hole (Fig. 10-161). This creates less operator stress since the half nut lever is disengaged after the bit is safely out of the hole, where you can see it. Both setups are based on simulating unscrewing the cutter while it cuts a right-hand thread; therefore the lathe must be able to run in reverse.

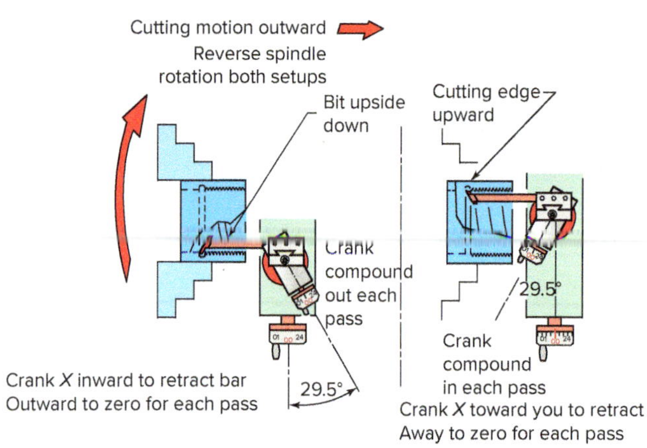

Figure 10-161 Safer setups—boring outward: option 1, bit upside down, in front; and option 2, bit up, on far side of hole.

Bit Upside Down

In option 1, the bit is turned over in the boring bar. The cutting tool is positioned on the near side of the hole.

Bit on Far Side Right Side Up

Here the bit is upright on the far side of the hole. In some cases this means that the machinist can see it at full depth, when it's parked in the relief—an added advantage.

> **KEYPOINT**
>
> For either setup, a thread relief is necessary. If a relief isn't possible, then neither method can be used since there is no place to park the cutting tool until the thread pass begins.

Cam Lock Chucks Only for Reverse

In order to achieve a right-hand helix, the lathe must rotate backward. That means a screw-on chuck cannot be safely used.

> **TRADE TIP**
>
> **Thread Stops** Many industrial lathes feature a special *X* axis setup provision that sets hard limits on the amount and direction the *X* axis handwheel can be cranked. They are very useful when making both internal and external threads. If the lathe has a thread stop, look for a small thumb wheel on the perimeter of the *X* axis collar, behind the crank. Ask your instructor to explain how they are used.

> **KEYPOINT**
>
> Thread stops correctly set up prevent accidentally cranking the *X* axis the wrong direction at the wrong time!

UNIT 10-7 REVIEW

Replay the Key Points

- Given the choice, always thread to a relief.
- On the first threading pass, you aren't cutting the thread but rather are making a scratch that can be tested to see that all the gear selections yield the pitch distance required.

- To cut an internal thread the compound is rotated 29.5 °in a clockwise direction to advance the bit's cutting edge into the work
- Chasing dial rules:
  Even number threads–on any of the eight lines
  Odd number threads–on the numbered lines only
  Fractional threads–on any two opposing numbered lines

Respond

Threading is complex to explain on paper but it will become clear with hands-on practice. Let's see what retention occurred here. On a sheet of paper, list the steps to setting up a thread and the actions you must take to make one. Questions A through D are critical thinking and there are no answers provided. Double-check yours against the reading, or better yet, talk them through with a partner.

A. The Setup
   1. 2. 3. 4. 5.

B. Operating the Lathe
   1. 2. 3. 4. 5.

C. Hand Positions

D. After positioning the bit for a pass what are the starting actions:
   A) Watch the ___________ ___________
   B) Push [ ] Up or [ ] Down on the ___________ ___________
   C)
   The stopping actions
   At the relief?
   A. B.
   Before the relief? (Usually an emergency stop)
   A. B.
   Positioning the tool for a new pass

Answer the following about threading. Answers found at the end of the chapter.

1. To what angle must the compound be turned for a Unified thread?
2. Explain why you swivel the compound in Question 1.
3. What is the width of a thread bit, nose flat for 16 threads per inch?
4. Engaging the lead screw is one of the five setup steps for threading. Is this statement true or false? If it is false, correct it to be true.
5. How would you make a left-hand thread? This is not covered in the reading–think it out.
6. To make 20 threads per inch, where can you engage the chasing dial? Where for 19 TPI? For $27\frac{1}{2}$ TPI?
7. What is the depth of thread for a UNF $\frac{1}{2}$-20 screw?

Unit 10-8 Measuring Threads

Introduction: When machining a screw thread, we make several light cuts, called **passes**, until the thread is at its full depth in the shaft. Each pass reduces the diameter of the thread until it reaches the full pitch depth. But since the deepest feature, the bottom crest of the triangle called the **root diameter *or* minor diameter**, includes a clearance amount to allow the screw to assemble with a mating nut, it isn't the key feature that must be measured.

Measure the Pitch Diameter The function being controlled by the dimensions and tolerances is the fit between the faces of the screw and nut. To do so, we need to measure somewhere within the thread form. So, we measure at a theoretical point (Fig. 10-162), very near to the center of the thread faces, called the **pitch diameter**.

KEYPOINT

The pitch diameter is actually the middle point of the nut and screw contact area.

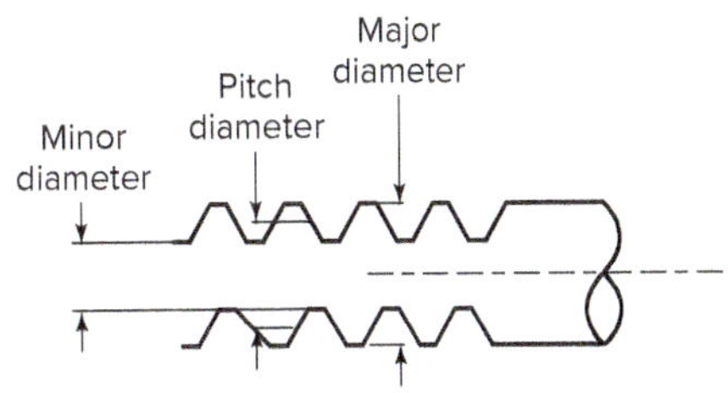

Figure 10-162 The *pitch diameter* compared to the other thread dimensions.

The pitch diameter is actually the center of engagement between the mated screw and nut. Therefore it isn't exactly at the center of either component. It occurs at the center of their interface. However, for convenience, it can be envisioned as the physical center of any given screw thread since it's within a few tenths of a thousandth of the middle for screw sizes smaller than 1 in.

Unit 10-8 introduces several options for pitch diameter measurement that might be found in a tech school. There are more sophisticated methods to be learned on the job. Methods beyond those presented here are industry specific, depending on the use to which the products will be put. They range from optical projection comparitors through electronic probing equipment and jigs.

KEYPOINT

The pitch diameter is a theoretical measure of the diameter of the overall thread.

Completely measuring threads can be a complex issue depending on the use to which they will be put and their tolerances. Not only is pitch diameter to be checked, but the thread form must be verified and the lead measured over a great distance, too. Here, we'll limit our investigation to ways of measuring the pitch diameter. These tests represent the larger part of thread measuring until the threads become very closely controlled. Check with your instructor to learn which methods are used in your lab.

TERMS TOOLBOX

Passes (thread) Each cut to deepen the thread toward the pitch diameter.

Pitch diameter The theoretical center of engagement between internal and external mated threads.

Pitch gage (snap gage) A horseshoe-shaped gage with two pairs of pitch rollers. If the first pair slips over but not the second, the thread should assemble.

Root diameter (minor diameter) The smaller diameter on the thread form–internal or external threads.

Thread ring gage A precision nut that can be set to test the pitch diameter of threads.

10.8.1 Measuring Pitch Diameter

As we've learned previously, dimensions can be measured or they can be functionally gauged. Of these methods, two are measurements and two gages of fit.

Gage 1 Testing Diameter Against a Nut Standard

The bottom-line, functional test finds if a nut assembles with the screw. If it goes on without dragging or excessive wobble, it's okay for some utility applications. This is the least technical method. But when a thread must be quickly produced for tooling or other noncustomer purposes, it works! Both this test and the next verify thread lead to some extent but only over a short distance–the width of the nut.

Gage 2 Adjustable Ring Nuts

A more technical version of the nut gage is the *adjustable* **thread ring gage**. Simple to use, sometimes one gage nut is set up at the smallest size allowable and a second is set at the largest size to become a go, no-go pair.

Measurement 1 Measuring with a Thread Micrometer

This is a special micrometer featuring pointed tips that fit into the Unified form thread grooves (or other thread form if the mic has interchangeable tips). To use this method, refer to the data chart in the mic's kit or in standard reference books such as *Machinery's Handbook*. It will list the size that the mic will read when the thread is at the correct pitch diameter.

This is a fairly accurate method of determining pitch diameter but doesn't actually verify thread form or lead. A thread can pass this test yet not fit to a mating part! However, if the thread form is under control and the lathe produces accurate leads (the usual situation), then thread mics are a good option.

Thread mics are more universal than test nuts, since they can measure any thread. There are reference tables to determine the pitch diameter for a given standard thread and a formula to calculate the pitch diameter for special threads not found in the chart.

Measurement 2 Thread Wires

This is an accurate, universal method but slow. Three equally sized precision wires are held in place as per Fig. 10-163. A standard micrometer then measures over the wire tops. A data chart (or formula found next) is needed to determine the exact measured distance over the wires when the pitch diameter is correct.

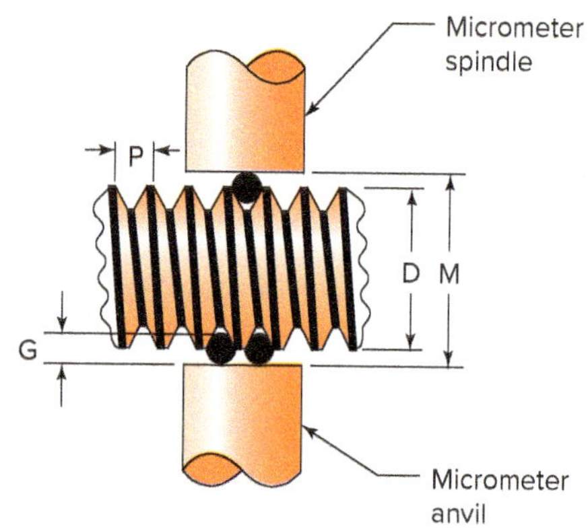

Three wires arranged correctly for measuring 60° threads

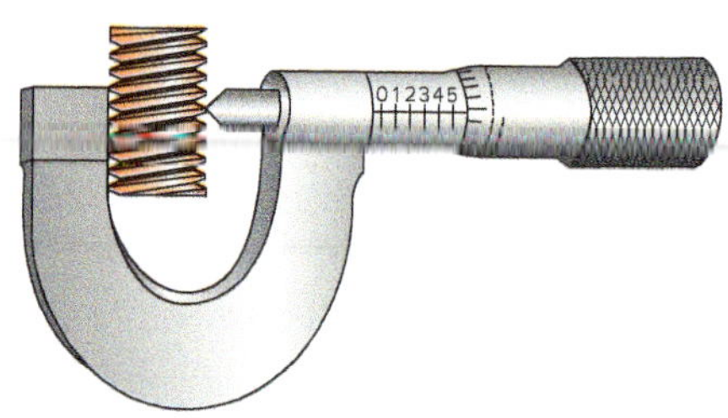

The screw thread micrometer measures the pitch diameter of a thread

Figure 10-163 Two common methods of measuring threads.

While they are supplied in handy sets (Fig. 10-164), thread wires can be any set of three round pins of the same diameter. They are selected such that the wire diameter contacts the thread form near but not perfectly at the pitch diameter for that thread's pitch. All that's required is that they protrude slightly above the major diameter to contact the micrometer face.

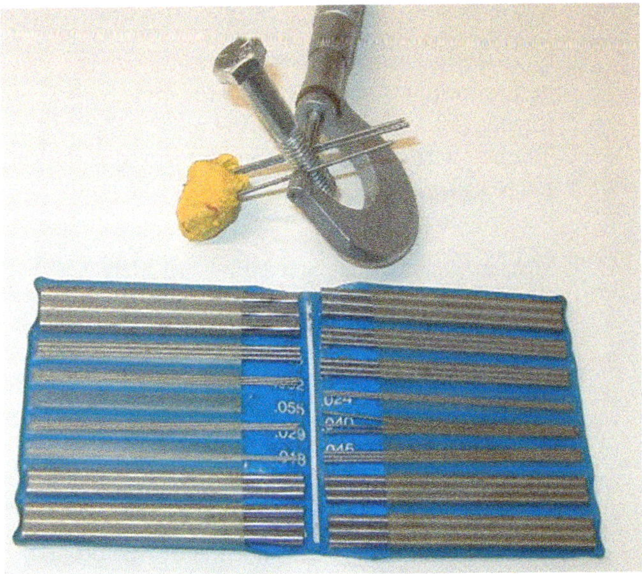

Figure 10-164 A set of wires—three of each diameter. Note the clay holding the wires until they are trapped by the micrometer.
Courtesy Mike Fitzpatrick

TRADE TIP

When measuring nonstandard threads or you do not have a three-wire set as shown in Fig. 10-164, here are three useful formulas from reference books:

To find M, the measurement over the wires at the correct pitch diameter, within a standard 60° thread form:

Major Diameter Known (from *Machinists Ready Reference*)

$$M = D + (3 \times W) - \frac{1.515}{N}$$

Pitch Diameter Known (from *Machinery's Handbook*)

$$M = E - 0.86603P + 3W$$

Best Wire Diameter The closest, standard wire size that contacts thread nearest the pitch diameter in 60° threads.

$$W = 0.57735 \times \text{PD}$$

where

M = Measurement over wires in thousandths of an inch

D = Thread major diameter (from a reference book)

W = Wire diameter

N = Number of threads per inch

P = Pitch diameter of thread

PD = Pitch distance = 1/TPI

These formulas do not compensate for the compounding of the measurement caused by the thread's lead angle. They are close enough for class 3 threads in everyday manufacturing. Closer formulas that do compensate plus a more complete explanation are found in ASME B1.2-1992. Read further in *Machinery's Handbook*.

10.8.2 Gage Test 3 Using a Pitch Gage Standard

The **pitch gage** (Fig. 10-165) is an adjustable go, no-go set of thread rollers. If the first pair can fit over the thread but not the second pair, the thread is within the calibrated tolerance. These gages offer one advantage over the previous methods—they functionally simulate mating with several threads; thus, check a small part of the lead as well as the pitch diameter.

Figure 10-165 A pitch gage slipping over the first rollers, but not the second, shows the thread is within tolerance.
Courtesy Mike Fitzpatrick

This gage is accurate and quick to use; thus, it is used most often in production machining. It's sometimes called a **snap gage** due to the way it drops on the test piece; it requires calibration against precision threaded standards.

10.8.3 Other Thread Forms

We will study technical screw threads as a subject in Chapter 14, but for now, you should recognize the other thread forms used in mechanical devices (Fig. 10-166). They are cut in a similar fashion as the Unified form and measured similarly, with the exception of the square form thread, which must always be functionally gauged.

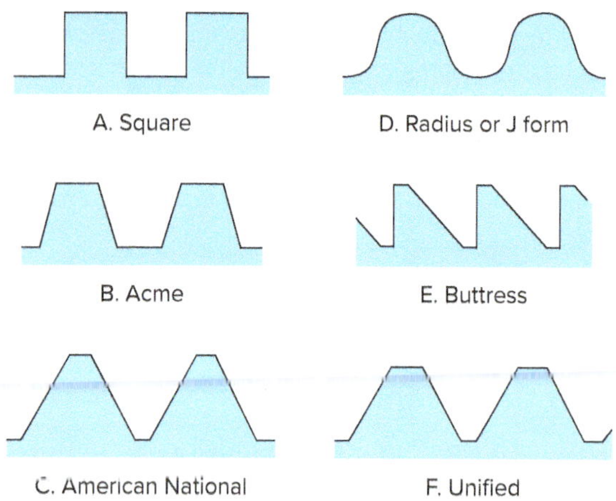

Figure 10-166 Other thread forms you might be called on to machine.

UNIT 10-8 REVIEW

Replay the Key Points

- The pitch diameter is a theoretical measure of the diameter of the overall thread.
- The pitch diameter can be measured or gauged.
- For measurements over given wire sizes, look in *Machinery's Handbook.*

Respond

1. Which methods of measuring pitch diameter use a micrometer?
2. What is the deciding factor for choosing the size of the three wires used to measure pitch diameter?
3. Describe the pitch diameter for a thread.
4. True or false? Pitch diameter can be verified by functional gage and measuring. If it's false, what will make it true?

Critical Thinking

5. Of the four methods we've just discussed, which verify thread form? If they can't, then how might you think thread form can be evaluated?
6. From *Machinery's Handbook,* what wire size is best for testing American Standard Unified threads (60°) with 16 threads per inch? (*Hint:* The best wire diameter contacts the thread at the pitch diameter but there is a range of wire sizes that will do the job.)
7. Measure an American Standard $\frac{1''}{2}$ – 13 UNC screw thread. From the *Handbook,* what formula would you use to determine the pitch diameter if you have the measurement *M* over three wires of known size *W*?
8. Can wires of 0.050-in. diameter be used to measure an American Standard Unified thread form of 13 TPI?
9. Using *Machinery's Handbook,* you have three wires of 0.050-in. diameter, to measure the pitch diameter of a $\frac{1}{2}$ – 13 UNC thread. What should the measurement *M* be if the thread pitch diameter is correct?
10. For the thread of Question 9 the measurement *M* (over 0.050-in. wires) is found to be 0.5235-in. diameter. How much more must be machined away to make it per specs?

CHAPTER 10 REVIEW

Unit 10-1

All of Chapter 10 is offered not only to learn how to recognize and safely perform operations on the manual lathe but to also get a feeling for them such that you can write repeatable program statements that do them right on the CNC machine as well.

Unit 10-2

To run and maintain a lathe correctly, it's critical to understand the way they are put together. While there are differences in the way they are made, they deliver the same range of functions. If you understand the range of functions a lathe must perform, then it's easy to adapt to the specific machine.

Unit 10-3

Do you clearly understand the differences in the six ways to hold work and how to choose the right one? In my career, I have seen two spectacular crashes involving misheld work on the lathe. In one, the operator was injured seriously. In the other, we were pretty white knuckled as the part went spinning throughout the shop causing all sorts of damage to things but not people–lucky that time! Education by disaster is unacceptable and inevitable if you don't understand work holding.

Unit 10-4

You now know about basic cutting tools to set up and perform the 15 operations. However, we've only seen the tip of the iceberg. Cutting tools are evolving as I write this book. Stay informed through supplier catalogs and seminars, by going to tool shows, and by subscribing to trade publications such as the *American Machinist, Modern Machine Shop Magazine* (and its online articles–both are excellent references), the *Machine Tool Digest, CNC Machining,* or several other fine magazines. Ask your instructor for a recommended publication for your area.

Unit 10-5

I hope I've made my case for extra caution when setting up and running a lathe. You must stand in front of them while they spin the work. Chips can fly out like bullets as they are flung off the spindle. Uncontrolled, they can be as sharp as a razor and strong, too. Then there are the heated chips and work and finally the misheld part flying out. All can be controlled with professional caution and skills.

Unit 10-6

Although CNC lathes have displaced manual machines on the front lines, to be a true journey-level machinist working in a tooling or job shop, you must be able to deliver the quick part made the good-old standby way. Manual machines will never be eliminated from the factory or especially from tool and die, repair, or other small shops.

Unit 10-7

Almost as difficult as rubbing your stomach with your right hand while patting your head with your left, the hand-eye coordination for threading takes practice! It is a good idea to watch a demonstration, then practice the actions with the tool pulled back a safe distance. This too is a skill a competent journey machinist must understand even though we make most modern threads using programs or automated equipment, not hand action.

Unit 10-8

The real trick here is the same as for all measuring tasks–determine the tolerance; then find the tool that delivers the repeatability and resolution needed. Then practice.

QUESTIONS AND PROBLEMS

1. Of the basic metal cutting operations on a lathe, which are performed from the tailstock? (Units 10-1 and 10-2)

2. Facing on a manually operated lathe presents a problem that is easily overcome on a CNC lathe. What is it? (Unit 10-1)

COMPLETE PART SETUP PROBLEMS

3. In Fig. 10-167, name the operations in sequence required to complete the spur gar shaft before the gear teeth are cut on it. It comes to your lathe as a 2.500-in.-diameter bar of CRS–cut to 8.0-in. lengths. (Unit 10-1)

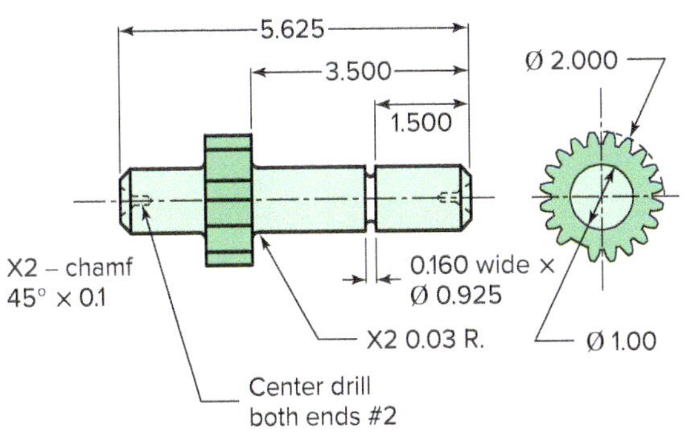

Figure 10-167 Write a plan to machine this spur gear blank before the teeth are milled on it.

4. For Fig. 10-167, what holding method would be correct? Explain. (Units 10-3 and 10-5)

5. Would a tailstock support be *required* when turning (Fig. 10-167)? (Units 10-2 and 10-6)

6. To complete the gear shaft on the lathe, which tool post would be ideal? Explain. (Unit 10-3)

7. To complete Fig. 10-167, which cutting tools should ideally be carbide insert? Which would probably be HSS? (Unit 10-4)

8. To machine Fig. 10-167, calculate the correct RPM to machine the 2.0-in.-diameter gear blank and the 1.00 shaft assuming carbide cutting tools. Use Appendix III, the SS chart, and the short formula. (Unit 10-6)

9. The RPM calculations were based on finished diameters. Would they be correct for the roughing passes? You have two options. Explain. (Unit 10-6)

CRITICAL THINKING

10. The pilot on the center drill is 0.125-in. diameter. What is the correct RPM for center drilling? (Units 10-5 and 10-6)

11. If there were 20 parts to be made per Fig. 10-167, is there a more economical way to machine them on the manual lathe? Explain. (Units 10-2 and 10-3)

12. Assuming you adopt the method suggested for making 20 parts in Question 11, is there a special safety consideration? There are two options. Explain. (Units 10-3 and 10-5)
13. True or false? The center gage is used to center threads on the uncut shaft. If it's false, what will make it true? (Unit 10-7)
14. True or false? When facing work it's a good idea to lock the *X* axis to avoid cutting tool slippage. If it's false, what will make it true? (Unit 10-1)
15. Explain why a face-grooving (trepanning) cutting tool must have a slightly different geometry than a tool meant to groove the side of the workpiece. (Unit 10-4)
16. Of the four thread-measuring tools and methods, which is the most accurate at measuring the thread pitch diameter and lead for several threads? (Unit 10-8)
17. What is the pitch diameter for a $\frac{7''}{16}$ - 14 UNC screw thread? (Unit 10-8)
18. What is the recommended wire diameter and measurement over the wires to check the $\frac{7''}{16}$ - 14 UNC thread of Question 17? (Unit 10-8)

CNC QUESTIONS

19. Why must you set the RPM so far below the normal recommended surface speed when threading on the manual lathe, yet they can be cut at the correct spindle RPM on CNC lathes? (Units 10-2 and 10-7)
20. Beyond the actual computer control of the various functions, what is the basic mechanical difference between a CNC lathe and a manually operated machine? (Unit 10-2)

CHAPTER 11
Mills and Milling Operations

Units

11-1 What Does a Mill Do?

- Given a drawing or part, you will be able to identify 12 basic operations by name and a bit about typical setups

11-2 How Does a Mill Work?

- Identify vertical, horizontal, and ram and turret milling machines
- Identify the functions and orthogonal axes of a milling machine

11-3 Setting Up Mill Cutters

- Identify and choose the right cutter mount for the job
- Calculate the best RPM for that cutter and work material
- Identify two common mill tapers: R-8 and ISO (American) standard

11-4 Avoid These Errors—Great Setups and Safety

- Use clamps and vises to their best mechanical advantage
- Recognize and prevent nine most common learner accidents

11-5 Mill Setups

- Follow the 10-task list to safe, efficient setups
- Use climb milling safely and correctly
- Do not use climb milling in two exceptions
- Calculate mill cutter feed rates
- Align work using DTI skills
- Coordinate machine axes to datum points on the work
- Square up work using datum concepts

Based in Rocky Mount, North Carolina, Ossid Corporation is a manufacturing company that specializes in tray overwrappers, weigh price labelers, case scales, and horizontal form-fill seal machinery. Ossid employs more than 12 machinists, many of whom have had training at Nash Community College. While at Nash, I worked 30-plus hours a week and went to school full time during the day. During my time there, they had only two HAAS VF4s, two manual lathes, and six manual mills—two of them with Acu-Rite 2 axis controls. When I left in 2003 to further my education, the machine shop had grown to four CNC mills and a CNC turning center. Almost 10 years later, Ossid continues to grow. It currently has more than seven CNC machines and is one of Rocky Mount's largest employers of machinists.

—Justin Cole, Nash Community College

Courtesy of Justin Cole, Nash Community College

Typically, most shoppers take food packaging for granted. That neatly packaged item didn't just happen. Look at the machines that produce them!

—Michael Fitzpatrick

Source: Michael Fitzpatrick

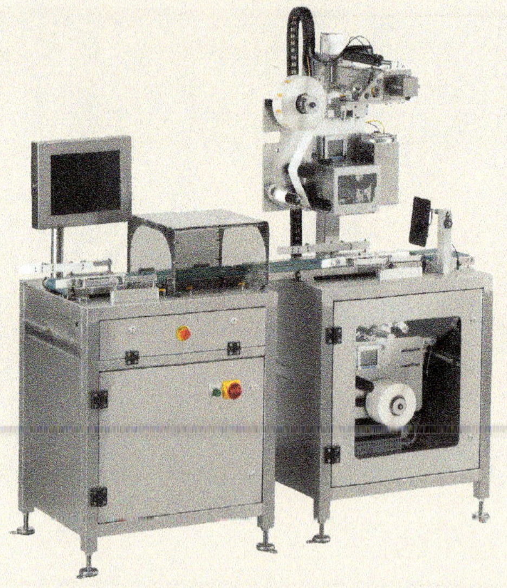

Courtesy of Ossid, Battleboro, NC

INTRODUCTION

When this text was compiled, the National Machine Tool Builders Association reported that 80 percent of new machining equipment sold in North America were milling machines. They shape metal by rotating a cutter and moving it through the workpiece. Clearly, this is an essential skill and a central issue in machining.

Although their prime duty is to create flat and square surfaces, mills can perform a wide range of tasks. Skilled machinists can cut angles, form radii, bore holes, and using attachments make circles or parts of circles and helices (spiral cuts—plural of helix). Using special cutters, mills can cut gear teeth and make dovetails and tee slots. The most amazing thing is that all of these shapes can be made on manual machines, without CNC control!

Now, add the computer to drive the feed axes, and the possibilities are almost limitless! Using CAD/CAM-generated programs, there's practically no 3-D shape that cannot be created. Today, if it can be drawn on screen it probably can be milled. Here are the units of learning in Chapter 11 to gain this valuable milling background.

Unit 11-1 What Does a Mill Do?

Introduction: Unit 11-1 provides a background in the operations expected of a beginning machinist. It gives you the ability to look at a part print or a finished item and then identify the basic operations required to make it. We will also discuss a few basics about the way the setup is made and which cutter might do the job. We'll explore setups and cutters in greater detail in Units 11-3, 11-5.

Like all others in this book, Chapter 11 assumes you'll be practicing these operations on manual mills, then transferring the experience to CNC machines. When differences are significant, they will be pointed out. Most of the 12 operations are common on programmed machines as well.

TERMS TOOLBOX

Arbor A precision shaft used to support wheel type cutters.

Ball nose end mill A cutter featuring a full radius at the tip.

Boring Machining precision holes by rotating a single point tool in a circle.

Bull nose end mill A cutter featuring radiused corners and a central flat face.

Climb milling Peripheral cutting that amplifies the cutter feed action–the work feed agrees with the cutter rim and, if not controlled, can lead to serious accidents.

Conventional milling Peripheral cutting where the cutter opposes the work feed direction.

Depth of cut (DOC) The distance from the work surface to the cut surface.

Fillet An inside corner radius between two surfaces.

Indexed (spaced) work Machining multiple features in a circular array–for example, a bolt circle.

Keyway (key) A square slot or groove in an object in which a key is inserted to prevent rotational slippage on the shaft.

Peripheral milling (profiling) Machining around the outside edge or surface of the work.

Plunge Drilling into the work with an end mill that is ground for the purpose.

Pocket milling Machining down inside a part using both face and peripheral cuts.

Radial engagement (RE) The amount of cutter diameter contacting the work compared to its full diameter; 66 percent RE is ideal.

Ramp A multiaxis movement of the cutter into the work.

Straddle (gang) milling Using two or more wheel cutters on a single arbor.

The upcoming operations divide into four families by the type of cutter and how it is fed into or past the work:

1. Facing–cutting one flat surface using the end of the cutter teeth
2. Profiling–cutting the outer rim of the work, using the side of the cutter
3. Drilling–pushing straight inward also called plunging.
4. Boring–creating round internal surfaces

11.1.1 Operation 1—Face Milling

Arguably the most basic operation of all is a face cut. The cutter is lowered below the surface of the work to the **depth of cut (DOC)**, then set to engage the work at a given portion of its diameter, called **radial engagement (RE**; Fig. 11-3**)**. The RE is expressed as a percentage of the cutter's diameter that is in contact with the work. The upper illustration in Fig. 11-3 has a 2 to 3 radial engagement (66 percent). The cut width is about two thirds of the total it could cut.

There are several reasons to set up a cut with less than 100 percent RE. We'll reason out why, as we go. However, there are times when we must use 100 percent RE because of the lack of a bigger cutter or from other machine limitations, but the better solution is to set up a bigger diameter cutter when a large area is to be faced, but why? Read on.

Step Cuts

It's common to make both face and peripheral cuts simultaneously as in Fig. 11-2, where a high-volume cut is progressing. Note the thick blue chips the cutter is taking from the steel workpiece using a 40 percent radial engagement but a very large depth of cut.

KEYPOINT

While it's possible to use a 100 percent radial engagement for a face cut (Fig. 11-3), for tool life and finish, during face cutting it's desirable to keep RE below 66 percent.

Critical Thinking

Study Fig. 11-3 carefully, and see if you can solve the puzzle of why there is the difference in tool life and finish for the 66 percent RE compared to the 100 percent RE. The answer won't be fully discovered until the end of Unit 11-5.

Removal Rate—In Cubic Inches per Minute

The combination of depth of cut, radial engagement, and feed rate create the overall removal rate–stated in cubic inches (or cubic centimeters) of metal removed per minute. The removal rate translates directly to horsepower required to make a cut. We'll delve into calculating these things in Units 11-4 and 11-5.

KEYPOINT

Mill feed rates are specified in inches of travel per minute (IPM) or millimeters per minute. That's different than drill presses and lathes, where it's in inches per revolution.

Three Cutters That Face

Face milling can be accomplished using three different type cutting tools. Learn their names and general characteristics.

Face Milling Cutters

These cutters are made for the purpose. Supplied in either HSS (Fig. 11-1) or carbide (Figs. 11-2 and 11-3), face mills remove the most metal per minute of the three types. The carbide insert cutters shown in Fig. 11-4 are by far the most common in industry due to their high removal rates, quick replenishment of cutting edges, and easy control of cutter geometry.

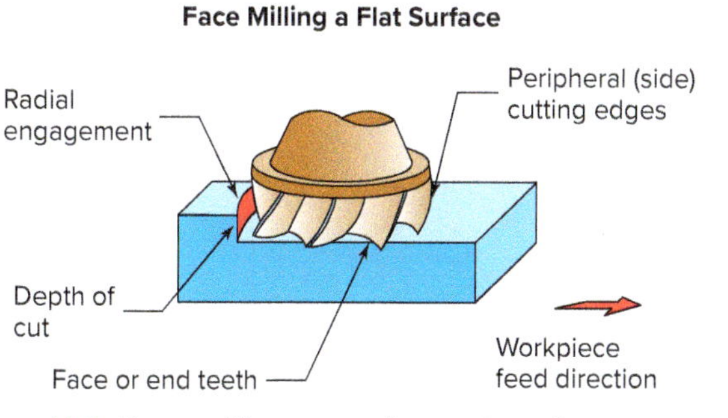

Figure 11-1 Face milling uses the end teeth to create a flat surface.

Face mills produce a medium fine finish around 64 microinches (μin.) or better under ideal conditions. While face mills are costly compared to the next two cutter types (a

10-in.-diameter, 12-tooth insert cutter can cost 2 weeks' wages for a journey-level machinist), remember they are the fastest metal removing tool of the three. Considering productivity, they often pay for themselves in the first month of use!

Figure 11-2 This high-tech, $1\frac{1}{2}$-in.-diameter carbide insert cutter blasts through this steel block.
Courtesy Iscar Metals, Inc.

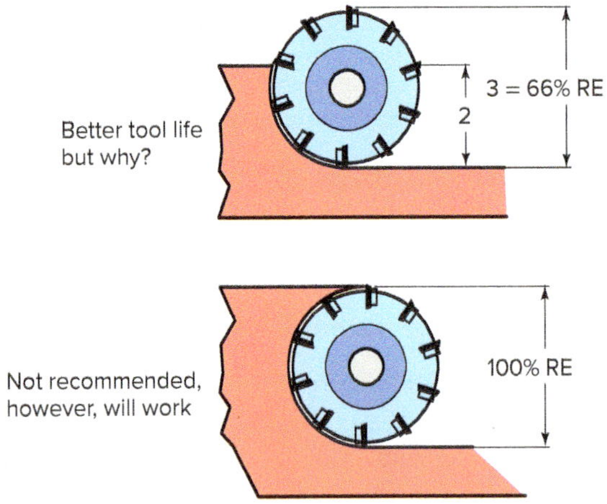

Figure 11-3 Radial engagement is an important part of the face mill setup.

Figure 11-4 A variety of insert tooth face mills.
Courtesy Kennametal, Inc.

End Milling Cutters (End Mills)

These cutters are also used to face cut because they have cutting teeth both on their end and sides. Also, when their face teeth are ground to do so, end mills can **plunge** straight down into the work, similar to drilling. Face mills cannot plunge; therefore, they must be set at the DOC when they are not in contact with the workpiece.

KEYPOINT

Not all end mills can plunge—they must have cutting ability to their face center. Only the plunging end mill can begin an inward hole or shaped pocket in a workpiece surface (Fig. 11-5).

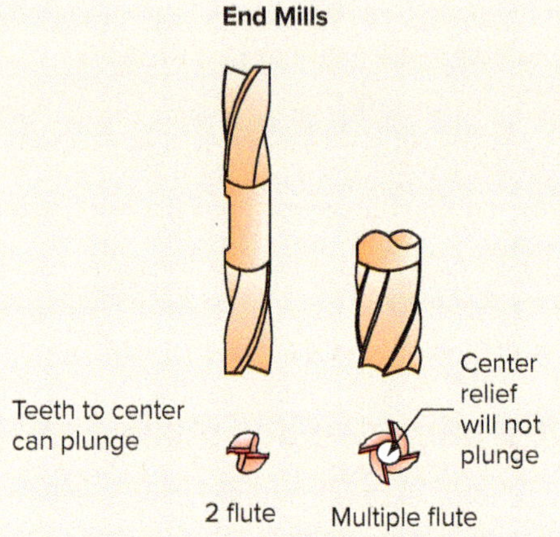

Figure 11-5 End mills feature both face and side teeth. Some are double-ended for cost savings; that is, they are one body with two cutters.

End mills with two or four teeth are common. Three-tooth cutters are also used in some free-cutting metals such as aluminum. Limited numbers are made of cobalt HSS, which is a bit harder and tougher than standard HSS, but cobalt cutters are more costly. The extra cost for cobalt HSS is not often justified, except in cases where carbide can't do the job because its cutting edge chips due to chatter but plain HSS dulls too quickly due to work hardness.

End mills are also supplied in solid carbide, which is more costly than HSS but cuts three times faster. They are also made as carbide insert tooth versions. The choice of which end mill to use is a basic skill that you'll explore in the shop—the right choice can make all the difference in accuracy, tool life, and production!

TRADE TIP

Which End Mill? If the cut requires plunging, the two-flute end mill is the best choice. Harder metals such as steel warrant using four-flute cutters because they feature more cutting edges to share the wear of the cutting action. To plunge, the end mill must have one or more teeth to its center.

End mills produce a medium fine finish and when facing, remove metal at a relatively slow rate, compared to face mill cutters. Using lower feed rates, end mills can produce a surface finish of around 64 μin.

KEYPOINT

To cut softer metals, choose the least number of flutes; harder metals warrant more teeth.

Fly cutters are a single-tooth facing cutter used more in tooling work than production due to very slow cutting (Fig. 11-6). They are correctly set up where a very large surface must be exceptionally flat or smooth and have an excellent finish of 32 μin. or better. The inserted HSS or carbide bit in a fly cutter is ground with a shape similar to a lathe bit.

Fly Cutter

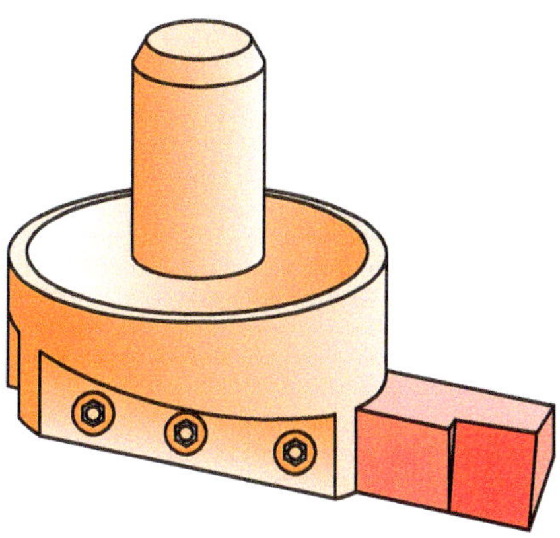

Figure 11-6 A fly cutter creates fine finishes.

Although fly cutters are inexpensive and can be shop-made, they are very slow metal removers with some serious safety considerations. Tool chatter and centrifugal force at the end of the extended fly bar are challenges for fly cutting (Fig. 11-7).

KEYPOINT

Caution! The out-of-balance nature of fly cutting makes maximum RPM, feed rate, and DOC all critical issues. A suggested limit might be 0.015-in. DOC, a 3-in.-per-minute feed rate, and an RPM under 700 for a 5-in.-diameter cut. *Check with your instructor when setting up a fly cutter—some have different limits.*

Figure 11-7 A fly cutter makes light face cuts only but creates a very flat surface with good finish.
Lake Washington Technical College, Kirkland, WA/McGraw Hill

11.1.2 Operation 2—Peripheral Milling

Peripheral milling (also called **profiling**) means cutting around the outside or inside surface of a part using the side teeth of the cutter. The most common cutter for profiling is the end mill (Fig. 11-8). The top on this block is being profile milled by a two-flute end mill.

Profile Milling

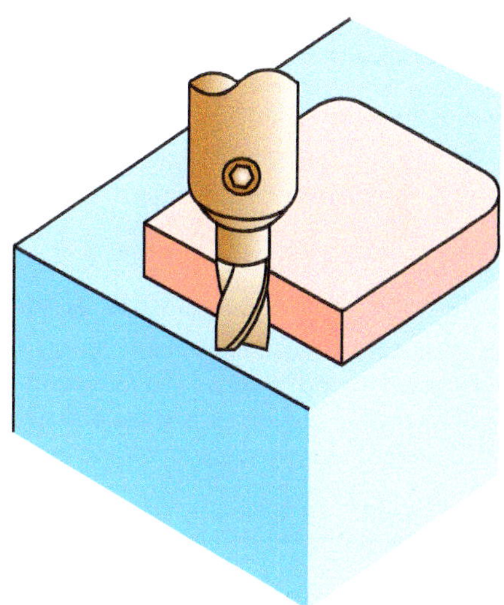

Figure 11-8 This block is being profile milled using the peripheral (side) teeth of the cutter.

Climb Versus Conventional Milling a Critical Issue!

Profiling has its own set of safety rules and challenges, which we will study in Unit 11-4, but the most important is knowing the differences in the forces, depending on the feed direction. When cutting with side teeth, the feed direction causes a multiplication of forces, with the cutter rotation and feed going the same way, as a wheel driven upon a road—called **climb milling**. During climb milling, the cutter rotation in the work-piece tries to pull the milling machine feed axis and work through the cut. Alternately, you can feed the cutter the opposite direction along the perimeter, with the teeth opposing the feed axis direction, called **conventional milling** (Fig. 11-9). It's critical for safety, tool life, and finish to clearly understand the difference. In Unit 11-5, we'll discover the real reason climb milling is the better process.

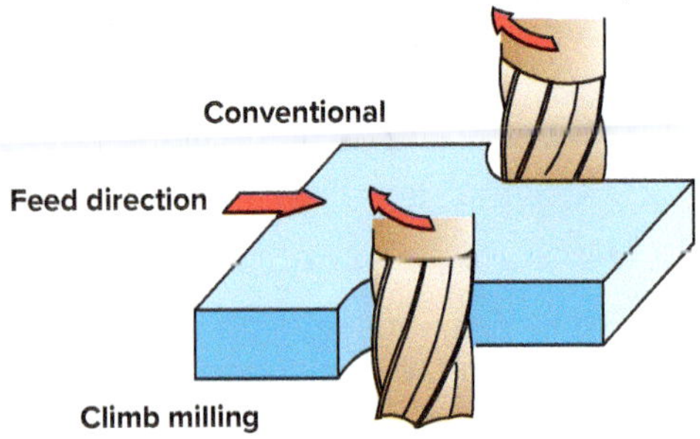

Figure 11-9 It's important to know the difference between climb and conventional milling.

Critical Thinking

Here are four questions to answer. Between climb and conventional cutting,

Which creates the best finish?

Why?

Which requires extra strong setups and backlash control?

Which provides the best tool life and why?

The truth is related to radial engagement. These questions will be answered soon, but reason them out now if you can.

KEYPOINT

There is a big difference in the forces created when climb milling compared to conventional milling. They make a difference in outcomes and especially in the setup. We almost always use climb milling in a CNC program.

11.1.3 Operations 3 and 4—Step and Pocket Milling

Often a combination of face and peripheral milling is required in one cut. A *pocket*, shown in cross section in Fig. 11-10, is a depression or cavity milled down into the metal. In CNC programming, a second pocket milled within the floor of another is said to be *nested*. These operations are called steps, or pockets.

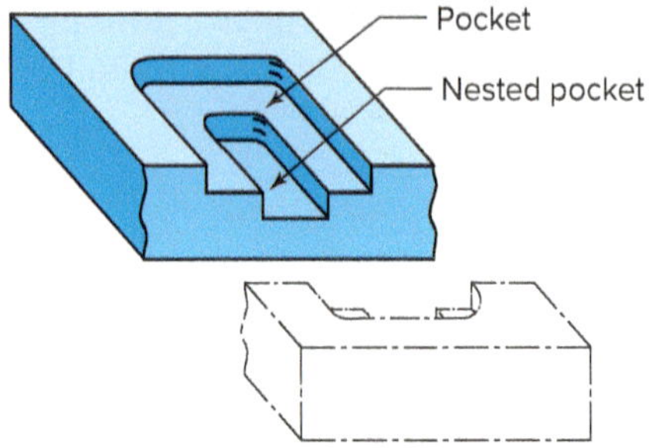

Figure 11-10 A cutaway view of two pockets with the second nested within the first.

Plunging to Cut Depth

To start the pocket operation, a hole must be opened to the rough pocket depth—a third operation similar to drilling—called *plunging* when it's done with an end mill. While it is possible to use a twist drill to start the pocket, it's faster to plunge the end mill to depth, then mill sideways to make the pocket. To do so, examine the end mill. It must have the teeth ground with one or more cutting edges completely to center—similar to a drill but with a zero degree point angle (Fig. 11-11).

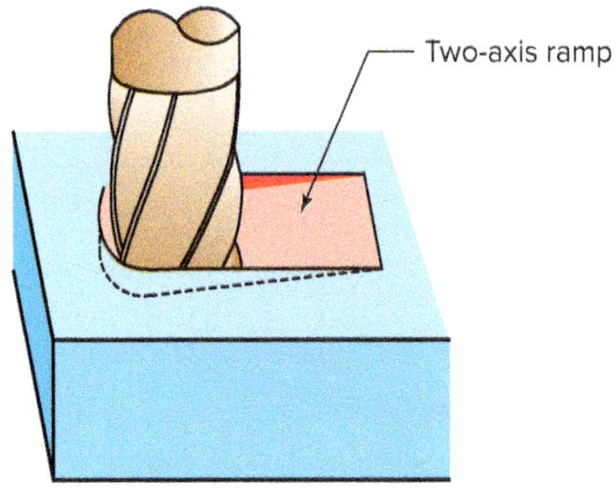

Figure 11-11 Ramping down to depth allows chips to escape and coolant to enter far better than plunging.

Three- and four-flute cutters can plunge if correctly ground with one or more teeth that cross the face center. Most two-flute cutters can plunge, but not all! Some feature hollow faces at their center, without cutting edges.

KEYPOINT

End mills must be ground with one or more teeth that cross the center face, such that they remove all metal during a plunge. If they are not, the unremoved center material will stop the plunge and probably burn or break the cutter.

Ramping to Cut Depth

See Fig. 11-11. Feeding the end mill sideways while plunging down is known as a ramp plunge, or shortened to just **ramp**.

This operation is not often used on manual mills, but it's common on CNC because it makes a slanted groove that releases chips better than a drilled hole; plus it allows coolant to reach the cut better. Programmers almost always choose ramping over plunging to cutter depth, while manual mill operators usually use simple plunging. Spiraling the ramp down, in a helix rather than in a straight line, offers a couple of advantages. First, it keeps the plunge area smaller, and it can terminate tangent to the wall of the pocket, thus causing fewer tool marks in the finish. That move is called a *spiral ramp.*

TRADE TIP

Pocket Secrets Pocket milling tosses a couple of serious challenges your way, both caused by backlash in the feed axes. When making the final cuts around the inside of the pocket, at each inside corner, the cutter can be pulled into the work surface by the cutting action when the clearance between the axis screw and nut is slack. It leaves an unwanted indent in the surface, often called an undercut. This problem exists on both manual and CNC milling machines, but it's far worse on manual ones due to the type of axis drive screws they have.

Changing directions in the corners, the cutter tends to dig (pull beyond the intended path) into the corners as the backlash is taken up on the axis. The second challenge exists only on manual mills lacking digital positioners, because the micrometer dials are valid only when cranking one direction. Yet you need to crank to locations on both sides of the pocket. If the machine has digital readouts, this problem is solved for cutter position, but not from the corner digging. Without a digital position, use layout and measuring tools to track progress.

To avoid the backlash corner digging, first rough within 0.015 to 0.030 in. of final size. Then take two light finish cuts around all surfaces. Second, always use climb milling as it tends to push the cutter away from the work rather than dig in. Third, on manual mills, always tighten the lock on the unused axes (*X* or *Y* and the *Z* quill). Finally, use digital readouts over mechanical dials whenever possible because they are not subject to backlash errors.

11.1.4 Operation 5—Angle Cutting

On manual mills, angular cuts can be set up several different ways, by either tilting the cutter or tilting or rotating the part. Angles can also be cut using form-ground cutters. Today the most common method of making an angular surface is to feed two axes simultaneously through CNC control.

Figures 11-12 through 11-17 show five manual mill setups that could be used to machine the angle on a drill gage blank. Each offers different mixes of accuracy and required setup time.

1. **Tilting the Part**

   Using a layout line and the parallel bar (see Fig. 11-13). Using the Trade Tip, repeat this setup at around 1° to 1.5° repeatability—good only for semiprecision work. Using a sine bar setup improves accuracy.

2. **Improving Layout Alignment** (see **Fig. 11-14)**

   To improve accuracy when tilting the work in a vise, lay a ground parallel on the top of the vise jaw; then level the layout to it. Remove the block before cutting begins.

3. **Tilting the Head**

   The spindle head of many smaller vertical milling machines features one or two axes that can be set at an angle with respect to their table (see Fig. 11-15). On their flange rim (see Fig. 11-16) they feature angular graduations with a resolution of either $\frac{1}{2}$° or 1°.

4. **Rotating the Part**

   There are a few ways angular work can be set up parallel to the mill table, then rotated either to the 59° angle or its complement, depending on the setup accessory: a rotating vise, a protractor, or slot block parallels (Chapter 7) and a sine bar. To finish the setup in Fig. 11-17, the work would be clamped to the table with a protective space between it and the table top. However, for the drill gage example, milling the angle with the side of a cutter presents a problem—the sharp corner will not be milled correctly due to the cutter diameter (see Fig. 11-17).

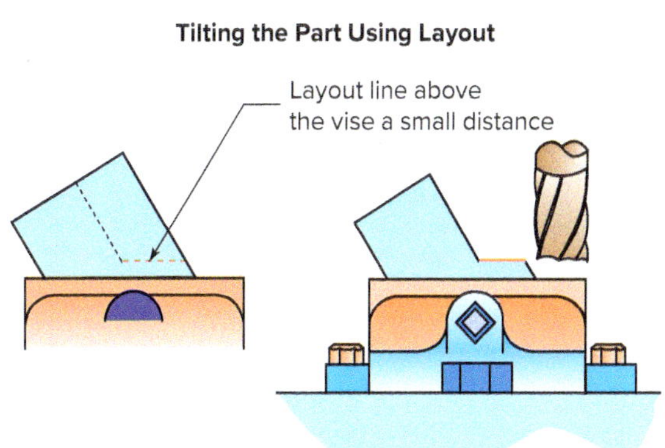

Figure 11-12 Tilting the part using layout is a simple method with repeatability around 1°.

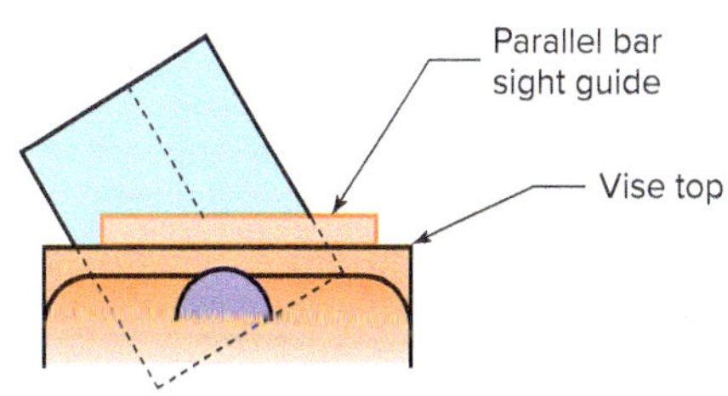

Figure 11-13 Improving alignment using a parallel bar.

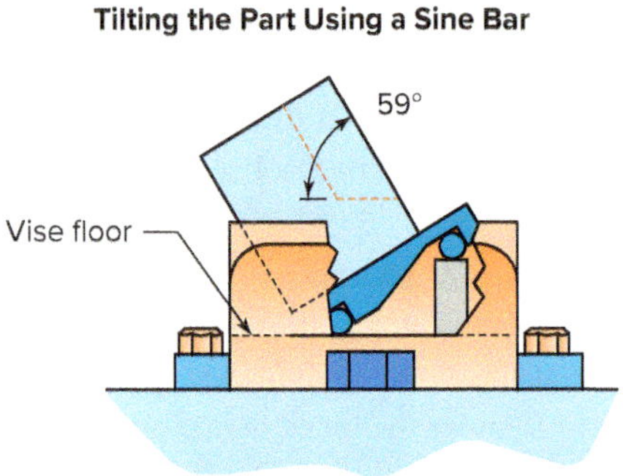

Figure 11-14 Tilting the part using a sine bar with repeatability around .01 degree.

Figure 11-15 Tilting either the *A* or *B* axis (shown) is a simple way to cut an angle.
Lake Washington Technical College, Kirkland, WA/McGraw Hill

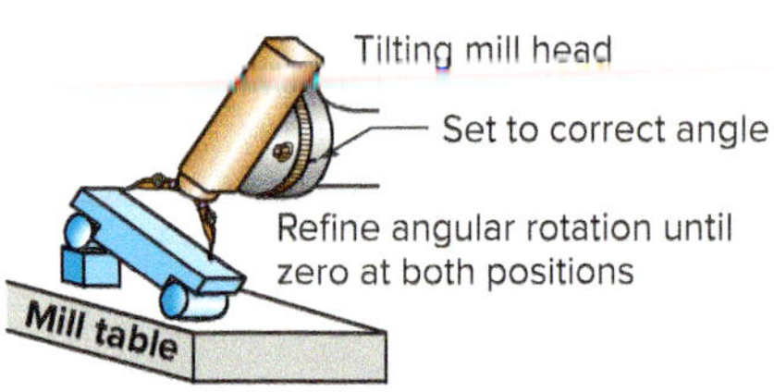

Figure 11-16 The tilt angle can be refined beyond the machine's scale using a sine bar and DTI.

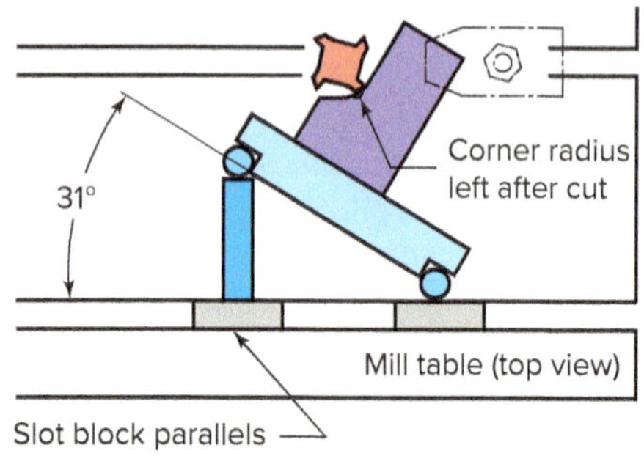

Figure 11-17 Rotating the work using slot blocks and a sine bar.

Critical Thinking

In Fig. 11-16, the setup is made to mill the 59° angle on the drill gage. So, what angle do you set on the sine bar? (The answer is found just before the Unit Review.)

KEYPOINT

Laying the work on the mill table and rotating it relative to an axis works particularly well for large parts too big to tilt. When doing so, never clamp the work directly on the table; always provide a space between the two with shims.

11.1.5 Operation 6—Forming Cuts

Radius Cutting

On all mills, inside or outside radius cutting (rounded corners) is accomplished using form-ground cutters. Here the pocket shown previously now has received several radii (plural of radius). Inside corner radii are also called **fillets** (rhymes with billets) on drawings (Fig. 11-18).

The teeth of radius cutters have either a convex curve or a concave curve ground onto their corner (Fig. 11-19). As with lathe tools, creating a shape in the work using a shaped cutter is known as *forming*. The two types of end mills ground to form concave corner radii are the **ball nose end mill** (full end radius) and **bull nose end mill** (a small flat on the end).

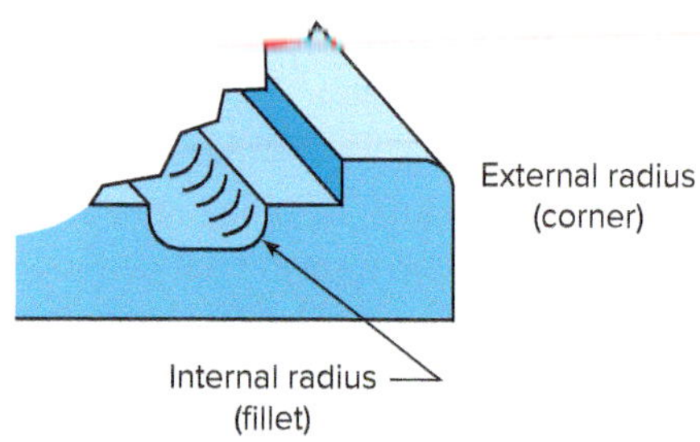

Figure 11-18 A radius corner has been added to the pocket and the outside corner. The concave radius is often called a fillet.

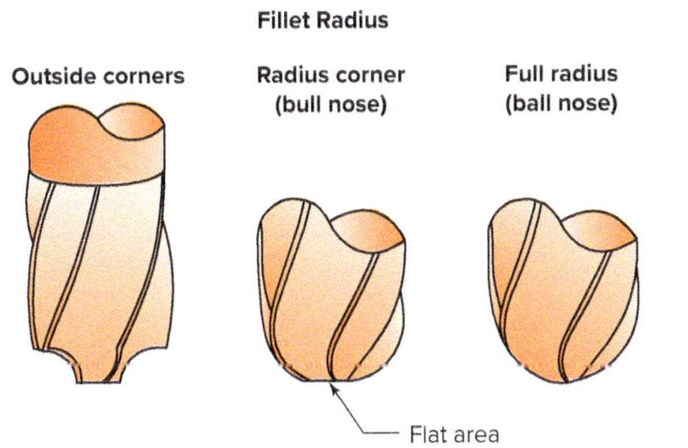

Figure 11-19 Three form-ground radius cutters: an outside corner radius, a bull nose, and a ball nose end mill.

Custom Form Shapes

When the shape is unusual, the simplest shop-made form cutter is the fly cutter. Its HSS bit can be custom ground to nearly any shape without special equipment. In Fig. 11-20, an angle and corner radii are put on this object. Figure 11-21 shows two types of purchased form cutters for special shapes, the tee slot and the dovetail.

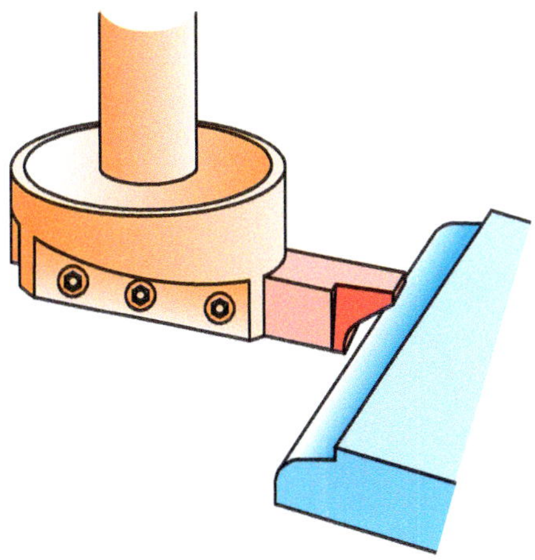

Figure 11-20 The simplest shop-made form cutter is ground into the bit for a fly cutter.

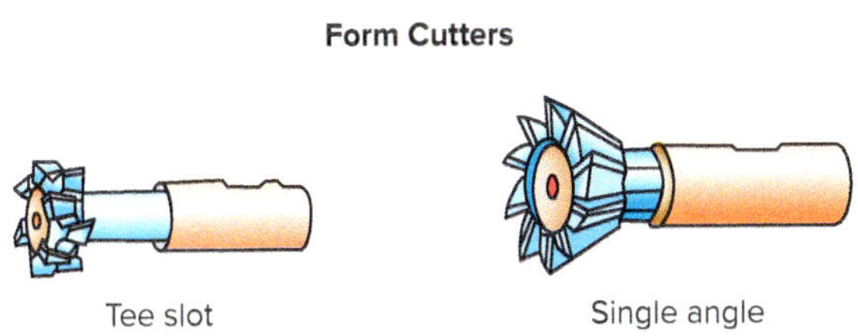

Figure 11-21 Many different angles and several other shapes can be purchased preground onto end mills.

11.1.6 Operation 7—Drilling/Reaming

We perform hole drilling on the milling machine when the hole's location tolerance is tighter than drill press accuracy, at around 0.015 in. or less. Using micrometer accuracy, manual mills can locate their tables from within 0.001 to 0.0005 in. in the *X*-*Y* plane, and with digital positioners do even better. The bits and reamers are held in a drill chuck or a taper adapter, but they can also be held in collets, which tend to hold the drill on the spindle axis with less runout.

Drill tools can be hand or power fed into the work. While mill spindles are far more rigid than the drill press, the drills can still wander off location as they start, so a pilot drill (or center drill) is required to aid location accuracy. **Boring** after drilling is the best location solution, which we'll explore next.

Hole Accuracy—Location Repeatability

Manual mill location repeatabilities of 0.001 are typical using *X*-*Y* micrometer dials and finer using electronic digital positing but only when the hole is bored on location.

> **KEYPOINT**
>
> To achieve both location and diameter accuracy, sometimes it's best to center drill on location, drill the rough hole, bore it within 0.015 in. of finished size, and then finish the size with a reamer.

Diameter Repeatability

Diameter accuracy of drilled and reamed holes on milling machines repeats no different than on a drill press. You can expect around 0.003 in. for drilling alone and 0.0003 in. for reamers. However, on mills we can add an adjustable boring head (coming up) to machine any diameter with 0.0002-in. repeatability, if a quality head is used along with good speeds and feed, and sharp boring tools.

Coolants, Accuracy, and Tool Life The constant challenge for tool life and the accuracy associated with sharp cutting tools is the delivery of coolants down to the cut area. When hole depths exceed three to five times the drill diameter, the chips being wedged up and out of the helical flutes (or plunging end mill) stop the coolant as it tries to penetrate down into the hole. There are three solutions.

Peck Drilling As discussed previously, there are two types. *Chip breaking* is used for medium-deep holes, whose depth is up to eight times diameter, whereby the infeed is halted and the drill pulled back from around 0.010 to 0.050 in., then sent back to drill. Beyond the 8:1 ratio, a second peck is called *chip clearing* because the cutter is pulled all the way out of the hole then sent back down to drill again. That allows most of the chips to get out and a burst of coolant to flow down to the bottom of the hole. CNC codes for these moves are G81/G82 drilling cycles, and G83 pecking, G84 tapping, and G85 boring cycles.

High-Pressure External Coolant Here a focused stream of coolant is sprayed down the tool flutes and can penetrate to depths of five or six times drill diameter. But it, too, becomes blocked by the outgoing chips. The high-pressure coolant system is an add-on feature for all manual machines and most CNC machines, but it's a good solution for many machining problems including medium-deep hole drilling.

Figure 11-22 The coolant delivery holes are shown in the blown-up end view of this carbide tipped deep hole drill.
Courtesy Aloris USA

High-Pressure Coolant Through the Tool The winner for chip clearing, accurate holes, and tool life features an injection hole through the tool center (Fig. 11-22). High-pressure coolant is forced down to the action area. Special sealed machine collars surround the tool shank and holder and force the coolant down to the point of application. Not only does this cool the cut, thus extending tool life while also aiding diameter accuracy; it also pressure forces the chips up and out of the hole, thus eliminating the need for pecking. Gun drills using this center coolant concept can drill to a nearly unlimited depth. This is an add-on accessory and requires special tools that feature the coolant galleries through which the stream is pumped (Fig. 11-22).

Drilling on Ram and Turret Mills

It's common to drill on the smaller universal mills (Fig. 11-23) called ram and turret machines, often called Bridgeport-type mills. Typical reasons might be close location and/or diameter tolerances within 0.003 in. or less or when no reamer is available for a special diameter so we must bore the hole.

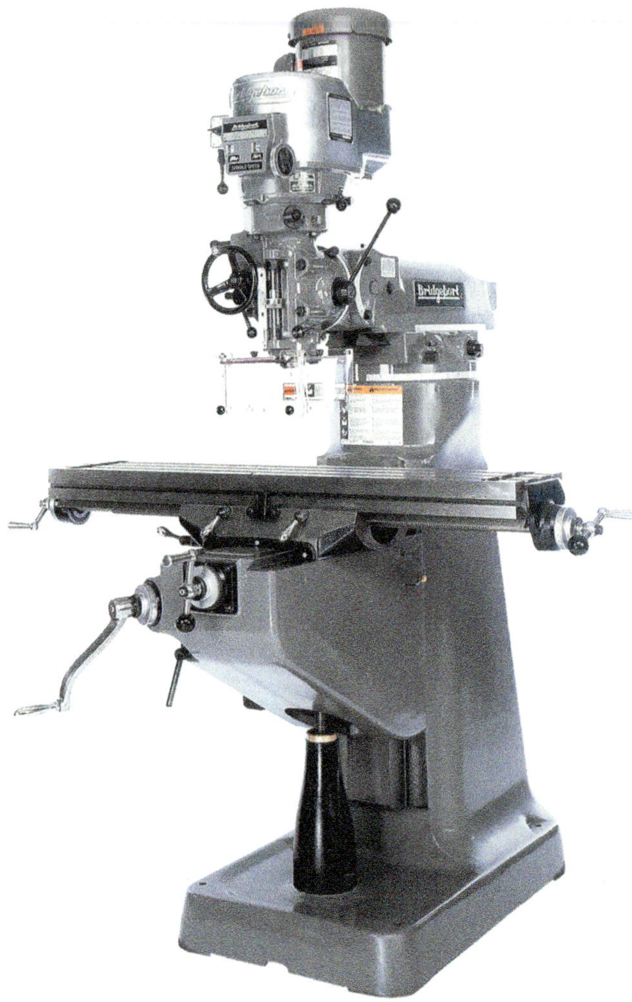

Figure 11-23 The ram and turret mill is popular for off-load support work, tooling, and training.
Courtesy Bridgeport Machines, Inc./Bridgeport, CT

KEYPOINT

Due to their flexibility and popularity in training labs, we'll be using Bridgeport mills in our examples throughout the manual milling chapter.

SHOPTALK

Although ram and turret mills (see Fig. 11-23) are made by many manufacturers, the early success of Bridgeport Machines of Connecticut made theirs the most popular brand in the past. So, like Crescent wrenches, known not by their proper name but by the name of a popular manufacturer, today we almost always call machines of this type by the originator. This is not an endorsement of a particular brand; there are many manufacturers of ram and turret mills.

CNC has changed much, and the Bridgeport mill is no exception. Manual versions are not commonly used for production work today, but the CNC versions are. However, today they remain a most adaptable machine in terms of configurations and for offload and tooling work.

11.1.7 Operation 8—Boring

Precision holes are bored on mills using an adjustable boring head. This specialized tool allows outward micrometer positioning of a single-point cutter (Fig. 11-24). A rough-out hole is first machined; then finish passes are taken by dialing the boring tool outward to a bigger radius.

Micrometer Adjustable Boring Head

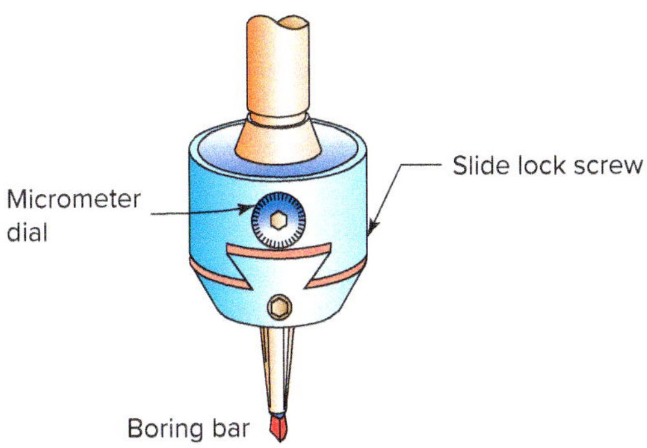

Figure 11-24 A typical adjustable boring head using the Swiss type boring bar.

Using the single-point tool and several passes to eliminate out-of-roundness and mispositioning of the rough hole, the boring process tends to machine the hole exactly on location, very round, and straight through the work. Boring creates a round, smooth hole of 32 μin. or better with size tolerances to 0.0005 in. (Fig. 11-25).

11.1.8 Operation 9—Arbor Cutter Work

Sometimes part features such as slots or tabs are cut using *arbor cutters*, sometimes called *wheel cutters* in shop lingo. They might also be called *plain mill cutters* (Figs. 11-26 and 11-27). These cutters are made to cut on their rim teeth, as shown in Fig. 11-26. Like a wheel on an axle, they are mounted on a precision shaft called an **arbor** that is held in the mill's spindle taper. The arbor features a square **keyway** along its entire length. The cutter has a similar square groove so that when a square steel key is inserted it positively rotates the cutter without slipping. Precision-ground spacers are used to position the cutter in place along the shaft. These cutters are either HSS or carbide insert tooth (Fig. 11-28).

Typical Operations using a Boring Head

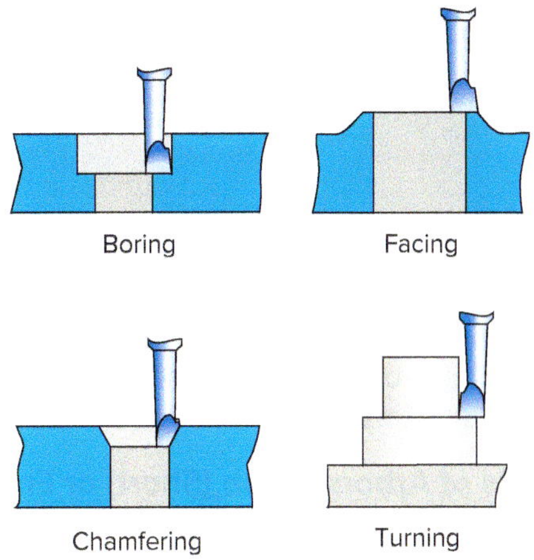

Figure 11-25 Several operations are possible with a boring head and the right cutting tool!

Arbor-Mounted Tools

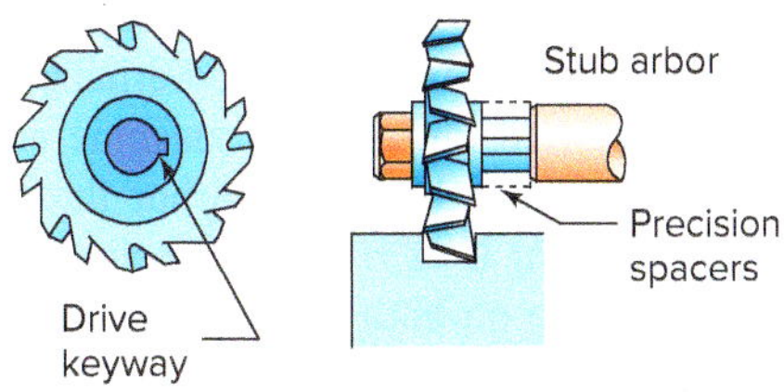

Figure 11-26 Wheel-shaped cutting tools are mounted on arbors.

Slitting Saw Cutter

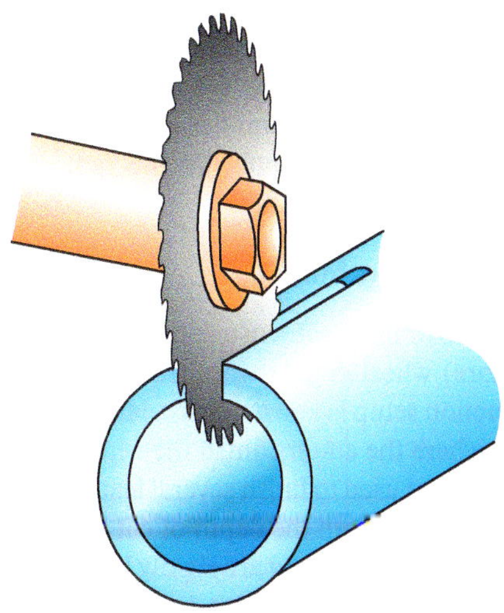

Figure 11-27 A thin wheel cutter is called a slitting saw.

The rotary table can be used both as an indexer (moving parts from position to position) and to move the part in an arc during machining. Used in that capacity, it becomes a feed axis that rotates by hand cranking. The machinist can work from layout lines on the part or by setting datum stop points on the graduated table. The precision rotary table may be moved as fine as 0.2° using a vernier scale adjacent to the handle.

Sine Bar Answer

The complementary angle is 31°. Return to Fig. 11-14, the same logic also applies there.

UNIT 11-1 REVIEW

Replay the Key Points

- Mill feed rates are expressed in inches of travel per minute (IPM), or millimeters per minute.
- Not all end mills can plunge.
- To cut softer metals, choose the least number of flutes; for harder metals choose more teeth.
- Safety–There is a big difference in the forces of climb milling and conventional milling–it makes a difference in the setup.
- Safety–The out-of-balance nature of fly cutting makes maximum RPM, feed rate, and DOC all critical issues. Check with your instructor for shop limits.
- When placing work on the mill table, never clamp it directly on the table; always provide a space between the two with shims.
- To achieve location and diameter accuracy, it's best to center drill on location, drill the rough hole, bore it within 0.015 in. of finished size, and then finish the size with a reamer.

Respond

1. Of those listed here, identify the operations that can be accomplished with an end mill.

   Facing

   Peripheral cutting

   Step milling

   Pocket milling

   Cutting angles

   Forming–inside and outside radius

   Drilling, reaming

   Boring

   Straddle and gang cutting
2. True or false? When drilling down to form the start of a pocket, a two-flute end mill must be modified such that it will nest. If this statement is false, what would make it true?
3. Identify the cutters shown in Fig. 11-34. For what operation would each be used?
4. Name the three types of facing tools for milling machines. Please note, there are more cutter types, but these are the basic three experienced in your first year.
5. Milling with two cutters mounted on an arbor that are spaced an exact distance apart is called milling.
6. Why might you move a hole-drilling operation from a drill press job to a vertical mill?
7. Explain and compare a ramp to a plunge cut.
8. Objective: Modify this brake disk in Fig. 11-35 by drilling cooling holes. Name the operation that will be used. What is the name often given to a pattern of holes such as shown in Fig. 11-35?

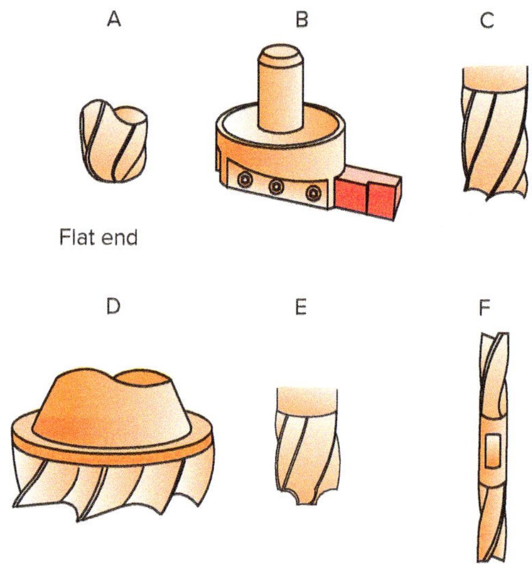

Figure 11-34 Identify the cutters.

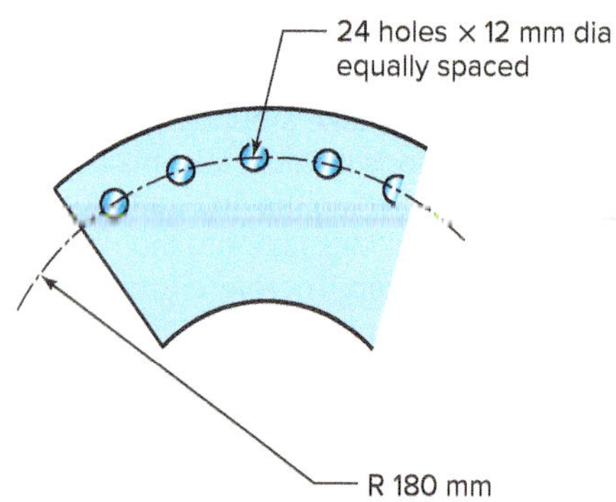

Figure 11-35 What operation is needed to drill these cooling holes in this disk brake? What is the name of the pattern produced?

Critical Thinking

9. When setting up to drill holes on a milling machine, the operator must first locate the spindle directly over a datum, then move it to each hole location. There is a vital step missing in this statement–what is it?
10. Describe the two advantages of using the operation shown in Fig. 11-36 over drilling and reaming a hole.
11. Using the operations we've discussed, identify the four used to make the part sketched in Fig. 11-37 and the possible setup/cutter combination to make it. There may be more than one way to accomplish the feature.

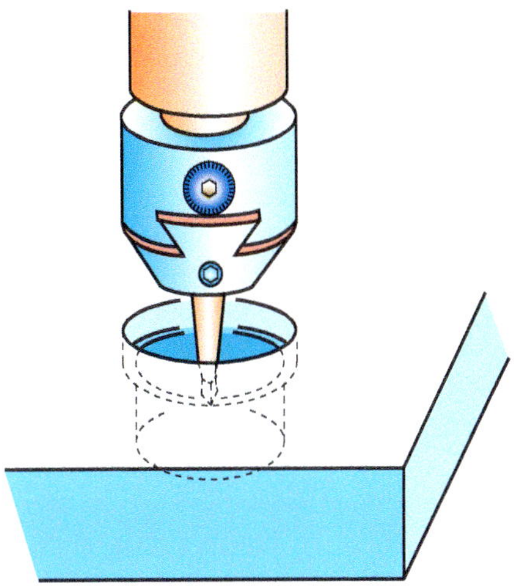

Figure 11-36 What is the name of the operation being performed and the advantages over drilling and reaming?

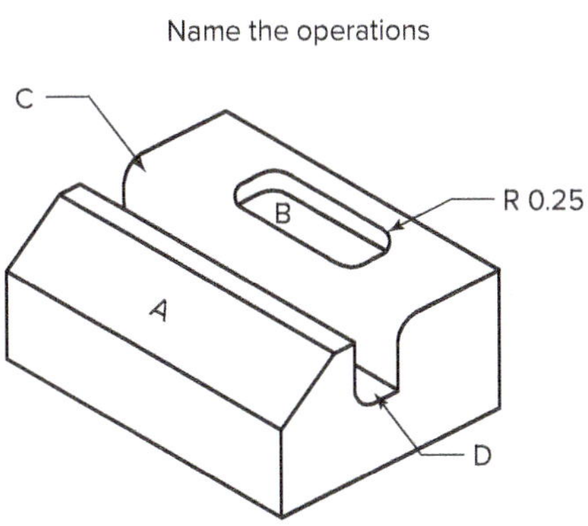

Figure 11-37 Name the operations and setup needed to make this object.

Unit 11-2 How Does a Mill Work?

Introduction: Similar to what we did with lathes, here we'll assemble, then examine a ram and turret mill, one feature at a time. But first, we'll take a quick look at the difference between horizontal and vertical milling machines.

TERMS TOOLBOX

Knee The second-largest component mounted vertically on the column. The knee provides the *Z* axis motion on a vertical mill.

Linear axis Straight-line motion. May be used as a motion for machining or limited to positioning during setups only.

Orthogonal axis set A 90° set of linear axes found on all CNC machines, which may have two or three primary linear axes. Their orientation to the world varies, but they always remain the same to each other.

Rotary axis A motion or position axis that moves in an arc about a pivotal axis (partial or full circle).

Saddle The mill component that mounts on the knee providing the *Y* axis motion on vertical milling machines. The lathe component that fits over the ways.

Work envelope The space where machining can occur bounded by axis limits.

11.2.1 Comparing Modern Vertical and Horizontal Spindle Mills

Vertical milling machines, which include the ram types, are very popular in industry and schools due to the ease of setup and use. Our early investigations focus on them. However, horizontal machines have their purpose especially in heavy industry. You should know when a job is better suited to a horizontal than to a vertical spindle mill. Let's briefly examine the differences.

Both machines shown in Fig. 11-38 have horizontal work tables, carried on a knee that runs vertically, up and down on a dovetail column. That forms one axis of movement for both–but it's the *Z* axis for the vertical, and it's the *Y* axis for the horizontal.

A second version of the horizontal mill features a vertical surface for mounting the work. The vertical table, horizontal mill is generally the larger of the two and is usually found only in shops that machine massive work (Fig. 11-39). But both horizontal machines are better suited to heavy or extra large work compared to the vertical. We discover why next.

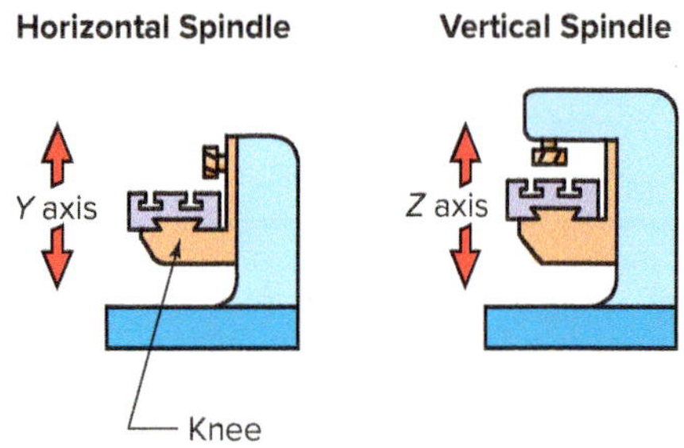

Figure 11-38 Vertical and horizontal milling machines are similar.

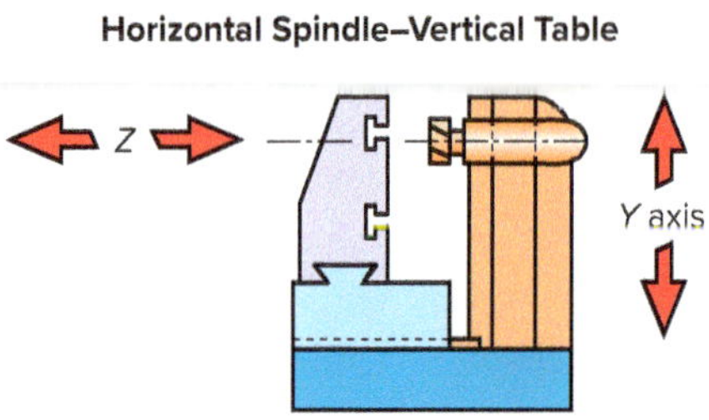

Figure 11-39 This is the larger version of the horizontal mill.

Horizontal Mill Advantages over a Vertical Mill

1. **Part Loading**

   Large parts can be suspended from above with a crane, then easily loaded directly down onto either type of horizontal table. No special crane rigging is required. In comparison, it's difficult to load a vertical mill with large parts, since its milling extends out over the table (Fig. 11-40).

2. **Chip Clearing**

   When large volumes of chips are made on the vertical table horizontal mill, they simply fall away along with coolants. They flush easily out from deep pockets with coolant injection.

3. **Larger Parts**

   Because there is no overhead frame, very large parts can be loaded onto either version of a horizontal mill, but the vertical table version can securely hold literally tons of work within its work envelope.

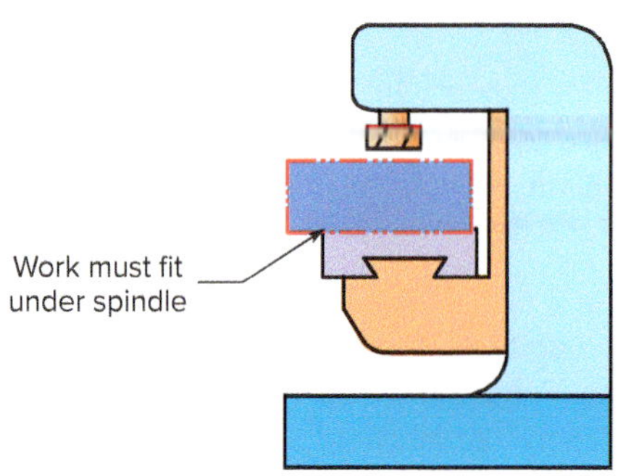

Figure 11-40 Placing the work in a vertical mill work envelope.

Work Envelope

The **work envelope** is a term used to describe the boundaries of space in which machining can occur (on any machine–not just mills). The envelope is bounded by the extent of spindle and axis travel. On mills, the work envelope becomes a cubic space bounded by the X, Y, and Z axis limits (Figs. 11-40 and 11-41).

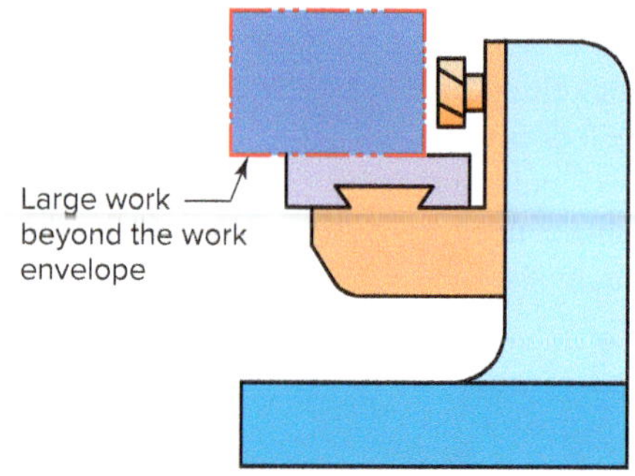

Figure 11-41 Placing the work in a horizontal mill's work envelope.

One of the greatest distinctions between vertical and horizontal mills is that on the vertical mill, the workpiece must be able to fit within the work envelope–the Z axis limits (Fig. 11-42). In other words, inside the machine. However, on horizontal machines, the work sits on or clamps to the table, but the work size is not limited to the work envelope (Fig. 11-43).

Figure 11-42 Chips fly in every direction as this horizontal-spindle hypermill rips through a large aluminum billet. Note the vertical table.
Courtesy Milacron LLC

Figure 11-43 A horizontal spindle machining center with horizontal table.
Courtesy of Haas Automation, Oxnard, California

11.2.2 Vertical Milling Machines

That said, in most of today's training shops the vertical milling machine is the more popular equipment (Fig. 11-44). It is easy and quick to set up, fast to turn around to another setup, and very versatile for work up to a maximum of 2 or 3 ft long. From here on we'll be discussing a vertical mill.

Figure 11-44 The MVP in most machine shops—the vertical machining center, milling machine.
Courtesy Okuma America Corporation

Column and Base

The spine of the mill is the solid metal column and base that compares to the bed ways on a lathe. It features a vertical, precision dovetail that guides and supports the knee—the up-and-down sliding axis that supports the table. The dovetail sliding upon the column becomes the Z axis (Fig. 11-45).

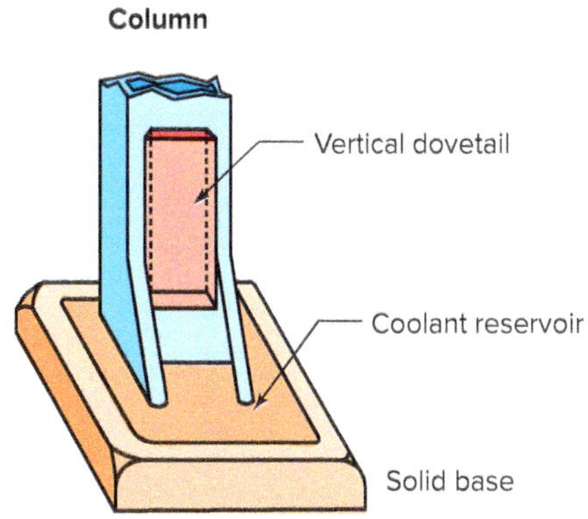

Figure 11-45 The base and column of a typical vertical mill.

Lubrication of the dovetail ways is essential for long life and precision. Your machine may have automatic or manual lubrication—always know which. The lube reservoir and pump are probably mounted beside the base; use them on a daily schedule if your machine is so equipped. Take a minute to look at the mill in your lab to determine which lubrication system is provided.

The Knee

The next most massive part of the axis system is the knee (Fig. 11-46). The **knee** mounts on the column dovetail using gibs similar to lathe axes. The knee on most mills can be hand cranked up and may feature a power feed also. The Z axis will feature a micrometer dial and an axis lock. The first of three axes is now complete; two remain to be added.

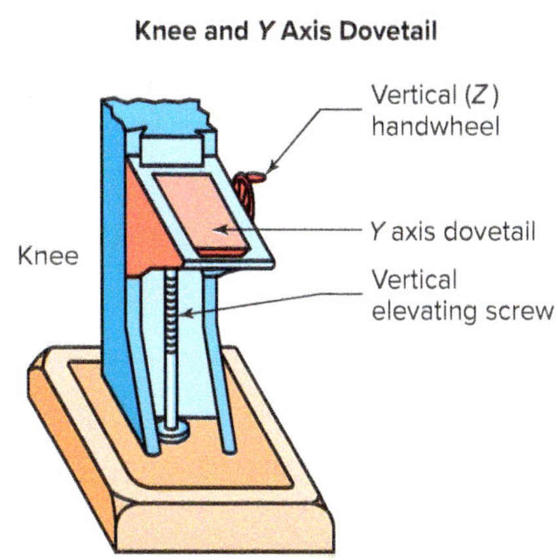

Figure 11-46 Fitting over the column, the knee slides up and down on dovetail ways. It's moved by hand and/or power feed. The knee forms the Z axis and supports the Y axis.

Table and Saddle

The mill now receives the *Y* axis **saddle** that moves in and out with respect to the operator. It slides upon a dovetail machined to the knee's top surface. Then upon the saddle, the table dovetail forms the right/left *X* axis (Figs. 11-47 and 11-48). Each of these axes has an adjustable micrometer dial and axis lock. Some feature power feed, while all feature hand cranks. Each also has adjustable gibs to minimize unwanted axis motion. Each axis has a given amount of backlash in the positioning system due to mechanical clearance between its drive screw and axis nut. Backlash must be accounted for by the machinist, as we've discussed previously.

X Axis Saddle Added over Y Axis Dovetail

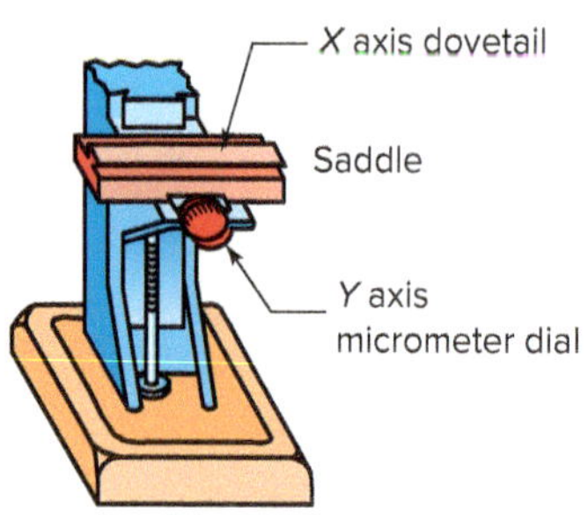

Figure 11-47 The mill has now received the saddle and the *X* axis.

The X Axis Table Completes the Set

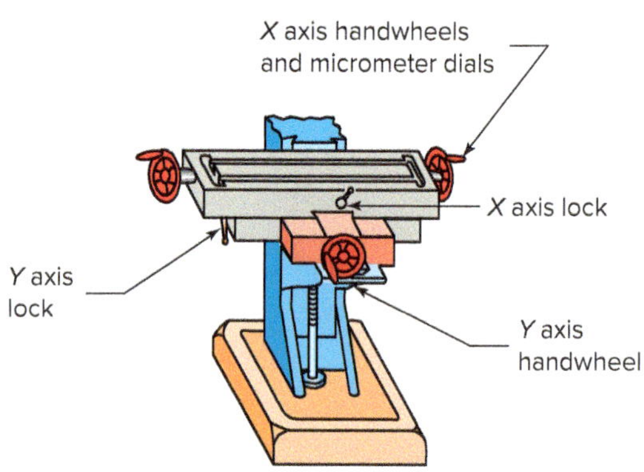

Figure 11-48 The table slides on the *X* axis dovetail to complete the axis set for a vertical milling machine.

11.2.3 Industry Standards for Axis Notation

Axis movement is either straight line—called **linear**—or in a circle—called a **rotary** motion. Some rotary axes can move during machining, while others are adjustable for setups only and are then locked in position during the cutting. These are called positioning axes.

When two or three linear axes are at 90° to each other, such as in the milling machines in Fig. 11-49, they are called an **orthogonal axis set** (the same root word as for orthographic projection). This 90° relationship of the axes is constant from machine to machine, worldwide. Depending on the machine type, the set might be in a different position relative to the floor—for example, the *Z* axis between horizontal and vertical mills—however, their orientation to each other within the set remains a constant.

Standard Axis Nomenclature

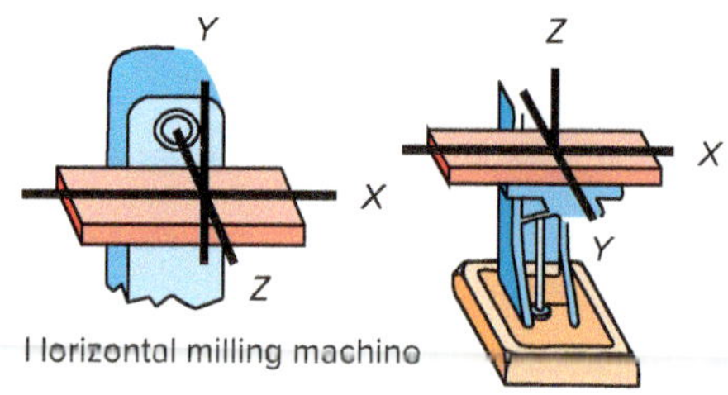

Figure 11-49 Compare the orthogonal set for a horizontal milling machine to a vertical.

In Fig. 11-49, compare the vertical mill's axis set to the set for a horizontal mill. The horizontal mill's set has been rotated such that *Z* axis is now parallel to the floor.

> **KEYPOINT**
>
> Note that according to convention, the *Z* axis will always be the linear axis parallel to the main spindle centerline.

Rotary Axis Identification *A*, *B*, and *C*

A relationship exists for rotary motion letters, too. If the center of the rotary motion revolves around a line parallel to the *X* axis, then it is an *A* axis, motion around *Y* becomes *B* motion, and that around *Z* is then designated *C*.

> **KEYPOINT**
>
> Rotation around a line parallel to *X* is the *A* axis, around *Y* is *B*, and around *Z* becomes *C*.

You will soon see that the ram and turret mill features six auxiliary movements. Two rotary motions of the head are for setup position only. Next, the quill is powered up/down or hand fed during machining. The ram rotates about the turret, and it slides in and out for setup positioning. So, other than the quill feed, all are limited to setup only—they cannot be used as machining motions.

> **KEYPOINT**
>
> On CNC mills, the *A* and *B* axes are common fourth and fifth machine rotary feed motions.

TRY IT

KEYPOINT

Although learning the axis nomenclature doesn't make a whole lot of difference for operating manual machines—right/left, in/out, and up/down would be okay—this is an essential preparation for working with CNC machines.

A. From what we learned in Chapter 10, is this statement true for lathes? *The Z axis is parallel to the main spindle axis.* Draw a sketch to assist—sketch both the *X* and *Z* axes.

B. On an R&T type mill (Fig. 11-50), the two rotary positioning movements of the spindle head can be called *A* and *B* according to the industry standard notation. In looking from the side of the mill table, what rotary axis positioning would this be, *A* or *B*?

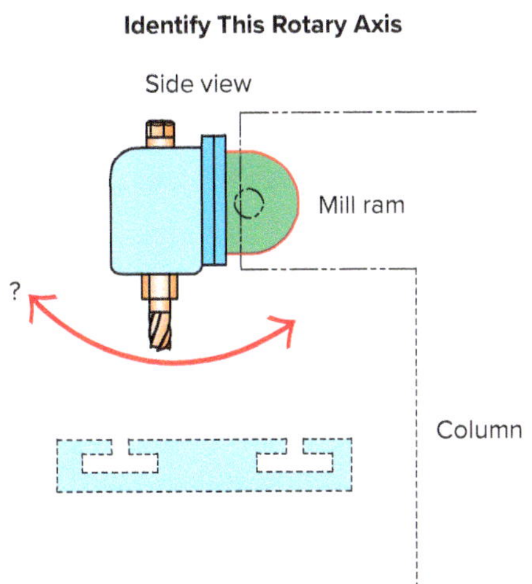

Figure 11-50 Name this setup rotary axis.

C. Then viewed from the front of the vertical mill, according to the industry standard, what rotary axis is shown in Fig. 11-51?

D. One variety of *horizontal* milling machine, known as a universal mill, features a horizontal rotating work table that is positioned, then locked. Looking back at the horizontal mill axis set, what rotary axis would this be?

ANSWER

A. Yes. The *Z* axis is the lathe saddle, which moves parallel to the spindle axis. That shouldn't be a surprise; these are world standards!

B. See Fig. 11-52.

C. The rotary axis is an *A* axis because it rotates around a line parallel to the *X* axis (Fig. 11-53).

D. The rotary table on horizontal mills would be a *B* axis because it rotates around a line parallel to the *Y* axis in the orthogonal set.

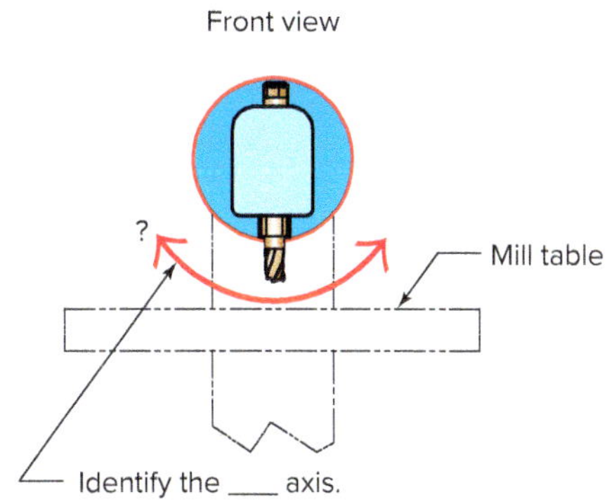

Figure 11-51 Name this rotary setup axis when the mill is viewed from the side.

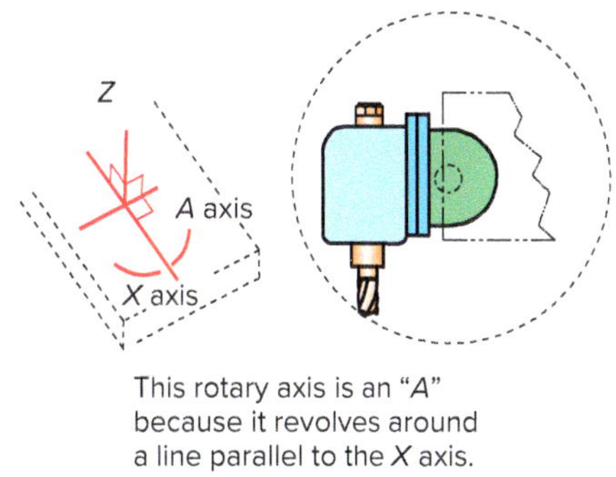

Figure 11-52 Name this setup axis viewed from the front.

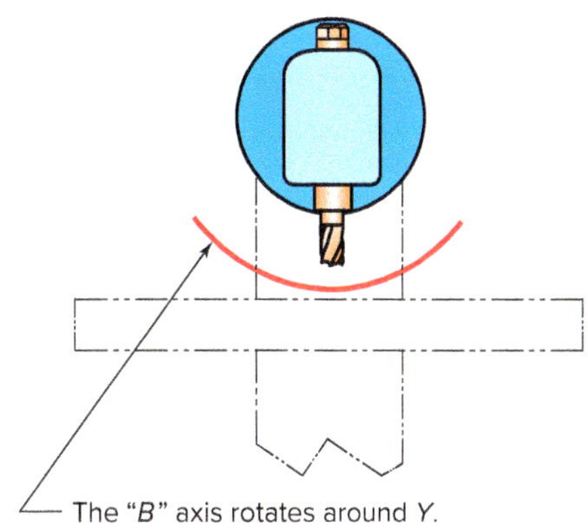

Figure 11-53 The *B* rotates around the *Y*.

11.2.4 Machine Table and Feed System

If the mill is equipped with a power feed, then the power feed gear case is found on one end of the table, as seen in Fig. 11-54. Feed selections for the *X* axis will range from below 1 in. per minute to 20 in. or more per minute (entry-level rates). Note, only a few small mills, such as the ram and turret, do not feature a power feed on all axes.

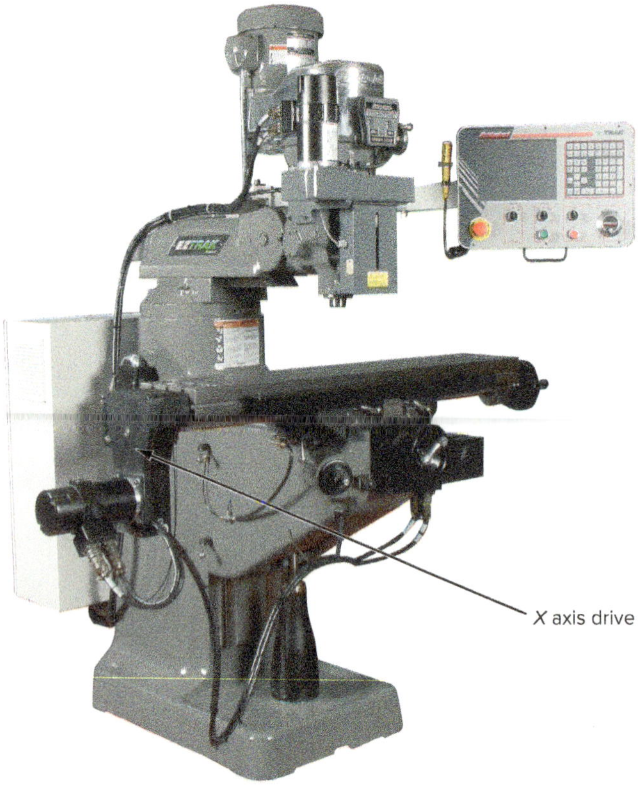

Figure 11-54 The *X* axis feed gear case can be seen on this CNC Easytrack mill.
Courtesy Bridgeport Machines, Inc./Bridgeport, CT

KEYPOINT

Feeds range from below 1 in. to around 20 in. per minute on small mills.

The mill table forms the *X-Y* plane and features *tee slots* used for bolting work or accessories (Fig. 11-55). The upper groove in the slots can be used to align work in several ways. Using the slot correctly can save a lot of extra time and add rigidity and precision to your setups.

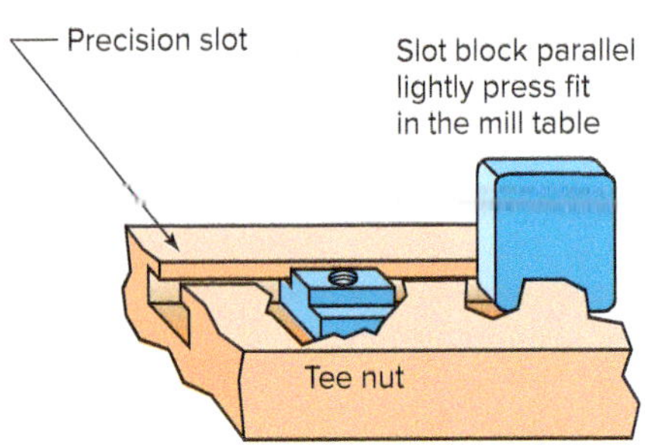

Figure 11-55 The tee slots are a *precision part of the machine.*

TRADE TIP

Using Tee Slots to Save Time Precision keys on the bottom of accessories such as vises will drop into exact engagement with the table slot grooves. This eliminates misalignment of the vise. Also, precision slot-block parallels fit perfectly into the slots. Material can be instantly aligned to them for machining (Fig. 11-56).

Figure 11-56 Work can be aligned to the *X* axis with slot blocks and clamped directly to the mill table. Is there a problem with this setup? See Fig. 11-57.
Photography by Prographics at Milwaukee Area Technical College/McGraw Hill

KEYPOINT

Tee slots are a *precision part of the table.*

Pro Table Care

In Fig. 11-56, something is very wrong! With all your setups, the table must be protected from dents, scratches, and especially from machine cuts! That includes the sides of the tee slots. Do not hammer on the table or store material or tools on it. Any nicks must be lightly honed away using a fine-grain flat stone.

KEYPOINT

Never machine close to the table with an end mill, face cutter, or drill bit. Use some form of spacer or vise to hold work up and away (Fig. 11-57).

Figure 11-57 Improving the setup in Fig. 11-56, parallel bars are under the work to keep the cutter away from the table.
Photography by Prographics at Milwaukee Area Technical College/McGraw Hill

11.2.5 *X–Y–Z* Digital Position Readouts (DROs)

A superhandy attachment is often added to milling machines to increase the accuracy and to help deal with backlash. The electronic display shows the position of the table with reference to the operator-chosen datum zero (Fig. 11-58). This is similar to adjusting the axis dials to zero at a datum.

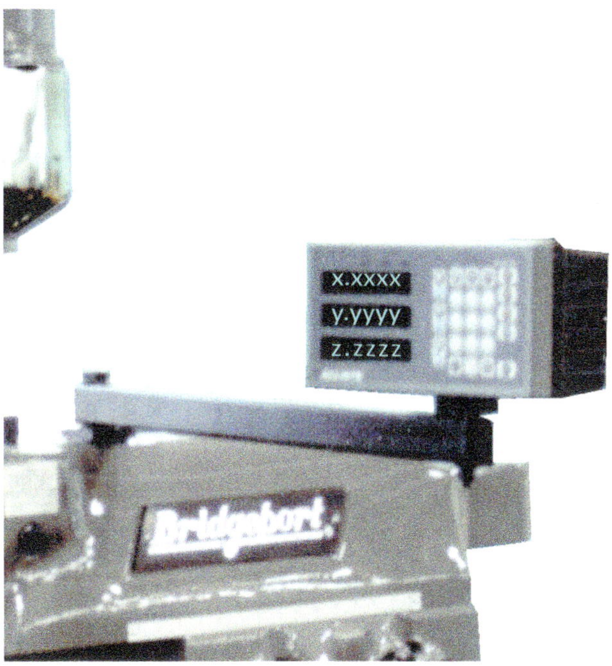

Figure 11-58 A digital readout displays the *X–Y–Z* position with reference to a movable zero point chosen to represent the datum basis of the workpiece.
Courtesy Bridgeport Machines, Inc./Bridgeport, CT

This ability to zero the axis position at any point within the work envelope is called *full floating zero.* You'll find the very same concept in CNC work. Modern digital positioners also convert between imperial inch and metric units, and most can store and recall significant points, then count down the distance to go to that point, thus making it easy to return to a location when machining several parts in a batch.

Using electronic location rather than micrometer dials, the axis positions are displayed independent of mechanical backlash. So, without the stall of a mechanical micrometer dial, drilling becomes more accurate. However, backlash still remains a challenge to be dealt with when moving the axis during machining due to the looseness of machine motions.

11.2.6 The Machining Head

The machining head features the following:

- Quill
- Quill feed rates
- Automatic and manual quill depth control
- RPM adjustments
- Feed levers and stops

Before looking at these features, let's examine the way in which the head is mounted on vertical machines.

Figure 11-59 Comparing the milling head on the standard vertical machine in the foreground to the ram and turret in the back, you can see it's made for heavier work, but it's less universal.
Photography by Prographics at Milwaukee Area Technical College/McGraw Hill

Standard Vertical Mill Head

On a heavy-duty vertical mill, the head is an integral part of the column (Fig. 11-59). It is positioned out from the column above the table. The work envelope is then limited to the area over which the head can be positioned.

Universal Ram and Turret Head

On these machines, the head is mounted on the ram. Following the minimum overhang principle, the machining head and ram must be withdrawn into its slide to the minimum that reaches the work. In Fig. 11-60, the ram has been extended to reach over a large part, but it may need to be withdrawn again for your setup. (Also see Fig. 11-61.)

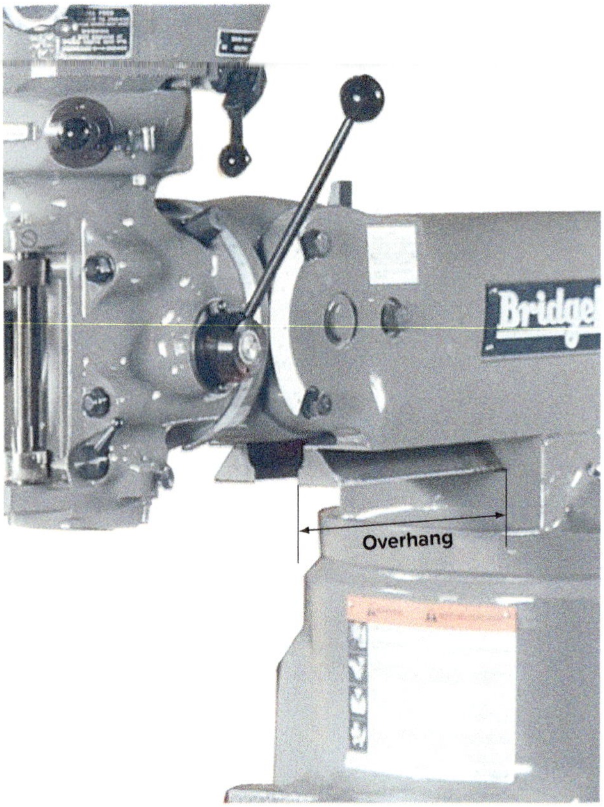

Figure 11-60 The head and ram are extended to reach a part further out on the table.
Courtesy Bridgeport Machines, Inc./Bridgeport, CT

KEYPOINT

Always draw the ram extension back to the column for normal work and lock it securely in position.

SHOPTALK

Your small vertical mills may be slightly different from those shown in the illustrations. There are several brands commercially available. However, the range of functions described will still be there; they are basic to this type of machine.

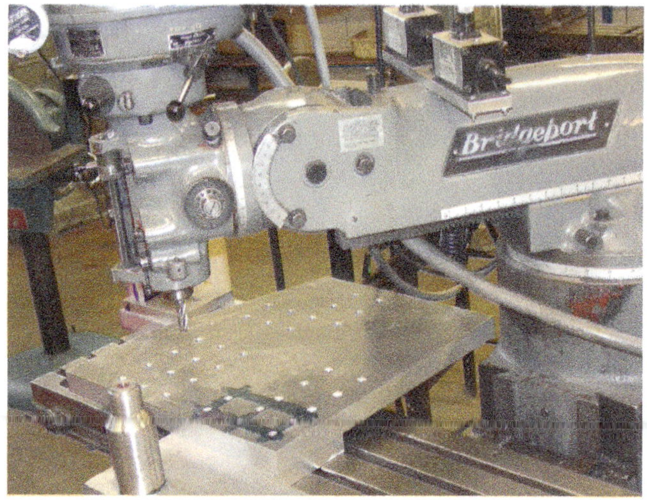

Figure 11-61 To machine this bicycle assembly jig we had to rotate the turret and extend the ram, extending the work envelope.
Lake Washington Technical College, Kirkland, WA/McGraw Hill

11.2.7 Features Found on the Machining Head

The R&T spindle head has the following features:

- RPM selection (step pulley or variable speed transmission)
- Spindle start switch
- Spindle brake and lock to stop rotation
- Quill feed by hand similar to a drill press, by power feed by engaging a clutch, and by manual powering of the power feed system. By shifting the motor drive gear into neutral, then engaging the feed clutch, one can turn a handwheel to power the quill feed system. This mechanical feeding action provides a very controlled method of drilling down.
- *A* and *B* axis tilt (position only, not for feeding)
- Quill lock and quill stops for production drilling and boring. These allow stopping automatic feed and provide a solid stop for quill movement.
- Quill feed rate selection in inches per revolution. Ram and turret mills commonly offer selection of 0.006 IPR, 0.003 IPR, or 0.0015 IPR feed rates, where IPR is inches per minute, as in drilling rates.
- Cutter holding drawbar
- Ram slide and turret rotation—lock bolts and crank for the slide.

You will need a demonstration and must learn the appropriate response for each feature from your instructor before setting up and operating the machine. Go to the lab if possible, and using Fig. 11-62, identify these functions on your machines.

Figure 11-62 Learn these features and functions on the ram and turret mill.
Lake Washington Technical College, Kirkland, WA/McGraw Hill

UNIT 11-2 REVIEW

Replay the Key Points

- The *Z* axis is always parallel to the main spindle centerline on all machines. Rotation around a line parallel to *X* is the *A* axis, around *Y* is *B*, and around *Z* becomes *C*. Rotary motion, *A* rotates around *X*, *B* rotates around *Y*, and *C* rotates around *Z*.
- Auxiliary axes are either for setup positioning axes, and must be locked during machining, or are feed axes that move during machining.
- Feeds on an R&T mill range from below 1 in. to around 20 in. per minute.
- On a milling machine, tee slots are a *precision part of the table.*
- Always draw the ram extension back to the column for minimum overhang setups; then lock it securely.
- *Caution! Never machine close to the table with an end mill, face cutter, or drill bit.*

Respond

Note that no answers are provided for these questions.

1. With a partner, identify the following milling machine components on a vertical mill in your shop. If you have a horizontal mill, find as many as you can on it also. Remember to put on safety glasses and avoid interfering with others working in the shop.

- ☐ *X* axis
- ☐ Spindle start
- ☐ Feed rate selector
- ☐ RPM selection (May be a set of levers, a dial, or a step pulley belt)
- ☐ *Y* axis motion lock
- ☐ Tee slots
- ☐ *Z* axis and *W* axis if it exists
- ☐ Feed start lever

Figure 11-62 *(Continued)*

- ☐ Lubrication system (May be automatic, in which case there will be nothing visible except an oil reservoir and an oil level indicator)
- ☐ Spindle brake (May be a function of the start clutch—when it is pulled down, it engages the brake. It may be a separate lever as on a Bridgeport mill. The separate brake should never be actuated while the spindle is turned on except in extreme emergencies.)
- ☐ Quill lever (Present on smaller ram and turret vertical mills only)
- ☐ *A* axis pivoting head—lock bolts and crank point, if so provided
- ☐ Ram slide and lock bolts
- ☐ Boring feed rate selector (Found on the head of Bridgeport machines and many others)
- ☐ Cutter drawbar
- ☐ *B* axis lock bolts

2. From the shop manual or from measuring, what are the limits of the work envelope of a typical machine in your shop? With the ram and turret centered?

Unit 11-3 Setting Up Mill Cutters

Introduction: In this unit, we look at mounting cutters on the machine and selecting the correct RPM. Both safety and quality depend on making the right choice.

TERMS TOOLBOX

Collet—DIN standard A small flat-nosed spring collet used in end mill holders for manual and CNC machines.

Collet—R-8 A popular mill spindle shape designated as type 7 by the ASTME.

Drawbar A long bolt that pulls the cutter into the mill spindle. There are two types: direct pull and tension nut pull.

ISO/ANSI mill tapers A world standard mill spindle taper featuring $3\frac{1}{2}$ in. per foot. Formerly called the "American Standard Mill Taper." Common sizes are #30, #40, and #50.

Shell end mill A hollow cutter that mounts on a pilot arbor.

11.3.1 Setting Up Cutters

There are five major methods of mounting and holding a cutter in a spindle. Each has a particular purpose and advantage.

Milling Machine Tapers

Machine tapers divide into shallow and steep versions. Shallow tapers such as the Morse tapers (studied in Chapters 10 and 11) are used for drill work. Once driven into their tapered receiving hole, they tend to stay in place, thus their name–self-holding. But they stay in place under loads from the end. NOT SIDE LOADS!

KEYPOINT

Self-holding tapers are not good at holding side loads, which is exactly what milling forces create. They will not stay in place. Therefore, Morse tapered tools are used only for drill work when used on mills, never for milling!

Steep Drawn Tapers for Milling

To resist milling vibrations and side-load forces as well as end forces, we employ standard machine tapers defined by the American Society of Toolmaking Engineers. *Critical fact–they are not self-holding.* Both types discussed next must be pulled into their receptacle with a draw bolt or draw-bar, which remains a permanent part of the setup.

Two kinds of milling tapers are commonly used on mills, the R-8 taper, used on smaller machines, and the Standard Mill Taper, used on heavier machines. You may hear veteran machinists refer to them as *American Standard* mill tapers.

1. **The ISO/ANSI Standard Milling Machine Taper**

   As shown in Fig. 11-63, most larger milling machines feature a spindle hole with a taper ratio of $3\frac{1}{2}$ in. per foot. Similar to the various Morse taper sizes, different sections of that $3\frac{1}{2}$ in. per foot cone are used for varying machine sizes. They run from #30 (small end around $\frac{5}{8}$ in.), #40 (small end 1 in.), to the #50 (small end around $1\frac{1}{2}$ in.). Figure 11-63 is proportionately drawn for three sizes.

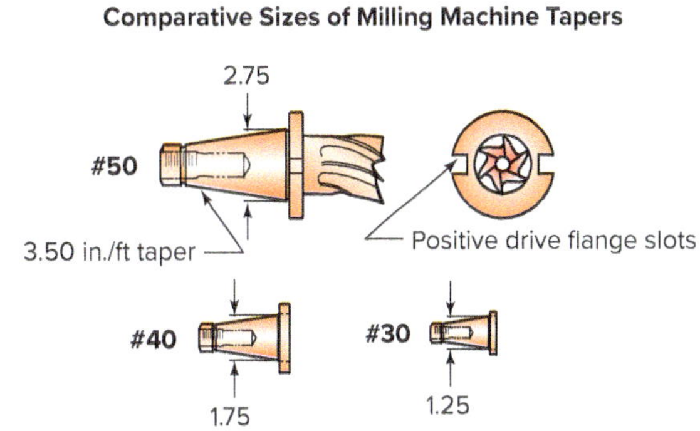

Figure 11-63 These three ISO/ANSI standard milling machine tapers all have the same taper, $3\frac{1}{2}$ in. per foot, but they are different portions of that cone.

2. **Torque Transfer**

   Spindle torque is considerable for mill cutters compared to drills. It is also an intermittent load because the cutter teeth enter and leave the cut with each revolution, causing a hammering action that would stall the rotation if they are not positively driven. For milling, spindle torque is transferred to the cutter through drive keys that protrude from the milling machine spindle flange (Fig. 11-64) and mate with slots on either side of the taper flange.

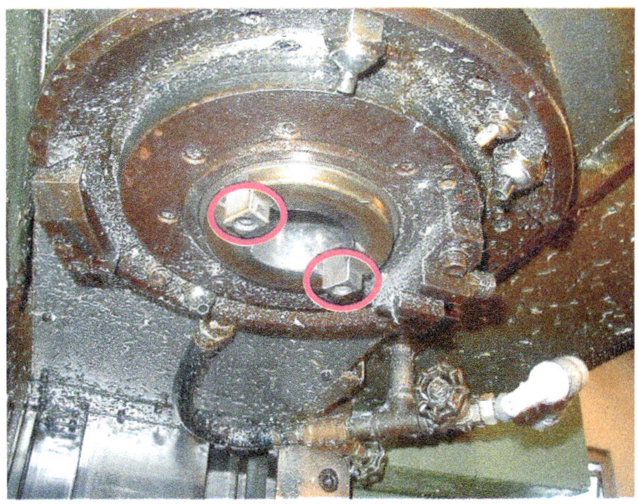

Figure 11-64 The drive keys show on the flange of this milling machine spindle.
Courtesy Tyee Aircraft, Everett, WA

While a few cutters are a solid unit (the taper is part of the cutter) as the upper end mill in Fig. 11-63, it's far more likely you'll find four other types in use on the manual and CNC mills (Fig. 11-65). They are the collet end mill holder (Fig. 11-66), the plain end mill holder, the pilot arbor (Fig. 11-67), and the stub arbor (Fig. 11-68). The pilot arbor, also called a C style arbor, can be loaded with hollow center cutters called **shell end mills** and with shell face mills.

Figure 11-65 A variety of different size end mill cutter holders with standard milling machine tapers.
Photography by Prographics at Milwaukee Area Technical College/McGraw Hill

TRADE TIP

Snap It In! Never put a collet into a **DIN** or Erickson type end mill holder (see Fig. 11-66) without first popping the collet into the extraction snap ring in the closing nut. If you fail to take this step, the collet will not be removable after tightening the collet on the cutter.

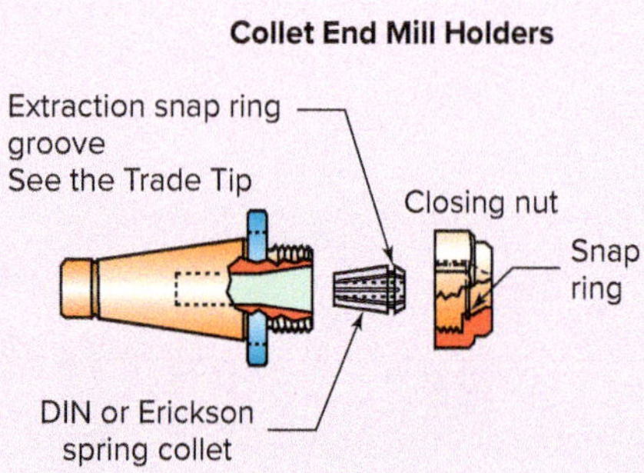

Figure 11-66 A collet end mill holder is useful on manual or on CNC machines.

KEYPOINT

If this step is forgotten, the only way to get the collet out of the hole usually requires destroying the expensive collet!

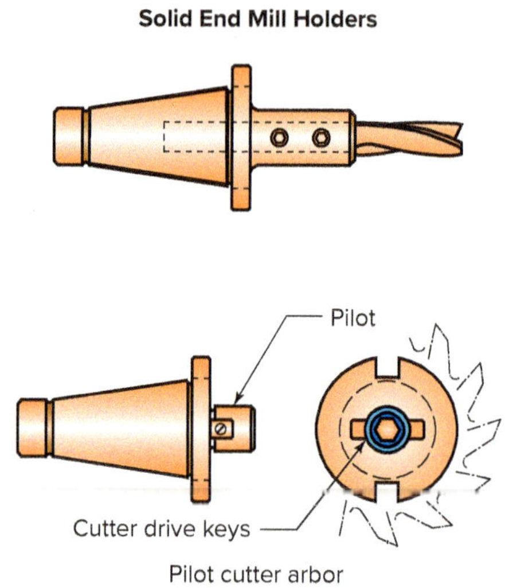

Figure 11-67 Pilot arbors mount hollow center end mills called shell end mills and face mills.

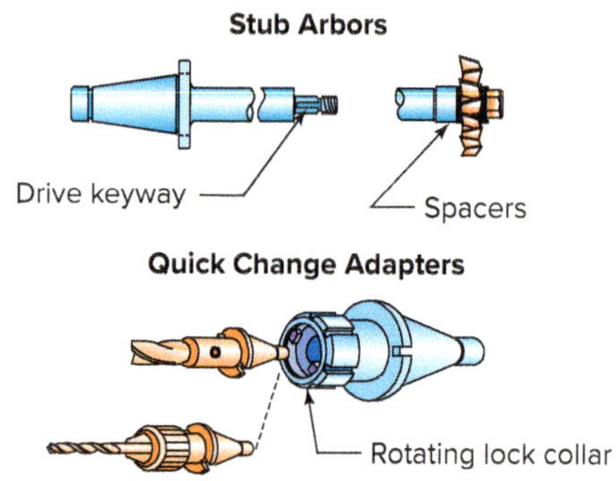

Figure 11-68 Stub arbors are used to hold slitting saws, wheel cutters. A quick change holder is useful on CNC machines that have no automatic tool changer.

R-8 Machine Taper

Most small mills feature a spindle nose taper designated style 7 by the ASTME, but it's commonly called an R-8 taper (Fig. 11-69). There are a wide array of tool holders and accessories featuring R-8 shanks (see Fig. 11-69).

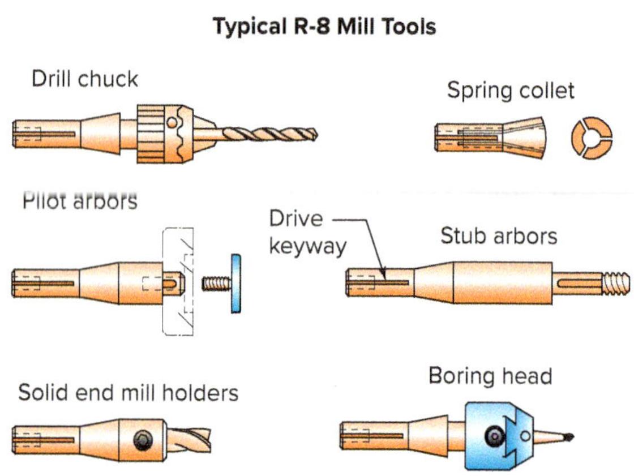

Figure 11-69 An array of R-8 cutter and tool options.

R-8 Max Capacity = $\frac{3}{4}$ in.

The largest standard size R-8 spring collet is 0.750-in. (19-mm) diameter. However, slightly larger collets can be special ordered.

KEYPOINT

- All collet type end mill holders transfer torque to the cutter by squeeze action alone; they are not meant for heavy machining. When loads increase, change to solid end mill holders.
- All mill tools use some form of precision taper. Always clean both components, the taper and the receiving hole, before assembly. The pressure created by a single trapped chip can permanently dent both the taper and, worse, the milling machine spindle.
- *Caution!* Larger tool diameter **R-8 collets** break very easily when sprung even slightly beyond their range capacity!

11.3.2 Drawbars

KEYPOINT

For safety you must understand the difference between direct pull and tension nut drawbars.

1. **Direct Pull**

   A direct pull drawbar is truly an extra long, high-strength bolt. Using your hand and then a wrench, it is screwed in until the right pressure is developed to pull and hold the cutter in the machine spindle (Fig. 11-70). About 40 pounds of pressure on the wrench is a good guideline, but have your instructor demonstrate the amount of wrench torque needed.

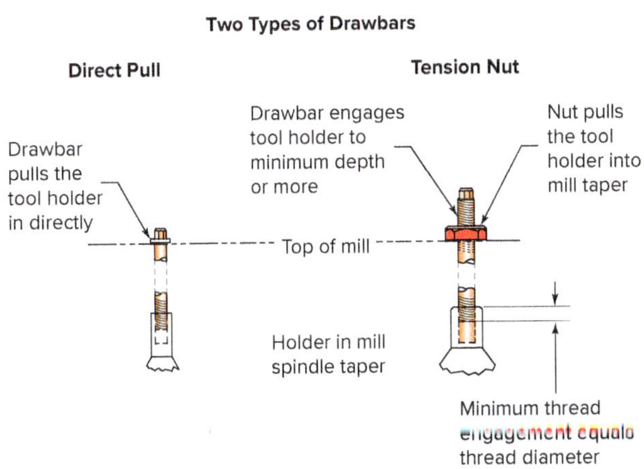

Figure 11-70 Cutters fly out violently when these two drawbars are not recognized. The nut type is used more often on larger mills.

2. **Tension Nut**

   The difference for this drawbar is that the final cinch is accomplished by the nut, not the drawbar. The nut pulls the tool into the spindle by shortening the drawbar. The nut type is more heavy duty and is found on larger machines. See the Trade Tip.

Drawbar Engagement

While a rule of thumb is to screw the drawbar into the tool holder, a minimum distance equal to the thread's diameter (e.g., a $\frac{1}{2}$-in.-diameter thread screws in $\frac{1}{2}$ in. deep), deeper is better to maintain a constant pull without stretching the drawbar threads. But to avoid an accident, be sure you do not bottom out the tension nut type. See the Trade Tip.

TRADE TIP

Tightening Cutters Right Two drawbar details, if missed, can cause serious accidents:

- First, if the nut type is mistaken for the direct pull type, and it is simply screwed in until it bottoms out, using it as if it were the pull type, then there is no tension on the cutter. The nut type must first be screwed into the holder a minimum distance equal to its diameter, but 10 turns or more is usual. Do not bottom it out—use the nut to pull.
- The second accident I've seen more than once is caused by not aligning the torque drive keys on the cutter taper with the flange slots on the mill spindle. From the top of the mill, the unseen cutter feels as though it is tightening since the flange is bottomed against the key tops, as would be the case in Fig. 11-71, if it were tightened in that position. However, with the first touch of the cutter to the work, the cutter spins, loses its hold, and a serious disaster ensues. Always check that the keys are in the drive slots!

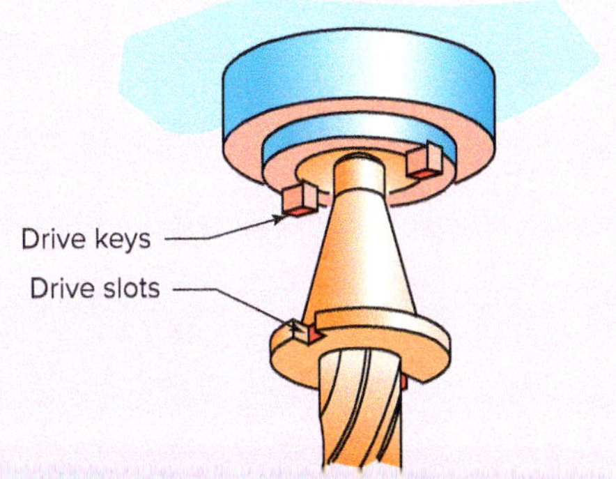

Figure 11-71 Caution, the keys can miss the slots!

- *When Tightening Set Screws, Do the "JOGGLE."* Sometimes, when tightening the set screws in solid-end mill holders, the flat spots on the cutter aren't perpendicular to the screw. Tightened just once, they seem okay, but they loosen instantly when the cutter hits metal. To avoid this potential accident, do the joggle. First tighten one screw; then release an $\frac{1}{8}$ turn or less. Now move the end mill in and out and rotate it around as you slowly retighten the screw. You can feel the cutter fall into alignment. But do it again. The second time, just loosen the set screw until the cutter can joggle a tiny bit. Then on retightening, it should be perfectly aligned to the set screw. Not clear? Ask for a demonstration of this simple but effective trick.

11.3.3 Drills and Reamers

Drill tools can be held in mills using three different methods:

1. Straight shank tool directly in collets.
2. In chucks drawn into the mill. The chuck may have an R-8 taper or it may be a straight shank itself and held in a collet.
3. Morse taper adapter—drawn into the mill spindle. This sleeve converts the mill spindle to accept standard drill tapers (Fig. 11-72).

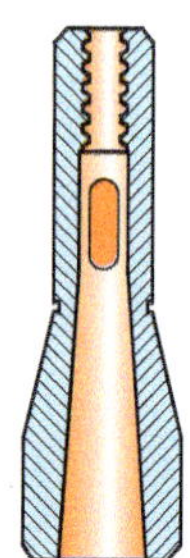

Figure 11-72 Morse taper tools must be used for drilling and reaming tasks only, never for milling, which creates side thrust and will dislodge the Morse taper!

KEYPOINT

Caution! Drills can vibrate or even pull out of Morse adapters during heavy use or during chatter.

11.3.4 Selecting Cutter RPM

Like lathes and drills, RPM depends on the combination:

1. Alloy and how hard it is
2. Operation
3. Cutter type

There are three convenient ways to learn RPM selection.

Chart RPM

There are tables in all machinist's reference books that provide mill RPM selections. They are useful but, as with any chart, have range and limits.

Calculated RPM

The short and long formulas we learned in Chapters 10 and 11 in the drilling and turning setups also apply to milling cutter speeds. The diameter variable is always the spinning object, so for milling it's the cutter. Let's review.

The long formula is

$$\text{General RPM} = \frac{\text{Surface speed} \times 12}{3.1416 \times \text{Diameter of cutter}}.$$

The variables are cutter alloy, cutter diameter, and work material.

The short formula is

$$\text{RPM} = \frac{\text{Surface speed} \times 4}{\text{Diameter of cutter}}.$$

Remember, using our short formula, the RPM will be within 5 percent of that calculated from the general formula. And similar to drill and lathe work, the recommended surface speed is found in a chart in Appendix III, in reference handbooks, and in cutter manufacturer supplied charts. The chart is entered with two arguments: the cutter type (carbide, HSS, ceramic, or composite) and the work material.

Example

This setup requires a $\frac{1}{2}$-in. HSS end mill. You will be cutting aluminum. Start by finding the surface speed in Appendix III.

$$\text{SS} = 250 \text{ feet per minute (HSS cutter in aluminum)}$$

$$\text{RPM} = \frac{250 \times 4}{0.5}$$

$$\text{RPM} = 2{,}000$$

TRY IT

Find the following RPMs.

A. For a carbide cutter of 6-in. diameter cutting mild steel, find surface speed and RPM.

B. When machining brass with a 2-in.-diameter HSS face cutter, find SS and RPM.

C. When a pocket must be milled into a carbon-steel forging using a $\frac{3}{4}$-in. HSS end mill, find SS and RPM.

ANSWERS

Keep in mind there are many other factors that affect your final RPM selection. For example, what speeds are available on the machine? Is coolant available, and how rigid is the setup? Factors such as these will suggest either more or less RPMs to the experienced machinist.

A. SS = 300, RPM = 200

B. SS = 175, RPM = 350

C. SS = 90, RPM = 480

Slide Rule Calculations and Smartphone Apps

Perhaps the most useful tool for RPM selection is the dedicated slide rule calculator. These are provided by cutter manufacturers. Slide rules make the long formula a snap to solve by printing the arguments and results on a sliding index card. Take a look at examples later in Fig. 11-97 in Unit 11-5. Also, go to your app store and look for cutting tool supplier applications for speeds and feeds–very handy!

11.3.5 Carbide Insert Mill Cutters

Similar to lathe insert cutters, the ASTME/ISO have designated milling machine standards to define cutter shape, size, and hardness of the insert tooth.

It's not necessary to learn all the facts now; however, at this time, please turn to the *Machinery's Handbook* section on indexable inserts and scan the information and illustrations. Also look over the industrial chart shown in Fig. 11-73. A similar reference will be used where you work. We'll return to it for a few insert selections in the unit questions and in the Chapter 11 Review.

Insert Carbide Tooling Advantages and Disadvantages

While cutters similar to that shown in Fig. 11-74 are popular in industry for both manual and CNC work, they aren't ideal in every situation. Knowing when carbide tools are right for the task and when they are not makes a big difference in successful setups. Let's compare them.

Carbide Insert Cutter Advantages

1. **Faster Metal Removal**

   Carbide insert cutters are usually three times faster in most steel-milling situations compared to HSS. They may increase speeds many times that in milling other metals.

2. **Faster Turnaround**

   Dull or broken inserts are easy to exchange and due to the cutter construction, the new insert nests back within 0.001 in. or better. Thus, all machining will resume at nearly the same size.

3. **Quick Change Geometry**

   By exchanging inserts, the machinist can change hardness, nose radius, and rake angle in just moments (Fig. 11-75).

4. **Long Tool Life**

   Correctly applied, carbide tooling extends tool life and reduces down time many times over HSS.

5. **Can Eliminate the Need for Coolants**

   Some materials cut better if no coolant is used when using the right cutter.

Carbide Insert Cutter Disadvantages

1. **High Initial Cost**

   Insert tooling costs many times that of HSS. However, carbide becomes quite economical when put into long-term industrial machining.

2. **Vulnerable to Breakage**

   Used incorrectly, carbide tooling lacks the forgiveness of HSS, and any excessive shock loading will chip and break the inserts. Often when an interrupted, hammering action breaks carbide cutters, HSS can withstand the punishment.

3. **Less Sharp**

   Because carbide is so brittle, it must have a tiny edge radius to avoid chipping. In comparison, HSS cutters can be honed to a very fine edge when extremely sharp cutters are required. Carbide cutters then cause more force and heat compared to HSS.

For all these differences, most training facilities teach students to use HSS cutters first.

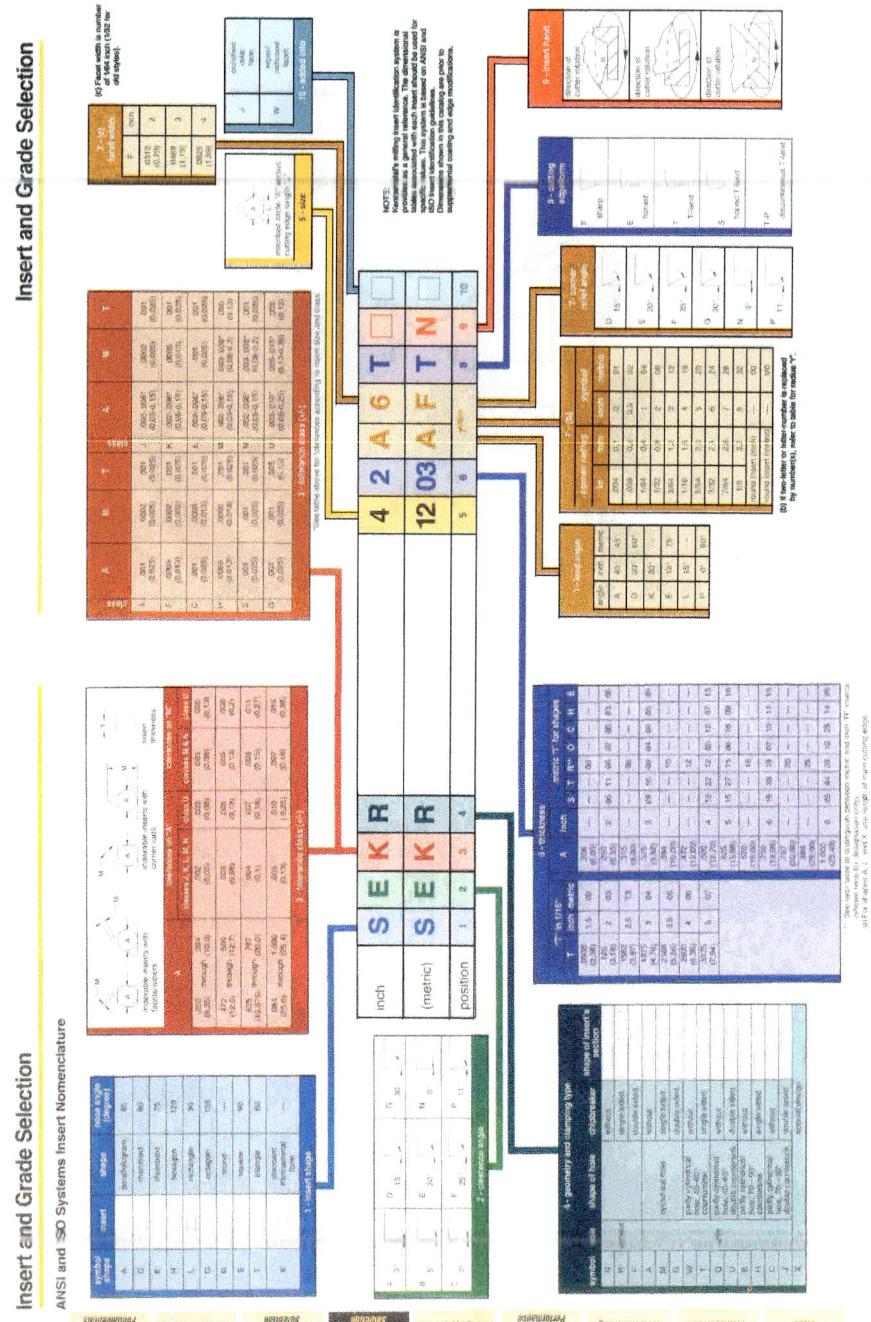

Figure 11-73 Industry standard carbide insert identification. Review this information and know where to find data of this kind when the job requires it.

Figure 11-74 An advanced design, Chip-splitter™ carbide insert, mill cutter, ripping through steel at high speed.
Courtesy Iscar Metals, Inc.

Figure 11-75 When exchanging inserts, it's critical to clean out the tiniest chips and dirt before installing the new tooth.
Photography by Prographics at Milwaukee Area Technical College/McGraw Hill

UNIT 11-3 REVIEW

Replay the Key Points

- *Be sure to snap the end mill collet into the snap ring. If this step is forgotten, the only way to get the collet out of the hole usually requires destroying the expensive collet!*
- All collet type end mill holders transfer torque to the cutter by squeeze action alone; they are not meant for heavy machining. When loads increase, change to solid end mill holders.
- All mill tools use some form of precision taper. Always clean both components before assembly. The pressure created by a single trapped chip can permanently dent both the taper and, worse, the milling machine spindle.
- *Caution!* Larger diameter R-8 collets break very easily when sprung even slightly beyond their range capacity!
- There are two kinds of drawbars–be sure you know the difference.
- When exchanging inserts, be sure to clean the insert seat.

Respond

1. From *Machinery's Handbook,* what does a number (single or double digit) in the fifth position in a carbide insert call out–for example, TNMG682 (assume a nonmetric insert)?

To get a feel for carbide selection standards, but using *Machinery's Handbook*, describe these inserts:

2. TNMG-432
3. DPMP-631
4. The top of a cast iron transmission case must be machined flat using a 4-in.-carbide face mill. What are SS and RPM?
5. A peripheral cut must be made to an aluminum bar using a 0.25-in.-diameter end mill. Identify SS and RPM.
6. Cut a slot in a stainless steel forging using an expensive solid carbide 1.0-in. end mill. What are the SS and RPM?
7. You have seen three different methods of selecting RPM; what are they?

Unit 11-4 Avoid These Errors—Great Setups and Safety

Introduction: Mills are generally safe machines, but they do have potential dangers, including fast rotating cutters, flying chips, and incorrectly held work that can shift in the setup, resulting in broken cutters and part ejection. Unit 11-4 lists the common errors that new machinists commit on milling machines, but more, it presents ways to avoid these mistakes. Similar to the "hit list" for lathe training, these potential accidents are gleaned from other's experiences (mine included). Each accident described here can be avoided.

The top item in terms of frequency is presented first. This list isn't a summary of accidents but rather a list of the *preventive skills* you'll need plus some Trade Tips to get it right! In fact, you might envision this entire unit to be a Trade Tip.

Unit 11-4 does not include dangers such as cut or pinched fingers, eye injuries from not wearing glasses, and so on. It highlights specific mill problems, but most important, it *offers solutions*.

TERMS TOOLBOX

Backlash compensation (compensator) A one-way brake that is set to hold back the *X* axis of some commercial duty mills. Compensators help avoid problems from aggressive feeding during climb milling.

Deflection (cutter) Bend in an end mill (or any log tool) due to a low rigidity ratio and machining forces.

Packing block The portion of the clamp setup that supports the back of the clamp.

Rigidity The relative ability to resist movement—can apply to the work, the setup, and the cutter.

Solid jaw (vise) The datum surface used for nesting work surfaces.

Tangential cut A movement of the face cut to one side such that the cutting force is lengthened and shifted.

11.4.1 Common Problems and Their Prevention

Solving Work Not Held Properly

During machining the part shifts.

Results

Cutter digs in and the part is ruined.

Cutter digs in and breaks into flying fragments.

The part is flung from the setup.

The Solutions

Use table clamps and vises the right way—they serve as a lever.

Use clamp leverage to your advantage—see the Trade Tip.

Use vises the right way—see the Trade Tip.

TRADE TIP

Using Table Clamps There are many versions and combinations of table clamps and the support **packing block** at the back. All must follow a few simple physical rules to hold right. Use your built-in sense of mechanical devices and compare the two clamps in Fig. 11-76 to determine what is wrong with the one on the right.

Clamp Answers (Fig. 11-77).

- *The clamp on the right has the bolt too far from the work*—the larger portion of the holding force is concentrated on the packing block rather than the work where it is needed.
- *The clamp on the right is tilted back; thus, all its force is concentrated on the corner of the work.* This not only damages the work; it also tends to tilt the workpiece up off the table. Plus it will push itself off the work when the cutting begins—its greatest danger. The one on the left is tilting slightly down to avoid this problem; even when it is tightened hard enough to bow the clamp down, it's level.
- *The left clamp is set up with a thin, soft metal shim to protect the workpiece. The clamp on the right has none.*
- The differing packing blocks are equally okay—the special step block and clamp on the left or the flat top block and plain clamp work fine for this application. You may need to shim the flat block to get the right height.

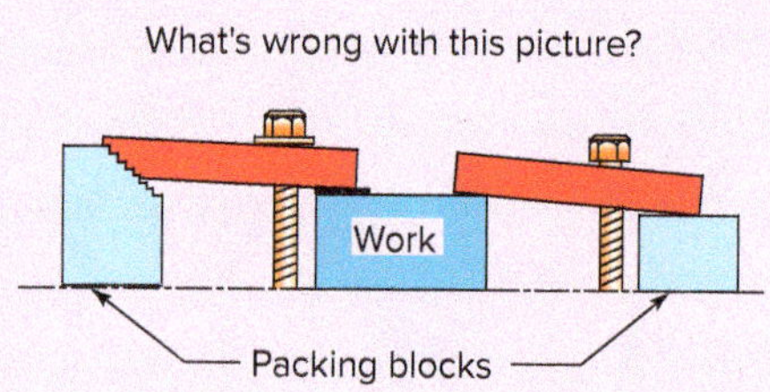

Figure 11-76 Compare these clamps. Can you spot four errors?

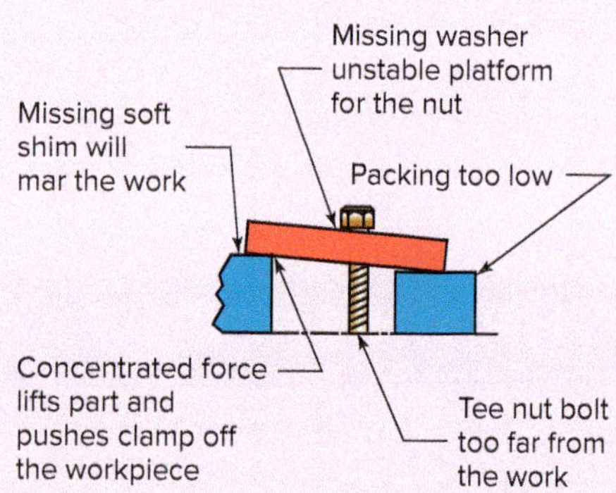

Figure 11-77 Did you find all these problems?

TRADE TIP

Correctly Using Mill Vises

Rule 1—Make sure the work is down low, close to the vise with plenty of grip.

Rule 2—Make sure it is true to the solid (nonsliding) vise jaw. An uneven part that is not parallel will not hold well because all the force will be concentrated in one location.

Critical Question: What's the best fix for the uneven part in Fig. 11-78? (The answer immediately precedes the unit review.)

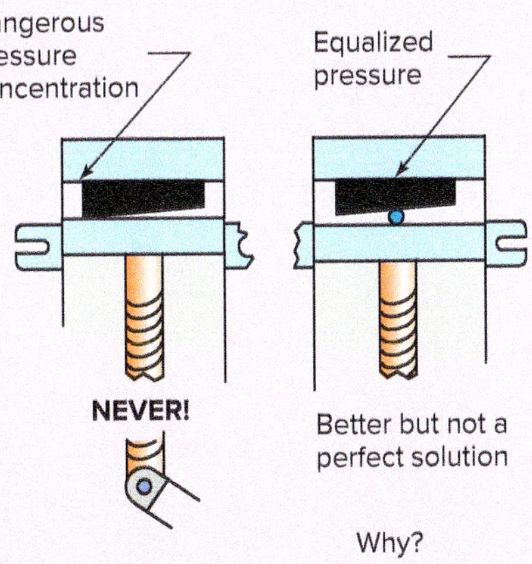

Figure 11-78 Sometimes a round pin can help to equalize work pressure against the solid jaw—but not this time. It might mar the work, but there's another reason. Why?

Using the Solid Vise Jaw The jaw that does not move on the mill vise is the gage surface. It forms an important datum for milling. *Use it correctly by placing any work requiring alignment to the vise, against the* **solid jaw** *always* (Fig. 11-79).

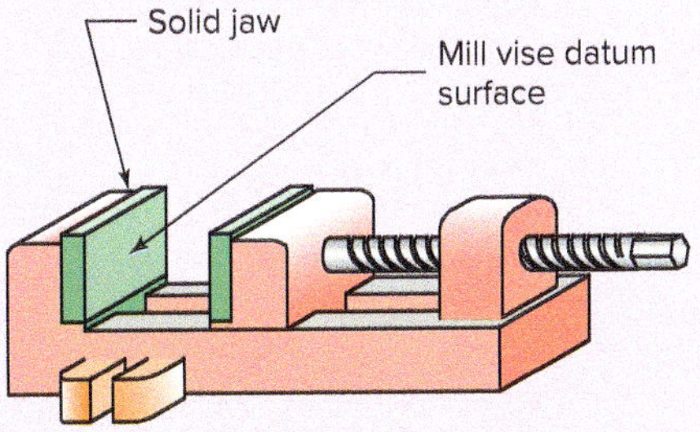

Figure 11-79 The solid jaw is the *datum surface*. Align it to the machine axes and always place important edges of the work against the solid jaw.

TRADE TIP

Using a Mill Vise Whenever parts are loaded into a vise, it's imperative to test how well they are seated down against the vise floor or supporting parallels. If, after closing the vise, the parallel bars supporting them can be moved, the part has lifted up due to vise closure. This is a predictable error that often causes undersize parts. One way to detect this problem is to place a piece of paper under the parts, on both sides of the vise. Then, if they are seated correctly, the paper will not move when tugged after the vise is tightened (Fig. 11-80).

Figure 11-80 Two paper feelers show whether the workpiece has lifted off the vise floor when tightened.

Photography by Prographics at Milwaukee Area Technical College/McGraw Hill

But if the test shows the parts have lifted up when the vise was closed, here's what to do. Release then snug the vise pressure about two thirds of final pressure. Now tap the part down with a deadblow (sand-filled) plastic hammer (Fig. 11-81). Now close it and give the part one last firm hit with the deadblow.

Figure 11-81 A deadblow hammer tends to cancel the rebound that happens when driving parts with a brass face hammer.

Lake Washington Technical College, Kirkland, WA/McGraw Hill

If the part won't seat, then check its shape. A part whose sides aren't square to its base can't be seated down to full floor contact. In this case, either accept side contact parallel to the floor but not fully seated or do not use a vise; use clamps and parallels to hold the work.

Vise Quality Another pro tip for vise work is that mill vises aren't all alike. Some work just fine for utility work holding, but they aren't precise enough for the job in Fig. 11-80. Their jaws aren't hardened and ground perpendicular to the vise's axis, their construction is loose, and they tend to lift parts up. Other vises are precision ground machine accessories (Fig. 11-82) featuring well-guided moving jaws that do not tip when they are closed. Due to the way they are engineered, the mill vise of Fig. 11-82 places $\frac{1}{2}$-lb down pressure on the work for every pound of closure pressure. Their moving jaw guides are adjustable to compensate for wear. Learn to recognize which vise is which in your shop; then use the better vise on the precision jobs.

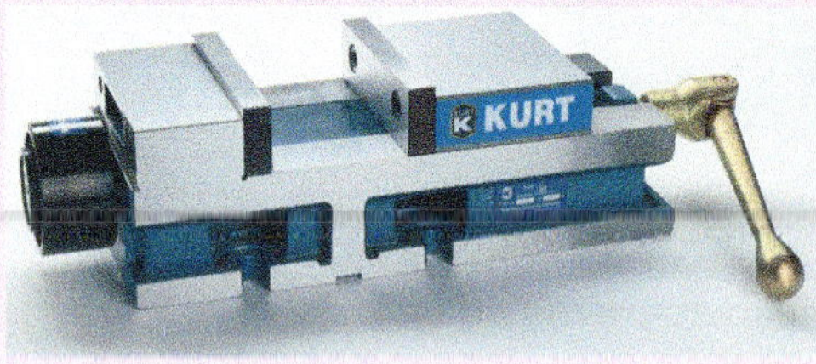

Figure 11-82 The AngLock precision vise is built to 0.001-in. tolerance for size and squareness. When it's closed, it puts $\frac{1}{2}$-lb down pressure on the workpiece, for each pound of closing pressure. It can deliver thousands of pounds of closing pressure.
Courtesy Kurt Manufacturing Company

Soft Jaws When work shape becomes impossible to hold in a flat jaw vise, we exchange the hardened jaws for soft jaws that can be custom machined to the right shape. Soft jaws also aid in holding batches of parts in just the right location. More coming up in Unit 11-5.

Solving Poor Cutter Choice

Work is too long, too weak, or poorly held. (Fly cutters can be a problem in this area.)

Results

The cutter spins in the collet, ruining the collet and workpiece.

The cutter chatters and breaks with flying fragments.

The Solutions

Always use the shortest, sturdiest cutter to do the job without **deflection** (bending; Figs. 11-83, 11-84, and 11-85). Long cutters tend to chatter and break.

Always use the joggle method of tightening set screws.

Double-check the drawbar during setups and often during operation.

Question: What is wrong with the setup of Fig. 11-83?

Solving Wrong Spindle Speeds

Speed is too high or too low.

Results

Too high—heat, chatter, cutter dulling, potential work hardening

Too low—poor finish, cutter digging, and breakage

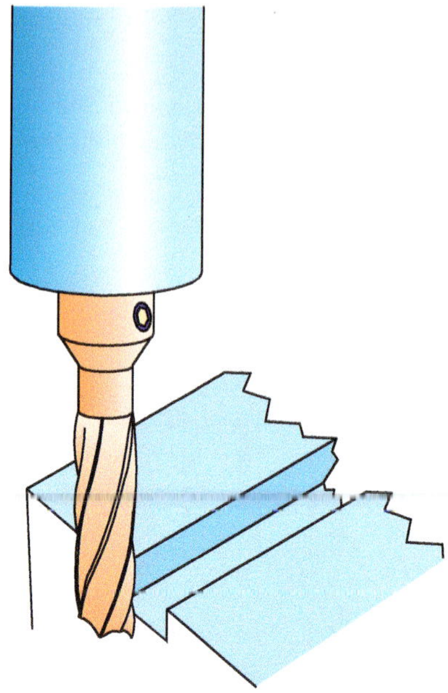

Figure 11-83 Explain what must be done to this setup to make it safe.

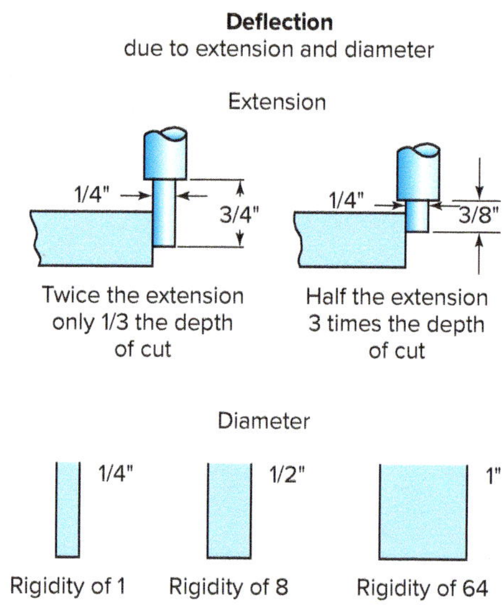

Figure 11-84 The rigidity of the cutter increases exponentially as cutter diameter increases.

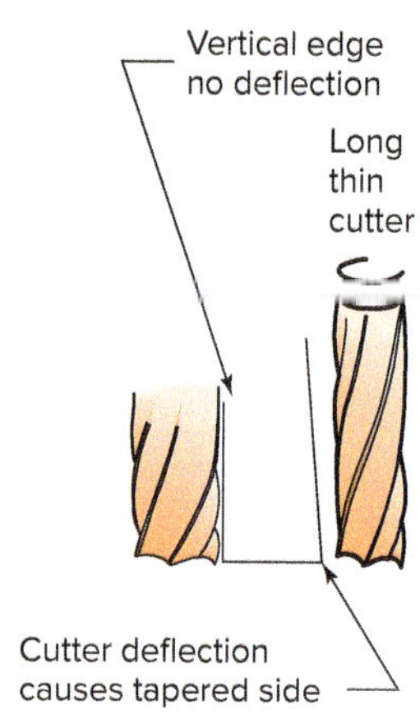

Figure 11-85 Poorly chosen, the cutter will deflect (bend) and can even break!

The Solution

Always compute the spindle speed and use coolant whenever practical.

Solving Dull Cutters

Either the student begins cutting with a dull cutter or it dulls during machining.

Results

High heat causes excessively hot chips, which burn the operator.

High heat burns up cutter or work hardens the material (steel, bronze, and other metals prone to hardening).

The cutter loads up on chips and breaks–common in aluminum (Fig. 11-86).

Figure 11-86 *Stop and solve it before proceeding!* Your cutter is *loaded,* probably from dullness or inadequate coolant, or possibly due to excessive spindle speeds.
Photography by Prographics at Milwaukee Area Technical College/McGraw Hill

The Solutions

Examine the cutter before mounting it in the machine.

Use coolant.

Use correct speeds and feeds.

Look and listen for signs of progressive dulling.

Solving Incorrect Operator Action

The operator initiates the wrong movement at the wrong time.

Results

The machine moves the wrong direction and the cutter contacts either the part, vise, machine table, or fixture, resulting in a ruined part, fixture, and cutter. Possible broken and flying cutter may also occur.

One common action is to start the cutter rotating backward–no kidding; it happens regularly!

The Solutions

Know and review the function of each machine control. Think out actions. *Know and practice how to stop all actions in an emergency.*

Solving Unrealistically High Feed Rates

Results

Rough finish on parts

Broken cutters or possibly shifting parts in setup

The Solution

Pay close attention to feed rate training coming up. Start low and increment up to a maximized rate.

Solving Climb Milling Aggressive Feed

Climb milling is the best process, but without proper part and machine axis control.

Results

Broken cutter

Part begins to self-feed and may be ripped out of setup

The Solution

Climb milling is a better way to peripheral cut but you must know and control the danger. It tends to pull the part along which can get out of control. (More training coming up.)

On manual mills, lock the machine axes not being used for feeding movements and set a drag on the axis being used. Note, using axis locks to hold back aggressive feeds causes wear to mill dovetails and to the lock pads, but sometimes it's the only solution to use climb milling safely. Larger milling machines feature **backlash compensaters**. They are one-way brakes designed to retard aggressive feed without excessive wear to machine parts.

Solving Leaving a Wrench on the Drawbar Nut While Starting the Spindle

Results

The wrench hits the machine frame, flies off into the shop.

The Solutions

Never take your hand off the wrench when it's on the machine.

Always check before turning on the machine that the wrench has been cleared from the drawbar (Fig. 11-87).

Figure 11-87 Dents in the motor housing show how dangerous leaving the wrench on the drawbar can be!
Lake Washington Technical College, Kirkland, WA/McGraw Hill

TRADE TIP

Work Entry Angle and Forces Milling is an intermittent action. Each cutter tooth enters and exits the work, thus causing a hammering action on both cutter and work. However, in Fig. 11-88 we see good and bad ways to make a cut. The lower cutter has simply been shifted off to one side of the part, to cause a longer shear action of the chip.

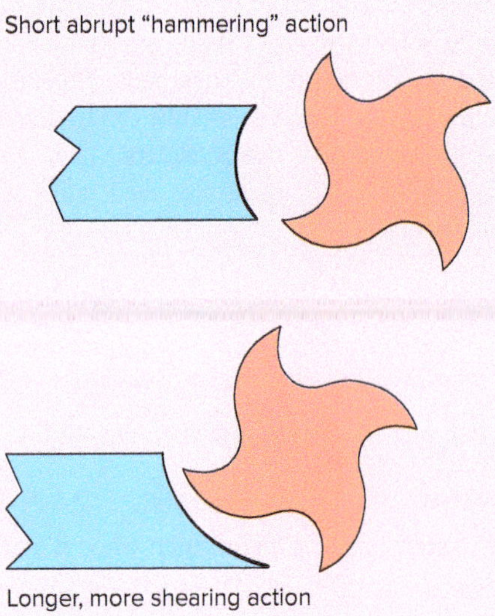

Figure 11-88 Which cut will vibrate less?

KEYPOINT

Both cutters in Fig. 11-88 are at 50 percent radial engagement, but there is a big difference in the cutting action.

The short hammering cut can be improved to a longer action known as a **tangential cut** (cutting on the tangent, not directly across). Much of the cutting force is transferred to the longer axis of the work, and these tend to vibrate less.

KEYPOINT

Controlling Chips Chip ejection can be aimed. Figure 11-89 shows two cuts of equal efficiency, but one sends the chips out at the operator, and one sends them harmlessly into the machine frame. A bonus is that the better setup forces the work toward the solid jaw of the vise—a good practice!

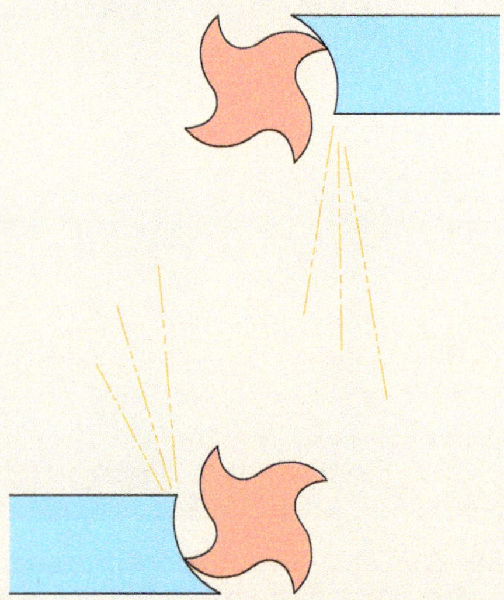

Figure 11-89 Top view. The direction of cut can control the direction of the chip ejection.

Answer to Critical Question

For the uneven part shown in Fig. 11-78, machine the block parallel, if possible, and then put it back in the vise. If it can't be machined parallel, then hold it in some other way; for example, with down pressure from table clamps.

UNIT 11-4 REVIEW

Replay the Key Points

- Setups are complex demonstrations of physical science. Be on the lookout for ways to make them more rigid and safer.
- Cutter actions can be rearranged to improve setups.
- There is a right and a wrong way to use machine vises.

Respond

1. On a sheet of paper, sketch a safe table clamp. There will be seven components to each clamp setup.
2. Why are there parallel bars under the work in Fig. 11-90? What is missing from the setup?

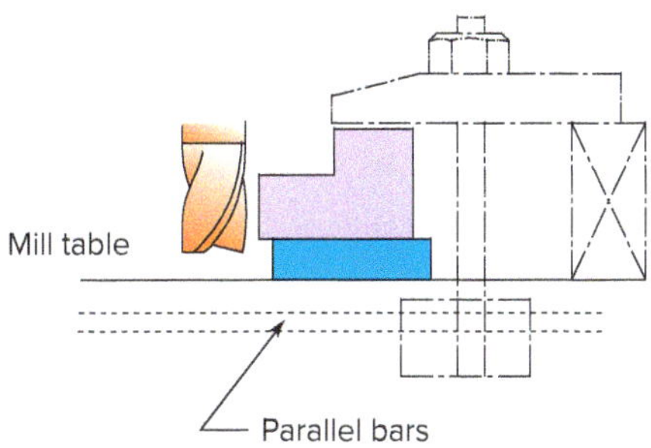

Figure 11-90 Why the parallel under the part? What's missing here?

Critical Thinking

3. To machine this pocket and drill the hole through the part, the plan calls for holding it in a vise (Fig. 11-91). Sketch your ideas for holding it against the solid jaw. Ask others in your training group how they would do it. This is a typical setup problem, although it is not directly covered in the reading.
4. What is good and bad about a *tangential* cut?
5. After cutting an aluminum bar with an end mill, for a short while, the cutter snaps off! What could have been the problem(s)? What is (are) the solution(s)? Identify at least five.
6. **A.** What is dangerous about *climb* milling?

   B. What must you do to control the machine during climb milling?

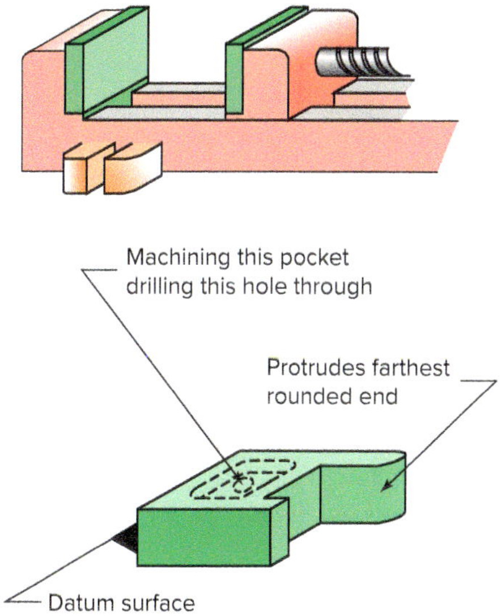

Figure 11-91 Sketch or describe your solution to holding this part correctly in the vise. Invent the solution from simple shop components.

Unit 11-5 Mill Setups

Introduction: This unit contains tasks that are critical to operating both CNC and manual milling machines. They involve computing feed rates and aligning work and work-holding vises and other accessories on the mill table.

TERMS TOOLBOX

Coaxial indicator A vertical orientation DTI for finding the axis of a hole by rotating its probe while the dial face remains stationary.

Coordinating the axes Positioning the spindle or cutter in an exact relationship with the work or holding accessories, then setting axis registers (dials or digital readouts) to zero. Creates the RZ or PRZ.

Edge-finder An alignment tool that wobbles until it comes into tangent contact with a surface, at which time the spindle's centerline is exactly 0.100 in. in from the edge.

Feed per tooth (FPT) (chip load) The recommended amount of advancement one tooth can withstand per revolution.

Jig bore indicator A vertical orientation indicator whose dial faces up when the probe is on an edge or within a hole.

Program reference zero (PRZ) A starting point reference for a program that must be coordinated on the workpiece or tooling where the axis registers are set to read zero.

Reference zero (RZ) A starting point reference on a workpiece where the dials and axis registers are set to read zero. All subsequent cuts will be made with dimensional reference to the RZ.

Wiggler A universal edge and center finding tool set that also includes an articulated indicator holder.

11.5.1 Ten-Point Checklist for Mill Setups

The following road map can be used to make certain you've not missed anything in your setups. Some points will be familiar, while some are new. It might be a good idea to copy, then check the items for your first few setups. There are a few more tasks needed for setting up a CNC that we'll add later.

- ☐ **1.** Mount the cutter.
- ☐ **2.** Calculate and select the RPM.
- ☐ **3.** Mount and align the work-holding accessory.
- ☐ **4.** Bolt the work to the table using clamps.
- ☐ **5.** Calculate and select the feed rate.
- ☐ **6.** Check all safety guards.
- ☐ **7.** Lock all unused axes including the quill lock/backlash compensator
- ☐ **8.** True the universal head (if equipped).
- ☐ **9.** Align the spindle with the part datums.
- ☐ **10.** Test the coolant before starting the cut.

11.5.2 Using Climb and Conventional Milling

We need to look more closely now at the chip being made to understand the differences in the two cutting methods (Fig. 11-92). The first fact we uncover is that *conventional* cutting isn't the convention in the modern world of CNC. There, the machine axis drives have nearly zero backlash, and the tiny bit they do have is controlled automatically. CNC machines don't aggressively self-feed out of control, due to the climb cut's force multiplication, as can happen on hand-cranked manual machines. This applies only when using the side teeth of a cutter on the outer perimeter of a surface, peripheral or sometimes called contour milling.

> **KEYPOINT**
>
> For reasons we'll explore next, climb milling is the better process, producing a finer finish and longer tool life, but it also produces a multiplication of forces that must be anticipated. In CNC programs, peripheral, pocket, and step cuts are all correctly programmed using climb milling.

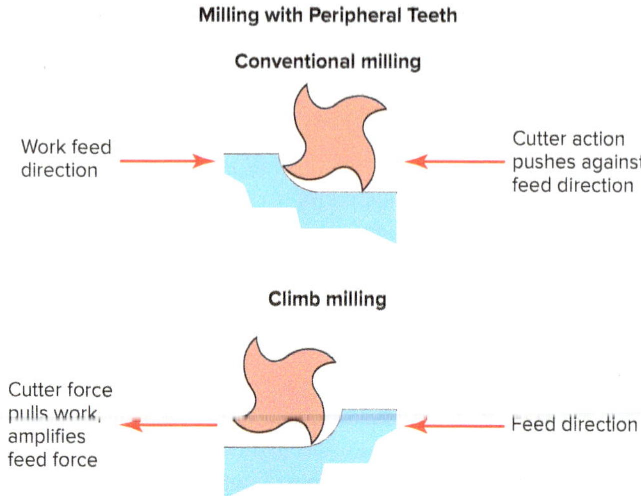

Figure 11-92 The forces of climb versus conventional milling.

When hand cranking a climb cut using a manual mill, the crank handle will move nearly on its own! But that can (and sometimes does) get out of control. If allowed to self-feed, the axis movement can accelerate until either the cutter breaks or the part is pulled out of the setup. Also, if the part isn't held strongly enough, it can begin to move into the cutter, which also spells disaster. These are the result of uncontrolled climb milling.

> **KEYPOINT**
>
> **Caution!** To safely perform climb milling, use either the feeding axis lock, snugged as a damper, or a compensator (recommended if the mill has compensation).

Compare the Action

Take a moment to compare Fig. 11-93, which shows climb cutting, to Fig. 11-94, where a conventional cut is being made. Note that the climb cut starts with a thick chip as each tooth enters the work. That's good; it makes a nonrubbing entry into the work. Then as the tooth exits the metal, the chip becomes progressively thinner. That's good, too. The machined surface is formed by a final, sheared slice, from inside the material toward the outside. Both work finish and tool life are excellent when compared to the way a conventional cut progresses.

Notice first off in Fig. 11-94 that during the conventional cut, the cutting edge doesn't start into the metal until some distance past the theoretical entry point. For a short distance, as the machine feeds forward, the tooth builds up pressure until the pressure is great enough for the cutting edge to dig in to start cutting. Yet, that rubbed zone is what will become the finish left behind! The tooth edge also suffers as it rubs on the material, especially if the material is hard, such as carbon

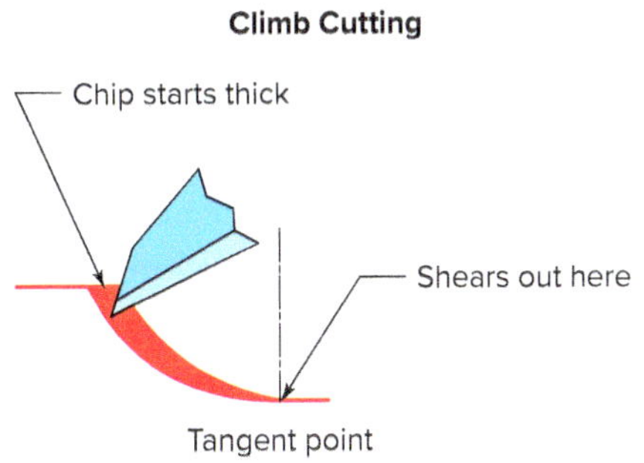

Figure 11-93 A climb cut produces a better finish and longer tool life.

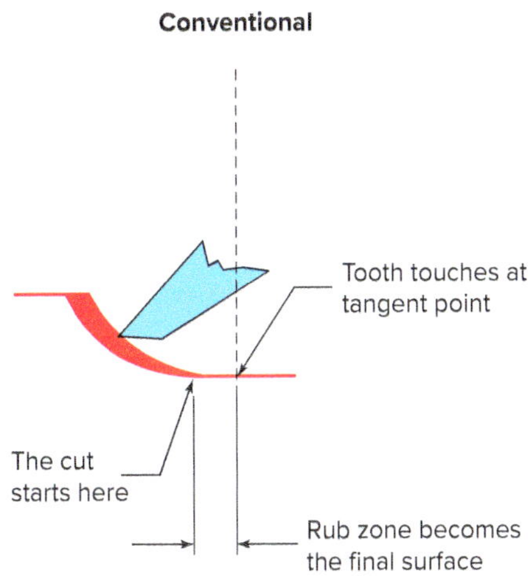

Figure 11-94 The conventional cut starts by rubbing until the tooth builds pressure to dig in, producing poor finish and dulling the cutting edge.

steel or bronze. Using conventional cutting, surface work hardening will occur faster in a material prone to it, and the dulling/work hardening accelerates as the edge begins to fail.

SHOPTALK

You might hear climb milling called *down cutting* and conventional called *up milling* due to the way the cutter pushes or lifts the material.

KEYPOINT

During a peripheral conventional cut, the cutting edge rubs for a short distance before it builds enough pressure to dig in.

TRADE TIP

Exceptions—When to Use Conventional Cutting On manual or CNC mills, it is acceptable to cut back and forth using first conventional, then climb milling when reducing a surface with peripheral cutting, especially on softer materials such as aluminum. Just keep in mind the different actions produced—snug the brake for the climb pass, then release for the conventional cut. However, to produce the best finish, choose climb milling for the final pass.

When to Use Conventional Cutting There are two times when conventional cutting is called for:

1. *Hard Crust* Cast iron often forms a hard thin outer surface, but it has a soft interior (Fig. 11-95). Here, by correctly choosing conventional cutting, the tooth starts under the crust until it breaks upward, pushing the crust away. In this case, climb milling would tend to dull the cutter more quickly because each tooth would start its cut in the hard crust.

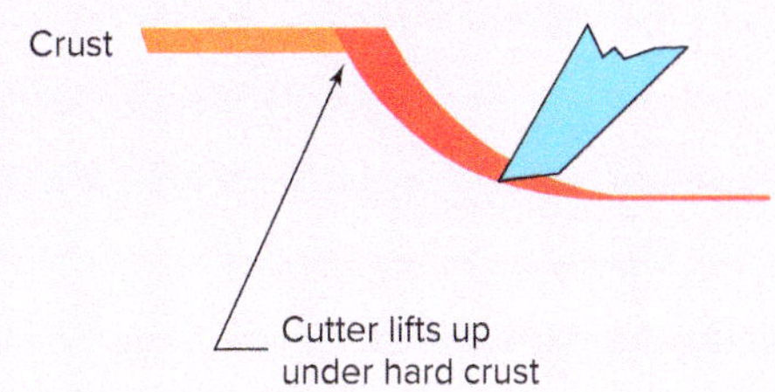

Figure 11-95 Conventional cutting is correct for work with a hard crust.

2. *When the Setup or Part Are Too Thin or Weak* Sometimes, the aggressive action of climb milling will flex, bend, or lift a part in a setup. In this case, try conventional cutting to stabilize the work and the setup.

11.5.3 Computing Mill Feed Rates

Recall that mill feeds are expressed in inches per minute (IPM) or meters per minute. Reasonable student rates vary from under 1 in. per minute to around 20 in. per minute depending on the work material, the setup rigidity, and the machine. Another big factor is your skill level; stay on the conservative side while learning. Industry rates are in the hundreds of feet per minute and are accelerating as machines and controls evolve. Feed rates of 100 IPM or greater are everyday stuff with CNC rates at 300 to 700 IPM and faster where the computer can control the action and still stay on course! (See Fig. 11-96.) There is a specific study of maximized machining, called high-speed machining, that we'll look at in Chapter 27.

To Find a Feed Rate in IPM

The process of finding a feed rate is a three-step multiplication.

Determine Feed per Tooth (FPT or F/T)

From a chart, find chip load for one tooth making one revolution–the recommended bite for one tooth.

FTP is also called the **chip load**.

Figure 11-96 This horizontal machining center is removing aluminum at 750 IPM.
Courtesy Milacron LLC

FPT is found on the chart by crossing two arguments: the cutter material (HSS or carbide) and the work material on the chart.

From experimenting, researchers have found minimum and maximum amounts of forward feed, for one tooth, making one revolution. The range provides good finish and tool life. It's then compiled into a recommendation chart.

FPT × Number of teeth = Feed per revolution

Multiply FPT times the number of teeth in the cutter.

This determines feed advancement for one single revolution of the cutter = FPR.

FPR × RPM = Feed rate

RPM has been precalculated. It's the first fact.

Quick Review—Calculating RPM

After finding the recommended surface speed from a chart, it is combined in either the long or short formula with the cutter diameter. The SS is determined with two arguments: work hardness and cutter hardness (carbide or HSS).

Example Problem

We'll be using *Machinery's Handbook* or CD for this exercise; however, commercial slide rules or your smartphone app designed just for milling are available and quick to use. Please turn to the section on "Speeds and Feeds–Feed Rates for Mills."

Calculate the feed rate for milling cast steel, using a $\frac{3}{4}$-in.-diameter, four-tooth, HSS end mill cutting to a depth of 0.250 in. with the RPM at 480.

Find the chart "Recommended FEED in inches PER TOOTH for HSS Milling Cutters."

Question?

What is the total range of tooth feeds for cutting cast steel?

The range is 0.001 in. to 0.012 in. depending on cutter type, depth of cut, and hardness of work material. We'll assume the steel is the least hard (top row). The rate is?

Answer

0.003 in. per tooth if at 0.250 DOC

Next multiply times 4 teeth, which equals 0.012 in. per revolution.

Now, multiply times the RPM at 480.

480 × 0.012 = 5.76 in. per minute feed rate

KEYPOINT

Feed rate = FPT × Number of teeth × RPM

TRY IT

From *Machinery's Handbook,*

A. Find the feed rate for milling plain carbon steel (lowest hardness) using a $\frac{1}{2}$-in. two-flute HSS end mill cutter at 0.050 in. depth of cut. Calculated RPM = 800.

B. Cutting gray cast iron (hardest this time), with an eight-tooth HSS face mill. RPM = 200. This time provide the range of possible feeds.

ANSWERS

A. Feed per tooth = 0.002 in.; 0.004 in. per revolution; *3.2 in. per minute.*

B. Feed per tooth = 0.002 to 0.008 in. per tooth (range of possibilities); 0.016 to 0.064 per revolution. From *3.2* IPM to finish cut up to *12.8 in.* per minute for roughing.

TRADE TIP

Speeds and Feeds With experience, you'll develop a "feel" for the right rate. Experienced machinists often select initial feed rates from their head, but not always!

There are times when a special material or unusual cutter is used; it's then that we all fall back on calculating.

The two pocket slide rule calculators shown in Fig. 11-97 are a convenient way to quickly find speeds and feeds. They are supplied by cutter manufacturers, usually free of cost. Search for Sandvik Speed and Feed or check your app store to get the free calculator. Check with your local tooling company or go online and request one. You'll find they follow the same formulas and input arguments we've already discussed for both RPM and feed rates.

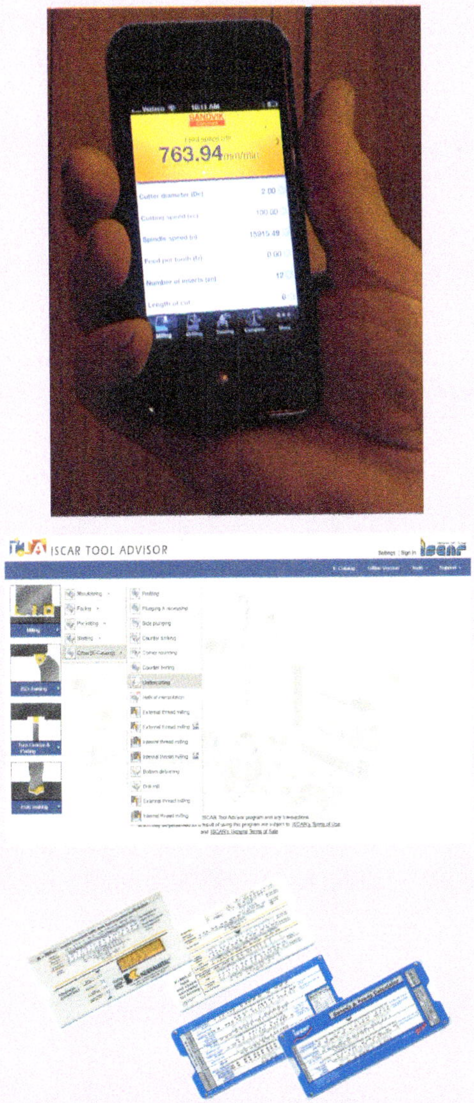

Figure 11-97 Speed and feed calculators should be a part of your toolbox.

(*digital calculator*) Courtesy Mike Fitzpatrick, (*manual calculators*) Courtesy Iscar Metals, Inc. (Blue) Courtesy Mike Fitzpatrick

11.5.4 Setting Up Milling Machine Vises

Now we turn our attention to the nuts and bolts aspect of the setup, literally!

Clean It First

When mounting vises, always wipe the bottom with a clean, lint-free rag; then give it a final wipe with your clean hand. Always clean the mill table with equal care. This palm trick will detect any chips or nicks protruding. The final hand swipe is an important practice.

Aligning

If the vise has slot alignment keys, do not set it directly on the mill table, then slide it across to location. Set the vise on a piece of plywood or other protective material then slide it into position–tilt the vise up to remove the wood, then lower the keys into the slots, making sure no chips or dirt has fallen on the table while you were doing so.

Indicating Vises

There are two cases when alignment keys are not used and we turn to a dial test indicator to set up the vise (Fig. 11-98).

The vise has no keys–yet the setup requires perfect alignment to a machine axis (common).

The setup requires the vise at some angle to an axis.

Naturally, there are a few trade tricks!

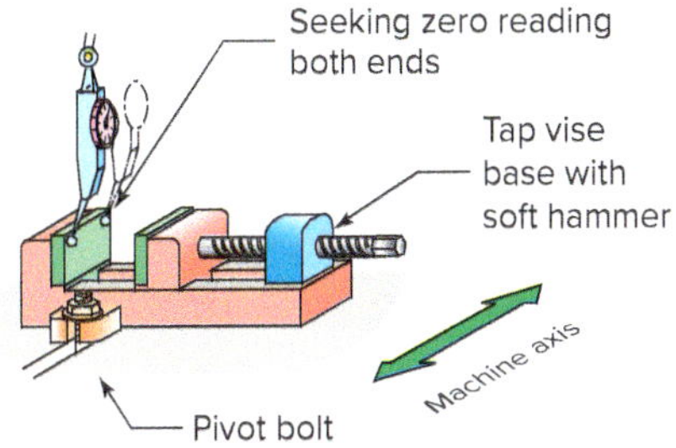

Figure 11-98 The indicator is against the nonmoving (solid) jaw.

ARMCHAIR FIELD TRIP

Be sure and watch this Haas Tip of the Day video about aligning vises.

In this video, you will learn how to properly mount a vise on a mill table. Go to: **https://www.youtube.com/haasautomation** and search for this video called: **How to Square and Indicate a Vise on Your CNC Mill—Haas Automation Tip of the Day**

KEYPOINT

Always indicate the *solid jaw* of the vise, never the moving jaw. The solid jaw will become the reliable datum for your part. A ground parallel held in the vise averages out jaw damage or irregularities.

Set Up a Pivot Bolt

To begin, lightly tighten one tee nut bolt; then snug the second bolt just beyond finger tight. That simplifies moving it as the vise pivots on the tighter bolt during final alignment. Now, with an indicator mounted on the mill frame or spindle, place the probe against the pivot bolt end, zero it, and then move the mill axis back and forth. While it moves, tap the vise side with a soft hammer until a zero reading is obtained on both ends of the vise jaw (Fig. 11-98).

Mount the indicator on the mill, in one of five possible ways (Fig. 11-99).

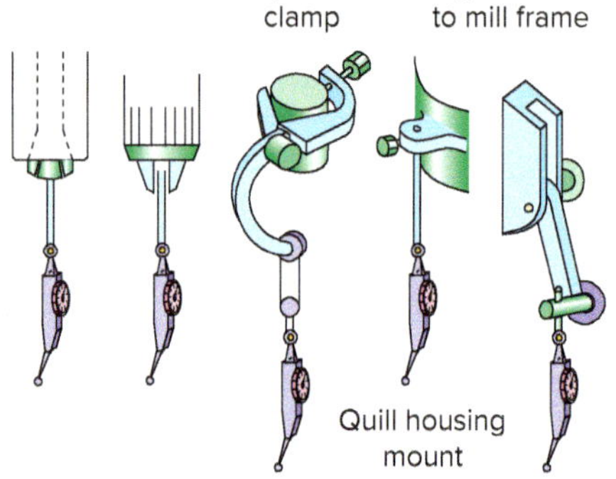

Figure 11-99 Five ways to hold your indicator on the mill.

KEYPOINT

Avoid the embarrassment of flinging your indicator across the shop!! Shift the spindle to neutral for the first three indicators shown in Fig. 11-99.

TRADE TIP

Positioning Mill Vises and Aligning Work Here are a few handy tips that simplify the task of positioning a vise in mill setup!

Read the Trend! Instead of cranking back and forth, end to end to indicate the vise, try this: With practice you can whip a vise into alignment in seconds! Start the power feed at a moderate rate around 2 IPM. As it moves away from the pivot bolt, watch the dial movement to determine which way the vise is misaligned. Start lightly tapping the vise as the indicator travels across the jaw. The indicator's needle movement will slow, then stop as the vise approaches alignment. Using this technique, one can often complete an alignment within 1 minute!

Balance It. Move It. Given a choice, never mount a vise at the extreme end of the mill table (manual or CNC). This creates an unbalanced load on the table and excessive wear on the gibs. However, it is good practice to move different setups to various locations somewhat near table center (Fig. 11-100), especially on CNC machines where moves repeat many times in production runs. That avoids long-term concentrated wear to the drive system.

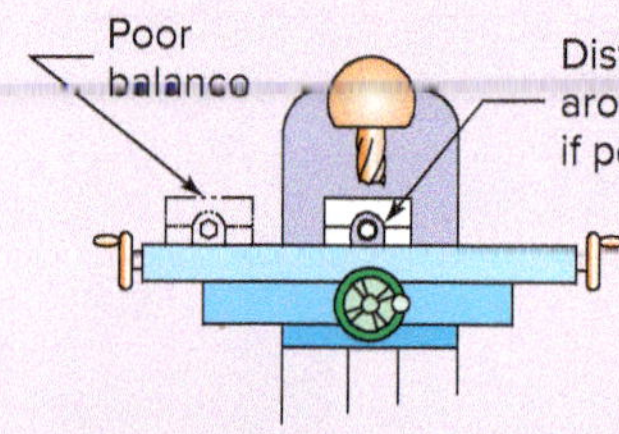

Figure 11-100 If possible, mount the vise near the table center to balance wear on dovetails and gibs.

Angle It. Here are three methods of positioning a vise at an angle to the *X* or *Y* axis. Each is progressively more accurate but requires a bit more time to perform.

1. *Rotating angular* vises with graduated bases are useful when angular tolerances are no tighter than 1° accuracy (Fig. 11-101).
2. *Indicating a protractor* is the most convenient method. Lightly clamp the base of the protractor in the vise then indicate the blade. Expect 10 minutes or better accuracy, which is 12 times as accurate as method 1. Don't forget to clamp the second ear down to the table as shown in Fig. 11-102.
3. *Indicating a sine bar* is the most precise method available (Fig. 11-103). Accuracy will be within 0.0005°—a major improvement!

Quick Alignment Here's another tip that will save you much time: Use the table slots and slot blocks to align a workpiece datum to the *X* axis (Fig. 11-104).

Figure 11-101 Some vises feature a swivel base. The better vises feature angular graduations as in this example.
Courtesy Kurt Manufacturing/Industrial Products Division

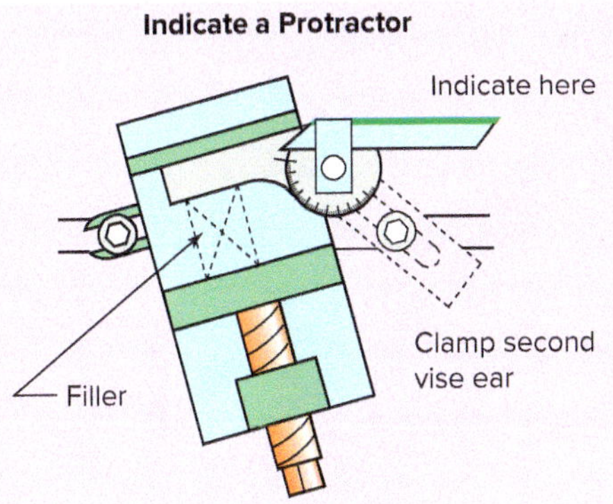

Figure 11-102 Setting up a vise to a precision angle using a *vernier protractor* and DTI.

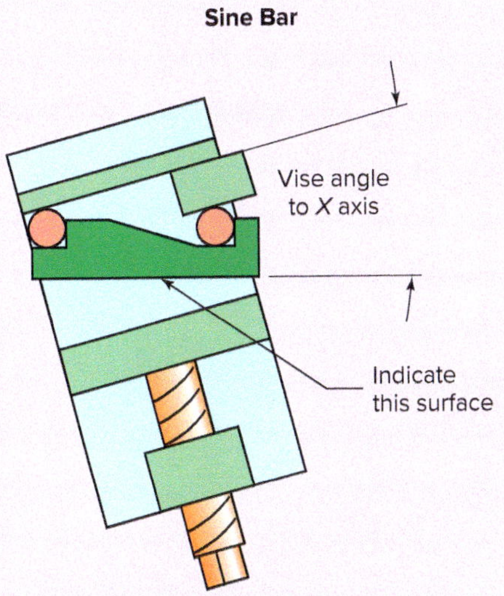

Figure 11-103 Setting up a vise to a precision angle by indicating a *sine bar*.

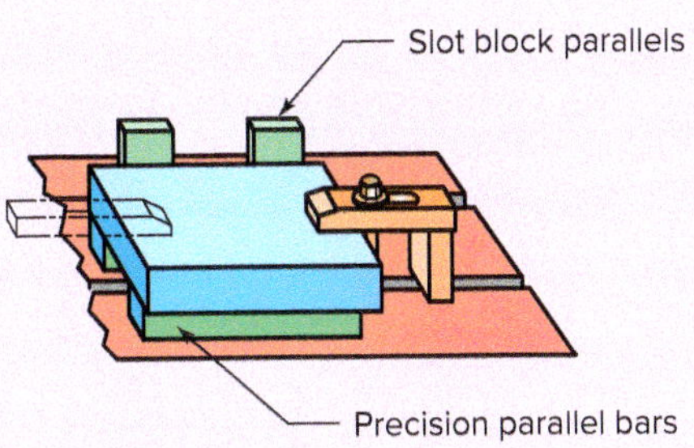

Figure 11-104 Saving time by using slot blocks to align a workpiece.

11.5.5 Indicating a Universal Head Machine

When drilling or boring vertical holes on mills, the head must be perpendicular to the table. If not, the hole produced will not be vertical. On rigid vertical mills, the head cannot be out of alignment. However, when using a ram and turret mill, or one with an *A-B* head auxiliary, the *A* and *B* axes must be aligned for precise drilling, boring, and many other operations.

Aligning the *A-B* axes perpendicular to the mill table is accomplished with an indicator fastened to the spindle. The indicator and spindle in neutral are rotated by hand, in a circle, against the machine table or a ground parallel to average irregularities and bridge tee slots. You will require a demonstration of this task to completely understand it. There are the three major phases.

KEYPOINT

Caution! Be sure the main power switch is turned off and the spindle is in neutral.

Phase 1—Testing and Roughing the Alignment

Graduations First

Look at the graduations on the head (Fig. 11-105). Both the *A* and *B* axes have them. Any gross misalignment will be apparent.

Using a Square and Alignment Notches in the Head

Now, using the small notches machined into the machining head, refine the alignment. As shown in Fig. 11-106, these notches are parallel to the quill and can be used against a master square sitting on the mill table.

TRADE TIP

Paper Feelers to Eliminate Guessing To get a good "feel" at both contact points between square and pad, hold a piece of paper at each contact point and tug on each. Any difference in pressure will be easily felt. If this is done correctly, the head error can be almost removed without using the indicator.

Phase 2—*A* Axis Alignment (Front to Back)

Leaving the *A* axis lock bolts finger tight, use the front-to-back swing of the dial test indicator sweeping the table to determine perpendicularity (Fig. 11-107). Your objective is to adjust the head until the indicator reads zero-zero, front and back. Once the *A* axis alignment is accomplished, lock the bolts a little at a time, and then retest to see if locking disturbed the position.

Figure 11-105 The *A* and *B* axes angular graduations are a rough indicator of alignment.
Lake Washington Technical College, Kirkland, WA/McGraw Hill

There is a complexity in aligning the *A* axis. Note in Fig. 11-108 that the indicator pivots at the end of a long arm. That means that as the head's tilt is adjusted, the sweep of the indicator moves in an arc at the end of the arm. Both the front and the back of the sweep move in the same direction. That fact makes a difference in how you adjust the head toward alignment. Here's an example to clear this up. In Fig. 11-109, the head is not perpendicular to the table. At position 1 the indicator shows a difference of 0.020 in. farther away from the table, in the front. From that, you can deduce that it's tilted outward.

So, with the indicator in the front, we crank the axis down 0.020 on the dial. But after doing so (Fig. 11-110), we still find a 0.009 front-to-back difference. The adjustment didn't bring the head into alignment due to the indicator being at the end of the long arm. Both the front and the back swung down closer to the table with the first adjustment.

KEYPOINT

The correction turns out to be a ratio of *amount taken* compared to *amount achieved* on the indicator dial.

The ratio is a function of the long indicator arm from the head's pivot point to the table contact of the probe. That distance changes from setup to setup, depending on indicator holder and quill position.

The ratio, in this example, is a little over half–in cranking out 0.020 in. on the indicator; 0.011 in. of error was removed. Discovering that with the first adjustment, reposition in the front and dial down say 0.017 in. to take out the 0.009-in. needed correction. The ratio changes as the probe approaches parallel, so a final adjustment is probably going to be required, but it will be a small correction. Have your instructor demonstrate and explain this concept. It's easy to understand when you are doing it, but difficult to put into words.

Figure 11-106 Equalizing the drag on paper feelers against alignment notches in the head roughly aligns the *A*–*B* axes.

Lake Washington Technical College, Kirkland, WA/McGraw Hill

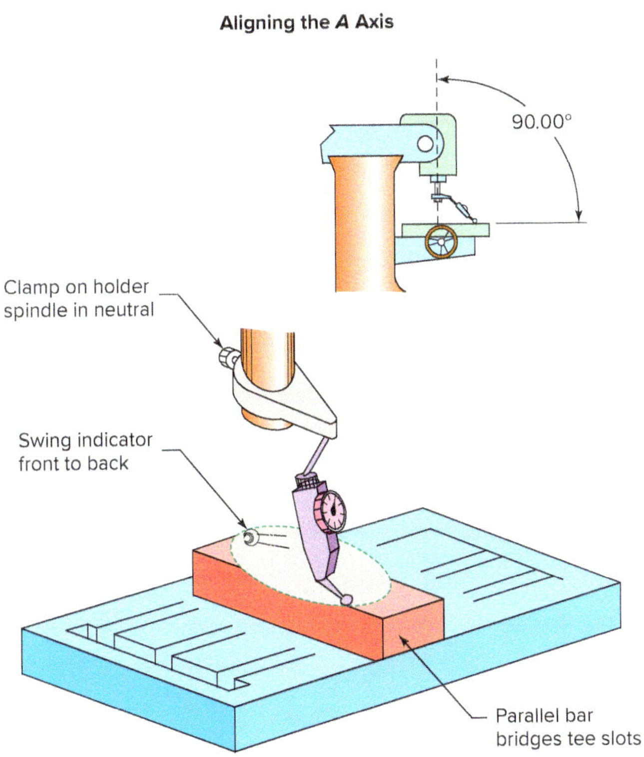

Figure 11-107 Correct the *A* axis error first.

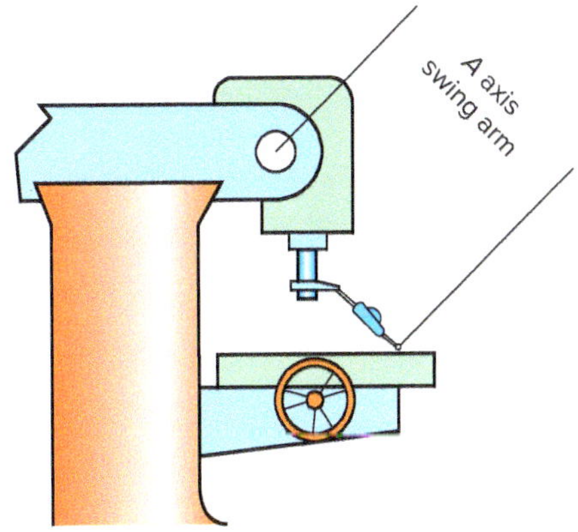

Figure 11-108 The indicator is at the end of a long arm.

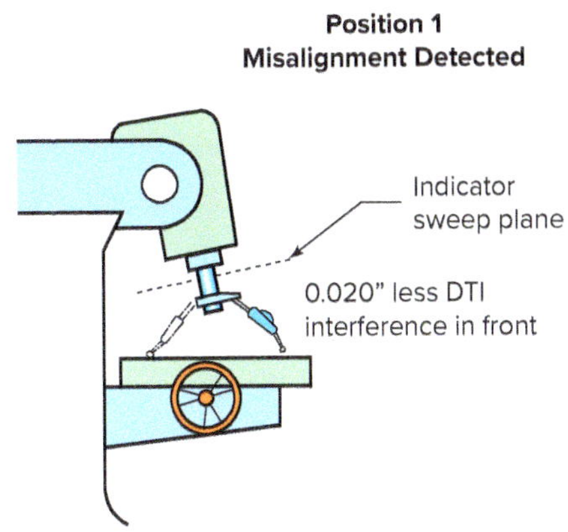

Figure 11-109 Position 1—0.020-in. error read on the indicator.

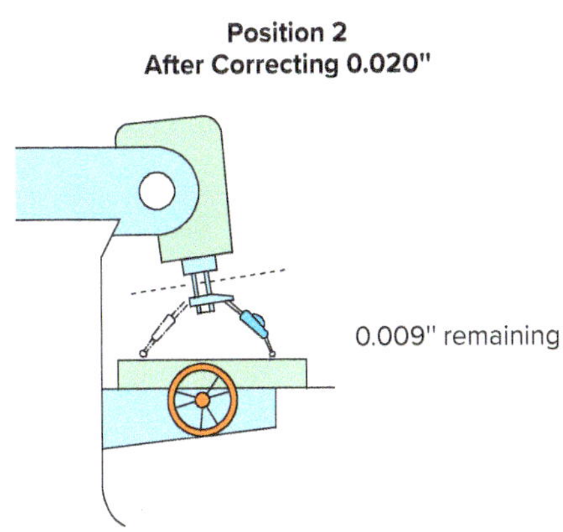

Figure 11-110 Position 2—0.009-in. error remaining.

TRADE TIP

Use a Ground Parallel to Smooth Out Table Irregularities When performing the DTI alignment of the universal head, place a large ground parallel on the table to account for the tee slots and any dents that might disturb the probe as it swings. A fine stone run across the table will remove protrusions but the DTI needle can still trip in dents. A parallel on the table solves that.

Phase 3—*B* Axis Alignment

The *B* axis is easier to align than the *A* axis (Fig. 11-111). This also is accomplished by sweeping the indicator in a circle on the table while the right-hand reading is compared to that on the left. The head is adjusted until these readings are equal, zero-zero. But this time there's no ratio challenge. All one must do is determine the right-left difference then adjust away half the amount needed. One side goes up while the other comes down.

Sine Bar for Angles For setting up precision angles of the *A* or *B* axes, after roughly adjusting the head using the graduations on the axis, use a sine bar and sweep it for zero-zero on a DTI (Fig. 11-112).

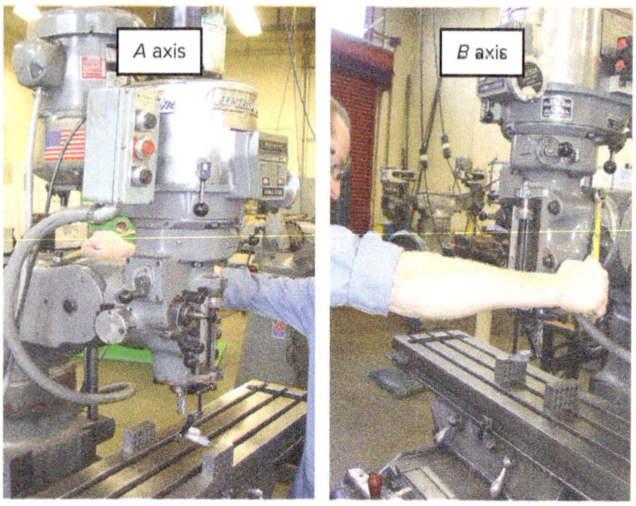

Figure 11-111 Rough align the *A* axis first, then the *B* axis. Then align them a second time and check the alignment after tightening lock bolts.
Lake Washington Technical College, Kirkland, WA/McGraw Hill

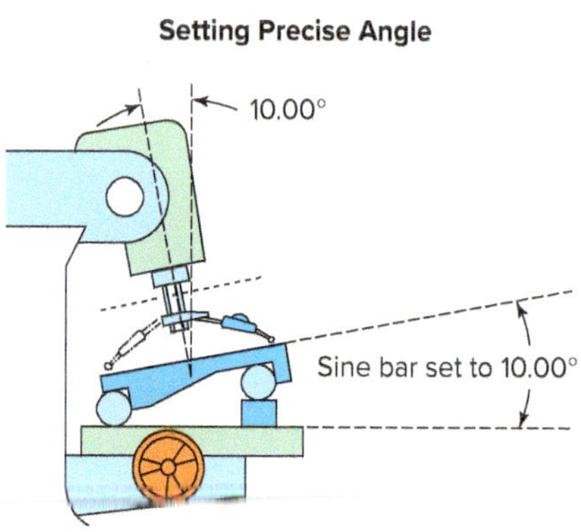

Figure 11-112 Use a sine bar and DTI to set precise angles.

11.5.6 Datum Locating a Milling Machine

For many manual mill and all CNC setups, it becomes necessary to locate the spindle over a unique place on the part:

- An intersection of two edges
- Centered over a hole
- On the intersection of layout line's–center-finders are covered in Chapter 9.

With the spindle located at that point, the position registers (micrometer dial or readouts) will be set to zero. This is a prime CNC setup skill.

KEYPOINT

That *X–Y* datum reference point is called the **program reference zero (PRZ)** in CNC programs and setups. On manual machines we call it the datum point, or the **reference zero (RZ)**. It's symbolized by a bull's-eye target (Fig. 11-113).

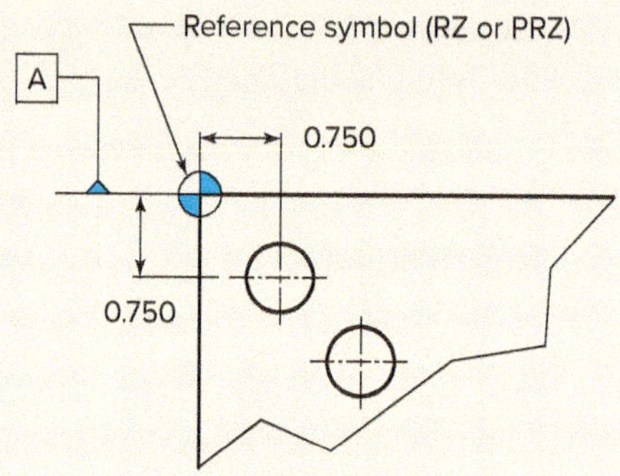

Figure 11-113 We need to drill this part to a 0.010-in. location tolerance.

There are three tools used to position the mill's spindle at a desired RZ.

Tool	Repeatability
Edge-finder	0.003 to 0.0005 (using light flash) (Trade Tip coming up)
Center-finder	0.005 to 0.003 (using colinear) (Trade Tip in Chapter 10)
Dial test indicator	0.0005 or better

Edge Finding

For this job (Fig. 11-114), we will be drilling five holes in a part that is already shaped. The hole location tolerance is plus/minus 0.010 in., so an **edge-finder** is appropriate. The RZ is at the intersection of datums A and B, the upper left corner. Once we position the mill spindle over that corner, we'll zero both micrometer dials (or *X* and *Y* readouts), then crank out, in turn, to each of the five locations and drill the holes.

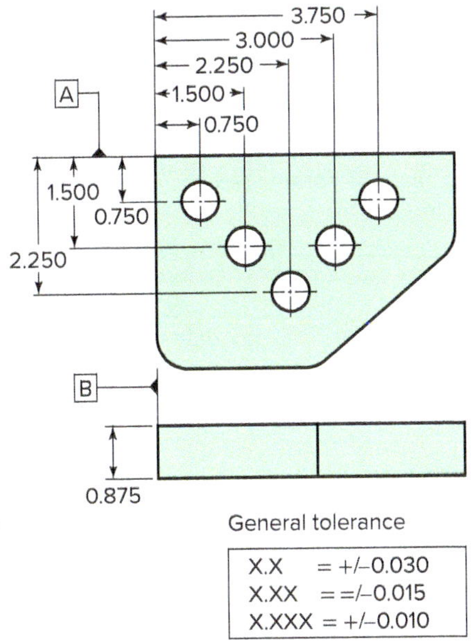

Figure 11-114 The intersection of datums *A* and *B* will become the RZ.

KEYPOINT

In Fig. 11-115, the RZ is at the intersection of two primary datums. The vise has been indicated true to the mill's *X* axis. The part is loaded into the vise with datum feature A against the solid vise jaw.

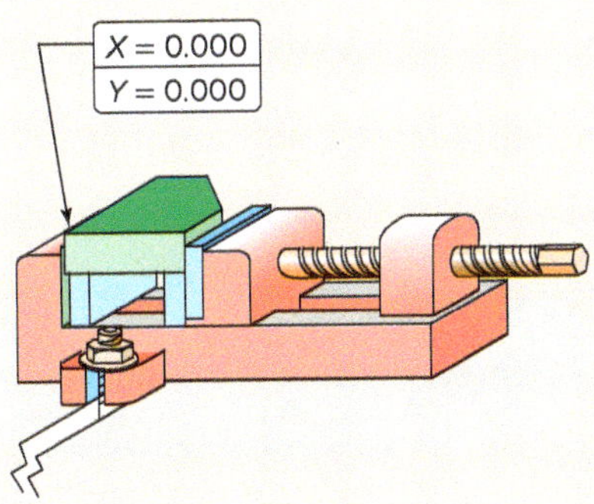

Figure 11-115 The symbol used to denote RZ or PRZ in setup documents.

Using the Edge-Finder

This handy little tool makes edge datum location a snap (Fig. 11-116). It's a priority 1 for your toolbox purchase plan. We'll be using the cylindrical probe. In theory, any round object of known diameter could be used for edge finding, except that there's always some runout in the way it's held in the spindle. But, due to the probe's ability to float on the end of the body, it centers on the spindle axis independent of spindle runout.

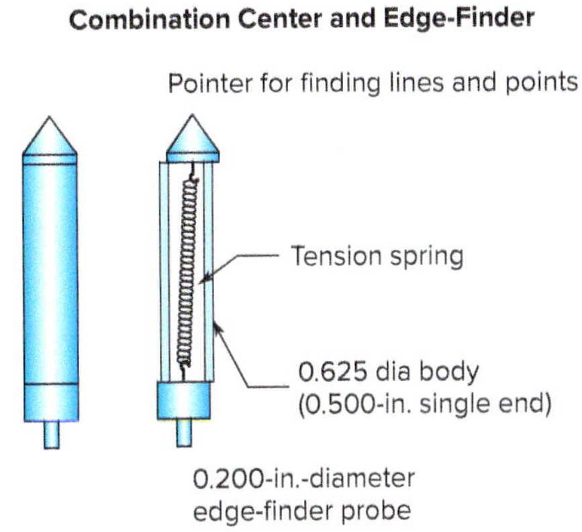

Figure 11-116 The edge-finder helps place the spindle axis over a part edge within 0.003-in. repeatability.

Mounted in a drill chuck or collet, it's spun at around 200 to 300 RPM but no more. The spring can be stretched if the RPM is set too high. The objective is to start it with a deliberate amount of runout, then, watching the wobble, approach the edge with the probe touching the work, until the runout disappears.

The objective is to stop moving the spinning probe (0.200 in. diameter) toward the edge at the instant that's it's perfectly tangent. At that location, the spindle axis is 0.100 in. from the edge (half the probe diameter). All that remains to do is set the position register at plus or minus 0.100, depending on which side of the edge the probe is relative to the work. In Fig. 11-117, it's to the left–in the minus direction compared to the edge.

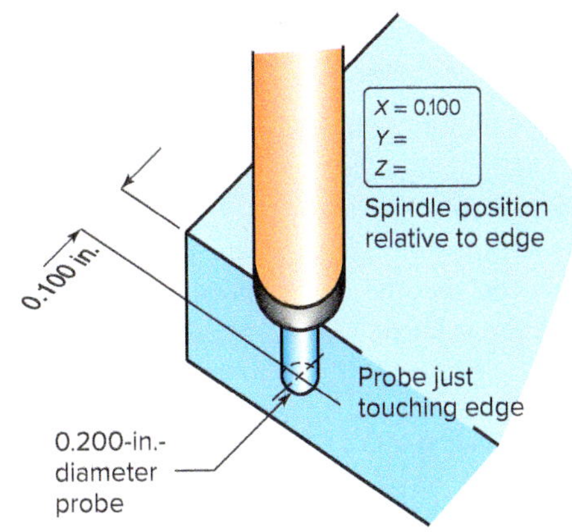

Figure 11-117 The spindle is 0.100 in. to the left of the work edge as the probe just touches.

KEYPOINT

The radius of the probe is 0.100 in. By moving the adjusted axis to zero in Fig. 11-117, the spindle would be directly over the *X* axis edge, and the backlash would be accounted for. It's now ready to position over the upcoming hole locations.

Setting the dials and registers to represent the spindle position relative to a RZ or PRZ is called **coordinating the axes**. After coordinating the *X* register, do the same to the *Y* register to complete the RZ.

Right and Wrong Edge-Finder Use

A less accurate method is to watch for the probe to sideways, then assume that location is the perfect tangent location–it is not. In order for the probe to snap, friction must be present, and that only happens once the edge-finder is in forceful contact with the edge–too far past perfect contact! That probe snap location will reproduce reliably at the same location every time, but it's beyond perfect alignment by around 0.001 to 0.002 in. It can be used for rough edge finding, but it doesn't produce the best results. Trade folks often assume that at that break-over point it's 0.003 past contact (assume?). See the Trade Tip.

TRADE TIP

Use the "Flash" To improve edge finding accuracy use the light coming through the flashing gap between the probe and work, rather than trying to see the probe settle into perfect tangency on the work. Just before it reaches perfect tangency, you'll see light flashing as the gap opens and closes due to the probe wobble. Slowly crank the table axis and watch the flash. The moment it stops, the probe is tangent to the work. To aid this effect, hold a piece of white paper behind the probe to enhance the light flash. Try it–it works wonders and improves accuracy!

Using a DTI to Coordinate an Edge

This is a vital skill for coordinating reference points when tolerances are finer than edge finding can deliver–below 0.001 in. The objective is to place the DTI on the edge, then hand rotate it in an arc (with the spindle in neutral). When the numbers seen on the dial are zero-zero on both sides of the edge, the spindle is located. You'll need to hold a ground parallel on one side for this test, as shown in Fig. 11-118.

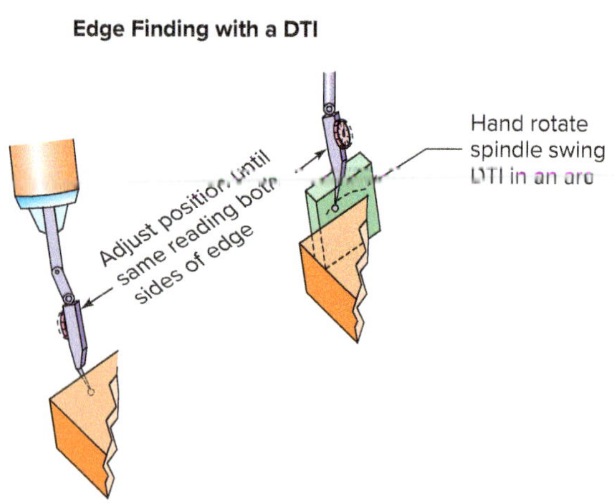

Figure 11-118 Locating over an edge using a DTI.

How Does It Work? With the indicator set just off the spindle axis, swing the spindle and indicator in an arc as you crank the axis toward the edge. As it touches the edge, the indicator readings will increase until a highest interference point (the *null* point) is reached, then drop off again.

Return to the null position, set the indicator to zero. Now pull it up to clear the edge, turn it 180°, and move the probe to the other side of the edge. Read the difference in this null value compared to the zero. Determine by the readings (plus or minus from the zero), which way the spindle is from the edge; then move one half the difference toward alignment. Test again as it usually requires a second and even a third fine adjustment.

The Wiggler Set

As shown in Fig. 11-119, the **wiggler** and its accessories are used in a number of tasks for alignment and indicator holding. You must have either an edge-finder or a wiggler in your toolbox. When picking up lines, wigglers aren't as accurate as center-finders because they do not reflect the line's image. The pointer is used for eyeball location.

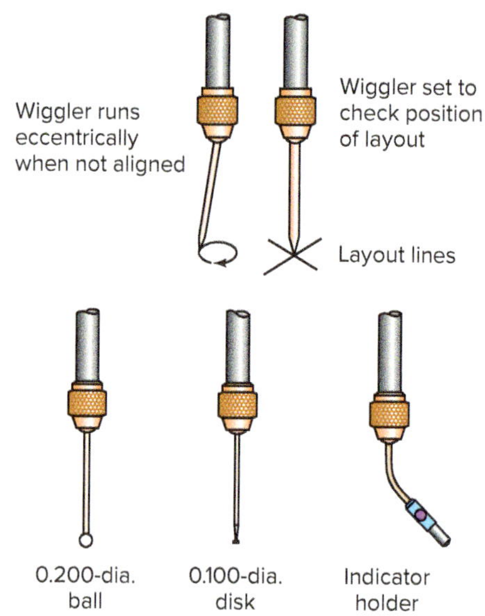

Figure 11-119 The wiggler set is used similarly to an edge-finder to place a spindle over an RZ.

11.5.7 Squaring a Blank Part Using the Datum Process

This is a basic skill for milling and very often the start of a job plan. We'll square up the drill gage blank to the print dimensions. The four steps are shown next, but not everything is explained. Questions will be asked at the end based on what you see in the drawings. The objective is to mill a 3.000-in. × 5.000-in. rectangle with 90 °corners.

We'll start with 0.100-in. machining excess on all four edges. The setup includes a vise true to the *X* axis of the machine with a $\frac{1}{2}$-in. end mill in a collet. *Hint:* Watch datum edge A as it progresses through the operations.

KEYPOINT

The only way this process will produce accurate results is to clean chips and remove burrs between each operation!

Step 1 Create the Datum Feature Edge

Based on the priorities on the print, start with the most reliable, longest, or most important edge of the part. In this case, the long edge is also Datum A on the print (Fig. 11-120), but that may not always be the case. It might be necessary to create an intermediate surface on the way toward making datum feature A.

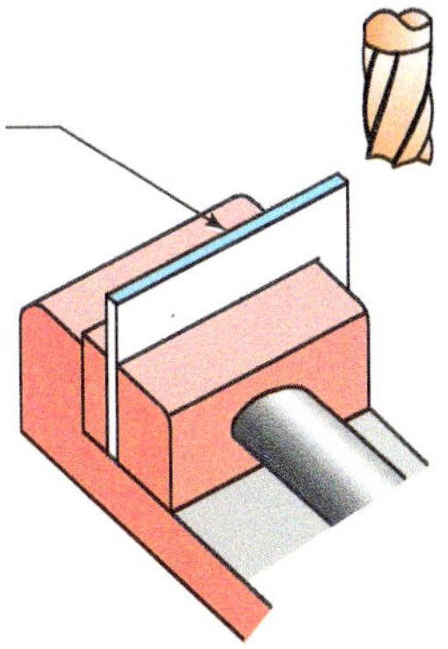

Figure 11-120 Machining Datum A—Minimum cleanup to produce a clean surface free of saw marks.

Step 2 Machine the Opposite Edge Parallel and to Size

Reversing the part in the vise and testing to see that datum feature A is fully seated on the vise floor, cut the opposite edge parallel to Datum A. Take a small test cut and measure the result. Coordinate the *Z* axis position (readout/micrometer dial) to your measurement, then take roughing passes, check it again. Make the final pass according to the coordinated reading.

After the coordination is completed, as long as the quill remains locked in place, subsequent face, step, and pocket cuts with that end mill will produce the same *Z* axis depth relative to the vise floor datum.

KEYPOINT

Do not skip this micrometer dial coordination, it assists and speeds precision. Note that just a bit of the part was left protruding from the vise, to facilitate measuring it (Fig. 11-121).

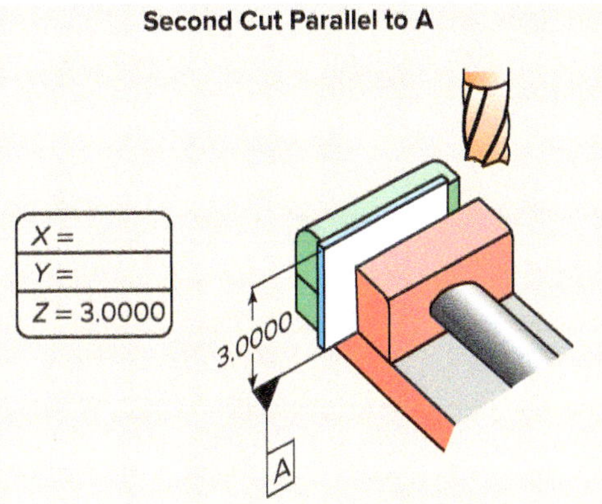

Figure 11-121 Machine the opposite edge to size. Measure the result on the part protruding from the vise.

Step 3 Machine One Corner Square

Lay the part down in the vise on parallels. Machine one edge for cleanup using the *Y* axis this time. *Notice that datum edge A is against the solid jaw of the vise* (Fig. 11-122).

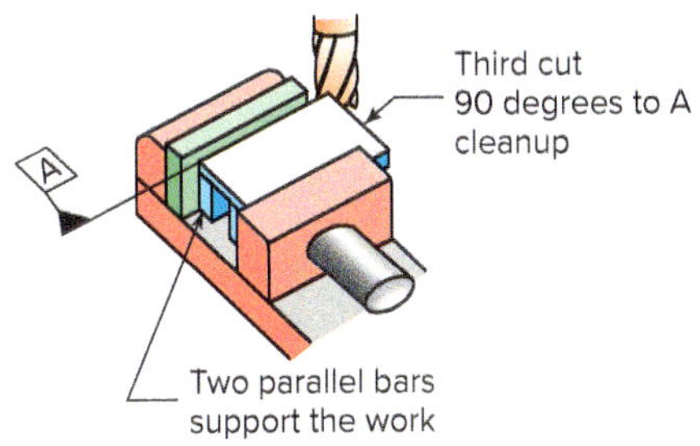

Figure 11-122 Note that the original Datum A edge is against the solid vise jaw.

Step 4 Machine Final Edge to Size

Slide the part such that the other end protrudes out of the vise but keep datum edge A against the vise jaw. Now machine the final end to size (Fig. 11-123). There are two other possible ways to make the final edge. It could be accomplished with the part standing up and Datum A on the vise floor, as in step 2. Or the part could be flipped over keeping datum edge A against the vise and cutting the part exactly as in step 3, with the cutter on the same side of the vise.

11.5.8 Indicating a Hole for Datum Point Location

Sometimes the RZ is a hole's center rather than intersecting edges. Often a CNC mill fixture will have a precision bored hole that represents the setup program reference zero for the fixture. Both the *X* and *Y* axes must be coordinated over this hole. Use an indicator in the mill spindle for this setup skill.

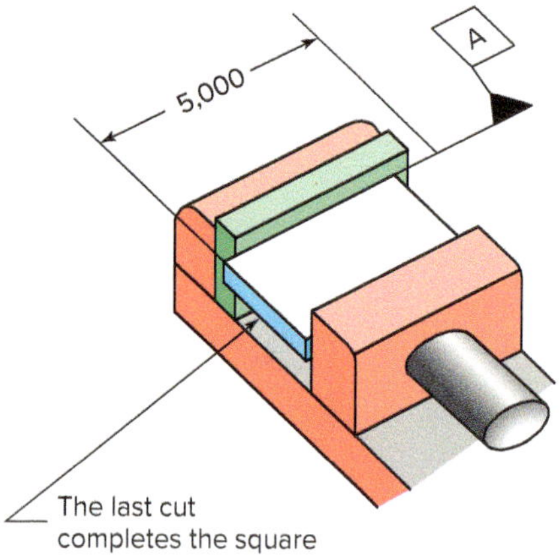

Figure 11-123 Machining the final end to size, square to Datum A.

Vertical Indicators

To locate over a hole, use an indicator, as shown in Fig. 11-124. Although an everyday DTI can be used for this task, there are two indicators made for the task of overhead location: A **jig bore indicator** has the dial facing up when the probe is vertical. They are also called *vertical indicators.* They speed up this operation by presenting their dial faces in such a way that they can be easily read without parallax.

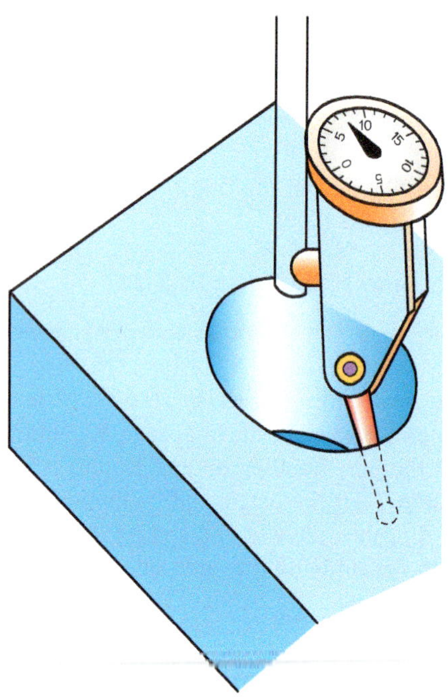

Figure 11-124 Locating over a hole using a jig bore (vertical) DTI.

Jig Bore Indicators

With the spindle in neutral, hand rotate the indicator while adjusting spindle position until a zero-zero reading occurs in the *X* and *Y* axis direction. A jig bore (or jib borer) indicator would be a priority 2 for your toolbox. You can get along without them, but they are so very handy to have when setting up over holes.

Coaxial Indicators

A **coaxial indicator** is also designed for this task. Its dial also faces up, but this time the machine revolves the probe under power at around 100 RPM or so, but the dial remains stationary by holding back on a handle.

First the *X* spindle position is dialed in by holding the handle parallel to *X*. In that orientation, the swings of the needle represent *X* misalignment. Adjust the spindle position relative to the hole while watching the swings diminish until no movement of the dial hand is detected–it's on-center for *X*. Then the indicator handle is rotated parallel with *Y* and the task repeated. This takes just seconds. These indicators are complex, precise mechanisms, and as such they are relatively expensive. They are normally shop owned.

11.5.9 Specialty Vise Setups

Vise work can be very creative. The challenge is to find and set up the right combination of holding tools to keep a complex part stable, repeatable, safe, and in the right datum relationship. Here are a few tips as to how to proceed.

Vise Soft Jaws

Often on CNC mills and occasionally on manual mills, it becomes necessary to mill special shapes into sacrificial aluminum (or steel) jaws. The blanks are mounted in the vise, then cut to fit one or more odd-shaped parts (Figs. 11-125 and 11-126). Note that in Fig. 11-126, the jaws on this fixture vise can be rotated to accommodate the round part.

Figure 11-125 The hard jaws are removed.
Courtesy Kurt Manufacturing/Industrial Products Division

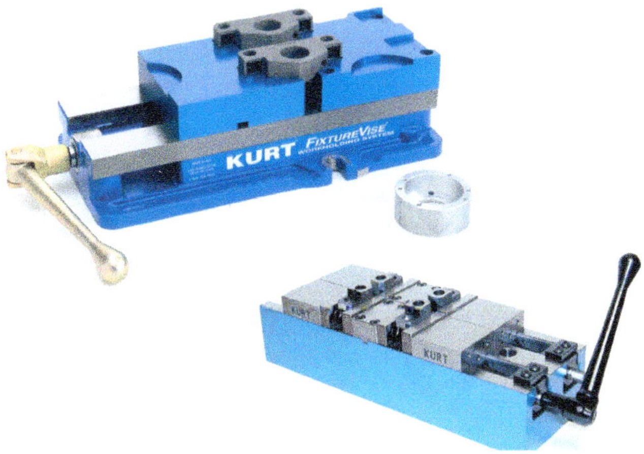

Figure 11-126 Two examples of custom-shaped vise jaws.
Courtesy Kurt Manufacturing/Industrial Products Division

Vise End Stops

Sometimes it's necessary to load a series of parts into the vise, with each stopping in the same datum relationship. Datum A is against the solid jaw, while Datum B touches the stop. If the vise isn't equipped with these kinds of stops (Fig. 11-127), then a stop can be bolted directly to the table.

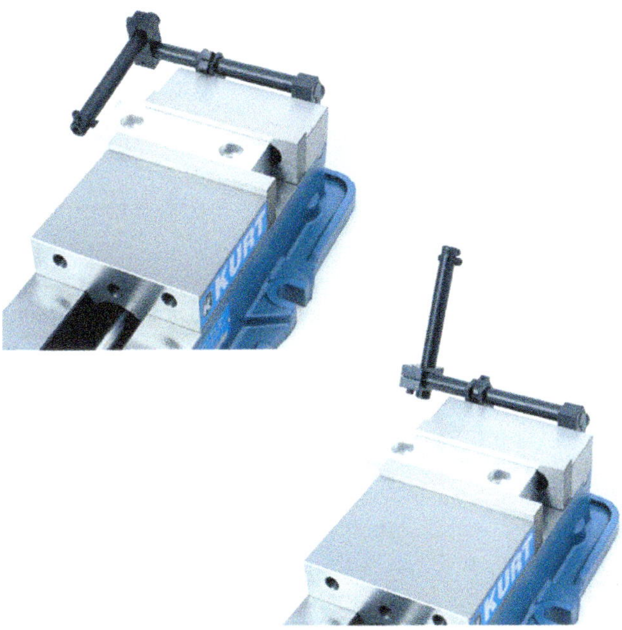

Figure 11-127 This vise is set up with swing-away end stops.
Courtesy Kurt Manufacturing/Industrial Products Division

Compound Rotate and Tilt Vises

The possibilities are limitless. Figure 11-128 shows just one of many. (I consider these kinds of challenges to be less like work and more like fun.)

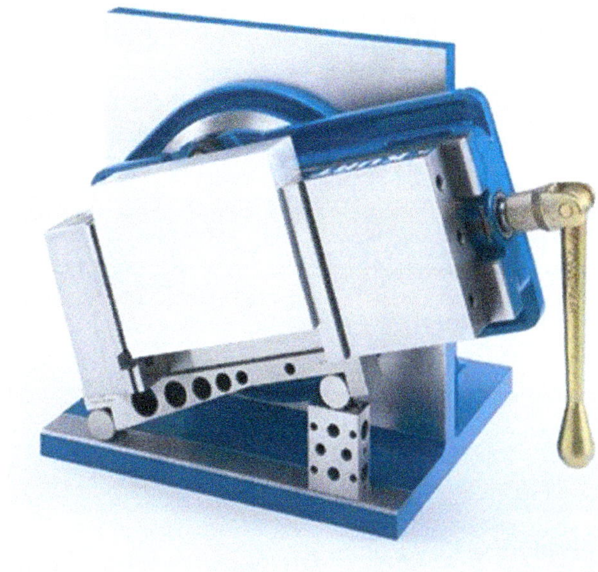

Figure 11-128 One example of the infinite possibilities.
Courtesy Kurt Manufacturing/Industrial Products Division

11.5.10 Cutter Question Answered—Why Longer Tool Life with 66 Percent Radial Engagement?

The answer divides into two parts: chip heat and cutting edge entry angle. To solve it, I need to ask another question: Study the two magnified views of tooth entry, and decide which is the better way to set up a face mill cut, positive or negative entry (Fig. 11-129). *Hint:* Remember, chip formation has two phases.

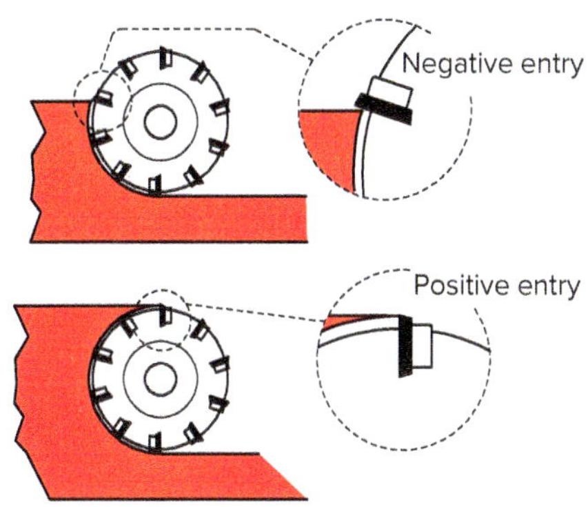

Figure 11-129 Compare entry angles of the teeth.

Next clue: The positive entry cutter starts the cut first; then, as the cutter feeds forward, the chip thickens and the rake begins. The negative entry cutter starts the rake before the cut.

Contrary to initial logic, negative tooth entry is better. Why? First off, it would seem that the face contacting the work before the cutting edge wouldn't cut, that this would break the tooth, especially if it's brittle carbide. But, in truth, there's little more pressure on the rake start-up than for any rake at any time in the cut (just a bit more until the plastic flow begins, but that's almost instantaneous). In fact, the recommended feed per tooth force is chosen such that the rake won't break the tooth. On the other hand, the positive entry cut contacts the delicate sharp edge first, which tends to chip the edge, and it wears until a sufficient rake builds up to move the boundary zone up the bit face. Heat builds up quickly, but the chip cannot conduct it away as it's ultrathin at the start of the cut and at the end.

KEYPOINT

Strange as it may seem at first, cutting tools last longer if a negative entry is used.

UNIT 11-5 REVIEW

Replay the Key Points

- Climb milling is a better process than conventional cutting, producing a finer finish and longer tool life, but it also produces a multiplication of forces that must be anticipated.
- During peripheral cutting, with a conventional cut, the edge rubs for a short distance before it builds enough pressure to dig in.
- Always indicate the *solid jaw* of the vise, never the moving jaw.
- Avoid flinging your indicator across the shop by shifting the spindle to neutral!
- Nearly all parts are loaded into a mill vise with datum feature A against the solid vise jaw.

Respond

Answers follow–check your understanding of milling.

1. In the square-up exercise in this unit, what two types of cutting were used?
2. When the peripheral cuts were being taken on the part ends (Fig. 11-122), which axis lock must be left snug during a climb cut but able to move?
3. Describe two advantages to climb milling compared to conventional. Explain.
4. True or false? Coordinating the milling machine (or lathe) requires positioning the axes to RZ, then testing the distance between work and cutter. If this statement is false, what would make it true?
5. Compute the feed rate for the following facts:

   Cutting aluminum with a two-flute, 0.750-in.-diameter HSS end mill

   Recommended FPT = 0.007 in.

   Recommended SS = 400 F/M

CHAPTER 11 REVIEW

Unit 11-1

The focus of this unit was to first identify and then to learn the nature of the basic milling operations. These are skills you'll probably be practicing at the manually operated milling machines. By doing so, you'll be safer and more efficient at learning on the CNC mills. But there is also a good chance you'll need to run manual mills during your first lab assignment or your first employment. That's because another theme in this unit was that milling rules the shop in the modern world.

Unit 11-2

We learned how the orthogonal axis system fits together and how these axes relate to all other machines–worldwide. We also explored the flexibility of work envelopes and that the ram and turret (Bridgeport) mills have an amazing range of setup configurations and capabilities that make them unique to the shop.

Unit 11-3

Setting up a mill means making lots of choices. One of the most basic is that of which cutter to use and how to mount it in the milling machine. The right cutter can make all the difference in quality, productivity, and tool life. Getting the RPM close to the optimum is another aspect of a good setup, but the point was made that there is a range of speeds that will do the job.

Unit 11-4

Why not learn from others' mistakes? It saves a lot of embarrassment. This unit presented a compilation of the most frequent problems we instructors see every semester. Raising your level of consciousness about them wasn't our main objective. Our goal was for you to know the solutions. It should start you on the way to being able to walk past a setup and know whether or not it's rock solid. No exaggeration, I can turn my back on the shop and know when things are going right or wrong on most of the machines, just by sound! I suspect most instructors would tell you the same.

Unit 11-5

This is one of my favorite subjects to teach. Every new part brings a set of puzzles to solve. How to hold the work? Which cuts to make first? What's the most efficient and safest way to plan the job? What tooling would be the best? Each unique part might be planned out and set up in at least 50 different ways—each offering some advantage and disadvantage. After pondering dozens more of the skills, the committee and I felt these were the essentials. They are the ones you need for the manual machines but transfer onward to the CNC.

QUESTIONS AND PROBLEMS

1. There are two types of plunging. Name and describe them. Which is the better entry process? (Unit 11-1)
2. The table tee slots are used for two purposes. Identify both and list the items that are used in the slots. (Units 11-2 and 11-5)
3. What vertical and horizontal milling machine component is held on the column dovetail? (Unit 11-2)
4. The feed rate for mills is expressed in inches per minute. Is this statement true or false? If it is false, what will make it true? (Units 11-2 and 11-5)
5. Tools, collets, and milling machine arbors are held in the spindle with what two devises working together? (Unit 11-3)
6. Although there are others used in industry, name the two common milling machine tapers shown in Fig. 11-130. (Unit 11-3)
7. Describe four necessary table clamping practices. Be complete. (Unit 11-4)
8. What is a normal feed rate range for a student machinist? (Unit 11-5)
9. When a pocket is milled within the bottom of another pocket, it is called a normal pocket. Is this statement true or false? If it is false, what will make it true? (Unit 11-1)
10. Briefly describe the advantages of *climb milling*. (Unit 11-5)
11. What is dangerous about climb milling? (Unit 11-5)
12. On a sheet of paper, draw a sketch of conventional and climb milling. Be sure to include the nature of the chip formation. (Unit 11-5)
13. When is conventional milling acceptable and when is it necessary? (Unit 11-5)
14. Compute the speed and feed rates for the following job. Use *Machinery's Handbook* for the feed rate and Appendix III for the RPM

    Cutting cast aluminum—as cast

    Using a $\frac{3}{4}$-in.-diameter end mill with four teeth

    Depth of cut is 0.250
15. Identify the tools and holders used on milling machines, as shown in Fig. 11-131. (Units 11-1 and 11-3)

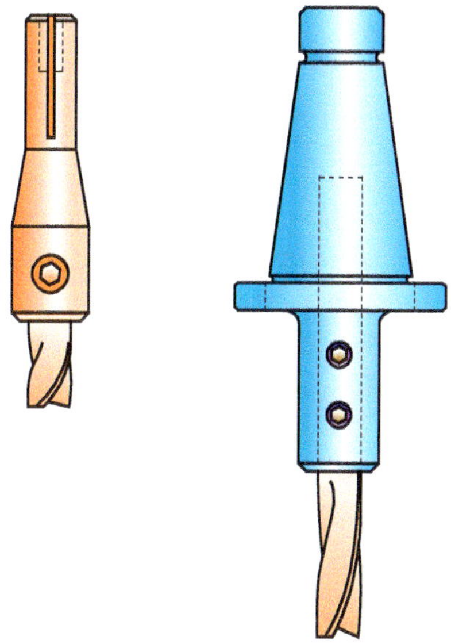

Figure 11-130 Identify these two mill spindle tapers.

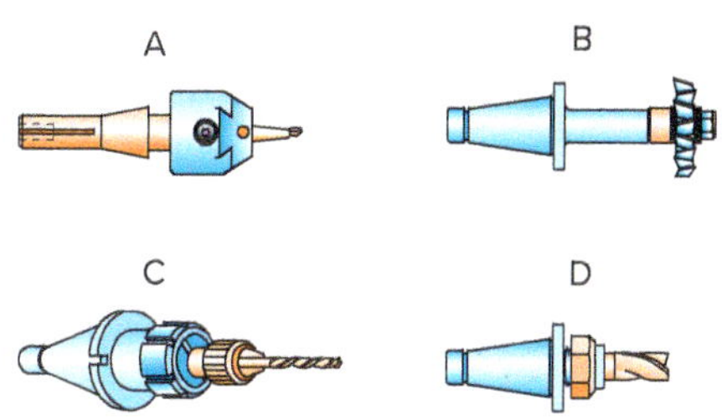

Figure 11-131 Identify these tools and holders.

CRITICAL THINKING

16. See Fig. 11-132. You need to machine eight $\frac{1}{4}$-in.-thick aluminum spacers to 4.000 in., removing 0.070 in. on one side only. You've discovered that they will fit in the mill vise either way (A) or (B). One way is safe, the other dangerous. Explain.
17. Describe how you could use an end mill to locate a spindle over the edge of an object. In doing so, what would be the major challenge to accuracy? How does an edge-finder overcome this problem?
18. How could you go about aligning a spindle over a hole that was not round? There are two ways. Describe each and the geometric result of each. This is a tough question.

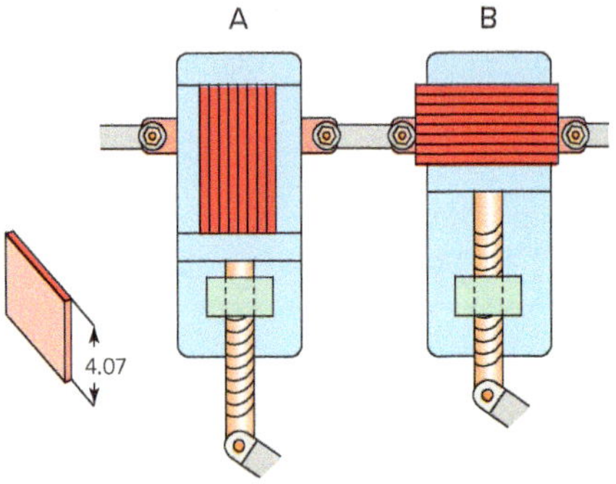

Figure 11-132 Which vise load is smart? Which is not?

CNC QUESTIONS

19. In Fig. 11-133, identify the *X*, *Y*, and *Z* axes.

20. How is it possible for a CNC mill to make a spiral ramp? Describe the axis moves.

Figure 11-133 Identify the three linear axes on this CNC vertical mill.
Courtesy Bridgeport Machines, Inc./Bridgeport, CT

CHAPTER 12
Precision Grinding Operations and Machines

Units

12-1 Selecting the Right Grinding Wheel for the Job

- Read industry standard grinding wheel labels
- Identify five different wheel characteristics and their effect on grinding wheel performance
    - Select the right general *abrasive* for a job
    - Select the right *bond* for the job
    - Select the best grinding wheel *structure*
    - Select the right *grit size* for the duty
    - Troubleshoot a poorly performing setup

12-2 Reciprocal Surface Grinding

- Set up and operate tool room type grinders
- Set up and operate power surface grinders
- Dress, balance, and mount grinding wheels
- Safely set up magnetic chucks
- Grind a plate flat and parallel

12-3 Setups That Work Right

- Make setups that safely hold work on a surface grinder
- Balancing, mounting, and dressing a grinding wheel on a surface grinder
- Setting and operating the surface grinder

12-4 Other Grinder Operations—Visual Inventory

- Do a visual inventory to recognize and describe industrial grinding machines
    - Form grinding
    - Tool and cutter grinding
    - Cylindrical grinding
    - Jig grinding
    - Rotary table grinding
    - Gear and thread grinding

INTRODUCTION

Precision grinding is an operation that usually follows turning, drilling, and milling. It improves surface finish and accuracy, plus it shapes metal that's hardened beyond the capability of standard cutters. Grinding produces finishes as smooth as 16 μin. with repeatable tolerances near 0.0001 in. (0.003 mm; Fig. 12-1).

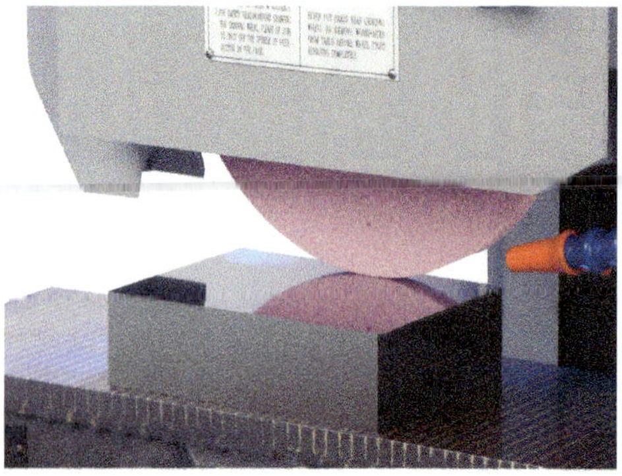

Figure 12-1 Mirror finishes require vibration-free spindles, the correct grinding wheel (Unit 12-1), correct balance and dress (Unit 12-3), and good setup parameters (Unit 12-4).
Courtesy Chevalier Machinery, Inc./Santa Fe Springs, CA

Removing metal with a grinding wheel seems different from cutting it away; it's even called an abrasive process. The Latin root *abrade* means "to scratch or wear away," not cut. The leftover *slurry* produced by grinding doesn't look anything like chips. It's a dust, or mud if coolant is used.

But examined closely, the waste particles reveal microscopic chips along with broken abrasive grains and bonding agent torn from the wheel. Before those grains broke away, they were actually tiny negative rake cutters, each with its sharp edge, making one chip per revolution as it whizzed by the work surface—but that also means external and internal heat. But in this case, the heat is so great that metals that oxidize ignite, resulting in sparks (Fig. 12-2).

The work heats, too. The solution is familiar: Add coolant and choose the right grinding wheel for the task. Stopping to resharpen the wheel before it's too dull is another heat control. So, in grinding work, coolants perform the same antifriction and heat conduction services they do elsewhere in machining. But with grinding, they also lubricate the chip and fling it out of the wheel.

Different from drilling, turning, and milling, grinding isn't performed on a single machine. Every machining process has its specialized machine that grinds. For example, cylindrical grinding machines take on some of the lathe's capabilities, producing internal and external round surfaces, tapers, and contoured shapes. But the manual versions can't cut threads, so the thread grinding machine assumes that task. On the other hand, CNC cylindrical grinders can cut threads. Jig grinding machines parallel the vertical mill or the jig borer's duties, precisely locating and finishing nearly perfect holes. They use small-diameter grinding wheels mounted on steel mandrels. Due to their small diameters, those wheels are driven by air turbines screaming up at an amazing 80,000 to 150,000 RPM and higher!

Figure 12-2 This spray of sparks is actually made up of microchips elevated above 2,500°F and reacting with room oxygen.
Courtesy Chevalier Machinery, Inc./ Santa Fe Springs, CA

In the past, grinders lagged behind in computerization, but today every grinding machine is supplied in a programmable form. Tool and cutter grinders and gear grinders are leading the race, but high-speed CNC cylindrical machines are now in service that rough cut in steps just like a lathe, dress their wheel, then make final cuts to shoulders, tapers, and contours. Their manual counterparts require time-consuming setups when going from one operation to another, but this is not so on the computerized versions (Fig. 12-3).

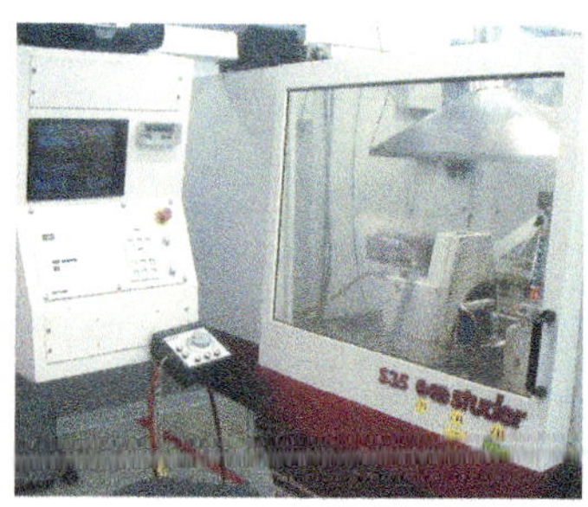

Figure 12-3 This third-generation cylindrical grinder is a true high-speed machine. Using advanced abrasives and CNC control, it can rough turn, face, and make finished contours as well as redress its own wheel, all at removal rates unheard of in the past.
Courtesy Smiths Aerospace-Actuation Systems/Yakima, WA

We're here not to learn how to run those machines but to understand how a grinding wheel removes metal, then to practice solid setups on a surface grinder.

Unit 12-1 Selecting the Right Grinding Wheel for the Job

Introduction: Just like changing cutting tools to achieve a different result, different grinding wheels return varying rates of removal, finishes, and tool life. Beyond the simple bench grinder wheels we've discussed previously, there are a lot of variations in the abrasives, the way they are held together, and how the abrasive grains and bonding agents are structured in the wheel.

TERMS TOOLBOX

Alumna The semirefined version of bauxite, aluminum ore, from which aluminum oxide abrasive is made or further refining produces the metal aluminum.

Bond (wheel) The material that glues the abrasive grains together.

Boron nitride (cubic boron nitride) A synthetic abrasive next to diamond in hardness. Its crystal structure looks cube shaped. It tends to stay sharp and to fracture well.

Friability The ability to shear away small flakes to self-refresh an abrasive's sharp edge before it shears out of the bond.

Glazing A grinding wheel condition where the grains have not self-dressed, yet they have no sharp edges left. The wheel must be dressed.

Grade (grinding wheel) A relative index of the hardness of the bond not the abrasive–the hardness of the wheel.

Grit size The number of screen squares within a square inch. The larger the number is, the finer the grain size.

Post (grinding wheel) The column of bond material linking individual grains.

Self-dress (self-refresh) New sharp edges are created by either the dull grain fracturing from pressure or the bond letting go.

Structure (grinding wheel) An index of the relative amount of porosity in a grinding wheel relative to grit and bond. Structure runs from 1 = dense to 16 = very open.

Truing (grinding wheel) Dressing the wheel to reshape it and to remove runout.

12.1.1 Adding CNC Control

Grinding wheel selection is an example of how much CNC has changed the way we plan and set up jobs. When working on manually operated equipment, wheel dressing robs production time to stop and redress the wheel, but we must do it for accuracy and efficiency. This is especially so when form grinding, where the wheel is shaped as a counterpart of the desired contour (Figs. 12-4 and 12-5). But depending on the process chosen, the wheel dress can take nearly as long as the grinding itself!

Figure 12-4 The wheel guard has been removed to show a typical form grind operation using a shaped wheel.

Courtesy Chevalier Machinery, Inc./Santa Fe Springs, CA

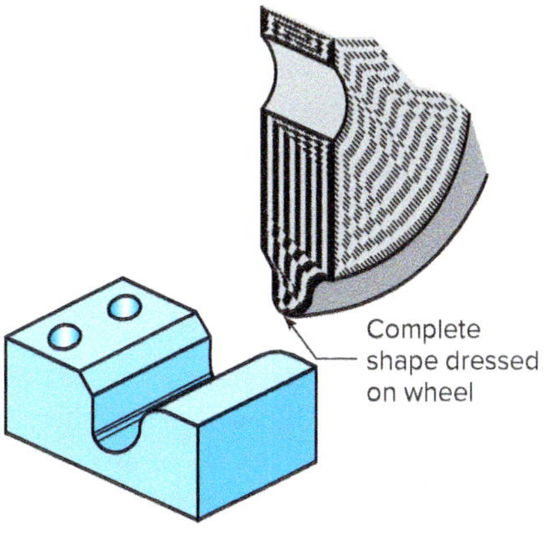

Figure 12-5 Manually dressed form grinding wheels must be the perfect counterpart to the desired shape.

So, for efficiency on the manually operated grinder, the wheel selected should hold its shape for the longest time. But in choosing a hard wheel for shape holding, we trade away the ability to remove larger amounts of metal–we'll see why in just a bit. That means the work has to be milled very close to the finished shape before grinding, due to the low removal rate of the finishing wheel, usually within 0.003 to 0.005 in. of the desired size. So, pregrind machining is a costly, time-consuming operation itself.

But CNC equipment dresses itself automatically and quickly. The dress time factor becomes far less critical (Fig. 12-6). A fast-cutting, softer wheel will work more efficiently without creating heat, so the extra cost of very close tolerance machining can be dropped, and more excess can be left on the work to grind away. And, even though the wheel will need dressing more often, it's not a problem. It can be scheduled into the program as often as needed.

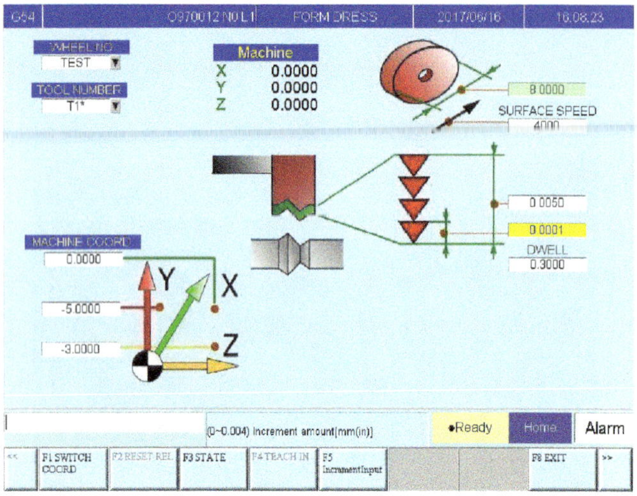

Figure 12-6 A CNC grinder dresses its own wheel faster and more accurately.

Courtesy Chevalier Machinery, Inc./Santa Fe Springs, CA

2-D Compensation for Wheel Shape

But beyond faster cycle times, there are more advantages gained with CNC control. First, we no longer need to dress the wheel into the exact shape. Using the CNC surface grinder shown in Fig. 12-7, the grinding wheel corners need to be radiused only with a known arc (Fig. 12-8). Then by keeping that radius tangent to the work surface, a complex shape can be ground. Also, that nonspecialized wheel can be used in many other ways–as is. In contrast, manual grinder shaped wheels must be completely reshaped to grind another job, or they must be put on a shelf to await a similar part!

3-D Compensation

The final ability brought by CPU-driven machines is to produce three-dimensional ground shapes. By controlling all three axes simultaneously, contours like that shown in Fig. 12-9 are possible. That's something we could not do previously.

Extreme Safety Planning–No Kidding! The theme throughout grinding must be based on the orbital speeds of the grinding wheels. They range from a bottom line of 4,000 feet per minute up to 12,000 feet per minute or more! That's bullet speed! So, while these speeds are right for the way a grinding wheel cuts metal, they also dictate the speed of an accident! Need proof? Go look at the left scatter shield on your lab's surface grinder–undoubtedly my case will be proved by chipped paint, craters, and dents!

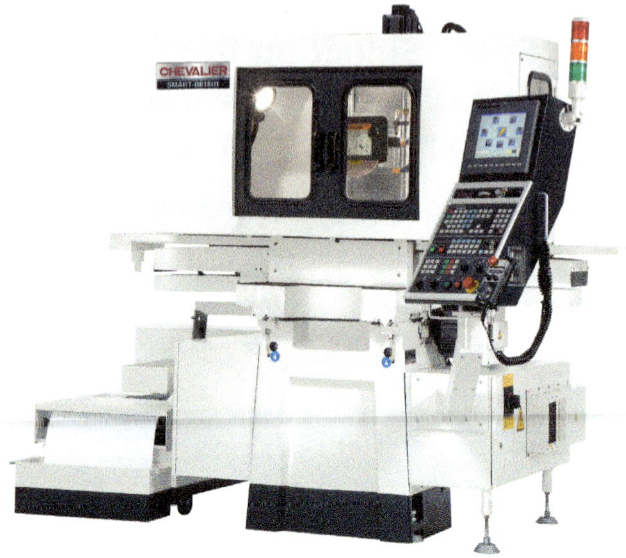

Figure 12-7 A CNC grinder can perform many tasks more efficiently and economically due to wheel selection options.

Courtesy Chevalier Machinery, Inc./Santa Fe Springs, CA

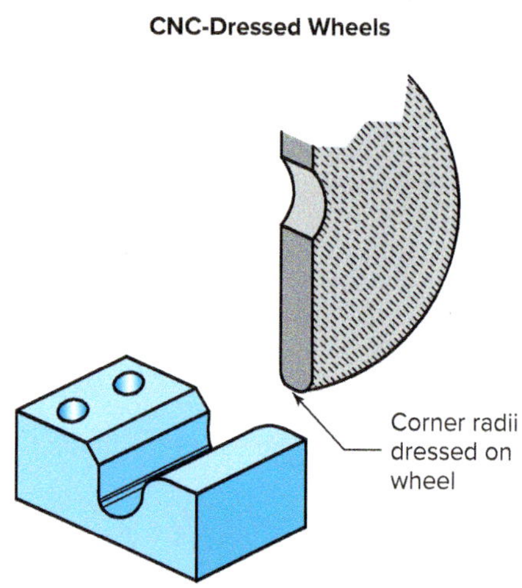

Figure 12-8 CNC grinders contour the work by keeping the known wheel radii tangent to the desired work shape.

KEYPOINT

Common grinding accidents occur more often due to poor setups that do not anticipate the dynamic forces and heat of grinding. Fewer are caused by incorrect operator actions at the wrong time.

Review We're going to learn a ton of new facts about abrasives and grinding wheels, but before we do, do you recall these facts?

Three-dimensional contours possible only with CNC

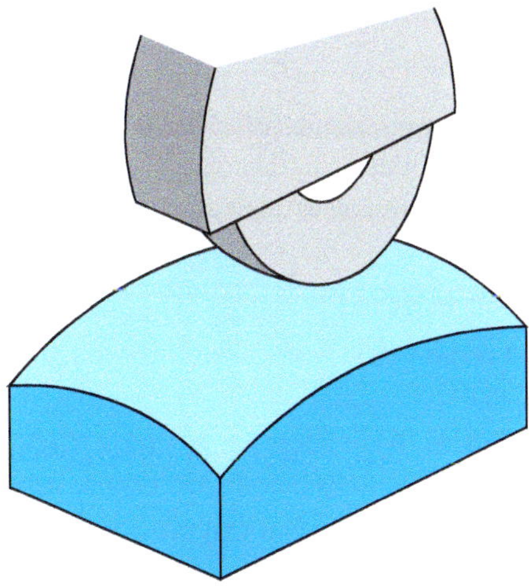

Figure 12-9 By controlling all three axes simultaneously, CNC grinders can create 3-D shapes.

- For normal day-to-day grinding, identify the two common types of abrasives.
- What are the two methods of bonding them together into grinding wheels?

Answer *Aluminum oxide* is the common, economical abrasive for everyday duties, and *silicon carbide* is used for harder work like grinding carbide cutting tools. You may have also experienced synthetic or natural diamond abrasives to grind extra hard materials or for finishing work finer than 32 μin.

The bonds are *vitrified,* clay fired to the hardness of ceramic, and *resinoid,* the low-heat process of gluing grains together with flexible plastic compounds and sometimes reinforcing them with fibers.

SHOPTALK

Like cutting tool selection, recommendations for applications of abrasives are a set of technical facts that are evolving quickly. Different sources may have different technical information.

For example, if you go online and find a recommendation chart supplied by a grinding wheel manufacturer and then compare that to recommendations in *Machinery's Handbook*, you will find differences between these two definitive sources. That is because the manufacturer's information is specific to the formulation of its product. *Machinery's Handbook's* information is more generalized.

12.1.2 Five Factors for Grinding Wheel Selection and Construction

Thus far, we've learned that wheels are changed after prolonged use and dressing that makes their diameter too small for efficient rim speeds or when they are possibly fractured from abuse. There's a lot of condensed technical information in this chapter–it's impossible to memorize, so for now scan it and know where to come back for reference when faced with a grinding wheel selection.

KEYPOINT

Due to the cost of maintaining an inventory of different grinding wheels, premounted and balanced on quick change hubs, a school budget may not allow many wheel change options, but in industry it's expected that the machinist includes the correct wheel selection during the setup of most precision grinders.

For example, one wheel is best for roughing, while another lasts longer for finish work, or one type is good for hard steel, but another is best for bronze or aluminum. One removes metal quickly, but a second creates a 16-μin. finish. Sound familiar? It's the trade-off game based on skillful use of knowledge and experience.

To assist in identifying the wheel's makeup, we use the industry standard ANSI code to tell us the wheel's makeup and purpose. It's on the paper gasket glued to the wheel, or it may be stenciled directly on the wheel. There will probably be six or seven columns, separated by dashes or spaces (Fig. 12-10). However, the first and last columns are manufacturer/product specific, so we'll investigate the five universal data in the order in which they are found. These are your control factors for all grinding wheels, no matter where they are used. There are a lot of technical specifications to understand.

1. Abrasive type
2. Grit size
3. Bond
4. Hardness structure
5. Bond type

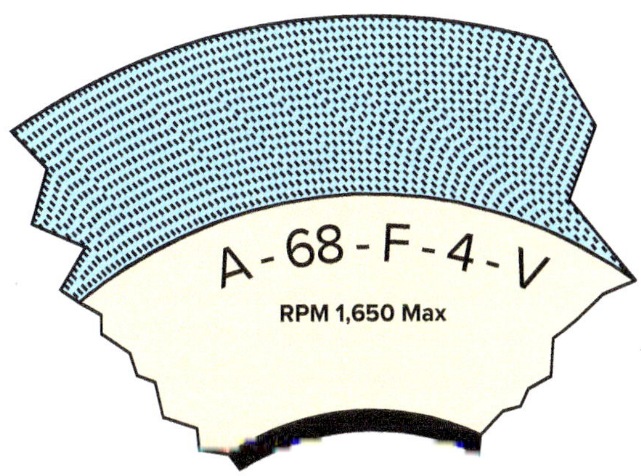

Figure 12-10 An identification number similar to this will be stenciled on the wheel or printed on its paper gasket.

KEYPOINT

The culmination of this unit is a cause-and-effect chart. It's there to begin grinding wheel selection skills.

Decision Factors for Wheel Choice

Choosing the right grinding wheel for the job will be based on

1. *Amount to remove*—Is it a lot (over 0.015 in.) or very little metal?
2. *Metal being ground and its hardness*
3. *Required finish*
4. *Batch size*
5. *Cycle times and tool life*

For each of the five wheel characteristics, I'll lay out a simplified baseline for correct application. See the Shoptalk.

On the practical side, unless your shop specializes in technical grinding, where wheel choices can be limitless, most shops don't stock a huge inventory of different kinds of grinding wheels. Your options will normally be two or three kinds of abrasives at the most, in several grits and structures. However, when those universal wheels don't deliver, then the following information must be clearly understood. Tech schools will probably have even fewer options due to budgets.

Abrasive Types—Column 1 (Fig. 12-11)

There are four basic kinds of abrasives in use today. The first three are produced by chemical reactions between two elements when elevated to high temperatures in an electric furnace:

1. **Aluminum oxide (A)**
2. **Silicon carbide (C)** (Also called carborundum, an early brand name.)
3. **Boron nitride (B)** or **cubic boron nitride (CBN)** (It is called cubic due to its box-shaped crystal structure.)
4. **Diamond (D)** Both natural and the synthetic.

SHOPTALK

Human-Made Diamonds Are a Machinist's Best Friend. They are created by superhigh compression of carbonaceous graphite in a high-heat environment. The synthetic version is five times stronger than the naturally occurring form due to its uniform microstructure.

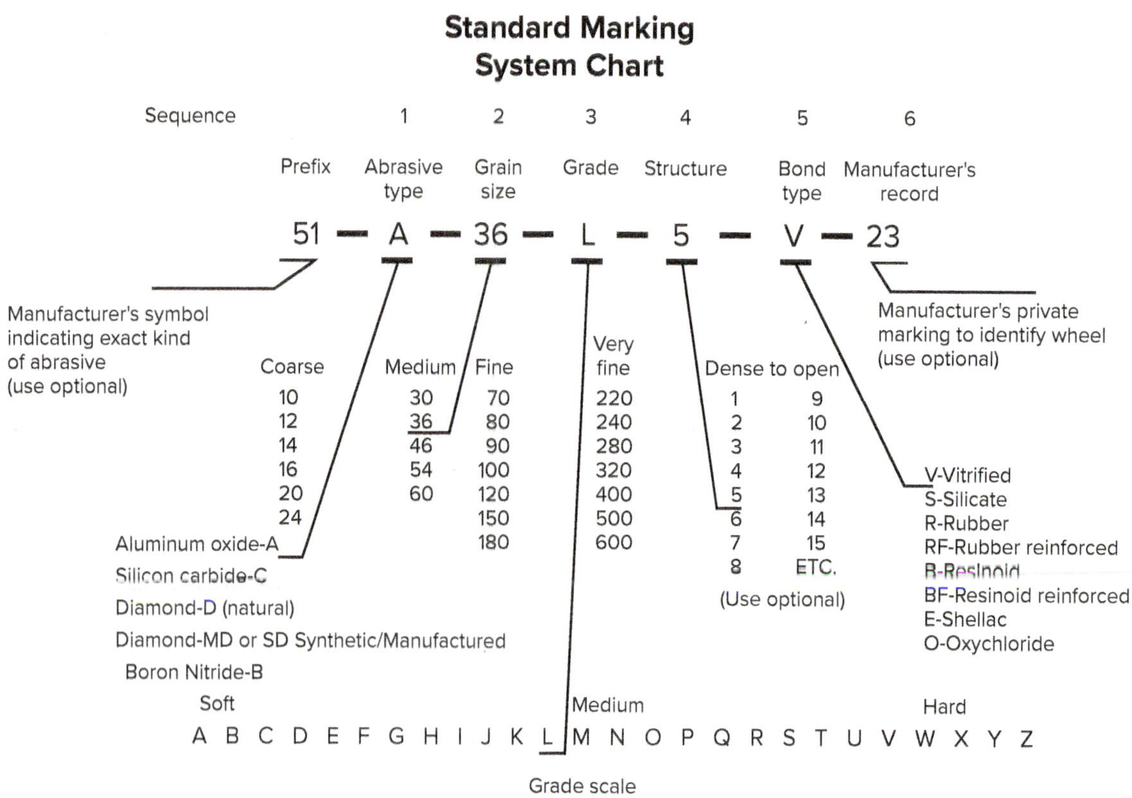

Figure 12-11 The ANSI numbering system for grinding wheel selection features seven columns of information.

Differences in hardness and crystal structure (which contributes to its ability to fracture away in small flakes–a desirable quality), overall grain strength, and resistance to edge dulling make each abrasive type best suited for a particular application.

KEYPOINT

Similar to cutting tool differences, the main abrasive property is hardness, but with extra abrasive hardness, purchase price also generally goes up. Machinists soon learn when grinding difficult materials that the least expensive may or may not be the economical choice for the job.

Superabrasives

Most advanced abrasives designed for specialized tasks are subcategories of these four. But there are also combinations of two or more along with ceramic/metallic compounds. For example, if we add an oxide of the metal zirconium to the aluminum oxide grain mix, the wheel gains efficiency in cut speed and grain durability. The ratio of zirconium oxide to aluminum oxide determines specific applications, but in general all combinations are faster cutting than plain aluminum oxide abrasives. These abrasives are beyond the scope of this unit. Learning to apply the four will provide a good working knowledge.

Abrasive Friability

One valuable abrasive property is its ability to stay sharp and to refresh its own sharpness between dressings. An abrasive that has good **friability** allows it to shear away tiny flakes, many times before the whole grain breaks completely out of the wheel. For example, the more modern versions of boron nitride produce flakes measured in millionths of an inch! This is known as the **self-dress** ***or*** **self-refresh** friability of the grain (Fig. 12-12). The wheel outlasts other abrasives several times over. But its life also depends on its hardness and how long each sharp edge lasts. We'll start with the softest abrasive and work toward diamond.

SHOPTALK

In reading technical articles and bulletins about abrasives, note that most generally compare their new product to aluminum oxide's properties. That's a sure sign that it's the standard. Aluminum oxide occurs in nature as the second-hardest natural mineral after diamond. It forms where aluminum deposits have been subjected to heat, but silicon is rare in the earth because the silicon bonds more readily to the aluminum.

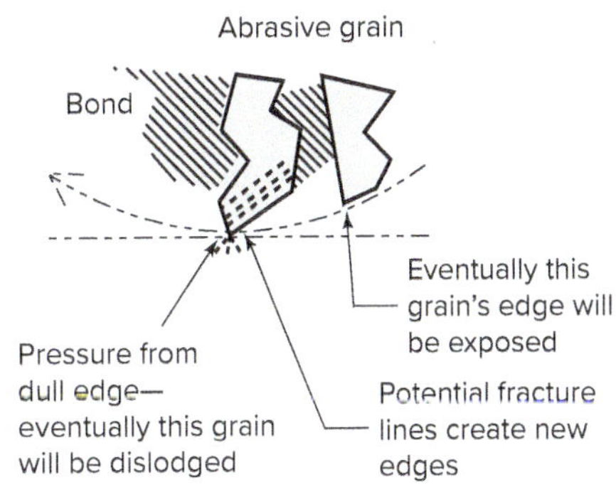

Figure 12-12 As one grain dulls it either fractures to create a new edge or the bond shears to expose another grain behind.

Aluminum Oxide

The workhorse of abrasive grains, aluminum oxide is made by combining **alumna** (semirefined aluminum ore) with oxygen at very high temperatures. It's the universal choice, providing good economical results over the widest range of work materials. Aluminum oxide stays sharp when grinding high-tensile alloy and tool steels. The chances are most of the grinding wheels in your lab are aluminum oxide.

Purity and Hardness. There is a range of hardness **grades** for aluminum oxide, depending on the purity of the process by which it's made. Purity relates to hardness. As the grain's hardness goes up, its edge lasts longer, but overall toughness goes down.

Dark gray 94 percent pure This is the tougher version. But its sharp edge doesn't last as long, especially when grinding hard metals and cast iron. While it's the abrasive of choice for pedestal grinder wheels due to its toughness, this abrasive is not used on precision grinders. It is shown in the center of Fig. 12-13 (yellow label).

Figure 12-13 A variety of abrasive wheels used in manufacturing. Not all are for precision grinding.
Saint-Gobain Industrial Ceramics, Inc.

Correct Application

- Good results when grinding plain and medium hard alloy steels.
- Not recommended for cast iron because aided by the wheel's impurities, the aluminum bonds to the iron, and the wheel dulls quickly.
- Can grind some nonferrous metals—not the best choice. However, when doing so, wax or oil helps chips release from the wheel when it has cleared the cut.

Light gray 94 to 97 percent purity. See the top yellow and orange label in Fig. 12-13. This is the harder version but is more costly. This popular abrasive represents an estimated 50 percent of all wheels on surface grinders and cylindrical grinding machines.

Correct Application

- Good for soft through hard alloy steels up to fully hardened tool steel.
- Acceptable for most other metals—will grind cast iron but not the best choice.

White (or stained) 97 percent. See Fig. 12-13 (upper right corner). The purest form of aluminum oxide is bright white, but it's often stained a bright color to distinguish it from lower-cost grinding wheels.

Correct Application

- Good results on extra hard tool and alloy steels.
- Can grind softer grades of carbide but not the best choice.
- Best for general tool and cutter grinding.

Silicon Carbide

This material is made by combining common silica sand with coke (refined carbonaceous coal) in furnaces purged of oxygen. The result is an abrasive that's about 33 to 50 percent harder than aluminum oxide. Silicon carbide's hardness combined with a tough but friable crystal structure means that it creates sharper cutting edges than aluminum oxide; thus, it cuts faster. But the crystals tend to shear in bigger flakes, thus grains shear completely from the wheel sooner.

Observed closely, grains of silicon carbide appear as a deep-colored background reflecting rainbow colors. There are varying hues of this abrasive running from black through green—each with slightly different properties. You've seen silicon carbide on high-quality wet or dry sandpaper and cloth–the deep black abrasive paper with a sheen.

Correct Application

Black and Gray SC (Fig. 12-13—blue/white label)

- Excellent for cast iron.
- Good over a wide range of steels, but the expense isn't usually justified over aluminum oxide.
- Good for grinding aluminum, copper brass, and bronzes.
- Acceptable results grinding some stainless steels and titanium.
- Acceptable results on tungsten carbide tools.

Green SC (Not shown)

- Preferred for grinding tungsten carbide cutting tools.
- See the Trade Tip.

TRADE TIP

Green Wheels Augmenting the greenish hue of the grains, the clay binder in silicon carbide wheels, made just for hand bench grinding of carbide tools, is stained green to note that it is reserved for that purpose only.

KEYPOINT

Green wheels are usually reserved for *nothing but carbide tools.* They are costly by comparison to aluminum oxide bench wheels, and due to the abrasive and binder's properties, they wear away too fast for everyday use.

Boron Nitride

Boron, a mineral somewhat like silicon, is combined with nitrogen to form an abrasive that's twice the hardness of aluminum oxide and the closest synthetic abrasive to diamond hardness. The process by which it's created is costly; however, boron nitride's edge-holding ability cuts extremely well compared to the two previous choices. Since it's best used for extra tough grinding jobs, the cost is often balanced against efficiency.

KEYPOINT

While it's a superior abrasive, in general, due to cost, we do not use boron nitride unless aluminum oxide or silicon carbide cannot do the job. However, its extreme hardness and fast yet cooler cutting action, along with less dressing of the wheel, sometimes combine to make it the best choice on a tough job.

Correct Application

- Best for form grinding of extremely hard metals.
- Excellent on very hard metals and other materials such as glass.
- Performs well on carbide tools but expensive if SC can do the job.
- Excellent on high nickel/cobalt superalloys.
- Excellent for long abrasive life and high-demand applications such as CNC high-speed grinding.

Diamond Abrasive (D) Natural (MD) Manufactured (SD) Synthetic (Fig. 12-14)

Now we're talking expensive! However, there are some tasks that only diamond can do. Its greatest advantage beyond hardness is that diamond cutting edges tend to remain sharp longer than all other abrasives; thus, it cuts with a great deal less heat. That property alone makes it the right choice for fine tool and cutter grinding where we cannot risk overheating thin cutting edges.

Figure 12-14 A diamond wheel is ideal for finishing these slots.
Courtesy Chevalier Machinery, Inc./Santa Fe Springs, CA

Correct Application

- Excellent, fine finish on carbide tools.
- The only choice for grinding exceptionally hard materials such as ceramic or glass.
- Excellent for hardest grades of tungsten steel tools.
- Best where heat-sensitive work may harden—diamond cuts with less friction and heat.
- Excellent for fine finishing.

Grit (Grain) Size—Column 2

Grit size is a simple issue but an important one for success. In general, the finer the abrasive, the finer the finish it will produce. But with small abrasive size, removal speed goes down and heat becomes a greater issue as the tiny cutting edges become dull. But there's a unique solution to these challenges found in the way the grains are bonded together–called the wheel **structure**. We'll look at that after bonds.

For now, let's get an initial feel for grit sizes. You've already seen grits specified as coarse (18–24 grit), medium (26–46 grit), and fine (46–80 grit), when selecting pedestal grinding wheels. That means that after crushing the large abrasive crystals down to small grains, they are sifted through a screen with the specified *number of spaces per square inch* (Fig. 12-15).

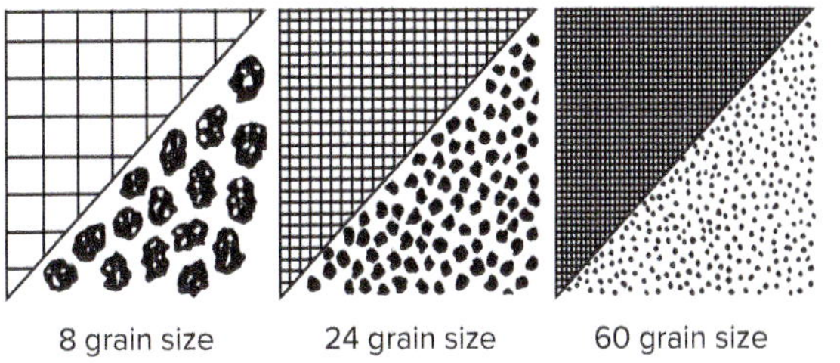

Figure 12-15 Grain size is based on the number of squares per inch, through which a single grain can pass.

Correct Application

For various grinding machines (surface, cylindrical, and tool grind), different grit sizes are typical for rough and finish work.

	Roughing	Finish	Form Grinding/Fine Finishes
Surface grinding	30–36	36–46	60 (very light removal)
Cylindrical	36–46	46–60	80 (heat can be a problem)
Tool and cutter	46–54	54–120	120+ (rarely required)

KEYPOINT

Note in the application chart that the three basic grinding machines are listed in an order. Each uses a slightly finer grain for its general duties. (There are some exceptions to this guideline.)

Grade—Bond Hardness—Column 3

In selecting the right wheel for the job, it's important to consider wheel hardness expressed as a letter grade, with "*A*" being the softest grade. Wheel hardness is a combination of the ratio of **bond** relative to abrasive grains and the air spaces between them and the hardness of bonding agent itself. The bonding material holding the grains in the wheel forms links between grains called **posts** (Fig. 12-16).

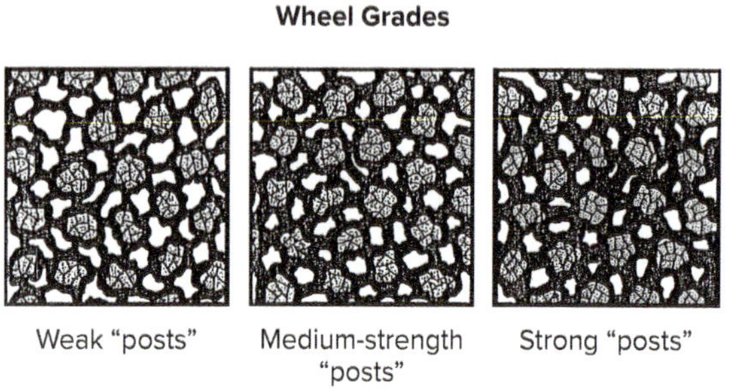

Figure 12-16 Wheel grade (hardness) is directly related to the wheel bond.

Varying the amount and hardness of the posts causes grinding wheels to perform differently, even though the abrasive grain hardness is kept a constant. More bonding material tends to hold the grains longer and stronger–a harder wheel. Sometimes a harder wheel is just right, when form grinding, for example, but it's not always a good thing. When grinding harder work materials, for example, grain edges dull more quickly. In these situations, it's best if the grains release sooner. If not, a condition known as **glazing** can occur.

A glazed wheel is one that has lost its cutting ability. It has been used long beyond the point where it should have been dressed. When this is done, the heat and friction create a glasslike surface that rubs and does not abrade away material. A glazed wheel will reflect light and is easily recognized. If not dressed to clear the problem, it burns the material rather than cutting it. Wheels glaze quickly if they are poorly chosen for grinding hard alloys.

KEYPOINT

Wheel glazing can be blamed on (1) a wheel grade that's too hard, (2) an abrasive that's too soft, (3) using it beyond the point where it should have been dressed (most likely), and (4) choosing grinder feed rates too low to cause the self-dress (not a common reason).

SHOPTALK

Dressing and Truing a Grinding Wheel, What's the Difference? Using a diamond or other wheel dressing tool to restore wheel shape and zero runout is called **truing** the wheel. Resharpening it to cut better is called dressing—in truth they are one and the same action. (More in Unit 12-4.) See *balancing* the wheel, coming up.

Self-Dress Property

Working together, the bond hardness (grade) and the abrasive's ability to fracture continuously refresh the wheel's ability to cut cleanly, between dressings, but only up to a point where it must be redressed to restore wheel surface runout and shape due to the random release of grains. The time between dressing the wheel is a function of several factors:

- Work material type and hardness
- Amount of material to be removed
- Type of grinding operation
- Coolant application
- Wheel grade
- Abrasive type

Knowing when to stop grinding and dress the wheel is an essential skill learned from experience.

Grinding wheel self-refreshing action is both good and bad. While it renews the cutting action for a limited time, it also means the wheel is eroding away. So, how much self-dress is right for the job? How hard should the wheel bond be?

KEYPOINT

When grinding a *hard metal,* use a *soft wheel,* and when grinding a *soft metal,* use a *harder wheel.*

In general, softer bonds are needed for harder material to cause more self-dressing, but abrasive hardness also affects this choice. Softer materials grind more economically with a harder bonded wheel.

Grades from A (Soft) to Z

Bond hardness varies from very soft grade A through very hard grade Z. Within a span of perhaps two or three grades, wheels deliver nearly the same result. For example, a wheel of F bond hardness will perform similarly to an H bond.

Correct Application—Bond Hardness

Bond Grade	Application
A through F	Carbide and extremely hard tool steel, glass Large contact area processes Rapid material removal
G through P	General hardened alloy steels (85 percent of all jobs)
Q through S	Nonferrous materials—CI with crust removed Small, narrow contact areas where wheel is fragile High-force-demand applications
T through Z	Nonferrous, very soft materials, extreme RPM

Harder bond wheels are also chosen for final grinding operations, where very little metal is to be removed or for form grinding where the wheel is dressed into a special contour shape. Here, it is desirable that the wheel hold that shape longer so the self-dress is less important–assuming, of course, that it's a manually dressed machine.

KEYPOINT

Wheel hardness is a combination of bond material and amount of bond within the ratio—it has nothing to do with abrasive hardness.

Wheel Structure—Column 4

Open space between grains also has a lot to do with wheel performance. A wheel is truly composed of abrasive grains, bond material, and *empty space.* Figure 12-16 also shows the effect of wheel structure. Space is necessary in every wheel for three reasons:

1. **Provide a Temporary Place for the Chip to Form**

   When the grinding wheel is in contact with the work, the forming chip must have someplace to stay until the wheel clears the work, to eject it.

2. **Create Coolant Induction Channels**

   Properly directed at the wheel-work interface, coolant is injected into the air space just before it contacts the work. The coolant is trapped in the air space as the wheel rotates onto the work. Thus, it is carried into the action area right where the chip is forming. Then on leaving the cut, centrifugal force ejects the coolant and chip.

KEYPOINT

That's a new coolant duty—it sprays out of the wheel as the cut finishes and tends to carry the chip out, too.

3. **Cools the Wheel**

   During noncontact while the wheel rolls over for another cut, air swirls in the space to cool individual grains. This is an especially important property when coolant is not possible and room air carried in and out of these spaces becomes the wheel's coolant. So, a dry setup might warrant a more open structure wheel compared to one where coolant will be injected.

KEYPOINT

In general, a wheel with more space is needed for heavy removal grinding while less space is required for finish work.

Structures from 1, Dense, Though 16, Open (More Space in the Ratio)

Within a range of perhaps two or three units, wheels perform similarly.

Correct Applications of Structure

1 through 5	Regrinding cutters, form grinding, removing less than 0.010 in. overall material from the workpiece
5 through 10	General grinding operations, removing from 0.010 to 0.020 in. overall
10 through 15	Rough grinding, or where heat is a problem, removing above 0.020 in.

Porous Wheels

The overall ratio of spaces compared to solids (grains and bond) can be created by larger individual air space holes or by multiple smaller holes between the grains. If a wheel is deliberately created with large, singular pores, its designation will end in a letter P. That's a wheel expressly made for high-volume, rough grinding.

A-48-G-12-V-P

TRADE TIP

Missing Structure Designation? Normally, when selecting a given grit size, the structure will be automatically correct. That's because most wheel manufacturers make wheels in the most commonly required ratios. Some wheels do not even feature structure variations—they are supplied in one structure only. In these cases, the structure number will be missing in the designation.

A-48-G-V

Bond Types—Column 5

There are three basic bonds used in conventional precision grinding wheels, with variations within each:

Vitrified Resinoid Metallic

We've previously discussed the first two, so that leaves the metallic nickel-copper-based alloys used to embed grains of diamond, CBN, or other expensive crystal grains onto tool steel disks, as shown on the diamond wheel in Fig. 12-14.

They are precisely assembled such that approximately one third of the grain's height is above the bonding agent and all are equally spaced, creating a highly aggressive and cool cutting wheel (Fig. 12-17). These wheels are made by electrodeposition (depositing metal with electricity) of the bond metal around the grains.

A Metallic Bonded Face Grinding Wheel

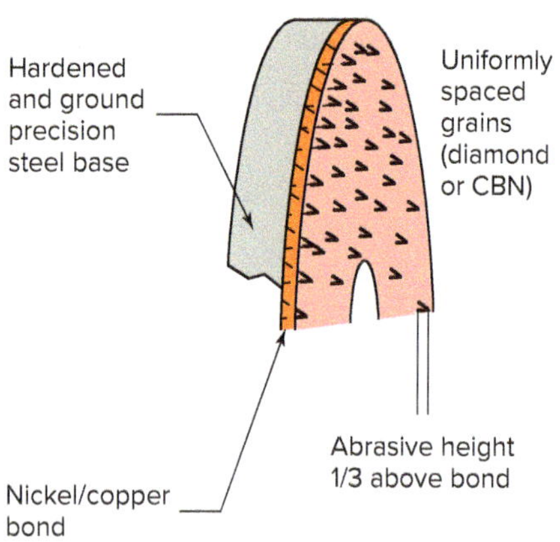

Figure 12-17 A metallic bonded, diamond, or CBN face wheel cuts very quickly and creates less heat than other types.

Correct Application of Bonds

Vitrified

- Used where coolants or oils are used—glasslike material is unaffected by water, oil, and most other chemicals.
- For general-purpose grinding where exceptional mechanical shock is not going to occur.
- Can withstand exceptional heat.
- Used where rim speeds do not exceed 6,500 F/M.

Resinoid

- Used where high-speed CNC grinding demands rim speeds above 12,000 F/M.*
- Used where mechanical shock may be a factor.*
- Not recommended for use around certain solvents and acids.

Metallic Bonds

- For extreme long life where dressing is impractical.
- To hold expensive grains firmly for a long time.
- For very high-speed grinding.
- For certain electrical and chemical applications where the wheel must conduct electricity.

*Requires fiber reinforcement for this duty.

12.1.3 Control Factors

The following is a compilation of basic problems associated with grinding, the target goal for improving, and the action needed to do so. But as usual, something gained often means something lost–most adjustments other than adding coolant are trades in time or cost or some other downside element. For example, changing from aluminum oxide to silicon carbide and then to boron nitride results in improved production but costs rise.

Memorizing these facts is not the objective. Understanding why each works is your goal. Note that when a sharper abrasive is indicated, it means a harder abrasive that holds its edge better.

Symptom	Target	Action	Trade-Off
Slow cutting	Increase production	Coarser grain	Rougher finish
		Harder abrasive	Wheel cost
		Softer bond	Increases wheel wear
		More open structure	Increases wheel wear
Dressing too often	Increase production Less downtime	Harder bond	Glazing/loading
		Harder abrasive	Expensive wheel
		Increase cutting	More force against work
		Pressure feed	Self-dresses better but requires strong setup
		Add coolant	None

Poor finish	Improve quality	Finer grain Dress more often Harder bond Rebalance wheel Slow cutting rate Sharper abrasive	Heat buildup Production time Possible glazing Downtime Extra time Cost
Parts overheating	Protect workpiece	Add coolant Dress more often More open structure Coarser grain Sharper abrasive Focus coolant	None Wheel wear Coarser finish Coarser finish Cost Overspray
Wheel glazing	Increase dress life	Coarser grain More open structure Sharper abrasive Add coolant Dress more often	Coarser finish Faster wheel wear Cost None Wheel wear and time
Early wheel breakdown	Increase dress life	Harder grade Sharper abrasive Less feed pressure Denser wheel	Potential glazing Cost Longer cycle time Glazing/heat
Wheel loading	Increase dress life and production	Inject coolant More open structure Dress more often Spray mist if no coolant Softer bond Sharper abrasive Take lighter cuts	No downside Fast wheel wear Wheel life shorter Breathing protection Breaks down faster Generally costlier Slower production

UNIT 12-1 REVIEW

Replay the Key Points

- Predictable accidents occur due to incorrect setups that do not anticipate the dynamic forces of grinding. Some happen from incorrect operator actions.
- Grinding wheels are exchanged to suit the job.
- Green, silicon carbide wheels are usually reserved for nothing but carbide tools.
- In general, due to cost, we use the softest abrasive that will do the job.
- Hard metal = Soft wheel and soft metal = Hard wheel.
- Wheel hardness is a combination of bond material and amount of bond within the ratio–it has nothing to do with abrasive hardness.
- A wheel with more space is needed for heavy removal grinding while less space is required for finish work.

Respond

1. Interpret this grinding wheel.

   A-46-H-7-V

   Describe what each data field means and what would be a typical application for this wheel. You may want to refer to *Machinery's Handbook* for this question.
2. List three reasons a grinding wheel might not cut the work material and a possible solution to each.
3. In general terms what might this grinding wheel be best suited to do?

   C-80-R-3-V
4. True or false? It is more economical to use the softest grade of aluminum oxide wheel that will do the job because as the hardness of the grains increases, so does the cost. If this statement is false, what will make it true?
5. On a piece of paper, fill in your selections for a roughing wheel cutting mild steel. You needn't be exact–provide a range for the characteristic needed.

SHOPTALK

Nuts! When a grinding wheel is produced, the powdered clay bond and abrasive grains are pressed into a mold, then removed. At that stage, it's easily broken with a light tap. That unfired wheel is furnace heated to vitrify (melt the bond) into a porcelain-like substance. On cooling, a precision lead or plastic hub is pressed in the center hole. Finally, after a factory dress, the wheel is carefully packaged and shipped to the customer. But, the question is, Given that the raw material was smashed into a mold under great pressure, how does the wheel end up with air spaces?

Interestingly, the original mixture starts out with three ingredients: abrasive, bond, and an organic filler. Among other things the filler can be made of specially ground and sized walnut shells! When the wheel is fired to create the bond, the filler turns to vapor and a tiny bit of ash; it disappears, leaving spaces in the wheel.

Unit 12-2 Reciprocal Surface Grinding

Introduction: In a lab or on the job, your first experience with precision grinding will probably be a small surface grinder. They create precise flat surfaces with finishes smoother than 16 μin., finishing parallel bars, for example. They're an indispensable machine for precision tooling work and many other tasks supporting production. With just a bit of setup and operation training, very rewarding results can be yours with these little gems.

Safety is a big item even on these machines. Though they are simple to understand and operate, there are several specific precautions because the work is often held by a magnet. While cylindrical or tool and cutter grinders may or may not be included in your first-year experience, it's certain that surface grinders and magnetic chucks will be. For that reason, they are the central focus of Unit 12-2 and Unit 12-3.

TERMS TOOLBOX

Electromagnetic chuck Grip can be switched on or off and residual magnetism can be reduced to zero.

Hydrostatic bearing A predictably thick oil film that separates sliding components.

Permanent magnetic chuck Sets of magnets either oppose or agree with each other inside the chuck. A lever rotates one set relative to the second set, which either cancels the pull or causes it.

Pick feed (grinder down feed) An adjustable ratchet feed that cranks the Y axis down at the end of each X stroke of a surface or cylindrical grinder.

Reciprocating A back-and-forth action.

12.2.1 Reciprocal Grinders

There are two different types of surface grinding processes used in industry. The first, **reciprocating** (back and forth), uses the rim of the wheel and moves the work in a straight line under it. By far, these are the more common grinders due to their simple construction.

Rotary surface grinders, the other type, use a spinning chuck whose face is horizontal to hold and move the work past the grinding wheel. The grinding wheel cuts with its face rather than its rim—similar to a face mill. In general, these high removal machines are a specialty process not found in every shop and rarely found in schools—we'll cover them in Unit 12-4.

Manual or Tool Room Surface Grinders

See Figs. 12-18 and 12-19. These machines grind by moving the work table back and forth with hand motion only. They also feature manually actuated motions of the wheel head Z axis and Y axis saddle, within a work envelope of approximately 18 to 24 in. long by 8 in. wide by 10 to 18 in. tall. As seen in the lower right corner of Fig. 12-18, most small grinders feature a fine movement for the Y axis—to enable removing a very small amount of material (#10 in the figure).

To impart the nearly friction-free action, the X axis glides on roller bearings. Differing from milling machines, the table is not held down with a dovetail. It sits atop the frame, suspended by antifriction rollers. The rollers are held together in sets called carriers. Note, the rollers are sealed from dirt and slurry, but the protective covers are removed in the drawing to illustrate concept only.

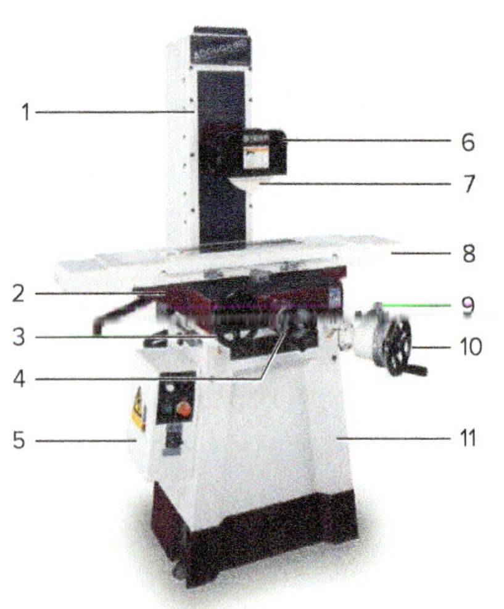

Figure 12-18 A small surface grinder has three movements, X, Y, and Z.

Courtesy Chevalier Machinery, Inc./Santa Fe Springs, CA

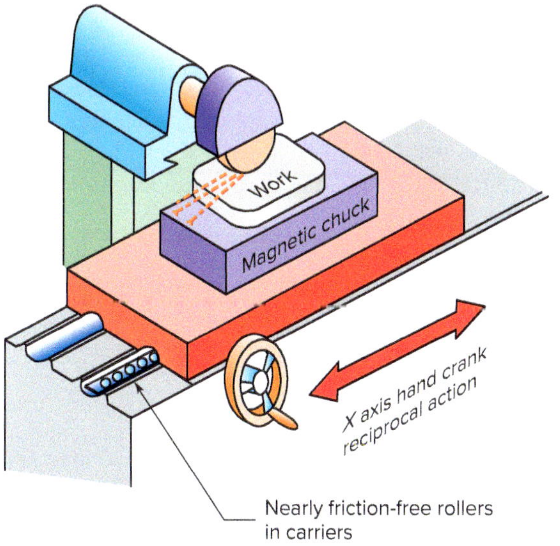

Figure 12-19 The X axis rests and rolls on nonlubricated, precision-matched sets of ball bearings.

As such, these machines are designed for light loads only. The table is usually not attached to the ways; it can be lifted straight up for cleaning and maintenance. The table X axis is moved left and right by a gear on a rack, using hand pressure only (Fig. 12-20).

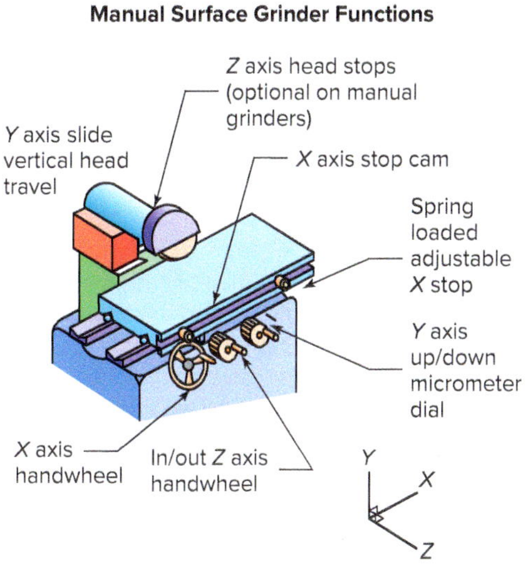

Figure 12-20 The surface grinder functions and setup options.

Two more hand-operated, dovetail slide motions move the wheel head up and down, the Y axis that controls work thickness, and the Z axis saddle in/out relative to the operator, stepping the grinding wheel sideways across the work.

KEYPOINT

Breathing Protection When Grinding The air around most grinding machines can be polluted with fine metallic and wheel dust particles and fine mist coolant flung from the grinding wheel (Fig. 12-21). This is especially true when dressing the wheel.

Figure 12-21 Always wear eye protection and a dust mask when dressing the wheel or running a grinder without coolant.
Prographics at Milwaukee Area Technical College/McGraw Hill

TRADE TIP

Learning to Use a Hand-Operated Machine Operating one of the small machines is like patting your head while rubbing your stomach with the other hand. The coordinated movements require practice to get them smooth and consistent. One must move the table back and forth with the left hand, while moving the Z axis in/out, at the end of each X stroke with the right. Then as the wheel comes to half its width off the work, the Y axis must be cranked down for another pass across the work.

Here's how. First, using your left hand, move the X axis back and forth against the spring-loaded stops. With no other movements, just repeat that motion for a while. Next, your right hand cranks the Z axis in and out, moving the grinding wheel sideways half its width each time, at the end of each X stroke.

Moving the wheel only when it is off the part at the end produces straight grinding marks. Once this two-axis motion is mastered, using your right hand, crank the Y axis down no more than 0.001 in. only at the inner or outer edge of the work. Do not crank down at the same edge each time. Alternating the down feed equalizes the wear pattern on the wheel and grinds the work flatter.

Safety Note

The predictable operator-caused accident results from confusing the Z hand crank with the Y axis "down." Both are usually actuated with your right hand. The accident occurs by forgetting which handwheel you are grasping! This can cause a big "down" when "across" was the intention. The unhappy result is a burned or undersize workpiece or a stalled wheel against the work, possibly breaking the wheel or gouging and dislodging the work.

When first learning, try placing a piece of tape or tying a short strip of rag to the Y axis handwheel. It's there to remind you that this is the down handle, just until you get comfortable with this three-movement operation. In Unit 12-3, we'll learn how to set up these machines.

Hydraulic Production Grinders

The big brother of the reciprocal surface grinders, these machines have work envelopes of from around 24 in. in X to over 20 ft or even more. For example, in Fig. 12-22, the table length is 6 m (nearly 20 ft).

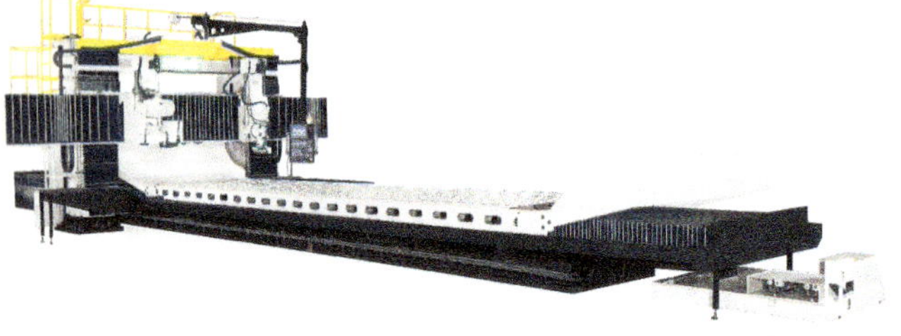

Figure 12-22 Larger surface grinders such as this and the one shown in Fig. 12-23 feature power-driven *X*, *Y*, and *Z* axes.
Courtesy Chevalier Machinery, Inc./Santa Fe Springs, CA

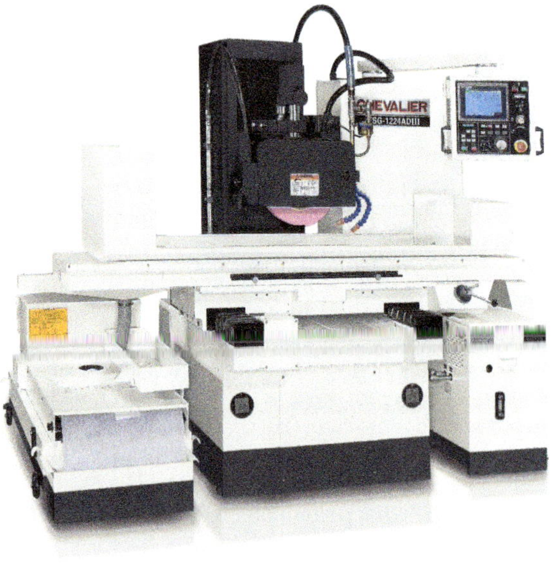

Figure 12-23 A typical size surface grinder.
Courtesy Chevalier Machinery, Inc./Santa Fe Springs, CA

Beyond its hydraulic actuated X axis, its Y and Z axes are also power fed. Most power-fed grinder functions can be operated automatically or by shifting gears, or by hand, too. The X axis on very large grinders does not have a hand motion as it's too big to move without hydraulic force. Using adjustable stops for all three axes, these machines often feature totally automatic or CNC operation.

Since removal rate and heat are constant challenges, these grinders always feature powerful coolant systems! Particulate masks are even more important here as well as very good safety glasses.

Safety Note

Amazing accidents on grinders happen faster than you can react. They are not a first-timer's experience. In my shop, students learn to operate a small grinder first.

Due to the heavy-duty nature, the table no longer rolls but rather is pushed along with a hydraulic cylinder. The table slides on a **hydrostatic bearing** (see Unit 1-3).

Warming Up a Hydraulic Machine

While any machine surface that slides usually requires a lube film between the moving and stationary components, true hydrostatic bearings force a predictable thickness of oil between the two. When they are warmed up to full pressure, no metal-to-metal contact occurs, and the components are separated by an exact gap filled with oil. The Trade Tip applies to manual, large automatics, and all CNC machines.

TRADE TIP

Exercising Most Machines Similar to a long-distance runner or weight lifter, it's important to "stretch out and warm up" hydraulic and CNC machines before heavy demands are placed on drives and bearings. After a long period of downtime, this is an operator's startup responsibility for accuracy and machine life. Here's a three-step procedure for a surface grinder, but other machines are similar.

After checking and servicing the hydraulic oil level, if needed, first start the pump (without moving the table) and let the machine idle for 2 or 3 minutes while the pressure builds up and the table lifts up on its oil film cushion.

Pressure Build—Exercise Warm-Up—Full Stroke Then lightly "exercise" the slides by hand crank or power feed. This entails short stroking the *X* axis several times—around 4 to 6 in. Finally, to distribute the oil evenly and to bleed any trapped air in the ram cylinder from a long period on nonuse, run the table to its full left and right limits. *Caution!* Your supervisor or instructor will need to show you how steps 2 and 3 are accomplished on the machine in your lab, especially the third step, where you will be taking the machine onto its end stops.

To exercise CNC equipment, I recommend a warm-up program that loops through several long and short axis moves. It must take into account any setup on the machine so as not to hit the grinding wheel or cutter against work or tooling. Load and run it while you open your toolbox, get the apron on, and read the instructions for the day.

12.2.2 Setting Up a Reciprocal Surface Grinder

Setting up a reciprocal surface grinder entails four tasks:

1. *Setting the stroke and feed limits–X* length, *Z* width, and stepover amounts; *Y* down feed if the machine has this feature.
2. *Setting safe work holding.*
3. *Coolant supply.*
4. *Mounting and dressing the grinding wheel* (Unit 12-3). Is the wheel right for the job and dressed true and sharp?

Correct Stroke Length and Width

Setting up a surface grinder requires three or four features depending on the type of grinder:

1. ***X* Stroke Limits Right/Left**

   Set beyond the workpiece surface by at least $\frac{1}{2}$ in.
2. ***Z* Axis Limits–In/Out Full Travel of the Wheel**

   Set to go beyond the edge by half the wheel's width.
3. ***Z* Axis Stepover Amount per *X* Stroke**

   Distance the wheel is moved sideways for each new *X* stroke.

   Set to step in/out half the wheel's width with each stroke.
4. **Amount of *Y* Axis Down Feed–Situational**

   From 0.0001 to 0.0005 in. depending on work, wheel grit, and structure and strength of the setup. These are conservative amounts for learning. Manual machines do not feature autofeed.

Surface grinder setup stroke and feed limits are set depending on the type of grinder–manual, automatic, or CNC. The setup rectangle (*X* length and *Z* width) must be bigger than the workpiece surface by a given amount. Figure 12-24 depicts a typical setup–a rectangle of stroke length and width.

All surface grinders feature adjustable stops for the *X* stroke length, and most also have *Z* axis width stop cams. On manually operated surface grinders, the *Y* axis down feed is then a hand function. The down feed on automatics is a ratchet that engages a tooth gear on the *Y* axis handwheel, called a **pick feed**. It can be disengaged or swung onto the *Y* handwheel when automatic feed is needed. The pick is adjusted to pull a given number of 0.0001- or 0.0002-in. clicks down at the end of each stroke. On CNC machines, all these functions are menu driven on the controller (see Fig. 12-23).

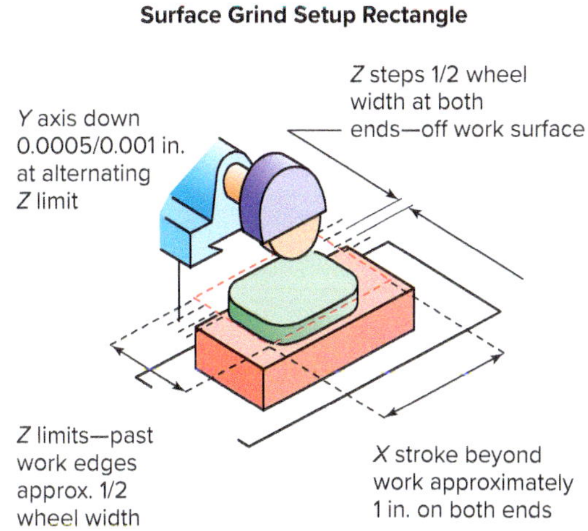

Figure 12-24 The typical setup stops and motions for any surface grinder, manual, automatic, or CNC.

Guidelines for Setup and Operation

Setting *X* and *Z* cam stops entails moving the grinding wheel beyond the work in each axis (see Fig. 12-24). It is there, when the wheel is not in contact with the work surface, that lateral and down movements occur. The right/left *X* stroke should clear the work by at least 0.5 in. minimum (Fig. 12-25).

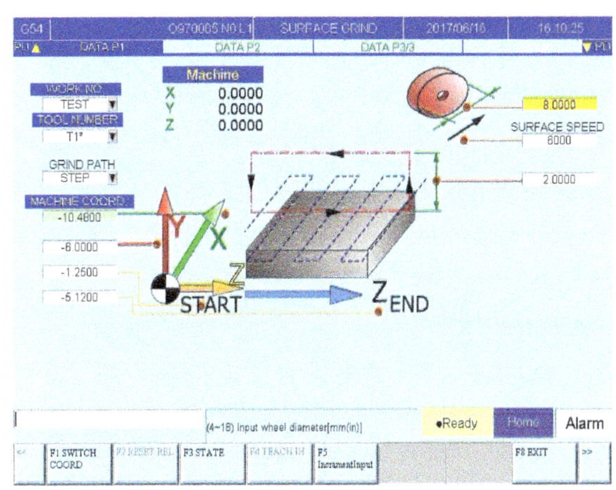

Figure 12-25 The same limit rectangle and steps become program entries (called parameters) on this CNC surface grinder control.
Courtesy Chevalier Machinery, Inc./Santa Fe Springs, CA

KEYPOINT

The extra *X* stroke length beyond contact allows good coolant flushing, but it also improves professional looking results. It is the zone where the wheel is moved sideways in *Z* and the down feed occurs. See the Trade Tip.

Z Axis Steps and Limits

The Z axis has two functions: sideways movement for each stroke and full limit at each edge of the work.

KEYPOINT

Z limits are set such that the wheel travels half its width with each stroke and off the work edge at each side.

TRADE TIP

A Bigger X Stroke Is Safer Other than cycle time, there's no problem in setting the X length much larger than the object. In fact, it's safer for your first few setups. However, in production the goal is to set the X limits such that the wheel clears just enough to avoid stepover (Z axis) tracks in the finish produced.

12.2.3 Coolants for Grinding Operations

Two Ways: Low-Pressure Flood and High-Pressure Injection

At the extreme rim speeds required for grinding, friction and heat become huge issues. On surface grinders equipped with coolant systems, in general, a low-pressure flood is used. On many surface grinders, the wheel guard contains the flood system or it can be a nozzle. Either way, it's a general stream of coolant that protects the work from overheating. While some of that flood also becomes trapped in the air spaces in the wheel, it's not enough, nor is it consistent enough.

Coolant Injection

To reach the point where the chip is being abraded from the work, a focused stream is directed at the contact point where the wheel begins its pass across the work. That traps coolant in the wheel's air spaces. Trapped coolant performs two services:

1. Lubes deformation
2. Chip ejection

What to Do When the Machine Has No Coolant

Most small tool room and manual grinders do not have flood coolant. If so, you have two options.

Mist Coolant

A coolant spray accessory works well. For best results, it should be aimed at the wheel contact area. The spray that doesn't become trapped within the wheel will help to cool the work, but the important aspect is to get coolant to the chip formation zone. Also, if a mist isn't available, then plain air can be directed at the wheel and can work to keep it cool, but it will have no effect on chip deformation.

Create a Cooler Setup or Operation

There are several things one can do when no coolant is available:

1. **Choose a Cooler Grinding Wheel**

   One with a more open structure, or larger grain size, or a harder abrasive that tends to hold its edge longer.

2. **Grind Less Material/Less Removal per Minute**

   This is the last option. Let the material cool naturally by conduction to room air.

A few materials can be successfully ground with no coolant (cast iron, for example), but if available, coolants always improve grinding in four ways:

1. **Reduce Heat Production**

   They lower friction, and there is thus less deformation heat.

2. **Remove Unwanted Heat from the Work and Holding Accessories**

   On precision grinders, this is a critical issue since tolerances are small. Thermal expansion of the part or work-holding device can be a real accuracy spoiler.

3. **Extend Wheel Life by Cooling the Abrasive Grains**

   Because the mineral grains are exceptionally hard, coolants extend their lives by stabilizing unwanted temperature fracturing from hot/cold/hot cycles. But that's only when an adequate supply of coolant is injected into the interface between wheel and work directly at the contact point (Fig. 12-26). Flooding the general area helps cool the work, but it can cause hit-and-miss cooling of the abrasive grains (Fig. 12-26). The best coolant injection on any grinder setup occurs by forcing a stream into the wheel as it contacts the work.

4. **Cause Chip Ejection**

   To cause this effect, coolants must be injected forcefully at the point of wheel-work contact (Fig. 12-27).

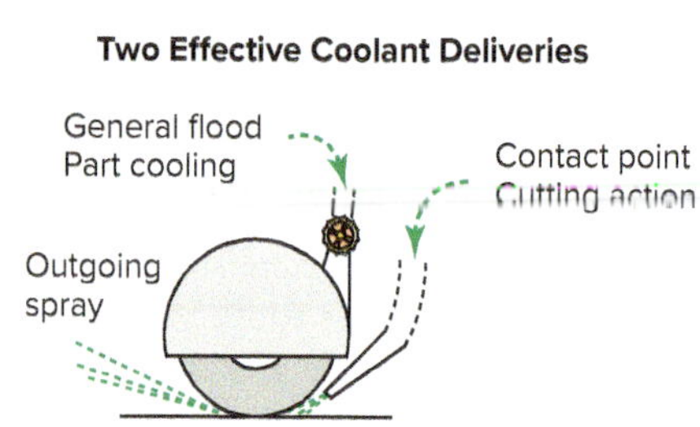

Figure 12-26 Flooded coolant reduces general work temperature, but it lubricates the cutting action only on a hit-or-miss basis. A better source is the nozzle on the right, focused on the wheel work interface.

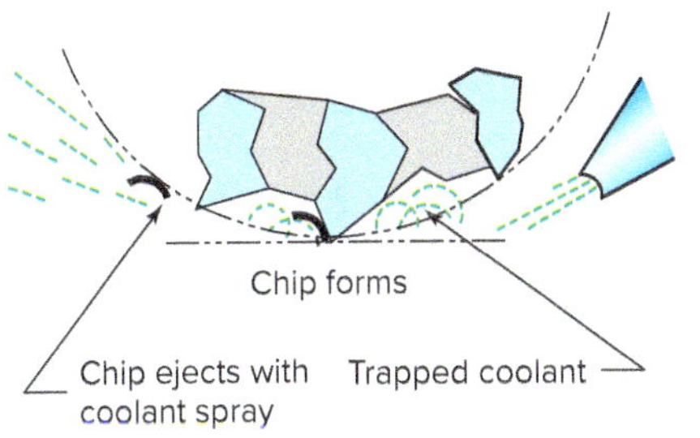

Figure 12-27 Correctly injected coolant is entrapped in the wheel pores until it is flung out, helping to clear the chip from the wheel.

KEYPOINT

Correctly used, grinding coolant creates a fine spray out the far side of the wheel, but that indicates the right use of the fluid (Fig. 12-28). This is true for cylindrical grinders as well.

Figure 12-28 Coolant is correctly injected into the wheel–work interface on this automotive cam grinder.
Courtesy Fives Landis Corp.

Coolant-to-Water Ratios for Grinding

Most shops use the same coolant syrup as that which is used throughout the shop. But for grinding, it's mixed in a thinner ratio of water to syrup, at or near 50 to 1. Read the directions for the product. Shops that do lots of grinding or where special alloys, such as titanium or magnesium, are ground use specialized coolants designed for abrasive work only. See the Trade Tip.

TRADE TIP

Mixing Coolant Always read mixing instructions for coolants carefully. They aren't all the same. Follow specific recommendations for the particular application. Ratios at or above 50:1 are common for grind work.

Beyond diminishing performance, the upper limit to the mix is not to exceed the coolant's ability to protect the machinery from overnight rust.

Also, many manufacturers recommend *adding water to the concentrated syrup while stirring constantly,* not syrup to the water. Why? It has to do with the way the molecules bond. A more complete mix is obtained in this way.

12.2.4 Six Work-Holding Methods

There are six options to hold the workpiece for surface grinding. We'll do an overview of the two most common first, then consider the rest in the next unit.

Precision Grinding Vises

Work can be held in a small vise, much the same as on a milling machine. But for grinding, there is a more precise version reserved for surface grinding and tool room operations only (Fig. 12-29; shown in use in Fig. 12-4). The vise skills you already have also apply in grinding.

- Always use the solid jaw to nest datum surfaces.
- Remember for every load of the vise, there's sure to be slurry dust left on the precision surfaces. Wipe them before loading the vise.
- Use a deadblow hammer to lightly tap the work until firmly seated.

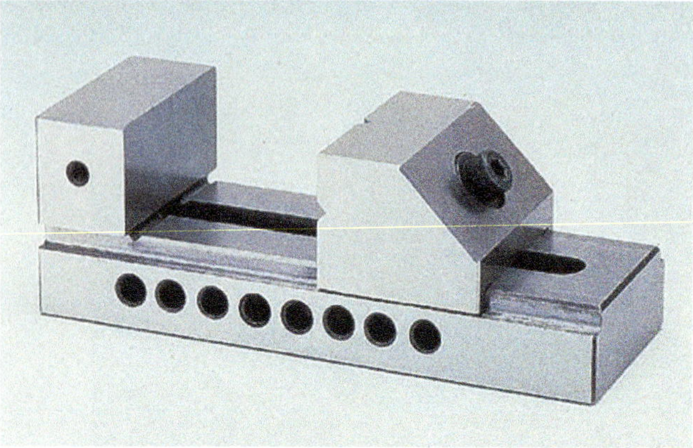

Figure 12-29 A typical precision grinder vise.
Courtesy Chevalier Machinery, Inc./Santa Fe Springs, CA

KEYPOINT

For any setup, in a vise or work held with some other method, the wheel's action forces the work to the operator's left (see Fig. 12-30). Be certain no person stands to the left of the grinder. This will become more important with more complex setups and for larger reciprocal surface grinders.

Magnetic Chucks

Work is often held on surface grinding machines with a magnetic chuck, which might be an **electromagnetic chuck** (Fig. 12-31), where the grip can be switched on or off. But more likely on a small machine, a **permanent magnetic chuck** is used (Fig. 12-32). No electricity is involved (Fig. 12-33).

Similar to a magnetic-base indicator holder, they are activated by a mechanical lever that either opposes or aligns magnets within their interior. When they oppose, the force is canceled, but when they agree, the attractive force can be great—great enough to pinch fingertips!

No—Danger

Yes—Safe

Figure 12-30 By incorrectly moving to the left of the wheel (left photo), the operator is in danger; the machinist on the right is in the safe zone.
Prographics at Milwaukee Area Technical College/McGraw Hill

Figure 12-31 A typical electromagnetic chuck.
Courtesy Chevalier Machinery, Inc./Santa Fe Springs, CA

Figure 12-32 Never have your fingers under an object as you put it upon a magnetic chuck. If the chuck is accidentally left on, the downward force can smash and cut them!
Prographics at Milwaukee Area Technical College/McGraw Hill

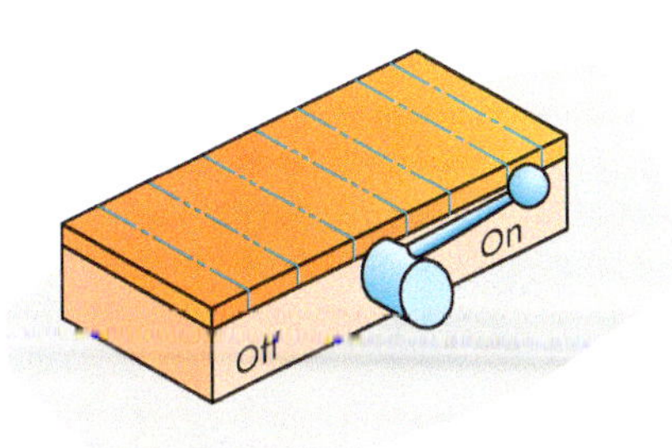

Figure 12-33 A permanent magnetic chuck.

KEYPOINT

Always check that the magnetic chuck is turned off before setting any object on it.

Ensuring a Safe Setup

When setting work upon a magnetic chuck, never place your fingers under the object! While it should be turned off, if the chuck is accidentally left on, when the work gets close to the chuck, the downward force can be great enough to smash and cut fingertips! See Fig. 12-32.

After the magnet is on, test it by grasping the work, then push and pull it to be certain there can be no movement sideways or any tendency to rock or tip. There is so much to learn about these devices for both safety and successful setups that Unit 12-4 is dedicated to setting them up. Have the supervisor or instructor check your first few setups.

The activities for Unit 12-2 are combined with those for Unit 12-3.

UNIT 12-2 REVIEW

Replay the Key Points

- *Breathing protection!* Beyond flying work and wheel fragments, grinders present a breathing danger from dust and flying coolants–wear a mask or filter.
- Coolants help to fling the tiny chips out of the wheel during the split second when the wheel isn't in contact with the work.
- Correctly used, grinding coolant is injected into the wheel-work interface. This creates a spray out the other side of the wheel but indicates the right use of the fluid.
- For any setup, in a vise or held by some other method, the wheel's force against the work is to the left. This will become more important with more complex setups and when using larger reciprocal surface grinders.
- *Never put your fingers under the object!* Accidentally left on, when the work gets close to the magnetic chuck, the downward force can be great enough to smash fingertips.

Unit 12-3 Setups That Work Right

Introduction: Safety and solid setups blend together, especially when using magnetic chucks and high-speed grinding wheels. I cannot stress this enough that using grinders requires individual instruction and guidance.

Using safety guards and eye protection, even if you are not actually removing metal, is a good start–why? *Wheels can and do explode even when they are revolving without touching the work.* We'll discuss how to rev them up safely in this unit.

Magnetic chucks add a new twist to setup safety. While magnets produce precise results, their relative grip compared to the positive gripping chucks, clamps, and vises of milling machines, make them a subject for special focus. When magnetic chucks are incorrectly applied, parts tip, slip, and sometimes fly out of the machine. That almost always leads to a broken grinding wheel–and flying objects!

TERMS TOOLBOX

Arbor (grinder) The precision hub, shaft, and flange mount for a grinding wheel.

Demagnetizing Removing residual magnetism from magnetic workpieces exposed to the chuck.

Dressing Reexposing sharp grains to unclog, sharpen, and true a grinding wheel.

Dynamic balancing Rotating grinding wheels and cutters at working speeds while removing all out-of-balance conditions.

Flange (grinding wheel) The wide sides of the wheel mount designed to spread the load over a large area. Must be one third the wheel diameter or greater.

Float grinding Holding and grinding nonmagnetic material using blocking and a magnetic chuck.

Glazed grinding wheel An overused wheel that's lost its cutting edges. Heat causes the bond to smear and forms a nearly glasslike surface that will not cut.

Nib A mounted diamond on a stick–used for dressing a grinding wheel.

Ring test Tapping the unmounted grinding wheel in several spots while listening for a clear bell sound. A broken wheel produces a thud.

Static balancing A gravity test of grinding wheels to shift weights to neutralize roll down.

Stop blocking Extra blocks that adhere to the magnetic chuck, used to guarantee the work will not move.

Strap clamp (magnetic chuck) Used when stop blocks aren't secure enough or they are too thick to clear the grinding wheel. This U-shaped clamp holds with set screws against the chuck side.

Truing a grinding wheel Dressing a grinding wheel to make it perfectly cylindrical and concentric to the grinder spindle.

12.3.1 Mounting, Dressing, and Balancing a Grinding Wheel

The first setup step is to mount the wheel on the grinder. Here's the step-by-step procedure.

SHOPTALK

Double Speed Option Many larger industrial grinding machines feature a high-speed spindle option to increase RPM for efficient use of grinding wheels that have been dressed down to a small diameter or for using CBN wheels. Never select the high-speed option until the wheel has been worn down to the half-size diameter. Do not use this option unless the wheel's original RPM limits for its full diameter meet the machine's RPM. Overspeeding and rim explosion can result if the RPM is exceeded. To help meet these requirements, some grinders place the double speed button in a place where it cannot be activated until the wheel is half worn.

Removal of the Old Wheel

Use a wood block or other protective material on the chuck to protect both the outgoing wheel and the chuck/table. Loosen the center spindle nut a quarter turn. Tap the spindle hub with a plastic hammer to break the self-holding taper; then remove the wheel and hub. (If the grinder features a simple **flange** and nut wheel mount similar to a bench grinder, then the tapping isn't necessary.) *Caution:* Grinders often use a lefthand thread for flange nuts!

After removing the outgoing grinding wheel, never roll it or lay it where other items might be stored upon it.

KEYPOINT

If you find it's a damaged wheel, show it to your instructor or supervisor; then take it directly to the garbage can—don't take chances that it will accidentally be reused at a later time. Break it up completely!

Check the Identification, RPM, and Size

Be certain the replacement wheel is the right makeup, diameter, and shape using the standard number selection for the machine and the job. Look for the maximum RPM label and check that against the machine on which it is to be mounted.

Task 1 Check for the Right Wheel and Condition

Visually inspect the wheel for cracks and chips and that it has two new paper gasket washers.

Paper washers perform two vital services for grinding wheels (Fig. 12-34).

Figure 12-34 Can you see the crack in this wheel? Always inspect wheels for visible signs of damage and check that it has paper washers in good condition or that new ones are ready to use.
Prographics at Milwaukee Area Technical College/McGraw Hill

1. **Equalize Flange Clamping Pressure**

   Without the washers, the clamping force is concentrated on individual, high grains, promoting wheel breakage if the flange is clamped too tightly. Remove a used washer to see the imprint of the grains on the paper. The soft paper spreads the load over a safer area of the wheel.

2. **Absorb Shock**

   They prevent wheel breakage by acting as a damper when a wheel is accidentally overloaded or some other accident occurs such as work slipping or tipping.

Never attempt to use a chipped or damaged wheel. While it's possible to trim dress a wheel beyond the damaged portion, that's to be done by supervision; you are not authorized to do so at the beginner level. Just don't use a wheel that looks anything other than perfect. But, as Fig. 12-35 shows, you can't always see damage, so don't skip task 2.

Figure 12-35 A ring test helps find hidden cracks.

Task 2 Ring Test the Wheel

A broken vitrified wheel sounds much the same as a cracked coffee cup: It will thud rather than make a clear ringing sound. To **ring test** the wheel, suspend it with any convenient hard object in the mounting hole (or your finger will work), then tap it in four places around the wheel with a plastic screwdriver handle or some other hard nonmetal object. Rotate it about a fourth of a turn and do it a second time (Fig. 12-36).

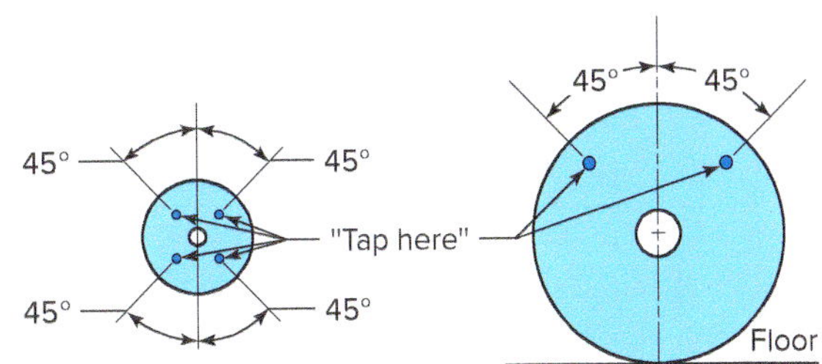

Figure 12-36 The ring test can find most hidden cracks if they are extensive.

But you can't always hear cracks that don't leave a long fracture in the wheel, so after mounting the wheel perform the acceleration test (coming up).

Task 3 Mount the Wheel on the Grinder

There are several details to see when mounting a precision grinding wheel. Grinding wheels mount on machines in one of two different ways: directly or on quick change hubs.

Wheel Flange Mount Styles–Direct or Quick Change. The direct mount grinder spindle is similar to a bench grinder's. The center hole in the wheel fits the machine's spindle shaft (Fig. 12-37). The wheel is held in place by a pair of steel flange washers and a flange nut (usually left-handed; Fig. 12-38) that screws onto the grinder shaft. This system is used on some tool and cutter grinders and small surface grinders. However, simple, direct mounted wheels have two problems. First, the machine's spindle shaft is subject to damage when an accident happens that is severe enough to break the wheel. Second, these mounts do not include wheel-balancing provision.

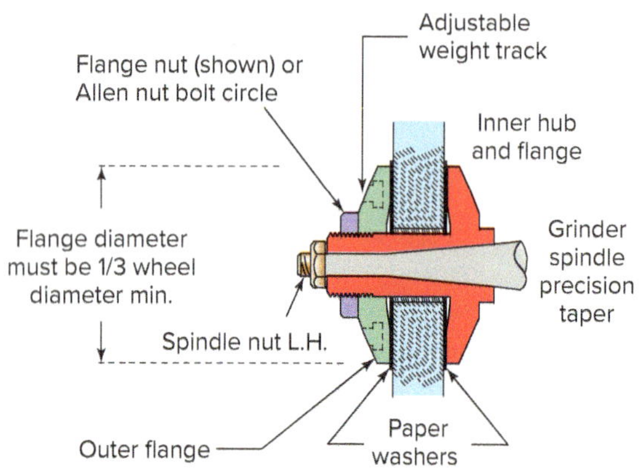

Figure 12-37 A cross section of a typical precision grinding wheel flange and quick change taper mount.

Figure 12-38 The guard has been removed to show this typical small grinder wheel flange.
Courtesy Chevalier Machinery, Inc./Santa Fe Springs, CA

The quick change hub is more expensive, but it is the better solution where setup turnaround time matters (Fig. 12-37). Grinding wheels mounted on a hub can be balanced using sliding weights.

Shops usually own many of these flange sets so they can keep an inventory of often used wheels mounted, balanced, and true–ready for quick turnaround.

KEYPOINT

Grinding wheel balance is a critical issue not only to achieve good finish and accuracy, but it's also necessary for long life of the grinder's spindle bearings. We'll see how in just a bit.

Task 4 Torque the Wheel Flange Nut

When mounting the wheel on the **arbor**, tighten the nut or the circle of Allen bolts to the manufacturer's specifications–if available. Someone will need to show you how much is acceptable. Too much is as dangerous as not enough! Larger grinding wheels require more pressure on the outer flange, but any wheel can be cracked from overtightening.

Torque Patterns. If the outer flange is tightened with a series of bolts in a circle as in Fig. 12-39, then use a cross-over torque pattern and procedure. Tighten slowly, a bit at a time in the order shown. Never tighten on one side first. Distribute the force equally.

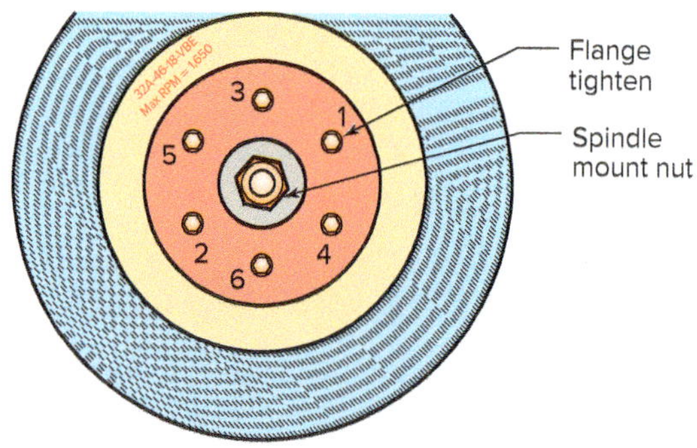

Figure 12-39 To avoid concentrating force in any one spot, bolt circle wheel flanges require tightening a bit at a time following a pattern, until they reach the specified torque.

Task 5 Do Primary Wheel Balance

Wheels placed on arbors require a balancing operation in one step (or a finer two-step process) for two reasons:

1. **Wheel Density Differences**

   Quality wheels are uniform in their density, but they are not perfect.

2. **Mounting Clearance**

   There is a necessary clearance between the grinding wheel hub and the spindle arbor, which will affect balance.

Static and Dynamic Balancing. There are two ways a wheel may be balanced. You've seen both processes in your local tire dealer's workroom.

TRADE TIP

Be on the watch for a stenciled arrow on some industrial grinding wheels. It will say "Up" or "Down" (Fig. 12-40). Mounted this way, the wheel should come closer to initial balance. That was the way the wheel was mounted, balanced, and trued at the factory.

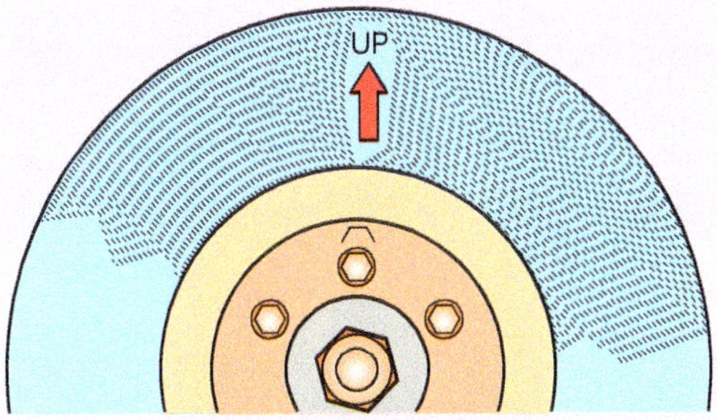

Figure 12-40 Mount the wheel according to the arrow if one is stenciled on the wheel or washer.

Static balancing is the simpler method, but it's not the best. It's a gravity test made by supporting the wheel and mounting the flange on a nearly friction-free set of rollers (Fig. 12-41). The heavy side of the wheel will tend to roll down. Weights are added and shifted until the wheel has no net tendency to roll. This is the method you will most likely experience in school. But static balancing doesn't eliminate all the problems of balance when it's spun up to cutting speed.

Dynamic balancing rotates the wheel at its normal operating speed then detects through sensors where and how the weights must be placed. To see why, consider a bicycle pedal. t should balance out statically, as long as both pedals weigh he same. Taking the chain off the sprocket to allow free rotaon, it shouldn't have a heavy side. But now suppose it's spun t 1,000 RPM. You can see that, although statically balanced, e weight isn't distributed about the axis of the pedal. It will ve a tendency to wobble, but the bearings prevent it, and at translates to vibration. Similarly, to balance a grinding eel, weights must be placed on both flanges of the wheel, ner and outer, to cancel dynamic balance problems.

namic balancing is the preferred industrial process for close rance grinding (Fig. 12-42). It's also necessary to preserve hine bearings and deliver the very finest finishes. Newer C machines include dynamic balancing as part of their rnal routines. We'll see this concept again when we get to -speed machining (see Chapter 25), where extreme RPM rs in their holders must also be dynamically balanced.

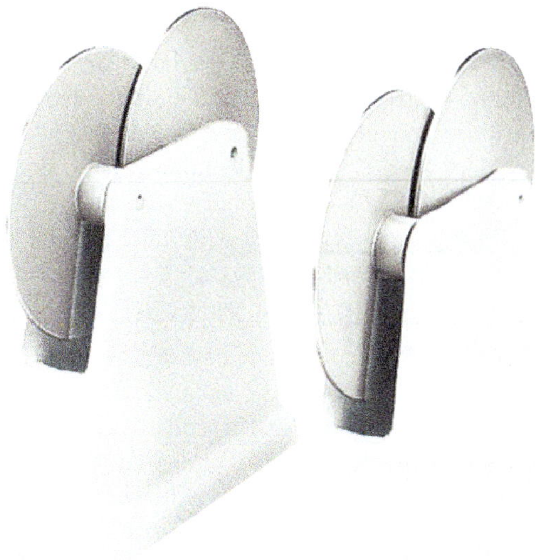

Figure 12-41 A static balance fixture consists of two pairs of knife-edge, hardened steel wheels with very free running bearings.
Courtesy Chevalier Machinery, Inc./Santa Fe Springs, CA

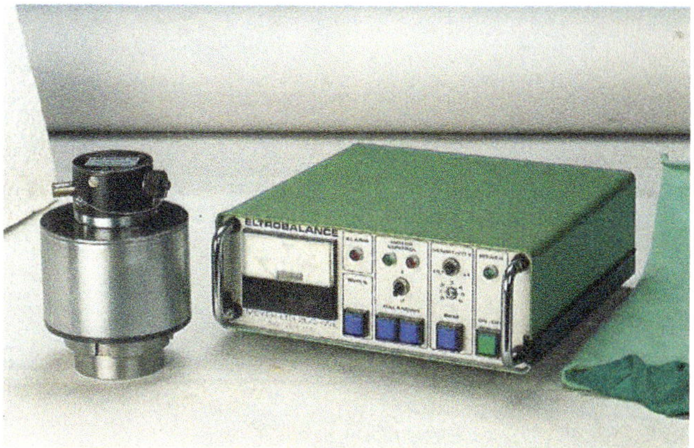

Figure 12-42 A dynamic balance sensing unit is attached to the wheel while it's spinning. A computer determines where and how big the weights must be.
Courtesy Chevalier Machinery, Inc./Santa Fe Springs, CA

Task 6 Mount and Dress/True the Wheel

Dressing the wheel is a resharpening of its edge by deliberately fracturing it and/or exposing the new ones. **Truing the wheel** is making its rim concentric to the machine's axis and making its grind surface perfectly cylindrical. They are done at the same time.

Grinding wheels must be dressed when they are loaded (see Fig. 5-49) and when they are **glazed**. Heat causes the wheel bond to smear with the resulting surface something like smooth glass. Normal wear has made the wheel face out of true.

Precision Wheel Dressing Tools

Precision Wheel Dressing Tools. Assuming we are dressing an aluminum oxide, or silicon carbide wheel, for precision surface grinder and cylindrical work we almost always use diamonds (Fig. 12-43). A single diamond crystal mounted in copper/nickel is called a ***nib***. It's not a gem-quality diamond, but nevertheless, it is expensive. Alternately, less expensive diamond chips can be used when they are embedded in soft materials such as copper or hard resins (Fig. 12-44).

As seen in Fig. 12-43, either diamond dresser can be mounted in a block that adheres to the magnetic chuck for surface grind dressing. It is placed in the end of a round rod for dressing cylindrical wheels. If the mounting block is a solid cube, the nib can be mounted on its side when it's necessary to dress the side rather than the rim of a grinding wheel.

Figure 12-43 A diamond nib in a surface grind block.
Courtesy Chevalier Machinery, Inc./Santa Fe Springs, CA

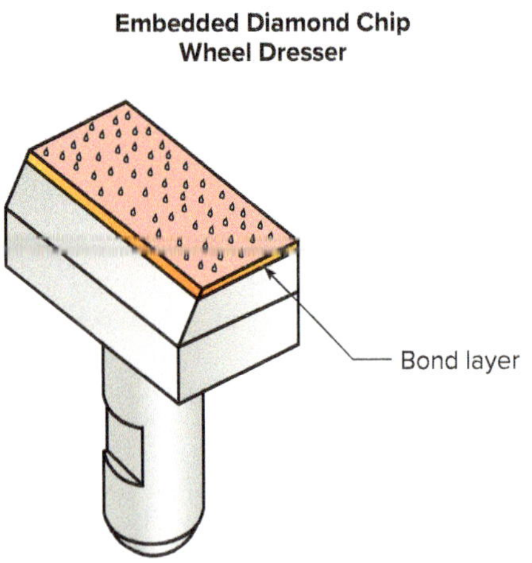

Figure 12-44 A diamond chip dresser works nearly as well (for most dressing situations), but costs less.

KEYPOINT

When dressing a wheel take one final pass slightly slower than when truing, but not too slow. Be sure the *X* axis can't move.

TRADE TIP

Dressing Here are a few tips on the right way to dress a wheel.

The diamond is set at about 15° to 20° beyond the wheel's center, as shown in Fig. 12-45.

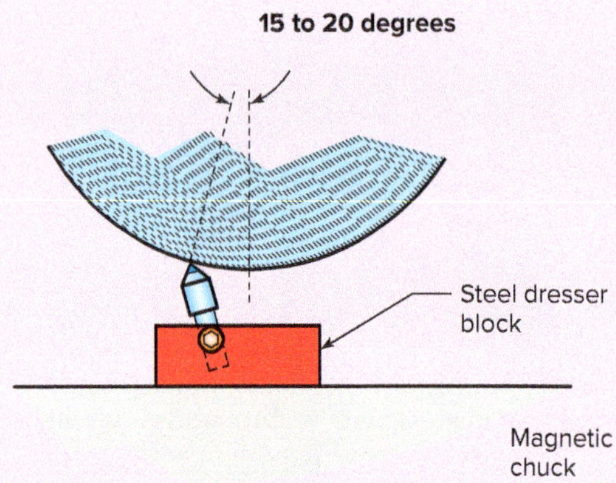

Figure 12-45 The dresser is set 15° beyond the wheel's center—can you see why?

Make sure the *X* axis is locked or stopped from moving.

Using the *Y* axis, the wheel is dropped until it just touches the diamond held on the magnetic chuck. Then crank it down about 0.005 in. more for the first pass.

Using either the *Z* axis handwheel or power feed, it is drawn across the diamond, moderately slowly at about 1.5 to 2.0 in. per minute.

To know that the wheel is true, look for consistent color across the face and listen for intermittent sounds turning to a solid sound. Bringing it to full true may take two or more passes of around 0.003 to 0.010 in.

Do not take a deep pass, for the possibility of grinding away the nib mounting material exists. The diamond is never found once it's ejected! (I've never found it! Good luck.)

The final pass is taken slightly slower at around 1.0 in. per minute. There are two mistakes made here, both of which only dull the new grains: going too slow such that the new sharp edge hits the diamond a second time on the next revolution or taking an extra pass back across without lowering the wheel again "just to be sure."

The final tip is to rotate the nib in its holder from time to time. That keeps it a uniformly pointed cone and prolongs its life in the mount by wearing it away on all sides.

Side-Wheel Dressing. Grinding a workpiece surface using the side grinding wheel (not the rim) isn't a beginner operation. But to finish steps in the work and for other situations that do arise, it's sometimes required. We often side-wheel when doing tool and cutter grinding. But there we use a wheel made for the task. On a surface grinder, where the wheel's rim is intended to do the job, heat is the greatest problem. When we contact the work along an arc, too much wheel surface is in contact with the work at any time. When side-wheel work is necessary, use Fig. 12-46 to guide the dress setup and seek help for the first time.

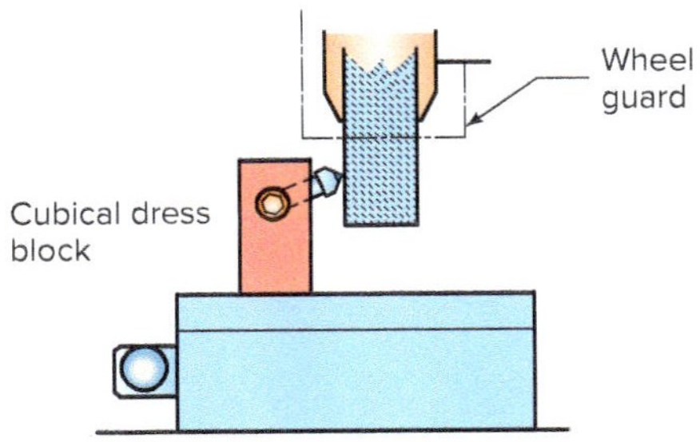

Figure 12-46 Side-wheel dressing requires extra caution.

The Fine Balancing Act

When using static balancing, to achieve the best possible finish and a long wheel life, after the first balance and initial truing, remove the wheel and hub from the grinder; then balance it a second time. You will find that the truing operation has slightly changed the initial settings in many cases.

TRADE TIP

Relieving a Side Wheel Here's a wheel dressing tip to reduce heat when performing side-wheel grinding. A relief is dressed into the wheel such that it contacts the work only along a narrow band near its rim. First dress it fully flat, up the side per Fig. 12-46. Then the diamond is moved up from the rim by about $\frac{1}{4}$ to $\frac{1}{2}$ in. and the *Z* axis carefully cranked into the wheel's side by 0.025 in. or so. It is then dressed up the side just beyond the depth of intended cut. Caution—this operation removes a lot of life from the wheel. Use it only when heat is sure to be a problem with a side-wheel operation.

Task 7 Warm Up the Wheel

With all guards in place, start the grinding wheel in accelerating stages. Turn on the spindle for a count of two then switch it off before it gets to full speed. Then turn it back on for a count of four and off again. Finally let it go to full speed but stand out of the path. Let it spin freely for a minute or two as a test. A wheel with hidden damage will normally blow during that time!

This acceleration test is a good practice after a shift change or when the machine has not been used for some time. On machines that use coolant, the fluid will collect at the bottom of the wheel in the air spaces as it slows to a stop, and this slow start method will sling out the unbalancing liquid.

Task 8 Always Inspect a Mounted Wheel

If the wheel is already on the machine, rotate it slowly and inspect it for cracks and chips before you use it. The ring test is not valid here because of the paper washer and flange pressure. Report any damage immediately–do not try to dress it away.

KEYPOINT

Mounting a Grinding Wheel

Remember these seven points:

1. Ring and visual test the wheel.
2. Double-check wheel size and RPM.
3. Always use paper washers.
4. Torque the nut—per instructor demonstration.
5. Balance the wheel, dress it, then rebalance it.
6. Warm up the wheel.
7. Always visually inspect a mounted wheel before using it.

12.3.2 Safely Using a Magnetic Chuck

The magnetic chuck can be used in countless ways. We'll look at a few basic methods, but you'll soon find other creative ways to combine lessons here, with parallels, vee blocks, and other items used to set up mills and other machines. It can be used to even grind nonmagnetic material such as aluminum or brass, which won't adhere to the magnet. To do so, the work is held by trapping it in place using **stop block** packing to completely contain the work (Fig. 12-47). (This is shown shortly in Fig. 12-50.)

KEYPOINT

Most surface grinder accidents are caused by an improper setup, where the workpiece can move or be tipped over.

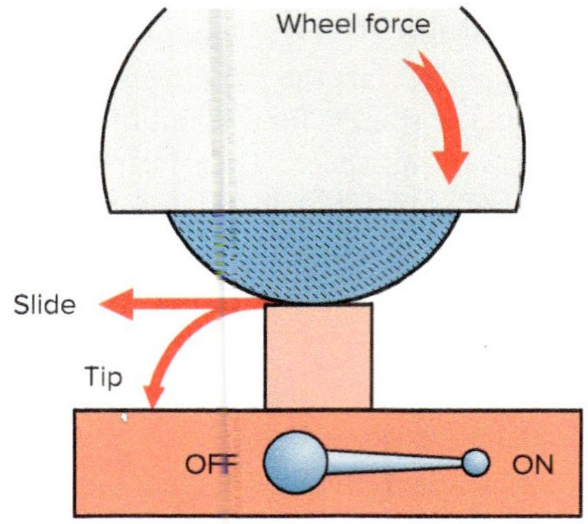

Figure 12-47 Anticipate and prevent movement based on the forces of grinding.

The bottom line for success is to analyze the forces and the work's potential for unwanted movement. Then set up stops, clamps, and other holding tools to stop it dead! *If the setup looks even slightly weak, it can't be used!*

Critical Safety

Potential movement due to the grinding wheel's force is to slide to the left and/or to tip over to the left (see Fig. 12-47).

Critical Question

Which movement is more dangerous, sliding or tipping, and why?

Answer: Both are totally unacceptable, but the tipping is worse because it pivots the block on the leading edge, which jams its back edge into the wheel.

Now, examine Fig. 12-48, where both blocks on the chuck are steel. *Which is dangerous?*

That answer is easy: The tall block will tend to tip over more readily. But here's a key point: *Even the short one isn't well set up.* It too will tend to tip or slide to the left. One key is to appraise the amount of hold area relative to the grinding forces and height. If you can tilt the work with your hand, it's definitely not safe!

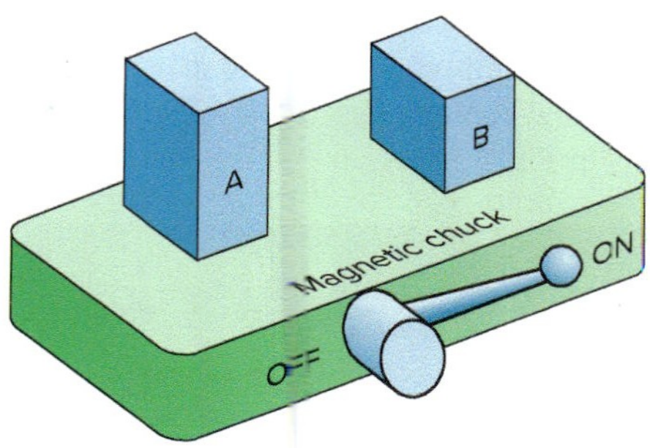

Figure 12-48 Which steel block is dangerous in this setup?

But How About This Pair?

Each is the same steel block. In Fig. 12-49, one is a better setup—why?

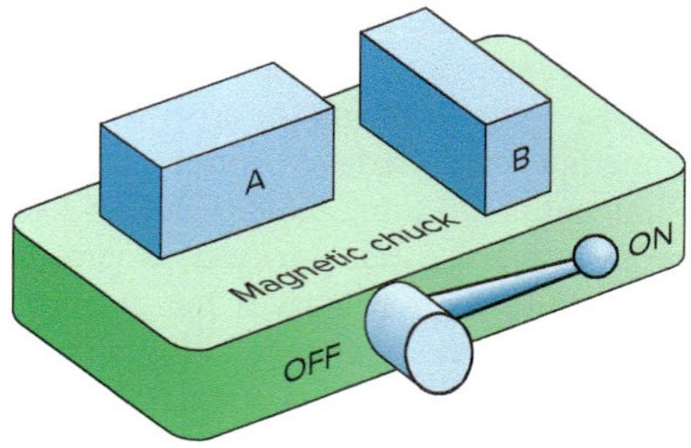

Figure 12-49 Placement can be right or wrong, too!

While both blocks will hold to the magnet with nearly the same attractive force, block A has its long axis braced against tipping. Because its longer face is held at right angles to the grinder force, it presents less resistance to tipping. But the important point is that both blocks could possibly tip. Neither is ready to grind! They must be trapped with stop blocks.

Stop Blocks

When setups include tall work such as in Fig. 12-48 or Fig. 12-50, tipping problems can sometimes be solved in a couple of ways. Stop blocks of steel are the more immediate solution. One can use parallel bars and burr-free steel bars of any size and shape as stops, as long as they are shorter than the workpiece. They can't be rough, or they might damage the chuck (Fig. 12-50). The larger surface area they have in contact with the magnet, the better they will hold.

Stop Blocks Prevent Movement

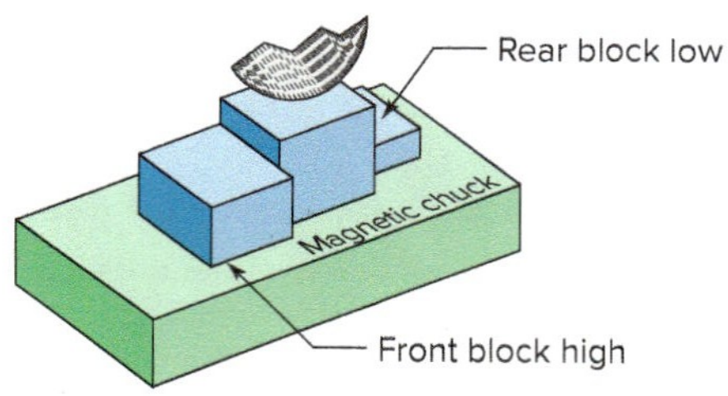

Figure 12-50 Correctly placed steel stop blocks prevent unwanted movement.

TRADE TIP

After setting the magnet to *on*, use a reboundless hammer to tap stop blocks firmly against the workpiece. If there's any doubt put paper feelers between the blocks and the work. By tugging on it after setting the blocks, the paper will show if they are firmly set.

Then just leave the paper in the setup after the blocks are tapped up tight. If the paper is well trapped between the work and the blocking, the coolant will not affect the paper fibers enough to change its thickness.

Using a firm push with your hand, always test the work to simulate the grinding action before starting the machine. Test to see that the work cannot move.

Full Containment—Float Grinding

In Fig. 12-51, the work is fully stopped on the chuck, on all four sides. If, using the hammer and paper trade tricks, the packing is tight against the work, it cannot move in any horizontal direction. This setup will allow grinding of nonmagnetic work. Think about it–once the wheel is directly over the work, it's totally trapped. It floats in the setup but cannot move–thus the name **float grinding**. However, heat expansion then contraction can loosen the packing. Grind slowly; use coolants if possible, take small down feed passes; and check it often to be certain it remains tight.

Note in setups similar to that shown in Fig. 12-51, that the front block must be nearly as tall as the work to prevent tipping–no less than 75 percent of the work's height. The back block need not be as tall.

12.3.3 Precision Angle Plates

Utility plates can be magnetically chucked then used to hold work using clamps or bolts. This is the best solution for work that is too tall or otherwise unsuitable for direct chucking (Fig. 12-52). But even though they adhere to the magnet very well, the angle plates can be tipped. Blocking them in place is necessary, too.

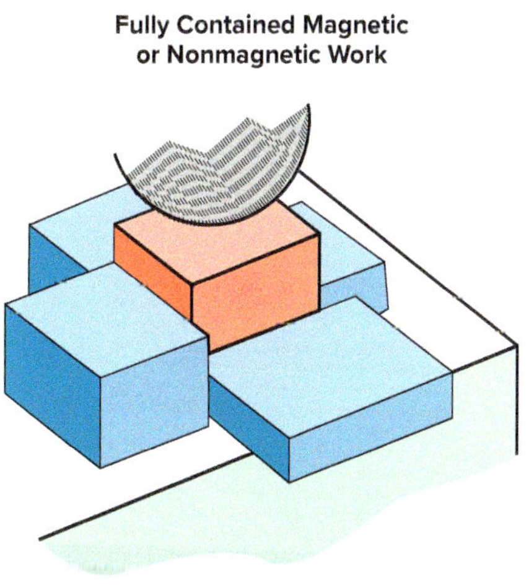

Figure 12-51 Four stop blocks fully contain the work—it cannot move laterally, and the wheel above stops it from coming up.

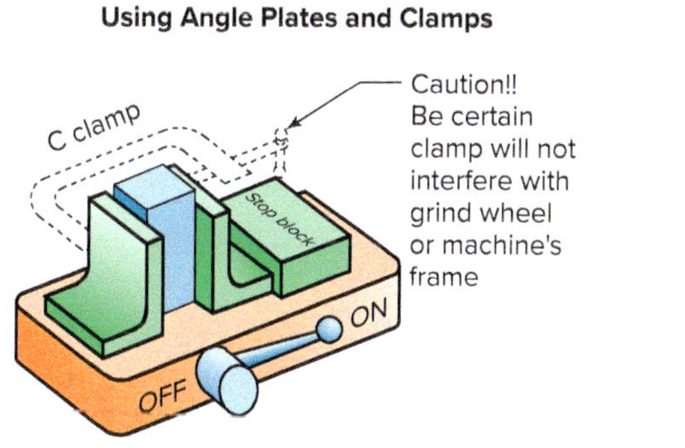

Figure 12-52 A combination of angle plates, clamps, and stop blocks is set up here.

Critical Question

Can you see another way to rearrange the setup of Fig. 12-52 to ensure squareness in two axes?

TRADE TIP

When using C-clamps on grinders, a bit of tape to hold the tee handle from bouncing into the wheel is a very good idea!

Answer

Move the rear (to the right) angle plate to the work's side so that it's trapped in a pair of angle plates that are 90° to each other. Then move the rear stop block forward (to the left) against the work, to trap it. Still use a C-clamp or other holding method to pull the work to the angle plates.

12.3.4 Magnetic Parallels

Odd-shaped work can be held using magnetic parallel blocks that transfer the force up to a part requiring some space from the table. We discussed them back in Chapter 7 but here's where they truly come to the forefront of setups. *Magnetic parallels* are made of alternating pieces of steel and any nonmagnetic material–brass, aluminum, or even plastic. This creates a series of spaced magnetic cores that pass the force of the chuck through to the object being held. But be aware, they weaken it to some degree.

In Fig. 12-53, a steel hinge base needs grinding on the bottom. It is easily held using two matched magnetic parallels to accommodate the hump. Even though this looks to be a pretty secure setup, I strongly recommend adding additional packing around the work to be certain it cannot move.

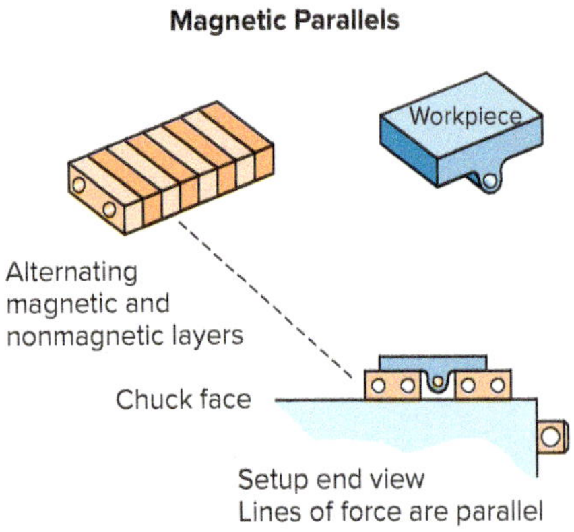

Figure 12-53 Magnetic parallels transfer the chuck's force, although they diminish it to some degree.

> **TRADE TIP**
>
> **Safety Note** The cores of magnetic parallels and other magnetic transfer tools must be placed *parallel* to the lines of force from the magnet below them or they will cancel most of the holding force! Try an experiment. Note the magnetic cores on a chuck in your shop. They run front to back in the *Z* direction. Now place a magnetic parallel one way with a steel object on top of it; then turn on the magnet. Try to move the object. Now, turn the magnet off, rotate the setup 90°, and try again. That's the proof you need.

12.3.5 Chuck Strap Clamps

Sometimes blocks alone are not secure enough or they aren't thin enough to clear the grinding wheel. Under these circumstances, a *chuck strap clamp* accessory can be used as a front and possibly as a backstop, too. They are a simple U-shaped steel bar with set screws in the front (Fig. 12-54). **Strap clamps** work best when the magnet is turned on to keep the central thin portion from bowing up due to the set screw force, but they can be used without the magnet.

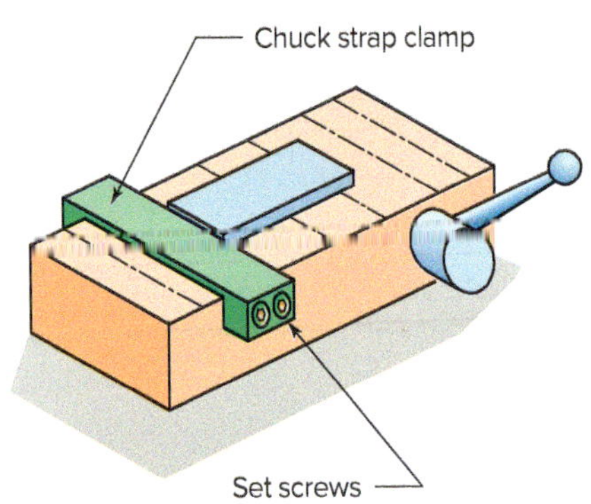

Figure 12-54 A chuck strap clamp can take the place of thin blocking.

> **KEYPOINT**
>
> Strap clamps are mainly used as very low front blocks.

12.3.6 Chuck Back Guides

To aid in alignment of work to the axis of the machine, when grinding slots of form grinding, for example, most chucks feature a back guide. Its height is adjustable within a small range (Fig. 12-55).

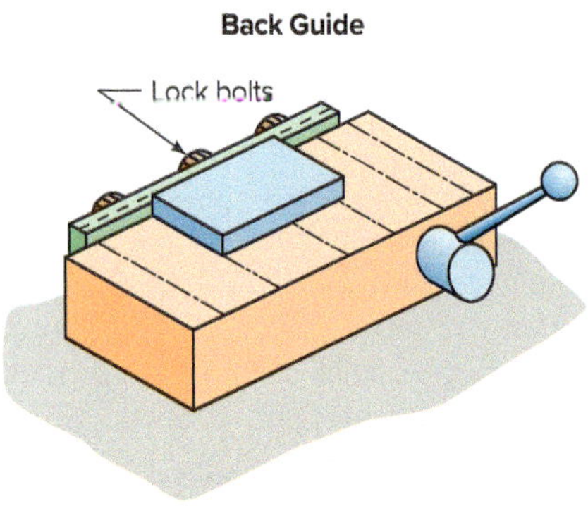

Figure 12-55 Most chucks are fitted with back guides to align work, the vise, or other accessories to the *X* axis and to block work against the back of the chuck.

> **KEYPOINT**
>
> Don't assume the chuck and back guide are aligned to the grinder's *X* axis. They only bolt in place and can be far from parallel. Accidents can also misalign them. Mount a DTI on the wheel guard and run along the length of back guide to test its alignment.

12.3.7 Angle Grinding Sine Plates

Expensive and precise, these holding tools are used to set up and grind angular surfaces (Fig. 12-56).

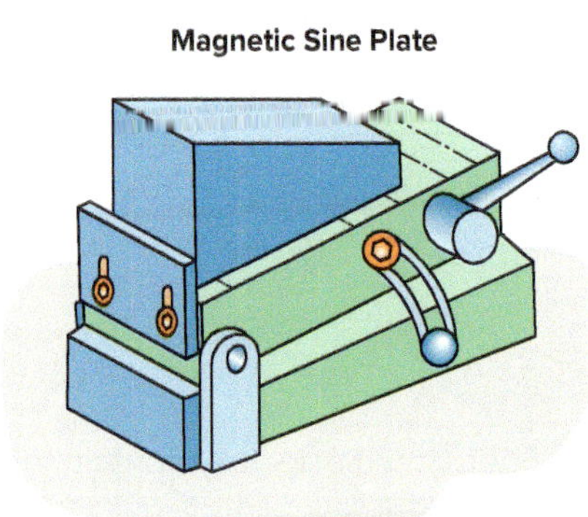

Figure 12-56 Magnetic sine plates tilt the work to a precise angle and hold it for grinding.

12.3.8 Maintaining Magnetic Holding Tools

At long intervals, the shop supervisor will choose to regrind the top of the chuck itself or other magnetic holding tools. Although regrinding is not an everyday maintenance operation, the chuck is made with enough top metal to regrind it many times over a long period of years. With professional use, the chuck should not need grinding more than twice a year maximum, and then only 0.002 to 0.005 in. will be removed.

Between grinding, small nicks and dents can be flat filed and honed away. Only after many of these imperfections have occurred is chuck grinding justified.

Never leave coolant sitting on the chuck at the end of a shift and be certain the magnet is turned off. If the electrical types are left on, they will heat up, and the permanent types will weaken over a long period of being left on.

12.3.9 Demagnetizing Work

Any work material that is drawn to a magnet (iron, steel, some stainless steels, some nickel alloys, and a very unusual alloy of aluminum-nickel and cobalt called Alnico), if left on the magnet for a time, will themselves become magnetized. Usually, that residual magnetism must be canceled to get the object off the chuck. And it must be canceled to send it to the next workstation. It's difficult to measure magnetized metals in some cases.

Demagnetizing on Electromagnetic Chucks

To cancel residual magnetism, electro chucks often feature some method of cycling the force down, thus automatically canceling the magnetism. Switching these magnets to the off position begins the cycle, which lasts about 1 minute. Using the chuck itself, the demagnetizer control creates alternating fields of slowly diminishing force that drain the work metal of its residual magnetism, thus **demagnetizing** it. An indicator will tell you when it's okay to remove the work (Fig. 12-57).

Demagnetizing on Manual Chucks

Manual chucks require a two-step process. First, switch the magnet off and move the work to a position that is 90° to its original if you can. Quickly throw the chuck lever two or three times–on, then off. This usually cancels most but not all the magnetism. If all residual magnetism is to be removed, the work is then placed on or in an electric demagnetizer accessory.

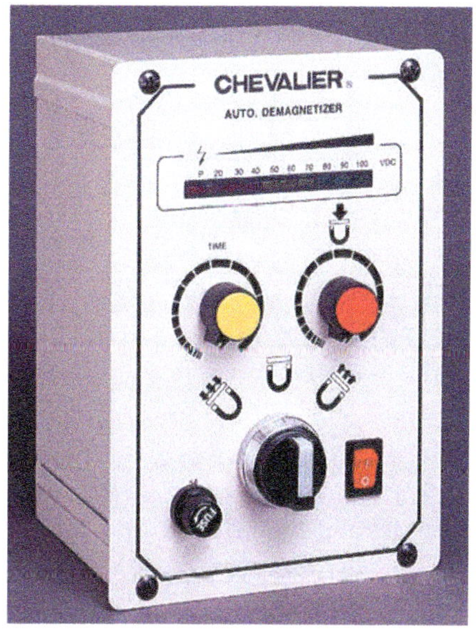

Figure 12-57 This chuck control includes automatic demagnetization. It also includes cycle buttons and a rotary switch to diminish the chuck's holding force—a useful feature when plates are warped.
Courtesy Chevalier Machinery, Inc./Santa Fe Springs, CA

TRADE TIP

If your shop doesn't have a demagnetizer to use with the permanent magnetic chucks, take your work to the electro chuck!

12.3.10 Grinding a Plate Flat and Parallel

The most basic operation on a surface grinder is to reduce the height of an object to a specific size. Other than setting the trip stops and motions, the actual grind procedure requires no discussion beyond your instructor's demonstration of the machine's features.

Thin and Warped Plates

But when the given plate is warped or flexible, it cannot simply be chucked down and ground. As it is released, it will spring back from flat on the chuck to its original pregrind shape. Here's how to solve that problem. The secret lies in a trade tip called "shimming and flipping."

The real tip is in this first step (Fig. 12-58). Lay the plate with the curvature upward and shim the middle such that the magnet does not pull out the warp when it is turned on–this is a critical step. With no shim, the plate would spring down under magnetic pull, then spring back after it is released.

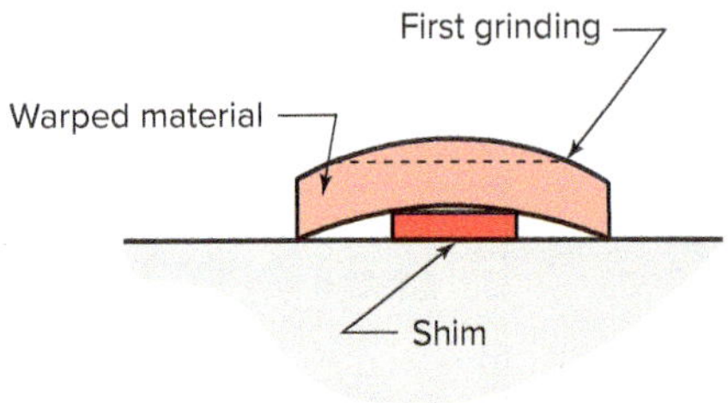

Figure 12-58 Step 1—Lay the work with the dome up, then shim it. Turn on the magnet and grind about 75 percent of the surface flat.

With the shim(s) in place, turn on the magnet and grind the plate for a minimum cleanup of about 75 percent of the surface. Next, release it, demagnetize it, flip it over, and shim the other side at the outer edges (Fig. 12-59).

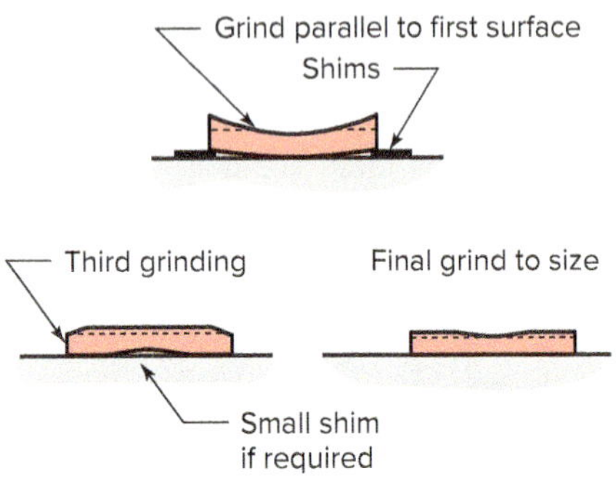

Figure 12-59 Release, demagnetize, reverse, and shim the work on the opposite side. Now grind it for 75 percent cleanup.

Again the shims prevent magnetic change from its natural position. Then grind about 75 percent cleanup on this side. Release, flip, and repeat steps 1 and 2 a second time and presto, the plate will be flat and parallel. Then it can be held directly on the magnetic chuck with no warp for final size grinding.

12.3.11 Operating a Surface Grinder

Each surface grinder is different. This instruction is general and will require specific information and a demonstration of the grinder in your lab. Here's a set of guidelines for success.

Always Clean All Surfaces

Loose grit is the enemy of precision! Always wipe the chuck surface with a rubber squeegee, then a clean rag, and then give it a final test using *your hand* (Fig. 12-60). Your hand is the best grit detector! One single piece of grit left on the table between the work and chuck will be 5 to 10 times the expected repeatability of the operation!

Figure 12-60 After cleaning with a squeegee, then a rag, do a final wipe with your clean hand. It's the only way to guarantee that no grit is left behind.
Prographics at Milwaukee Area Technical College/McGraw Hill

Depth Feed—How Much?

A common error made by beginners is to attempt to remove too much metal per stroke. Each setup, work, wheel, and machine combination will be different. Similar to milling and turning, there is no hard-and-fast rule, but here are some guidelines to remember for setup and operation of the small surface grinder.

Cutting	Down Feed for Pass Across the Work	Suggestions
Steel mild	0.0005 to 0.003	Use coolant
Steel hard	0.0005 to 0.001	Use coolant and dress often.
Cast iron	0.002 to 0.006	Coolant not required—but it helps keep dust under control. Cast iron dust will contaminate coolants.
Aluminum	0.0005 to 0.005	Wheel is open structure and coarse. Be on the watch for loading and heat buildup, especially when no coolant is available.
Nonferrous metals	0.0005 to 0.003	
Stainless	0.0005 to 0.001	Use coolant.

12.3.12 Removing Work from the Chuck

Removing work from a magnetic chuck can be challenging because of both the air seal created against the chuck and the residual magnetism.

Safety Always

Never attempt to take any work off the chuck before the wheel completely stops turning.

Save That Chuck!

To avoid gouges in the precision chuck surface, do not slide the work if you can tilt, then lift, it. A paper gasket placed under the work at setup time will prevent scratches when trying to remove work. The secret here is an old telephone book: Its thin paper is perfectly parallel and will stay so even when the coolant soaks it. Okay, maybe telephone books are old technology—but use something else!

UNIT 12-3 REVIEW

Replay the Key Points

- Grinding wheel balance is a critical issue not only to achieve good finish and accuracy, but also for long life of the grinder's spindle bearings.
- When dressing, take one final pass slightly slower than when truing, but not too slow. But do not go back over the wheel to be sure. Be sure the *X* axis can't move.
- When mounting a grinding wheel
  - › Ring and visual test the wheel.
  - › Double-check wheel size and RPM.
  - › Always use paper washers.
  - › Torque the nut(s) per specs or instructor demonstration.
  - › Balance the wheel, dress it, and then rebalance it.
  - › Warm up the wheel with a short run before grinding.
  - › Always visually inspect a mounted wheel before using it.
- Potential movement due to the grinding wheel's force is to slide to the left and/or to tip over to the left. Tipping is more dangerous.
- Appraise the amount of hold area relative to the grinding forces and height. If you can tilt the work with your hand, it's definitely not safe!

Respond

1. List the seven steps to mounting a new wheel on a grinder.
2. How is a grinding wheel balanced?
3. Which of the two equal sized steel blocks shown in Fig. 12-61 is safest for grinding the top? Describe how to improve the setup of both.
4. On a sheet of paper, sketch the correct blocking for the cast iron object in Fig. 12-62 to be ground safely. Or describe it in words.
5. Would blocking and grinding aluminum on the surface grinder be correctly called float, flat, or finish grinding?
6. Name at least five other options beyond blocking that can be used to secure a steel workpiece on a surface grinder.

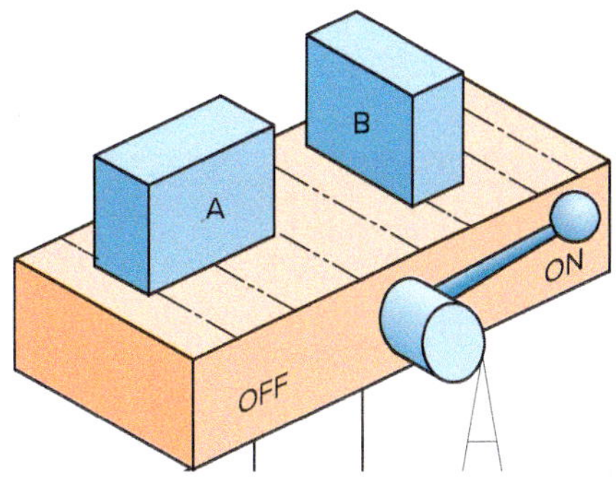

Figure 12-61 Which is safer and why? How can this setup be safer?

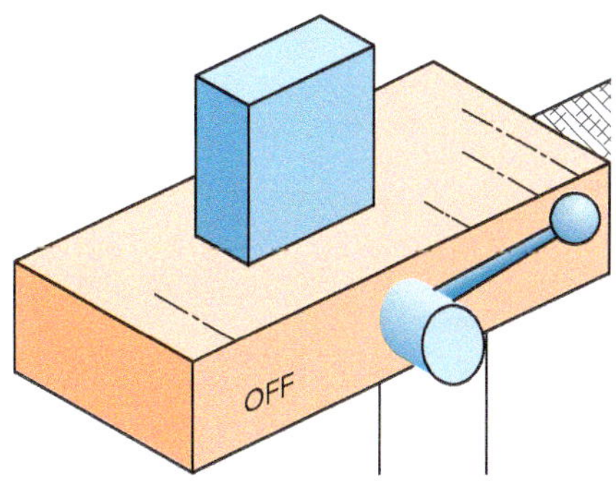

Figure 12-62 What is needed to make this setup safe?

Critical Thinking Questions

7. When dressing the grinding wheel, it is important to place the diamond at least 15° to the left of the wheel (beyond center; see Fig. 12-45). We didn't discuss this; however, you should be able to answer why.
8. When must a demagnetizer be used?
9. Name two times when a demagnetizer would not be needed.
10. For what three reasons is dressing with a diamond performed?

Unit 12-4 Other Grinder Operations—Visual Inventory

Introduction: Our objective here is to provide a background idea about other types of grinders found in industry that may or may not be in a technical school curriculum. They will most likely be introduced in an industrial apprenticeship or on-the-job training. When any new grinder is to be learned, the skills gained in wheel selection and metal removal by surface grinding will apply.

Unit 12-4 is offered so you can describe and recognize these technical processes. As such, it does not contain safety or skill training. We'll just take a brief look at the machine and the kind of work it can do.

TERMS TOOLBOX

Blanchard grinder The common name for a rotary table, vertical-axis-wheel-head, surface grinder featuring segmented wheels.

Centerless grinder A cylindrical grinder that does not hold the work but rather contains it in a three-point suspension among a grinding wheel, support blade, and rubber regulating wheel.

Cylindrical grinders One of two kinds of grinders that create round objects: the center and the centerless types. Only the centerless type can grind the entire length of a shaft. Only the center type can grind a taper.

Front feeding Putting cylindrical work through a centerless grinder by tilting the regulating wheel. This grinds the work along its full length.

Regulating wheel (centerless grinder) The wheel that rotates and controls the workpiece in a centerless grinder.

Through feed Feeding work from front to back on a centerless grinder.

12.4.1 Manually Operated Grinders Live On

Although each machine described is found in both manual and CNC versions, the manuals are still doing their job in many shops. These have been slow to be replaced by CNC versions because, in the past, grinding machines were envisioned to be single-operation equipment–the part is put in, and a single surface is ground. For a long time, job planning was written with that image in mind. Each machine does its particular specialty very well and won't be pushed out of the shop for some time to come. But it will happen someday.

That one-at-a-time image is changing for grinders in general due to CNC control. It has already changed for tool and cutter and gear grind machines. In Fig. 12-63, a CNC tool grinder is finishing a worm gear on its OD.

Figure 12-63 A CNC tool and cutter grinder finishes the OD of this worm gear.
Saint-Gobain Industrial Ceramics, Inc.

CNC grinders can perform multiple operations, dress their own grinding wheel, change grinding wheels, dress the new wheel into a contour if needed, and make complex shapes not possible by manual grinding (Fig. 12-64). We even see grinding heads attached to CNC machining and turning centers and tool grinders as part of the machining center's capability. Here are a few selected industrial grinders. To win the game and sound "in the know," learn the trade terms!

12.4.2 Cylindrical Grinders

There are two general types of machines for grinding precision round objects: the center type and **centerless grinders**. Each machine grinds round objects but in a radically different way.

Center Type

These grinders are similar to a lathe. They hold and rotate the work by the same means as a lathe: collets, chucks, and between centers.

Basic Construction

Much of your lathe training will be invaluable here especially for work holding. But there is a difference between the grinder frame and the lathe.

Figure 12-64 Looking closely, the aluminum oxide wheel shows when the safety door is opened on the CNC cylindrical grinder. Otherwise, it might be taken for a turning center.
Courtesy Okuma America Corporation

Pivoting Bed

When adjusting to cut a straight cylinder shaft on a lathe, the tailstock is moved laterally until the shaft runs parallel to the lathe's Z axis. The lathe frame remains stationary. Not so on a cylindrical grinder. The entire bed pivots at its center, which includes the headstock and the tailstock. That flat bed supporting the headstock and tailstock is clamped at each end to a saddle that sits atop bed ways. The bed and ways reciprocate in the Z (right-left) axis under hand cranking or power feed (Fig. 12-65).

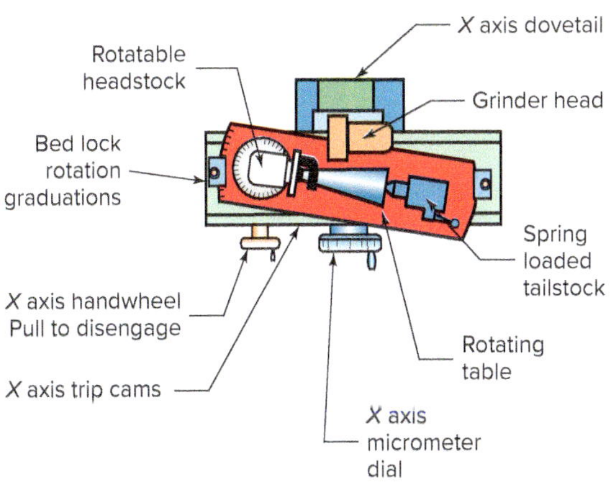

Figure 12-65 Overhead view of basic manual cylindrical grinder.

In the overhead view, the bed is set to grind a long taper with the work held between centers. To adjust away taper, the locks at each end are released, and the entire bed with head and tailstock are rotated relative to the saddle.

Other Setup Options—Grinding Steep Tapers

The cylindrical grinder can cut short tapers and grind shoulders in two different ways (Fig. 12-66).

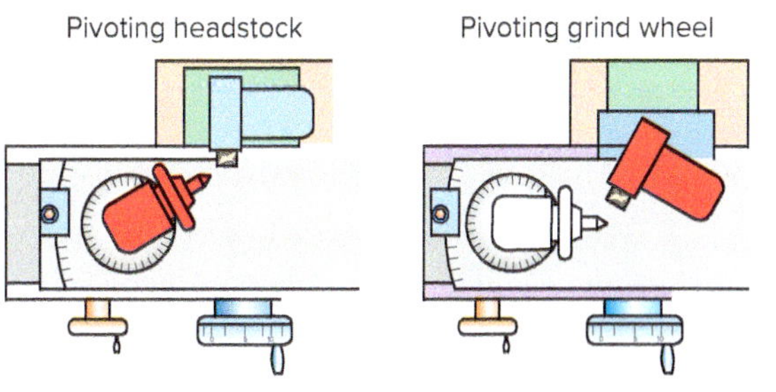

Figure 12-66 An overhead sketch of two different kinds of taper setups. Both functions can be used together to achieve special operations.

Pivot the headstock or the grinding wheel The headstock can be moved on the bed ways–forward and back similar to a tailstock. And it can be turned and locked relative to the bed, as shown on the left in Fig. 12-66, or the grinding wheel can be pivoted, as shown on the right. In both setups, the bed X axis is locked, and the work is held and rotated in a front-closing collet. Figure 12-67 shows a typical use of these pivoting functions. Here the grinding wheel has been pivoted, then dressed parallel to the machine's axis again. This allows grinding up to a sharp corner shoulder without touching the face of the work–only the diameter.

Universal internal attachment Some cylindrical grinders feature a second grinding head that allows internal grinding operations (Fig. 12-68). Here, the work is held in a chuck or on a faceplate. The external wheel and guard have been removed. The internal head is mounted and driven by a belt from the spindle motor. A mounted stone is being spun in the internal head. These stones rotate at high rates–from 10,000 to 30,000 RPM on cylindrical grinding machines. But, on dedicated internal grinding machines, air turbines drive them to 150,000 RPM or higher (Fig. 12-69).

Why Choose a Center Type?

Cylindrical grinders can hold repeatable tolerances of 0.0003 to 0.0001 in. with finishes easily within 16 μin. or finer. The advantage of these machines over the centerless type is that they can grind work that is not initially round and can round areas on work of any size or shape that can be swung within the work envelope, for example, a bearing on a camshaft, a crankshaft, or a shoulder diameter on a hex bolt (Fig. 12-70).

Figure 12-67 This cylindrical grinding wheel has been angled to widen its face and to reach the work shoulder (top view).

Courtesy Fives Landis Corp.

Internal Grinding Setup

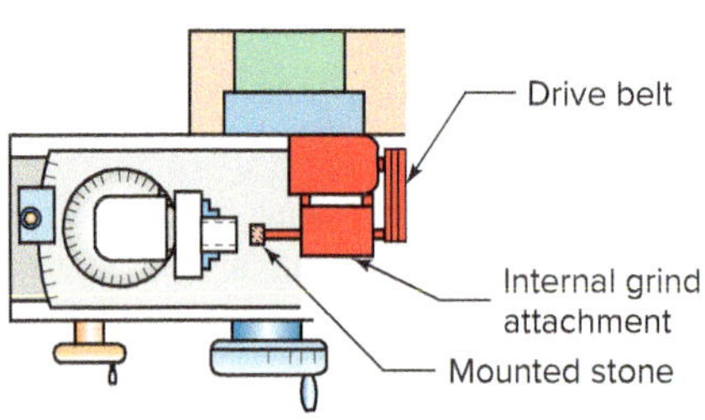

Figure 12-68 The internal head is set up on this cylindrical grinder.

Inside-Diameter-Mounted Grinding Wheels

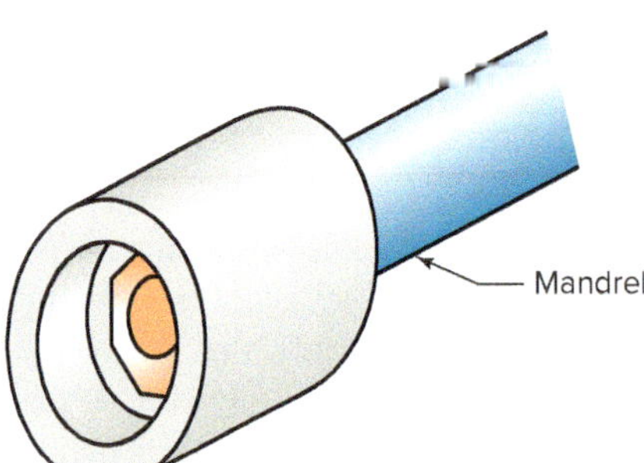

Figure 12-69 High-RPM grinding wheels for internal diameters are mounted on mandrels.

Figure 12-70 Center type cylindrical grinders can hold and grind objects that aren't round.

Courtesy Maximum Advantage-Carolinas, Precision Grinding Machinery/Tega Cay, SC

Centerless Cylindrical Grinders

The unique **cylindrical grinder** can grind a round shaft along its full length–something a center type cannot do. Work is not chucked. It floats and is trapped in a three-point tunnel between the steel support blade, the rubber regulating wheel, and the grinding wheel (Fig. 12-71). The centerless grinder's main purpose is to reduce diameters. They do not finish tapers or shoulders. A centerless is best suited to pins and shafts, but they grind them accurately and very fast. Repeatable tolerances of 0.0001 to 0.0002 in. with finishes of 16 μin. are common (Fig. 12-72).

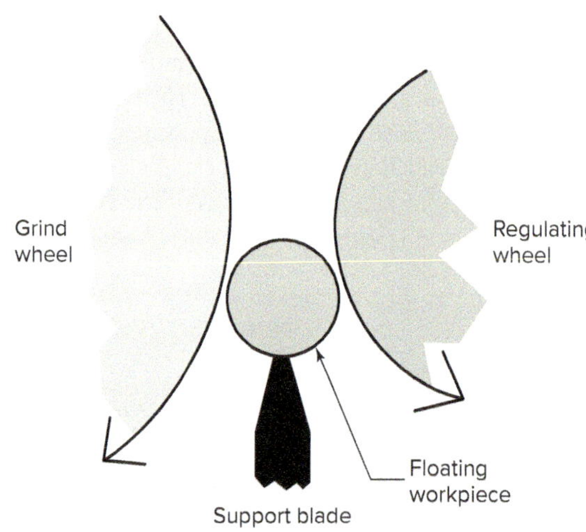

Figure 12-71 The three components that support the work on a centerless grinder.

KEYPOINT

Due to the unsupported work and the nature of inserting it into the tunnel, it is unlikely that you'll encounter a centerless grinder as a beginning machinist. They require experienced operators.

To use a centerless process, the work has two requirements:

1. It must be round and smooth initially.
2. Its diameter must fit into the three-point work tunnel.

Through Feed or Front Feed

There are two ways the centerless grinder is set up to grind round work.

Through feed for large batches By slightly tilting the rubber **regulating wheel** axis relative to the grind wheel, work can be inserted in the front to eject from the back into a part-catching tray. The grinding wheel is set at the correct diameter. It is not fed inward, so a very limited amount of material can be removed in this setup (0.003 to 0.005 in. maximum).

A **through feed** setup is made where a large number of parts are to be ground to a specific diameter, without taper, along the entire length. Figure 12-73 is a simplified diagram of a through feed setup. This is a very fast method. The work can be loaded, one after the next, as fast as the operator can insert them! They fall out to a catch pan.

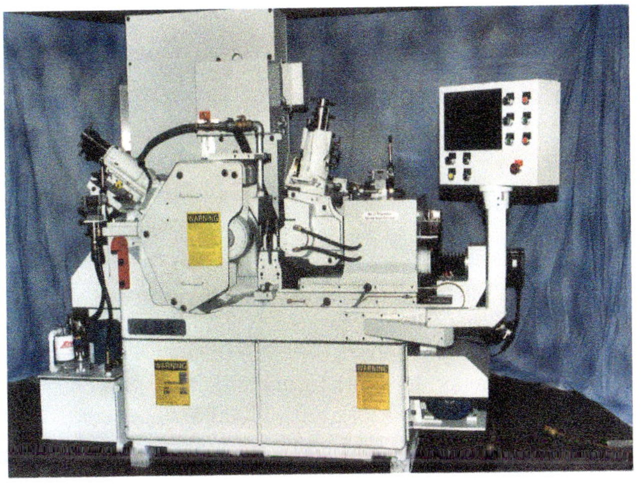

Figure 12-72 Both CNC and manually operated centerless grinders are found in industry.
Courtesy RSS Grinders & Automation, Inc.

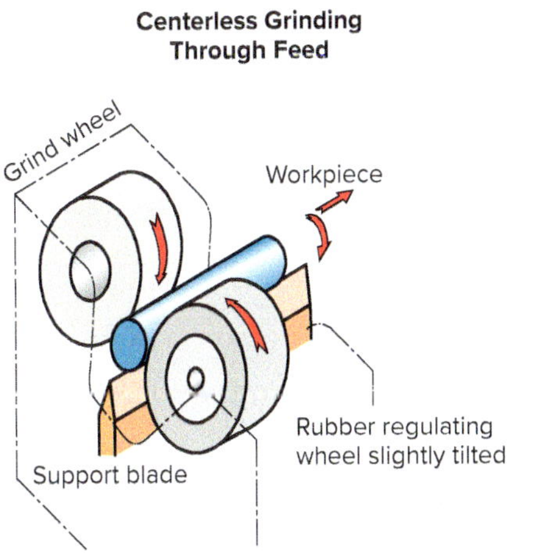

Figure 12-73 A simplified sketch of through feeding on a centerless grinder.

Front Feed (Insert and Remove)

In **front feeding**, the regulating wheel is parallel to the grind wheel, and the work is inserted in the opened tunnel, bigger than finished size. A combination stop gage and part pusher rod is at the far end of the tunnel to ensure the work goes in to a preset depth and that it stays inserted at that depth (Fig. 12-74).

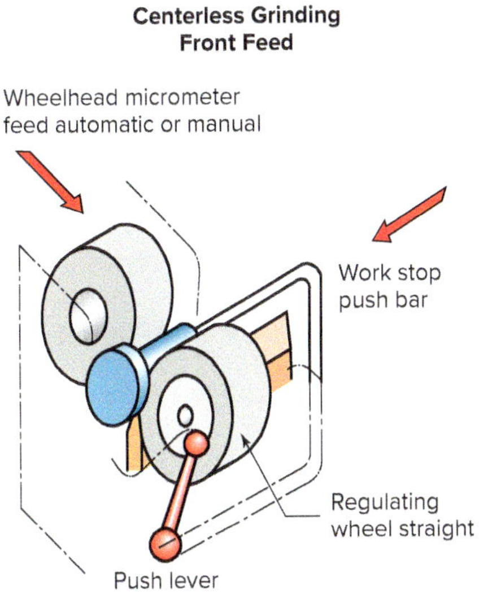

Figure 12-74 A simplified sketch of front feeding on a centerless grinder.

Then the grinding wheel is dialed toward the regulator until it reaches final diameter. This action is automatic or hand cranked. Upon completion, the grinding wheel is pulled back to open the tunnel. The part eject lever is pulled, pushing the work out of the tunnel, toward the operator where it's removed. This setup allows grinding to a step, which cannot be done with through feed.

KEYPOINT

Centerless grinders do not hold the work, but rather, they trap it in a three-way suspension between support blade, grinding wheel, and regulating wheel.

12.4.3 Tool and Cutter Grinders

End mills, reamers, lathe tools, face mills, and all other cutting tools are sharpened on a universal tool and cutter grinder. These machines are exceptionally flexible in their setup configurations. The manual machines feature a fully tilting and rotating head plus a horizontally rotating table. It's likely that your school will have one of these machines (Figs. 12-75 and 12-76).

Figure 12-75 Sharpening the spiral peripheral teeth on an end mill requires screwing the cutter forward by resting it on a thin support finger and pushing it past the grinding wheel.
Courtesy Get an Edge-Tool Grinding Service/Mukilteo, WA

Figure 12-76 Sharpening the end teeth on an end mill.
Courtesy Get an Edge-Tool Grinding Service/Mukilteo, WA

On the manual versions, there are literally dozens of accessories available to hold, rotate, and grind tools. Additionally, many kinds of grinding wheel shapes are available (Fig. 12-77) as well as different kinds of straight-line and radius dressing tools. Learning to operate a tool and cutter grinder is a trade in itself.

Grinding Wheels for Tool & Cutter Grinding

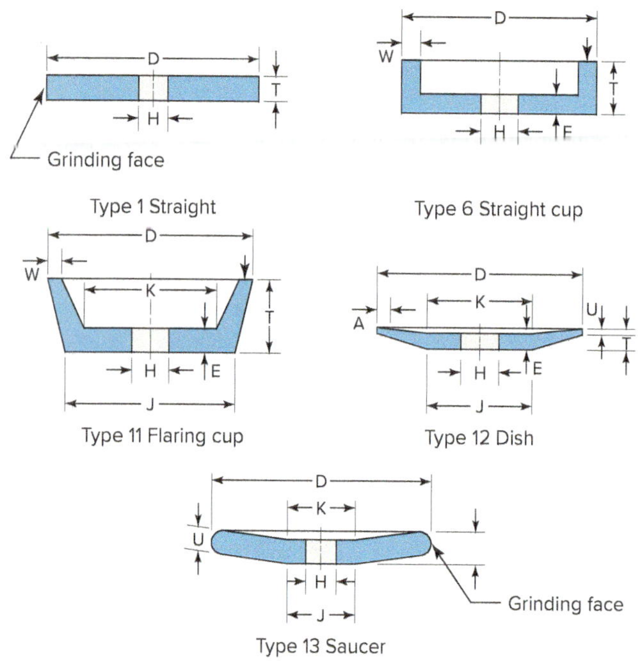

Figure 12-77 Several different grinding wheel shapes are used on tool and cutter grind machines.

12.4.4 Internal Cylindrical Grinders

Three varieties of machines finish the inside diameter of work. First are jig grinders (Fig. 12-78) which locate and grind holes to exact size and position. Next, internal diameter (ID) grinding machines (Fig. 12-79), which are similar to cylindrical grinders with internal heads, rotate the work while stroking the mandrel and stone in and out of the hole. They can grind large objects. The third is an internal attachment for cylindrical grinders that we've already seen.

KEYPOINT

Mandrel Wheels Must Achieve Extreme RPMs In order for these relatively small diameter grinding wheels to achieve the 4,000 minimum F/M surface speed required for efficient grinding, their RPM must be between 10,000 and 150,000 RPM, depending on the diameter of the wheel. Be extra careful and double-check RPM setting on machines that feature multiple setting.

Figure 12-78 A jig grinding machine features a very accurate X–Y locating table and the ability to dial in hold diameters within 0.0002 to 0.0001 in. accuracy.
Courtesy Smiths Aerospace-Actuation Systems/Yakima, WA

Figure 12-79 An internal diameter (ID) grinder resembles a cylindrical grinder, but it's designed to rotate larger work and the smaller grinding wheels must rotate at much faster RPM.
Courtesy Mike Fitzpatrick

Jig Grinding Machines

The objective of a jig grinding machine is final machining of holes to the best possible cylindricity, finish, size, and position tolerances. They are considered extra precision equipment. Operation of this machine is often considered a specialty within the machinist's role.

Jig grinders feature an ultraprecise positioning table to locate holes within 0.0001-in. repeatability. The grind head functions something like an adjustable boring head, in that the radius of swing of the mandrel grinding wheel can be dialed out to any radius, in 0.0001-in. increments. Constructed somewhat like a vertical mill, jig grinders feature a vertical spindle usually powered by an air turbine (see Fig. 12-78).

ID Grinding Machines

Inside diameter grinders are much like an OD machine in purpose but not in construction (see Fig. 12-79). Having short ways and extra heavy headstocks, they are made to hold and rotate large and heavy work while finishing a single hole. Their purpose is extra precise control of the diameter and its taper.

12.4.5 The Rotary Surface Grinding Processes

These machines are high-volume metal removers compared to reciprocal machines. The large grinding wheel rotates in a horizontal plane (vertical axis). Different from regular grinding wheels, they are composed of replaceable segments bolted to a base plate, called a composite. It's similar to a large, insert tooth, face mill cutter. The advantage of segments is that one is easy to handle at a time.

Work is held on a rotating, magnetic, horizontal table. These two disks, the chuck and wheel, are rotated off-center to each other while the distance between them is closed to control work thickness. They both spin the same direction (counterclockwise), which means at their contact point the wheel action opposes the chuck and work (Fig. 12-80).

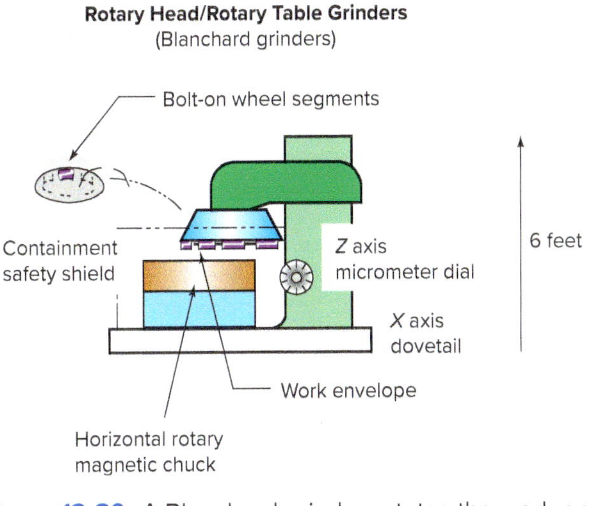

Figure 12-80 A Blanchard grinder rotates the work on a horizontal table while grinding it with an overhead vertical axis wheel.

It doesn't take too long to see that only an expert ought to run one of these machines. The forces are tremendous.

SHOPTALK

By Any Name Like the Bridgeport mill, these machines are often referred to as **Blanchard grinders** due to a popular manufacturer (Fig. 12-81). These brutes can often remove more metal per minute than a milling machine! That's not an exaggeration. Using a 48-in. chuck Blanchard, I have regularly made fairly large "C"-shaped blue chips when grinding steel! Rotary machines are especially good for premachining where one or two sides of a group of parts must be made flat or to size. The accuracy is within 0.001-in. repeatability or better.

Figure 12-81 A Blanchard grinder can remove metal efficiently using a high powered, horizontal grinding wheel (yellow with stone segments below) while holding close tolerances.
Courtesy Blanchard Division of DeVlieg Bullard II, Inc.

UNIT 12-4 REVIEW

Replay the Key Points

- Cylindrical grinders can hold repeatable tolerances of 0.001 in. with finishes easily within 16 μin. or finer. The advantage of these machines over centerless is that they can grind work that is not round initially.
- Centerless grinders do not hold the work, but rather they trap it in a three-way suspension among support blade, grinding wheel, and regulating wheel.
- Be extra careful and double-check RPM setting on machines that feature multiple setting.
- Mandrel-mounted wheels need to run at very high RPMs to achieve correct cutting speeds.

Respond

1. Name three machines that grind the inside hole diameters.
2. Beyond the fact that it cuts with a grinding wheel, how is a standard cylindrical grinder different from a lathe?
3. True or false? Only the jig grinding machine can finish several holes on one part. If it's false, what will make it true?
4. Which machine uses segmented grinding stones? What is its shop lingo name?

Critical Thinking Question

5. Approximately how fast (RPM) must the rim of a $\frac{1}{2}$-in.-diameter, mandrel-mounted grinding wheel go to equal the rim of a 14-in.-diameter cylindrical grinder doing 1,714 RPM? What RIM speed is that? (*Hint:* The short RPM formula helps solves this.)

CHAPTER 12 REVIEW

Unit 12-1

Like so many other skills we've discussed, we concentrated on the basics of wheel selection to begin your education. I'm sure you realize there will be dozens of new abrasives types and wheel combinations found out in industry–it's going to be far more complex. But the skills you've already gained on lathes and mills, combined with surface grinding experience, will be enough preparation to transfer to many different kinds of grinders. The only missing piece lies ahead in the CNC sections and lab practice.

Unit 12-2

Upon completing Unit 12-2 you should have gained a fairly good idea of how a reciprocating surface grinder is constructed and how it works. You are ready to see the demonstration and try a small job on the tool room version. You are ready to proceed to Unit 12-3, to make setups that overcome these forces and safely grind most materials.

Unit 12-3

An understanding of magnetic chucks along with grinding wheel skills of Unit 12-1, is the heart of Chapter 12. You now know how to analyze the forces and to anticipate how to put together setups that guarantee safe grinding.

Unit 12-4

There are more grinding machines than the ones we've seen in Unit 12-4. Other than thread and gear grinders, which require a great deal of training, they all follow the same principles we've learned: Choosing the right grinding wheel, anticipating the heat and forces, and making setups that overcome them will be fairly simple.

QUESTIONS AND PROBLEMS

1. Interpret the grinding wheel shown in Fig. 12-82. How will it be best used? (Unit 12-1)
2. What variable does the bond hardness control? Explain the grade selection when grinding a fully hard tungsten steel tool bit (HSS). (Unit 12-1)
3. True or false? In general, coarser grit wheels with less air space are needed to perform heavy-duty removal grinding–rough grinding. If it is false, what will make it true? (Unit 12-1)
4. True or false? In general, coarser wheels with less air space are needed to grind soft metals such as aluminum. If it is false, what will make it true? (Unit 12-1)
5. What duties does coolant perform for grinding? Identify the extra duty beyond those it performed for chip cutting. (Unit 12-2)
6. What are you seeking during a ring test? (Unit 12-2)
7. What does a *paper washer* do? Name two things. (Unit 12-2)
8. True or false? A mounted diamond used for wheel dressing is called a *nib*. If it is false, what will make it true? (Unit 12-2)

Figure 12-82 In general, how might this grinding wheel be used? For what duty and material?

CRITICAL THINKING QUESTIONS

9. While we didn't cover it in the reading, why is a diamond nib set 15° beyond wheel center when dressing a wheel? (See Fig. 12-45.) Why not place the diamond on the other side of the wheel? (Unit 12-2)
10. Using Fig. 12-83 as a guide, describe or sketch the chuck accessory that would prevent slipping of this thin steel object. (Unit 12-3)

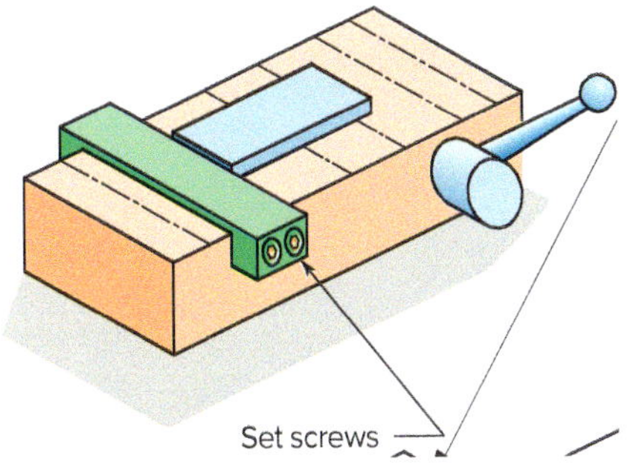

Figure 12-83 The setup.

11. Describe and explain the final pass when dressing a grinding wheel. (Unit 12-2)
12. On a sheet of paper, sketch or describe in words, the stages of grinding a warped plate–to become flat and parallel. (Assuming it is thin enough that the magnetic chuck will pull it down flat to the chuck.) (Unit 12-3)
13. Describe briefly or sketch the difference in the two kinds of cylindrical grinders. (Unit 12-4)
14. On a sheet of paper, sketch the work tunnel in a centerless grinder. Include the regulating wheel, the grinding wheel, and the support blade. Indicate the direction each wheel rotates. (Unit 12-4)
15. Explain or sketch the difference between lathe frames and manual cylindrical grinders. (Unit 12-4)

CRITICAL THINKING PROBLEMS

16. What's wrong with the surface grind setup shown in Fig. 12-84?
17. The aluminum workpiece shown in Fig. 12-85 is not ready to grind. Why? Describe or sketch the improvements needed in the setup.
18. You are surface grinding a particularly hard tool steel die component with a white aluminum oxide grinding wheel. After a short time, the wheel begins leaving a dark band of burned metal as it strokes across the work. List as many reasons as you can think of that might have caused the problem.
19. You dressed the wheel from Question 18, made sure the coolant is directed correctly, and then began grinding again. It starts grinding well, but the burn marks soon return. In sequence, what might you do to troubleshoot this setup?

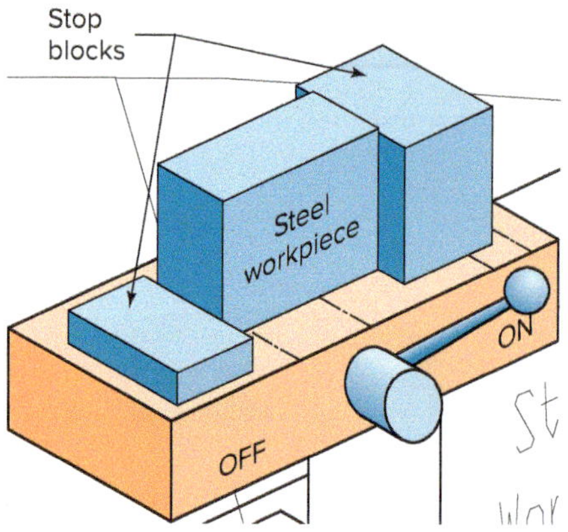

Figure 12-84 How can this setup be improved?

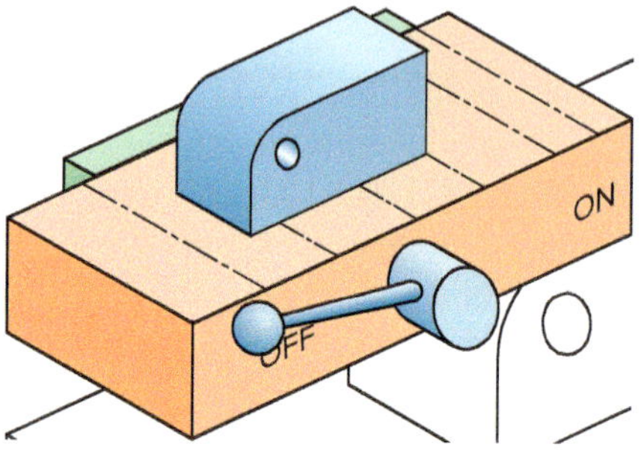

Figure 12-85 What can be done to improve the grind setup of this aluminum workpiece?

CNC PROBLEM

20. Identify the advantages that CNC surface grinders have over manual ones.

Design element (gear): ©Design Pics/PunchStock

CHAPTER 13 Screw Thread Technology

Units

13-1 Technical Thread Variations

- Define the function and form of acme, buttress, and square threads
- Calculate the lead of multiple threads
- Draw a sketch of thread variations—left-hand and multiple start
- Recognize pipe threads on technical drawings or on real parts
- Identify various grades of bolts by their markings

13-2 Getting and Using Thread Data from References

- Use the index in *Machinery's Handbook*
- Gain a working knowledge of the kinds of thread information found in the *Handbook*
- Find and calculate thread facts for a setup

13-3 Manual Lathe Setups for Unusual Threads

- Set up the manual lathe to machine a *left-hand* thread
- Set up the lathe to machine a *multiple-start* thread
- Set up the lathe to machine *tapered* threads

ATC Automation, based in Cookeville, Tennessee, since 1977, is a company that builds custom assembly automation and test equipment for a variety of industries, including the medical devices, consumer products, and automotive sectors. The company applies a wide range of assembly technologies and engineered solutions in all facets of discrete part assembly and test systems. It focuses on areas such as lean manufacturing and robotic assembly. The company has effectively designed and developed the machines that build many products we commonly use today, including the equipment used to manufacture headlamps in automobiles, the electrical components of electric toothbrushes, and numerous surgical devices.

A lean automated assembly and a fully automated assembly system assemble and test products for the life science industry at ATC Automation.

Courtesy of ATC Automation, Cookeville, TN

INTRODUCTION

In the mechanical world, threads perform a wide range of duties beyond fastening. They drive aircraft control surfaces or quickly move and accurately position CNC axes. Threads also join pipes that carry high-pressure fluids or hold scaffolding together—each a similar but different duty. Threads are used to lift objects as heavy as your car or to secure the lid on a mayonnaise jar. Each function has its unique thread, different from the standard versions we've thus far studied. Knowing how to machine them is one indicator of journey-level skills, because you'll need to find and use data.

After a discussion on the technology of specialty threads, we'll practice using information found in *Machinery's Handbook*. You'll need a copy available either on CD or hardbound to complete problems in this chapter. When scanning the *Handbook*, you'll quickly see that we're touching only the highlights. Any edition or version of a machinist's reference should yield the same info, but it might be found in a slightly different place than described in the questions.

When prepping manually operated lathes for threading, you'll need to find the specs to grind or select the tool bit's shape and size. Also set the lead selectors, and to make other setup changes. When programming CNC lathes to cut any thread, we use prewritten routines called *canned cycles* or *fixed cycles*, a generic template capable of cutting any thread, except it's missing the numbers (parameters). It's a matter of filling in blanks or leaving them off when they don't apply (Fig. 13-1). **The point is that operating a manual or CNC lathe, you'll need to understand thread formulas.**

Figure 13-1 A CNC thread routine makes thread cutting a snap.
Keith Smith

Today, few shops make batches of threaded parts by any method other than an automated or CNC lathe. Still, when the job is *hot* and your shop needs one or two specialty threads made right away, it's the top machinist who's capable of whipping them out on the old reliable manual lathe. Here's the background you need to be that machinist, making specialty threads on any lathe, CNC or not.

Unit 13-1 Technical Thread Variations

Introduction: Threads perform three very different task families (Fig 13-2). Some designs require a thread to perform a combination of duties. Based on the target function, there are subtle differences. Those small details make a difference in the machining and measuring of a specific thread. Before we can discuss the hard data, let's look briefly at the ways threads are put to work. Knowing the terms matters when threading–give extra attention to learning them.

Three Task Families

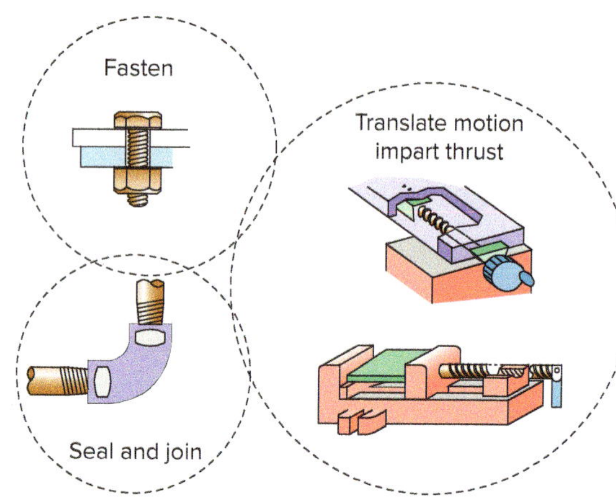

Figure 13-2 Screw threads have three different functions that sometimes overlap.

TERMS TOOLBOX

Acme form A thread built on a 29° triangle.

Axial thrust (force) The lateral force of a screw on a nut, parallel to the screw's axis.

Ball-screw (recirculating) Split nuts with a strong preload forcing them apart on the screw. Round balls fill the hollow space between the nuts and screw. The ball-screw has zero backlash and very little friction.

Buttress form A thread designed for thrust in one direction only.

Galling (rhymes with "falling") Deformation of metal under heavy loads. Causes smearing and lockup of threaded fasteners.

Helix angle (threads) The angle created in a thread, whereby the triangle has its opposite side equal to the thread's lead and its adjacent the circumference of the thread.

Lead (pronounced "leed," long *e*) The distance a thread translates in one full revolution.

Multiple start Threads featuring more than one helical groove.

Radial thrust (force) The outward push of the screw against the nut, tending to keep the nut locked in position on the screw when tightened.

Square form Designed for multiple translations–not commonly used today.

Stripping Shearing away of the threads from excess force.

Stud bolts Bolts with no head–threads on both ends.

Thrust The aspect of threads that is the *push* along the axis of the thread.

Translation The function of a thread that moves an object.

13.1.1 The Three Task Families

Task 1 Fasten and Assemble

A common bolt fastening, a component on your car's engine, is a one-time-only assembly. It may be removed later, but it's designed to be permanent. On the other hand, bolts that locate and fasten an inspection cover on a hydraulic fluid reservoir must be screwed down tight to prevent leaks. But they are going to be removed and reinserted many times throughout their service lives. They will be made with more precise threads and be of higher carbon steel–perhaps hardened to an SAE Grade 8.

Still others are fasteners on which lives depend, so they must not vibrate loose and/or must be made of extra strong metal (e.g., the bolts and nuts holding wheels on trains and airplanes). Their *root* (bottom of the thread shape) will be rounded to distribute stresses better. They, too, would be made from hardened steel but the Grade 8 would be too brittle–perhaps a slightly tougher but a little softer bolt, a Grade 7, would do the job better. Using the wrong bolt in this application could lead to disaster.

Task 2 Impart Motion and/or Thrust (called *Translation*)

The feed screws that move the milling machine table in *X*, *Y*, or *Z* are bidirectional **translation** screws (**lead** screws). They are designed to move their axis hundreds or even thousands of times per day, for years. They are machined with one of four very different kinds of threads, depending on how much force is imparted, called the **thrust** *load*, and on how fast the translation must occur. Another great divisor between translation screws is the amount of acceptable backlash (discussed later in this unit).

As you've discovered in the lab, manual mills and lathes have a given amount of lead screw backlash–reversing the axis crank causes a stall in motion until the clearance between the nut and screw reverses. However, that kind of action is unacceptable in precision CNC machinery. Requiring zero backlash, CNC axis drives make use of a highly engineered thread called a ball-screw. Then there are thread and nut combinations that fall between the simple lead screw and the ball-screw, with just a bit of backlash, called a preload antibacklash screw (as discussed later in this unit).

On the less accurate side, the screw in a bench vise or on a screw jack for lifting an automobile are both translators, but the force is in one direction only and backlash is of no consequence. But between them there are subdifferences. A screw jack for lifting your car must translate its load farther with each turn of the crank than a bench vise; otherwise, it would take too long to lift it. So, we modify the thread for each function. The vise must also remain tightly closed during vibration and cutting forces so it fastens temporarily as well as moves the vise jaw. Those two duties require two different versions of a form called an **Acme** (see Unit 13-2).

Task 3 Form a Seal and Join

Threads joining pipes where leakage is not permitted may carry low-pressure water, gasoline, or high-pressure gasses. Each requires a similar but slightly different pipe thread.

Some pipe threads are tapered so they seal when tightened. As the tapered surfaces tightly contact each other, we call that bringing the thread *home*. The everyday pipe thread is less precise. It needs a sealant compound between the two joining pieces to make a tight seal when it's home. A second version is more precise in its form, because it must seal without compounds, to avoid chemical reactions with the material carried within the pipe–called a dry-seal pipe thread.

KEYPOINT

Threaded mechanical devices *fasten and assemble*, *translate and thrust*, and *seal and join*. Some threads are called upon to perform a combination of these duties. Each specialized duty has a specific modification for the task. Threads fail in two ways—by **stripping**, whereby the threads break off completely, or **galling**, where they deform to the point that they jam and cannot be used.

13.1.2 Thread Forms for Various Functions

Unified Threads

Before discussing other thread forms, we need to take a closer look at some of the terms and formulas that define the Unified standard thread. Take a moment to study Fig. 13-3. Throughout Unit 13-1, we'll be comparing other threads to the standard 60° triangular form to highlight functional differences.

Unified Thread Terms

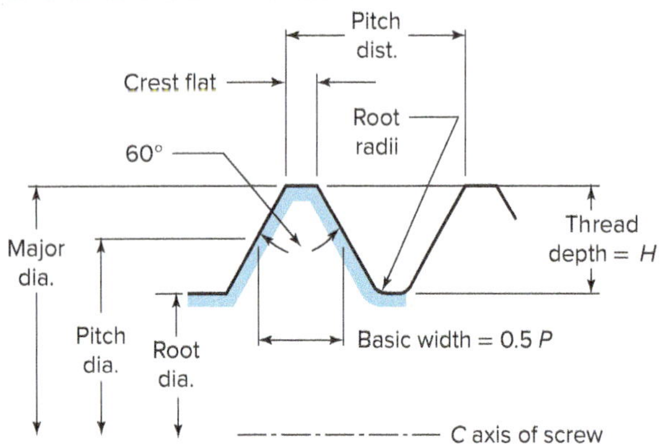

Figure 13-3 Basic screw thread terms found in ANSI/ ASME B1.7M (revised).

Radial Versus Axial Thrust

There are two kinds of forces acting on a thread: when it is turned (called torquing the thread) along the threaded shaft's long axis–**axial thrust (force)**. The other is outward, known as **radial thrust (force)**. Much of the difference between threads lies in how much radial thrust is needed to maintain fastness.

> **KEYPOINT**
>
> Increased radial thrust means the thread will remain fastened better, meaning it won't loosen or vibrate loose. But radial thrust robs axial thrust from the total input torque put upon the thread.

The next three forms are designed to push laterally. In other words, compared to a 60° triangular thread, they convert more of the rotary torque into axial thrust.

Radial thrust is the characteristic of a screw thread to translate some of the torque put upon it, into force outward against the nut (Fig. 13-4). For a fastener, radial thrust is a good thing. When a standard nut is tightened on a bolt, it locks on the thread due to the wedging effect of radial thrust (force), such that it resists loosening under vibration, changing loads and temperatures. On the other hand, radial force becomes lost power and undue friction in a translating screw.

Screw Thread Forces on the Nut

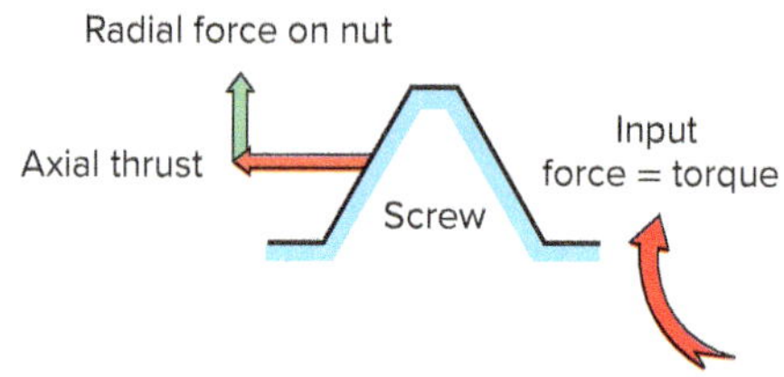

Figure 13-4 When the face of the screw pushes on the nut, two forces are created: The axial force is parallel to the screw's axis, and the radial force is outward against the nut.

> **KEYPOINT**
>
> In all screw (and gear) systems, the input force never perfectly equals the output forces due to friction loss. But the objective is always to use as much of the force in a useful direction as possible, known as effective force.

Acme Threads

Evolving from the 60° thread, the Acme thread (Fig. 13-5) is also triangular and truncated. But its triangle isn't an isosceles, it's an equilateral with an apex angle of 29° (Fig. 13-6). When we observe their shapes, it's obvious the Acme threads are better suited to axial thrust.

General-Purpose Acme

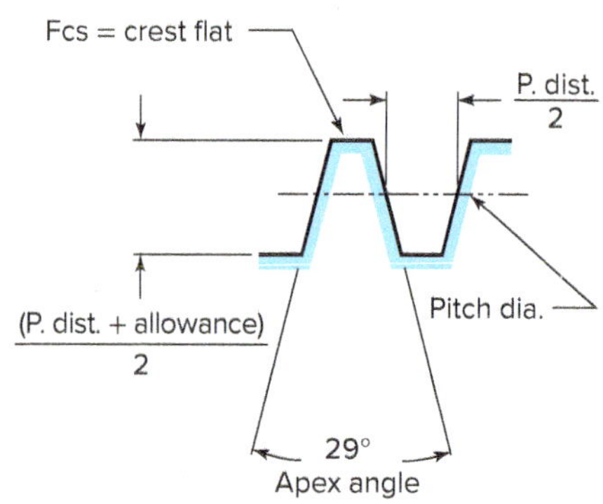

Figure 13-5 Acme screw threads impart bidirectional thrust and moderately accurate translations (ASME/ ANSI B1.5 [revised]).

Comparing Forms

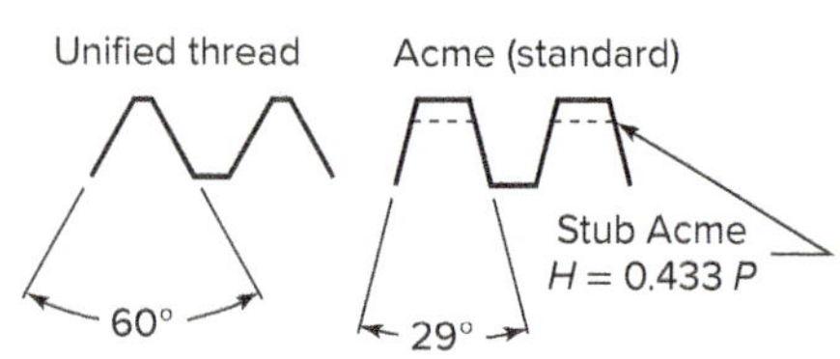

Figure 13-6 A comparison of Unified and Acme forms reveals their functional differences.

Acme threads are designed to push harder—their faces are more vertical with larger contact areas; thus, they tend to wear less and convert more torque to axial thrust. They last longer when constantly translating, as a milling machine's lead screw would.

> **KEYPOINT**
>
> Functionally, Acme threads resist stripping and wear better during constant use. They are specified for devices that must repeat translations and thrusts—in both directions.

But, due to steeper faces, Acme threads do not stay fastened as well as 60° threads due to reduced radial force. The 14.5° faces (29°/2) on acme threads would make poor fasteners. They loosen due to temperature changes and vibration.

There are several subforms and classes of Acme thread, depending on the specific application. Figure 13-6 depicts two: the standard and the stub (truncated farther down the triangle, making shorter, less deep threads). The choice depends on the size of the shaft upon which they will be cut. To avoid stripping, the stub thread must deliver a slightly diminished thrust force compared to the taller version. But the stub doesn't weaken the shaft on which it's machined nearly as much. It creates a more compact assembly.

> **KEYPOINT**
>
> For the same reasons as an Acme, based on function, most other thread forms can also be specified as stub threads.

Buttress Threads

These threads perform similarly to Acme threads, delivering repeated translation thrusts, but here, in *one direction only.* They feature an even sturdier form and vertical face compared to both Acme and Unified forms (Fig. 13-7). It's obvious which face is the thrust direction. They efficiently translate torque into nearly 100 percent axial thrust (less friction loss).

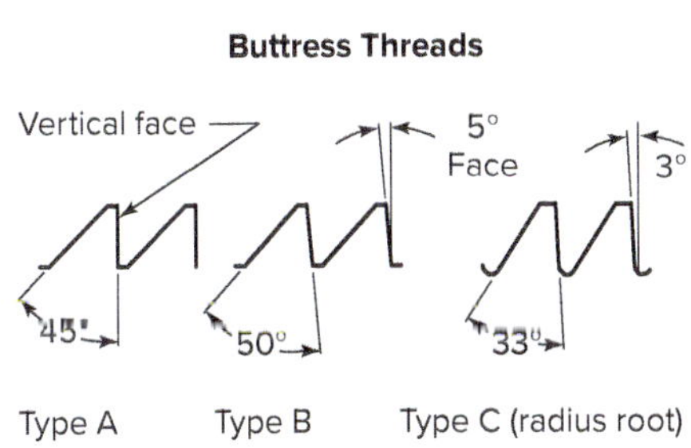

Figure 13-7 Buttress threads are designed to thrust axially—in one direction.

The screw for a drill press vise shown in Fig. 13-8 or a screw type car jack might be made with a buttress thread. Since there is a momentary thrust in the opposite direction to open the vise, the Acme thread might also be specified, especially for higher demand vises such as in milling. The car jack's thrust will always be up, and it's a perfect application of the **buttress form**.

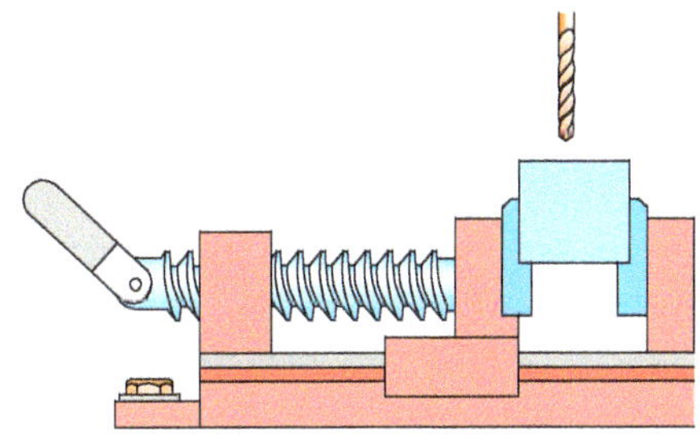

Figure 13-8 A drill press vise screw might be made with a buttress thread.

> **KEYPOINT**
>
> Buttress threads are designed to translate and thrust in one direction.

Keeping to our theme of subtle differences, there are three subcategories of buttress thread forms: Type A, straight vertical faces, and Type B, 5° slope to the load face, causing a bit of radial thrust and a radius root for extreme duty applications machined upon hardened screw shafts.

Square Threads

These are the strongest threads for multiple translation and thrust loads in both directions. But, totally lacking radial thrust, they work loose the easiest of all the thread forms, so they're never used where permanency (semi-secure fastening) is of concern. **Square form** threads offer the advantage of large surface contact area between nut and screw; thus, they last a long time during repeated use (Fig. 13-9).

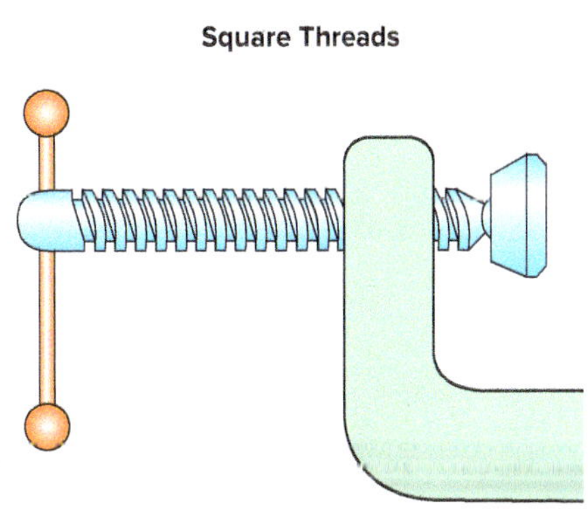

Figure 13-9 Square threads are the best at imparting thrust, but it's difficult to measure their effective diameter.

Not Used in Engineered Devices

Even though they offer translating advantages (strong, long lasting, and bidirectional), square threads are the least popular for precision applications due to the difficulty in measuring them as they are machined. Their straight sides make it impossible to gauge pitch diameter (PD) other than measuring the *root diameter*, then calculating PD based on it. Also, due to their broad, flat *crests* (the top of the thread form), they are more difficult to produce with a tap than all other thread forms. There is no practical measuring device (other than optical comparitors, which require removing the work from the lathe or by functionally screwing them into a comparison nut) that tells when they are machined to the right diameter.

The bolt is tested against an acceptable nut. Conversely, while machining a nut, an acceptable screw is the functional test. If the gauge assembles, not tight or loose, the threads are complete. Due to this lack of size control, square threads are usually found on crude devices such as the clamps in Fig. 13-9.

> **SHOPTALK**
>
> **Not Quite a Complete Metric World!** The complexities of changing to a metric pipe size would be a big headache for construction applications, especially remodeling. So, inch standard pipes remain in effect worldwide. Therefore in many metric nations, although their pipe threads are actually in inches, they are specified in the metric equivalent size, which creates very odd metric pipe size numbers.

Pipe Threads

Pipe threads fall into two major types: mechanical fastening and sealing types. Each of these types is then subdivided into two categories that could truly be called loose and precision fitting. Radial thrust becomes important again, since these threads must not only stay permanently assembled but also seal in many of its forms. So most pipe feature tapered internal and external threads–they tighten as they are screwed in (Figs. 13-10 and 13-11). Therefore to gain radial thrust, pipe threads use the familiar 60° triangle on both straight or tapered pipe threads.

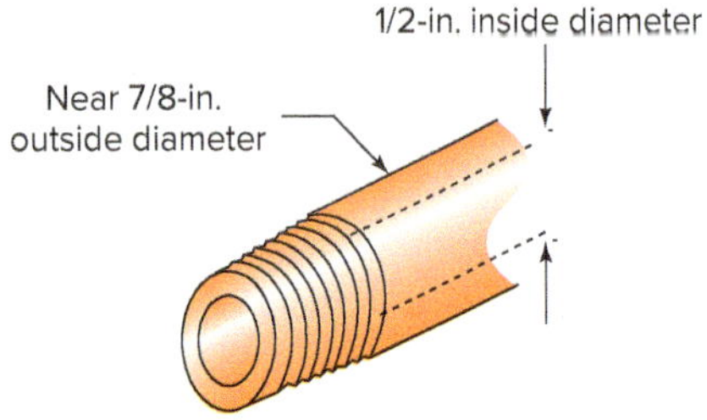

Figure 13-10 Because pipe threads are designated by the ID of a standard wall pipe, they are physically much bigger than their nominal size.

Tapered Pipe Threads

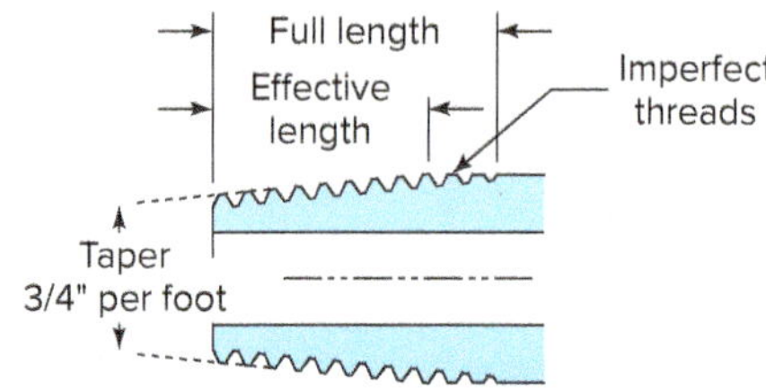

Figure 13-11 Tapered pipe thread terms.

Nominal Differences for Pipe Threads

Pipe threads share one unique difference from all other threads. Their measured physical size is much larger in diameter than their nominal size. (Recall *nominal* means the target or design size.) In the past, when pipe pressures were low due to the technologies of the day, pipe threads were based on the inside diameter of the pipe, not the major diameter of the thread. So, a $\frac{1}{2}$-in. pipe thread was close to $\frac{7}{8}$ in. in diameter because the $\frac{7}{8}$-in. hole in the pipe is $\frac{1}{2}$ in., the determining factor (Fig. 13-10).

That relationship remains true for standard wall thickness, wrought steel pipes today (the metal and plastic pipes you would find at the hardware store). However, for heavy or extra heavy wall pipes, the difference between outside diameter and thread size is even greater.

> **KEYPOINT**
>
> Pipe sizes are based on capacity, not physical diameter. Pipe diameters are bigger than the nominal size.

13.1.3 The National Pipe Thread Designations

In North America, pipe threads fall within the American Society of Mechanical Engineers (ASME; formerly ANSI) standards B1.20.1-revised. Similar to the unified system, their designations look similar to other threads. Here are the three most common, but as seen in *Machinery's Handbook*, there are a bunch of subcategories.

National Pipe Tapered (NPT)

Example: $\frac{1}{8}$-27 NPT. This tapered thread is commonly used for joining pipes carrying low-pressure fluids. The seal between the male (outside thread) and the female thread is accomplished when the conical thread comes home with a sealant compound between them. Taps and dies are available to cut these threads (Fig. 13-12).

Figure 13-12 A set of tapered pipe taps and dies.
Photography by Prographics at Milwaukee Area Technical College/McGraw Hill

KEYPOINT

Pipe thread pitches are different from common fasteners such that no other thread might accidentally be assembled with them.

National Pipe Dry-Seal (NPTF)

Example: $\frac{3}{4}$-14 NTPF. A tapered pipe thread that seals without compound or sealing material. This is the more precise thread used for gasses and high pressures. The F is used to designate fit-up dry.

National Pipe Straight Mechanical (NPSM)

Example: $1\frac{1}{4}$-$11\frac{1}{2}$ NPSM. Designed for general assembly where no internal pressure is required. Notice that the pitch is $11\frac{1}{2}$ threads per inch to prevent assembly to other threads, which could lead to disasters, as in scaffolding struts and bars, for example.

Critical Application Threads and Fasteners

When the thread must be rigidly controlled such that it will not fail, we must turn to standardized quality. These parts are made according to very rigid rules and specifications *not* found in *Machinery's Handbook* but in government and professional association documents such as the ASME-ANSI or military specifications (Mil-Spec).

SHOPTALK

The Bad and Good News About Tapered Threads!
Without a die, outside tapered threads are not commonly made on the manual lathe. However, they can be machined using a tapering attachment. To find out more about setting up a manual machine to cut tapered threads. Since that setup is very time-consuming compared to making them on a special pipe thread lathe using dies, tapered threads are not considered a normal manual lathe task. The dedicated thread lathe is a portable machine that can be moved to a job site to quickly produce straight or tapered outside diameter pipe threads. When they are occasionally requested in the machine shop, they are usually produced with tapered dies and taps, if available (Fig. 13-13). But, as I'll bet you've already guessed, CNC lathes machine them just by changing a parameter to taper while threading!

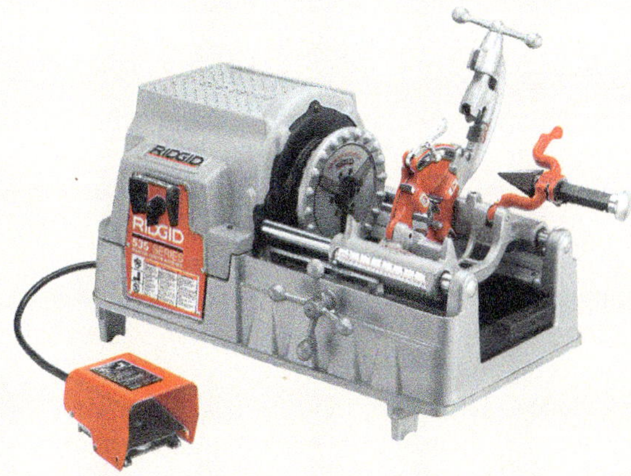

Figure 13-13 A portable pipe threading lathe is used at construction sites.
Courtesy RIDGID Tools

These documents are published at several levels, national government (Mil-Spec) or International Standards Organization (ISO), professional association standards (Society of Automotive Engineers, American Society of Mechanical Engineers), and individual companies publish their own specifications.

Where a national or international standard is in existence, lower authority specs such as company standards may expand on but not override or change them. They can supplement the specification to tailor it to the individual industry, but that expansion can never dilute a higher authority.

These closely controlled specifications are not typically given on the drawing–but they are noted usually with flags that tell the reader what specification to use (Fig. 13-14; flag 11 on the drawing).

KEYPOINT

If a thread spec is shown on the drawing but the information is not (the usual condition), then you must obtain the data and guarantee that your work follows them exactly. To be legal and certified, the parts must meet those specs.

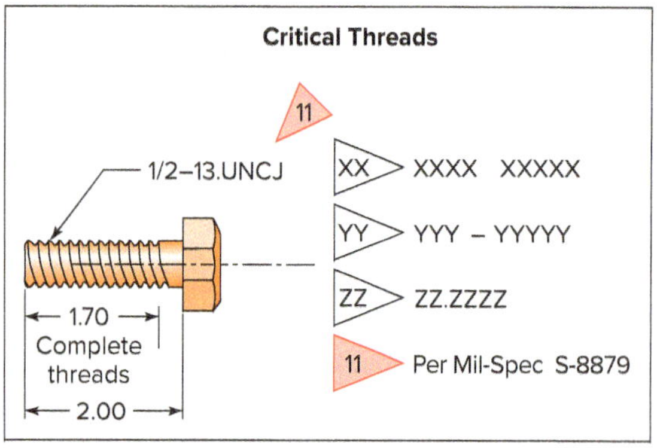

Figure 13-14 When threads must meet standard specifications, the control document will be noted on the drawing. You must meet the specs if the thread is to be legal.

Controlled Radius Root Threads

J form radius root threads are a good example of a standardized thread for a critical application. They are called out where fastening threads must endure vibration and other extreme duties. They must not fail. To help meet these extreme demands, the straight-sided 60° triangular form must be modified due to the hard intersection of the thread face and root diameter. The sharp corners of the truncation create points where forces can concentrate and start a failure in the bolt. The bolt's strength is improved by distributing the forces over a rounded corner, by adding closely controlled root radii. Note the J designation in Fig 13-14. Radius roots can be called out on any thread form, but when they are critical, a specification will be noted.

Where that radius must be controlled for exactness, the thread callout on the drawing will include the letter "J" (e.g., $\frac{1}{3}$-13 UNCJ). You must be able to determine the exact radius on the bolt or nut. There are challenge problems coming up where you can test your skill at this kind of data access.

Threaded Inserts

Standard tapped internal threads work fine in softer metals such as magnesium or aluminum, for many applications. But when great demand is put upon bolts screwed into these softer materials or where they must be assembled repeatedly without failing, standard tapped threads can strip or deform.

So here, too, when the assembly must not fail, we create a stronger, harder internal thread than the parent material by inserting 60° cross section, stainless steel or other hard alloy wires wound into a helical coil (Fig. 13-15). The insert wire is threaded into specially tapped holes to create a standard threaded hole that will accept a standard bolt.

Figure 13-15 A Helicoil® thread insert creates strong, stainless steel threads (or other alloys) within a softer metal workpiece.
Courtesy Emhart Fastening Technologies

KEYPOINT

Threaded inserts are used in soft and hard metals and even in plastics, to create stronger and more reliable threads.

Installing Threaded Inserts

To install a threaded insert (Fig. 13-16), first a specific diameter hole is drilled. Then a special tap is driven into the material, creating the thread that accepts the insert. It has the right pitch distance but a larger diameter than the standard size thread.

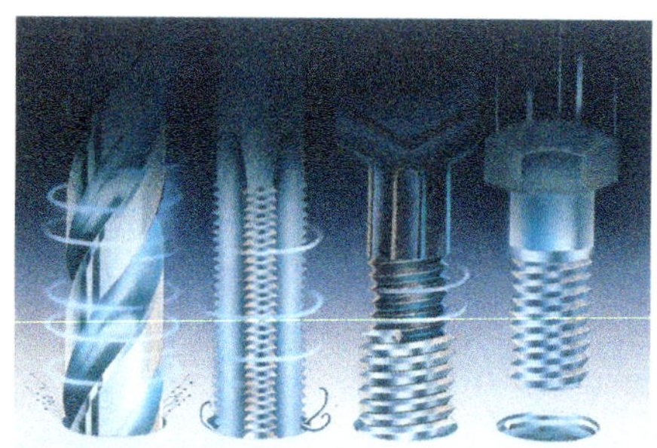

Figure 13-16 Installation is a four-step process: drill, thread the hole with special tap, screw in the insert, and then break off the drive tab.
Courtesy Emhart Fastening Technologies

To insert the coil, an installation tool is placed over a driving tang that protrudes from the coil's outer end, to drive it into the tapped hole. Once installed, the tang is removed from the coil by striking it with a rod.

Advantages

After insertion, the coil expands outward with a springlike action, to anchor it permanently in place. The key advantage to threaded inserts is that the coil distributes the assembly load over the full length of the thread, thus providing greater overall strength. They not only resist stripping; they also wear better for repeated use. Threaded inserts are common in aircraft, automotive, and many other manufacturing industries. The most common manufactured variety is HeliCoil®.

13.1.4 Fastener Grades

Steel bolts are classified by their strength grade. Hardness determines strength. Bolt grades span from 1 (softest) to 8 (hardest; Fig. 13-17).

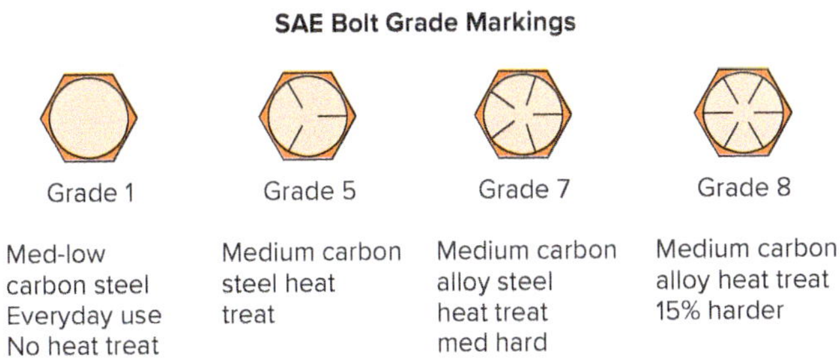

Figure 13-17 SAE bolt grade markings indicate the material from which it's made, the heat treatment (if used), and relative hardness of a bolt.

But as the bolt's tensile strength (rated strength during stretching) increases, the ability to flex and bend without breaking goes down. The hardest bolt isn't always the best choice. Here's a baseline for application.

SHOPTALK

Home and Automotive Applications of Threaded Inserts Accidentally stripped spark plug holes in your lawnmower or car engine can be retapped with specially made tools found at automotive supply stores. They come in kits with the right insert to fit the original spark plug. When tapping the hole, be extra careful to not let any of the chips fall into the cylinder, which would destroy the engine. The best solution is to remove the cylinder head. With small engines, you can sometimes plug the hole with a rag and then turn the entire engine upside down during tapping, such that chips tend to fall out—downward. Careful removal of the plug afterward helps prevent unwanted chips from entering the engine.

Correct Application

Grade 1 = Everyday applications–the bolts and nuts found at the hardware store. Assembling a backyard barbecue or child's wagon.

Grade 5 = Medium strength–fastening noncritical but demanding items such as a universal joint or a wheel on your car. Often found in automotive stores.

Grade 7 = More demanding applications–a cylinder head on your car or a jet engine part. Might be at the hardware store but usually bought through companies specializing in fasteners. May need to be certified to meet specs.

Grade 8 = Highest demand applications–milling fixtures, dies, and a part on a CNC lathe. Usually purchased through certified suppliers if they will be assembling critical items where safety is concerned.

ARMCHAIR FIELD TRIP

Don't miss this Haas Tip of the Day video about identifying bolts.

This video discusses how to tell the difference between metric and inch screws and much more! Go to: **https://www.youtube.com/haasautomation** and search for this video called: **The Nuts and Bolts of Screws: Bolt ID—Haas Automation Tip of the Day**

KEYPOINT

Bolt grade is determined by code marks on the bolt head. Grade is normally specified in the drawing notes—not in the thread callout.

13.1.5 Thread Variations

Now let's examine two unusual "twists" that can be added to any of the thread forms we've studied: left-handed and multiple-start threads.

Left-Handed Threads

These threads tighten in the opposite direction to those we've thus far studied. They have a functional purpose: sometimes the right-hand helix spirals the wrong direction. They might loosen under load and vibration.

For example, the nut holding the grinding wheel on the left side of a pedestal grinder would loosen from the grinding forces if it were a right-hand thread, but if we make it a left-handed spindle and nut, it remains tight. Studs holding the wheels on some cars are left-handed on the left wheels. You can see the LH stamped in the end of such **stud bolts** (bolts without a head–threaded on both ends).

SHOPTALK

In addition to the Society of Automotive Engineers, the American Society of Tool Making Engineers (ASTM) also publishes bolt grade standards that contain a few more classes for specific applications. The SAE bolts are the more common in everyday life. To see a comparison chart turn to *Machinery's Handbook*, under "Grade Identification and Mechanical Properties of Bolts."

A third example is a turn-buckle joiner for a cable (Fig. 13-18). Its purpose is to tighten the joint between two cables or rods, when the center buckle is rotated. As it is turned, the right-hand thread pulls from one end while the left-hand thread also pulls inward on the opposite end (Fig. 13-19).

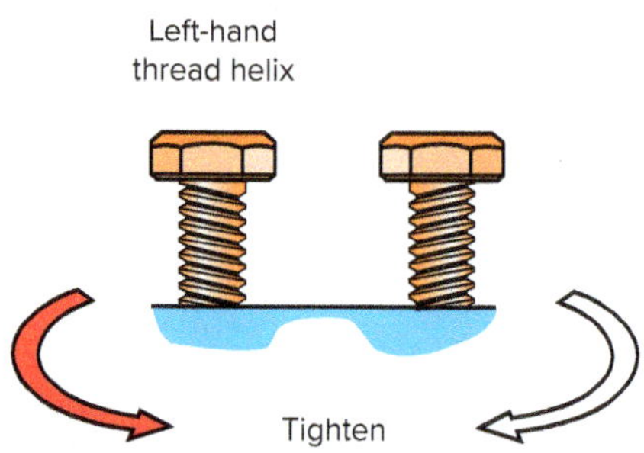

Figure 13-18 Left-hand threads tighten in the reverse direction.

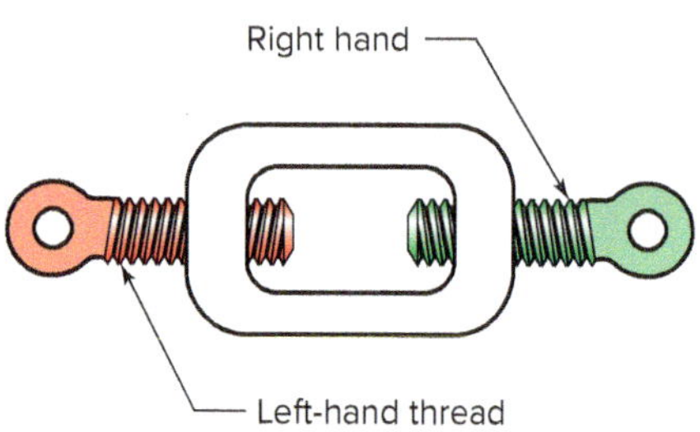

Figure 13-19 A turn-buckle is an example of left-hand thread use.

Any thread can be specified as left-hand spiral. For example,

$$\tfrac{3}{8}\text{-24 UNF LH}$$

or

M12-1 LH

Left-hand taps and dies are available but it's not common to find them in most shops unless they make many parts requiring left-hand threads. In Unit 13-3, we will explore two manual lathe methods of making these threads. Naturally, left-hand threads are a snap to produce on CNC lathes, but the trade tips we'll be learning will be useful.

Multiple-Start Threads

These threads are something of a mystery to beginning machinists. They feature more than one helical groove on the screw and on the mating nut. At first, they might be difficult to envision—more than one thread groove on the same shaft? Their function is to increase the rate of translation per revolution.

KEYPOINT

Multiple-start threads move the nut farther in one revolution compared to single-start threads.

The term **multiple start** means there are two or more individual thread grooves in the object. To explore them further, we need to define the term *lead* (long *e*) and compare it to pitch distance.

Lead is the distance translated by the nut in one revolution.

Pitch distance is the repeating distance between threads.

KEYPOINT

Lead and pitch distance are the same on single-start threads but different on multiple starts.

A double-start thread translates twice the pitch distance in one revolution compared to a single-start thread. Look at Fig. 13-20 to see a double-start thread not yet complete. As yet, it has only the first thread groove completed. Notice the blank space between grooves, ready for the second start. Then in Fig. 13-21, it's finished with both grooves. Placing an imaginary pencil in a groove, then tracing it one full turn of the helix, it would lead twice as far as would a single-start thread of the same thread pitch—it has a steeper (larger) **helix angle**.

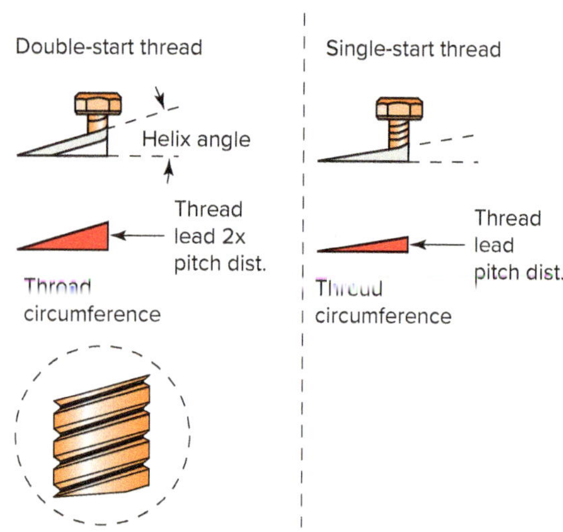

Figure 13-20 A double-start thread has two separate helical grooves. Only one groove has been machined in the magnified view.

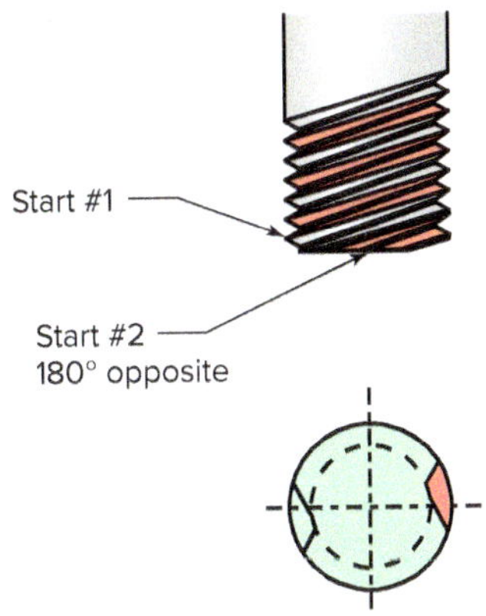

Figure 13-21 Both thread helices are now machined on this double-start thread.

The relative stripping strength of the double-start thread remains about the same as a single-start thread of the same pitch; the finished thread has the same number of threads per inch as a single start, and it has the same amount of thread surface carrying the load. The difference is the steeper angle of the individual threads.

Examples of multiple-start threads can be found all around you (e.g., the screw-on top of a glass mayonnaise jar or on a quality writing pen cap or your gas cap on your car). With one-quarter turn of your wrist, it's assembled or removed. They are either three- or four-start threads.

Calculating Lead

The lead of any thread is equal to the pitch distance multiplied by the number of starts.

KEYPOINT

Lead = Pitch distance × Number of starts

For example, what is the lead for this thread: $\frac{1}{2}$-20 UNF 2S (double start)?

Find the pitch distance.	1.000/20 = 0.050 in.
Multiply PD × 2	0.100 in.

In Fig. 13-22, it follows that, if the same thread were a triple start, what lead would it have? Three times the pitch distance at 0.150 in.

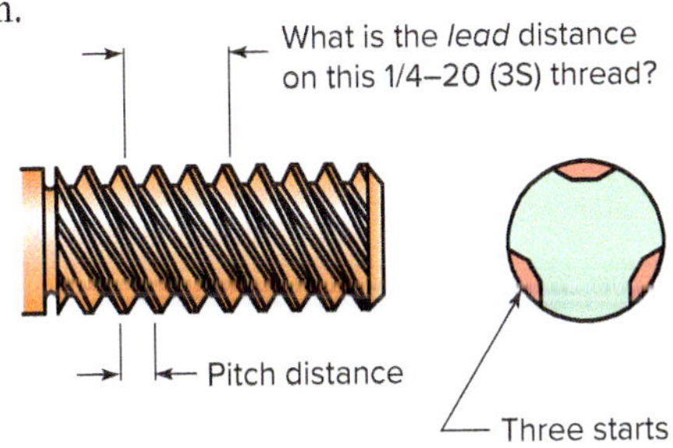

Figure 13-22 Calculate the lead.

Trading Mechanical Advantage

Using multiple-start threads in a device creates a trade-off of mechanical force for distance traveled, compared to single-start thread. With the inclined plane (wedge effect) becoming a steeper helix angle, the amount of mechanical force created per unit of torque applied is directly reduced as the lead increases. In other words, it takes a harder twist on the wrench to achieve the same tightening force as the multiple starts go up.

SHOPTALK

Multiple-"Lead" Threads You might hear multiple-start threads called multiple-lead threads as well. While not exactly correct usage, it's a common shop expression understood by most machinists. Truthfully, the thread has only one lead accomplished by several starts.

Backlash Control Threads

In some mechanical devices, backlash must be controlled to a minimum or eliminated completely, based on function. There are three common kinds of screw-nut systems that control backlash: split nuts, loaded split nuts, and ball-screw split nuts. Each is a design evolution toward less backlash.

An axis drive on everyday machinery can be as simple as an Acme screw and a suitably well-fit bronze nut. That system must have some mechanical clearance, even when new, otherwise it will bind. The best fit screw and nut will wear over time, even when well lubricated. There is nothing to do but replace the nut when backlash exceeds practical limits. The nut is generally made of a softer material (brass or bronze) to deliberately wear out before the more costly lead screw. But there are better systems.

Split Nuts—Adjustable

Most industrial-quality machines are fitted with bronze split nuts–two nuts running on the same screw shaft–with a small space between them. We saw an example in Chapter 10 on the X axis cross feed on most lathes. Backlash is adjusted to a minimum by moving one nut relative to the other–then locking its position on the pair's carrier. Adjusted correctly, it can have very little backlash. If adjusted to zero, the system would have too much friction and binding would be a problem (Fig. 13-23).

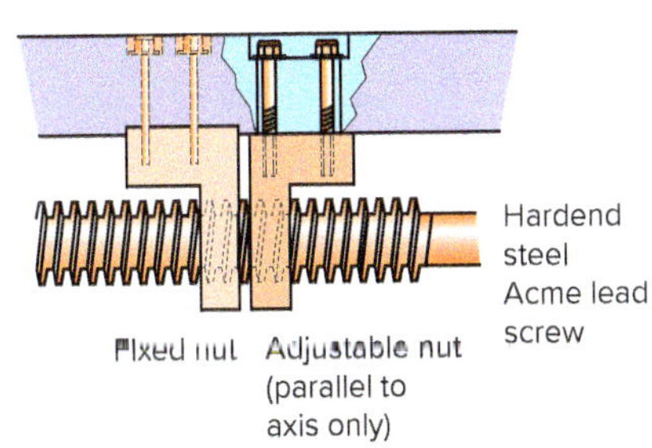

Figure 13-23 A simplified adjustable split nut backlash screw and nut.

Manual mills feature similar devices on their *X*, *Y*, and *Z* axes. These systems require periodic adjustment to compensate for wear.

Preload Split Nuts—Self-Adjusting

The next higher backlash design also uses the split nut concept, but now the two nuts have a strong spring (or hydraulic piston) placed between them. It allows the adjustable nut to move laterally (within a very small distance) relative to the other. The spring (red) pushes the sliding nut (orange) away from the fixed nut. In this system, pressure is exerted against both sides of the thread face, creating zero backlash for thrust loads up to a given amount (Fig. 13-24).

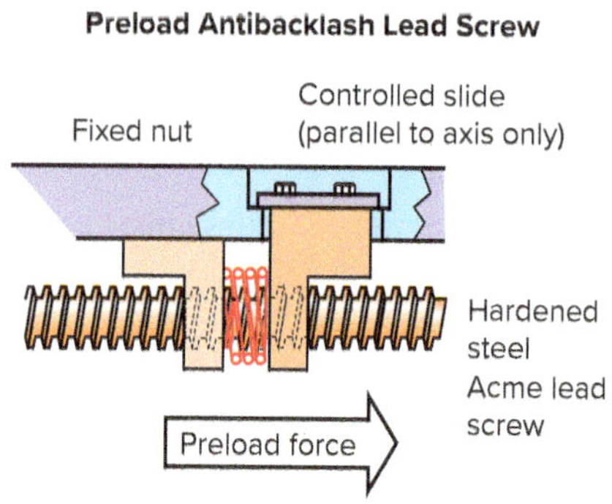

Figure 13-24 A simplified preloaded split nut system uses spring or hydraulic pressure to force nuts apart.

They continuously compensate for wear in the screw, even when the wear isn't uniform. The problem is that the screw often wears more in its center than at the two ends. The spring load nut system can compensate over these changing wear patterns for backlash, because it constantly adjusts. But accuracy is lost.

The preloaded split nut has two limitations. First, the spring action accelerates wear on the nut. The nuts must be replaced. The other is that the spring can be overcome by extreme machining loads, making the axis jump or bounce; not a good thing during a cut. As such, spring-load axis screws are limited to small, light-duty CNC machines only. One solution to the wear problem is to use Delrin® or Teflon® nuts or some other high-density plastic that slides extremely well on the metal screw. But they too cannot withstand heavy machining loads.

The final solution for the big gun, industrial-strength machines is the ball-screw.

Ball-Screws—Continuous Feed

While other axis drive devices have been tried for CNC equipment, we keep coming back to **ball-screws** as the translation system of choice. Again, a preload is used between two split nuts, but this time the pressure is so strong that no machining force can overcome the parting force on the nuts. Under that much side load, the friction would be so great between the nut and the screw that the system would bind or the nuts would wear out almost immediately. The friction solution is to forsake the Acme thread and machine and precisely grind thread grooves into half-round spaces on both sides–nut and screw (Fig. 13-25). This creates a round space between the two components.

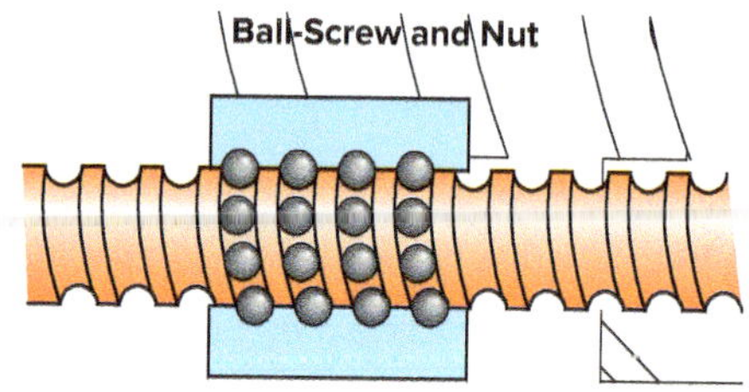

Figure 13-25 A ball-screw and single nut.

Then precision balls are introduced between the nut and screw. They roll and take the thrust load while reducing friction close to zero. But, in designing a ball-screw, one more problem had to be overcome–if the axis were to move farther than the width of the nuts, the balls would roll out of the nut. So, a recirculation channel surrounds the nut pair (Fig. 13-26). It can be as simple as a tube leading from the opening on one end of one nut to the opening on the far end of the other nut. As the screw translates over a great distance, the balls roll out into the channel to be pushed into the far end, thus continuously recirculating the balls.

Single and Multiple-Start Ball-Screws

To take advantage of the fast central processors in newer CNC controls, ball-screws feature multiple-start threads. Older CNC machines and other devices such as truck steering systems use single-start ball threads.

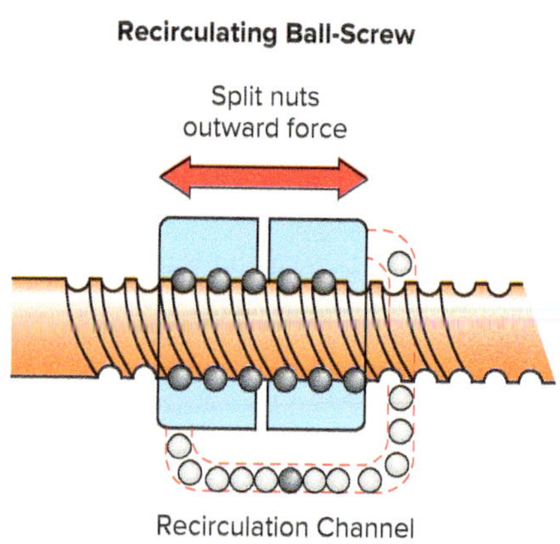

Figure 13-26 A ball-screw is the most popular CNC axis drive translating screw.

UNIT 13-1 REVIEW

Replay the Key Points

- Lead of a thread is equal to the pitch distance times the number of starts. These two numbers are the same on single-start threads.
- Pipe threads are specified with pitches different from common fastener threads such that no other nonpipe thread might accidentally be assembled with them.
- The Acme thread form is a truncated 29° triangle offering more face area compared to 60° threads.
- Buttress threads are specified when translation or thrust in one direction is the function of the device.

Respond

1. True or false? A *buttress* thread is designed for thrust in one direction only but has less contact surface between the nut and bolt, compared to a standard 60° triangular thread. If it is false, what will make it true?
2. Identify the thread form shown in Fig. 13-27.

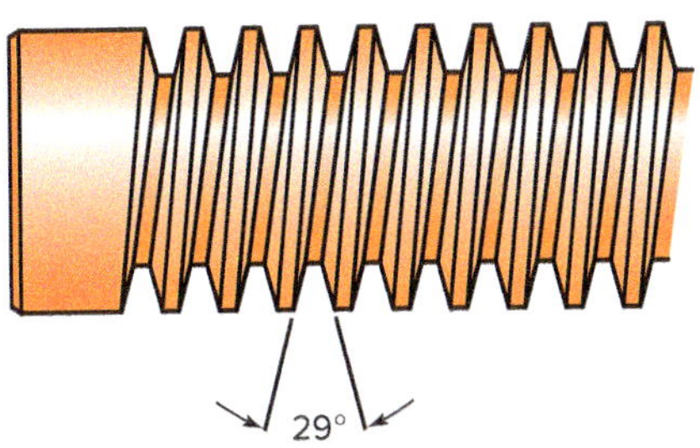

Figure 13-27 Identify this thread form.

3. In Fig. 13-28, the screw has 8 TPI. What is the lead?

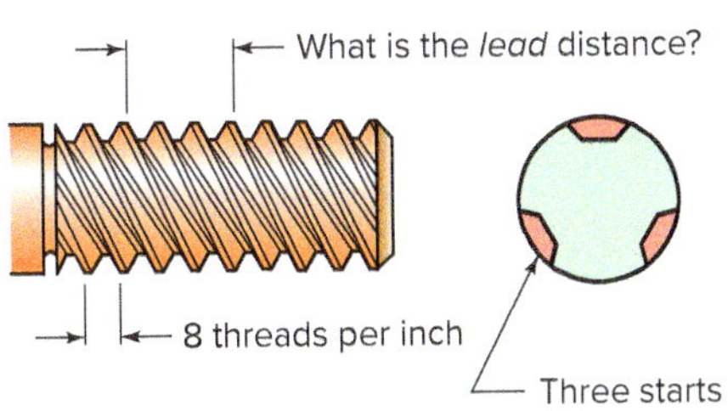

Figure 13-28 Calculate the lead.

4. Write the correct drawing callout for the pipe threads in Fig. 13-29. Requires looking in the *Handbook*. What thread form is used on this thread?

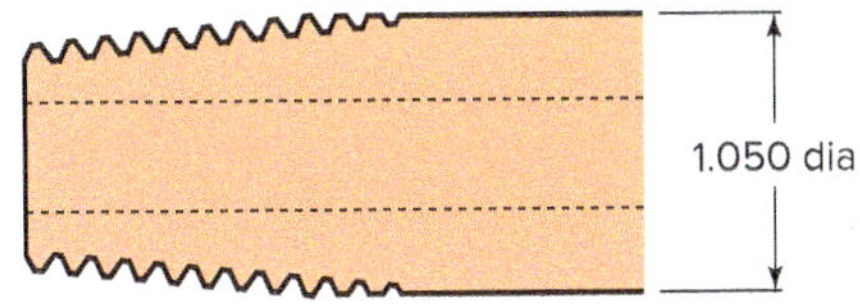

Figure 13-29 Write the thread callout for this pipe thread (standard wall pipe).

5. From *Machinery's Handbook*, what is the correct *pitch* for this tapered pipe thread?

$$2\frac{1}{2} - \_\_\_\_ \text{ NPT}$$

6. How would the callout change for the pipe thread of Problem 5 if it were to seal without a sealant compound?
7. How big is the inside diameter of the pipe shown in Fig. 13-29 in Problem 4?
8. Describe the three major functions of screw threads.
9. True or false? The difference in helix angles between a single- and double-start thread of the same diameter and pitch is that the double-start thread is greater (larger). See Fig. 13-20. If it is false, what will make it true?
10. A. Name two reasons a designer might specify a left-handed thread.

    B. Identify one functional reason to use multiple-lead threads over single-lead threads.

Unit 13-2 Getting and Using Thread Data from References

Introduction: You will need to make access data when setting up to machine, program, and measure a specialized thread. They will be based on reference books, charts, and company data. Here we'll concentrate on *Machinery's Handbook*, but many other sources will yield the same results. Keep in mind that this unit intends to provide a working knowledge of the kinds of information needed and how to find it when you need it—not to memorize the data.

13.2.1 Using Formulas and *Machinery's Handbook*

There is little instruction needed; only practice is required. Next are a few examples with answers found at the end of this chapter. Check them as you go.

UNIT 13-2 REVIEW

Respond

Solve the following puzzles that typify a custom thread setup in industry.

Problems 1 and 2 refer to this general-purpose thread:

ACME 1.0-8

1. What will the *smallest* possible *major diameter* be for the bolt (external thread)? This is the lower size limit of the shaft before threading. (*Hint:* Look under screw threads–Acme, in the index or search words if using the CD version.)
2. You are to set up and bore an internal Acme thread to the specification shown here. How wide should the crest flat be when this thread is to its finished size?
3. You must spotface a minimum cleanup flat area on a drill jig to accept a $\frac{1}{2}$-in. standard wing nut. What diameter counterbore should you obtain? See the hint in the answers if you get stuck.
4. You have been assigned to turn a type B buttress thread. To do so, you will need to grind the included angle of the bit to __________ degrees.
5. You are to rethread a spark plug and need information about the various dimensions. You suspect it is a metric thread. Where will you look?
6. There are actually two common types of American Standard Acme threads; what are they?
7. *Finalize Your Scan* From scanning the *index*, list the various major topics in which you could find information about threads. See if you can find more than I did in 3 minutes or less.
8. Turn to two subjects from your list or mine in which you have no information at all, such as rolling threads or Aero threads. Read just enough of the technical data there to satisfy your curiosity; then stop.

There are no new terms nor key points to review in Unit 13-2. Go to the Online Learning Center located at www.mhhe.com/fitzpatrick4e to find information on how to machine technical threads.

CHAPTER 13 REVIEW

Unit 13-1

After looking through *Machinery's Handbook,* I'm sure you realize that we've only touched the high spots on technical thread variations. In my apprenticeship, we studied them for an entire semester! The rest of your education will happen on the job when the nature of your shop's business determines which specialty threads you'll need to understand and make. An aerospace machinist in California will be making radically different threads compared to a tool and die maker in Detroit.

Unit 13-2

Although working with charts and specification books, and all the numbers they contain may seem boring, management programmers, inspectors, and engineers must have that deep understanding of their usage. Assuming you are climbing a career ladder, having a working knowledge of how to use them is an investment in your future!

QUESTIONS AND PROBLEMS

1. Other than fastening, name the two other functions of threaded devices. Describe each. (Unit 13-1)
2. Which thread form is built on a 29° triangle? (Unit 13-1)
3. Which thread form features a radius root? (Unit 13-1)
4. For what engineering reason was the radius root thread invented? (Unit 13-1)
5. For what functional reason do we sometimes machine left-hand helix threads? (Unit 13-1)
6. Briefly describe four ways a left-hand thread can be made. (Unit 13-3)
7. A steel hex head bolt has three lines on its head, equally spaced. What class thread is it? (Unit 13-2)
8. For what function is the buttress thread designed? (Unit 13-1)
9. What advantage does an Acme thread offer over a 60° thread form? (Unit 13-1)
10. Identify two possible ways to index a three-start thread on a manual lathe. (Unit 13-3)
11. When single-point turning a tapered thread, there are two possible setups. Identify them. (Unit 13-3)

PROBLEMS

12. The pitch distance of a given thread is 0.07692 in., but when a nut is turned upon it, it advances a little over 0.300 in. Explain. (Unit 13-2)

13. What is the number of threads per inch pitch for the screw in Problem 12? (Unit 13-2)

14. Explain what is different about the following thread compared to a standard unified form thread: $\frac{7''}{16}$-18 UNCJ. (Unit 13-2)

CRITICAL THINKING

The two bolts shown in Fig. 13-30 both have a problem in their manufacture.

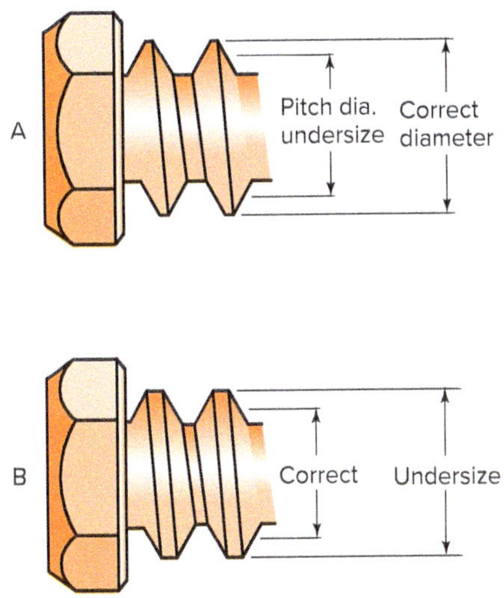

Figure 13-30 How will a correctly machined nut fit on each problem bolt?

On Bolt A, the *pitch diameter* is smaller than it should be by 0.010 in. However, the major diameter is correct. In other words, the round rod from which the bolt was made is the correct diameter but the threads themselves are too small.

On Bolt B, the major diameter is undersize by 0.010 in., but the pitch diameter, of the thread checks out to be correct.

15. When these bolts are assembled with a correctly machined nut, how will they fit?

16. How will each bolt perform when used? What are its weaknesses?

17. To set up and machine a $\frac{1''}{4}$-20 UNC thread, you must know the pitch diameter. From *Machinery's Handbook,* identify the minimum and maximum PD value for a Class 2 screw thread.

18. A $\frac{7}{16}$-18 UNC screw is measured and all dimensions are found to be within tolerance including the pitch diameter. Then the outside diameter of the shaft and the thread crests are damaged. To repair it, the diameter of the entire shaft length is turned down by 0.015 in. Then a chasing die is run back over the threads (see Chapter 5). How much did that rework change the pitch diameter?

CNC QUESTIONS

19. Briefly describe the most popular CNC axis feed screw and nut. Why is it used?

20. Name three or more reasons to choose CNC lathes to produce specialty threads whenever possible.

21. What is the one thread a CNC lathe might make that a manual lathe cannot?

CHAPTER 14 Metallurgy for Machinists—Heat Treating and Measurin Hardness

Units

14-1 Steel and Other Alloys

- Identify the various kinds of alloy numbering systems
- Use the AISI system for machining data
- List the four steel groups
- Identify the various tool steel numbers

14-2 Heat Treating Steel

- Describe the three major states of steel
- Quench steel to full hardness
- Temper steel to a lower controlled hardness
- Define annealing steel to a full soft state

14-3 Case Hardening Steel

- Identify the professional, case-hardening methods
- Explain the shop surface hardening process
- After a demonstration, perform a case-hardening process using the solid pack method

14-4 Aluminum Alloys and Heat-Treated Conditions

- List the conditions of aluminum
- Identify alloys of aluminum including heat treat conditions

14-5 Measuring Metal Hardness

- List the three common hardness testing methods
- Identify the destructive and nondestructive tests
- Relate hardness to function and to machinability
- Perform a Rockwell hardness test
- Convert units between Rockwell, Brinell, and Shore hardness

14-6 Physical Properties of Metals

- Define ductility, malleability, and elasticity
- Compare yield strength to tensile strength of metals
- Using *Machinery's Handbook,* identify physical properties and differences in given alloys

INTRODUCTION

This section rounds out your background in metals. Its primary goal is to understand why heat treatment of metal is often necessary in a job plan, and to learn the basics of do-it-yourself heat treating of steel and how to test hardness in metals before machining or after hardening, tempering, or annealing.

There are five reasons the various implements in Fig. 14-1 might have their physical conditions changed before, during, or after machining, forming, and welding. To

Make it harder for better wear or strength characteristics.

Make it softer to increase machinability or to improve long-term durability.

Relieve stresses created during its original formation at the steel mill, or to relieve stress created by machining, welding, and other processes.

Change chemical properties, which then permits further heat treatment.

Stabilize the metal from future changes in shape.

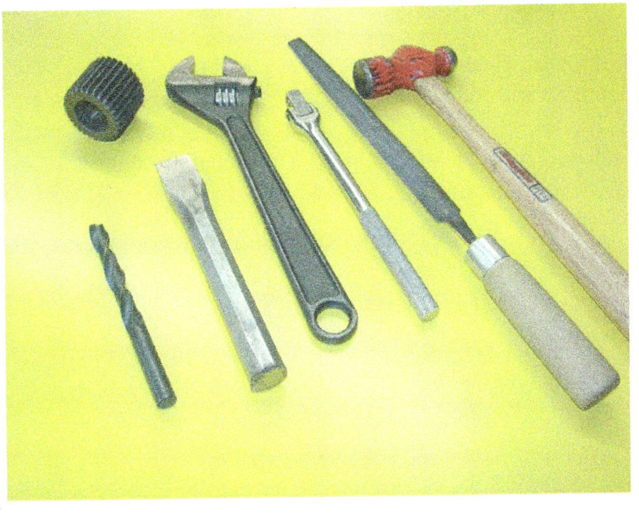

Figure 14-1 Various items are heat treated to various states of hardness and tempered durability.
Courtesy Mike Fitzpatrick

LEVEL OF ACCURACY

Similar to machining, heat treating can be performed to close tolerance using exacting equipment and standardized processes. But it can also be done to less demanding requirements. When traceability, product quality, and function are critical, processes are performed by specialists (Fig. 14-2). But utility heat treatment of shop tooling is another matter. In many shops, there's a need to harden or soften jig and fixture components and other items such as steel vise jaws. These are things we make in-house to keep the work moving along, but they won't be sold to customers. Using a small electric furnace or a handheld torch, the tradesperson is able to perform the basics.

Figure 14-2 Today, most shops contract their certified heat treating to outside specialty shops.
Courtesy Procedyne Corporation

Figure 14-3 Machinists with the ability to do their own heat treatment of shop tooling, short run production, or repair work deepen their career potential.
Prographics at Milwaukee Area Technical College/McGraw Hill

Learning to perform in-house heat treatment is a major goal in Chapter 14 (Fig. 14-3). But you'll also gain an appreciation of the skills of specialists. They must understand the use of powerful high-heat gas and electric furnaces (see Fig. 14-2) and other thermal equipment. Once those energy-hungry beasts are up to temperature, they stay energized! So, unless a shop has a constant need for heat treatment, it's not practical to maintain a facility. Today, most shops send their certified work to an outside contractor that serves an industrial area.

Measuring hardness is a basic requirement for machinists. You should be able to take your work to the hardness tester to determine if it's right according to the job plan, or to test mystery metal to determine correct cutting speed.

Unit 14-1 Steel and Other Alloys

Introduction: As we learned in Chapter 5, steels are supplied in large families known as the alloy steels, tool steels, and stainless steels. These groups could also be called consumer production steel, technical application steels, and noncorrosive steels.

There are many subtypes within each. Adding to the list, steel manufacturers offer proprietary (their own patented formula) alloys. So, similar to other specifications we've studied, memorizing data is not the objective. Knowing where to find the information such that you make good setups for a given metal is your goal here in Unit 14-1.

Some references create a fourth classification for steels, *plain carbon steel* or also called the *carbon steels*. These are the simple formula steels used for car bodies, lawnmower handles, sheet roofing, and many other applications. Depending on their carbon content, those above a given percent (see Unit 14-2) can be heat treated to improve physical hardness. These are steels with nearly no extra metals or minerals added to create superior characteristics other than hardness. Carbon steels don't bend, flex, or take changing loads as well as those steels alloyed with further ingredients. In this text, we will consider plain steels to be the base material among the alloy steels.

SHOPTALK

The Word *Alloy* The word *alloy* has two meanings. First, as a noun, it denotes a specific combination of ingredients; a formula for a particular steel or other metal. For example, the alloy 4340 is a chrome-moly steel (nickel carbon steel with chromium + molybdenum). But as a verb, the word also means to combine metals. For example, when we alloy chromium and nickel with steel we obtain stainless steel.

When performing heat treatment, it's critical to identify the exact alloy. This unit is an introduction to the way steels (and other metal alloys) are designated. Based on a given steel's formula, heat treating temperatures, times, and cooling media vary.

TERMS TOOLBOX

Alloy A specific formula for a metal.

Alloy steel (SAE steel) A group of steels tailored to production part making.

Corrosion resistant steel (CRES) Stainless steel used in medical, food, and other applications where rust is a problem.

Quench Rapidly cooling a metal from high to lower temperature to create hardness.

Tool steels A family of custom-designed steels used for extreme duty.

14.1.1 The Numbering System

In the past, steels (and most other commercial metals) were identified with one of several numbering systems, depending on the branch of manufacturing in which it was being used. Each professional organization or association tailored a system to be useful to them.

AISI	American Iron and Steel Institute (used most for machining)
SAE	Society of Automotive Engineers
ASTM	American Society for Testing Materials
ASME	American Society of Mechanical Engineers
ANSI	American National Standards Institute
AWS	American Welding Society

Plus there are other military and government organizations that classified metals.

Combining systems, the ASME created the Unified Numbering System (UNS), which we'll look at in a moment.

TRY IT

Please turn to *Machinery's Handbook,* under "AISI, SAE and UNS Numbers for Plain Carbon, Alloy and Tool Steels." Note the similarity in these systems. Note, too, that the UNS adds a letter to help identify alloy steel's group, similar to tool steels. Keep a marker in that section–problems are coming up next.

KEYPOINT

Due to popular use in the machining field, from this point onward in Chapter 14, we'll use the AISI classification system.

Steel Classifications—Main Modifying Ingredient First

Within the AISI numbering system, the major family is identified by the first digit or first two digits of the series. They denote the second-most common ingredient beyond plain carbon steel. That provides a general idea of how the **alloy** will perform and how it should be heat treated.

For example, all steels having the heavy metal molybdenum as a secondary ingredient begin with a number 4. Therefore, 4340 steel is a particular kind of alloy steel found in mountain bicycle frames, for example. The 4130 is also a molybdenum steel but with slightly different properties. A bit less costly, 4130 is used for high-demand automobile parts such as axles.

To start your vocabulary, here are four useful groups to learn:

Number	Type
1xxx	Carbon steel (including mild steels)
2xxx	Nickel steels
3xxx	Nickel chromium steels (stainless)
4xxx	Molybdenum steels

TRY IT

To get a feel for the system and where this information is found, answer the following questions from the *Handbook*. Quickly scanning the section named "Standard Steels," you will see that there are three kinds of information about the various alloys:

Numbered alloys with ingredients
Word descriptions of group characteristics
Recommended applications

The following three questions will take you to those three areas of the handbook:

A. You are given a work order to make snap rings (a circular retaining pin). They must be made of the correct steel. Which steel alloy would be recommended?

B. As you will see in Unit 14-2, carbon is the major factor for selecting heat treating temperatures. What is the amount of carbon in the specific alloy of Question A?

C. The print calls for using a stainless steel of the 30316 alloy. Can it be hardened by heat treatment? If you can't find this information, see the hint in the answers.

ANSWERS

A. 1060 steel

B. 0.55 to 0.65 percent carbon. (*Hint:* Look under **steels—corrosion resistant** abbreviated to **CRES**.)

C. No, it cannot be made harder. This group of stainless steels is very resistant to corrosion but cannot be hardened by heat treatment to change their characteristics. They can be stress relieved, however.

SHOPTALK

We tend to think of metals as the durable and beautiful things we use: the steel in our cars, aluminum in windows, or brass and gold in works of art, jewelry, and coins. But there are dozens of metals you'll never see. In fact, of all known elements, there are more metals and semi-metals than any other group! Some are rare and very difficult to extract from their chemical bonds with other elements, a few more are far more valuable than gold ounce for ounce, and some have weird and wonderful properties. The semi-metals have most but not all of the properties of a metal. Look up *semi-metal* in your browser, then metallic elements—very interesting!

The metal mercury melts at −38°F, while tungsten melts at 6,170°F (almost three times the melting temperature of steel). But even stranger, the pure form of the metal sodium (the stuff in table salt) not only melts at normal body temperature (98°F), but at that temperature it also reacts with oxygen so quickly that it will ignite! There are metals that melt in your hand but freeze when poured on a table. There are metals that dissolve in water! In fact, sodium will nearly explode in water!

Metallurgy is one fascinating science!

14.1.2 Tool Steel Identification

The second group of steels is custom designed for extra demanding applications. In general, they are more expensive than **alloy steels**, but there are exceptions. Due to cost and their extreme properties, they aren't often specified in consumer products but are used more for tooling applications. The tool steels are generally higher in carbon content, alloying minerals and metals. They are custom tailored for uses such as dies, punches, wood lathe tools, drills, and end mills. Oddly shop tools such as hammers, files, and chisels are not usually made from tool steel but rather from the AISI alloy steels.

Tool steels are identified by a letter followed by a number. The letter not only identifies the alloy family, but it also identifies the general procedure for heat treatment. When steel is cooled rapidly, from a high temperature to a much lower level, it becomes hardened. That's called **quenching** the steel. So, the W class of tool steels is quenched in water. But water would be too severe and would shatter some steels so a few are correctly quenched in oil (the O group), which removes heat a little slower than water. Still others are supersensitive to the quench rate and they harden by exposure to moving room air, the A tool steels.

For now, scan the following types with a more complete chart in the handbook:

Tool Steel Applications

Designation	Application	Quench Media
A	Die components for medium duty	Air (very slow quench)
D	Heavy die components	Oil/salt (medium quench)
M	Cutting tools (HSS)	Melted salts (slow quench)
M, T	Tungsten metal cutting tools	Melted salts
W	Pins, bushings, and shafts for medium duty	Water
O	Pins, bushings, and shafts requiring more toughness	Oil

We will look at tool steels again after some discussion of heat treating processes coming up next.

KEYPOINT

A. Each type requires a given kind of quench and each alloy within the type requires different temperatures before and after the quench.

B. Alloy steels for general production applications are designated with a series of four, five, or six numbers. The first one or two digits identify the group's major added ingredient(s) and its specific formula. Further digits identify specific alloys and carbon content.

C. Tool steels for heavy-duty applications are identified by a letter followed by a number. It tells not only the type but also something about the needed heat treatment.

UNIT 14-1 REVIEW

Replay the Key Points

- Due to popular use in the machining field, in this book we'll use the AISI steel classification system.
- Each alloy requires different heat treat temperatures and procedures.

Respond

Questions are deferred until Unit 14-2.

Unit 14-2 Heat Treating Steel

Introduction: Heat treating steel requires changing workpiece temperatures, in three stages, to permanently alter its hardness up or down. Some transitions are slow, and some are very fast. Some require high temperatures—to 2,700°F—while some are relatively low—from 275° to 500°F. Performing any of these operations requires the handling of superhot metal! Safety must be considered! (There's a very well-defined set of heat treat protective gear coming up in Fig. 14-18.)

KEYPOINT

Safety

Regular shop safety attire is not adequate to perform heat treating of metals. Your instructor will show you exactly what equipment is required in your lab.

Here in Unit 14-2, we'll learn to perform three in-shop processes on heat-treatable steel—quench hardening, thermal tempering of hardened steel, and thermal annealing. The objective is to change its hardness and physical condition.

TERMS TOOLBOX

Annealed The softest state of steel—100 percent of the hardness has been drawn out.

Annealing Drawing 100 percent of the hardness from a steel, taking it back to its softest state.

Austinite A fine-grained solution of carbon and iron that is not stable. It will revert back to rough layers of soft metal if left to cool slowly.

Carbon stripping Surface disturbance of steel whereby carbon molecules have migrated to the surface to form a black crust. Caused by overheating too long.

Critical temperature The point at which carbon begins to dissolve into the iron crystals and austinite forms in steel. See *Transformation temperature.*

Drawing (heat treat term) To reduce hardness down from a full hard, brittle condition by heating to a temperature below the critical point.

Martensite A desirable, permanent crystal structure created by freezing austinite structure through quenching.

Normalized (steel) Steel that has been heated just below the critical point and cooled slowly to leave it stress free but to retain a bit of its physical hardness as well. A subset of annealing.

Thermal shock The fast reduction of heat through quenching. Each alloy requires a given level of thermal shock to lock the crystal structure into full hardness.

Transformation temperature See *Critical temperature.*

14.2.1 Three States of Steel

A little bit of carbon makes a big difference in steel! To be heat treatable (the ability to be hardened), steel must have a minimum of $\frac{3}{10}$ of 1 percent (0.3 percent) up to a maximum of around 3 percent. Some exotic steels may contain 4 percent. But in general, carbon above the 3 percent has little or no effect on hardness.

KEYPOINT

Steel must have from 0.3 percent up to 3 percent maximum carbon. The carbon content has everything to do with the heat treatment process.

Mild (low-carbon) steel below the 0.3 percent threshold cannot be hardened without first changing its carbon content—which we'll look at in Unit 14-3. Heat-treatable steel can be placed in four physical states.

1. **Annealed**

   The softest, most machinable state—the easiest to machine.

2. **Full hard**

   The hardest condition of steels capable of being hardened. Most steels are too brittle to use at this point. And often too hard to machine in any way except grinding or electrical discharge erosion (see Chapter 26).

3. **Tempered**

   A lower, tougher useful hardness state. The condition the product will be put in for its life as a useful object.

4. **Normalized**

   Not fully annealed—used to relieve stress from processing and welding assemblies.

14.2.2 The Three-Step Hardening Process

To perform your own hardening of steel, you must take the metal through the three steps. Here's a quick run-through. We'll discuss each in more detail right after.

1. **High Heat**

   Heating thoroughly to a specific prequench temperature of 1,300° to 1,900°F, depending on the alloy (700° to 1,000°C). Temperature varies by alloy. It's called the **critical temperature** or also the **transformation or conversion temperature** (Fig. 14-4).

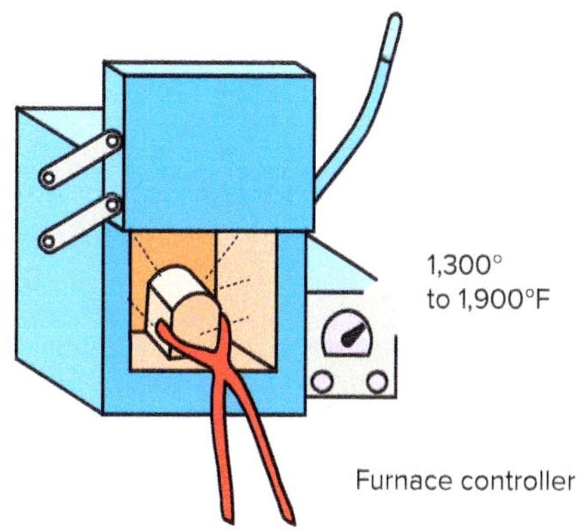

Figure 14-4 Heat to the recommended conversion temperature.

2. **Quench**

   A rapid cooling to create desired full hardness. Plunge the metal into water, saltwater brine, oil, air, or melted salts. Once it's cooled, the metal is too hard for practical use. It tends to crack if used at that hardness. It can even crack during the quench, or soon after, due to internal stresses created during the quench (Fig. 14-5).

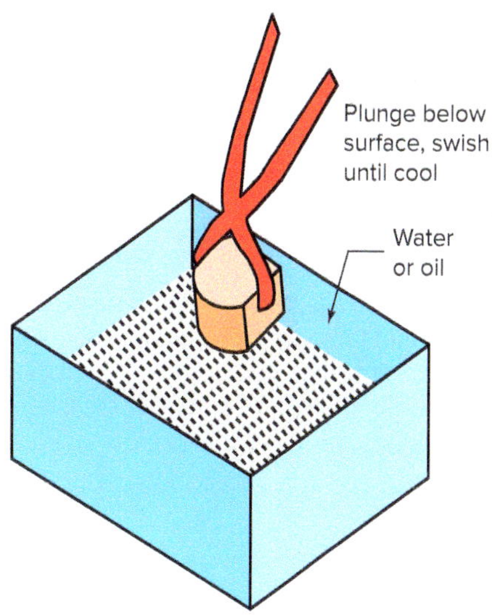

Figure 14-5 Quench to freeze the structure, creating the desired full hardness.

3. **Temper**

   Reheat and hold at a controlled lower temperature to reduce the brittleness of full hard metal hardness to a usable condition, tougher but reduced hardness

(Fig. 14-6). Nothing is then done to the steel other than letting it cool slowly in room air or even more slowly by wrapping it in an insulating blanket for some tool steels.

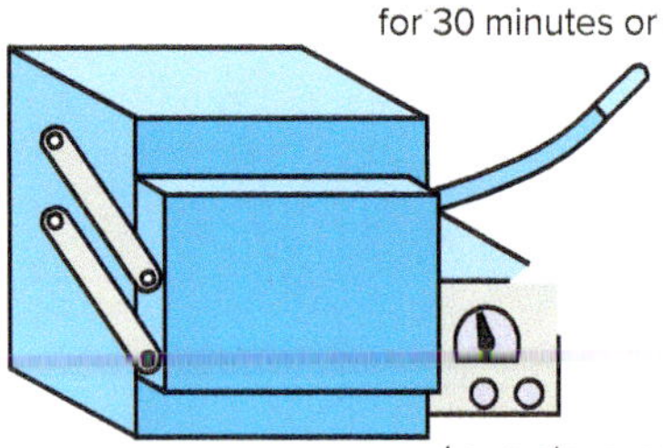

Figure 14-6 After quenching, the parts must be returned to the furnace as soon as possible at a specific preset lower temperature.

Step by Step

Here are the expanded facts you'll need to harden steel:

Step 1 Elevating to the Hardening Temperature Elevate the metal slightly above (50° to 100°F) the *transformation* state. At that point, the steel's crystals change from coarse layers separated by nonuniform layers of carbonaceous metal, into fine crystals with the carbon in uniform solution. This state is called the **austinite** *structure, and the steel is called austinitic.*

When it is held at this elevated temperature until the metal is uniformly hot (called soaking), the internal structure forms a desirable, fine-grained, well-distributed condition. The change in structure isn't instant; it takes time. So, the steel must soak at that elevated temperature for a while. A rough guideline is about 1 hour for every inch of thickness.

KEYPOINT

Slightly exceeding the transformation temperature and holding it at that temperature to soak creates austenitic structure: fine, well-mixed grains. But that state is not stable; it reverts back to a coarse, separated structure when left to cool slowly.

Austinitic steel is a good condition, but it's not stable. If left to cool slowly, the austenite grains separate back into layered materials that are not hard. But if frozen quickly enough with the right quench media, the grains remain fine and locked into full hardness. Then in this state, the steel is known as **martensite**–the objective of the process.

Here are a few general conversion temperatures to memorize, to impress your friends! Note the inverse relationship of carbon to temperature.

Carbon Percentage	Transformation Temperature
0.65 to 0.80	1,450° to 1,550°F
0.80 to 0.95	1,410° to 1,460°F
0.95 to 1.10	1,390° to 1,430°F
1.10 and above	1,380° to 1,420°F

Step 2 Quenching To create the necessary locking action, different alloys require varying amounts of **thermal shock**, the rate at which heat is conducted away from the hot metal. Too little shock and the steel doesn't reach full hardness. But the hardness that does occur is nonuniform and nearly impossible to predict. It's softer than full hard and has soft and hard spots within–a very undesirable condition. It will tend to break along demarcation lines between the areas. Thin sections harden more. The only sure way to achieve complete, uniform hardness is to hold the metal at the prescribed high heat; then quench it at the prescribed rate (Fig. 14-7).

Figure 14-7 To avoid warping or stressing the work, plunge it completely into the liquid, then stir it to quench all sides evenly.
Prographics at Milwaukee Area Technical College/McGraw Hill

Various quench media conduct away heat at various rates. For example, cold water absorbs heat far faster than hot water. Between the two quenches in terms of shock rate, we use brine, water mixed with salt. All the water quenches are too fast for many high-carbon steels, so heat treat oil is commonly used in the small shop or even exposure to room air for air hardening steels.

KEYPOINT

Different alloys require different rates of quench—not too fast or too slow.

To bring steel to the full hard state, you must know:

What is the right transformation temperature for this alloy?

How long must it remain at the transformation state?

What is the correct quench media?

To quench right is to immerse the steel in the quench media, then move it about until it stops boiling the liquid. Do not remove it from the quench just to see if it's cool—hard and soft spots can result.

Too much shock! Quenching too fast can be harmful to metal. Some alloys shatter if given too much shock. Charts are supplied by the manufacturer of the particular steel. Many steel companies supply a complete reference book for their line of steels.

TRADE TIP

Quenching *The Right Way to Plunge Hot Parts* (Fig. 14-8) Plunge the hot workpiece into the quench media with its broadest surfaces vertical, which then surrounds the workpiece with cold liquid as close to simultaneously as possible, called a vertical plunge. Then keep it well below the surface and move it constantly in a figure eight pattern to flush the boiling gas and expose all sides of the object to fresh liquid. Also keep moving it up and down in the tank. Do not just drop it to the tank bottom, as the side that's down will not be exposed to fluid.

It's important that no area of the workpiece cools at a different rate due to the boiling action of the media around the hot part or in the way it was plunged into the quench. Not plunging right and swishing results in warpage, cracks, soft spots, and unnecessary internal stress.

Combo-Quenching Sometimes parts with thin or complex sections will crack or warp excessively when quenched in water even though it's the right media and the plunge is performed right. If so, try to quench them for a few seconds in the oil, then transfer them quickly into the water while still hot. If you do so, the sharpest edge of the shock will be taken away. *Caution!* When moving the part from the oil to the water, oil on the surface will vaporize and may flare up. Wear a complete set of heat treat protective clothing, use long tongs to hold the work, and work only in a fire-preventive ventilated heat treat facility. Do not try a double quench without supervision.

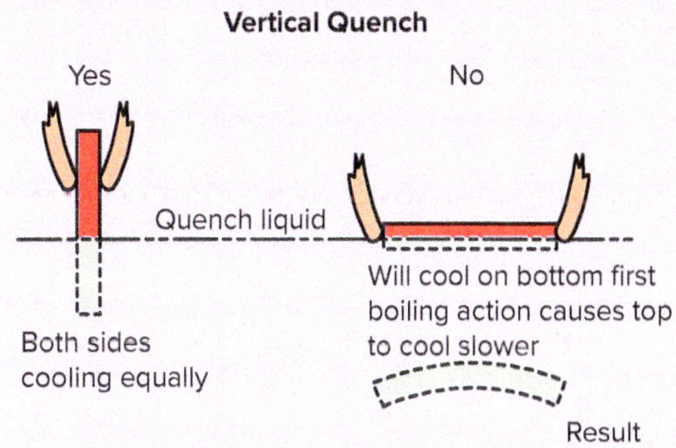

Figure 14-8 The right and wrong way to quench a part.

Step 3 Tempering the Quenched Steel Okay, the steel is at the full hard state, but that's too hard to be useful. Tempering the steel down to a useful hardness where it won't crack under load involves heating it to an exact temperature well below transformation. It's done by placing the work in the furnace (or the torch flame), then holding it there until it is uniformly well heated throughout, usually 30 minutes or more per inch, in an electric furnace, depending on the total volume of metal, the desired temperature, and the alloy.

SHOPTALK

Tempering is also called **drawing** because it draws out the brittleness.

KEYPOINT

The higher the draw heat, the lower the resultant hardness will be. Higher heat = lower hardness. Draw temperatures range from around 400°F to over 1,200°F. To avoid the formation of stress cracks, full hard steel should be tempered immediately after quenching.

Here, too, it's critical that all sections of the work achieve uniform temperature. After the work is heated and soaks, it is usually placed on a wire rack so that air can circulate around all sides, then allowed to cool slowly. Everyone understands to not touch work with their hands when it's in that location. Some sensitive steels might be placed in an insulating blanket or steel box to safely cool them.

KEYPOINT

After the draw is finished, there is little difference in the way the work is fully cooled. The metal needs to cool naturally to room temperature without a steep transition.

Draw Temperatures and Hardness

In Unit 14-5 we will learn how to measure hardness and what the resulting numbers mean. For now, I need a scale to illustrate the inverse relationship of draw temperature to hardness. One of the more popular systems for comparative hardness is the Rockwell index. When a Rockwell test is made on a typical hard, tool steel block for example, it will rate as high as 63 to 65 at full hardness. At its annealed (softest) state, it will be near 20. For reference, mild steel, which cannot be hardened, would measure very close to 0 (zero). Draw charts (Fig. 14-9) or draw temperature tables like the one shown next provide half the information needed for doing your own heat treat.

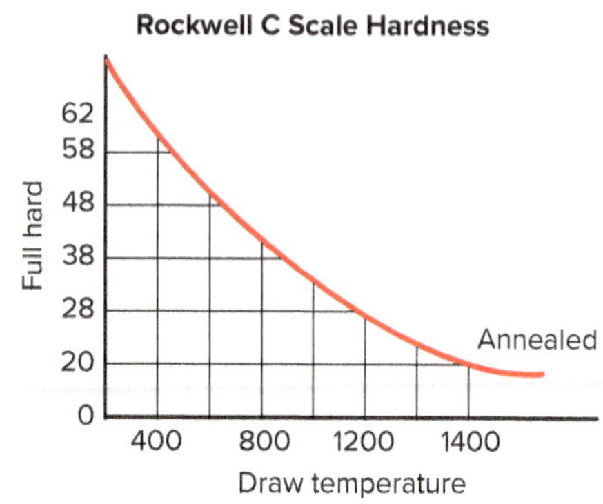

Figure 14-9 A typical hardness curve as draw temperature is raised.

Typical Draw Temperatures and the Application

Draw Temperature (degrees Fahrenheit)	Hardness Rockwell Index	Typical Application
375 to 400	58 to 62 (highest useful without brittleness)	End mills
450	56 to 58	Slitting saw
500	54 to 56	Punches
550	52 to 54	Bending dies
600	48 to 52	Drill bushing
1,300 (1 hour—slow cool)	Normalized (discussed next)	
1,400 (slow cool in furnace)	Annealed	

14.2.3 Annealing, Stress Relieving, and Normalizing Steel

All three of these heat treat processes are subsets of tempering. They are all accomplished by putting the metal into a very high draw temperature. Each process softens the metal, but each has a different objective.

Annealing Takes It to the Softest State

Annealing is a 100 percent draw back to the steel's softest state. It's done to make the metal machinable or to remove work hardening spots. To anneal steel, the work is put in the furnace set at *just above transformation temperature,* then left to cool very slowly (usually by turning off the furnace without opening it). But it can also be wrapped in an insulating blanket, buried in dry sand, or put in a closed container.

Stress Relieving

Internal stress can be induced into metal in several ways: by machining, welding, through bending and forming, and even at the rolling mills where it is originally made. To relieve that stress the metal is heated close to but *just below the transformation point.* It's held there for 1 or 2 hours, depending on thickness; then it is cooled on an extra slow schedule of furnace temperature reductions.

Normalizing

Also a stress relieving process, this time the top temperature is *farther below the transformation point prescribed for each alloy*—slightly less than the stress-relieving temperature. During normalizing, the metal must be cooled on a schedule. Normalizing might be seen as very long, very severe draw. It relieves stress but doesn't quite take all the hardness from the object, leaving it nearly stress free but slightly hard. The difference between normalizing and stress relieving is that normalizing can be performed only on heat-treated metal whereby plain steel unable to be hardened can only be stress relieved to become **normalized steel**.

KEYPOINT

All three processes require a slow, controlled cooldown.

A. Annealing occurs above transformation temperature.

B. Stress relieving occurs just below transformation temperature.

C. Normalizing occurs at a prescribed temperature farther below transformation.

TRADE TIP

High Heat *Caution!* This unit is not about correctly using an oxygen/ acetylene torch. You must be trained to use it correctly. Ask for a demonstration and safe operation test from your instructor.

Uniform Heating While the best results are accomplished in a controlled furnace, you might be called on to perform a heat treatment using a torch. If so, it's important to heat the object uniformly. Concentrate the flame on the thicker parts while "playing the flame" on other parts (Fig. 14-10). Apportion the heating (where you aim the flame), based on volume in the work. This helps create an even heat throughout the part.

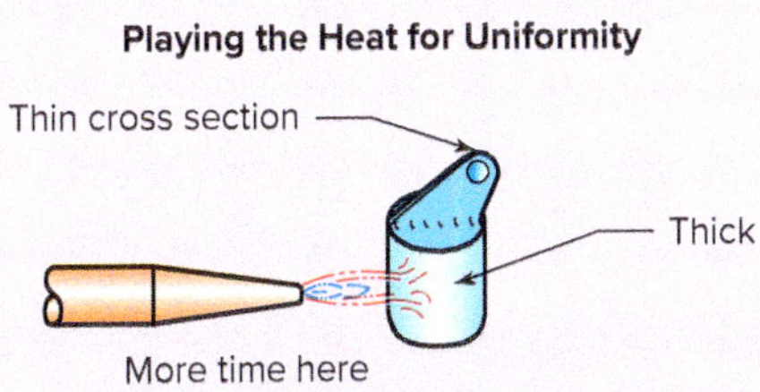

Figure 14-10 When using a torch, be sure to play the heat by concentrating on thick sections.

Carbon Stripping At the high temperatures required for heat treating, carbon in the steel can move due to the high molecular kinetic energy. Near the transformation temperature, carbon can migrate to the surface of the workpiece—called **carbon stripping**, a highly undesirable condition (Fig. 14-11). It's seen as dark spots on the white-hot steel surface. After cooling, the stripped carbon lies in flakes on the surface. Beneath those flakes, the surface is rough. The destroyed soft shell can be as deep as 0.060 in. Without reintroducing more carbon, it cannot be hardened due to its lack of carbon.

Carbon-Rich Flame To help prevent stripping, a torch flame can be adjusted richer (higher) in carbon than a normal welding flame. To do so, reduce the oxygen-to-acetylene ratio in the mixture (Fig. 14-12). The rich flame appears less focused, more orange, without the blue cone of the neutral flame on the bottom. It isn't hot enough for other duties, but during heat treat work it will strip less carbon due to the higher levels surrounding the hot steel.

Figure 14-11 Dark carbon flakes can be seen on this overheated steel as it passes the critical temperature. The result shows more clearly after it cools—free carbon stripped from the steel.
Courtesy Mike Fitzpatrick

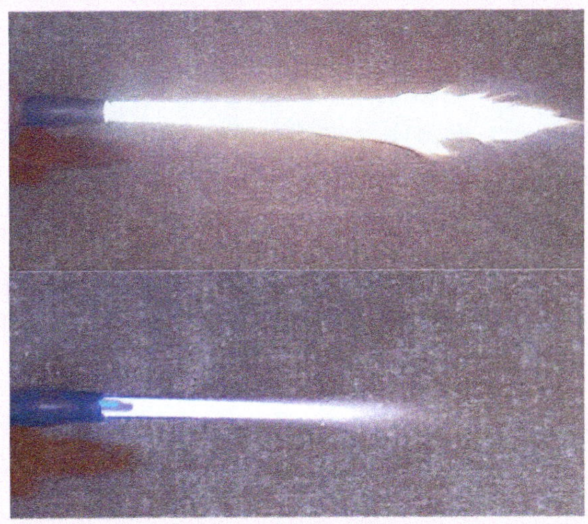

Figure 14-12 The rich flame above isn't as efficient, but it tends to not strip carbon during heat treat operations.
Courtesy Mike Fitzpatrick

Time and Temperature Also, to control stripping, do not overheat the work as was done intentionally in Fig. 14-11. Bring it uniformly up to the prescribed high temperature (dull red through glowing orange, depending on alloy); then quench it immediately. If you hold it at or above the critical point for extra time, carbon will continue to migrate to the surface.

Figure 14-13 A commercial furnace can control the heat within 2 percent accuracy.
Courtesy Engineered Production Systems/Division of Engineered Product Sales Corparation

KEYPOINT

Stripping Is a Function of Temperature and Time

To avoid stripping, do not overheat and do not hold parts at high temperature any longer than required for uniform heating.

14.2.4 When You Must Use a Torch—Utility Temperature Controls

Commercial furnaces can be set to any temperature between 300°F and approximately 2,600°F. They hold temperatures within 2 percent of the target and graph and record temperature profiles in order to create a certificate of the process (Fig. 14-13). For a general shop heat treat, the electric furnace controls we use range from 5, 7, to 10 percent, depending on the quality of the control.

But, when controlled furnaces aren't available, or the work is too big to fit in the chamber, the next best temperature control is the *pyrometer* (Fig. 14-14). A pyrometer is a high-heat measuring device. In industry, there are two types: those that must touch the part with a probe, similar to a meat thermometer, and those that measure optically and thus do not need to touch the hot part. They are used mostly where the heat is too great to come near the parts.

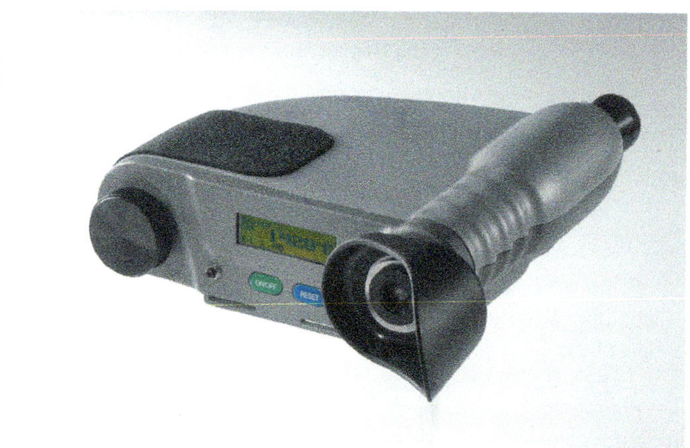

Figure 14-14 A pyrometer can tell the temperature of hot metals.
Courtesy Spectrodyne, Inc.

Pyrometers are common in heat treat shops, but it's unlikely you'll have one in a small shop. When both furnace and pyrometer aren't available, we fall back to a set of trade tips to control temperatures. They aren't accurate enough to perform a critical heat treat, but they work well for utility purposes.

High Heat Control Using Color Brightness—Steel Glow

As steel molecules (and those of some other metals) heat, they begin to radiate energy as visible light, called iridescence. Beginning with a dull red, the metal glows brighter as the energy builds toward a higher temperature. With practice, that color brightness can be used as a rough gauge of temperature (Fig. 14-15). This sequence in Fig. 14-15 was shot as the part heated.

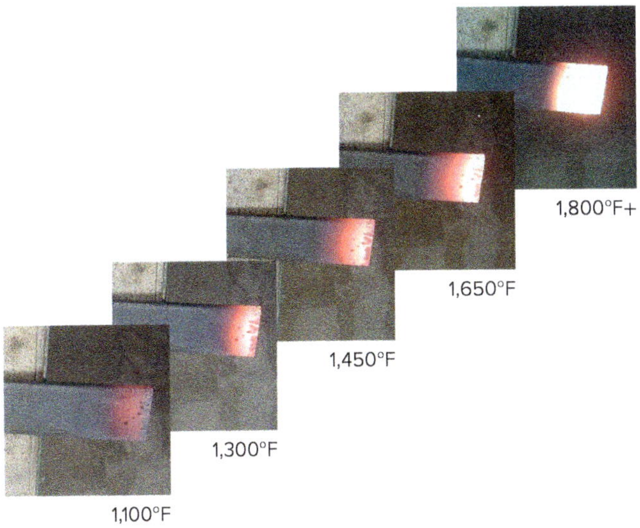

Figure 14-15 Learn to identify temperature by glow.
Courtesy Mike Fitzpatrick

Color Brightness for Different Temperatures

To Gage Steel's Temperature	Approximate Temperature
Dull red	1,100°F
Bright red (called *cherry red*)	1,300°F
Orange	1,450°F
Bright orange	1,650°F
White orange (*white hot*)	1,800°F

Note that carbon steel melts at 2,000 to 2,200°F.

KEYPOINT

I estimate temperature discrimination by glow to be no better than ±200°F, and only after some guided experience.

The only way to get a grip on this skill is to work with a journey-level craftsperson and see a demonstration.

Low-Temperature Controls for Drawing

There are two ways to know how hot metal is becoming before it glows: the melt gauge crayon and surface oxide colors.

1. **Melting Gages**

   These handy little pocket crayons are specially formulated to melt at different temperatures (Fig. 14-16). They can be purchased in 50° increments from 250°F up to 1,200°F. When the draw temperature is below 700°F, several lines are drawn on the cold metal, at various places so different cross section masses can be tracked as they heat. For temperatures above 700°F, have three crayons bracketing the desired temperature ready. Swipe each on the hot surface. The middle temperature should melt but not the higher temp.

2. **Avoid Overheating**

   Not exceeding the draw temperature, especially on thin sections, is the main challenge. Once it's melted, the crayon indicates no further information.

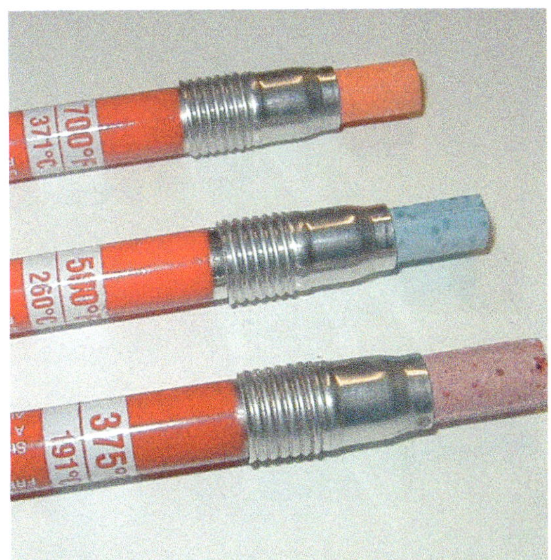

Figure 14-16 A Tempilstick® melting crayon gage can be used to indicate draw temperature from 250° up to 1,200°F.
Courtesy Mike Fitzpatrick

TRADE TIP

Surface Oxide Color and Temperature Here's another tip that helps determine draw temperature when no other low-temperature gauge is available. Clean, shiny steel will react with room oxygen as it is heated. The oxides change color as the temperature increases (Fig. 14-17). The shiny steel in the figure was heated on one end. Starting from room temperature, it shows a light straw yellow at around 400°F, and then it slowly progresses up through a series of color changes, to a light bluish color before it reaches the dull red glowing point. The problem here is that the surface temperature doesn't always tell the core temperature. Also, different alloys vary slightly in their color spectrum, so this hint, although useful, must also be regarded as a rough estimate. Still, oxide color can be helpful when no other temperature control is available. Clean the metal surface to bright steel, where you need to determine temperature. It's smart to error on the cool side of the temperature. If the finished product is tested and found to still be too hard, there is no serious harm in retempering it to a slightly higher temperature, causing a lower hardness.

Figure 14-17 Various oxide colors can be used to indicate temperatures of steel.
Courtesy Mike Fitzpatrick

However, heating it too hot the first time is a one-way trip! There's no way that the overtempered part can't be increased in hardness a bit! It must be completely hardened, then redrawn back.

Oxide Color for Various Temperatures (Carbon Steel)

Color	Temperature
Pale yellow	430°F
Straw yellow	450/460°F
Dark yellow	480/490°F
Brown-yellow	500°F
Brown-purple	520°F
Light purple	530°F
Purple	540°F
Dark purple	550°F
Blue	560°F
Dark blue	570°F
Light blue	600°F and above

14.2.5 Safety Equipment Requirements

Heat treatment of metals requires the right protective equipment and clothing (Fig. 14-18).

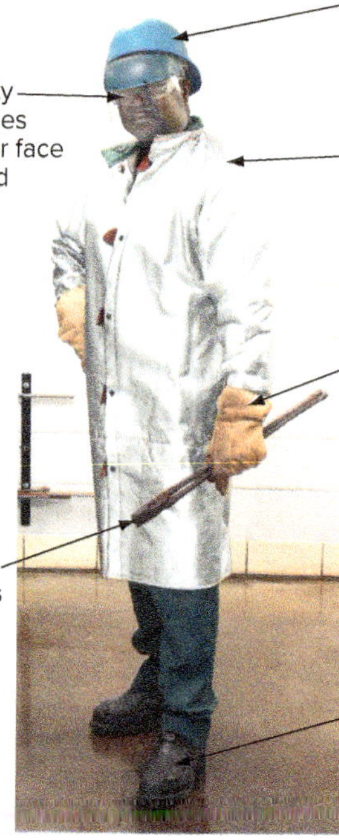

Figure 14-18 This craftsperson is correctly dressed for safe heat treat work.
Prographics at Milwaukee Area Technical College/McGraw Hill

1. **Extra Eye Protection**

   Safety glasses plus a full-face shield made for high temperatures. It may also be partially mirrored to reflect heat.

2. **Fire Extinguisher**

   Immediately available.

3. **Protective High-Temperature Coat**

   Made for heat treating and with long sleeves.

4. **Gauntlet Gloves**

   Long to overlap coat sleeves, insulated for high temperatures. *Caution!* You will not handle hot parts with these gloves. *Hot parts are always handled with long gripping tongs.* Always have two pairs of *tongs* available.

5. **Steel Shoe Protectors**

6. **Other Precautions**

   Be certain there is someone within voice range when heat treating. Memorize where the nearest *eye bath* is located.

SHOPTALK

Hot Parts Become Slippery! Highly heated parts above the transformation temperature become slippery on their surface. Handle them with extra care.

SHOPTALK

Liquid Salt Molten salts are used in several heat treat applications, at both low and high temperatures. They are efficient media for both heating and quenching. Thus far, we've discussed liquefied salt as a quenching media, but it can also be used for high temperatures (Fig. 14-19). But a common question arises—is it the same salt I use on my food?

Figure 14-19 A high-heat, liquid salt bath at 1,300°F.
Courtesy Fluid-Therm Corporation

The answer is no. The definition of a salt is a neutralized combination of an acid and an alkali base. Before melting, the granulated salt does resemble table salt. To identify different kinds of heat treat salts, note that they are sometimes stained various colors. They are definitely not the familiar spice on your potato chips!

Additionally, it's wise to have a backup person, correctly dressed for heat treat, standing by during transferring of hot parts. A second pair of tongs can often save the day.

Quenching Media for Heat Treating

In professional shops many different liquid materials are used to quench work. But in the tooling and small shop, there are three:

Plain cold water (can be heated to prevent overshocking).

Brine, water with salt added to slow its conductivity.

Heat treating oil, a slower, gentler media–less thermal shock.

Heat treating oil is specifically formulated to not ignite unless the entire vat is heated beyond its flash point. But it will flare when the hot part is accidentally removed before fully cooling.

Melted Salt, Air, and Compound Quenching

To further minimize shock, heat treat shops use liquid salt quenches. Molten quench salt becomes a liquid at around 350°F. Maintained in a electrically heated tank, this quench is very gentle in terms of thermal shock.

KEYPOINT

Safety Using Quench Oil

Flame Control Once a hot part is plunged into the heat treat oil, never remove it until the boiling action ceases unless performing a double quench. Otherwise, if the hot oily part contacts room oxygen, it will flare up. If you accidentally do remove it, plunge it back, and the flame will extinguish. Never heat treat with any oil other than one specified for quenching! Regular lube oil cannot be used.

Do Not Overheat the Oil Vat Do not use a small volume of quench oil to heat treat many parts. If the vat is small, with each quench, it will become hotter, which reduces its shock value progressively, and it increases the danger of flare-up. If using a small furnace as in Fig. 14-3, this isn't usually a problem as not enough parts can be loaded and heated to overheat a normal-sized oil vat (4 or 5 gallons).

Air Hardening Tool Steel—Air Quenching

In terms of gentleness, moving air provides enough heat removal for one family of high nickel tool steels known as the air hardening tool steels (Fig. 14-20). Designated by the letter A, examples commonly used are A-2 and A-6 die steels. Once heated to their transformation temperature, which is generally higher than that of other tool steels, they are removed from the furnace and set on racks that allow air movement to surround the hot parts. This provides enough freeze rate to lock the austinitic crystal structure to martensite.

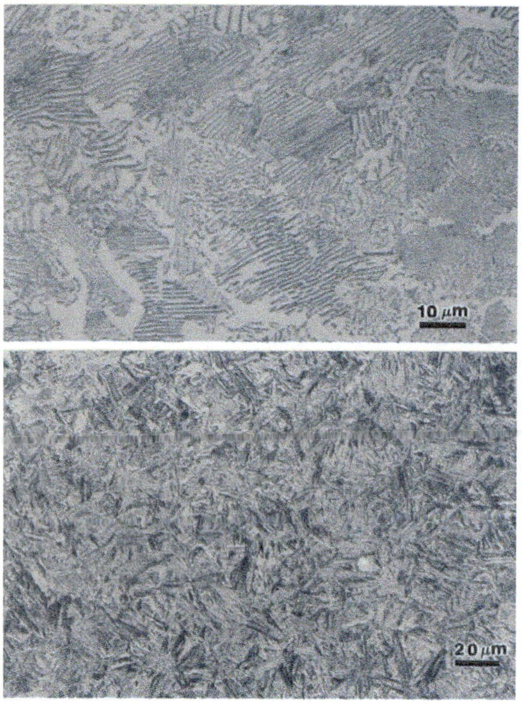

Figure 14-20 Compare microphotographs of 4140 steel in its annealed state, coarse and layered (upper photo) to the same alloy at 54 Rockwell C scale (R_C 54), where the structure is fine and well distributed.
Courtesy George VanderVoort, Buehler, Ltd.

UNIT 14-2 REVIEW

Replay the Key Points

- Heat-treatable steel runs from 0.3 percent up to around 3 percent maximum carbon.
- Slightly exceeding the transformation temperature creates an austenitic structure: fine, well-mixed grains. Quenching it freezes the grains into desirable martensite.
- To harden a particular alloy you must know the specific hardening temperature for this alloy and the correct quench media.
- The higher the draw heat, the lower the resultant hardness will become. It is critical that all sections of the work achieve a uniform temperature.
- Normalizing, stress relieving, and annealing are similar processes, requiring elevating the work at or near the critical temperature, then cooling slowly.

Respond

From the reading and from *Machinery's Handbook*

1. List the steps required to bring a steel to full hardness.
2. List the various quench media used to heat treat steel
   - A. For the in-shop heat treating
   - B. For professional heat treating
3. An optical pyrometer is one method of monitoring the temperature of steel for high heat. Is this statement true or false? If it is false, what will make it true? Name two more methods.
4. In order for steel to be heat treated, it must have a carbon content above 3 percent. Is this statement true or false? If it is false, what will make it true?
5. Define annealed steel.
6. To temper steel what must be done to it?
7. Why is steel not used in full hard state?
8. Name the three missing reasons to heat treat metal: Harden it, soften it, ______, ______, and ______.
9. Describe the transformation state for steel.
10. After tempering a steel part at 500°F, it is tested and found to be too hard for the application. What must be done?
11. Starting with annealed steel, what must be done to temper it to a prescribed hardness?
12. List the four temperature monitoring methods for *tempering,* in order of their exactness. Which are in-shop methods?
13. As the tempering temperature is elevated, the temper becomes harder in most tool steel. Is this statement true or false? If it is false, what will make it true?

Unit 14-3 Case Hardening Steel

Introduction: Mild steel can be modified with extra carbon, such that its surface can be made nearly as hard as tool steel. This process creates a shell of high-carbon steel (called the *case*) from 1 to 2 millimeters deep. After adding the carbon case, the steel's surface can then be hardened in the normal fashion.

Like other forms of heat treating, there are methods for exact case depth control and shop utility processes. In this unit I'll briefly discuss other methods but concentrate on how you can do it in the shop.

Steels Are Case Hardened for Two Reasons

To harden mild steel This is usually done for the economy of not using expensive alloy steel for a product yet having a surface hardness able to withstand abrasion and wear.

To change the properties of alloy steels Others need to be very hard on the outside and tough on the inside. For example, an automobile axle, made of heat-treated and tempered alloy steel, can withstand flexing and heavy loads, but it must also be very hard at the bearing surfaces (Fig. 14-21). In other words, the steel has a double requirement for surface hardness and interior toughness. This is possible using the case process over a medium high-carbon alloy steel.

Figure 14-21 The axles on this muscle car must be heat treated for toughness to withstand 800 horsepower without snapping, yet they must also be hard on their outer shell for bearing purposes. (See Fig. 14-22.)
Courtesy Mike Fitzpatrick

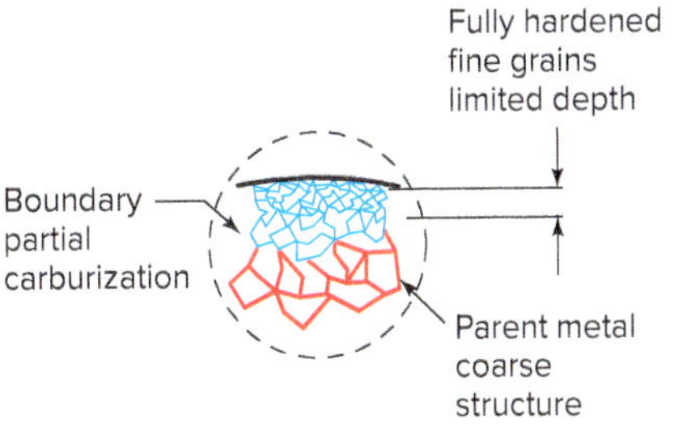

Figure 14-22 A simulated microphotograph of a cross section of an axle shows fine, high-carbon crystals on the surface, then a gradual increase toward coarse structure below the case.

TERMS TOOLBOX

Carburizing Adding carbon to the outer surface of steel. Accomplished with liquids, solids, and gasses.

Case hardening Several processes that modify the surface of steel by adding extra carbon, then hardening that shell.

Nitriding A gas hardening process that adds nitrogen rather than carbon to a thin shell around specially formulated steels.

14.3.1 The Case-Hardening Process

If we were to cut open the axle shown in Fig. 14-21, it would appear as the drawing in Fig. 14-22. The outer surface has been modified with extra carbon, then quenched to lock the fine grains. The case-hardened surface creates a set of characteristics for the axle's function. The core must be flexible and not prone to stress cracking, but its outer surface must be hard to resist wear. This is possible using case or surface hardening.

The case zone extends down from 0.015 to 0.200 in., depending on the process used (shop or commercial). A small boundary zone occurs where carbon content falls off rapidly, leading to the core of the original unmodified structure. That core can be alloy steel that is also heat treated but to a much lower hardness. This process can also be used to modify low-carbon steel incapable of heat treatment otherwise.

Before hardening, the carbon content increases by elevating the steel's temperature from 1,500° to 1,650°F. That's at or above what would be the transformation point assuming the steel had enough carbon to convert. At that point, due to thermal energy, carbon molecules begin to migrate outward—we see it as the carbon crust that forms on the outside of the work (Fig. 14-10). But once the carbon moves outward, it can also migrate inward if there is a carbon-rich material on the outside of the metal.

KEYPOINT

Once the steel is heated slightly above critical temperature, it is then immersed in a carbon-rich material—solid, gaseous, or liquid. See the Shoptalk.

14.3.2 Shop Case Hardening

For everyday shop tooling work, we use the pack-hardening process by surrounding heated metal in powdered carbon made just for utility **case hardening**. It produces a shell depth of from 0.015 in. to 0.045 in. maximum, depending on the number of times the steel is heated and coated with fresh compound. Compared to commercial processes, carbon cannot penetrate very far. But, other than shallow depths, it offers several advantages:

It can be performed with torch or small furnace.

It is a relatively simple, quick way to harden a steel item.

Specific areas can be spot hardened while others are left soft.

The compound is inexpensive and if used conservatively, will harden many parts.

Heating, then packing the object in a case-hardening compound produces from 0.015 in. (one hot immersion then quench) up to around 0.045 in. case depth (immersion, clean, reheat, and reimmerse repeatedly up to three or four times, then quench). This more thorough process also produces about a 20 percent harder surface.

SHOPTALK

In the past, metal workers used charcoal, powdered coal, and even charred wood chips to induce carbon into hot metal. But you can see the problem with those materials—fire. The work had to be buried far enough within it that room oxygen didn't make it to the hot interface. Today, we use specially formulated prepared compounds (Fig. 14-23) that look like coarse black sand. They are stabilized carbon and will not ignite or give off dangerous vapors.

Figure 14-23 Case-hardening compound is rich in carbon. It's made to transfer carbon but to not flare up.
Courtesy Mike Fitzpatrick

Three-Step Process to Raise Carbon Content

Adding carbon to the steel's surface is not complex. It involves heating the metal then exposing it to the compound.

Step 1 Heat to 1,500°-1,650°F

Step 2 Surround with carbon compound (Fig. 14-24)

Step 3 Absorption time—sustain at temperature This last step has two forms, depending on whether a furnace or torch is used.

Figure 14-24 After removal from the furnace nearby, using long-handled tongs, the heated part is quickly buried in case-hardening compound.
Lake Washington Technical College, Kirkland, WA/McGraw Hill

Torch process If the shop has no furnace or if a shallow case of less than 0.5 mm is required, simply burying the heated work in the carbon compound, then leaving it to cool naturally will produce results. After cleaning the spent compound from the work, this process can be repeated to add extra depth but never to the depth of the next method.

Controlled furnace A deeper case is gained by burying the heated workpiece in a steel tray filled with compound, back in the electric furnace and holding it at the absorption temperature for 1 hour or more.

Step 4 Quench Quenching of the case should be performed soon after the case has been created.

If using a torch, the metal is removed from the compound, then cooled enough to handle. Then the crust of spent compound is removed. The metal is reheated to 1,500° to 1,650°F and plunged in cold water.

If using the furnace absorption method, it's possible to move the hot metal directly from the furnace to the quench tank. The problem is the crust of spent compound will contaminate the water. Unless you plan to change the water, it's better to remove the work from the furnace, cool it, clean it, then reheat it and quench it. But to avoid surface cracks, this should be done soon after the work is cased.

14.3.3 The Professional Surface-Hardening Methods

While these are not machinist tasks, you should know a bit about them.

Gas Carburizing

Gas **carburizing** is a highly controlled case-hardening process where a deep uniform layer is required, up to 3 mm. This process does not distort the surface of the work as much as pack hardening. After placing the parts in a sealed furnace, the furnace interior is purged of oxygen by displacing it with nitrogen. Then the furnace and parts are heated to the critical temperature. At that point preheated natural gas is flooded into the furnace to displace the nitrogen. The carbon-rich gas is kept at a positive pressure for a length of time, depending on desired case depth. The gas contributes the carbon but does not burn since there's no oxygen.

The depth of the case is controlled by pressure and exposure time.

Liquid Salt Case Hardening

Using salt compounds heated electrically to a liquid state, similar to salt quenches, but to the critical temperature required to move carbon, an enriching environment can be created by adding carbon compounds to the liquid. This process brings several advantages for case hardening over other methods.

Faster cycle time—The heated liquid transfers heat and carbon to the work fastest of all the methods.

Less distortion

Selective hardening—See the copper Trade Tip.

TRADE TIP

Using Copper Plating as a Carbon Barrier Copper will not melt at the temperatures required for heat treating steel. Copper plating on steel will stop carbon movement in or out of the surface and is used to protect the surface during regular heat treating of high-carbon steel. It is also used to selectively mask off areas of the workpiece that are not to be cased.

This tip can be used to protect precision parts from carbon stripping during normal heat treating. For example, a complex-shaped, finished tool steel die can be preplated with copper, then heat treated. Once the copper is stripped back off the work after heat treat, the finish is near the original! The copper formed a barrier to carbon stripping.

Copper plating is also used during liquid carburizing where selected areas of the work are required to remain unmodified—no carbon. Only the raw surfaces without the copper shield will absorb carbon from the salts; similar to stenciling, the copper shields selected areas of the work.

Nitride Gas Hardening

This method is unique in that it modifies the steel in a different way. Steels formulated for this process will produce a shallow hardness of a few thousandths when their structure is modified by nitrogen, not carbon. The process is similar to gas carburizing except the heated gas is ammonia. Ammonia gas requires extra precautions, therefore this process is performed in specialty heat treat shops only. **Nitriding** is relatively quick to perform compared to the other surface-hardening processes and produces a very hard but thin case on the work.

UNIT 14-3 REVIEW

Replay the Key Points

- Steel is given a case to modify and harden the surface of mild steel and to change the properties of alloy steels.

Respond

1. Using the solid pack method, the hardness of the case is dependent upon the length of *time in the furnace,* while the metal in question is exposed to the carbon compound. Is this statement true or false? If it is false, what will make it true?
2. List the steps required to create a case hardened steel workpiece.
3. Using the solid pack method, when would you not replace the packed work in the furnace?
4. Name and briefly describe the professional case-hardening processes.
5. Describe why we can case harden steel.

Unit 14-4 Aluminum Alloys and Heat-Treated Conditions

Introduction: Due to its low weight-to-strength ratio, and its availability in our earth's crust, aluminum is a very common metal in modern manufacturing. It too is heat treated to improve physical characteristics, but the way heating and quenching are used is exactly opposite that for steel.

Some aluminum alloys can be hardened for strength and durability. They can also be softened to increase malleability, to easily form or stretch it without cracking. Like steel, aluminum can also hold internal stresses caused by casting, mill rolling, machining, welding, and other processing. Aluminum can be relieved using thermal heat treatment, but it is accomplished with a very different set of processes compared to steel.

SHOPTALK

Aircraft Aluminum The 5000, 6000, and 7000 aluminum series are often referred to as *aircraft alloys* or military aluminum due to their superior strength and ability to be heat treated for improved properties. However, they are also used in many other applications and products.

No aluminum alloy will accidentally work harden, nor can it be deliberately hardened by heat treatment, beyond the machinability range. Thus, annealing isn't used to bring it back to machinability. But it is annealed to better bend, stretch, and form it.

SHOPTALK

Napoleon III of France was instrumental in financing early research into the refinement of aluminum. Aluminum was known before this time, but due to the great difficulty of extracting and refining it, it was one of the rarest of metals known. In 1855, when important foreign dignitaries visited his court, they ate from the everyday gold dinnerware. But when royal visitors came, they brought out the aluminum! Aluminum was the best!

Unit 14-4 provides a background about the possible heat treat conditions of aluminum. When the engineering drawing specifies a given heat-treated aluminum, the material must meet that callout. But it's not a how-to-do-it unit. Aluminum heat treating is a specialty usually not performed by machinists. I'll bet you usually skip the reviews at a chapter's end. But if the fascinating facts surrounding metal is getting through to you, read them this time; you'll learn a lot more strange facts.

KEYPOINT

Aluminum specifications usually include not only the alloy identification but the heat-treated condition, too.

Because the heat treat process has many variations, a complete designation will include not only the alloy but also its exact heat treat condition and often even the way in which it was put in that condition.

TERMS TOOLBOX

Condition The heat treat hardness level, and process for aluminum.

Wrought Any process that forms (but does not cut) a metal when it is not in the liquid state.

14.4.1 Aluminum Alloys

Similar to steels, the alloys of aluminum are specified by number (Fig. 14-25). The first digit identifies the major alloying ingredient. The remaining numbers identify specific ingredients in the subgroup. For example, the alloy 5052 is modified with the metal magnesium.

Figure 14-25 The alloy, hardness, and method of tempering can be identified on this aluminum bar.
Courtesy Mike Fitzpatrick

Aluminum is supplied in cast alloys and mill rolled bars and sheets. All alloys that are cold or hot worked after they are formed at the smelter are called **wrought**, referring to the working of the metal after it is solidified. Here are the major groupings:

Type	Major Added Ingredient (Metal or Mineral)
1000	Iron and silicon
2000	Copper
3000	Manganese
4000	High silicon
5000	Magnesium
6000	Magnesium/silicon
7000	Zinc/magnesium/silicon/copper
8000	Lithium

14.4.2 Heat Treat Conditions

The temper (called **condition** for aluminum) is indicated by a letter and a number, or often a series of numbers that follow the alloy number–separated by a dash. For example, 5051-T6 is hardened to condition T (fully hard) using process 6 (found in the *Handbook*).

This is a magnesium aluminum and it has been put in the tempered condition T6. Each letter denotes a certain physical state. These you might try to remember if you work with aluminum.

Letter	Condition
F	As fabricated including work and weld stresses (not often used in this condition).
O	Annealed—softest state usually created for upcoming forming operations.
H	Hardened and will be a specific process; for example; H1 = strain hardened and annealed.
W	An unstable soft condition whereby the aluminum will be formed, then hardened.
T	Heat treated to a useful hardness by a specified process. This is the common category for finished products. Numbers following indicate how hard, then which process was used. For example, 5051-T65 is hardened to level 6 using process 5.

TRADE TIP

For more information on aluminum conditions and alloys, go to http://en.wikipedia.org/wiki/Aluminum or use your search engine.

UNIT 14-4 REVIEW

Replay the Key Points

- Aluminum specifications include the alloy, heat treat condition, and the process by which it arrived at the condition.
- When a specific aluminum is called out on the drawing, all criteria must be met, including the process by which it was placed in a given condition, if it is to be a legal part.
- Aluminum heat treating requires hot and cold baths, thus the specialty equipment isn't normally used as an in-shop process.

Respond

Using *Machinery's Handbook* as a reference:

1. Describe an aluminum alloy denoted to be 7075-T6-51.
2. List the various temper conditions of aluminum.
3. Which conditions would be soft and stable?
4. How much copper (abbreviated CU) and silicon are in a typical 6061 aluminum alloy?

Unit 14-5 Measuring Metal Hardness

Introduction: Beyond verifying our in-shop heat treatment, testing hardness is sometimes necessary for production work as well. Even though it's bad planning, occasionally a job arrives at our machine with an unknown alloy or maybe its composition is known but the hardness isn't. It is possible to use a file to roughly test the machinability of that metal (if you can file it you can machine it), but the best way to select cutter types, speeds, and feeds is a true hardness measurement.

TERMS TOOLBOX

Brinell Testing hardness by reading the diameter of a ball penetrator mark.

Elasticity The property used by the Shore system to test hardness.

Rockwell Testing hardness by reading a penetrator depth.

Scleroscope The device that performs a Shore test, which measures the rebound of a diamond-tipped hammer off the test piece.

Shore Testing hardness by measuring the bounce of a diamond-tipped hammer.

14.5.1 Properties Used to Test Hardness

As metal becomes harder, or softer, its physical properties change. We will study these properties further in Unit 14-6, but for now, we are interested in two. The three tests we'll look at use these two properties. Which test you will use depends upon a given product line, hard tooling, medium hard work such those found in automotive parts or specialty work that cannot have its surface dented! It's an industry-by-industry issue.

1. **Malleability**

   This is metal's ability to be deformed without tearing or cracking. From our study of chip making, you'll recall that the harder the metal, the more it resists deformation. Gaging this property can provide a relative hardness value.

2. **Elasticity**

Elasticity is the ability to stretch without breaking. It can be measured by bouncing an object of known hardness off its surface. The harder the work is, the higher the test object will bounce.

14.5.2 The Rockwell Hardness Test

The **Rockwell** is a widely accepted method for both soft and hard metals. This system gauges malleability by measuring the ***depth*** a pointed probe of known shape and size will penetrate into the material given an exact amount of force upon it. Due to Rockwell's range it is the most popular test in tooling and small production shops and training labs.

Rockwell Numbers

There are several different scales within the Rockwell system, which we'll look up in the *Handbook* later. For this discussion we'll use the Rockwell C scale, correctly used on hardened steel. The C scale can be said to start at 0 (annealed steel) and runs up to 68, harder than a HSS tool bit, near that of a carbide tool. It is symbolized by a larger R with the scale subscript.

$$\mathbf{R}_C$$

KEYPOINT

Rockwell C numbers rise as the hardness rises. Rockwell tests cover a wide range of material hardness.

The Two-Step Rockwell Test

To perform a Rockwell test, a part with a clean smooth surface is placed in what could be described as a very strong, large micrometer frame or precision arbor press (Fig. 14-26).

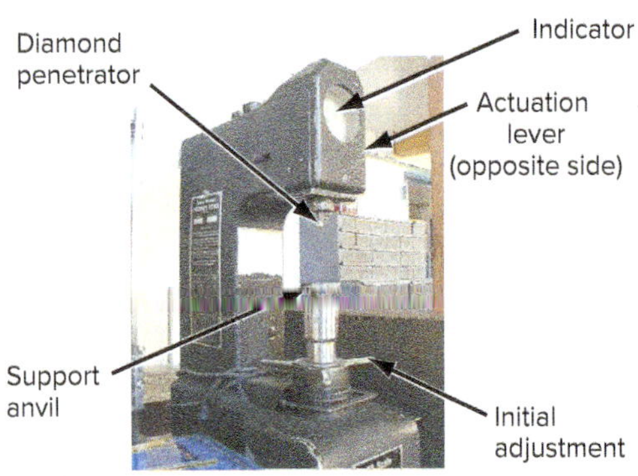

Figure 14-26 A Rockwell tester has a strong frame with a dial to indicate diamond penetration depth into the workpiece: deeper penetration indicates softer metal.
Courtesy Tyee Aircraft/Paine Field, WA

Step 1 Calibrate Load The test object is set upon the lower anvil such that it's stable and won't move when pressed down from above. Next, a cone-shaped diamond penetrator is brought into contact, then driven into the metal to a predetermined pressure of 10 kilograms (22 pounds). That causes the conical point to sink into the metal from 0.003 to 0.006 in. This is the initial calibration load. At that time, a large dial indicator is rotated to read zero (Fig. 14-27).

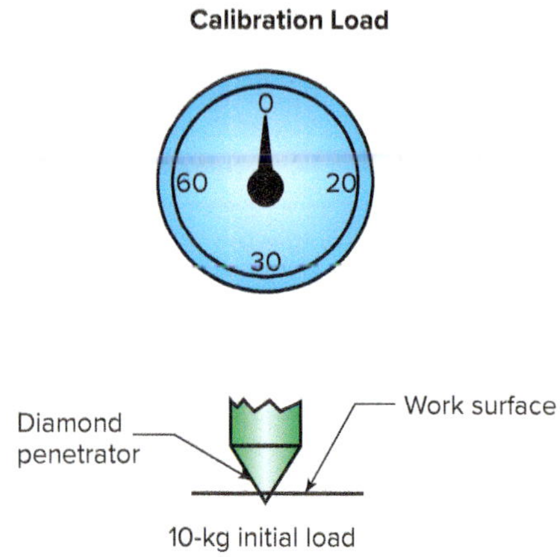

Figure 14-27 The initial load is placed on the penetrator, then the dial set to zero.

Step 2 Test Load Then with the calibration pressure upon the penetrator and indicator set to zero, a second, additional 10-kg test load is added. As the diamond sinks farther, its added depth is translated to the dial, but in an inverse relationship. The deeper the diamond penetrates, the softer the metal tests, therefore the lower number that must appear on the dial face. Inversely, when the point can't go very deep, the metal is hard and registers higher on the dial face (Fig. 14-28).

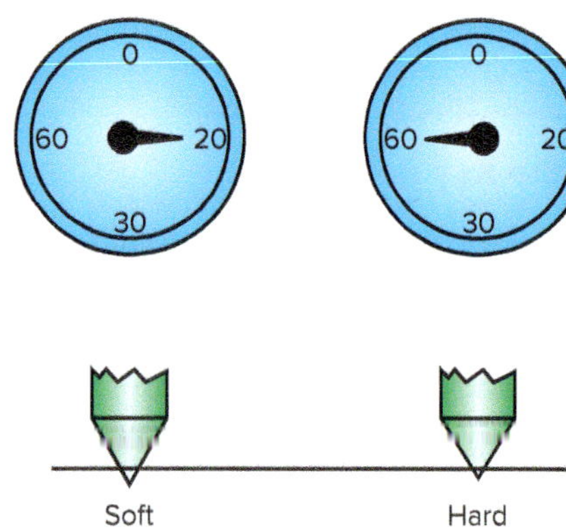

Figure 14-28 Then by hand or by automatic action, the test load is added to the penetrator. Shallow penetration shows hard metal. Increasing depth shows softer metal.

KEYPOINT

Step 1 Touch surface, initial load, and set dial to zero.
Step 2 Add the second load—measure penetration depth.

14.5.3 The Brinell Hardness Test

A second popular hardness rating, the **Brinell** test, is very similar to the Rockwell system in that a *penetrator* is forced into the sample; however, here the measured gauge is the *diameter* of the dent made by penetration of a hard steel ball of known size into the workpiece surface. Hardened tool steel balls are used for testing softer material, while a carbide penetrator ball is used to test harder metals (Fig. 14-29).

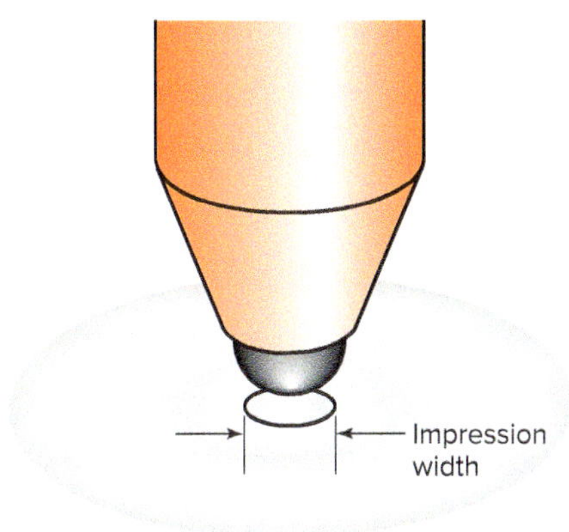

Figure 14-29 The Brinell hardness test measures the diameter of the round impression made by a hard steel ball. Brinell discriminates well in softer metals but isn't applicable for very hard metal.

Due to the upper hardness, limiting factor of Brinell balls, this test is correctly used as a test of soft to medium hard metals–it is used more for customer products than for hard tooling.

KEYPOINT

Brinell is correctly used for relatively soft to medium hard metals. As such, most production drawings often specify hardness ranges in Brinell.

Brinell Scale Numbers

The scale runs from 160 for annealed steel up to approximately 700 for very hard steel. They offer three advantages over Rockwell:

1. Brinell hardness numbers relate to the metal's tensile strength (which we'll investigate in Unit 14-6). Briefly, tensile strength determines how much force is needed to permanently deform one square inch of the metal, by pulling it. Tensile strength is expressed in pounds per square inch, required to deform one square inch of metal by pulling.

KEYPOINT

Brinell readings are approximately twice the tensile strength in 1,000-pound units. For example, a metal reading of 260 Brinell means that a given square inch would require 13,000-lb pull to deform it permanently.

2. Brinell numbers are greatly expanded compared to Rockwell numbers, allowing more discrimination between softer metals of only slightly different hardness. This can be useful where a small difference in hardness can make a big difference in performance of the material.
3. The Brinell system has a single scale covering its entire range of hardness, while the Rockwell system must be switched between ranges as softness range varies.

TRADE TIP

Test It, But Don't Scrap It! The Rockwell and Brinell systems are a ***destructive test.*** They create a permanent mark in the metal. It is a common error for entry-level people to make the test in the wrong location on the work. Tests should be performed only on nonfunctional surfaces or in the exact location noted by a flag on the drawing.

14.5.4 The Shore Hardness Test

The **Shore** test is very different from the previous two: it gauges elasticity, not malleability. If a precision test probe is dropped from a known distance, upon the work surface, its bounce height can be measured. The harder the metal, the higher the probe will bounce. For this test, a precisely rounded, diamond-tipped hammer of exact weight is dropped 255 mm to the work's surface. It bounces back up a scale that retains the highest reading.

The Shore test has two advantages over Rockwell and Brinell:

There are no permanent deep marks, and this test is nondestructive.

The **scleroscope**, the instrument that makes the test, is small and portable (Fig. 14-30). This test can be taken to the work, while the work must come to the test for both Brinell and Rockwell systems.

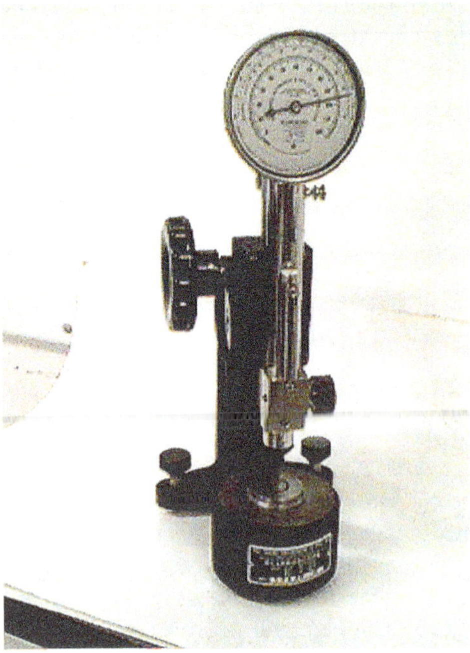

Figure 14-30 Shore scleroscope, a portable hardness tester that measures rebound height of a diamond-tipped drop hammer.
Courtesy Mike Fitzpatrick

The Shore test has two requirements:

1. **Horizontal, Flat Test Surface**

   The work to be tested must have a flat surface big enough to support the scope, and it must be able to be positioned horizontally such that the scope tower is oriented vertical and perpendicular to the work.

2. **Solid Sample**

   The work must be solid enough to cause a true reading of the bounce. Any superficial case-hardening, work movement, or flexible surfaces will falsify the reading, as will round or unusual shapes which absorb some of the bounce.

14.5.5 Hardness and Your Machinist Work

Just today a student brought a piece of stainless steel to the lab intending to make a model jet engine rotor. The piece of steel was purchased from a local scrap dealer who didn't know the alloy or hardness. The student asked, "What cutting speed do I use?" A hardness test provided the setup information.

Recall that in finding the recommended cutting speed for any setup, two arguments were crossed:

Tool bit hardness   Work hardness

Below is a baseline chart of Rockwell C and Brinell hardness samples. When in doubt, always look it up.

Critical Question

The student's stainless did adhere to a magnet and tested R_C, 40. We then assumed it was in the 400 series (martensitic stainless steels), probably 416 as there was a lot of marine shipbuilding in the region. The student was in the Engineering I lab and HSS cutters were the only available option. Based on the chart, what cutting speed did we set up?

Cutting Speeds Based on Work Hardness

			Speeds in ft/min	
R_C/Brinell	Typical Product	Machinability	HSS	Carbide
20/226	Soft carbon steel	Good	100	210
30/286	Drill shank	Fair	80	180
40/371	Axe head	Poor	50	120
45/421	Chisel, wood	End of HSS range	X	80[#]
50/475	Chisel, metal	End of carbide range	X	40
60/625[*]	HSS end mill	Grind only	X	X

[#]Considered the end of practical machinability—tool life and work hardening become major issues below this line. Extreme machining conditions indicate grinding.

[*]Average number—600 is beyond Standard Ball Brinell Test.

Answer

Fifty feet per minute using coolant, HSS tools, and watching the cutting action like a hawk. The hardness of the metal was close to the end of the HSS machinability range. One incident of dull tool, and work hardening would occur, and the part would be scrapped or need to be annealed. The student knew the tools would dull far faster than normal and must not be allowed to go beyond the first stage of breakdown. Gaining this kind of awareness is a major objective for the following problems.

UNIT 14-5 REVIEW

Replay the Key Points

- Rockwell C numbers rise as the hardness rises. Rockwell tests cover a wide range of material hardness.
- The Rockwell test: step 1–touch surface, initial load, and set dial to zero; step 2–add the second load, measure penetration depth.
- Brinell hardness is correctly used for relatively soft to medium hard metals. As such, production drawings normally specify hardness ranges in Brinell.
- Test it, but don't scrap it!

Respond

1. What two properties are used for metal testing?
2. List the three common tests for metal hardness and describe their method.
3. Of the three hardness tests, which do not mar the part?
   - ☐ Rockwell
   - ☐ Brinell
   - ☐ Shore
4. Briefly summarize the cutting speed table with respect to work hardness. (From *Machinery's Handbook,* hardness testing sections.)
5. A workpiece tests to be 42 on the Rockwell C scale. What would its Brinell hardness be?
6. Using the workpiece of Problem 5, what tensile strength would the workpiece exhibit?
7. From the *Handbook* we find that there are several Rockwell scales. Identify the correct Rockwell scale for
   A. Extremely soft metals
   B. Very hard metals
   C. Medium hard metals
   D. Case-hardened work
8. A Brinell test is performed and read to be 344. What is the equivalent Shore reading?
9. A milling machine is already set to cut a pocket using a high-speed end mill. The work material tests at 328 Brinell. The RPM for the $\frac{3}{4}$-in. cutter is 400. Is this correct? If not, what is the correct speed? Use the Cutting Speeds Based on Work Hardness chart in this unit; then convert to RPM.

TRADE TIP

Note, each of these scales uses a different amount of test pressure and might require a penetrator change.

Unit 14-6 Physical Properties of Metals

Introduction: In addition to cutting or abrading metal from a workpiece, most metals can be formed by bending, hammering, pushing, rolling, pulling, casting of liquid metal, and extruding. There are subset methods within each category. Before any process is chosen to machine or form a metal, the planner must know the way it "behaves" while being worked. The master property that predicts its cooperation or refusal to be shaped is **malleability**, its ability to be permanently moved from one shape to another without breaking. But malleability is only one of several characteristics a given metal will exhibit. You should know something about each.

There are three groupings of properties attributed to metals: physical, chemical, and electrical. Each has significance to manufacturing. Chemical and electrical properties are important for design work; however, in Unit 14-6 we'll limit exploration to physical characteristics since they affect machine setups.

In choosing a metal for a design, engineers balance factors of hardness, wear, and corrosion resistance, and toughness against cost, machinability, availability, and **weldability** (can it be welded). All tied to physical characteristics. Terms matter here, as always–test your understanding in the vocab Terms Toolbox game!

TERMS TOOLBOX

Ductility The measure of the ability to be drawn out to thinner cross section, in one axis, without breaking.

Elastic limit (elasticity) The measure of the ability to be stressed by bending, pulling, twisting, etc. without permanent deformation.

Eutectic An alloy that melts at a lower temperature than any of its constituent ingredients.

Malleability The measure of the ability to be deformed in multiple directions without breaking.

Newton The SI standard unit of force = gravitational pull of 1 kg.

Tensile strength (ultimate strength) The final strength of a metal in lb/in.^{2} or Newtons/cm^2. Metal fractures at this point.

Weldability Some metals can be welded, and some cannot. Some can be welded but only using special materials and processes.

Yield strength The point at which the elastic limit is exceeded.

14.6.1 Metal Characteristics

Malleability

A metal is malleable when it can be deformed by pushing, rolling, or hammering without cracking, separating, or breaking. A metal that exhibits good malleability can be deformed in all directions. Gold is so supermalleable that it can be rolled or hammered out to sections of an amazing 0.00001 in. (a hundred thousandth of an inch!). Aluminum is close to gold in this trait. We use the foil rolled from aluminum every day in our homes.

In other words, harder metals of the same alloy cannot be deformed as much without cracking. This ratio is the major contributing factor to the machinability of a metal. A less malleable metal resists deformation more and thus creates more heat and requires lower cutting speeds.

However, metals that are very malleable are sometimes also difficult to machine. For example, lead is downright difficult to machine to a dimension because it's so soft that it "squirms" while being cut. It won't stand up against the cutter force. It is similar to cutting a sponge with a spinning cutter. Softer alloys of aluminum also tend to be "gummy." Instead of forming clean chips that eject from drills and end mills, they smear across the tooth face and clog flutes, stopping further cutting action. We've discussed this gummy property previously.

KEYPOINT

Malleability is the ability to be deformed in *all* directions, and it is inversely related to hardness.

Ductility is a subset property of malleability. A metal that is ductile can be drawn out by pulling or rolling—but in a *single direction*. Soft copper is a ductile metal. It is easily drawn into wires and tubes.

KEYPOINT

Ductility is the property that allows metal to be deformed in one direction only. It is a subset of malleability.

SHOPTALK

Amazing Alloys Combining metals yields some bizarre results. For example, a **eutectic** alloy is one that oddly melts at a much lower temperature than any of the ingredients from which it was made. Solder is a eutectic. Made of lead melting at 621°F and tin melting at 449°F, solder melts at the far lower temperature of 350°F! Combining different amounts of each metal affects the melting temperature and that effect can be graphed. The eutectic melting point is at the bottom of a funnel-shaped line; the lowest melting point combination is called the eutectic point.

Another and even stranger fact arises when adding the metal bismuth to the solder combination. Bismuth is a pinkish metal melting at 520°F, which further lowers the eutectic point. The new alloy not only melts far below the temperature of boiling water, a truly amazing 149°F for the eutectic, but in a particular ratio of bismuth to the lead and tin, it also exhibits the totally unique property of remaining a constant volume from liquid to solid. No other alloy does this. This characteristic makes it useful in complex, technical machining.

Commercially named Cerro-True, it is used when delicate materials must be supported to be machined without bending, flexing, or vibrating. Liquified, this amazing metal is cast around and inside delicate workpieces, to support them during machining. Due to the zero shrink factor, the liquid metal remains supportive of the workpiece after it cools to a solid. Both work and supporting Cerro-True are machined together. After the cutting is complete, the chips and work are both put in hot water where the eutectic melts away and can be drawn off the bottom of the tank to be reused!

A second custom-designed version of this metal is called Cerro-Bend. Using slightly different proportions, it expands just slightly when it solidifies from the liquid state! This property is used to fill then support metal tubing to prevent wrinkling and collapse when bending them around relatively small radii. After bending, the removal of the solderlike metal is easy with the hot water tank.

Greater Than Their Parts! Another startling fact: Many alloys achieve greater properties than any of their individual ingredients. An exceptional example lies in the metal beryllium-copper, commonly used for electrical conducting springs. While it is 96 percent soft copper, with only 2 percent beryllium and 1 percent cobalt, and 1 percent zirconium it is possible to heat treat the alloy to a R_C 50, hard enough to machine soft steel! No kidding.

Tensile, Yield, and Elastic Strength

These three properties are measures of the ability to withstand stress (push, pull, bend, and so on) without damage. Each is directly proportionate to the heat treat condition of the alloy.

The Three Stages of Measurable Strength

To understand the different stages, let's perform a test on 1 in.^{2} or 1 cm^2 of a given metal. This is a **tensile strength** test—meaning we'll put tension on the metal by pulling it apart. We'll increase tension progressively until it breaks and three stages of strength are observed (Fig. 14-31).

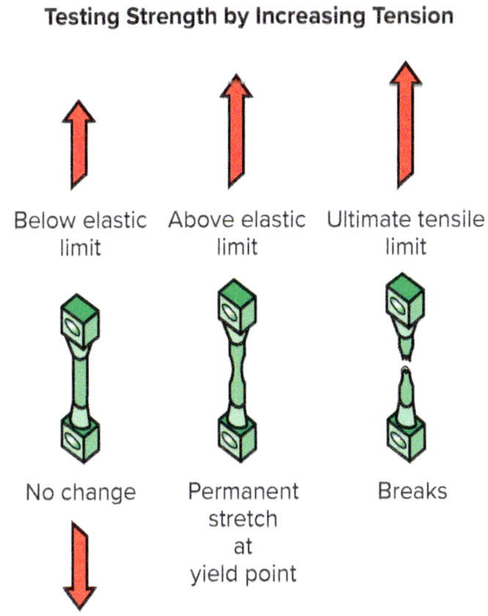

Figure 14-31 A tensile strength test finds three stages of stressed metal: below, then above the elastic limit, then at the ultimate limit, where it breaks.

Stage 1 Elastic Strength Until a given pressure is reached nothing permanent happens to the test piece. It will stretch a tiny amount but will return to its original length when the force is removed. The force remained within the **elastic limit** of the test metal. A safe design ensures that the alloy chosen will perform well within a safe margin below its elastic limit.

Stage 2 Yield Strength Then at a greater force, which is different for each alloy and heat treat condition, it would start to stretch permanently. It would not return to its original shape if the strain was ended. It has exceeded its elastic limit. That permanent deformation point is noted in pounds per square unit, as its **yield strength**. The yield point must be known by the toolmaker if building a die that forms by bending a given metal. It indicates in pounds per square inch how much force will be required to cause a permanent formation of the metal.

Stage 3 Tensile Strength Then as the pull continues to increase, the resistance to stretching begins to decrease on an accelerating curve. It starts to pull apart rapidly, then snaps. It reached its *ultimate strength* expressed as its tensile strength in pounds per square inch or **newtons** per square centimeter (see the Shoptalk).

SHOPTALK

Force Units Tensile, yield, and elastic strengths are expressed in one of two possible units: Imperial (pounds per square inch) or SI (newtons per square centimeter).

A newton is the international basic unit for force. To reproduce a newton, you would measure the amount of gravitational pull on 1 kg of mass. In other words, 1 newton of force is how much pull a 1-kg item exerts toward the earth's center.

Remember: A newton is a standard unit of force equal to just over 2 lb. It is the force of gravity on 1 kg of mass.

It should be noted that some alloys have little or no ability to stretch, thus they go from the elastic limit directly to the breaking point. Many cast alloys exhibit such a trait.

UNIT 14-6 REVIEW

Replay the Key Points

- Malleability is the master physical property. It is the ability to be deformed in all directions. It is inversely related to hardness.
- Ductility is the property that allows metal to be deformed in one direction only.
- There are three stages within tensile strength test: elastic limit, yield point, and ultimate strength.

Respond

1. Using 10 words or less, define malleability, ductility, and elasticity.
2. A newton is a unit of mass that describes the tensile strength of metal. Is this statement true or false? If it is false, what will make it true?
3. Which will be the greater values for every alloy, the tensile or yield strength?
4. How are stress forces expressed for various alloys?
5. You need to make a shaft for a food processing machine and think that a stainless steel would be the best material for the application. It must withstand at least 40,000 pounds per square inch tension with a 5 percent safety margin. Would 30302 stainless steel be acceptable?
6. Many car enthusiasts brag about their "Mags" (magnesium wheels). But in actual fact, most of those wheels are made from aluminum! Is aluminum's tensile strength stronger or weaker than cast magnesium?
7. Of the two alloy steels 4130 and 4340, which has the stronger yield strength?

CHAPTER 14 REVIEW

Unit 14-1

Every time I read about metals, I find surprises. In doing the research to write this section, I learned perhaps 10 new things about some of the oddball metals and their properties.

For enrichment only, see where you might obtain lab samples of some of the metals and alloys you might be called upon to machine. Compare their weight to steel. Find out how rare their ingredients are in the earth's crust. Are they an alloy or a pure element? Research what it takes to extract them from the earth. Some metals are actually abundant but so tightly bonded to other elements that they are too expensive to use. Where are they found, and what nations control them? This is a big issue, as many are needed here in North America to keep manufacturing going, yet they are mined thousands of miles away where the supply is controlled by other governments. A few are so critical to national defense that they are stockpiled within our borders! Here are a few suggestions for your list:

Magnesium	Beryllium-copper
Pure beryllium (lighter than aluminum and more costly than gold)	
Titanium	Tin
Tantalum	Molybdenum
Hastalloy®	Cobalt
Nickel	Pure tungsten
Chromium	Manganese
Inconnell	

Unit 14-2

Similar to the special ability to cut technical threads using a manual lathe, the ability to do your own heat treatment of tool steel is a skill that separates the machine operator from the journey-level machinist. Today, it isn't required of the CNC machinist. However, if your career plans include progressing to tool making, heat treating then becomes an essential skill. A thorough knowledge is also needed to become a job planner, engineer/designer, or programmer.

Unit 14-3

All the other heat treat processes change the physical properties of metal. Case hardening changes steel's chemical composition by adding more carbon to the alloy. However, using the shop process of packing the carbon compound around hot steel, the carbon can migrate only so far between the hot austinetic crystals even if the process is repeated two or more times. After quenching, it forms a shallow shell that's 0.045 in. thick at best. To *push* carbon into the steel further requires technical processes performed by a specialist.

The area where this process most affects the machinist is that of planning to grind a job after it has been case hardened using the packing process. The challenge is to plan rough machine cuts that leave just the right amount of excess to be finished after case hardening. If the case is not deep (0.030 in. average), it can easily be ground completely away!

Unit 14-4

Heat treating aluminum is a very different process from heat treating steel, and a good example of the strangeness of metallurgy. It's exactly backward in every respect when compared to steel. When aluminum is heated, then quenched, usually in very cold water, it enters its softest state, not the hardest. But if left alone, over time it will return to a harder state by a process called aging. But using a lower heat, similar to tempering steel, the process can be speeded up. So here, too, its opposite, heating it after quenching, speeds up and increases aluminum's hardening.

Although it's the third-most abundant element in the earth's crust, behind oxygen and silicone, aluminum is a difficult metal to extract because it forms tight bonds to other minerals and metals. Found in many compounds in the ocean and on most continents, there is only one form, bauxite, that can be economically refined. But even then the process requires huge amounts of electrical energy.

Perhaps one of its more interesting properties is that in its pure state it forms an oxide that's nearly impervious to further reaction or corrosion. Pure aluminum literally creates its own anticorrosion coating. But aluminum without alloys is far too malleable to use in products. So, to take advantage of the oxide property, aluminum manufacturers coat strong alloys with a thin layer of pure aluminum, called *cladding.* It then oxidizes and stops further reaction to outside elements that would corrode the alloy without protection.

Unit 14-5

Unlike heat treating, measuring hardness and understanding Rockwell and Brinell index numbers aren't added skills for the modern machinist. They are an integral part of the job. There are other systems and methods, but the two, Rockwell and Brinell, are by far the most popular ways we test and rate hardness. And they are the way we communicate how fast a metal is to be machined. This is important information for high production today, but as computers and the machines of the next generation speed up, using hardness information will become ever more critical.

Unit 14-6

It's useful to understand the various physical properties of metals when machining, heat treating, and measuring hardness. This becomes especially true when we maximize production and cutter life during CNC machining.

However, if you're entering the tool and die field, an in-depth understanding of those properties is critical. I know this for a fact–I'm a journey tool and die maker. The tools must be based on calculating forces and they are expensive. For example, a three-station, progressive die (one that shapes and cuts metal in stages) can easily cost $15,000 or far more. The die maker must calculate the exact shear property of the metal being formed in order for it to work right. If not the punch doesn't fit the die opening (both too much clearance or too little won't work), and it will fail catastrophically or constantly need resharpening.

QUESTIONS AND PROBLEMS

1. What facts are found in the AISI numbering system to help indicate how a given alloy must be heat treated? In other words, what makes the difference in critical temperatures and quenches? (Units 14-1 and 14-2)
2. Identify and describe the three general groupings in which we divide steels. Some references include a fourth group; what is it? (Unit 14-1)
3. In terms of carbon, what is the dividing line between mild and heat-treatable steel? (Unit 14-1)
4. What AISI family includes the mild steels? (Unit 14-1)
5. Using *Machinery's Handbook,* what is the critical temperature required for heat treating for AISI 1019 steel? (Units 14-1 and 14-2)
6. What is the critical temperature for AISI 1095 steel? (Units 14-1 and 14-2)
7. Identify and describe the three stages of heat treating steel. What must be done to put the steel in each state? (Unit 14-2)
8. When the steel is heated above the critical temperature, then quenched, what occurs? (Unit 14-2)
9. Following quenching, why must we then temper the hard steel? (Unit 14-2)

CRITICAL THINKING

10. Why don't we save time by heating steel to a temperature just below the critical point, then quench it? Wouldn't it save the tempering step? (Unit 14-2)
11. What is another name for tempering steel? (Unit 14-2)
12. What are the correct quenching media for these tool steels: A-2, O-1, and W-2? (Unit 14-2)
13. What is the objective of heating steel above the transformation temperature (critical temperature), then quenching it? (Unit 14-2)
14. Identify two reasons we might case harden a steel part. (Unit 14-3)
15. Describe the case-hardening process you can perform in your shop. (Unit 14-3)

REFERENCE BOOK PROBLEMS

16. Completely describe the information in this aluminum specification: 7075-T651. (Unit 14-4)
17. What is the tensile strength of the 7075-T651? What is the expected Brinell hardness index for this aluminum in this temper condition? (Units 14-4 and 14-6)
18. Name the two most common hardness tests in North America and breifly describe each. (Unit 14-5)
19. List and describe at least five physical properties of metal. (Unit 14-6)

CNC QUESTION

20. True or false? Other than when performing CNC HSM (high-speed machining), we normally do not look at the heat treat condition or alloy of the aluminum. If the statement is false, what will make it true?

CHAPTER 15 Job Planning

Units

15-1 Planning Jobs Right

- Answer the seven prime questions leading to successful plans
- Write a job plan for efficiency and quality for four jobs

15-2 Troubleshooting

- Find and correct operation sequences that are unsafe
- Find and fix sequences that lead to poor quality

Even making a single workpiece with the simplest of shapes requires careful choices before putting it into work. More to the point, not planning the simplest or any job, easy or complex, is a surefire guarantee it will be difficult to complete, of poor quality, unprofitable, or even dangerous!

One of the highest skills within our trade is job planning, and the machinist who excels at this has great opportunity for advancement. But it's a hard-won competency that's different from previous subjects we've discussed, for four reasons.

First, planning solutions are often matrix-like in that any single job could take many different routes through the shop on its way to completion. Some sequences are faster, while others lead to more repeatable accuracy, some toward cost or time savings, and on it goes. Every job presents a puzzle with a nearly infinite set of solutions, each with trade-off advantages and disadvantages. There is no single or perfect answer.

Second, planning often requires inventive thinking beyond conventional solutions. Some new designs create a new challenge: how to hold an oddly shaped object as it nears completion or how much material is just enough for roughing but not too much for economical manufacturing.

Third, plans are never quite finished. Even the best plans may change after the work is in progress. As safer ways are discovered or as new ideas occur we edit them as part of normal operations to upgrade capabilities. In industry, I've seen the same job go through the shop dozens of times, and before each run, improvements are edited in based on the last time we ran it. CNC programs experience the same evolution toward perfection, too.

Fourth, writing plans requires a full understanding of all the machine types and the operations they can perform—the big picture for the shop's capability. If you've followed the book outline, you are at that point now. You now see the bigger picture—that the shop functions as a unit.

Here in Chapter 15, after a few hints for success, we serve up six part-making puzzles for which our solutions (reviewed by several instructors) are offered. Four jobs have not been planned yet, and two are already done, but they might need improvement. You'll need to compare your solution to ours, discuss it with others, then decide for yourself if your plans are the better solution. It's quite possible that your plan is better than mine!

THREE KINDS OF PLANNED SUCCESS

In the shop, there are three different kinds of planning. They work together to keep everything running smoothly:

1. Job flow
2. Shop flow
3. Operation sequence

Job Flow This is the route any single part or batch takes through the shop. For example, do we surface grind the work before or after heat treat? Does the job go to the mills first, then the lathe? These plans must occur first.

Shop Flow This is the day-to-day operation of the entire shop. It schedules and moves jobs through all workstations simultaneously. Unplanned, they arrive at critical workstations (example: CNC mills or surface grinders) at the same time. When they clash, one must be given priority while the others are sidetracked, halted, rerouted, or sent to another machine shop. Keeping the total workload moving efficiently might also be called *shop load* or *shop floor planning*. As a management task, shop flow is beyond the objectives of Chapter 15, but it's a big part of operations.

Shop flow is vital. Most large companies employ specialists in the field. They maintain job boards and wall charts where they move tags to track it all. But even the experts often turn to management software called (Enterprise Resource Planning [ERP]) when visual aids can't solve the matrix. PC-based management predicts bottleneck points and schedules and adjusts sequences to avoid them. A big challenge happens when priorities change—one job becomes HOT. Every other job in its way must move over!

Shop flow is a perfect description with a river of jobs flowing through. New ones enter upstream daily. Each has its own priority and potential to affect those already in work. Often, a job's priority will change halfway through the loop. Suddenly it goes from normal to *hot*—the customer needs it now! That changes everything in its routing. It causes turbulence, which then disturbs other lower priority work. You can see that only a computer with specific programming can reschedule the flow when things don't go as planned—which is almost always!

Absences and Breakdowns The two worst dams in the flow occur when the machinist doesn't make it to work that day or a machine breaks down, especially if that station is one of the bottlenecks! If available, another machinist can be substituted or the job could be moved to "bump" another job. But then, a whole series of disturbances ripples out.

Operation Sequences The third kind of planning is the machinist's problem, which is logically sequencing operations. They're step-by-step instructions, aimed at achieving the drawing dimensions and specs. Chapter 15 starts your education in this kind of job planning.

One of the highest skills within our trade, the machinist that excels in job planning has great opportunity for advancement.

Unit 15-1 Planning Jobs Right

Introduction: Learning by doing—it's the only way to learn to plan. With some guidance, in Unit 15-1 you are to plan out four jobs, each more complex. Then upon compiling your best ideas, compare yours to mine. That's the key here—trying, then comparing. Although I consider this kind of planning to be the "challenge and fun" stuff of machining, check with your instructor, as you may not need to do all four. But even if you skip a couple of planning tasks, read the solutions I offer—it's entirely possible yours will be better!

TERMS TOOLBOX

Grip excess Excess used to hold a bar in a lathe chuck beyond that which is needed for the part.

Temporary datum Planning a cut sequence that creates a reference to be cut away when no longer needed.

15.1.1 The Seven Prime Questions

A solid plan must answer six or seven questions, depending on whether or not it's to be machined by CNC methods. All must be answered at the same time or nearly so. They are intertwined, each affecting the others. Use the following as a checklist for writing your plan:

☑ *Which machine types first?*
The answer to this will depend on the number and complexity of parts in the batch and the kinds of machines available for this job. Job priority affects this decision, too, since a hot job can bump others from certain machines.

☑ *How is the work to be held and located during machining?*
Often overlooked by beginners, this must be solved right after you decide which machine gets the job first. An incorrect choice can lead to needless variation being introduced into the job.

☑ *Does the job need work-holding and reference excess material?*
Well-planned excess material to hold onto or to create temporary reference surfaces can save money and time.

☑ *Do the first cuts lead to reliable reference?*
The first vital cuts must create references that will be used for moving the parts between setups.

☑ *Is the overall sequence efficient?*
Can several operations be accomplished in one setup?

☑ *Are specialized holding or cutting tools justified?*
This decision is based on two factors: batch size and the probability of completing the job using only standard holding and cutting tools. For example, are soft jaws for the vise or chuck justified? This decision is often based on part quantities–more parts in the batch justify specialized tooling. But if you decide to use special tools, they need to be ordered and made right NOW, to be available when the job hits the shop floor!

☑ *Reference point (CNC jobs only).*
Since our plans will be limited to manually operated machines in this chapter (CNC planning comes later), we won't be using this factor. But it must be in the list of questions when planning a job that's to be machined from a program. There must be zero point chosen, a place where all coordinates are set to $X = 0.0$, $Y = 0.0$, and $Z = 0.0$.

That reference point applies to the program, to the holding tooling, and to the machine setup. Everybody in the loop must know where it is located on the part or on the holding method. Choosing that point wisely is a prime decision that must be made at the time when the other questions are decided.

Other than a reference point, the upcoming challenges focus attention on these issues. Here are a few tips about how to answer them.

15.1.2 Excess Material Planning

Planned for Holding and Intermediate Operations

Although excess must be made into chips, which consumes money and time, sometimes by planning the right amount in the right place, the added material costs are outweighed by improved safety and quality, or the excess provides a far more efficient way to hold and cut the part. Thus far in this text, excess has been presented as a minimum allowance to completely machine a surface. But there are two more aspects.

KEYPOINT

There are times when it is very cost effective to plan excess metal beyond a cleanup cut minimum. It's used for *work-holding of difficult objects* and *a planned temporary datum to start the job right.*

Work-Holding Excess

Here's an example of work-holding material for a lathe job (called **grip excess** *material* for lathe work).

You are to turn five brass surveyor's plumb bobs (swinging pointers to indicate a vertical line; Fig. 15-1). They're made from $1\frac{1}{2}$-in.-diameter brass bar stock, a relatively expensive material. This stock doesn't fit through the lathe's spindle hole, so bar feeding and parting off aren't an option. That leaves two other options:

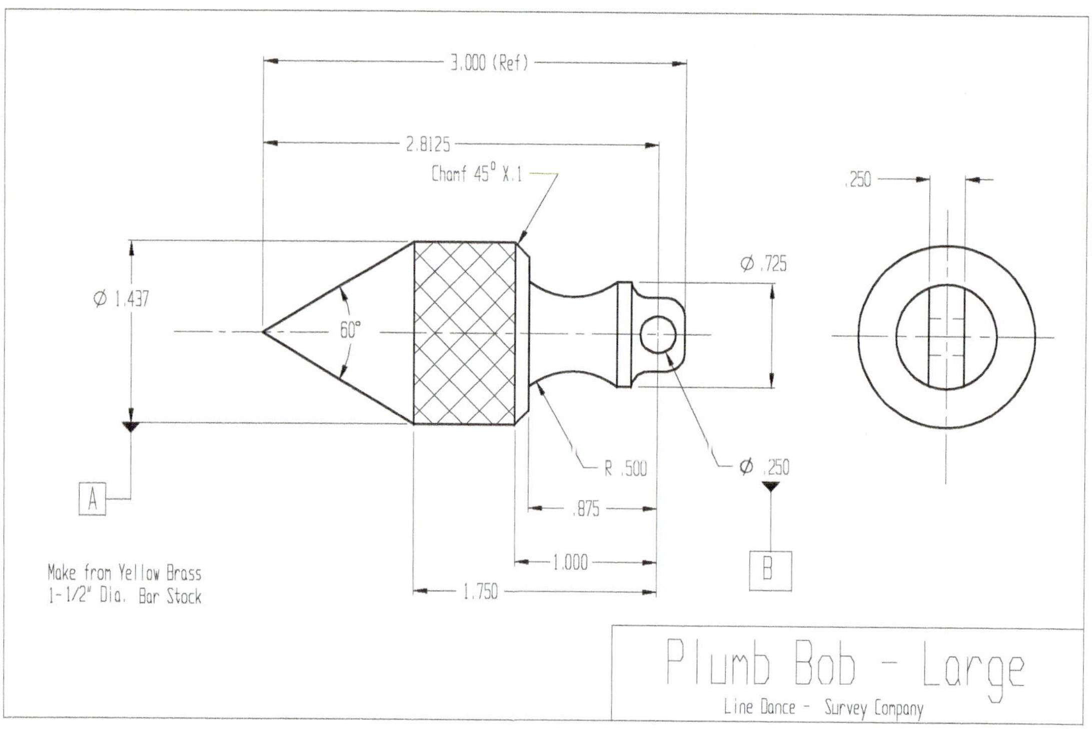

Figure 15-1 There are several different ways to plan operation sequences for this plumb bob.

1. **Sawing Blanks to Length** (Fig. 15-2)
   Turn half the shape, reversing to finish the job.
2. **Sawing with Grip Material** (Fig. 15-3)
   Holding an extra length, machine the entire shape in one setup.

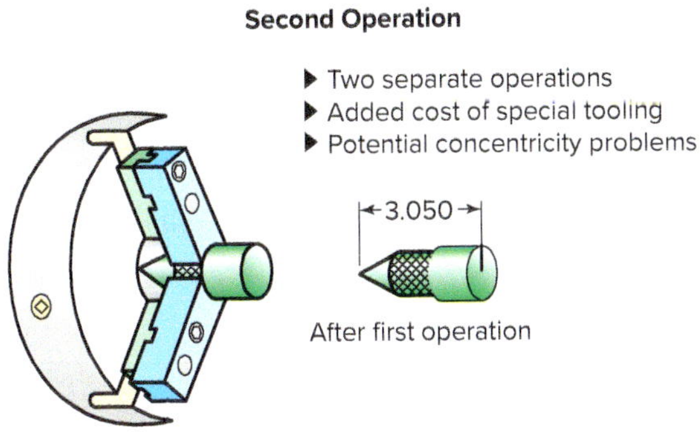

Figure 15-2 Option 1—The second of two setups.

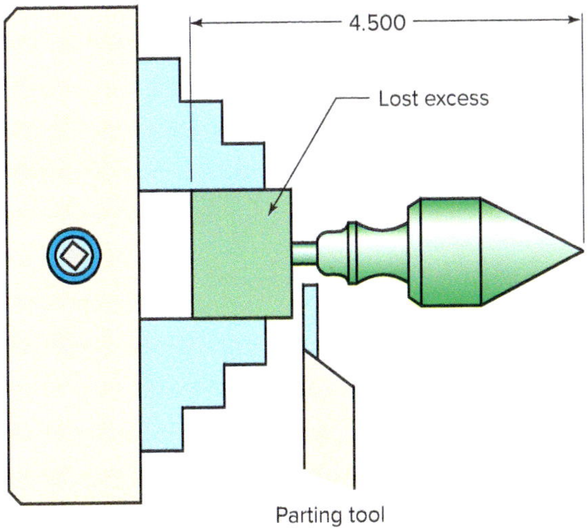

Figure 15-3 Option 2—Cut all features in one setup; then part off the finished work.

Both plans would require the bar to be cut into lengths before machining. But the grip excess (plan 2) would include 2 in. extra material per part length. Let's compare.

1. Cut to 3.100 in. long. We'll first turn the point and knurled section, leaving a small shoulder between the bar stock and 1.437 in. diameter. This could be done in a standard three-jaw chuck. Then reverse it by pushing the shoulder against soft jaws (see Fig. 15-2). The soft jaws would help ensure concentricity of the second cuts and to not mar the finished knurls. With the temporary shoulder nested against the jaws, it's possible to finish the shape.
2. On the other hand, how about cutting the bar into 5-in.-long pieces? (See Fig. 15-3.) That leaves a minimum of $1\frac{1}{2}$ in. stock to grip in the chuck so the entire shape can be turned, then parted off at the final operation. No special tooling is required, and the whole profile can be turned in one setup.

Critical Question

Is plan 2 more economical? Probably, even though a length of brass is lost for each part. It's going to go faster and be safer, and the shape is going to be more concentric end to end. But the extra brass stock is costly, too. The final answer would be a time study *versus* current material cost. (See the Shoptalk.) But how about the stability of the part as the cutoff blade nears center? Is this a safe plan?

> **SHOPTALK**
>
> In plan 2, a small but significant fact to be considered when batches become large is that the excess will be parted off as a solid chunk. A solid piece of metal commands a better price when it's recycled, and it will offset some of the added cost of excess.

> **KEYPOINT**
>
> In this example, plan 2 is likely to be more economical, but there are times when plans similar to plan 1 might work better.

15.1.3 First Cuts Based on Design Datum Basis

Related to excess material is the decision of how to create a reliable reference datum(s) at the start of the job. Unless the parts can be finished in one setup, as plan 2 was, the parts will need to be moved from one setup to the next, or to another machine. The shoulder in plan 1 that allows us to reverse the parts and be able to locate them in the soft jaws with reference to the pointed end was a **temporary *reference* datum**. It is slower with concentricity challenges but far less dangerous than plan 2.

Temporary Reference

So the third reason for planned excess is to create a temporary basis for the next cuts. That extra step is used when one can't immediately cut the actual datum to begin the job. It's machined early in the plan, then sequential setups or cuts are referenced to it until an actual high-priority datum can be cut. The temporary surface simulates a high-priority datum until one can be created. Then, after shaping the part, where it's no longer needed, the temporary surface is removed. Of course, the objective is always to minimize that planned amount.

KEYPOINT

Good plans begin with vital, first cuts that create sequential control throughout the remaining operations. That first set of operations might be as simple as squaring two sides of a block or a lathe face cut, but it is the starting point reference from which all sequential operations originate.

Recall squaring a part on the milling machine (see Chapter 11). After establishing a reliable datum feature, all sequential cuts were made by placing that surface against the solid jaw of the vise.

15.1.4 Holding Tooling and Method

The next prime question is how will the job be held on that machine to make the first cuts. This must be answered before writing cut sequences or a program. The obvious reason is safety and efficiency. But the subtle reason is to control variation. Does your choice of holding method eliminate or cause variation in the job? Rather than discuss it here, one problem was expressly created to challenge your thinking in this aspect of planning.

KEYPOINT

Always analyze the holding method and ask whether it causes needless variation.

Note, too, that each problem asks you to machine different numbers of parts. The quantity in the batch changes some of the thinking, especially on holding tooling. Larger part runs may justify special tooling (soft chuck or vise jaws or fixtures), where small batches are more economically solved using standard vises and chucks, even though they take longer to use.

Problem 1 Cast Iron Angle Plates

Okay, it's your turn. Plan and write a set of operations to machine these four precision angle plates including semi-finish milling and final grinding on all surfaces marked with a 64-μin. callout (Fig. 15-4).

Instructions

- Machine only features that have surface finish dimensions.
- 6-mm excess has been cast on every face of the casting to be machined.
- Leave 0.4-mm grind excess after milling on every face.

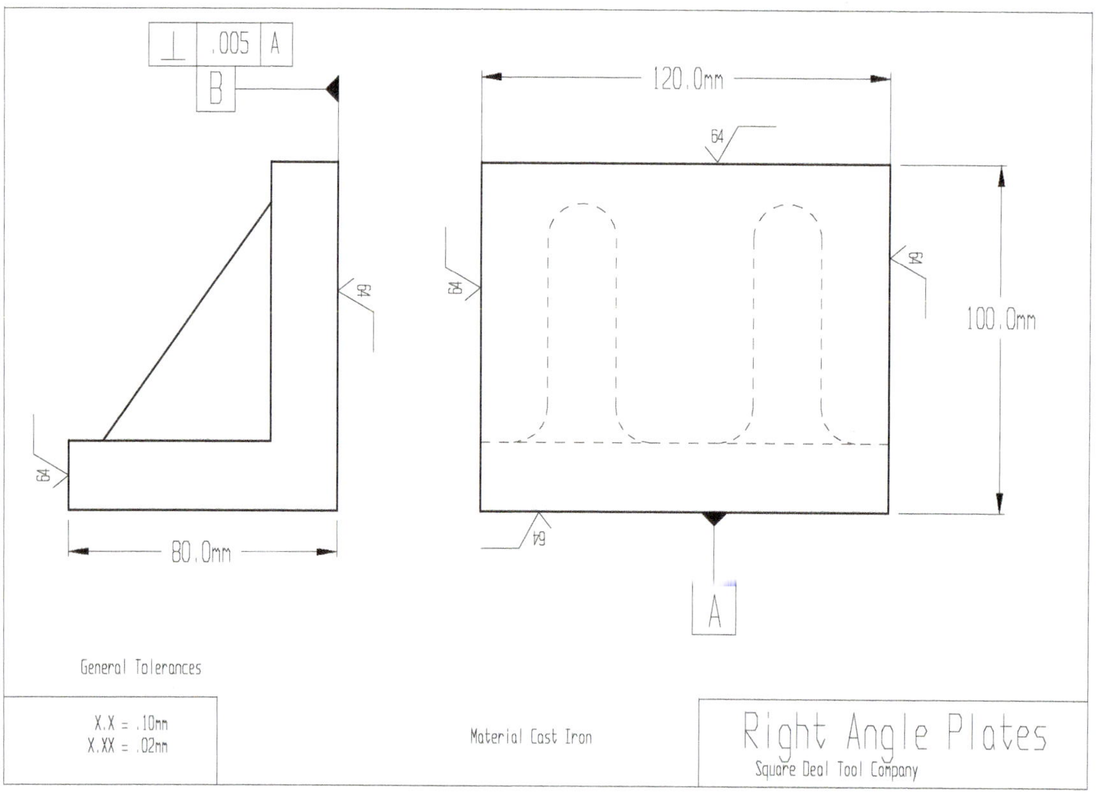

Figure 15-4 Planning Problem 1—Right-angle plates.

Using the model here, write a complete machining plan on a sheet of paper. Add more sequence numbers if required. Then compare your final version to mine.

Job Planning Sheet

Job Name: Right-Angle Plate **Number:** 4 Machined in Matched Set
Material: Cast Iron
Special Tooling or Holding:

Sequence	Operation	Comments
005		
010		
015		
020		
025		
030		
035		
040		
045		
050		

Problem 2 Valve Domes

These valve domes (Fig. 15-5) are to be machined from precision brass forgings. Each has 0.10 in. excess on all dimensions to be machined. Include operations to make all holes and slots. On completion, turn to the answers and compare your plan to mine.

Instructions

- Write your plan to produce a quantity of 150.
- Plan this job without the benefit of CNC equipment.
- How would you change the plan if CNC mills and lathes were available?

Problem 3 Tap Wrench Nuts

You are to machine five of these tap wrench nuts from a 1.00-in.-diameter 4130 steel bar preheat treated to Rockwell-C 28 (Fig. 15-6). Study the print and write your plan on a sheet of paper.

Instructions

- Make a quantity of five of these nuts.

Problem 4 Locator Bolts

These special fixture bolts (Fig. 15-7) might need more than one setup to complete, or do they? Note that they're made from mild steel but that no raw material size or shape has been specified.

Instructions

- Make 25 in this batch.
- *Choose the raw material* shape and size.
- You may need to look up standard steel shapes and sizes in *Machinery's Handbook*.
- Problem 4 is a time-versus-quality challenge.

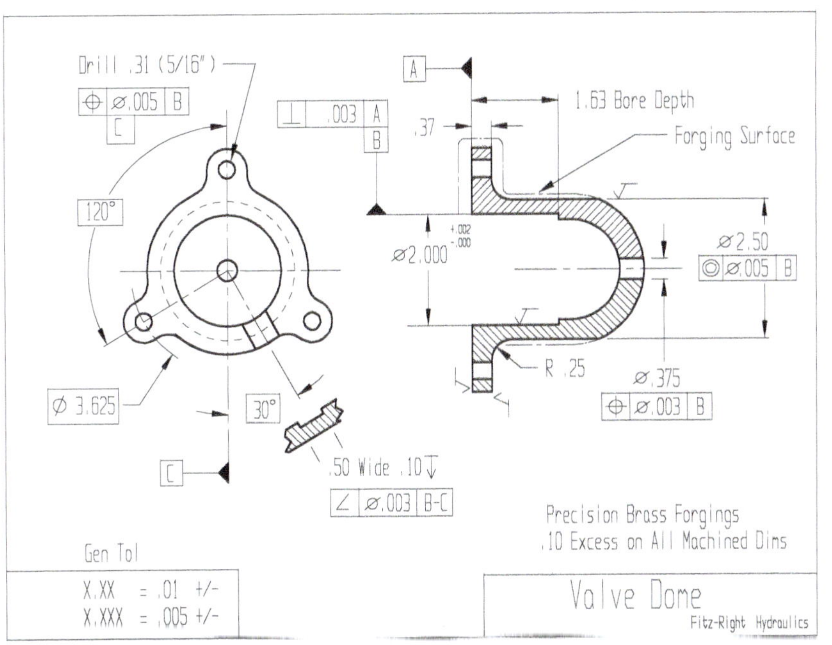

Figure 15-5 Planning 2—Valve dome.

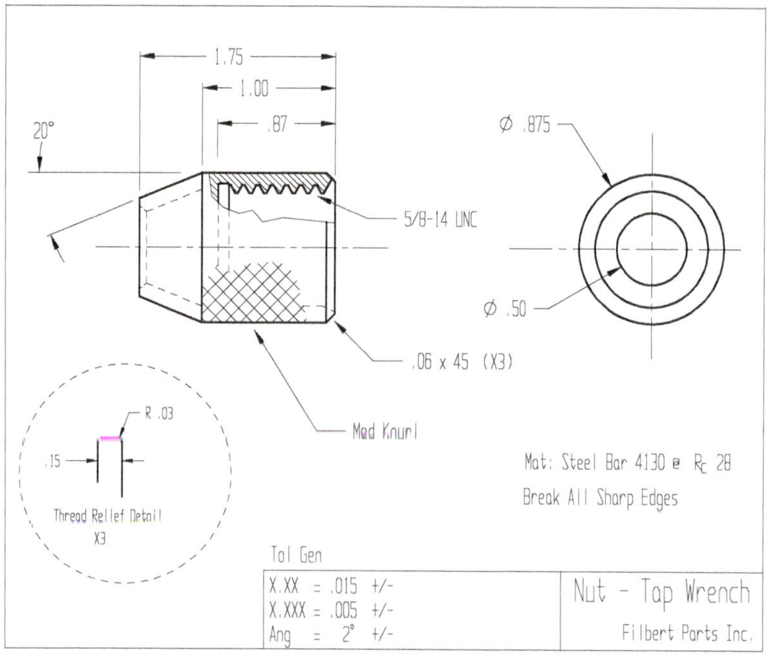

Figure 15-6 Planning Problem 3—Tap wrench nut.

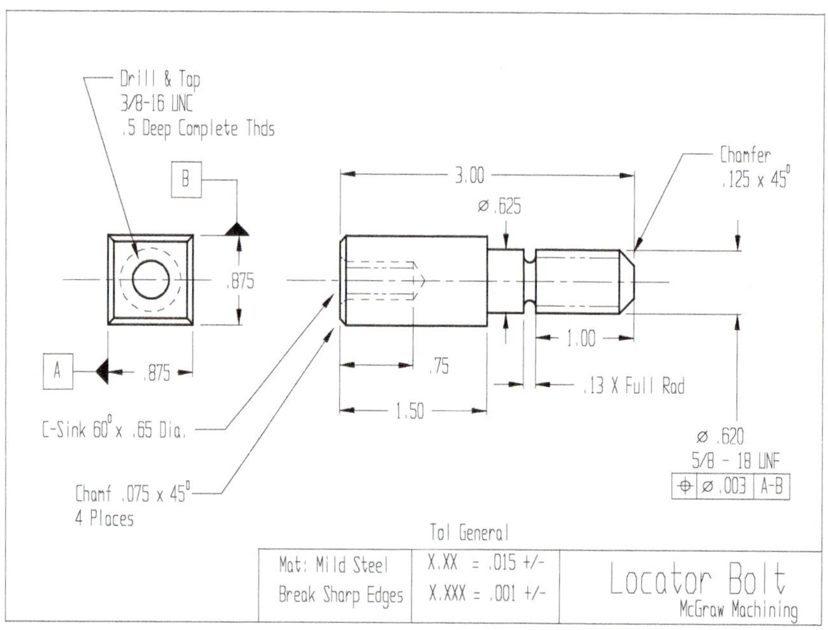

Figure 15-7 Planning Problem 4—Locator bolt.

15.1.5 Troubleshooting

The most skilled planners cannot foresee every trap that arises as the machining commences. Plans are improved regularly as part of normal operations. However, as your instructor in print, I'm compelled to add a bit of advice: As a new person, exercise restraint in criticizing a plan or instructions. If safety and quality are at risk, then go ahead, but do it with finesse. You are the "new person," and they did just fine before you got there!

Seeing problems with work orders and improving them is a skill needed for competency in both manual and CNC machining. Problems that cause part errors and safety hazards must be eliminated when they are found. Here's an example (Fig. 15-8).

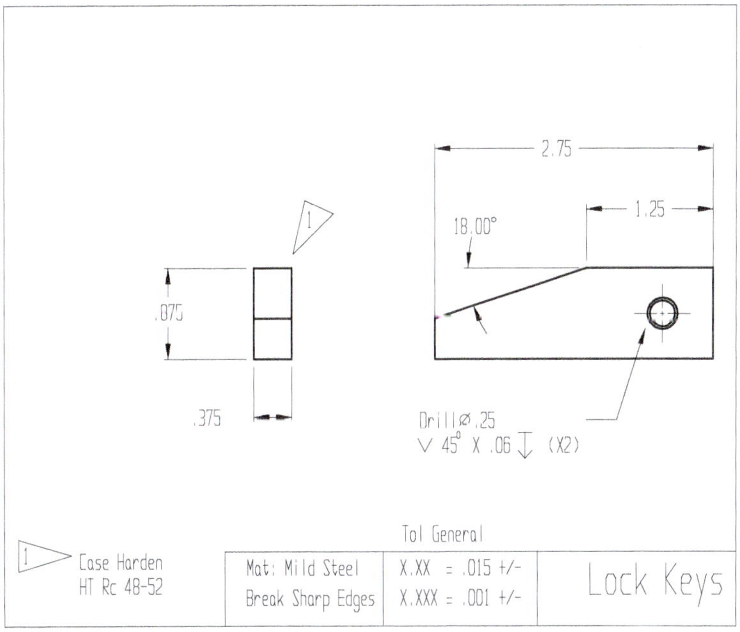

Figure 15-8 Planning Problem 5—Lock keys.

KEYPOINT

In every shop, there will be an official revision procedure that keeps everyone that it affects informed of the changes. Never write on or alter a work order/plan without permission.

Problem 5 Lock Keys

After reviewing the plan offered, edit it or reroute it altogether to solve sequencing problems if they exist. On completion, compare your improvements to mine. (*Hint:* There are at least four big errors in this plan.)

Problem 6 Double-Arm Bracket

You are to make a batch of double-arm brackets (Fig. 15-9) from a solid block of aluminum. After reading the original plan, edit it with improvements or scrap it altogether and start over, if you feel a justification.

Now write your own plan for the double-arm bracket. It need not take the path I've chosen here.

Job Planning Sheet 15-5

Job Name: Lock Keys
Number of Parts: 15
Material: 1.0 by $\frac{3}{8}$ in. CRS Bar 45 in. long
Special Tooling or Holding: None

Sequence	Operation	Comments
005	Saw to lengths 3.0	0.125 in. excess each end
010	Layout angle	
015	Rough saw angle	Leave 0.030 machining excess
020	Mill 18.0° angle	Tilt mill head
025	Mill 2.75 × 0.875 rectangle	
030	Case harden R_C 48 min	
035	Drill $\frac{1}{4}$-in. hole and chamfer 2 sides	

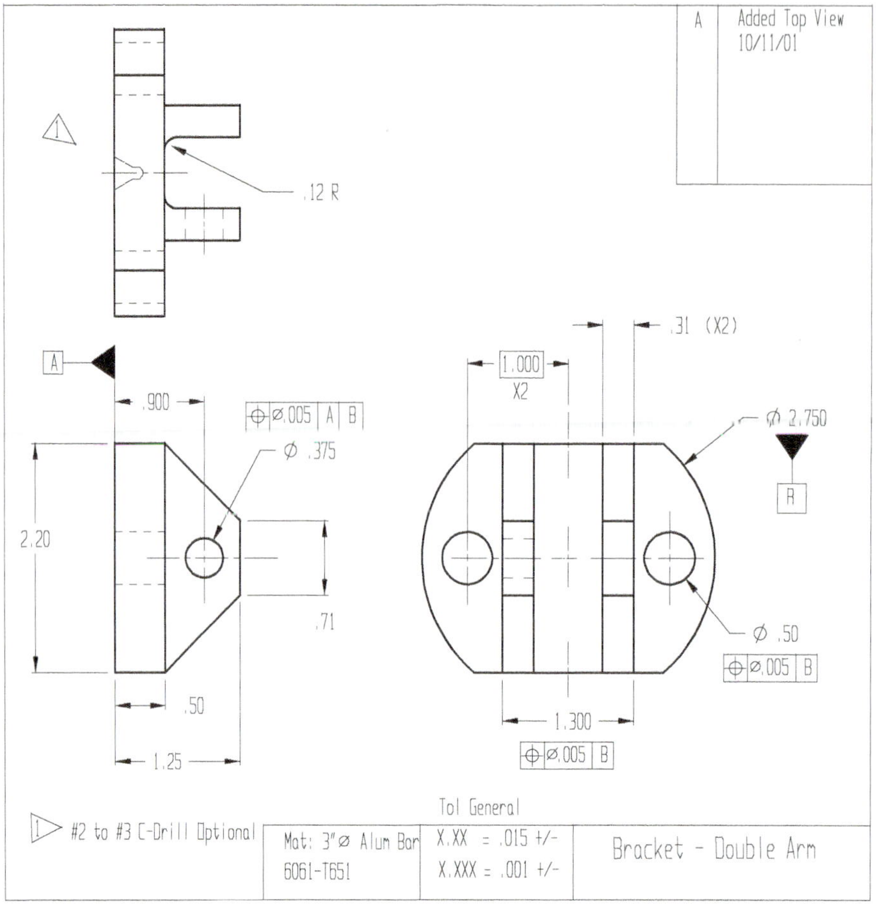

Figure 15-9 Planning Problem 6—Double-arm bracket.

Job Planning Sheet

Job Name: Bracket Double Arm
Material: Aluminum: 6061-T651 Aluminum
3.0-in.-Diameter Bar
Sawed to 15-in. Lengths
Qty 17
Tooling or Holding Soft Jaws Lathe/Mill Vise

Sequence	Operation	Comments
005	Turn 2.750 dia (2 bars)	4 jaw chuck & tailstock center 14 in. long turned portion Excess for grip & facing parting
010	Face and part off 1.5 lg	Soft three jaw chuck—tailstock support creates datum feature A excess for milling
015	Mill 2.20 × 0.50 deep	Soft vise jaws on (Fig. 15 10) 2.750 dia—Datum B
020	Reverse part mill 1.3 × 0.25 R	Standard vise on 2.20 dim 1.300 dim across ears
025	Mill 0.31 ears	
030	Drill 0.375 through	Vise with end stop
035	Layout ear profile	
040	Rough saw	Leave 0.060 excess
045	Mill angle sides	
050	Mill 1.25 dim	
055	Remove all burrs and sharp edges	

Operation 015

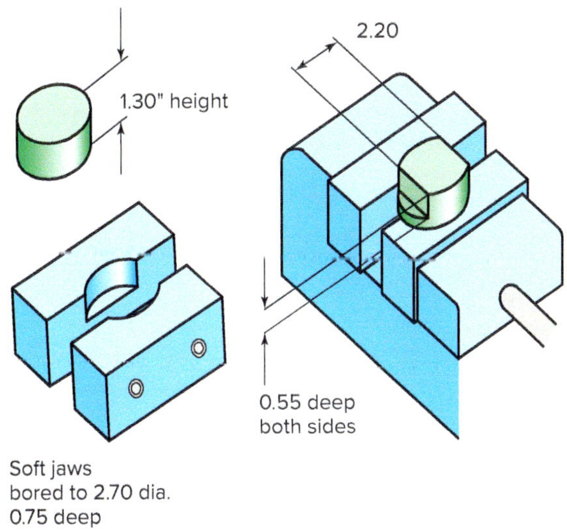

Figure 15-10 Operation 015 for the double-arm bracket.

PART 3

Introduction to Computer Numerical Control Machining

To be a master, yesterday's machinist needed many skills and traits. But even though the knowledge bank was large and took years to gain, once journey level was reached, it remained relatively stable. Sure, processes and technologies did change over time, but the rate of change was nothing compared to today!

For a career to flourish today, machinists must add two more skills to their list. We'll start with *adaptability*. It's more than a willingness to change; it's about looking forward to evolution, seeking it, and embracing it. Adapting will surely be your greatest challenge and your biggest career asset.

In that early machinist's world about three quarters of the price of an average manufactured item was in the labor cost to make it, but not today. Due to the efficiency of computers and software, and to fierce international competition, the wage portion is being driven down and down. Today, any business wishing to have a future foresees and adapts to technologies that deliver faster, better, and less costly products. That goes for the machinist as well.

The workplace is evolving, too. Previously, programmed machines were used for long production runs. Their great cost, slow setup, and programming time justified purchasing them only when the shop expected to make hundreds or thousands of similar parts. But today, these machines are so easy to get going that batches of 5 or 10 parts are normal and economical.

Progressive shops today consider one part to be a batch! Instead of completing a large group of parts before moving the job to another station, these shops move single parts or small batches through in a continuous flow, from machine to machine, and from station to station. It's a flash-quick turnaround environment.

After conferring with planners and lead machinists, on the weekend, they reconfigure the machines to match the most efficient part flow for the upcoming workload (Fig. P3-1).

Today's toolmakers are using CNC to quickly and accurately make one-of-a-kind molds, dies, and other implements. Not only are they using computers to do their work more quickly; they can also create unique items that were impossible to machine previously.

Figure P3-1 These machines are not monoliths.
Courtesy Contour Aerospace/Everett, WA

As I write this, the tasks required to process a normal job from the engineer's drawing to final inspection are those of the

Designer/engineer

Estimator (cost to manufacture, thus price out the door)

Programmer

Planner

Setup/lead

Toolmaker

Machine operator

Quality assurance (previously called the inspector), possibly using programmed coordinate machines

But in the next generation, these tasks will merge, evolve, or even disappear!

Better communication skills are the second survival trait. Yesterday's craftspeople had a print, a machine, and perhaps a planning sheet. For the most part, they made and executed most of the decisions as to how and when the part would be machined. But today, the planner, programmer, toolmaker, lead machinist, and operator are all involved in the final outcome. In the new shop, they need to share program data, design planning, and setup information. Most if not all that data a solid model!

Without communication much of the advantage of automation can be lost and the process can even become dangerous! Today conservative feed rates are 300 in. per minute with mill spindles spinning at 8,000 RPM—double or triple those figures for high-speed machining! Many accidents can be traced back to misunderstood directions or missed setup statements. We'll weave the ability to read and write concise instructions into many of the chapters and units.

So, welcome to the world of programmed machine tools. It's an adaptable, communicative place. Today is an excellent time to be an entry-level craftsperson.

CHAPTER 16 Coordinates, Axes, and Motion

Units

16-1 World Axis Standards

- Identify *X*, *Y*, and *Z*—the primary and secondary linear axes of CNC
- Identify *A*, *B*, and *C*—the primary rotary axes
- Identify *U*, *V*, and *W*—the secondary linear axes

16-2 Coordinate Systems and Points

- Identify geometry points using absolute Cartesian coordinates
- Identify geometry points using incremental Cartesian coordinates
- Select the correct coordinate value based on drawing dimensions

16-3 CNC Machine Motions

- Define rapid travel and linear and circular interpolation
- Compare two-, two-and-a-half-, and three-axis motion

16-4 Polar Coordinates

- Identify drawing features defined with polar dimensioning
- Use polar coordinates to define unique points

INTRODUCTION

The bulk of any CNC program involves point coordinates that refer distances to axes. Along with statements that determine how the machine is to use them, those *X*, *Y*, and *Z* coordinates are used either for tool motion or positioning tools on the work, or for reference. The axes they refer to are the same ones we've already learned on mills and lathes. But here in Chapter 16, we'll formalize them.

The four units in this chapter are bedrock knowledge for everything in CNC and most of the remaining textbook.

Unit 16-1 World Axis Standards

Introduction: There are nine standard axes universally used in CNC machining. Three are the familiar *primary linear* (straight-line) movements *X*, *Y*, and *Z*. Three *primary rotary axes* (*A*, *B*, and *C*) are used to identify arc or circular movements such as a programmable turntable, lathe spindle, or an articulating, wrist action milling head (rotary motion).

Finally, we have three secondary, straight-line motions called the *auxiliary linear* axes (*U*, *V*, and *W*).

SHOPTALK

EIA-RS267-B Axial motion and position are standardized by the Electronic Industries Association (EIA) here in North America in its Recommended Standard EIA267-B. There is also a parallel standard through the ISO (International Standards Organization). These standards actually include 14 defined axes for motion and position; however, using the 9 described here covers all you'll need for training.

TERMS TOOLBOX

Articulate A wrist type action that moves in an arc but not a full circle. Mill heads articulate.

Discrete planes One of three unique planes defined by the axes that lie on it: *X-Y, Y-Z,* or *X-Z.*

Five-axis mill A vertical or horizontal machine with a head that articulates in the *A* and *B* axes.

Orthogonal axis set Three axes lying at mutual 90° angles.

Right-hand rule Used to determine the letter identification of axes within an orthogonal set.

Rule of thumb Used to determine the sign value of rotary axes.

Three-space envelope A three-dimensional work envelope defined by *X, Y,* and *Z* coordinates or by spherical polar coordinates.

World orientation The relationship of a machine's axis set with respect to the floor and operator.

16.1.1 The Primary Linear Axes *X*, *Y*, and *Z*

The basic axes used to define a **three-space** (three-dimensional) **envelope** lie at 90° to each other and as such are called an **orthogonal axis set**. Using the same root word as orthographic projection, the set is comprised of axis lines at mutual 90° angles, intersecting at a common reference point (Fig. 16-1). The limit of machine's axis travel is defined as the *work envelope.*

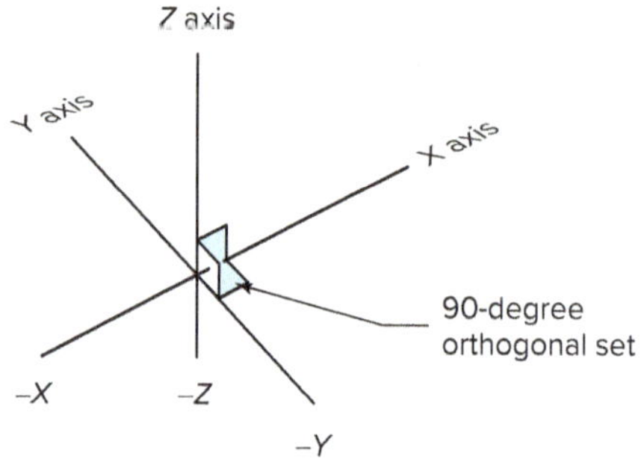

Figure 16-1 The three primary linear axes of *X*, *Y*, and *Z*.

Three Primary Planes

Combining any two primary axis lines defines a flat plane. There are three planes: *X-Y, X-Z,* and *Y-Z* (Fig. 16-2). For example, when viewing a part placed on a vertical milling machine, the table represents the *X-Y* plane, while a lathe object is viewed in the *X-Z* plane–usually from overhead. Here are the code words for each.

G17 = *X–Y* Plane
G18 = *X–Z* Plane
G19 = *Y–Z* Plane

Three Discrete Planes

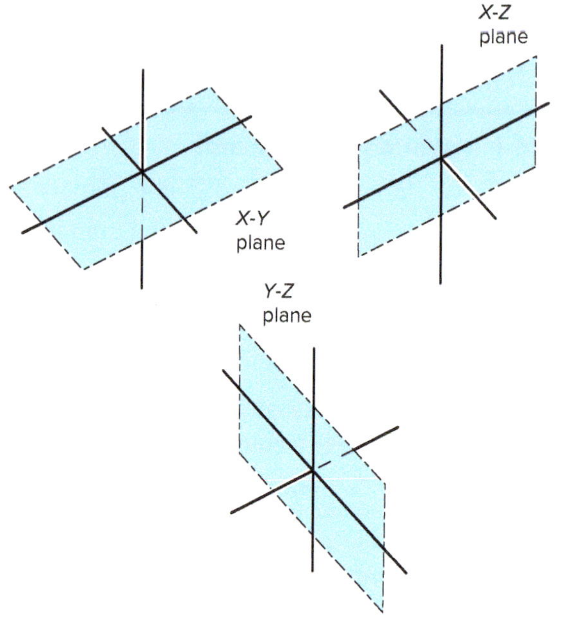

Figure 16-2 The three primary planes *X-Y*, *X-Z*, and *Y-Z*.

When the machine control is capable of cutting curves in more than one of these three **discrete** *(unique)* **planes**, the programmer must add a code word to define in which plane the motion is to occur. We'll learn to use the codes later, but they are G17, G18, and G19.

KEYPOINT

For now, understand that changing to a two-axis curved cut within a different primary plane requires a new code to be entered to signify which plane is wanted.

Axis ID on a CNC Machine

When facing a new machine for the first time, the **world orientation** of its axis set (relationship to the floor and to the operator) can often be identified this way, in this order (Fig. 16-3).

Z The axis parallel to the main spindle

X Usually the longest axis, *usually* parallel to the floor

Y The axis perpendicular to both *X* and *Z*

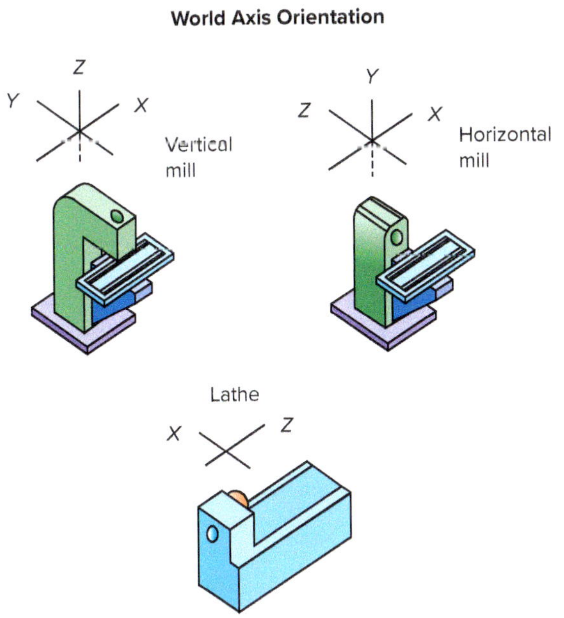

Figure 16-3 The primary axes as they apply to three familiar machines.

These are conventional guides, not a standard. For any given CNC machine, the set needn't be in any given relationship to the world. While the axis set remains orthogonal and the axes are in the same position to each other, the axis set can be rotated to any world position.

Try it yourself. Use the right-hand rule for a mill in your shop (Fig. 16-4). Depending on the world perspective of the axis set, you may find your hand in any position, but it will be found to fit the rule. There are almost no violations to the orthogonal set, but it can be found lying in some odd positions relative to the world, especially on advanced CNC equipment and robots.

TRADE TIP

When faced with an unfamiliar CNC machine, always look for the *Z* axis first as it will be the easiest to identify. The *Z* axis carries the tool to the work, as with lathes, or the work to the spindle or, vice versa, on mills. Then with the *Z* axis under control, apply the **right-hand rule** to identify the other two axes (Fig. 16-4).

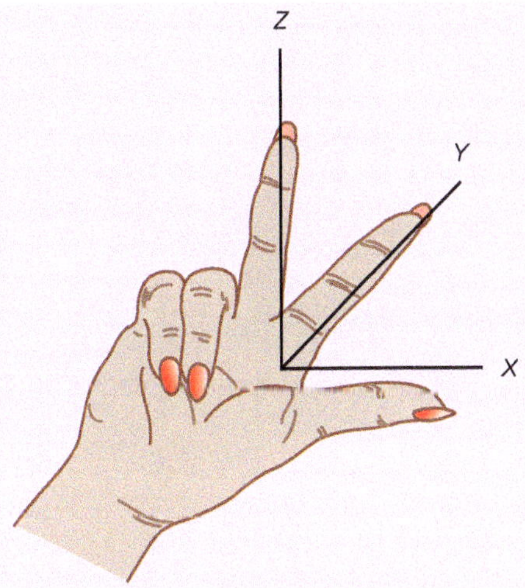

Figure 16-4 The *right-hand rule* helps identify the machine axes.

KEYPOINT

The Right-Hand Rule

Pointing the thumb of your right hand along the positive *X* axis direction, your first finger points out the positive *Y* axis direction. Finally, the raised middle finger points out the positive *Z* axis.

The most common example of skewed world orientation, shown in Figs. 16-5 and 16-6, is a *slant bed lathe,* where the *X* axis has been tilted relative to the floor. This modification makes chip and coolant ejection more efficient and improves operator access for setting up tools.

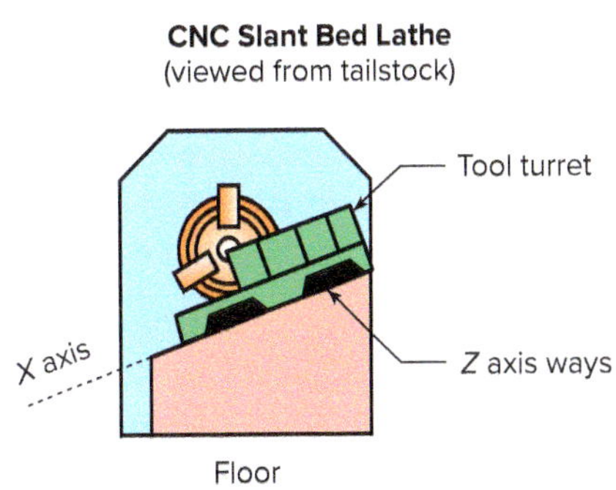

Figure 16-5 A slant bed lathe features a tilted *X* axis relative to the floor to improve chip ejection and operator access to tools.

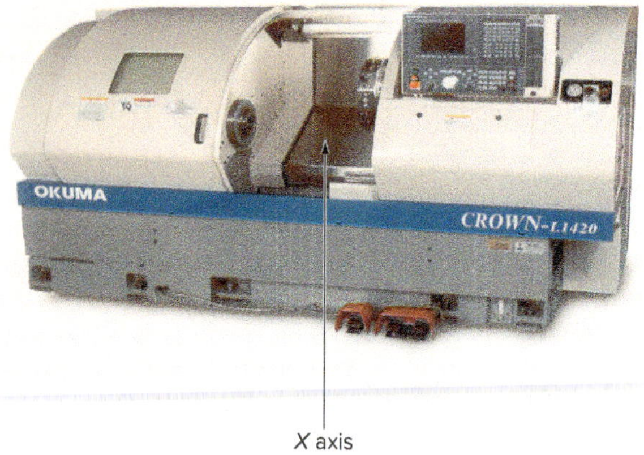

Figure 16-6 The *X* axis of slant bed lathe is tilted relative to the floor to improve coolant and chip ejection and to aid operator access during setup.
Courtesy Mike Fitzpatrick

SHOPTALK

Oops! Some time ago, a few early programmed lathes were given reverse axis sign values for their *Z* axis only. The grand idea was to eliminate the negative sign on most *Z* axis coordinates and thus create shorter programs. But this "improvement" led to so many serious crashes due to the nonstandard axis set that they were never produced again.

16.1.2 The Primary Rotary Axes *A*, *B*, and *C*

Some CNC machines feature programmable axes that rotate or **articulate**. According to the EIA267-B standard, there are three primary rotary axes:

A, *B*, and *C*

Each is identified by the central primary linear axis around which it pivots. The

A axis rotates around a line parallel to *X*.
B axis rotates around a line parallel to *Y*.
C axis rotates around a line parallel to *Z*.

KEYPOINT

The primary rotary axes are identified by their central axis *X*, *Y*, or *Z*.

Question

To identify a rotary axis, first find its central axis. For example, a lathe equipped with a positioning spindle that can stop at any given rotation angle facilitates drilling holes spaced around the perimeter of a turned part (Fig. 16-7). What is this axis called?

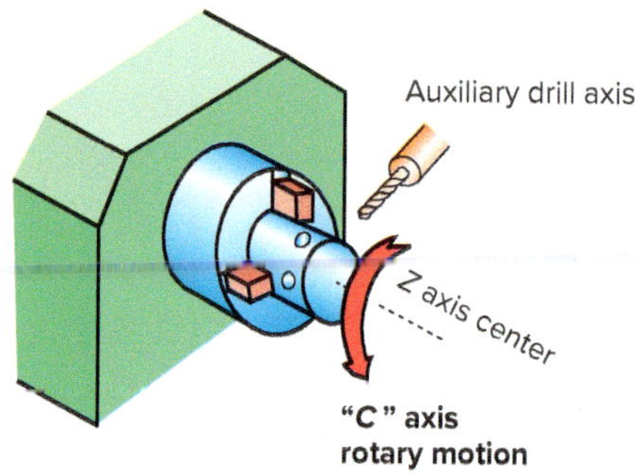

Figure 16-7 A lathe spindle capable of positioning to an exact location would be designated to be a "C" axis according to the EIA standards.

Answer

The indexing chuck turns about an axis parallel to the lathe's *Z* axis; therefore it's a *C* rotary axis. If it can stop only at a given location, not feed through an arc during machining, it is a positioning axis.

That same *C* axis could also be capable of rotating at programmed, slow feed rates, which would then enable milling operations as well, assuming a milling head was added. It is then known as a *feed axis*. To facilitate the milling operation, a powered tooling head is added to the tool turret (Fig. 16-8; coming in Chapter 17).

Figure 16-8 Multitasking lathes feature powered mill cutters in the tool turret, often called live tooling.
Courtesy Tyee Aircraft/Everett, WA

Mill Rotary Axes

When angled or warped surfaces are required, but they can't be cut with standard milling equipment, we turn to rotary axis machines. There are two ways a CNC mill might employ a programmable rotary axis: by rotating the part or by rotating the cutter head or a combination of the two.

Articulating Mill Heads

These machines are equipped with a spindle head that can rotate in one or two planes during a cut (*A* or *A* + *B*). The articulation is similar to the two Bridgeport type mill head rotations, but here it's empowered to feed in an arc (Fig. 16-9).

A-B Articulating Mill Head

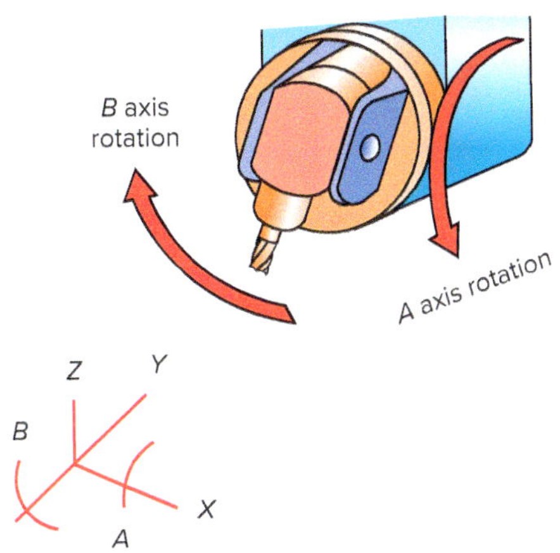

Figure 16-9 A five-axis mill features two articulating motions—*A* and *B*.

Five-Axis Cutter Heads Machines with this capability are known as **five-axis mills** with axes *X*, *Y*, and *Z* and *A* and *B* (Fig. 16-10).

Figure 16-10 Five axes machining a complex part.
Courtesy of AMT Senior Aerospace

Plus or Minus Rotary Motion—Rule of Thumb

To define which direction a rotary motion is to occur, clockwise or counterclockwise, we use a plus or minus sign on the coordinate. Try it yourself. Use the **rule of thumb** (Fig. 16-12) to solve the following questions about cutting a warped surface.

Rule of Thumb—Rotary Axis Sign Value

To identify whether the rotary axis direction is positive or negative (clockwise or counterclockwise), use the rule of thumb (Fig. 16-12).

First, identify the positive direction for the central axis around which the rotation occurs (+X, +Y, or +Z). *Then, pointing the thumb of your right hand along that positive direction, your fingers will curl in the positive rotary axis direction. Negative rotary motion would be against your fingers.*

TRY IT

Warped Part Critical Questions: The part shown in Fig. 16-11 must be profile milled by climb cutting (right to left on the page). To make this warped surface, an *A/B* articulating head must follow the changing warp as the mill moves on the *X* and *Y* axes. Use the rule of thumb (Fig. 16-12) to determine which direction the head must rotate, in the plus or minus *A* and *B* directions.

A. Before cutting, is the *A* axis tilted in the plus or minus direction from vertical? (This is shown on the right of Fig. 16-11.)
B. Which way does *A* move during the cut?
C. The *B* axis starts at a zero-degree tilt; does it end in a plus or minus tilt at the end of the cut? (See the left side of the figure.) Answers found in the Chapter Review.

Milling a Curved Warped Surface

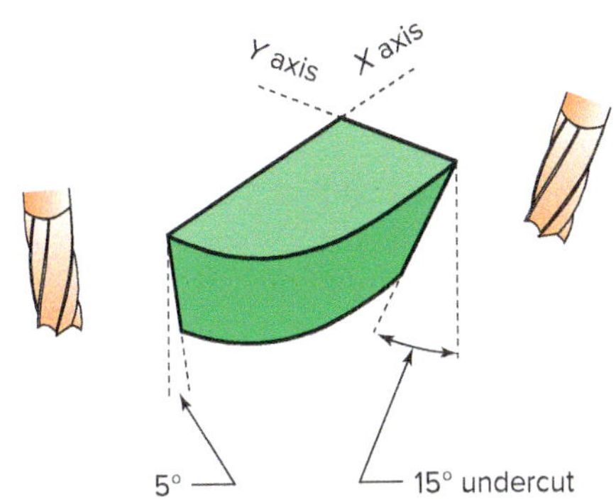

Figure 16-11 To climb mill this part requires both *A* and *B* axis articulation. What direction are they (plus or minus)? Use the *rule of thumb* to solve it.

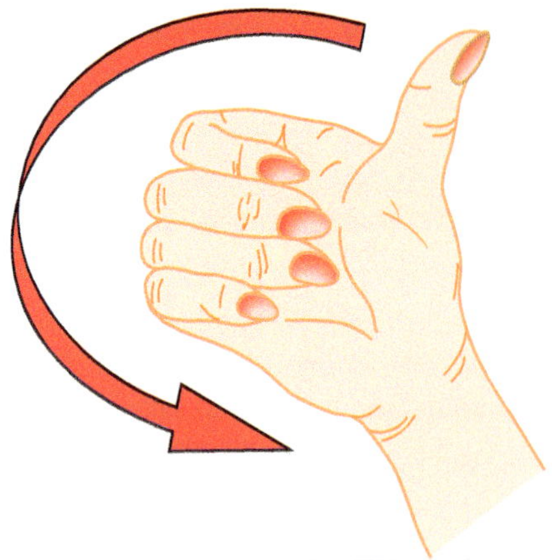

Figure 16-12 The *rule of thumb* helps to identify rotary axis sign value, direction sense.

Part Rotating and Indexing

The second way a machining center or CNC mill might achieve rotary axis cutting and indexing is by moving the part rather than the cutter head. This can be accomplished on a programmable *A* axis accessory (more common), as shown on the right of Fig. 16-13, or a built-in axis in the middle of the table. There are also several other versions of programmable part index and rotate accessories. The name for this addition is a *Trunion.*

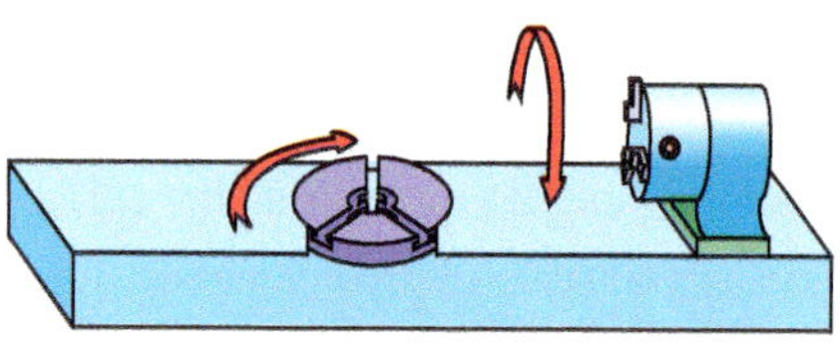

Figure 16-13 Programmable rotary axes can be add-on accessories or built-in parts of the machine.

Part Indexing

A very efficient way to index parts is a *four-sided, tombstone table* (Fig. 16-14). Here parts are loaded on four sides of a sturdy, vertical work-holding tower. Each side can be indexed and locked toward the spindle. For example, the part receives all face milling cuts at one station then the tombstone indexes and peripheral, and side cuts are made on the second face. Drilling and boring are done on the third. Three sides of the part can be machined in one setup. Another advantage of this kind of holding and indexing tooling is that the parts can be exchanged safely, on the fourth side, opposite the spindle, while the machining is in progress. The third advantage is that whole towers can be preset with the correct holding fixtures such that when *turning the machine around* (setting it up for a new job), it takes just minutes to remove one tombstone and replace it with a second, completely ready to go with parts preloaded! This is a good example of ways to keep the chips flying for profit!

Figure 16-14 Production tombstone fixtures rotate the work to create safe, efficient access to more than one side of the workpiece.
Courtesy Smiths Aerospace Actuation Systems, Yakima, WA

> **KEYPOINT**
>
> For program purposes, part rotation accessories and some index accessories must be assigned an axis letter identification. When this is done, the identification follows the rule of thumb, based on the right-hand rule.

16.1.3 The Secondary Linear Axes *U, V,* and *W*

The lathe drilling attachment in Fig. 16-7 illustrates another axis set. CNC machines occasionally receive secondary, straight-line axes to add auxiliary tool slides or boring quills and other machining functions to their capability.

Secondary Linear Rule

To identify the secondary linear axes, determine the primary linear parallels (*X, Y,* or *Z*). If the secondary axis is parallel to

X, it is the *U* axis

Y, it is the *V* axis

Z, it is the *W* axis

For example, the small vertical milling machine shown in Fig. 16-15 has its programmable quill designated as the *Z* axis; therefore the knee, which also moves vertically, becomes the "*W*" axis according to EIA267-B.

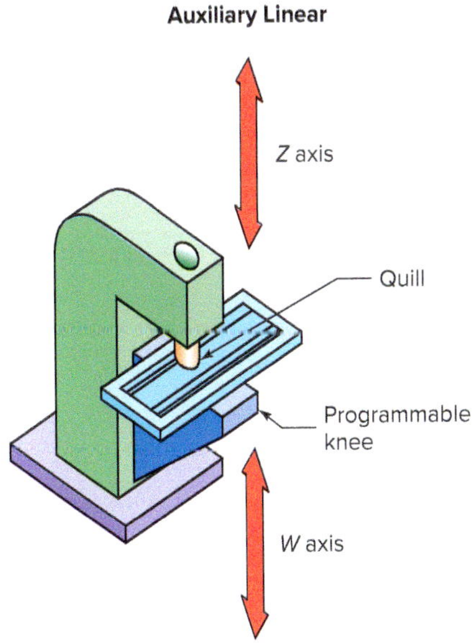

Figure 16-15 The programmable milling machine knee would be called a *W* axis since it moves parallel to the *Z* axis.

TRY IT

Warped Axis Answers

A. The program starts with a +15 °(positive) *A* axis tilt.

B. Then the head starts pivoting in the negative *A* direction.

C. The *B* axis ends with a negative tilt.

Did you miss this one? Remember, the thumb points along the positive central axis–opposite each *X* and *Y* part edge.

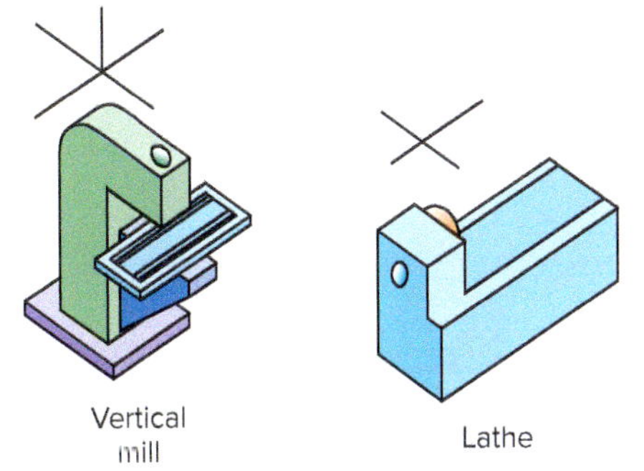

Figure 16-16 Identify the axis set using the right-hand rule.

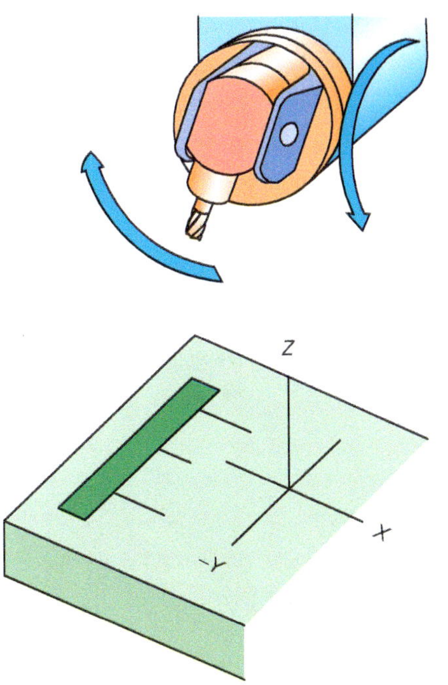

Figure 16-17 Identify the two rotary motions on this five-axis, articulating mill head.

UNIT 16-1 REVIEW

Replay the Key Points

- To change from one discrete plane to another, a code word must be included in the program.
- The rotary axes are identified by their central axis.

Respond

1. On a sheet of paper, using the right-hand rule, complete the orthogonal set for the sketch in Fig. 16-16.
2. Which primary linear axis is parallel to the main spindle axis? Which axis is central to a *B* axis move?
3. Identify the rotary axes on this five-axis mill in Fig. 16-17 and the direction they are moving (+ or −).
4. A multitasking CNC lathe capable of cutting both sides of the work features two programmable tool turrets. Each can move individually in two directions. Identify the missing axes on this machine and their sign value (+ or −: Fig. 16-18).
5. The primary axes correspond to each other. Fill in this chart.

Primary linear		Y
Secondary linear	U	
Primary rotary		C

6. This five-axis mill in Fig. 16-19 is rotating its cutter head in the direction indicated. Is this a positive or negative *A* or *B* axis movement according to the rule of thumb?
7. Complete the orthogonal set for the CNC horizontal mill in Fig. 16-20.

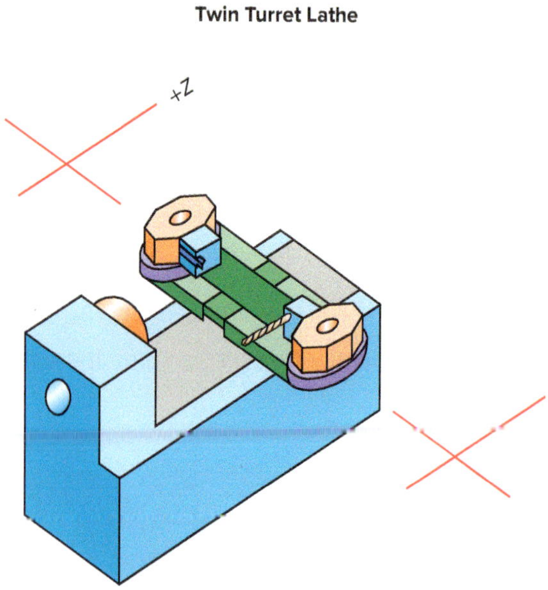

Figure 16-18 Identify the primary and two auxiliary axes of this lathe.

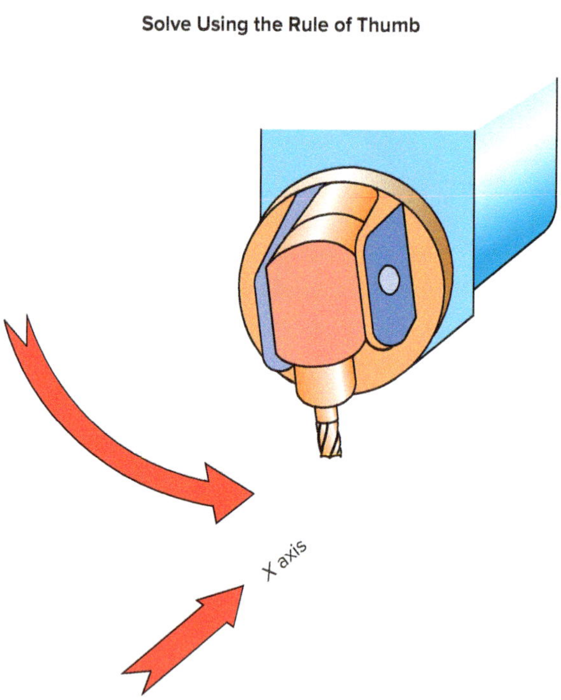

Figure 16-19 Is this a positive or a negative *A* and *B* axis motion?

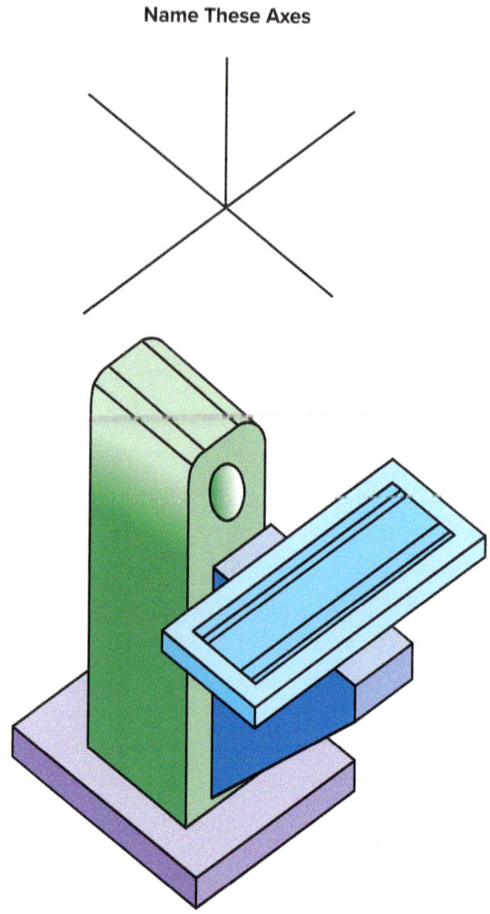

Figure 16-20 Complete the axis set for this horizontal mill.

Unit 16-2 Coordinate Systems and Points

Introduction: The ancient mapmaker René Descartes used a grid of lines crossing at 90° to identify any point on his geographical maps. Four centuries later we're still using them on street maps. Way beyond his intention, mathematicians and machinists adapted them to programming and called them *Cartesian coordinates* to help us make CNC machining possible.

Cartesian coordinates don't serve all the point identifications required. Sometimes information comes to us on an engineering drawing that's not in *X*, *Y*, and *Z*. Like a rocket leaving the earth, we sometimes think of point identification in terms of how far it has traveled from the origin (radius = *R*) and at what angle (*A*). These are known as **polar coordinates**. For example, a bolt circle is much easier to define using polar coordinates rather than *X* by *Y* locations.

KEYPOINT

In CAD drawing and CNC programming by hand or by CAM software, we mostly use rectangular coordinates, but every so often it's easier to switch to polar coordinates. They can save a lot of math when drawing dimensions are provided in terms of radius and angle relative to datums.

Unit 16-2 is about three kinds of coordinates used in CNC. It's also about the way we assign values to them to determine direction.

TERMS TOOLBOX

Absolute coordinates Values based on the origin of the axis set–the PRZ.

Cartesian coordinates Rectangular coordinate set that refers points to reference axes.

Coordinating (CNC axes) Physically moving the cutter to a known position relative to the part geometry, then setting axis registers to represent that position.

Default (CNC condition) An expected or preset value or mode.

Entity A straight line or curved arc with a start and end point coordinate.

Full floating reference The ability to place the PRZ anywhere within the work envelope (and sometimes beyond in special situations).

Homing (the CNC machine) Driving the machine to a fixed position on the machine; machine home.

Incremental value (relative value) Coordinates based on the previous entry. Jump-to-jump values. Sometimes called relative coordinates because each entry is based on the last.

Machine home (M/H) The never-changing utility position used for safe parking, accuracy refreshment, and setup purposes.

Null entry A command, mode, or coordinate that has been set previously and can be optionally omitted from a program statement.

Polar coordinates Point identification using radial and angular displacement from the origin.

Program reference zero (PRZ) Master origin for program and part geometry. Chosen by the programmer, PRZ is coordinated by the machinist during setup. It can be placed anywhere within the envelope on modern machines.

Quadrant One of four possible 90° segments lying on a flat plane, created by the intersection of the two axes that define the plane. Absolute coordinate values depend on the quadrant in which a point lies.

16.2.1 Absolute and Incremental Value Coordinates

There are two different ways to use both Cartesian and polar coordinates. How we use them depends on whether the coordinate refers to the origin (absolute–more common) or take reference from the previous point called incremental coordinates. Figure 16-21 shows examples of two different kinds of dimensioning. Note that in the real world these would not be mixed, as in the example drawing, because the tolerances would conflict.

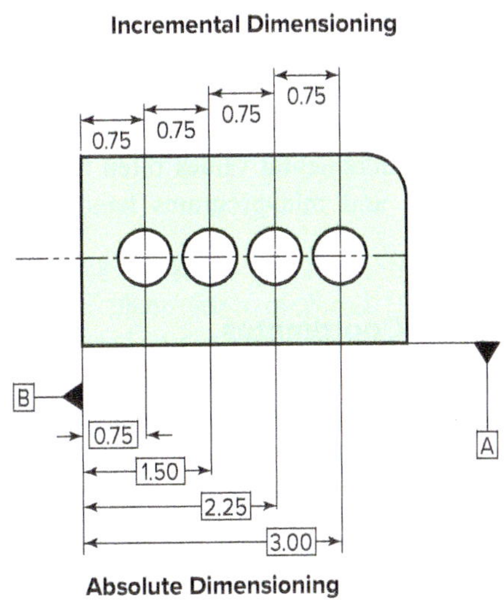

Figure 16-21 Examples of absolute and incremental (also called relative) dimensioning.

KEYPOINT

Normally, the designer would use one type or the other—absolute or incremental dimensioning.

SHOPTALK

Invented in the United States, CNC technology is leading our trade, but it's far from a new subject! Machinists have been using some form of hard-programmed, numerical-controlled machinery since 1950. Craftspeople have been right at the front of the crowd surging along the information highway, but we did more than move data—we moved chips and made complex helicopter and jet parts with our programs and computers.

Diameter Programming

A second lathe convention is illustrated in Fig. 16-25. In nearly all lathe programs, the *X* axis coordinates are based on diameter. Example: A change from X2.125 to X2.000 would result in a diameter change of 0.125 in.

KEYPOINT

For lathe programming, the *X* values are conventionally written as diameters.

Unconventionally, on some controls and CAM systems, it is possible to program in radius values, but only after specifying radius-valued *X* coordinates; *X* = diameter is a **default** condition for most turning centers. (*Default* here means the fall-back value or condition.)

KEYPOINT

The datum priority of the design is the overriding factor for PRZ placement.

16.2.2 Points for Geometry and/or Reference

Coordinate points fall into two categories by the way they will be used in the program as position or reference identification. Some points can do double duty. In Fig. 16-22, the PRZ was also on the part as a geometry point to be machined.

Geometry Points

Geometry points are those you intend to use for position or motion–a location to be machined. They are the start and end of a given line or arc (Fig. 16-26).

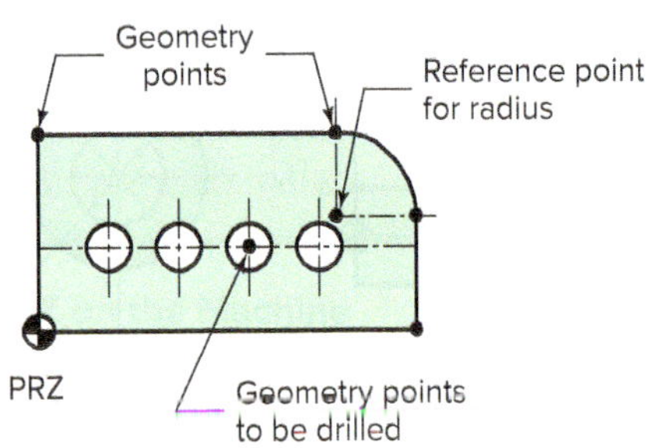

Figure 16-26 Geometry points occur at significant locations.

Identifying Geometry Points

An entity is a familiar term for those with CAD training (Fig. 16-27). They are individual straight lines or arcs. Each has a start and end point coordinate. Where any two entities join, become tangent, intersect, or cross, there is also a unique geometry point common to both.

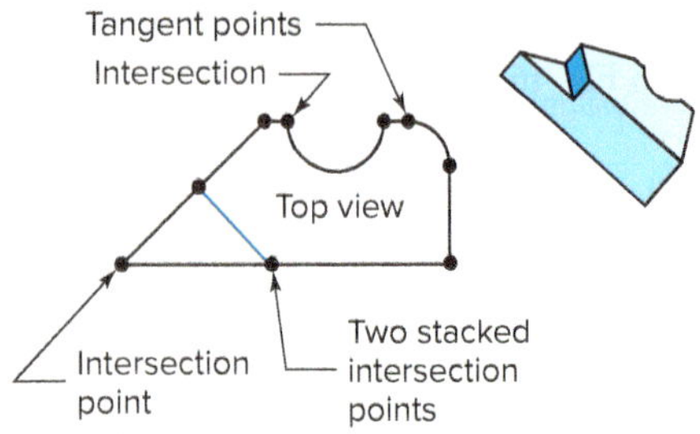

Figure 16-27 Points can be coaxial (stacked on top of each other along a primary axis).

Reference Points

In addition to the PRZ, there are two other kinds of reference points used in CNC. One not normally used for program execution is **machine home (M/H)**. M/H is not a floating point, it lies at an exact fixed (repeatable) location within the machine's envelope, usually with all axes fully withdrawn away from the chuck or table. Driving the machine to M/H is called **homing**.

It's a utility location used mostly for setup functions and for accuracy coordination in older machines lacking the ability to retain their position when they are turned off. We'll set M/H aside for now.

Local Reference Points

When writing a program it's possible to switch from the PRZ to a temporary *local reference zero (LRZ) point* that lies at a known distance from the PRZ. Once the features are machined that relate to it, the LRZ is canceled to again refer to the PRZ. Local references are used for a couple reasons:

1. A feature, or group of features, refers to a reference other than the PRZ (Fig. 16-28). Many calculations would be required to write absolute value coordinates for each cutout referring back to the PRZ for each coordinate. But setting a local reference point, at the center, then referring to it for that group saves math and programming time when writing the program without CAM. Using a CAM system like Mastercam (see Chapter 26), this step wouldn't be needed–it can make all the calculations, one of the many advantages to computer-assisted programming.
2. An LRZ can also be used to repeat the entire program or part of a program (called a subroutine) in another location within the work envelope. The repeat could be on one part as the instrument cutouts, or we could be machining two of the same parts in two vises. The second part would use a local reference, at a given distance from the first (Fig. 16-29). Another way to accomplish this is called fixture offsets. We assign a new PRZ in a new location with a G54, G55, G56 through G59, or even higher. Each vise would have its own reference point. So G54 X10.0 would go to vise 1 at X10.0, but G55 X10.0 would go to vise 2. We'll see more of this in program planning.

KEYPOINT

In CAD drawing and CNC programming by hand or by CAM software, we mostly use rectangular coordinates, but every so often it's easier to switch to polar coordinates. They can save a lot of math when drawing dimensions are provided in terms of radius and angle relative to datums.

Unit 16-2 is about three kinds of coordinates used in CNC. It's also about the way we assign values to them to determine direction.

TERMS TOOLBOX

Absolute coordinates Values based on the origin of the axis set—the PRZ.

Cartesian coordinates Rectangular coordinate set that refers points to reference axes.

Coordinating (CNC axes) Physically moving the cutter to a known position relative to the part geometry, then setting axis registers to represent that position.

Default (CNC condition) An expected or preset value or mode.

Entity A straight line or curved arc with a start and end point coordinate.

Full floating reference The ability to place the PRZ anywhere within the work envelope (and sometimes beyond in special situations).

Homing (the CNC machine) Driving the machine to a fixed position on the machine; machine home.

Incremental value (relative value) Coordinates based on the previous entry. Jump-to-jump values. Sometimes called relative coordinates because each entry is based on the last.

Machine home (M/H) The never-changing utility position used for safe parking, accuracy refreshment, and setup purposes.

Null entry A command, mode, or coordinate that has been set previously and can be optionally omitted from a program statement.

Polar coordinates Point identification using radial and angular displacement from the origin.

Program reference zero (PRZ) Master origin for program and part geometry. Chosen by the programmer, PRZ is coordinated by the machinist during setup. It can be placed anywhere within the envelope on modern machines.

Quadrant One of four possible 90° segments lying on a flat plane, created by the intersection of the two axes that define the plane. Absolute coordinate values depend on the quadrant in which a point lies.

16.2.1 Absolute and Incremental Value Coordinates

There are two different ways to use both Cartesian and polar coordinates. How we use them depends on whether the coordinate refers to the origin (absolute–more common) or take reference from the previous point called incremental coordinates. Figure 16-21 shows examples of two different kinds of dimensioning. Note that in the real world these would not be mixed, as in the example drawing, because the tolerances would conflict.

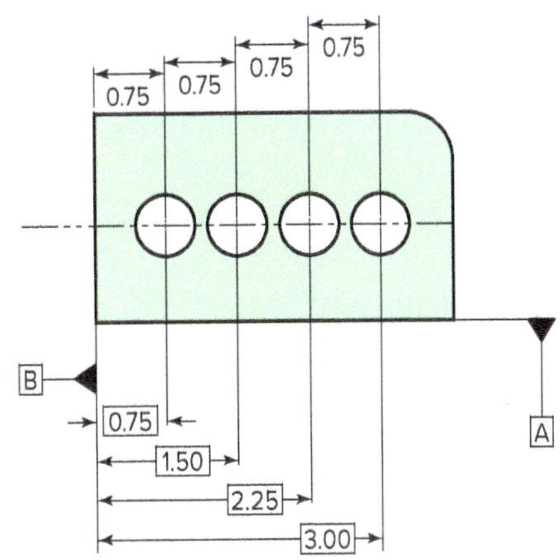

Figure 16-21 Examples of absolute and incremental (also called relative) dimensioning.

KEYPOINT

Normally, the designer would use one type or the other—absolute or incremental dimensioning.

SHOPTALK

Invented in the United States, CNC technology is leading our trade, but it's far from a new subject! Machinists have been using some form of hard-programmed, numerical-controlled machinery since 1950. Craftspeople have been right at the front of the crowd surging along the information highway, but we did more than move data—we moved chips and made complex helicopter and jet parts with our programs and computers.

But even earlier than machining, other industries used punched cards to control weaving looms to make flawless, intricate patterns in fabrics.Programmed automation has been in manufacturing for more than 60 years!

Definitions

Absolute Value Coordinates

An **absolute coordinate** is one where each entry represents the point's distance from the master origin: $X = 0$, $Y = 0$, and $Z = 0$. We use absolute point identification more often in CNC work because prints usually refer features to origins on the print (datums A and B, for example). They are conventional for CAM-generated programs.

Incremental Value Coordinates

These refer to a coordinate set where each entry represents the identified point's distance from the previous point. These points can be envisioned as jumps from the present location to the next. Using incremental values often saves math and time during setups and mini-programs handwritten at the control.

Absolute Value Coordinates

Now let's develop a working ability using coordinates. We'll start with absolute value rectangular.

Referenced to the Origin—Program Reference Zero

All absolute coordinates refer to a singular starting place—the origin. In CNC work it's called the **program reference zero (PRZ)**, *the origin of the grid.* The PRZ is the master reference point on which the program and the setup are based. Selecting the PRZ location relative to the part's geometric features is among the first critical decisions for planning.

KEYPOINT

The PRZ must be based on geometric priorities whether or not the print is GDT.

Setting the PRZ on the Machine

When the machine is set up, the PRZ must be located at the same position on the physical part as it was selected for the program. This is a critical setup task on both mills and lathes. We'll use a mill example.

The action is similar to setting something up (Fig. 16-22) to be drilled on a manual milling machine. First, the vise is indicated true to the machine's axis and bolted in place. It does not matter where the vise is located on the table because the PRZ can be set at any position within the work envelope on modern equipment. This is known as **full floating reference**.

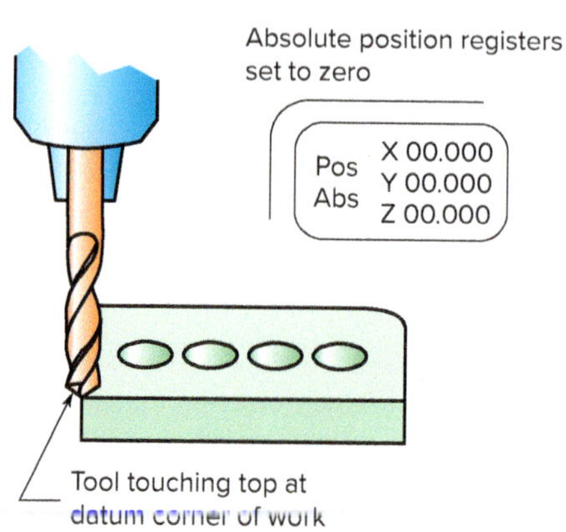

Figure 16-22 Coordinating the PRZ to the physical part requires setting the axis registers to zero when the tool is at the PRZ position.

When the machinist sets the PRZ relative to the part, vise, chuck, or fixture, the action is called **coordinating** the machine. Using edge finders and indicators the spindle is located over one corner of the work in this example. The next action is to set the digital readout and/or micrometer dials to read zero at that position on the manual mill. On the CNC it's the same; the control registers are set to $X0.0$ and $Y0.0$. Then, after coordinating the X and Y axes, the cutter would be touched to the top of the work and its register set to $Z = 0.0000$, assuming that's where the Z axis PRZ has been chosen for the program (a common practice).

KEYPOINT

The tool just touching the work surface is the simplest way to illustrate the Z zero position. It works fine for programs that use just one cutting tool. Later we'll learn that Z zero in a setup with several cutting tools may be at machine home or some other height off the part to allow for different length tools. This is called cutter length offsets. Each tool has its own compensating offset such that they all touch $Z0.0$ as if they are all the same length.

With the PRZ coordinated, all absolute moves of the machine's axes would then refer to that point. On a mill part, the PRZ is commonly placed on a corner of the part as in Fig. 16-22, while on a lathe part the PRZ is often placed at the outer tip and centerline of the work.

The PRZ can sometimes be placed outside the work envelope, under special circumstances. First the spindle or tool is positioned on the part at a known location. But this time the axes aren't set to zero; rather their position relative to PRZ is written into the workpiece registers to represent where the tool is at relative to PRZ. Not all controllers will permit this action. However, it is possible since the machine refers coordinates to it but needn't actually move the spindle or cutting tool to it.

By engineering convention, the PRZ is shown by the bull's-eye circle shown in Fig. 16-23.

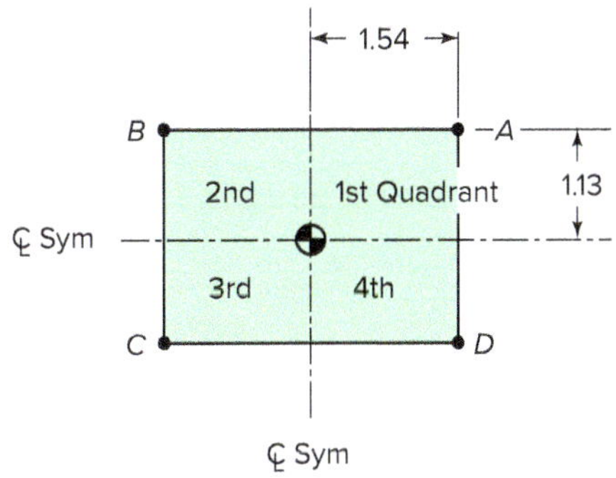

Figure 16-23 Absolute values in the four quadrants.

SHOPTALK

Depending on local custom, you might also hear the PRZ called *program home, program datum point, program zero*, or *X–Y–Z zero*. Whatever you call it, it's the point from which all absolute coordinates originate or to which they refer.

Absolute Value in the Four Quadrants Code G90

Depending on which side of the PRZ a coordinate lies, its value will be either positive or negative. Using the *X*-*Y* plane for the example, the intersection of the two axes creates four possible zones called **quadrants**. They are numbered from the upper right in a counterclockwise direction: 1 upper right, 2 upper left, 3 lower left, and 4 lower right (see Fig. 16-23). Each ordinate (*X* or *Y*) combines to make a coordinate and each has a positive or negative value depending upon in which quadrant they are located.

KEYPOINT

Slightly different from math statements, for CNC coordinates, we place the minus sign after the letter ordinate to indicate a negative value. We do not use a plus sign. No sign indicates a positive value.

X - is negative *X* is positive

Each *X* and *Y* value is made unique by the quadrant in which it is located. If, as shown in Fig. 16-23, point *B* is within the second quadrant, then its coordinates are

B = *X*-1.5400, *Y*1.1300

In Fig. 16-23, what are the absolute coordinates for points *C* and *D*?

Answer:

C = *X*-1.5400, *Y*-1.1300

D = *X*1.5400, *Y*-1.1300

In Fig. 16-24, the drill tip position becomes negative when it is below the *Z* axis zero.

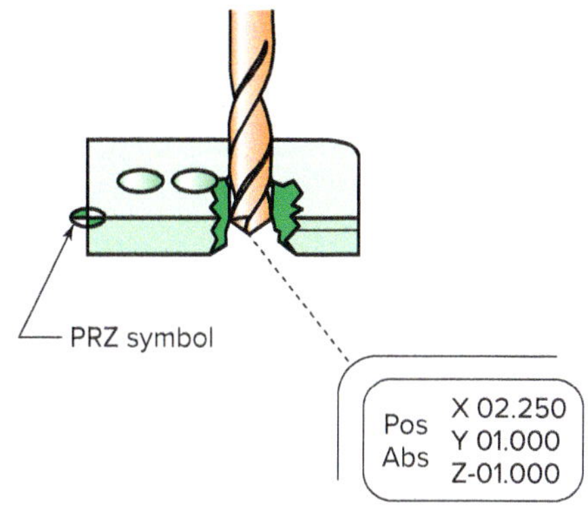

Figure 16-24 *X* and *Y* are positive absolute value, but *Z* is negative.

Lathe Quadrants

When programming work for the lathe, the workpiece is viewed in the *X*-*Z* plane–from overhead (Fig. 16-25). The PRZ is usually placed at the outer end centerline (but not always). If it is located in this easily coordinated place, the entire path to be machined lies within the second quadrant, making all *Z* values negative but all *X* values positive, as seen in Fig. 16-25.

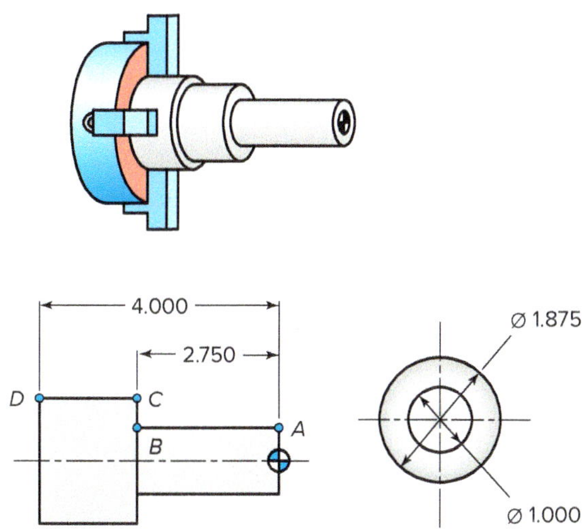

Figure 16-25 The quadrants for *X* and *Z* axis values on lathe work.

In Fig. 16-25, the coordinates for point *A* are

A = *X*1.0000, *Z*0.0000

What are the coordinates for points *B*, *C*, and *D*?

Answer:

B = *X*1.000, *Z*-2.7500

C = *X*1.8750, *Z*-2.7500

D = *X*1.8750, *Z*-4.0000

Diameter Programming

A second lathe convention is illustrated in Fig. 16-25. In nearly all lathe programs, the *X* axis coordinates are based on diameter. Example: A change from X2.125 to X2.000 would result in a diameter change of 0.125 in.

KEYPOINT

For lathe programming, the *X* values are conventionally written as diameters.

Unconventionally, on some controls and CAM systems, it is possible to program in radius values, but only after specifying radius-valued *X* coordinates; *X* = diameter is a **default** condition for most turning centers. (*Default* here means the fallback value or condition.)

KEYPOINT

The datum priority of the design is the overriding factor for PRZ placement.

16.2.2 Points for Geometry and/or Reference

Coordinate points fall into two categories by the way they will be used in the program as position or reference identification. Some points can do double duty. In Fig. 16-22, the PRZ was also on the part as a geometry point to be machined.

Geometry Points

Geometry points are those you intend to use for position or motion–a location to be machined. They are the start and end of a given line or arc (Fig. 16-26).

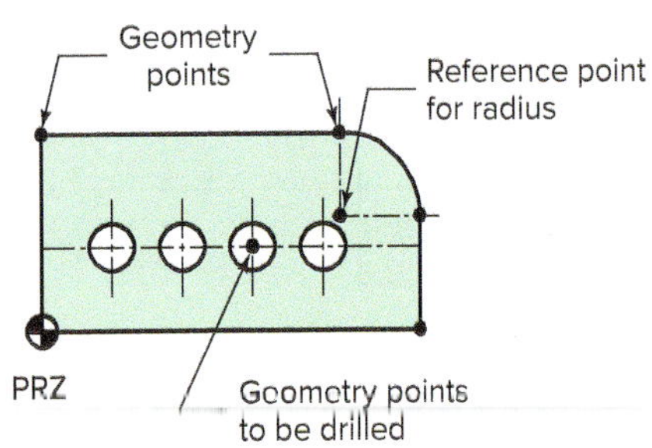

Figure 16-26 Geometry points occur at significant locations.

Identifying Geometry Points

An entity is a familiar term for those with CAD training (Fig. 16-27). They are individual straight lines or arcs. Each has a start and end point coordinate. Where any two entities join, become tangent, intersect, or cross, there is also a unique geometry point common to both.

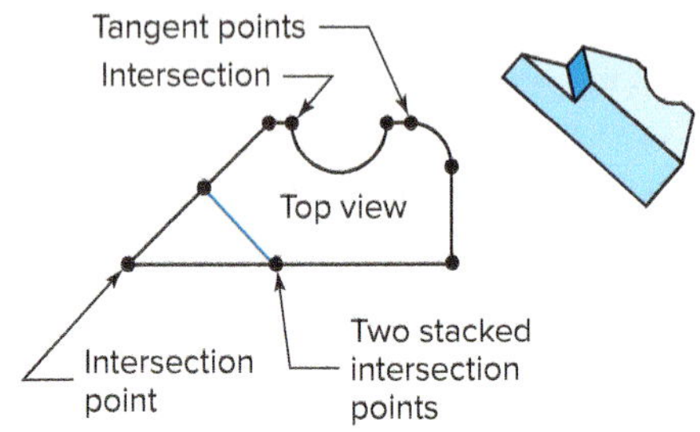

Figure 16-27 Points can be coaxial (stacked on top of each other along a primary axis).

Reference Points

In addition to the PRZ, there are two other kinds of reference points used in CNC. One not normally used for program execution is **machine home (M/H)**. M/H is not a floating point, it lies at an exact fixed (repeatable) location within the machine's envelope, usually with all axes fully withdrawn away from the chuck or table. Driving the machine to M/H is called **homing**.

It's a utility location used mostly for setup functions and for accuracy coordination in older machines lacking the ability to retain their position when they are turned off. We'll set M/H aside for now.

Local Reference Points

When writing a program it's possible to switch from the PRZ to a temporary *local reference zero (LRZ) point* that lies at a known distance from the PRZ. Once the features are machined that relate to it, the LRZ is canceled to again refer to the PRZ. Local references are used for a couple reasons:

1. A feature, or group of features, refers to a reference other than the PRZ (Fig. 16-28). Many calculations would be required to write absolute value coordinates for each cutout referring back to the PRZ for each coordinate. But setting a local reference point, at the center, then referring to it for that group saves math and programming time when writing the program without CAM. Using a CAM system like Mastercam (see Chapter 26), this step wouldn't be needed–it can make all the calculations, one of the many advantages to computer-assisted programming.
2. An LRZ can also be used to repeat the entire program or part of a program (called a subroutine) in another location within the work envelope. The repeat could be on one part as the instrument cutouts, or we could be machining two of the same parts in two vises. The second part would use a local reference, at a given distance from the first (Fig. 16-29). Another way to accomplish this is called fixture offsets. We assign a new PRZ in a new location with a G54, G55, G56 through G59, or even higher. Each vise would have its own reference point. So G54 X10.0 would go to vise 1 at X10.0, but G55 X10.0 would go to vise 2. We'll see more of this in program planning.

3. We also use an LRZ on very large parts such as wing spars, for example, where the master PRZ might be 100 ft away. A temporary point can be established that's far more convenient for setup work. The bed of the gantry mill in Fig. 16-30 is 200 ft long.

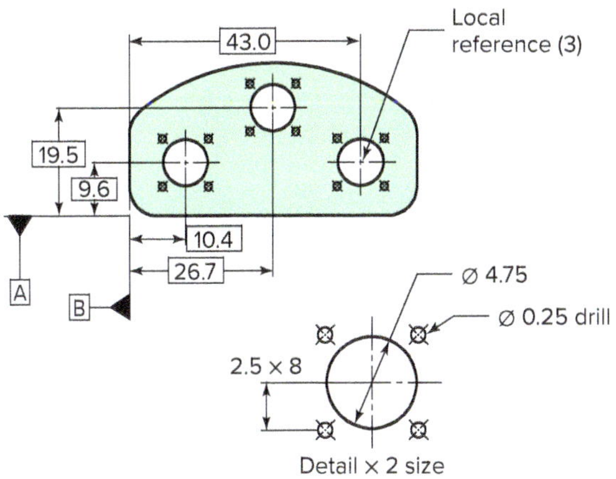

Panel instrument	Suitable for the use of local references

Figure 16-28 The math required to write coordinates for this instrument panel's features could be simplified by using a temporary local reference for each cutout pattern.

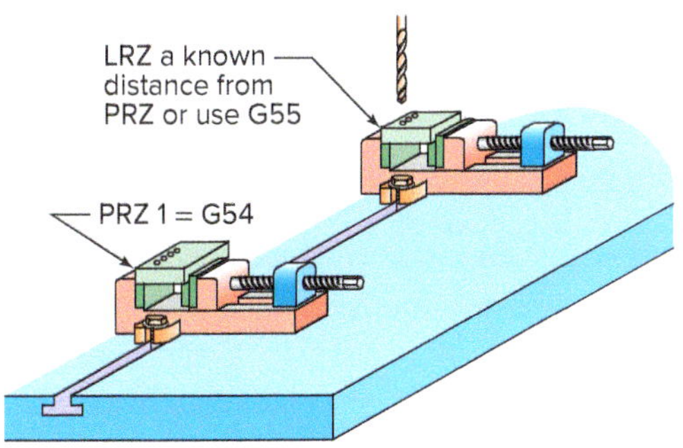

Figure 16-29 An LRZ might be used to run the same program in two vises.

Figure 16-30 Machine home and PRZ could be 20 ft apart (left) or 200 ft on this giant, three-spindle, five-axis, gantry mill.

Left: Courtesy Aerospace Manufacturing Technologies/AMT/Arlington, WA.

Right: Courtesy of Contour Aerostructures, Corp/Everett, WA

TRADE TIP

In CNC work there's almost always more than one way to do a job using the power of the codes.

Subroutines that contain all the commands for one instrument cutout (see Fig. 16-28) can be used to simplify the instrument panel program. Such would be written using incremental coordinates (coming up next). Then we need only move to the center location to begin the subroutine. We'll learn about subroutines in logic commands later.

Most controls store from 6 to 10 fixture locations with reference to the machine home. Haas controls can store many more. The codes commonly used are G53, G54, G55, G56, and so on, depending on the manufacturer. Once the new fixture offset code is read, the machine begins acting on the reference location stored on the fixture offset page of the control. Many shops load G53 with all zeros so it's an easy way to return to (or refer to) machine home. G53 then cancels any fixture offset. G53 is usually global on most controls—adding, for example, .030 to the *X* value shifts all other fixtures .030 in. to the right.

16.2.3 Incremental Coordinates Code G91

Incremental **Cartesian coordinates** are a useful, second method of identifying a point. They might be called *relative* coordinates because each movement of the tool is related to the last spindle or tool position. The coordinate is the distance from the present position to the next. **Incremental values** are sometimes chosen for several reasons. For example, when

The *drawing,* or a portion of it, is incrementally dimensioned and the program will be written without CAM software. In that case, using absolute values will require lots of math. On the other hand, the incremental values are all there—no calculations to make.

Manual *keyboard entries* were used, for example. A brief tooling program written at the control to shape soft jaws, or to move an indicator probe an exact amount using keyed in entries, would both be simplified using incremental coordinate entries.

Incremental solutions are needed. When a solution to a programming calculation yields an incremental value. In other words, solving for the next position or local reference, creates the incremental distance. To compute the absolute distance, further math would be necessary. In this case switch the program from G90 Absolute, to G91 incremental values. The control continues to track tool location absolutely but displays the incremental distance to go.

Incremental values are required. A few specific commands must be in incremental values. We'll see curve commands where the center of the circle must be identified with incremental coordinates.

Extra moves are to be edited in. A written program could require editing. It's often easier to slip in incremental moves to fix something in a program otherwise written in absolute values.

KEYPOINT

Using Incremental Coordinates

There is no difference in accuracy using either incremental or absolute values, as the control tracks progress internally, always absolutely. However, when bad entries are made, incremental values can create greater problems. Examples are coming up.

As long as it doesn't violate shop policy, it is possible to switch freely between absolute and incremental values inside a program and even inside a single command line. But the right command code must be provided to the control so it knows whether absolute or incremental values are being used. We often use incremental coordinates for editing where we must insert an extra move or to write quick setup or tooling program at the control.

Incremental Motion + / – Direction

The axis value, plus or minus, describes the direction in which the tool movement is to be made from the present position to the next. The center of Fig. 16-31 should be envisioned as the present location of the tool. For example, at this position, the coordinates *X*1.0 *Y*0.0 take the tool to position 1. However, an *X*-1.0 *Y*0.0 would take it the opposite direction.

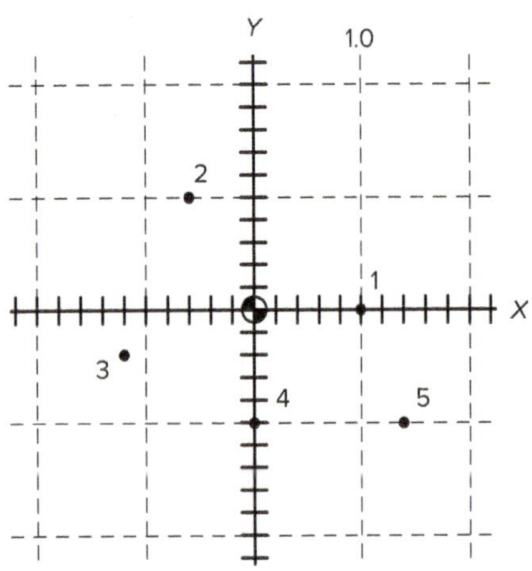

Figure 16-31 Identify these points incrementally, in sequence.

TRY IT

Critical Question *Starting at Point 1,* determine incremental coordinates to move from Point 1 to Point 2, then onward to Points 3, 4, and 5 in sequence. Note the 1-in. scale on the drawing.

ANSWER

To Point 2	*X*-1.600	*Y*1.000
Point 3	*X*-0.600	*Y*-1.400
Point 4	*X*1.200	*Y*-0.600
Point 5	*X*1.400	*Y*0.000

KEYPOINT

Note that the Point 5 coordinate *X*1.400, *Y*0.000 can be shortened to *X*1.400 with no *Y* entry since there is no movement of the *Y* axis from Point 4 to Point 5.

Metric Coordinates G21 EIA or G71 ISO

Metric coordinates are as simple to enter as those in inches. But here, too, the right code must be provided to alert the control whether the units being entered are metric or imperial (inches). Although it would be poor practice, both units can be mixed in the program. That would only happen if a wacky design mixes them!

As you know, a millimeter is 25.4 times smaller than an inch. So, metric coordinates are usually carried only to three decimal places; for example, *X*3.750.

KEYPOINT

There is a code for metric-valued coordinates and one to tell the control the program is in imperial (inch) units. It can, for example, be G20 = Imperial and G21 = Metric, or G70 = Imperial or G71 = Metric.

G21 X2.000

This line sets metric values and the entry is 2 mm.

G20 X2.0000

This line tells the control to return to imperial values and that the entry is 2 in.

Modern controls allow complete flexibility of units and values. Most can accept either absolute or incremental and metric or imperial values with ease, as long as they are accompanied by the right code. It's risky but the units and values (absolute = G90 or incremental = G91) can be changed within the program or even within a single command with no loss of accuracy.

TRADE TIP

Incremental coordinates can be useful when several parts of nearly the same size or shape are programmed (Fig. 16-32). Using one base program, the modification can be inserted where needed. For example, the three handles could be made from a single program with two additional *X* moves added to adjust the part length to machine the Dash 1, 2, or 3 version. This wouldn't work using absolute coordinates because the next sequence would go right back to the original shape. For example, by adding a couple of incremental moves of 0.400 in., the Dash 1 is changed to the Dash 2 part.

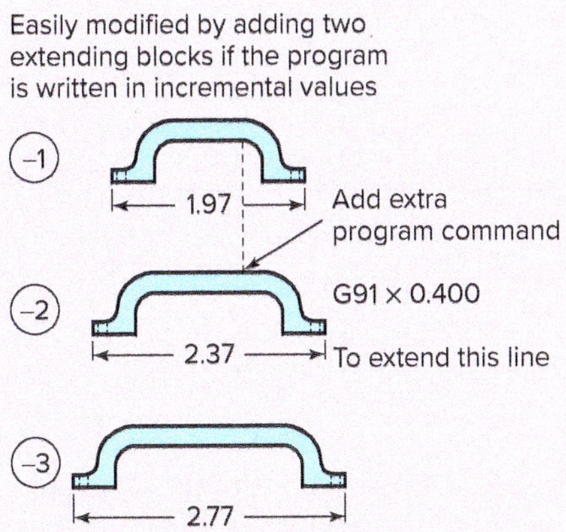

Figure 16-32 Using incremental coordinates, the Dash 1 program can easily be edited to machine the Dash −2.

16.2.4 Conventions in Program Commands

Eliminating Nulls (Repeated Absolute Coordinates)

While it's correct to represent each geometry or reference point in a program with a complete *X*, *Z*, or *X*, *Y*, and *Z* coordinate, not every coordinate must contain all entries to be complete data.

Absolute Null Rule

If an absolute coordinate value has been entered in a previous command statement, then that axis ordinate entry might be skipped. Called **null entries**, the duplicated portion can be left off for convenience and to create shorter programs.

For example, compare the two sets of coordinates shown next. Relate the full coordinates to the abbreviated set that follows where the nulls are omitted. The machine actions remain the same.

Complete Set		**Nulls Removed**
From start to PRZ	*X*0.0 *Y*0.0 *Z*0.0	*Z*0.0
To point *A*	*X*2.5 *Y*0.0 *Z*0.0	*X*2.5
To point *B*	*X*2.5 *Y*−2.0 *Z*−1.0	*Y*−2.0 *Z*−1.0

TRADE TIP

There are times when nulls are better left in. As a beginner, there is a chance your work will require some debugging. Nulls help with this. There are other times to keep them in the program. For example, when programming curves nulls occur in the references and geometry points, yet they are best left in for good record keeping of your calculations.

Entry Order

Modern controls can be programmed in any *X*-*Y*-*Z* order with acceptable results. The coordinate *X*1.0, *Y*2.0, *Z*3.0 could be entered *Y*2.0, *X*1.0, *Z*3.0, or ordered *Z*3.0, *X*1.0, *Y*2.0, too. Most controls do not require data order as long as it is complete within one command. However, ordering it alphabetically is the norm.

Leading and Trailing Zeros

On modern controls, decimal number entries need only have their decimal in the right place, and the significant digits correctly arranged. For example, *X*.37 and *X*0.37000 are both acceptable. However, there are three reasons that program entries should be ordered conventionally. The program is often easier to edit. Compare these:

Unstructured

*X*1.37 *Y*1
X.256
X.037 *Y*1.125 *Z*.5
Z−0.2500
X.037 *Z*−.05

Structured

*X*1.3700	*Y*1.0000	
*X*0.2560		
X.03750	*Y*1.1250	*Z*0.5000
		Z−0.2500
*X*0.037		*Z*−0.0500

Both sets of coordinates convey the same data. However, in the second set, the columns align, and they extend to the accuracy of the machine. All the zeros aren't necessary, but they set a uniform, easy-to-scan format. Ordering the program is a matter of choice, shop policy, and how the CAM software creates them.

KEYPOINT

When you are assigned to a new control find out what conventions are required for data entry. *Older controls often require a fixed format* with rigid ordering. Modern Haas controls do not have any special conventions.

When Whole Numbers Are Required

Some entries must be entered as whole numbers without a decimal. For example,

Line numbers never have a decimal point

N001 cannot be N001.0

Tool numbers

T10 cannot be T10.0

But they might be T0010 if the control holds many tool definitions.

Number of passes to complete a shape

P12 cannot be P12.0

These situations will be learned on an individual control basis. Read the control manual.

UNIT 16-2 REVIEW

Replay the Key Points

- Absolute coordinates always refer to their *X*, *Y*, and *Z* distance from the PRZ.
- Incremental coordinates identify the distance and direction to the *next* point using the present point as its reference.
- Coordinate points are used either for geometry or reference, and one point can be used in both capacities at times.
- The PRZ is chosen from its datum basis on the print.
- The PRZ also exists on the physical envelope of the machine and setting it is a setup task.
- If the machine reads *X*0.0, *Y*0.0, and *Z*0.0, and the cutter is positioned at PRZ, it is coordinated.
- Each primary plane has four quadrants, which determine absolute coordinate sign values.
- Null values may or may not be omitted by the programmer's choice.
- Modern CNC equipment offers certain entry flexibilities which can be used on a choice basis.
- There is no difference in accuracy using either absolute or incremental or metric or imperial values as the control tracks progress internally and always absolutely.
- We can switch freely between absolute and incremental as long as the right command code is given to the control so it knows which is being used.
- We often use incremental coordinates for editing where we must insert an extra move or to write quick setup or tooling programs at the control.

Respond

Resolve any misunderstandings or get assistance with answers that are incorrect or unclear. Do not go beyond this unit without 100 percent comprehension.

Absolute Coordinates

1. Coordinates that refer to the PRZ for their values are known as absolute rectangular coordinates. Is this statement true or false? If it is false, what will make it true?
2. Name in their order of importance the two guideline rules for selecting the PRZ.
3. Draw a sketch of the PRZ symbol on a sheet of paper. Where should it be chosen for mill job shown in Fig. 16-33?

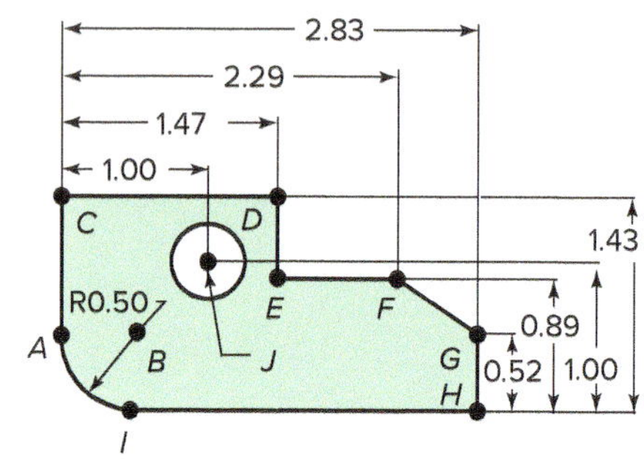

Figure 16-33 Mill questions and problems.

4. Sketch the correct location for the PRZ on the lathe job in Fig. 16-34.

   Student note: Often only half of the profile is shown on lathe work.

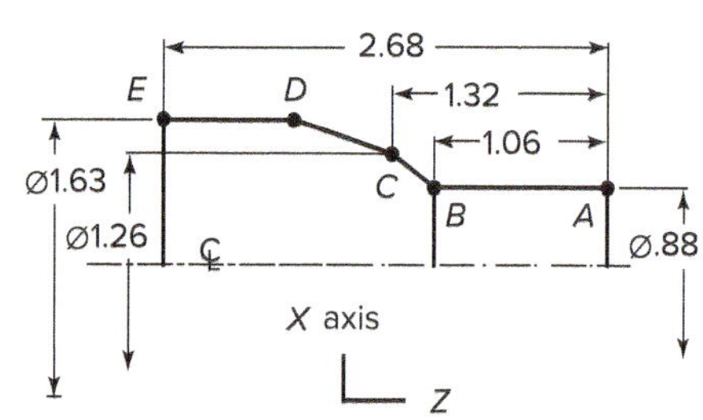

Figure 16-34 Lathe questions and problems.

5. Write a set of *X*-*Y absolute value* coordinates for the indicated points in Problem 3 (see Fig. 16-33). Record them in the sequence shown, then check your answers against those provided. Note, *Z* values are not needed for this exercise, but would be necessary in a program.
6. Write a set of absolute value coordinates for the lathe job in Problem 4 (see Fig. 16-34). Note, typically half the part is shown for symmetrical lathe work. The *Z* axis is horizontal on the page and the *X* is vertical.

7. On graph paper, draw a graph similar to that shown in Fig. 16-35; then plot these points and label them in order. Then connect the dots to form a familiar shape. All lines will be straight. Some may be diagonal on the page.

Point	Absolute coordinates	
A	X2.00	Y0.83
B	X0.83	Y2.00
C	X0.83	Y2.00
D	X-2.00	Y0.83
E	X-2.00	Y-0.83
F	X-0.83	Y-2.00
G	X0.83	Y-2.00
H	X2.00	Y-0.83

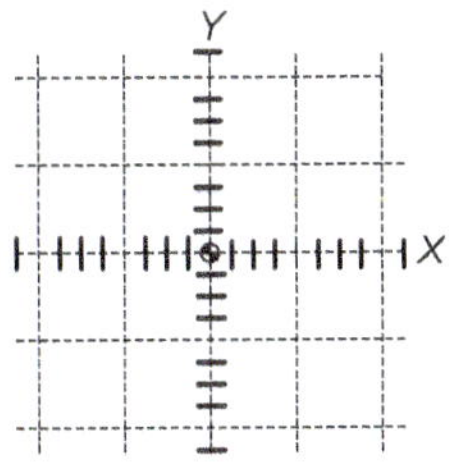

Figure 16-35 Use graph paper to create a program grid like this.

Critical Thinking

8. This problem (Fig. 16-36) illustrates a major difference between absolute and incremental coordinates. We will rerun it a second time in the incremental problem set. The set of *X*- *Y*, absolute mill coordinates in Fig. 16-36 has a deliberate error that *does not produce the shape shown.* Using engineer's graph paper (0.1-in. or 0.2-in. grid), plot and connect them in sequence using a contrasting color to see what happens. Save your drawing.

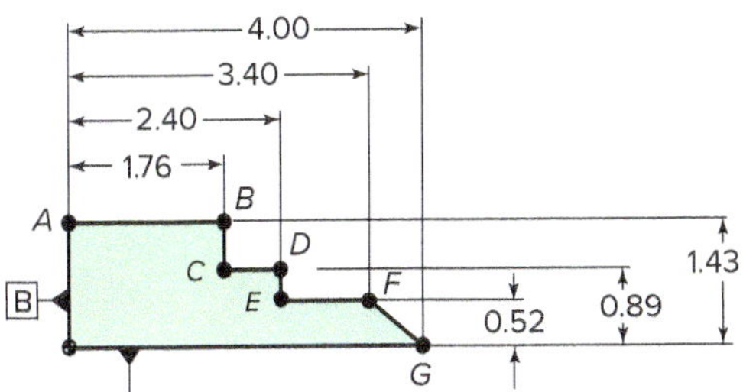

Point	Abs. coordinates	
A	X 0.00	Y 1.43
B	X 1.76	Y 1.43
C	X 1.76	Y 0.89
D	X 2.40	Y 0.52
E	X 2.40	Y 0.52
F	X 3.40	Y 0.52
H	X 4.00	Y 0.00
PRZ	X 0.00	Y 0.00

Figure 16-36 Plot these points to see what the error does to the shape. Save your drawing for later analysis.

KEYPOINT

To enrich this exercise, plot the points on a CNC simulator software or using the geometry creation part of a CAD/CAM system.

9. Find the null values in the coordinate set of Problem 5. Copy the page; then use a highlighter or circle those that could be dropped from the program.

Problems in Incremental Coordinates

10. Write a set of incremental coordinates for the shape shown in Fig. 16-37. Note that the drawing scale needn't be actual size. Assume the tool is positioned at the PRZ (lower left corner) and proceed to point *A*, then on around the part clockwise. Close the shape from point *I* to point *A*. No *Z* axis movements are needed for this exercise.

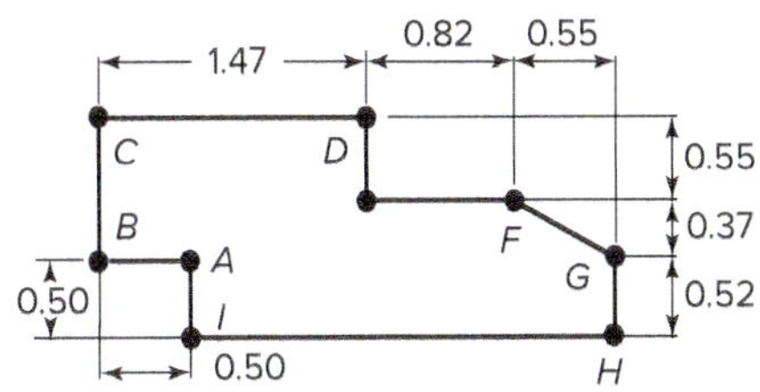

Figure 16-37 Write a set of incremental coordinates for this part.

11. Plot the following incremental coordinates with a pencil, in sequence. Choose a scale on graph similar to that shown in Fig. 16-38. The program begins with the tool (pencil in this case) 5 mm above the paper at the PRZ. It touches to make a line when it's at *Z*-5.0 mm. Nulls have been removed.

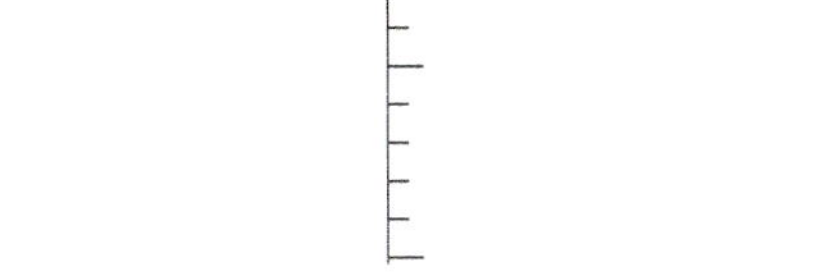

Figure 16-38 Your plot grid needn't be any particular scale or actual size. Choose a scale that fills the page with the part dimensions.

Sequence Number	Coordinates
N005	*Z*–5.0 (touches the paper to make a line)
N010	*Y*10.0
N015	*X*–5.0 *Y*10.0
N020	*Z*5.0
N025	*X*10.0
N030	*Z*–5.0
N035	*X*–5.0 *Y*–10.0 (makes diagonal line)
N040	*Z*5.0
N045	*X*–0.0 *Y*10.0 (diagonal)
N050	*Z*–5.0
N055	*X*–10.0
N060	*Y*–10.0
N065	*X*2.5
N070	*X*–2.5 (retraces the previous line)
N075	*Y*–10.0
N080	*X*10.0
N085	*Z*5.0
N090	*X*10.0
N095	*Z*–5.0
N100	*X*10.0
N105	*Y*10.0
N110	*X*–10.0
N115	*Y*10.0
N120	*X*10.0
N125	END

TRADE TIP

In the coordinate set of Problem 11, sequence numbers have been added. They are an organizational tool you should include in most programs. However, in many cases, they are optional. Numbering them five apart leaves space for future editing.

Sequence numbers are sometimes required when certain programming logic tools called loops, subroutines, and fixed cycles are used. These statements jump or bridge the program forward or backward out of the normal inline sequence. This is called *branching logic*, which we'll study later. In these instances, sequence numbers become a necessary address for the jump and/or for the return back to the normal program sequence on completion of the branched commands.

Critical Thinking

12. After verifying your understanding of Problem 11, return and modify only the first five lines by inserting the null coordinates that have been omitted. Note, this is a critical thinking point to teach yourself. Read the answers carefully.

13. The set of *X*-*Y*, *incremental* coordinates in Fig. 16-39 have an error that does not produce the shape shown—once again. Start the PRZ. Plot and connect them in sequence using a contrasting color, to do the same problem in absolute value. What happens to the shape? Solving this puzzle will point out a big difference between the two kinds of coordinates when an error is made.

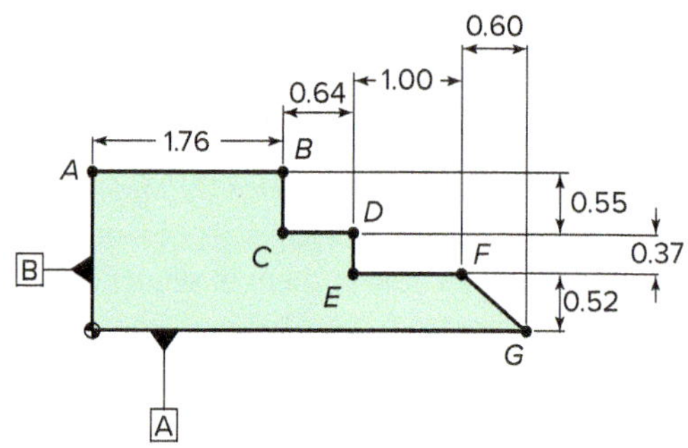

Point	Incr. coordinates
A	*Y*1.44
B	*X*1.76
C	*Y*-0.55
D	*X*0.64 *Y*-0.37
E	*Y*-0.37
F	*X*1.00
G	*Y*0.60 *Y*-0.52
PRZ	*X*-4.0

(Sum 0.52 + 0.37 + 0.55)

Figure 16-39 Once again, the program errors and does not produce the shape shown. Analyze the difference in absolute and incremental errors.

Unit 16-3 CNC Machine Motions

Introduction: Whether setting up, operating, or programming a CNC machine, the various kinds of movements they are capable of making must be understood. We will look at CNC machines from the simplest to the more complex.

For safety, it's important to know how your CNC machine is going to respond to various types of motion commands. There can be a difference, depending on the age and processor speed. Older controls with slower computer power move differently in some cases, when driven by up-to-date high-speed computers. Where it will matter, these differences will be pointed out.

TERMS TOOLBOX

Circular interpolation Machining an arc in two or more axes, at the specified feed rate.

Default (value or condition) A fallback number to which the machine will refer when entries exceed axis or controller limits.

Linear interpolation Machining in one or more axes in a straight line at the specified feed rate.

Rapid travel (rapid) The fastest axial speed that the machine can produce. Used as a positioning move for efficiency.

Retract height (R) The safe location in the *Z* axis that the tool is moved to before rapid repositioning.

2-D control A control capable of machining a circle using two axes at one time only.

$2\frac{1}{2}$-D control A control capable of circular interpolation in any two axes while performing linear interpolation in the third axis.

3-D control Performing circular interpolation in three axes at a specified feed rate. Not limited to three planes for arcs.

16.3.1 Axis Moves

There are several ways tool motion occurs on lathes, mills, and combination turning-milling machines.

Rapid Travel

Rapid travel, usually shortened to simply *rapid*, is the fastest speed the machine can produce. Rapid is an efficiency move to reposition for another cut or to change tools or parts. Common speeds range from a slow 100 in. per minute (IPM) on training machines to a blazing 1,000 IPM on industrial equipment or more. Rapid is a one-speed entry—no velocity is specified.

KEYPOINT

Rapid travel can be manually adjusted below full speed using the *rapid override* function (Fig. 16-40). Rapid cannot be adjusted above the 100 percent level (in comparison, feed rate moves can be adjusted above 100 percent).

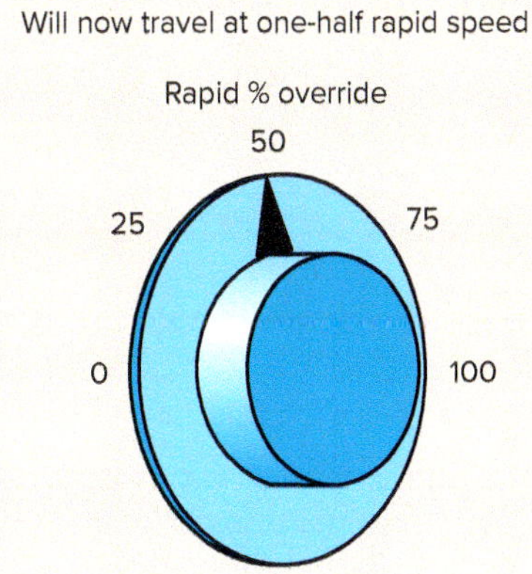

Figure 16-40 A typical override control can reduce rapid travel below the machine's maximum velocity.

Rapid Height or Distance Off

In planning a program, one must determine the closest rapid approach movement toward the work. The tool must be slowed to a feed rate at that distance from the work material. The safe distance must take into account clamps, vise, chuck jaws, and other holding tooling. As a beginner programmer, stop cutting tool rapid travel when it's 12 mm or $\frac{1}{2}$ in. from the work surface. This is large for industry but safe. It can be modified later when you are more confident in your programs.

That safe travel height above the work/tooling is called the **retract height** on milling machines (abbreviated to **R**) where the tool must pull back several times, then move over the work surface as in drilling a series of holes, for example. On lathes it might also be called the *departure point* in certain commands where the tool must retract back to a safe starting place before taking another pass. Retract moves away from the work are usually made at rapid travel.

Linear Versus Nonlinear Rapid

The path a CNC machine takes while performing rapid travel falls into two versions depending on the speed of the microprocessor in the controller (Fig. 16-41). It's important to know which path your machine follows. Older machines take a dog-leg path. If the odd path goes unaccounted for in a program, the cutter can run into the part or the holding tools.

KEYPOINT

All newer machines with 16-bit or higher microprocessors follow the true linear motion. Haas machines follow the true path.

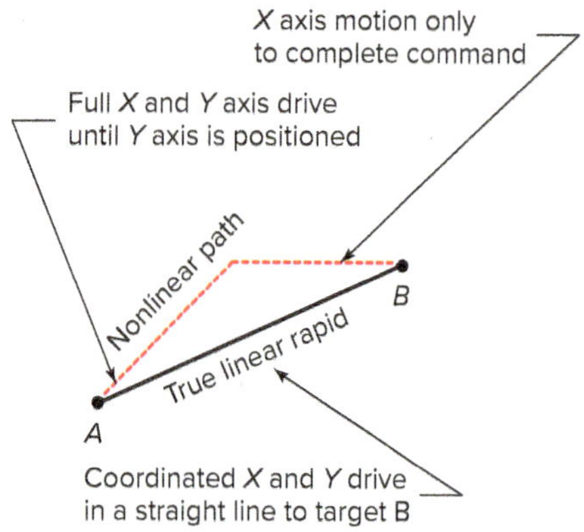

Figure 16-41 Older controls do not follow a true straight line path from point to point while in rapid travel mode. *It's critical to know what path your control will take!*

In the illustration, two controls reposition the tool from *A* to *B* at rapid speed. Note the difference between the slower CPU dash line and the newer solid line. The slower CPU cannot output motor control data fast enough to portion the two drive motors to move in a straight line from *A* to *B* in rapid travel.

When given a rapid command, the older control rotates both axis drive motors at their full speed until one of the axes reaches its destination, producing a 45 °angle for a distance. Then the other axis continues on at full speed to its destination.

The faster CPU control portions out the motor control speeds such that both *X* and *Y* arrive at the destination at the exact same time–producing a straight line from *A* to *B*. That true linear movement requires more computer speed.

Linear Interpolation—Straight-Line Motion at Feed Rate G01

Linear motion occurs at the specified feed rate in the program. Like twiddling the dials on an Etch-A-Sketch® or both *X* and *Y* handles at differing rates on a mill, you could almost produce the diagonal line shown in Fig. 16-42. The result would not be very smooth or accurate. Additionally, achieving the desired feed rate would be an impossible challenge.

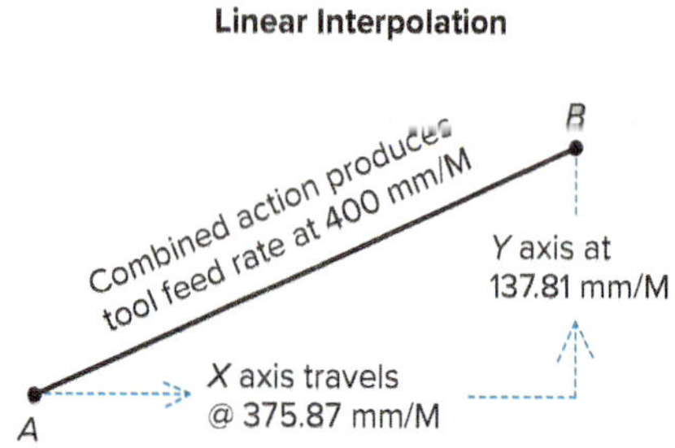

Figure 16-42 The combined action of differential *X* and *Y* feed rates creates the linear path at the programmed feed rate.

As in Fig. 16-42, when the CNC machine is commanded to move in **linear interpolation**, one, two, or three axes must move at coordinated speeds to arrive at the destination, in a straight line, at the tool feed rate specified in the program.

From the drawing you can see that when more than one axis motor is driving the tool, each must be turning at some subrate below that which is specified. Their combined effect produces tool movement at the specified rate. This is known as *interpolation* (interpolation means to find a value between two others). For example, to linear interpolate a straight line at 20° going from *A* to *B*, the *Y* axis must rotate at 36.3 percent of the *X* axis speed (the tangent ratio of the line's slope).

If the programmed feed rate is 400 mm per minute (mm/M), for example, the *X* axis must drive at 375.87 mm/M while the *Y* axis turns at 136.81 mm/M. The CPU controls each axis drive to achieve the result.

Feed Rate Override

Similar to rapid movement, feed rate movements can be overridden by the operator to fine-tune the cutting action. But there's a difference between rapid and feed override control. The usual adjustment range for feed rate is from 0 percent up to 150 percent of the programmed rate. Unless the programmed rate is near or at the machine's fastest rate already, the feed can be increased above the programmed rate.

KEYPOINT

On most CNC controls, feed rate can be increased above the program value to a maximum increase, usually 150 percent of the programmed rate.

Circular Interpolation: G02 Clockwise, G03 Counterclockwise

Circular interpolation is the next higher CNC motion. It produces a full circle or a part of one (arc), at the programmed feed rate. Circular motion adds an additional challenge for the computer beyond linear interpolation. When producing a true circle, the control must continuously change the ratios between the drive motors for each increment of arc.

Each axis is interpolating below the target feed rate, but their combined action creates the correct tool velocity. Keep in mind that in addition to the axis motor control calculations, the CPU is also comparing resultant feedback to keep everything on track (Fig. 16-43).

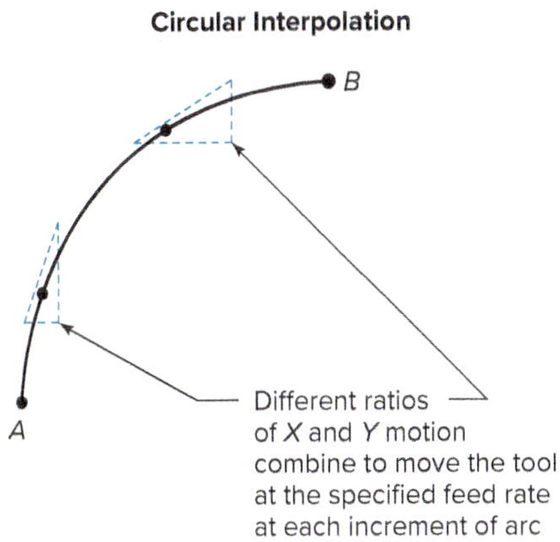

Figure 16-43 By constantly varying the X and Y rates, a true circle can be produced at the programmed feed rate.

Default Rates

During circular interpolation, due to the high number of calculations per second, some older (slower CPUs) must limit feed rates to a maximum **default** *rate.* These controls cannot produce circles at the same speed as straight lines. For example, an older control can perform linear interpolation at 200 IPM but must slow to 140 IPM for arcs. Even though the program is at a higher rate, the control will fall back or default during the circular movement. Newer 32-bit (or higher) CPUs have no such limits until extreme feed rates are encountered.

The term *default* will be used elsewhere in CNC. It means a fallback or preset bias built into the control. A default also occurs when a choice must be made by the control when other information is not available.

16.3.2 Axis Combinations for Milling Machines

CNC lathes can perform the tool motions of linear and circular interpolation as already described in the primary plane of *X*-*Z* only. This discussion is about mills only. Three-axis milling machines are further categorized by their ability to perform various circular moves:

Two-dimensional motion

Three-dimensional motion

Two-and-one-half-dimensional motion

The difference is significant.

Two-Dimensional Motion

The lower level **2-D control** is restricted to moving two axes while producing a circle. That means arcs in the primary planes of *X*-*Y*, *X*-*Z*, or *Y*-*Z* only—one plane at a time (Fig. 16-44).

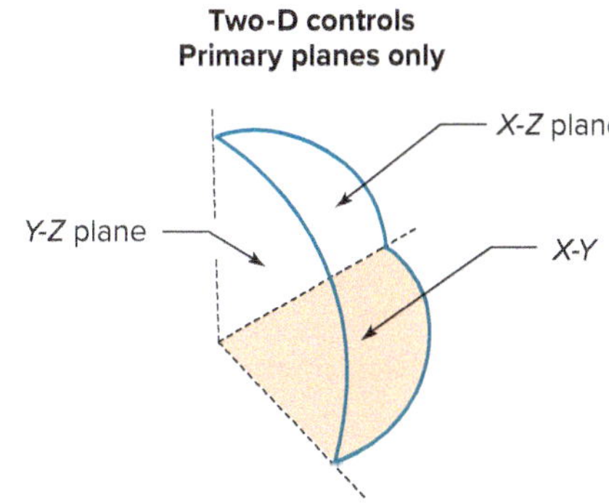

Figure 16-44 A 2-D control can produce a circle in one primary plane at a time only.

The program must include the command G17 to indicate circular interpolation in the *X*-*Y* plane, G18 to produce a circle in the *X*-*Z* plane, and G19 for *Y*-*Z* circular interpolation.

Three-Dimensional Motion

A 3-D mill requires a fast processor or very slow feed rates. It must interpolate and drive three axes simultaneously to produce arcs. Perfect circular tool motion involving *X*, *Y*, and *Z* axes at the programmed feed rate is a great challenge for the processor. The **3-D controls** are not common in industry (Fig. 16-45).

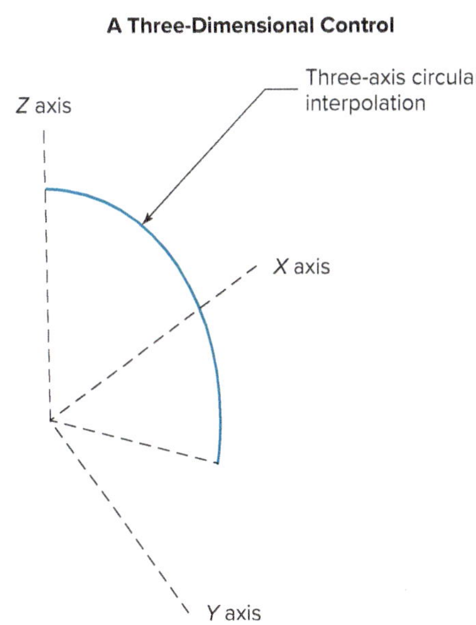

Figure 16-45 A 3-D control can produce an arc using three axes simultaneously. That means arcs in any plane can be machined.

Two-and-One-Half-Dimensional Controls

Many controls today fall between the preceding two, producing circular motion in one of the primary planes *X*-*Y*, *X*-*Z*, or *Y*-*Z*, while moving the third axis in a straight line at the same time. That motion, called a $2\frac{1}{2}$-D move (two axis circular while one moves linear), produces a spiral similar to a thread–always parallel to a primary axis (Fig. 16-46).

A 2.5-Dimensional Control

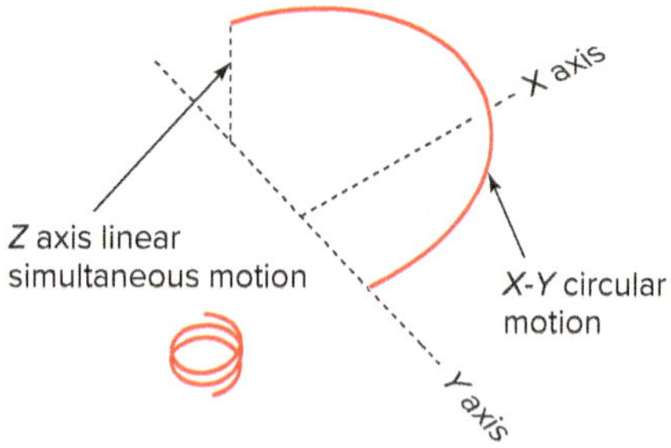

Figure 16-46 Most CNC controls can produce $2\frac{1}{2}$-D motion.

Machines with **$2\frac{1}{2}$-D control** are able to machine the spiral ramp pass we've discussed previously in milling. Today, in many cases we do not need CNC controls that feature high-level interpolations as long as a capable CAM system backs it up for program commands (Fig. 16-47). The CAM system can approximate complex surfaces using either 2-D or **$2\frac{1}{2}$**-D movements.

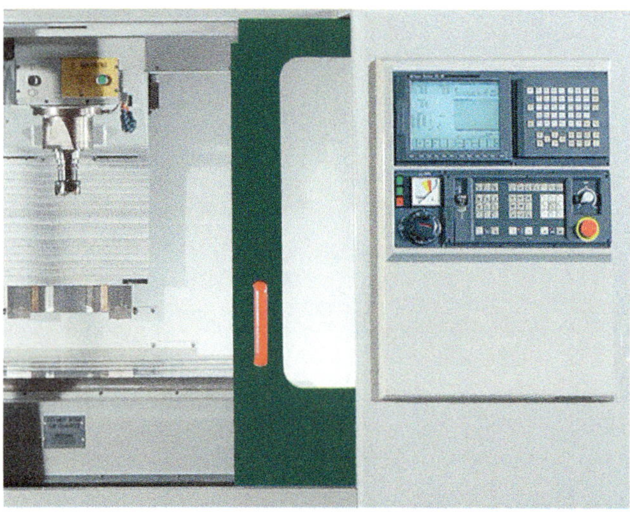

Figure 16-47 Most CNC controls are capable of $2\frac{1}{2}$-D motion.
Courtesy Bridgeport Machines/Bridgeport, CT

TRADE TIP

Conic Interpolations Not all regular curves are parts of circles. The regular curves shown in Fig. 16-48 are called the conic sections. When a cone is cut at angles other than 90° across its axis, first an elliptic cross section is produced; then tilting the slice further, a parabola and then a hyperbola are produced. Each is a curve with different characteristics. Some controls can interpolate these curves using the same 2-D and 3-D rules outlined.

Conic Section Curves

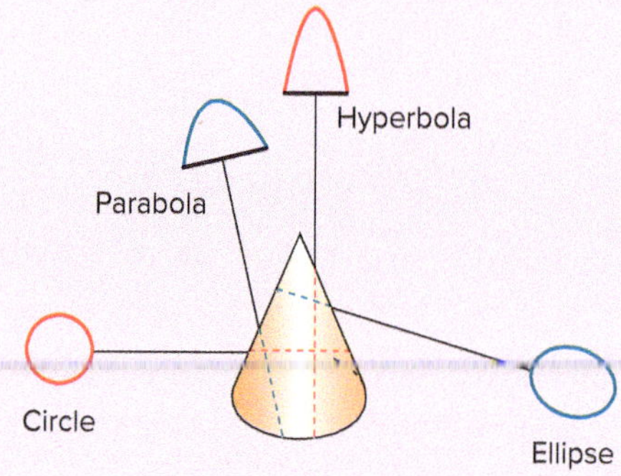

Figure 16-48 Some advanced controls are able to perform conic interpolations.

16.3.3 CAM-Generated Shapes

Because most machine controls are limited to linear and circular motions, CAM software cannot simply generate an irregular surface. Instead, the CAM software breaks the surface into tiny arcs or straight lines, then links them to approximate the contour.

The length of these approximation lines is defined by a set tolerance away from the perfect shape, and how steep the curvature is. Advanced software such as Mastercam, our example CAM program, can reprocess the program once it has been generated to create more efficient arcs that fit the shape better, but there is still a lot of data compared to a true 3-D process.

However long the data might be, the CAM process for complex shapes is the method of choice today. Down to the microscopic level the difference between true 3-D generated shapes and CAM-approximated is indistinguishable. Additionally, using CAM, the CNC machine is far less costly and complex.

UNIT 16-3 REVIEW

Replay the Key Points

- Rapid travel can be manually adjusted below full speed using the *rapid override* function that allows reducing velocity only.
- Rapid cannot be overridden above the machine's 100 percent level.
- Older slower CPUs may produce a dog-leg kind of rapid travel, which must be accounted for in programs.
- Feed rate can be increased above the program rate up to the maximum rate of the machine, usually to 150 percent of the programmed rate.

Respond

Critical Thinking

1. When might a shop need a higher level control than 2-D for a lathe?
2. During a circular interpolation, the feed rate is specified with the code word G21 F600, meaning the programmer has set the metric values (G21) and the feed rate at 600 mm per minute. However, during actual operation the machine is observed to cut at 400 mm/M under that command. Offer two possible explanations.
3. A hemisphere must be machined on the top of an odd-shaped casting. The shop has a CNC lathe and 2-D CNC mill. Describe ways in which it might be machined.
4. The move illustrated in Fig. 16-49 is called a *helical ramp plunge or a spiral ramp*. It is used in CNC milling to plunge down into the work to start milling a pocket. What type of control is needed to perform such a move? Referring to your manual milling machine skills, why would the illustrated move be superior to straight plunging and to a linear down ramp? This is a critical thinking question not directly covered in the reading.
5. Name a CNC mill type and action that can produce a smooth version of the hemisphere of Question 3.
6. Describe the actions of the three levels of mill controls: 2-D, $2\frac{1}{2}$-D, and 3-D.
7. Two control limitations described in this unit were default feed rates and types of circular interpolations. To what are both due?
8. What type machine movement performs the ramp down into a pocket shown in Fig. 16-50?

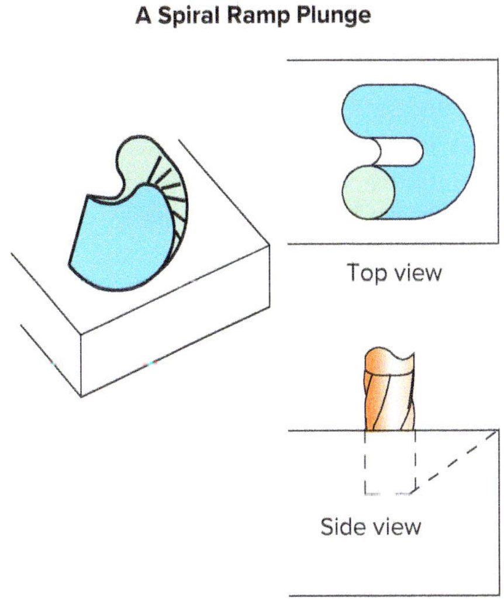

Figure 16-49 What kind of CNC control can make a spiral ramp, 2-D, $2\frac{1}{2}$-D, or 3-D?

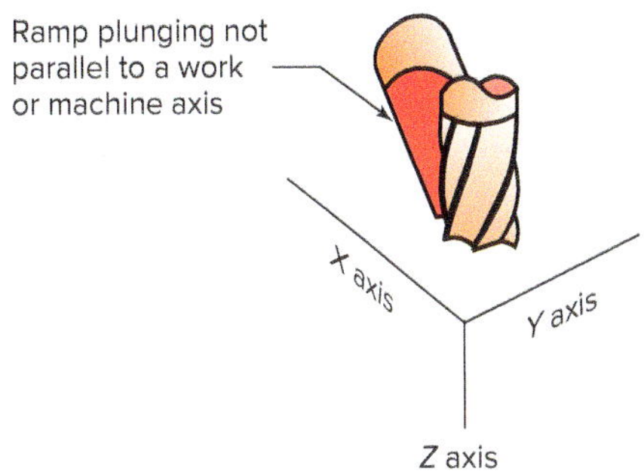

Figure 16-50 What control level can produce this ramp plunge?

CHAPTER 16 REVIEW

Unit 16-1

Machine evolution is underway. Whole new species are on the market that have blurred the distinction between cutting and grinding, turning and milling; they are complete, one-stop machining centers. That trend will continue since it reduces the cost of manufacturing, which means that being able to identify machine axes on new machinery will become more complex and necessary.

Those multitasking machines will feature many slides and rotary motions. All must be assigned axis names and directions. It's predictable that we might exceed the nine EIA standards we've studied.

Unit 16-2

I don't foresee a time when the ability to write keyboard-entered tooling programs or to edit existing CAM-generated programs won't be a job requirement for a lead machinist. To do so, a good working ability with both absolute and incremental Cartesian coordinates and command codes will remain essential job skills.

Unit 16-3

The ability to perform in-depth data manipulations at the machine is one of the major factors separating operators from machinists. The machinist can control and improve the process, while the operator controls the running of the machine.

Therefore an understanding of the ways a machine is able to interpolate is needed for the same reasons as already outlined, to improve and try out programs and to write mini-tooling routines for your machine.

KEYPOINT

While they are optional in CNC work, when drawing part geometry in CAD software, polar coordinates are an indispensable tool.

QUESTIONS AND PROBLEMS

1. We've studied nine axes that define motion and position in CNC work. Describe them. (Unit 16-1)

2. True or false? When an axis moves with a wrist action, not in a complete circle, it is said to be interpolating. If it is false, what would make it true? (Units 16-1 and 7-2)

CRITICAL THINKING

Problems 3 through 10 refer to Fig. 16-51, and the following planning.

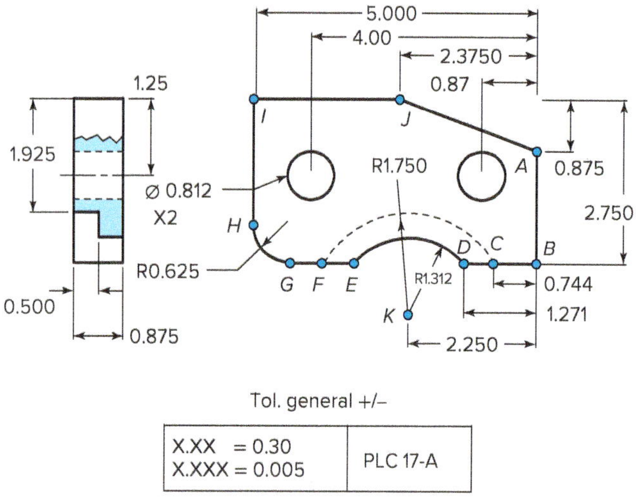

Figure 16-51 Problems 3–10.

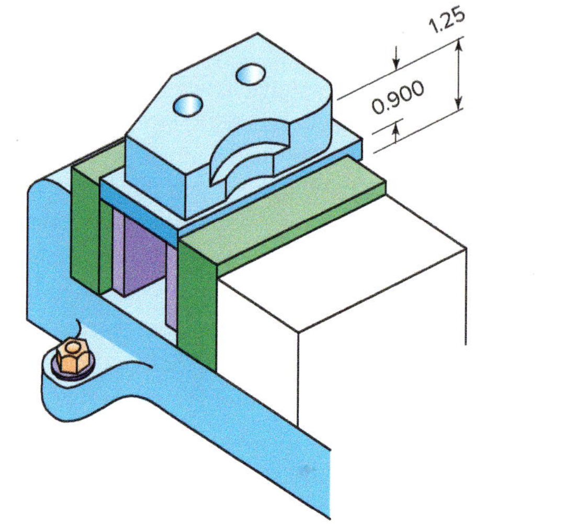

Figure 16-52 Here's how we'll hold and profile mill the part.

Job Planning Part Number PLC 17-A

005 Cut aluminum blanks 5 × 2.75 × 1.25 (thickness excess for holding in vise)

010 Profile mill all details × 0.900 total depth (0.875 + 0.025 extra)

015 Reverse parts and fly cut the thickness excess to 0.875 in. dimension.

The final operation will be to turn the parts over and hold in soft jaws and cut away the excess. This allows the parts to be 100 percent profile milled at one time (Fig. 16-52).

3. In Fig. 16-51, note that part PLC 17-A is not dimensioned geometrically. Where should the PRZ be placed on the drawing to write a profile milling program? Explain. (Unit 16-2)

4. Figure 16-53 depicts four possible positions we could use to set up a vise to hold the part for milling of the PLC 17-A. In which quadrant does each setup lie? (Units 16-1 and 16-2)

5. How many circular interpolations will be required to mill the profile of the PLC 17-A? (Unit 16-3)

6. Of the four possible setup orientations of the PLC 17-A (see Fig. 16-53), which could be used to machine the part, A, B, C, or D? Which would be the most efficient, A, B, C, or D? Explain. (Unit 16-2)

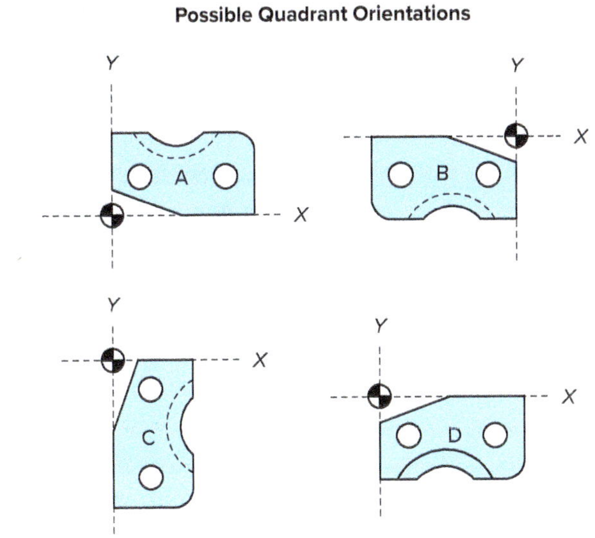

Figure 16-53 Setup options in various quadrants.

7. With the part in position D (Fig. 16-53), what are the absolute coordinates for the center of the two 0.812-in. holes? (Unit 16-2)
8. Using the drawing orientation (Fig. 16-51), what are the incremental coordinates of point *K* (arc center) using point *D* as the present reference location? (Unit 16-2)
9. What are the absolute coordinates of point *K* with the part in the *D* orientation (see Fig. 16-53)? (Unit 16-2)
10. With the part in the drawing orientation, write the absolute coordinates of the geometry. Figure 16-51 points in the order given. Leave null entries off. You are not writing curve commands, just listing the coordinates. (Unit 16-2)
11. To perform a spiral ramp plunge, what level machine motion must the control be capable of performing? (Unit 16-3)
12. How do CAM programs produce three-dimensional shapes? (Unit 16-3)
13. True or false? An incremental polar coordinate uses the PRZ as its center reference and a local line as its angular reference. If this statement is false, what will make it true? (Unit 16-4)
14. In Fig. 16-54, where should the PRZ be placed to turn this pin chuck collet? In what orientation should it be machined? Explain. (Unit 16-2)
15. Are there any circular interpolations required for the pin chuck (see Fig. 16-54)? Why? (Unit 16-3)
16. Since the 50° dimension is angular, would polar coordinates be used to define the tapered surface (see Fig. 16-54)? (Unit 16-4)
17. Assuming the part is 180° rotated, what are the geometry point, absolute coordinates, for turning this part? Start from PRZ, then turn toward the chuck. You are not writing a program, just listing the significant geometry points required to turn the OD. (*Hint:* See Problem 18; then check the answers.) (Units 16-1 and 16-2)
18. The second geometry point for Fig. 16-54 was the diameter at *X*0.6840, *Z*0.0000. It's not on the drawing. How was it calculated?
19. Figure 16-55 depicts a part requiring two wrench flats milled on the end of a round inspection cap. Would the coordinates be more efficient in Cartesian or polar? Explain. (Units 16-2 and 16-4)
20. What are the absolute polar coordinates of points *A*, *B*, *C*, and *D*, at the ends of each wrench flat (Fig. 16-61)? (Unit 16-4)

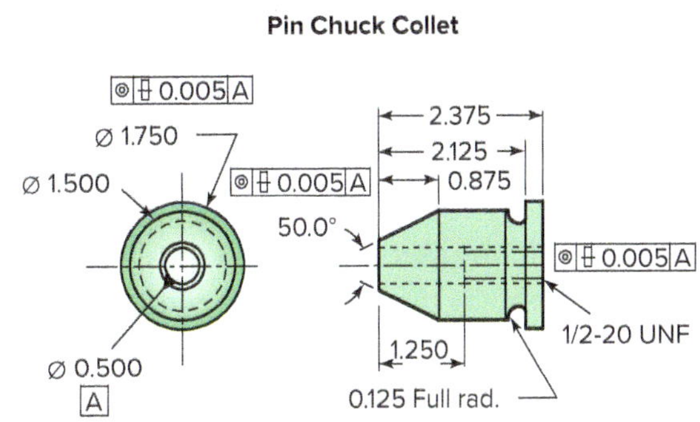

Figure 16-54 Turning problem.

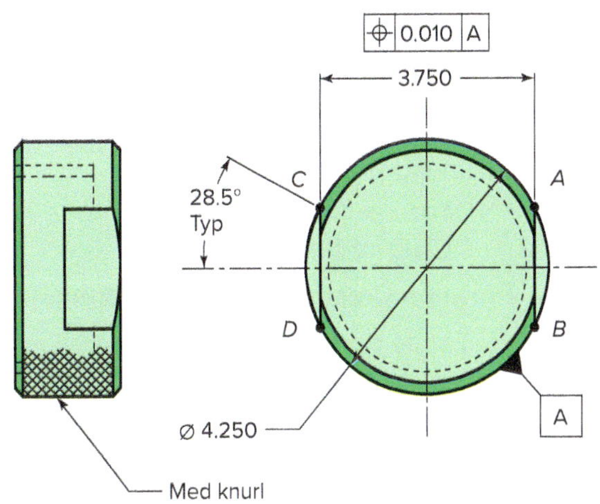

Figure 16-55 Inspection cap.

CHAPTER 17
Advanced CNC Systems

Units

17-1 A CNC Axis Drive

- Define and sketch the five components in a CNC axis drive
- Describe a kinetic sense motion (closed loop)

17-2 Industrial Machining and Turning Centers

- Based on a timeline, list the three branches of changing technology
- Identify the capabilities of a modern vertical machining center
- Identify the capabilities of a turning center

17-3 Multiaxis Machining

- Define the 3 axis of rotation
- Explain the differences between a 4 axis and a 5 axis machining center
- Locate the machine rotary zero point on a multiaxis machine
- Describe how a rotary axis works on a turning center

17-4 Adaptive Process Control

- Explain how a touch trigger probe is used to measure parts during production
- Describe how a tool setting probe helps define tool offsets in a CNC machine

17-5 Program Creation and Data Management

- Define part geometry
- Describe the way CAM software creates a machine-readable program
- Use CNC data management terms correctly

At Morrison and Marvin Engine Works, employees work to replicate the form, fit, and function of antique engines in exact quarter scale. They build the prototypes, and then they sell casting sets and plans to model builders and home shop machinists in the United States.

For example, they have replicated the 1895 MERY Explosive Engine, a double acting, six-cycle engine built in California, and the 1903 Kansas City Lighting Balanced Engine, an opposed single cylinder, double piston engine specifically built to power a hay press.

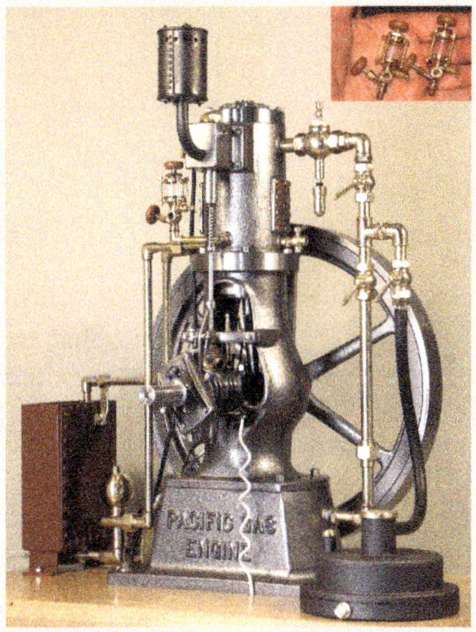

Courtesy of Morrison and Marvin Engine Works

A favorite engine is the 1889 Pacific vertical engine that was built in California. The fuel/air mixture was ignited by a pair of insulated points inside the cylinder. At near top dead center, the rising piston closed these points, and on the beginning of the downward stroke, the points opened and ignited the fuel charge. This replica is pictured above.

A new addition to our inventory is our ModelBuilder Vise, a set of 4140 steel castings which allow the builder to construct a high-quality miniature machinist vise. We also sell miniature spark plugs that are made in America. We have several small engine casting sets and are always developing new projects.

INTRODUCTION

Chapter 17 provides the background concerning the machines and the data they use necessary to manage CNC equipment. We'll look at three aspects of CNC systems:

Axis Drive Mechanics We'll learn how CNC lathes and mills can make a near-perfect surface when cutting into and out of changing material loads. Even with varying excess from one part to the next, every part comes off the machine a near-perfect duplicate of the last!

Machine Variations and Evolution While today's machines are far superior to anything in the past, it's a sure bet they aren't the ultimate. So, our goal is an understanding of the drive behind machine evolution such that the future is easier to see. Reading the timeline carefully, you might foresee a big change just around the corner!

Process Control Probes are used to help machinists set up CNC machines and to control part dimensions during the machining process. These tools can reduce machining costs by monitoring part feature dimensions in "real time" and making adjustments before the part is unloaded from the machine.

Programs We'll also examine program creation and data management. Unit 17-5 isn't about writing programs; it's about where they come from and how they are managed.

Unit 17-1 A CNC Axis Drive

Introduction: A CNC axis shares a lot of similarities with using your arm to move and place an object. Your brain sends signals through connective nerves to the synapses that trigger muscles. It also receives reports back that compare the command to the result—called a kinetic sense. Without looking, you always know where your arm is located and how fast it is moving. Then, if the item doesn't budge, you can add more power to get it going or at some point decide it's beyond your ability. If the object moves easily, you can back off on the force.

TERMS TOOLBOX

Absolute positioned system A CNC drive system that retains its grid position even when turned off, then on. These machines feature linear feedback devices.

Ball-screw A backlash-compensating linear drive system in which a split nut is forced laterally both directions against a screw. Balls roll through the circular channel between nut and screw.

Closed loop A circuit that includes monitors to send a progress signal back to the CPU.

CPU (central processing unit) The component that decodes the program, sends the drive commands to the servo motors, and monitors and adjusts progress.

Feedback The signal returning from the drive system to the CPU.

Hard limits Physical switches that verify the location of an axis position.

Homing the machine Sending all axes to limit switches in order to achieve a zero base for the encoders. Generally required only on older machines.

Initialization Turning on the control, depending on sophistication, routines must be followed to prepare for machining. Newer machines load software and ready themselves; older units need more preparation.

Machine home A never-changing position where the machine must be driven to refresh its feedback signal at initial start-up.

MCU The master control unit; see *CPU.*

Open loop A budget drive system that does not feature any provision for feedback signal processing.

Servo error An unwanted condition in a closed-loop system where the axis progress exceeds the tolerance from the expected result.

Servo motor A highly controllable motor with predictable power, speed, and acceleration curves based on input energy.

Soft limits Optical sensors that verify proximity to hard limit switches.

Stepper motor A drive motor that moves a given part of a rotation, given one pulse of energy.

17.1.1 Five Axis Drive Components

Using five electrical/mechanical components, a CNC drive performs the same tasks as using muscles to move and position your arm. The controller can keep a cutter hugging the true programmed shape from within 0.001 or 0.0005 in. for heavy cuts down to an amazing 0.0002 in. on light loads, depending on machine construction and the nature of the shape and volume of metal being removed.

The control constantly compares actual tool position to the expected position. If the cutter's progress doesn't fall within a preset lag tolerance, the controller has one or two options, depending on basic architecture within its operating system.

More commonly, it can wait a preset time, until the physical cutter location catches up to the expected position before issuing another axis drive command. Second, on more powerful machines, the control can increase the axis drive force when the cutter enters heavier cuts and starts to lag behind, and it can back off on the axis drive as the cutter breaks into lighter work sections. The system performing these marvels is a **closed-loop**, *linear servo axis drive.*

That mouthful is shortened to a CNC axis drive! It's responsible for the cutting motions and for the positioning accuracy. Like your muscle movements, five components are required (Fig. 17-1):

CNC Drive	Muscle Motion
1. Controller unit	Brain–processing in/outgoing data
2. Axis drive relay card	Synapses–triggering muscles
3. Translation axis screw	Tendons pulling
4. Controllable drive motor	The muscle
5. Feedback device	Sensory nerves reporting progress

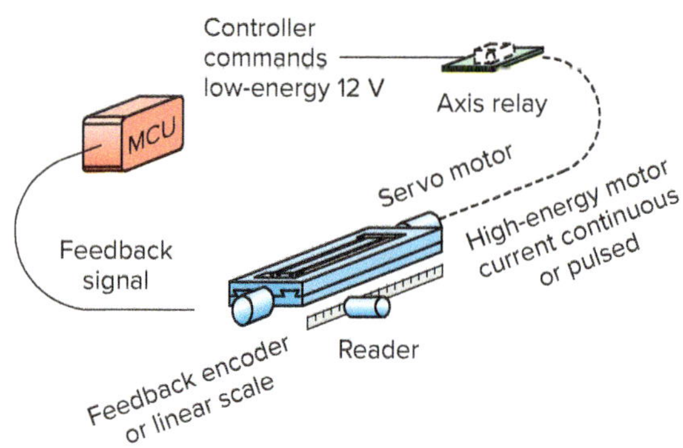

Figure 17-1 A closed-loop CNC axis drive system.

To achieve the maximum results and understand limitations, the action of each component should be generally understood by the machinist, which is our goal for this unit.

17.1.2 Three Kinds of Repeatability

When buying a CNC machine, one of the first facts to inquire about would be its axis repeatability. We begin with resolution (Fig. 17-2).

First Machine Resolution

This is the smallest movement the axis drive is able to make, almost always 0.0001 in. or 0.02 mm. Either it's a mechanical limit, based on the finest movement the axis drive motors are able to make, or it can also be a sensor limit of how fine the machine can read movement and send a signal back to the controller–one or the other, usually the mechanical limit.

Second Axis Repeatability

Next up we discuss the capability of the axis drive to move the spindle or tool away from a specific location, then drive it back to the same place within a close tolerance. The approach test must be made from every part of the work envelope and at varying speeds. But, the return test is never under a cutting force load. CNC axis repeatabilities are usually around 0.0002 in., again with exceptions. But repeatability doesn't tell the whole story. By itself, repeatability can be misleading.

Figure 17-2 Super precision CNC machines such as this Super-Mini Mill have resolutions below 0.0001 in.
Courtesy Haas Automation, Inc.

Third Shape Repeatability—The Final Factor

Machinists sometimes think of axis repeatability as the tolerance limit for holding to a curved path; however, that isn't always so, especially as removal rates increase. Many factors combine to create the shape repeatability. Depending on the overall strength of the machine's axis drives and the nature of the machining, a few machines are able to reproduce shapes within the specified repeatability while most tolerate more deviation.

SHOPTALK

There are extreme accuracy machines designed for specialty work that feature 0.00005-in. resolution (fifty millionths inch). These machines can reproduce parts within 0.0001 in. each and every time.

Physical Servo Errors

Similar to discovering your arm hasn't the strength to push the work along, when a program demands more than the machine or the cutter can deliver, or when a cutter dulls or loads up, the machine can become out of sync. When its physical path exceeds a preset deviation limit, it enters a condition known as **servo error**. On screen and possibly with flashing lights, the control will signal the operator and, at the same time, halt all progress when it cannot bring the axis back into coordination within a given time lag.

SHOPTALK

Shape repeatability is more difficult to quantify than location repeatability due to the wide variation in machine cuts and forces. Close tolerance shape repeatability is often provided as a selling point when choosing new heavy-duty CNC equipment. However, there is no standard test to determine it, as there is for axis repeatability. We often find ourselves comparing apples to oranges, one machine to the next.

TRADE TIP

As the operator, you have three main weapons against servo error. First, use the feed and speed override controls to fine-tune cutting forces (editing in permanent changes to the program if allowable—ask first!). To do so, machinists must understand the forces and physics of chip making and the limits of their machine. Closely monitoring cutter sharpness and coolant delivery can also help prevent a lot of servo stalls.

There are also programming methods that can prevent servo and shape errors. They are the same skills learned on manual machines, like taking a light final pass with a full-sharp cutter reserved for finish work. Your manual machine experience really counts here!

17.1.3 Mechanical Axis Translation

The heart of the precision drive is mechanical: the *backlash eliminating, recirculating ball-screw, and preloaded split nut.* Its purpose is to move, position, and reverse axis motion with no backlash. For a reminder of how they look see Fig. 17-3.

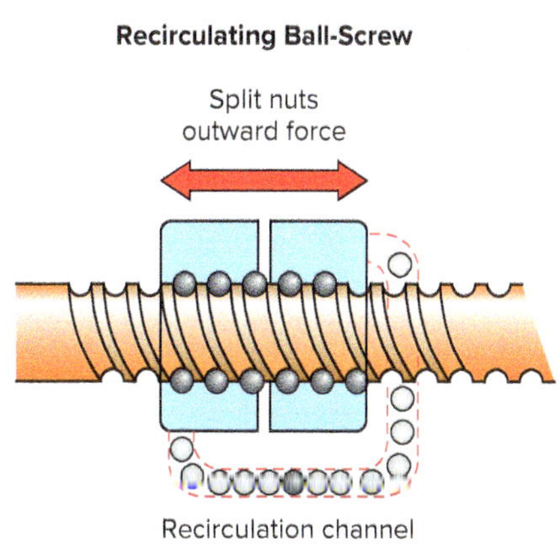

Figure 17-3 Ball-screws are the mechanical heart of the drive system.

17.1.4 Controller

The CNC brain is sometimes called the *master control unit* **(MCU)**, or the **central processing unit (CPU)**. This component has four or five functions, depending on whether or not it is connected to an outside computer. For our study of axis systems, the controller's CPU converts stored program commands or keyboard entries by the operator into drive commands, then sends them to the axis motor relays.

> **KEYPOINT**
>
> Axis drive commands do not go from the CPU to the drive motors; they go to intermediate relays.

The CPU issues a low-energy command (usually 12 V DC) to relay devices that work as electricity amplifiers and switches. At the same time, it is also processing incoming feedback from the axis drive sensors to determine what action is required to maintain the programmed feed rate.

17.1.5 Axis Drive Relays

An individual CNC controller can be used to drive a variety of different CNC machines. To do so, the controller doesn't directly command the motors to move. It commands an axis drive relay that's custom tailored to the specific drive motor. Drive relays make it possible for a single control model to move any kind of motor, big or little, servo or stepper (studied next). One control is able to operate a small offload mill, or it can be wired into a giant horizontal mill with over 100 hp per axis, because of the axis relays.

Wired between the controller and the motor, the intermediate relay card receives the low-energy control command, then amplifies it, and configures it to the motor's needs. Then the relay triggers high-energy current as pulses, or a stream depending on the kind of motor being driven.

17.1.6 Drive Motors—Servo or Stepper

CNC drive motors are different from manually operated machines that often use one motor to drive several functions through gears, belts, and clutches. CNC drive motors are highly controllable and predictable: one motor per axis. Given an exact amount of input energy, they reliably move the same way each time. They accelerate and decelerate on an exact time/load curve.

> **KEYPOINT**
>
> CNC machines use one drive motor per axis, and each motor is very different from those on manuals.

There are two kinds of popular drive motors, **servo motors** and **stepper motors**. (A few machines, such as precision grinders, use hydraulic motors powered by pumps, but they are far less common.)

Direct Current Servo Motors—More Power for Industrial Machines

Electric servo motors are stronger but also more costly. They produce accurate RPM based on input energy. As the input energy increases, servo motors spin faster or drive harder when pushing axis drives.

> **KEYPOINT**
>
> The DC servo motor is by far the most common.

Steppers—Moving in Small Increments

Steppers are the second most common drive muscle. They are generally used in smaller machines. Steppers do just that; they move in tiny jumps rather than spinning as servos do. Grasping the shaft of an unmounted stepper and spinning it, you'd feel it click or pulse as the motor rotates through small magnetic arcs.

> **SHOPTALK**
>
> Precision ball-screws are so critical to the manufacture of CNC machines that a certain foreign machine building company wished to make high-quality machines here in the United States. Before they even considered building a factory, they sent a study team to determine if there was more than one ball-screw manufacturer that offered the accuracy required. Not until they knew that single fact would they commit to thinking about the investment (Fig. 17-4).

Figure 17-4 This giant ball-screw is one of the largest ever made in the United States.
Courtesy Wedin International, Inc.

Using the CPU's clock function to coordinate timed pulses to the drive relay, after reading the programmed feed rate, the controller determines how fast the pulses must be sent. The faster they are issued, the faster the motor rotates (up to its maximum speed).

Each pulse from the axis relay rotates the motor an exact arc. For example, common steppers rotate through one full circle with 15, 30, or 45 pulses per revolution, depending on their construction. So the motor moves 24°, 12°, or 8° per pulse. That, in turn, translates to an exact rotation of the ball-screw.

Since the movement of a stepper is predictable, light-duty machines can be constructed with no feedback system, called **open-loop** *circuitry*. Open-loop machines are budget equipment only. With no feedback, they do not sense or control variables, nor do they repeat shapes very well. In fact, an open-loop axis can be stopped completely, yet the control will continue blindly issuing program commands, with no movement resulting. No kinetic sense; the control doesn't know it's out of sync!

KEYPOINT

Open-loop machines are best used for detail and off-load work or for woodwork, but they aren't practical for heavy machining.

17.1.7 The Feedback Device

The fifth component is the watchdog in the system. Monitoring the axis progress and location, the **feedback** device is responsible for detecting and reporting position in real time back to the processor. There are two general families:

1. **Direct Linear Feedback**

   Sometimes called **absolute positioning**, these systems make use of various kinds of calibrated distance sensors. Although more sophisticated, the scales are similar to digital positioning systems on manual machinery. The controller tracks axis position using signals from the scale. Direct feedback provides the advantage of acquiring position instantly during initial powering up. Machines equipped with direct feedback reacquire their axis and tool positions when the control is first turned on.

SHOPTALK

Blueprinting the Drive Screws The objective is to make a perfect **ball-screw** with zero lead variation throughout its length. Machining them requires highly controlled cutting, grinding, heat treating, finish grinding, and lapping. Although close to perfect, they simply cannot be absolutely free of lead errors. So, the power of the computer is put into action on certain high-quality CNC axis drives. Using laser sensors, lead variation in the screw can be detected, mapped, and stored in the controller memory. Corrections are then applied to compensate for the tiny errors in the screw. But as the machine warms up and over time, the error pattern changes. From time to time, it can be updated using portable laser technology. However, for some precision situations, periodic calibration isn't good enough. So the next, level machine has the laser feedback system built into the machine's axis drive. It continuously checks and updates its own blueprint—now that's engineering!

2. **Indirect Feedback**

   These devices monitor the amount of drive screw rotation as shown in Fig. 17-5, called *encoders (counters)*. About the only advantage of these encoding devices is their low cost compared to absolute positioning systems. The biggest disadvantage of encoders is that they can become lost either from depowering or from heavy machining loads. Since the CPU must count pulses to track position, it can lose the count when the control is switched off or when the cutter encounters sudden heavy cuts. The system that's built in this way must be homed by parking it on relay switches **(hard limits)** or optical sensors **(soft limits)** to regain a sense of where it is. Machines of this type are set up such that they will not operate until they are homed after being powered up in the morning, so they have a zero base from which to begin positioning.

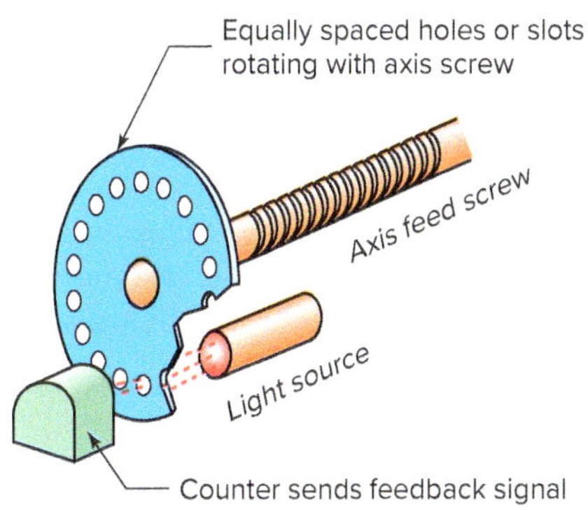

Figure 17-5 An indirect encoder counts pulses as the axis screw rotates.

17.1.8 Initializing (Turning on a CNC Drive)

When **homing the machine**, or **initialization** (powering up the system) of encoding machines, each axis must be driven up to a reliable start point, usually against axis limit switches. That position is known as the **machine home** or *axis home*. Once parked at that never-changing hardwired zero base location, usually at the far end of each axis, the controller, then has a datum origin for its axis count.

TRADE TIP

On older encoding drives, it's a good idea to insert a home command into the program, after a long or difficult roughing cut, before taking the finish passes. To simplify the action, homing can usually be performed with a program command, rather than manually by the operator. Then with the axis coordination refreshed, the final pass can be taken with improved accuracy. It's often convenient during homing to program a tool change to a finish cutter since the tool change position is often at or near the home position on both mills and lathes. Presto—the machine has a sharp cutter and a refreshed axis position.

There are other kinds of linear and rotary feedback drive mechanisms, but scales and encoders (see Fig. 17-5) represent most of the possibilities for common equipment.

KEYPOINT

No matter what kind of device, feedback's purpose remains the same: sense the exact position of the machine in *X*, *Y*, *Z*, *A*, *B*, and *C* axes; then feed that information back to the CPU.

UNIT 17-1 REVIEW

Replay the Key Points

- There are five components in a modern CNC axis drive system: controller, axis relay, servo motor, ball-screw, and feedback device.
- Servo errors usually happen when cutting forces suddenly build beyond the axis drive's ability to push–for example, when an end mill loads up and stalls.
- Open-loop machines are best used for detail and offload machining or for woodwork but never for industrial machining.
- The direct current servo motor is by far the most common.
- The feedback device senses the exact position of the machine in *X*, *Y*, *Z*, *A*, *B*, and *C* axes, then sends that information back to the CPU.

Respond

1. On a sheet of paper, roughly sketch the five components of a CNC drive system–identify each component by name. Check your sketch against Fig. 17-1.
2. Describe in fewer than 10 words:
   A. Direct feedback
   B. An open-loop drive (critical thinking question)
   C. A servo motor
   D. An axis drive relay
3. Describe in fewer than 20 words how a ball-screw aids CNC accuracy.

Critical Thinking

4. When during a cut an axis drive is suddenly loaded, the MCU senses the lagging progress and adds extra energy to the drive to compensate. Is this statement true or false? If it is false, what will make it true?
5. Some budget controls lack ______________; therefore, commands are not verified by the CPU.

Unit 17-2 Industrial Machining and Turning Centers

Introduction: In Unit 17-2 we'll take a look at the kinds of machines you might find on the job. We'll look at the features and accessories of modern machinery, many of which may not be present among tech school lab equipment due to their expense and because they consume far too much material per hour to fit into a training budget! Just a few examples; I could not keep up with advancements if I tried.

TERMS TOOLBOX

Bar cutter/loader A turning center accessory that saws bar stock into blanks ready to load, then turn.

Bar feeders A turning center add-on that contains a long spinning bar and advances it when a new length is required.

Bar loaders A turning center accessory that places new bars of turning stock into the bar feeder or bar cutter.

Chip-to-chip (CTC) cycle time The time between cutting with one tool, then a second tool.

Downtime Nonproductive time when no work is being done.

Live tooling A lathe tool that can spin, making milling and drilling operations possible on turning center jobs.

Pallet changer A machining center with twin workstations–each able to rotate into the work envelope while the other is outside being loaded with new parts to be machined.

Part-to-part cycle time The time between finished parts in a multipart run.

Process control robot (PCR) An in-process inspection robot that can measure size during a program. It can also feed back data to the control to adjust for variation.

Qualified tooling Cutters inserts and holders that can be replaced with very little variation from one to the next.

Random tool storage The ability to store an outgoing cutting tool into any tool drum position, with its tool number and new position being tracked by the control. This improves CTC time.

Tombstone A vertical square work-holding fixture often used on pallet-changing machining centers. The tombstone allows access to three sides of the workpiece in a single setup.

17.2.1 Machine Evolution

The drive toward the perfect machine is motivated by four factors:

1. **New products** Not possible previously
2. **Error elimination** Remove human intervention variability; thus, reduce scrap
3. **Increased production** Reduce labor cost and lead time
4. **Higher quality** More reliable reproduction

To clearly see into the future, we take a brief look backward. While it's enriching to know our trade's family tree, this information has a greater purpose: to deepen understanding of the reasons behind machine evolution.

The Big Clues

The first stride up from manual machines began because there "had to be a better way"! But at a certain point, the drive changed from doing what couldn't be done to doing it better and ever more quickly. Also notice in the timeline that soon after the machines were born a big branch of the changing technology had to do with programming systems–where and how they were written. Note, too, that the person, place, and method of writing a program have changed several times in the timeline, and will predictably change again in the future! The next phase might lay the job back in your hands, skipping the programmer. Here's an overview of unfolding dates.

1946 **Aerospace spin-offs following World War II require complex shapes**. To machine helicopter rotor blades, John Parsons's Corporation experiments with tables of *X–Y–Z* movements called out to one human operator at each axis crank of a manual mill. You can guess how that went?

1949 **Parsons set up a demonstration of his early solution for the U.S. Air Force**. It showed it was possible to automatically generate complex shapes using his *X–Y–Z* data as a program that drove the machine axes.

1952 **First true programmed machine**. Joint effort put forth by the Air Force, Parsons, and the MIT Servo-Mechanisms Lab yielded the first true programmed machine tool. It worked! Driving a Cincinnati Hydrotel®, the electrical relay cabinets were larger than the machine tool itself (Fig. 17-6), but it worked!

1955 **Tape drive machinery is born**. The Air Force awards a $35 million contract to produce 100 numerical control (NC) machines for aerospace manufacturing. Machinery is reserved for military purposes. Note, there's no "C," just NC.

Figure 17-6 The first programmed machine tool had larger electrical cabinets than the machine!
Courtesy MIT Museum

KEYPOINT

These machines have no memory; they follow numeric programs on punched tape or cards, a lot like a music box.

1960 **Private industry gets NC machinery**. Machine tool manufacturers begin supplying simple two-axis tape drive drills, lathes, and milling machines, to nonmilitary industry. Machines begin the ascent toward the "best machine." Programming remains a matter of calculating and entering commands by hand.

KEYPOINT

At this stage, the time required to write the program is usually many times longer than the time needed to run the program! Math challenges and entry errors are ever-present challenges.

1960 **Standardized code word vocabulary for programming machine tools becomes necessary**. Organized by the Electronic Industries Association (EIA), RS-267-B code vocabulary is born. Programs are written by highly skilled specialists with an in-depth math ability.

1963 **Word routine, automatic programming language (APT) is born**. Using subroutines in a language called APT, programming can be done by writing word statements such as "*goto PT 1 [point 1]*" or "*goaround R2.5 [2.5 radius]*," then feeding that into large mainframe computers to calculate the coordinates.

Owning a mainframe is beyond most budgets, other than the largest corporations! So, if using APT, we use teletype machines (tape-reading typewriters that can also punch new tapes) to send raw APT programs to the mainframe located several states away, which then returns the finished program commands! *There has to be a better way to program machines!*

1970 **CNC is born**. Machines with RAM (random access memory) become commercially available. Programs can be edited and written right at the CNC machine, but that means it isn't running while it's being programmed.

1975 **Desktop PC computers become affordable and programming becomes a simpler, more affordable issue**. Programs can be written for many different machines, using a single PC workstation. Desktop APT becomes a reality and programming gets a big efficiency boost. CAD drawings begin. **CAM begins**. Experiments begin by creating software that solves some of the programming math for *X–Y–Z* coordinates. As the math solution software becomes ever more capable, it becomes obvious that the software might be able to write programs directly from the part CAD image template. However, due to computer software expense, CAD drawings are rare. Programs are still written by specialists, but far less math is required.

KEYPOINT

Note that at this point, evolution is on three ladders, each affecting the other: CAD/CAM software systems, PC chips, and the physical machines, each on their own path each driving the others and improving the process.

1980 **Full-functioning graphic programming**.

Affordable CAD desktop software.

Chip technology increases calculation speed. Multitasking CNC machines.

Graphic programming systems become common on machine controllers. Some programming is performed on the shop floor on the control by the machinist. Onboard graphic programming systems add expense to CNC controllers. Most new machines have their own CAM system. "There has to be a better way."

KEYPOINT

Now, the physical machines begin to change rapidly due to the CPU's ability to multitask.

1990 **PC-based CAM systems move most programming away from operators to the office environment**.

Solid CAD drawings become more common (Unit 17-3).

1995 **CAM systems operate on solids**. Wireframe drawings (Unit 17-3) are still used, but solid drawings begin to make gains in industry.

2000 **Automated CAM software takes solid drawings into programs automatically**.

Today **PC-based controls and lightning-fast programming based on solid models make shop floor programming by the machinist a possibility once again**.

17.2.2 Machine Efficiency Today—Challenges

From the mid-1980s, machine evolution has been mainly driven by the need to improve efficiency (make more parts per hour). To compute manufacturing costs, machine time is divided into two broad categories: *uptime* (making chips), also called *WIP* (work in progress), and **downtime** (not making chips). Due to their great cost and potential for profit, CNC machines must deliver uptime production in the area of 90 to 95 percent, to pay for themselves, plus wages, overhead, and profit. Uptime is challenged by five kinds of nonproductive downtime that drive machine evolution and the shop team to solve.

1. **Part-to-Part (PTP) Turnaround (Cycle Time)**

   This is the time required to remove the finished part, then reload a second. Efficient cycle times include cleaning the setup of dirt and chips and carefully reinserting and clamping or chucking the next part. Well-thought-out operator actions are a key element in this battle, but the machines themselves have been changed to minimize PTP time, too, as we shall see. CNC evolution has created a whole group of part-loading options: bar loaders and feeders for lathes, pallet-changing systems for mills, and robot part-loading arms for both kinds of machines; plus in Unit 17-4 we'll look at how the machines now inspect their own work.

2. **Ongoing Maintenance Downtime**

   This involves replacing dull cutters, maintaining lube and coolant levels, and performing other tasks that do not make parts but must occur, even though they slow production. If safe, many of these tasks are best dealt with while the machine is operating but not all. Some can be performed safely only with the machine stopped. There are some computer solutions to this downtime, too. When machines have the tool capacity, two tools with the same shape and size can be put into the tool storage drum. Then after using the first for a given number or cycles, the machine automatically switches to the second sharp tool. At that time, the first can be removed from the setup to be sharpened or replaced. This is known as *scheduled tooling*.

3. **New Setup Turnaround**

   These are tasks required to get a new setup up and running. Writing and inserting a new program; unbolting the old, then aligning and bolting the new work-holding tools; and inserting new cutters in the turret and trying out the program on a first part run are just three of the many machine nonproductive turnaround times.

4. **Chip-to-Chip (CTC) Tool Change Cycle**

   This is the time required to change cutting tools during a program. It includes stopping the spindle, rapidly traveling the machine to a tool change position, exchanging tools, starting the spindle again, and repositioning it back at the work with chips being removed once again. The Haas SS machines feature tool changes at lightning speeds.

5. **Inefficient Programs**

   This category is complex. It includes unneeded traveling of the cutter without touching metal (an "air cut") or changing tools or moving parts between setups too many times. Later we'll examine some ways to find these time bandits and how to get them locked up through smart editing of programs.

KEYPOINT

Today's machine evolution is driven by the need to maximize WIP.

So, now we investigate the turning and machining center functions and accessories that trim away nonproductive time.

17.2.3 Vertical Machining Centers

Unarguably the backbone machine in industry today, the vertical machining center (Fig. 17-7) is capable of changing its tools and parts, and as the word *center* indicates, multioperational work can be finished on them. They can thread internally and externally, bore, drill ream, and mill round objects that would have been machined on a lathe in the past.

Figure 17-7 The backbone of today's CNC shop, a vertical spindle machining center (VMC).
Courtesy HAAS Automation, Inc./Oxnard, CA

Machining centers are able to machine three-dimensional shapes when they are programmed with CAD. If we add an A-B head and/or a rotary table (4 and 5 axes), vertical CNC mills can machine nearly any shape that can be drawn on a computer screen.

Tech-Specs for Vertical Mills—Machining Centers

Here are the capabilities these machines bring to the workplace.

Tool Changing

Most machining centers can store, sort, and automatically change a minimum of 16 or 18 cutting tools, up to approximately 36 within standard equipment. Some feature optional tool carriage extensions or conveyor systems that hold an unlimited number of cutting tools.

Chip-to-Chip Time, CTC Cycle Time, or Sometimes Called Tool-to-Tool (TCT)

The speed with which these machines can withdraw from cutting metal to change then return to fully making chips with a second cutting tool, is called the chip-to-chip tool change time. Cycles of from 2.5 to 10 seconds are common. The cutting tool flange is gripped by a robotic arm shown in Fig. 17-8. This efficient tool arm is known as an end-for-end changer. It has the incoming tool in one gripper, on one end of the arm, and has gripped and pulled the outgoing tool from the spindle at the other end.

Figure 17-8 An end-for-end robot tool change arm pulls the outgoing tool and has the incoming on the opposite end ready to load into the spindle.
Courtesy Mike's Machines/Arlington, WA

CNC Tool Management

Most modern CNC tool management systems perform **random tool storage** (older CNC machines do not). Random storage means that any tool may be restored at the closest storage drum hole, or tool storage conveyor position, at any time during the program run (Fig. 17-9). At the first insertion, the operator informs the control that a given tool number is in a particular drum position (Fig. 17-10). Then the control keeps an accounting of where each tool is located in the drum after each tool change. The point is, the drum needn't rotate to put the tool away in the same hole each time. After use, it places the tool in the first available open hole.

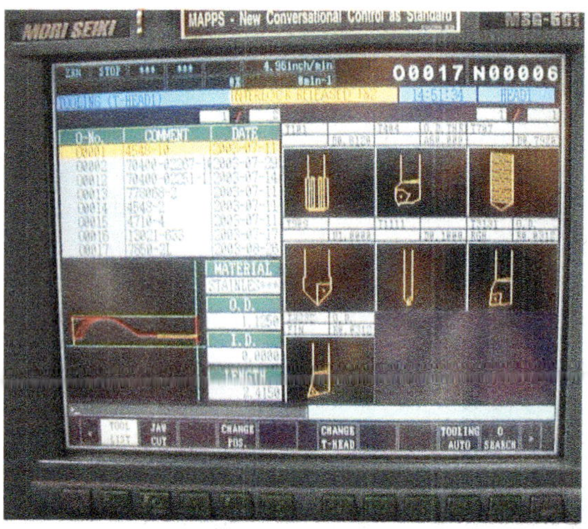

Figure 17-9 Tool information stored on one page of the controller tool size, tool offsets, and current location.
Courtesy Tyee Aircraft/Everett, WA

Figure 17-10 A typical tool storage drum on a machining center stores up to 36 tools (varies with machine).
Courtesy Bridgeport Machines/Bridgeport, CT

KEYPOINT

In setting up autoload cutting tools in a machining center, there are two numbers that must be coordinated: the *tool number* assigned by the programmer (found in the setup document) and the *tool position number on the machine's tool drum,* chosen by the operator at setup.

Once this tool load coordination is performed, the computer tracks and moves the tools for efficiency.

KEYPOINT

Older CNC machines with tool changers may require the tool number to match the tool storage drum position, called an assigned location. On these machines we try to put tool 1 in hole 1 if it's possible.

TRADE TIP

Drum Rotation as Soon as Possible Smart programmers write a command to rotate the tool drum or conveyor to the next machining center tool to be used before it's needed. For example, while cutting with tool 3, the drum is commanded to rotate to tool 4. Then, when a tool change command is issued, the tool is in the change position ready and waiting to improve chip-to-chip time.

CNC Cutter Holders

If the CNC mill has no automatic tool changer, then standard mill tapers and flanges can be used to hold cutters in the spindle. However, to facilitate autotool changing, cutters must be mounted in holders with three modifications compared to manual mill tool holders (Fig. 17-11).

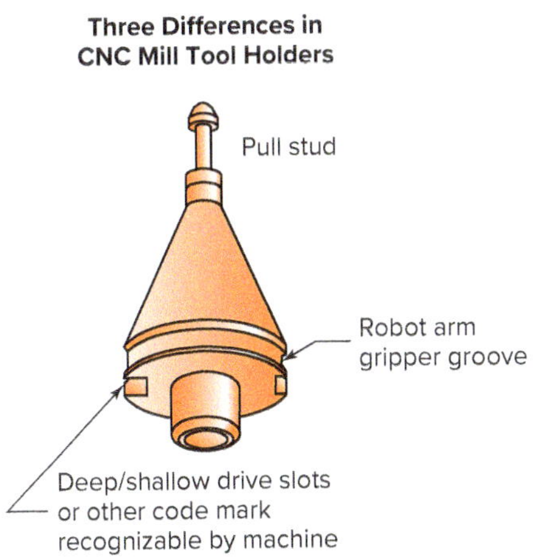

Figure 17-11 CNC mill cutter holders have three modifications from manual mill holders.

In many cases, tool holders are interchangeable between machines. But there are some differences between the following tool-holder features. Although undesirable, some shops must use a variety of tool holders based on the tool changer arms on various machines. In these situations, it's dangerous to mix holders between machines.

1. **Spindle Orientation Coding**

   Spindle orientation has two purposes: to ensure that wide tools don't interfere with each other in the storage drum and to stop the spindle in a given position during machining. Spindle codes can be a dot or bar code on the flange, seen by the robot. It is often a difference in the drive slots and lugs, one being deep and the other shallow (see Fig. 17-11). Stopping the cutter at a specific rotational orientation during a program can be useful. For example, withdrawing a boring bar from a finished hole can scratch the surface if it's pulled straight up. However, if the programmer commands the machine to orient the cutting tool tip with respect to the hole's axis, he or she can add a tiny *X* or *Y* axis retract away from the wall, then pull the boring bar up with no mark.

2. **Automatic Draw Studs**

   The CNC tool change cycle includes an automatic tightening into the taper. The CNC holder is fitted with a draw stud as shown in Fig. 17-11. Many modern CNC machines include an interlock sensor that will not allow the spindle to start unless the tool is correctly nested in the spindle.

3. **Robot Grip Slots**

   The flange rim is modified with a groove machined to nest the gripper claw of the robot arm. There are a couple of different patterns for this and not all tool changers require this modification.

KEYPOINT

There are variations of each of these three modifications. Be certain the tool holder you intend to use fits the specific CNC mill tool changer.

17.2.4 Work Pallet Changing

To reduce cycle time, machining centers are fitted with twin rotating work surfaces on which *pallets* are located. A pallet is similar to an interchangeable mill table or **tombstone** (a vertical work mounting surface). This feature may not be present on school machines as it is an expensive option. However, in production machining, a **pallet changer** greatly reduces part-to-part cycles, plus improves operator safety. The ingoing part is preloaded and ready to rotate in when the current part being machined is done. Pallets are usually exchanged in 15 seconds, chip-to-chip time (Fig. 17-12).

Figure 17-12 A pallet-changing machine speeds part-to-part cycle time.
Courtesy of AMT Senior Aerospace

Preset Pallet Fixtures (Tombstones)

Additionally, to reduce setup time, CNC shops might own several pallets that can be readied before they are needed. The new setup can be assembled on a pallet off the machine, then quickly exchanged when the new program is loaded.

17.2.5 High-Speed Accurate Spindles

Most small to medium CNC machining centers feature spindles capable of higher RPMs than standard milling machines. From 10,000 to 15,000 RPM is common, but high-speed machining (HSM) spindles achieve revs at 20K or faster. These spindles accommodate the high RPM demands of small cutters and/or high-volume metal removal of materials such as aluminum.

Temperature-Controlled Spindle Oil

To maintain stable accuracy and long life lubrication, many machining centers feature temperature stabilized bearing oil. As the work forces heat the spindle oil, it is circulated through a cooler/filter before reentering the bearings always at the same thickness (viscosity). This not only stabilizes accuracy, but it also imparts long life to exceptionally high-quality bearings (Fig. 17-13).

Figure 17-13 Spindle oil is cleaned, cooled, and recirculated by this stabilizing unit.
Courtesy Smiths Aerospace-Actuation Systems/Yakima, WA

17.2.6 Coolant Systems

Chip ejection and cutter lubrication become even more important at the high removal and spindle rates used in CNC milling. To meet this challenge, machining centers include high-volume coolant systems. Some feature more than one programmable outlet such that coolant can be directed from two or more sides of a cut. Coolants are often "blasted" more than flooded over the cut area (Fig. 17-14).

Programmable Coolants

Some industrial CNC machines feature coolant nozzles that can swivel in two axes. Using a joystick or two axis knobs, the operator aims them during the first run of the program. The coolant positions are attached to the tool number memory, then recalled during subsequent runs of the program.

Figure 17-14 Twin coolant streams flood this end mill cut.
Courtesy Haas Automation, Inc.

Mist Coolants

A second type of coolant is the air-driven mist. Combining both air and coolant streams, mist coolant not only blows chips away, but it also often reaches deep into pockets and bores where flood coolant is driven back by flying chips (Fig. 17-15).

Figure 17-15 Mist coolant is often more efficient than flood.
Photography by Prographics at Milwaukee Area Technical College/McGraw Hill

At the correct time the programmer might call on mist or flood coolant. But more likely, these decisions will be the machinist's.

KEYPOINT

If program editing is allowed by management, coolant improvements can be edited into the program at tryout time.

High-Pressure Coolants

Usually an added accessory, many machining tasks can be improved if the coolant can be injected forcefully at or into the cut. Figure 17-16 shows a high-pressure coolant system that injects a strong enough stream to remove wire burrs from aluminum!

Figure 17-16 High-pressure coolant can be forced into deep pockets and holes or through the tool.
Courtesy ChipBlaster, Inc.

Through the Tool Coolants

Technology borrowed from the gun drilling department, high-pressure coolant can be injected through special cutters featuring central galleries to channel the coolant right to the cut area. They help force chips up out of deep holes and get to the action where they are needed (Fig. 17-17).

Figure 17-17 Using special cutters HP coolants can be injected through the tool to the point of greatest need.
Courtesy ChipBlaster, Inc.

17.2.7 Automatic Safety Features

Autodoors

A common feature often found on CNC turning centers as well as machining centers is the automatic closing safety door (Fig. 17-18), many with sensors to detect and protect anything in their path. These doors are more prevalent on pallet-changing machines since they require larger front space to pass the rotating pallets, but they are also found on standard vertical machining centers and turning centers.

Figure 17-18 Automatic safety doors are found on many newer *lathes* and *mills.*
Courtesy Contour Aerostructures/ Everett, WA

KEYPOINT

Containment screens and doors found on machining and turning centers not only stop flying chips; they also retain expensive coolants in the machine.

Remote Monitoring

Using replaceable video cameras rather than human eyes, the operator can display the machining action on the control screen (Fig. 17-19) as opposed to putting their face in harm's way. Here we see a split screen with program and distance to go screens plus the actual machining. This is useful where safety guards, flying chips, and coolants obscure visibility from the operator side of the machine.

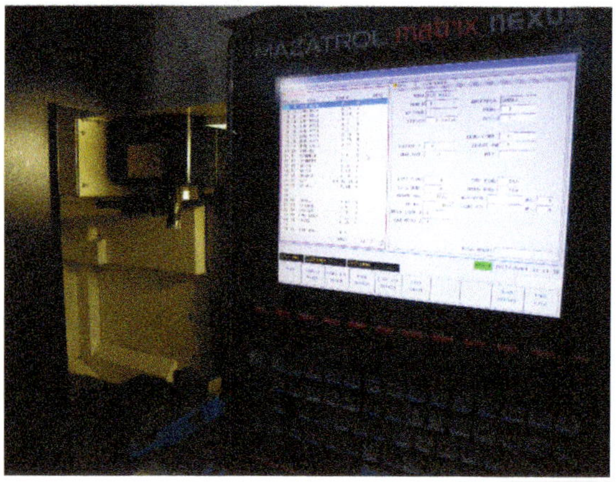

Figure 17-19 Many monitoring control options can be arranged on-screen.
Courtesy of AMT Senior Aerospace

Automated Chip Conveyors

High chip volume, coolant containment, and fast-moving machinery make it difficult to remove chips by hand. Automated conveyors are added to machining and turning centers to serve the purpose. A chip conveyor not only aids ongoing maintenance but also improves operator safety by making it unnecessary to reach into the machine to clean away chips. As shown in Fig. 17-20, the chips are sent to a cart that can be taken to recycle bins when it's convenient. Larger systems send their chips directly to dumpsters, to then be handled by forklift trucks as part of the factory's recycling program. Where extreme chip volumes are common, conveyors take them away to "puckers," machines that compress them into tight bundles to be sent to the recycle facility.

Figure 17-20 For high-volume machining, chips are conveyed away by automatic systems.
Courtesy Okuma America Corporation

Gated Control for Peripheral Devices

The next two advancements aren't machine accessories, but rather, they are programmed machines themselves. They work in conjunction with one or more CNC machines. They load or inspect parts at appropriate times, using their own internal controllers and programs. When it's time for their assistance, the CNC control halts progress, then signals them to proceed–it opens the gate for the go-ahead. Completing their task, they signal the CNC control, which then closes the gate switch to be certain of coordinated activity. The CNC control is the master control for gating.

Automated Part Loading (Robotic)

Where work is heavy or the turnaround demand is constant and/or fast, part-loading robots can improve cycle times, reduce operator strain and boredom, and improve safety. When placed in a cell of two or more machines as shown in Fig. 17-21, they can keep groups of machines working without human intervention. NOTE: The human barrier has been removed in this photo, which prevents people from entering the danger zone. Some shops use beams to halt robot action when someone crosses into the danger area

Figure 17-21 This robot efficiently and safely moves heavy engine parts between CNC machines.
Courtesy Okuma America Corporation

Automated Part Inspection

When in-process part quality is critical and inspection must be performed at special times within the PTP cycle, the **process control robot (PCR)** also greatly enhances PTP cycles and quality (Fig. 17-22). Results from the PCR are fed back to the control to enable automatic adjustments to cutter offsets to maintain tolerances.

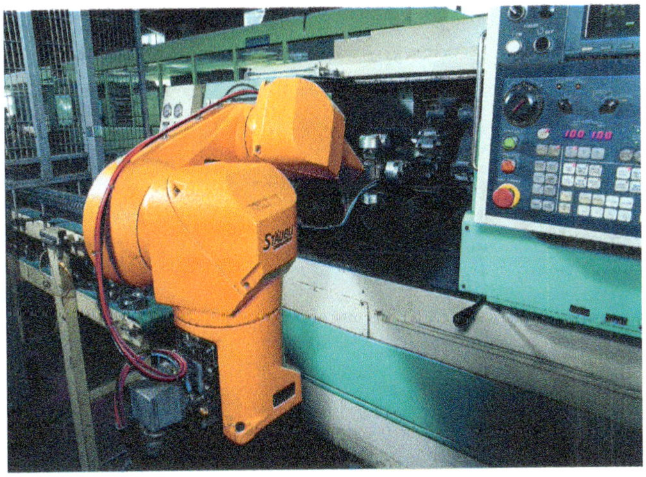

Figure 17-22 This production control robot (PCR) automatically inspects critical operations during the cycle.
Obuljen & Stäubli Robotics

Automated Cutter Offsets

This popular accessory has become a near standard on both machining and turning centers. When they are first loaded into the machine, the exact cutter diameter and length must be entered into the control's tool memory.

While a program is written with a certain cutter length and diameter in mind, the actual tool needn't be an exact match, just close to the one for which the program was written. If the capable control is informed of the real cutter size and shape, it can compensate for the difference between programmed and actual size.

These differences are called tool offsets and loading them into the control is an operator responsibility. Using offsets saves time and money since any good, resharpened cutter can be used. The offsets must either be determined by hand measurement and then the data entered by hand or can be automatically determined by probes or optical sensors.

Operator Warning Lights—Yellow Announcement and Red Warning

To aid operator monitoring, many CNC controls feature bright flashing lights placed high for visibility. If enabled, the yellow signals the need for operator intervention such as for a clamp change or the end of the program. This feature allows operators to manage more than one machine within a cell. The red light indicates something is seriously wrong such as low coolant, a program fault that cannot be processed, or a servo error (Fig. 17-23).

Figure 17-23 Operator beacons warn and inform of major events.
Courtesy Mike Fitzpatrick

17.2.8 Tech Specs for CNC Turning Centers

Since lathes had already evolved into very sophisticated automatic versions of turret lathes, multiple-spindle lathes, and screw machines, long before programmed machine tools came on the scene, the CNC version represents a lesser degree of evolution than milling machines, up from their manually operated ancestors. Still, they are changing rapidly today. Many of the cycle time and safety improvements discussed for machining centers are standard or available on turning centers as well:

- Chip removal conveyors
- High-volume coolant systems with low-level warnings
- Automatic safety door closure with interlock switches
- Video monitoring systems on split-screen controllers
- Automatic cutter offsets where, by touching the cutting tool to a sensing probe (Fig. 17-24) or passing it through a laser vision system, the control stores the size and position relative to the setup
- Individual part-loading robots

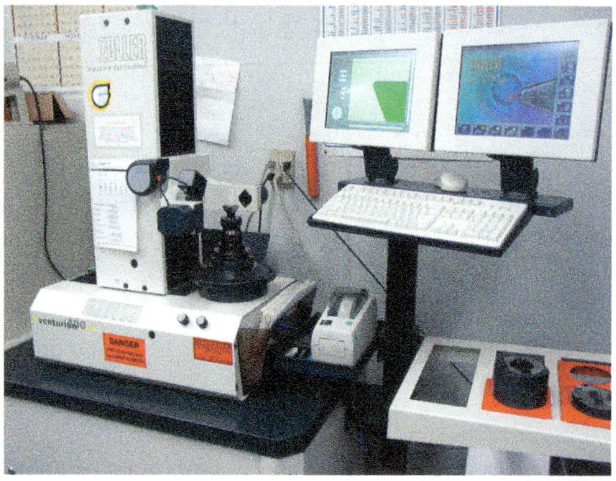

Figure 17-24 Tool offsets can be determined by laser vision or touch probes.
Courtesy of Aerospace Manufacturing Technologies (AMT)/Arlington, WA

17.2.9 Automatic Bar Feeding

For machining many parts from a single long bar of material, the **bar feeder** pays for itself, often within a month or two. As shown in Fig. 17-25, the bar feeder supports and safely contains a rapidly spinning bar, up to 20 ft long.

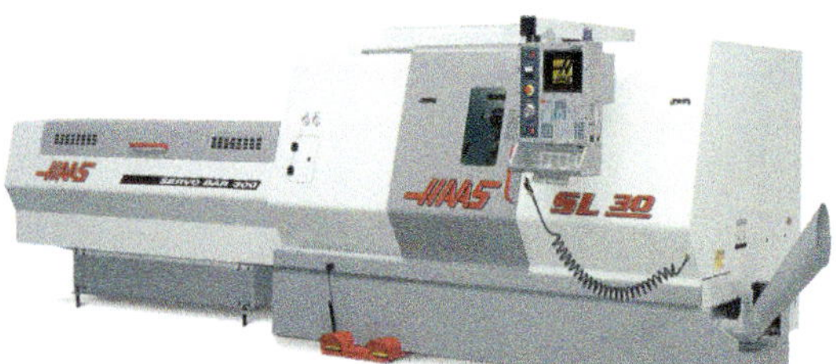

Figure 17-25 A bar feeder (left side of the machine) safely contains and advances the long bar.
Courtesy Haas Automation, Inc.

Upon parting off the finished work into the part catcher (another automated accessory), the bar feeder pushes the bar forward an exact amount to make another part.

Bar feed is accomplished in two different ways. On better feeders, the bar is located forward an exact distance by the feeder advancement. On less expensive versions, the bar feeds forward until touching a programmed stop mounted in the tool turret.

Nonstop Feed

If collets are used to hold the work, using both types of bar feeders, the spindle need not be stopped to advance the work unless it protrudes very far from the spindle where part whip would be a problem.

Through the program, without stopping the spindle, the collets open and the feeder advances the material. This ability allows long unattended operation, freeing the machinist to perform other duties such as writing future programs or burring parts.

Changing Acceleration Loads

Bar feeding is by far the more common way to automate long runs but it has one problem: If the job requires RPM changes, the entire bar must be accelerated and stopped. The answer is the bar cutter changer (Fig. 17-26).

Figure 17-26 A bar cutter loader takes the strain off the spindle motor by placing shortcut parts in the chuck.
Courtesy Tyee Aircraft/Everett, WA

Sometimes when the work is closely toleranced, the differences in spindle acceleration of a long bar that become continuously shorter with each part can make a difference in final shape, finish, and size. Speeding up and stopping long bars also load the spindle motor. To help with that, some bar feeders offer part rotation drives to assist the spindle but even though the bar is rotating in an oil bath, all vibrations caused by the long bars in the feeder cannot be damped. The bar cutter is the answer.

Automatic Bar Cutter/Part Loaders The **bar cutter/loader** is an automatic accessory that is a saw and loading robot combination. The saw cuts the work into slices, then automatically loads them into the stopped spindle. The controller opens the chuck or collet, then sends a gate signal to the loader to pull the present part and load a new blank.

Automatic Bar Loaders If the production run requires many lengths of bar to be consumed, in both a bar feeder and a bar cutter, an automated **bar loader** is the answer (Fig. 17-27). It holds and inserts one bar at time into the bar feed system. Using this feature, it can be literal shifts before the machine requires human attention.

Multiple-Tool Turrets Most new CNC turning centers feature multiple-station tool turrets (one or two opposing turrets). Many different cutting tools can be mounted on the turret as well as individual coolant nozzles (Fig. 17-28).

Programmable Tailstock Many but not all turning centers use a programmable center support. The addition of a tailstock depends on the kinds of products being made (Fig. 17-29). If they are long, an autotailstock is a good addition.

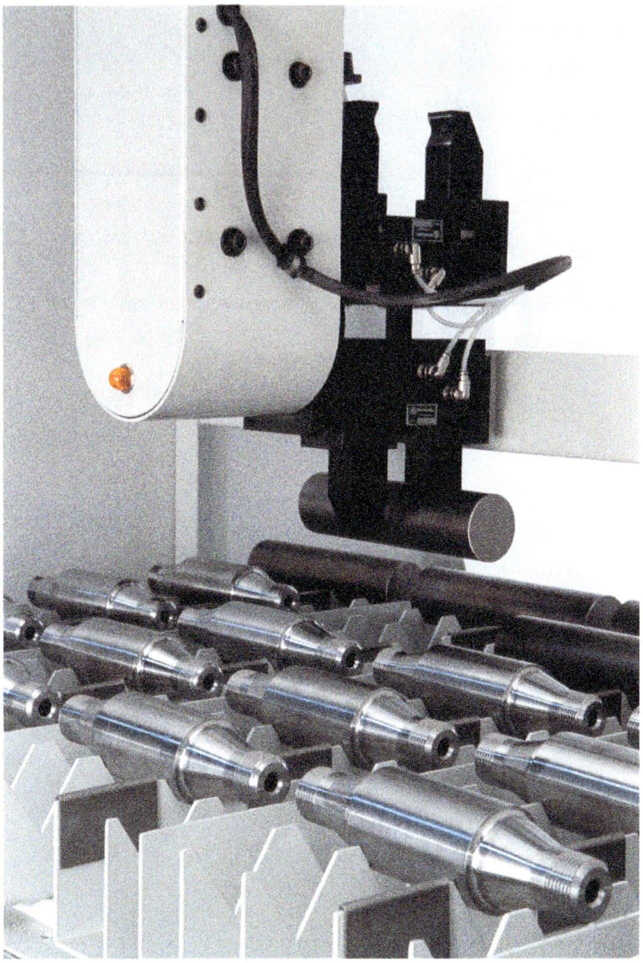

Figure 17-27 An automated part-loading arm.
Courtesy Haas Automation, Inc.

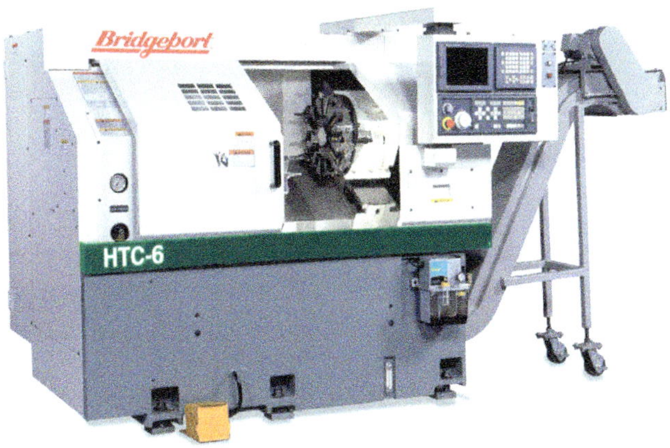

Figure 17-28 This standard tool turret can accept from 10 to 18 tools depending on size.
Courtesy Bridgeport Machines/Bridgeport, CT

Live Milling and Drilling Tooling When specially equipped, some CNC lathes can spin mill cutters, drills, and reamers from the tool turret. This is accomplished using hydraulic powered motors. Figure 17-30 shows a lathe drilling a hole using **live tooling**.

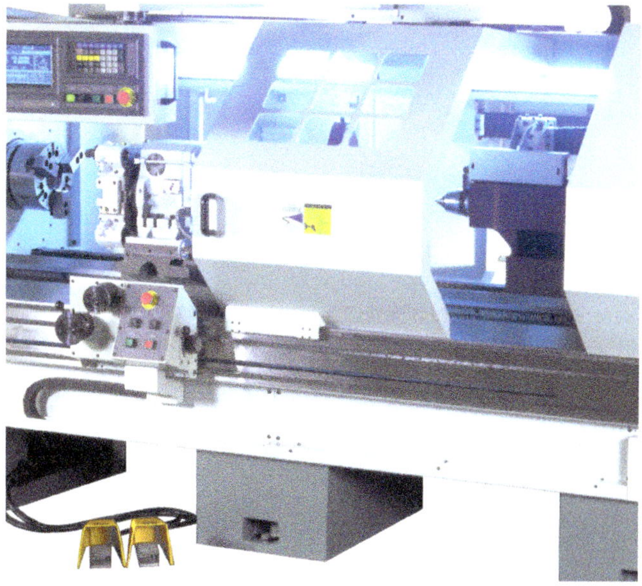

Figure 17-29 A programmable tailstock (right end) is needed when work is long.
Courtesy Bridgeport Machines/Bridgport, CT

Figure 17-30 Live tooling accessories allow milling type cuts using lathes.
Courtesy Tyee Aircraft/Paine Field, WA

Some newer lathes feature a third axis of motion (*Y*) for the turret. That allows milling of shapes off center to the workpiece. These machines can complete an amazing array of part shapes (Fig. 17-31).

Hydraulic Chucking

CNC turning centers include automatic collets and chucks that speed up PTP turnaround time. Hydraulic-powered chucking ensures fast, consistent grip from one part to the next when compared to manually tightened chucks. The amount of grip pressure is adjustable.

Figure 17-31 This CNC turning center features a *Y* axis to allow curved mill cuts.
Courtesy Tyee Aircraft/Everett, WA

TRADE TIP

The close-open-close chuck joggle trade tip learned in manual lathe training will be useful here, too. When just placing the work in the three-jaw chuck and closing it, the work can dangerously be pinched off center between only two jaws. To avoid this problem, after gripping the work, the chuck is opened a bit, then reclosed while jiggling the part. It will usually find the true center.

Safety Foot Pedal Switches

For safety and convenience, hydraulic chucks are usually activated by a foot pedal, so the operator can use both hands to support the work. Using a combination toe lift plate, then a down-activated switch farther back beyond that, the chuck cannot be accidentally opened at the wrong time (Fig. 17-32). Safety features are also built into the program and control to ensure that the chuck cannot open during a cut.

Qualified Tooling

Toward the goal of more WIP time, the CNC lathe takes quick-change tooling to new levels of efficiency and accuracy with **qualified tooling**. Improvements in machines and cutter tooling combine for faster cycle and setup times.

The turret is made such that sturdy tool holders can be quickly mounted and any replacement holder within the same system will position in exactly the same position relative to the work and the turret. Additionally, the tools themselves feature close tolerance, replaceable inserts. These features mean quick exchange of tool holders and cutting tip inserts with little disruption to production. The alignment pins of this turret are shown Fig. 17-33.

Figure 17-32 Safety chuck foot switches are interlocked through control gates and double trip foot action to prevent accidental opening at the wrong time.
Courtesy Okuma America Corporation

Figure 17-33 Qualifying pins on the turret ensure cutter tooling can be replaced with minimum variation.
Courtesy Mike Fitzpatrick

Tool Interference and Turret Capacity

Since lathes must hold in the turret all the tools required to finish a job, two issues arise that affect both program planning and program tryout. First, placing long tools such as drills and reamers next to each other (for example, T001 = Drill and T002 = Reamer) can create a situation where during drilling, the reamer also touches the workpiece, near its rim if the work's diameter is large. It's the machinist's job to test for these situations during program tryout. Be certain the turret withdraws far enough to rotate safely to the next position. Then after indexing to the next tool, slow the rapid approach back to the workpiece to be certain that adjacent tools do not touch the work.

The second issue arises when the program plan requires more cutting tools then can be loaded into the turret. Many cutting tools can be loaded into large turning centers (16 to 24, depending on turret size and the size of the cutting tools) since CNCs feature big turrets. However, on smaller turning centers, where capacity is limited, programmers can turn to multipurpose tools, such as shown in Fig. 17-34. They can face, turn, and part off, as shown, in making this control rod ball end. Multipurpose tools also reduce turret rotation and total cycle time, but they aren't as strong for heavy cuts compared to standard tooling. The decision to use multipurpose tools is the program planner's and must be built into the program and tool numbers must be assigned.

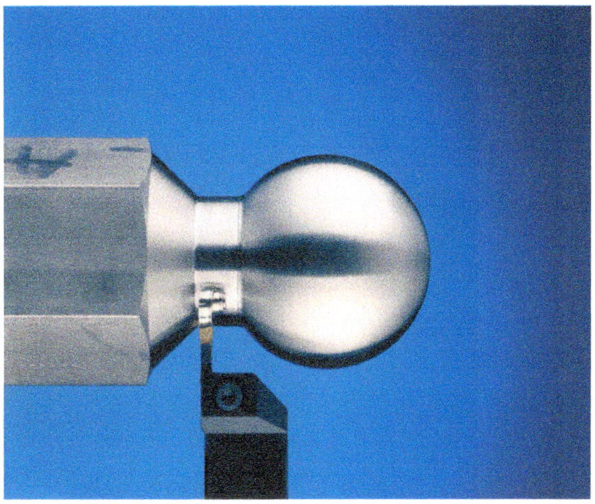

Figure 17-34 Multipurpose cutting tools can accomplish several operations without indexing the turret. They reduce tool inventory and cycle times.
Courtesy Carboloy, Inc./Mesa, AZ

UNIT 17-2 REVIEW

Replay the Key Points

- *NC evolution* began because skilled machinists using manual machinery could no longer deliver advancing aerospace designs.
- *CNC machine evolution* toward faster, more accurate production continues today.
- *CAM software evolution* now carries the torch of programs that machine creative new shapes impossible to machine otherwise.
- When loading tools into a machining center, there are two numbers that must be coordinated: *tool number* in the program and *tool position number* on the machine.
- Accidents equate to downtime as well as hurt workers and disabled machinery. CNC machines feature several improvements to avoid them.

Respond

1. When was the first true programmed machine tool invented? And who were the three partners in the consortium?
2. List as many improvements as you recall, without looking back, that evolved milling machines into CNC machining centers.

   Identify those that require computer direction.
3. Name three major differences between a CNC turning center and a manual lathe.
4. On a turning center, name the accessory that contains long bars and pushes them forward when another length of stock is required.
5. What was the original, main reason programmed machine tools were invented? What drives machine evolution today?

Critical Thinking

6. Extending the timeline forward, what advancements in CNC turning and machining centers might we see in the future for mills and lathes?

Unit 17-3 Multiaxis Machining

Introduction: A multiaxis machine uses rotary axes. So far in this book we have concentrated on machines that move in a linear fashion. In other words, moving from point A to point B. Whether the motion has been in a straight line or a curved arc, the tool path has passed through a series of points along the *X*, *Y*, and *Z* axes. These points have been identified by their distance from an absolute point located on the part. Multiaxis machines, on the other hand, rotate the work or cutting tools in circular motions that are described by rotation around one or more axes. The tool motion is described both by movement in the *X*, *Y*, *Z* axes, as well as by an angle of rotation in one or more rotary axes.

TERMS TOOLBOX

***A* axis** The circular rotation around the *X* axis.

Axial tools Tools that orientate parallel to the *Z* axis

***B* axis** The circular rotation around the *Y* axis.

***C* axis** The circular rotation around the *Z* axis.

Gouging Unintended removal of material.

Inverse time feed rate A feed rate that commands a move to be completed in a specified amount of time.

Linear axes The straight line motion of moving in the *X*, *Y*, and/or *Z* axis.

MRZP (machine rotary zero point) The centerline of the rotary axis.

Multiaxis machine A machine that has both linear and rotary axis motion.

Multiaxis machine configuration The mechanical method of creating rotary axis on a CNC machine.

Radial tools Tools that orientate perpendicular to the *Z* axis.

Rotary axes The circular rotation around the *X*, *Y*, or *Z* linear axes.

Trunnion table A rotary fixture mounted to milling machine table.

17.3.1 Rotary Axis

In Chapter 16 we discussed the six major axes of motion. There are three **linear axes:** *X*, *Y*, and *Z*. In addition to the linear axes there are three **rotary axes:** *A*, *B*, and *C* (Fig. 17-35). The *A* axis the rotation around the *X* axis; the *B* axis rotates around *Y*, and the *C* axis rotates around the *Z*.

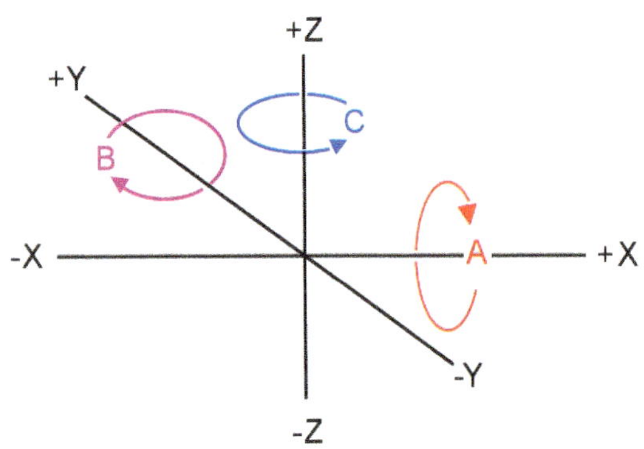

Figure 17-35

For most machining processes, moving in linear motion has been adequate. But in order to gain access to machining features on the sides of our work, we have had to stop the machine, unload the part from the fixture, reposition the part so the side we want to machine is up, and then restart the machine. Each of these steps takes time and risks having the operator misload the part.

Multiaxis machining using rotary axes allows us to program the machine to automatically reposition the work in relationship to the tool. This allows access to the sides of the work that we can't reach when working only in linear axes. For example, in a 3 axes mill, you can only access the work from the top. You can mill the sides with the side of the cutter, but you wouldn't be able to drill a hole in the side without repositioning the work in the vise so that the side to be drilled is facing up. What you'll see in this chapter is that by adding rotary axes, we will be able to rotate the fixture without stopping the operation.

17.3.2 Four and Five Axis Machining

Four Axis Machines

Rotary axes are nothing new. Take the old-fashioned indexing head used by machinists on conventional milling machines. (Refer back to Fig. 11-31.) When installed on a milling machine table, the indexing head becomes a rotary axis. The hex for a bolt head can be milled by setting up the blank in the indexing head, milling one of the flats, then rotating the indexing head 60° to mill the second flat and so on until all six sides are cut (Fig. 17-36).

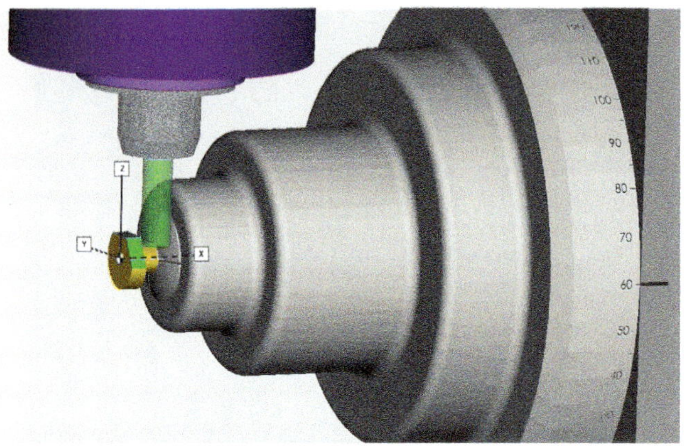

Figure 17-36 Milling flats for a hex bolt

Even though the tool path moves through the identical *X*, *Y*, *Z* locations for each cut, the results are different as we rotate the indexing head between cuts. If the indexing head's centerline is parallel to the *X* axis then we can think of this as rotation around the *X* axis, or on a CNC mill this would be our ***A* axis**.

SHOPTALK

In a machine shop, a 3 axis mill with an additional axis of rotation is called a 4 axis mill. The first three axes are the linear axes, *X*, *Y*, and *Z*, the axis of rotation becomes the fourth.

Look at this example in Fig. 17-37. Here, two holes are being drilled in this shaft. Even though the tool has been drilling identical *X*-*Y*-*Z* coordinates, the location of the two holes are different. The first hole was drilled at A0° (or *A* axis 0 degrees); then the second hole was drilled at A90° (or *A* axis 90 degrees).

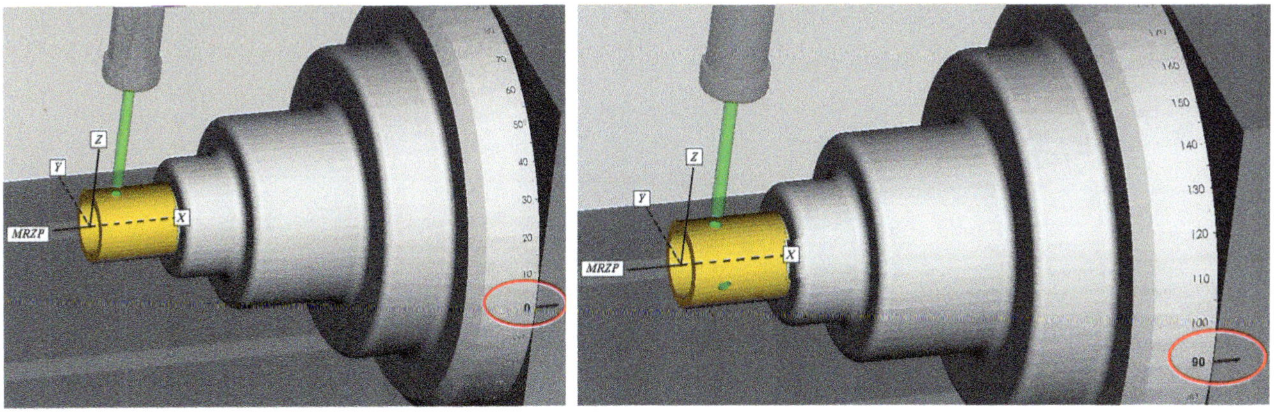

Figure 17-37 First hole drilled at A0° on left. Second hole drilled at A90° on right.

Centerline of Rotation

When programming a linear axes CNC machine, the programmer must decide where the *X, Y, Z* axes originate to establish the program reference zero or PRZ. When programming or setting up a multiaxis CNC machine, the centerline of rotation must be taken into consideration. To better understand the center of rotation we'll use the indexing head example again. Think of the centerline of the indexing head in the same way as the centerline of a lathe chuck. As the stock rotates in the indexing head, the centerline remains in the same place while the part rotates. This centerline is fixed by the mechanical action of the index head and can't be altered. This is the **machine rotary zero position (MRZP)**.

KEYPOINT

The MRZP remains at the centerline of the rotary axis. The part will rotate around the MRZP.

Five Axis Machining

Adding a fifth axis to a milling machine is simply mounting another rotary axis to the *A* axis. In a 5 axis milling machine, there are the three linear axes and two axes of rotation. The two rotary axes can be achieved using different **multiaxis machine configurations**. Here are the three possible ways to create two rotary axes working together in unison:

- Table-to-table: In this machine configuration, there are two rotary axes mounted to the table. One rotary axis is mounted to the other.
- Head-to-table: Another possibility is to have the spindle head rotate around one axis while another rotary axis is mounted to the table.
- Head-to-head: The third option is to have a spindle head that rotates around both rotary axes.

To illustrate how two rotary axes work together, we'll use this example of machining a cube. Figure 17-38 shows a six-sided cube with engraving on each side. The job is to mill the cube and engrave the letters on each side.

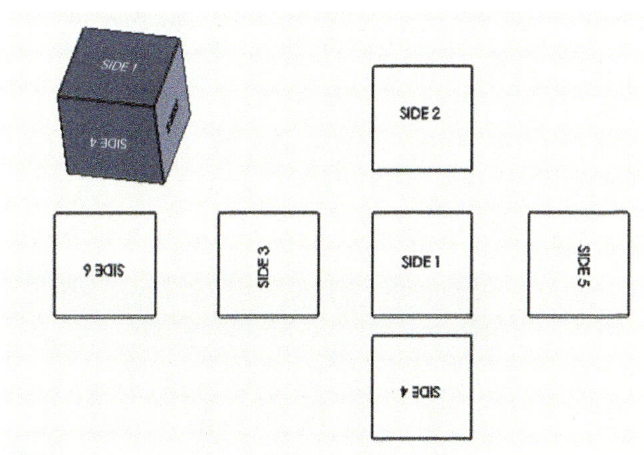

Figure 17-38 Six-sided cube drawing.

We will use a 5 axis mill with a table-to-table configuration. Mounted on the table is a **trunnion table** that will serve as the *A* axis. The trunnion table will rotate around the *X* axis. Think of the trunnion table as a playground swing. The swing rotates around the attachment points or the centerline of the *A* axis. The trunnion table in Fig. 17-39 is rotated 45° in the *A* axis.

Figure 17-39 Trunnion table mounted on a 5 axis CNC mill.
Courtesy Mike Fitzpatrick

Mounted to the trunnion table is another rotary axis that will rotate around the *Y* axis. This will be our ***B* axis**. (Note: Some manufacturers may call this the *C* axis.) These two axes will work in unison to present different sides of the cube to the cutting tool. The MRZP is where both the *A* and *B* axis centerlines intersect (Fig. 17-40).

So let's look at the example of machining the cube. "Side 1" will be up and ready to machine when the *A* axis and *B* axis are at 0°. To machine "Side 2," the *A* axis will be rotated a positive 90°; the *B* axis will remain at B0°. "Side 3" can be machined when the *A* axis is at 90° and the B axis is at 90°.

Figure 17-40 MRZP is where the two rotary axis intersect.
Courtesy Mike Fitzpatrick

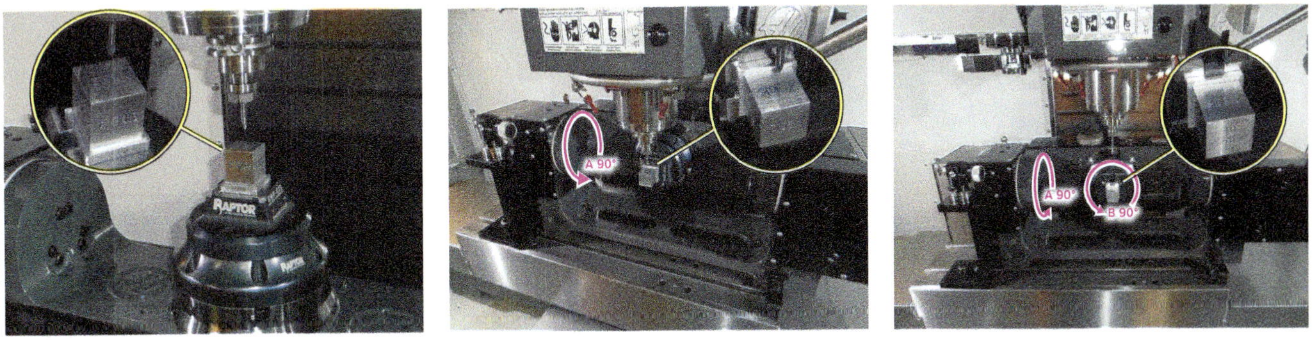

Figure 17-41 Side 1: A0° B0° / Side 2: A90° B0° / Side 3: A90° B90°.
Courtesy Mike Fitzpatrick

Five Axis Machine Rotary Zero Point

Although the part rotates around the MRZP, the distance from the part's features and the center of rotation never changes. However, the *X*, *Y*, and *Z* coordinate locations of the part's features do change as the part moves. Take the analogy of the playground swing: As the child swings, the distance from the child to the attachment point doesn't change, but his or her location in space does. Look at the cube in Fig. 17-41. At A0°/B0°, in relationship to the MRZP, the center of the top face, or "Side 1," is at X0, Y0, Z4.870. The top face is 4.870 in. from the MRZP. When the cube rotates to A90°, the center point of "Side 1" is now at X0, Y-4.87, Z0. Even though the face has changed location in XYZ, the distance to MRZP is still 4.870 in.

KEYPOINT

The MRZP on a 5 axis machine is where both the *A* and *B* axis centerlines intersect. The relationship between the part and the MRZP doesn't change as the part rotates.

Dynamic Work Offsets

Until recently, programmers needed to know where the part would be located in relationship to the MRZP before they could write a program for a multiaxis machine. That information was fed into the CAM software program and the computer calculated the changing coordinates of the part as it was rotated. This process demanded expensive fixturing to maintain this relationship. Any changes to the fixturing made on the shop floor required the programmer to reprogram the part.

Today, because of increased processing power, modern controllers on multiaxis machines can accept an offset between the program reference zero (PRZ) the programmer used and the MRZP. The newer Haas controls, for example, use a dynamic work offset (DWO; Fig. 17-42). When setting up the job, the machinist determines the difference between the PRZ and the MRZP. The difference is entered on the DWO page. During the initial setup, adjustments can be made to the DWO eliminating the need to involve the programmer.

17.3.3 Multiaxis Lathes and Live Tooling

At first it can be confusing to think of lathes having rotary axes. After all, in a simple axis lathe, doesn't the work already rotate in the spindle? The answer to that is yes the work rotates but not in a locational way. The work will rotate at the programmed RPM but will stop randomly. In a multiaxis lathe, the rotation of the spindle will be controlled by an axis of rotation called the ***C* axis**. The *C* axis is an axis that rotates around the *Z* axis, or in this case the centerline of the spindle.

The cutting action in a lathe usually occurs when a cutting tool moves by a spinning piece of work that is mounted in a fixture such as a chuck. For example, the tool is positioned at different *X* axis coordinates and moved along the *Z* axis to cut different diameters.

Multiaxis lathes have the added capability of using live tooling. Live tools are tools mounted in the tool turret that rotate independently of the main spindle. They can be programmed very much like a milling tool using spindle speed and feed per inch commands. This capability is useful when drilling holes that are not on the centerline or when using an end mill to machine features such as flats or slots. When using live tools, the main spindle will be programmed to come to a halt, then programmed to rotate to a desired degree of angle in the *C* axis (Fig. 17-43). The live tool will then rotate at a given RPM and move along the *X* or the *Z* axes.

KEYPOINT

With live tool capability, we can machine features that are off centerline in a CNC lathe.

Figure 17-42 Dynamic Work Offset page on a Haas controller.
Courtesy Haas Automation, Inc.

Figure 17-43 A live drill is preparing to drill a hole cross hole on a C axis lathe.
Courtesy Mike Fitzpatrick

C Axis

Adding a programmable axis of rotation around the centerline of the *Z* axis, in a CNC lathe, allows machinists to drill holes that normally would have needed another operation on a milling machine. Look at the print in Fig. 17-44.

The four holes on the 1.75 bolt circle, shown in Fig. 17-44, would be impossible to machine on a 2 axes CNC lathe, but we can install these holes on a multiaxis lathe. On a multiaxis lathe the diameters and faces are still machined in the traditional way. After these features are machined, the program will stop the spindle and the *C* axis rotates to 0°. The live drill is positioned at X1.750 and moves along the *Z* axis. When the first hole is drilled, the *C* axis will then be rotated so the next hole can be drilled until all four have been completed.

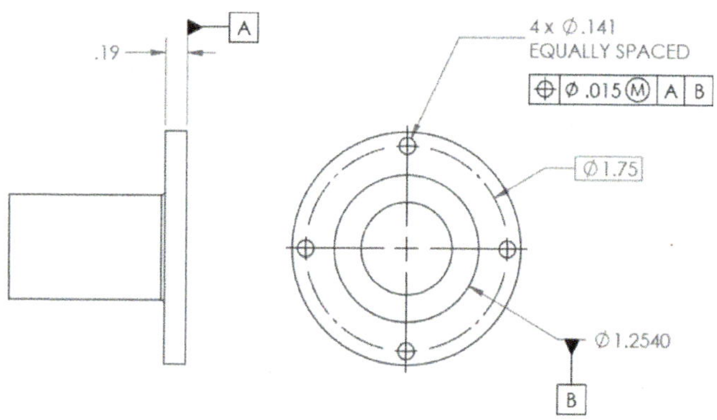

Figure 17-44 Print showing 4 equally spaced holes.

As was mentioned in the beginning of this chapter, two major benefits have occurred in this example. The first is that the need for an additional mill operation to drill the four holes has become unnecessary, saving time and money. Second, because the drilled holes and the turned diameters were machined at the same time in the *C* axis lathe, the relationship between the turned diameters and the drill holes are guaranteed. If the holes had been drilled in a milling operation, the relationship between the hole pattern drilled in the mill and the centerline of the features machined in the lathe would have been dependent on the precision of the fixturing and the skill of the operator loading the part.

TRADE TIP

Although the holes will be guaranteed to be concentric to diameters turned during this operation, the calibration of the live tool's X geometry offset will need to be set correctly for the holes to be on the correct bolt circle. As with any tool, small adjustments are usually necessary during setup.

Axial versus Radial Tools Live tools in turning centers are either axial or radial. The difference is that **axial tools** are parallel to the centerline of the main spindle, and **radial tools** are perpendicular. The drill in Fig. 17-43 is a radial tool. Its centerline is perpendicular to the axis of the spindle. The part in Fig. 17-37 could have been made complete in one operation on a *C* axis lathe. The diameters could have been turned; then radial tools could have drilled the cross holes.

The centerline of axial tools, on the other hand, are parallel to the axis of the spindle. Drills normally can only drill on the centerline of a part because they depend on the main spindle turning. Live tools, however, can drill off center because they spin while the work remains stationary. To drill the holes shown in Fig. 17-44, you will need an axial tool.

17.3.4 Simultaneous Rotary Axis Machining

So far we have discussed rotary axis machining in terms of indexing. In other words, we have described programming a CNC machine to index to a set of axis degrees and then began machining. In the machining industry this is called 3 + 2 machining. Most multiaxis work is done this way, to reduce the number of operations and fixturing needed to complete a part.

But if we need to machine complex sculpted surfaces with a free flowing tool path, the machine would have to move the rotary and linear axes at the same time. Many of you have seen videos of the machining of impellers or complex sculpted shapes on 5 axis machines. A powerful CAD/CAM programming system such as Mastercam is needed to generate these types of tool paths. Often, complex surfaces such as that shown in Fig. 17-45 use ball end mills that travel along parallel paths stepping over very small amounts for each cut.

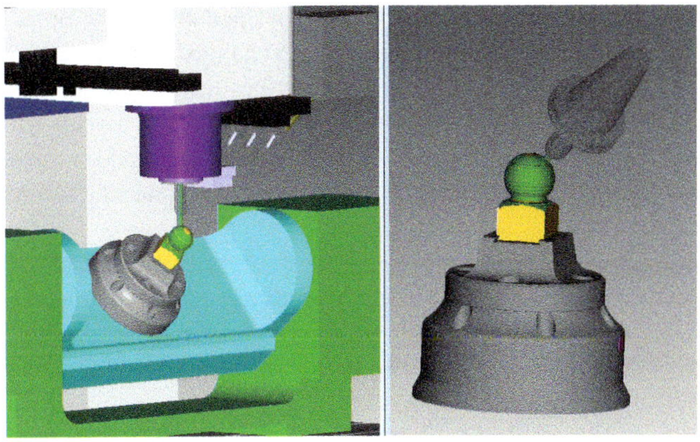

Figure 17-45 5 Axis mill with trunnion table

Heavy machining is usually avoided during simultaneous multiaxis machining due to loss of rigidity of fixtures and tools. Fixture rigidity is lost due to the fact that the braking mechanism of the rotary axes must be off during simultaneous rotary machining. Tools must be long enough to clear the fixtures during the entire routine. One strategy when large amounts of material is required, is to use a 3 + 2 approach first to rough the part out. Then switch to simultaneous machining for the finish passes. In this way shorter tools can be used at any particular index position and the rotary axis brakes are also applied. When taking light finishing cuts the lack of rigidity does not have as much impact during simultaneous motion.

Inverse Time Feed Rate

Finally, let's look at feed rates. On linear axes machines the default description of motion is expressed in units of travel per minute; which would be inches or millimeters per minute. When using rotary axis movement, the units per minute are described as degrees per minute. If the line of command moves the tool along the axis in only linear or rotary distances, that is fine. But if the tool path needs to move in both rotary and linear distance at the same time, it would be possible that either the rotary or linear move is completed before the other. This leads to **gouging** or the removal of too much material in localized spots. Look at this line of code, for example:

```
G1 G94 X2.0 A20. F10.
```

The G94 commands the machine to feed in "units per minute," which is the normal default in most CNC machines. So, the feed rate along the *X* axis will be in *inches* per minute, but along the *A* axis, the rate will be in *degrees of rotation* per minute. Moving at the rate of 10 in. per minute, the *X* axis motion will be completed in 2/10 of a minute, or 12 seconds. Motion along the *A* axis is at the rate of 10° per minute so it will take 2 minutes to complete. The tool path will not flow as the programmer intended and will lead to an undesired contour.

One way to deal with this problem has been to use **inverse time feed rate**, or G93. This type of feed rate commands the machine to complete a move in a specified amount of time. If the G94 in the line of code above is changed to G93 (inverse time feed rate): the 10 in F10. describes the number of times this move could be made in 1 minute. Using the inverse time feed rate means both the linear axis and rotary axis moves will be completed in 1/10 minute or at the rate of 10 times per minute.

KEYPOINT

Inverse time feed rate ensures that both a rotary and a linear move will be completed at the same time.

Tool Center Point Management

Modern controllers have a different way of handling the problem of synchronizing the feed rate with axis rotation. Controllers now can record the machine rotary zero point (MRZP), and then calculate the correct rotary and linear speed based on the feed rate expressed in inches per minute. This eliminates the need for inverse time feed rates. The controllers on newer Haas machines, for example, have dynamic work offset and tool center point control options, which allow the controller to determine the location of the tool tip in relationship to the center of rotation. Based on that understanding the controller can calculate the speed of rotation based on the inches per minute feed rate.

To clarify this point, consider this command:

G1 G94 A360. F50.

Notice that inverse time feed rates are not used in this command. The newer controllers will base the speed of rotation on the length of tool travel around the circumference of a circle with the MRZP as its center to maintain a feed rate of 50 in. per minute. Because the controller knows the distance of the tool tip to the MRZP, it can calculate the distance that will be traveled in 360° of rotation. So, if the diameter of the contour is, say, 4 in., the circumference of the circle will be pi times the diameter or 12.6 in. To maintain a feed rate of 50 in. per minute, the controller will complete the rotation in 0.25 minutes or 15 seconds.

UNIT 17.3 REVIEW

Replay the Key Points

- Rotary axes are moved in degrees of rotation around a linear axis.
- There are three major axes of rotation: *A*, *B*, and *C*.
- Five axis CNC mills have three linear axes and two rotary axes.
- A multiaxis machine's rotary zero point, or MRZP, is the centerline of the rotary axis.
- Live tooling on multiaxis lathes turns independently of the main spindle.
- The *C* axis on a multiaxis lathe rotates the work around the *Z* axis. *C* axis on a multiaxis lathe rotates the work around the *Z* axis.
- Simultaneous machining describes a multiaxis machine moving in rotary axes while the tool is engaged with the work.
- Inverse time feed rates help solve the problem of having both linear and rotary axes move on the same command line.
- Tool center point management strategies used on modern controllers eliminate the need for inverse time feed rates.

Respond

1. Look at Fig. 17-36. When cutting the flats for the hex head bolt, why is 60° chosen for the amount of rotation to cut each flat?
2. If the part in Fig. 17-37 had been machined in a *C* axis lathe, what amount of rotation would you program between the two cross holes?
3. Given that the indexing head in Fig. 17-36 is aligned parallel with the *X* axis, which axis would be rotated to cut each flat?
4. If the Z0 is at the MRZP in Fig. 17-36, what would be the Z height of the flat if the hex bolt has a dimension of 0.75 across the flats?
5. You are programming the six-sided cube shown in Fig. 17-38. Side 3 was machined at A90°/B90°. What would be the position of the *A* and *B* axes to machine Side 5? Side 3 was machined at A90°/B90°. What would be the position of the *A* and *B* axes to machine Side 5?
6. Refer to Fig. 17-41. How many degrees of rotation are necessary to index between holes? How many degrees of rotation are necessary to index between holes?
7. In Problem 6, which axis should you rotate?

Critical Thinking

8. Fill in the blanks for the following program code to complete the slot in the cylinder drawing provided. The slot for the cylinder is cut on a 4 axis mill. Tool 1 is a 3/8 end mill.

```
%
N10 T01 M06
N15 G0 G94 G17 G40 G49 G80 G90
N20 G0 G54 X0.62 Y0.0 S7500 M03
N25 G43 Z 1.0 H1
N30 G1 Z .75 F25.
N35 X________
N40 A________ F150.
N45 G0 Z1.1
N50 G91 G28 Z0
N55 M30
%
```

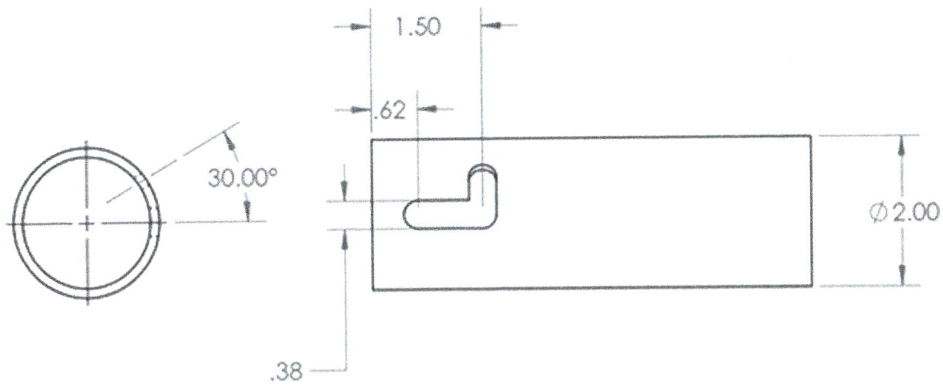

Unit 17.4 Adaptive Process Control

Introduction: Machine processes are becoming more capable, speeding the production process. One such improvement is the ability to inspect their own work while it's being made.

Most CNC production is **proactive** and **reactive**. It is proactive in that the programs follow carefully crafted plans to ensure the best possible outcomes. After first articles are produced, inspectors check parts to verify they are produced to specification using precision measuring equipment, comparing it to engineering prints. Then during the production run, as variation is found, machinists react by adjusting the machine offsets or changing tools. Machinists have had the ability to anticipate problems and react to problems but no other way to control variation. This unit will discuss how manufacturers are using probes during the machining process so that machines can adapt to conditions as they change (Fig. 17-46).

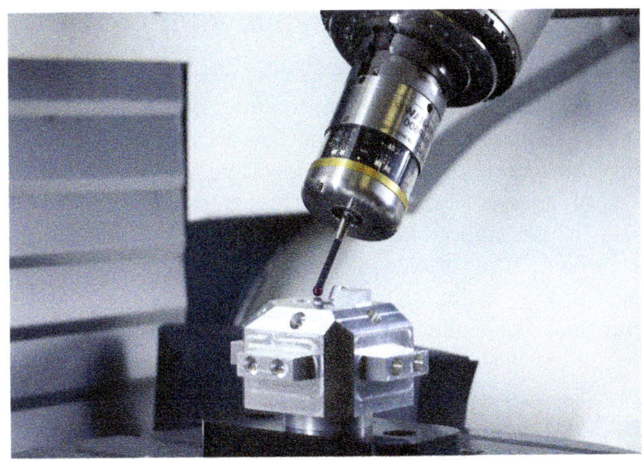

Figure 17-46 Touch Trigger Probe

Courtesy Renishaw Inc

TERMS TOOLBOX

Adaptive machining Manufacturing process where machine settings adjust to changing conditions during the machining operation based on information received from probes or sensors.

Fixture offsets Coordinates that are entered into a CNC controller to establish the program reference zero (PRZ) on the part.

In-process inspection Inspecting parts before they are removed from the machine.

Lights out operation Running production when no one is in the building. (Lights then become unnecessary.)

Macro Saved sequences of commands that when called upon set the machine into motion to complete a task.

Proactive Preparing for possible problems.

Reactive Compensating for problems that have already taken place.

Stylus The part of the touch trigger probe that makes contact.

Thermal expansion The lengthening of machine or tool components due to heat.

Tool diameter offset Define the diameter of a tool to the CNC machine's controller.

Tool length offset Establishes the tool tip's relationship in the Z axis to the program reference zero during the setup of a CNC machine.

Tool setting probes Electronic devices that can set tool offsets and detect tool wear or broken tools.

Touch trigger probes Probes that are mounted in a tool holder that can detect components when contact is made.

17.4.1 In-Process Inspection (IPI)

Today, manufacturers are integrating the use of **adaptive machining**. The most popular improvement is the use of probes mounted in the machine that actively monitor tooling and inspect dimensions during the production process, known as **in-process inspection**. Probes can monitor accuracy in real time, and then send data to the machine's controller. The machine's controller in turn can make adjustments during the process based on information it receives. The ability of the machine to adapt to current conditions leads to less rework and downtime. Adaptive probing enables less active monitoring and even **"lights out" operation**.

There are two categories of probes used in CNC machine production. The first is **tool setting probes** (Fig. 17-47). They can be used in the initial setup of **tool length** and **tool diameter offsets**. Tool setters can also detect a tool's condition during production and alert the machine controller when there are problems such as tool wear or when the tool has become broken. These probes use laser, camera, or touch feedback.

The other category includes **trigger touch probes** that can measure part features while the parts are on the machine (see Fig. 17-46). These probes can determine feature size and location. They are used during setup and in process inspection.

Figure 17-47 Renishaw Tool Setting Probes.
Courtesy Renishaw PLC

ARMCHAIR FIELD TRIP

Watch this Haas Tip of the Day video about the use of probes.

Can your CNC machine really measure parts? To find out the answer to this question and more, go to: **https://www.youtube.com/haasautomation** and search for this video called: **In—Process Part Measurement; Probing in 5 Simple Steps—Haas Automation Tip of the Day**

17.4.2 Tool Setting and Broken Tool Technology

Tool setting systems (such as shown in Fig. 17-48 later in the chapter) can improve production efficiency in two ways. First, tool probes help with both the preprocess steps of tool setting and with monitoring tools during production. Automatic tool setting routines can establish the tool length offsets as well as the diameter offsets of tools being prepared for a machining routine. Even though many shops today use preset tools, where the geometry and length offsets have been established in the tool room before they have been sent out to the floor, things can go wrong. Mistakes can be made as the tool data are entered into the machine controller, or tools can become mixed up. Either way, these errors can lead to disastrous results causing loss of production time. Tool setting probes can ensure that the offset information given to the controller is correct.

Figure 17-48 Checking for broken tool.
Courtesy Renishaw PLC

Tool probes can also gather information about the tool's condition during the machining cycle and send it the controller. Here are three adverse tool conditions that commonly occur during the cutting processes that can lead to scrap parts or machine downtime:

- Tool wear: The normal wear that occurs during cutting operations can cause variation in feature dimensions.
- Broken tools: For various reasons a tool can become broken or can lose a cutting insert during machining. When this happens, too much material can be left on the part.
- **Thermal expansion**: The spindle can expand due to the heat generated during the machining operation. When the spindle expands significantly, the tool tip's relationship to the part Z zero will change.

When the controller receives information about tool wear or thermal expansion during the production cycle, it can adapt to the changes and alter the tool offsets. In the case of a broken tool, the machine can be programmed to call for a "sister" tool to stand in for the one that is broken. The operator will be alerted to attend to the broken one.

SHOPTALK

Train wrecks. When a roughing tool is broken during production it can leave significant unintended amounts of material on the part, which will cause the tools that follow it to break as well. Sometimes CNC machinists refer to this type of situation as a train wreck, because once one train car leaves the tracks, it causes the following cars to derail.

ARMCHAIR FIELD TRIP

You don't want to miss this Haas Tip of the Day video about setting tool offsets.

This video shows how to make tool offsets in a CNC mill both manually and by using tool setters. Go to: **https://www.youtube.com/haasautomation** and search for this video called: **Tool Offsets and Manual Tool Setters—Haas Automation Tip of the Day**

17.4.3 Touch Trigger Probes

Touch trigger probes are usually mounted in a tool holder (Fig. 17-49). The part of the probe that makes contact is called the **stylus**. The stylus is a stem with a round ball at the end. The stylus ball can be made of ruby, silicon nitride, zirconia, ceramic, or tungsten carbide and has been manufactured to exacting dimensions. When the stylus contacts a component in the machine, a signal is sent to the machine's controller and the machine position is recorded.

Probing during the production process can be used to

- Determine part or stock orientation
- Set or adjust fixture offsets
- Measure features before they are completed
- Inspect the part after the cycle is finished

Setting Program Reference Zero

After the machine operator loads the stock or part into the holding method, a probing routine is sometimes used to check to make sure the part has been located properly. If the part material is not centered, the probe routine can be designed to automatically update the **fixture offset** or to alert the operator. This allows manufacturers to use a minimal amount of stock because the probing routine will ensure that the offsets are centered on each piece of stock regardless of variation.

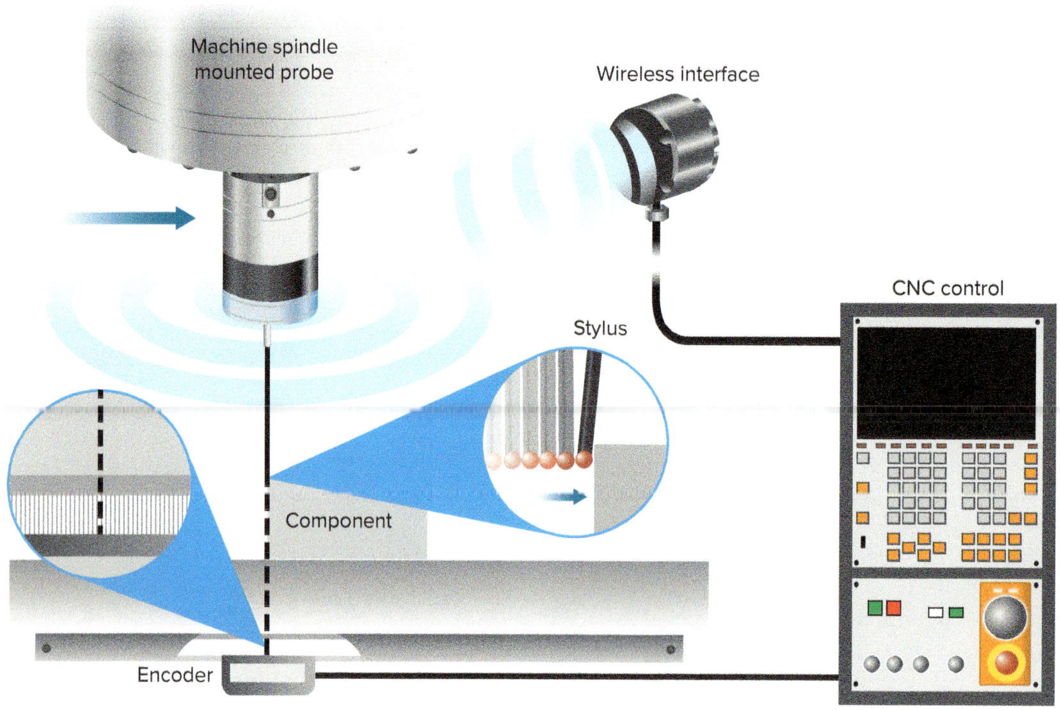

Figure 17-49 Transmitting data to the controller.
Courtesy Mike Fitzpatrick

Where there are tight locational tolerances between features machined in previous operations and those that are about to be machined, the updating of the fixture offsets can prove critical to the success of the part. The probe can be programmed to locate features previously machined to update the fixture offsets.

Using Probes to Measure

As a machinist, you will most likely use probes on CNC machines for setting up tools and finding fixture offsets. If the machine you are working on has been fitted with a Renishaw probe, macros have been preloaded into the controller's memory to perform certain probing tasks. **Macros** are saved sequences of commands that will set the machine into motion to complete a particular task. Many controllers have a page that shows templates for common probing tasks such as setting tools or finding fixture offsets. Some machinists are not aware that these same routines can be used to measure part features.

Take, for example, the probing routine to find the center of a bored hole to establish a fixture offset zero point. In this example a 0.750 hole machined in a previous operation becomes the fixture offset for the current operation.

If you are using one of the controller's probe routine templates, a couple of things will happen. First, you will be instructed to move the tool to an approximate location where the probing routine will be initiated. Second, the controller will ask you a few questions and a program will be created and sent to MDI for execution. Questions will include

- Which fixture offset will be used?
- What is the approximate size of the bore?
- What is the depth to be probed?

If the answers to these questions were

- Fixture offset: G54
- Approximate diameter: .750
- A depth of minus .500 in. from where the tool is now

the resulting lines of code would look something like

G65 P9901 M2 A1. D.75 W-.50 S54 (See Fig. 17-50.)

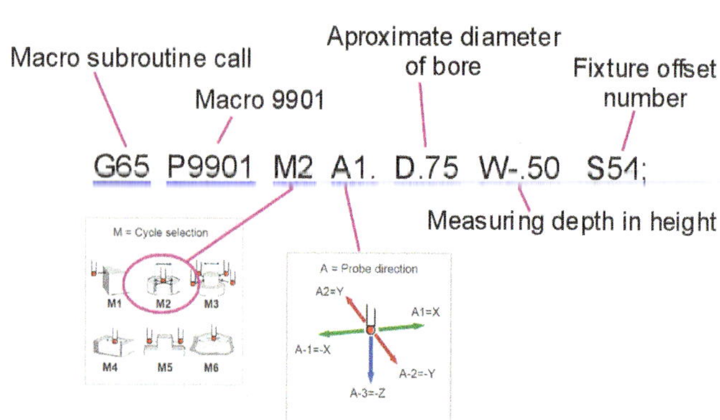

Figure 17-50 Macro definitions.

Even though the primary reason for the probing operation was to find the fixture offset, other information is also stored in the controller, including the size of the feature that was probed. Somewhere in the controller, information about the size of the bore is also found (Fig. 17-51). On the Haas controller, Macro Variable 188 contains the information about the actual measured size of the probed feature. In this case, it was 0.7528 in.

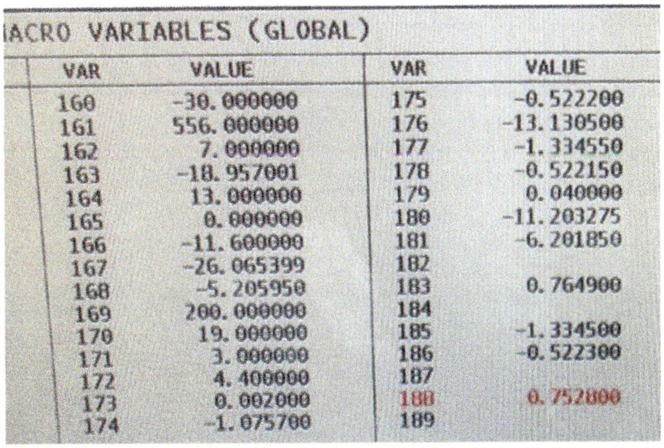

IACRO VARIABLES (GLOBAL)

VAR	VALUE	VAR	VALUE
160	-30.000000	175	-0.522200
161	556.000000	176	-13.130500
162	7.000000	177	-1.334550
163	-18.957001	178	-0.522150
164	13.000000	179	0.040000
165	0.000000	180	-11.203275
166	-11.600000	181	-6.201850
167	-26.065399	182	
168	-5.205950	183	0.764900
169	200.000000	184	
170	19.000000	185	-1.334500
171	3.000000	186	-0.522300
172	4.400000	187	
173	0.002000	188	0.752800
174	-1.075700	189	

Figure 17-51 Probing the bore and recording the size
Courtesy Mike Fitzpatrick

ARMCHAIR FIELD TRIP

Your CNC machine gathers information that you might not be aware of. Watch this **Haas Tip of the Day** and find out how your CNC controller gathers important information that you can use to measure part features Go to: **https://www.youtube.com/haasautomation** and search for this video called: **Output Your Machine Data for Analysis—Haas Automation Tip of the Day**

TRADE TIP

Machinists can take measurements by employing this probing method. The only caution is that the measured reading is only as accurate as the calibration of the machine.

17.4.4 In-Process Control

In the case of very tight size tolerances, the program can purposely leave a predetermined amount of material on that feature to be checked by the probe and then finished. Consider this example:

- A 2.500 in. bore with a tolerance of + or − .0003 is to be machined.
- The bore is roughed leaving 0.040 in. on the diameter.
- A finishing end mill is programmed to machine the bore leaving 0.020 in. on the bore diameter.
- The inspection probe is then called up to inspect the feature and to verify that precisely 0.020 in. has been left on the diameter. If not, the probe routine can automatically adjust the finish tool's diameter offset for the finish pass.

These measurements and adjustments take place without human intervention. This kind of routine is not used on all features on a part because inspection routines do consume machine time. They are performed on features that have very tight tolerances or those that are known to be difficult to dimensionally control.

KEYPOINT

Probes can help CNC machines adapt to current conditions and make adjustments without human interaction by automatically updating tool and fixture offsets.

Inspection Probing

Probing routines can be programmed to perform the final inspection after the part has been completed and before it is removed from the machine. Inspection probes can be programmed to act very much like a coordinate measuring machine, or CMM. The CMM is discussed in more detail in Chapter 28, but basically a CMM is a computerized measuring machine that is normally used by inspectors after part production. The touch trigger probe tool on the CNC machine can measure all the critical features on a part and check to see if they meet specification. This can eliminate the need for inspection later. The biggest advantage is that the part is known to be good before it's removed from the machine. This saves time by eliminating the need to reload parts that need to be reworked because their dimensions are out of tolerance.

After the probe has run through its inspection routine, a report is created and can be associated with the part that was inspected. Although inspection routines do consume machine time, the benefit of knowing the part meets specification before it's removed from the machine can greatly outweigh the costs of the lost production time.

Programming complex operations such as inspection routines requires the use of software. Renishaw has its own programming software called Active Editor Pro, which assists programmers to embed probing routines in machining operations. Renishaw has teamed up with other companies to integrate Active Editor Pro directly into CAM software such as Mastercam.

UNIT 17.4 REVIEW

Replay the Key Points

- Adaptive process controls allow the machine to actively react to changing conditions during machining without human interaction.
- Tool probes are used to set tool length and diameter offsets.
- Probes can verify correct information has been entered into the machine's controller.
- Broken tools, thermal expansion, and tool wear can be detected by tool probes.
- Touch trigger probes are used to determine fixture offsets and stock position.
- Machine offsets can be adjusted by information sent to the controller from the probe, to control part feature dimensions.
- Final inspection of a part can be conducted by a probing routine before the part is removed from its fixture.
- Machinists can use probing routines to take measurements.

Respond

1. How does thermal expansion affect dimensions of a part?
2. In the example given in 17.4.3, the bore feature was programmed to machine the bore at 2.460 before coming to a nonoptional stop (M00). If the diameter turns out to be 2.452, how much cutter *radius* compensation will need to be made?
3. In Problem 2, will the compensation change to the cutter compensation be positive or negative?
4. Why do you think the stylus probe tips are made of materials such as ruby, silicon nitride, zirconia, ceramic, or tungsten carbide?
5. What problems might occur if there is variation in the size of the stock used for a production run, and why would probing avoid these problems?

Unit 17-5 Program Creation and Data Management

Introduction: This unit provides an overview of the way in which program data are created and managed. It will help to clear up some of the data management terms specific to CNC work. It's not about writing programs.

TERMS TOOLBOX

Bulk load A version of DNC whereby large packages of program code are sent to the controller. Then when these are nearly consumed, another package is sent down.

DNC (direct numerical control) The host computer assumes the role of RAM sending program data to the control as needed. *Used mostly where CAM programs exceed the CNC control's RAM.*

Download Sending a program from an offline PC to a CNC control.

Drip feed A version of DNC whereby small packages of data are sent from PC to controller in just enough data for the present command and the next.

Hand load Bringing a completed program to a CNC control via Wi-Fi or portable storage device.

Imported image Bringing a part geometry file from a CAD system into a CAM system. Or bringing any computer file from one software to another.

Part geometry The image of the part to be machined. Can be drawn online or imported from a CAD system.

Postprocessing Converting a generic tool path program into a machine-readable version.

Tool path A generic program used by the CAM computer on which a postprocessed machine-readable program is constructed.

17.5.1 Three Types of Part Geometry

CAM and graphic interface programming systems create the code or conversational words based on an accurate drawing of the intended job, called the **part geometry**. The programming software will follow the drawing lines or surfaces to create the program codes (Fig. 17-52).

Figure 17-52 Graphic interface (GI) programming at the CNC control allows the operator to create a program while the machine is cutting another part.
Courtesy Tyee Aircraft/Everett, WA

Part geometry falls into three categories, depending on whether it is going to the lathe or mill and how complex the milled shape might be.

Two-Dimensional Wireframe Line Drawings

For simple parts, a flat pattern works fine for both lathe and mill work, called a wireframe since it's composed of line segments. Similar to a pencil drawing, this geometry is a profile, two-dimensional accurate image composed of straight lines and arcs. Flat patterns are the only kind of drawing needed for *X*-*Z* lathe work (Fig. 17-53). Both the full and half part profiles are commonly used.

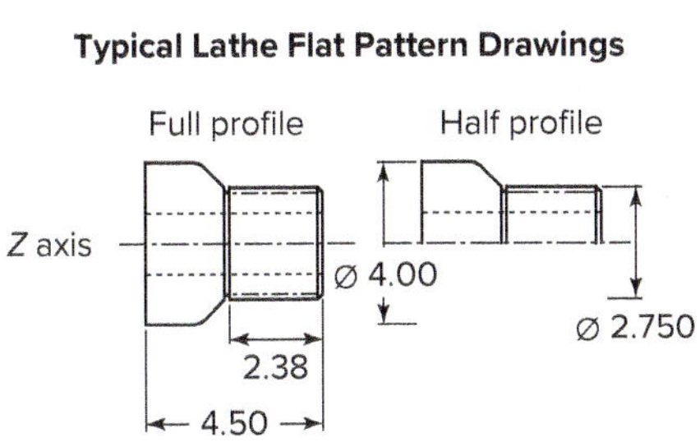

Figure 17-53 Lathe part geometry can be two dimensional.

To employ a flat pattern for an *X*-*Y*-*Z* mill program, at the time of program creation the user must assign *Z* axis depth to various cuts (Fig. 17-54).

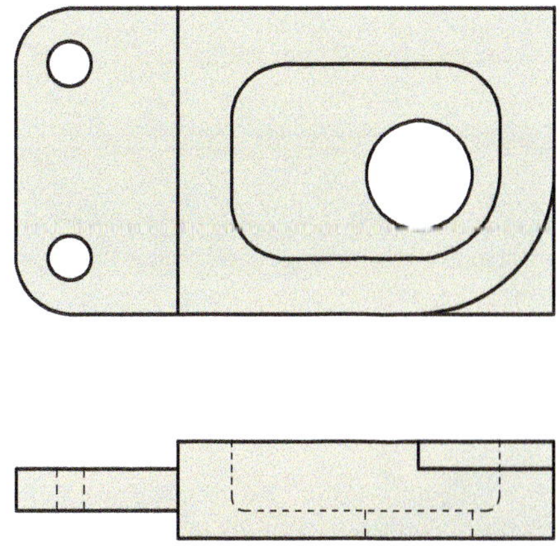

Figure 17-54 The top view, if drawn correctly (per Fig. 17-55), can be used as a flat pattern geometry.

Three-Dimensional Wireframe Geometry

Used for mill work, a wireframe is still a set of lines, but they have depth, as shown in the perspective view of Fig. 17-55. If the geometry is in wireframe form, then the CAM system can assign *Z* axis cut depths automatically but *Z* depth can be overridden when needed. Wireframes also help to show part features not easily visualized with a flat pattern drawing. But wireframes do not indicate surfaces, just lines. The user must still tell the CAM system to which side of the line the cutter must go. When shapes are complex, that becomes difficult. Then the next higher geometry in terms of usefulness for CAM programming is the solid model.

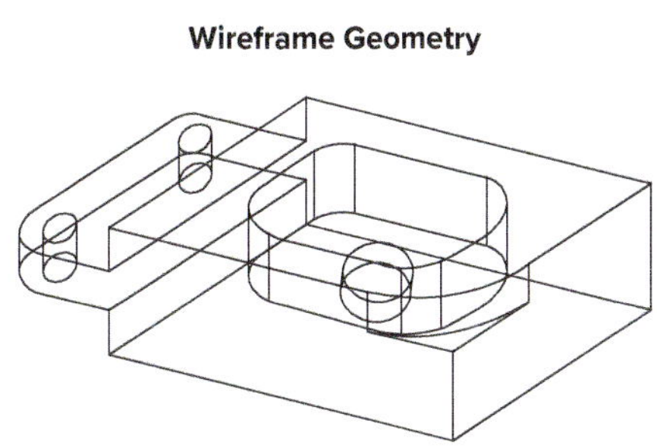

Figure 17-55 A wireframe consists of lines and arcs in *X*, *Y*, and *Z* dimensions.

Solid Model

A solid model looks as though the part was made from a piece of clay. To create a solid part geometry, one begins with a large rectangle (or other billet shape), then unwanted pieces are removed just as though it was machined from raw stock. Shaped pieces can also be added to the geometry as it takes shape (Figs. 17-56 and 17-57). Solid model part geometry offers several advantages:

Easy pictorial visualization of all part features. Note that features are not easy to see in the wireframe form.

Near automatic programming capability on complex shapes with compound curves (software dependent)

Complete control of every side of the object

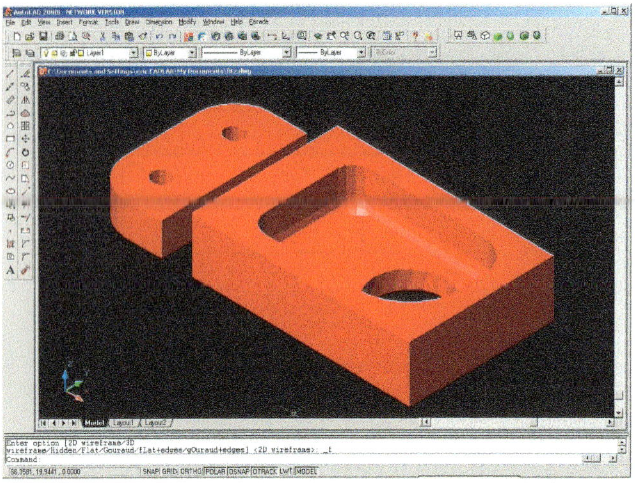

Figure 17-56 A solid model starts as a machined product would, with a billet of raw material.
AutoCAD software

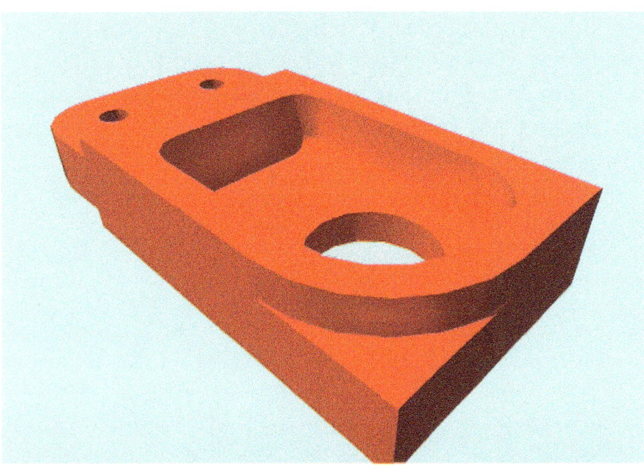

Figure 17-57 Because it represents the finished product, the solid model can be viewed from any side.

17.5.2 Program Creation

Importing Geometry

Once the geometry is drawn in the CAD, it's ready to **import** (bring the image into the CAM software as a readable file). It may need to have dimensions and other details that clutter the drawing removed. If the drawing was correctly drawn for CNC, those details will be on layers that can be removed or made transparent (masked) while the program is being created.

Next, the user begins the process of turning geometry into program codes. This procedure can be grouped into two general types.

> **KEYPOINT**
>
> We'll devote much training to actually writing CNC programs later, both by hand and by CAM. This material is an overview of how programs come into existence.

Flat Pattern or Wireframe Element Selection

Starting with a flat pattern or wireframe of the geometry, individual elements are pointed out by the user using the mouse, in the sequence and direction they are to be machined. The programming software does not know they define a surface; it only sees the elements as lines and arcs, so the user must indicate to which side of the line the cutter must go.

Then, either before or after element sequencing, the user assigns machining data such as cutter size and shape and number of roughing and finishing passes to the sequence. Then with the elements linked together in a sequence, the software uses the machining data to create a complete **tool path**—or sometimes called an *element chain.*

Differences in Software

Adding Machining Data Before or After Path Selection

Depending on the software, machining data such as cutter size and number of roughing passes are either added before the path is selected, or some require the data to be entered as the tool path is created one chunk at a time.

Pre- or Postcustomizing into Codes for a Given Control

All programs must be custom translated into a specific machine's conventions. Control to control, there are differences in format as well as usage of the unassigned code words. The CAM software has a library of these control specifics. The various templates for each control are called *posts* (**postprocessing** *templates*) or sometimes *filters.*

When a CAM system has the ability to create the custom program for a given control, it is said to *support* it (write specific codes for a specific control).

> **KEYPOINT**
>
> Some CAM software requires that the specific template be selected before element selection while others like our example Mastercam require cut parameters be chosen and entered during cutter path selection.

Geometry Associative Programs

When the design changes, so must the program. Most modern CAM software is able to automatically update the program if any element in the geometry is changed. Some less sophisticated software requires the user to go back and redo the program, then repost the program.

Geometry Drawing Accuracy

To be of use for a CNC program, the person doing the drawing must ensure absolute accuracy of elements (Fig. 17-58). Lines must be drawn in the exact length desired on the work and line intersections must be trimmed exactly. Solid models can't simply look like the part; they must be an exact size and shape replica of the real part. If these rules aren't followed, the CAD drawing might look right to the eye, but it will not generate the right part in CNC. That accurate picture can be the original CAD drawing, or it can be developed within the CAM system.

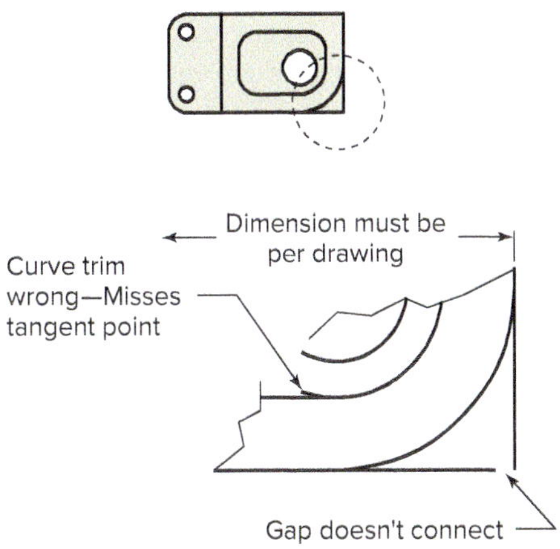

Figure 17-58 For programming purposes, CAD drawings must be accurately sized and have each intersection or tangent point trimmed perfectly.

17.5.3 Sending Programs to the Controller

Data Transfer

Once a machine-readable program is ready, it is sent to the controller. This act is known as **downloading** (Fig. 17-59).

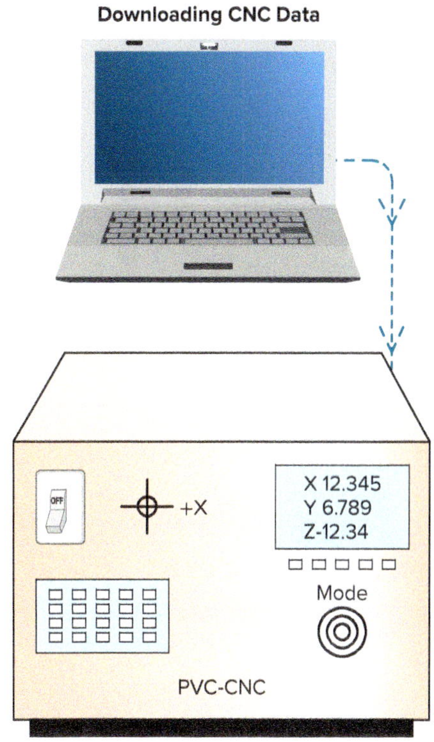

Figure 17-59 Downloads send entire programs to the controller's permanent memory.

Active or Permanent Memory

The CNC controller stores programs in one of two locations: active memory ready to machine or permanent storage for later use. Permanent storage is where the downloaded program goes initially. The total capacity of a control to hold programs in both locations is an issue.

Transfer Types—Hand, Drip, and Bulk

The program can be transferred to the CNC memory in one of three ways:

1. **Hand Loading**

   The program is copied to a standard computer storage media then wifi, or hand carried to the controller. This is sometimes called the *sneaker net* (*using your shoes to transfer the data, or a foot-shake routine*).

2. **Direct Download**

   Through a communication wire, the PC host sends the entire program to the controller. The word *download* implies that the program is coming from a higher or master source. *Uploading* occurs in the reverse direction, program data are sent to the host or PC from the control.

3. **Direct Numerical Control (DNC)**

   The third download method is used when program data exceeds the memory capacity of the control. CAM-generated programs use many small lines to approximate a surface, thus programs for complex shapes become exceptionally large. The solution is to send it in packets–just in time to be used. There are two ways in which this direct numerical transfer can occur (Fig. 17-60).

Drip Feed Transfer. The data are sent in small units just in time for the control to translate the codes and make the machining movements.

Bulk Load Transfer. The data are sent in large packages that nearly fill the controller's RAM. As the data are consumed, the host computer sends the next package of data at the needed time.

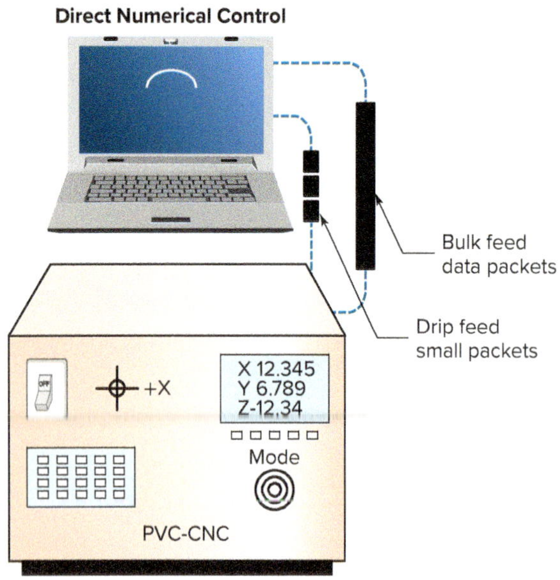

Figure 17-60 Direct numerical control (DNC) sends packets to the control's active RAM just in time for execution.

KEYPOINT

DNC is used mostly for the times when the CNC control's RAM is inadequate to hold an entire CAM-generated program. Working together, the host supplies the data, and the controller continues to drive the machine.

UNIT 17-5 REVIEW

Replay the Key Points

- Within the first 100 G and M words, many will be specific to the machine. Words 100 and above are also machine specific.
- For a CAD drawing to be of use to the CAM system, it must be drawn to perfect accuracy and must be trimmed with perfect intersections between elements.
- A download occurs when a CNC program is transferred from a master computer to a dedicated CNC controller.
- A DNC occurs when a host computer feeds a program in small or big packets to the control to use as needed.

Respond

1. Test your data vocabulary use. How many of these CNC data terms can you define in your own words?

Bulk load	Download	DNC (direct numerical control)
Drip feed	Hand load	Imported image
Offline	Tool path	Postprocessing
Taxonomy		

2. In your own words, describe the two ways CAM software creates a program using part geometry.
3. Define solid model part geometry.
4. Extending the terms you have thus far learned, can you describe a CNC *upload?*
5. Given the timeline, who might be responsible for programs in the future?

CHAPTER 17 REVIEW

Unit 17-1

We keep coming back to ball-screws. Other drive mechanisms have been tried for CNC axis work. The linear drive is one experiment that held some hope. A linear is somewhat like a stepper motor unwrapped into a straight line. Given a pulse of energy, the sliding component moves forward one magnetic section along the stationary driver. They are fast and accurate. However, the problem is there is no mechanical advantage as with screw drives. To develop the amount of thrust needed, the linear magnets must be huge and consume too much energy!

Unit 17-2

In addition to cutting metal, as we do on CNC lathes and mills, we'll soon see machines grinding in conjunction with cutting metal. The technical challenges to integrate them into classic machinery aren't great.

Many other operations are CNC driven and can be added to most machines, too (for example, laser cutting, waterjet cutting, electrical discharge eroding, and deposition of metal using focused lasers and powdered metal). These and several other metal cutting and shaping processes are also CNC driven. Look for them to be added to the complete machining center sometime soon.

Unit 17-3

Rotary axes increase machining efficiency by reducing the number of operations needed to complete a part. This is important because it saves the costs of setting up and loading and unloading parts. In addition, every time a part needs to be handled and loaded into fixtures, we run the risk of introducing errors.

Simultaneous rotary axis machining is often necessary to produce complex surfaces used in aerospace, automotive, and mold making industries.

Unit 17-4

The use of probes in CNC machining has become indispensable. They help the machinist to quickly and accurately set up jobs. Touch trigger probes can also monitor production while it is occurring and adjust machine offsets to control part feature dimensions that are challenging to produce.

Probing routines can also be used to inspect parts before they are removed from the machine. This can save time lost due to reworking parts that don't meet engineering specifications.

Unit 17-5

While the software war rages on as to the leading CAD and CAM systems, we the machinists can only benefit. As software becomes more automatic in its ability to master programming, the responsibility may be returned to the machinist.

QUESTIONS AND PROBLEMS

1. Which component in a complete CNC axis drive sends low-energy commands to make the servo motors move? (Unit 17-1)
2. True or false? The axis encoder makes it possible for a CNC control to drive several different kinds of axis drive motors. If this statement is false, what would make it true? (Unit 17-1)
3. What kind of axis drive motor is the most popular for heavy-duty CNC machines? (Unit 17-1)
4. What condition can occur within a closed-loop feedback system when the program demand exceeds the ability to thrust the axis? With what result? (Unit 17-1)
5. What factors and controls do the operator have to prevent servo errors? (Unit 17-1)

CRITICAL THINKING

6. Why can't a stepper motor machine with no feedback go into servo error? (Unit 17-1)
7. True or false? CNC machines lacking encoders must be homed when initialized. If this statement is false, what would make it true? (Unit 17-1)
8. True or false? CNC machines with direct reading scales need not be homed when first powered up. If this statement is true, then why? (Unit 17-1)
9. For what reason do manufacturers use stepper motors in CNC axis drives? (Unit 17-1)
10. Why are CNC machines evolving today? (Unit 17-2)
11. Why were NC machines invented originally? (Unit 17-2)
12. Name the three separate branches of evolution that intertwine to improve CNC production. (Unit 17-2)
13. Identify the feature that speeds up automatic tool changing by not requiring a given tool to be placed in a specific drum location. (Unit 17-2)
14. When setting up a CNC mill, what two tool numbers must the control know initially? (Unit 17-2)
15. Identify the kinds of part geometry, and describe each. (Unit 17-3)
16. In order for a CAD drawing to be of use in CNC work what must be done? (Unit 17-3)
17. Describe DNC data transfer. Why is it necessary? (Unit 17-3)
18. On a sheet of paper, sketch the five CNC components in a closed-loop axis drive. Briefly describe each. (Unit 17-1)
19. Which component in Question 18 is responsible for reporting position back to the control? (Unit 17-1)
20. What is one main reason for DNC capability? (Unit 17-3)

CHAPTER 18 CNC Controls

Units

18-1 Initializing a CNC Machine (Powering Up)

- Describe and use the machine home cycle
- List the generic startup procedure

18-2 Controls for Setup and Operation

- Identify the *mode* switch on a CNC control and list the various mode purposes
- Be able to use both hard and soft keys on a control panel
- Define and perform a *homing cycle*
- Identify and perform safe *manual movements* of CNC machines
- Make safe program halts by using one of four possible methods

INTRODUCTION

Okay, you know how CNC machines make such amazingly accurate locations and motions. Now it's time to learn to use the various screens, buttons, and dials to perform them. The best advice is to take it one step at a time. CNC control panels look complex—but taken one function at a time, they are manageable. Chapter 18 will explain the range of functions a control must perform. For examples we'll use two, the Haas standard control and a generic control. Transferring to another will be easy once you understand the range of functions. Then it's simply a matter of discovering how that control performs it.

Keep two things in mind as you learn: Never "try" a function—be certain you understand what will happen when you touch the button. While it might be actuated differently, every function described in this chapter and the next will be on the CNC control panel in some form. It may be a dial, a button, or a soft key under the screen—but it will be there. All machines require the things you are about to learn for setup and operation.

Unit 18-1 Initializing a CNC Machine (Powering Up)

Introduction: Unit 18-1 is about performing a cold start. Depending on the machine's ability and its age, powering up may be automatic or it may require operator actions. But first, you must check "under the hood," including vital supplies such as air pressure, lubes, and coolants, then be sure safety guards and emergency buttons are in the operate position.

Now with the main breaker switch on, usually on the back of the machine, touch the Power On button (Fig. 18-1, green button). It may take a while as servos power up and the OS loads. Next touch the blue Power up button (Fig. 18-2). The rest of the procedure can be automatic as on your Haas mill or lathe where the cutting tool will pull away from the work slowly in the *Z* axis, to home on the limit switches; then the turret will index to Tool #1.

Some industrial machines do not need to home. They wake up knowing their axis locations, which tool number is in the active position, and where the others are in the tool changer or turret. All controls load their operating system automatically. In just a few moments, green lights and a video message indicate the machine is ready to exercise its axes and spindle bearings before it's put to work.

On the other hand, other machines will require an **initialization** procedure (a formal start-up sequence). Generally older or budget machines require axis registers to be manually driven near limit switches where they can find and zero base their position. This is necessary if the machine is open loop. It must know where machine home is located. That means they do not have sensors telling the controller where the axes are located until they are zero based. These machines should have their axis coordination refreshed often if the machining is relatively heavy or if you have experienced a stall/crash! Initializing may be complex, requiring several steps, in a defined order, or it might be semiautomatic. Accuracy and safety will depend on performing all steps correctly. All machines disallow full use or program execution, rapid moves, or spindle operation until the initialization is done right.

TERMS TOOLBOX

E-stop button The big red button that depowers all machine functions. May require several actions to restart after touching this button.

Initialize (initialization) A process through which some CNC machines are brought up to full operational readiness. May be semiautomatic or may need intervention.

Lockout key switch A multilevel switch that, with the key removed, allows or disallows certain control functions and authorities.

Midcycle stop (midauto halt) A way to quickly stop a program run without losing physical position or program data.

Overt (action) A deliberate and often difficult move to ensure it isn't done by accident.

RIS (reduced instruction set) controller Controls using icons and double function buttons to create a super-compact control pad.

Figure 18-1 A standard Haas mill control.
Courtesy Mike Fitzpatrick

Figure 18-2 The standard Haas keyboard.
Courtesy Keith Smith

18.1.1 CNC Controller Examples

In planning this text we chose Haas controls (Figs. 18-1, 18-2, and 18-3). These we compare to a generic panel (Fig. 18-4): the plain vanilla control (PVC)!

The functions found on the PVC are universal. They may have a different name or be actuated differently on a given unit, but they'll be there in some form because they're basic to setup and operation of the machine.

Figure 18-3 The Haas NGC control.
Courtesy Haas Automation, Inc.

18.1.2 Kinds of Controls

CNC control panels vary widely. Here are the three most common examples.

Standard View Screen, Dials, Keys, and Switches

Figure 18-5 shows a group of typical CNC controllers. They feature a video screen, control dials, switches, buttons, and indicator lights to show conditions and modes. Representing the largest share of control types, these units are made specifically for industrial machines.

PC and RIS Controls

Figure 18-6 shows a graphic **RIS *(reduced instruction set)* controller** touch pad. While it takes a bit of training to know what all the symbols mean, once learned, these highly compressed control panels work with simple beauty. The pad can be taken in hand to any viewpoint around the machine. Some are even wireless!

RIS units are dedicated to setup and operation modes only. To keep the panel palm sized, they have no keypad. Editing and program writing tasks are performed on an interfaced PC located nearby. Some RIS panels feature double or even triple functions per button, similar to a scientific calculator.

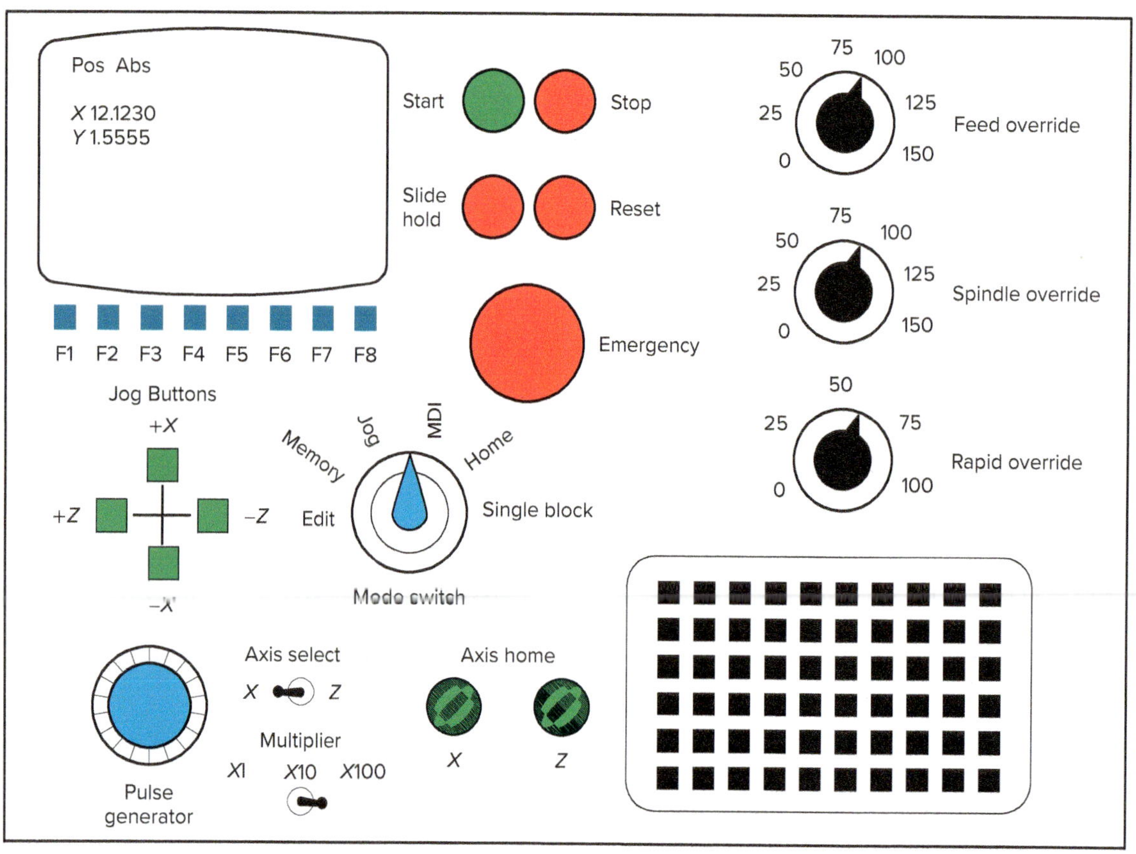

Figure 18-4 A generic CNC control panel.

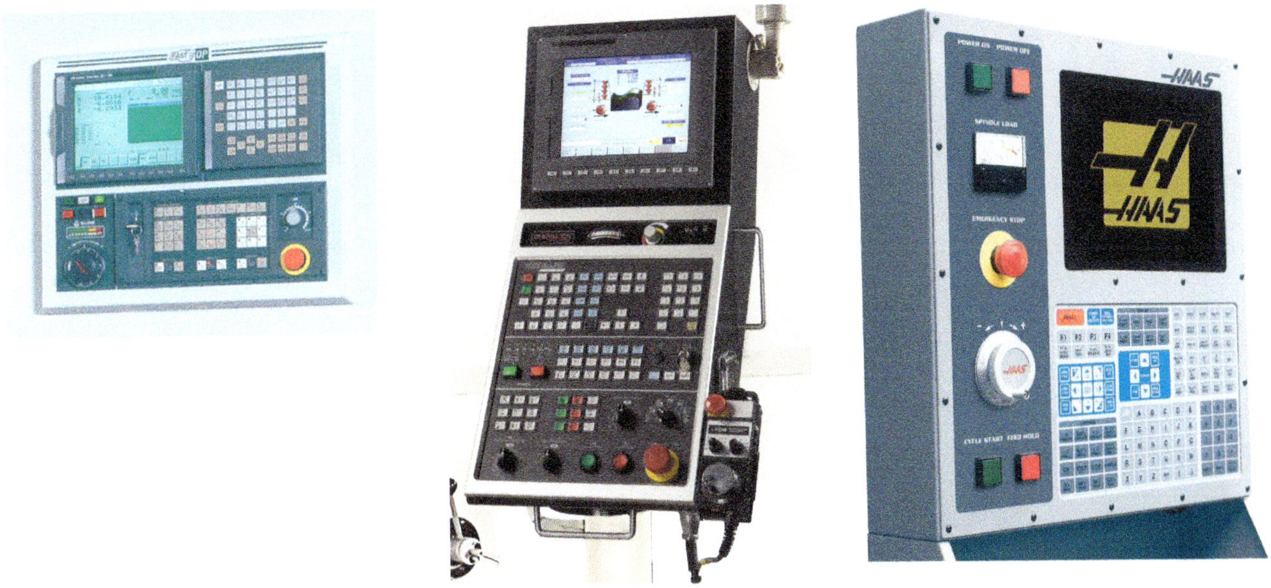

Figure 18-5 (left) Bridgeport, (middle) FANUC, and (right) Haas control panels.
(*left*) Courtesy Bridgeport Machines/Bridgeport, CT; (*middle*) Courtesy Chevalier Machinery, Inc.; (*right*) Courtesy Haas Automation, Inc.

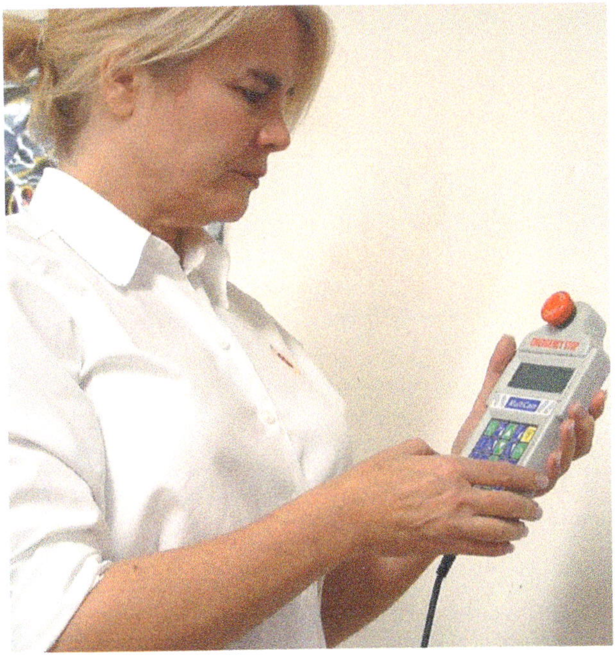

Figure 18-6 An RIS (reduced instruction set) controller performs all operation functions but not program editing.
Courtesy Mike Fitzpatrick

KEYPOINT

Operator Pendants

On many industrial controls, a handy portable pendant with just the manual setup functions is an accessory added to the main control unit. With the ability to take the portable pendant to the setup, this allows easy viewing of dial test indicators and other alignment items not easily seen from outside the work envelope. These pendants are not RIS controls. They cannot manipulate setups or access data or run programs. Pendants are for setup motions only (Figs. 18-7 and 18-8).

Figure 18-7 An operator pendant is often added to high-end industrial controls.
Courtesy Okuma America Corporation

Figure 18-8 A pendant provides manual axis movement from any view perspective.
Photography by Prographics at Milwaukee Area Technical College/McGraw Hill

PC-Based Controls

Finally, there is a controller based solely on a standard PC. Any fast processor and hard drive can store and read a program, then send a signal to the axis drive relays.

Familiar Operating Systems

Using everyday operating systems such as Windows® or Unix®, PC-based CNC controls offer the following:

Huge hard drive capacities to hold long programs

Standard hard drive and computer components that bring interchangeability and easy repair when necessary

CAD/CAM that can be performed at the CNC machine if the software is loaded

Part inventory using barcode scans

General shop communications such as e-mail, work order, or engineering drawing downloads

The ability to perform statistical process control charting and all sorts of other tasks directly at the CNC machine

Most PC-based controls include an operator pendant to assist with setup and program-run functions.

18.1.3 Lockout Switches for Safety

Not to be confused with the main power switch that is locked out during maintenance, this switch is on the control panel or near to it. To protect machines, setups, programs, and people, many controls feature a multiposition key switch. With the switch set and the key removed, various levels of responsibility are allowed while others cannot be accessed. Figure 18-9 shows such a **lockout key switch**. None of these settings prevents start-up functions, however:

1. **Operate** (lowest responsibility level)
   - *Allows* all program run and halt functions
   - *Disables* edit and setup information changes
2. **Setup** (medium level)
   - *Allows* operate functions plus process changes such as tool diameter, length, or PRZ location
   - *Disables* program edit and load capability
3. **Edit** (full control responsibility level)
   - *Allows* access to all functions, including changing setup information, writing new programs, and changing existing programs in storage

Figure 18-9 The key-operated protect switch on a control allows various levels of operator responsibility.
Courtesy Tyee Aircraft/Everett, WA

18.1.4 Prechecks

There are three or four "under the hood" checks to be performed before powering up:

1. **Air Pressure**

   Many but not all machines require clean compressed air. It's found on machines that blast tool holders to clear chips during tool changing or to power lubrication systems or to activate clamps and other auxiliary functions.

2. **Lube Oil**

   This is required by *all* machines; on most, it's monitored, and if found to be below the machine's requirement, an alarm will show on the panel and perhaps with a flashing red light! But there are a few machines that won't tell you that the oil is low! Remember, there will be more than one product: spindle oil and way lube. Many retrofit (manual machines made into CNC) or older model CNCs might run without lube oil, damaging the machine. In general, newer machines will not operate if the lube reservoirs are low.

3. **Hydraulic Oil**

   Usually lathes with hydrochucking require this. Check the oil level before initialization, and after the machine is up and running *check* that the correct pressure is set for the chuck. There will be an adjustable dial on the back or side of the turning center with a gage. Check the setup document or ask the lead machinist for the correct pressure for the specific job. Too much chuck pressure mars or even crushes work and can bend tooling; too little holds the work weakly and dangerously.

4. **Low Coolant**

   Nearly all machines require coolant, but it may or may not cause an alarm if it's low. It can be a real disaster if coolant runs out during a heavy cut. Many controls sense when coolant is at the add mark but will not stop the machine until a current cycle completes.

The next section provides a short list of things that may need to be seen to before the green button is activated.

ARMCHAIR FIELD TRIP

Watch this exciting video Haas Tip of the Day about machining coolant.

How do you mix cool and add coolant to your CNC machine? Watch this video to find out.

Go to: **https://www.youtube.com/haasautomation** and search for this video called: **Coolant Simplified—4 Steps to filling the tank on your Haas CNC machine—Haas Automation Tip of the Day**

18.1.5 Emergency Stop

We're almost there; the green button is next. But when you hit it, nothing happens? Why? Chances are you need to check the emergency-stop button or buttons and put them in the run position (Fig. 18-10). Every CNC machine has one or more. They are only used to stop during a crash or to lock up the machine after turning it off.

Emergency-Stop Buttons

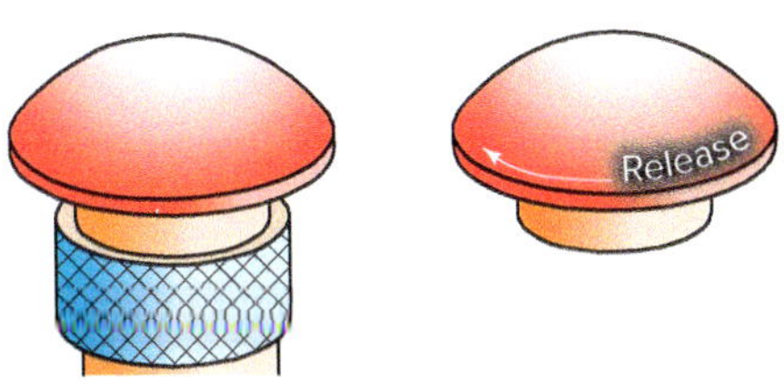

Figure 18-10 The largest red button is for final shutdown and emergencies.

Plunger Safety Switches—The Big Red Button

As seen in Fig. 18-10, these buttons are universally used as an emergency safety stop, **E-stop**. When depressed during normal power-down or in a real emergency, the button stays locked off until the base collar or the button itself is hand rotated to unlock it. This **overt *(deliberate)* action** creates a reminder of clearing the problem, if one existed, before running the machine. Depressing *Ctrl-Alt-Del* on your PC is an example of an overt action.

Never use the E-stop button midcycle unless all other halt actions would be too slow to be safe. In addition to stopping all axis moves, this button stops the spindle, coolant, and any other auxiliary functions being used. On older CNC machines with encoders, the PRZ location is lost because this action removes power to the feedback system, resulting in a lost register count. Later we'll learn several halt and **midcycle stop** techniques, all of which will be a better choice for situations that could become emergencies but aren't yet.

KEYPOINT

E-stop is for all-out emergencies. It may disable all safety brakes, allowing the spindle to coast to a stop. It is also used to power down at the end of the week or day. It helps guarantee that the machine can't be accidentally started.

Multiple E-Stop Locations

For quick access on larger machines, there might be more than one E-stop button. Machining centers might feature one on the front panel and one at the tool turret. Lathe operators might find one at the bar feeder and one by the coolant/chip system as well as the one on the main control panel. An E-stop is often part of the hand pendant, too. Check them all. If any single switch is depressed, the machine won't start.

KEYPOINT

Safety

As an operator, it's a good idea to review the location of all E-stops occasionally so you can react quickly in an emergency.

If, during machine use, the E-stop button is hit, you may need to reinitialize the machine, repower the axis drives, then manually jog (see the Terms Toolbox in Unit 18-2) the cutter away from the work or danger area, then start the program over. Okay—everything is ready to go. Push the green button to power up the control!

UNIT 18-1 REVIEW

Replay the Key Points

- Know and review often where the emergency-stop buttons are located.
- Depressing the E-stop button during a production run nearly always has a marked effect on setup and program operation.
- To start certain machine index tool turrets to position 1, close chucks and safety doors, and start hydraulics at a specific time—before starting controls, home the axis drives to reset the position registers, and coordinate the cutting tool inventory.

Respond

1. Describe why some CNC machines must be homed before they can be used to cut material.
2. __________ a machine may be fully automatic, requiring a single button touch, or it may require several steps in a given order.
3. Why does the E-stop button remain depressed when it is touched?
4. You need to change a tool diameter in a vertical mill setup. You *cannot* do that from which modes?
5. How is an operator pendant different from an RIS controller?

Unit 18-2 Controls for Setup and Operation

Introduction: Good job, the machine is up and ready to go! Unit 18-2 explains the five basic controls that all operators must know to exercise the machine's axes and spindle, load a program, set up the machine, and get the whole thing going. Again, the controls may be slightly different from machine to machine, but each must be present in some form. Before proceeding, Haas machines require you to perform a spindle run-up program stored in memory, or you can write your own.

TERMS TOOLBOX

Distance-to-go (DTG) screen A video display of the amount of travel yet to occur to complete the current command.

Hard keys Dedicated entry keys on the control panel.

Hard limit The emergency end of axis travel.

Jog A manual movement of a CNC axis, instituted by the operator.

Manual pulse generator (MPG) A rotating dial that has many finite positions (clicks) representing one jump of the axis selected.

Multiplier controls (rate control) A three- or four-position switch that sets the amount of jog feed per jump or pulse.

Override A dial control to fine-tune feed rate, rapid travel, and spindle RPM.

Slide hold The little red button that halts all axis motion with no consequence to other functions or the program.

Soft keys Control entry keys positioned just below the video screen that assume different functions depending on the mode and page selected.

Soft limits The switch that stops the normal range of axis travel. There is no consequence for touching this switch other than axis stoppage.

18.2.1 Mode Selection

Operation modes are used to simplify and make tasks more efficient and safer by putting the functions needed at the forefront. Each mode allows some functions while banning others. Usually, this prime control is a rotary switch as in Fig. 18-11, but it can also be a screen selection. Selecting the mode of operation is the first action to take.

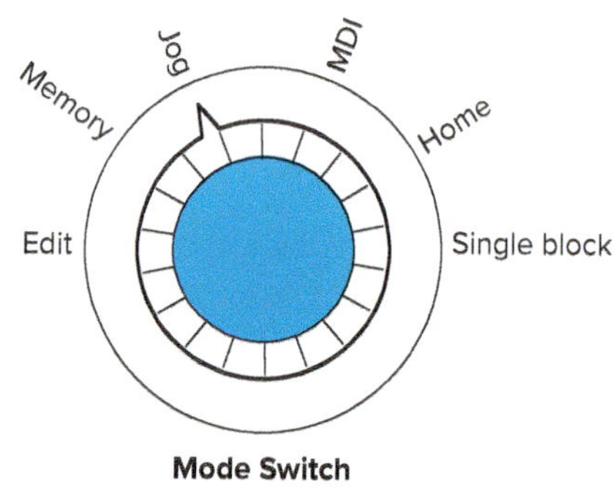

Figure 18-11 A rotary mode selection switch is common on a CNC control.

Modes and Submodes

Refer back to Fig. 18-2, and note that the dark gray arrow keys on the upper right are the six modes on a Haas control. The keys to the right are the functions available in that mode. Note, MDI means manual data input, which is where you write small programs at the controller. That mode might be Edit on another control.

Haas Modes

- Edit—Editing existing active programs
- Mem—Operating the machine automatically from a program
- MDI DNC—Various functions
- Handle Jog—Manually moving the machine's axes
- Zero Return—Homing the machine
- List Programs—Manage and select saved programs

Possible Other Modes on CNC Controls

Home

Home provides access to the automatic homing cycle or to manual performance. Home may or may not be present on newer machines with constant positioning capability.

Jog

The **jog** mode allows the operator access to manual motion functions discussed next. For safety, this mode disallows automatic program functions as the operator is usually observing indicators, tapping vises, and generally within the work envelope. This mode must be selected to enable the operator pendant.

CNC or Autorun

This is the program run selection. May also be called *auto*. When executing a full program run, the mode will be set here.

Single Step (Single Block)

This operates one command line at a time, then waits for another touch of the (green) cycle start button. A single step is used mostly during program tryout.

Manual Data Input (MDI)

MDI allows one or more command lines to be written at the keypad. They can then be executed, if so desired, using the cycle start button. The functions here might also be part of edit.

Edit

This mode enables full program memory and storage access except to the program currently in the active buffer (the one being used at the time).

TRADE TIP

Some controls allow multitasking. One can run a program while editing another, but not the same program at the same time. To work around editing an active program, a copy is made, then renamed. When editing any program, remember to add a note about the revision level in order to not confuse Rev levels.

Memory

This mode is for file management, calling a program from permanent storage, or performing a download or upload from a host server. Memory is also used by some controls when writing a program from the keypad.

Hard Keys Versus Soft Keys

There are two types of entry keys on a control panel (Fig. 18-12).

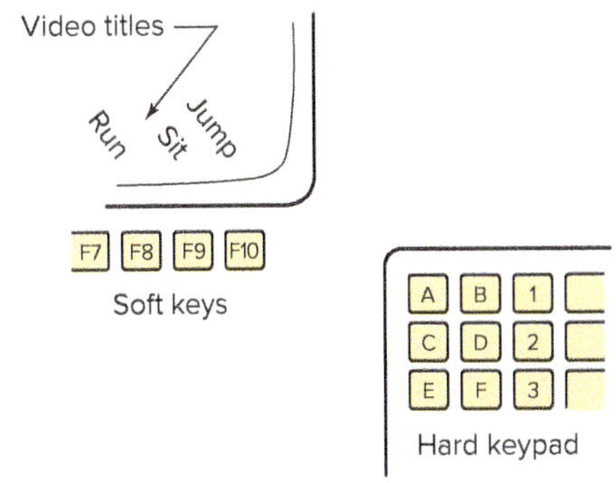

Figure 18-12 Hard keys have a single or possibly a dual purpose; soft keys have multiple purposes based on the mode and video screen selected.

Hard Keys

These are normal keys associated with computers. **Hard keys** may be plastic (as on your PC), or they may be membrane keys (as on cell phones) under a sealed layer to protect them from coolants and chips. Each hard key has a dedicated primary purpose and possibly a selectable second or even third function similar to those on calculators.

Soft Keys

Depending on the mode, the selection of **soft keys** expands the possible uses of function keys at the top of the keypad. They work like the F-1 through F-9 or F-12 keys on your PC; by adding video titles above them as a mode or screen page is selected, they assume new roles. Note F1 through F4 on the Haas. Each has a different effect depending on the mode.

18.2.2 Manual Movement of CNC Axes

Homing Axes

There are various reasons to home almost all CNC machines for setup and tooling purposes. Here's how: After selecting the home mode (Zero Return; see Fig. 18-1), the operator drives all axes to their home position, either manually using controls or with an automatic cycle, but only after the cutter is jogged clear of setup obstacles. Then the axis is slowly driven upon the limits, and the position registers are set to zero. Doing so requires an understanding of limit switches. See if you can find it on the Haas.

Limit Switches

The work envelope of a CNC machine is protected and contained by limit switches that stop the axis drives before hitting the physical end of travel. Damage would occur if an axis were driven onto its full mechanical limit–the end of the ball-screw. There are two kinds of limit switches you must understand.

Soft Limits (Normal Operation End)

Figure 18-13 depicts a simplified version of a **soft limit**. It's an optical cam device that signals the control to slow down, then stop. On most CNC machines, the soft limit switch can be contacted at full rapid travel speed with no consequence to program or position due to its deceleration function. The axis will simply stop traveling fast when the soft limit is reached.

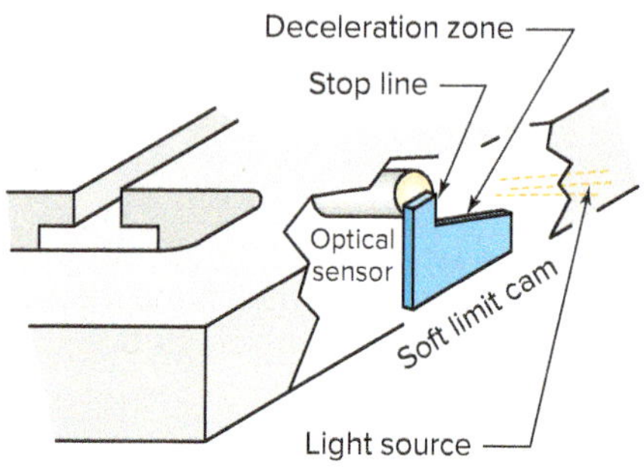

Figure 18-13 A simplified soft limit stops axis movement with no consequence to the program or setup.

> **KEYPOINT**
>
> Once an axis is on the soft limit, a command in the reverse direction requires no special prefix or action. Soft limits are the boundaries of the normal work envelope.

> **TRADE TIP**
>
> **Parking on the Soft Limits** A programming trick is often used when parking a machine for a part change. The programmer or operator commands the *Z* axis to rapid away 100 in. Although an easy figure to write, most machines haven't that much *Z* axis envelope. The objective is to simply drive the machine upon the soft limits, where it stops.
>
> When machining extremely heavy loads, on older indirect feedback machines, it may be necessary to refresh the coordination of the axis registers due to mechanical slippage and mathematical rounding in the control during rough cuts. After roughing is done, add a homing command into the program; then do the finish cuts with refreshed axis coordination.
>
> **Caution!** Do not try this parking tip without determining if the machine features dual soft/hard switches. Hitting hard limits at rapid speed is a crash.

Hard Limits (Crash Stops and/or Axis Refresh)

A **hard limit** is a physical contact switch beyond the soft limits but still located before the full mechanical limit of the axis travel. This switch is the absolute end of powered axis travel. The hard limit is a never-changing position used to zero axis registers. Contacting the hard limit during feed or rapid movement stops axis motion and puts the machine in an alarm condition.

Backing Off Hard Limits Once on the hard limits, the operator must make some overt action such as touching the jog and slow buttons at the same time, or the reset and jog together, to allow the machine to back off the limit toward the open work envelope again. The action will be different depending on the control, but it ensures that the axis is moved slowly and correctly away from the physical end of travel.

> **KEYPOINT**
>
> Hitting a hard limit during program operation is considered a minor crash.

18.2.3 Manual Axis Setup Movements

Your first lesson and an important one! There are four ways for moving machine axes for setup work, for homing them, and for driving a cutter (with no program) through material to machine a fixture or chuck jaws. One or two methods must be part of every control.

> **KEYPOINT**
>
> **Warning**
>
> The force behind the axis is great and a very clear understanding of the following controls is necessary before attempting to use them. Access these methods under Handle Jog on the Haas control.

1. **Continuous Jog**

   This is a steady axis movement as long as the jog button is depressed or Jog Lock is chosen (Fig. 18-14). On our PVC control the rapid feed override selects the speed at which the axis moves (discussed later). On the Haas it's the increments buttons. To use it on your Haas mill after selecting a slow increment, 0.001 in. for example, then touch the Jog Lock button, then the direction you wish to move the spindle relative to the table. CAUTION: First move away from the work to be certain there are no surprise movements.

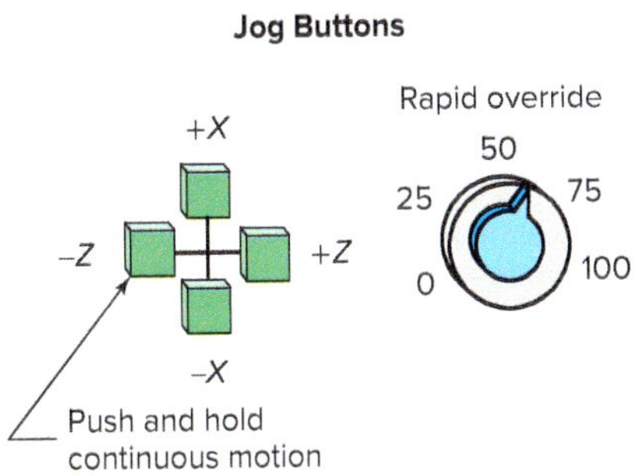

Figure 18-14 *Continuous jog* moves the axis selected while the button is depressed. The rapid override is used to determine the speed at which the axis moves.

2. Incremental Jog

The machine will jump a given amount with a single push of the button at the speed selected on the override (numbers by Handle Jog). A selector switch, called a **multiplier** (Fig. 18-15), is used to select the distance covered in one jump. For example, if the machine's resolution is 0.0001 in., then select

Multiples of the resolution
*X*1 = 0.0001 for each push
*X*10 = 0.001
*X*100 = 0.01
*X*1,000 = 0.1

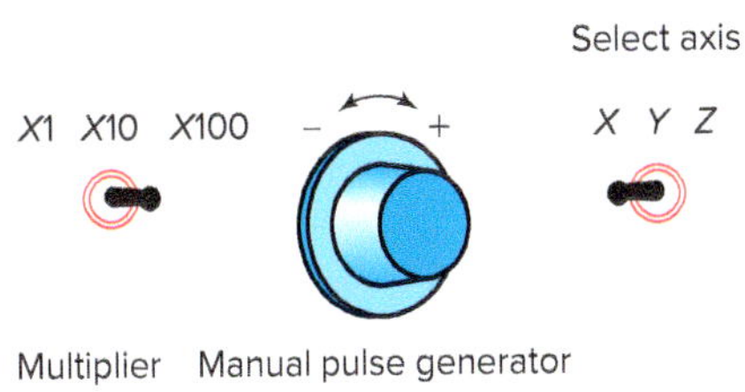

Figure 18-15 The resolution *multiplier* applies to either *incremental jog* or to *handwheel (MPG)* movements. For either movement, the axis and multiple of the resolution must be selected.

Incremental jog is used during setups to move the machine an exact distance in multiples of the machine's resolution. If the control has this feature, the same buttons are often used for both continuous and incremental jog. Not all CNC controls feature incremental jog.

Manual Pulse Generators

A common method of moving an axis, this control might also be called a "handwheel." As the relatively large dial is rotated by hand, small clicks can be felt. Each click, or pulse, represents one jump of the selected axis at the rate selected (see Fig. 18-1). Handle Jog, the **manual pulse generator (MPG)**, is a handy combination of incremental and/or continuous motions. If the dial is rotated continuously, a fast motion is obtained, but one jump is also possible by turning it one click. The jump amount is determined from the multiplier switch and the speed by the override dial.

In Fig. 18-15, the *X* axis will move 100 times the resolution (normally 0.0001 in.) per pulse = 0.010 in. Rotating the dial counterclockwise produces positive *X* motion. On machining centers, to aid in setups, the MPG is often placed on the pendant to be taken into and around the machine to improve viewing angles.

TRADE TIP

MDI (Manual Data Input) Axis Movement Here's a handy fourth method of moving one or more CNC axes a given exact distance. It's used by handwriting one-time-only, mini-programs at the controller. Full edit capability must be unlocked to use it!

After selecting the correct mode, the operator keys the needed commands. Then with a push of the cycle start button, the machine will obey that single command block, once only.

All jog actions described thus far cause single axis, straight-line moves, while this method expands the possibilities. Using an MDI statement, any motion within the capability of the control is possible: multiaxis linear and even circular. So if you wanted to move 1.375 in *X* while moving 0.250 in *Z*, you could enter: G91 G0 X1.375 Z.250. By entering that block then touching cycle start, the machine would execute the two axis movements at rapid travel incrementally (G91).

18.2.4 Machine Halt Controls

Making the Right Halt Choice

When an emergency happens, the need for practiced action is critical. Many crashes get worse due to inaction or worse, the wrong action. Make sure to know your stop and halt options. It's important to be prepared to make a quick decision between the big and little red (feed hold; see Fig. 18-1).

Remember, the big button ends the program and possibly disturbs the setup PRZ. However, never hesitate to hit it if safety is an issue. Reviewing the possibilities for stopping and halting, from time to time, is a good idea.

In Chapter 19 we'll investigate your operator actions when stopping the machine in situations ranging from in-process adjustments to an all-out emergency. Here in Chapter 18, we'll discuss the various controls used to halt the machine during a program.

KEYPOINT

There is a difference in stopping and halting. Halting implies that the action is temporary and normal: to measure a key dimension, to move a clamp, or to add coolant or brush away clogged chips. After halting, machining will resume. Halting is an operator responsibility, and the emergency-stop button shown in Fig. 18-10 is *not used in this capacity.*

1. ***Slide Hold***

   This control may also be called *feed hold.* It is present on most CNC controls (Fig. 18-16). The red slide hold button is smaller than the E-stop button. It immediately halts all axis motion with no other consequence to the program. Spindle and coolant and any other secondary functions such as clamps, lights, and so on continue as programmed or set from MDI. With a second touch of this button or possibly the cycle start button (depending on control), the program resumes normal motion.

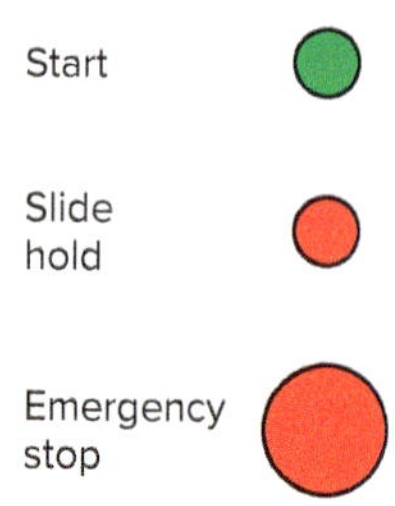

Figure 18-16 The three main operator control buttons are color coded: start (green), slide hold (red), and emergency stop (red).

TRADE TIP

Halting While there are no data, setup, and/or program consequences to halting, there may be physical machining effects. With the spindle continuing to spin, the cutting tool might be dwelling on the work. This can lead to work hardening, dull cutters, or unwanted marks in the material. Most controls allow a spindle stop during a slide hold.

2. **Turn the *Feed Rate Override* to *Zero***

   Here the halt is made by reducing feed during a cut to nothing. Because the spindle or milling cutter will continue to rotate with the cutter possibly in contact with the work, this method is not recommended. But small adjustments such as aiming the coolant can be quickly made in this way. See rapid override for another method of stopping the machine for small adjustments.

3. **Switch the *Mode Control* to *Single Block* or *Cycle Stop***

   The machine will finish the current command, then halt. The cycle start must be pressed to resume activity. Leaving the control mode in single block, it is possible to allow one command at a time using the start button, until the cutter is clear of the work; then fix the problem.

4. **Program Interrupt**

   Many controls feature the useful ability to halt machining, then take further axis actions such as moving a dull cutter off the work or stopping the spindle, then replacing it. Using program interrupt allows all jog and manual motion actions plus other functions normally disallowed during slide hold.

   With the problem resolved, with a touch of the green button the control will reset all machining and logic conditions within the program, then slowly move the cutter back to the exact position where the interrupt took place. It will continue normally.

18.2.5 Video Screen Selections

Depending on mode choices and tasks the video screen can be set to display a wide range of information. Most allow operator selection of split screens with more than one selection. Here are the four basic screens often selected for setup and operation.

Tool Path Graphics Screens

This selection (Fig. 18-17) can be used in three very different ways and is often selected for one half of a split screen:

1. **Program Monitoring**

   This process tracks the position of a cutting tool using a picture of the progress. When coolant, chips, and safety shields obscure the machining, this is one way of tracking progress.

2. **Program Dry Run**

   The first test of a program can be made on the video screen with all axes locked out–using the slide hold or axis lock switches. The graphics will provide a rough picture of the program path with no material being machined.

3. **Graphic Programming**

   An option on some advanced controls, the program is developed by drawing elements on the screen similar to CAD drafting.

Figure 18-17 When you cannot see the cutting tool, monitor progress with screen graphics.
Courtesy of Shoreline Community College

KEYPOINT

Verify a clear understanding of the following position screens. This knowledge is absolutely necessary for efficient, safe setup and operation.

Coordinate Position Screens

There are three kinds of position screens:

1. **Machine Coordinates Screen (Mach Coords)**

   This screen shows the position of the spindle or tool with respect to machine home independent of any program reference position that's been set. Machine coordinates are mostly used for setup functions, locating the PRZ, and aligning holding tools. The mach coords screen is available even on machines that do not require a homing operation. Machine coordinates are used to check major items such as clearance to tail stocks on lathes or approaching full travel range of an axis. It is unlikely that you will use machine coordinates to monitor a program in progress because the distances refer to the machine position rather than the tool to work position.

2. **Work Coordinate Position Screen (Work Coords)**

   This is one of two screens most likely to be displayed during a program run. The current position of the tool is shown relative to the selected PRZ. This, along with the **distance-to-go (DTG) screen**, are the most common screens.

3. **Distance-to-Go Screen**

   The amount of travel left to complete the current command line. This screen is extra useful when trying out a new program. Used in conjunction with the single block mode and feed and rapid override controls, a block of commands can be read and started, then halted at a questionable position. Then with the remaining distance to go displayed, the operator can physically measure the gap between the tool and the chuck, for example.

Split and Combined Screens

Various combinations of these screens may be selected by the operator. For example, during a proven program run, operators will often display machine and DTG and work coordinates along with graphic progress.

18.2.6 Override Controls

There are three functions on a CNC control panel that allow fine adjustment to spindle RPM and axis movement rates. We've already discussed the rapid **override** which adjusts the jog rates and also the rapid travel rate during a program. On the Haas control they are the lower-left dark gray buttons. Once the function is chosen, Feed, Rapid, or Spindle Speed, the handwheel is then used to control the override.

Notice in Fig. 18-18 the rapid override can be set all the way to *zero* (a trade trick method of stopping a CNC machine when it encounters the next rapid movement command in the program). Turning the override to zero, the machine will complete the current feed rate commands until it encounters a rapid travel command, at which point it stops moving until the override is turned up again (see Fig. 18-18).

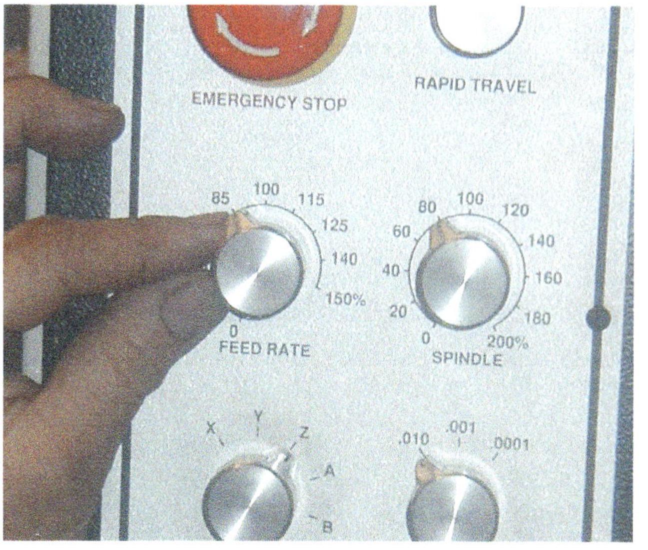

Figure 18-18 The *rapid travel override* control is used to slow programmed rapid moves and manual job commands.

Lake Washington Technical College, Kirkland, WA/McGraw Hill

KEYPOINT

Caution

In Fig. 18-19, stop the spindle before attempting to measure physical distance to go.

Spindle Speed Override

This control works during manual use of the spindle and during programs (Fig. 18-19). Notice that it not only decreases RPM below the programmed rate, but it can also increase it from 150 percent to 200 percent above that which has been entered.

KEYPOINT

Rapid Override

While the fastest possible rapid moves are desirable to trim cycle times, the rapid speed is slowed during a first tryout of a program. During program tryout, turning it to zero—the controller will finish the current feed rate command, then halt (a quick way to evaluate the next rapid move)—is it safe? If yes, then turn it up and let it continue.

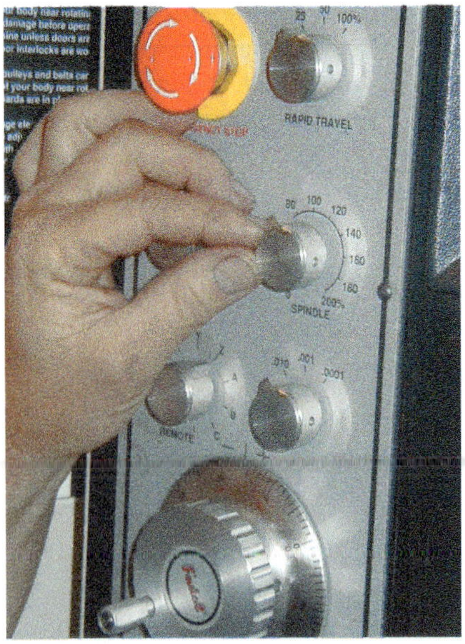

Figure 18-19 The spindle speed override switch can reduce RPM from 0 to 10 percent of the programmed rate or increase it up to 200 percent.
Lake Washington Technical College, Kirkland, WA/McGraw Hill

Feed Rate Override

This dial works similar to the spindle RPM control. It can adjust down to *zero* and up to *150 percent* of that which has been programmed (unless the machine is at its limit feed rate already).

UNIT 18-2 REVIEW

Replay the Key Points

- On some controls, edit mode is used when creating a new program from the control keypad. On others, program entry is done in MDI mode.
- Once an axis is on the soft limit, a command in the reverse direction requires no special prefix or action.
- Soft limits are the boundaries of the normal work envelope seen in Fig. 18-13.
- *Caution!* Do not try moving an axis to the limits without determining if the machine features both soft and hard limits. Hitting hard limits without the deceleration of soft limits at rapid speed is a crash!
- The force behind the axis drive is great and a very clear understanding of the controls is necessary before attempting to use them.
- There is a difference between stopping and halting. Halting implies that the action is temporary and normal, for example, measuring a key dimension, moving a clamp, or adding coolant.
- Knowing the difference between the three coordinate sets in a CNC control–machine, work, and distance to go–is absolutely vital to setup and operation.
- Always stop the spindle before attempting to measure physical distance to go.
- While the fastest possible rapid moves are desirable to clip cycle times, the rapid speed is slowed during a first tryout of a program.

Respond

1. Describe the difference between the big and little red buttons.
2. What does an MPG handwheel do?

Critical Thinking

1. Other than running a stored program, what is the only axis movement method that can make curves or move two or more axes simultaneously?
2. Identify the master control that allows certain functions while banning others.
3. Of the three override controls, which cannot go beyond 100 percent? Why?
4. Why do some machines not require an initialization?
5. A program is being tried out for the first time on a vertical machining center. During a rapid move, the slide hold is touched, and the axes are read on the DTG screen *X*1.5065 *Y*0.0055 *Z*1.0030. With the spindle stopped, the cutter is found to be 1.030 in. above the work surface. Is it safe to allow this command to continue?
6. What does the *mach coords* screen tell the machinist?
7. Your machine refuses to start up. There is no action at all when touching the start button. What might be the problem?
8. Manually jogging an axis can be in three forms. Name them.

CHAPTER 18 REVIEW

Unit 18-1

While I've tried to describe the various controls in a general way, every machine is different. Even two from the same manufacturer might have different Rev levels in their operating systems. As unlikely as that might seem, I've been in just such a situation. Moving from one machine to the next, I had to remember which of the two I was operating. One required a separate hydraulic start-up that had to happen after initializing the control. One started its hydraulics automatically. The first had no low-lube warning, while the second did. There were other differences, too, even though they were identical machines outwardly.

Unit 18-2

Problem 15 coming up will ask you to go to the shop and identify the basic controls on a unit in your lab or shop. This is more than an academic exercise. You'll be doing the same quick orientation on new equipment throughout your career.

While these universal controls are sure to become even more compact than the RIS unit shown previously, they have carefully chosen functions that must be present on any machine in the foreseeable future. Learn to find them first; they are the controls you'll need to be safe and efficient as soon as possible.

The best policy is to first read the start-up and control sections of the operator's manual. Then ask your lead or another machinist who is familiar with the machine, if you don't know for sure.

QUESTIONS AND PROBLEMS

1. On a sheet of paper, identify the CNC controls A through H in Fig. 18-20 and describe the function of each. (Unit 18-2)

2. Describe hard limits and soft limits in 10 words or less. (Unit 18-1)

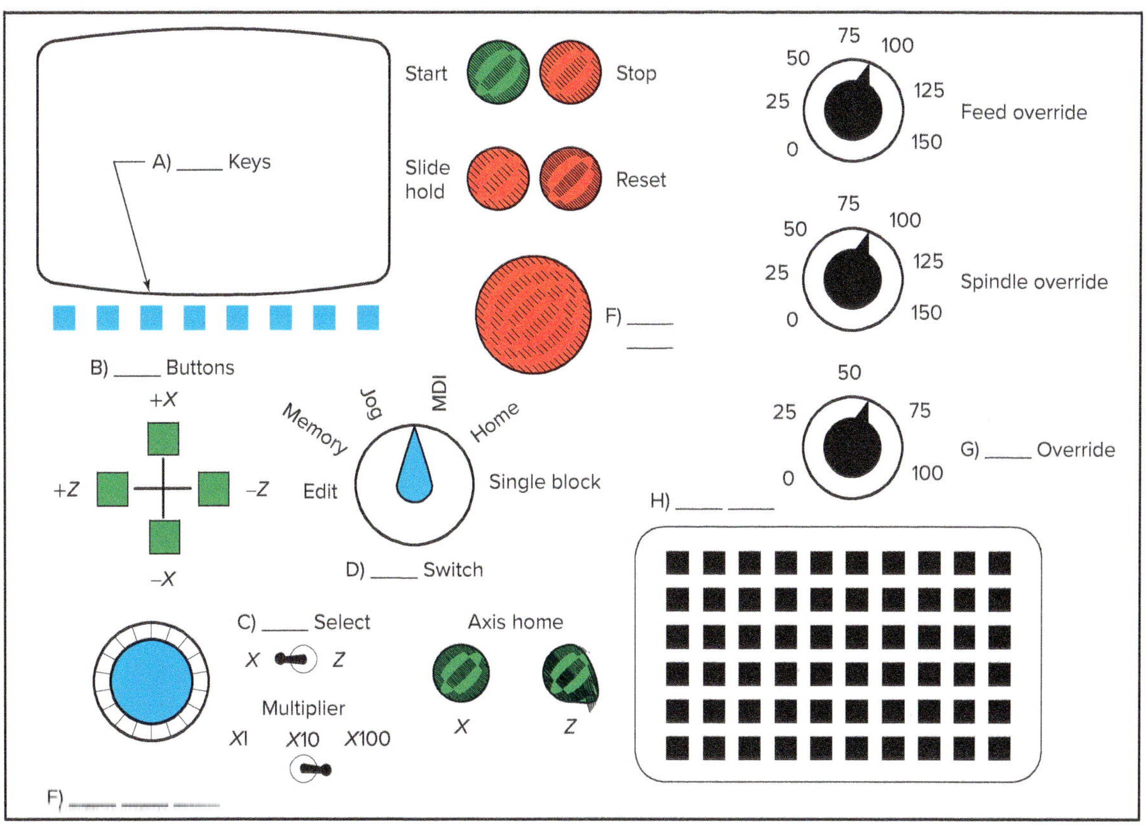

Figure 18-20 Identify the lettered controls and features.

CRITICAL THINKING

3. True or false? A CNC milling machine must be homed in order to set the video *X, Y, Z* work coordinate position to zero. If it is false, what will make it true? (Units 18-1 and 18-2)
4. Before starting a CNC machine name four items that should be checked. (Unit 18-1)
5. To perform a program *edit,* name two tasks that must be accomplished. (Unit 18-2)
6. Without looking back, name the other mode selections beyond edit. (Unit 18-2)
7. Name five possible methods of moving a CNC machine other than using a stored program. (Unit 18-2)
8. What button stops all axis motion with *no consequence* to the program? (However, there may be a consequence to the work.) (Unit 18-2)
9. What would happen during a cut in progress if you push or turn
   - **A.** rapid override to zero?
   - **B.** feed rate override to zero?
   - **C.** spindle speed to 125 percent?
   - **D.** emergency stop?
   - **E.** single-step mode? (Unit 18-2)
10. What do hard limits prevent? (Unit 18-1)
11. True or false? In using the MPG on a machine with a resolution of 0.0001 in., you wish to increase the rate from 0.001 to 0.010 in. per click. You would select the *X*10 multiplier from the mode switch. If it is false, what will make it true? (Unit 18-2)
12. What CNC functions may be overridden by the operator? (Unit 18-2)
13. Identify the three CNC controls in Fig. 18-21 that work together and explain their purpose. (Unit 18-2)
14. Not directly covered in the reading: You are to try out a new program. From what you have learned, name the panel controls and video screens you will use. (Unit 18-2)
15. Wearing correct eye protection, observe one or more CNC machines in your lab and list the controls and functions you recognize as compared to the generic control. Note, there may be different names and other controls not discussed here. (Unit 18-2)

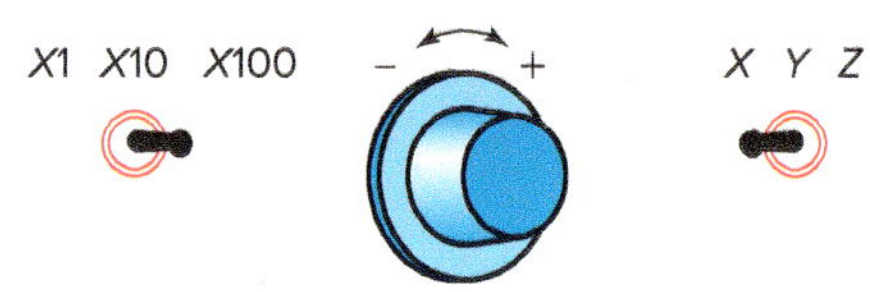

Figure 18-21 Identify the *name and purpose* of these three CNC controls that work together.

CHAPTER 19

Operating a CNC Machine

Units

19-1 CNC Planning

- Identify the PRZ location instruction
- Determine holding tooling based on datums and PRZ
- Select cutter tooling within the job parameters
- Define two kinds of cutter-compensated programs
- Read special job requirements

19-2 Controller Alarms

- From operator's manuals, classify alarms
- Take corrective action

19-3 Operator Actions (During Production Runs)

- Make a three-stage crash plan—practice emergency stops
- Look for ways to improve part-to-part turnaround time
- List the ways to monitor a CNC run
- Develop proactive operation by remaining alert

19-4 Monitoring and Adjusting

- Select and use the best monitoring method for the machine and control type
- Use tool memory pages, enter corrective adjustments

INTRODUCTION

Taking command of a production machine is a big responsibility. It means producing quality work at a rate that creates profit. But there's more: You must keep yourself safe, keep records of production, and maintain the machine and work area. This chapter and the next provide the information and skills needed. But they are also designed to teach more than the knowledge an operator needs; they are aimed at the upcoming craft master!

Typically in industry, new people are given the job of putting parts into a setup already proved by a lead person. You'll touch the green button to wait until the program ends to replace the workpiece with another. That's known as the *button-pusher* phase of your career. While it may not be the greatest challenge, don't underestimate the potential for learning; there's plenty of valuable knowledge and experience to be gained.

For three decades, the SME Education Foundation (www.smeef.org) has dedicated its resources to inspiring, preparing, and supporting our nation's youth toward careers in manufacturing education. It provides connections to K–12 STEM education programs, scholarship money for those interested in an advanced manufacturing career, and supports youth awareness camps every year for middle school children. The foundation is supported by industry and private contributions and is continuing the effort to change the image of manufacturing, inspiring the next Henry Ford or Steve Jobs.

Courtesy of SME Education Foundation

Unit 19-1 CNC Planning

Introduction: In a small shop or tech lab, you'll plan and execute your own program. That means you'll know all the facts about the setup and the tool path when trying it for the first time. From the beginning, you'll know what cutting tools will be used and what their tool numbers are, where the PRZ is located, and the sequence of events. Before writing the program, you've planned it all. The point is that there are no facts to find out.

But in industry, before the job reaches the shop floor, several people work on it. The planner, the programmer, and the setup and tooling people all make choices that affect the job; all that communication must be clearly understood by you the operator, or disaster is sure to follow.

In most shops, the setup and operation facts are recorded in a document similar to a work order. In some shops, the CNC document can be part of the work order but it's usually separate because it changes far more often than a normal work order.

TERMS TOOLBOX

Compensated program A program prepared to look for offsets.

Cove An internal detail that sets the upper limit for cutter radius.

Cutter centerline (cutter path) A less common compensated program, where a given cutter radius has been built into the path but can be compensated with offsets. See *minus compensation.*

Flank interference A compensation challenge on lathes whereby the control must keep the sides of the tool from touching the workpiece.

Generating A curved cutter path that creates a part radius larger than the cutter radius. The motion preferred to forming.

Minus compensation A negative offset number applied to a cutter centerline program to bring the cutter closer to the part geometry.

Offset A variable number entered into controller tool memory by the operator. The program refers to the offset by the tool number or from a code word in the program.

Optional stop (opt stop) Given the right halt code in the program, the operator can choose to halt at certain places or to switch them off and run through the halt.

Part path program A program based on positive compensation offsets away from the shape of the part. The more common program type.

Radius offset The tangent distance away from the part geometry for lathe or mill cutters.

Remarks Notes embedded within the program.

Tool bias/approach vector Defines the lathe tool's orientation in the setup, thus the way it must move toward the workpiece and away during compensation.

Tool orientation The direction the lathe tool is pointing.

19.1.1 Vital Setup Information You Must Have

A CNC document conveys two kinds of information:

Machine setup facts	**Program run instructions**

Those critical instructions are found in two possible places:

Hard-copy document

Embedded remarks within the program

Differences from a Standard Work Order

First, you'll notice that the step-by-step operation sequences are gone. Those facts exist within the structure of the program. In their place, you'll find details about the overall sequence in which sets of cutter operations are to occur and there will be a specific cutter that will carry them out. For example,

T01 = 6-in. carbide insert tooth face mill

Opps #N015 through N240

Face cuts to top surface

In addition to the usual work order information found in the heading of the document–print, part and revision numbers, material type and heat-treat condition, grain and type of material (castings, forgings, bars, etc.), and part counts–the CNC document lists the following:

- *Program identification code with revision level* (see Trade Tip)
- *Program location*–central computer, disk, tape, and so on
- *Specific machine for which it is processed*
- *Special jigs/fixtures* with ID numbers for this job
- *Raw material size, alloy, and heat-treat condition*
- *Standard tooling required* along with their tool number in the program
- *Custom tooling required* and their tool numbers
- *The PRZ location in the setup*
- *Specific location and kind of chucking/holding/clamping tooling*–This is a critical factor because the program expects them at a given place and height above the work.
- *Event sequence*
- *Any special instructions particular to the job*, for example,
  - › *Clamp changes* at midprogram halts
  - › *Warnings* about a large piece of excess material that will fly off the work at a given point in the program
  - › *That the part must be rotated* or *repositioned* in the chuck at a certain halt point
  - › A key dimension that must be measured a given halt point

KEYPOINT

The CNC documentation foresees potential problems, eliminates guesswork, and gets the job going right with the least amount of danger and experimentation.

19.1.2 Operation Remarks Within the Program

Most dedicated controls and all PC-based controllers allow **remarks** to be embedded in the program where they can be emphasized in bold or flashing characters (if commanded to do so) on screen, at the right time. In addition to making them appear at the best time to be noticed, embedded remarks can easily be updated by the shop as better ideas are found. (Check policy to edit them.)

TRADE TIP

Program Revisions Editing and tracking program updates are huge issues in CNC work. We often improve a program several times when it's run for the first few times. There will be a very well-defined policy about who edits them and when they are changed, and especially how the data are managed. Small edits such as feed rate changes might be allowed from the machinist or the lead person. But those changes must be communicated with the programming department, and the revision level for the program must be updated if they are incorporated. *Large changes are usually done only by the programmer, and they must be formally documented.* Due to potential data mix-up, some companies do not allow machinists to edit.

When a program is edited, the revision level must be included in the program header and also within the setup document. The revision level of a program can be tracked by date and time or by a Rev number/letter as on drawings, or it can be a serial number.

KEYPOINT

Matching the program's header Rev level to the CNC documentation, data crossing can be avoided.

A. When a job and program are complete, there will also be a policy about what to do with the old program in permanent storage in your machine's memory. Leaving it there is risky! It is most often required to be erased or returned to the program department.

B. Program revisions must be tracked in the program header to be certain that the data are the version required.

Each control has its own way of setting these messages aside without attempting to read them as program commands. Here are the two most common methods:

1. **REM: This is a special note _____ _____ END REM:**

   The control displays but ignores characters between REM and END REM. The control recognizes the REM as a bracket.

2. **(This is a special note _____ _____)**

   The control displays but ignores all characters between the parentheses.

The remarks are put in the program as warnings or flags of special events or they might describe better or safer ways to run the program (Fig. 19-1). For a simple job, a complete CNC document may be nothing more than a few lines of remarks in the program, perhaps accompanied by a sketch similar to Fig. 19-2. But the documentation can also be several pages of critical information. Now, let's examine each category.

```
Again Buffer Copy Delete Find - find Get Insert Jump          --space--
-----
:%
(PROGRAM 256-56b - HYDRAULIC FITTING - PROGRAM 2)
(REV JAN 12 - 2006)
(PRZ @ LLC, TOP OF WORK)
G80 G90 G40 G20
S2000 F20.0
N005  T01 M6 (.5 EM - 2 FLUTE)
N010  M3  M7
N015  GO  X1.00  Y2.500 (TO TOP OF RADII)
N020  Z.25
N025  G01  Z - .250
```

Figure 19-1 CNC instructions can be embedded and emphasized in the program.

Hill Machining - CNC Documentation			
Dwg # *106-54-309*	Rev *A*	Prog # */106-54*	Prog Rev *22*
Part Count *50*	Mat *7775 Alu*	Programmer *Fitz*	Date Orig *2-2004*
Holding Tools/Fixtures		Notes	
T01 *.5 ⌀ EM-2FL*	T02 *5/16" Drill*	T03 *3/8" Reamer*	T04
T05	T06	T07	T08
T09	T10	T11	T12
T13	T14	T15	T16
Special Instructions	*Parts are pre-cut with datum edges finished Program profiles contour*		
PRZ Setup Sketch			

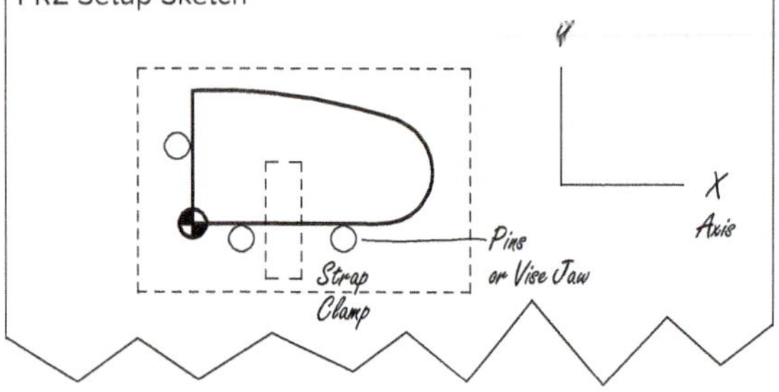

Figure 19-2 Typical simple CNC setup and run document.

19.1.3 The PRZ Location

At the top of the list, in pictures and/or in words, the PRZ location and axis orientation will be noted. It's your job to make it so on the machine. Also noted is the method of locating multiple parts in a batch, such that each nests in the setup with its datum basis and the PRZ in a constant repeating position.

The specific setup that creates the PRZ may be part of the instruction, or it may be left up to the machinist. Figure 19-2 shows two pins on one side of the part, and a third on the other along with a clamp location. If you miss those details, one of two problems might happen:

1. **Caution!**

   An accident can occur as the cutter hits the wrong location on the part or even tries to cut the holding tooling. Because this is an obvious danger, it is an uncommon mistake.

2. **Introduced Variation**

   This is a far subtler and common problem for beginners whereby needless tolerance is consumed. To understand this potential, answer the critical questions.

Introduced Variation Questions

The job is to CNC drill three holes in a batch of parts as shown in Fig. 19-3. On the print, note that the three holes are to be located relative to two datums and their intersection should be set up as the PRZ location for this job. For this exercise, the work has been premilled square, and the radius is already machined on the left corner.

The 15.0-mm and 20.25-mm positions are basic to the hole location. The overall length of 75.0 mm has a loose tolerance of 1 mm.

The setup in Fig. 19-3 will accomplish the goal. The solid vise jaw provides Datum A, while the end stop establishes Datum B. However, by incorrectly setting up a material stop on the wrong end of the work, as shown in Fig. 19-4, the operator has misplaced the B locator relative to the PRZ. The Datum A locator is correct, but Datum B is now on the wrong end.

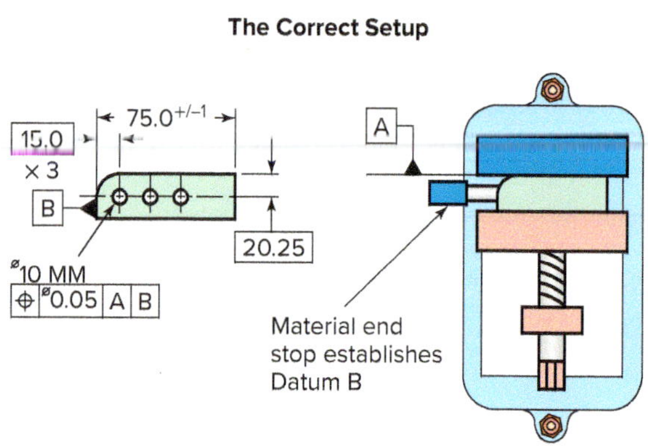

Figure 19-3 The part datum locators are placed right. Every part will reproduce the PRZ position.

The Wrong Setup

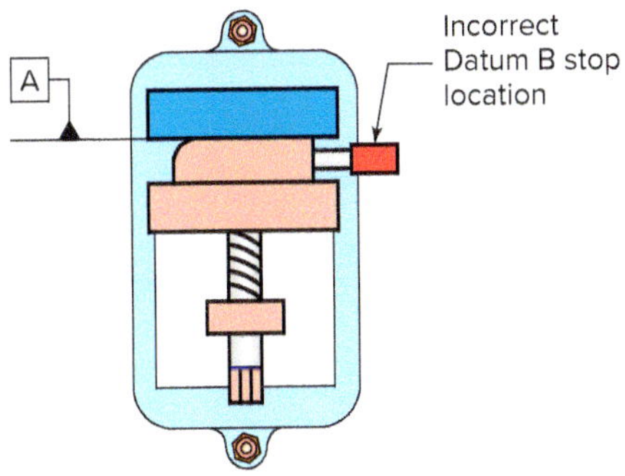

Figure 19-4 What variation does this cause?

Question

What are the consequences of this error? No accident will occur because the part is nearly symmetrical. But in this setup, by how much could the 15-mm end space be missed with respect to Datum B?

Answer

The end spacing to Datum B can be shifted by as much as the overall length tolerance for the bar, of plus or minus 1.0 mm. That's greater than the hole end spacing tolerance. The result is that the hole pattern will shift left or right with the overall length variation of the parts.

> **KEYPOINT**
>
> With the Datum B locator incorrectly placed, the holes will now be referenced from the right end, not the critical left, datum feature B.

But that kind of error is obvious; this next one is not. In an attempt to continue using the wrong setup, the length of all the blank parts was premachined to the 75.0-mm length within an overly tight 0.01-mm tolerance before the CNC drilling operation.

Question

The raw parts are now very close to 75 mm. So now, can't that right end locator setup be used? Think this through and you'll have gained an important CNC concept.

Answer

No matter how close the raw parts were machined to the 75-mm dimension, length variation still exists. Each part loaded starts with introduced left-end variation whereby, if the PRZ was correctly located on the left side, the only variation for the hole positions would be due to the drilling alone.

In trying to make the wrongly planned setup work, the cost per part was needlessly driven up by extra operations while the quality was reduced! Some parts will be potential scrap due to the combined effect of drilled and induced end variation. Normally, these PRZ issues will have been resolved by the programmer and toolmaker before an operator receives a job.

> **KEYPOINT**
>
> PRZ location and holding method are the most basic facts.

19.1.4 Selecting Cutters for Compensated Programs

For mills and lathes using modern CNC controls, the cutter tooling outlined in the CNC document is a guideline for size and shape. The actual tool selected can vary within small limits. After some shop talk about the savings gained from this ability, we'll explore the operator's role in cutter selection based on the CNC document.

Cutter Selection Based on Program Type

Compensated Programs

When a program is written, the programmer has a certain cutter in mind for turning or milling. But in most cases, that which is set up in the machine need not be the exact diameter, radius, or length originally chosen.

To be able to use this adaptive ability, the program must be correctly prepared as a *tool* **compensated program**. For now we're concerned with reading the CNC document and selecting cutters that will work right from those available. Later we'll look at finding those compensating factors and entering them in the control. We'll look at mill examples, then lathes.

Two Kinds of Tool Compensation

When a program is written it is in one of two possible types: cutter centerline or part path.

> **KEYPOINT**
>
> It's essential that you know which program path is used, since the cutter value you must enter into tool memory will be quite different depending on the type.

19.1.5 Program Type I Cutter Centerline—Built-In Compensation

A specific cutter size is built into the program either by the CAM system or by computing it then handwriting the program codes. The **cutter centerline** program generates the dotted line shown in Fig. 19-5. This path is calculated at the time it is written.

Cutter Centerline—No Compensation

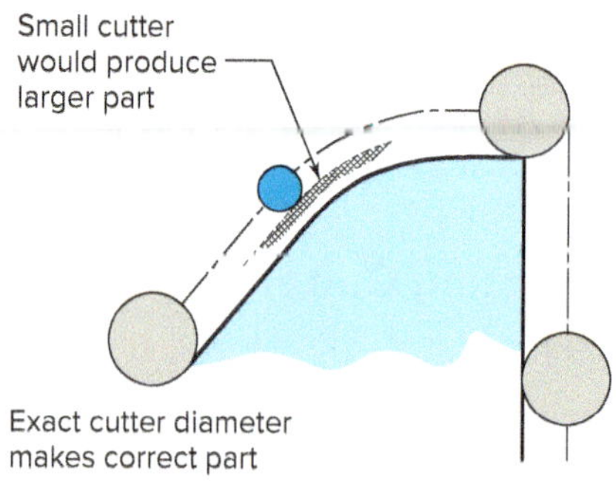

Figure 19-5 A noncompensated cutter centerline program will produce the correct part only if the exact cutter size is used.

KEYPOINT

A cutter centerline program moves the tool on a path that is the tool radius away from the part geometry.

Compensated or Uncompensated

If the cutter centerline program is written with compensation commands, any cutter can be used that's close to the size for which it was written (as long as it can fit into all coves in the part shape).

If the cutter path isn't set up for compensation, then it will produce the right size and shape only if the exact cutter for which it was written is used. Uncompensated, if any other cutter radius is used, the resulting part will be either larger or smaller than expected–the blue cutter is undersize and makes a part too large, by the radius difference in the two cutters.

SHOPTALK

In shop lingo, we often shorten the term *cutter radius compensation* to *cutter comp* or just *comp*.

Furthermore, using the correct cutter, as it wears, the uncompensated program subsequently makes part surfaces that become progressively larger due to deflection. There are two fixes: change to a new, full-size cutter or edit the program to include compensation.

Writing Centerline Programs—Point Shifts

Writing a manually compensated program such as in Fig. 19-5 is a good exercise for beginners. To do so, each coordinate point on the part geometry must be shifted out to the center of the cutter's radius (Fig. 19-6). To write the cutter path using a 1.00-in. end mill, point *A*, on the part geometry, had to be shifted up to point *A*′ (*A* prime) to compensate for the 0.50 radius distance off the part geometry.

Computing the Tool Path Centerline

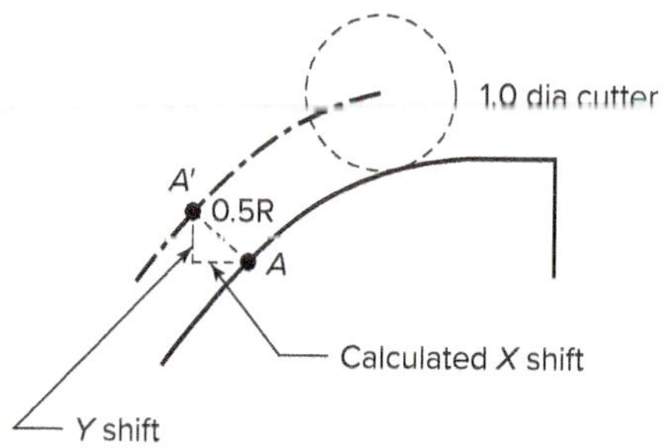

Figure 19-6 Point *A* on the part geometry becomes point *A*′ on the cutter centerline. It is shifted a small amount both in *X* and *Y*.

KEYPOINT

All program compensations, whether mill or lathe, are based on keeping the cutter radius tangent to the part geometry.

Minus Compensations to Centerline Programs

The centerline program is adjusted by entering the difference in the intended cutter radius for which it was written, and the actual radius of the real cutter put in the machine. As such, it's sometimes called a **minus compensation** or *minus comp program* since the real cutter is almost always smaller, due to resharpening, than the full-diameter cutter chosen by the programmer.

If the exact cutter size for which the program was written is used, there is no adjustment needed. But for example, in Fig. 19-7, if a 0.725-in.-diameter end mill is used for a program written for a cutter of 0.750-in. diameter, the compensation, based on the radius, is entered in the control as a *minus* 0.0125 in. (the required radial change to come closer to the geometry line by 0.0125 in.).

Question

For a compensated cutter centerline program, what adjustment must be entered in the control for a 1.060-in.-diameter cutter, if the program was originally written for a 1.00-in.-diameter cutter.

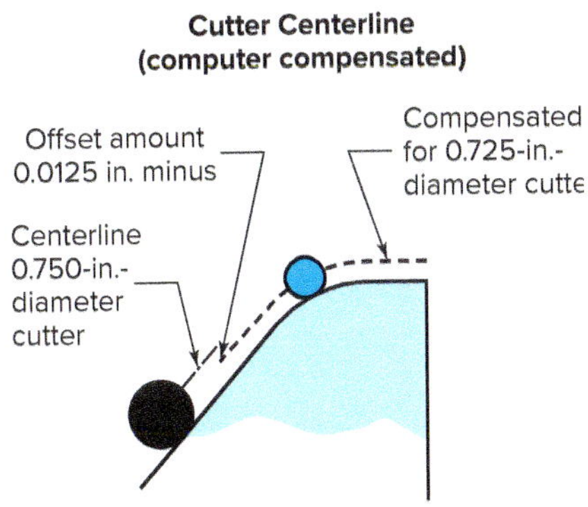

Figure 19-7 A 0.75-in.-diameter cutter centerline program using a 0.725-in.-diameter cutter is compensated 0.0125 in. toward the part geometry.

Answer

The actual cutter is bigger than the basic cutter–not the usual condition. So, it must be adjusted by *plus 0.030 in.* based on the radius. The new cutter must back away to allow for the larger diameter.

> **KEYPOINT**
>
> An uncompensated cutter centerline program can be edited to create a compensated version. But when the value of the work is much greater than the cutters, modern industry uses full size cutters only. This simplifies cutter compensation.

19.1.6 Program Type II Part Path (Part Geometry Coordinates)

Part path programs are based on the part shape. As shown in Fig. 19-8, the part shape coordinates are programmed directly from the print without any cutter radius shift off the shape.

Positive Compensation

To use the part path program, the operator enters the actual cutter radius (or sometimes the diameter depending on control type) into the control tool memory–as a positive compensation to the program line.

> **KEYPOINT**
>
> Using programs based on part path, the CNC control creates the cutter path by moving away from the geometry a distance equal to the cutter **radius offset** it finds in the control's tool memory. The machinist enters that offset.

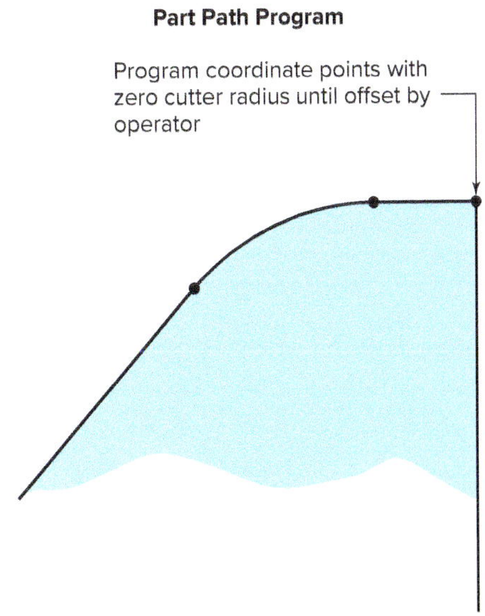

Figure 19-8 A part path program creates a cutter path of zero offset—zero diameter until told differently with a compensating tool radius offset.

In effect, compensated centerline and the part path programs are the same except a part path program is written for a zero radius cutter. It is subsequently compensated to some other real radius when used on the shop floor.

> **KEYPOINT**
>
> A **part path program** is composed of coordinate points shown on the print. While it can be a different size, there is a cutter size listed in the CNC document. It is the target choice to produce best results.

Part Geometry Programs Are Common

If the program is not a part path program, that fact should be noted in the CNC document. Otherwise, assume a positive offset away from the part geometry.

> **KEYPOINT**
>
> To set up a CNC machine, you must know what kind of program is used—cutter centerline or part path.

> **SHOPTALK**
>
> **Offsets** **Offset** is a CNC term used for the variable that adjusts a program for real conditions impossible for the programmer to know. Offsets are used to account for variance in end mill diameters and lengths and for lathe tool nose radius and the tools shape.

Offsets are entered in the control memory by the machinist. Offsets are not in the program; they exist in the control's tool memory. The correctly prepared program calls up these saved variables to compensate.

The offset being discussed is for tool radius compensation. You'll see others as well—for example, tool length offsets (Fig. 19-9). For now, remember, an offset is a controller-stored variable that adjusts a program for a specific condition. Using offsets adds great flexibility to CNC work.

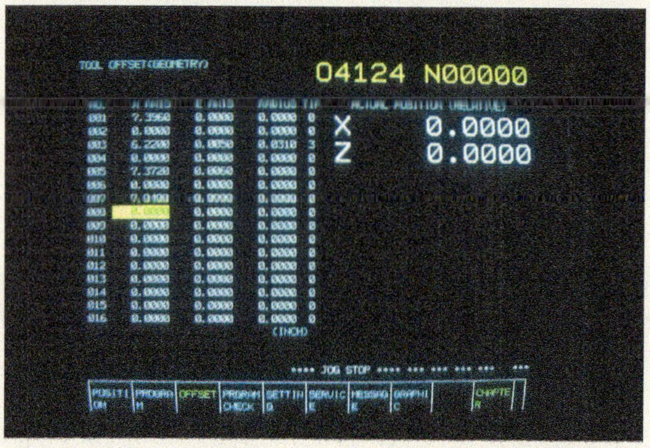

Figure 19-9 Tool length and radius offsets are written into their tool number.

Photography by Prographics at Milwaukee Area Technical College/McGraw Hill

For safety reasons, most shops use one or the other, most likely a part path program since they feature less math, but uncompensated full-size cutter programs are becoming more popular. Shops normally do not mix the two kinds of programs; however, some shops accept jobs that come with the program.

19.1.7 Cutter Radius Considerations

Lower Limits on Cutter Variation

1. **Minimum Lathe**

   Tool nose radius is a matter of *good cut qualities* up to and including zero radius. Here is another area where manual lathe practice is valuable. By now you should have experimented with varying nose radii to get different finish and antichatter results.

2. **Minimum Mill Cutter, Based on Cutter Strength**

   The lower limit for end mill diameter is strength. The cutter must hold up to the demand, but otherwise, a smaller cutter will always produce the correct shape if it is correctly compensated and long enough.

Upper Limits for Cutter Radius

To understand the upper limits of cutter radius, we need to define two new terms.

Generating a Curve Versus Forming

When the cutter moves in an arc, producing a curved surface, it is called **generating** the shape. Compare that to forming in Fig. 19-10, where the cutter simply leaves behind its own shape.

Forming and Generating Curves

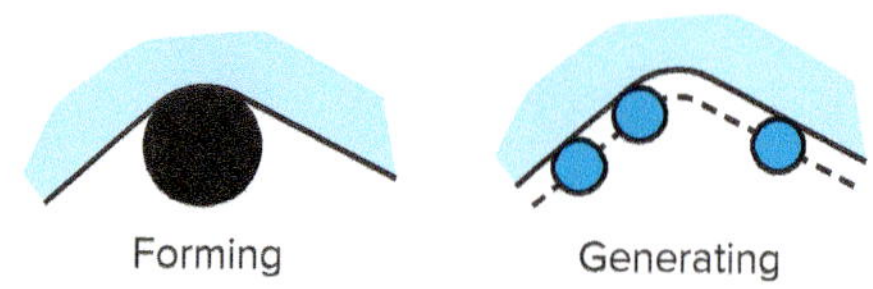

Figure 19-10 *Generating* versus *forming*.

> **KEYPOINT**
>
> In most cases, generating is the superior machining process in terms of finish and avoidance of undercut corners but requires a cutter with a radius smaller than the part geometry.

Coves

Borrowed from carpentry, the term **cove** means any internal detail that must be produced by profile milling or turning. A slot is a good example of the internal shape, as seen in Fig. 19-11. The cutter chosen must be small enough to enter and exit the cove without touching the opposite side.

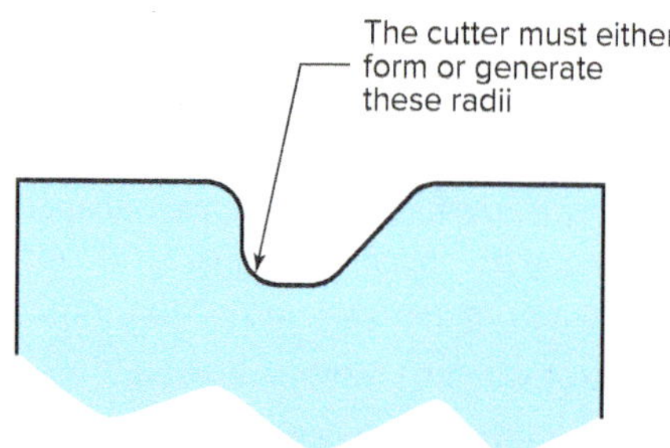

Figure 19-11 An internal cove is a set of details requiring the cutter to enter and exit without touching opposite walls.

Cutter Diameter/Radius Rule (Mill/Lathe)

When selecting a cutter size or shape, its radius cannot be larger than the smallest inside (concave) radius, or cove, on the part geometry. Of the three cutters shown in Fig. 19-12, only two will be able to machine the correct geometry. One will form, while one will generate the shape. The third will not work! But the result (scrap or uncut part) will depend on the capability of the control to see the problem.

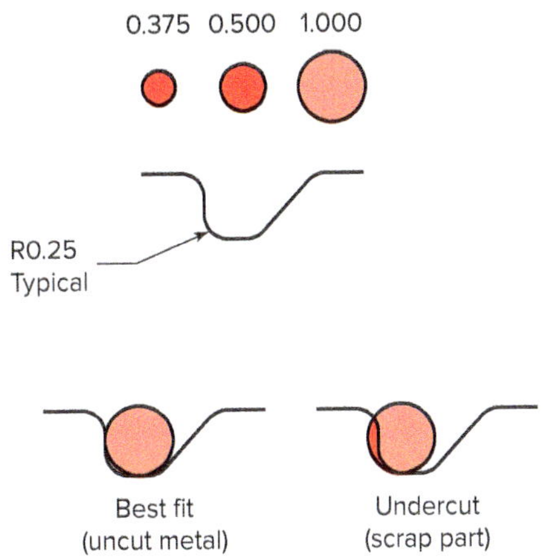

Figure 19-12 Only one cutter can generate the corner radius, one will form it. The third will either cut a *best fit*, leaving uncut metal, or the control will alarm and halt with a (control-dependent) geometry error.

19.1.8 Controls with Various Levels of "Look Ahead"

Various controllers react differently when we try to use a cutter radius too large for the cove–for example, using the 1.000-in.-diameter cutter in Fig. 19-12.

Using a very low-level control, when the 1.0-in. cutter follows the slanting line down, it will overrun the upcoming vertical surface and undercut it. A cutter that is too large to fit the cove will make an incorrect part.

Most CNC controls today can see that upcoming interference event by examining the part geometry and the cutter radius. Depending on the level of software in your control it will do one of the following:

Error out–some will cause an error code and stop machining.

Be best fit–some will take the larger cutter as far into the cove as possible, then out, without undercutting the opposite side of the cove. The part won't be the right shape, but it won't be scrap either.

Be scrap–lower-level controls won't see the problem at all. They would produce a scrap part if the 1.000-in. cutter was used in Fig. 19-12.

TRADE TIP

Avoid Forming If Possible When selecting cutters larger than the programmed size, it is poor practice to change up to a size that's equal to a cove radius on the part. If the cove radius is the same as that of the cutter, then the generated curve exists, but it has a zero radius. The cutter forms the cove radius. The control will be able to deal with the change. It will produce a tool path curve with "zero" radius, as shown in Fig. 19-13, but forming tends to pull the cutter into corners, leaving undercuts, and to produce tool chatter.

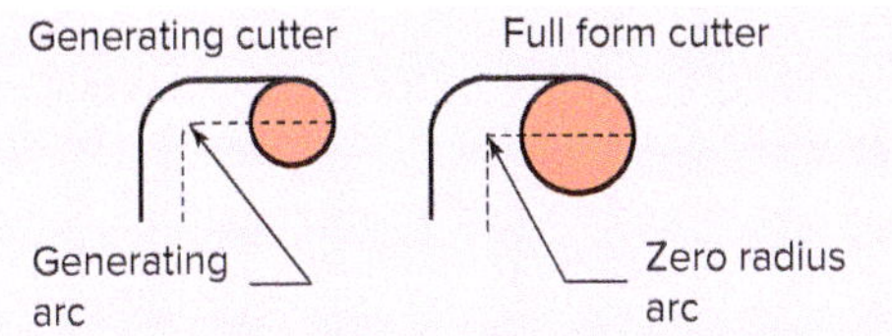

Figure 19-13 The compensation software calculates a curved path for the cut on the right, but the path has zero radius.

Let's sum this up:

- The lower limit is cutter weakness for mills, and finish produced for lathes.
- If there are no internal coves within the part shape, any cutter larger or smaller will work (Fig. 19-14) as long as the program is cutter compensated.
- The best choice is close to the original size for which the program was written.
- If the program is not written for compensation, the cutter must be the exact diameter for which it was written, but that program can be edited for comp.

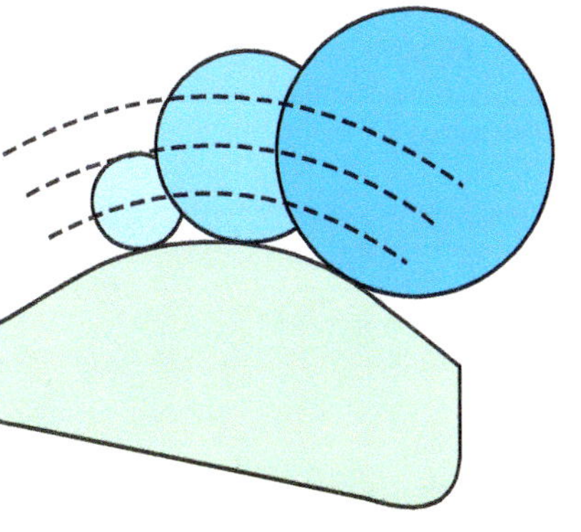

Figure 19-14 For outside shapes with no coves, any cutter diameter is able to produce good results if the program is compensated.

19.1.9 Lathe Cutter Selection

When selecting lathe tools from the CNC document, then loading tool information into the control, one additional issue that arises is tool **flank interference**.

Tool Flank Interference

Compensated cutter paths for lathe work must keep the cutter tangent to the part geometry as on mill compensation. But lathe controllers must also consider the sides of the cutting tool. The control must know which direction the tool is pointing and how wide the tool is in order to create a useful tool path following a part geometry.

In Fig. 19-15, the 30° diamond-shaped cutter is set up in the straight position. It could make the thread relief, while the

45° insert in the 45° holder would drag on the heel as it cuts the cove slope. CNC terms used for this lathe **tool orientation** are **tool bias *or* approach vector**. There are two ways in which this tool bias information can be entered in the control's tool memory:

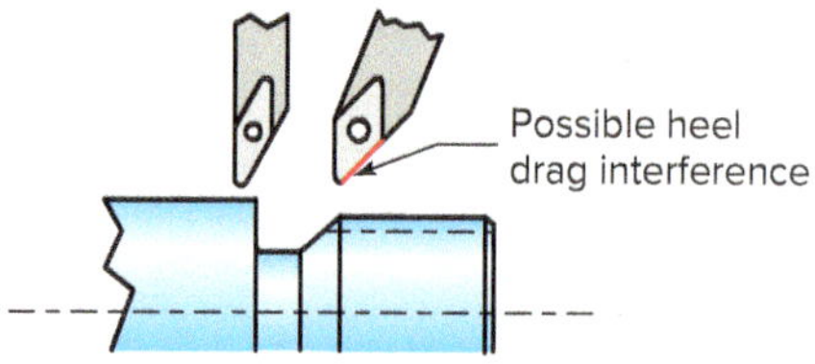

Figure 19-15 Lathe tool *shank size*, *orientation*, and *cutter shape* must be known to the control to avoid approach interference problems in the compensated path.

Tool Library Many modern CNC lathe controls contain a standard library of the common tools. During the setup, when a tool number is loaded into the tool memory, it is attached to a screen-selected image that represents the correct tool shape and the direction in which it's pointing in the setup. For example, after selecting turning tools from a menu, the image shown in Fig. 19-16 might appear. The machinist selects the tool shape and orientation that suit the setup. Then the shank size and nose radius offsets will be attached to that selection. Most tool libraries of this type are customizable. New tools are supplied on disk by manufacturers, or they can be custom drawn using control definitions.

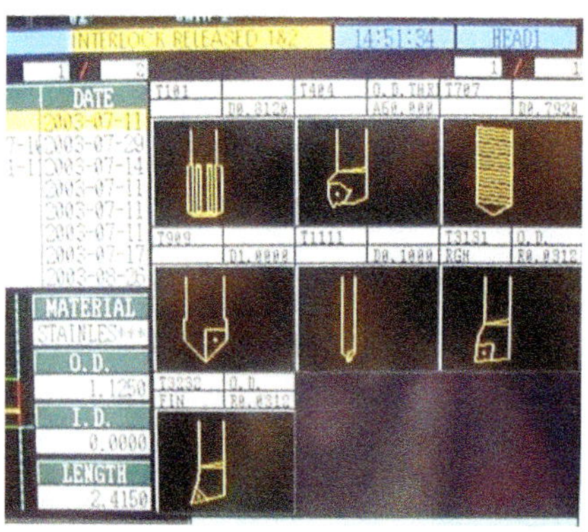

Figure 19-16 A standard tool library selection.
Courtesy Mike Fitzpatrick

Tool Vector Selections Without the tool library function, controls attach an approach vector to a tool selection. The vector numbers are determined from a chart, such as shown in Fig 19-17. For example, after selecting *facing tools,* an entry from 1 to 8 will be requested. The vector number represents the direction in which the grooving tool is oriented. During tool path compensation, it tells the controller how to approach and pull away from part surfaces.

It's your job to avoid these problems during tool selection for the setup and tryout. Call on your manual lathe tool selection experience to avoid this problem. Vector tool bias selections do not guarantee interference problems are solved. Try out new programs carefully.

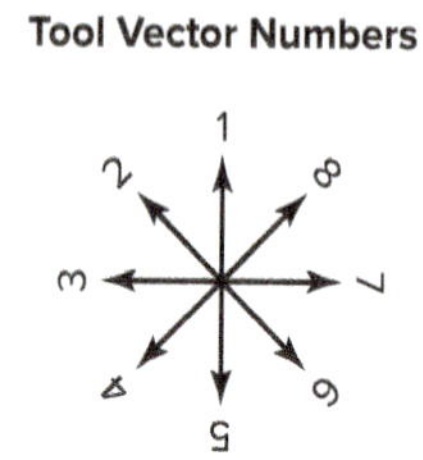

Figure 19-17 This control features eight possible *approach vector* numbers.

KEYPOINT

Because vector controls know only the direction the tool points, not the full shape, potential shank and heel drag problems can arise.

Phantom Tool

A puzzling problem occurs when using nose-radius-compensated programs for turning but the wrong radius is stored in tool memory. Similar to a mill program with the wrong cutter diameter in memory, a distorted part will be produced. But with lathes an odd twist occurs. Even with the wrong radius entered, or none at all, as in the example (Fig. 19-18), straight diameters and face distances are machined right, yet curved and angular surfaces are machined wrong!

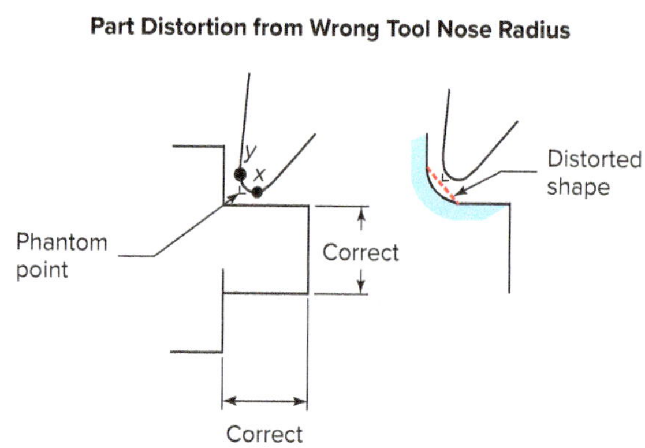

Figure 19-18 Without the nose radius offset in the control, a curved surface is altered.

Think this through–it's a bit of a brain teaser. Lathe compensation programs follow the same logic as a mill, in that the control keeps a round cutter tangent to the programmed shape (Fig. 19-18). If no nose radius is entered, the control assumes a zero radius, thus keeping a pointed tool tangent to the geometry. That works fine if the tool really is pointed.

In this example, the control has a zero nose radius in memory. However, the radiused tool has been coordinated for both *X* diameter and *Z* face distance from PRZ. Its radiused nose is touching the work at tangent points *X* and *Z*. As long as the tool cuts only straight diameters and perpendicular faces, it will make correct dimensions. It will form the tool nose radius at internal corner intersections.

However, when cutting curved surfaces, as shown on the right in the figure, the control perceives that it has a pointed tool, known as the phantom tool point; thus, it pulls back to keep the phantom point on the geometry. That makes a distorted shape. This occurs only when the operator forgets to associate a nose radius with a cutter memory. Given the wrong nose radius or no entered radius, distortion occurs on angular surfaces as well.

KEYPOINT

When mounting a lathe tool, both the tool radius and the tool shank shape must be entered into the control's tool offset memory.

19.1.10 Understanding Cutting Tool Numbers

Tool numbers in the CNC document must be correctly identified in the control (Fig. 19-19). If the machine is a lathe, the task is simple: Tool 01 goes in turret position number 1. Tool 02 goes in position 2, and so on. Some mills require the same: T01 goes in storage drum hole 1, T02 in hole 2, and so on. These are generally older mills lacking random tool storage ability. More modern CNC mills allow the storing of any tool in any drum hole. But the tool-loading task is a bit more complex. For example, the mill operator is storing T01 in open drum position 08 in Fig. 19-20.

Figure 19-19 On some older machines, tool position must match the turret position.
Photography by Prographics at Milwaukee Area Technical College/McGraw Hill

Figure 19-20 After storing random tools, write the tool number (T01) into the turret position (hole 8) on the tool storage page.
Photography by Prographics at Milwaukee Area Technical College/McGraw Hill

The next action will be to immediately go to the control and enter Tool 1 at position 8 on the tool tracking page.

Tool numbers are arranged in sets of two-digit data fields, in one, two, or three columns. Various controls use different tool number sets. Some take up the cutter offset information when reading the tool number, while others require a separate code to get and use offsets.

Separate Code Example T01 H43 D1

Use tool 1 and apply length offset (H43) = to the number stored in D1 (diameter 1)

Single Column T01

Tool number (Dropping leading zero, can also be T1)

TRADE TIP

Double-Check Mill Tool Numbers On older controls lacking random tool storage, Tool 1 must be placed in storage position 1, similar to lathe tool numbers, and it never changes throughout the running of the program. But no matter what kind of tool storage system is used, whether or not random, to be certain the tools are correctly loaded before running the program, handwrite and execute an MDI (manual data input) command for each to be placed in the spindle, one tool at a time. Then check each against the tool list in the CNC document.

For mills, the command is T01 M6. Here, M6 is the mill command to change tools. At that time, the tool listed as tool 1 should be in the active position. If it's not, something is wrong with your setup.

Tool Radius Attachment—Two-Column Format

T0101

Tool number/Offset number

Using this format, radius offset is the second column. In the example, radius offset value 1 is applied to tool 1. For consistency, offset 1 is usually attached to tool 1, but there are times when a second offset might be placed on tool 1 by calling it T0102.

For example, a roughing program uses T0102, which has a deliberate false offset bigger than the tool radius, which then roughs out the part, leaving a bit of finish metal. Then the offset is exchanged for the true compensation and the program is rerun with T0101 to finish the shape–the same cutter but different offsets.

Tool Radius and Length Numbers—Column 3

On some controls, the tool number includes a third data field to include tertiary offsets. The most common is the length offset used on machining centers:

T010203 = Tool number/Radius offset/Length offset

19.1.11 Understanding Mill Tool Length Differences

Tool length offsets make both the programmer's and the operator's jobs much easier. We'll study how to determine and enter them in Chapter 23. For now, you need to know that they apply to real tools of varying lengths. You also saw tool setting probes in Chapter 18, which also can determine tool differences.

KEYPOINT

Without a tool probe, during mill setup, the operator must determine and enter the tool length offset differences as determined by gage measuring the tools.

When the program is being written, the programmer assumes that all tools in the setup are the same length (it doesn't matter how long, just that all are the same). Programmers do not adjust Z axis moves to account for differing tool lengths.

During setup or tool replacement, the operator determines length differences in the real tools by touching a reference surface with each cutter, then reading the Z position on the screen. Next, the difference in each is stored as the tool's Z length offset (Fig. 19-21).

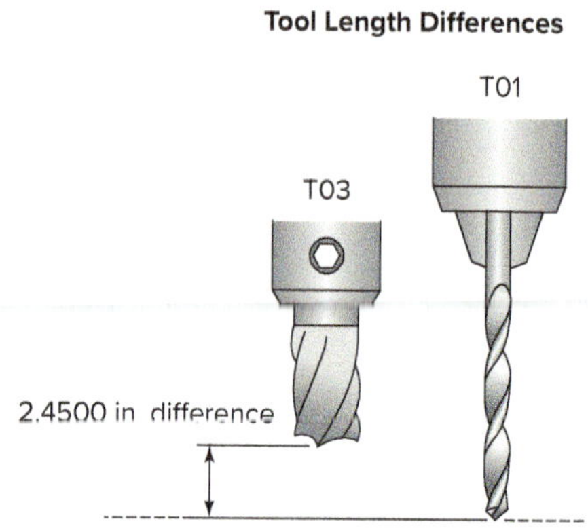

Figure 19-21 If the drill is set to touch the work surface at $Z = 0.00$, then the end mill requires a *length offset* of plus 2.450 in. in order to also go to $Z = 0.00$.

So, using the example of Fig. 19-21, under tool number T03, length offset is a plus 2.450 in. for the end mill compared to tool 1, the drill. Anytime, under length offset 03, the program calls the end mill to a Z axis position, it combines the axis move with the offset distance. The Z axis must move an additional 2.450 in. to reach Z zero using the end mill, when compared to the drill.

Figure 19-22 shows an operator using a feeler gage to test the cutter length. There are several versions of this task that will be covered later.

Figure 19-22 Touching the bit to the work surface is a common way to coordinate Z axis reference.
Lake Washington Technical College, Kirkland, WA/McGraw Hill

Presetting Tool Lengths

Many high-production shops set tool lengths such that every tool number 1, the drill in Fig. 19-21, is the same length before it is given to the mill operator. That way, a dull tool can be quickly replaced with no length testing by the operator. Another solution is the automatic probe shown in Figs. 19-23 and 19-24.

Figure 19-23 The *Z* axis position of cutting tools can be coordinated using an auto probe that communicates contact to the control.
Courtesy Mike Fitzpatrick

Figure 19-24 This CNC lathe features an automatic probe to determine tool position and offset.
Courtesy Tyee Aircraft/Everett, WA

While there are manual methods of determining length offsets, both at the machine and at the preset bench, as seen in Fig. 19-22, the automatic probe has some advantages in that it senses, then automatically loads the offset for the operator.

KEYPOINT

Using one of several methods, before a multiple-tool program can be run on a milling machine, length offsets must be determined and stored for each cutting tool.

19.1.12 Cut Sequence and Special Instructions

Sequence of Operations

When trying out a new program the operator absolutely must know the following facts from the CNC document:

Which cutter will be used in which sequence?

What machining operations occur in what order?

Are any special interventions required (list coming up)?

Sequence information may be listed in the document or embedded in the program as comments, often both. For example,

T02 [Tool 2–$\frac{3}{4}$ in. EM will rough three pockets and Datum C]

T03 [Tool 3–$\frac{1}{4}$in. drill–deep drill watch coolant stream]

If we know this information, adjustments such as feed and rapid travel override can be used as they are needed.

Special Instructions

Events that would otherwise be a "big surprise" to the operator should be made clear in the documentation beforehand. Here's a list of the unusual events that can happen.

Lathe surprise events

1. **Cutoff Operations**

   The part or excess material becomes unattached and may become a potential projectile. Instructions for avoidance might be included.

2. **Extreme Machining**

   The coolant stream is critical such as heavy roughing cycles or deep drilling.

3. **Tooling Challenges**

   Experience shows that the cutter might fail at a particular point in the program; thus, extra attention must be given to be able to halt operation instantly.

4. **Part in Chuck Changes**

5. **Optional Stops**

   These are critical places where the programmer has provided a halt if it is needed. For example, every 10 parts must be measured on a critical dimension. Between halt times, a toggle switch, such as the one shown in Fig.

19-25, is set to ignore the **optional stops *or* opt stops**. There may be more than one opt stop switch such that various intervention places can be turned on or off, at the operator's discretion.

Figure 19-25 Optional stop switches recognize or ignore halt points in the program by the operator's choice.
Courtesy Tyee Aircraft/Paine Field, WA

Typical mill surprise events

1. **An Exceptionally Deep Operation**

   Coolant and close monitoring are required here.

2. **Vibration and Chatter May Occur**

   The operator may need to override speed and feed controls as the machine cuts through a critical area.

3. **Midprogram Clamp Changes or Additions**

   Sometimes as machining progresses, the part becomes progressively weaker and requires extra support. Or to machine an area that was shadowed by a clamp, at a given point it must be shifted.

UNIT 19-1 REVIEW

Replay the Key Points

- The CNC documentation foresees potential problems, eliminates guesswork, and gets the job going right with the least amount of experimentation.
- By matching the program's header Rev level to the CNC documentation, data crossing can be avoided. There will also be a policy about dumping old programs from permanent memory.
- When a datum locator is not placed correctly, needless variation will be introduced into a production run.
- PRZ location and holding method are among the most basic decisions.
- Program compensations (mill or lathe) are based on keeping the cutter radius tangent to the part geometry.
- An uncompensated cutter centerline program can be edited to create a compensated version.
- When using programs based on part path, the control creates the cutter path by moving away from the geometry a distance equal to the cutter radius offset it finds in the control's tool memory. The machinist enters that offset.
- A part path program is composed of coordinate points shown on the print.
- You must know what kind of program is used—cutter centerline or part path.
- In most cases, generating an arc is the superior process compared to forming it, in terms of finish and avoiding undercut corners, but generation requires a cutter with a radius smaller than the part geometry.
- Vector tool bias selections do not guarantee that interference problems are solved. Use caution when trying new programs.
- When mounting a lathe tool, tool radius and the tool flank shape and size must be entered into the control's tool offset memory.
- During mill setup, the operator must enter the Z axis tool length offset differences as determined by gaging and comparing each tool against a reference.
- Using one of several methods, before a multiple tool program can be run on a milling machine, length offsets must be determined and stored for each cutting tool.

Respond

1. List the three kinds of compensated programs and give a brief description of each. Which is the most common, and why?
2. As parts are run on a lathe, their outside diameter becomes progressively bigger and their length increases. What can the operator do?
3. Using Fig. 19-26, to profile mill this part select the best mill cutter from these diameters: 0.25, 0.375, 0.500, 0.625, 0.750, and 1.00 in.

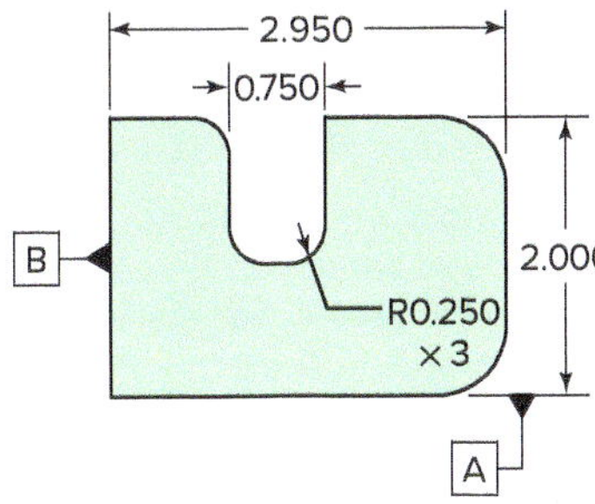

Figure 19-26 Select the best cutter from the list of Problem 3.

Critical Thinking

4. You are to set up to CNC mill the internal pocket and two drilled and tapped holes (Fig. 19-27). Sketch the setup tooling if the part has been premilled to a finished square and the 0.750-in. slot has also been machined into the block.
5. Don't look back. List as many important items as you can that might be listed in a CNC document. The book identifies 14.

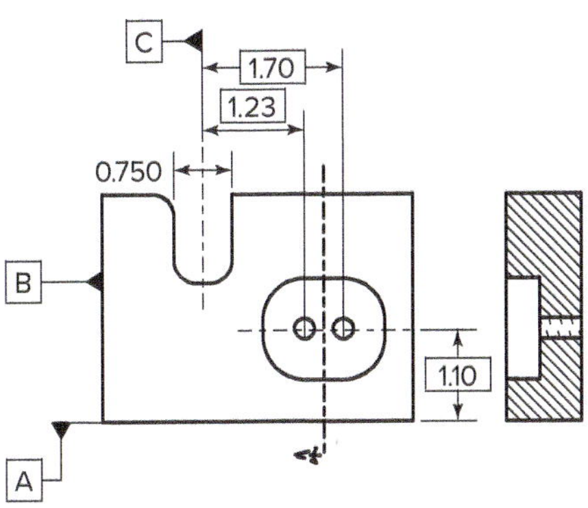

Figure 19-27 Describe the setup holding and alignment tooling for this part per Problem 4.

Unit 19-2 Controller Alarms

Introduction: Many potential problems are detected by an alert operator. But others are announced by the control itself, in codes or words, called an alarm condition. When the control finds a problem, it takes certain actions. It may keep on cutting but flash a screen error message or a yellow warning light or both. It also might just stop. Something must be corrected. Depending on the control type, the error message can be easy to read or sometimes it's cryptic and requires a bit of detective work to solve.

These messages are never welcome, because they not only announce something requiring operator (or programmer) intervention, but they also often halt the machine until solved. We'll cover a few of the more common ones. On the job, you'll need an operator's manual that lists them all. They differ from control to control.

TERMS TOOLBOX

Branching errors The program's logical sequence is in error. It wanders off and can't return to the original address from which it branched.

Look ahead The control's ability to analyze and compensate program commands ahead of the active line. This ability can cause some frustration when looking for the cause of an alarm.

Servo errors The machine has become uncoordinated between physical position and recorded position.

Syntax error Any entry that doesn't appear in the right order or form for the control to understand.

19.2.1 Alarm Classes

A few controls provide alarms in readable words. But many flash alphanumeric error codes (Fig. 19-28). The operator is then left with the task of decoding what went wrong. Given a little experience, it's usually easy to find and fix, but sometimes an alarm can be a real headache to sort out.

Figure 19-28 Control alarms can be simple or a real challenge to solve, but they are never welcome!

Take a moment to scan the following example alarms to get a feel for the various ways a control can stall. The main objective is to understand the kinds of alarms within a group; thus, you'll know where to start looking for the problem, inside the program data, in the machine setup, or in the machine itself.

KEYPOINT

Alarms fall into general groups. Recognizing what the group indicates helps speed the solution.

Physical and Hardware Errors

These errors can occur when running any program new or old. Here the problem lies outside the program!

- Machine supplies

  Air pressure, hydraulic, coolant, and lube oil

  Each may have its own code or all will be grouped into a single alarm condition. For example, alarm 01 = low fluids.
- Hard limit overtravel
- Violating a safety switch

  Door open, chuck switch, bar feeder, part loader, and so on
- **Servo errors**

  Due to exceptional action, extra heavy machining, or a crash, the machine has lost its position count on the feedback system.

Syntax and Program Errors

These problems lie within the program—but only within new programs. They are fixed by solving math or logical order errors, then editing them out. See the Trade Tip on look-ahead controls. They are also found in a "dry run" where the control reads but does not perform the commands. Last they are found in a program verification such as Vericut (see Chapter 26).

- **Syntax errors**—switching a letter *o* for a 0 (number zero), incorrect wording of statements, incorrect ordering of a rigid set of variables
- Missing data—no feed rate or RPM (very common for beginners)
- Math errors—curve or other computations do not make sense
- Extra data—two, mutually exclusive, actions are requested in the same command. For example, both rapid travel and feed rate commands are given in the same line. The machine might alarm, or it might accept only the last command it reads on that line.
- Limited power—two switching commands have been given in the same line, but, due to exceptional power drain, only one is allowed on any given line. For example, turning on the coolant and work light and spray mist all at one time might be disallowed on older controls.
- Radical commands—the programmer has asked for a rate beyond the ability of the machine. For example, 200,000 RPM when 2,000 was the intent.
- Branching logic—a loop, bridge, or subroutine is used (program tricks to save extra writing), but they aren't addressed such that after completion, the control can return to the main program and keep going. Control becomes confused by such a **branching error**.

TRADE TIP

Look Ahead Controls On modern controls with **look ahead** capability, one cannot assume the problem is on the line where the control has stalled. Depending on the level of intelligence, most controls look ahead from a minimum of three command lines up to the entire program, as they attempt to perform tool compensation. They read and compensate several lines ahead of the active command line. If not, tool interference such as that shown in Fig. 19-29 can occur.

The control is machining line *A–B*, but it has foreseen a problem coming up as it rounds the corner, it will hit point *E*; the cutter won't fit into the cove. It has detected the *problem five command lines ahead in the program*. It errors and halts. In this example, after finding out where the problem might be, the fix is to install a smaller diameter tool.

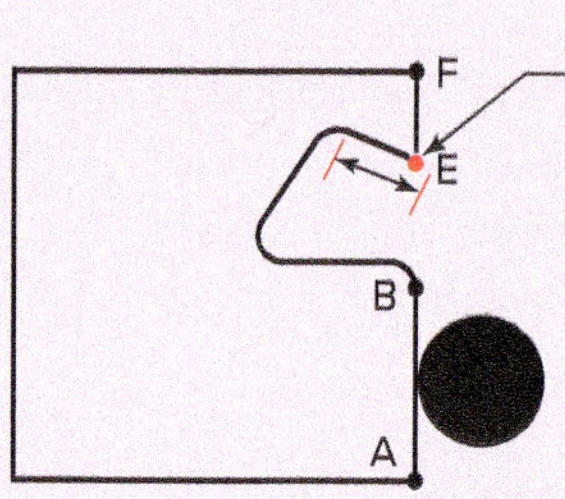

Figure 19-29 Looking ahead, the control may see the interference problem depending on how many blocks it analyzes ahead of the current command from *A* to *B*.

Operator/Controller Errors

Review of previous actions or entries.

- Compensation impossible—also common for beginners. Here, a cutter has been loaded such that the control cannot use it to compensate the part geometry. See Fig. 19-29.
- Erroneous offsets—impossibly big or small offsets, usually a decimal point error.
- Machine not initialized correctly—the homing sequence that has not been performed.

KEYPOINT

The alarm may be for a problem several lines ahead of the place where the control has stalled.

UNIT 19-2 REVIEW

Replay the Key Points

- Alarms fall into groups. Recognizing them helps find and fix the problem.
- Due to look ahead capability, the problem may be several lines ahead of the line on which the control has stalled.
- An alarm may be in flashing words that are easily understood, or it may be a code number that must be decoded from a list in the operator's manual.

Respond

The activity for alarm codes is to take a moment to scan the manual of the machine you are about to operate. Or your instructor may have a handout prepared on the same subject. Notice the types of errors and the way they are communicated on your specific control. Be aware that the explanations are often translated from the language of another country and as such can be difficult to understand.

Unit 19-3 Operator Actions (During Production Runs)

Introduction: Everything is set up and tested. Production is ready to go. This brief unit describes the activities that a pro-manager performs to be in full control. It's been put into a unit because alertness is such a vital subject for safety.

It's easy to sit back and do little or nothing during a run when everything *seems* to be going right. But it's just then, when you are in "potato mode," that something goes wrong suddenly. The solution is to be ready, but I admit, sometimes that's not easy to do when the run is long and has worked perfectly for some time.

19.3.1 Practicing Emergency Plans

By now you have seen or heard an accident on a manual machine. And you also know that crashes occur at fantastic speeds in CNC, especially on turning machines.

KEYPOINT

The best operator action during emergencies always results from preplanning and practice. Review and practice emergency moves at the beginning of your workday.

When starting a machine, do a quick mental review of the following three-level occurrence list. The real skill objective is to analyze quickly at which level the problem is and take the right action–at the right time. Some problems can wait until a natural division in cutting action while others must be halted, right now! There isn't much time to sort it out sometimes. Another challenge is that different machines feature varying kinds of halt controls.

Notice that as the emergency level increases, the action options decrease, yet they must be performed ever more quickly!

Level 1—Something Isn't Right

The Event

As yet this is a nonemergency that might get worse. Examples include a sound that hasn't been present or a finish that seems to be deteriorating.

The Action Plan

Memorize the methods that halt the machining at a convenient time, with no consequences. Some of these actions are not immediate; the machine will continue to run for a while, then halt at a natural time when the cutter is not touching the workpiece.

- **Switch mode to single block–**
  Will stop upon completing current command.
- **Turn rapid override to zero now–**
  Will stop as the next rapid travel code is read.
- **Turn rapid override to zero after a rapid move commences–**
  This halts the action with the cutter not touching the work.
- **Touch halt/slide hold at a tool change–**
  Will stop with tool off the work.
- **Midcycle halt–**
  Will stop right now, but then the cuter is dwelling upon the work. The spindle must be stopped.

ARMCHAIR FIELD TRIP

To find out more about using single block on your CNC machine, watch this **Haas Tip of the Day.**

You'll stay out of trouble when starting up a new program by watching this video! Go to: **https://www.youtube.com/haasautomation** and search for this video called: **Avoid Disaster Using the Haas Single Block Function—Haas Automation Tip of the Day**

Level 2—A Problem Is Obvious and Will Degrade—Immediate Action

The Event

A low-level emergency such as chip buildup exists. It must be solved efficiently before something much worse occurs.

The Action Plan

Memorize the controls available to stop the machining quickly:

- Slide hold
- Feed rate override to zero
- Midcycle halt

Level 3—An All-Out Crash—Fully Dangerous to Machine and Operator

The Event

Parts of the setup and cutter shards are flying out; axes are diving into and stalling on machine tables and chucks; it's the big one! Stop everything right now!

The Action Plan

Above all else, be certain you are out of danger! Only if it's safe to do so, hit the emergency stop with lightning-quick reflexes.

> **KEYPOINT**
>
> **Urgent!**
> Level 3 situations occur too fast to think about your actions. You must have a preplanned action and have practiced it enough that it's instantaneous. Know the location of all E-stops. If the machine has more than one E-stop button, practice turning toward it and away from the danger.

19.3.2 Problems and Emergencies Caused by Operators

Here are the four main actions of operators that cause problems. They are organized in their rough order of occurrence (author's opinion) within a long production run.

1. Not loading the part right.

   Misaligned, loose clamping pressure or part upside down or backward.

   Parts shift, breaking cutter. Or parts fly from setup.
2. Not removing chips well enough during cut.

   Keeps coolant away from the cutter–loading and breakage.

   Work hardening and poor finish.

   Causes chip loading in mill cutters.

   Can catch and destroy work finish on lathes.
3. Not keeping coolant streams aimed at critical cuts (mills only).
4. Using cutters too long beyond their time.

   Avoidance of changing dull cutters too long.

   Results in vibration and poor finish with a possibly oversize part.

Boredom—The Enemy of Efficiency (and Promotion)

If you allow it, long or repetitive CNC work can lull you into a less alert state. Here are a few suggestions as to how to avoid doing so:

Monitor the progress.

Improve part turnaround time (information next).

Remove burrs from finished parts.

Measure a critical dimension.

Record measured results in computer database.

Stack it in the outgoing pallet or tub.

Pick up next part from the "in pile" ahead of time to use it.

Blow off, deburr the part, ready to put it into the machine.

File and remove burrs on new part.

Place prepped part ready for next load.

Turnaround Time

Until given the responsibility of editing the program, you can't do much about inefficient movements in the program. But you can analyze your own actions. Many new operators seem to be working furiously for a short time, then doing nearly nothing for the remaining cycle time. Look for tasks that can be accomplished while the machine is running.

It may seem trivial at first, but 10 seconds cut off the cycle time can make a big difference in delivery schedules and job costing. Remember, one of the main advantages of CNC is the rapid cycle times.

> **KEYPOINT**
>
> Your job is to refine cycle times without compromising quality or safety. Don't rush; work efficiently.

> **SHOPTALK**
>
> **Debugging Was Once Real** Finding and correcting errors is called debugging the program. We assume the problems are called bugs—right? History proves differently.

Instead of silicone chips and transistors, the first monster computers controlled electronic pulses using vacuum tubes roughly three times the size of your thumb. The control mechanism within each was a pair of flat metal plates. Heating one changed the conductivity of the other, but the heat needed also caused one of the plates to glow with a small orange light.

Literally hundreds of these vacuum tubes were required to operate the computer, but their collective, happy glow was just too much for congenial insects! Every so often, technicians had to open them and using a brush and vacuum truly debug the system of suicidal guests.

UNIT 19-3 REVIEW

Replay the Key Points

- Know the location of all E-stops. If the machine has more than one E-stop button, practice turning toward it, away from the danger.
- The best operator action during these times always results from preplanning and practice. Practice every so often.
- Your job is to refine cycle times without compromising safety. Don't rush; work efficiently.

Respond

1. Describe the nature of a level 1 problem and give an example using a mill cutter.
2. Identify several ways to stop machining to investigate a level 1 situation.
3. Describe a level 2 situation and give an example that might require action.
4. What must be done in a level 2 situation?
5. Describe an example of a level 3 situation and what you must do.

Unit 19-4 Monitoring and Adjusting

Introduction: This unit explains several ways to stay in control for safety and efficiency. After exploring machine monitoring, we'll look at compensating and changing tools.

TERMS TOOLBOX

Control chart A statistical method of detecting trends requiring offset adjustments.

Critical dimension (key dimension) One or more dimensions chosen by function that represent all dimensions made by a single tool.

Position page (screen) The control monitoring page that displays positions and distance-to-go information.

19.4.1 Monitoring Production

Absolutely the number one method of knowing how the cut is progressing is the sound it makes. That's why personal earsets are almost always banned in machine shops, while a few permit radios on benches if not turned up too loudly (Fig. 19-30).

Figure 19-30 Ears are the best method for detecting an early problem. Personal music is a bad idea.
Courtesy of Keith Ellis—*Northwest Metalworker Magazine*

Next, we keep an eye on cutter progress, the finish produced, chip ejection, and coolants. But due to safety shields and to flying coolants and chips, visual monitoring isn't always possible, so we turn to technical solutions.

Alternate Monitoring Methods

There are three options for tracking what's going on inside your machine when you can't see the action directly:

Camera Controller position page
The graphic monitoring page

Video Camera

An add-on accessory, video cameras are not featured on all machines; however, they work well where machining is obscured behind screens covered with coolant and chips.

Controller Position Page

This is arguably the choice most often selected during normal program monitoring and unarguably the best screen for program tryout. The **position page** can be arranged on most controls to show one or all three displays:

1. The present coordinate position and kind of motion being made
2. The destination coordinates
3. The distance-to-go (DTG) to complete current command

Controller Graphics Page

Modern CNC controls offer varying levels of graphic sophistication. The latest show color-differentiated cutters, raw work, and finished cuts in animated real time. Older graphic screens depict the tool path similar to wireframe geometry.

Monitoring by Measuring

An essential way to monitor production is to measure certain carefully selected **critical dimensions called key dimensions** and surface finish. They indicate the condition of the cutting tools. If this information is fed into a computer program capable of creating a chart called a **control chart**, as shown in Fig. 19-31, the operator can get a good picture of tool wear and other factors such as heat buildup in the setup. For more information on this ability, read Chapter 27.

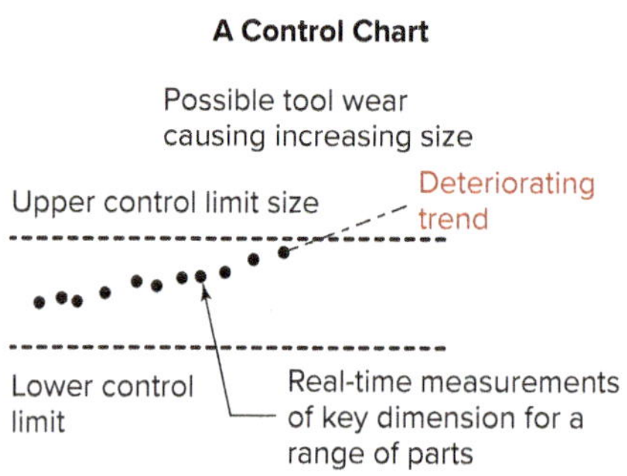

Figure 19-31 Control charts track progress and highlight trends that need operator intervention.

From the example chart, it's easy to see that if the operator takes no action, soon the parts will go beyond tolerance. Using these kinds of data, an operator can also predict how many parts might be made before it's time to change cutting tools.

Finding Key Dimensions

Often in CNC work, when a single tool machines several details on one workpiece, the operator need not measure every dimension it produces. For example in Fig. 19-32, a representative diameter and length can be chosen such that if they remain in control, all others normally follow if machined by the same tool.

If the object has varying tolerances based on function, then the key dimensions should be chosen from the tightest.

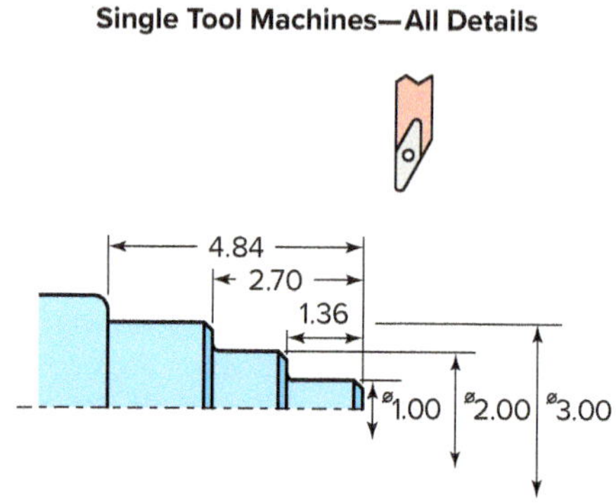

Figure 19-32 Key dimensions represent many features as shown here, or can be the most critical features of the part.

KEYPOINT

Sometimes the engineering drawing will note which dimensions are to be used for this tracking purpose and there may be a symbol for which dimensions are key to the function.

Maintaining Size and Shape

Offsets As shown in the control chart (see Fig. 19-32), if a trend shows in key dimensions produced by a single cutting tool, the likely cause is tool wear. You'll have two options:

- Adjust the tool offset (plus or minus depending on type of program)

Experience and SPC will tell you how many adjustments can be entered before changing tools (usually no more than two). Still, it is a standard practice.

- Exchange the tool for a sharp cutter

Solving Dimensional Error

When monitoring progress in an established run that was producing good parts, we often discover dimensions that begin to wander away from nominal. While changing cutter offsets will usually bring the dimension back to acceptable, you may be chasing a symptom of a problem, not attacking the cause. Before you can solve the root problem, you need to sort out what caused the dimension change. Here are seven general causes. There are others based more on specific situations.

- Dull tool
- Tool deflection—dulling weak bending tool
- Part or setup flexing
- Mislocated PRZ
- Misaligned holding tooling or misaligned in the holding tooling

- Incorrect information in tool memory (wrong radius, offset, or length)
- Incorrect program

TRADE TIP

Settling Down—Initial Tool Break-In Veteran machinists know there is an initial period when cutting tools wear away the fresh new edge, then stop deteriorating for a long period (Fig. 19-33). This is especially true for new carbide tools. When the change stops, the operation is said to "settle down." Then, at a given time, the changing dimension will begin again as the tool dulls to a given point. The real trade trick is the ability to predict how many cycles that tool will last before unwanted events occur.

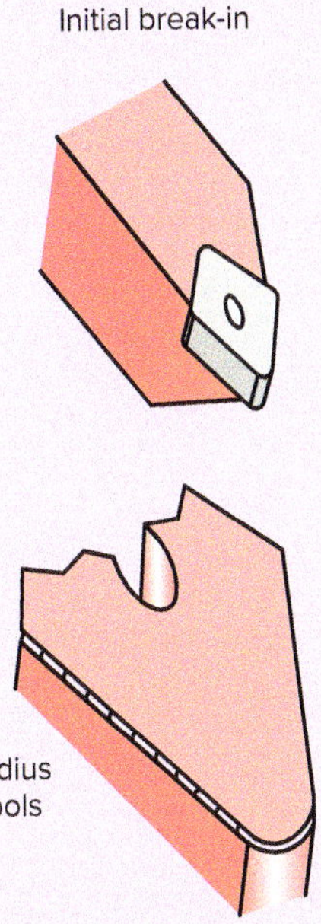

Figure 19-33 Initial size changes as carbide cutting tools break in. Then it stabilizes for an extended period as wear slows.

Using commercial inserts, an edge prep hone operation has already been performed during manufacture. But on shop-ground carbide tools, you can stabilize the break-in period by skillful use of a diamond hone. Remove from 0.003 to 0.010 in. from the cutting edge. As strange as that seems, this dulling improves tool stability because, using correct feeds, the tool's cutting edge is being depressurized by the rake face lifting the chip away from the parent metal. The extremely sharp edge tends to chip, while the slightly rounded edge stands up well to the action.

The final analysis might show a combination of causes. The upcoming activities present an opportunity to test your detective work in this area. Use this list to solve situations where parts are not coming out right.

19.4.2 Calculating Offset Change Amounts

Before adjusting tool nose radius, the following must be clearly understood.

For Lathes

1. **Reducing Nose Radius Values**

   Using smaller tool nose radius adjustments moves the cutter closer to the part geometry to remove more material on the workpiece. It produces smaller outside diameters and bigger bore diameters.

2. **Increasing Nose Radius Values**

   Adjusting nose radius values upward moves the cutter away from the workpiece to create larger OD dimensions and a smaller ID.

But, you can only use nose radius offsets to solve dimensional errors when *all diameters* produced by the same cutting tool are universally in need of adjustment.

KEYPOINT

Adjusting the nose radius cannot correct *Z* axis distance problems (critical thinking problem coming up).

But when you do change the nose radius, it's similar to changing the *X* axis micrometer dial on a manual lathe–the amount of change required is determined by dividing in half, the difference in the nominal dimension and the measured result.

For example, by entering a 0.003-in. smaller nose radius, the tool moves in 0.003 in. closer to the work but produces a 0.006-in. reduction in diameter.

Nose radius change value =

$$\frac{\text{Actual dimension} - \text{Target dimension}}{2}$$

Positive or Negative Nose Radius Change

The formula just given produces a positive value when the work is too big compared to the nominal target–therefore the stored nose radius value must be reduced by the result to bring the tool closer to the work. If the formula produces a negative value, the actual size is smaller than the target and the tool nose radius value must then be made larger in the control by the calculated amount.

Mill Offset Change—Critical Question

For profile milling cutters, it takes a bit of detective work to think out whether the correction needed is positive (away from the geometry line) or negative (toward the geometry line). Figure 19-34 shows three critical features, on three different parts: a tab, a slot, and a step. Each feature measures to be 0.017 in. too large. What radius offset correction is needed for each? Compare your answer to the response found at the start of the Respond for this unit.

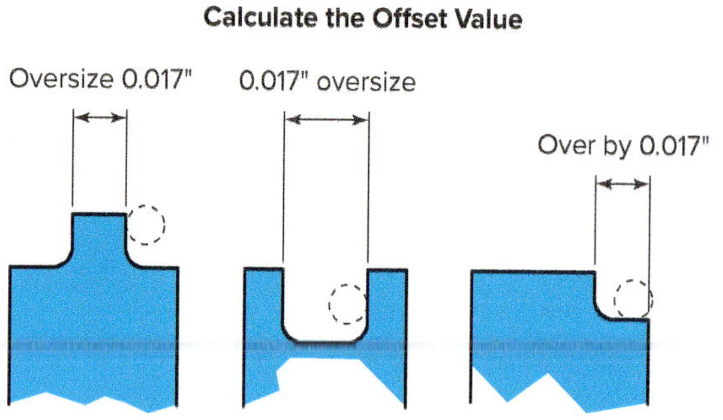

Figure 19-34 Calculate the offset.

UNIT 19-4 REVIEW

Replay the Key Points

- Personal music machines are a bad idea when running a CNC machine.
- Offset changes must be calculated by measuring the key dimension and comparing them to the nominal target.

Critical Question Response

The tab requires a negative correction of 0.0085 in.–toward the line–half the amount measured oversize. The slot requires 0.0085 in. away–positive offset. *No offset can correct the problem on the third part.* Think this through; the problem can only be a mistake in the program. Changing the offset would move the step right or left depending on whether it's a positive or negative correction, but it would not change the wall-to-wall dimension.

Respond

1. You are operating the CNC turning center on a first part run. After measuring four dimensions, you see a need for offsetting two tools. In Fig. 19-35, the actual measured results are shown in brackets. T01 cuts the 1.00- and 2.000-in. diameters and faces between them but does not face the end of the work. T02 cuts the larger face out to the dimensional limits of the stock size. Using Fig. 19-36 as a guide, enter your calculated offsets, a simulated tool memory page. Notice in Fig. 19-36 that since this is a first part with fresh tools, there are no initial offsets–only the nose radius for compensation.

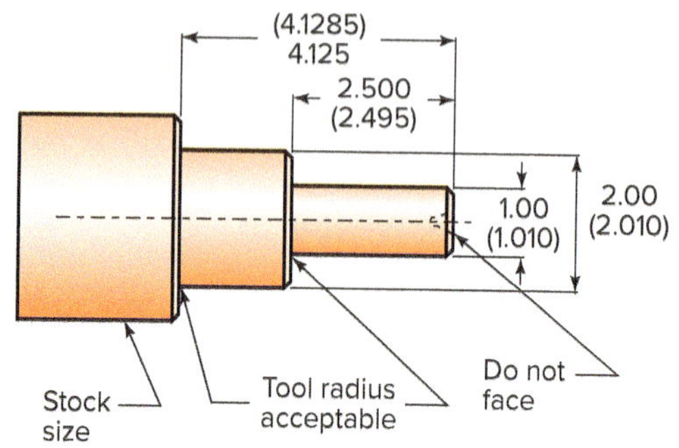

Figure 19-35 Problem 1.

Lathe Tool Memory

Tool	Name	Tool Radius	Radius Offset
T01	Turn –1	0.06	
T02	Turn –1	0.06	
T03	Part		
T04			
T05			
T06			
T07			
T08			

Note, the –1 behind the turning tool name indicates that the tools are both type 1 straight turning tools with a 15 lead.
This ID number comes from the tool library

Figure 19-36 Typical tool memory page.

Critical Thinking

2. In monitoring a long run on the CNC vertical mill, you find the condition for a nominal 0.750-in. slot width as shown in Fig. 19-37. It is being profile milled with a 0.5000-in. end mill, T0101. This is a mature setup and has already been offset, as shown in Fig. 19-38. Calculate and rewrite the radius offset required to bring the dimension to nominal again. Explain your calculated answer.

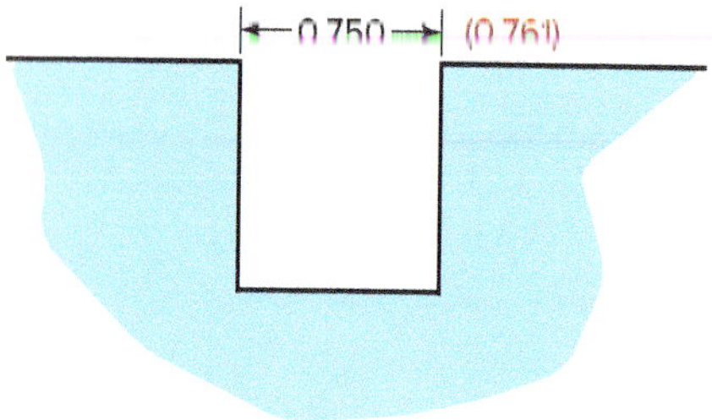

Figure 19-37 Problem 2.

Mill Tool Memory Page 1

Tool	Name	Tool Diam.	Drum Location
T01	EM	0.500	13
T02			
T03			
T04			

Note, this number changes as the control tracks random tool storage management

Mill Tool Memory Page 2

Offset	Amount	Offset	Amount	Offset	Amount
01	−0.007	09		18	
02		10		19	
03		11		20	
04		12		21	

Figure 19-38 Note the initial 0.007-in. offset on this tool memory page.

3. Explain why one tool might use two different offsets in a single program, for example, T0101 and T0102.
4. What does the DTG screen tell the machinist?
5. How is the DTG screen used for program tryout?

CHAPTER 19 REVIEW

Unit 19-1

I've worked in a shop that produces airplane wing spars. The blank aluminum is several inches thick and 150 ft long. Each billet is worth over $5,000 before the job begins! It weighs over 2 tons when it's laid down on the big gantry mill bed. At first, due to the great weight, it needs no clamps to keep in place for the first few program runs. The raw material goes onto the machine by overhead crane with special lifting devices, but it comes off by three people picking it up at the ends and in the middle! The finished part weighs less than 200 pounds. To get to that stage, it took several programs and setups. The point is that work like that takes good documentation and communication. One error caused by misunderstanding a direction can cost thousands of dollars!

Unit 19-2

Troubleshooting a problem is simplified if you become familiar with the classes of alarms. Then step back and look at the facts. Was the program CAM generated or by command entry at a keyboard? Is the part geometry complex or simple? Then apply PEP (predictable error point) logic. For example, if the program was written by hand key entry, always suspect a letter *o* was accidentally substituted for the number 0. If the geometry is complex and the program is CAM generated, suspect cutter shape interference. However, if it's handwritten, suspect a calculation mistake and so on. Before jumping right into scanning the program lines to find the problem, try a little logic on what kind of errors a particular program might have.

Unit 19-3

If you remember only one safety tip from this chapter, don't forget to practice what to do in level 2 and 3 events. Many times, I've seen operators turn a level 2 into a level 3 by their incorrect actions or by paralysis (no action). It's not enough to know the dials and knobs on the control. When a problem develops suddenly, you must move almost automatically. To be ready, you should practice the moves at the start of a shift or during a long job. Review how you'll halt and how you'll stop the machine.

Unit 19-4

The chances are that as an entry-level machinist you won't be changing offsets for a while. However, when the lead machinist does so, ask questions and ask if you can do it while he or she watches.

QUESTIONS AND PROBLEMS

1. Safety question: Why is it so vital that the PRZ location be clearly documented by the programmer and understood by the machinist? (Unit 19-1)
2. Why are program revision levels important to verify using the setup document and work order? Where is the program revision level found, in the document or in the program header? (Unit 19-1)
3. In what three ways are program Rev levels documented? (Unit 19-1)
4. Identify the two kinds of information a CNC document conveys. That information can be found in two different places—where? (Unit 19-1)
5. Describe a tool-compensated program. (Units 19-1 and 19-4)

6. Name the two kinds of programs that can be compensated. (Units 19-1 and 19-4)
7. Describe the nature of a level 2 incident and actions you must take. (Unit 19-3)
8. What operator actions are required for a level 2 incident on a CNC machine? (Unit 19-3)
9. True or false? The operator actions for a level 2 incident can also be used for a level 3, except they must be taken immediately rather than waiting until a natural place to halt. If it's false, what would make it true? (Unit 19-3)

CRITICAL THINKING

Use this list of possible factors that might be at fault to solve the following problems in a CNC part run. (Unit 19-4)

Clue List (partial—there may be other causes, too)

Dull tool

Tool deflection

Part or setup flexing

Mislocated PRZ

Misaligned holding tooling or misaligned in the holding tooling

Incorrect information in tool memory (wrong radius, offset, or length)

Incorrect program

10. Using a compensated, part path program and a 60° diamond insert cutting tool with a 0.060-in. nose radius, shank setup at vector position 4 as shown in Fig. 19-39. After coordinating the setup and turning one part, the following dimensions are measured. (Units 19-3 and 19-4)

Print	Actual
1.5000	1.4995
2.7500	2.7505
0.500 R	0.565
45° × 0.125 chamf.	45° @ 0.190 each leg
2.3125	2.3123

Is there a problem? If so, what might it be?

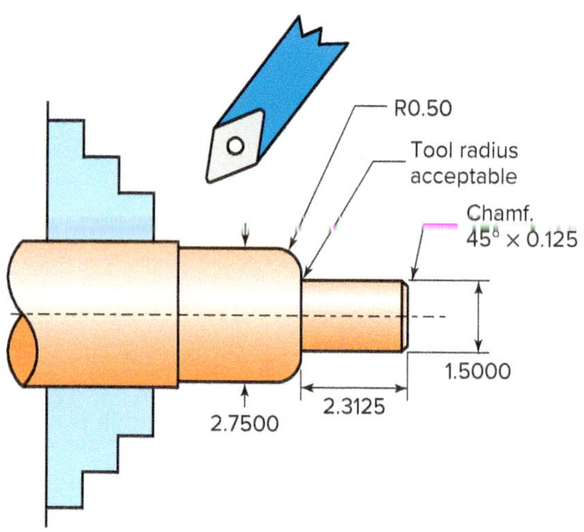

Figure 19-39 Problems 10 through 12.

11. Using Fig. 19-39, the part is machined and the 1.5000-in. dimension is measured at 1.507 and the 2.7500 one comes out as 2.7568 in. The radius, chamfer, and the 2.3125-in. dimensions are all measured to be right on target. What correction or corrections are called for? (Units 19-3 and 19-4)
12. In Fig. 19-39, the first run of the program shows the following turned dimensions. What's the problem, and how can it be corrected? The tolerance for diameters and length is ±0.003 in. (Units 19-3 and 19-4)

Print Size	Actual
1.5000	1.5003
2.7500	2.7503
0.500 R	0.501
45° × 0.125 Chamf.	45° @ 0.123 each leg
2.3125	2.2135

13. Per Fig. 19-40, the part shows the drawing measurements on the 1.500-in. straight diameter after making 74 parts. List the possible problems in their order of likelihood.

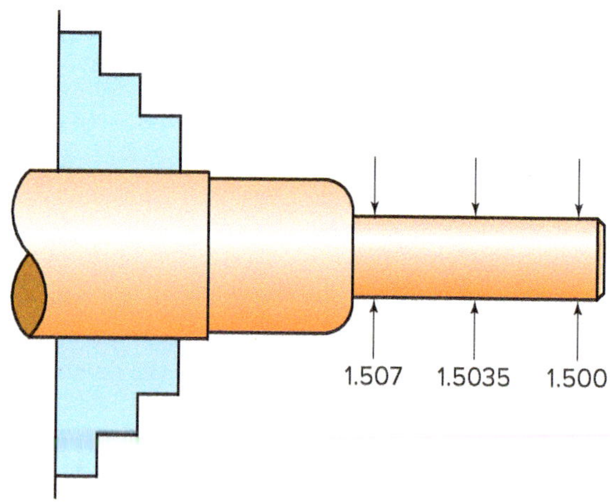

Figure 19-40 What is the most likely cause?

(*Hint:* Testing one part in every five, 40 parts made were found to be okay. Then the setup was assumed to be settled so no further measurement was done until the 75th part. All other measurements are found to be okay on that part.) (Unit 19-4)

14. See Fig. 19-41, Horseshoe Fitting 204-56B. A setup is made according to the document sketch (Fig. 19-42) of the fitting blank.

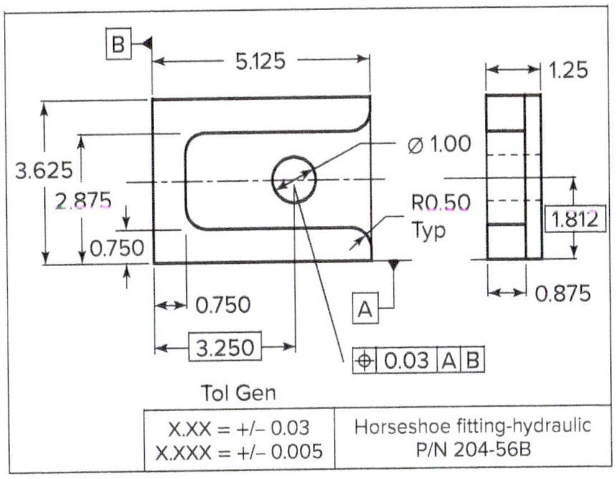

Figure 19-41 Problems 14 through 17.

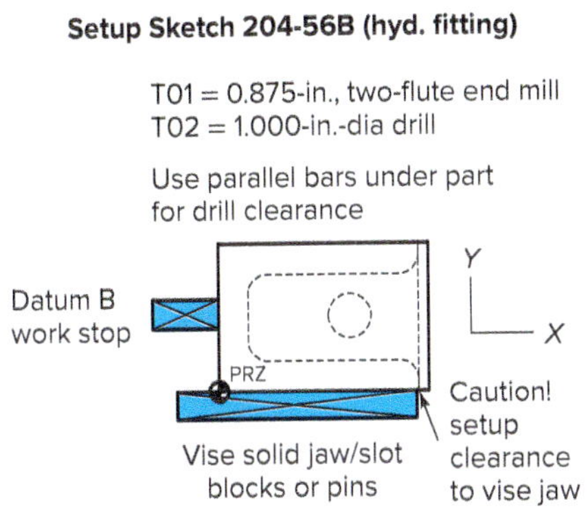

Figure 19-42 CNC setup sketch part number 204-56B.

Instructions

It has been precut 3.625 × 1.25 × 5.25 (excess for machining right side)

The program is a compensated part path

Using T01

End mill the 0.875 deep pocket including the 0.50 radius

Profile mill end to 5.125 dimension

T02

Drill the hole

On the first part, the condition shown in Fig. 19-43 is found to be true. What is needed to bring the four key dimensions exactly on target? (Unit 19-4)

15. Using the tools as outlined in the setup document (Fig. 19-42), will the program for the 204-56B Hydraulic Fitting form or generate the 0.50 radius at the left side of the pocket? What radius will the cutter path follow? (Unit 19-1)

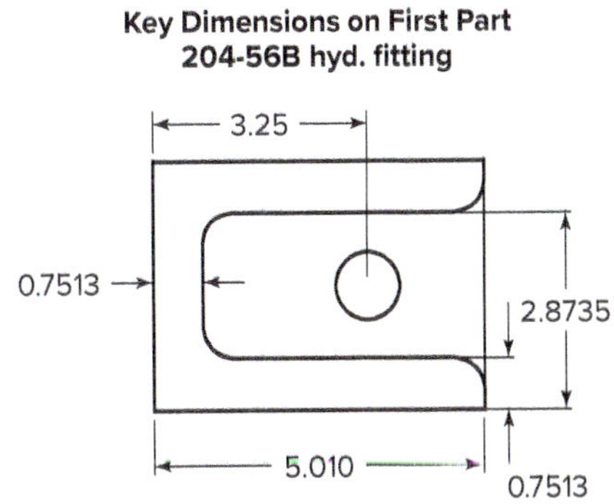

Figure 19-43 Is there more than one reason behind these measurement errors?

16. In machining the 204-56B, you have set up a 1.50-in.-diameter end mill in place of the 0.875 and reentered the cutter diameter offset for it. What will be the problem if any? (Units 19-3 and 19-1)

17. On the program tryout for the 204-56B fitting, the condition shown in Fig. 19-44 is discovered. What are the problem and fix? (Unit 19-1)

18. Monday morning, you touch the green button to start your machine. What's wrong in each situation? (Unit 19-2)

A. It will not start at all.

B. It starts to a certain point, the red light flashes, and further initialization is impossible.

C. It comes fully up and is a machine that does not require initialization. But now it will not move in jog or automatic modes.

19. Why would the control alarm out (stop machining) as you touch the cycle start button to make the 20th part after things have been going well? (Unit 19-2)

20. What is a tool path *look ahead?* Describe how it might complicate finding an alarm condition. (Unit 19-2)

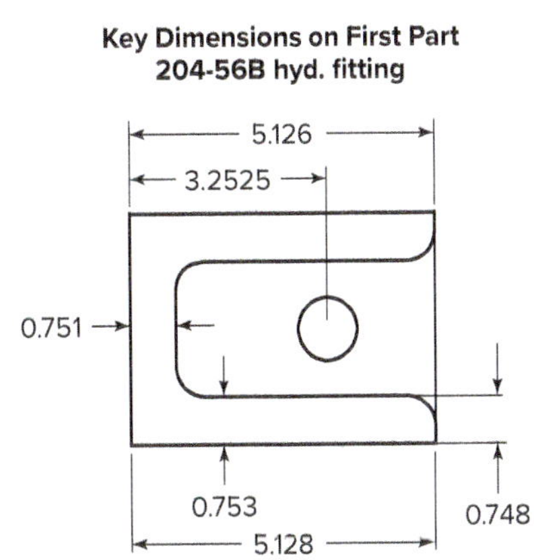

Figure 19-44 Why might this part be tapered on the edge?

CHAPTER 20 Program Planning

Units

20-1 Selecting the Origin, Quadrant, and Axes

- From part functional priorities choose the right location for the PRZ for mill or lathe work
- For mill work choose the axis orientation and quadrant

20-2 Selecting the Holding Methods and Cut Sequences

- Select holding tooling for the vertical CNC machining center
- Select holding tooling for the turning center
- Organize the cut sequence
- Plan intermediate cuts that lead to final results

INTRODUCTION

For me, writing and executing a successful CNC program is the ultimate computer game! I can still remember the first one I wrote by hand compiling codes—and to answer your question, yes it had flaws, but I eventually got a good part out of it. So I assure you, we instructors understand that students are more than anxious to do the same—create their own, then test it on metal. But experience proves over and again that while it's fun to do, starting there is a giant mistake!

Unguided, beginners often compile the program, then try to solve the details of carrying it out afterward. Here in Chapter 20, we learn that a modest investment in planning before writing returns big dividends in productivity, quality, and safety. Along with organizational skills, this chapter is also about communicating instructions to others. To do both, you will call heavily upon your manual machine experience. There are two vital task families to learn; then you can plan your first CNC job!

THE FINAL CHALLENGE

In the review we'll tackle a couple jobs, planning them from start to finish. The last problem has two answers: one provided by myself and then a second by the students and instructor of the Machining Technology class at Bates Technical College in Tacoma, Washington. When I wrote my job plan I had no idea if our plans would agree or diverge, but I expected they would diverge. Both plans will make good parts. The point is, although a geometric science, this is very much a creative endeavor (they didn't match at all).

Unit 20-1 Selecting the Origin, Quadrant, and Axes

Introduction: Selecting the axis orientation, the PRZ location, the holding methods, and the cut sequences are inseparable decisions. Each choice affects the other, and they must all be made at the same time. We study each as individual units only for book organization. The order in which these choices are made is so intertwined that there is no clear order in which to make them.

In some simple cases, the choice of holding tools can be made later at setup time (standard chucks, vises, and clamps). But many times in real CNC work, they must be preselected and sketched. Why? To avoid situations where your program accidentally machines through vise jaws when bar clamps would have been the right choice, or to avoid throwing parts from the wrong kind of lathe chuck. Clearly, everything must be well thought-out before writing the program, and those choices must be documented.

TERMS TOOLBOX

Coordinate shift A programming technique used on both mills and lathes to temporarily shift the PRZ for safety or convenience.

Tooling reference–set point An alternate method of establishing PRZ using a block attached to the fixture.

Touch method (touch off) A physical method of setting a *Z* axis PRZ at the machine by touching the cutter lightly to the work surface.

Work stop/spindle stop An adjustable stop to locate raw material in the same place with each part.

20.1.1 PRZ Selection for Lathe and Mill

The choice of PRZ is usually a relatively easy decision, but it's critical.

KEYPOINT

The PRZ must be based on functional priorities and geometric datums if they are within the design. If you are not using a geometric design, then the most reliable and easily located surfaces should be used.

In some cases, PRZ selection is based on temporary surfaces aimed at machining the higher priority functions and features as soon as possible. Some of the PRZ choices involve work and/or axis orientation. We'll start with lathe work, the simpler issue, since it's a two-axis machine, then look at mills. Here are two guidelines for PRZ selection for either.

Choosing the PRZ Location—Lathe or Mill

Primarily, PRZ must reflect the datum/dimensioning basis of the design.

Second, choose a location that's easy to physically locate during the setup–a large surface, a well-machined hole, and so on.

PRZ Location for Turning Work

X Axis on Centerline

To control diameters, the program reference zero always lies on the work center for the *X* axis. Diameters are always referenced from the lathe's centerline for turning work.

Z Axis PRZ

While the *Z* axis location is often at the far right tip, as shown in Fig. 20-1, away from the headstock, occasionally function will dictate *Z* location other than the tip, as shown in Fig. 20-2, where it's obvious where the PRZ should be. In this case, it's easy for the operator to locate it–2.00 in. inward from the outer tip, plus some excess material for machining, say, 2.050 in. (0.050 in. extra for a face cut).

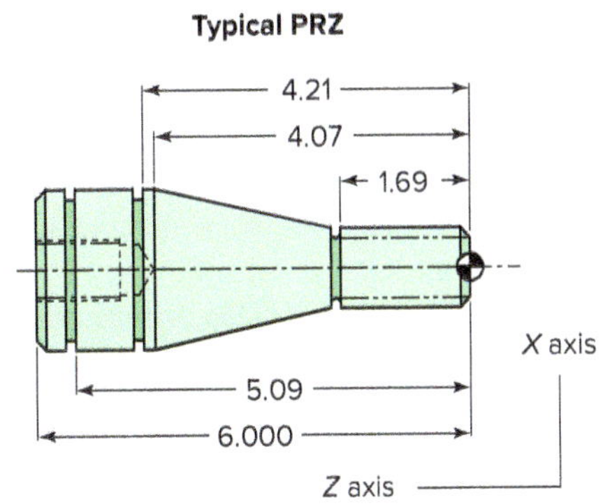

Figure 20-1 A typical PRZ location for turning work.

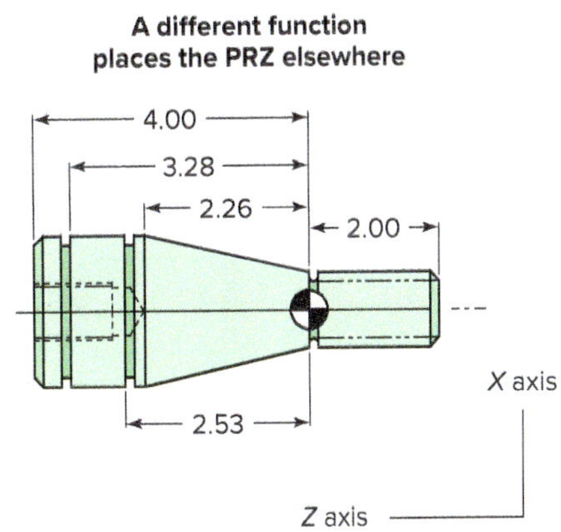

Figure 20-2 Function occasionally places the PRZ in some location other than the work tip.

KEYPOINT

For Fig. 20-2, the shifted PRZ is a critical fact (the PRZ is not at the "usual" location) and must be absolutely clear in the CNC document.

In the case shown in Fig. 20-2, during setup the machinist will touch the facing tool to the right outer tip of the work, then set the controller position registers at *Z*2.050–a plus *Z* distance from PRZ. That means the *Z*-PRZ is 2.050 in. farther into the part. If that setup detail is missed, the program run would be a disaster.

20.1.2 Work Reversal

Often, work must be machined on both ends, as shown in Fig. 20-3. Bar feeding can make the external details and the internal features on only one end. For example, it's going to require a second setup to machine the bore on the left. There are two ways to solve this:

1. **Machined Temporary Datum**

   Machine features in from one end, to a selected *Z* axis position. Then remove, reverse chuck the part nested against your temporary reference, and complete the final features.

2. **Spindle Work Stop**

   If the work is being machined from cut stock with no chucking excess, one solution to controlling a reverse *Z* axis position is to use a **spindle stop** or **work stop**, as shown in Fig. 20-4. After machining the features onto every part in the batch, each is reversed and rechucked. The adjustable stop sets the *Z* position of the work based on the far end.

   When we use a work stop, the PRZ remains the same physical location on the work, but it's now inside the machine spindle. The machinist must set the *Z* position registers to represent a PRZ that the tool cannot reach!

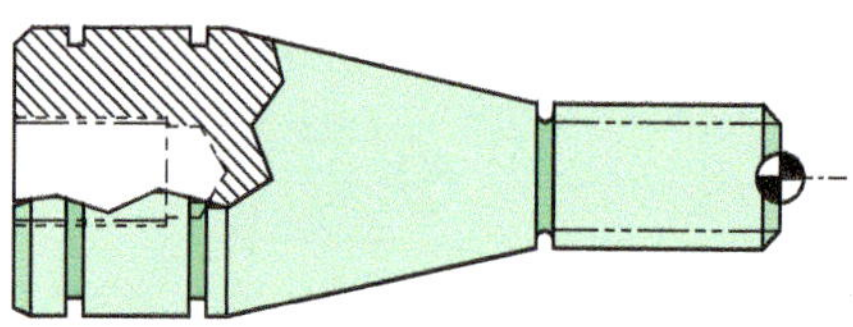

Figure 20-3 This job requires machining on both ends.

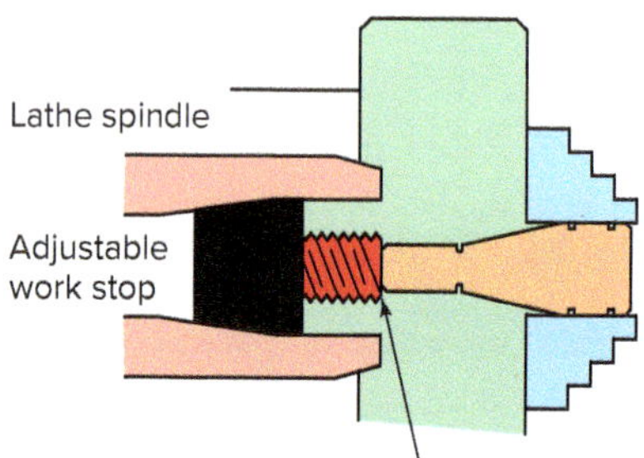

Figure 20-4 A spindle work stop provides a *Z* axis locator.

KEYPOINT

Once again this must be crystal clear in the document. A sketch is usually the best solution.

TRADE TIP

Chuck Jaw Faces A simple work stop can be the face of the chuck jaws in many cases. Often when machining custom jaws, a step can be faced into the jaws to function as a work stop.

To coordinate such a setup, with the PRZ inside the lathe headstock, the operator writes where the *Z*-PRZ is located relative to the cutting tool, even though the tool cannot travel there.

Caution! It's common programming practice to rapid travel the tool back to Z0.0 or close to it, when the next roughing or finish pass is to be taken. Forgetting that the PRZ is inside the headstock would result in smashing the tool into the chuck! There is a convenient planning fix for these situations using a programming tool called a **coordinate shift**. (See Trade Tip next.)

TRADE TIP

Shifting Coordinates There are times when the physical location for the PRZ on the work puts it in an unreachable location to coordinate it. It falls inside a chuck or beyond the work envelope on large parts, for example. In these cases, we can use a coordinate shift command. This program command sets up a temporary reference location, based on the master PRZ location. For example, in Fig. 20-5, positioning the tool 6.5 in. from the master PRZ, the programmer has moved the tool 0.5 in. away from the far tip of the finished work length—6.50 in. from the master work stop. All programmed *Z* axis moves are still relative to the original PRZ so no variation will be introduced. They now include the 6.5 in. The programmer can write the program using print dimensions, then shift the reference point out to a location that's useful and safe.

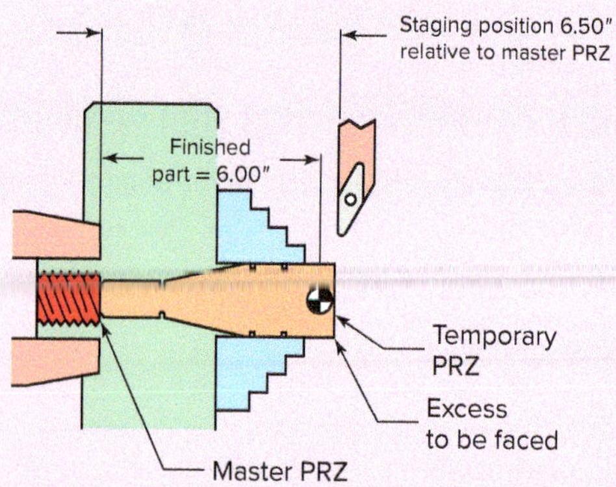

Figure 20-5 Shifting coordinates can be a time saver.

From that command onward, all positions relate to the temporary reference (TPRZ), yet that reference fully represents the original, 6.0 in. farther into the spindle.

This saves tons of extra programming math and potential accidents. You will see many possible applications of this time-saving tool in both lathe and mill work.

Mill Program—Reference Zeros

With mills, we add one more guideline for PRZ selection, beyond the datum basis and easy position to set up on the part:

> Try to use the first quadrant if possible. This is a secondary issue of rotating the work such that it lies in the first *X*-*Y* quadrant with the best axis advantage (Fig. 20-6).

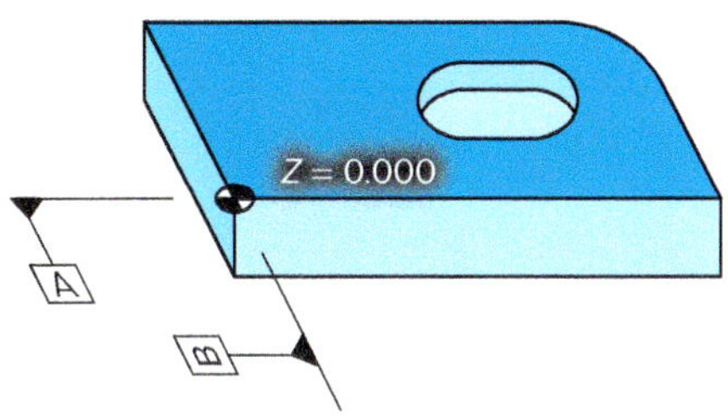

Figure 20-6 Top left corner is a common PRZ location.

Top Surface—Lower-Left Corner

The most common selection for mill PRZ is the top surface, lower-left corner shown in Fig. 20-6. We try to position the work such that the longer side is parallel to the *X* axis, but that's only for convenience. Located that way, all absolute value coordinates on the geometry will be positive (it's not a big deal, but it eliminates minus signs).

But that's not the only choice, as shown in Fig. 20-7. In each orientation, the datum basis of the design is preserved on the work. The only difference is the axis orientation and positive or negative values resulting when absolute coordinates are used.

Rotating the work to some other quadrant is often chosen to accommodate holding tools or work envelope space available, as shown in Fig. 20-8, where the vise has been turned sideways such that it will completely fit within the chip guard with the door closed. The long axis of the work is now parallel to the *Y* axis on the vertical mill.

KEYPOINT

Axis orientation and PRZ are preprogram decisions. Once the program is written, the setup must be made as shown in the document.

Question

In Fig. 20-8, in which quadrant is the work placed with the vise rotated as shown? (See the caption for the answer.)

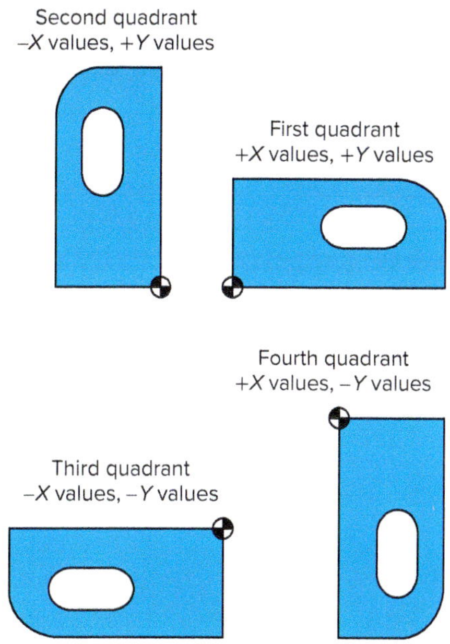

Figure 20-7 The same workpiece with the same PRZ but located in each quadrant.

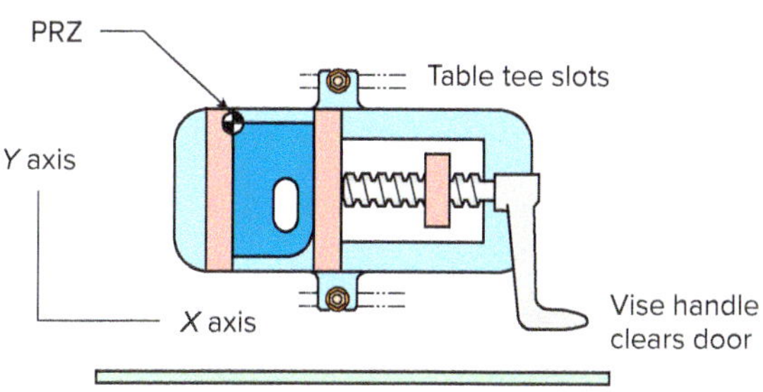

Figure 20-8 This part orientation was chosen to fully contain the vise within the safety shield. This places the work in the fourth quadrant. Watch that vise handle!

Touch Method for Setting Tool Reference

Selecting the top surface of the work for the *Z* axis reference is a common method. It simplifies the task of coordinating the *Z* axis position of tools and for determining length offsets. It's done by just touching the top of the work with the tool tip–lightly with no hard contact. Better yet, the *Z* reference can be set by staying a given distance off the work, then setting the *Z* axis position register to represent that height off.

This is known as the **touch-off method**. It's used on machines without tool probes or camera vision. We also touch the tool to the work on lathe setups. Touching a cutting tool to the work by jogging is not without risk. To do so, you must drive the tool toward the work using machine controls that have no sensitivity. Servo drives produce hundreds of pounds pressure without any sense of contact, which mars the work and breaks the tool! Don't miss the Trade Tip!

TRADE TIP

Feelers for Tool Touching Here are two methods that improve touching off when the machine has no tool probe (see Chapter 17).

Use a paper feeler (Fig. 20-9). Slipping it back and forth as the nonrotating cutter is slowly jogged toward the work, contact will be obvious when the paper drags. The thickness of the paper is then accounted for in setting the tool position; usually 0.003 in. must be added to the *Z* axis register position.

Figure 20-9 A paper feeler is a good practice when touching tools to work for *Z* axis touch methods.
Lake Washington Technical College, Kirkland, WA/McGraw Hill

An even better contact detector is a precision setup feeler (Fig. 20-10). They are usually made of a soft material such as plastic or aluminum to protect the work and cutter. Make your own to some easy math size such as 10 mm or 1.00 in.

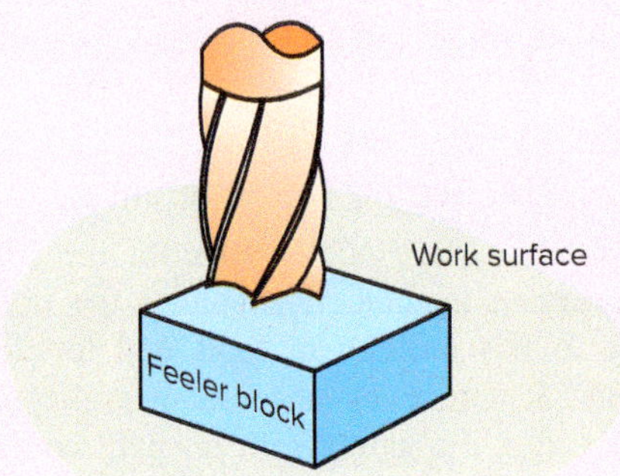

Figure 20-10 A precision, homemade feeler block of aluminum or plastic will protect cutting tools and work surfaces.

Tooling References—PRZ Not on the Workpiece

Fixtures are custom made holding devices that

- Hold odd shapes that cannot be held safely or reliably using standard methods.
- Provide quick part-to-part turnaround time.
- Reduce variation in production due to reliable loading and holding.

For now, we're concerned with a special type of reference provided by the fixture. Figure 20-11 is an example of an odd part shape that requires a fixture. In these cases, the raw stock is usually a casting, forging, or weldment, with no previous machining. In this starting condition, none of the surfaces provide a reliable location on which the physical PRZ can be located.

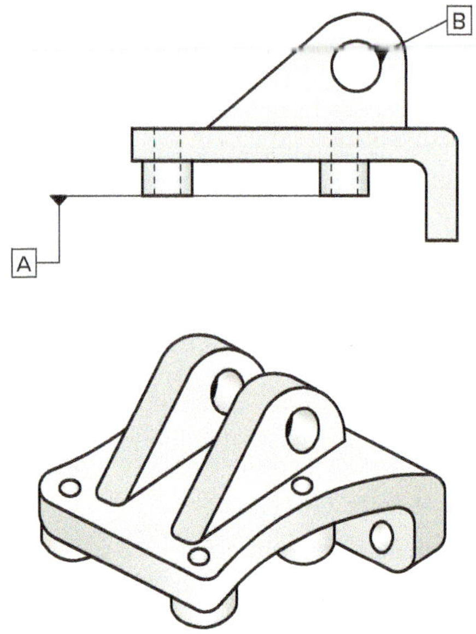

Figure 20-11 The design has a datum basis, but until the rough casting is machined, there is no reliable place to indicate a PRZ. This part is ideally held in a mill fixture.

Shown in Fig. 20-12, to cut the bottom pads to create datum feature A, and bore datum feature B, this part is held in a fixture. (Clamps not drawn.) The PRZ reference is planned and located on the fixture, called a **set point** or *fixture reference point.*

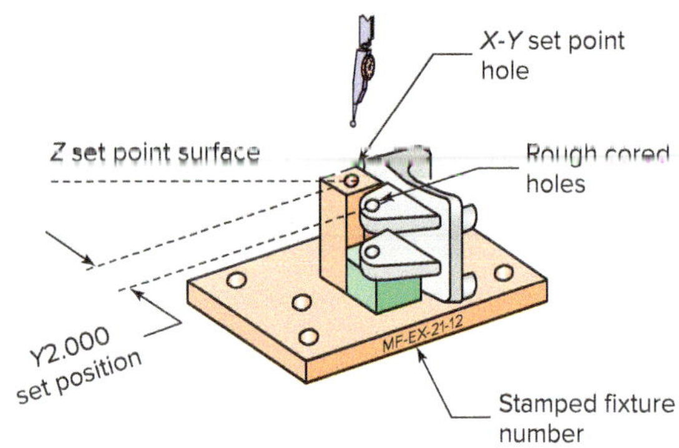

Figure 20-12 A mill fixture solves both holding and reference point location. Note the set point on the fixture is relative to the PRZ—it is not the PRZ.

KEYPOINT

The PRZ is not the set point. The PRZ is a known distance from the set point at datums A and B. In Fig. 20-12, when the spindle is directly over the set point hole, it is then at *Y*2.0 *X*0.0 relative to the PRZ.

Using a DTI to position the spindle over the hole in the set block, the operator then writes the location into the axis registers. This information must be clearly sketched in the document. The set point is usually a precision hole bored into a tooling block on the fixture.

TRADE TIP

Soft Jaw Vises and Chucks Custom-machined soft jaws are handy shop-made fixtures. For Fig. 20-11, can you see a way to hold the part in custom-machined soft jaws? They could be machined to custom fit the work. If the part shape is simple, the tooling program is written by the operator or toolmaker. But there's another trade tip coming up in Unit 20-2. A part program can be edited and used to cut the soft jaws by switching cutter compensation to the other side of the geometry line. One program can be modified to make both the workpiece and a fixture shape that holds the part!

UNIT 20-1 REVIEW

Replay the Key Points

- When choosing the PRZ location
  - Primarily, it must reflect the datum/dimensioning basis of the design.
  - If possible it should be some physical location that's easy to reproduce in the setup—a large surface, reliable machined hole, and the like. But sometimes that's not possible, so tooling reference becomes necessary.
  - When a PRZ is in a location other than what might be expected, it must be made crystal clear in the setup document.
  - Keep in mind that axis orientation and PRZ are preprogram decisions. Once the program is written, the part must be placed as shown in the document.
  - With a set point, the PRZ is normally still on the part, not on the fixture. The set point represents a known distance from the PRZ.

Respond

1. What is the main guideline for choosing a PRZ location?
2. What is the secondary guideline for choosing a PRZ location?
3. Why would you choose to plan a set point for establishing PRZ?

Critical Thinking

4. What could happen if the PRZ was not coordinated correctly on the machine relative to its program position? (Two potential problems.)
5. Explain why, if possible the PRZ is chosen such that the entire part to be milled lies within the first quadrant.

Unit 20-2 Selecting the Holding Methods and Cut Sequences

Introduction: Unit 20-2 builds upon setup skills gained in lab assignments using manual machines. But with CNC work, the method chosen must be rock solid to withstand the thrust and heat. This unit supposes you know how to use dial indicators.

Single-point threading is a perfect example. On manual lathes, threading occurs at an artificially low rate to accommodate the physical drop of the half nut lever. That results in light cuts because the speed is so low.

But on CNC lathes, the computer starts and stops the cutting tool in perfect coordination to the spindle. The CNC spindle can rotate near normal cutting speeds, therefore the cutting tool moves at amazing rates! The chuck grip and work rigidity must be able to withstand that force.

ARMCHAIR FIELD TRIP

Watch this Haas Tip of the Day video about threading inserts.

Watch a CNC lathe turn a rough piece of stock into a threaded part in no time!

Go to: **https://www.youtube.com/haasautomation** and search for this video called: **Rough—Finish—Thread Using HaasTooling.com Inserts—Haas Automation Tip of the Day**

TERMS TOOLBOX

Sacrifice plate A metal plate bolted above mill tables to protect them and provide a utility surface for drilling and tapping threaded holes for clamps and grooves for parallels and pins.

Subplate Same as sacrifice plates.

20.2.1 Machining Center Holding Tooling

We start by investigating the differences in CNC holding on mills. We'll limit our investigation to vertical CNC mills, where there are three common holding selections for your plan:

- Precision vise or vises–with hard or soft jaws
- Direct bolt down–over a sacrifice plate or set up on parallels
- Fixture–shop or professionally made

Precision Vises

We need say nothing about hard jaw vises at this point in your training. However, with CNC work, soft jaw vises are used more often because we need to hold odd-shaped work and also provide a locator for a PRZ.

The vise jaws are custom-machined to fit the part using the CNC control. Often, the finished soft jaw set (Fig. 20-13) is given a tooling number and saved for future applications of the program. This fact must then be added to the CNC document under comments–that tooling already exists and need not be made each time the program is run.

Figure 20-13 A shop-made soft jaw vise is custom-machined to fit an odd-shaped part, difficult to hold by any other means.

Courtesy Universal Aerospace/ Arlington, WA

TRADE TIP

Cutting Both Sides of a Line Cutting a set of soft jaws on either the lathe or the mill is simplified if you know this trick of the trade. By modifying the cutter radius compensation commands in the program written for the workpiece, the operator can create a reverse image program. Here's how.

In machining the box cutter cams shown at the top of Fig. 20-14, for the second set of operations the need arises to hold their perimeters to machine the pocket and drill the hole in the bottom. To cut the vise jaws, the outer profile program can be modified by switching the cutter to the opposite side of the geometry line (shown at the bottom of the illustration).

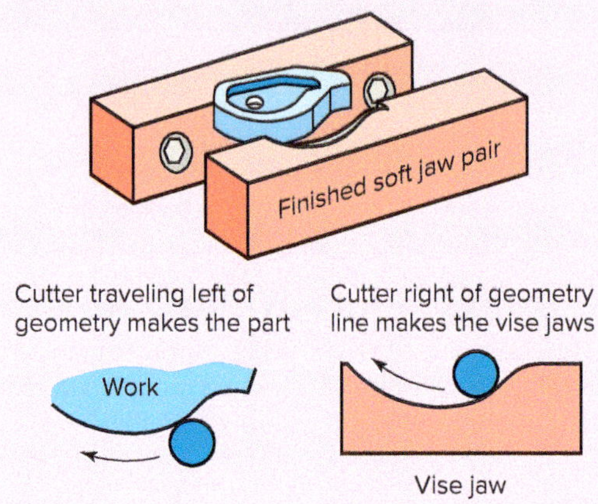

Figure 20-14 Using first left-hand, then right-hand compensation, both the work and the vise jaws can be machined using the same program.

The cutter is traveling along the left side of the geometry to cut the workpiece. But it's switched by editing to the right side to machine the reverse image vise jaw. Using program data might save some time!

ARMCHAIR FIELD TRIP

Making soft jaws made easy! You'll want to watch this **Haas Tip of the Day** video about creating soft jaws for odd shaped parts. Go to: **https://www.youtube.com/haasautomation** and search for this video called: **Mill Soft Jaws: The Proper Way to Make and Use Them—Haas Automation Tip of the Day**

KEYPOINT

Caution!

This is an advanced editing challenge that requires more training to perform. There are several modifications to be made to the program to eliminate unwanted tool motions, tool changes, and so on. Also climb cut profile cuts will be turned into conventional passes.

Direct Bolt Down—Sacrifice Plate

Bolting directly to the machine table is less common in CNC work mostly because it's slow to replace production parts but to also keep fast-moving cutters away from the machine table. We use bolt-down holding when the work won't fit into a vise or to gain a bit more *Z* axis work envelope for an extra tall workpiece.

For protection during bolt-down holding and for other setup utility reasons, a protective flat plate called a **sacrifice plate** or sometimes a **subplate** is usually bolted to the CNC mill table. These thick, aluminum plates (shown in Fig. 20-15) serve two purposes in CNC work:

- Protect the machine table
- Provide a soft utility surface for creating setups

Figure 20-15 This *sacrifice plate* on the machining center protects the table and adds useful setup features.
Courtesy Morgan Branch CNC/Arlington, WA

High-density, reinforced plastic is occasionally used as a sacrifice plate, too.

These plates often feature premachined, precision slots to fit vises or slot blocks, saving alignment time. Precision round dowel pinholes are sometimes bored to align work directly. They also feature a matrix of tapped holes, usually $\frac{1}{2}$-13 UNC, the common setup stud size. They are spaced regularly in *X* and *Y* directions.

Custom Machining the Subplate

A quick fixture can be made on top of the plate, as shown in Fig. 20-16.

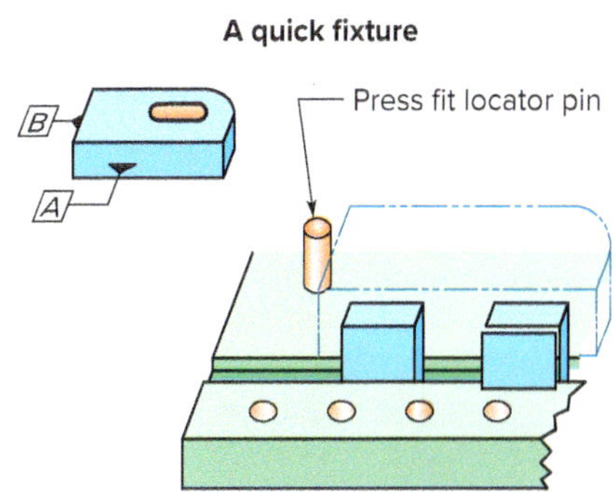

Figure 20-16 A typical quick, shop-made fixture features a slot fitting a pair of slot block parallels, a bored end alignment hole, and an array of tapped clamp holes.

KEYPOINT

Not all shops allow machining the subplate to create a special setup. Ask first!

While they are considered consumable, these utility plates are costly to replace. Make every effort to use features already machined on them. As cuts are made to the plate, it becomes an unstable platform, so there comes a time to resurface them with a fly cutter or replace them when they become too thin.

Machine Fixtures

Discussed previously, fixtures are a commonly used work-holding tool in industry (Figs. 20-12 and 20-17). These custom-made holding tools range from inexpensive shop-made to the designed, professionally made fixtures. Costing from a few hundred to thousands of dollars, fixtures are a good solution for some CNC plans but not for others.

Figure 20-17 A shop-made fixture improves safety and quality.
Courtesy of AMT Senior Aerospace

KEYPOINT

When using a designed fixture in your job plan, be sure there is no other less costly way to hold the work—and still deliver the needed repeatability. Custom fixture cost is usually not justified unless it is to be used more than once or there is absolutely no other way to hold the work safely or efficiently.

SHOPTALK

Creative Solutions Observe solutions in your shop. Collect creative ideas and think up others. Mandrels, rotary tables, and rotary chucks are all used—use the building blocks gained in manual machining. I've even been forced to use double-back tape to hold a very thin workpiece on a mill table!

Vacuum Chucks One useful holding tool for mill work is the vacuum chuck, as seen in Fig. 20-18. The mill sacrifice plate just mentioned can often be used to make a vacuum chuck by plugging the bolt holes, then machining the sealing details into the plate. A vacuum pump made for the purpose is used to hold the work tightly to the seal O-rings cut into the plate. For CNC router woodwork, vacuum holding is the universal choice (Fig. 20-19).

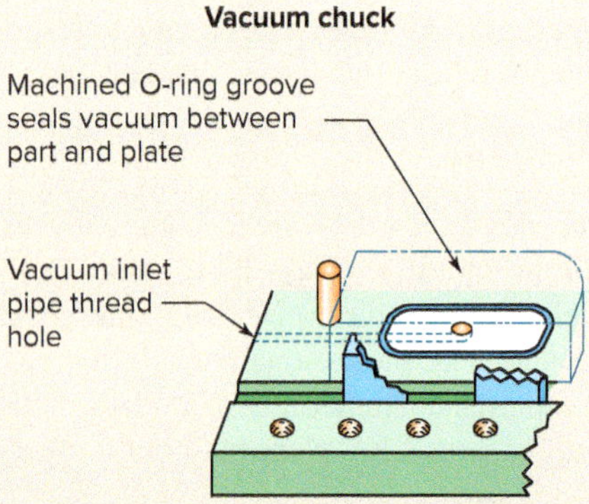

Figure 20-18 A vacuum chuck is made by machining an O-ring groove in a plate with a vacuum channel to distribute the holding force. A vacuum pump supplies the energy to hold the work to the plate.

Figure 20-19 This giant gantry mill is producing three wing spars at one time, all held down with vacuum chucks. Notice the large chip volume.
Courtesy Contour Industries/Everett, WA

20.2.2 Turning Center Work-Holding Tools

Similar to manual lathe work, there are the familiar selections of four-jaw chucks, mandrels, and faceplates, but they are normally used for one-off work or when a three-jaw or collet chuck cannot be used. The three-jaw and collet chucks are used for production. With both, there are differences from those on manual lathes. In all cases, long work is supported by a manually set tailstock center or the faster programmable tailstock.

- Common

  Three-jaw chucks–hydraulic with hard or soft jaws

  Collets–hydraulic
- Less useful

  Faceplates/turning fixtures–bolted to face plates or chucked up. With these any out-of-balance condition is very dangerous; thus, the speed must be kept low.
- Mandrels

Hydraulic Three-Jaw Chucks

There are five differences in three-jaw chucks made for CNC turning centers, compared to those on engine lathes. First, the chuck is made for extreme duty including much higher RPM. Second, the scroll plate is replaced by a hydraulic pull yoke, inside the spindle housing. CNC chucks aren't tightened with a wrench, because doing so would be too slow and the grip would vary depending on the operator. We've previously looked at the safety foot switch that actuates the chuck–the third difference. It's interlocked such that once the chuck tightens and the program commences, the chuck cannot be accidentally opened.

KEYPOINT

Reminder

The force developed by a hydraulic chuck can crush hollow parts and must be adjusted up or down by the operator.

The fourth difference is the jaw travel range from open to closed. Limited by the hydraulic action on CNC chucks, the open-close range of about 5 mm is adjusted to fit the work diameter. That's accomplished by bolting the jaws in place along a serrated carrier (Fig. 20-20). Commonly, each serration represents 1 mm of radial change for the jaw.

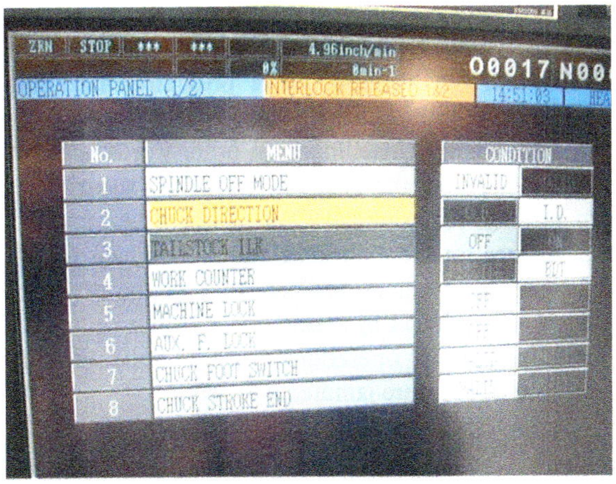

Figure 20-20 The hydraulically powered chuck's grip radius is adjusted through precise serrations on the back of the jaw.
Courtesy Tyee Aircraft/Everett, WA

KEYPOINT

Position the jaws such that they grip the work at about half their travel range to allow for raw part variation.

Closing Sense

The fifth CNC chuck difference is that on most industrial turning centers a closing sense for inside or outside gripping action must be preselected by the operator. Set for outside grip, they close inward and interlock upon reaching the full pressure set by the operator. Only then, with the green grip indicator on, will the spindle start (Fig. 20-21).

Safety Note!

To set up a three-jaw hydraulic, the operator must place all jaws at equal radii such that they will close on the work at about *half their full travel.* If not, when there is an unusually small diameter part within a batch, it might not be held as the carriage runs out of travel.

Figure 20-21 A grip sense switch tells the machine which orientation is the safety close position. If no closed feedback exists, the machine will alarm out and not start.
Courtesy Tyee Aircraft/Everett, WA

Hard or Soft Jaws

Soft aluminum jaws are very common, but we also use steel jaws for CNC production work. Correctly bored, these jaws can serve the function of a work stop while gripping the odd-shaped part. Another advantage arises when tight concentricity is an issue. By boring the jaws in the setup in which they will be used, the work centers on the lathe's axis with zero runout.

When custom jaws are unbolted, then reused at a later time, they are normally lightly rebored due to the impossibility of refastening them in the exact location of the original bore.

KEYPOINT

Consider Soft Jaws

For many reasons, soft jaws are a creative solution for CNC work holding.

Hydraulic Collets

In our manual machining discussions, we considered collets a medium-grip work holder. While CNC collets don't grip as securely as a chuck, they hold strongly enough due to hydraulic closing to do moderately heavy roughing of work up to around 1.5-in. diameter.

CNC spring collets are heavier than their manual counterparts and stronger, but they can be broken. Be certain the bar stock to be gripped is within the range of the collet. Work that is too small will bend or break the collet leaves when flexed beyond their working range. Be certain to adjust the hydraulic pressure based on the size and strength of the work. Like chucks, collets can crush tubing. CNC collets are also supplied (or custom machined) to fit odd-shaped work.

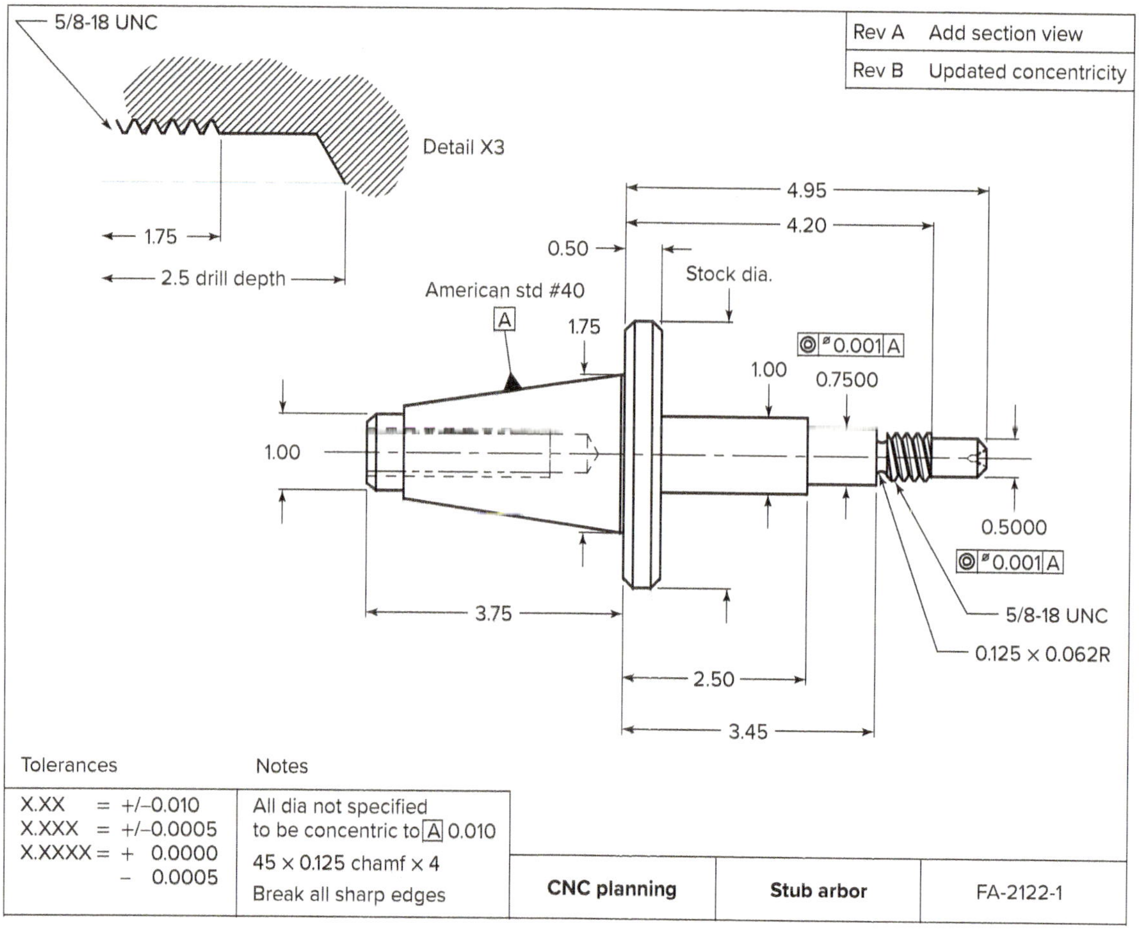

Figure 20-22 Problem 9.

> **Work Order—Stub Arbor**
>
> Make Qty-50 Part Number FA-2122-1 Rev B Sht. 1 of 1
> 4340 Bar 2.50 dia stock—Cut to length plus 3.0″ chucking excess
> Prefaced one end and center drilled—both ends
> Stress Relieved and Heat-treated to R_C-24/48
> Lot 1 of 50 lots*

*More parts to be made over many months.

Step One—Organizing the Major Factors
From the print, make notes about the five job characteristics:

- ☐ Critical tolerances and functions
- ☐ Overall shape and condition
- ☐ Material available
- ☐ Number of parts
- ☐ Value and complexity

Step Two
Select the type of turning tools needed for the job and assign them tool numbers on the document (see Fig. 20-23).

Step Three
Now, with a pretty good feel for the part and program, plan out the PRZ, holding tooling, and cut sequences on the document. If, in your opinion, another operator might not understand printed word instructions, draw sketches. Assume this may be a long-term instruction used repeatedly by others in the shop.

10. Valve Body (Units 20-1 and 20-2)
This is a product typical of small parts encountered in high-tech, electromechanical devices. Use Fig. 20-24 and the work order.

Hint: The end mill that will fit between the pockets, walls, and the $\frac{3}{4}$-in. post is $0.3125\left(\frac{5}{16}\right)$. See the following Shop-talk for some vital background information. Note that the cutter must also have a 0.125-in. corner radius to form the internal radii between walls and pocket bottom.

Divergent Answers
To illustrate the different ways products can be made, I will offer my solution and for comparison, so will another instructor (Mr. Bob Storrar). Neither is superior. Compare yours to ours. Have your instructor or another journey-level master review your solution.

KEYPOINT

Caution!

This is an advanced editing challenge that requires more training to perform. There are several modifications to be made to the program to eliminate unwanted tool motions, tool changes, and so on. Also climb cut profile cuts will be turned into conventional passes.

Direct Bolt Down—Sacrifice Plate

Bolting directly to the machine table is less common in CNC work mostly because it's slow to replace production parts but to also keep fast-moving cutters away from the machine table. We use bolt-down holding when the work won't fit into a vise or to gain a bit more Z axis work envelope for an extra tall workpiece.

For protection during bolt-down holding and for other setup utility reasons, a protective flat plate called a **sacrifice plate** or sometimes a **subplate** is usually bolted to the CNC mill table. These thick, aluminum plates (shown in Fig. 20-15) serve two purposes in CNC work:

- Protect the machine table
- Provide a soft utility surface for creating setups

Figure 20-15 This *sacrifice plate* on the machining center protects the table and adds useful setup features.
Courtesy Morgan Branch CNC/Arlington, WA

High-density, reinforced plastic is occasionally used as a sacrifice plate, too.

These plates often feature premachined, precision slots to fit vises or slot blocks, saving alignment time. Precision round dowel pinholes are sometimes bored to align work directly. They also feature a matrix of tapped holes, usually $\frac{1}{2}$-13 UNC, the common setup stud size. They are spaced regularly in X and Y directions.

Custom Machining the Subplate

A quick fixture can be made on top of the plate, as shown in Fig. 20-16.

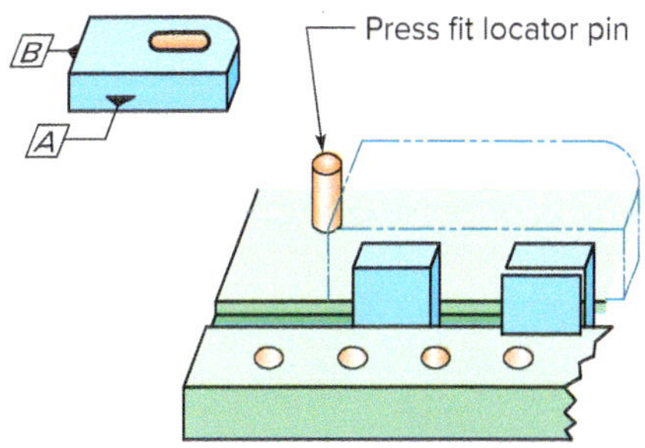

Figure 20-16 A typical quick, shop-made fixture features a slot fitting a pair of slot block parallels, a bored end alignment hole, and an array of tapped clamp holes.

KEYPOINT

Not all shops allow machining the subplate to create a special setup. Ask first!

While they are considered consumable, these utility plates are costly to replace. Make every effort to use features already machined on them. As cuts are made to the plate, it becomes an unstable platform, so there comes a time to resurface them with a fly cutter or replace them when they become too thin.

Machine Fixtures

Discussed previously, fixtures are a commonly used work-holding tool in industry (Figs. 20-12 and 20-17). These custom-made holding tools range from inexpensive shop-made to the designed, professionally made fixtures. Costing from a few hundred to thousands of dollars, fixtures are a good solution for some CNC plans but not for others.

Figure 20-17 A shop-made fixture improves safety and quality.
Courtesy of AMT Senior Aerospace

KEYPOINT

When using a designed fixture in your job plan, be sure there is no other less costly way to hold the work—and still deliver the needed repeatability. Custom fixture cost is usually not justified unless it is to be used more than once or there is absolutely no other way to hold the work safely or efficiently.

SHOPTALK

Creative Solutions Observe solutions in your shop. Collect creative ideas and think up others. Mandrels, rotary tables, and rotary chucks are all used—use the building blocks gained in manual machining. I've even been forced to use double-back tape to hold a very thin workpiece on a mill table!

Vacuum Chucks One useful holding tool for mill work is the vacuum chuck, as seen in Fig. 20-18. The mill sacrifice plate just mentioned can often be used to make a vacuum chuck by plugging the bolt holes, then machining the sealing details into the plate. A vacuum pump made for the purpose is used to hold the work tightly to the seal O-rings cut into the plate. For CNC router woodwork, vacuum holding is the universal choice (Fig. 20-19).

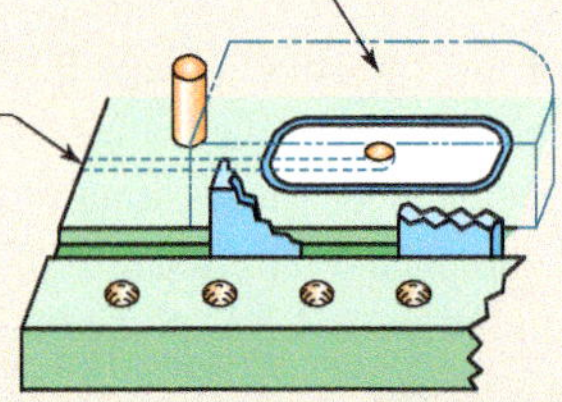

Figure 20-18 A vacuum chuck is made by machining an O-ring groove in a plate with a vacuum channel to distribute the holding force. A vacuum pump supplies the energy to hold the work to the plate.

Figure 20-19 This giant gantry mill is producing three wing spars at one time, all held down with vacuum chucks. Notice the large chip volume.
Courtesy Contour Industries/Everett, WA

20.2.2 Turning Center Work-Holding Tools

Similar to manual lathe work, there are the familiar selections of four-jaw chucks, mandrels, and faceplates, but they are normally used for one-off work or when a three-jaw or collet chuck cannot be used. The three-jaw and collet chucks are used for production. With both, there are differences from those on manual lathes. In all cases, long work is supported by a manually set tailstock center or the faster programmable tailstock.

- Common

  Three-jaw chucks—hydraulic with hard or soft jaws

  Collets—hydraulic
- Less useful

  Faceplates/turning fixtures—bolted to face plates or chucked up. With these any out-of-balance condition is very dangerous; thus, the speed must be kept low.
- Mandrels

Hydraulic Three-Jaw Chucks

There are five differences in three-jaw chucks made for CNC turning centers, compared to those on engine lathes. First, the chuck is made for extreme duty including much higher RPM. Second, the scroll plate is replaced by a hydraulic pull yoke, inside the spindle housing. CNC chucks aren't tightened with a wrench, because doing so would be too slow and the grip would vary depending on the operator. We've previously looked at the safety foot switch that actuates the chuck—the third difference. It's interlocked such that once the chuck tightens and the program commences, the chuck cannot be accidentally opened.

KEYPOINT

Reminder

The force developed by a hydraulic chuck can crush hollow parts and must be adjusted up or down by the operator.

The fourth difference is the jaw travel range from open to closed. Limited by the hydraulic action on CNC chucks, the open-close range of about 5 mm is adjusted to fit the work diameter. That's accomplished by bolting the jaws in place along a serrated carrier (Fig. 20-20). Commonly, each serration represents 1 mm of radial change for the jaw.

Figure 20-20 The hydraulically powered chuck's grip radius is adjusted through precise serrations on the back of the jaw.
Courtesy Tyee Aircraft/Everett, WA

KEYPOINT

Position the jaws such that they grip the work at about half their travel range to allow for raw part variation.

Closing Sense

The fifth CNC chuck difference is that on most industrial turning centers a closing sense for inside or outside gripping action must be preselected by the operator. Set for outside grip, they close inward and interlock upon reaching the full pressure set by the operator. Only then, with the green grip indicator on, will the spindle start (Fig. 20-21).

Safety Note!

To set up a three-jaw hydraulic, the operator must place all jaws at equal radii such that they will close on the work at about *half their full travel.* If not, when there is an unusually small diameter part within a batch, it might not be held as the carriage runs out of travel.

Figure 20-21 A grip sense switch tells the machine which orientation is the safety close position. If no closed feedback exists, the machine will alarm out and not start.
Courtesy Tyee Aircraft/Everett, WA

Hard or Soft Jaws

Soft aluminum jaws are very common, but we also use steel jaws for CNC production work. Correctly bored, these jaws can serve the function of a work stop while gripping the odd-shaped part. Another advantage arises when tight concentricity is an issue. By boring the jaws in the setup in which they will be used, the work centers on the lathe's axis with zero runout.

When custom jaws are unbolted, then reused at a later time, they are normally lightly rebored due to the impossibility of refastening them in the exact location of the original bore.

KEYPOINT

Consider Soft Jaws

For many reasons, soft jaws are a creative solution for CNC work holding.

Hydraulic Collets

In our manual machining discussions, we considered collets a medium-grip work holder. While CNC collets don't grip as securely as a chuck, they hold strongly enough due to hydraulic closing to do moderately heavy roughing of work up to around 1.5-in. diameter.

CNC spring collets are heavier than their manual counterparts and stronger, but they can be broken. Be certain the bar stock to be gripped is within the range of the collet. Work that is too small will bend or break the collet leaves when flexed beyond their working range. Be certain to adjust the hydraulic pressure based on the size and strength of the work. Like chucks, collets can crush tubing. CNC collets are also supplied (or custom machined) to fit odd-shaped work.

KEYPOINT

Most lathes require the grip switch set to outside when using collets.

Turning Fixtures

Due to safety issues and expense, lathe fixtures are considered a last-resort decision. You would choose a fixture when the work could not be held any other way and the job justifies the expense.

Due to centrifugal forces combined with cutting forces, CNC lathe fixtures must include bulletproof clamping and extragood balancing. Fixture making and use are beyond the scope of this unit. However, their care is something we must discuss.

Fixture Care

Toolmakers work for days on fixtures–they are precise and costly. The tolerances used are below one-third that of the critical dimensions of the intended work. *They are expensive and must be treated with the greatest respect.* Here are the seven commandments of fixtures:

1. Never modify them without permission. If modified to any great degree, a note in the CNC document is necessary.
2. Keep all clamps, pins, and other attachments with the fixture when you are through with it.
3. Oil all steel components before long-term storage.
4. Always clean fixtures before mounting on the CNC machine and after use.
5. When loading parts, do not use hammers, pry bars, and so on. Fixtures are tough but can be damaged!
6. *Know the fixture before using it!* If the part can be inserted wrong, determine which way is right. Also study the clamping. It might be possible to distort the work with excess clamping force.
7. Read and check all instructions provided and be certain the part and fixture numbers match the work order!

20.2.3 How to Get Started Writing Your Plan

To begin, review the engineering drawing for these five major factors:

1. **Critical Features and Dimensions**

   Look at tolerances and datums.
2. **Scan the Overall Work Shape and Special Features**

   Do these details create special cuts or holding considerations?

   They may require special cutters or holding tools.
3. **Know the Material Type, the Raw Billet Cut Size, and Consider Excess**

   Remember, excess material may save money in the long run.
4. **Number of Parts**

   Doing this may justify certain decisions about tooling.

   Do future runs of the same part justify better tooling costs?
5. **Raw Material and Finished Part Value**

   May justify cutting and holding tooling decisions.

As you solve the problems coming up, consider these five factors. It's rare that a perfect answer shows itself. It's entirely possible that your plan will have improvements over mine. With each, there will be a blank CNC document to copy, then fill in.

20.2.4 Don't Dismiss Manual Machining

Due to book organization, we've presented machining as though it were either manual or CNC, while in industry, machining is simply machining. In other words, a job might efficiently start on the manual machines, squaring up or rough turning, for example. Then it might be moved on to the CNC area for the main machining, then back for offload operations. For example, the bar stock of the first problem has been premachined by facing one end plus center drilling on both ends. Those operations could be done on a manual lathe, if the CNCs were jammed with work.

If they can deliver the work at the required rate, using manual machines often improves the price of the final product. The manual machines are far less costly to own and maintain. Use them to allow expensive CNC equipment to do the work only they can do.

UNIT 20-2 REVIEW

Review the Key Points

- Not all shops allow machining the subplate to create a special setup.
- Use a designed fixture in your job plan only if you are positive there is not a less costly way to hold the work.
- The force developed by hydraulic chucks can crush hollow parts and must be adjusted up or down by the operator.
- Always consider soft jaws in CNC work.
- Depending on the need, be sure to select the interlock inside or outside grip.

Respond

1. When would the machinist choose a fixture for a CNC plan?

2. Based on Unit 20-1, what program reference is most likely when using fixtures?
3. What is different about a three-jaw chuck on a CNC lathe compared to a manual chuck? (There are two facts.)
4. True or false? When setting up a CNC three-jaw chuck, one must set the jaws such that they close to the middle of their grip range.
5. In addition to the midtravel range, what other consideration is needed for setting CNC three-jaw chucks?

CHAPTER 20 REVIEW

Unit 20-1

On the Space Shuttle now retired, there are liquid oxygen booster turbines that cost in excess of $15,000 each, and you could hold them in your hands! The program that made them had to be right! But no matter whether it's a high-value part like that or an assignment for training, all programs need to run right. Yet for any given part geometry there are probably over a hundred different plans that all lead to a finished part. How do you know yours is among the best? There are three gages in this order:

Operator and machine safety
Part quality—variation is minimized
Cost of manufacturing
Overall cycle time
Value- and non-value-adding time (no chips) in the cycle
How much handling—more than one setup?
Does it cause offload processing and burring?
Cutter tool life and cutters costs
Excess raw material consumed

Unit 20-2

Real experience is the only teacher of this skill. Observing others will help, but doing a plan that matters is the real test. To get it going, after a few questions, there are two challenge problems in the review, one for turning and one for milling. Neither is easy. Go at them as though the job is real! In the workplace, I always consult others on a difficult plan. Do the same here—bounce your ideas off another student. Because the CNCs can do so many more operations in one setup, compared to manual machining, every job plan for a new shaped part is radically different. The possibilities are nearly unlimited.

QUESTIONS AND PROBLEMS

1. Before planning out a job there are five factors that should be considered. Name them. (Unit 20-2)
2. When would a coordinate shift be used? Why? (Unit 20-1)

CRITICAL THINKING

3. A milled part is very large. Its end must protrude out beyond the area where the spindle can be brought over the physical PRZ to be coordinated. Would a coordinate shift or a fixture reference solve the problem? (Unit 20-1)
4. True or false? In setting up a part you discover the vise protrudes out beyond the safety door. It won't close. But you can turn the vise 90° (with the part) and it closes. You can edit a coordinate shift into the program now to make everything work. If the statement is false, what would make it true? (Unit 20-1)
5. Name at least three important details that are different about setting up a CNC hydraulic three-jaw chuck compared to a standard three-jaw chuck. (Unit 20-2)
6. Name an improvement to aid in the touch-off method of coordinating cutting tools. (Unit 20-1)
7. Is it desirable to place milled parts within the first quadrant? Why? What is the advantage? (Unit 20-1)
8. What are the major factors in PRZ selection before writing a program? (Unit 20-1)

CNC PLANNING PROBLEMS

9. **Turned Stub Arbor** (Units 20-1 and 20-2)
Per the work order instructions and the part drawing in Fig. 20-22, plan the holding methods, cut sequences, PRZ location, and special instructions such as rechucking or part reversal. Be sure to sketch any details you feel the operator will need. A blank CNC document (Fig. 20-23) is supplied to be copied, then used for this and the next assignment. There is also a blank master in the Instructor's Guide for this book.

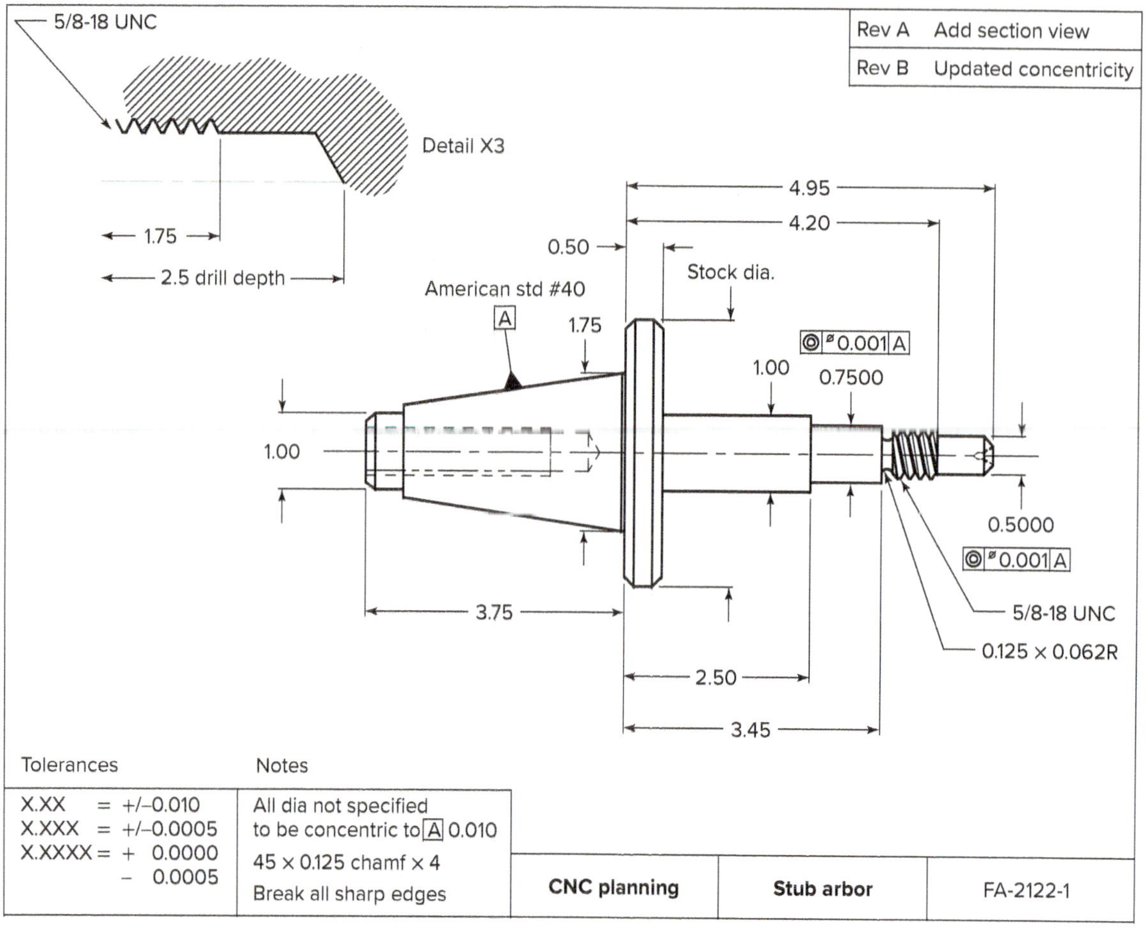

Figure 20-22 Problem 9.

> **Work Order—Stub Arbor**
>
> Make Qty-50 Part Number FA-2122-1 Rev B Sht. 1 of 1
> 4340 Bar 2.50 dia stock—Cut to length plus 3.0″ chucking excess
> Prefaced one end and center drilled—both ends
> Stress Relieved and Heat-treated to R_C-24/48
> Lot 1 of 50 lots*

*More parts to be made over many months.

Step One—Organizing the Major Factors
From the print, make notes about the five job characteristics:

- ☐ Critical tolerances and functions
- ☐ Overall shape and condition
- ☐ Material available
- ☐ Number of parts
- ☐ Value and complexity

Step Two
Select the type of turning tools needed for the job and assign them tool numbers on the document (see Fig. 20-23).

Step Three
Now, with a pretty good feel for the part and program, plan out the PRZ, holding tooling, and cut sequences on the document. If, in your opinion, another operator might not understand printed word instructions, draw sketches. Assume this may be a long-term instruction used repeatedly by others in the shop.

10. Valve Body (Units 20-1 and 20-2)
This is a product typical of small parts encountered in high-tech, electromechanical devices. Use Fig. 20-24 and the work order.

Hint: The end mill that will fit between the pockets, walls, and the $\frac{3}{4}$-in. post is 0.3125$\left(\frac{5}{16}\right)$. See the following Shop-talk for some vital background information. Note that the cutter must also have a 0.125-in. corner radius to form the internal radii between walls and pocket bottom.

Divergent Answers
To illustrate the different ways products can be made, I will offer my solution and for comparison, so will another instructor (Mr. Bob Storrar). Neither is superior. Compare yours to ours. Have your instructor or another journey-level master review your solution.

McGraw Machining-CNC Documentation			
Print Number FA-2221	Part Number - 1	Rev Level B	Sheet 1 of 1
Date	Part Count 50 of 500	Material 4340 HT	
Holding Tools/Fixtures		Name-Stub Arbor	
T01 T03 T05 T07	Cutter Tooling T02 T04 T06 T08		
Event Sequence and Special Instructions			
PRZ Setup Sketch			

Figure 20-23 Photocopy and use this blank for your planning sheet.

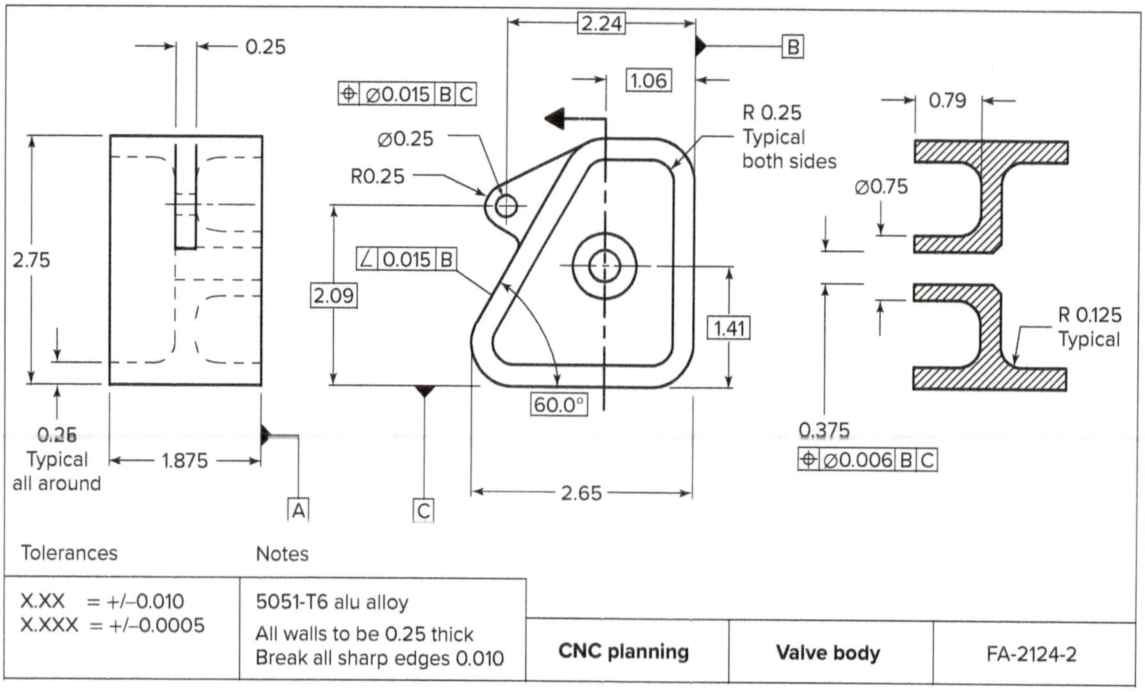

Figure 20-24 Problem 10.

> **Work Order—Valve Body**
> Make Qty—20 Part Number FA-2124-2 Rev NEW
> Sht 1 of 1
> Aluminum Blocks—Grain direction not critical
> Saw and premachine flat and square to
> 3.0 × 3.0 × 2.0
> A one-time-only order—not to be repeated

SHOPTALK

Islands On the valve body, there is a pocket with a $\frac{3}{4}$-in. post protruding from the floor of the pocket. A trade term for that protuberance to be machined is *an island.*

There is a tight spot between the island and the angular wall on the left side. It limits the diameter of end mill that can be used because the cutter must fit between the island and the wall. I determined that gap, using the CAD database, to be greater than 0.312 in. but less than 0.375 in.; therefore, the only standard cutter that would fit between the island and the wall is the $\frac{5}{16}$-in.-diameter cutter.

Using the $\frac{5}{16}$-in. cutter to rough out the entire pocket floor and island won't take too long as there are only 20 parts and they are small. But there are times when you might choose to rough out most of the pocket with a larger cutter even though it won't fit into the tight spots. This programming technique saves lots of time over making the many cuts required using the smaller cutter. Then with most of the metal removed from the pocket and islands, go back and profile mill the features to final size with the smaller cutter that clears all tight spots.

CHAPTER 21

Level-One Programming

Units

21-1 Code Words and Program Conventions

- Name six word prefix letters
- Explain groupings of codes
- Start to form a vocabulary of alphanumeric words
- Define a command line compared to a code word

21-2 Outer Program Structure

- Start programs by setting safe and necessary modes
- Cancel unwanted modes
- End programs with safe tool movements
- Organize data

21-3 Manually Compensated, Linear Cutter Paths

- Calculate coordinates for the tool path
- Write a manually compensated mill program

21-4 Writing Arc Commands

- Write arc commands using the radius method
- Write arc commands using the center or "*IJK*" method

21-5 Writing Compensated Programs

- Use the G41/42 compensation codes in a program
- Write on ramps for the cutting tool
- Write off-ramp commands

Made Right Here

Whether it's your smartphone or your mountain bike, someone had to design and manufacture these products. Did you ever stop to think about the people behind the product? Edge Factor produces films that tell the stories of the people and technology in the manufacturing world. Think extreme mountain bikes, 3-D camera equipment, rescue missions, and prosthetic limbs . . . you get it?

Manufacturers build the world you see all around you. The best part is that machinists get paid to interact with advanced technological toys and to see ideas transform into real-life products. To watch the latest Edge Factor production, visit www.edgefactor.com.

Courtesy of Edge Factor

INTRODUCTION

In this book, we approach programming in stages, writing codes by hand first, then working up to CAM programming. Here in Chapter 21 we study arranging code words into complete statements that prepare for or cause a machine action, called commands or command blocks. Then we will put the blocks together to write two kinds of programs:

Manually Compensated Creating a cutter centerline tool path without assistance from CAM software or graphic interface controls.

Tool Radius Compensated Writing a part path program with commands that tell the control to compensate for a given cutter size and shape.

To complete the shape assignments, you'll learn to write arc commands.

Completing Chapter 21, you should be able to write programs for most of your basic school assignments, with the exception of branching logic statements, special routines, and complex surfaces only possible with CAM assistance, all subjects we'll study later.

WHY MANUALLY COMPILE PROGRAM CODES AT ALL?

Today in industry nearly all programs are generated using CAM software and, to a lesser degree, graphic interface or conversational input right at the CNC control. It's a matter of efficient use of your time and staying up with the competition. CAM programming brings more than turning hours of calculating and code writing into minutes, it also programs shapes not possible any other way.

So the question is valid. If it's time-consuming and math intense, why not skip writing codes manually and go directly to CAM? The answer is fourfold.

The first answer is to be in control of a CNC machine one must be able to read the program! Being able to read code means you'll know what the machine is about to do before it does it! It's also necessary to read code to solve what's wrong when an error message appears on-screen. Writing codes yourself is the fastest way to gain these skills.

Second, the CAM program (or any program) is rarely a perfect work. There are places where it can be edited for safety or efficiency. With code and program structure training you will have the skill to do them.

Third, during a machine setup the need often arises for a quick mini-program to make soft jaws, a fixture, or some other provision for holding the work. These tasks you can do at your keypad if you can write code. Called manual data input, the MDI program is only possible if you know codes.

The last reason is that code writing is a backup skill. When a programming system isn't available or the PC goes down, the competent journey machinist can still get the machine up and running. Beyond that, my personal reason number five—it's fun and rewarding to be able to write code!

The units of learning in this chapter provide the fundamentals of brain-assisted machining (BAM!).

Unit 21-1 Code Words and Program Conventions

Introduction: To begin, we'll take a look at a code word vocabulary. It's not a full set, but these words are a good beginning. Then we'll investigate data conventions: how to make entries the control can accept.

The main objective is not to memorize codes, although doing so will accelerate mastery. Here, we'll be focusing on the possibilities of the words. If you know the function exists, you can look it up when you need it. Haas Automation provides great tutorials:

Enter: ***Haas Customer Resource*** Center in your browser.

TERMS TOOLBOX

Alphanumeric code A combination of a prefix letter with trailing numbers.

Code groups Similar words that are mutually exclusive within any single command line.

Command (command line) A complete statement causing a machine action.

Command word (word) A part of an instruction that causes action or prepares for action or some utility function. The basic building blocks of programs.

Conversational A CNC language based on spoken words.

Default A fallback or a preset condition built into the control.

Machine assembly language (MAL) The underlying operating commands that your program causes. This language works behind the scenes and is caused by CNC programming.

Modal command A word that stays on until canceled or superseded by any other word from its group.

One-shot codes Words that affect the line in which they are written but have no further effect.

Preparatory (Prep) codes (G codes) Codes that cause or prepare for a machine action.

RS274-D (EIA Recommended Standard 274-D) The standardization of code words for CNC programming.

21.1.1 Programming Languages

Three Ways to Compile Programs

When the control reads commands, it translates each word into a more basic set called **machine assembly language (MAL)**. MAL works at the on/off level to process and send signals to actuate drives and start or stop various machine functions. Although MAL is the universal language working within all machines, writing programs in it would be very slow and clumsy for day-to-day use.

To speed and simplify the task, friendly languages have been invented. Similar to computer programming languages, each word in the vocabulary is actually a routine in itself. The MAL understands and breaks each code word down into machine-readable signals.

Today there are three different methods of compiling a machine-readable program:

1. **Graphic Interface**

   This method generates commands from screen pictures drawn on the CNC machine's control. Many control manufacturers offer this feature. While graphic programming is similar to CAM programming and is easy to learn, in fact it's designed with ease in mind–each system is designed for a given control. Transfer of knowledge is simple, but there's no universality between systems. Therefore we'll not study it here. Today, most graphic controls also convert their programs into EIA codes to facilitate editing and data exchange to other machine systems.

2. **Conversational**

   In this second method the machinist writes the program statement on the CNC control. These command words are based on the spoken language of the country in which it is being used. If it's English, for example, then command words might be *go-to*, *go-around*, and *spindle on and arc*. Like graphic programming, conversational input is also easy to learn. But it too is controller dependent. What you know on one control will transfer to another manufacturer's product, but not verbatim. Even in the same language, they don't speak the same "dialect"! It is less common for a conversational control to convert programs to EIA codes since editing is easy in the original form (Fig. 21-1).

3. **EIA Codes (ISO)**

   This is the closest we come to a universal language set. The basic vocabulary is standardized by the **Electronics Industries Association, EIA–Recommended Standard RS274-D**, here in North America. Worldwide it's overseen by the International Standards Organization (ISO). ISO and EIA codes are nearly the same at the beginner level, and there are only a few differences at any level.

   While it's close to universal, you will find code differences from control to control. Being over 30 years old, the code language was originally written for lower-level machine tools than those of today (reading punched cards and tapes, no computers in the control—called NC). But the originators foresaw the need for future expansion. They interspersed defined words with blank words that had no meaning, to be defined by the manufacturer of the control. Thus, the language had room for growth.

Today, every manufacturer uses the originally recommended core of 50 motion statements (G codes) and 50 miscellaneous statements (M codes) of a possible 100 within each set (one set for turning and one for milling). But we outgrew the total 200-word vocabulary some time ago.

Industry then created new words, starting with the 50 unassigned words within the basic set. Then as computers grew in power and speed, many more beyond the original 100 level were added.

Many of the first 50 new words have become universally accepted for CNC work because they became popular on FANUC CNC controls, an early leader in control distribution. These words could also be argued to be core words. Haas controls use the same core words.

So the core remains a near constant throughout the industry, with a few differences that are easily understood. The best news is that core commands comprise nearly all programs except the most complex of shapes and program structures.

KEYPOINT

While it's not perfectly universal, the EIA code language comes as close as any we have. It is the standard in CNC work.

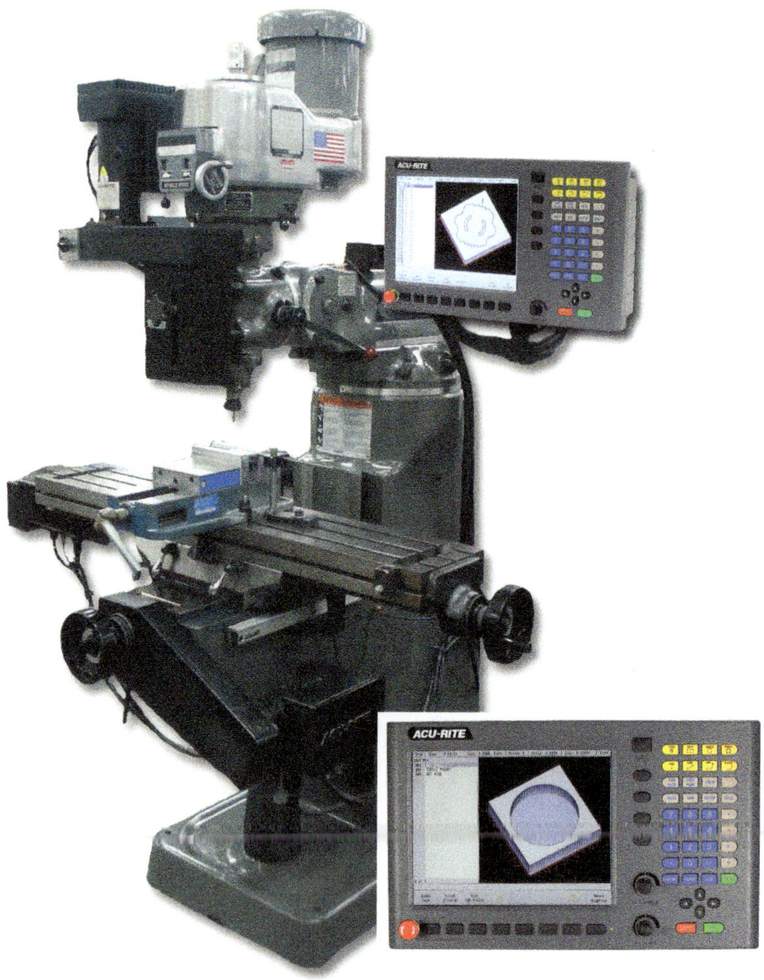

Figure 21-1 A typical conversational control is programmed using machinist language.
Courtesy HEIDENHAIN Corporation

Input Conventions

The flexibility of today's microprocessors and text editors makes it easy to write programs. Many rigid rules and conventions of the past have been relaxed.

For example, while it's good policy to put complete numbers into the program along with correctly placed decimal points, you needn't worry about leading and trailing zeros on entries:

> 10. or 10.0 are both acceptable input as much as 010.0000 (past practice).

Without the capabilities of today's CPUs, previously you would have needed to write:

> 0010060 */tab/*005000 */tab/*0200090 */end of block/*

While today you can write:

> *X*1.006 *Y*5.0 *Z*20.009

Or you could also write:

> *X*1.0060 *Y*5.0000 *Z*20.0090

If all necessary code and math components are present and the decimal point is in the correct place in each, the statement will be understood by the control. That said, flexible input and edit conventions may or may not be present on the machine you run. It depends on your machine's age and level of technology. Find out by reading the program manual. Haas machines use full functioning editors in Edit Mode.

21.1.2 Program Organization

Sometimes it's useful to arrange data in a form that makes reading it easier. Scan the two example side-by-side programs that convey the same data. Don't be concerned with what these programs do; they both do the same thing. Look at the data organization that follows.

Notice in the outline form that all motions using T01 (tool 1) are set up so it's easy to tell how many axis moves and codes pertain to that tool. Also the tool is noted in brackets. It's easy to see how many lines T01 covers: from N020 down to N055 where the program changes to T02. Also notice that the same applies to each feed or rapid travel mode. N030 is a rapid move, and so too is N035. Then on line N040, the G01 mode covers lines 40, 45, and 50. On line 45, it's easy to expect no *Y* tool motion.

KEYPOINT

The outline form is a suggestion only. The line-by-line program works equally well.

When given an outlined program, some controls will rearrange the data into line-by-line form as the program is downloaded, depending on the sophistication of the control's text editor. Others accept the data as is and leave them in outline form. Mastercam color codes their output (Chapter 25).

CAM software generates line-by-line programs. However, tool and operation notes can be inserted, and they can be arranged into outline form by editing.

Positive and Negative CNC Values

The value sign occurs differently than it would in standard math notation. Place the minus sign between the axis letter and the significant numbers. That order first indicates the axis or other code to which the negative value applies; then the numbers follow.

Line by Line		**Outline**				
N005	G80 G40	N005	G80 G40			
N010	G94 F20. G70	N010	G94 F20. G70			
N015	G17 G90 S1000	N015	G17 G90 S1000			
N020	T0101 M6	N020	T0101 M6 (0.5 End Mill)			
N025	M3	N025		M3		
N030	G0 *X*.5 *Y*0.0 *Z*.5	N030	G0	*X*.5	*Y*0.0	*Z*.5
N035	*X*.1 *Y*–0.1 *Z*0.0	N035		*X*.1	*Y*–0.1	*Z*0.0
N040	G1 *X*0.0 *Y*–0.5 *Z*–0.5	N040	G1	*X*0.0	*Y*–0.5	*Z*–0.5
N045	*X*3.0 *Z*–1.5	N045		*X*3.0		*Z*–1.5
N050	*Z*–3.0	N050				*Z*–3.0
N055	G0 *X*100. *Y*10. *Z*100	N055	G0	*X*100.	*Y*10.	*Z*100.
N060	T0202 M6	N060	T0202	M6	(#3 C'Drill)	

21.1.3 Code Words and Commands

A code word is a single statement that does the following:

- Sets a mode
- Causes an action
- Prepares for an action
- Starts or stops a function (coolant or spindle, for example)

Complete Commands

Program words are arranged into sets of **command lines** (shortened to **commands**), or *command blocks.* Whatever you call them, a command is a complete instruction containing words and/or coordinates for one action. If the action includes axis movement, it's sometimes called a *motion statement.* The command might be as short as G00 *X*1.0, which has one word and one coordinate, causing a rapid positioning to *X*1.0. Or the command may contain several words, numbers, and coordinates longer than one entire screen line.

Controller Difference—End of Block

When the command line is complete, most controls accept the ↵ (Enter on a standard keyboard) to signify the end of that command. For example,

G01 *X*1.5000 *Y*-0.375

After the 0.375 entry, I touched the same enter key, and the text editor moved down to the next line. The CPU inserted a transparent end-of-line code that signifies a finished command that the MAL understands. However, some controls like Haas, for example, require an *end-of-block symbol entered at the end of the command:*

G01 *X*1.5000 *Y*-0.375:

> **KEYPOINT**
>
> The colon (might be semicolon) tells the control that the block is complete and is to be acted upon. On these controls, the colon must be present or the control thinks the following line is part of the present line, which can get ugly depending on what it reads there!

21.1.4 Baseline Vocabulary

Code Word Prefixes

We've already seen a few words in previous chapters comprised of a prefix letter followed by numbers (called **alphanumeric code**). Your objective for now is to become familiar with six common word prefix letters plus a set of selected words. You'll find the entire vocabulary in Appendix V of this book.

> **KEYPOINT**
>
> The six prefixes are G, M, T, S, N, and F.

Question

While scanning the code prefix words, notice that they are in groupings. See if you can figure out why they are in groups. Given these few words, you could compile a fairly complete program for a multifeature part.

G Codes

These are the prep and motion words. G words with axis coordinates form the heart of the program.

> **ARMCHAIR FIELD TRIP**
>
> **Be sure and watch this Haas Tip of the Day** video program code.
>
> This video discusses the different type of motion commands used in CNC programming
>
> Go to: **https://www.youtube.com/haasautomation** and search for this video called: **G00, G01, G02, G03—EVERY PART YOU'VE MADE USED THESE CODES!—Haas Automation Tip of the Day**

M Code Words

Miscellaneous functions or actions are the second-most common words used in the program. They cause utility functions such as tool change, spindle on or off, and coolant.

Control Differences

On many controls, including Haas, to prevent current surge and CPU data overloading, only one M code is permitted per command line–it's become a convention. For example, entering M3 and M8 on the same command line would cause a program syntax alarm. Many controls have no problem with multiple M functions. But that's not the reason for the groupings–keep guessing.

Notice that nearly all of this short list, other than M14, M15, and M30, can be abbreviated to a letter and a single number.

T Code Words—Tool Numbers and Time

T words have two applications. By far, the more common is to denote tools. For example, T0101 M6 rotates the mill storage drum to tool 1, then changes to that tool, and takes up offset 1 from tool memory. T words are also used for time functions in dwells.

Control Differences When Using T for Time

Most controls accept G4 T10.5 to indicate a dwell of 10.5 seconds. On them, when the T follows the G4 (or some other command requiring a time span), the control accepts the entry within the same command line, to indicate time. However, others require some notation to indicate the T word is time, not a tool: M4/T10.5 (for example, a slash is added to denote time). Using the code word list (or from memory), interpret the following commands A and B. The answer is in the Trade Tip–Inserting Line Numbers.

F Code Words—Feed Rates

A. G94 G70 F20.0

B. G95 G71 F.5

Controller Differences for Metric Feeds

Note, most controls accept metric feeds in parts of millimeters, which would change the last example to F0005. Read your control manual for the specific form.

Prep Codes—Cause or prepare for an action

Group 1		
	G00	Move at rapid travel—one or more axes
	G01	Linear movement at feed rate—one or more axes
	G02	Circular motion—clockwise at feed rate 2 axis limit
	G03	Circular motion—C-clockwise at feed rate 2 axis
	G04	Dwell—a stall for a specified time
Group 2		
	G17	Machining in the *X–Y* plane—mills only
	G18	Machining in the *X–Z* plane—mills
	G19	Machining in the *Y–Z* plane—mills
Group 3		
	G70	Inch input (ISO = G20)
	G71	Metric input (ISO = G21)
Group 4		
	G40*	Cancels tool radius compensation
	G41	Compensate cutter radius to left of geometry line
	G42	Compensate cutter radius to right of geometry line
Group 5		
	G80*	Cancels previous routines
		See "Modal and One-Shot Code Words"
	G81	Drilling cycle
	G82	Drill cycle with pecks
	G83	Drill and bore cycle
Group 6		
	G90	Absolute value coordinate input
	G91	Incremental value coordinate input
Group 7		
	G94	Feed at inches/millimeters per minute—usually used for mills but lathes understand G94
	G95	Feed at inches/millimeters per rev (lathes and drilling)

*On some controls M02 or M30 also cancel compensation and routines.

M Codes (miscellaneous functions)

Group 1

M00	Program halt—utility stop for operator actions
M01	Optional stop—operator selected or ignored
M02	Program end
	Can also use M30—stop and rewind tape—from days when we used punched tape as data storage media. But this sends to control back to line 1, so we still use it.

Group 2

M03	Spindle start forward (lathe or mill)
M04	Spindle start reverse
M05	Spindle stop
M13	Spindle forward + coolant on
M14	Spindle reverse + coolant on

Group 3

M06	Tool change

Group 4

M07	Coolant on—spigot B (often assigned to mist coolant)
M08	Coolant on—spigot A
M09	Coolant off

SHOPTALK

Because so many commands contain G code words in a program, you will hear machinists refer to the entire language as "G codes." For example, "Is that control programmed only in G codes, or can it do graphics, too?"

S Code Words—Speeds

Causes the spindle (lathe or mill) to rotate at the specified XXXX RPM. For example, S2000 M03 starts spindle at 2,000 RPM forward rotation.

N Code Words—Line Number Addresses or Number Entry

N code words are used in two ways:

N001 = Line Number Addresses

Line numbers are used for program organization. In most cases, they can be omitted, although most programmers use them. If using logic commands, line numbers are a must for branching statements and loops to begin or end–this will be studied later.

N20 = Numeric Value or Repeat in a Routine

For example, in a milling pocket routine, the cutter is to cut down N20 times by dividing the total depth into 20 small plunges downward.

KEYPOINT

There are times when using certain routines and branching logic statements that line number addresses must be in the program to define where to start and stop routines and to reenter the normal program again.

Where the correct entry is always a whole number (N20) no decimal point is entered. *That is a rigid convention that must remain.*

TRADE TIP

Inserting Line Numbers You can choose to place line numbers where they are needed and leave them off the program where there is no need. You can also spread them apart for future editing, for example:

N005
N010
N015

Then a line can be inserted at N006, for example, without disturbing the overall sequence.

Why Codes Are Grouped?

Mutually Exclusive Words—Code Groups

It's common and often necessary to have several **code groups** on the same command line. Each is needed in combination to complete the command. However, some G and M words are incompatible with each other and cannot be within the same command. They are *mutually exclusive*, meaning each overrides, conflicts with, or cancels the others from the same group.

> **KEYPOINT**
>
> **Standard Convention**
>
> By now you have the answer to groupings, that no two words from any group can be entered in the same command line.

For example, you cannot ask a controller to rapid travel and feed an axis all in the same command line or command it to turn the coolant on and off simultaneously!

Controller Differences Safety Warning!

Accidentally writing two exclusive G or M codes in the same command line will cause one of two possible events depending on the control:

Not dangerous–The machine will hang up on a syntax error code for you to sort out.

Dangerous!–A few controls will ignore all but the last code of that group, on that line, which could be a disaster if it wasn't the intended action!

Modal and One-Shot Code Words

Called **one-shot codes**–some codes affect only the line on which they appear. For example,

N005 M01
N010 G01 *Y*1.0000

At line N005, an optional stop has no bearing on line 10 following. These are called *one-shot or one-time-only* words. We'll see others when we scan the EIA set in Appendix V.

Modal commands work as follows: Upon reading some command words, the control retains their settings. They stay in effect (turned on) until they are canceled by another word in a future command. The cancel word can be

- A program end
- Another word from the same group in a following command line
- A specific cancel word–The master cancel words are G80 (end all canned cycles) and G40 (end tool radius compensation), and some use G49 (with nothing following) to cancel tool length compensation.

These *modal* words are similar to a two-position toggle switch: If it's turned on, it stays on until turned off (canceled), thus their other common name: *toggled* commands. For example,

N005 G00	*X*20.0000
N010	*X*30.0000
N015	*X*45.0000

After line 5, the rapid travel is still in effect. All three moves occur at rapid, but

N005 G00	*X*20.0000
N010 G01	*X*30.0000
N015	*X*45.0000

would be quite different. Here, the G01 on line 10 overwrites the rapid mode of line 5. So, what action will occur on line N015? The machine will feed in a straight line to *X*45.0. The G01 is the mode.

Default Conditions

The word **default** has two similar but slightly different meanings in CNC work:

1. **Best-Guess Value**

   This is a fallback value that the control chooses when incomplete or impossible information is supplied, for example, single-point threading on a turning center. Suppose a previous RPM was set at 3,000. The next command line initiates threading of a large shaft with a fast lead. The machine cannot possibly coordinate the starting and stopping of the tool at the high RPM so it will fall back, or "default," to a lower RPM at which it can control the threading action.

2. **Preset Value**

   When first powering-up some machines, certain conditional values or modes will be assumed. They are the "default values" for that control, a bias to one condition or another, built in. For example, a given control always starts in absolute value mode until commanded differently. Each control exhibiting default values has its own list.

Most newer controls have no default values. They start up in the modes and values set in the program. Therefore, make it your policy to set all expected modal conditions such as feed type, metric/inch values, and absolute/incremental coordinates at the start of all programs no matter if the control has defaults or not. We'll learn to do this soon. It's called the safety lines or program start, studied next.

Outside Research

This activity will expand your list, but again, it's for exposure to the possibilities. Ask your instructor for the handout listing codes of the machines in your lab, if available. If not, ask for permission to use the programming manual.

First, compare the codes in your lab to the short list in this unit. If you find differences, note them and mark them on the real vocabulary you will use in your lab. That's what any machinist needs to do when moving to a different machine: be certain of the vocabulary of non-core words.

Now, go to Appendix V to see the full EIA list of code words. Please notice that columns 3 and 4 contain check marks as to whether the code is "Function Retained"–modal, or "Affects One Line Only"–one shot. Compare that list to the codes used in your lab. They will be used for future reference as you move between controllers.

UNIT 21-1 REVIEW

Respond

1. List the six code word letter prefixes and describe their usage.
2. What actions will occur on these three lines?

   N025 G01 G94 F19. X30.0000

   N030 Y1.0000

   N035 G00 X0.0 Y0.0
3. What will occur given this command? N001 G0 X1.000 G01 Y15.0
4. Working with a partner machinist, without looking back, list as many G code words as you can remember.
5. Now list the M words you remember.
6. Why are CNC codes arranged into groups?
7. True or false? In the example lines of Problem 2, line N030 is not a command, it is a word only. If it's false, what will make it true?

Unit 21-2 Outer Program Structure

Introduction: This unit shows how programs start and end. There are a few commands needed to ensure that unwanted actions do not occur and that your program launches safely and ends in a parked location that's safe for part changing. In industry, there will most likely be a specific template for start and end program structure that will be pasted into every program. When CAM programming, you can compile the program header and finisher, then save them to be pasted in at the right place.

Special Communication and File Symbols Before any safety codes are written, a few controls may require a special symbol set as the first two lines or a program start command word. Where program data start symbols are needed, they alone must precede any remarks or code words, for example, on a Haas or Fanuk control

O2345

For both controls the % (program start) is followed by the letter *O* and program number (O2345). To determine if your control requires a file symbol, read the program manual or scan any program stored in memory.

M99 (program start code word for another control in my lab)

TERMS TOOLBOX

Air cut (shop lingo) An inefficient time when the cutter is moving at feed rate but not making chips.

Departure point The safe place from which a program starts and then parks the tools at completion. Also the safe place from which certain cycles begin and end.

Retract (R) position The safe distance off the work beyond which no rapid moves toward the work should occur.

Start-up commands (safety commands) The first few command words that set expected modes and cancel unwanted actions or values.

21.2.1 Program Start Words

There are two categories of **start-up command** information that must be in the lines after the program start and file name:

Cancel words Mode-setting words

As stated, not all programs need all these preconditions, but there is no harm in including them unless they are not wanted. For example, I usually start my programs with the G20 code to signify it's in inch values or G21 if it's metric. Normally, the control would not be set to metric mode–except that one time when I assumed it was inches! Big surprise, my incorrectly headed program tried to start in millimeter mode. So you see, there are exceptional "trap" situations that can be disastrous without the right start-up codes.

Here is another example. Suppose a clogged cutter has caused the operator to halt mid-program to clean the chips out. Then the operator chooses to restart the program from the beginning. If, somewhere within the original run through the program, a mode had been set just before the halt, the control will restart in it, unless cancel words begin the program.

KEYPOINT

A complete program start, including safety cancel lines and mode settings, helps avoid exceptional traps.

Setting and Canceling Modes

Required modes can actually be written into the program anywhere, as long as they are entered before or on the first line in which they are needed. For example,

Sample 1

N010 G94 G20 F20

N015 G1 *X*2.00

But

Sample 2

N015 G94 G20 F20. G1 *X*2.0

will work equally well. The feed method G94, G20 inch values and feed rate were both available at the time they were needed, in each sample.

Suggested Start Codes and Modes

Here is a list of those modes and values that should be set at the start of your program:

General Modes

G94 or G95	Feed type (per minute or per revolution)
G20 or G21	Inch/metric values (may be G20 or G21)
G90 or G91	Absolute/incremental values (usually absolute to start)
G17, G18, or G19	Mills only—working plane *X-Y* or *X-Z* or *Y-Z*

On both lathe and mill start-up commands, most programmers spread out the commands over two or three lines, and they might also write a bracketed note to explain.

Specific Machining Data

FXXX	Initial feed rate
SXXX	Spindle RPM

The feed rate and spindle speeds must be entered before any machining can occur—but they need not be in the start section of the program. Again, there would be no harm in placing them there.

Canceling Modes for Safety

To ensure that a program starts cleanly without trying to execute some unwanted mode from a previous command, we put cancel codes in the start-up. There are two for now. Others will be needed as your programs become more complex.

G80	Cancels all previously used canned cycles
G40	Cancels any previously used radius compensation

KEYPOINT

Cancel commands should have been entered in the previous program where the mode was ended. They are placed at the start as double protection from traps.

Here's a suggested generic start-up template.

N005 G80 G40	(Cancel modes)
N010 G20/21 G94 (or 95) FXXX	(Initial feed)
N015 SXXX G90	(Coord system [or 91] abs./incr.)

Then, if it's a mill program

N020 G17/18/19	(Milling plane)

Secondary Start-Up Commands

If a given tool is to be exchanged, it would be the next command: T01 M6 for a mill and T01 for a lathe, since it only needs to rotate, then spindle, and coolant starts M3 M7 but not on the same line. Next are the physical tool motions. Every program has a staging area called the **departure point**. It is the place where your program parks the machine to end the previous cycle and where it also begins the next cycle. On a mill, it is probably the tool change position. On a lathe it is a full *X* and *Z* retract, which is also machine home.

Yesterday in our lab, a student learned the hard way about setting a homing command as part of program startup. A bolt held work to the subplate. The mill program start departed from PRZ with the cutter 1 in. above *Z* zero. But the student had jogged the cutter away from PRZ to tighten the bolt. When the program, the cutter rapid traveled right through the bolt on its way to the first cut! If there had been a homing command in the start-up, that wouldn't have happened.

TRADE TIP

Homing the Machine to Park Many controls feature a specific command that drives the machine in rapid travel to home or tool change positions. This homing command is not a core word but will be found on your handout for the control. Rather than putting axis commands in your program, a simple code word will do it all for you. Check your programming manual for the "Machine Home" command. Using this command, always visualize the path a rapid traveling tool will take. It may be necessary to first move it to a safe place from which to start the homing. (See safe rapid travel next.) On Haas use G28 for this, and if you need a single axis move before heading to machine home use G28 Z4.0. On that command the machine will rapid up to Z4 inches, then to machine home.

21.2.2 Safe Rapid Travel Toward or Above the Work—The Retract Position

The next item after starting the spindle is to rapid travel the cutter toward the work. The rapid travel destination is called *retract height* or *retract distance* (Fig. 21-2).

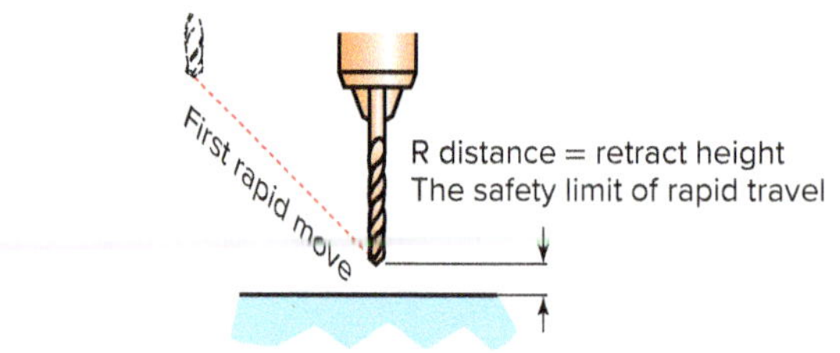

Figure 21-2 A program typically starts from a *departure point,* then moves at rapid travel speed to the safe *retract height* (or retract distance on lathes).

This **retract (R) position** is the safe coordinate distance of the cutting tool from the work beyond which it should not rapid travel. At the R distance, you will switch to feed rate to complete tool contact with the work.

A drill cycle on a mill is a good example. One of the data fields in the G81 drill routine is the R coordinate. It is the jump height of the drill between holes. It is the safe distance above the work for a rapid travel move to a new location. The letter R is used in many cycles to indicate this safe distance from the work (see Fig. 21-2).

> **TRADE TIP**
>
> **Coolant Starts** It's good practice to turn on the coolant at the end of (or during) the rapid motion to R position. Doing so ensures full flow before tool contact with the work. Turning the coolant on right when the cutting begins means the flow will sputter for a second or two. In some low-machinability metals, that can be a problem.

Safe R Distance Versus Program Efficiency

For student programs and even your first few professional programs as well, I strongly recommend R heights of 0.5 in or 12 mm (or more, as your shop policy sets). While that's safe, it requires relatively long **air cuts** with the tool, making its final approach at feed rate, without making chips. Later, these distances can be shortened for efficiency if shop policy allows it.

Lathe tool changing falls into this safety concept as well. Professionals retract the turret just enough to safely rotate it without interference from any tool in the turret. That said, as a beginner pull it all the way back then change the tool turret (until you get more experience!).

> **KEYPOINT**
>
> Depending on which direction the lathe turret indexes, the safe retract distance to change tools may need to be far enough from the chuck to clear every tool in the machine. (See the Trade Tip.)

But close tool changing is not a good idea for beginners. For student lathe programs, pull the turret as far back as it will go, usually to the home position, the farthest possible distance from the work; then change tools.

> **KEYPOINT**
>
> For beginners, it's a very good idea to run the turret all the way back in *Z* and out in *X* (home position) before changing to another tool.

With experience both lathe and mill retracts and lathe tool change positions can be shortened for efficiency.

> **TRADE TIP**
>
> Some lathe tool turrets rotate only one direction when changing tools. However, many newer lathes turn either clockwise or counterclockwise, depending on which is the shortest distance to the tool called out in the program. When retracting the turret back for a tool change, it's critical to think that through. The longest tool may or may not swing past the work and chuck. Don't take a chance—retract all the way back for your beginner programs.

21.2.3 Program End Commands

Ending the program is a simpler issue, but it must still have a few vital features to be safe and efficient. Modes that are in effect should be canceled, *even though they will be canceled by the start section* when the program is run again.

This is especially true for tool compensation programs. As the tool completes the final cut, it's pulled off a safe distance from the work, then rapid traveled back to the departure point at which time the compensation is ended, when the tool is safely away from the work surface. That way, when it moves from the tool radius distance away from the geometry to center over the geometry line, it won't accidentally touch the work or chuck or vise. Tool compensation is canceled with the G40 code. We'll explore this more thoroughly coming up.

As the tool is pulled off the work, stop the coolant M9. The M30 or M2 command will also turn it off, but it's poor practice to do so at the very end. That generally slings coolant everywhere! Not only does it cause a slippery mess; the cost of coolant lost to the shop floor robs profit as well. This problem occurs mostly on smaller mills without full enclosure. Turn it off at the end of the final feed movement. Another convenient program end is to put a blank tool in the spindle or rotate the lathe turret to a blank position (see the Trade Tip).

TRADE TIP

Blank Tool at the Departure Park Position Sometimes it's a good idea, for operator convenience and safety, to exchange a large or very sharp tool for a blank—no tool at all. This small operator concession is appreciated for part changes where hands need to be near the tools.

21.2.4 Safe Rapid Travel Away from the Work

A potential accident can occur when pulling the cutting tool away from the work for a tool change or to park the machine at program end. The total path must always be considered.

In Fig. 21-3, a drill has completed drilling to depth. To simply home the machine from the bottom of the hole would be a double crash. It would start moving in *X*, *Y*, and *Z* axes at rapid travel. It must be pulled straight out of the hole, then sent to home. But when homing it right from the R position, it could also hit a vise jaw or clamp if not retracted far enough above the work first. The drill must be retracted straight up in the *Z* axis, above the vise, then sent to home.

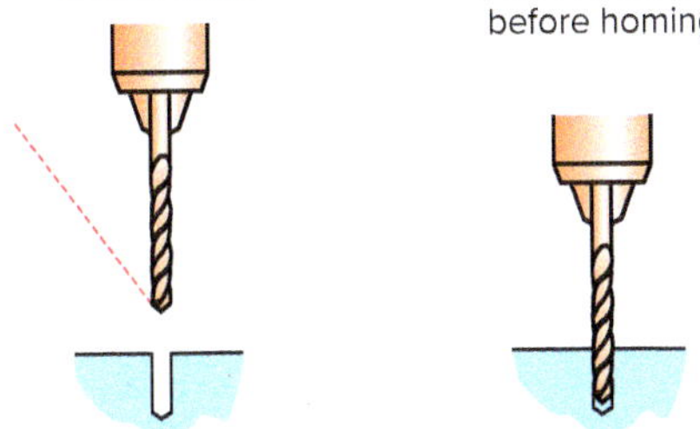

Figure 21-3 Rapid moves must always be made to and from a safe location.

KEYPOINT

For lathes or mills, pull directly out of the hole or off the work, to a safe distance; then analyze the path back toward the departure point before homing the cutting tool!

UNIT 21-2 REVIEW

Replay the Key Points

- A complete program start, including safety cancel lines and fixing needed modes, helps guarantee success.
- For beginners, it's strongly suggested that all axis moves toward the work stop rapid traveling at least 0.5 in. from the work.
- For beginners, it's strongly suggested you withdraw all tools to their farthest (safest) position before changing parts or changing tools.
- Always pull directly off the work, at feed rate, to a safe distance, and then analyze the path toward the departure point before homing the cutting tool!

Respond

1. A program begins with
   %
   O5432
   What does it mean?
2. True or false? Modal commands are those that cancel or override all previous modes set in the program. If it's false, what would make it true?
3. Why do we add the word *G80* to the beginning of a program?
4. Explain the two elements of good program start.
5. Identify at least two modes that should be set at the start of a program.
6. Why must a tool retract path always be analyzed before sending the tool back to the departure point?

Unit 21-3 Manually Compensated, Linear Cutter Paths

Introduction: Okay, it's just about time to give program writing a go. At the end, I'll offer a solution program in the answers. Check it against yours. If your lab has CNC simulation software this is the time to learn to use it. If so, write your program in the text editor; then define the raw material size, cutting tools, and the PRZ location on the material. Then machine the cyber-blank using your program.

KEYPOINT

You can use the student version of Vericut in MDI mode to simulate your program (see Chapter 26).

Not only is it rewarding to see the codes turned into cyberparts, but it's also the fastest way to test your program. At this stage simulators are by far the best learning tool. If you have no Haas simulators or vendor software, you can also test a tool path by backplotting in Mastercam, by using the lab's CAM system, and by using controller graphics but with the machine safely locked up in dry run mode. (You'll need some guidance for that test.)

After a discussion of how to calculate the tool path, we'll make a CNC mill part, start to finish. You will write the plan, then the codes. I'll provide a few math hints for coordinates. Notice that there are no arcs. Curve commands are coming up in Unit 21-4. Note, you will need trigonometry and the Pythagorean theorem to complete some of these activities.

21.3.1 Calculating the Tool Path

Here we will take a look at *X*-*Y* shifts for a profile mill tool path, but they could also be *X*-*Z* shifts for turning a workpiece using a radius nose cutting tool on a lathe.

Sketching the Tool Path

You are going to calculate coordinates, then write a tool path program for the latch lock (Fig. 21-4). To do so, this program will climb mill from point *A* to point *I* using a 0.500-in.-diameter cutter (0.25 distance off the geometry). See Fig. 21-5–this is known as a cutter centerline program.

Latch lock (0.25 thick MS)

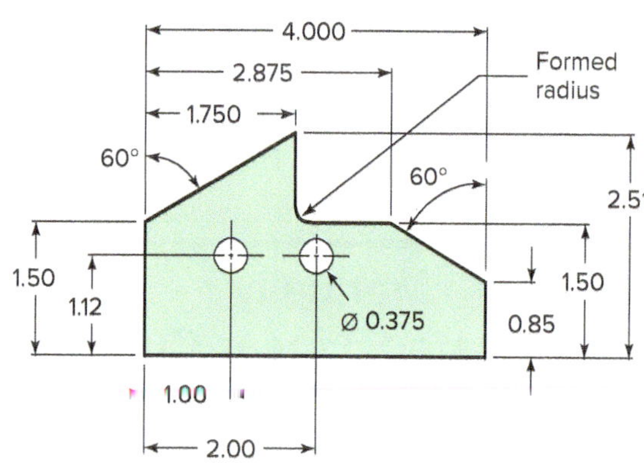

Figure 21-4 The latch lock to be programmed.

The first step is to draw the tool path at the needed tool radius offset distance. The centerline is blue in Fig. 21-5. We don't do this on a day-to-day basis, but it helps at the start. Use a contrasting color to draw a dotted line to represent the cutter centerline. A circle template often assists with this process.

Sample tool path sketch

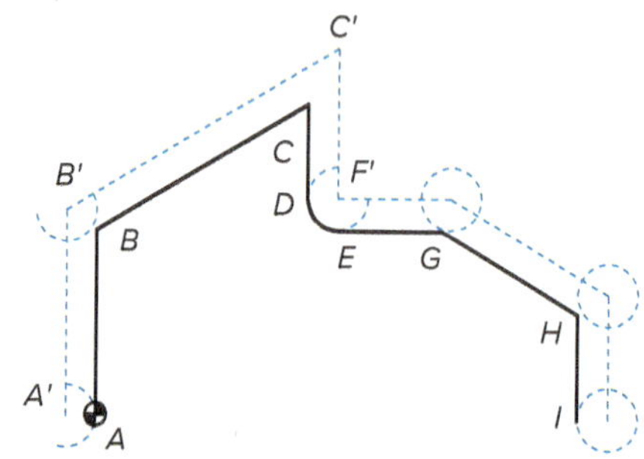

Figure 21-5 Sketching the tool path.

KEYPOINT

Keep in mind that while the circles are easy to envision as end mills profile milling a part, they could also be the nose radius on a lathe tool. The concept is the same—a radius tangent to a geometry line.

Four Kinds of Compensation Shift

Now, label each significant point on the part geometry, then look for its counterpart on the tool path. Points lie at each end of lines, at line intersections, and, when using arcs, at their tangent points and centers.

KEYPOINT

The coordinate that will be entered in the program is not on the geometry. It is the cutter center, on the cutter path.

In this book, I'll refer to the shifted coordinate point on the cutter path as a prime of the point on the part geometry. For example, Point *A* on the part geometry becomes Point *A'* on the cutter path. There are four distinct situations illustrated by the latch lock and Fig. 21-5:

1. *Straight Shift Point* A' Because the cutter is tangent to a vertical line, point *A* shifts to the left by a minus 0.25 cutter radius to become point *A'* in the tool path.
2. *Compound Shift Point* B' Point *B* shifts in both *X* and *Y* to become *B'*. This requires a triangle solution. There are others on the part geometry.
3. *Tangent Point* F' Point *F* is a special situation. Points *D* and *E* on the geometry disappear because this particular tool path will form the 0.25 part radius. The cutter will cut in a straight line, to point *F*, then turn 90° to cut to point *G*. If the cutter was smaller in diameter, we would need to write a curve command based on points *D* and *E*.
4. *Centers Don't Shift* Center references, such as drilling holes, are not shifted.

The Shift Triangles

Now sketch the shift triangles where they are needed, as shown in Fig. 21-6. Then solve for the X' and Y' distances. It will be useful to blow them up to some greater size. Continue the color scheme from the parent sketch.

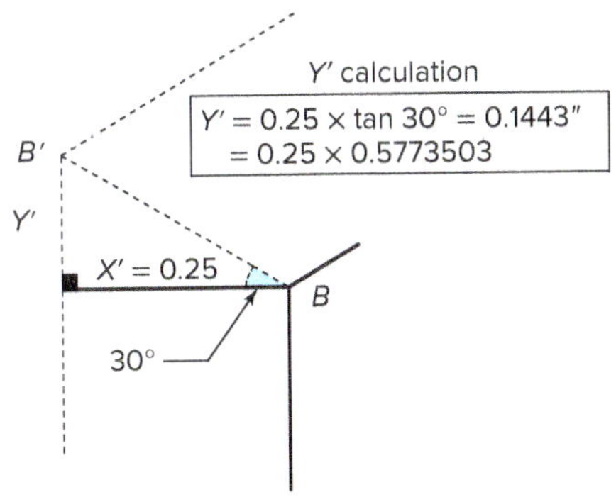

Figure 21-6 Solving for the X' and Y' shifts. Each triangle is different based on the part geometry.

> **TRADE TIP**
>
> **Finding and Solving the Coordinate Shift Triangles**
>
> 1. Label the cutter radius first. It will be the distance the centerline lies from the part line.
> 2. Label the 90° angle.
> 3. Using basic geometry, look for lengths or angles that are known. In the example, the 60° angle on the print.
>
> Or, if you do not have the math skills to solve these problems, try drawing the geometry and tool path accurately in CAD; then analyze the elements. This is how the CAM system derives point coordinates.

When these coordinate shift calculations are complete, it's time to write the program codes.

UNIT 21-3 REVIEW

Respond Program Problem—Latch Lock

Calculate and write a cutter centerline program to mill Fig. 21-4, the steel latch lock. Use the following program sheet as a model. If you get stuck on any math, you will find hints in the answers.

Job Instructions

1. Work in absolute values.
2. Calculate spindle speeds with 100 F/M surface speed. Conservatively feed the cutter at 10 in. per minute for milling and 0.003 in. per revolution for drilling.
3. Upon completion your program may not be in the exact order of the answer, but they are okay as long as all data are correct and present. The simulator will show if the program works or not.
4. Use a 0.50-in.-diameter four-flute cutter and a 0.375-in. drill. For this first program, center drilling is not necessary.
5. Cut the profile using climb milling—just one finish pass around (although a rough and finish pass would produce more consistent results if this were a real part).
6. The lower edge has been premachined. Place your clamps along that edge and start machining along the left-hand edge, which has 0.030-in. machining excess.
7. Use absolute value coordinates.
8. If your machine uses codes that differ from the book list, use them.
9. Work on a program sheet similar to the following.
10. The end mill and drill may cut into the sacrifice plate.

Program Sheet

N005	N100	N195
N010	N105	N200
N015	N110	N205
N020	N115	N210
N025	N120	N215
N030	N125	N220
N035	N130	N225
N040	N135	N230
N045	N140	N235
N050	N145	N240
N055	N150	N245
N060	N155	N250
N065	N160	N255
N070	N165	N260
N075	N170	N265
N080	N175	N270
N085	N180	N275
N090	N185	N280
N095	N190	N285

Unit 21-4 Writing Arc Commands

Introduction: Programming arcs is easy with a bit of training. There are actually three methods, two of which we'll consider here.

Radius Arc Method

Center Point Arc Method (*IJK*)

Polar Arc Method (Not studied)

An arc is a polar entity; thus, writing arc commands is simple enough that it requires little instruction. However, not all controls feature this ability. Unique to this method, the entries include the radius and number of degrees in the arc, plus or minus (clockwise or CCW).

Depending on how the dimensions are provided, using one of the three methods eliminates nearly all math required to program arcs.

TERMS TOOLBOX

Center point arc method (*IJK*) The arc command that requires curvature code; end point coordinates in *X*, *Y*, or *Z*; and center point coordinates in *I*, *J*, or *K*.

Curvature codes G02 CW and G03 CCW The code words that cause clockwise or counterclockwise arc motion.

Polar arc method The arc command that requires curvature direction, radius of arc, and number of degrees in the arc.

Radius arc method An arc command that requires curvature code, end point, and radius entries.

21.4.1 The Five Arc Facts

To draw any defined arc on paper using a compass, there are five parameters required. They are the same facts the controller will need (Fig. 21-7).

1. Arc radius
2. Center point
3. Start point
4. Direction clockwise or counterclockwise
5. End point

Default Facts

However, just as with a compass on paper, all five facts are not needed in each command. Given any three, two will be automatically known. For example, if the start point and the center points are known, then the radius is the distance between them–it is automatically available.

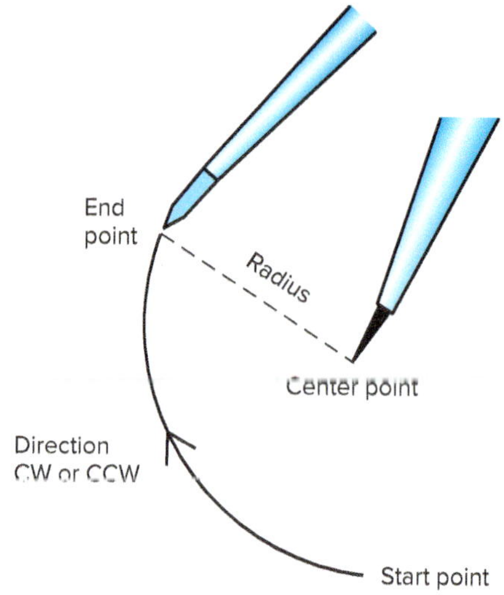

Figure 21-7 The five arc facts.

Or, given the arc's start and end points, the center point will lie on a line perpendicular to a line connecting them (Fig. 21-8). The control can derive the center point in this case if it's given the radius and direction, clockwise (CW) or counterclockwise (CCW).

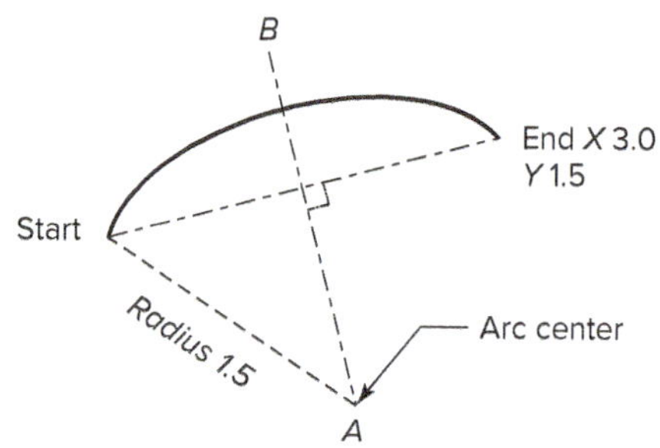

Figure 21-8 Given the *start* and *end* points, the control can derive that the center point lies along line *A–B*.

KEYPOINT

The difference in the two arc commands, center identification, or radius method is which facts must be entered and which are determined automatically.

Current Tool Position—Always the Start Position

In all arc methods (polar included) one fact is always automatically known: the start point. So, with the start point known, only two more facts must be provided in either command.

21.4.2 The Radius Method for Arc Commands

The radius arc method is the simpler of the two nonpolar commands, as long as the arc's radius is known. Factors in the command are

Curvature Code	End Point Coordinates	Radius
G2	*X*3.0 *Y*1.5	R1.5

This example moves the tool clockwise (G2) from its present position to *X*3.0 *Y*1.5 with a radius of 1.5 in.

The end point coordinates may be absolute or incremental per the information available but either the G90 or G91 mode must be in effect (Fig. 21-9).

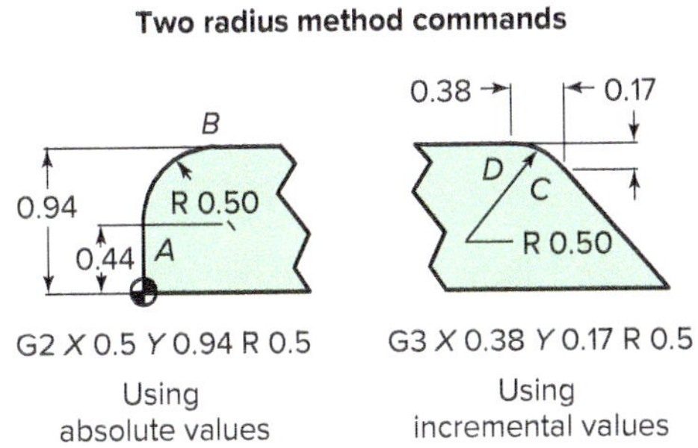

Figure 21-9 Two example radius method arc commands.

Arcs Larger Than 180°

There are actually two possible arcs that could connect the start to the end point, given the factors outlined. One is less than or equal to 180° and the other is greater than 180°. In Fig. 21-10 both arcs move clockwise from *A* to *B* and both have the same radius. The solid blue arc is the default entity, less than 180°. To program the arc greater than 180° using the radius method, the control must be given an additional symbol.

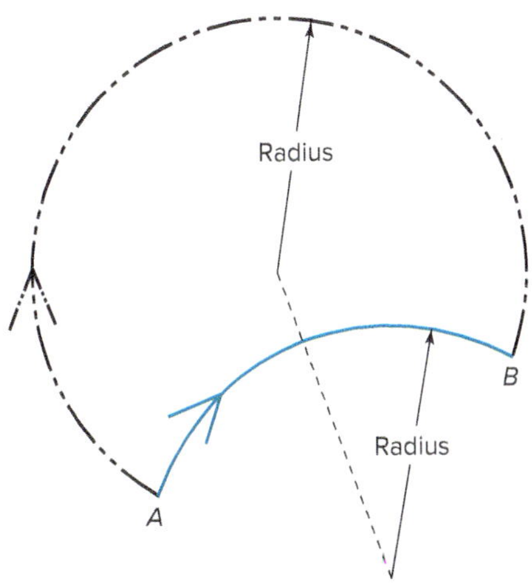

Figure 21-10 Two possible arcs given start, curvature, end, and radius factors.

KEYPOINT

All controls (of which I'm aware) default to the lesser arc unless a special symbol is added to the command, to indicate that the arc is the larger segment.

Controller Differences—Greater Arcs

The symbol that signifies the arc larger than 180° varies from control to control. Here are two examples:

Default arc: G02 *X*14.0 *Y*12.0 R16.0

Greater arc: G02 *X*14.0 *Y*12.0 R−16.0

The minus sign placed on the radius indicates that the arc is greater.

/G2 *X*14 *Y*12 R16

The slash indicates the larger arc on our generic control. Read your manual.

KEYPOINT

On Haas controls for arcs greater than 180°, add a negative sign to the R−.

The Math

When the arc is 90° on square part edges, the radius command is ultrasimple; just write the **curvature code**, end coordinates, and radius. However, when the arc is not a simple corner, there are some computations to be made. Each geometry presents its own set of conditions.

Make a sketch; then fill in what you know. Use the Pythagorean theorem and trigonometry to solve for the unknown end coordinates. The upcoming activity will give you some practice at calculating and writing arcs using both methods.

KEYPOINT

The radius method is the simpler of the two choices as long as the data are available.

21.4.3 The Center Point (*IJK*) Method for Arc Commands

For the center point arc method (*IJK*), this command is the one CAM systems use because it can denote an arc of any number of degrees (zero up to a full circle), with no confusion between lesser and greater arcs. Factors in the command are

Curvature Code	End Coordinates	Center Coordinates
G2	*X*.75 *Y*1.75	*I*.75 *J*0.0

The example moves the tool from the present position to the *X-Y* position using a center point that's 0.75 in. from the start in the *X* direction and right on the *Y* line (*J* = 0.0).

For this command method, the end point coordinates may be absolute or incremental depending on information at hand; however, the *center point coordinates are normally incremental* on all controls with which I'm familiar (except one wood router), using the start point as their reference. Read that last statement again—center point identification is incremental from the start point's present location. And it is identified with alternate axis letters *I*, *J*, and *K*.

Why Use *I*, *J*, and *K*?

When working in an *X-Y* plane on a mill, for example, there are two sets of coordinates within the arc command, the *X* and *Y* end point coordinates and the *I* and *J* center coordinates (Figs. 21-11 and 21-12). Note that *I*, *J*, and *K* are alternate forms of *X*, *Y*, and *Z*.

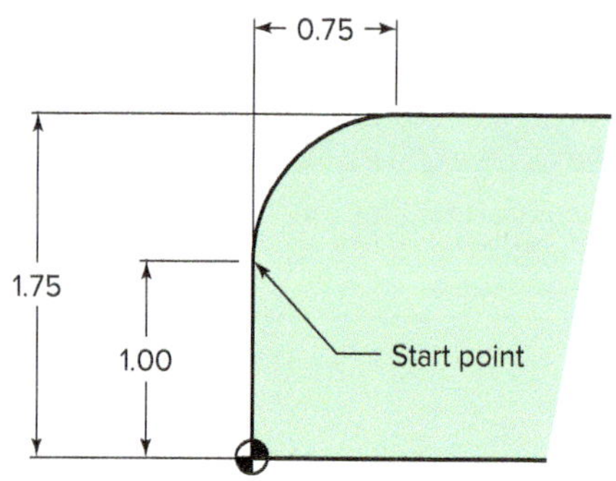

Figure 21-11 An arc command using the center point method.

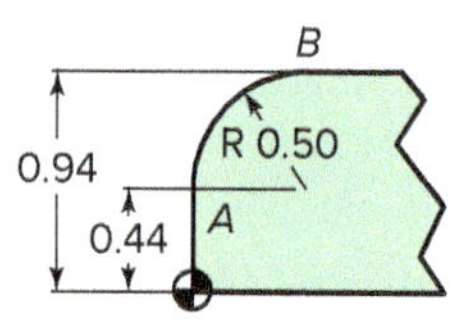

Figure 21-12 A second example.

> **SHOPTALK**
>
> Using the *IJK* arc method, the center point coordinates are incremental on most controls I've experienced other than two models. However, that's only due to *I*, *J*, *K* conventions in CNC programming—there is no mathematical reason the center could not be absolutely identified on all controls.

> **KEYPOINT**
>
> To differentiate end point coordinates from the center point, the center is written in alternate forms of *X* and *Y* to become *I* and *J*. According to convention, *I*, *J*, and *K* are usually incremental values in CNC work.

If a lathe program is being written, then the center is identified with *I* and *K*.

Controller Differences—Other Uses of *I*, *J*, and *K*

Depending on the control, other uses are made of the *I*, *J*, and *K* prefix in commands other than for arcs. Again, they are used as alternate *X*, *Y*, and *Z* coordinates, usually as incremental values. For example on a given lathe control, *I* and *K* are used as incremental moves of the *X* or *Z* axis without changing to incremental mode.

FANUC Lathe Controls and Incremental Coordinates

When you wish to use incremental coordinate values on many older FANUC lathe controls, there is a special convention: G91 is not required nor used to set the incremental mode. These controls default to the absolute mode and have no incremental mode. When you wish to use an incremental coordinate, the *X* and *Z* are replaced with a *U* and *W*. So, on most other controls G91 *X*10.0 would be an incremental jump of 10 units in the plus *X* direction, However, on FANUC, *U*10.0 would cause that action.

UNIT 21-4 REVIEW

Replay the Key Points

- In all arc methods (polar included), one fact is always automatically known, the start point.
- Between the *IJK* and radius methods, the radius is the simpler to calculate and enter as long as the given data support it.
- There are two arc direction codes, G02 = CW and G03 = CCW.
- Arcs less than 180° are the default value. Arcs greater require some symbol.
- In CNC work *I*, *J*, and *K* are used as incremental, alternate forms of *X*, *Y*, and *Z*.

Respond

Calculate and write a command for the given arcs. For the first two problems, assume *the control is in incremental mode.*

1. See Fig. 21-13.

A. Write the default arc command using the radius method.

B. Now write the same arc command using the center method.

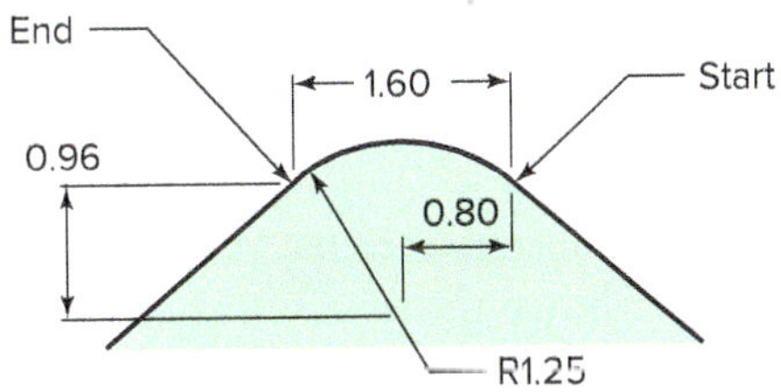

Figure 21-13 Problem 1—Write a radius method command, then a center method command.

2. Using the information in Fig. 21-14, write three different arc commands: the default command from *A* to *B*, the arc greater than 180° from *A* to *B* (use the minus radius symbol for greater arc), and the default arc from *B* to *A*.

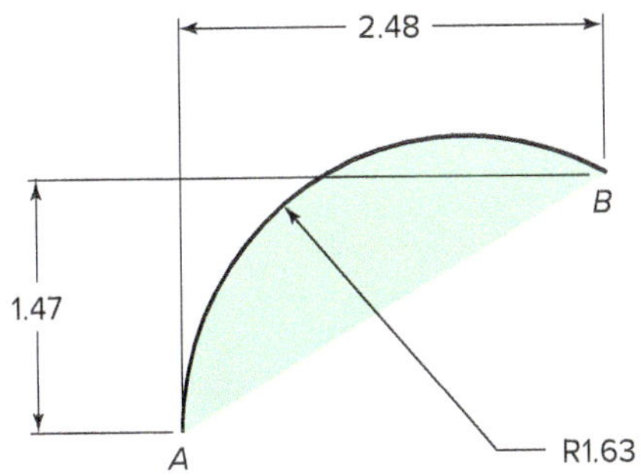

Figure 21-14 Problem 2—Write the radius method command.

3. See Fig. 21-15. Write an *IJK* arc command for this turning job using absolute value coordinates. Then write the radius arc command.

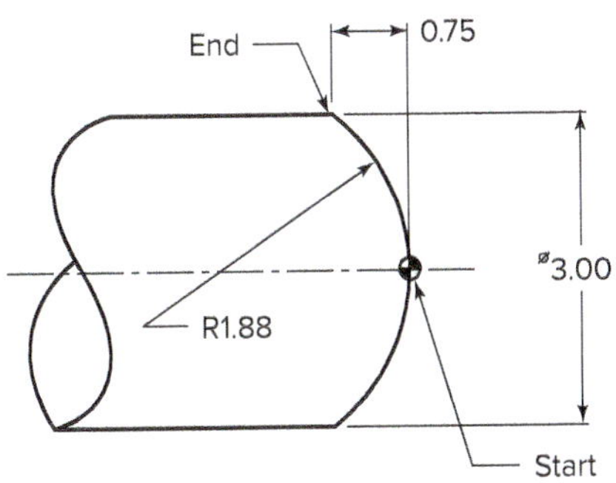

Figure 21-15 Problem 3—Write the radius method command.

4. Now write a metric program to turn (in one pass), then part off the aluminum bullet nose pin in Fig. 21-16. This is a bar feed job held in collets. Do not turn the stock diameter. Use a 2-mm nose radius lathe tool and write your program in absolute values. You may use either arc command method—both will be shown in the answers along with some math hints if you get stuck. Do not write roughing passes—this is an exercise to compile codes and arc commands only. Use 1,000 RPM and feed at 0.1 mm per revolution.

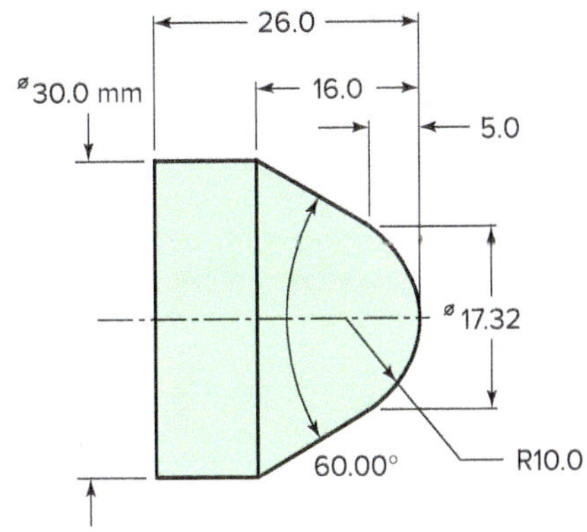

Figure 21-16 Write a manually compensated program to turn this bullet nose pin using a 2-mm nose radius tool.

Unit 21-5 Writing Compensated Programs

Introduction: One of the most powerful advantages of CNC: with "tool radius comp" we can save calculation time and use any cutter that's not a specific size. Now we write part path programs (part dimension coordinates from the part drawing) with a couple more commands that cause the control to calculate the tool path. That means no more *X*-*Y* shift calculations to make! The programmed coordinates are part definitions taken from the drawing. They may still need to be calculated, but they are easier to determine. From the previous activities, you can see this skill is a real time saver.

Also, a compensated program allows some latitude as to which cutter will be used. With all factors correctly written into the program and the cutter radius offset stored in the control, the cutter's rim for peripheral milling or nose radius for turning will remain tangent to the work along the entire part geometry. In other words, the cutter's outer radius will follow the part path. The cutting tool can be exchanged for another with a different radius and still produce an accurate part as long as the new tool radius has been correctly entered in the control, called cutter comp offset.

TERMS TOOLBOX

Dummy move A line of any convenient length that transitions to the ramp.

Lock on The cutter has arrived at tangency to the line and will remain so until told differently.

Off ramp The short line used to smoothly transition off the work and to sometimes create a phantom buffer before turning off compensation.

On ramp A short added line used to lock on to the part geometry before chips are made.

21.5.1 Preparing a Program for Compensation

Three items are required in the program to initiate compensation:

1. **Which Side of the Geometry Line—G41 Left or G42 Right?**

   The compensated program defines to which side of the upcoming geometry line the controller is to move the cutter. That happens on the ramp.

2. **The On Ramp**

   Similar to an automobile starting onto the highway, compensated programs require a distance to get started—sometimes called the **lock on** *distance*. This ramp is an added geometry line or series of lines that begin away from the workpiece but lead onto it in a logical place. The cutter is first sent to a departure point at the beginning of the ramp to get a compensation sense which side of the next line command. Then as it follows the ramp, the cutter becomes tangent to the line before cutting metal.

> **KEYPOINT**
>
> The cutter becomes tangent to the geometry before touching the metal to be cut.

3. **How Far Right or Left—Radius Offset**

   The control will move the cutter left or right of the first ramp line, by the distance stored in the tool radius offset memory. The G41 or G42 is entered in the program, on or just before the first ramp line where tangency is to begin.

> **KEYPOINT**
>
> During compensation, the distance the cutter stays right or left is determined by the radius offset stored in the tool memory. It's not in the program; it's in tool memory in the controller!

The complete command to compensate includes G41 or G42 plus an entry to tell which offset to use:

G41 D01 (cutter left, use offset stored in D1)

G42 T0101 (cutter right, use tool 01 with offset 01)

G41 D1 (cutter left using cutter radius 1)

Cutter Right or Left G41 or G42

The tangency designation is as though you were behind the cutter looking in the direction of the cut (Fig. 21-17). If the metal is on your right (to the right of the cutter) it is cutter left and vice versa.

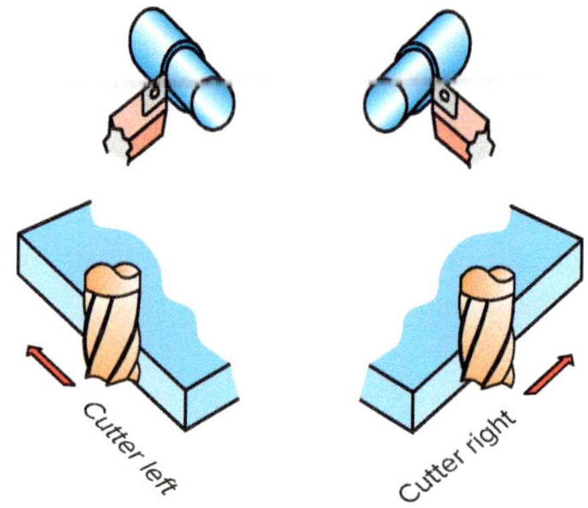

Figure 21-17 Compensating left and right.

> **KEYPOINT**
>
> The terms *left* and *right* refer to the cutter's relationship with the geometry, when viewed *in the direction of the cut.*

In the illustration, the mill cutter is climb cutting a profile—cutter left and conventional cutting when on the right. Adding either word to the program causes the cutter to move to a tangent relationship with the part geometry *on the next axis move after the code, in the current program plane.*

> **KEYPOINT**
>
> If the program is for a mill, the cutter becomes tangent in the current programming plane after the G41 or 42 code is given.

The command varies slightly with different controls. Here are two examples in a CNC mill program:

G17 G41 D1

X-*Y* plane, cutter left using stored tool offset 1

T0101 G17 G42

Tool 1 offset 1, *X*-*Y* plane, cutter right

Current Program Plane on Mills

The G17 code tells the control to G41 or G42 compensate in the *X*-*Y* plane. Under the G17, any *Z* axis motion has no effect on compensation on modern controls such as Haas.

> *Controller Differences* There are a few older CNC controls that will unlock from compensation, if given a *Z* axis move.

The G17 should already be present in the program start lines on mills. No plane code is needed for lathe programs, since they operate only in the *X*-*Z* plane.

Then, once the cutter radius is correctly locked on to the geometry (tangent and following connected lines), it will remain so along the entire part path until the G40 command is given to end compensation or the program is ended.

KEYPOINT

G41 and G42 are modal commands. Once entered they are in effect until canceled.

Designing Ramps

The on ramp can be a single line that leads onto the geometry, or it may need to be multielement (Fig. 21-18). On the left side of the drawing, the cutter is first positioned at point *A*. It is uncompensated, centered over the line. Then the comp codes are read, and the cutter commanded to point *B*. As it moves, the cutter becomes tangent to the line at point *C*. Following that angular path, there's a chance the cut path will nick the corner of the part.

So, one improvement is to preposition the cutter back a bit and change the approach angle as shown on the right side of the drawing. Another improvement involves two or even three connected lines of any convenient length (Fig. 21-19). The multielement ramp joins with the geometry better, to begin the cut.

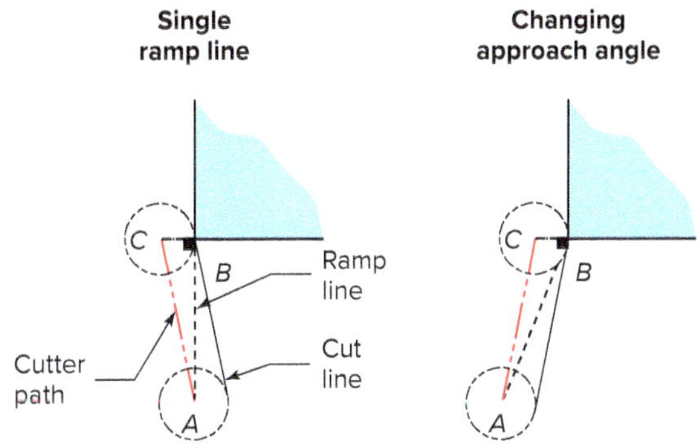

Figure 21-18 The ramp line provides a direction sense for the control to lock on to the shape. There's a slight chance it might nick the corner with the first ramp.

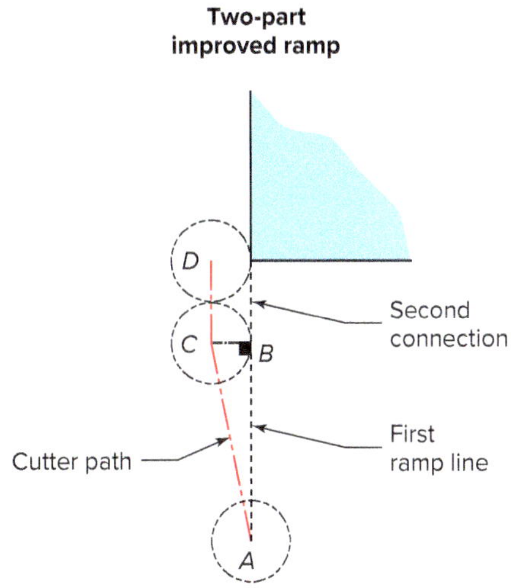

Figure 21-19 The double ramp line reduces chances of nicking the part.

They provide a smooth transition into the cut to avoid cutter marks and nicks in surfaces and corners. The cutter is fully tangent to the line before it touches metal, at point *C*.

Program Example Codes

Here's a typical set of commands to write a tool compensated program:

(Demo Mill Comp Start)

N005 G80 G40	
N010 **T0101** M6	(Control has radius offset 01)
N015 S1000 G94 G90 **G17**	(The program is in the *X*–*Y* plane)
N020 M3 G0 *X*–0.5 *Y*–0.5 *Z*.25	(Rapid positioning—point *A*)
N025 G41	(Cutter left on next *X*, *Y*, or *X*–*Y* move)
N030 G1 *Z*–0.25	(Not *X*–*Y* plane—no comp yet)
N035 ***X*0.0 *Y*0.0**	(Planar move, becomes tangent at ramp end)
N035 *Y*1.0	(Now ready to follow remaining geometry commands)

Ending Compensation

When the cutter has completed the shape, it's very important that the path moves off the work smoothly, again with little or no abrupt transition. The off ramp is easily created by adding an extension line to the geometry. It usually continues in the same direction as the cut as long as chucks and clamps are not in the way.

> **KEYPOINT**
>
> Only after the cutter is well clear of the work, turn off compensation with the G40 code. *Warning:* Only under very special conditions do we ever turn off compensation while the cutter is in contact with the work! Normally the tool is withdrawn safely off the surface.

Controller Differences—Safety Notes on Compensation

As CNC processors became faster and competition forced features on manufacturers, many of the bugs in compensation have been fixed. However, in industry or tech school, there's every chance you'll be running two different machines that behave very differently during the lock on, while staying locked on, and during the unlock phase of compensation.

Lock On Modern controls commanded to become tangent will gradually move away at an angle to the line until they are tangent at the far end, as shown in Figs. 21-18 and 21-19. But on a few older controls, they move all the compensation distance before starting along the ramp–at point *A*. Given the command to move from point *A* to point *B*, they move directly sideways, 90° to the upcoming line, then start along the ramp. Read the manual or ask before starting a real compensation program! In general, that same control is incapable of initiating comp during a rapid travel move as all Haas controls are.

Third Axis Moves Mills Only As mentioned already, some lower-level controls will unlock even though not commanded to do so, if an axis move outside the program plane is given. That bug made it complex to drop the cutter down for sequential passes when roughing a pocket, for example. That was fixed very early on, and few controls today have this handicap.

> **TRADE TIP**
>
> **Comp Tricks** Here are a couple tips that might be useful and a warning.
>
> *Ramp Lines Can Be Ultrashort* There are times when the cutter can't be positioned very far from the surface where the cut will begin—a vise jaw, tight bore, or opposite wall of a narrow pocket, for example. There are two solutions: On a mill, begin the *X–Y* ramp above the surface in the *Z* axis; then after tangency to the ramp is established, drop the cutter down to a short line leading onto the surface to be cut. Modern controls will remain locked on to compensation during a motion not in the present plane. So in G17, a *Z* motion has no effect on the comp.
>
> A second solution drops the cutter to the *Z* level surface at the start of the ramp line but just a short distance off the geometry. Then a very short first ramp line is programmed (for example, 0.001 in. long). The control follows it while becoming tangent. This is known as a **dummy move**. What actually occurs is that the cutter moves forward 0.001 in. but also sideways by the radius amount. It's still following line *A–B* 0.001 in. long but going to *C*. With the cutter now locked on the dummy, it connects to a second line that leads onto the part geometry. Voilá, no nicks.
>
> **KEYPOINT**
>
> Ramp lines can be very short if need be.
>
> *Rapid Lock* Most modern controls are capable of locking onto the ramp line while in rapid travel! It depends on CPU speed. To save time, you can initiate compensation while moving from the program departure point toward the work. Be careful to make the rapid motion continue onto the work surface in the same general direction; otherwise the direction sense might be reversed and the cutter end up on the wrong side of the line (Fig. 21-20).
>
> *Warning! Do Not Reverse the Cutter Direction While Locked On* If a program is written to move from point *A* to point *B* while in G41, for example, then commanded back to point *A*, a big problem occurs. Think this through. On the return move, the cutter must jump back across the line to stay on the left as shown in Fig. 21-20. This jump usually results in a crash. The control does not "see" a metal part on one or the other side of the line; it simply sees a line connecting points *A* to *B*, then a new line from *B* to *A*, and it will go to the left side of each!

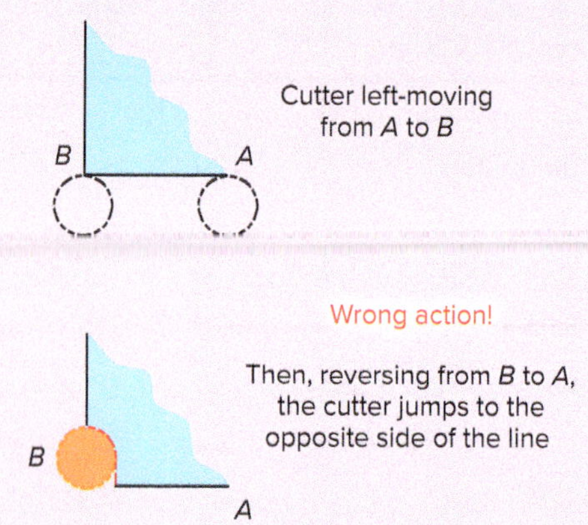

Figure 21-20 Wrong action! Reversing cutter direction while locked on results in a jump across the line.

Unlocking Seeing the G40 cancel code, most new controls gradually move away from the off-ramp line on the next physical line command within the current plane. But an older control will unlock the same as it locked on–90° directly sideways, then follow the line!

Buffer off ramps Some controls require a compound **off ramp** that has at least three line elements beyond the actual part geometry. Since the control is looking ahead at the program line, that control must not see the G40 command within three off-ramp moves ahead, while still making real moves on the part. If the three-part off ramp is not present to act as a buffer, the control will stop compensating while still on the part geometry–a disaster in many cases. These extra buffer elements can be dummy length.

UNIT 21-5 REVIEW

Replay the Key Points

- During compensation, the distance the cutter stays right or left is determined by the radius offset stored in the tool memory. The command to compensate is in the program, but the distance is not.
- After the compensation code, the cutter becomes tangent at the far end of the next line (for most modern controls).
- Do not reverse cut direction while locked on to geometry–the cutter will jump across the line.
- Turn off compensation with the G40 code *only* after the cutter is well clear of the work. Never turn off compensation while the cutter is in contact with the work!

Respond

Note: We'll write compensated programs in the review problems. Answer these questions to determine if you understand the concept.

1. What advantages are there to using cutter compensation in a program?
2. True or false? To compensate to the left side of the geometry line one must enter M41 on or before the next move within the current plane. If it's false, what would make it true?

Critical Thinking Problem

(Not covered in this reading but mentioned previously.)

3. I have a previously written *cutter centerline* program that's long. I don't want to start over to write a part path program. It was specifically written for a 0.750-in.-diameter end mill, exactly. But I have only a 0.735-in.-diameter cutter.
   A. What happens if I use that cutter?
   B. Can I edit the program to include compensation for the different cutter?
4. Describe at least two things the on ramp does for compensation.
5. Why must you clear the cutter well off the part geometry before entering G40?

CHAPTER 21 REVIEW

Unit 21-1

How many code words do you now know? Day by day, you'll add more to your vocabulary. But occasionally we all need to look one up in the index for a given control!

Unit 21-2

Programs aren't unlike a painting. All artists have their own techniques that work for them, as do programmers. That said, don't take too much time trying to make the data look like a masterpiece of organization. Most CAM systems generate a line-by-line arrangement, so it's the standard for data.

Unit 21-3

One item probably overstressed in Unit 21-3 is to know the control's way of doing things when initiating compensation. In general, most of the odd differences are gone in today's controllers. They all operate pretty much the same, locking on and off gradually at rapid travel if needed and staying locked until told differently.

Unit 21-4

Unless one calculates and writes a few arc command lines, they remain a mystery, especially the *IJK* version. You should be able to do it, but today most of us solve our computations through the CAM system. CAD drawings can also be analyzed for distances. That's how I checked the problems here–I drew an accurate version of the problem, then analyzed the database for distances and coordinates. Today, however, arc coordinate computation is a backup skill I wouldn't want to be without.

Unit 21-5

Program verification is a big issue in CNC work. You'll be comparing yours to mine in the review coming next. But in the real world, that's not the way it's done. It's just not possible to read all the lines of even a modestly long program to check it. Nobody can do it unless it's short and simple. So that means verification by another means. We have three options.

In a CAM system like Mastercam®, one of the utilities is called *Tool Path Verification.* Higher up the test scale, programmers use surface verification programs (Fig. 21-21) that create complete cyber parts along with tool marks and errors if they exist. Note the yellow rapid travel mistake in the program. This software compares the CAD drawing surface to the projected machined results.

Test Part the Final Proof Finally, if the part is very expensive, we machine material of lesser value, then measure it.

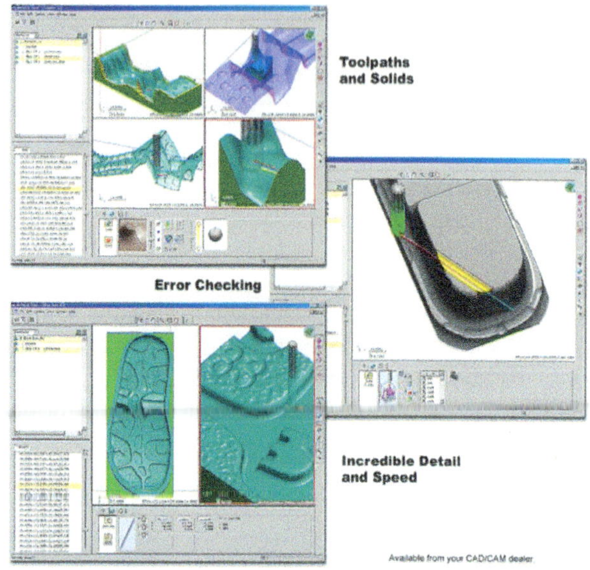

Figure 21-21 The tool path and surface are analyzed by the software and an error is found—see the yellow rapid travel flag where the cutter ruined the work.
Courtesy Northwood Designs, MetaCut View ® Software @ HTTP://www.MetaCutView

QUESTIONS AND PROBLEMS

1. Identify the six common code word prefixes. (Unit 21-1)
2. If you saw an entry of T10.5, would you assume it to be a tool or time word, and why? (Unit 21-1)
3. True or false? CNC codes are grouped together for easy reference. If this statement is false, what would make it true? (Unit 21-1)
4. True or false? An entry of *G71 X-042.32* would be rejected by most modern CNC controls due to the leading zero before the 42 in. and the trailing zero beyond the 0.32. (Unit 21-1)
5. After the file number and communication symbol (if needed) your program should have two kinds of safety start data–what should they do, and why? (Unit 21-2)
6. When pulling a drill back to a safe distance beyond the work, to what level is it withdrawn? (Unit 21-2)
7. Write a cutter centerline coordinate for point *A'*, using Fig. 21-22 and assuming 0.125 radius on the tool nose. Use absolute values. *X* axis values are diameters on the drawing. See the blow-up sketch. (Unit 21-3)
8. See Fig. 21-22. Complete the centerline cutter path coordinates (not the program) from PRZ to point *E* using absolute values and *X* diameters. Do not write the curve command (see Problems 9 and 10). (Unit 21-3)

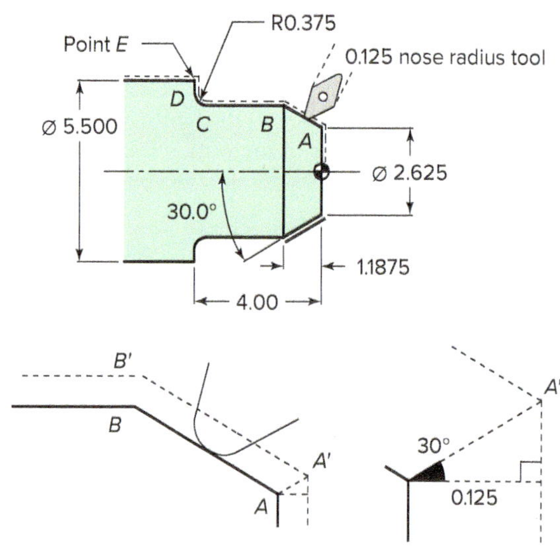

Figure 21-22 Review Problems 7 through 10.

9. See Fig. 21-22. Write the curve command from point *C* to point *D* using the radius method and absolute values. (Unit 21-4)
10. See Fig. 21-22. Write the curve command using the center point method (*IK*). (Unit 21-4)

Problems 11–16

Latch lock—cutter compensation program

See Fig. 21-23. Write a complete tool radius compensated program to profile machine the latch lock. (Unit 21-5)

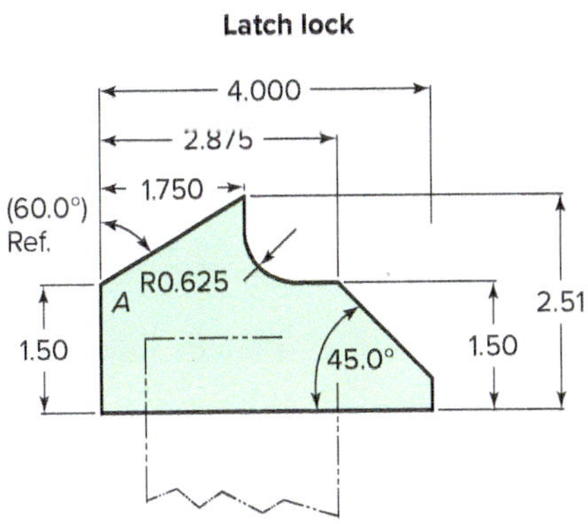

Figure 21-23 Review Problems 11 through 16.

- Planning:
  - Climb mill from the left corner of the 60° angle (point *A*) to the far end of the 50° face. One pass only.
  - Be sure to include on and off ramps for compensation. The on ramp is a double line.
  - The stock is $\frac{3}{16}$-in.-thick steel plate. It has been sheared with 0.030 in. excess on the 2.51-in. dimension. It is 4.00 in. wide.
  - The stock will be clamped over the lower edge on a sacrifice plate that will be slightly cut by the end mill.
  - Use absolute values and the radius method command for the 0.25-in. inside corner radius. (Unit 21-5)

Problems 17–20

Write a tool radius compensated program for the familiar bullet nose pin (Fig. 21-24). (Unit 21-5)

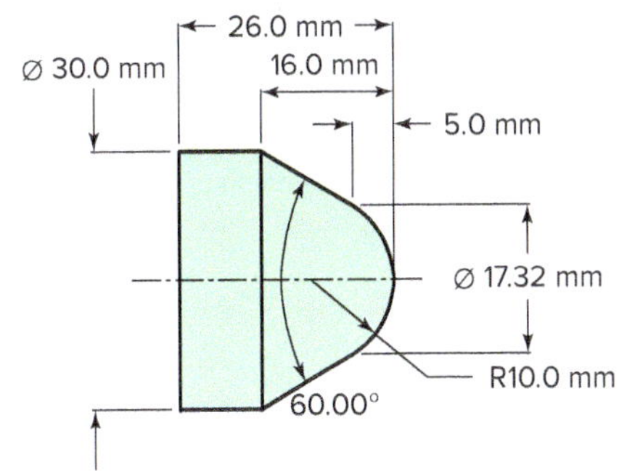

Figure 21-24 Review Problems 17 through 20.

CHAPTER 22 Level-Two Programming

Units

22-1 Fixed Cycles

- Identify and use parameters in fixed cycles
- Find specific cycles in the programmer's manual in your shop
- Begin to form a vocabulary of standard fixed cycles
- Write a threading fixed cycle for a turning center

22-2 Branching Logic—Loops and Subroutines

- Draw a logic diagram of a branching program
- Define loops, nested loops, and subroutines
- Use subroutines on a control in your lab

22-3 Special CNC Programming Tools

- Write a constant surface speed program command
- Write an RPM limit command
- Describe an axis-scaling program command and how it is used
- List the special capabilities of your control
- Describe axis mirror imaging and how it is used

INTRODUCTION

Chapter 22 is about the artful aspect of CNC programs compiled by hand, using logic statements. With today's popular CAM-generated programs, which make less use of these tools, one might be tempted to skip this chapter. But you'd be missing the real art of programming. Using these techniques one can simplify math and reduce total data. But more, it's just plain fun to do these things! As a bonus, in learning the logic commands coming up, you'll gain insight into all programming languages based on a branch of math called Boolean logic.

We'll start with fixed cycles. Used for milling and turning, all CNC controls contain a few within their capability. Many feature a library of *canned cycles* (everything is contained within the command including the formulae that make the calculations to carry out the routine).

CUSTOM CYCLES

Many controls allow custom-written (or purchased) specialized cycles. They support the manufacturing of a particular feature or unique product. For example, a shop makes custom chain sprockets. While each sprocket is different, they vary in just three characteristics. So a tailored cycle that profile mills the chain teeth is a real time-saver.

After entering the code to begin sprocket milling, let's say it's G151, the cycle would require a pitch parameter for the sprocket *P* (which sets the diameter), *W* the width of the chain (depth to plunge the end mill), and the *N* parameter to determine the number of teeth, plus the *X*–*Y* center location.

G151 *P*.5 *W*.1875 *N*23 *X*1.500 *Y*0.000

Presto! Supplied with just those variables, using the formula within the cycle, the control calculates the size and shape of each tooth at the correct angle with respect to the axis of the sprocket, then makes the correct axis moves. So, with no other programming, the G151 roughs, then finish mills the sprocket. The magic happens transparently within the controller after it reads the G151.

In comparison, the same program generated by CAM software would be hundreds or even thousands of command lines long. The CAM software does not use formulae or algorithms to generate shape; it breaks the graphic image of every tooth profile into tiny segments. It then calculates the best fit arc or straight line for each segment. The program then writes axis commands for each. Down to the microlevel, the profile is identical to the canned cycle shape—it's just the data to make it that are different. We won't investigate writing these custom routines (sometimes called *parametric cycles*), as they are beyond the beginner's level. Here we will explore the standard cycles all machinists should understand.

Then after canned cycles, we'll look at some higher programming functions:

- Branching logic with loops and subroutines
- Mirror imaging
- Scaling
- Constant surface speed

Most are program tools that CAM software doesn't use but that the hand programmer can employ to great advantage. Here are the units that come close to an art form and are fun.

Unit 22-1 Fixed Cycles

Introduction: A fixed (canned) cycle begins with a code such as G81, which starts the drilling cycle. The code is followed by several variable entries called *parameters*. By entering the parameter once, the *modal cycle* retains them until changed or canceled. Recall that a modal command means it's a mode to be switched on by the G81, and it stays on until canceled by the G80–cancel cycle code–or by another word from the same group such as G83, the peck drill cycle. M2 (end of program) also cancels cycles on most controls.

TERMS TOOLBOX

Bolt circle A pattern of holes or other features distributed about a central datum.

Canned cycle A second common name for a fixed cycle.

Departure point The initial position from which a given routine is started. Often designated with the letter *I* in programming manuals.

Parameter A blank requiring an entry to customize a fixed cycle.

Shape address numbers (shape definition) The line numbers often defined by *P* and *Q* parameters in canned cycles.

Tiled routines Overlapping and connecting mill routines to machine away a given area of a complex shape.

SHOPTALK

Same Language, Different Dialects! Keep in mind as we look at various cycles for mill and lathe, and at their parameter designators, that these are evolved words that came into existence after the original assigned code words. While the cycles operate similarly from control to control, the code and the parameter letters may be different than those used here. For example, a unit in our lab uses *D* for the hole depth parameter, not *Z* as discussed. The parameters and codes cited herein are those I'm familiar with; it's impossible to know them all!

22.1.1 How Fixed Cycles Save Time

Follow this example G81 for drilling a series of holes on a vertical mill:

N019 T01 H1 G43 M6 (G43 = tool length offset, H1 = apply length offset 1)

N020 G81 *X*1.00 *Y*2.00 *Z*-0.5 R.375

N025 X2.0

Line N019 uses G43, a common code to initiate tool length offsets. Command line N020 tells the control to go to the *X*-*Y* location and drill a hole to a depth of *Z*-0.5, then retract at rapid travel to 0.375 above the surface of the work. On line N025 a new location was entered; the drilling routine would still be the same: move to the new *X*-*Y* and drill another hole. This will continue for each entry until the fixed cycle is canceled with a G80 command.

KEYPOINT

The G81 code is modal. It stays in effect until it is canceled.

ARMCHAIR FIELD TRIP

Don't miss this Haas Tip of the Day video drilling cycles used on CNC machines.

How are canned drilling cycle commands structured? What are some of the problems with using canned drilling cycles? Go to: **https://www.youtube.com/haasautomation** and search for this video called: **Don't Waste Cycle Time; Peck Drilling Essentials—Haas Automation Tip of the Day**

Critical Question?

So on an inserted line, if I enter

N021 *X*1.75 *Y*2.5

what will happen? Don't read the answer; think it through.

Answer

The control will move in rapid travel to the new location and repeat the modal cycle because it's still in effect. Now, what happens on line 22?

N022 *X*3.00 R.75

It goes to the next location moving the *X* axis only (*Y* was a null entry) to drill the hole to *Z*-0.5 but this time retracts to a different R height. You can see, a modal **canned cycle** can be useful. All one needs to do is enter the parameters that change, hole to hole.

Fixed cycles are used in two different but closely related ways:

1. **Abbreviate Repeating, Sequential Commands**

   The preceding drill routine is a good example. To program 50 holes without a cycle would require writing all the *X*, *Y*, and *Z* axis movements 50 times!

2. **Simplify Otherwise Complex Operations**

   Some operations would be hard or nearly impossible to write with standard commands. The threading cycle on the lathe is a perfect example (coming up). In some cases, there are no standard commands that produce a perfect thread. But by changing **parameters** within the various threading cycles, a variety of threads can be easily programmed, including internal, left-handed, and tapered versions–as easy as entering a code word, then filling in five parameters!

22.1.2 Lathe Cycles

Threading Cycles

To understand CNC threading we must first compare manual to CNC lathes. There's a major mechanical difference. On a manual lathe, when the half nut lever is engaged, a geared connection between spindle rotation and tool lead is locked in. But on a CNC lathe, there is no mechanical connection. The spindle has its drive motor, and the tool turret has one drive motor for *X* and one for the *Z* axis.

The computer determines the moment to start the tool moving and at what axis speed it must move to produce the thread lead. It does it perfectly, hitting the mark every time.

KEYPOINT

But that creates four critical checks for CNC operators and programmers using a threading cycle.

Any Lead Is Possible

With computer control, any thread lead can be produced: metric or imperial, left- or right-hand, fractional, tapered, and even nonconstant lead threads.

Acceleration Distance Required

Without gearing, the tool cannot instantly start moving in the correct relationship to the spindle. The *Z* axis must accelerate over a short distance before it's up to the correct lateral speed. Depending on the power of the drive system, some acceleration distance must be built in by starting a short air cut–not contacting the workpiece. Otherwise imperfect threads will be made during the acceleration. Read the manual for that machine to see how much acceleration zone might be required.

On a few older machines with less responsive drive systems (Fig. 22-1), there is also a deceleration distance at the end of the thread as well, but it is always shorter than the acceleration zone.

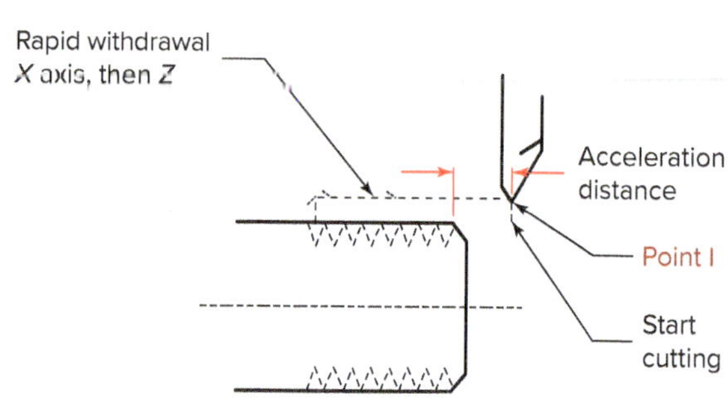

Figure 22-1 The *thread cycle* requires a departure point outside both the *X* and *Z* positions relative to the finished thread.

Lack of Operator Control

During threading, the demand on the control to keep everything in coordination is intense. In order to govern the RPM-to-lead ratio, once the thread cycle begins, the control defaults to a specific RPM and locks out these operator functions during threading:

Single-step mode	Spindle speed override
Feed override	Rapid override

KEYPOINT

Because the control must maintain rigid control of the tool movement to spindle rotation, once the thread cycle begins there is no altering of speeds, stopping, or even slowing the lathe except with the emergency stop—it goes on autopilot!

That makes for some nervous first part tryouts when a thread cycle is in the program!

The Departure Point

The prepositioning of the tool before the control reads the thread cycle command is critical. The tool must be parked outside the diameter (or inside for internal threads) and beyond the start length of the thread in *Z* because that's the position to which the tool will rapid return to begin successive passes. The departure point, sometimes labeled "I" for initial position, creates the retract position in the cycle. This prepositioning is also important for several other lathe cycles coming up.

But it should not be too far away either. It's that position that determines where the first pass begins. Positioning the tool too far away in the *X* diameter can create several air passes the full length of the thread before metal is cut. To see this effect, experiment with changing departure points then viewing the results on a screen graphics dry run or on a PC simulator or in a CAM toolpath backplot.

There are four or five parameters for the thread cycle depending on which cycle is chosen–single-pass or multipass threading. For example, a $\frac{3}{4}$-10 UNC thread:

G33 X.75 Z-1.5 H.866 A60.0 P.1

X = Thread diameter 0.75

Z = Final length 1.5 from PRZ

H = Thread depth (single depth–might be a *D* parameter on your lathe)

A = Thread form angle–60° (This parameter determines infeed angle similar to rotating the compound on a standard lathe.)

P = Pitch distance accurate to five decimal places (Your lathe might accept number of threads per inch for this parameter.)

Ta = Parameter varies, but it equals taper amount (Used with G33 and G34 nonconstant thread leads. If this parameter is left off, the thread will be straight.)

The Multipass Threading Sequence (Fig. 22-2)

1. The tool will start from the initial departure point. That's the preposition of the tool from the previous command.
2. It will rapid to the first *X* diameter for the first pass. Note that you do not make this calculation, the control will do it.
3. Then cut to full *Z* length.
4. Withdraw in *X* axis rapid, to the initial *X* diameter; then rapid to the *Z* departure point.
5. The control then repeats all these steps, in coordination such that the tool cuts the same groove each time, progressively deeper until a complete thread is made.

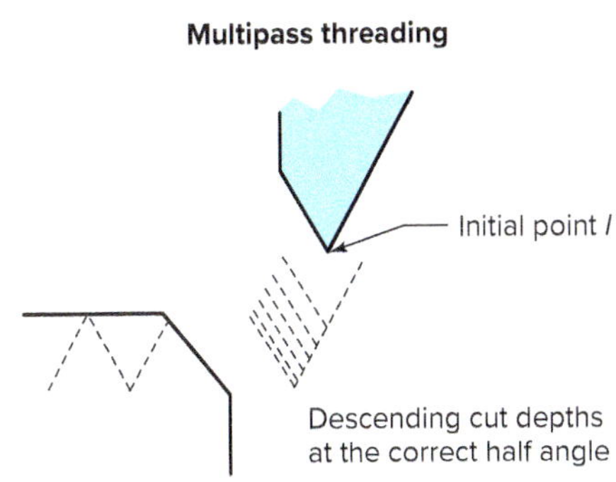

Figure 22-2 The sequence of passes.

TRADE TIP

Most lathes will feature more than one thread canned cycle. One will be a simplified, one-pass cycle, while a second divides the total infeed thread depth into equal chip volumes—not by equal infeed amounts. Based on a formula, this multipass cycle takes progressively smaller infeed amounts as more of the bit contacts the work—this cycle produces the better-quality thread but requires more time to complete.

Some lathe controls feature internal (nut) threading cycles, which simplifies departure point position and how the control will rapid the bit away from the thread. Other lathes use the same cycle for both internal and external threads. On the controls using one cycle for both internal and external threads, the departure point in relationship to the thread diameter determines the direction of infeed and rapid withdrawal of the tool—important information coming up.

Material Roughing and Finishing Cycles on a Lathe

These are exceptionally useful programming tools for removing a large volume of metal. Without them, many roughing passes would need to be written. These three cycles (Fig. 22-3) are commonly numbered as

G70 Finish Cycle

G71 *Z* direction roughing (turning)

G72 *X* direction roughing (facing)

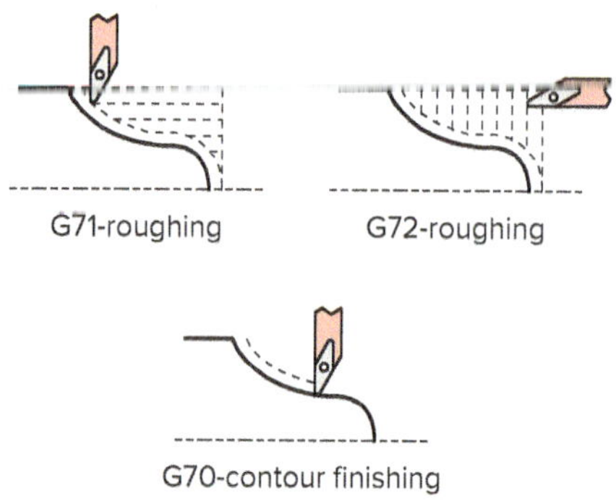

Figure 22-3 The roughing and finishing cycles efficiently remove volumes of metal.

Combined, they first rough out in efficient straight-line passes, leaving a stepped, rough surface. The G70 cycle then finishes the shape in a smooth contour cut, based on your parameters. Often it's more efficient to rough cut using a straight line and then switch to contouring to remove the final material.

These cycles work similar to tool compensation in that they follow the part geometry.

When one of the roughing cycles is read by the control, using the *D* depth of cut parameter, the control divides the removal zone into straight-line passes that stop short of the finished surface. That shell of excess left for the finish pass is determined by the *I* and *K* parameters (Fig. 22-4). *I* equals distance away from the final shape along the *X* axis while *K* equals distance off in *Z*.

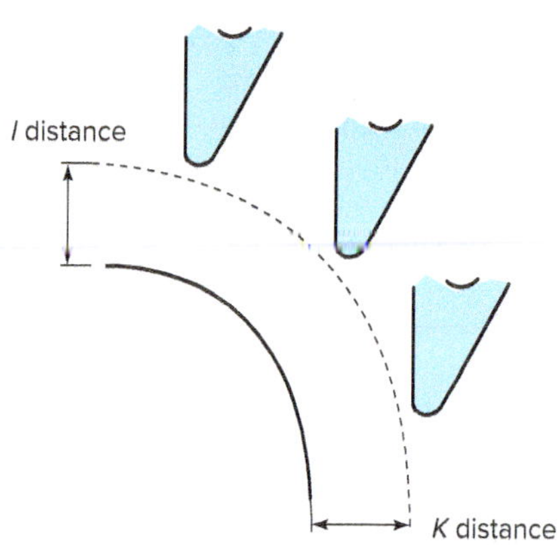

Figure 22-4 The *I* and *K* parameters determine how much *excess* is left in the *X* and *Z* directions after roughing.

These cycles will make their feed passes either inward or out, depending on where the programmer places the tool before evoking them. In other words, if the tool must progress from its departure point, toward the chuck as the cycle begins, the control interprets that to mean the passes are inward. All feed passes will be inward, and all rapid returns will be outward back to the departure point.

Shape Definition Addresses

To use these three roughing/finishing cycles, the parameters include a program line number on which the cycle is to start working and a line number on which it is to stop. These are called the **shape definition addresses**. The program entries between the start address *P* and end *Q* are the part geometry program lines that the cycle will follow.

For example,

N0015 G71 *P*0020 *Q*0035 *I*.5 *K*.3 *D*2

This *X* axis roughing command will start cutting from the departure point written on the previous line, close to (but leave excess) over the contour between line number 0020 through N0035. It will divide the total removal into *X* axis passes of 2 mm (*D*2) depth and leave 0.5 mm excess away from the finished shape in the *X* direction and 0.3 in the *Z*.

KEYPOINT

The finished shape of the object lies between these two line numbers in the form of regular program commands and coordinates.

ARMCHAIR FIELD TRIP

Watch this Haas Tip of the Day video about canned roughing cycles.

How to solve common problems when using canned roughing cycles on CNC lathes.

Go to: **https://www.youtube.com/haasautomation** and search for this video called: **Troubleshoot your lathe G71 and G72 roughing cycles quickly—Haas Automation Tip of the Day**

Controller Difference

Note, your control may require a slightly different form for the *D* parameter as in *D*2.000 equals 2 millimeters depth.

After the roughing cycle is finished, the control moves to the next line where it reads

N0016 *G70* *P*020 *Q*035

Here on line 16 the control reads the finish cycle command and will contour cut the shape between lines N0020 and N0035.

KEYPOINT

A. These roughing and finish cycles are the first program commands in this book, where line numbers are essential to implement the command. The cycle must include parameters for the line number to start and to end the contour.

B. Roughing and finish cycles are nonmodal. Once the control branches back from the shape end address, it continues on into the program with the cycle no longer in effect. (See the next paragraph for branch definition.)

Controller Differences—Order of Entry

On most CNC controls, the shape definition (line numbers 20 through 35 in the example) must be placed beyond the command line in which the cycle is implemented, not before. The search for definition then occurs forward into the program structure–that's called a *branch*. However, a few controls will also branch backward to find the needed line number addresses, that's often called a *loop*. As usual, read the programming manual to find out; however, if the program is written with nothing but forward branches, it will work in all cases.

TRADE TIP

Using Roughing and Finishing Cycles *Writing the Definition* It's difficult to know the exact *P* and *Q* numbers until the entire program is written. To solve that small dilemma, write the program with the roughing and finishing cycle commands in their correct location within the program, but leave the "*P*" and "*Q*" parameters as XXX and ZZZZ (or blank). Then when the program is finished and you know exactly which line numbers contain the final shape definition, go back and fill in the shape addresses. But don't forget to do it!

Two Cutting Tools For better surface finish and longer tool life, rough with one cutter under G71 or G72; then change tools to a finish cutter before commencing the G70 cycle. Using one finish pass, sometimes the square corners of the roughing cuts leave disturbances in the final shape as the cutter hits large and small cut depths. To solve it, take two passes with the finish cutter. On the semifinish pass, set the radius offset such that the cutter backs away and leaves 0.5 mm or 0.015 in. for the finish pass. Return and cut again with the correct offset, but now there are no bumps in the excess.

Modal Feed Rates

In reading the program manual, you will see that a provision for a feed rate is made within most canned cycles. If an *F* parameter is given within the cycle command, that rate is modal along with the cycle. It is in effect as long as the code is in effect, but as soon as the G80 cancel cycle command is read or another cycle implemented, the original feed rate, before the mode was commanded, is reinstated. This applies to lathes as well except the thread cycle is given in pitch or lead distance not inches per minute feed rate. It too is canceled as soon as the cycle is canceled.

The Departure Determines Quadrant Sense

For many cycles the departure point is a critical consideration in your planning. It must be slightly outside the contour in both *X* and *Z* directions to provide a quadrant sense, as shown in Fig. 22-5. There are four cases, with each determined by the relationship of the initial departure *point I* with respect to the work. Each quadrant sense then causes the positive direction for *I* and *K* excess as well. Quadrants 3 and 4 would be applied for internal roughing with a boring bar.

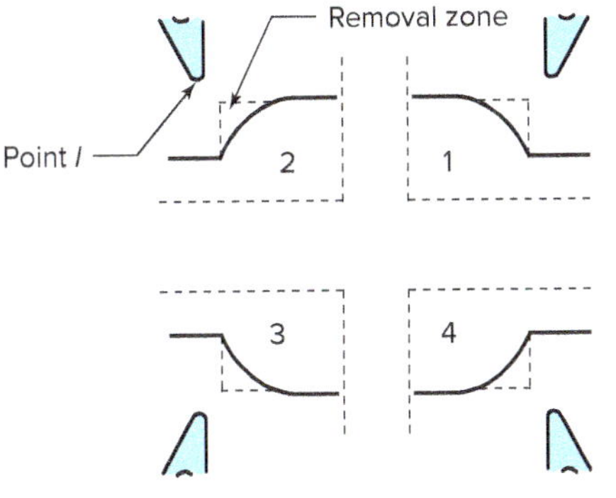

Figure 22-5 Four cases for roughing cycles are determined by the departure point *I*.

TRADE TIP

Combining Cycles With both lathe and mill roughing cycles, it is often necessary to use more than one cycle to completely rough the work (using them in overlapping blocks to machine the entire shape). This program trick is known as tiling, which rhymes with *smiling* (laying routines next to each other as you would floor tiles) so that we have **tiled routines**.

KEYPOINT

If a G70 command follows either the G71 or G72, the control will first rough, then contour finish the work with no extra programming other than the shape definition and line numbers.

22.1.3 Mill Cycles

Similar to lathe finishing cycles, these are time-savers. However, there are two notes to consider.

Simple Shapes Normally

A few high-end controls do feature complex shape routines but they are not universal control to control. We can look at only three canned cycles that are close to common leaving full investigation for outside research.

No Island Geometry Within Pockets

Due to the complex shapes encountered in milling, they are usually provided as rectangular and round pocket clearing routines on all controls but not normally will they finish a contoured geometry with islands nested within. However, as the Trade Tip describes, both rectangular and circular clearing cycles can be combined to machine a given area.

Listed next for three fixed cycles are the common parameters. Their function will be the same on all controls. For example, each must be provided a total depth, number of passes, and overall shape size, but the letter designator might be different in your lab.

Rectangular and Square Pockets (Fig. 22-6)

- *X* and *Y* coordinates for one corner of the pocket
- *L* and *W* parameters for length and width
- Plus or minus values to designate direction from the reference corner
- *D* parameter for the total depth
- *H* parameter for the distance to the top of the pocket from PRZ
- *N* parameter for the number of vertical rough passes
- *P* or *R* parameter for the number of lateral stepovers

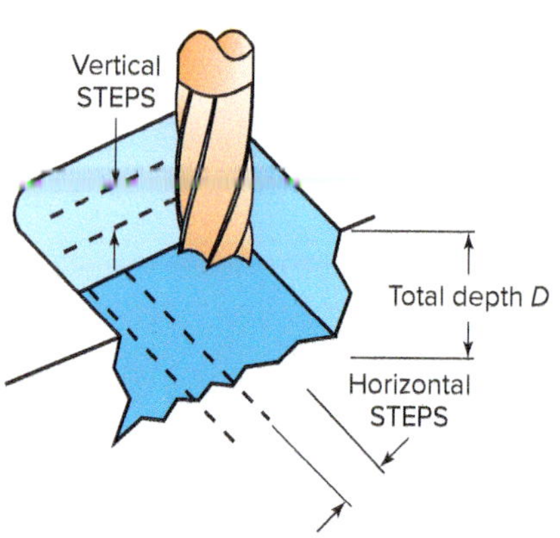

Figure 22-6 A rectangular pocket routine.

Note, this routine mills away all the material from the top to the total depth of a rectangle in both vertical and horizontal steps. It can be applied to the outside of an object as well as a pocket. You can even use this routine to face down the top of a part completely—all *X*-*Y* passes and *Z* steps down—one command, not dozens!

Round Pocket Routines

- *X* and *Y* coordinates of the center
- *D* parameter for total depth
- *H* parameter for the height of pocket top from PRZ
- *R* parameter for the pocket radius
- *P* or *R* for the number of radial movements inward or outward
- *N* parameter for the number of vertical passes

Note, these round area routines are often capable of both clockwise or counterclockwise circular passes and may start in the center and work out or on the rim and work inward.

Bolt Circle Routines (Fig. 22-7)

A third common mill cycle provides a quick way to calculate and drill holes in a pattern distributed about a datum, called a **bolt circle**. While there is no standardization, this routine is fairly common on most controls.

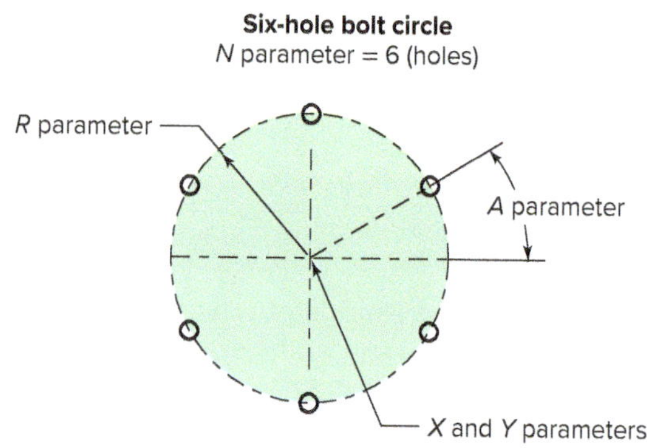

Figure 22-7 A bolt circle fixed cycle.

The **departure point** is usually the center of the pattern. It is either set as a local reference, or there is a *X*-*Y* parameter for the pattern center within the routine.

G88 *N*12 A12.5 *D*-0.5 *P*.125 *R*.5 *H*0.0

- *N* = Number of holes in the pattern
- *A* = Absolute angle from the horizontal reference line to the first hole. This establishes the pattern relationship to the *X* and *Y* axes.
- *D* = Total hole depth
- *P* = Parameter for pecking action
- *R* = Parameter for the retract height between holes
- *H* = Parameter for the top of the hole from PRZ

UNIT 22-1 REVIEW

Replay the Key Points

- Many cycles are modal. They remain in effect unless canceled (G80) or a program end (M2) is read.
- Once the thread cycle begins there is no altering of speeds, stopping, or even slowing the lathe except with the emergency stop!
- Roughing and finishing cycles require a line number address.

Respond in Your Lab

1. Why is the departure point and/or initial point critical for many cycles? Give three reasons.
2. Are canned cycles appropriate for complex shapes on milled parts?
3. Ask for the handout describing the lathe and/or mill fixed cycles in your training lab. If this is unavailable, ask to read the programming manual for the control. The object here is to know the possibilities. Look too at the parameters and compare them to those given here.
4. Write a program that uses fixed cycles to machine an object. If you have graphic simulators, try out the program. A program using cycles and tool compensation is fairly high level, and it is anticipated that there will be bugs to work out.
5. When using a thread cycle on the lathe, why must we park the tool slightly back in the *Z* axis before the thread starts?

Unit 22-2 Branching Logic—Loops and Subroutines

Introduction: In Unit 22-1 when we used the *P* and *Q* addresses in both the rough and finish routines, the control had to leave the command line, where the roughing cycle command is read, and move forward in the program to the line number address designated by the *P* parameter. Then after completing the shape definition, it returned to the branch origination. It would then continue onward in the normal program.

This nonlinear program flow can be shown in a graphic form called a **logic diagram**. If you have any experience in computer programming, many of the logic manipulations described will be familiar.

We will investigate two useful techniques, but there are others beyond the scope of this unit. Within those advanced techniques left for outside research are the ability to

Measure the work using interfaced equipment and automatically adjust offsets up to a limit.

Sense the cut pressure and change to a new sharp cutter when the pressure exceeds a preprogrammed limit. This ability is called *adaptive control.*

Count and record the number of parts produced per hour and per day and the total run count; then stop at the number required–this is common on bar feed turning centers.

Communicate with part loading robots and outside control databases.

TERMS TOOLBOX

Branch A departure from a linear format whereby the program flow jumps to a designated address to create a loop or to find and operate a subroutine, then return to the origin of the callout.

Logic diagram A graphic tool used to visualize the syntax of branching logic.

Nesting The use of a loop or subroutine within a loop or subroutine.

Smart loop A programming tool that turns the program flow back upon itself a given number of times.

22.2.1 Subroutines

Miniprograms Called Within Programs

The purpose of a subroutine (often called sub) is to avoid repetitious programming or to put aside groups of commands to be used again at a later time. Clever programmers sometimes save a number of subs, then paste them together to create programs quickly. For example, an instrument shop makes custom panels that hold various instruments, customized to each request. For each instrument one could save the panel cutout shape as a subroutine. Then, in the program locate the mill spindle over the needed hole and call the sub.

A subroutine is nothing more than another program. It has a start, a body, and an end. But different from the programs we've been studying, it is stored in memory, then called up, and used by an active master program. The actual commands within the subroutine are those we have studied including fixed cycles and tool compensation. Nearly any program command can be used within a subroutine.

Controller Differences

Logic programming was introduced long after codes were invented; thus, there's a great deal of variation in the codes and parameters used to start and end subroutines. Some controls call a subroutine from the main program using a G code, while others use an M98 code and still others use the logic statement *Gosub.*

To simplify explanations, the words GOSUB *NNNN,* SUB *NNNN,* and ENDSUB are used to represent the **branch** commands along with their line number addresses (*NNNN*). These are words borrowed from conversational controls (also from computer programming languages); however, they convey the meaning easily.

Outside Research—In Your Lab

To use subroutines, you will need your control's equivalent words that do the following:

1. **Call the Subroutine from the Main Program (GOSUB *NNNN*)**

   Causes the program to branch out to the stored file name given in the command.
2. **File Identification Name (SUB *NNNN*)**

   The control searches memory for this identification, then, upon finding it, operates it as a program.
3. **End Subroutine (ENDSUB)**

   The last line of the subroutine. This causes the control to end the subroutine and branch back to the next command line in the master program, *after* the call subroutine command. Several controls use M99 for this command.

These can be shown in logic diagrams (Fig. 22-8).

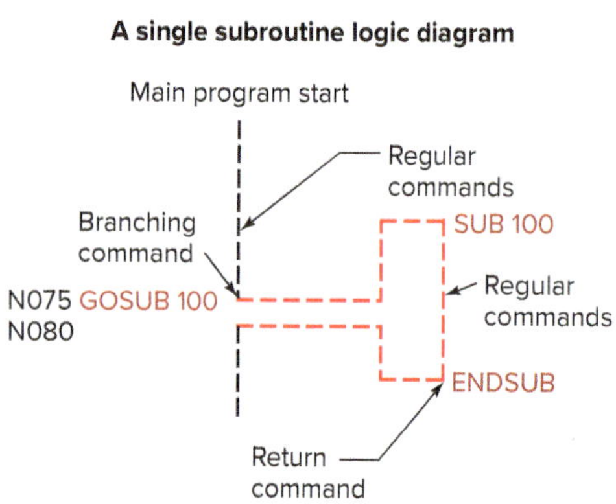

Figure 22-8 A simple logic diagram of a subroutine.

Critical Thinking Challenge

A subroutine could be used to program the drilling of the 16 holes, 0.250 in. diameter (Fig. 22-9). Look at the detail view and see if you can think of how this would be accomplished. Diagram or write the program structure. Then read the program provided just before this unit's Review to see one possible solution. The real trade secret lies in the subroutine.

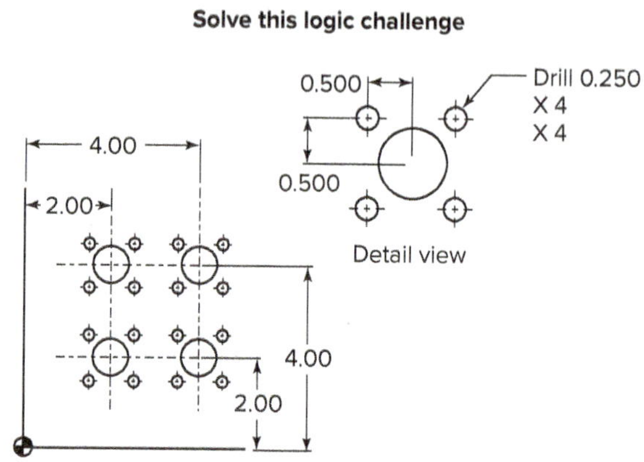

Figure 22-9 This instrument plate has four repeating patterns.

KEYPOINT

Caution!

When swapping modes (abs/Incr) as in this example, be certain to reinstate the original mode (G90) before leaving the subroutine. Otherwise the next move after the last subroutine has run might be a big surprise!

Nested Subroutines

It is possible to have a subroutine call out a second subroutine. That second branch is then said to be **nested** within the first. For example, the circular cutout in Fig. 22-9 could be a second subroutine, called out in the drill pattern routine. The second subroutine is also stored in memory and called when needed.

Some controls limit nesting of subroutines (and loops—discussed next) to three levels deep, a subroutine within a subroutine within a subroutine. Other more capable controls feature a higher nest limit, all the way up to no limits. However, nesting much beyond three deep usually ends in a tangled mess, especially when something goes wrong or editing is required!

SHOPTALK

Loops The origin of a CNC loop can be traced back to the early days of numerically controlled machine tools where a punched Mylar® plastic or paper tape was fed through the machine to cause the various machining actions—like a music box. To save rewinding time, a loop then was created by actually gluing the back of the tape to the front. It ran in a continuous loop through the CPU.

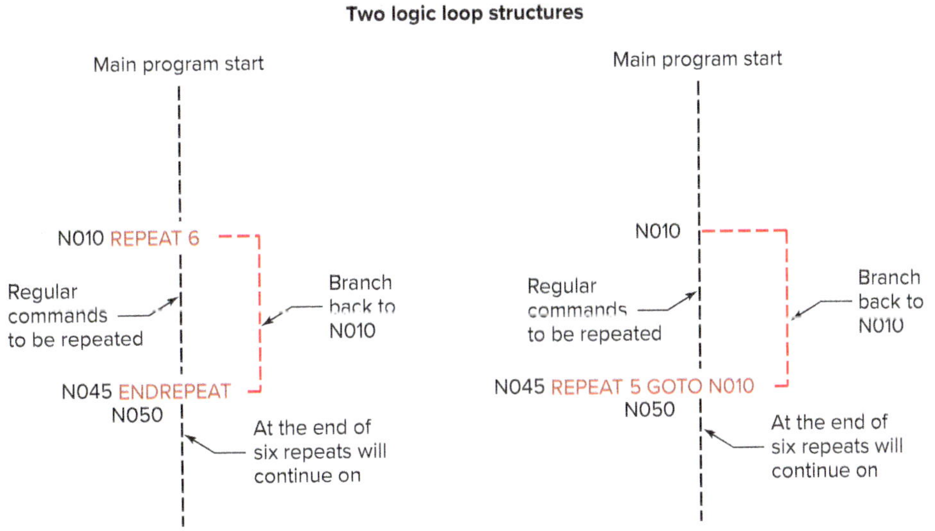

Figure 22-10 Both loops would repeat the commands between line numbers 10 through 45 six times, then skip forward in the program.

22.2.2 Loops

A logic loop is a useful technique that also makes efficient use of repeating data, but this time it's located within the main program. There are two kinds of loops, the *counted* or **smart loop** and the *unending, dumb loop.* The counted loop is of prime interest as a programming tool, while the dumb loop has limited use.

Smart Loops

A smart loop includes the branch statement and a count function. It branches back to repeat a selected group of command lines within the program, a given number of times, then, completing the loop, moves on to the next command line after the loop start.

There are actually two different structures to these loops, as shown in a logic diagram (Fig. 22-10). Dependent on the control type, the branching statement is placed at the front of the loop or at the end. This is a controller difference in words only.

The overall effect is the same: The program branches backward to repeat a group of commands a given number of times (see Fig. 22-10). Both forms repeat the indicated commands six times, until the count is satisfied, then move forward onto line N50. Notice the difference in repeat requests based on the loop structure. The second structure needed five more repeats to complete six times through the loop, because it has already completed the commands one time as the flow reaches the branch back.

> **KEYPOINT**
>
> If on your control the loop repeat statement occurs at the *end of the loop*, then the *requested* repeats will be one less than the number desired, $N = R - 1$.

Dumb Loops

A dumb loop is a logic branch that does not count but turns back on itself forever if not stopped by the operator (Fig. 22-11). The only way to stop a dumb loop is to include a program halt or optional stop within the command lines or by an operator intervention such as feed hold, cycle stop, or emergency stop. An uncounted loop is actually a syntax error. It never reaches the end of the program. Dumb loops have only two applications.

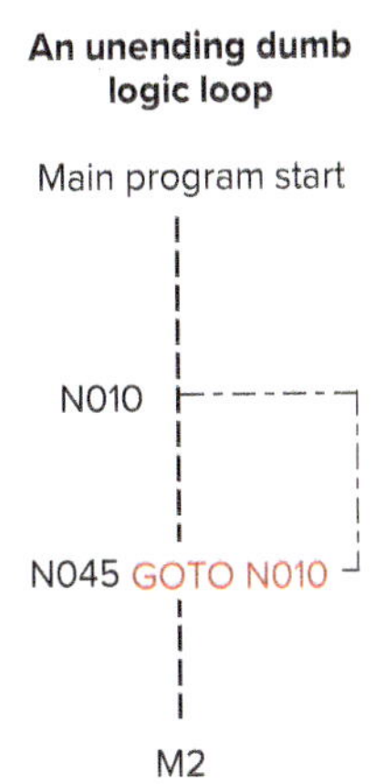

Figure 22-11 A dumb loop diagram.

Warmup Routine

They can be a stored program that can be quickly called up to "exercise" all axes of a cold machine before it is put into production.

(Exercise Loop—Departure from Machine Home)

```
N005 G0 X50.   (run to limit switches or end of travel)
N010 Y50.
N015 Z-50.
N020 X0.0      (or a machine home statement)
N025 Y0.0
N030 Z0.0
N035 GOTO N005
N040 M2
```

Never-Ending Program

Dumb loops can be a program that runs continuously, whereby the operator places the part in the machine, then depresses "cycle start," which overcomes the halt command on line N025.

(Simple Drill and Repeat Loop)

```
N005 M3 S2000 G95
N010 G0 Z.5
N015 G1 Z-0.5 F0.003
N020 G0 Z100. (limit home)
N025 M0   (program halt)
N030 GOTO N005
N035 M2
```

Neither dumb loop ever reaches the M2 command, so it's actually not needed to run it. However, most controls will make a syntax check as the program is loaded into memory. Without the correct program end code, the control will error out.

Answer to Critical Thinking Challenge

To simplify, most of the usual start codes and modes have been eliminated; study the logic and subroutine only.

The trick was to position the tool at the center of each pattern using absolute coordinates, then switch to incremental values within the sub. In effect, that makes the center position a mini-PRZ or local reference for the first move within the subroutine.

Instrument Plate Main Program (from Fig. 22-9)	Instrument Plate Subroutine 01
T01 M6 ($\frac{1}{4}$ drill)	SUB 01
G90 M3 S3000	G91 (incremental mode)
G0 *X*2.0 *Y*2.0 (center of 1st group)	G81 *X*.5 *Y*.5 *R*.25 *D*−0.5
GOSUB 01	*X*−1.0
G0 *X*4.0	*Y*−1.0
GOSUB 01	*X* 1.0
G0 *Y*4.0	G80 G90 (absolute mode again)
GOSUB 01	ENDSUB
G0 *X*2.0	
GOSUB 01	
G0 *Z*100 *X*100 *Y*100 (park at limits)	
M2	

UNIT 22-2 REVIEW

Replay the Key Points

- When changing modes within nested routines or loops, be certain to reinstate the original mode before leaving the routine; otherwise the next move afterward might be a big surprise!
- For loops, the *n*umber of *r*epeats will be either *N* = *R* or *N* = *R* − 1 depending on where the repeat statement is located in the loop.

Respond

1. Can you invent a way to drill the grid plate shown in Fig. 22-12 using a loop? Try it yourself but use the commands of your equipment this time, if available. If not, use the words shown in the logic diagrams. There are several solutions to this problem including subroutines, but for now, use only one loop.
2. Now, can you see a way to improve your program using a nested loop?

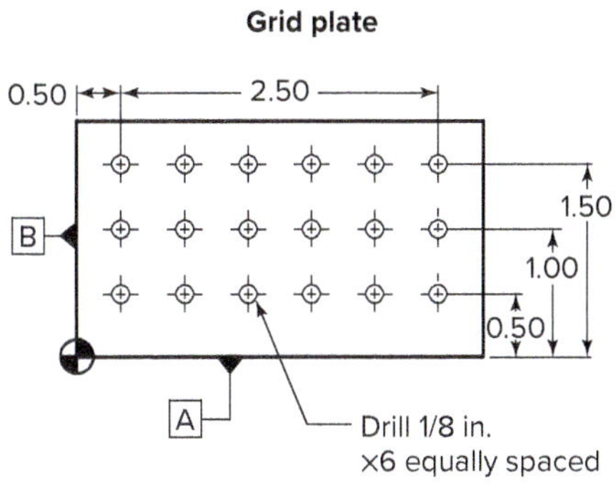

Figure 22-12 Program this *grid plate* using loops.

Unit 22-3 Special CNC Programming Tools

Introduction: This unit lays out a small collection of programming tools of varying degrees of usefulness. The goal here is to introduce the concepts. Each will require a bit of reading within the programming manual to implement. They are all highly specialized computer functions; therefore none are standard code words.

Unquestionably, the most useful is the constant surface speed (CSS) command, which allows the CNC lathe control to maintain a selected cutting speed on operations where the cut diameter is constantly changing, as when facing the work.

TERMS TOOLBOX

Mirroring A program code or draw utility whereby geometry is reversed.

Obverse (shape) The counterpart shape required to mold a correct feature.

Scaling A program factor (or draw utility) whereby dimensions are expanded or contracted by a percentage.

Shrink factor The amount of percentage change of part geometry.

22.3.1 Turning Center—Constant Surface Speed

Probably the most useful of this unit, this command is easy to write and demonstrates a powerful difference in CNC and manual machining where, due to changing part diameter, the selected RPM is always a compromise when face cutting or parting are done. As the cutter progresses inward or out, the RPM should be adjusted, but without this tool it is a compromise.

Before instituting CSS, the tool is first moved to an initial point. That *X* diameter will then serve as a model for the surface speed. The RPM is set at that point, and the CSS code is given:

G0 *X*2.0 *Z*.5	(Rapid to Initial Point)
G96 S2000	(G96 is a common code used for CSS, but the code on your control might vary)

After the model surface speed is set with the G96, any change in *X* axis diameter will cause a corresponding increase or decrease in the RPM according to the familiar RPM formula.

The RPM Limits

You can see that, using CSS, as the lathe cutting tool faces inward as it approaches the work center, the calculated RPM will approach infinity–super high! For example,

X Axis Diameter	RPM
2.00	2,000 (the model)
1.00	4,000
0.50	8,000
0.25	16,000
0.0125	32,000 (No way!)

The G50 RPM Limit Code

A common code used to stop RPM at a safe ceiling is G50. When the command line combines G50 with a spindle speed, the spindle speed represents the fastest the spindle will be allowed to go. (See the Trade Tip.)

N015	G0 *X*2.0 *Z*.5
N020	G96 S2000
N025	**G50 S4500**

The RPM in the preceding constant surface speed program will not exceed 4,500.

TRADE TIP

One reason to "cap" the spindle speed with a G50 command is to make sure the spindle RPM does not exceed the machine's capability. But another important reason is to make sure the spindle speed does not exceed the safe ceiling of a large hydraulic chuck or an unbalanced fixture that has been mounted to the spindle. In these cases, the spindle cap may need to be lower than the machine's capability.

Milling Cutter—Constant Feed Rate

CSS is very common in lathe programs but it also exists for profile milling. As an end mill cuts around an object (Fig. 22-13), the feed rate at the contact rim varies slightly depending on whether it is producing an outside or inside curve, as shown in Fig. 22-13. That's because the rate is predicated on the cutter centerline, the dotted line.

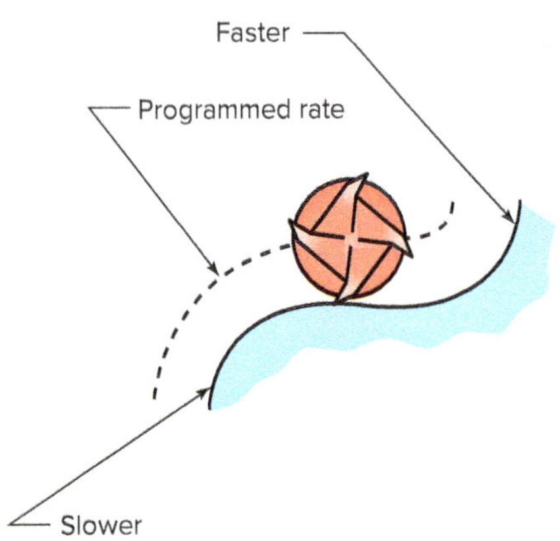

Figure 22-13 As the end mill cuts around curves, its feed rate varies at the rim.

Under normal milling conditions this velocity differential has little effect, and we need not do anything about it in the program. However, when milling metals with a narrow machinability window such as hardened stainless or phosphorous bronze, for example, or when performing high-speed machining, which pushes limits and creates a very narrow window for surface speed, CSS can be useful. An uncommon code, the word and application vary from control to control.

KEYPOINT

For now, remember that this command exists for mills as well as lathes and that in certain critical machining situations, it can extend tool life.

Axis Scaling

This programming tool is used to expand or shrink machined results of a program. It can be applied to all axes or to a single axis, depending on the desired effect. It's used where a complex geometry and drawing would make it difficult to calculate the expansion or contraction of the workpiece, so the print dimensions are used for program coordinates–then the shape is scaled up or down using the power of the computer. If you are familiar with CAD drawing, then you've probably already used **scaling**.

One good example is that of milling a precision, plastic injection mold cavity. On these molds, the plastic is forced at elevated temperature to flow in and conform to the mold. But as it cools, it also shrinks. The mold must be slightly larger than the finished part by a predictable ratio.

That **shrink factor** is calculable and simple to edit into the program. After writing the finished part geometry, a positive factor is inserted at the start. After the code word (we'll use G51 used by Haas controls), a percentage of increase or decrease per axis is included with each axis:

G51 *X*1.05 *Y*1.05

That would cause all *X* and *Y* axis movements beyond that command line to be 5 percent larger. In this example, the *Z* axis commands would be unaffected. But they too can be scaled, as well as radii and other parameters. Look up G51 in the programming manual or on the Internet.

SHOPTALK

Right- and Left-Hand Parts In some production designs, a right- and left-hand part are needed. This can be accomplished with the single program and a mirror command of the *X* or *Y* axis. That axis reversal quickly produces opposite hand parts.

But there's a technical hang-up in this procedure: Mill cuts that were originally climb milled then become conventionally cut in the mirrored version. One solution is to use an opposite hand end mill (left-hand rotating and spiral). These are expensive special-order cutters.

The better solution for making an opposite hand part is to simply write a new program using the CAM computer to reverse the part geometry before producing the cutter path. In Fig. 22-14, I have done exactly that; the drill gauge on the right is as drawn. Then it was mirrored across the *Y* axis vertical line, to become the left-hand part. That new geometry could then be used to write a new program using CAM.

Another use of scaling is found in high-volume (big chip loads, not HSM) machining where deformation heat expands work. Then as it cools, the parts are found to be undersized.

Using part scaling, a slightly expanded part can be made that accounts for thermal expansion, such that the cool part is the right size. If the problem occurs on one axis of the work because it is long, then the scaling can be applied to only that axis. The ratio can be adjusted by the operator as conditions change. For example, a new tool causes 1.5 percent expansion but as it dulls, 2 then 3 percent, at which point the cutter must be changed to a sharp tool again and the ratio is returned to normal in the program.

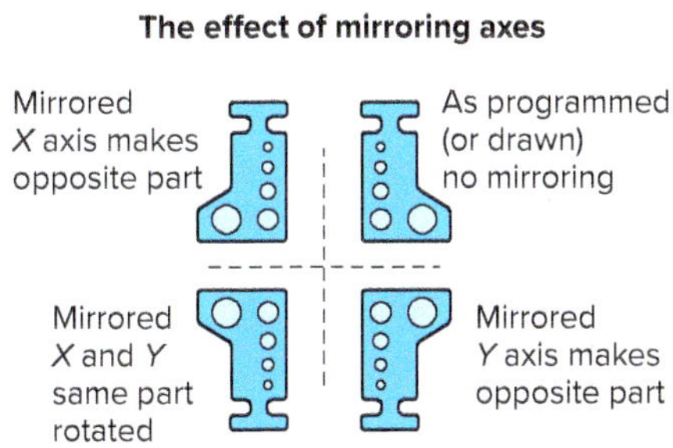

Figure 22-14 Results of mirroring.

CAD/CAM Scaling and Mirroring

The scaling and mirroring commands (next) are used when editing a previously written program or handwriting a new program. However, CAM and CAD software also have these capabilities. One can easily expand or contract the part geometry by a given ratio before creating a program. It can also be turned into a mirror image just as easily.

> **TRADE TIP**
>
> **Cutter Offset Scaling** This is more of an operator's just-for-now trick than a programming technique; it isn't recommended as a planned-in move other than for roughing and finishing with the same tool. However, by falsifying cutter radius offsets, there are a few times when part size scaling can be accomplished at the machine. For example, a mill part is heating up during machining and found to be 0.005 in. too short when it returns to room temperature. If the problem can't be fixed with good cutter geometry or better coolant flow, then add 0.0025 in. to the cutter radius offset. The cutter backs away from both sides of the geometry. It will expand work perimeter sizes. But be aware; it also thickens walls and reduces internal features by the same amount!

Axis Mirroring

Similar to scaling, **mirroring** modifies all axis movements beyond the command, until it's canceled by a flip/flop of the same command–one time turns on mirroring, a second time turns it off.

Given a mirror axis command, the axis drives the opposite direction to the programmed command. One or more axes may be mirrored depending on the effect desired. Mirroring is another unassigned code; it's G100 (mirror on) and G101 on Haas controls but may be a far different code word on your control!

A good example of mirroring is that of machining a mold or die. In this case, it's making plastic letters (Fig. 22-15). The mold must be the mirror image of the finished part. Often, it's far easier to program the letters or other features from the print, in their normal form, and then edit and insert an axis mirror command before the geometry begins.

G100 *X*

Presto, that's all the command required to reverse the *X* axis, but leave the *Y* and *Z* alone.

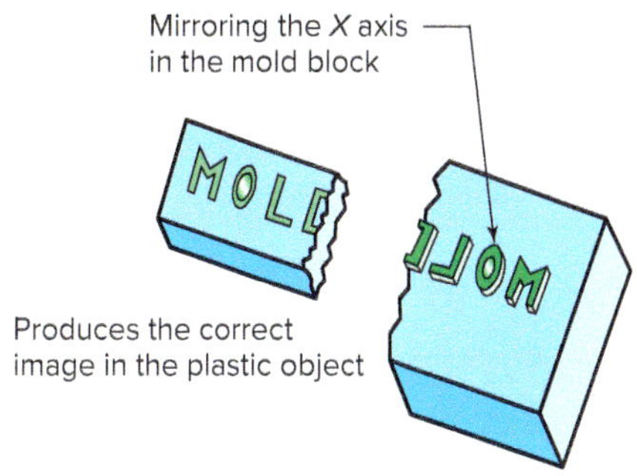

Figure 22-15 Mirroring a letter mold.

> **KEYPOINT**
>
> Mirroring only *X* is the needed command to perform the letter reversal of Fig. 22-14, called an **obverse** mold. Mirroring *X* and *Y* axes wouldn't produce a mirror image part; it would change its location only.

Double Axis Mirroring

Be aware that mirroring both axes does not produce the obverse effect. Shown in Fig. 22-15, mirroring both the *X* and *Y* axes on a mill program will only serve to shift the part quadrant.

UNIT 22-3 REVIEW

Replay the Key Points

- The objective was to learn the possibilities of these commands, not to learn how to use them. Each requires more study in the individual programming manual.
- Mirror only one axis to obtain an obverse geometry.

Respond

1. Describe a *subroutine* compared to a *loop*.
2. True or false? Constant surface speed is a lathe programming tool used to maintain a model cutting speed as diameters vary. If it is false, what will make it true?
3. When G96 is used in a turning program, what safety factor should be included?
4. Describe canned cycles. How are they used? What is another name for them? What are the blank variables called within?
5. Compare a smart loop to a dumb loop.
6. Describe the difference in G71 and G72 roughing cycles for the lathe.
7. In the following command lines, what purpose does the G80 code perform?

```
N075   G81   X1.0 Y1.0 D-0.5 R0.2
N080         X1.5
N085   G80
N090   G0    X0.0
```

8. Describe axis mirroring and axis scaling.

Outside Research

From a handout or from the programming manual,

9. How many threading cycles does your lathe feature?
10. Determine the recommended acceleration distance for a thread cycle on the lathe in your lab.
11. What codes, if your control features them, are needed to cause RPM limits, axis scaling, and axis mirroring?
12. Identify the area-pocket roughing fixed cycles found in your milling machine programming manual.
13. Does the CNC milling machine in your lab feature constant feed rate?

CHAPTER 22 REVIEW

Units 22-1, 22-2, and 22-3

All of Chapter 23 was introductory, to whet your appetite for the higher functions when the opportunity arises. We only tipped the iceberg as far as the special things that this generation of CNC controls can do, especially the PC-based versions!

QUESTIONS AND PROBLEMS

Problems 1 through 6 refer to the following command:

N0195 G71 *P*0155 *Q*0185 *I*0.03 *K*0.03 *D*0.05

1. Describe the machine type and the action. (Unit 22-1)
2. How much material will be withheld for a finish pass? (Unit 22-1)
3. How much material will be taken with each advancement of the *X* axis? (Unit 22-1)
4. Write a finishing command to follow the example command. (Unit 22-1)

CRITICAL THINKING QUESTIONS

5. What does this command tell you about the logic of the control on which it is to operate? (Unit 22-1)
6. How could you write a set of program commands to semirough the shape leaving 0.015-in excess, then return a finish the shape? (Unit 22-1)
7. When performing a threading routine on the lathe, which operator control functions are disabled? Explain. (Unit 22-1)
8. Write a threading routine (use the code your lathe uses or G33) for a 12-mm, 1.75-pitch, 24-mm-long standard 60° thread form. Unit 22-1)
9. True or false? A loop is actually a syntax error because the control never gets to the end of the program. If it is false, what would make it true? (Unit 22-2)

For problems 10 through 12, the following looped program is written to drill the holes for Fig. 22-16. Assume a single tool program ($\frac{1}{4}$-in. drill) with Z0.0 at the work surface. The tool is parked above PRZ 5.0 in. The material is 0.375 in. thick, and the work is clamped over a 0.50-in.-thick sacrifice plate that will be drilled.

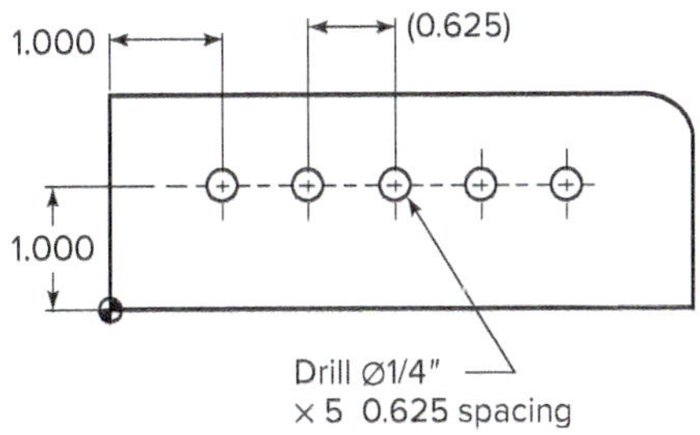

Figure 22-16 Review Problem 10.

10. Does the loop drill the five holes? (Unit 22-2)

11. Is there any danger with the program? (Unit 22-2)

12. If the program needs editing, repair it to run correctly. (Unit 22-2)

(Final review loop program—Problems 10 through 12)

```
N005   G80 G40
N010   S2000 G94
N015   T01 M6            (1/4-in. drill—no offset)
N020   M3
N025   G0 X1.0 Y1.0
N030   Z.25 M8           (rapid to retract height)
N030   G91 G1
N035   Z-0.700           (drill into sac. plate—0.1)
N040   Z0.700
N045   REPEAT 005 GOTO N030
N050   G0 Z5.00          (park spindle)
N055   X0.0 Y0.0
N060   M2
```

13. Define a subroutine. How would it be used? (Unit 22-2)

14. What is different about a subroutine compared to a master program? (Unit 22-2)

15. Define a nested loop subroutine. How deep can one normally nest branches? (Unit 22-2)

Critical Thinking—Not Covered in the Reading

16. Can mirroring be used to machine soft jaws to fit a contoured part? In other words, can you edit mirroring into an existing program to machine a counterpart of the object to be held into the soft jaws? (Unit 22-2)

17. Describe the purpose of G96 for turning. What precaution must be used? (Unit 22-3)

18. True or false? An axis-scaling code would not be found in a program created by CAM. If this statement is false, what would make it true? (Unit 22-3)

19. Describe when and why constant surface speed might be used for milling profile work. (Unit 22-3)

CRITICAL THINKING PROBLEM

20. **A.** In the following lathe commands, what surface speed (not RPM) has been set as a model?

B. At what diameter will the RPM stop accelerating?

Hint: The solutions require manipulating the long RPM formula:

$$\text{RPM} = \frac{\text{SS} \times 12}{\text{Circumference}}$$

The needed formulae can be found in the answers.

```
N005   G0 X2.500   (rapid tool to model diameter)
N010   S1000 M3
N015   G96
N020   G50 S3000
N025   G1 X0.0
```

CHAPTER 23 Setting Up a CNC Machine

Units

23-1 Aligning and Coordinating One Cutting Tool in Machining Centers

- Align parallel to a machine axis
- Align to an *X–Y* datum point
- Enter *X–Y* PRZ into a controller
- Set fixture offsets
- Determine and enter cutter offsets
- Learn the features of turning centers
- Determine *X* and *Z* axis positions for a single tool
- Set PRZ locations

23-2 Multiple Cutting Tools—Determining and Setting Offsets

- Compare master tool, dummy tool, and master position methods for offsets
- Determine and load tool length and diameter offsets for mills (machining centers)
- Illustrate or describe the offset procedure for lathes

23-3 Proving a Program—Tryout

- Based on the situation, determine the level of testing required
- Select correct tryout methods based on the situation

> Tesla Motors, headquartered in Palo Alto, California, has been making electric cars since 2008. Recently, I had the distinct pleasure of driving one of these slick vehicles; what a ride!
>
> Mike Fitzpatrick, author

INTRODUCTION

In Chapter 23 the planning and program are finished but untested. So it's time to focus on getting that first CNC job up and running safely and efficiently. It's time for that white-knuckle, first part run!

How white-knuckled you get depends on an understanding of the goals of Unit 23-3, where we'll discuss 12 different ways to ensure success—9 of which occur before touching the cutter to the metal. The number of precut tests proving it works will vary by situation. But clearly, recognizing the tricky part runs, then taking the best tryout action, is a worthwhile skill to have!

Chapter 23 is about managing a CNC machine. While much of it is deep, study it to the max! In my opinion, knowledge of setting up a CNC machine is the center of this book.

MACHINE AND SHOP DIFFERENCES

While the setup tasks of Chapter 23 have a common objective, the way each shop and CNC control performs them varies. Your instructor might have a particular way of setting up a PRZ, while your shop supervisor may set a policy that's completely different. The key point is that we'll discuss the range of possibilities, but it would be rare for a shop to mix them; it would be too confusing. They'll have one procedure.

The following goals teach you how to get your machine going and perhaps get a career up and running right as well!

Unit 23-1 Aligning and Coordinating One Cutting Tool in Machining Centers

Introduction: First we'll make the work-holding setup; then we'll synchronize its physical location within the work envelope to the axis registers in the control. We'll start with the simplest case, a single cutting tool, first in a CNC mill, then in a lathe. In Unit 23-2 we'll study setting multiple tools, where a lot of the variation between procedures arises.

This chapter assumes previous experience with dial test indicators and edge/center-finders to align tooling to machine axes. If you don't have that experience, be sure to review these skills with your instructor or in this text. Previous practice with digital axis positioners on either lathes or mills will also be useful.

SHOPTALK

For brevity I often state that a concept applies to a "mill," which means a vertical machining center probably equipped with automatic tool changing. A lathe means a turning center with the cutting tools mounted in an indexing tool turret.

TERMS TOOLBOX

Axis preset command A code that writes machine coordinates into work coordinate registers based on the present tool position.

Calc (position) Modern controller function whereby axis values can be written into work coordinate registers or offset memory.

Coordination (coordinating) Physically aligning holding and cutting tools, then writing axis registers to represent that position as the PRZ.

Fixture offset A method of setting PRZ. A code word (usually G54 through G59) that refers to axis position relative to PRZ, using stored data in memory. Fixture offsets can be changed by editing memory.

Master tool method A setup method whereby one cutting tool is used as a model for all other tools in the setup. All other cutting tools are given offsets so that they perform as though they were the length of the master.

TIR (Total indicator readings) When moving an indicator across a surface, TIR is the total change of the indicated reading.

Touch-off method Bringing a cutting tool to the work until it touches a feeler of known thickness.

23.1.1 Using Five Different Data Screens

To perform CNC setups, you'll need to clearly understand the difference in and how to access and write data into the following control screens (sometimes called pages) on your control. Here's a brief summary of what each represents and how it's used for setup and operation.

Screen 1–Machine Coordinates Screen

The absolute *X-Y-Z spindle position* with reference to machine home (even for machines not requiring homing at start-up).

Used for

Alignment and setup reference locations and holding tools, to machine tooling such as vise and chuck jaws. When setting PRZ you will store that physical location using machine coordinates in G54 through G59. Then turn to Work Coordinates next.

Not used for

Machine coordinates are absolute and fixed, they cannot be changed, nor will they be used during program operation. They do not directly show a relationship with the program PRZ. They work transparently behind work coordinates.

Screen 2–Work Coordinates Screen

Both Lathes and Mills The *position of the cutting tool with reference to the current PRZ* that has been established during the setup.

Used for

Basic data for *monitoring program tryouts and during normal program runs.*

Not used for

Work coordinates are not used until the setup is synchronized to the physical location to the PRZ. They are useless until that final step since they represent a previous setup no longer on the machine.

A main objective for CNC setup is to coordinate the *X* and *Y* (and *Z*) axes to represent the actual tool position in these coordinates. You will be reading this screen to represent the setup during the part run.

KEYPOINT

Work coordinates aren't complete until the setup is finalized.

Screen 3–Distance-to-Go (DTG) Screen Data

These are countdown incremental values that tell the operator how far the cutter will move in each axis, ***to complete the current command.*** They are used when clearance is in question during cuts, either during program or MDI commands. Often we halt the motion with override or slide hold functions, turn off the spindle, measure how far the cutter is from danger, then compare that to the DTG values. If it looks okay, restart the spindle and continue.

Used for

Program tryout; for example, a cutter rapids toward the vise. To be certain it won't hit it, touch slide hold, ***stop the spindle rotation***, then measure the distance the tool can travel. Then we compare that to the DTG screen.

Not used for

Machine or PRZ reference–they report how far the cutter is going to move to finish the current axis command. *DTG values cannot be changed* by hand data entry.

Screen 4–Tool Memory Page

Used for

Discussed previously, this screen compiles the tool number and its tool changer location on mills, along with its compensating offsets for diameter/radius and length. Lathe tool shape or approach vector are also stored here.

Screen 5–Fixture Offset—Mills

Fixture offsets *locate the PRZ relative to machine home* (usually on mills but lathes can use them, too). They are used to locate one single PRZ or multiple PRZs. Common codes for fixture offsets are G54 and G55, up to G59. On Haas machines G53 is commonly reserved to identify machine home. It is a global offset, so changing it will change all others, too.

Used for

In industry, several vises and fixtures can be prelocated to reduce setup time. As they are mounted, keys or pins fix them in an exact location on the sacrifice plate. In one quick step, it's aligned and at a precise, known machine reference location within the work envelope. Then saved fixture offsets can be implemented to establish the PRZ. On most controls, a minimum of seven locations can be set. The Haas control adds G110 through G119 making a total of 16 possible unique fixture offsets. In other words, you could have 16 different aligned vises on the table.

TRADE TIP

A quick method to return reference to machine home, and to move the tool to M/H, is to write G53 *X*0 *Y*0 *Z*0, where the G53 offset has been preloaded with *X*0, *Y*0, and *Z*0—It has been reserved for machine home.

Do you need more than the base-6 offset? Haas uses G110-G129, G154 P1-P99, and G154 P1-P20. Most modern controls use similar extra offset numbers. Finally, G52 is a global command on Haas controls, so whatever value is stored under G52, it applies it to ALL offsets when entered into a program.

23.1.2 Vise Alignment and Coordination

One of the more basic setup tasks is to locate and coordinate a single vise on the table. With practice, you should be able to complete this step in less than a half hour. Here's how.

SHOPTALK

In one shop where I worked early in my career, the supervisor used a vise setup as a test to see if he would hire someone who reported being a journey-level machinist. The supervisor took the applicant to a mill, gave him or her what was needed, and then watched to see if the applicant first cleaned everything before assembly and how long it would take him or her to complete the task. Upon completion the supervisor would go over and sweep the indicator over the vise jaw to see if it was within 0.001 in. If the applicant passed all that within 15 minutes, then he or she usually got the job!

Work Orientation First

If you didn't do the planning, start by determining how the raw stock must lie relative to the machine. That is, determine its axis orientation, then find the holding tool(s) to make it so. In most cases, the vise, clamps, or fixture can be placed near the center of the table, but not always. Long or odd-shaped parts sometimes need the setup made elsewhere.

Next, clean everything; the vise and sacrifice plate or mill table must be free of chips.

X–Y Alignment Step by Step

- ☐ **Load the Indicator Holder—Manually or MDI.** Many shops keep an indicator holder as a permanent part of the tool drum load. For example, in our shop it's always T16 = indicator (Fig. 23-1).
- ☐ **Shift the Spindle to Neutral or Turn the RPM Override to Zero.**
- ☐ **Select the Manual Axis Movement Method.** Pulse generator—or handwheel or jog buttons.
- ☐ **Alignment Indication Procedure.** At this point, the procedure is the same as on manual machines. Use either the end-to-end or trend method of aligning the datum surface, solid jaw, to the machine axis. Remember, after the final bolt tightening recheck the alignment.

Figure 23-1 Many CNC mills maintain a permanent indicator holder as Tool Number 16.
Lake Washington Technical College, Kirkland, WA/McGraw Hill

KEYPOINT

Caution!

Lock out axis motion before reaching in to place the indicator in the holder.

KEYPOINT

Caution!

During early experiences with manual movements, turn the rapid and feed rate override controls to below 25 percent.

TRADE TIP

Reminder: For better alignment of vises, a ground parallel held in the jaws averages imperfections and provides a smooth datum surface upon which to run the indicator probe.

X–Y PRZ Coordination

The vise is bolted in place and its solid jaw is parallel to the X or Y axis. The next task is to synchronize the PRZ in the axis registers, called **coordinating** the setup. We'll look at two alignments—PRZ over a hole and PRZ at an intersection of two surfaces. For now you'll be setting X and Y zero points. The Z axis position can be coordinated only after cutting tools are in the machine (see Unit 23-2).

Parking the Spindle Over a Hole

Either the part has a precise hole that's at a known distance from the PRZ or it might be the PRZ, or the setup includes a setpoint hole bored in a block that's at a known relationship with the work PRZ. In other words, the PRZ is at the hole, or it's at a known distance from the PRZ.

ARMCHAIR FIELD TRIP

Finding the precise location of a hole can be kind of tricky. **Watch this Haas Tip of the Day** to find out how to use a test indicator to "sweep" in a hole quickly and accurately.

Go to: **https://www.youtube.com/haasautomation** and search for this video called: **Manually Pick Up a Bore or a Hole with an Indicator—Haas Automation Tip of the Day**

Step by Step (Fig. 23-2)

- ☐ **Shift Spindle to Neutral to Rotate the Indicator by Hand.**
- ☐ **Rough Alignment.** With the mounted indicator probe just above the hole, switch to manual movement, then jog

until the probe swings visually concentric around the hole's rim, through 360°. Try to get it within 0.015 in., so the indicator will be able to read the full amount when it's put in the hole.

- ☐ **Hand Tilt the Probe Safely Beyond the Hole's Edge, Then Drop the *Z* Axis.**
- ☐ **Final Alignment**. Hand rotate the spindle while jogging until **TIR** = 0.0.

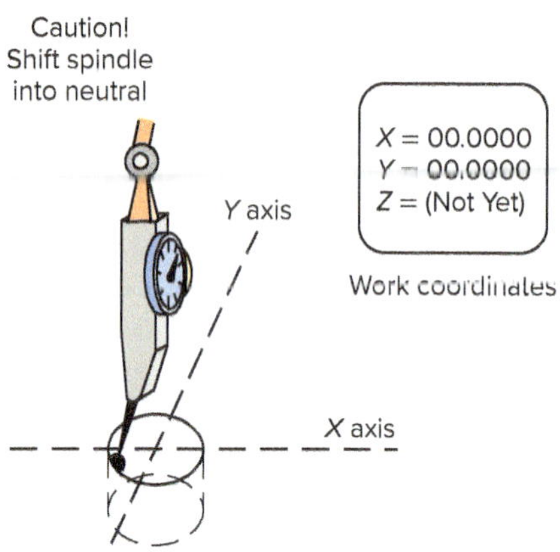

Figure 23-2 Aligning the indicator over a set point hole.

Note, after alignment it's a good idea to drop the probe deeper into the hole, then test the *X*-*Y* alignment a second time just in case there are roundness or angularity issues. The second test will detect them. If the two tests don't agree, try placing a precision pin in the hole or alternately try averaging the two locations.

KEYPOINT

A. **The spindle is now at the exact *X*–*Y* PRZ location.** With the fixture offset page open and the cursor on the right line number, G54, for example, touch Part Zero Set on Haas controls.

B. **The spindle is at a known set point position relative to the PRZ, according to the planning.** *But the axis registers do not represent that position yet.* Write the value into the MDI command line, then touch F1 on Haas controls. That writes it into the *X* or *Y* register. (See upcoming discussion.)

Loading Registers (see Fig. 23-2)

- ☐ **Switch to Work Coordinates Screen**. There are three ways in which a number can be entered into an axis register in the control.

1. **Axis Zero**

   The simplest is to use a reset button that zeros the registers. However, on some controls that also resets other data. You may not want to lose it all. Some controls don't have a reset function, but where they do, it's useful at times if one knows how it affects all other data.

2. **Manual Data Input (MDI)**

   The universal method is to place the cursor over the column and write the digits, one at a time, from the keypad. On Haas controls use F1 to set the written value in the register while F2 writes the negative value in the data field.

3. **Calculate and Code Functions**

   Beyond the MDI input, there are differences in axis loading protocols from control to control. Following are a couple of representative examples.

Preset Code Word

Not all controls feature this code word, especially those with fixture offsets. It is a word that sets the axis registers to a given value. All the data are in the command; none are stored in memory as with fixture offsets.

KEYPOINT

An **axis preset command** writes whatever value it contains into the axis registers at the time it reads it in the program. The current location becomes the *X*–*Y* PRZ.

To use this feature, the tool/spindle must be at the EXPECTED known position relative to the PRZ. If by accident it's not at that expected position, then the PRZ is lost until reestablished by homing the machine and starting over.

Suppose our generic control uses G60 as its axis preset code, and we're parked at a set point 2 in. positive, from the PRZ along the *X* axis. With the work coordinates page displayed, turn to MDI mode and key in G60 *X*2.00 *Y*0.0

>cycle start< in single block mode.
The registers will now read *X*02.0000 *Y*00.0000.

(Calc Function) Calculating Position

Most modern controls provide a way to load register axis values directly into registers or to use math in data entries. Go to the work coordinates page and touch the "CALC" soft key below the screen.

If the present location is the PRZ, enter the calc function, and then touch it a second time, indicating that all the numbers should be zero. The register then reads *X*0.0000, *Y*0.0000. (Note: There may be control differences in how this function works on your machine.)

TRADE TIP

Many controls allow the use of math functions when entering data in the calc mode. For example, suppose an edge-finder is just tangent to a fixture surface that's a plus 2.625 in. from *X* axis PRZ. Enter 2.625 − 0.1 (the minus 0.1 accounts for the radius of the probe). The register will then read 2.525.

Parking Over a Corner Intersection *X*–*Y* PRZ Datum Point

This alignment procedure is a bit more complex since you'll be aligning and coordinating axis registers to two edges, one at a time. Using an indicator, one can position the spindle at an intersection point between two edges, near or at the repeatability of the indicator.

One datum edge for mill work is usually the solid vise jaw or a parallel held against that jaw, to become Datum A. The second is probably the raw stock held against a material stop to establish Datum B. It does not matter which is coordinated first. In Fig. 23-3 we see such a setup requiring *X*-*Y* datum alignments.

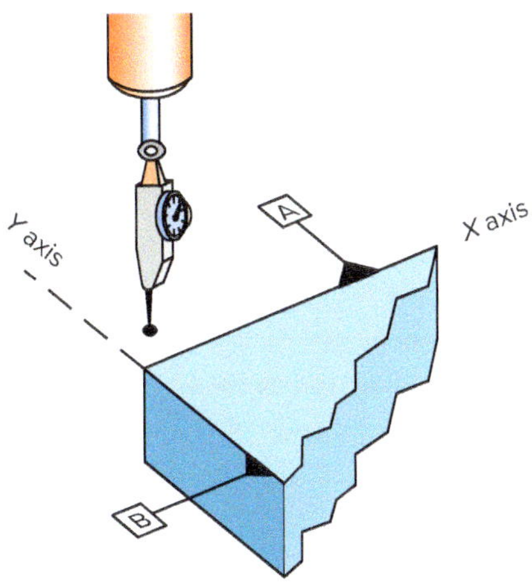

Figure 23-3 This setup requires physically parking the spindle over the intersection of edges A and B to establish PRZ in the *X*–*Y* plane.

Step by Step

☐ **Rough Align**. Begin with an edge-finder or center-finder. Jog the spindle into eyeball location over the *X*-*Y* point. At that time, it can be useful to zero out the work coordinates and use them for final alignment. However, machine coordinates can also be used until the alignment is finalized.

☐ **Final Indicator Alignment, One Axis**. We'll start by placing the spindle directly over the *X* axis line (vise jaw), which *zeros the* Y *axis.*

☐ **Indicator Null Over Vise Jaw—Parallel to *X* Axis**. The objective is to jog until the same indicator reading (usually zero) shows on both sides of the line, as shown in Fig. 23-4. It's done just like on the manual machines, jogging the axis in small amounts until it's centered over the edge.

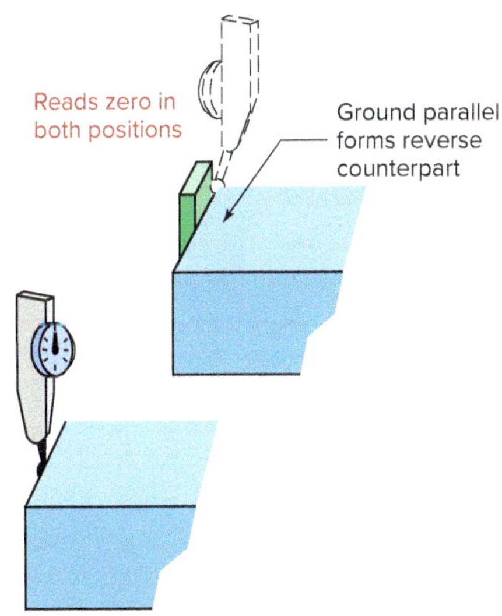

Figure 23-4 Establishing the *Y axis zero*. The indicator reads *zero* on both sides of the *X* axis, edge A. To do so requires a ground parallel held against the edge as shown.

☐ **Work Coordinates Page, Set *Y* Axis Register to Zero**.

☐ **Null Over the *Y* Axis Edge—Set *X* Zero**. The operation is repeated against the surface parallel to the *Y* axis, thus establishing the *X* axis zero. PRZ in the *X*-*Y* plane is then complete.

KEYPOINT

Aligning the spindle over the edge parallel to the *X* axis establishes the *Y* axis zero. You can shortcut this step using only an edge-finder, but it's not as accurate as indicator edge finding. It depends on how tight the tolerance is for the job.

To zero the second axis, do not use the reset function as that will erase the previous axis coordination.

TRADE TIP

There are three hints that simplify edge alignment:

1. *Rough Alignment* Using a precision ground pointer or better yet a center-finder, one can bring the position quite close to alignment (Fig. 23-5). Sometimes, if there's enough rough stock on the work, this may be accurate enough to position the spindle. If not, this operation saves time by prealigning for the more accurate methods.

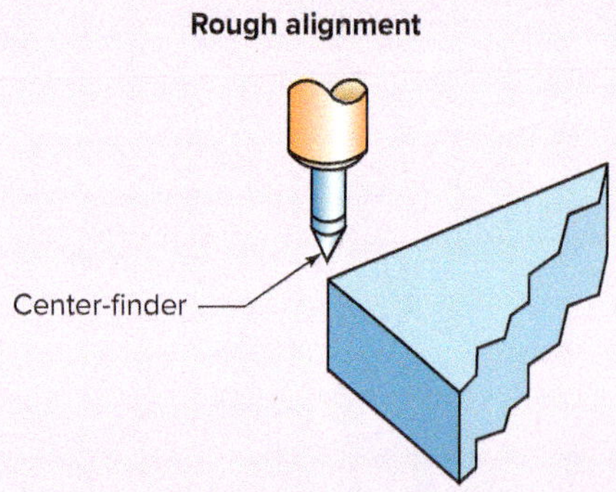

Figure 23-5 A center-finder (or even a precision pointer) saves time.

2. *Datum Hole Accessory* This little homemade gem can save time during precision datum alignment. It features a hole precisely bored directly in line with a ground test edge. When clamped against the edge in question, indication is simplified by spinning the probe in the hole. You can make this tool as a shop project! (See Fig. 23-6.)
3. *Edge-Finders* More accurate than center-finding but less than the indicator methods, edge-finders are often employed to align axis registers in CNC setup work. In one step per edge, the test cylinder is brought to the exact tangent relationship with the edge in question. At that position, the spindle is 0.100 in. if the probe is 0.200 in. diameter (most common). The axis register is then set to represent that coordinated position at plus or minus 0.1000 in., or the calc function can be used to account for the probe radius.

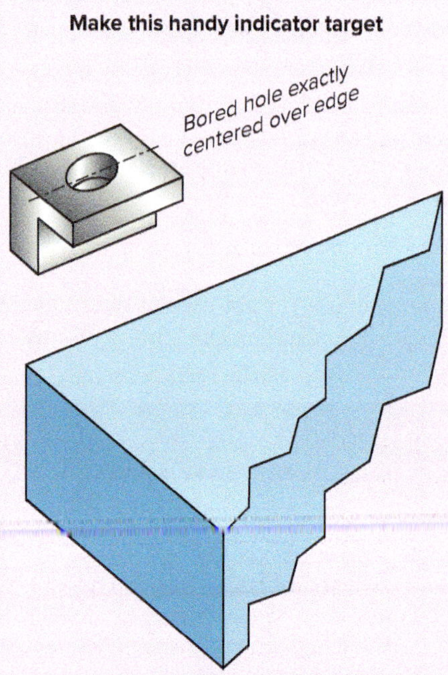

Figure 23-6 An indicator target speeds the use of DTI over an edge.

Using Fixture Offsets Instead of Setting Work Coordinates

The preceding procedure involved a single vise with a single *X*-*Y* PRZ coordination, which required changing *work coordinate* entries to represent that PRZ datum point. But there is another way the setup is synchronized: by loading the *machine coordinates* of the PRZ into a given fixture offset memory.

> **KEYPOINT**
>
> To use fixture offsets for PRZ synchronization, the program must contain a fixture offset command (G54 through G59). Then you must ensure that the correct offset memory is loaded with the *machine coordinates* of that offset—you will determine those values during setup.

To use fixture offsets we begin the same way, by locating the spindle at the PRZ. But here we read the *machine coordinates* of that location. The machine reference values are then written into fixture offset memory under G54 through G59 (most controls). That can be done manually or some controls will load them automatically using the calculation (calc) function. On Haas it's the Part Zero Button (on the upper left).

Then when the control reads a fixture offset command in the program, all subsequent axis moves are shown in work coordinates relative to that location as PRZ.

> **KEYPOINT**
>
> From the moment a fixture offset command is read in the program until it is canceled, work coordinates will represent the new PRZ.

These programming tools can be used in three different ways:

1. To set up a single PRZ
2. To set up and use more than one machining station in a single program—multiple vises, for example
3. To set up and save several different PRZs for different programs

In some shops, to maintain consistency, fixture offsets are used to set up every PRZ. Every program header starts with the fixture offset code, where the coordinates of the reference point are stored. Always begin your setup by reading the document provided and looking at the program to see if the fixture offset code is there.

Most newer controls feature offsets starting at G53. Most keep G53 loaded with *X* = 0.0 and *Y* = 0.0. So entering G53 cancels any offset and refers to M/H. G53 is global on most controls–changing it changes all other offsets by the same amount.

Many controls have a specific fixture offset cancel code. G28 is one popular code, but there are others.

Edge Intersection Step by Step

☐ **DTI Position at the PRZ Location**. Using the methods already outlined, position at PRZ 1.

☐ **Go to *Machine Coordinates* Page**. Write down the absolute position of the PRZ with respect to M/H. Then switch to the fixture page and enter the location under the particular fixture memory line (G54 through G59).

Auto-Loading Most CNC controls feature autoentry of machine positions into fixture offset memory. Once the spindle is parked at the PRZ position, a button touch will load that position into the highlighted fixture offset line. If so, use it to avoid transposing digits from hand entry. Those controls also provide calc entry when the set point location of the PRZ is different than the spindle's position.

TRY IT

Got It? Test Your Understanding: In Fig. 23-7, vise 1 PRZ is stored under G54 and vise 2 under G55. The dimensions are the machine coordinates discovered with an indicator during setup. They are relative to M/H. Each was entered into fixture offsets. Next the program calls

N025 G54, G0 *X*0.0 *Y*0.0
(no *Z* axis movement)

A. Where is the cutter going to go?
B. After the move, what will the work coordinates then read?
C. What will the machine coordinates read?

Okay, that was easy but now, command line 30 is read:

N030 G55 (no axis movement)

D. What then will the *work coordinate* registers read?

ANSWERS

A. Above the PRZ for vise 1.
B. Work coordinates: *X*0.0 *Y*0.0.
C. Machine coordinates: *X*13.3767 *Y*-06.5566
D. Work coords: *X*-5.3240 *Y*5.9567
The cutter won't move, only change references. It's the distance difference in the two vises, *X*-*Y* location.

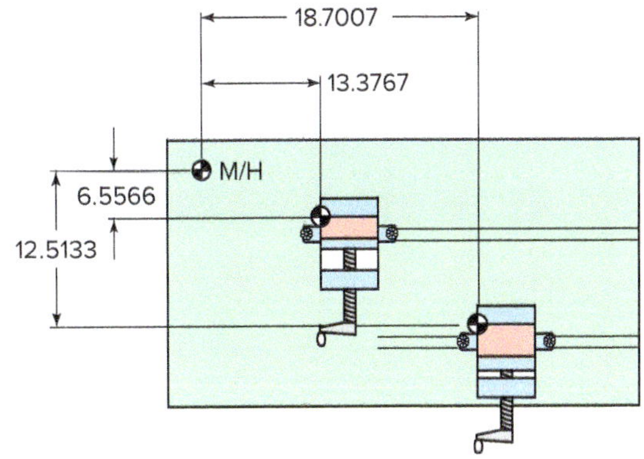

Figure 23-7 These machine coordinates are stored under G54 and G55.

KEYPOINT

Fixture offsets are memory-stored locations relative to machine home. They are called by the program, but the numbers are not in the program. Fixture offsets are modal until canceled usually with a G53 or G28.

23.1.3 Setting *Z* Axis Coordination

Okay, the *X*-*Y* PRZ is finished; now the single cutting tool is loaded into the spindle and it's time to coordinate the *Z* axis zero. Continuing with our simple case, it's an end mill. The plan states it's at *Z*0.0 when it is just touching the workpiece surface. Here, too, there are several variations, so it's critical to understand the range of possibilities.

KEYPOINT

Caution!

Not setting *Z* height correctly is a common cause for tryout crashes!

Step by Step

☐ **Load Cutting Tool**. Either draw the cutter and holder into the spindle manually or if it's in the tool drum, switch to MDI, enter the tool number and M6 code; then execute the tool change in a single block.

☐ **Change to Manual–Jog Mode (without autoprobe)**. Jog the *Z* axis toward the work until the end mill's tip is at an exact relationship with the planned PRZ location–the top surface of the work, in this example. Again, read the setup planning–the *Z* datum point could be elsewhere.

Approaching the surface slowly, the objective is to hand sense contact, known as the **touch-off method**. Touching a cutter to work material isn't the best policy. The tool is manually rotated while being jogged toward contact with the work–just until it drags a bit!

KEYPOINT

Do not rotate the cutter backward to avoid cutting edge damage.

Gaging *Z* Height An improvement over touching the work surface with the cutter is the gage block or paper feeler method previously discussed (Fig. 23-8). Either object will be of known thickness and prevents the tool from touching the work directly. It also acts as an improved sensitivity gage.

☐ **Set *Z* Axis Register in Work Coordinates** Upon physically positioning the *Z* axis, the control is then set to the correct positive value relative to the tool position above the work or to *Z* = 0.0000 if it's touching.

Figure 23-8 Using a tool-setting probe to establish *Z* axis position.
Courtesy Mike Fitzpatrick

Probes and Machine Vision

Discussed in Chapter 19, the safest, most reliable, and fastest method of setting *Z* axis coordination is an autoprobe (see Fig. 23-8) or a tool vision system. Easy to use, the tool probe has a known thickness saved within the control. Once the tool probe cycle is commenced, it slowly drives the tool toward the probe lying upon the reference surface. Then at the exact location that it touches, it halts motion and sends a signal to the control to set *Z* zero.

Controller Differences for Tool Sensing and Setting

Probe sensors range from small electronic pads, hand located between the datum surface and the tool, up to sophisticated laser vision cameras on swing-away arms (Fig. 23-9). After the probe senses the tool's length, loading the axis registers varies, too. At contact, the position might be hand loaded using the probe's known thickness. But more likely on these sophisticated controls, the *Z* position will be automatically loaded into work coordinates. If there ever was a time to read the manual, this is it!

Figure 23-9 This probe is sensing cutter position.
Courtesy Haas Automation, Inc.

23.1.4 Coordinating PRZ on a Turning Center

Here, too, we start with a single turning tool, then build upon that with offsets and hints in Unit 23-2. Similar to the mill, the cutting tool must be positioned in a known relationship to the intended PRZ in both *X* and *Z* axes. Then one axis at a time, the PRZ is loaded into the work coordinate registers.

Coordinating the *X* Axis

Normally, it matters not which axis is coordinated first. We'll start with the *X*.

TRADE TIP

Keep in mind, all *X* values on the controller are *diameters* under normal circumstances.

Differing from the mill procedure, it is not possible to accurately position a turning tool exactly at *X* zero–the work centerline. The procedure is to position it at a known diameter, then load that into the *X* axis position. This is especially true when using a tool with a nose radius, since it's difficult to detect when the tip is exactly at the center of the work. Assume the work-holding device with a length of material is already in place.

X Diameter Coordination

Touching a Known Diameter

Using the touch-off method, the cutting tool is slowly jogged into contact with any utility, round stock of known size, held in collets or a chuck (Fig. 23-10). To facilitate a feel sense, the spindle is hand rotated in neutral and a paper feeler is used. (See the Trade Tip.)

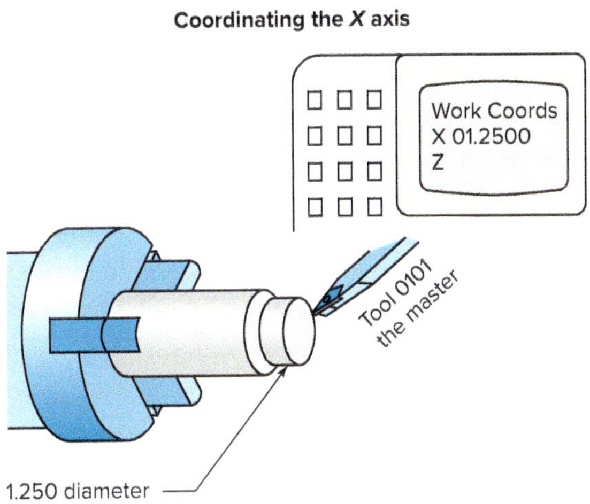

Figure 23-10 Coordinating the *X* axis involves making or touching off a known diameter.

Better Accuracy—Machining a Test Diameter

Here we use manual movement of the cutting tool to machine a short test cylinder. The setup bar or workpiece is machined to any cleanup diameter.

> **KEYPOINT**
>
> That measured diameter becomes the *X* position of that turning tool.

Creating this temporary diameter with the turning tool eliminates potential runout and feel inaccuracies of touching the workpiece, and it reduces fine-tuning of the setup due to work and tool deflection experienced during actual machining. But it doesn't eliminate them altogether, as the manual cutting won't perfectly match program movements and metal removal—you may need to refine the *X* PRZ during the first few parts to be made with that cutting tool.

> **TRADE TIP**
>
> **Precaution** Like a mill, when touching the lathe cutting tool to the workpiece while hand rotating the spindle, never rotate the spindle backward, especially when a carbide tool is touching the work. It will chip the cutting edge.

> **ARMCHAIR FIELD TRIP**
>
> **Watch this exciting video Haas Tip of the Day** about setting offsets on a CNC lathe.
>
> Manually setting offsets on a CNC lathe can be confusing. Watch this Haas Tip of the Day to help you get started with setting up CNC turning centers. Go to: **https://www.youtube.com/haasautomation** and search for this video called: **Set Your Lathe Offsets Manually—Haas Automation Tip of the Day**

Step by Step

Because metal will be cut in this procedure, there is danger of a crash. *Be certain you understand the control and have instructor assistance.*

- ☐ **Select Manual Mode.**
- ☐ **Index the Tool Turret–Mount Cutting Tool.** Be certain that the turret is fully withdrawn from the chuck to ensure tool clearance as it rotates.
- ☐ **Select RPM–Moderate.**
- ☐ **Start the Spindle.** Check the workpiece first to be certain it is tightly held.
- ☐ **Select Jog or MPG Mode.**
- ☐ **Move the Cutting Tool $\frac{1}{2}$ in. From the Work.**
- ☐ **Slow the Axis Motion.**
- ☐ **Move *X* Axis to Make an Estimated 0.015-in.-Deep Test Cut.**
- ☐ **Using Moderate Jog Rate or Continuous Feed, Machine the Test Diameter.** Double-check which axis is selected when you begin the machining pass. You will be moving the *Z* axis perhaps 0.25 in. or so into the workpiece to create a test diameter.
- ☐ **Withdraw the Cutting Tool in the *Z* Axis Only–Stop the Spindle.**

> **KEYPOINT**
>
> **Warning!**
>
> *Do not alter the X position when withdrawing.* Then, before reaching in to measure the test diameter, lock out accidental axis movements.

- ☐ **Measure the Test Diameter.**
- ☐ **Load the Test Diameter into *X* Work Coordinates.**

Z Axis Coordination

Setting the *Z* axis PRZ value is a one-step touch-off procedure performed on the work material, not on a test bar. Or for improved accuracy, a test cut can also be made.

Step by Step

- ☐ **Chuck the Work Material.** Place it in the position in which is to be machined.
- ☐ **Jog the Turning Tool into End Contact.** Or using MDI/JOG, machine the end for a minimum cleanup.
- ☐ **Withdraw in the *X* Axis Only, Then Stop.** The tool is withdrawn in *X* to preserve the tool's *Z*0.0 position.
- ☐ **Load *Z* Axis Registers.**

Probes and Tool Vision Systems

Similar to the mills, the most accurate and fastest method of coordinating a lathe setup is to autoprobe the cutting tool's position relative to the machine coordinates (Figs. 23-11 and 23-12).

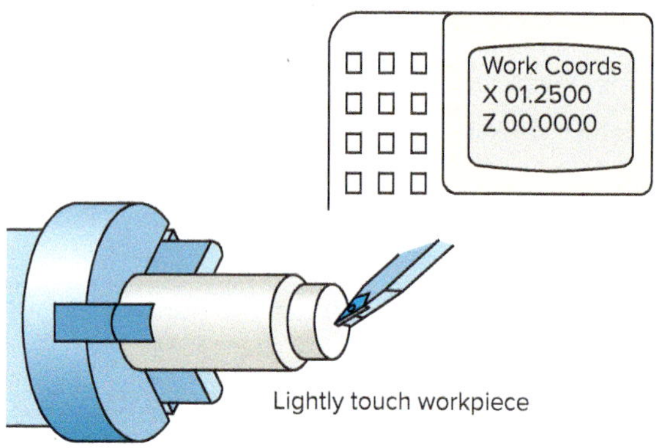

Figure 23-11 Coordinating the *Z* axis.

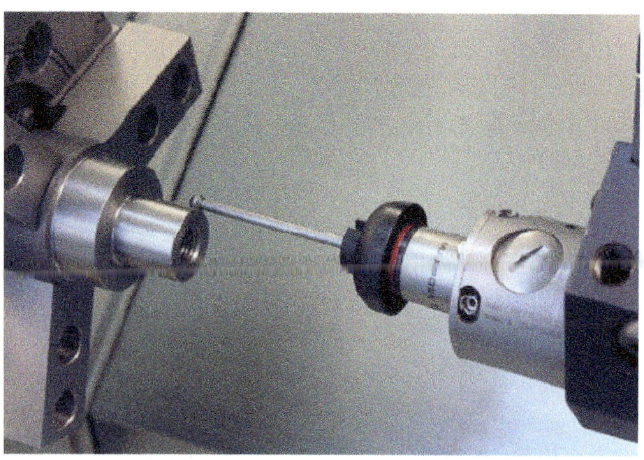

Figure 23-12 *X* axis measurement or tool coordination is easy using a probe.
Courtesy Haas Automation, Inc.

23.1.5 Master Tool Method

Now begins the great argument among machinists–follow it closely here and in the next unit. This discussion is critical for a deep understanding of CNC setup procedures. So far, examples of setting up a single cutting tool at the physical PRZ location (or at a known distance from it), then writing that location into work coordinates is known as the **master tool method**. It's simple and easy to explain for a first experience–move the tool to PRZ, then set the registers to represent PRZ. But a master tool brings complexities to making setups with several cutting tools.

Length Offsets Compare to Master

Using a master tool, when additional cutting tools are used they undoubtedly have different lengths on mills and both *X* and *Z* differences for lathes compared to the master. Those differences must be accounted for as tool length offsets in tool memory. It works fine as long as the master tool remains in the setup. Each tool's offset makes it perform as though it had the same length as the master. All replacement tools must be offset against the master so they touch the workpiece when commanded to go to *Z* zero in a mill program or some *X* diameter or *Z* position on a lathe.

> **KEYPOINT**
>
> Using the master tool method, each cutting tool is given a length-compensating value, but the original master's value is zero until it must be replaced. All other tools are length-compensated relative to the master.

Calling out tool 2, you would add on a command to use length offset 2, which either holds back or adds *Z* movement to compensate for the difference in it and the original master.

Example

Let's use a Haas mill example: Length offsets are commanded with a G43 on our mill using *H* lines in tool memory. Suppose tool 2 is 1.5375 in. longer than the master. Its *H* value would be -1.5375 in offset memory, not in the program. (It's longer, so it must be held back by the difference in it and the master.)

T02 M6

G43 *H*2 (use length offset *H*2 = -1.5375)

G0 *X*0.0 *Y*0.0 *Z*0.0

Holding back by the value found in *H*2, it then moves to *Z* PRZ exactly as would the master. At the moment the offset command is read both work coordinates and DTG screens represent true tool-to-work distance.

Replacing the Master

The big problem arises when the master becomes dull or broken. Its replacement must be exactly the same length, or values must be changed in the control or the following happens.

Critical Question

In a mill setup where the master cutting tool (an end mill) touches the work top surface at *Z*0.0, after setting the PRZ, you then exchange a dull master with a *longer replacement*! Then without changing values in the control, command:

T01 M6

G0 *X*0.0 *Y*0.0 *Z*0.0

What will happen?

Answer

The longer tool will try to rapid travel below the work surface, at the PRZ location, a serious crash!

To solve this problem, there are four options if using master tool.

Option 1 Offset the New Master

Here the programmer put a length-compensating command in the program for all tools, including the master originally, or it can be edited in. But until the master is replaced, its compensating offset value remains zero.

At replacement the machinist uses the touch method to find the difference in the first and second masters, then stores the new compensation in tool length memory. Using this solution, the new master tool is compensated to perform like the original master.

The problem now is that all other tools have no physical master for comparison and replacing them becomes a problem. It can be done, but it's a complex process. To do so, one must first take up the *Z* axis offset for the master so that as the machine is moved, it represents the work coordinates of the new master tool. Next, touch off the replacement tool and read its work coordinate position relative to the new master. That becomes its offset (but with the sign reversed).

Option 2 Reset All Tools

The next solution is to set up again the new master's PRZ. Touching off the new master, a new *Z* PRZ is written into the control. Then all other tools in the setup must be offset relative to it again. It's starting the setup over again with a new master tool—very time-consuming unless there are only two or three tools! This option is not recommended, but it does work.

Option 3 Preset Duplicate Tools

Some shops use fixtures to set the cutting tool's *gage distance*—the distance it protrudes from the mill spindle taper. This requires special cutter holders that allow adjustment of tools with respect to their length. Using preset tools, since each replacement is the same length, no length offset change is needed for master or other tools.

Using a preset replacement, the new tool can even be loaded into the storage drum before it's needed, then switched into the production based on a replacement schedule. Or at a predictable time, the switch can be triggered by either part count (common) or due to increasing tool pressure beyond a set limit (uncommon). With both options, there is zero downtime. Presetting tool length is the best solution for accuracy, efficiency, and operator simplicity. But it requires a big investment in presetting equipment and tool holders. Note that presetting is also used with master position methods coming up.

Option 4 Use a Dummy Master Solution

A final solution for tool replacement under the master tool method is to plan and load a dummy tool in the drum or turret. It's touched to the workpiece, then the *Z* PRZ is set to it. All other tools are offset to it. Not used as a cutter, it never wears out or loses its master status. Usually when a dummy master is used, its tool number is one of the higher numbers in the machine so that real cutting tools occupy the lower numbers, T18 or T20, for example. No reference is made to the dummy in the program—it is not called up in the program at all.

Okay, the single tool setup is ready to go on both machines, lathe and mill. Review carefully, since we'll be making more complex setups in Unit 23-2.

UNIT 23-1 REVIEW

Replay the Key Points

- Work coordinates aren't complete until the setup is finalized (coordinated).
- When setting the PRZ, the spindle or cutting tool must be at the exact *X*-*Y* or *X*-*Z* PRZ location (or it might be at a known setpoint position relative to it per the planning).
- The axis preset command writes whatever values it contains into the axis registers.
- When aligning over the intersection of two edges in mill setup, zeroing over the edge parallel to the *X* axis establishes the *Y* axis zero.
- Fixture offsets are memory-stored *machine coordinate positions* of individual PRZ locations—G54 through G59—called by the program, but the numbers are not in the program modal until canceled usually with a G53.
- Setting a primary cutting tool's tip at the PRZ is known as the master tool method of setting *Z* axis coordination.

Respond

1. Identify and briefly describe the controller screens needed for setup work.
2. True or false? When aligning over a set point hole, you cannot use work coordinates since they do not represent the PRZ as yet.
3. If Question 2 is true, then what position screen can you use?
4. List the main steps to coordinating a mill setup, in their order of occurrence. Use one cutting tool using the master tool method.
5. List the main steps to coordinate a lathe setup using the master tool method and one cutting tool.
6. What do machine coordinates tell the setup person?
7. In the planning you are told that a mill job has a fixture set point that is *X*-3.00 from PRZ and *Y*-2.00 in. What is the setup procedure after bolting and aligning the fixture to the mill table?
8. True or false? The touch-off method is one way to coordinate *X* and *Z* tool position relative to the PRZ, for turning setups. If false, what would make it true?
9. Beyond eliminating feel inaccuracies, what advantage does turning a test cylinder bring to the touch-off method?

Critical Thinking Question

10. Using the master tool method of setting length offsets, suppose the master tool T01 breaks and is replaced by a new end mill. What can the operator do to get the mill setup running again (not a program change)?

    Facts: The replacement end mill is 0.0377 in. longer.
    The program uses three cutting tools: T02, an end mill, and T03 a drill.
    The master tool had no length offset at the time it broke.

In Your Lab—CNC Setup Problem—The Mystery Block Challenge

Here's a test to perform on your CNC mill that verifies understanding. The objective is to align and coordinate a ground parallel on a CNC mill. But there's a special twist to the game.

Instructions

After coordination is complete, using only the indicator probe and axis position registers, determine the length of the block to within a half thousandth! Your instructor may have a set of blocks prepared for this activity. However, any parallel of a size not known to you will be okay (Fig. 23-13).

Mystery block challenge

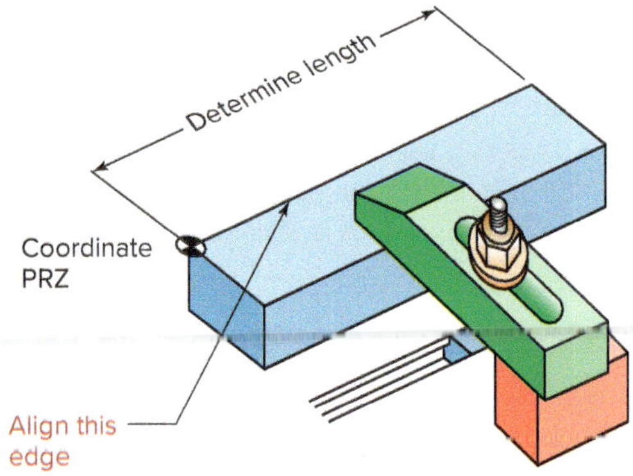

Figure 23-13 Align and coordinate the block using DTI. Then determine its length using only the indicator and the work coordinates.

> **KEYPOINT**
>
> **Caution!**
>
> If this is your first manual movement of this machine, be sure to reduce all axis override functions to 25 percent or less and lock the spindle in neutral for this activity.

Step by Step

After lightly clamping one end, align the parallel block using a test indicator. Aim for an end-to-end accuracy of 0.0005 or less.

Use an edge-finder first, then DTI to establish the PRZ following the previous edge intersection procedure.

Load the PRZ position into the work coordinates.

Critical Thinking Now, see if you can think through how to use the axis position in work coords and the DTI to measure the length of the block.

Upon completion, check your result with a micrometer or against the instructor's answer sheet. You should be within 0.0005 in. or closer.

Unit 23-2 Multiple Cutting Tools—Determining and Setting Offsets

Introduction: Now we investigate setting up several cutting tools in a tool-changing machining center and in a turning center with an indexing turret. We're going to determine and enter two kinds of offsets for mills.

Tool radius for keeping the cutting tool tangent to the geometry—assuming the program is compensated

Tool length differences—for both turning tools and milling

For lathes we add the tool shape and direction vector such that the tool approaches the work correctly.

TERMS TOOLBOX

Mach coords (abbreviation) Shop lingo for machine coordinates with reference to machine home.

Material shift offset Allows a small shift in the PRZ to accommodate abnormal workpieces.

Work coords (abbreviation) Shop lingo for work coordinates.

23.2.1 Determining Tool Offsets

Tool Radius

Tool radius offset is a simple issue for both lathe and mill. All that's required is to know the kind of tool path (cutter centerline or part path), then enter the right offset in tool memory.

Radius Values Entered in Memory

For mill tool path compensation, tool radius is the standard protocol over tool diameter. The reason is that changing radius distance away or toward the geometry simplifies the effect. The change in radius is the amount the cutter moves closer or farther away from the geometry—therefore, it equals the difference in size change. However, I have seen controls that used tool diameter, then divided that value in half for compensation. Lathe tool nose offsets are always expressed in radius values.

Why Change Radius Comp for Profile Milling or Turning?

Radius offsets might be adjusted by the operator to compensate for the following:

1. **Tool Wear**

   Wear occurs up to a limit; then a new tool must be mounted.

2. **Machine Warmup**

   Heat and bearing characteristics can affect size.

3. **Tool Replacement from Failure**

   While qualified insert tools index within a guaranteed tolerance from one insert to the next, they all have some variation. Sometimes it's a good idea to back the tool path out just a bit (0.0003 in. or so) when putting a new insert in the holder.

4. **Change a Dimension**

   Remember, this will change all dimensions made by that cutter.

5. **Adjust for Deflection**

   Also adjust for other factors such as vibration.

Which Kind of Path?

Even with just one or two cutting tools in the setup, it would be unusual for an industrial program to be written without tool compensation. However, a comped program is even more certain if multiple cutting tools are used. But even though the chances are near 100 percent that it is compensated, it's a good idea to verify it. Look for a check block on the plan:

[ ] *Tool Comp*

Or Look for a Note in the Planning

If the path type isn't indicated in those ways, scan the program for G41/42 and comp codes in the first few lines after the first tool change.

Part or Cutter Path Program?

Next, verify whether the program is a *cutter centerline* version, in which case the cutter radius offset will be the small difference between the recommended cutter envisioned by the programmer and the actual cutter put in the machine:

[X] *Cutter Centerline Path* [ ] *Part Path*

For example, the planning states T01 = $\frac{1}{2}$"End Mill and the program is written for cutter centerline, as checked earlier. You determine that the actual cutter is 0.4950 in. diameter. The radius offset entered will be a -0.0025, shifting the slightly smaller cutter's path closer to the part geometry. The offset value is the difference in tool radii.

SHOPTALK

Where parts are expensive, many shops today are choosing cutter centerline programs and use ONLY full-size mill cutters. They feel that the cost of always using a new cutter outweighs the possible problem of entering the wrong offset.

TRADE TIP

In using Mastercam, the program supplied with this text, you will be given a choice of either *[ ] Comp in Computer* or *[ ] Comp in Program* when writing a program.

Choosing "**comp in computer**" generates a cutter centerline program based on a cutter radius or diameter selected by the programmer. This path requires small negative offsets for undersize cutters.

Choosing "**comp in program**" generates a part path program requiring large positive offsets equal to the real tool's radius.

Suppose the program is a part path:

[ ] *Cutter Centerline Path* [X] *Part Path*

Then the tool radius offset value will be the whole cutter radius that's required to keep it tangent to the geometry. Using the same 0.495-in. end mill, the offset will be plus 0.2475 in.

SHOPTALK

To avoid confusion, machine shops normally do not mix program types regarding tool compensation. All programs will be either part path or tool centerline—*but that's when they write their own programs.* However, if it's a shop that contracts outside work, many jobs are accompanied with a drawing and a program! In these shops, be on the watch for tool path differences.

23.2.2 Tool Length Compensation—Mills

Using three cutting tools in a milling machine, we'll compare the two methods of setting the length offsets: *master tool* and *master position.* Each protocol has its own particular advantage and disadvantage. Most shops use master position. With the tool just touching *Z* zero, press **Tool Offset** on Haas controls

KEYPOINT

The choice of length offset method is a setup decision only—not a planning item. The difference lies in how you set the values in tool memory.

Possible Algebraic Controller Difference

We assume in these examples that the control combines the commanded axis move with the length offset by the rules of algebra (algebraically). That is, it adds plus value offsets and subtracts minus values from the command distance in the program. That's the common usage, but it's possible that a given control may combine the offset opposite from this. Read the programming manual to be certain or ask someone.

Mill Tool Length Offsets—Master Tool

In Fig. 23-14, the master just touches the work surface at *Z*0.0 (PRZ zero). Then, T02 and T03 length offsets represent the physical length difference between them and the master.

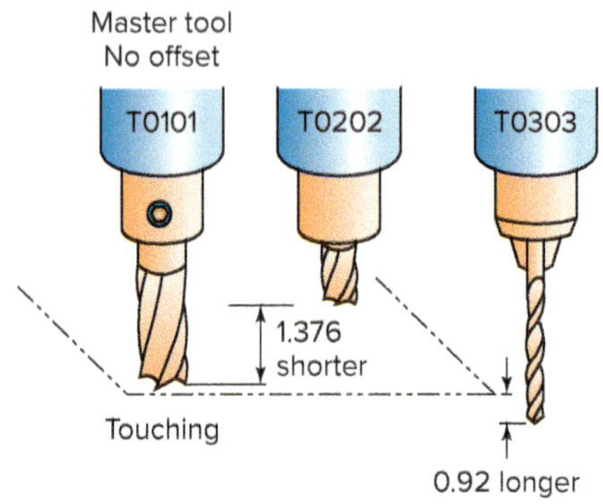

Figure 23-14 Under the master tool method, every other tool receives an offset equal to the difference in length except the master tool. But what if you break T01? See Fig. 23-15.

Finding and Storing Length Offsets

Under the master tool method, first the master is set to read *Z*0.0 when it's touching the reference surface.

KEYPOINT

Caution!

Before setting up the master tool's *Z* zero position, go to tool memory and be sure it has no length offset.

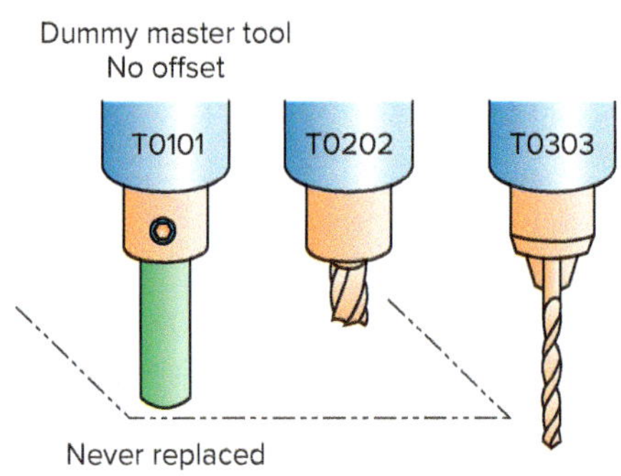

Figure 23-15 One solution is to use a dummy master that never gets broken or worn.

Then each successive tool is brought to the touch point (or shim or probe), and its *Z* axis value read in *work coordinates.*

Critical Question

If after setting the *Z*0.0 for T01, with no length offset in effect, then T02 is loaded into the spindle and touched to the work surface in Fig. 23-14, what *Z* axis position will be displayed on the *work coordinate screen*?

Answer

Z-01.3760–The shorter tool had to move 1.376 beyond *Z* zero to touch the work.

TRY IT

Got the Idea? Here are two critical questions (see Fig. 23-14):

A. With T01 just touching the reference surface and the *Z* axis register zeroed, with no offset in effect, the work coordinates read *Z*00.000. Then the following command line is read.

G43 H2 (No tool change or axis motion—just the correct offset for T02 is read and used.)

What will the work coordinates read? Plus or minus 1.376 in.?

B. Okay—Tool 2 is loaded into the spindle, then brought down to just touch the work surface. No change in offsets, and *H*2 is still in effect. What do the work coordinates then display?

ANSWERS

A. *Z*01.3760. Why? Because the *Z* axis quill position will be 1.376 above where T02 would reach zero.

B. *Z*00.000. The tool is touching and its length offset is in effect, so it's now at *Z* zero.

TRADE TIP

Autoentry of Length Offsets Some controls feature an automatic method for entry of tool length offset. For example, tool 2 is touched to the work surface, and the cursor placed in the correct data field in tool memory; then the *SET* button is touched (a soft key). The control automatically calculates the difference and enters it *with the reversed sign value*. Read the manual to see if this math-saving feature is available.

Controller Differences—Example Commands

In addition to the *H* word as on Haas controls, other controls use the tool ID line to denote the length offset. The following command lines would compensate tool 2 to use offset 2:

T0202 M6 (Load tool 02 using offset line 02 from tool memory.)

G43 G0 *Z*.5 (Now under length comp 02, rapid this tool to a position 0.5 above the *Z* axis PRZ. For Fig. 23-14, this tool will move 1.376 in. more than it would without compensation.)

23.2.3 Master Position Method for Length Offsets

Here's the method that's more common in industry for multiple tools. In this procedure, all tools are assigned an offset from the beginning; *there is no master.* Offset values are each tool's distance from machine home to the datum reference (Fig. 23-16).

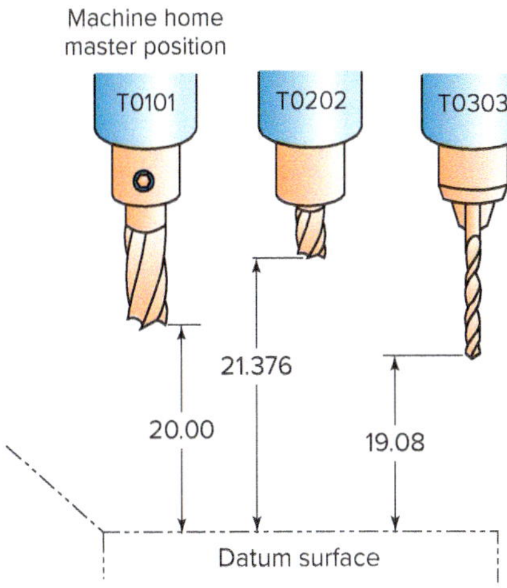

Figure 23-16 Compare the master position method to the master tool. Every tool receives a positive offset from machine home to the datum surface.

KEYPOINT

Under the master position (MP) protocol each offset represents every tool's distance from machine home to the datum surface—*Z* zero. They will be large numbers.

Using the same tools from the previous example, note that the offset values still represent the physical length differences, but this time they are large numbers.

Using the MP setup method, when any tool is commanded to a *Z* position, the control must always *add the offset* to the movement. An uncompensated *Z* axis move would actually be no movement at all. No motion would occur until the length offset is added on. Using this protocol, length offsets in tool memory would now look like this:

Offset Number	Length Offset	Radius Offset	Drum Position
01	20.000		
02	21.376		
03	19.080		

The length differences between the three tools are still there, but they represent their distances down from machine home (master position) to the datum surface. The offset then is the positive correction placed upon the real tools such that each will drop to the work surface (Z = 0.0) by their offset distance when commanded G43 *HXX* G0 *Z*0.0.

Another way to envision master position is to compare it to a master tool setup. In a master position setup, envision that there is an imaginary dummy tool, but it's so long that it just fits between the work surface zero with the spindle retracted to machine home. That imaginary tool is at Z zero when the spindle is all the way retracted at machine home. Because the real tools are all much shorter, they all require positive length offsets. So in reality, that very long master is actually replaced by a master position (Fig. 23-17).

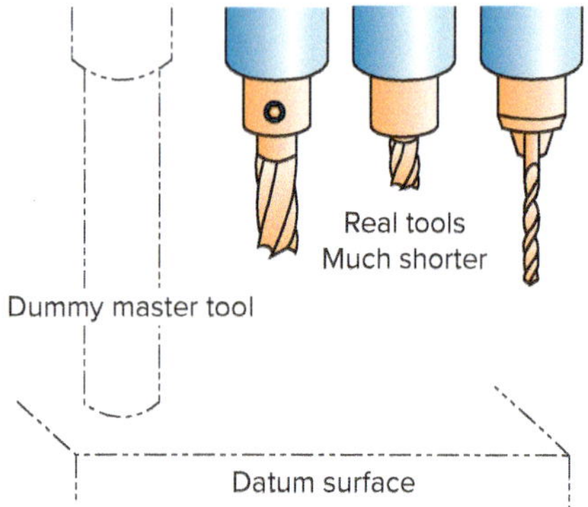

Figure 23-17 Imagine a very long dummy tool—all other tools would need a positive offset to equal its length.

Confused? Don't worry. Machine experience will help clear it up. And remember, no shop mixes the two methods–and they usually choose master position since it simplifies replacing cutting tools.

23.2.4 Lathe Multiple Tool Setups

Like mills, lathe offsets and PRZ setup can be made using master tool or master position. Differing from mills where only length varies, between different lathe cutting tools there are both X and Z axis differences.

Both *X* and *Z* Differences

If the turret is rotated with no axis motion, the turning tool, the facing tool, and the drill all index into a different position relative to the workpiece (Fig. 23-18). Those X and Z differences must be stored as tool offsets–either as master tool or master position numbers. And again this is a setup decision only–either method will work.

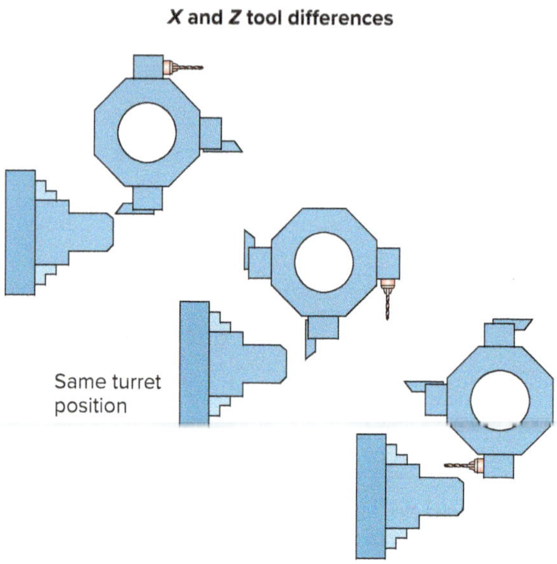

Figure 23-18 Three cutting tools show X axis and Z location differences when the turret is indexed but not moved relative to the spindle.

X Axis Offsets and PRZ

We'll start with the X axis since it's a fairly simple issue for most setups with multiple tools. When cutting tools such as drills, reamers, and taps are used, X axis PRZ and offsets are normally based on them. They are tools that must be positioned on the spindle centerline to cut correctly–called centering tools.

X Axis Constant—Turret Position on-Center

As shown in Fig. 23-19, centering tools are mounted in bored mounting blocks. Each block places the tool the same X distance from the turret center. Therefore, the X axis machine coordinate position of the turret, when it centers these tools, becomes a constant for X setup.

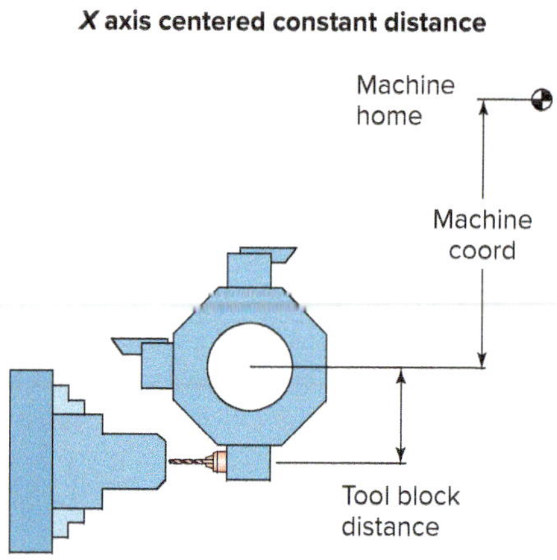

Figure 23-19 The X constant is the sum of two diameters.

How that crucial machine coordinate position is used depends on whether it's a master tool or master position method, regarding *X* axis offsets.

Master Tool Normally all center cutting tools have no *X* axis offset. They are *all* the master for *X* position. Replacing a centering tool has no effect on its *X* offset since the replacement goes right back on-center.

If there are no center cutting tools, you have two options: use a dummy centering tool or select some other master tool.

Master Position Every cutting tool is offset from M/H to the *X* PRZ. It will be the sum of the turret and tool distances shown in Fig. 23-19.

KEYPOINT

Keep in mind for CNC lathe work, *X* axis numbers are diameters, even machine home. Think of machine home as the largest diameter the turret can reach.

Finding the Centered Position If you do not know the *X*-centered position, it can be found by placing any cylindrical object in a center cutting tool block (a 1-in. test pin in the drawing). Then using the touch method, bring its rim into tangency with a turned test cylinder of known diameter (Fig. 23-20).

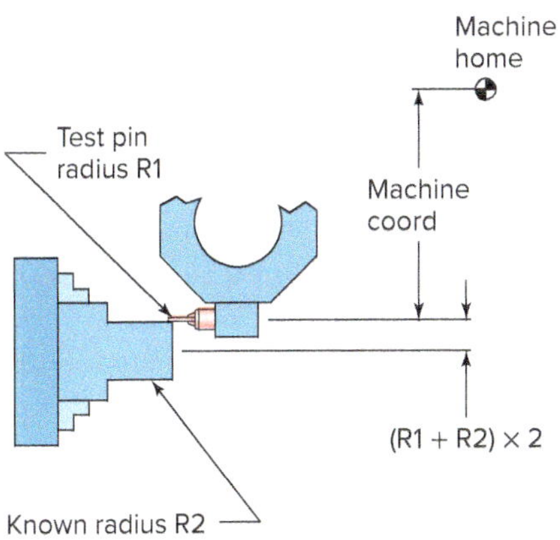

Figure 23-20 The constant *X* position is the sum of three numbers.

KEYPOINT

At that tangency location, the turret is at the sum of the two *diameters* away from the center of the spindle, or as shown (R1 + R2) * 2, double the sum of the two radii.

Critical Question

Suppose in Fig. 23-20 that the *machine coordinate* is -18.303 in. with the test pin touching the test cylinder. The test pin is 1.0 in. diameter, and the turned cylinder is 3.500 in. diameter. What is the *X* center constant on that CNC lathe?

Answer

It will be 18.303 + 1.000 + 3.500 = 22.803 in. The sign of the 18.303 was reversed to make it a positive distance.

That machine coordinate of the centered turret position never changes. It's usually hardwired into the machine home routine. That *X*-centered machine coordinate will be recorded in one of these two places:

In the operator's manual

In the tie-in book

TRADE TIP

I've even seen it steel stamped on a noncritical surface of the lathe—but that's not a recommended practice because with time and crashes, it can and does change slightly.

Z Axis Offsets

Compensation for tool differences with respect to the *Z* axis PRZ is very similar to mill setup and depends on whether it's a master tool or master position method.

Z Comp Master Tool

For a master tool setup, after setting the *Z* axis zero in work coordinates each subsequent cutting tool is brought up until touching the *Z* reference (usually the work face), then the **work coordinate (work coords)** position is noted. That's the *Z* axis difference. The sign value is reversed to become the saved offset value for that tool (assumes algebraic combining). The offsets will be either positive or negative depending on whether the tool in question is longer or shorter compared to the master.

Z Comp Master Position

To offset tools in the *Z* direction, each is touched to the reference surface and its **machine coordinate (mach coords)** position relative to machine home is noted and saved as its offset—always positive values.

Material Shift Offsets

Most modern CNC lathes (and mills, too) offer an offset screen that allows the operator to shift the PRZ a small amount but not lose the original PRZ location to provide a **material shift offset**. This handy operator control is designed to deal with variations in raw stock, castings, and forgings, or when previous machining leaves the work close to not having enough excess on one side, to clean up, but there is plenty on the other side. Sometimes the stock is slightly different from the norm, a little longer or shorter or maybe warped, for example. The part can still be made but only by shifting the PRZ just a bit to equalize excess or keep the machining within the odd-shaped part. It's probably not going to be a permanent shift but applied to a limited number of parts, then canceled.

UNIT 23-2 REVIEW

Replay the Key Points

For milling:

- *Master tool* Using the MT protocol, the length offset value of a given tool is found on the *work coordinate* screen, as the physical difference in the master at Z zero and the current tool touching the work surface (or feeler or probe).
- *Master position* The length offset under master position for each tool is the *machine coordinate* Z position when the tool touches the datum surface—but with the sign reversed to positive. Remember that the offset is determined from the machine coordinate screen.

For turning:

- *Master tool* Normally all center cutting tools have no X axis offset. They are *all* the master for X position. Breaking one has no effect on X offsets since the replacement goes right back on-center.
- *Master position* Every center cutting tool receives the same X axis offset—to position it from machine home to X center.

Respond

The following critical thinking problems involve setup of PRZ and offsets:

1. A rectangular workpiece that is faced on its top and a pocket milled into its top surface is measured. The pocket is the right depth but the part is too tall by 0.005 in. In this program, a single end mill faces the part and cuts the pocket. The program is written with tool length compensation. Is the problem with the program or the setup? How can the problem be corrected? (There are two methods.)
2. See Fig. 23-21. After machining and measuring the profile of this object, the machinist finds the two dimensions to be 2.7536 and 1.6286 in. The program is a compensated cutter path that was written for a 0.500-in.-diameter cutter. What offset correction is needed?

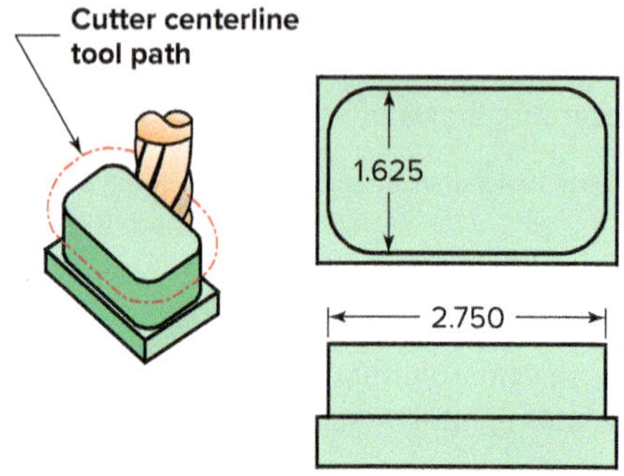

Figure 23-21 Compute the offset change needed.

   A. What will the new offset be for the cutter path program?

   B. What would the offset need to be if this was a part path program and this part was produced with a 0.500-in.-diameter cutter offset?

3. Is this statement true or false? In a setup as in Problem 1, one end mill with the program written with tool compensation, the control does not require a length offset. If it is false, what will make it true?
4. Describe the coordination of the PRZ on a vertical mill CNC machine, when the alignment is at the intersection of two datum edges on the work or tooling. Assume the rough position has already been performed.
5. When performing the PRZ coordination of Problem 4, the controller will be on the tool memory page. Is this statement true or false? If it is false, what will make it true?
6. After turning, an object shows two diameters to be oversize by 0.05 mm while two are exactly right. All Z axis dimensions are too short by 0.1 mm. What must be done to correct this single tool setup and part run? The program is *not written with tool comp offsets.*
7. Describe the master tool method of setting up a lathe or mill.
8. Describe the master position method of setting up a lathe or mill.
9. How are material shift offsets used?
10. Which tool compensation method is less complex when replacing work or broken tools? Explain.

Unit 23-3 Proving a Program—Tryout

Introduction: Okay, the setup is ready to go. The objective is to get the program proven and working as soon as possible, but safely. There are 13 testing methods presented here in a hierarchy, each exposing a bit more about the program. We begin the process tests that present zero risk, then work toward those that move machine axes without cutting material so there's a bit of risk. Finally, the ultimate tests are those that make chips then real parts. There are three different ways to cut metal, with lessening degrees of test control. Which of these 13 methods are selected depends on the following:

The number and cost of the parts to be run. When making many parts each with a relatively low value, the first needn't be perfect. Called a *tryout part*, it can be measured and corrections made to produce exact dimensions on the next part. However, when making only one or two parts from expensive forgings, for example, more testing is required before cutting metal.

The history of the program (and programmer). Has it been run before? If so, you might skip most steps and do a single step run until the cutter seems to be going okay, then turn to full autorun and let it go but still keep a close eye on the progress.

The nature of the object. A complex part will predictability need more analysis.

Your experience level. First time? Use every trick available!

TERMS TOOLBOX

Associativity Linked elements in a program. Each is associated with the others. Highlighting or changing one changes all linked windows.

Axis disable A tryout method whereby the operator can lock an axis but the control displays as though it is moving.

Dry run Any one of several program testing methods whereby the machine does not make chips but goes through all or part of a full autocycle.

Optional stop (opt stop) A controller function whereby certain codes may be recognized or ignored within the program.

Single-step tryout Evaluating a program by fully running it one code at a time. The last test before a full autorun.

Tryout material Soft or inexpensive material used to machine a shape without the danger of wasting real material or breaking cutters.

23.3.1 Program Tests

It would be unusual that all these options would ever be used in any program test. However, each is a tool that should be understood.

Safest Testing

Test 1 Reading Programs

Not an effective method for testing an entire program. Trying to decipher the data has little value especially when a typical CAM-generated program is many lines long, or the shape is complex. Program scanning is mentioned here for the rare situations when there is no graphic capability for evaluation.

Tests 2, 3, and 4 Graphic Evaluations

One or more of these tests are a must-do in modern machining. They are found in the shop in three varieties:

Test 2	CAM tool path evaluation
Test 3	Specific tool path utility software (Vericut in this text—lessons coming up)
Test 4	CNC control graphics

CAM tool path evaluation, also called a *program plot or backplot, is a graphic test of the toolpath—not the code* (Fig. 23-22). Most CAM software offers the ability to display the work material and the cutting tools, too, and to view the results progressively, from varying angles to see cut depths and other programmed movements not visible from a simple top view. Figures 23-22 and 23-23 show two views of the same toolpath. One is a wireframe of the toolpath with blue lines showing G01 moves, while G00 moves are in yellow.

Depending on user selections, the backplot can display the cutting tool (Fig. 23-23) and the work object and view it from many different perspectives during and after the screen-machining is complete. Where there is a need to zoom in close, details of the machining can be displayed.

But a graphic backplot cannot tell the user if dimensions are correct or if the program has danger points where the cutter might hit bolts, vise jaws, or other objects in the setup unless they are included in the material designation. Testing for these things is yet to come.

TRY IT

The best way to learn this test option is to do one. We'll be using Mastercam and Fusion 360, but if your class has different software it will work similarly. In Chapter 25 you will use trial versions of either Fusion 360 or Mastercam to write a toolpath. The following is the simple way to do a backplot in Mastercam but on completion experiment with different views and display options.

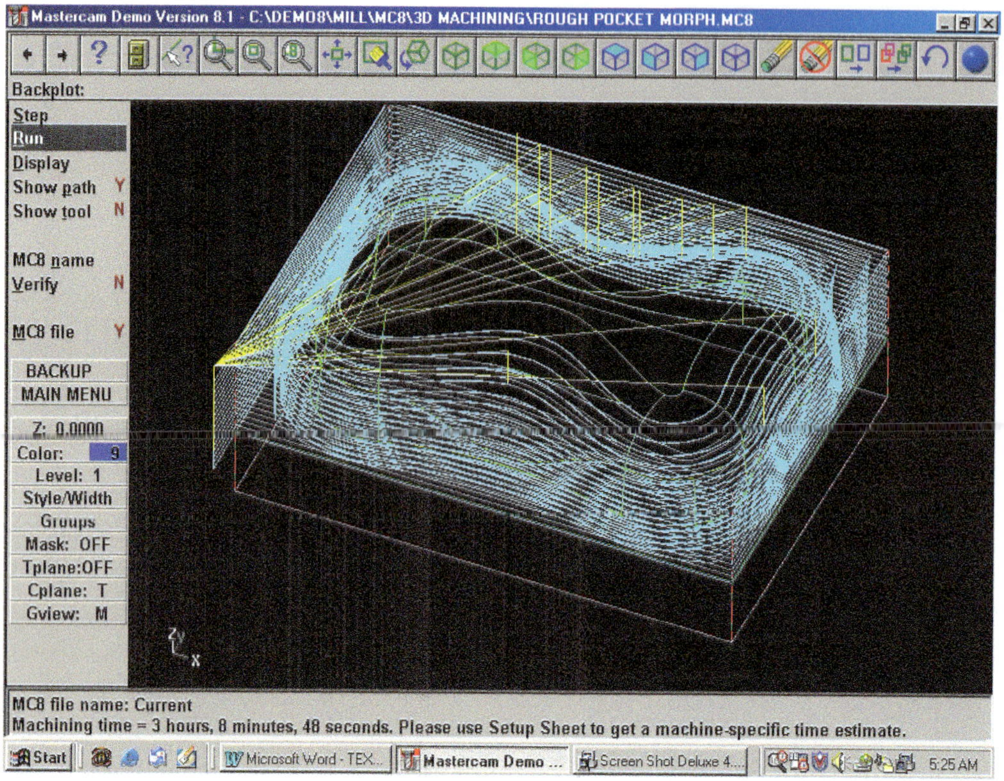

Figure 23-22 A sample Mastercam backplot shows a wireframe tool path in blue and rapid moves in yellow.

Source: CNC Software/Mastercam Software

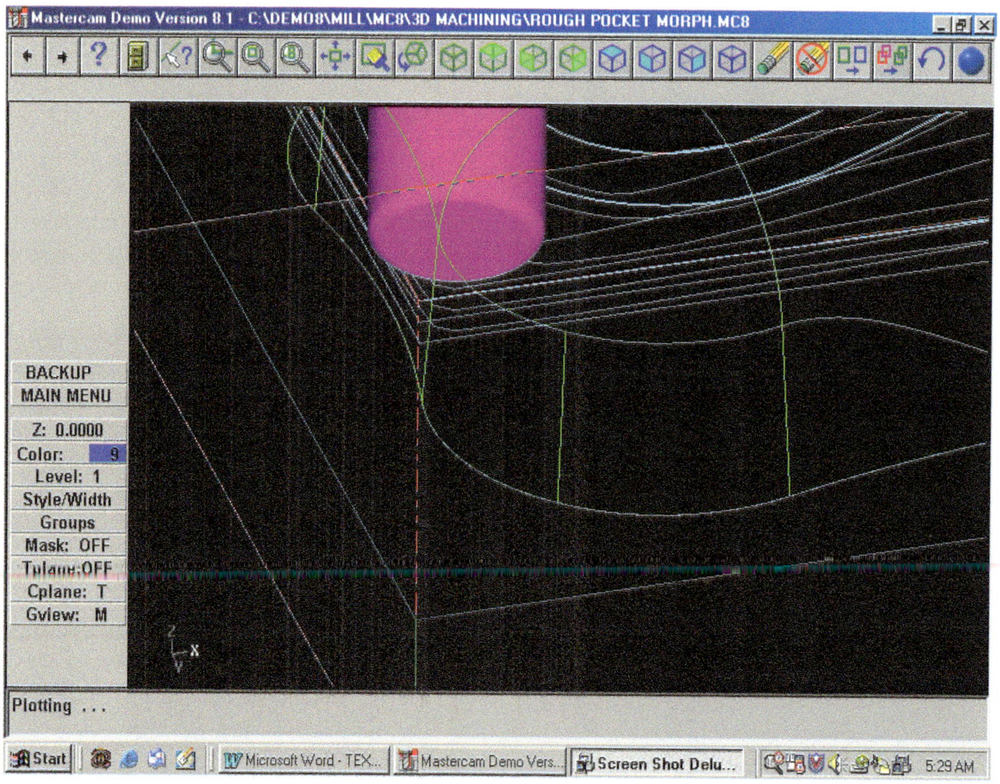

Figure 23-23 Testing further, the view has been zoomed in where a problem is suspected, with the cutter displayed.

Source: CNC Software/Mastercam Software

Mastercam Toolpath Verification

To test run a toolpath after starting MC-X:

[ ] **Files—Open (any demo** file)

[ ] **Operations Manager Tab (Fig. 23-24 left side)**

[ ] **Toolpath tab**

[ ] **Verify or [ ] Backplot**

Older Mastercam Backplot Step by Step

Start Mastercam Mill

[ ] **Files,**

Get,

2-D or 3-D

Click any file—The part geometry will appear on-screen.

[ ] **Main Menu**

[ ] **NC Utils** (Utility tools for programming and testing)

Backplot,

Run

Blue lines are cutter paths; yellow are rapid moves.

or Step

Click to read each command block.

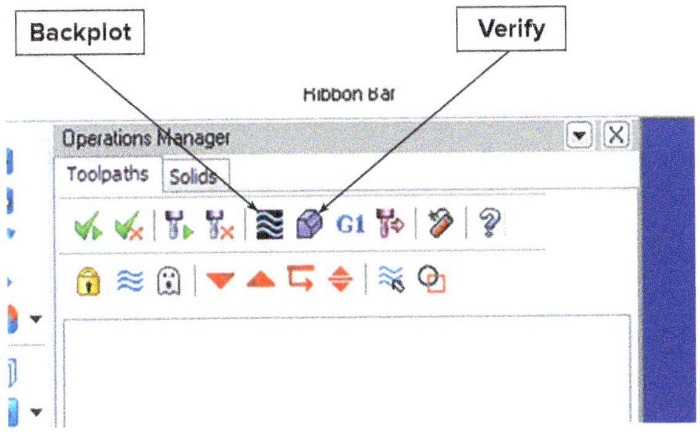

Figure 23-24

Source: Courtesy Mastercam Software

Try Different Options Now change the various backplot options to see different view perspectives, and different cutting tool displays.

In Fig. 23-23, I made four changes. First, using G-View-Dynamic (Get View at the bottom of the control window), the geometry was mouse rotated to show the corner of the part better, then zoomed in to a specific feature. Next, the tool display was set to shaded. Then I single-stepped through until the cutting tool reached the windowed area in question.

Change tool display.

Single-step to see one block at a time.

Change perspectives and zoom in.

Test 5 Toolpath Utility Software

The ultimate graphic test, these programs test the program code and simulate most of the real conditions, too. Vericut is a utility program by CG Tech that simulates all kinds of machines and work-holding devices. It can compare the finished object to the original solid model to see where undercuts or material is left to machine. You'll be using this program after writing your toolpath in Chapter 25.

> **TRADE TIP**
>
> No modern CNC shop should be without some form of tool path and surface analysis system. It can save time and improve both safety and quality and avoid potential scrap work. Training coming up in Chapter 26.

Similar to CAM, Vericut can show a wireframe of the cutter path or a shaded solid model of the product (Fig. 23-25), from any view perspective. Zooming in to see small details during the cyber cut, or after, makes it possible to analyze small details such as mismatches and uncut transitions. It even displays tool marks. It can detect events that are crashes—for example, when in rapid travel mode, the cutter touches the surface.

Total Screen Associativity and Surface Analysis

For evaluating, debugging, and editing programs, a most useful feature is forward/reverse **associativity**. This means that all windows of the program are linked. Cutting tools, program code lines, and cut elements in each window can be highlighted to show their part in the program.

For example, as you click on a line of code, the cut element that it makes is highlighted, plus the cutting tool that made it will be highlighted. Or if you click on a cut area in the graphic window, or on a single cut line, then the line(s) of code that made it will be highlighted as will the tool. Finally, if a tool is highlighted, the playback shows those cuts made by it along with the command codes. Highlighting one, in any window, instantly connects it to the other two windows. That makes program investigation and editing a snap, especially when it's many lines long as are most CAM programs.

Test 6 CNC Controller Tool Path Graphics

Once the program has been downloaded into the control, it's a good idea to give the program a dry run test with all axes locked out. This test can determine if the program works in the particular control and that the toolpath is that which you expected—not a wrong program.

Machine Motion Tests

That's the end of graphic methods, so it's time to use the CNC machine to further test the program. There are four levels to which one might go.

A Dry Run

A **dry run** is an actual run of the program with one or more machine functions disabled—no chips will be made. There are several different ways to perform a dry run.

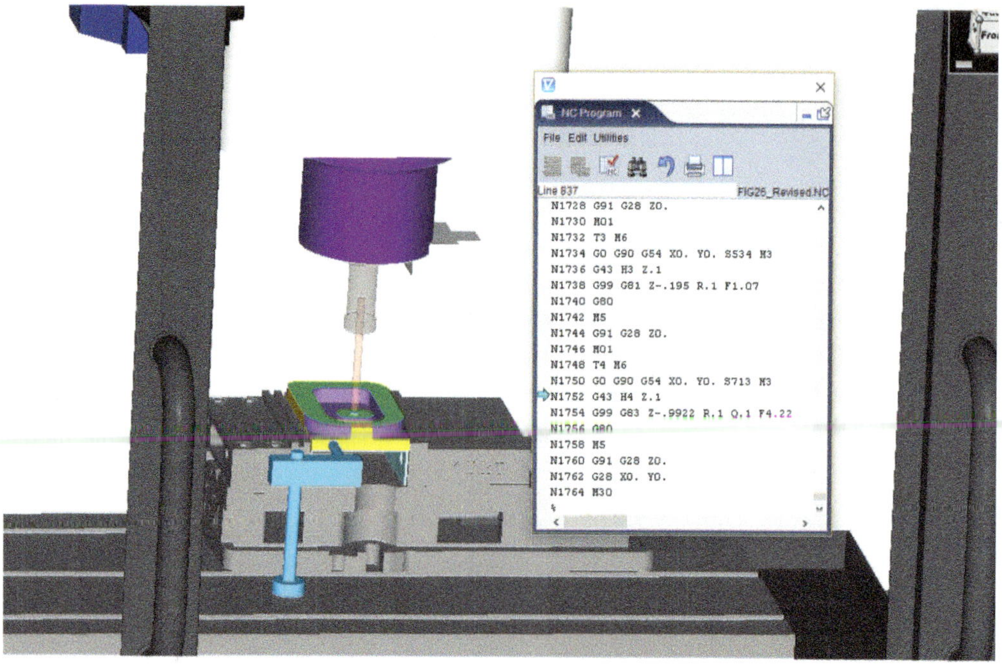

Figure 23-25 Evaluation software simulates run conditions and program code.
Courtesy Mike Fitzpatrick

Test 7 Axis Disable

If the control offers an **axis disable**, you can first lock up all axes, then run through the program in dry run status. This test shows that tool changes are okay and that the program syntax is acceptable. If the control has screen graphics it will also show the tool path and the real times required for each cut (Fig. 23-26). During the run, the work coordinates will change as if the machine were moving, and all modes and active codes will also show on-screen. It's a rough outer test but it's quick and has zero risk.

Figure 23-26 Many CNC controls can display a tool path analysis from many different angles.
Courtesy Keith Smith/Shoreline Community College

Next, using a mill for the example, unlock *X* and *Y* axes, but withhold the *Z* up at machine home (or withhold *X* or *Z* if it is a lathe). Now the *X*-*Y* spindle motions can be analyzed for danger points where a rapid move might hit a clamp or the vise. You might move the *Z* axis a little farther toward the work, then lock it again, and dry run it again.

Another addition to the dry run is to allow tool changing during the run but only after a lockout test. Here, the tools index into position and out at the correct time in the program, but the axis slides remain locked. This feature may not be available on all machines.

Test 8 Cutting Tool Withheld

Using this method, all the axes and tool changer are allowed to move, thus all motion can be observed for danger points. A disabled tool can be set up in three different ways:

1. Remove the cutting tool from the machine.
2. Use offsets to force the cutting tool back a great distance. Don't forget how far back so it can be reversed for the real program run.
3. Better suited to multitool setups, shift the PRZ in the *Z* axis, up away from the work. Again don't forget the amount of shift so it can be returned afterward.

KEYPOINT

In each of the cutting tool withhold tests, there is some risk as axes will be moving.

Test 9 Missing Material

This is another form of dry run, but it has more risk since the cutting tools will be coming to their intended height with respect to the vise, chuck, or clamps even though the work material is removed for now. You can use this method in conjunction with the PRZ shift up away from the setup.

Test 10 Tryout Material

Here everything is enabled–cutting tools, tool changing, and all axis motion. However, the part is not the real material. Instead, a **tryout material** that is less expensive and machines more easily is used. This test creates the first-dimensional examination of the product and is usually performed because the work material is expensive or in scarce supply–for example, if the job requires machining three expensive titanium blocks and no more can be had! Here the first part must be right as well as the following two! A test material run always follows a series of dry runs.

Test material can be special wax designed just for a CNC tryout. It, too, isn't cheap, but the chips can be remelted and cast back into a block! See the Trade Tip.

Special, high-density foam that is also made for the purpose can be used for tryout parts (Fig. 23-27). Sometimes when nothing else is available, we use wood or even aluminum. Foam and wood, however, can be dust problems.

Figure 23-27 Tryout material is a safe way to obtain real part dimensions.
Photography by Prographics at Milwaukee Area Technical College/McGraw Hill

TRADE TIP

Reusing Wax Before a wax tryout, first clean the machine's chip bin so as to not mix metal chips with wax. But 100 percent cleanup isn't necessary if you follow the upcoming hint. After we make the cuts, the chips are picked up with a small battery-powered handheld vacuum that we reserve just for reclaiming wax. Then melt the reclaimed wax in a pan over a hot plate. The liquid is then poured through a tea strainer to remove stray metal chips, when it's recast in the mold. After several recycles the wax's properties begin to degrade, so a special conditioner is added to the liquid before recasting. It restores dimensional stability.

Test 11 Oversize Radius Offsets

This test only works if the program is tool radius compensated. Here, radius offsets are faked such that the cutter backs away from the work by a large amount; thus, little or even no contact is made on profile cuts.

KEYPOINT

Be sure to record the offset changes in a multitool setup to guarantee they are written back correctly when the test is complete.

Using radius offsets, one can machine the work into an oversized version, then measure it to help ascertain that the true offsets will produce a good part! Be aware, the next part after the offsets are restored may not be of correct size due to tool deflection differences in the light test cut and the full removal cut.

Test 12 Single Step

Okay–it's showtime! Here we make chips but not at the rate they would be made under autocycle. This test is universally used on all CNC controls with which I am familiar–old or new. It's a full run of the program with all offsets correctly entered. It will make a good part, one command at a time. But it can dull tools when allowed to dwell on the work too long.

For a **single-step tryout**, first select the single-step mode with the setup complete, program loaded, and work material ready. It's a complete setup and part run with only axis motion slowed. For this test have the distance-to-go (DTG) screen active. Then, with rapid travel turned down to 25 percent or lower, and the feed rate backed down, too, touch the >cycle start< button. The control will complete the given command block, then await another push. The problem is the spindle will dwell while the control awaits a second touch of the start button. Dulling the cutter against the work is a potential problem here.

KEYPOINT

The real trick to using this test is to analyze the program block under the cursor. Then watch the position registers and the distance-to-go screen. *Caution! Look and think before touching the start button.* Ask yourself, "Is the next move reasonable?" Don't assume anything–you are especially vulnerable to this mistake if you wrote the program yourself.

Using DTG During a single-step test, when it's questionable that a cut or rapid move is safe, allow most of the command to occur, then turn the override function to zero (rapid or feed depending on the command), stopping the axis move midway through the command. Next, stop the spindle and use a rule or some other means of measuring how far the tool can go safely. Then compare that to the DTG screen on the control. *Caution!* Do not put your hand near a spinning cutter or lathe chuck to make this test! Turn off the spindle.

In Run

Test 13 Optional Stops

To use this test, the codes to cause an optional stop must be built or edited into the program. Optional stops halt the program at specific points when the operator selects them, such as key measurements that might be performed on every twentieth part, for example. But between, when switched off, no halting is desired; therefore, the switch is left in the open position, and the **optstop** code is ignored (Fig. 23-28).

Figure 23-28 Optional stop 1 has been enabled by the machinist while the others remain disabled.
Courtesy Tyee Aircraft/Everett, WA

> **KEYPOINT**
>
> This test is used for ongoing part evaluation, but it can also be used for a first part run if the stop points are well thought-out.

Many CNCs indicate an optional stop by a yellow flashing light.

Notice in Fig. 23-28 that there are multiple switches. Various levels of intervention may be created; for example, one optional stop is for a problematic dimension or supercritical finish requiring a lot of operator attention, then a second for a larger range SPC sampling.

23.3.2 Comparing the Methods

The objective is to choose the best combination from each category to match the job. Here's a quick overview of the possibilities:

Prerun Tests

Program reading–for spot checks of error points

Graphic evaluation–a must-do if available

CAM PC graphic evaluation–if available (best choice)

Dry Runs

An axis disable

Tool or material withheld

Oversize offsets–no cutter contact

Test Runs

Tryout material

Fake offsets–reducing until final size achieved

Single-step testing

Halt Run

Optional stop–One or more custom intervention points. *Must be edited or written into the program.*

Single-step cycle–Switching to this mode the control will complete the block it is currently reading, then wait.

Feed/speed override–Correctly used only during a non-contact movement of the tool with work. Any other time, this is a low-level emergency-stop procedure.

UNIT 23-3 REVIEW

Replay the Key Points

- Tryout material can be used to make a test part that can be measured.
- Shifting the PRZ away from the material only works if the setup is a master tool PRZ.
- Using oversize radius offsets, one can machine the work into a blown-up version of the part, then measure it to help ascertain that the true offsets produce a good part!
- To best use a single-step test, analyze position and distance-to-go. Ask yourself, "Is the next move reasonable?"
- Optional stops are used for ongoing part evaluation but can also be used for a first part run if the stop points are well thought-out.

Respond

1. Which program tests have no risk?
2. Of the graphic program tests with no risk, which is the best for program testing? Why?

3. What are the main issues a backplot can reveal?
4. What cannot be determined by a backplot?
5. How would you go about moving all three axes of a vertical mill yet not make chips?
6. In order to change the PRZ in a multitool mill setup, to try out a program, what kind of setup must be made?

Critical Thinking

7. Your assignment is to CNC mill an expensive stainless steel forging (Qty 1) into a complex, close tolerance shape. Which program tests would you perform?
8. True or false? All graphic tests of programs have zero risk.
9. Describe a single-step test through a program. At what stage of testing would you perform it?
10. True or false? Optional stops are more useful for part checking during the run than for program testing.

Special Research

If you haven't already done so, it's time to install the Machining Technology disk and run the tutorial for the Metacut Utilities. Note the following:

- All aspects of the screen image are associated.
- Changing any line of program code changes the part shape when the screen is refreshed.
- You may mouse select any element or code line and they are strung together.
- As the program is run, each element and its direction are indicated by a chain snake.

CHAPTER 23 REVIEW

Unit 23-1

When making a setup on any machine, it's critical to consider the forces that will be brought to bear on the work. By now, you've experienced these forces in your lab and perhaps even seen an accident occur. (I hope not, but . . . !)

That's not news, and neither is the fact that when it's to be a CNC job, the safety issue is multiplied. All work must be held in the very best grip possible. Be certain that studs, bolts, teenuts, and clamps are in good condition and that they are grade 5–made for machine work. All bolts must be torqued down but not overstressed. All chucks must be tested and retested for grip. Leave nothing to chance before the program begins.

Unit 23-2

The chances are, you'll work in more than one shop in your career. When moving from one to the next, you'll experience differences in the way they do things with regard to setups, offsets, and PRZ setting. For example, in teaching in an overseas shop, the standard policy was that PRZs on lathes were set up using the same command codes we would fixture offsets on a mill. Their reason was they felt it was consistent with the way the mills were set up.

In particular, you'll find people who insist that either master tool or master position is the *only way* to set up a PRZ. It depends on the way they were taught originally. The savvy craftsperson learns to adapt through understanding the possibilities. You'll also find combinations of the preceding three methods: master tool, master position, and fixture offsets. If you understand the differences between them, then adapting will be easy. That was the theme of Units 23-1 and 23-2.

Unit 23-3

There are dozens of combinations of the 13 possible testing methods. I wouldn't begin a program run without at least 2 tests no matter how proven it was in the past. Beyond training situations, you'll always be expected to get the machine up and running as soon as possible. "Time truly is money." However, the time cost of an accident is far greater than the testing beforehand. Program testing, especially at the beginner level, is a time investment that pays dividends.

OVERALL CNC REVIEW

We're at the end of the CNC-specific chapters. Next we take a look at CAD Design Using SolidWorks, then CAM programming, then other advanced technologies. As such, the following problems and questions are a final review of your understanding of the whole subject, Chapters 16 through 23: foundation knowledge, monitoring production, editing, operating CNC machines, programming, and machine setup and testing. Each question is keyed to the chapter from which it was taken. Restudy might be indicated if you miss the target on two or more from any given chapter.

1. Based on Fig. 23-29, identify the five axes of a CNC vertical mill and the sign value for each (plus or minus). (17)
2. If the spindle of a CNC turning center is programmable–that is, it can make controlled angular movements as would an indexing head–it is identified as the axis. (17)

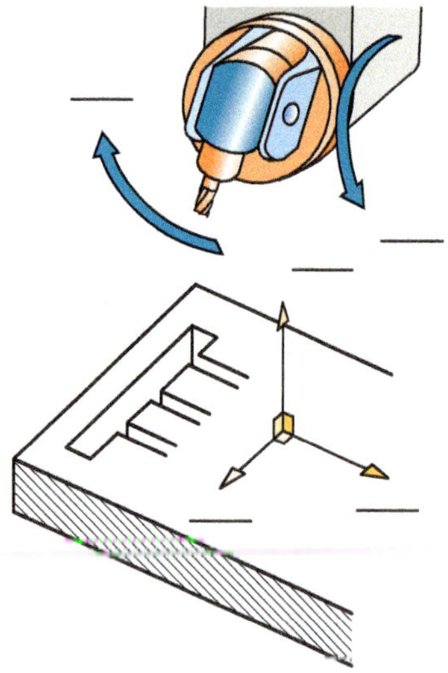

Figure 23-29 Identify these CNC axes.

2. Is this statement true or false? According to the *rule of thumb,* by pointing the thumb of your right hand along the positive axis of rotation, your fingers will curl in the positive direction of rotary motion. If it is false, what will make it true? (17)

3. In handwriting the program coordinates for Fig. 23-30, what would be an easy method of avoiding calculations for points *E, F,* and *G*? (23)

4. In programming Fig. 23-30, the correct answer was to switch to incremental coordinates at point *D*. What EIA code made that change? (22)

5. The modal code required in Question 5 for incremental values will remain in effect until. (22)

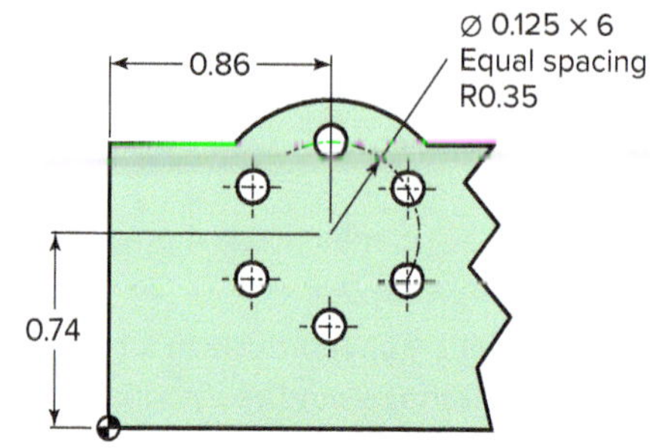

Figure 23-30 Review Problem 4. How can these points be identified?

CRITICAL THINKING

7. In handwriting a program for Fig. 23-31, the bolt circle holes must be drilled. Name at least three possible methods of writing this portion of the program other than graphic/CAM. What is the fastest of all three? (17, 23)

8. Is this statement true or false? Cartesian coordinates that refer to the PRZ for their values are known as absolute Cartesian coordinates. If it is false, what will make it true? (17)

9. Name in their order of importance the two guideline rules for selecting the PRZ. (17)

10. Where should the PRZ be chosen for the mill job shown in Fig. 23-32? (17)

11. A hemisphere must be machined on the top of an odd-shaped casting. The shop has a CNC lathe and 2-D CNC mill; describe at least two simple ways in which it might be machined. (17)

12. Identify and describe three electrical CNC drive components. (18)

13. What is the function of a ball-screw in a CNC machine? (18)

14. Is this statement true or false? DNC can be accomplished in bulk or drip feed modes from the CNC control to the host computer. If it is false, what will make it true? (18)

15. Some CNC machines require a *home cycle* when initially turned on, while others do not–why? (19)

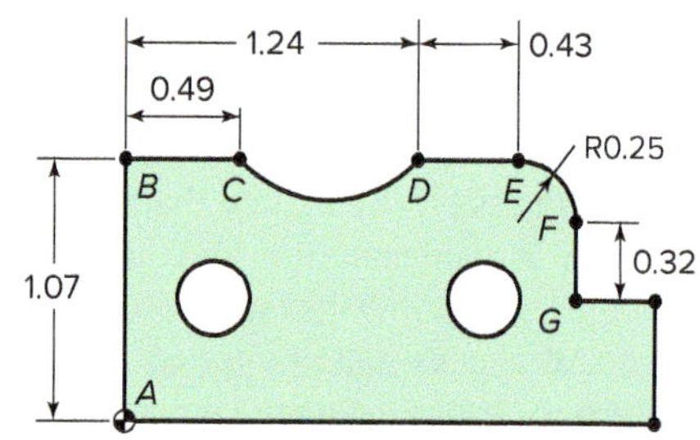

Figure 23-31 Describe several ways to write the program code without CAM software.

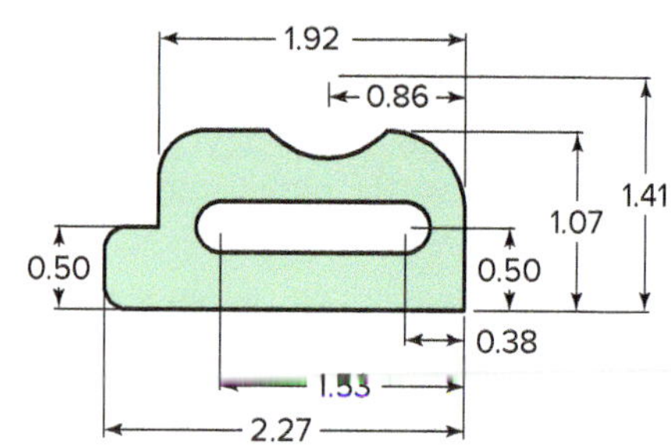

Figure 23-32 Problem 10. Where should the PRZ be located?

16. There are four methods of *halting* a CNC autocycle during a program with no consequence to the run other than the tool might be idling upon the work. With each method the autocycle may be restarted with a simple action. Name them and the restart action required. (19, 24)

17. During a setup, you wish to position the tool turret of a turning center at an exact position within the work envelope with respect to machine home zero. Name three operator methods of accomplishing this move. (19)

18. Name and describe three position screens commonly found on a CNC control. (19)

19. To correctly use the MPG, one must place the control in mode and use two other operator controls. Name and describe them. (19)

20. Label the CNC controls A through H on the PVC controller shown in Fig. 23-33, and describe the function of each. (19)

21. What would happen during an autocycle if you push or turn (19)

A. rapid override to *zero*?

B. feed rate override to zero?

C. spindle speed to 125 percent?

D. emergency stop?

E. single step?

22. List as many categories of vital information found in a CNC setup document as you can recall. These are vital to both operator and setup person. (19)

23. List two kinds of compensated profile mill programs and give a brief description of each. Which is the most common, and why? (20)

24. During an autocycle, several errors can cause alarms. List and describe them. (20)

25. Safety question—*vital*. Name and describe the three levels of operator intervention for problems. Under each level, name one or two methods of stopping the run to make the fix. (20)

26. Name at least three or four methods of monitoring a production CNC machine run. (20)

27. You have produced the part shown in Fig. 23-34 on a CNC machining center with T0101 = 0.500-in. diameter, two flute end mill for the pocket, and T0202 = 0.375-in. drill. Actual measurements after machining are shown in red (brackets). What must be adjusted to bring the part into target nominal, PRZ, or offsets? (Note, no tolerance is given for this problem.) (20)

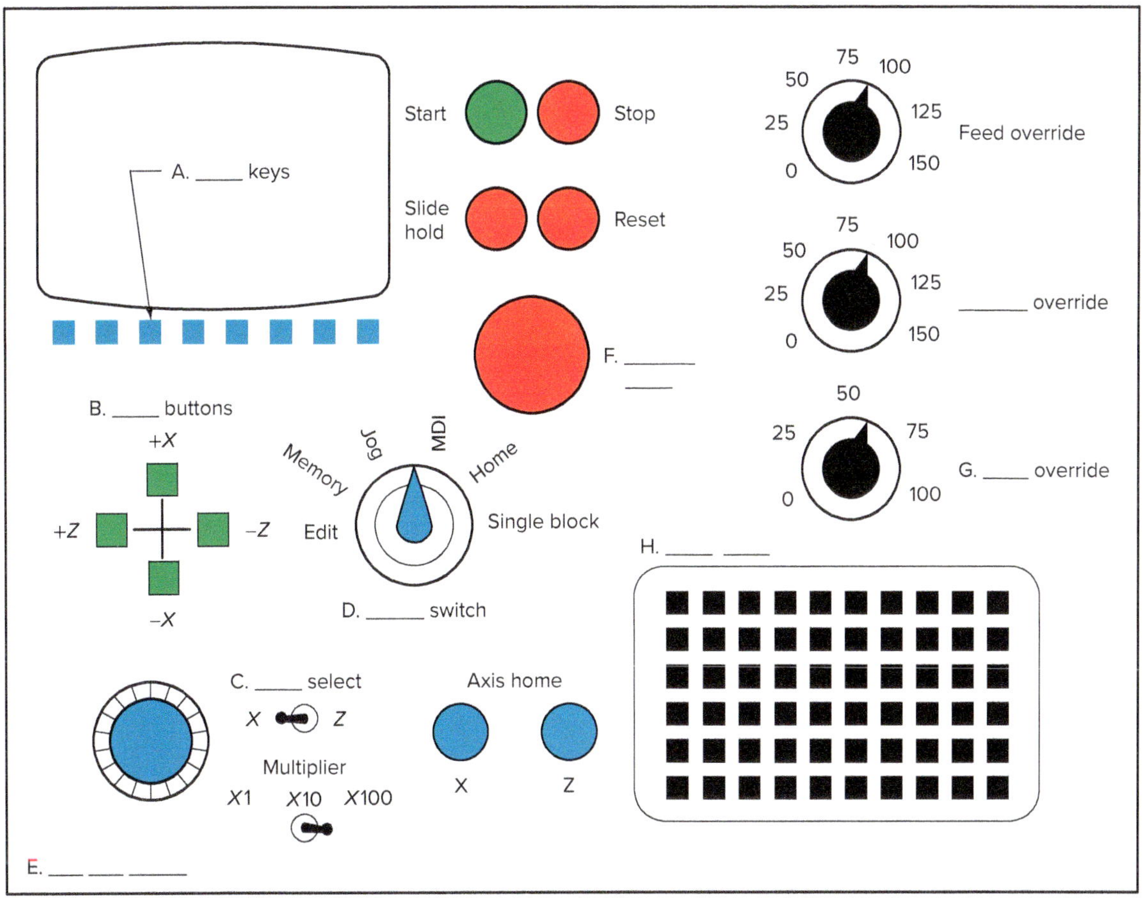

Figure 23-33 Problem 20.

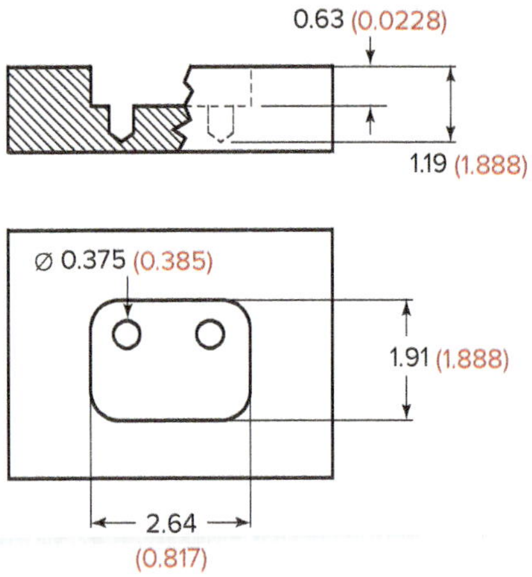

Figure 23-34 Problem 27. What can be done to correct these dimensions?

28. There are many factors to consider when planning out an efficient, safe CNC program. Beyond machine type, name the three most critical. (21)

29. Define the following planning and setup terms: (21)
coordinate shift
tooling reference–setpoint
tooling points
touch method

30. Is this statement true or false? Soft jaws are a problem for machinists. Made of the wrong material, both vises and chuck jaws deteriorate from use. If it is false, what will make it true? (21)

31. Name the three input methods of generating a CNC program. Which is more common? (22)

32. Fill in and describe as many of the following codes as you can remember. Note, the object was never to memorize codes but to learn the possibilities. See how many you do remember. (22)
G00
G01
G02
G03
G04
G17 (mill code)
G18
G19
G70
G71 (might be G21)
G40
G41
G42
G80
G90
G91
G94
G95 (mostly on lathes and drilling)
M00
M01
M02 (can also use M30)
M03
M04
M05
M06 (mill only)
M07
M08
M09

33. Why are G and M codes grouped into blocks? (22)

34. What actions will occur on these three command lines? (22)
N025 G01 *X*30.0000 G94 F19. G90 G70
N030 *Y*1.0000
N035 G00 *X*0.0 *Y*0.0

35. What will occur given this command? (22)
N001 G0 *X*1.000 G01 *Y*15.0

36. You are in G17 on a vertical CNC mill. Write a part path, *IJK* method (center ID) command block for the arc of Fig. 23-35. Then write the same arc command using the radius method. (22)

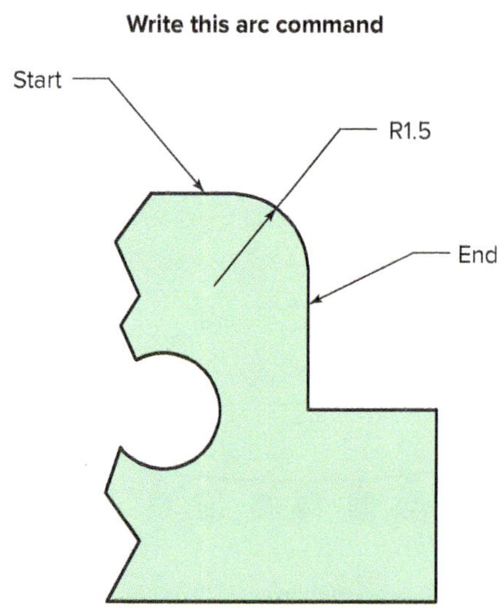

Figure 23-35 Problem 36. Write these arc commands.

37. To initiate cutter radius compensation, what considerations must be (22)
A. in the program planning?
B. in the program?
C. in the controller?

38. How are *fixed cycles* used? What is another shop term for these programming tools? (23)

39. Safety question. What is the singular danger of using the threading cycle on a CNC lathe/turning center as compared to all other fixed cycles? (23)

40. Is this statement true or false? A bolt circle is a threaded fastener that is bent into a round shape. If it is false, what will make it true? (23)

41. With respect to fixed cycles, define the following CNC terms: *departure point*, *parameter*, and *tiled routines*. (23)

42. Define the following setup terms: *coordinating*, *PRZ–datum point*, and *master tool*. (24)

43. A machinist wishes to evaluate a CNC program *graphically* but the CNC control has no graphic capability. Name at least two of three possible solutions. (*Hint:* Think globally.) (24)

44. What is the best test before loading the program into the control and performing a dry run? (24)

45. How many program tests evaluate part dimensions? Describe them. (24)

46. What are the factors that determine how much program testing is advised for a given job? (24)

47. You wish to machine a cone-shaped hole in an aluminum block. It has been predrilled to 1.00-in. diameter (Fig. 23-36). List the various ways it can it be machined, both manual and CNC. Which would be the most accurate with the best machine finish?

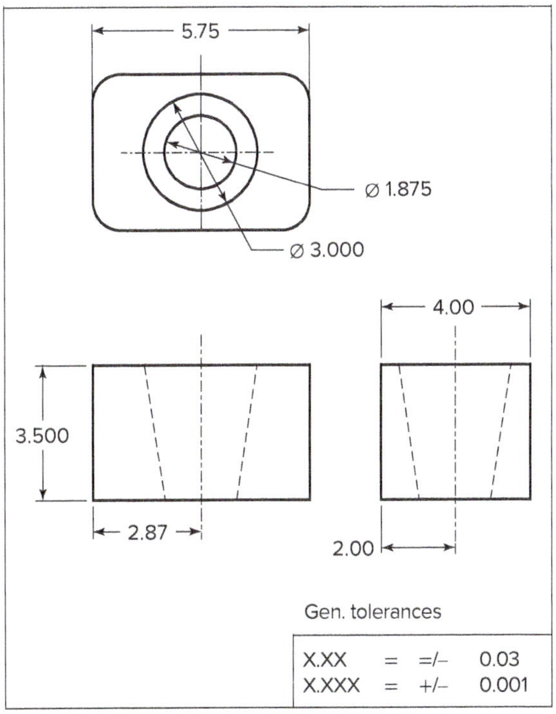

Figure 23-36 How many ways are there to cut this tapered hole?

48. A program has machined castings correctly for several hours and made 25 good parts. The 26th part is removed from the fixture, and it shows that one side hasn't cleaned up all the way. What could be the problem?

49. You are attempting to do a dry run, graphic evaluation of a handwritten program for a lathe control. It halts on the highlighted line (*), with error code #34B. Looking that up in the operator's manual you find it means a syntax error has occurred. What's wrong? How do you fix it? See Fig. 23-37.

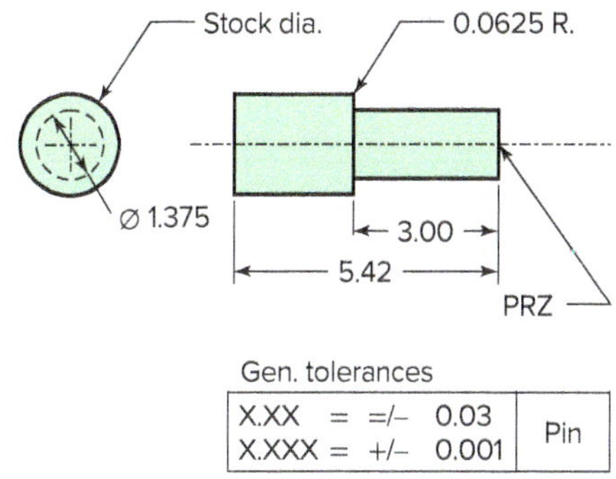

Figure 23-37 Problems 49 and 50. Compare the program to the drawing and edit out errors if they are found.

```
:O12345
(Final CNC Quiz—Turned Pin)
(2.000 CRS Stock—Cut to length—PRZ outer Tip—Per
Fig. 24-37)
N005  G80  G70  G90  G40
N010  G0 X100.  Z100.
(Home turret to soft limits)
N015  T0101
(0.03NR, RH turning—offsets 01)
N020  S650  G95    F0.005
N025  G0   X2.100  Z0.250  M03          (Rapid to departure)
N030  G41  X2.00                        (First ramp)
N035  G1   X1.750  Z0.050               (Ramp on)*
N035               Z–3.000              (First pass)
N040  G0   X1.800  Z0.250               (Rapid back for second pass)
N045  G1   X1.375  Y–3.000              (Second pass—finish 1.375 diameter)
N050       X1.875                       (Face shoulder)
N055  G3   X2.000  Z–3.0625  R.0625     (Corner radius)
N060  G1   X2.125                       (Pull off contour)
N065  G0   X100. Z100.0
N070  G40
N075  M2
```

50. You now have the syntax problem repaired in the program. It dry runs okay, but the graphics show an odd shape at line N045—not per drawing. What's the problem? How do you fix it?

PART 4

Advanced and Advancing Technology

Fast technical evolution is the very reason for this book, but it's also your greatest challenge as a machinist. The subjects in Part 4 are typical of the manufacturing world to which you are aiming your career. Some have been with us for a while. For example, Mastercam, Solidworks, and CGTech Vericut (new for this edition), but they evolve every year, reacting to market needs. Others are proven technologies, while a few are just now on the scene.

Part 4 has two major goals: the first is to introduce the following high-tech subjects:

Chapter 24 Design for CNC Manufacturing—Solid Modeling Instruction and a web-based activity in the Instructor Resources. You also qualify for a cost-free Solidworks Student Design Kit.

Chapter 25 CAM Mill Programming for CNC Machinists Instruction and web-based activity. You get the Student Mastercam Home Learning Edition HLE Download.

Chapter 26 Advancing Technology Sandvik Coromant discusses their latest research in dynamic tool paths.

Chapter 27 Statistical Process Control (SPC) Not a new technology but a powerful tool every machinist should know how to apply.

Chapter 28 Computer Coordinate Measuring A major area of advancement. New high-tech measuring tools appear regularly.

So taking them all as a whole, the second goal of Part 4 is to keep your eye on future technical developments. Learn when to adopt and adapt—always looking forward but not leaping too soon. But clearly, in your career you will face ever-more change.

CHAPTER 24 Design for CNC Manufacturing—Solid Modeling

Units

INTRODUCTION

Now we explore solid modeling. Solids are the way parts are engineered today and they are the basis for many program creations. Chapter 24 is partly for entry-level machinists But it's also here for three valid reasons:

1. You should understand them as a basis for your work.
2. Engineering and programming might be a second career move.
3. Even the best of designs need a tweak or two due to production needs, planned holding cuts, for example, or a set of tabs to keep a large piece from flying off the work as it is profile milled away—called a ***disconnect***. Another example: a planned temporary datum must be machined on the part to be removed later—it's not on the solid model but must be added before programming. These production edits require some ability in manipulating part geometry.

Changes can be made to the solid within the draw utility inside Mastercam. But they can also be done in the CAD software.

What You Won't Learn

We've partnered with SolidWorks due to its major support of technical training and worldwide distribution in industry. It's a great product capable of designing, assembling, and testing the simplest to the most complex part or complete assembly. But Chapter 24 is only for exposure to SolidWorks (and a fun challenge), because to become competent requires a lot more study. The SolidWorks tutorials are an excellent place to start, and a formal course is another. Since our focus is baseline knowledge, we will skip many powerful techniques and functions, leaving them for further study.

Chapter 24 is divided into two parts:

Unit 24-1 The Solid Model Advantage for Industry, which requires some background understanding of drawing a solid model

Unit 24-2 Drawing Your Solid Model, which is a web-based activity to create a solid model of the part shown in Fig. 24-1. Before proceeding, you will either need access to a class-licensed SolidWorks or will need to download a student set of the free **SolidWorks Student Design Kit** for home use. Your instructor will provide a password for the download. **The student software is not for classroom use.**

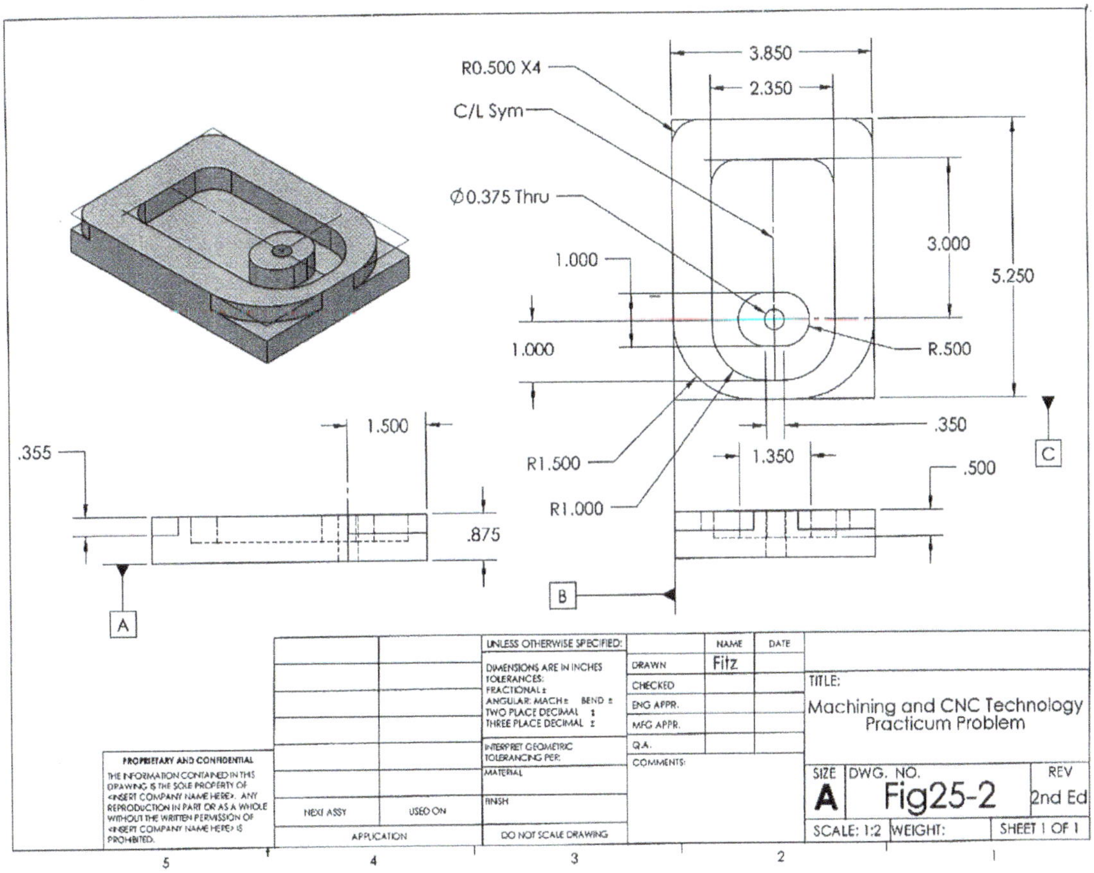

Figure 24-1 The master drawing for Chapters 24 and 25.

Source: Courtesy of Mike Fitzpatrick

The final file you will produce would normally be transferred directly to Mastercam for programming. Because the SolidWorks student version will not export a file, I have provided the finished solid file found in the student side of our learning center.

The web assignment is found in the Instructor Resources. This is an Abobe PDF file; thus, you can print it if you prefer to have paper to reference or mark as you go.

Unit 24-1 The Solid Model Advantage for Industry

Introduction: I really enjoy creating designs in SolidWorks and suspect you will too. Solids bring several fantastic advantages to the manufacturing process—not possible using previous CAD wireframe drawings. Then if you wish to make a paper orthographic projection it, too, is a breeze once the solid has been created!

TERMS TOOLBOX

Base extrusion The start-up volume.

Boss Any added material (solids term).

Extrusion Changing volume—positive or negative.

Extruded cut Removing volume.

Features mode Extruding and changing volume.

Normal (view perspective) Looking directly at a surface or plane.

Sketch mode Creating a flat-shaped template to be extruded in features mode.

24.1.1 Creating and Testing Working Assemblies

After drawing a set of solid parts, users can assemble them as if they were real. The assembly can be tested for function and fit. Time studies can be performed to see how the assembly coordinates in use. If interference problems arise, the software shows this in red, or it can even be made to halt at interference points as it's tested on screen! Making an assembly is one of my favorite aspects of SolidWorks!

Testing Designs

Assigning a given material to the design (aluminum in this exercise), a finished solid part or assembly can be weighed and stressed to see how it will perform. A visual display tells the designer locations where forces concentrate; thus, it might need more material or a redesign of the shape.

Safety Margins Ensured

SolidWorks also determines if the design's safety margin is low, good, or too great (overdesign). Based on test results, the designer can upgrade or modify the design before the prototype is produced–saving time and money by creating a dialed-in design earlier. Less testing and redesign!

24.1.2 End-User Advantages (Machinists, Programmers, and Inspectors)

Dimensional Accuracy

Previously, I often found myself missing crucial dimensions on a wireframe CAD drawing. However, with SolidWorks, if it's created correctly and fits together or functions right, then all dimensions are available by querying the solid–one can actually measure it for needed data! ***Nothing can be missing; it is the part!***

Fast Visualization

The pictorial view on the left in Fig. 24-1 provides an almost instant understanding of the object's general shape. It was drawn by SolidWorks with just a few mouse clicks! But even better, in SolidWorks, the on-screen object can be grasped with the mouse and rotated as though it were in your hand, and it can even be sectioned to see interior details, too.

Drawings Follow Solids

When shop drawings are required that differ from previous technologies where we once began with a CAD drawing, today after the solid model is complete, tested, and perhaps assembled with others, only then do we create a drawing. But we use the power of the program to do that, too! Simple, accurate, and fast, SolidWorks places the various views where you drop them–it draws the flat drawing by pointing with the mouse! That's how I drew Fig. 24-1. I first created the solid object on the left, then clicked in place the various views. That drawing can actually be paperless!

In this chapter we will not explore all these many wondrous abilities, because it all begins with learning to create a solid model–our single objective. However, nothing is stopping you from experimenting afterward! In fact, I encourage it–this is awesome software!

24.1.3 Launch Pad Facts, Guideline Rules, and Concepts

Extruding Base Volume

Working in solids is a lot like using modeling clay to make a shape–one must start with a lump! It's called a **base extrusion**–a term borrowed from metal forms. It all begins with a flat sketch that is then extruded outward from the sketch to the full size to become the base volume.

Adding and Subtracting Volume

Continuing to build the design, material can be added, called a **boss extrusion**, extending from further sketches drawn upon three places: (1) selected surfaces of the base; (2) basic planes, top, front, or side; or (3) auxiliary planes you create, to be extruded outward to join and increase volume. But volume can be taken away, too, called an **extruded cut**.

> **KEYPOINT**
>
> In the build stage, you are either working in **sketch mode**, making the pattern, or in **features mode**, where volume is added or subtracted based on a sketch.

Factoids:

- Solids can be modified after extrusion–for example, a fillet can be put upon a face or a hole drilled.
- At any point in the process, your solid can be assigned a material that will allow characteristic testing.
- This extrusion process is similar when using Mastercam Solids as well.
- You might experience different icon and menu arrangements than mine–SolidWorks allows many desktop options. Look for the icons as I show them, or use the dropdown menus if you can't locate them. You may need to customize your toolbars to display the buttons you need.
- To keep a data box or many other functions locked, notice the push pin. Clicking it pins the function until it's clicked again and unpinned.
- There are often a *minimum* of two ways to accomplish the same objective–one of my favorite aspects of SolidWorks!

Fully Defined Entities

An entity is anything that can be seen, modified, or measured: a line, a surface, or a hole, for example. It is fully defined when it has size/shape and *location* within the drawing window. An entity is also defined if it joins or intersects other defined entities.

> **KEYPOINT**
>
> Colors indicate progress. Blue entities are undefined; black are defined. Before being defined, an entity can be changed by changing its input value, but after definition, it can only be edited.

No Unit Review

Since most of this chapter is about creating a solid model, there are no Unit 24-1 review questions. Final credit for this chapter will be gained:

1. By showing your finished solid to your instructor.
2. By answering the Chapter Review questions at the end of the chapter—but only after completing the solid modeling activity.

Unit 24-2 Drawing Your Solid Model

24.2.1 Downloading SolidWorks for Your Personal Computer

To download your Student Design Kit for completing the activity at home go to www.solidworks.com/how-to-buy/get-educational-trial

Understand that the kit will expire and is not intended to be used as an in-school program. If you are a student, when your SDK license expires, you can purchase a student copy of the program. See your instructor or learn how on the SolidWorks website.

When you get to the Solidworks website, enter your educational email address and follow the instructions for receiving your free trial. You will receive an email that includes the serial number for your software license, which should be entered when you first open the program. (*Hint:* It's easy to copy and paste big numbers!)

If your lab has one or more seats of SolidWorks, it would be best if you use it to create the part. That part can be done in school, and it can be transferred to Mastercam for programming, or you can still use the final image found at our learning center.

Now it's time to go to Instructor Resources. Then, when your solid is finished, return to Chapter 24 Review.

CHAPTER 24 REVIEW

Replay the Key Points

Complete after your solid model is finished:

- In the build stage, you are either working in "sketch mode," making the pattern, or in "features mode," where volume is added or subtracted based on a sketch.
- Colors indicate progress. Blue entities are undefined; black are defined.
- Before being defined an entity can be changed by changing its input value, but after definition it can only be edited.
- To edit either a feature or a sketch, expand the feature to be edited in the manager window at the left, by clicking the plus sign, then right-click it, opening a dialog box. Now click "edit sketch" or "edit feature" (hand-and-pencil icon) to change anything!
- After editing some aspect of your solid, sketch, or extrusion, it's a good idea to rebuild the object (a little traffic light icon).

Respond (answers found at the end of the web-based activity)

1. Under what conditions would only one feature icon show on the features toolbar?
2. Can you extrude a feature directly? What is needed?
3. Name at least two end-user advantages when using a solid model.
4. Name the two modes used to create a solid block.
5. Identify two ways a hole can be created in a solid.
6. What is the name for a view perspective that looks directly into a surface?
7. What does the push pin do in SolidWorks?
8. What is a BLIND extrusion?
9. If an entity is blue, in what condition is it?
10. What step is necessary before testing a design for stress?

CHAPTER 25 CAM Mill Programming for CNC Machinists

Units

25-1 **Understanding the Possibilities**

- What is a graphic interface? (wireframe and solids)
- Standard tool paths
- Dynamic tool paths (new thinking)
- Feature-based machining
- Rest milling

25-2 **Creating a CAD/CAM Solid Model (web-based instruction)**

- Common draw icons and rules
- Sketching the wireframe
- Extruding Your CAD solid

25-3 **Creating a Tool Path (web-based)**

- Planning the tool path
- Work coordinate shift
- Part orientation
- Creating the tool path
- Postprocess the tool path into machine code

25-4 **Importing Part Geometry (web-based)**

- SolidWorks files for CNC programming
- Prepping outside files for CNC programming

Made Right Here

Based in Post Falls, Idaho, Buck Knives uses the best materials and state-of-the-art heat-treating and tempering processes to create high-quality knives.

Visiting Buck Knives, I was impressed with three outstanding traits: first, the blending of CNC and other technologies with an art as old as civilization; second, the absolute pride the company takes in its products; third, and the most heartwarming, the respect that management gives to its skilled workforce. If I was beginning again, I would choose Buck as a place for my machinist career to unfold! Everybody I met was engaged, proud to show me their work, and professional.

Courtesy Buck Knives, Post Falls, ID

INTRODUCTION

Computer-assisted machining is a power tool for programming that's also the major change engine in machining today (in my opinion). For sure, evolution is always about machines and their controllers, as they grow faster with each generation. Cutters are also evolving, which we'll discuss after this chapter. But today the change engine is the new tool path thinking! It's a totally new way to cut metal with a spinning cutter—which is only possible when an intelligent CAM program drives the tool motion!

You might think that with 250 years of machining behind us, we'd have put this subject to bed (i.e., how to move an end mill through a cut). Carried forward from manual machines, we held to the idea that milling is a point-to-point thing—even into the CNC era, it was the way we thought milling tools ought to move. Not so today! Driven by the possibilities of dynamic CAM tool paths, much of what we've held as "standard" milling is morphing into something brand new. Learning about these possibilities is one of your goals for Chapter 25 and for your career.

When I first wrote this book a few years ago, it was all about high-speed machining (HSM) with faster spindles and feed rates using **trochodial** tool paths, invented to take advantage of those extreme machine tools (which we'll explore here in Chapter 25). But today's **dynamic tool paths** are even better for HSM, and the best news is that *dynamics apply to all CNC milling, whether it's HSM or not!* Even the lightest mills can employ a dynamic tool path to remove more material than they could with standard cuts. As has been true since my apprenticeship, I am as much a student as I am a teacher, learning these new ideas to bring them to you. Truly understanding the following terms really matters, because many are new to our trade.

WHAT YOU *WON'T* LEARN IN CHAPTER 25

This chapter is about two things: an exploration of the possibilities and a first experience in CAM programming. But it is NOT a complete lesson in CAM programming. Mastery requires a lot more training in a technical school or online or in a factory training course or apprenticeship. Self-study and job experience are other avenues to mastery. But like me, with technologies such as these, you'll forever be a student. In this chapter we will be discussing two CAD/CAM software packages: Mastercam and Fusion 360, but keep in mind that there are many more makers of this kind of software. I stand in awe of the people who write programs such as Fusion 360 or Mastercam; like SolidWorks, these are awesome software packages!

Having said that, you will have an opportunity to try creating your own CAD/CAM projects. At the end of this chapter, you will find two exercises that will lead you through the process of getting started with CAM programming.

Unit 25-1 Understanding the Possibilities

TERMS TOOLBOX

Axial engagement Using the side teeth of an end mill.

Dynamic tool path Milling in adaptive, small circular whittles using more axial and less radial engagement.

Feature-based machining (FBM) Mastercam recognizes and creates a tool path based on window-selected solid features.

Optimized roughing (Opti-Rough-Mill) Adaptive auto-routines that remove the most efficient amounts of material per pass.

Postprocessing Generating program codes using a CAD/CAM system.

Radial engagement The amount of the mill cutters side engagement–how much of its cutter radius is used.

Rest milling Removing the rest of unwanted material after roughing.

Spiral entry A spiral end mill plunge using axial engagement for better cutting, chip ejection, and coolant entry.

Trochodial (tool path motion) A continuous series of arcing *whittles*.

25.1.1 What Is a Graphic Interface?

Wireframe and Solids

Compared to hand-compiled programs, CAM programming is faster, eliminates human math errors, and turns hours into minutes. But, of the greatest import, CAM makes possible the programming of shapes and tool paths that can't be achieved by any other means. They could never be handwritten, no matter how capable the programmer might be. It's called graphic interface (GI) programming.

A GI program can be based on three different kinds of part picture definitions:

1. A wireframe drawing (still effective for 2-D work)
2. Surfaces
3. But evermore the template is a solid model for good reason (Fig. 25-1)

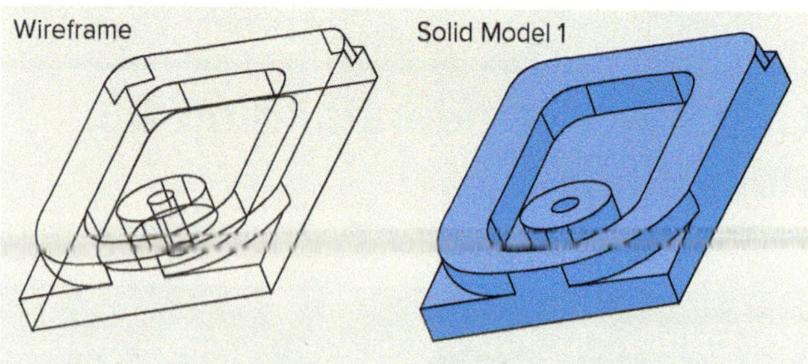

Figure 25-1 Part geometry can be a wireframe, but solids work better—especially when part shape is complex.
Source: CNC Software Mastercam Education Division, Gig Harbor, WA

Solids bring more automation to the process because they *are* the part, with complete feature and dimensional information built in. They can even have material characteristics. A solid provides information the CAM system can use with less intervention from the programmer. When we come to feature-based machining (FBM), the grand master of efficiently using solids, you'll see how much work can be automatic. FBM turns CAM hours into minutes!

25.1.2 Standard Tool Paths

In just a bit, you will go to the *Machining and CNC Technology* website for the programming challenge based on Fig. 25-2. But first let's take a quick tour of the possibilities that you might use to program a part.

Standard Possibilities

There are a set of feature programming options that have been with us for some time and continue to be useful—we'll call them *standard operations*. They often form a lot of programs, especially when using a 2-D wireframe template for milling or turning.

Pockets

Pockets are internal areas to be removed. When working with solid models, they can be defined by selecting a floor face or its bottom edges (Fig. 25-3) or sometimes by the surrounding walls. If using wireframe geometry, pockets are defined by the rim lines. When selecting a wireframe pocket, Mastercam will recognize which side of the line the cutter must go to because it's a closed shape–it will machine the material inside the shape. But here's our first template difference: When an object such as the oval boss surrounding the hole lies within the pocket (see Fig. 25-2), that boss must also be selected by the programmer as an area to be avoided in a wireframe. Otherwise the software will not recognize the boss and will cut the pocket without it. On the other hand, when a solid is used, boss recognition is automatic.

KEYPOINT

Using a solid model, it's best to select the floor face or the bottom wireframe of a pocket; by so doing, depth will be autoloaded to the operation depth parameter.

Contour (Also Called Profile or Peripheral Milling)

This operation uses the cutter's side teeth to mill a wall on the part, using either solids or wireframe lines. Now here's a huge difference: When selecting a contour from a wireframe, the user must tell the program to which side of the definition line the cutter must machine. Without indicating to which side the machining is to occur, a CAM cannot determine which side is metal and which is not–lines are all that exist, no material. With a solid this issue goes away–there is metal and not metal (air). If the wireframe is a flat drawing without depth, that must also be provided by the programmer–but not when using a solid. Contour depth is on the part.

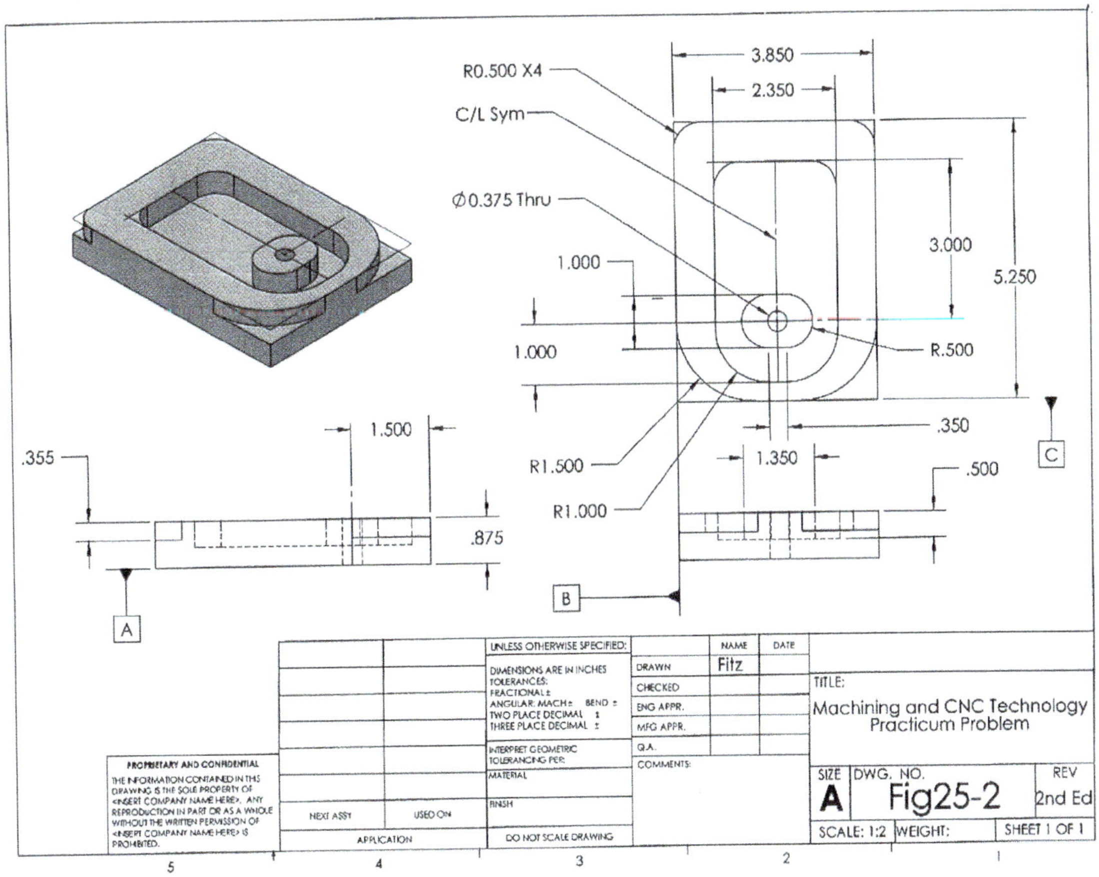

Figure 25-2 Put a marker here—we'll be using this part drawing.
Source: CNC Software Mastercam Education Division, Gig Harbor, WA

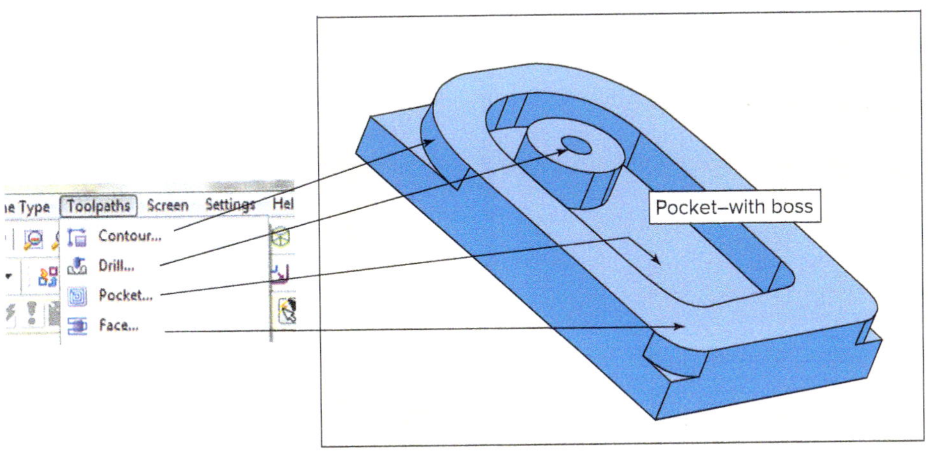

Figure 25-3 Standard milling tool path options applied to our example object.
Source: CNC Software Mastercam Education Division, Gig Harbor, WA

Better Billet Definition

Toward efficient programs, rapid cutter approaches should stop rapid travel within a minimal but safe distance from the raw stock, regardless of which side the cutter approaches from. If the part is simple like our part, that bounding box can be defined as an expanded version of the raw stock when using wireframes, surfaces, or solids (Fig. 25-4). One can define the raw stock billet large enough to prevent rapid travel crashes, but suppose the raw material is a casting or a forging–preshaped close to the finished product? Not bricklike?

Preshaped Definitions

Enclosing a preshaped object, like the casting in Fig. 25-5, using a standard bounding box would require enclosing the entire casting–causing a lot of air cuts before reaching metal on some approaches. But when the solid is the starting stock definition, Mastercam can define the material as a swollen version by a user-defined amount of machining excess. Then, when the cutter rapids toward the material, Mastercam knows exactly where to switch to feed rates.

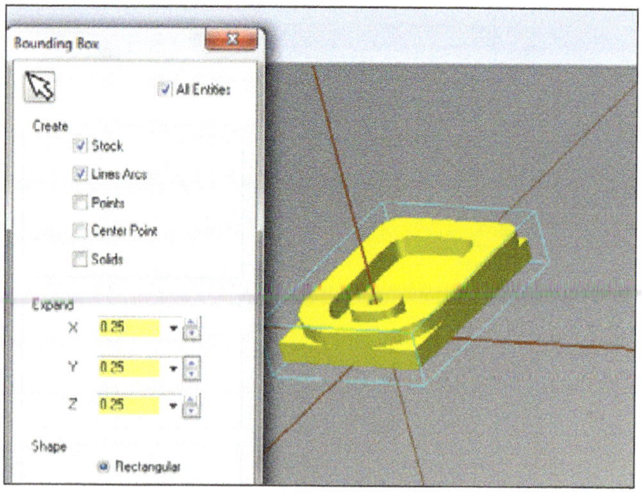

Figure 25-4 A bounding box defined in blue tells Mastercam the stock definition.

Source: CNC Software Mastercam Education Division, Gig Harbor, WA

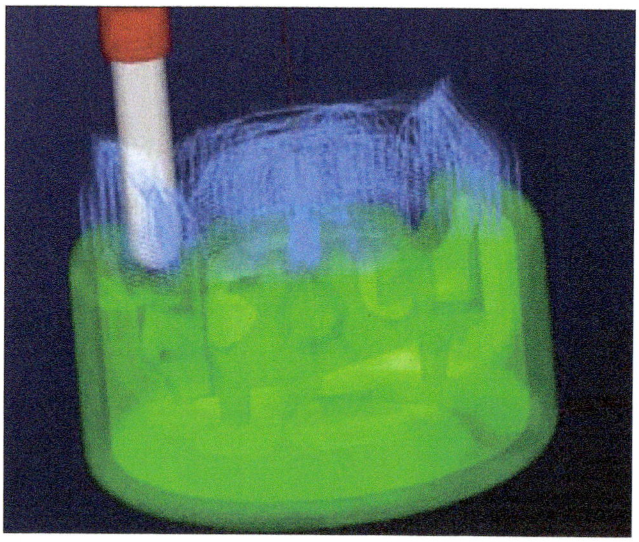

Figure 25-5 A complex bounding definition based on a solid model is far more efficient.

Source: CNC Software Mastercam Education Division, Gig Harbor, WA

25.1.3 Using Surfaces for Part Definition

Landing between wireframe geometry and solids–surfaces are shaped skins with no volume–there is not metal on both sides (Fig. 25-6). They, too, have been with us for a while:

- **Ruled:** Transitioning one shaped wireframe to another along a straight line
- **Revolved:** Sweeping a shape around an axis–like a soda bottle
- **Swept:** Transitioning one shaped wireframe into another along a third shaped defining path
- **Net:** Blending multiple curves together into a surface

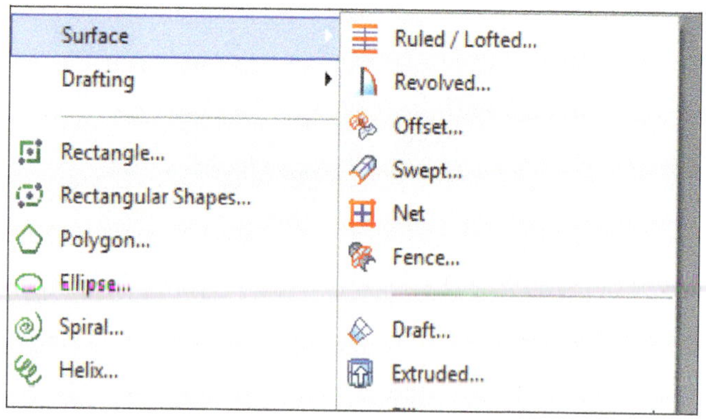

Figure 25-6 Surfaces have their place in CNC.

Source: CNC Software Mastercam Education Division, Gig Harbor, WA

Multisurface Part Geometry

Just like it sounds: Instead of a single surface, the part is defined by many connected surfaces, such as this welding machine case. Each surface is being driven by a wireframe boundary. Surfaces can fully describe the outside of a part, but they are still only a skin with no volume. They can also be designated as *Check Surfaces* in Mastercam; when creating a tool path, if selected as a Check Surface, the cutter must avoid it while machining other features to prevent gouging. Solid models are rapidly making surfaces a little-used technology.

TRADE TIP

Each part geometry has its best application. Think of all of them—wireframe, surfaces, and solids—as tools in a toolbox.

25.1.4 Dynamic or Adaptive Tool Paths

Now for the advancing solid model tool path technology! As we discussed in Chapter 17, numerical control machining has come a long way since its beginnings in the 1940s. When the first CAD/CAM programs were developed, they produced tool paths that mimicked the type of tool paths that machinists historically used as manual machinist (see Fig. 25-7). They were toolpaths that plodded along, taking heavy radial engagement (see Fig. 25-13) and shallow axial cuts. These tool paths moved from point to point in a straight or curved line. But, as our machining and computing technology grew, so did the science of metal removal. Today, CAM systems can produce tool paths that use light radial engagement and deep axial cuts (see Fig. 25-11) while using higher spindle speeds and faster feed rates. *But if we can plant only one idea here, the subject isn't just about speed; it's also about quality and tool life! And it's about protecting the machine itself.*

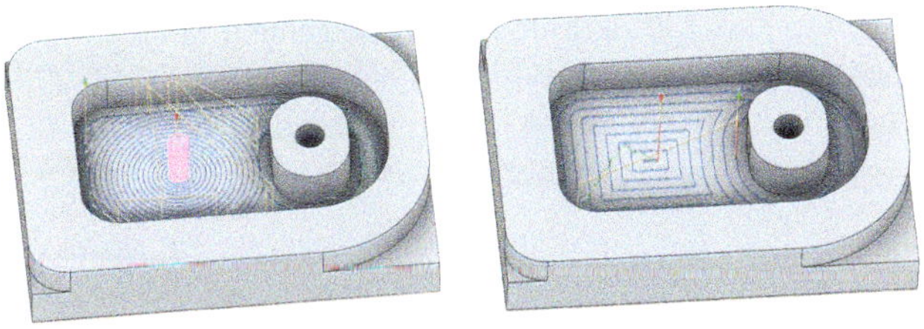

Figure 25-7 Adaptive tool path (left) vs traditional pocketing tool path (right)

Evolution Begins—Protecting the Machines

High-speed machining can be defined as taking shallower depths of cuts at higher federates (600+ in./min) and spindle speeds (10–20,000 RPM or higher). While this allowed parts to be cut faster with a better surface finish, when machine builders started producing those machines, tool path strategies had to be created to protect both the machine and the cutter!

The first problem to be solved was to make sure the tool path had no right angle turns, because the extreme feeds created staggering physical loads at hard corners. They could literally shake a fast machine apart. So one definition of a high-speed tool path (HST) is "rounding all corners, and having smooth lead-ins and lead-outs on surfaces" (Fig. 25-8).

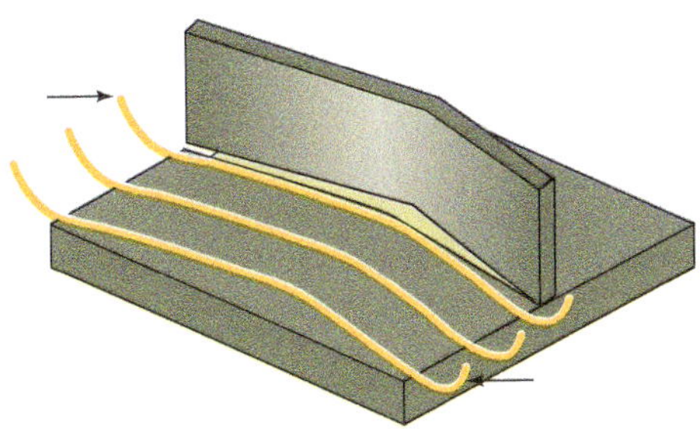

Figure 25-8 Smooth leads and corners.
Source: CNC Software Mastercam Education Division, Gig Harbor, WA

Protecting Cutters

The next two issues to overcome were the worst-case scenarios for shocking or stressing cutters, with the worst offender being plunging and 100 percent tool burial, then bashing forward in a linear path–a sure way to break the cutter!

Spiral Entry

Cutter shock and stress were significantly reduced by incorporating an elegant spiral motion while plunging, rather than hammering down like a drill press (Fig. 25-9).

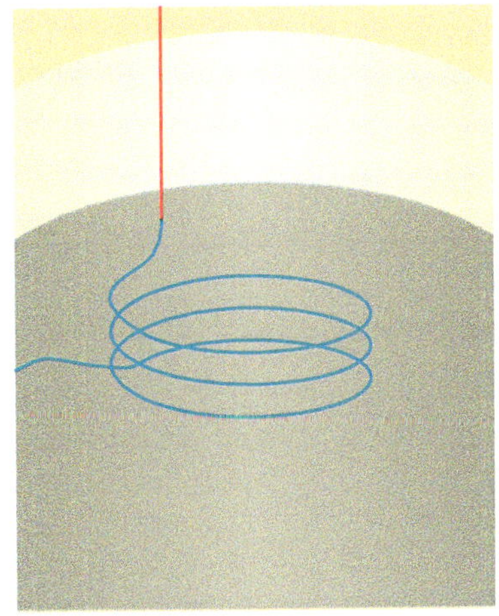

Figure 25-9 Spiraling compared to plunging in saves end mills.
Source: CNC Software Mastercam Education Division, Gig Harbor, WA

Trochodial Motion

Trochodial motion was the greatest change from straight-in cutting to advancing in a circular motion, taking a little bite/shave of the material with each cutter arc into and out of the work. At high feed rates, one cannot expect the cutter to survive when buried: That first cut in a pocket after the plunge is the worst for the cutter, as it is deep in the cut, and the trapped chips have nowhere to escape and are recut. Trochodial motion solved this; now with a spiral plunge and then moving immediately into little circular contact with the work, feed rates can remain high yet not break the cutter (Fig. 25-10). This circular motion shaves away evenly in that first full engagement, constantly climb cutting and maintaining a uniform light load. This holds true not only for the first cut but also for any cut that would bury the cutter.

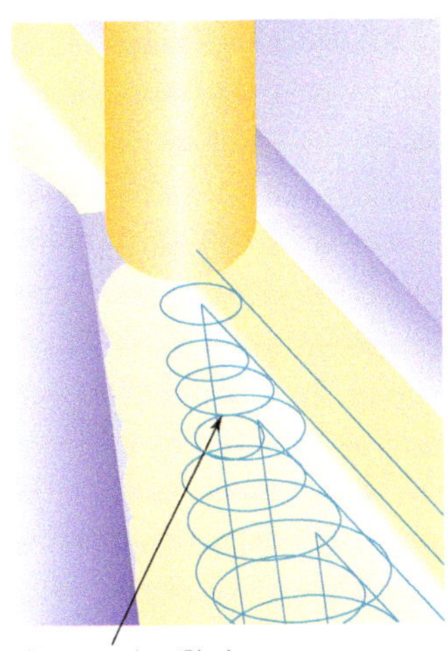

Progressive Circles

Figure 25-10 This cutter is advancing in trochodial motion.
Source: CNC Software Mastercam Education Division, Gig Harbor, WA

> **SHOPTALK**
>
> Trochoids are not true arcs. In order to truly understand them, check Wikipedia.

Dynamic Tool Paths Come Online

Evolving out of trochodial tool paths, the newest thinking is that for the most efficient contour milling we use the fullest possible length of the cutter's side teeth–called maximum **axial engagement**–while whittling away tall, thin chips using the longest end mill side teeth possible. The chips are thin but of constant volume as they trace the part contour (thus differing from trochodial, where they vary). This engages the best cutter geometry, as opposed to the previous method of mostly cutting with the end teeth in successive down layers, where the end of the cutter does most of the work. With dynamic tool paths, cutters last longer, since they use better cutter side tooth geometry, and longer cutting edges; thus, heat goes out with the sheared chip. Cooler cutters, better finishes, longer tool life, faster removal–yet lighter machines can perform them. They work for any mill operation.

> **KEYPOINT**
>
> Dynamic tool paths apply to any operation—HSM or standard!

Radial Versus Axial Engagement

As with trochoids, for dynamic milling we use 10 to 15 percent **radial engagement** (side tooth contact). We back away from deep radial engagement (end tooth contact) to around 10 to 15 percent of the cutter's diameter (Fig. 25-11).

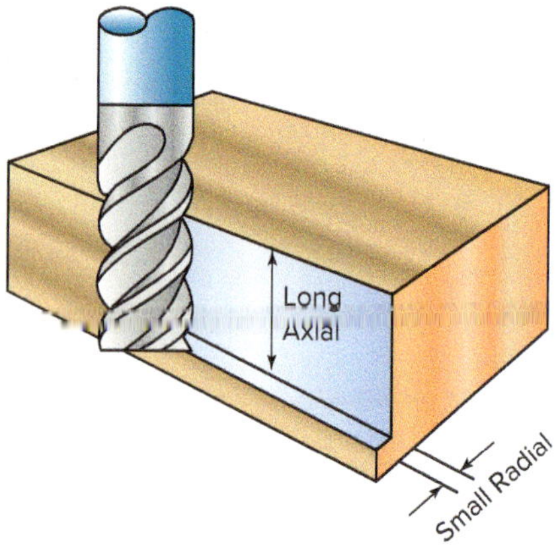

Figure 25-11 This cutter has a high axial and low radial engagement.

Source: Sanvik Cormant CNC Software Mastercam Education Division, Gig Harbor, WA

25.1.5 Compensating for Chip Thinning

Looking up a feed per tooth (chip load) for a given cutter and setup, you'll find numbers that represent feeding a cutter such as this with a large radial engagement (Fig. 25-13). For example, the feed rate for this cutter is based on 0.006 in. per tooth per revolution. Each tooth advances 0.006 in. into the work.

When we back away to a small radial engagement, as we've been describing, the chip is made thinner (Fig. 25-12). This means that the feed rate should be increased to keep the chip load constant. That upward adjustment maximizes cycle times, but it's also necessary to correct the thin chips to keep heat leaving with the chip and not overheating the cutter. This is true for standard milling and especially so for HSM, where heat removal is crucial. That adjustment is made automatically using Mastercam dynamics. We'll cover the chip thinning formula in a later chapter.

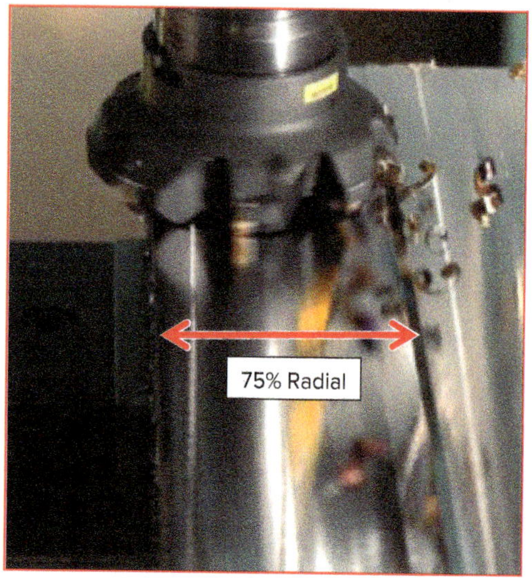

Figure 25-12 This cutter is at a 75 percent radial engagement.

Source: CNC Software Mastercam Education Division, Gig Harbor, WA

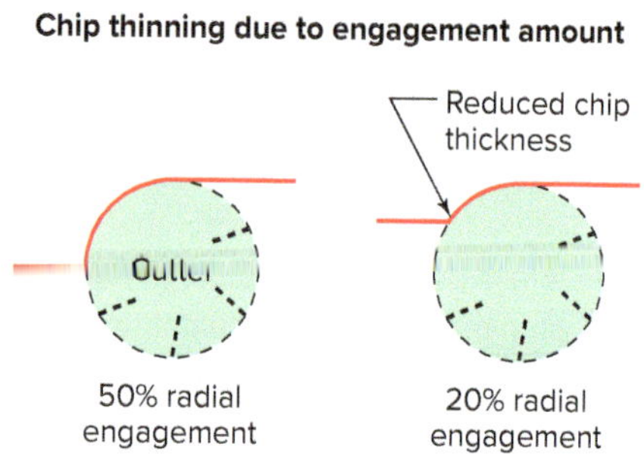

Figure 25-13 Less radial engagement requires increasing feed rates to keep chips thick enough to carry away heat and for efficient cut rates.

Source: Lake Washington Technical College—Kirkland, WA

Summing a Dynamic Tool Path

The dynamic tool path constantly adapts to the remaining material, with a constant chip load and volume. Each successive cut will be made up of the following components (Fig. 25-14):

- Large axial engagement
- No sharp corners
  - › Spiral entry
  - › Smooth entry/exit
- Climb cutting always
- Micro-lift rapid return reengage

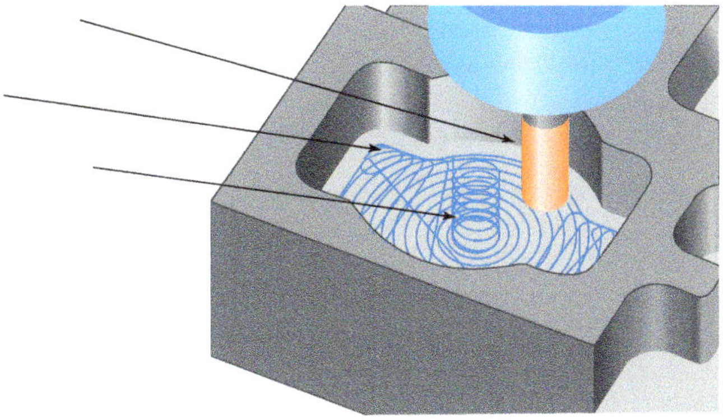

Figure 25-14 All dynamic tool paths share these features.
Source: CNC Software Mastercam Education Division, Gig Harbor, WA

Dynamic Tool Paths Benefits

From this brief description, the following are only possible through highly intelligent CAM software:

- Cycle times reduced by 30 percent or more compared to traditional cutting techniques
- Machines only where material exists
  - › Consistent chip load allows for maximum feed rate/material removal
  - › Accounts for chip thinning, keeping feeds maximized
  - › Better finishes due to constant climb milling
  - › Better accuracy due to light cutter loads–less deflection of work and cutter
- Improved tool life
  - › Full flute utilization for longer tool life
  - › Side teeth are better geometry for cutting/shearing the chip compared to end teeth
  - › Deformation heat leaves with the chip; far less cutter heat
- Economical
  - › Smaller tools can be used with similar results–fitting into tighter spots, too
  - › Less carbide needed, since cuts are not brute force
  - › Smaller, lighter-duty machines can accomplish more work
  - › Fewer spindle loads
  - › Less demand on setups due to reduced forces

25.1.6 Feature-Based Machining (FBM)

This is another aspect of the new tool path capability using solid model geometry. Basically **feature-based machining (FBM)** does for the programmer what he or she would do given enough time–selecting bounding definitions and cut parameters on parts with many features. In this example, all inner features can be selected for FBM with one window selection! Programming them one at a time would take time; however, with FBM you would window select all features, and the tool path would be created automatically in minutes!

FBM uses only solid models for part geometry. With a single operation, FBM analyzes the model, detects all machining definition features in a specified plane, and automatically generates all of the 2-D milling tool paths necessary to completely machine the selected features (Fig. 25-15).

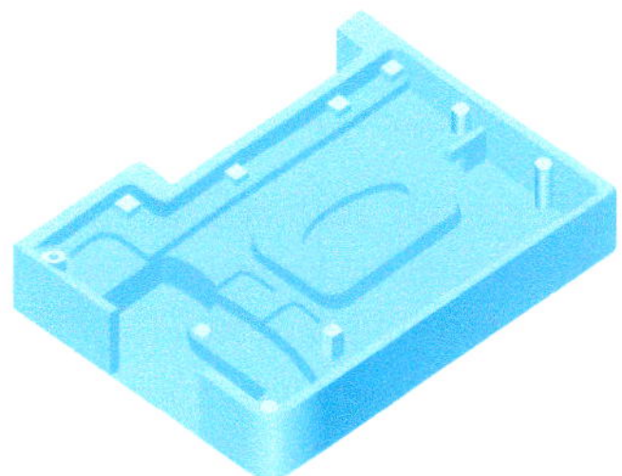

Figure 25-15 Programming these nested pockets would be a snap with FBM.
Source: CNC Software Mastercam Education Division, Gig Harbor, WA

Seen on this part, FBM auto selects closed, open, nested, and through pockets, with both vertical and tapered walls. For complex nested pockets (one within another), Mastercam creates a separate zone for each depth and creates the boundaries required to machine it. In the following example, six zones were created and machined separately (Fig. 25-16).

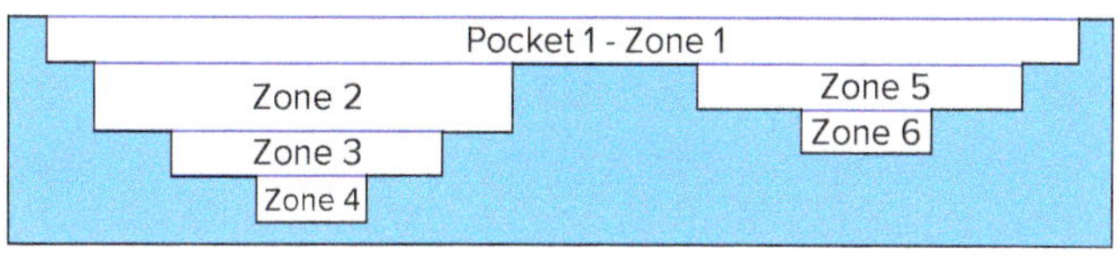

Figure 25-16 Mastercam creates zones to be completed in stages.
Source: CNC Software Mastercam Education Division, Gig Harbor, WA

25.1.7 Rest Milling

Previously, for efficiency, we usually roughed out a pocket with a larger cutter. But that cutter could not get into corners or tight spots–leaving material uncut. Thus, we had two options to finish the work (Fig. 25-17):

1. Re-create a second duplicate tool path using a smaller cutter able to finish the shape—but that meant it made 90 percent air cuts! This takes little programming time but adds lots of wasted machine time.
2. Find and draw fences to contain leftover metal, creating small pockets for another set of operations—taking a lot of programming time to do.

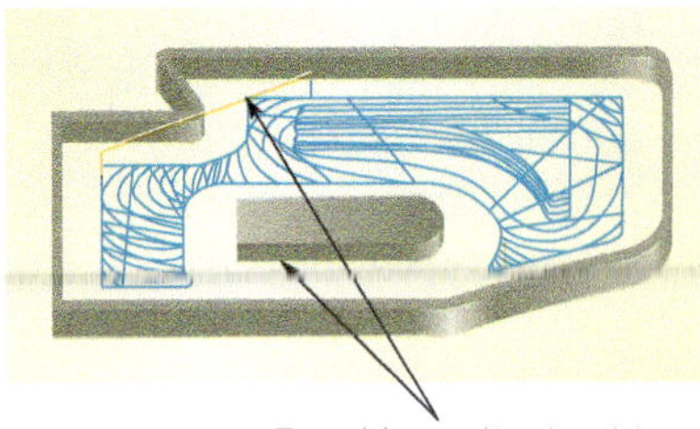

Figure 25-17 After roughing this pocket, several corner areas and one tight spot remain uncut.
Source: CNC Software Mastercam Education Division, Gig Harbor, WA

Neither solution was efficient.

Based on a solid model, **rest milling** compares roughing results to the solid model to identify uncut metal, then creates a dynamic tool path that removes the "rest" of the uncut material (Fig. 25-18). Efficient for the programmer and the machine cycle time, dynamic rest milling solves the issue.

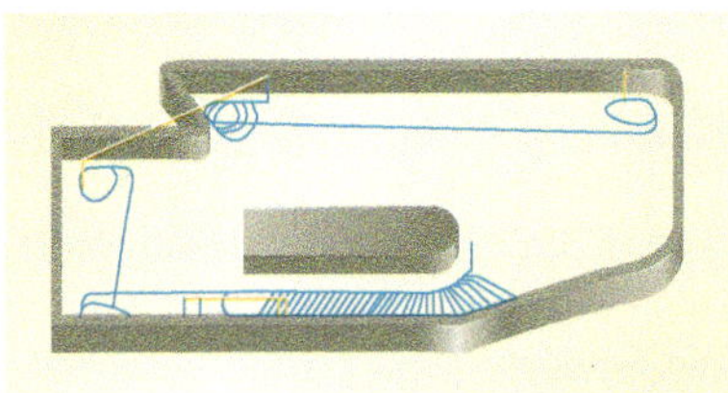

Figure 25-18 Rest mill creates an efficient follow-up tool path.
Source: CNC Software Mastercam Education Division, Gig Harbor, WA

Optimized Roughing and Peel Milling

Our final dynamic tool path subjects also make use of larger axial engagement, rethinking how we rough and finish features.

Optimized Roughing (Opti-Mill)

In standard roughing, we tend to make several down passes in stages, as shown in the Xcursion *film–standard versus dynamic tool paths*—just seen, where the V-9 cutter used that older down pass thinking (passes 1-7 in the Fig. 25-19 drawing).

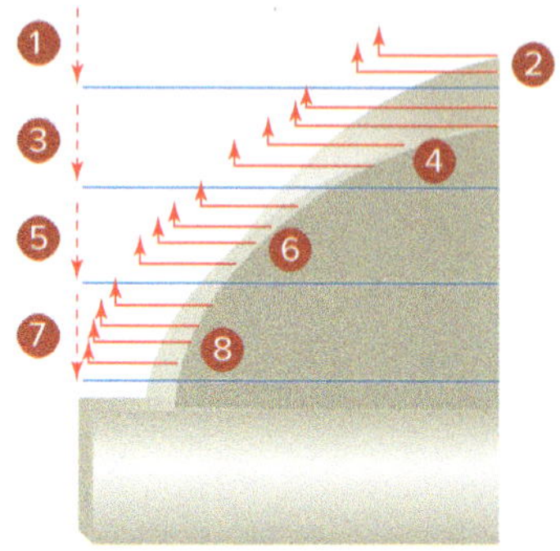

Figure 25-19 Down passes (1 to 7) compared to up passes that are far more efficient.
Source: CNC Software Mastercam Education Division, Gig Harbor, WA

Today, however, a dynamic roughing tool path is just the opposite: We spiral down as deep as possible, then rough with the cutter's side teeth as it is stepped up, with each pass cutting the fullest axial engagement (passes 8 up to 2).

As with all dynamic milling, this is far better chip removal and cutter usage! Mastercam's 3D, OptiCore, OptiArea, and OptiRest tool paths support cutters capable of machining very large depths of cut. It uses a fast, aggressive, intelligent roughing algorithm based on 2-D high-speed dynamic milling motion (Figs. 25-20, 25-21, and 25-22).

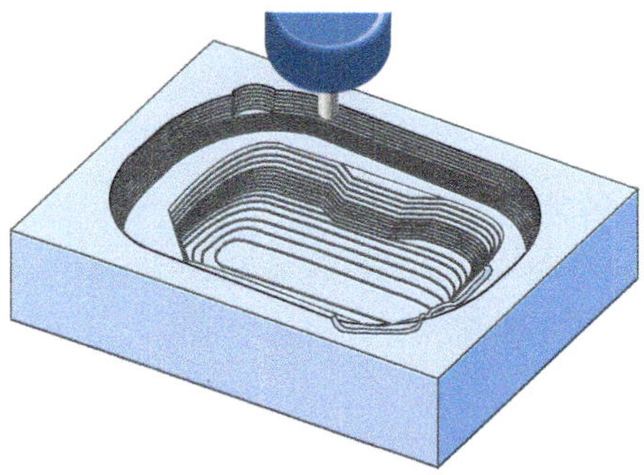

Figure 25-20 Plunge deep, then rough outward using more axial engagement per pass.
Source: CNC Software Mastercam Education Division, Gig Harbor, WA

Peel Mill 2-D

High-speed peel mill tool paths allow for efficient constant climb milling between two selected contours or along a single contour. It uses a trochoidal motion with accelerated back feed moves when the tool is not engaged in material. Peel mill can machine more than one contour by selecting them together (see Fig. 25-22).

Two Steps to Get Ready to Program

Okay, that's it—those were many of the possibilities of modern tool paths. Now it's time for a fun challenge.

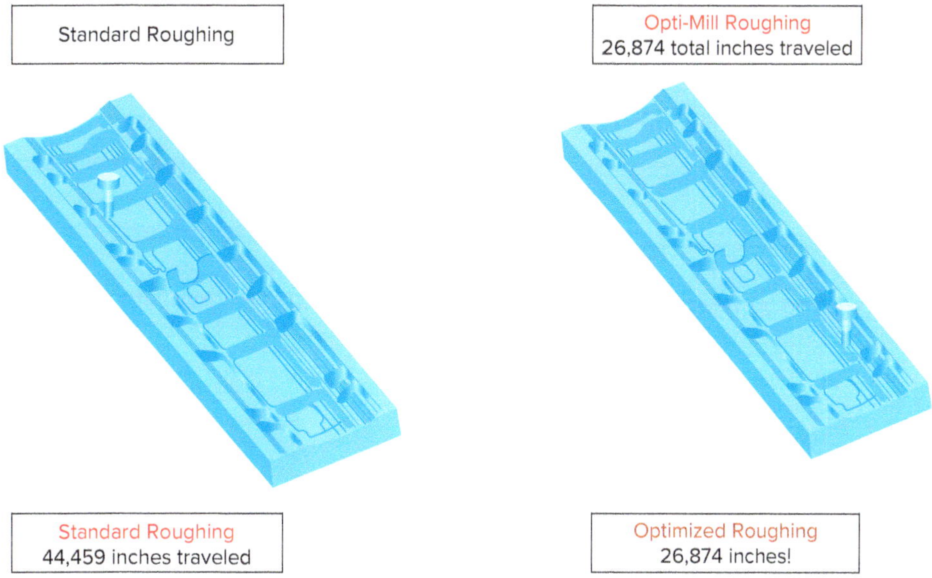

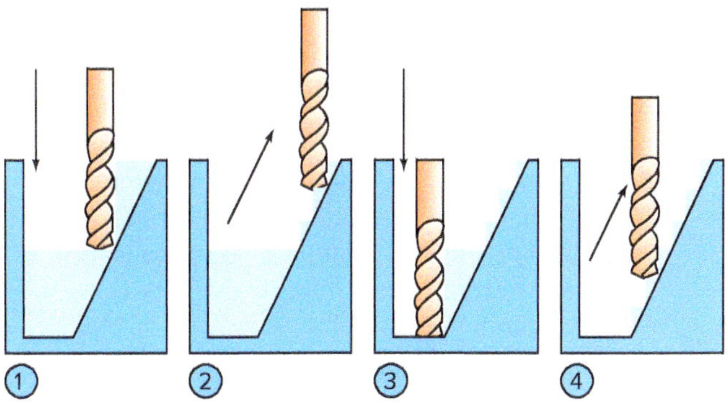

Figure 25-21 Up passes are far more efficient.
Source: CNC Software Mastercam Education Division, Gig Harbor, WA

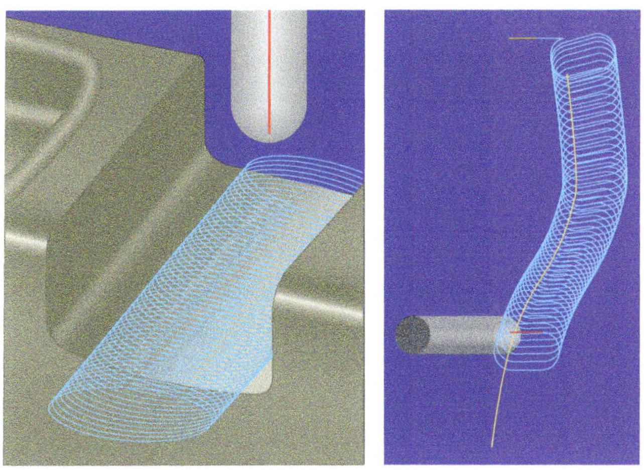

Figure 25-22 Peel mill can connect complex surfaces into a single tool path.
Source: CNC Software Mastercam Education Division, Gig Harbor, WA

Step 1: Gain Access to Mastercam or Fusion 360. If your lab has either one of these CAD/CAM software packages, go ahead and use it. If not, you can download a temporary copy to your home computer. For Mastercam, visit: https://www.mastercam.com/downloads/ and request a copy of their Home Learning Edition. For Fusion 360 by Autodesk, visit: https://www.autodesk.com/education/students and follow the instructions to access the Autodesk Education plan.

Step 2: Build Your Solid and Tool Path. Go to mhhe.fitzpatrick5e to find the step-by-step tutorial on how to create a solid model. Then, continue with the instructions that describe how to design tool paths upon that solid. There are separate tutorials for Fusion 360 and Mastercam, depending on which one you are using. For a final activity, you'll import the SolidWorks file drawn in the last chapter and build a tool path upon it, too. The benchmarks for Chapter 25 will be your completed solid and tool path. Additionally there is a 25-question test at mhhe.com/fitzpatrick5e–see your instructor for the *1st Experience CAM/CAM Certificate*. Your instructor and download issue this certificate to you once you complete the 25-question test and present your CAD/CAM project.

UNIT 25-1 REVIEW

Don't forget to test your skill in the terms game based on SolidWorks and Mastercam or Fusion 360–they truly matter with this fast-changing technology!

Replay the Key Points

- Graphic interface programs can be based on a wireframe, surface, or solid model.
- Using a solid model, many parameters can be automatic, compared to the other two part geometry models.
- Each part geometry has its best application–think of all of them as tools in a toolbox.
- Using dynamics, cycle times are reduced by 30 percent or more when compared to traditional milling.

- Dynamic tool paths improve tool life and are more economical in terms of cutters and the machines that can perform them.
- Dynamic tool paths apply to any operation, HSM or standard!

Respond

The following review questions are typical of a 25-question downloadable exam, which your professor or instructor can access from the Learning Center Instructor's Manual. Passing it to their satisfaction, they can then print a portfolio certificate of completion to show to employers—also found in their Instructor's Manual.

1. HSM means

A. High-speed milling

B. Helical special milling

C. High-shear machining

D. High-speed machining

E. None of the above

2. Axial engagement is

A. Using the side teeth on a cutter

B. More efficient for chip removal

C. A major objective of dynamic milling

D. All of the above

3. Radial engagement is

A. Using the end teeth of a cutter

B. Plunging into the work

C. Milling with the cutter's radius

D. The percentage of the cutter diameter that is engaged with the cut

E. Answers A and D

4. Identify the three kinds of part geometry used in graphic interface CAM programming:

A. Wireframe, engineer's drawings, and simple models

B. Solid models, surfaces, and wireframes

C. Solid models, feature-based models, and wireframes

D. Wire-based geometry, features, and solid models

5. A dynamic tool path. (You may select more than one answer.)

A. Is more efficient compared to standard tool paths

B. Runs cooler than a standard tool path—heat leaves with the chip

C. Runs hotter, since it only applies to HSM

D. Puts less force on the work, cutter, and machine

6. During HSM, why must we not make sharp cornered tool paths? (You may select more than one answer.)

A. They make poor finishes on the work.

B. They can't get into corners of pockets.

C. They can break cutters.

D. They put extreme force on machines.

7. Of the three types of part geometry, only one has volume. Which one?

A. Surfaces

B. Solid models

C. Wireframes

8. Question 7 was easy, but this one isn't: When programming with a wireframe or a surface-defined part model, what is the most challenging aspect in terms of crashing into the work with the cutter in rapid travel?

A. Mastercam X (or any CAM) does not recognize metal from not metal.

B. The bounding box, billet, or bounding model cannot be defined from them.

C. Neither part model has shape definition.

D. None of the above applies.

9. When plunging an end mill to depth, what path is dynamic to get to depth?

A. Pecking in

B. Ramming down

C. Spiraling in

D. Spiraling up and down

10. One problem associated with roughing out a pocket with a large diameter cutter is that it cannot get into corners and tight places.

A. True

B. False

CHAPTER 26 Advancing Technology

Units

26-1 Preparing for Success—Program Verification

- CAM software evaluates its own tool path but not the code on a machine.
- Vericut, our chosen example, does the final evaluation with nearly all production machines and setups. It comes as close as can be possible to the real program run in metal.

26-2 Better Tool Paths to Extend Tool Life and Improve Productivity

- Compensate for chip thinning for tool life and productivity
- Improve your tool paths based on Sandvik research

26-3 Beyond Chip Making—Three Different Ways to Remove and Cut Metal

- Be able to define the three processes, waterjets, lasers, and electrical discharge machining (EDM)
- Use your basic knowledge for OJT training in any of these three technologies.

26-4 Putting Metal Together—Direct Deposition

- Not unlike a 3-D printer, understand how we can make metal parts from digital files and lasers bombarding metal powder.

INTRODUCTION

Chapter 26 is loaded with new ideas and technology—some you can use today while others are preparation for your advancing career.

TRADE KNOWLEDGE OBTAINED

Unit 26-1 Program Verification

Knowing that a new program will be successful before it hits the shop floor shortens setup times and reduces machine collisions which can lead to extensive downtime. This unit will include an exercise that will allow you to test the program you wrote in Chapter 25.

- Set up a piece of stock in a verification project.
- Place a PRZ on a part based on your Chapter 25 CAM project.
- Import your program code into a verification project and test it.

Unit 26-2 Better Tool Paths

We're not done learning new ideas in machining. Following the dynamic tool paths we've just explored in Chapter 25, researchers in Sandvik Coromant's labs have discovered milling techniques that when applied to any tool path, dynamic or not, extend tool life and at the same time make better finishes, more accurately, and with improved cycle times!

- Apply chip thinning calculations to milling feed rates
- Modify your CAM tool path for better tool life and efficient, accurate milling
- Plunge milling for roughing
- Investigate the wealth of machining resources at Sandvik Coromant (www.sandvik.coromant.com/us)

Unit 26-3 High-Energy Cutting Without Chips

Next we'll explore three technologies that remove metal with super-pressurized water, high-wattage laser, or electrical spark erosion! All are fully implemented in manufacturing today. Using CNC control they continue to evolve and become ever more popular. They each have their place in cutting metal—with the right and the wrong application. Chances are you will work in a shop that uses one or more and may be challenged to run one in your career. Materials for each technology have been generously contributed by leaders in their field. Upon completion you'll know the basics and be ready to be trained to use them:

- Waterjet cutting
- Laser applications:
  Part marking
  Cutting metal
- Electrical discharge machining (EDM):
  EDM process
  Sinker electrode
  EDM wire feed EDM

Unit 26-4 Direct Deposition—Creating Metal Parts

The technologies of Unit 26-3 were once considered new ideas but not today: They grew from experiments to fully implemented tools. Now in Unit 26-4 you'll see the one that's at that leading edge stage—creating amazing new alloys and shaped parts using laser beams and powdered metal—not unlike a 3-D printer but a lot hotter!

- Laser-engineered net-shaping LENS Process

Unit 26-1 Preparing for Success—Program Verification

Introduction: Every time you write an NC program, there is that moment when you think, "I hope I've got this right!" In the past, there has been no sure way of knowing in advance if a new program would work. A machinist could spend time setting up a program only to find it unworkable or, in the worst cases, the new program might cause a collision. Having software that would show precisely what would happen in advance would give you a tremendous advantage in terms of time and safety and is the primary reason manufacturers use verification software.

So far we explored advanced options that manufacturers use for generating program code to run on CNC machines. We learned how SolidWorks can help with part design and how a CAD/CAM program, such as Mastercam or Fusion 360, can help develop the machining process and produce program code. In this chapter, you will see how industry uses verification software to test programs and to develop detailed setup plans before a job is sent to the shop floor. Later, you'll find out how you can check the program you wrote in Unit 26.1.

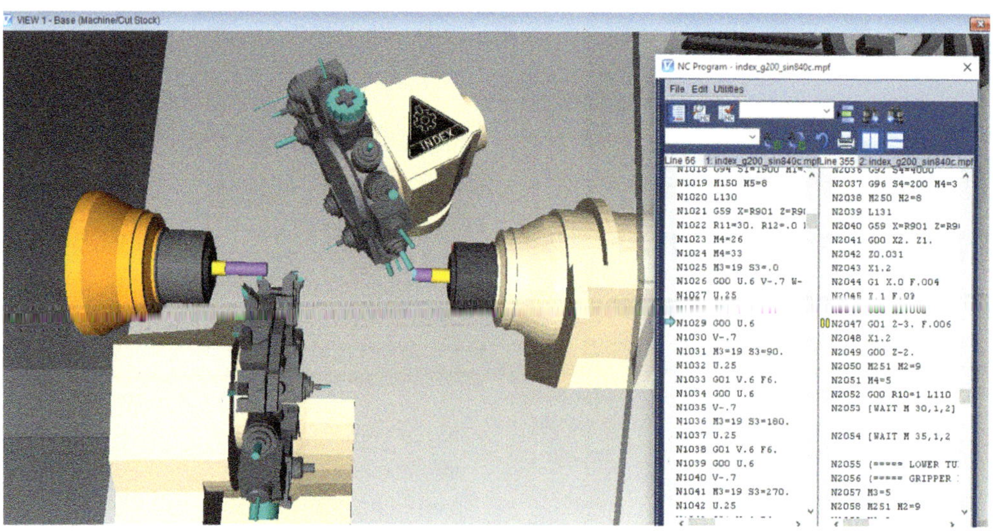

Figure 26-1 Using Vericut to test program code.
CGTech VERICUT

TERMS TOOLBOX

Collisions Unexpected contact between tools and machine components.

Cut stock An "in process" piece of material that has been partially or completely machined.

Design model A digital representation of a solid model that is created in a CAD program such as Solidworks.

Fixtures Work-holding devices used to accurately locate and capture stock to be machined.

Gouging Describes excessive material removed from a part during a machining operation.

Proving a program Running a program on a machine for the first time.

STL models (stereo lithography models) Digital solid models created in CAD software.

Tool definitions Describe the tool holder's profile, type of tool used, and tool extension from tool holder.

Virtual machine A digital replica of the physical machine used in computer simulations.

26.1.1 What Is Verification Software?

Introduction: Using verification software can be a lot of fun. It's similar to a computer game, but in industry it's serious business. Basically, you run an animated simulation on your computer that is based on the program code you created before it heads for the shop. Did you make a mistake and program a **collision**? The simulation will show you this before the program leaves your computer. Helping to develop programs that run safely is just one reason programmers use this software. Verification software is also used to help design tooling and fixturing.

Verification Simulation

Verification software, such as Vericut by CGTech, is used to test program code on your computer in a virtual environment that closely replicates the actual machine on the shop floor. You import an NC program into a verification project and a simulated animation is played on your screen. The animation will stop at any problem points in the program. One example of a problem point is where a collision occurs. In Vericut, the problem area will appear red, alerting the programmer that the program will need to be fixed (Fig. 26-2). Another example is when the program syntax, or language structure, is wrong. By testing a program in this way, the programmer can check to make sure the **program code is correct** while at the same time **testing the design of the setup**. Making sure that there won't be collisions between tooling and **fixtures** during a program operation is an important aspect of verification software.

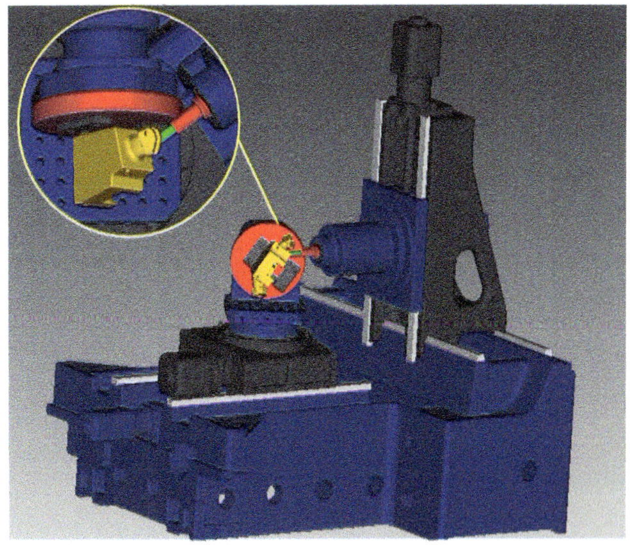

Figure 26-2 Red is bad! Collision between tool holder and fixture.
CGTech VERICUT

CAD/CAM and verification software programs both use animations and though they might look somewhat similar, the animations have different purposes. The CAD/CAM animations show you what the tool paths of a correctly written program would look like and their purpose is to help you write a program. Verification software simulations are based directly on the code and show exactly how the program will work on the actual machine.

The web-based activity that is discussed later in this chapter will walk you through the steps of putting together your own Vericut simulation. To do so you will download a home study copy of Vericut.

KEYPOINT

CAD/CAM animations help you create program code. Verification software bases its simulations on the code that has been written.

Proving a Program

Without using verification software, the testing of a new program can be a nerve-racking process! Without the ability to prove out a program before it comes to the shop floor, a machinist must employ other strategies such as changing a controller's setting to use single block, to use the rapid travel override, or to halt the machine at optional stops. Even when you are careful and take every precaution, it is difficult if not impossible to anticipate all the possible collisions that may occur, especially during the motion of multiaxis mills and lathes.

Proving out a program on the shop floor can be time-consuming and expensive. A new program with a 20-minute cycle time may take a couple of hours to test out. Verification software reduces the tryout time by eliminating predictable program conflicts.

Setup Sheets

Everything that is learned during the verification process will need to be clearly communicated to the shop floor. Detailed information about how to set up the fixture, stock, and tools are transmitted using setup sheets. The machinist works from these setup sheets, paying close attention to every detail. If these instructions are followed precisely, the shop floor results will match the programmer's verification simulation (Fig. 26-3).

Checking for Part Discrepancies

There are times when a machine operation will not have the capability to produce a part to print specifications due to the following:

- Problems with the program code
- Incorrect job planning
- Programmer error

CAD/CAM programs can sometimes generate code that will cause unintended overcutting (**gouging**) or undercutting (leaving excess material) a part surface. Another problem occurs when the choice of tools cannot produce a desired result. Finally there's just plain human error. No matter what the cause, it's disappointing when the inspector tells us that a hole has been placed in the wrong location or that a part feature is missing.

Figure 26-3 Setup sheet generated in Vericut.
CGTech VERICUT

But what if we could find these issues before real chips fly? Vericut can detect these types of problems. After running a simulation, Vericut can analyze the simulated **cut stock** and measure it against the **design model** to verify whether the part produced will meet the engineering specifications. By using verification software, the programmer can be confident that the program and operation design will be successful before it leaves the office.

> **KEYPOINT**
>
> Verification software predicts whether a machining process has the potential to produce a part that will meet engineering specifications **before** the job reaches the shop floor.

26.1.2 Components of a Vericut Simulation Project

So now that you have an understanding of what verification software can do, let's look at what goes into the creation of a simulation using Vericut. Setting up a simulation is not unlike setting up an actual machine. What would you need to build a part on a CNC machine? A machine, of course, plus tools, stock, a fixture to hold the stock, offsets, and a program. You need digital versions of these same components to build a simulation in Vericut.

Building a Virtual Machine

Vericut allows programmers to create an exact copy of your shop's machine including clamps, fixtures, stock, and tools in a virtual environment. The programmer starts by importing a machine template into the Vericut project. This template will include

- A machine profile: These are the physical components of a machine, the enclosure, worktable, spindle, tool changer, and so on.
- The machine's controller: There are many different manufacturers and models of CNC controllers. Each controller has its own way of interpreting program code. **For a machine simulation to be accurate, the Vericut project must use the same controller as your machine uses.**
- Cutting tools: The tools used in the Vericut project will need to have the same definitions as the tools the actual shop machine will use. This includes properties such as tool-holder profiles, tool lengths, and so on. It is important that the tools used in the Vericut simulation closely match the tools you will use on the actual machine. **Collisions caused by improper tool planning are one of the major causes of machine crashes.** The Vericut animation will alert you to possible collisions by stopping and highlighting the problem area in red, but it can't do this if you give it incorrect **tool definitions** (Fig. 26-4).

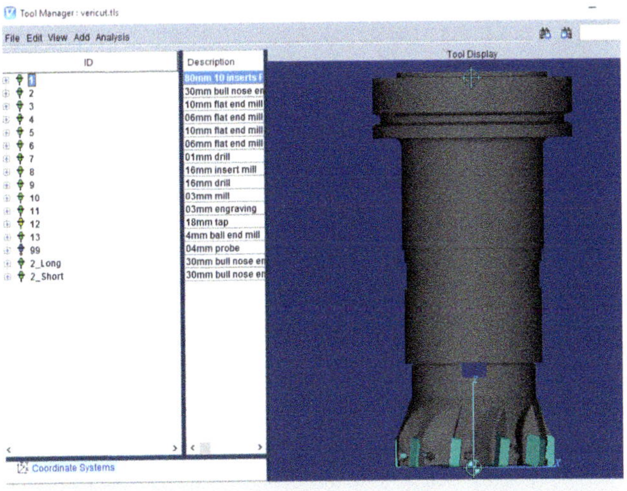

Figure 26-4 Defining tool definitions in Vericut.
CGTech VERICUT

Fixture Design

In the next step, the programmer imports a CAD model of the fixture to be used for a particular operation into the project. The fixture may be as simple as a 6-in. vise or a dedicated fixture designed to capture a partially machined part with an irregular shape. Either way, **the fixture can be designed in a CAD system**, such as SolidWorks, and then exported as an **STL model** to the Vericut project. Once the fixture has been proved out in Vericut, the fixture can be built by a toolmaker to the same design specifications. See Fig. 26-5.

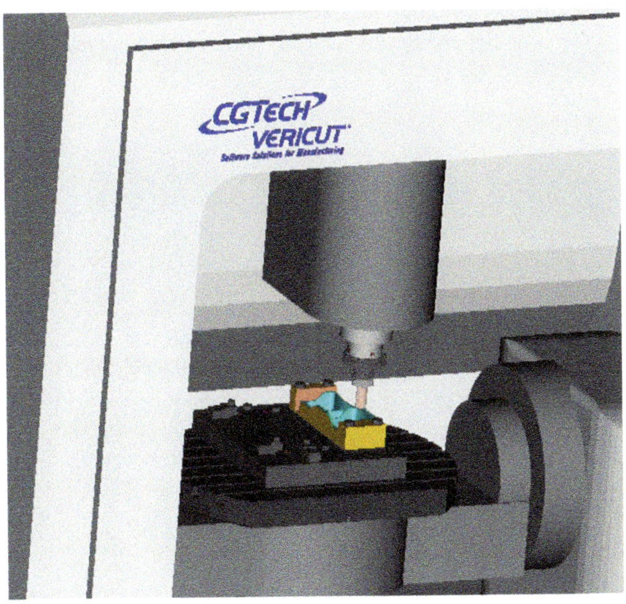

Figure 26-5 Simulation of a part being machined on a fixture plate.
CGTech VERICUT

Stock

The model of the stock used in the Vericut project must be a replica of the stock that will eventually be used on the shop floor. It may be raw stock such as a block, or it may be cut stock from the previous operation. **The PRZ will be established on the stock in the Vericut session just as you would when setting up a job on the actual machine.** It will also be necessary to use the same PRZ that was used by the CAD/CAM system that produced the NC program.

KEYPOINT

The reliability of the simulation depends heavily on the attention to detail you give to the setup of the virtual machine project.

26.1.3 Testing the Program from Chapter 25

Now it's your turn to try program verification. Go to instructor resources and download the *Machining and CNC Technology* Chapter 26.1.3 Student Activity. You will be provided a temporary license to Vericut, to load on your home computer.

If you used a home edition of Mastercam to program the 25.1 activity, you wouldn't have been able to generate a program code. You can download the NC code provided in the instructor resources website to practice this Vericut exercise.

For this activity, you will open a template Vericut project, create stock, establish the PRZ, load the NC program, and run the verification simulation. A video link will be provided that will walk you through the steps for this activity.

UNIT 26.1 REVIEW

Replay the Key Points

- A verification machine simulation animation is based on the NC program.
- Using verification software saves time by shortening the proving-out time.
- Proving out of programs is done on computers rather than on the shop floor, which cuts down on the number of unproductive work hours.
- Fixture and tool design is tested in the verification process.
- Eliminating machine crashes due to errors in programs or fixture design is an important function of verification software.

Respond

1. Where would the PRZ be placed in a Vericut project?
2. True or false? A Haas controller interprets an NC program exactly as an Okuma controller does.

3. Why would the CAD/CAM animation results differ from that of a verification simulation?
4. Name three ways verification software can reduce overall production time.
5. In Fig. 26-6, would this ½ end mill be capable of producing a part to print requirements? Why or why not?
6. How would Vericut alert you to a machine collision during simulation?

Questions 7 through 9 are based on the Vericut web activity:

7. What is a CSYS?
8. What kind of controller was used in the Vericut project?
9. Name two possible solutions to the vise stop problem.

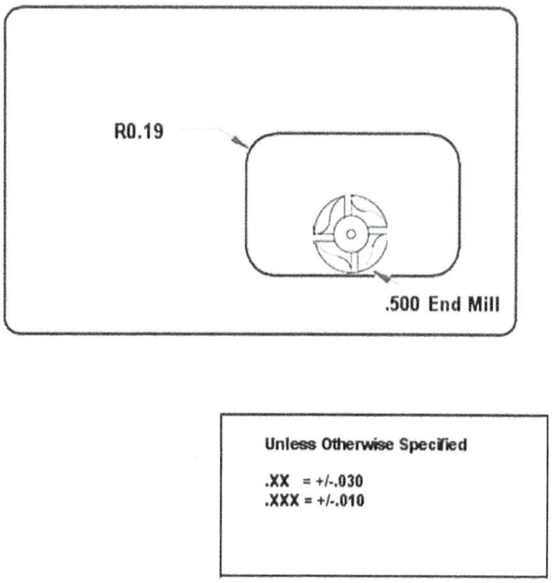

Figure 26-6 Problem 5 print.

Unit 26-2 Better Tool Paths to Extend Tool Life and Improve Productivity

Introduction: Unit 26-2 has two overreaching goals:

1. To apply several new tool path ideas for efficiency and longer tool life.
2. To direct you toward investigating resources–there's more to learn than I can put in this book! Plus it will evolve quickly. Many useful facts can be found in milling handbooks, free catalogs, and YouTube videos. Attend seminars and stay tuned–we're just getting started on tool path evolution!

TERMS TOOLBOX

Chip thinning Not achieving the chip load due to shallow radial engagement.

Cutter geometry immersion Amount of the tooth shape that engages the cut.

Radial engagement (RE) The percentage of the cutter's full radius that makes the mill cut.

26.2.1 Chip Thinning

Milling cutter feed rate calculations begin with one single data point found in a recommendation chart–the amount each cutting tooth should advance in one revolution, called either "feed per tooth" or "chip load." Setting the feed rate to achieve that feed per tooth is important for tool life, heat removal with the chip, and efficient machining.

Chip load numbers range from a conservative 0.001 in./rev up to about 0.045 in./rev or more, depending on the combination of material being cut, cutter and spindle speed, and horsepower.

Let's suppose that in *Machinery's Handbook* or *Sandvik's Milling Handbook* or smartphone app, the given F/T should be 0.020 in./rev. Multiplying that times the 12 teeth times the RPM, we arrive at 72 in. per minute feed rate.

The chip will be 0.020 in. thick at that feed rate as long as the cutter is at 50 percent radial engagement or greater (Fig. 26-7).

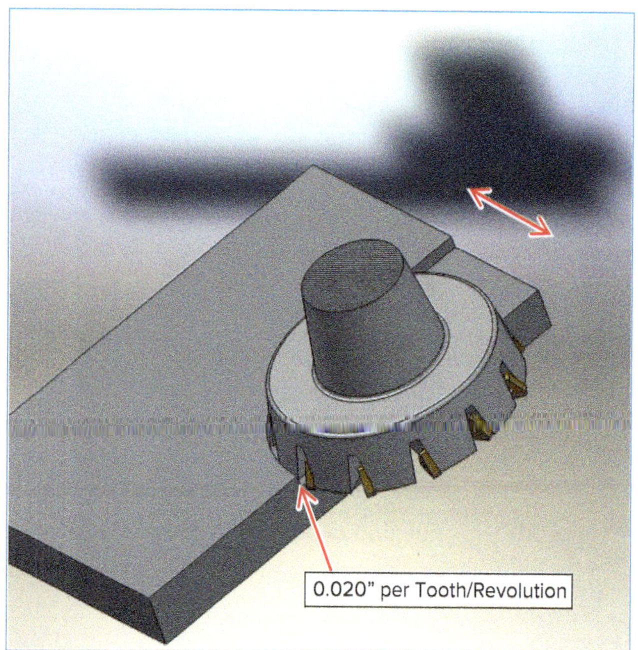

Figure 26-7 At 50 percent radial engagement or greater, the chip load is achieved with an uncompensated feed rate.

KEYPOINT

The mill cutter must be at a radial engagement of 50 percent of its diameter or greater to achieve the chip load at the calculated feed rate.

But when performing dynamic tool paths, we back away to a 10 to 20 percent **radial engagement (RE)**, and the chip is thinned by a progressive ratio (Fig. 26-8). This must be compensated–otherwise heat builds up and the program is inefficient.

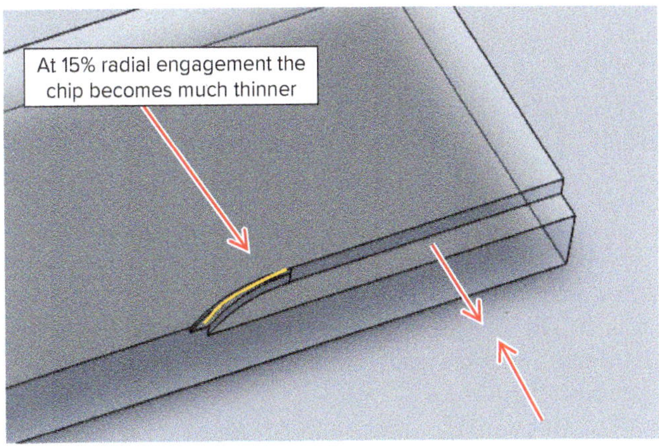

Figure 26-8 A small radial engagement creates a thin chip unless compensated with a faster feed rate.

Use the following formula to increase feed rate to achieve the desired chip thickness. This is true for any milling, but it is a critical step for HSM, where heat must be removed in the chip–otherwise meltdown is imminent.

Note on line 1 of this Sandvik chart (Fig. 26-9) that when the cutter is radially engaged at 25 percent of its diameter, the calculated feed rate must be increased by a factor of 1.2 (120 percent faster) to achieve the desired chip thickness for both face milling and profile cuts. At line 4, with 10 percent RE, the factor zooms up to nearly double (1.7 × calculated feed rate). You can extrapolate between these numbers for other adjustments to feed rates.

Tooth Shape Considerations (Fig. 26-10)

There is a second factor that can cause chip thinning–the **cutter geometry immersion**–which includes the lead angle on straight cutter teeth and, for round inserts, the total amount of the circle involved in the cut. Both can thin the chip by another factor. See approach angle (1) and radius insert (2). This is not new–we already know that lead angle has the effect of spreading the chip over a longer distance; thus, it's thinner. The needed upward adjustment can be calculated from a milling handbook formula or chart.

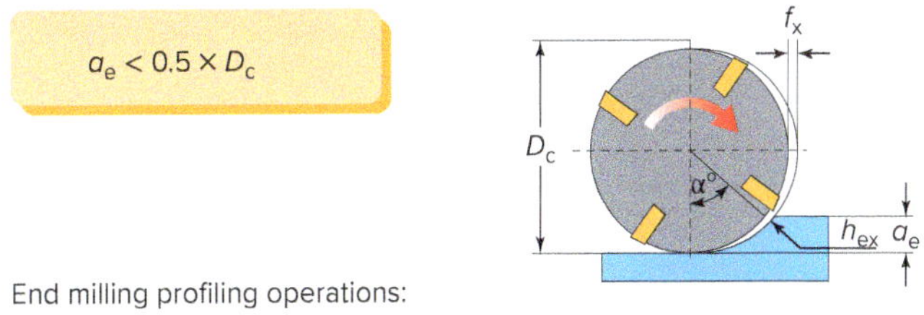

End milling profiling operations:

– CM390	– R390-11T308M-PL S30T	– h_{ex} = 0.003"
– Plura	– R216.24-xxx50-AKxxP 1620	– h_{ex} = 0.0016"

Radial depth of cut/diameter ratio a_e/D_c	Feed modification value	f_z inch for given chip thickness in inch	
		Plura - h_{ex} 0.0016	CM390 - h_{ex} 0.003
0.25	1.2	0.0019	0.0036
0.2	1.25	0.0020	0.0038
0.15	1.4	0.0022	0.0042
0.1	1.7	0.0027	0.0051
0.05	2.3	0.0037	0.0069

Figure 26-9 Sandvik chip calculation chart

Chip thinning
Entering angle effect on feed rates

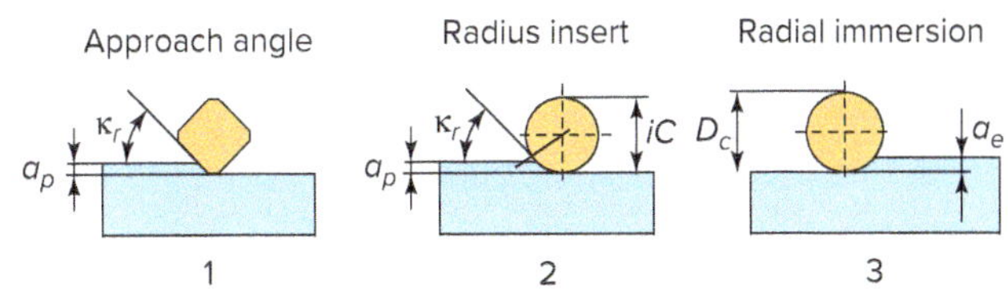

Figure 26-10 The tooth shape with lead and radius can also thin the chip.

KEYPOINT

Similar to RE chip thinning, *lead angle* on mill cutter teeth or round inserts *not buried* to their full IC will also cause chip thinning and should be compensated from data charts.

26.2.2 Modify Your Mastercam Tool Path for Better Tool Life and Efficient, Accurate Milling

Now we'll investigate a couple of logical innovations you can add to a tool path–one is added by selecting the right linking parameters in Mastercam (Fig. 26-11). The others you will need to add to the part geometry definition. They all involve curved tool paths when possible.

Rounding Corners and Rolling In and Out, to Prevent Tool Failure

The objective is to not shock or hammer the insert teeth by adding small modifications to your tool path. When facing an area in Mastercam, you have the option of turning on *corner rounding* in your tool path.

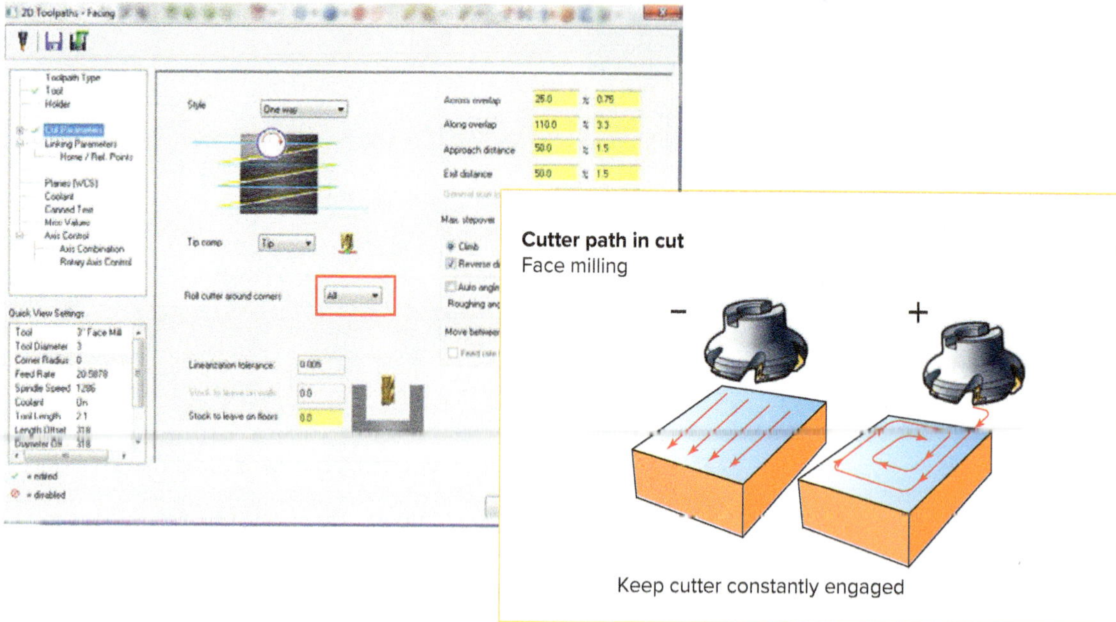

Figure 26-11 Mastercam cut parameter page
Source: Courtesy Flow International, Corp.

> **KEYPOINT**
>
> Tool life can be greatly extended by adding round corners in tool paths.

Editing Roll-In Approaches

Notice the added clockwise ¼ circle ramp onto the part geometry in the tool path (Fig. 26-11). That small but significant added wireframe curve can extend tool life by a big factor as the cutter teeth begin cutting smoothly. You can even hear this modification as the teeth begin with a thin entry building to full chip load, and the exit from the cut is smooth, too.

26.2.3 Plunge Milling for Roughing and Spiral Hole Milling

Our next example has been with us for a while, but it still works wonders in many applications. Cutters can often withstand an end push into deep cuts better than they can hold up to side motion (Fig. 26-12).

Plunge milling is a roughing technique that can often remove the bulk of the material in far less time than profile milling. It can be used to mill all metals but works especially well on hard metals.

Circular Pocket and Hole Milling

Similar to plunge milling, when we need to dig down into a pocket cavity, previously we ramped or spiraled to the first depth, then began facing. Here again researchers discovered an improved tool path. Using curved or circular tool paths, teeth aren't shocked and last longer; plus the removal rate increases (Fig. 26-13)

> **SHOPTALK**
>
> **Two Hints to Stay Informed**
>
> 1. Go to the Sandvik Coromant website to download a free smartphone/iPad app machining calculator.
> 2. Type "Sandvik Tool Paths" or "Dynamic CNC Milling" into your browser or directly into YouTube.

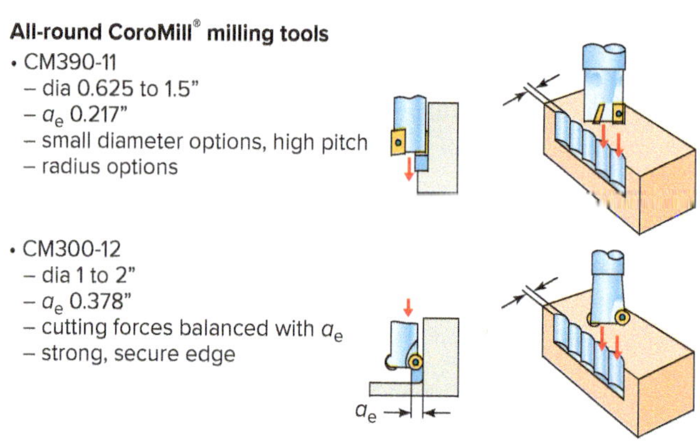

Figure 26-12 Using the right cutter—plunge roughing can be more efficient than profile milling.
Courtesy Flow International, Corp.

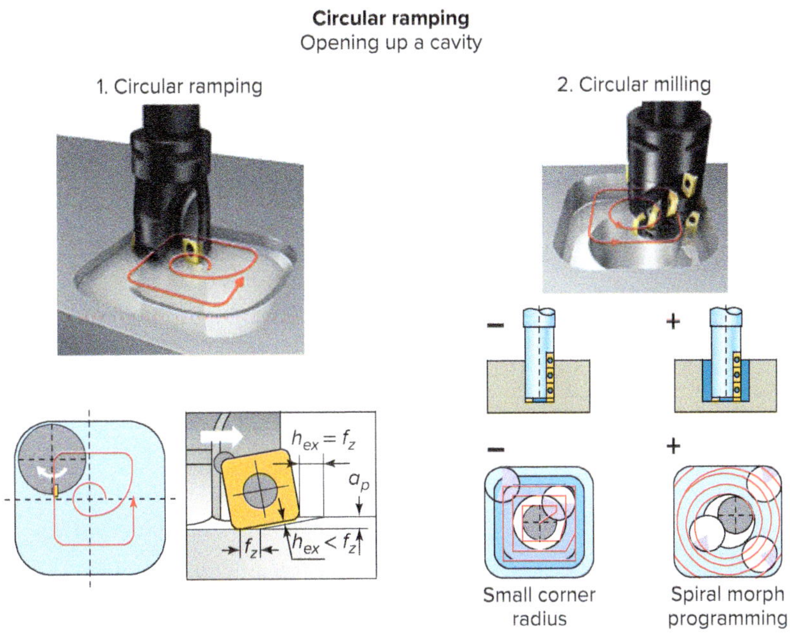

Figure 26-13 Tools last longer in curved tool path ramping.

Two Final Key Points—Based on Lab Results

1. As much as possible, plan climb mill tool paths so that each tooth exits each rotation with the smallest chip possible. This prevents exit shock and tooth chipping.
2. To avoid insert hammering and failure, *never set the facing step-over parameter at 50 percent radial engagement.* Choose either 25 percent or 75 percent steps with climb milling; thus, they follow the guidelines provided.

UNIT 26-2 REVIEW

Replay the Key Points

- Mill cutters must be at a radial engagement of 50 percent of its diameter or greater to achieve the recommended *chip load at the calculated feed rate.*
- Lead angle on mill cutter teeth and round inserts not buried to their full IC will also cause chip thinning.
- Tool life can be greatly extended by adding round corners in tool paths and on approaches to the work.

Respond

1. What facing tool path modification can be added for improved tool life using linking parameters in Mastercam?
2. What tool path modification must be added to the part geometry for better tool life?
3. Which statement is true?
   A. Mill teeth last longer if the chip is thick upon exiting the cut.
   B. Mill teeth last longer if the chip is thin upon exiting the cut.
4. You are using a radial engagement of 20 percent of the cutter's diameter. To compensate for chip thinning, what factor must be applied to the feed rate?
5. Okay, let's see if you got it! Calculate the feed rate compensated for chip thinning:

Cutter:	6.0-in. diameter–10 carbide insert teeth
Material:	Half hard bronze with a recommended SS of 280 ft/min chip load 0.020 in./rev
Tool path:	Step-over parameter = 1.5 in. per pass

Critical Thinking

1. Your tool path steps over 75 percent per pass in a facing operation. How much chip thinning compensation must be applied?
2. How can you add the small clockwise entry to a ramp on for a face cut?

Unit 26-3 Beyond Chip Making—Three Different Ways to Remove and Cut Metal

Introduction: In making a product, planners can choose from a list of possible methods to cut, shape, and form it. An informed machinist should build a working vocabulary of these processes and know where they fit within the manufacturing loop. They cut material using three very different media–plain tap water, high-energy laser light, and electric current. Each brings a unique advantage to the shop and a right and a wrong application:

1. Waterjet cutting
2. Laser cutting and surface marking/finishing
3. Numbered list

TERMS TOOLBOX

WATERJET CUTTING

Closed-loop recovery Reusing water where contamination or scarcity are issues.

Heat-affected zone (HAZ) Disturbed and modified material immediately surrounding many cutting processes, not present with waterjet cutting.

Intensifier pump The hydraulic piston that creates ultrahigh pressure.

Orifice The tiny hole in the nozzle that focuses the energy.

Ultrahigh pressure Pressure above 35,000 PSI up to 60,000 PSI.

LASER CUTTING

Emitter Atoms, molecules, ions, or semiconducting crystals that produce light as they return down from an excited state to a normal energy level.

Exciter Triggers the process by pumping energy into the medium, thus raising the energy state of the emitter.

Gain medium/mirrors A material that either supports or forms the emitter and will not release the beam until it is perfectly directed.

EDM

Current The amount of energy flowing in the arc, measured in amperes.

Detritus The EDM waste composed not only of mostly cooled work particles but also of electrode and destroyed fluid, especially with oils.

Dielectric (fluid) The insulator path between the electrode and work. The absolutely essential component of EDM–water or oil.

Duty cycle The ratio of on time compared to overall cycle time.

Heat-affected zone (HAZ) The thin, undesirable, modified surface that is a by-product of EDM cutting.

Ion/ion path An energy state whereby electrons are shifted and become electrically biased, resulting in a greater tendency to conduct current.

Overcut The result of the arc gap beyond the electrode. Overcut can be regulated by varying voltage to control size. Similar to kerf in sawing operations.

Recast layer The thin surface modification caused by metal refreezing to the parent.

Voltages In EDM, voltages are the potential to bridge a gap. The higher the voltage, the larger the gap it will jump. EDM voltages range from around 300 at open gap to 35 at the maximum cutting action during the on time.

White layer or white metal The thin upper layer, surface modification of carbonaceous (oil dielectric) or oxidized (water dielectric) metal that was metallic vapor within the plasma bubble when it collapsed. In collapsing, it solidifies and refastens to the parent above the recast layer.

26.3.1 Waterjet Cutting

The name is waterjet–no hyphen. Originally invented for the lumber industry, waterjet cutting has evolved from a whiz-bang technology to become the fastest expanding specialized machine tool cutting application in the industry (per Flow International, Fig. 26-14). An *X*-*Y* CNC-directed vertical waterjet can cleanly cut, without burrs or burn marks, nearly anything from soft paper, foam, and baby diapers to extremely hard materials such as titanium, marble, glass, and even ceramics! Cutting is the main use, but there are several other ways a waterjet can be used. Using rotating multijets, surface preparation and paint stripping are accomplished, too. Every day, the list of applications expands. The jet can also pierce through or to a given depth in some applications. Especially note that waterjets cut without burning the material (author).

The Basics

The first part of the process is to filter plain water in stages. Then it's pressurized from 35,000 to 60,000 PSI in a set of **intensifier pumps** (Fig. 26-15). The **ultrahigh pressure** water is then delivered through highly specialized plumbing, to be jetted through a tiny **orifice** from 0.003 in. up to 0.018 in. The tiny stream looks like a silver thread when light reflects from it, but make no mistake; it packs a big wallop!

Figure 26-14 A CNC vertical waterjet with water absorption tank below.
Courtesy Flow International, Corporation

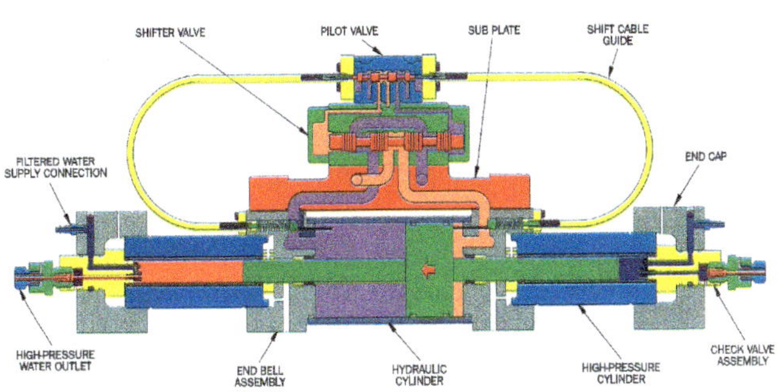

Figure 26-15 The intensifier pump creates the ultrahigh pressure required to cut material.
Courtesy Flow International, Corp.

The jet is usually directed by *X-Y* CNC axis drives as seen in Fig. 26-16, a vertical cutting machine, the more common form of waterjet. Or it can be mounted as the terminal implement on a robotic five-axis arm (*X-Y-Z-A-B*) to cut and pierce nearly anything from flat stock to complex 3-D shapes.

For safety and efficiency, the cutting head contains an instantaneous, off-on valve. Rather than stopping and starting the pump, this valve action is faster and safer. Opened with shop air pressure, should the supply fail, it quickly stops the stream and remains closed.

Adding Abrasives to Increase Cutting Power for Hard Material

When cutting hard material such as metal or glass, an abrasive is injected into the stream only *after* it exits the nozzle, for if we added it before passing even the hardest of nozzles, the precision orifice would be instantly blown out. Injecting the abrasive requires a merging and directing process to not diffuse the water stream's tight focus. This is accomplished using an acceleration tube, acting like a gun barrel, to join the grains to the water without disruption to the stream's focus. Typically, about 1lb of 80-grit abrasive is consumed per minute in this application (Fig. 26-17).

Pressure Pumps—The Process Heart

There are two ways water is brought up to the level that it can cut material: Intensifiers and gear displacement. The piston intensifier is the better in terms of peak pressure and overall life span, but it is more complex and costly as well.

Figure 26-16 A double-head CNC waterjet making large, decorative metal wagon wheels.
Courtesy Flow International, Corporation

Figure 26-17 A close-up shows the nozzle, valve, jet stream, and the abrasive induction equipment.
Courtesy Flow International, Corporation

Intensifier Pumps

With two or more double-action pumps working in tandem to eliminate flat spots in pressure, water is delivered, surge free, at pressures and volumes varying by the application. A double-action pump puts out pressure on both the outward and inward strokes of the piston, but working by itself, there would be a turnaround stall as the piston reverses. So the second pump is added to pump 90° out of phase from the first. That creates a continuous stream. The pumps work on the intensifier principle:

$$\text{Primary pressure} \times \text{Piston area} = \text{Secondary pressure} \times \text{Piston area}$$

So, large hydraulic pistons are powered with moderately high-pressure hydraulic fluid, which in turn pushes a much smaller piston against the water. The pressure gain, or intensification ratio, is 20:1.

Positive Displacement Gear Pumps

Developing enough pressure to do many jobs (but not all), liquid can also be pressurized by passing it through a set of precisely fitted gears. Gear pumps have a shorter life span, compared to intensifiers, but they are quick and economical to replace.

Waterjet Advantages

Using either pump method, due to the velocity of the jet, amazingly, nearly zero water is left behind on the product. It stays dry even when cutting absorbent diapers! Another bonus is that no heat is produced, a great advantage over other jet cutting methods. The lack of a **heat-affected zone (HAZ)** means no mechanical stress around the cut, no surface hardening, and no crack propagation.

KEYPOINT

Remember the term *HAZ*; it's a significant comparison item we'll see again in other articles and processes coming up.

Due to the fast cutting action on softer materials such as sheet rock or paper, a waterjet can be used to slice a moving slab of material at a perfect 90°, without stopping the slab. This is accomplished by moving the cutting head at an angle relative to the material speed, resulting in a square cut "on-the-fly" (Fig. 26-18). The final advantage is the availability of the water and its nonpolluting nature; a bonus, it can even be used over and over if necessary.

Figure 26-18 A multijet, paper-cutting machine uses waterjets without wetting the paper.
Courtesy Flow International, Corporation

Environmental Issues for Waterjets

Since the jet is so tiny, the narrow kerf carries away very little target material *swarf*, so cleanup of the tiny particles suspended in the water is simplified and economical before releasing it back to the environment. Furthermore, since an amazingly small volume of water is actually used in waterjet cutting, it may often be harmlessly returned to the environment as is, as long as the small amount of suspended material is nonpolluting. For light contamination or where abrasive cutting must be used, the water is directed to a settling pool where clean water flows over the top to be released. However, for toxic materials or where water is scarce, the stream may be efficiently recycled through the system again and again—called a **closed-loop recovery** system.

Safety Issues

Depending on the type and thickness of material being cut, and pressure used, as much as three-fourths of the original energy may still be in the stream after exiting the cut material. Stream catchers must be placed beyond the work. There are two types: For straight down cutting, a water pool with from 24- to 30-in. depth is required to absorb and dissipate the energy. The second method is a moving container that features a small hole for the jet to enter. It's filled with steel balls that deflect, dissipate, and absorb the excess energy with only 6-in. penetration. Obviously, operator protection is an absolute, and the plumbing must be of the highest quality to withstand the pressure and movement of the nozzle head.

Industrial Applications of Waterjets

Due to their versatility and efficiency, waterjets can and have replaced routers, saws, cutting torches, mill drills, and many other processes within the factory. Here's an amazing partial list of places you might see waterjets used to cut material. Flow International Corporation states that growth has far exceeded even its expectations as new and exciting applications for waterjets arise.

Aerospace: Cutting of expensive metals or metal-plastic fiber and graphite-epoxy composites

Automotive: Gaskets, plastic upholstery, floor mats, headlines, or any soft interior item

Building and construction: Building tiles, insulation, and wood; also to pierce (drill into) and remove concrete or other hard materials for shaping or for rework

Food industry: Chicken, frozen pizza, fish fresh, and frozen candy and vegetables (Fig. 26-19)

Job shops: Metal, glass, stone, plastic

Figure 26-19 The tiny jet neatly slices this candy bar without disturbing its structure at all!
Courtesy Flow International, Corporation

26.3.2 Cutting with Laser Light

Today, beams of high-energy laser light are doing an expanding array of tasks. Within the manufacturing world, lasers are used to heat in exact locations, to etch surfaces, and to weld (Figs. 26-20 and 26-21). They also mark parts and cut material, the two applications you might be called on to use as a machinist. Lightweight lasers can be the terminus on a five-axis robot arm, where in the auto industry they trim sheet metal after it is formed and neatly cut other complex shapes.

Figure 26-20 Computer direction of the moderate power laser, these machines can precisely and permanently mark a wide variety of items.
(Both) Courtesy Synrad of Mukilteo Washington State

Figure 26-21 A high-power, industrial CNC vertical laser cutting machine making intricate details in sheet metal components.
Courtesy Trumpf Incorporated/Farmington, CT

But, beyond that there's also a rapidly expanding market for lower-powered lasers. They're used for wildly diverse applications such as surveying, measuring, communications, tech-art displays called laser shows, delicate eye surgery, and marking extremely delicate or valuable objects such as serial numbers on diamonds and many more, but they are out of our arena.

Part Marking and Vertical Cutting

These are the two applications we'll investigate here. For both, CNC programs either direct the laser beam across the work surface or axis drives move the work.

The highly controllable nature of laser cutting makes it usable in unique applications not possible with other cutting methods. For example, the "flying optic" version involves mirrors that move in three axes to direct the beam thus expanding work envelopes without making the machine bigger. Here, both laser and work do not move, but rather, computer-positioned mirrors direct the energy to the work. Another redirection method is used in part marking where the mirror pivots to project the image to be etched onto nearly any 3-D work shape with a close control of penetration.

The controllability of laser means that different beam generation types with varying outputs can remove from only a fraction of a thousandth of the surface (Fig. 26-22) down to great depths. Varying kerf widths from a few thousandths and upward are possible with laser. Different from other thermal processes, due to the high density and small, focused spot size, extremely intricate contours may be achieved with close tolerances and exact repeatabilities (Fig. 26-23).

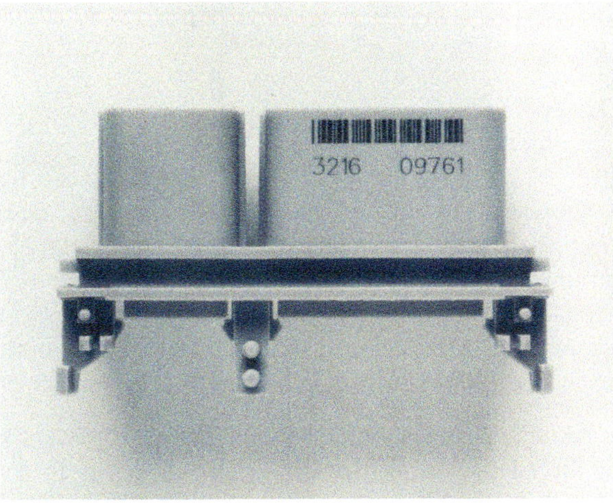

Figure 26-22 Note the tiny, precision barcode and part number etched upon the surface of this electronic component.
Courtesy Synrad Corporation, Washington State

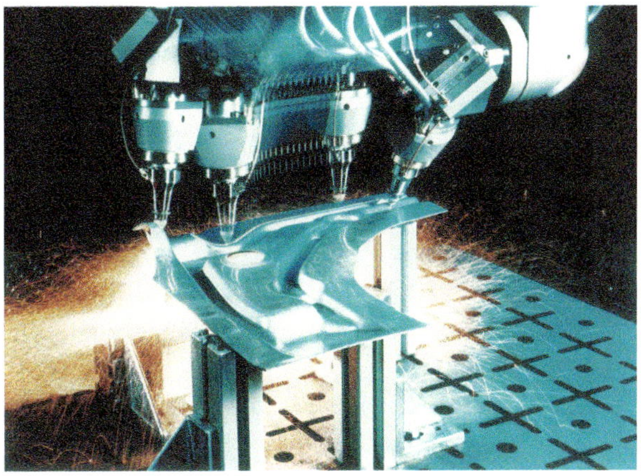

Figure 26-23 A five-axis robot moves a laser cutter to trim this complex sheet metal stamping.
Courtesy Trumpf Incorporated/Farmington, CT

Laser Advantages

Due to the narrowness and intensity of the beam, laser cutting can be quickly accomplished with little HAZ as compared with *conventional* thermal cutting methods (oxy-fuel flame or plasma cutting–author). Combining laser cutting and surface etching, designers can produce work not possible with any other method.

No tooling required
Easy to program
High accuracy
Can be projected to any surface image
Excellent sheet utilization (due to narrow swarf)
Quick part delivery
Quiet operation
Fast, permanent marking

KEYPOINT

Since a laser cut is made with a light beam, clamping/holding, work thickness, and turnaround time become nonissues, therefore this process is ideal for just-in-time manufacturing.

What Is a Laser?

An acronym for light amplification by stimulated emission of radiation, a laser beam is nothing more than light, but it's created in a special form. In 1960, the first laser shot out from a mirrored artificial ruby, as a pale pink light. The inventor, Theodore Maiman, explained why laser light can tightly pack more useful energy into the beam compared to other sources. There are two reasons:

1. **Uniform Frequency–Purity**

   Lasers emit in one single frequency. Various **emitters** produce different frequencies, but the uniformity of each means far more usable energy in one place compared to everyday light. Depending on the emitter, they range from infrared through visible light up to the ultraviolet regions of the spectrum. In comparison, everyday light, like that from the sun or a flashlight, contains many rainbow-like frequencies. Thus, it's difficult to focus its energy into a supertight beam.

2. **Uniform Direction**

   Due to the generating process, all elements of the beam emit along an exact parallel axis. This means far less beam separation as it gets farther from the source–very different from other light sources.

Coherent Light

Both properties combine to make laser beams usable as a process. The result is called *coherent* light; it stays closely focused over a long distance.

Laser Power Ratings

Manufacturing outputs are rated in watts. Cutting fabric or paper requires 200 to 250 watts while part marking occurs down around 10 to 25 watts, and there are literally hundreds of other uses below that, with outputs as low as a fraction of a watt. Cutting metal may require outputs from 1,000 or 2,000 watts up to 30,000 watts or more depending on material and thickness.

3 Basic Parts to Any Laser

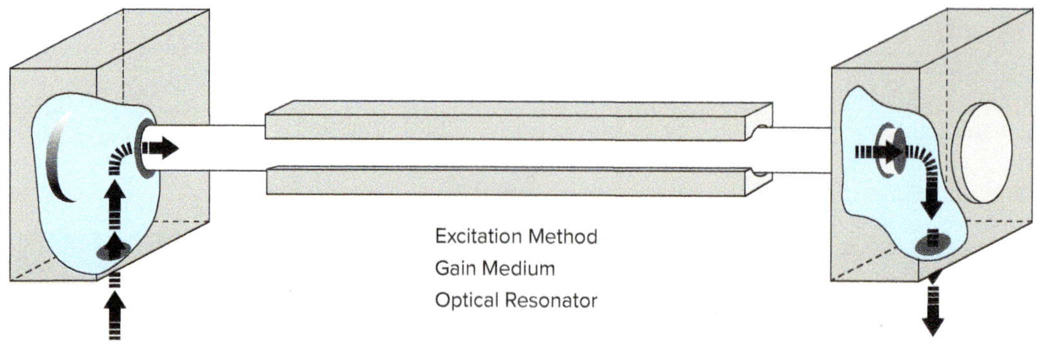

Figure 26-24 The three components of a laser.

Three Basic Components to Any Laser (Fig. 26-24)

1. **Exciter**

   Triggers the process by pumping energy into the gain medium, thus raising the energy state of the emitter.

2. **Emitting Entity**

   Embedded within the gain medium, atoms, molecules, ions, or semiconducting crystals first become excited by the pumped-in energy; then they produce light to return down from the excited state to a normal energy level.

3. **Gain Medium/Mirrors**

   A material that either supports or forms the emitter and will not release the beam until it is perfectly directed.

The Four Gain Media Types

1. Gas (CO_2 or argon, for example)
2. Solid (glass, ruby, semiconducting diode)
3. Liquid (with suspended dyes)
4. Vapor (He-Cd)

Two Differences Within the Laser Process

Similar to everyday lights that come from one of three sources, are iridescent (glowing element) or fluorescent (glowing gas) or LEDs (light-emitting diodes) that glow when current opposes their resistance path. Due to electrical energy put into the laser's emitting element, the laser process also begins by raising the emitter to a higher energy state (Fig. 26-25). In returning to the normal state, light photons are emitted by the excited element, but the big difference for lasers is they give forth their radiation at a uniform single natural frequency. Everyday lights emit in a range of color frequencies.

The second difference is that the photons cannot escape until they are all pointed in the exact same direction. To ensure uniformly focused beams, photons can pass through a semireflective mirror, but only if they are at exact perpendicularity. Otherwise they bounce back and forth within the medium, thus colliding with other charged molecules, which in turn begins a cascading effect, emitting further photons. Thus, they collide and bounce within the medium, until reaching exact perpendicularity to escape in a near-perfect beam. This bouncing action is known as a *laser oscillator* (Fig. 26-25).

KEYPOINT

Light cannot leave the gain medium until it reaches the required angle with the mirror, therefore the parallel nature of the beam.

Laser Output (Radiation)

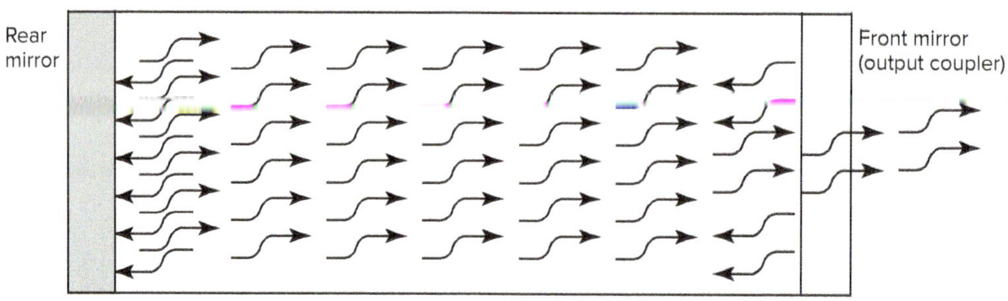

Laser beam exits resonator through an optical coupler

The output optic is 60% reflective, 40% transmissive

Figure 26-25 A laser oscillator features one fully reflective mirror and one at 60 percent reflectivity.

Environmental and Human Issues

It's vital that users understand that while laser is a form of radiation, it presents no special danger above other high-energy cutting methods. Other than direct contactwith flesh or eyes as concentrated beam heat or ultraviolet light, it has no effect on people or equipment nearby. Laser is just light, but it can and will cut and burn.

Since the beam is invisible in most cases, training and operator safeguards are an absolute must. Different types of protection are appropriate based on the frequency, cutting action, and output power. Also, when cutting certain materials, heated vapors and fumes must be recaptured, filtered, or recycled; plus latent heat in the product can burn hands. Otherwise, laser has no other danger or impact to people or environment.

26.3.3 Electrical Discharge Machining

Since its invention over 50 years ago, this metal cutting technology has expanded outward from hand-fed, broken tap removers into tool and die applications where it has become a central utility. But modern CNC controls and updated power supplies for the cutting action have now made EDM a practical solution in nearly all aspects of production manufacturing from aerospace, electronics, and automotive to medical equipment manufacturing.

Pacing machine tool evolution in general, CNC-driven EDMs now feature tool changers, palletized part loading and auto-tool-set changing, and autoslug removers. Some even have adaptive controls that adjust to varying cut conditions and autothreading of wire electrodes. Today, developments are coming fast, and wire machines (explained next) have increased their cutting speeds by an amazing 30-fold in the last 10 years alone. Manufacturers see greater removal rates with better surface integrity and improved finishes coming in the near future.

Why Use EDM?

This process is chosen when the work is too hard to machine conventionally or where the target shape is impossible to produce by other means. Finish is another reason. Modern equipment can produce mirror finishes with repeatabilities well below a thousandth of an inch.

KEYPOINT

Any conducting metal can be cut by this process, disregarding its hardness, and finishes can be as fine as in single-digit microinches.

Two EDM Processes—Wire Feed and Sinking Electrode

There are actually two distinct ways EDM is used to shape metal (Figs. 26-26 and 26-27). The wire machine, which performs similar to a continuous jewelers band saw (cutting a narrow kerf in any X-Y direction) but using a tightly stretched, slow-moving brass or alloy wire as a cutting electrode. The second type is the vertical ram or electrode sinker, which makes pockets and holes by burning downward with a shaped electrode into the material. The sinker machines make use of shaped electrodes that can be easily machined into many complex shapes. The electrode is made to mirror the final shape (reverse of the desired finished shape).

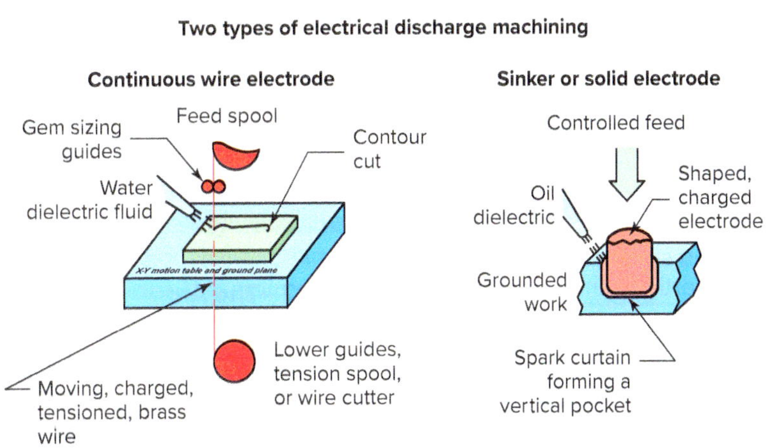

Figure 26-26 The two types of CNC EDM—wire feed and vertical ram/sinker.

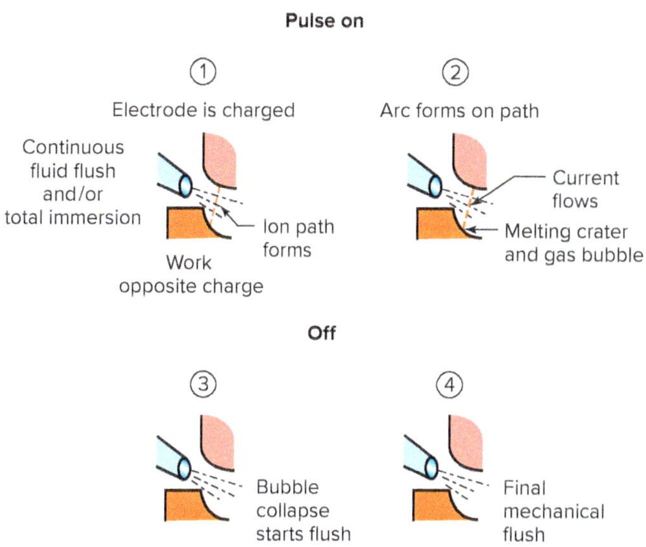

Figure 26-27 The four phases of one EDM arc.

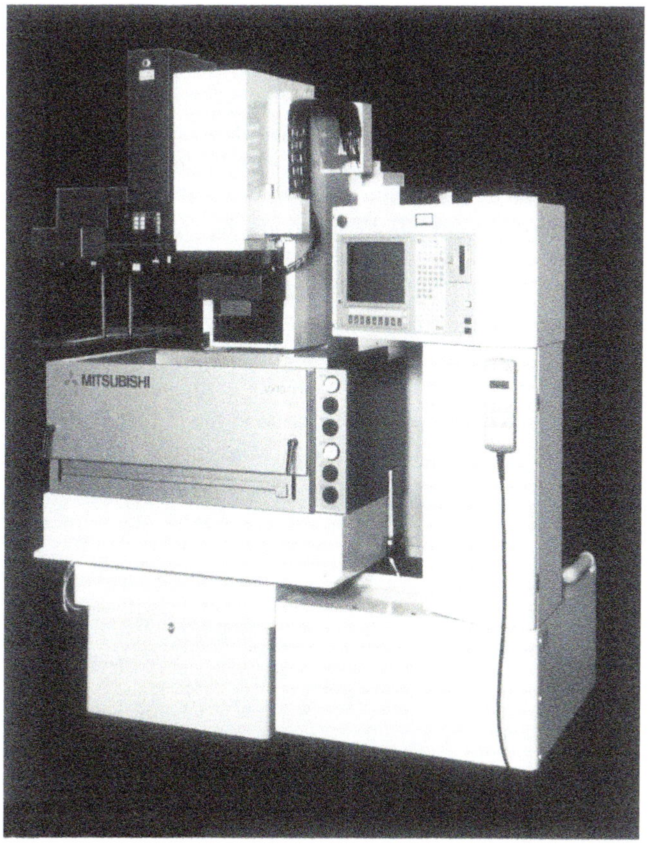

Figure 26-28 A CNC-controlled sinker EDM showing the oil tank, CNC control, and vertical ram. Note the tool changer on the left.
Courtesy Mitsubishi Division/MC Machinery

Additionally, by moving laterally with a CNC axis drive and control, many other operations can be achieved with both types: wire or sinker. Since both machine types cut with spark erosion, we'll explore the EDM cutting action before looking at the machines themselves. Refer to Fig. 26-28 to see an example of an industrial CNC EDM sinker and Fig. 26-29 for a CNC wire feed EDM.

The EDM Process

The "cutting teeth" in this process are millions of tiny electric arcs leaping between electrode and work, each melting and vaporizing metal to become a microcrater in the parent metal. The arc is an on-off pulse.

A fluid component between the work and electrode is necessary to not only cool but to control the action and flush away particles removed. This fluid, called the **dielectric**, performs several duties beyond cooling. Each is a separate but critical task. Truly, without the fluid no controlled EDM could occur. Early tap burners used plain air as the dielectric material supporting the arc between the electrode and workpiece. But due to air's inconsistent control of the arc, they were incapable of accuracy.

One Arc Through One Cycle

There are four distinct phases to a single on-off cycle. Fig. 26-27 shows them in sequence. This occurs thousands of times per second, over an area that looks to be a spark curtain to the eye. For simplicity, we'll envision it as a single arc.

The On—Cut Time

Phase 1—*Ion Path* Forms

During the on time, a charge builds between the electrode and grounded work, and atoms in the dielectric fluid within the gap between "line up electrically" in what is called an ***ion path*** or *flux tube*. Each atom, in that shortest path, shifts an electron, which causes a bias to conduct electricity.

No arc has started yet–just the path in the fluid, but within a nanosecond, the arc will follow. The fluid is supporting the path, but it's also the insulator preventing premature arc formation. Without it, the arc would be unstable and jump uncontrollably as it does in the air-supported tap burner.

The definition of a *dielectric* is a material that resists electrical **current** flow to a given point but through which an electric field may pass–thus holding off the arc until at a useful high-energy state.

Now, we've covered two of the three fluid duties: It supports the path, and it acts as an insulator, stabilizing the process to avoid premature arc formation.

Phase 2—The Arc

The arc forms and current flows, doing the real work of melting one tiny crater. During the spark time, the target metal first vaporizes into a gas bubble. Then as the crater forms, due to the expanding spherical crater surface absorbing the energy of the arc, the metal melts rather than vaporizing. Remember that aspect–we'll discuss it next. At this stage, molten metal lines the crater and some metal exists as vapor in a gas bubble surrounding the arc/crater.

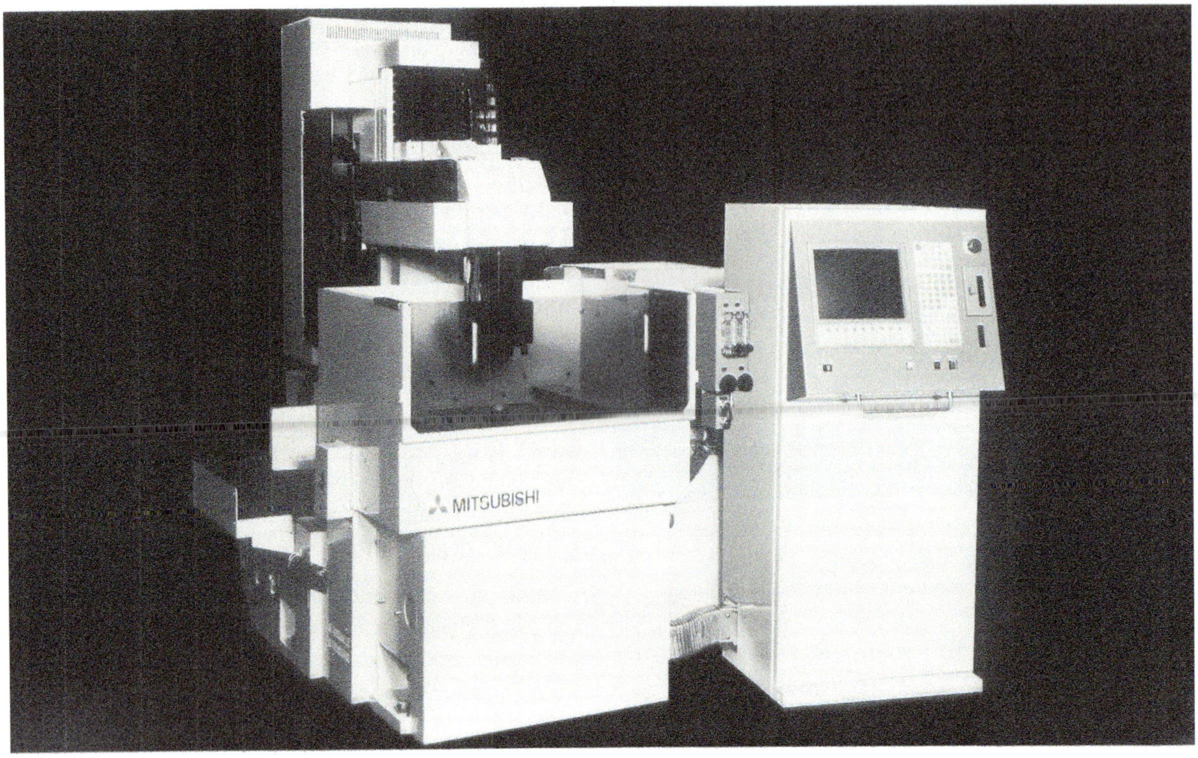

Figure 26-29 A wire EDM machine showing the water tank and CNC control.
Courtesy Mitsubishi Division/MC Machinery

Now two things occur that begin to progressively slow the action. First, particles of not only the work but also unfortunately destroyed electrode and burned fluid build up in the gap. As the crater expands outward from the original point of contact, the arc become less focused and stable. Second, the energy is being spread over an expanding area as the crater becomes larger; therefore temperature drops to a minimum useful level. So, due to both factors, at a given portion of the cycle, the current must stop to flush the area of particles to ready it for the next arc within fresh fluid.

Off Time

Phase 3—Bubble Collapse

Now, at a controlled time based on removal rates, desired finish, depth of cut, and other factors, the controller stops the arc for a microsecond. Here, the plasma-gas bubble (highly charged gas) formed from the intense heat collapses, which in turn forcefully pulls fresh fluid into the crater area. At this phase, the molten metal that wasn't flushed away freezes back to the parent. Its chemical makeup is not changed, but its structure has been modified. We'll discuss this again as the **recast layer**.

Phase 4—Mechanical Flush

The bubble collapse starts the flush, but a forced continuation is needed to completely sweep away particles and spent fluid. So, during off time flushing continues with fluid being pumped mechanically through the gap. The removed swarf is sometimes called **detritus** (total waste particles) in the EDM process. Contaminated dielectric fluid will be filtered for particles larger than 1 to 5 microns depending on the application, then sent back to be used again.

Phase 4 is complete. Now, with the gap swept clean, a single cycle is over, the control switches on again, and a new ion path forms.

Control Options and Factors

Since removal only occurs during the on time, the objective is to extend the burn period relative to the overall cycle time, known as the **duty cycle**. But faster removal means big craters and rougher finishes. Other factors, such as depth of cut with decreased flushing ability, come into play, too.

SHOPTALK

Marketability of EDM Skills Because EDM has been in the industry for some time and nearly all die and mold shops use it, operator skills are a valuable asset for employability.

KEYPOINT

As with all cutting operations, there are trade-offs when adjusting EDM cutting: Operators must choose from balances of removal rate, accuracy, finish, and tool wear. Sounds familiar, doesn't it?

Overcut and Electrode Wear

Because there must exist a gap between work and electrode both for flushing and to create voltage potential, the electrode must be smaller than the expected sinker pocket or wire kerf. A zero gap for either process would become a dead short! One prime function of the control is to sense and regulate the advancement of the electrode to maintain the correct gap.

Overcut is the EDM term applied to the difference between the electrode size and the cut area. It is a factor to be calculated before machining an electrode or selecting a wire. It must also be regulated during operation.

KEYPOINT

Due to the EDM arc length, the resulting cut is larger than the electrode (Fig. 26-30). By varying voltage, amperage, or duty cycle, fine adjustments of the finished size and surface finish can be achieved.

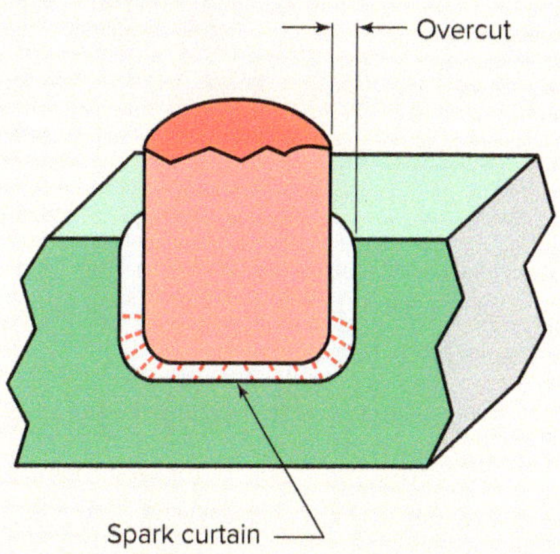

Figure 26-30 Cavity size is a function of overcut due to the arc gap between the electrode and work.

So, in addition to using the original electrode size to create final work size, overcut can be adjusted to fine-tune dimensions by varying voltage, amperage, and duty cycle times.

Roughing and Finishing on Sinker EDM

Due to the unavoidable erosion of the electrode as well as parent metal, it is often necessary on sinker processes to use two or more electrodes to rough, then finish the work. Since the wire machine continuously spools new electrode wire through the cut, electrode wear becomes a minor factor.

However, by misregulation of the wire speed or of the cut speed/power, the wire can be burned through, which brings the operation to a halt until it can be rethreaded. During wire work, where extreme accuracy of finish is a requirement, a second skim cut is often taken.

Surface Modifications to the Work

Heat-Affected and Recast Zones

Due to the nature of the process, two thin surface layer modifications are created and must be expected to varying degrees. The outer layer, known as the **white metal**, is actually material that was removed in the vapor but refastened to the parent as the bubble collapsed. Carbon from the oil dielectric, or oxygen from the water, combines with the metallic vapor. Due to intimate contact with the dielectric fluid, the white metal tends to be hard and carbonaceous on sinker machines and soft and oxidized on wire machines.

Then directly beneath is a second layer known as the recast metal. It was molten as the arc ceased, but since it had not been removed, it quickly froze to the solid parent beneath, during the flush. Its physical structure has been affected but not to the degree of the white layer.

Combined, these undesirable surface modifications are called the **HAZ (heat-affected zone)**. Both can cause cracking and other mechanical surface problems. But the good news is they can be controlled to varying degrees, depending on the operation. The HAZ can be almost eliminated completely in some cases with skilled machine settings. It can also be removed with secondary operations such as grinding or honing, where surface modification cannot be tolerated. Research is being conducted into methods of reducing and eliminating these effects. They are more prevalent on roughing cuts, shown in Fig. 26-31.

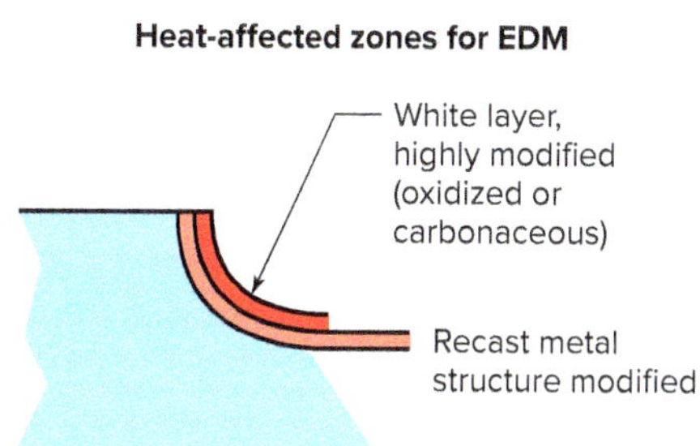

Figure 26-31 The two thin surface modifications result from EDM removal of material.

Two EDM Machine Types

Now, with an understanding of the process, let's take a second look at the two machine types: sinkers and wire feed.

Sinker Machines

The vertical ram holds and advances the electrode (Fig. 26-32). On many verticals, a CNC control can move it in any number of axes of motion (*X*, *Y* are common and *A*, *B* on fewer machines) to produce shapes beyond vertical pocket sinking.

Figure 26-32 The mold is submersed below the dielectric as this vertical EDM cuts precision injection mold cavities and plastic distribution runners.
Courtesy Jones Plastic and Mitsubishi Division/MC Machinery

Wire Feed Machines

Using plain water that has been processed to be electrically neutral (deionized) as the dielectric fluid, and a brass wire as an electrode, these machines also feature advanced CNC controls. With extra *A-B* axis control beyond *X-Y* motion, the wire angle can tilt with respect to the work, as it profile cuts. This results in a continuous draft angle effect shown in Fig. 26-33. The tilting axis machines can cut a cone from flat bar stock, for example.

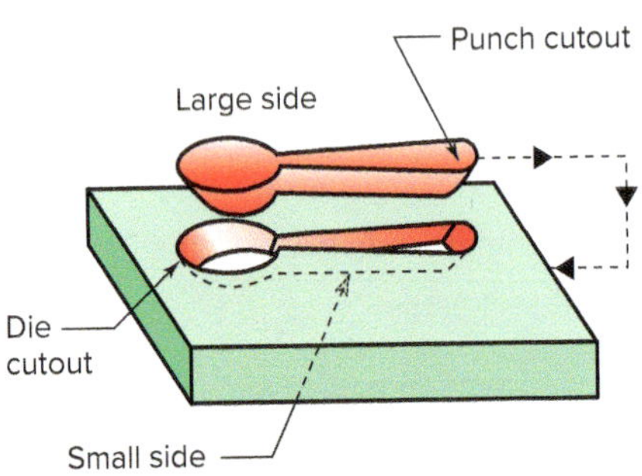

Figure 26-33 Continuously tilting the wire to create a draft angle, two parts can be made at the same time and can be sheared together for a perfect fit between punch and die!
Source: Courtesy Jones Plastic and Mitsubishi Division/MC Machinery

In the example, a die set is being made to blank (cut out by punching) a flat metal spoon. Both the die and punch are EDM hewn from the same piece of prehardened metal. The final operation will be to reverse their order, such that the wide part of the punch faces the narrow part of the die, then forcefully shears them together for a near-perfect fit. The draft is carefully precalculated such that there is a minimum amount of interference between the components; thus, they shear easily. The advantage is, the punch and die are made from a single piece of metal, in a single operation. The shearing removes the HAZ. Since EDM can cut hard metal as well as soft, EDM was the perfect technology to make these tiny detail punches. These blades (Fig. 26-34) were wire EDM cut from prehardened metal. They would have been very difficult to produce (or taken an impossibly long time) using any other process.

Figure 26-34 Impossible to make any other way using a wire feed EDM machine.
Courtesy of Mold Rite Inc

Wire Electrodes

Thin brass or brass alloy wire is used as the conductive electrode. Similar to a band saw, the cut must start at the edge of the work or have a predrilled starter hole through which the wire is threaded for internal cuts. Modern EDM machines have now been developed that will thread their own wire, unattended, thus putting them into the production line making use of robotic part changing and cell technology.

To control wire tension and exact size, the brass wire is stretched tightly between two tension rolls and pulled through a gem quality, sizing die before entering the cut. This ensures electrode uniformity. The wire is a one-time-only cutter. Therefore electrode wear becomes a near nonissue. After exiting below the cut area, it is either wound up on a take-up spool or cut into short sections and put into a bin, for recycling. Stratified wire with longer-lasting outer shells, and special alloys within to withstand tensioning, are available to improve finish, accuracy, and performance.

UNIT 26-3 REVIEW

Replay the Key Points

- There are six major components in a waterjet system: water filtration pump, special plumbing, nozzle head, the motion equipment, catchers, and waste equipment.
- A waterjet can cut nearly anything soft to hard.
- There is no HAZ with waterjet cutting.
- Laser is just light, so it poses no special danger over other high-heat processes other than ultraviolet light for some and invisibility. For these, specialized safety training and protection is critical.
- Due to the narrowness and intensity of the beam, laser cutting can be quickly accomplished with a little heat-affected zone (HAZ) as compared with conventional thermal cutting methods.
- Some designs can only be produced by combined laser cutting and etching.
- EDM removes by melting, then flushing small craters into the parent metal.
- Generally speaking, low frequencies are used for roughing and heavy metal removal.
- Any metal regardless of hardness may be cut by the EDM process. There are five major components in an EDM cutting tool: *electrode*—brass wire or solid carbon/copper alloy; *power supply*—creates and controls the on-off energy; *dielectric fluid*—water for wire machines and oil for sinkers—cools, paths the arc, and flushes away waste; *drive mechanism (machine tool)*—CNC drives the electrode causing vertical action and sideways cutting; *filter system*—cleaning the dielectric fluid in closed-loop recovery.

Respond

1. Of the three processes, laser, waterjet, and EDM, which affects the material the least? Explain.
2. Define *electrical discharge machining*.
3. Why must an EDM pulse the arc rather than make the arcs continuous?
4. True or false? Because it uses only plain water as the cutting medium, waterjet cutting is far less dangerous compared to laser. If it's false, what would make it true?
5. There are two kinds of EDM; describe them briefly.

Unit 26-4 Putting Metal Together—Direct Deposition

Introduction: From the first gut-wrenching moment when an ancestral machinist realized they had made their part undersized to this day, we've all wished for some way to put metal back on the part! Today, that wish has come true, granting us several processes that add rather than take away material—known as **additive manufacturing**. Search the Internet for the key words **direct deposition** *of metal* to read about the variety of ways it can be done. But as you'll discover, most of these crude repairs just add layers to existing parts; they do not create usable parts with complex shapes. One special combination of technologies can make parts, some impossible to produce by any other method.

It's a magic brought to us through advancements in four disciplines, all coming together to create solid or hollow, fully functioning metal parts starting with powder! The product can even be made of different metals combined into one part! These technologies are

Moderately high-powered lasers

Powdered metals

Exact CNC computer direction of the laser beam

CAD/CAM tool paths

TERMS TOOLBOX

Additive manufacturing Putting material together to make a part rather than removing material as in machining (subtractive).

Direct deposition Adding or building small deposits of material together to create a model or real part. See *additive manufacturing*.

Sintering An additive process which binds powdered metal together to form semihard parts.

Wire Electrodes

Thin brass or brass alloy wire is used as the conductive electrode. Similar to a band saw, the cut must start at the edge of the work or have a predrilled starter hole through which the wire is threaded for internal cuts. Modern EDM machines have now been developed that will thread their own wire, unattended, thus putting them into the production line making use of robotic part changing and cell technology.

To control wire tension and exact size, the brass wire is stretched tightly between two tension rolls and pulled through a gem quality, sizing die before entering the cut. This ensures electrode uniformity. The wire is a one-time-only cutter. Therefore electrode wear becomes a near nonissue. After exiting below the cut area, it is either wound up on a take-up spool or cut into short sections and put into a bin, for recycling. Stratified wire with longer-lasting outer shells, and special alloys within to withstand tensioning, are available to improve finish, accuracy, and performance.

UNIT 26-3 REVIEW

Replay the Key Points

- There are six major components in a waterjet system: water filtration pump, special plumbing, nozzle head, the motion equipment, catchers, and waste equipment.
- A waterjet can cut nearly anything soft to hard.
- There is no HAZ with waterjet cutting.
- Laser is just light, so it poses no special danger over other high-heat processes other than ultraviolet light for some and invisibility. For these, specialized safety training and protection is critical.
- Due to the narrowness and intensity of the beam, laser cutting can be quickly accomplished with a little heat-affected zone (HAZ) as compared with conventional thermal cutting methods.
- Some designs can only be produced by combined laser cutting and etching.
- EDM removes by melting, then flushing small craters into the parent metal.
- Generally speaking, low frequencies are used for roughing and heavy metal removal.
- Any metal regardless of hardness may be cut by the EDM process. There are five major components in an EDM cutting tool: *electrode*–brass wire or solid carbon/copper alloy; *power supply*–creates and controls the on-off energy; *dielectric fluid*–water for wire machines and oil for sinkers–cools, paths the arc, and flushes away waste; *drive mechanism (machine tool)*–CNC drives the electrode causing vertical action and sideways cutting; *filter system*–cleaning the dielectric fluid in closed-loop recovery.

Respond

1. Of the three processes, laser, waterjet, and EDM, which affects the material the least? Explain.
2. Define *electrical discharge machining*.
3. Why must an EDM pulse the arc rather than make the arcs continuous?
4. True or false? Because it uses only plain water as the cutting medium, waterjet cutting is far less dangerous compared to laser. If it's false, what would make it true?
5. There are two kinds of EDM; describe them briefly.

Unit 26-4 Putting Metal Together—Direct Deposition

Introduction: From the first gut-wrenching moment when an ancestral machinist realized they had made their part undersized to this day, we've all wished for some way to put metal back on the part! Today, that wish has come true, granting us several processes that add rather than take away material–known as **additive manufacturing**. Search the Internet for the key words **direct deposition** *of metal* to read about the variety of ways it can be done. But as you'll discover, most of these crude repairs just add layers to existing parts; they do not create usable parts with complex shapes. One special combination of technologies can make parts, some impossible to produce by any other method.

It's a magic brought to us through advancements in four disciplines, all coming together to create solid or hollow, fully functioning metal parts starting with powder! The product can even be made of different metals combined into one part! These technologies are

Moderately high-powered lasers

Powdered metals

Exact CNC computer direction of the laser beam

CAD/CAM tool paths

TERMS TOOLBOX

Additive manufacturing Putting material together to make a part rather than removing material as in machining (subtractive).

Direct deposition Adding or building small deposits of material together to create a model or real part. See *additive manufacturing*.

Sintering An additive process which binds powdered metal together to form semihard parts.

26.4.1 Assembling Metal—Here's How

Although there are several approaches to additive manufacturing, in this unit we will investigate one of the emerging technologies called the sintering process. This process begins with powdered metal and an epoxy binder. The printer lays a layer of powder then the laser scans the part shape to create a greenware version of the part. It then must undergo a second process where a binding metal is introduced as it melts the epoxy.

While the results are like magic, each process is not difficult to envision. They both begin with powdered metal and a printer. It's something like a 3-D printer making polymer prototypes but differs in that it lays down real metal, not plastic.

Sintering Process

The **sintering** process has several advantages over other additive processes:

- Many parts can be made at one time.
- No base start-up metal is required.
- Large batches can be produced in 48 hours start to finish (Fig. 26-35).

Figure 26-35 Dozens of small parts being printed in the powder layer.
3DX Industries – Ferndale Washington

Step 1—Build the Greenware

The machine deposits an even, fine layer of powdered metal then overprints it with fine droplets of binder. The print head then draws the part shape complete in the powder (Fig. 26-36). Next comes another layer and laser definition that binds the shape over the last. As each layer is printed the 3-D parts build into an uncured greenware part.

Using solid model files called stereo lithography (STL) files, the software divides the object into fine layers to be built one layer at a time

Figure 26-36 Large batches of parts can be produced at one time.
Courtesy Mike Fitzpatrick

Step 2—Interim Cure

Next the soft parts are heated to a temperature of 385°F to set the binder. This is necessary so that the parts can be handled without breaking in step 3.

Step 3—Introduce Metal Binder

Starting at 800°F the temporary binder gasses off, then elevates to the prescribed high temp that varies depending on alloy but not over 2,048°F. This step uses capillary action to melt and send a binding metal into the semihard part. The result is a fully functioning part.

The metal binder is in the fixture bar shown in Fig. 26-37. The metal binder can be brass or bronze (Fig. 26-38). This will go into the furnace along with the semihard parts.

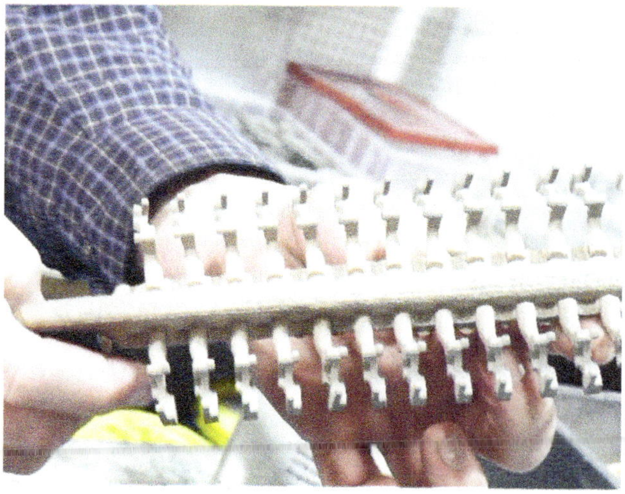

Figure 26-37 Ready to go to the high-heat furnace to induce the metal binder into the powder.
Courtesy Mike Fitzpatrick

Figure 26-38 The brass or bronze binder material.
Courtesy Mike Fitzpatrick

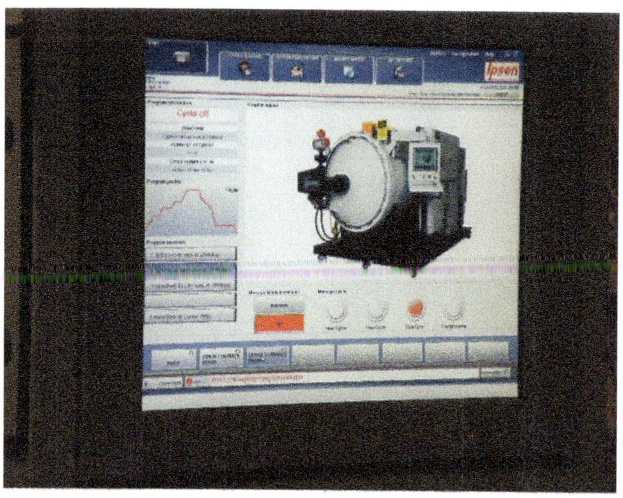

Figure 26-39 The final stage before any postprocess machining.
Courtesy Mike Fitzpatrick

The most critical aspect in terms of quality is the absolute control of the high heat in an atmosphere-purged furnace (Fig. 26-39).

As seen in Fig. 26-40, complex parts can be made. Many can be used as is, while others require postprocess machining to finish close tolerance features.

Figure 26-40 Impossible to make any other way. A bit of finish machining and this part is done.
Courtesy Mike Fitzpatrick

Replay the Key Points

- Direct deposition creates metal parts from powder directly, with no mold.
- The sintering process makes many duplicate parts in large batches.
- Sintering may require postprocess machining.

Respond

1. What three advantages does direct deposition offer?
2. Define additive manufacturing.
3. What kind of metal is used as a binder in the sintering process?
4. During the sintering process, how is the metal binding material introduced?

CHAPTER 26 REVIEW

Units 26-1, 26-2, 26-3, and 26-4 Paradigm Shifts

A paradigm is a perceived model of how we do things–a set of mutually and widely accepted rules. We machinists live under a set that tells us how we cut and shape metal. Milling tool paths is the perfect example–they used to go from point *A* to point *B*.

A shift in paradigm isn't an advance in the way things are done. It's a giant, and usually sudden, leap. It changes the rules. The problem is when any group or organization is entrenched in a paradigm (technology and procedures, in this case), the new way will be made invisible by those it challenges. The Swiss example shows this.

In the past, the phrase "Fine Swiss watch" was a part of the world's vocabulary. Swiss watchmakers have refined and kept secret the manufacture of microgears, escapements, springs, shafts, and other mechanical parts. They were unarguably the best at the craft. The industry was a large part of their nation's economy and of their image as the finest craftspeople worldwide.

But then the electronic digital timepiece appeared. Simple, with few moving parts, it was seen as an inexpensive novelty by the watchmakers but nothing more. Other nations saw the potential and began producing the new watch. In less than 5 years, the Swiss watchmaking industry was devastated. A final twist to this illustration: The digital watch was a Swiss invention in the first place! I'm sure you see the point: Dynamic tool paths and direct deposition are quantum differences in the way we do things.

QUESTIONS AND PROBLEMS

1. At what pressure range does a waterjet cut?
2. What dangers are there to waterjets?
3. Is this statement true or false? The gear drive, intensifier pump is the better but more costly means of creating ultra-high-pressure water for waterjets. If it is false, what would make it true?
4. What is the size of a waterjet stream?
5. What must be done to cut harder materials with a waterjet?
6. In your opinion, what are the main advantages of waterjet cutting?
7. Define EDM.
8. Define the phases of one EDM arc making one crater.
9. Describe briefly the two different kinds of EDM machines.
10. What materials can an EDM cut, and why?
11. Each type of EDM uses a fluid. Describe each and list five different purposes for the fluid.
12. Which fluid is used on a wire feed EDM? Which is used on a sinker?
13. What does the acronym LASER mean?
14. What are the special dangers of laser cutting? Is there a radiation danger?
15. Which of the three processes–EDM, lasers, and waterjet cutting–leaves behind a HAZ of two distinct layers? Describe each layer.
16. Why do we roll into a face cut?
17. Identify three possible sources of chip thinning.
18. What advantage is there to plunge milling?

CRITICAL THINKING

19. What facing tool path modification is made by changing a Mastercam linking parameter?
20. Describe how four technologies make the LENS process possible?

CHAPTER 27 Statistical Process Control (SPC)

As the world's most trusted source for 3-D measurement technology, FARO Technologies Inc. develops and markets portable coordinate-measuring machines (CMMs) and 3-D imaging devices to solve dimensional metrology problems.

Technology from FARO permits high-precision 3-D measurement, imaging, and comparison of parts and compound structures within production and quality assurance processes. The devices are used for inspecting components and assemblies, production planning, documenting large-volume spaces or structures in 3-D, and more. FARO's 3-D measurement technology allows companies to maximize efficiencies and improve processes.

Courtesy FARO Technologies, Inc., Lake Mary, FL

Units

27-1 What Is and Why SPC?

- List and describe the two objectives of SPC
- Define normal variation and real-time control
- Define the Cp ratio
- Compare a histogram to a control chart
- Define the two types of variation—special and common cause

27-2 An SPC Experiment

- Create a control chart and derive a Cpk ratio
- Provided a number of manufactured items; select a critical dimension
- Measure variation and produce your own histogram and control chart

INTRODUCTION

Statistical process control (SPC) equips users with an early warning radar that detects variation *before* it can cause problems. It works on any process with a quality target but especially well for machining. With the data picture SPC provides, we can learn how close to the target we should be able to machine a given feature, a difficult thing to deduce otherwise. Then, putting that powerful knowledge to work, we monitor, detect, and reduce variation when it exceeds norms. The goal is improvement and control of our overall process, with the ultimate goal that we make all excellent parts!

We know that within any batch of parts, features are going to vary in size, position, and other characteristics. Some of that unwanted variation can be reduced without changing the process, but to control or eliminate other sources, we must change the process itself (new cutting tools, new machines, and reworked plans or programs, for example). The skill is to know the difference and take the right actions. SPC is a data tool that helps us do just that: identify the two kinds of variation and then take steps to control each.

An understanding of the *nature of variation*. Arguably this is the most empowering aspect of SPC and a major goal of Chapter 27. Learning these concepts will help you make better decisions, even if you don't apply SPC to your work.

A clear picture of what we should *expect to happen*. After producing a number of parts (or operations), the capability of the process becomes known. That benchmark shows us how much variation is normal when everything is going right. That clear picture of "what is normal" in our process becomes the basis of all we do, called the Cp.

A clear picture of *exactly what is happening during real-time machining*. Using process capability as our guide, rather than job tolerance, we measure parts as they are made, called the Cpk.

A VISUAL DATA TOOL

While SPC can be performed with paper and pencil, it's a lot more efficient on PCs or CNC controllers, where the software crunches measurements of critical part features into graphs that show capability and control charts (run charts) that are used as real-time guides during production.

KEYPOINT

Knowing and using capability as a guide, rather than job tolerance, we better concentrate attention on reducing variation rather than inspecting, detecting, and perhaps reworking after it has occurred.

TOTAL OVERSIGHT

The same data are used in many other ways throughout the shop by management. With hard facts showing capability, they then bid or plan work that can be delivered with existing machines and tools. It also clarifies where equipment and tooling dollars should be spent to improve capabilities or pursue new business.

FINAL TRAINING

In Unit 27-1, we'll discuss the *what and why* of SPC for machinists. Then in Unit 27-2 we will present a lab activity. A great deal of the *how* depends heavily on the shop's application methods, the size of an average part run, general tolerances typical of the shop, and, to a lesser extent, on the specific software chosen. An employer survey showed they wanted entry-level machining graduates with an understanding of the benefits and a grasp of basic concepts and terms of SPC. Beyond that, they wanted to finish specific training themselves. Chapter 27 is written to provide that platform.

Unit 27-1 What Is and Why SPC?

Introduction: All machinists, whether or not they use **statistical process control (SPC)**, must understand the difference between the two kinds of variation: *assignable cause* and *common cause*. SPC helps us differentiate the two more clearly and early, so we can take the right action. If we can assign a reason for feature differences, then we can reduce or control them. But some variation is built into the process; thus, the process must be improved or we must accept the results as they are.

KEYPOINT

It's important to know the differences in variation because each is dealt with differently.

TERMS TOOLBOX

Assignable cause variation Variation specific to a factor outside the process.

Bell curve The natural graphical shape of variation, tending to centralize about a target result with diminishing results to each side.

Centering (a process) Adjusting such that results distribute equally around the target dimension.

Common cause (normal) variation The unwanted but ever-present spread of results in a process.

Control limits, upper and lower (UCL and LCL) The boundaries of normal variation in a process–they are not the drawing tolerance.

Key dimension A dimension assigned for SPC tracking. May be noted on the drawing with brackets or notes. May be assigned in the work order.

Range The amount of dimensional variation found within a group of parts.

Real-time control Tracking production of key dimensions, in sequence, as they are made.

Spread (variation) The natural bell-shaped tendency of variation to be domed with tapering to each side.

Statistical process control (SPC) A math/graphical method of seeing normal variation in any process with a quality outcome.

Trend A process that is starting to deviate out of the normal control limits.

27.1.1 Two Kinds of Variation

Common Cause Variation

Common cause variation is that which results from the process. It can be quantified (given a numeric range) by SPC and then can be reduced but only by process improvement. It cannot be eliminated altogether. Examples include cutting tool wear, work-holding flex or runout, and tool holder or machine rigidity.

Assignable Cause Variation

Assignable cause variation originates from special events that do change results but are not integral parts of the process. They often show as bumps or waves on production graphs. Examples might be a forklift driving past during a delicate operation, every 15 minutes, or the daily heat buildup in the shop, in a machine, or in the coolant. Using SPC, variation of this type can be identified as outside normal occurrences; then with a bit of detective work, the source *can be* eliminated. According to employer requests, if you get a good grasp on these terms you are well on your way to meeting their stated goals for SPC background.

27.1.2 Setting Control Limits—The Boundary of Normal

Let's clarify an important key point:

> SPC detects and helps sort variation, but it does not show the reason—identification of the source and what to do about it remain the machinist's job and are based on skill and experience.

Okay—here's how it's done. The first step is to compile how much normal variation is part of the process, the *common cause.* Those data are converted by the software into **upper control limits (UCL)** and **lower control limits (LCL)**. Control limits set the bounds for production norms. Then as we produce ongoing parts using those natural control limits, we can see and distinguish variation.

KEYPOINT

The UCL and LCL boundaries are not related to drawing tolerance; they are the *natural limits* of normal variation within the process when everything is going right.

For example, a drawing shows a tolerance of ±0.001 in., applied to the diameter of a 1.000-in. hole. After drilling and measuring, say, 100 parts, then entering the size, the SPC software shows us how well we can control its diameter—consistently around 1.0002-in. diameter, with the variation spread at ±0.0003 in. Through statistical formulas, those data are converted to the control limits for diameter around 0.0006-in. total range. The UCL/LCL spread won't come out at exactly 0.0006 in. We'll see why in just a bit.

Okay, we have the first ability—we know how well the process should go. For now, let's suppose there are no special causes—all variation is due to the nature of drilling this hole. That's the usual; most variation is common cause.

Provides Real-Time Comparison to Normal

Then with control limits set at 1.0005 UCL max diameter (1.0002 + 0.0003) and 0.9999-in. LCL (1.0002 − 0.0003) naturally occurring, the machinist continues to make parts and the **key dimension** feature is compared to the normal control limits. To be under control, the job must stay within the UCL/LCL limits.

To further improve, we could center the process on the design target of 1.0000-in. diameter using the 1.0003-in. UCL and 0.9997-in. LCL. But that step would be possible only once we have a clear understanding of the natural variation in this drilling process—hard data that are provided by the SPC software. To close the range we would need to change the process—use more coolant, implement a reamer following a smaller drill, or change the cutting speed, for example.

KEYPOINT

SPC isn't a voluntary tightening of tolerances but rather functions like a dipstick to show us how much variation should be occurring if everything is going right.

Properly executed, SPC empowers the machinist to predict future process outcomes, thus reducing the risk of poor quality.

Because SPC makes the process more predictable, budgets can be wisely spent on prevention, as opposed to detection and reworking.

Real-Time Control Charts

With some of the specific details left out, we now have a grasp on variation, and we come to the graphic tool used at the machine. By tracking progress graphically, they focus attention on trends in the operation as they occur, and indicate when the job is within control limits or show that intervention is needed. In Fig. 27-1, a key measurement is being tracked as parts are made. The key feature's target size was 0.6257 in.

You can see in Fig. 27-1 that things were going along pretty well until around 4:30 p.m. Until then, the diameter was coming out just a little big, but well within bounds, then a trend started. The measured key feature began deviating toward the upper control limit.

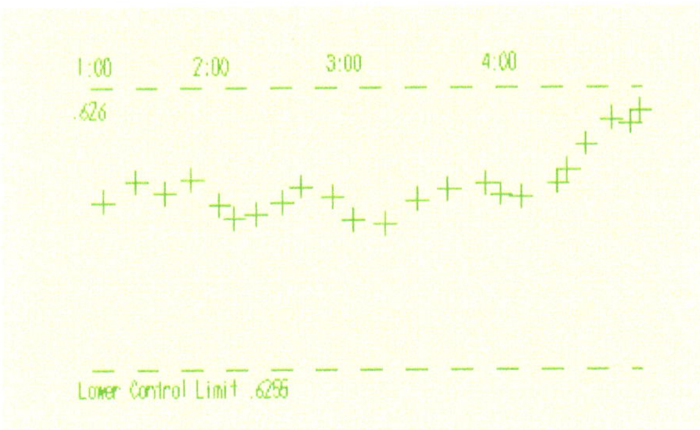

Figure 27-1 You'll be working on charts similar to this at your workstation.

Critical Question

If that trend continued across the UCL, would the parts be scrap?

Answer

Not yet; remember, the UCL is the boundary of the normal. The job tolerance still lies beyond that. But something needs to be done now!

The cause of the problem might be cutting tool wear, or the setup may need to be tightened up, or heat might be building in the machine itself, but it might also be that it's at shift change and a new machinist just took over and is doing something different! The graph provides clues but not solutions.

27.1.3 A Mind Experiment—Finding Normal Variation

All the principles are illustrated by shooting two different rifles at two targets of the same size. The objective is to hit the bull's-eye (a nominal target dimension), but there will be some variation. In this example, the rifles will be bench mounted and the ammunition standardized so only the rifles' ability to shoot straight is in question.

Each rifle could also represent two different CNC machines drilling a hole at a given geometric location. Both are attempting to locate the hole's axis at the same position, within the same nominal dimension. The bull's-eye represents the hole's true geometric position. The diameter of the target is the drawing tolerance for location. As long as the hole's axis hits somewhere on the target, the work is acceptable.

Okay, we fire each rifle one time and see what the results show (Fig. 27-2). Based on this initial information, it appears rifle A is a bit more accurate than rifle B, although that isn't much to go on. If you were to take corrective action at this time, what would you do? Move A to the right and B to the left and down. But do we have a complete enough picture?

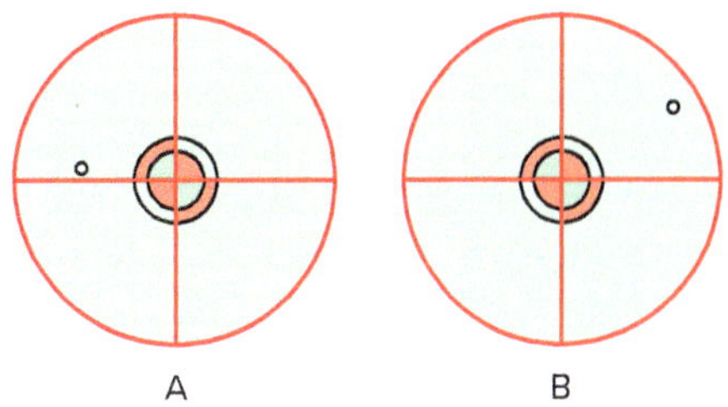

Figure 27-2 These targets illustrate two different machines performing the same process.

> **SHOPTALK**
>
> We're discussing the use of SPC to track key features within a long run of parts. However, it can be applied in two other ways: short-run production and operation tracking. It's accepted that it does work for short-run production, but opinions vary among users as to its effectiveness.

Not yet. So we shoot a few more times (Fig. 27-3), and something very different begins to appear. Process A is not very well suited to the job. It seems that normal variation for A consumes more than the job tolerance–it's predictably going to make a lot of scrap unless it's improved. The small crosses indicate that some shots missed the target completely.

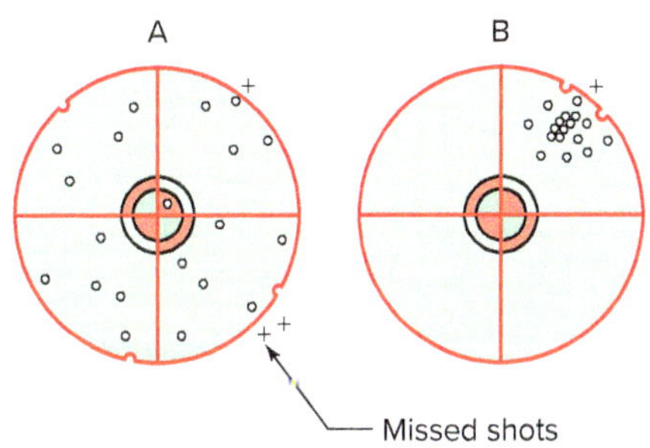

Figure 27-3 After more data are collected, a clear picture forms.

The two graphs show us three pieces of valuable information:

1. Rifle A is not a good process for this job, and rifle B has the capability to be very good.
2. Rifle B isn't centered–it's not aimed at the bull's-eye.
3. Rifle B has missed the target one time, too.

Capability Shows on Histograms

Note that at this stage it truly did not matter in which order the shots were fired. Given enough shots, the histograms show us the capability of both processes–rifles A and B. Those total capability charts are called histograms, and they show overall capability in a process.

Control Limits

Next, a small circle can be drawn around rifle B's spread (Fig. 27-4). That's representative of its capability, and it becomes the control limits for shooting B. Still we'd need to shoot and record many more times to determine an exact capability, but it gives us useful data. In fact, that's what the SPC software does: It continuously updates as more data are collected.

> **KEYPOINT**
>
> The capability plot indicates we can correct the process such that it will not make scrap unless new variation is introduced! Rifle B shows a good process that isn't centered on the target.

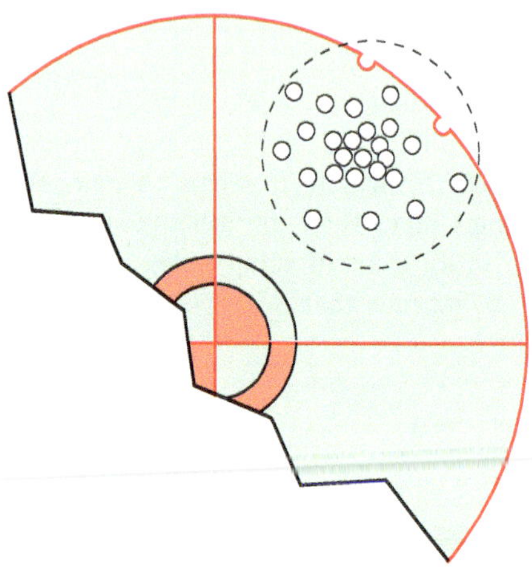

Figure 27-4 Rifle B's record shows it has the *capability* to hit very near the bull's-eye once the process is centered.

There are several kinds of histograms that we'll leave for factory training. But no matter what the form, they all compare process variation to the drawing feature tolerance. In Fig. 27-5, I've put the data into SPC software, and it has created a bar graph histogram of B's hits.

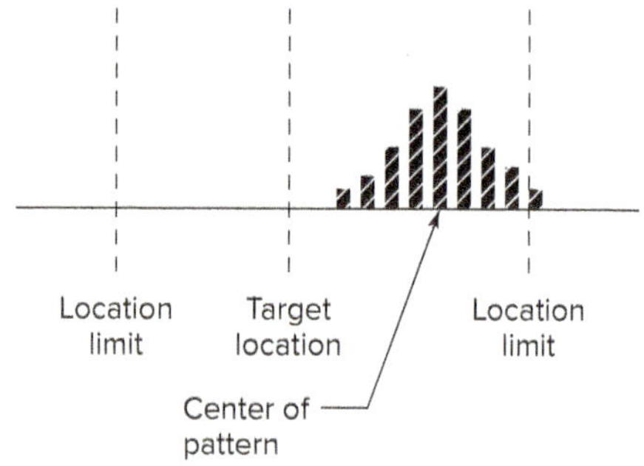

Figure 27-5 A histogram in bar form shows the spread of B's variation.

Bell Curve Distribution

Notice the way rifle B's holes were distributed–most were near the center of a tight pattern, and a diminishing few hit farther out. Notice that they concentrate at a central location, then thin out from their natural center. That's called the variation or **spread**. The spread of variation normally assumes the same shape, concentrating at a point, then thinning in a flattening curve to each side of the process center–this is called a bell-shaped curve or a **bell curve**.

Figure 27-6 is a training aid that shows the tendency of any process with a quality outcome to form a similarly shaped curve. If we dropped balls into the top, they would fall at random, bouncing off pegs. The drop point can be moved, to simulate a process that is not centered, as our example rifle B is. If enough balls are dropped, they naturally form a bell-shaped curve around a central location.

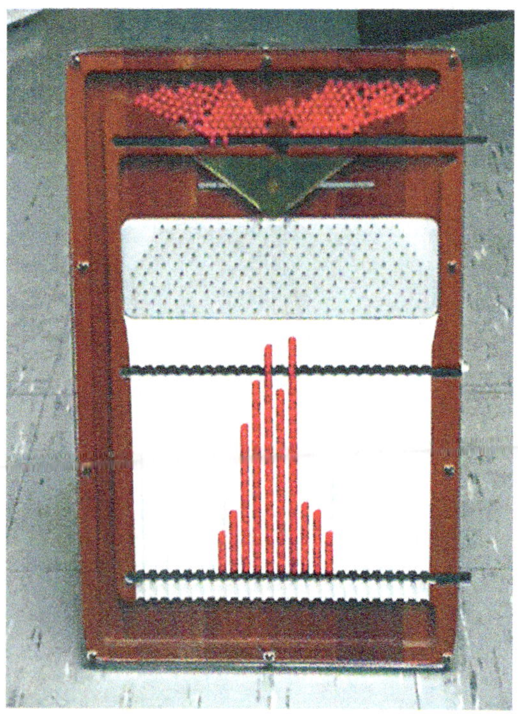

Figure 27-6 The bell curve is starting to form. If all balls are dropped down, a full curve would appear.
Courtesy Tyee Aircraft/Paine Field, WA

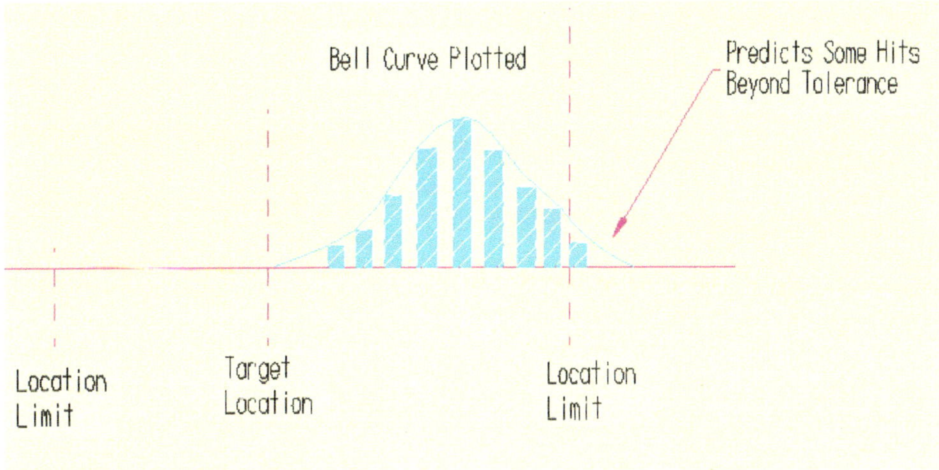

Figure 27-7 The SPC software can display a histogram of the variation.

When enough data are present, the SPC software can create a well-defined definition of the variation in this process, to the point that it can predict future events. Figure 27-7 graphs the spread; notice that the curve goes out beyond the collected data. That's why control limits are a bit wider than early data. SPC software can predict how wide the spread will be, based on the bell curve it produces.

Looking at Fig. 27-7, notice that the plotted curve extends out beyond the data we've collected. It shows that a significant number of shots will inevitably hit beyond the samples we've taken. That's why the circle drawn around the first few shots of process B wasn't exactly equal to its control limits. The extended bell curve shows that when a much greater number of parts are made, a predictable few will fall a bit farther from the pattern center. But they need not make scrap parts if we center the process.

Based on this distribution graph, the software can predict once the rifle is aimed correctly (**centering** the process) how many parts will exceed the job tolerance (become scrap or require a rework) within a batch of 100, 1,000, or 10,000 parts. That's powerful information!

Actions Based on Information

The target examples illustrate an important concept in SPC—*complete knowledge equals better control.*

Critical Question

Going back to Fig. 27-2, right after the first shot, what would have happened had you corrected each setup before shooting a second time? After the early correction, where would the bullet hit on the second shot?

The answer, of course, is that you would have mistakenly aimed A to the right. But that misinformed correction would have then gone far off the target on the plus side of the tolerance, to the right—a scrap part. Truthfully, once a clear picture came to be, A's histogram showed us that it's actually aimed right—its process is centered on the target, but it's a crummy gun!

That questionable action of adjusting without knowing is a lot like a machinist adjusting the machine before having a clear understanding of the capability of the process. The machinist might be "chasing normal" rather than refining the process. That said, however, in reality, based on skills and knowledge, when it can be anticipated that the process will be within reasonable control, the machinist would still make early adjustments because no scrap is acceptable under any circumstances, even to get the data picture.

A Is Out!

The problem isn't that the adjustment was wrong, it's that rifle A is just not up to the task. So, from here on, we consider A as invalid. The histogram proves it. We'll continue with B only. Rifle B is more like the usual situation—a process able to deliver but requiring some refinement.

Critical Question

Okay, what correction would you have applied to rifle B after the first shot?

Making the early correction (Fig. 27-8), to the left and down, brought the process closer to the target, but that preliminary information would not perfectly center it. If the big picture was known (a complete histogram), then the best correction would have been from the *center* of B's pattern, to the bull's-eye center. But at that point, with only one shot, the pattern center was unknown.

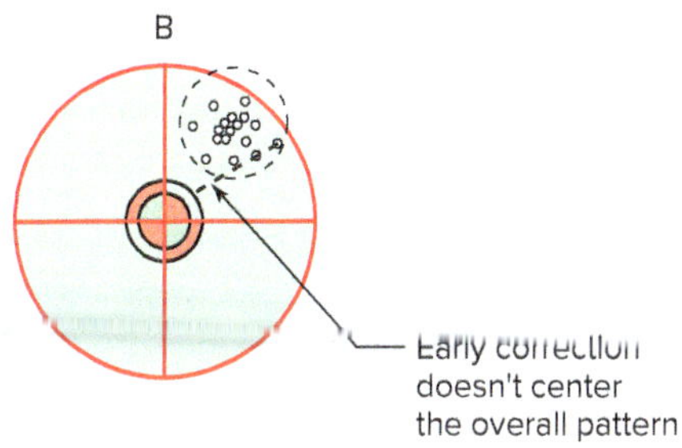

Figure 27-8 A process capability (Cp) ratio compares the normal spread to the target tolerance.

KEYPOINT

Making that perfect correction would have been possible only after collecting enough data, then centering the normal process. However, that first correction is not wrong, but it does invalidate the data, which must be started again for a better process capability.

Those perfect centering data are available only if you can leave the process alone until the normal variation comes into focus on the histogram. However, it's impossible for a skilled machinist to make a bunch of questionable parts at the onset, to collect data. For if the machinist did, one piece of data would be the final paycheck!

In real machining, that first correction would be made if experience told the operator that the process was known to exhibit tight repeatability as on rifle B, as we do when we know our machine's ability. But in making early corrections, the data collected would no longer be valid in the histogram database. After the initial correction to B, the parts would be reproducing very close but not perfectly centered on the target. At that time, the operator might make one more adjustment, then let the machine run for a while, producing high-quality parts but not the absolute best possible, with little (or no) further corrections for a time. That will then result in an accurate histogram with perhaps one last correction.

So you see, just having the SPC understanding of variation helps with decisions about corrections–even without the graphs. But with clear data, corrections can be spot on!

The Cp Ratio Describes Capability

To rank the potential of a process, a ratio is formed by comparing the overall feature tolerance on the drawing, to the normal variation found by the software. Process capability (Cp) equals

$$\frac{\text{Feature tolerance}}{\text{Variation normal}}$$

In this example, it's determined by dividing B's spread diameter into the target's rim diameter, yielding the Cp. It is the capability of the process expressed as a decimal ratio.

Rifle A had a very poor Cp below 1.0, since, predictably, many of its hits were going to be off the target. Its normal variation consumed more than the tolerance. Even before centering B, the histogram showed it had a great Cp.

KEYPOINT

The Cp ratio shows us what we can expect from the process—not what it is doing at any given time. Centering is unimportant—it's part of a histogram.

Left alone, aimed as it is, rifle B will make a bad part or two, but the potential to make 100 percent all good parts was evident. Centered or not, rifle B shows a good Cp well above 1.

KEYPOINT

Cp Ratio

The Cp ratio compares the drawing tolerance to the variation. It shows *potential* capability disregarding centering the process. From the Cp of rifle B, we see it *can* make good parts once we center the process (re-aim the rifle).

A Cp above 1.3 is considered okay; 1.5 is better. Bigger is better yet.

27.1.4 The Second Step—Control During Production

We now return to rifle B, starting with a clean target but one whose outer rim is not the location tolerance. This time, it represents the control limits. During this phase, we're producing parts. We track each shot in sequence–that's known as **real-time control**. Since we have the data needed, we center on the target.

SHOPTALK

New? SPC is far from a new science. It's accepted that it was invented in the United States in 1920 at Bell Laboratories and was widely in use by 1950.

As Fig. 27-9 shows, all goes well for a while, and the process is behaving normally. Then in Fig. 27-10 we see that a problem begins to develop, called a **trend**. The process is wandering off-center. No scrap has been made as yet. However, left uncorrected, it soon will be. The process isn't performing normally. Now the machinist earns his or her pay. Okay, now we have the basics. Let's review.

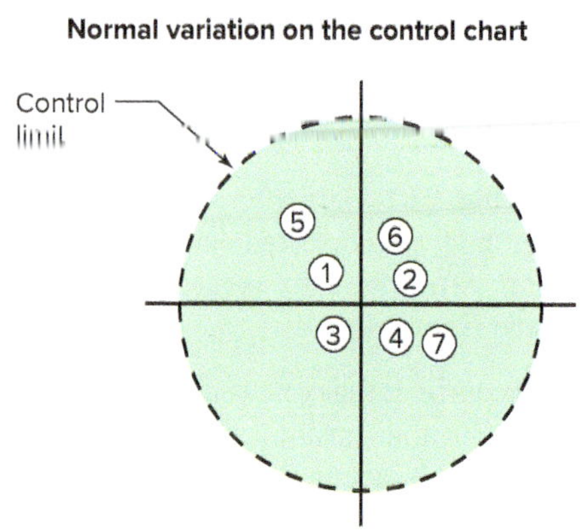

Figure 27-9 Using the histogram, we've now centered the process.

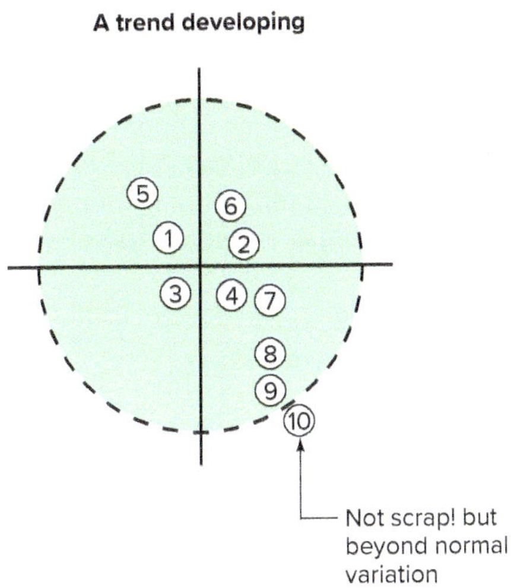

Figure 27-10 A trend is starting, but no bad parts have been made yet.

UNIT 27-1 REVIEW

Replay the Key Points

- SPC can be applied to any process with a measurable quality target, but it works particularly well for precision machining.
- SPC signals clearly and sooner that a trend is occurring away from normal.
- SPC doesn't identify the root cause of a trend or excess variation. It isn't a voluntary tightening of tolerances–it highlights natural capability.
- A Cp above 1.0 is better–the larger the Cp, the better the chances of making all good parts.
- A histogram detects capability and normal variation.

Respond

1. What does SPC tell the machinist in real time?
2. What is the graph that shows the Cp of a process?
3. True or false? Because a histogram analyzes the spread of many measurements of key features, those measurements need not be in any order. If the statement is false, what would make it true?
4. True or false? A Cp of 1.2 is closer to 1.0, therefore better than a Cp of 1.3. If the statement is false, what would make it true?
5. Using 10 words or less, explain why SPC is useful in machining.

Unit 27-2 An SPC Experiment

Introduction: In Unit 27-2, we proceed in two stages: Based on a batch of sample parts, you set up a histogram to determine UCL and LCL. Then you incorporate those data to simulate production using a control chart of your own making.

TERMS TOOLBOX

Control chart A real-time evaluation of ongoing production sometimes called a run chart that shows the unavoidable differences from one part to another.

Cpk ratio The ratio comparison of the present production variation to the control limits.

Using Software

If your training lab has SPC software for student use, then by all means use it instead of paper and pencil as suggested here. Alternately, see the Shop-talk in Unit 27-1 about Internet resources for SPC downloads. But if you wish this can be a mind experiment, too.

27.2.1 Making a Histogram

After the walk-through explanation, find 30 or 40 items that *should be* exactly the same size on a key dimension that can be easily measured with a micrometer. The number sampled doesn't matter in this exercise, but there should be enough to create a database. I'll use the inside diameter of stamped steel washers as the key feature based on function–you could use any key dimension in a group of parts (Fig. 27-11). The nominal hole size should be a bit oversize–let's say it's 0.510 in. with a ±0.005-in. feature tolerance.

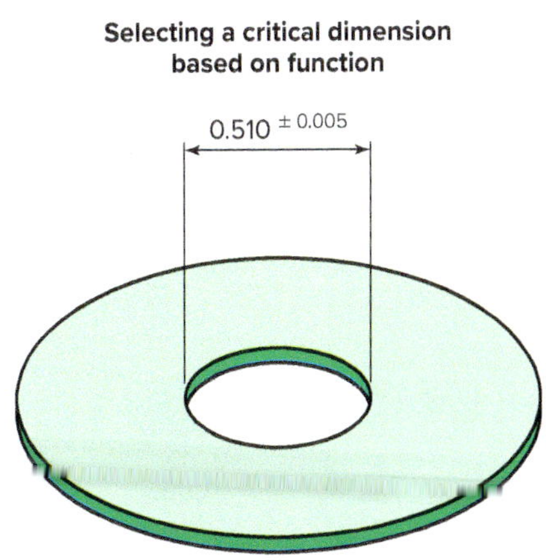

Figure 27-11 We'll use washers for the example parts.

In the graph in Fig. 27-12, height represents the number of parts that measure a certain size. The horizontal graduations are 0.001-in. size difference. You may need to change the scale on yours to see the variation more clearly if the range of variation is tight. Remember, for a histogram, the order in which the washers are measured doesn't matter. We're only interested in the spread. Here are my results:

Number	Size
12	0.511 in.
8	0.512 in.
7	0.510 in.
5	0.513 in.
4	0.509 in.
2	0.508 in.
2	0.514 in.

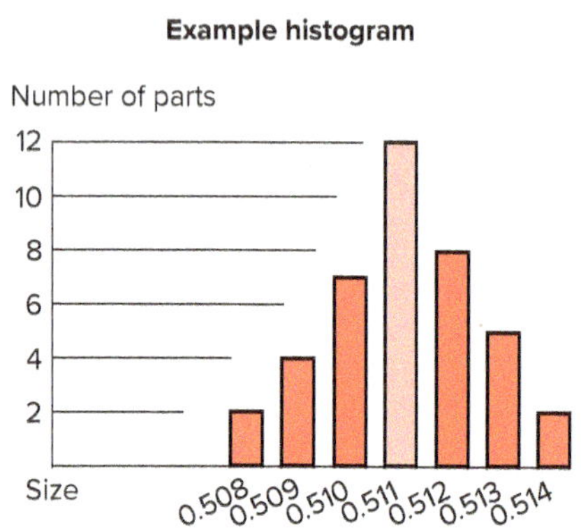

Figure 27-12 Measure and record the key feature on all parts; then create a bar graph similar to this.

Based on this small sample of data, the software can draw the bell curve (Fig. 27-13). From it we're reminded that the Cp for a given process isn't exactly the final spread until a large database (DB) is created. The curve extends beyond my data to predict that a few parts will vary more than those we've detected thus far.

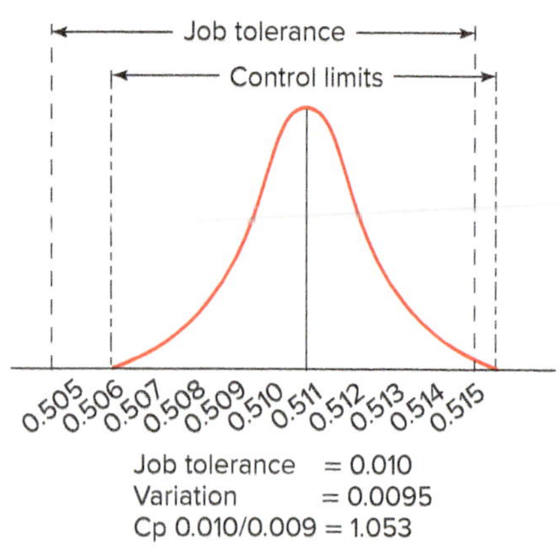

Figure 27-13 Plot your bell curve.

Sample Range—The Dipstick

The larger the number of overall parts represented in the DB, the more accurate will be the histogram. If, for example, on a run of 4,000 washers, 5 out of every 100 are removed and measured, then the histogram represents a great deal more data. Taking 5 to represent every 100 parts is known as a *sample range.*

What Does the Graph Reveal?

The bell curve shows us that the process isn't centered. Diameters center around a slightly large 0.512 in. Looking at the diameter tolerances (vertical lines), it also shows it will predictably make a few parts with the hole too big at 0.5155 in. and left off-center. There's not much chance of making an undersize part as the curve intersects the horizontal axis just before the lower size limit.

The Cp for my process isn't very good! It should be improved before going into production, but we'll continue with it as-is for the example. The print tolerance range adds up to 0.010-in. total diameter difference. The observed spread is ~0.0095; thus, the Cp is 1.053:

$$\frac{0.010}{0.0095} = 1.053$$

Centering the process will slightly reduce the number of predictably bad ones, but still, this Cp shows the process will make reject parts!

KEYPOINT

If the process has a Cp of 1.5, then it can be statistically shown to be capable of making only 3 parts beyond tolerance, in 10,000! A Cp of 1.3 is the minimum acceptable for production; 2.0 shows no scrap will be made ever, as long as the machinist pays attention to trends and keeps the process centered.

TRY IT

Now, make your own histogram based on a batch of sample parts of your choosing that are easily measured with a micrometer.

INSTRUCTIONS

1. First, draw a horizontal line near the bottom of the paper to represent the scale.
2. Then divide that line into equal segments representing the resolution of the measuring tool. For this experiment, label them in 0.0001 increments.

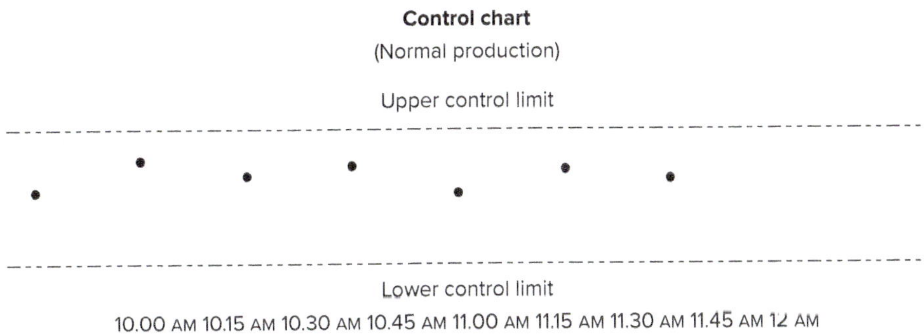

Figure 27-14 Measure each part as though it was coming off in real time; then make a control chart.

3. Make sure to spread the horizontal graduations apart enough to discriminate between data points–the total range expected should consume about two-thirds the width of the paper.
4. It does not matter what scale you use for the vertical spacing for this exercise as long as they are consistent. Perhaps each tick should be 0.2 in. higher on the stack.

Measure, then mark all 30 or 40 parts on the chart. The bell curve may be a bit rough using a small sample set. Then, if not working on SPC software, roughly sketch the bell curve. The width of the curve at the baseline can represent the control limits for this experiment for now.

FIND THE PROCESS CAPABILITY

Now calculate the Cp for your samples. Remember, the Cp is derived by dividing the total tolerance available by the range of variation found under the bell curve.

Based on the feature you've chosen, what would be a reasonable feature tolerance? Choose 0.030 for this experiment. Use the measured spread on your chart as the variation range for this experiment.

SIMULATE A CONTROL CHART

Instructions This time, we'll measure the same parts but in a sequence to simulate real-time production.

The most common **control chart** is the paired horizontal lines introduced back in Fig. 27-1, representing the upper (UCL) and lower control limits (LCL) (Fig. 27-14). The space between represents normal variation. Just like driving on a highway, the idea is to keep it centered between the lines!

Draw your control chart using the two derived control limits from your bell curve. Let's assume this time that each washer represents 5 measured from a range of 100 parts coming off the machine. So each dot represents the average of 5 washers taken from the run, every 15 minutes.

27.2.2 The Cpk Ratio—Capability Applied

Now the need for a numeric index arises to summarize how well the process is performing in real time. This ratio is called the **Cpk ratio**. It's the capability of the process–applied. It compares process capability to present results. It's derived by dividing the Cp by actual variation results. And again, a Cpk of 1.3 is good, but 1.5 is better.

$$\frac{\text{Cp}}{\text{Present variation}}$$

Cpk is a moving value. It changes as we go along. If the operation is working well, the Cpk will remain above 1.3. If variation leaves the control limits, the Cpk falls below 1.00, and that's bad. The tighter the variation can be controlled in the process, the higher the Cpk will be. On the control graph (Fig. 27-15), the Cpk is dropping off toward a trend.

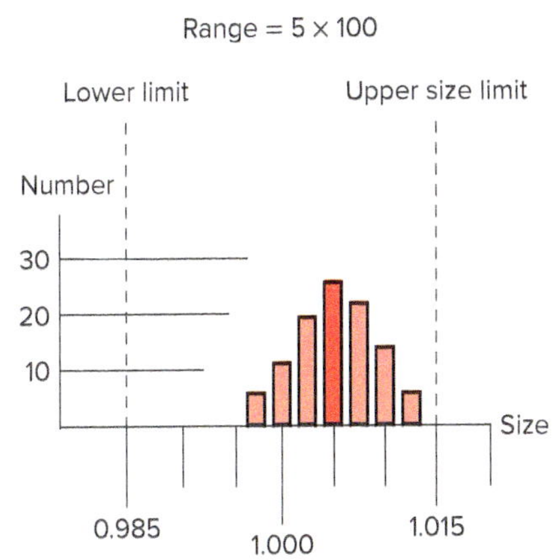

Figure 27-15 What does this histogram show?

KEYPOINT

Differing from a Cp ratio, a good Cpk depends on keeping the process centered as well as controlling variables in the setup and outside it.

Detection, How We Earn Our Pay!

There's a possible problem developing. I need to solve a diminishing Cpk by investigating for the assignable cause. There are four general categories:

1. **Common**

   Not a cause for adjustment–key features can and will vary in repeating patterns or irregular patterns but they do not show any trend toward the control limits.
2. **Trend**

   A steady up or down change toward the control limits.
3. **Exceptional Repeating**

   An odd bump in production on the control chart–that will happen again if not corrected. It is probably assignable but will require some thought to sort out.
4. **Exceptional Nonrepeating**

   An odd bump in the control chart that investigation proves will not happen again. In this case, the bad parts are not entered into the database–they become nonevents.

KEYPOINT

Trends account for 95 percent of all our challenges. It's the daily things like tool wear, chatter, and heat. Generally, finding and fixing these things are easy.

Exceptional problems add a little mystery to the challenge. They fall into two categories. First are events that follow a pattern and can be assigned and corrected, but they don't happen very often (see the Shoptalk). For example, every day at 4:00 p.m. a freight train goes by the factory. When it does, I make five washers that aren't the same as the rest. Or each Monday, I make several that are too big, before the machine warms up. The second cause can be from needed improvements. Perhaps the carbide grade isn't right and breaks down or chatters too often. This is not routine tool wear, but, for example, we experience catastrophic failure at around 250 parts. That requires a bit more investigation to find and fix.

SHOPTALK

Here's my best example of SPC detecting regular–exceptional variation. We were operating an ultraprecision jig boring lab making aircraft assembly fixtures. The lab was on the side of the shop near a large river. To our frustration, we found inexplicable variation on a day-to-day basis, yet the jig boring machines were the most precise in the factory and the machinists all the most experienced. We set up control charts and watched the trends. Soon after, we saw that hole locations were moving apart then back by about 0.0002 in. or so, in about a 6-hour swing. It took us awhile to realize we were about 10 miles upstream from Puget Sound and we were looking at a tide chart! The river backed up enough to slightly change the floor shape! Knowing that weird fact, our machinists consulted the tide tables each shift, then took corrective action. No kidding, tide charts!

Nonrepeating exceptions are once-only events. They could be a sand pocket in a casting, or my machine goes offline due to a power failure, so it loses data. These events are assignable, but won't happen again. Since they would affect your Cpk, yet you can't fix them, it's not cheating to eliminate their key data from the DB.

KEYPOINT

Summary

If the cause is common, it must be corrected.

If it's regular, exceptional, and controllable (I press slide hold, while the train goes by), then I take action.

If the exceptional cause is nonrepeating, but assignable, then I can eliminate its data from my collection of Cpk data.

Okay–that's all you need to be ready for specific training on how to apply this super tool to your new career. Review to be sure you have the shop-ready concepts.

UNIT 27-2 REVIEW

Replay the Key Points

- A Cpk of 1.5 has the ability to limit bad parts to only 3 within 10,000.
- An initial Cp can change over time even without refining the process.
- Differing from a Cp ratio, a good Cpk depends on keeping the process centered as well as controlling variables in the setup.
- Sample order does matter when computing a Cp.

Respond

1. Suppose a trend develops that takes a critical size just over the lower control limit line. This part must be thrown away as it is scrap. Is this statement true or false? If it is false, what will make it true?
2. What must be done assuming the trend continues in Question 1?
3. Is the compiled comparison of the overall results of a group of parts to the job tolerance the Cp or Cpk ratio?

4. The Cp ratio compares a group of machined results to.

5. Three companies are seeking your business. They each present their normal Cpk ratios under which they historically perform production. That of company A is 1.0, company B's is 1.33, and company C's is 1.65. If the price is similar, which company would get your purchase order?

Critical Thinking

6. A production run starts up with the goal of 2,000 parts. The early period shows a Cp of 1.5 and a Cpk of 1.3. After 1,000 parts, the Cpk is higher at 1.55. What might have happened?

CHAPTER 27 REVIEW

Unit 27-1

When I was researching this section, a local shop manager told me, "We'd do SPC training for our people even if we didn't actually use it on our shop floor. Just knowing the concepts, our skilled people become better decision makers." The concepts of SPC create powerful control oversight. Job experience will prove this, but take it from the experts, the smart machinist can prevent scrap, improve quality, and be in better control using SPC.

Unit 27-2

We've left some details out of this discussion. But most focus on statistical methods and terms. If you now understand and can compare Cp and Cpk ratios and can define histograms, normal variation, control limits, and charts, then you are ready for the workplace. Now from this end of the lesson, you can see that SPC is really a matter of common sense.

QUESTIONS AND PROBLEMS

1. A *histogram* is compiled as shown in Fig. 27-15, for a hole that's nominally 1.000-in. diameter with a ±0.015-in. tolerance.
   - **A.** What's the approximate Cp? If in doubt how to proceed, see the hint in the answers.
   - **B.** Based on the histogram, what should be done to this process?
   - **C.** Can this process make all good parts? Is it an acceptable Cp?
   - **D.** This histogram represents how many parts?
2. What does the control chart in Fig. 27-16 tell us?
3. Does the variation shown in Fig. 27-16 need to be corrected?
4. Identify some possible causes for the variation in Fig. 27-16.

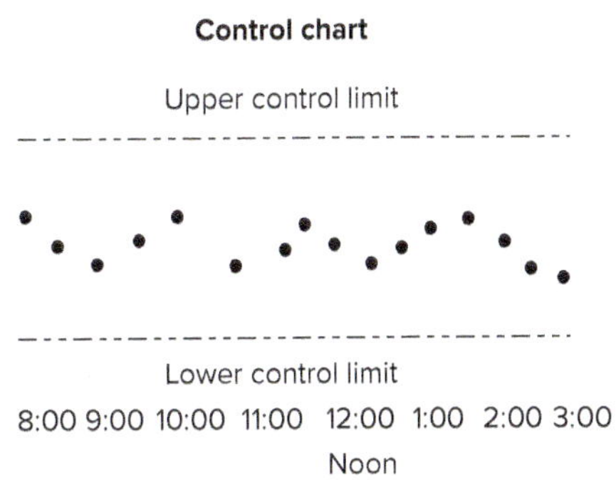

Figure 27-16 Control chart.

CRITICAL THINKING

5. I have started a job and am recording on a control chart *starting with part number 1 off the machine.* How could I do that? Can my control chart limits vary over time?
6. The feature tolerance is ±0.001 in. After running 1,000 parts, I've found no more variation than 0.0005-in. total spread. What is the Cp of the process, and what does it tell you?
7. Is the result of Problem 6 strictly correct—why or why not?
8. Identify the three kinds of causes of variation and your actions for each.
9. Define the four terms employers want you to know:
   - **A.** Cp
   - **B.** Cpk
   - **C.** Histogram
   - **D.** Control chart
10. Sum up what you believe SPC brings to the shop, the machinist, and the factory

CHAPTER 28 Computer Coordinate Measuring

Units

28-1 What Does a CMM Do? How Does It Work?

- Define the floating work axis system
- Describe the three different types of CMMs
- List the advantages of computer inspection compared to standard methods
- Compare data collection methods and know which is right for the task

28-2 Set Up a Manual CMM for Inspection

- Plan a complex CMM inspection routine
- Describe and perform the three phases of setup: qualifying the machine—before inspection, aligning the workpiece—first part of inspection, and menu selections for the inspection process

INTRODUCTION

Although this is the last chapter in this book, it could have been at the front. Its importance to industry is at the forefront of improving quality, with two overall goals:

1. To impart enough background to use any generation CMM on the job (with some instruction)
2. To show the entry-level machinist how this technology fits and functions within today's manufacturing plan

In Unit 28-2, we'll be setting up and using a manually positioned CMM, since they are likely to be in a training lab. We'll also investigate the higher-level machines and how they collect and process inspection data automatically.

As much as anywhere in our trade, geometric understanding plays a big role in CMM success. Often there's as much creativity in planning and writing a CMM program as there is in the machining program that made the part. Without the insight of geometric dimensioning and tolerancing, writing inspection programs to evaluate complex features is nearly impossible. Just like cut sequences, there are usually several ways to solve a given geometric question, but only one or two yield best results or the functionality being sought.

CMM skills are a must-have competency for today's machinist. But that wasn't always so. In the shops of the past, the CMM was considered a specialist's job even among inspectors. It was programmed and run by a white-lab-coat person, not to be touched by shop people!

Today, in progressive shops, the quality philosophy is moving toward access. Two factors combine to make it so:

- The person most responsible for quality, the machinist, should also be armed with the best measuring tools to achieve it.
- The equipment is evolving toward user-friendliness; thus, less training and handling are required, and it's also better suited to the shop environment. It does not need to be in a perfectly clean and temperature-controlled lab. Not all but some CMMs now deal with workplace dirt and temperature swings. We see these *shop-hardened* versions conveniently placed within

cells of machines. Portable CMM arms are meant to be taken to the work and machine.

So, if I've made my case, you can see that while CMMs are the last chapter of this book, they are by no means an add-on subject for the modern tradesperson; they are a central competency.

Unit 28-1 What Does a CMM Do? How Does It Work?

Introduction: Without experience, most assume computer measuring does what the manual instruments and processes do, but they do it faster and more accurately. And that's true in many cases (but not all), but it's only part of the reason they improve quality and profitability. In Chapter 28, we're focusing on computer coordinate machines; however, similar arguments could be presented for other computer-directed measuring equipment as well.

KEYPOINT

CMMs verify geometric requirements in ways no other process can. We have the proof coming up.

TERMS TOOLBOX

Conformal axis system Axes are assigned to the work independent of machine axes; also called a floating axis set.

Digitizing Programming by hand probe motion–see *teach-learn.*

Hit (CMM) Shop lingo for one recorded touch of a data-collecting probe.

Nontactile measurement Using optical methods to measure surface elements.

Scanning CMMs Two types both of which collect entire surface element data. The tactile type maintains its probe in constant contact with a surface. The non-tactile type collects the entire surface element without touching.

Shop hardened An instrument built to withstand the temperature swings and dirt of a shop environment.

Tactile measurement Measuring by touching the object with a probe.

Teach-learn (read-learn) Programming by recording hand probe movements and menu picks.

28.1.1 Computer Inspection Advantages

Cures Choke Points

CNC machines often make parts faster than they can be measured with everyday measuring tools. Without a CMM, there are three options.

One solution is to load the cutting tool drum with a touch trigger probe and then, if the CNC controller is equipped to do so, write inspection routines into the machining program itself (Fig. 28-1). In previous chapters, we've seen probes testing cutting tools to autoload offsets and measuring parts–in process measuring. But not all feature inspection lends itself to *in-process inspection*, nor can all CNC machines perform it. There are many times when the part must still go to the stand-alone CMM.

Figure 28-1 In-process inspection helps check key features as they are made and keeps real-time SPC up to date.

Courtesy Haas Automation

CMMs Evaluate Complete Shape Envelopes

Using CAM-generated programs based on solid models, there's almost no shape that cannot be cut accurately. But, without a CMM, we are often put in the odd position of having confidence that the shape is probably okay. However, there's no definite way to prove it using standard measuring equipment! A tool path evaluation showed the program was right and no errors occurred when we cut the part, but we are unable to conclusively prove the part is right, to ourselves or, more important, to the customer!

KEYPOINT

A CMM can compare a CAD database or digital shape definition to the actual result.

Geometric Reports of True Values

The main reason these reports are so vital to part evaluation is that other methods of measurement don't always report data in a way that's useful, in two related ways:

1. **Element Analysis (Fig. 28-2)**

   A simple question is how far the hole location is from the datum edge. There are three possible answers depending on the function being sought.

Edge to Datum The answer on the left is easy to get with a caliper, but simply measuring the hole diameter, then adding the radius to the edge measurement doesn't account for the datum and, due to hole irregularities, can yield a low-reliability answer.

Reserved Circle from Datum The middle test in Fig. 28-2 is accurate if the question is how far the reserved round space is from the datum. To find it, the height of a snug fitting test pin is measured above the datum table. Here, a fairly accurate answer is obtained using standard tools. A typical example of this functionality would be that this hole is to receive a bolt in an assembly to another plate.

Mean Hole to Datum But one functionality can only be determined by computer analysis. Suppose the hole is a precision laser aperture or a spray nozzle that must exactly center its total output on a target location. The functional question then becomes how far the mean hole is from the datum. The mean hole accounts for all form irregularities. Only the CMM can sum the irregularities and then determine the location of its mean round diameter relative to the datum. Think about it: That distance will be slightly different from the pin test if the hole is not perfectly round. This is a simple example of the concept. But the conclusion is that when features and shapes become complex, only computer analysis can solve their total shape envelope before location or orientation questions can truly be answered.

2. **Geometric Value Opposed to Acceptance**

   The second justification for CMMs, with their ability to analyze elements, is that they report the value of the measurement rather than an acceptance within a tolerance range, which is often the result when using standard testing. Acceptance occurs when we measure the part to be within tolerance, but do not know the exact value of the variation.

A simple example is using a micrometer to measure a diameter and assuming the roundness is okay as well. We already know what's wrong with that assumption. Another acceptance example is illustrated in Fig. 28-3. Using shop methods, the test on the left determines that the TIR is 0.003 in. It's flat enough, and within the 0.005-in. tolerance, but truthfully the exact flatness is not known.

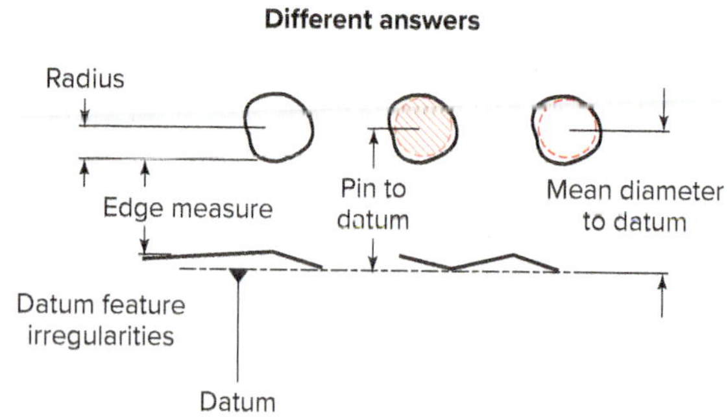

Figure 28-2 Three answers to the question of how far the hole is from Datum A.

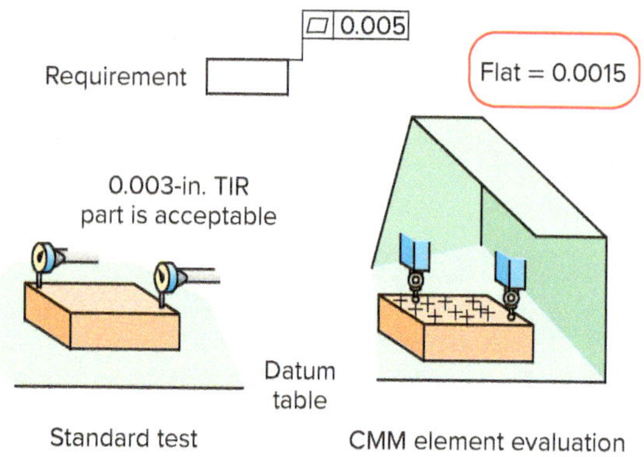

Figure 28-3 Only the CMM can detect the true flatness value filtered from tilt.

In fact, this part could be perfectly flat or slightly unflat, but also tilted to cause the 0.003-in. TIR. Part of the problem is that the height gage/indicator test is actually a parallelism test. Flatness is embedded, so we can say the part is good enough but not exactly how flat it is. When requested to report flatness, the CMM filters out tilt and finds true flatness to be 0.0015 in. The part in Fig. 28-4 illustrates this concept.

KEYPOINT

The CMM can filter out and report the value of the required characteristic from other variations—that is, flatness from tilt, for this example.

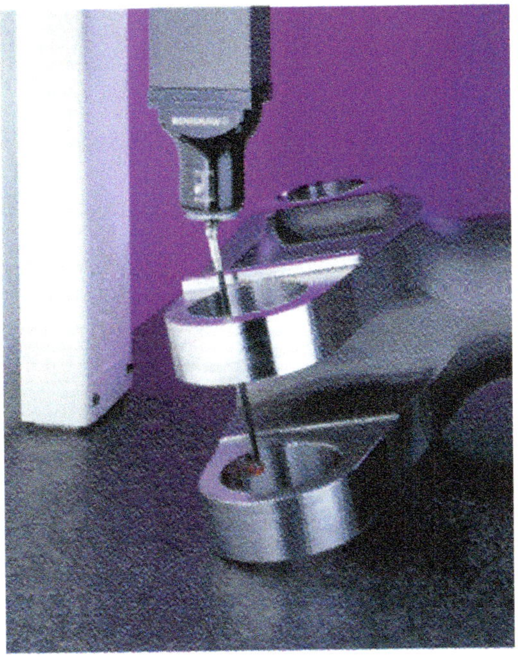

Figure 28-4 Near-perfect inspection of size, shape, and orientation of this tilted hole requires collecting much surface data. A scanner is the right CMM.
Courtesy Renishaw Inc

If the true variation of any process is not known, then the capability of the process (Cp ratio, Chapter 27) will be wrong!

Reliable Accuracy

An obvious advantage of computer inspection is that it eliminates all the classic inaccuracies of standard measuring. We could go back to the list in Chapter 6 and show how each—heat, pressure, alignment, and so on—are eliminated altogether or reduced beyond any significant effect by CMMs. The repeatability of today's CMM is in the microvariation zone below 0.0001 in. or 0.003 mm.

Data Management

Another advantage of all electronic and computer testing equipment is direct input of data to the SPC database. All math is instant and to the tenth decimal place! It also reports in metric or imperial values with equal ease. Finally, a CMM provides printed inspection reports that take zero operator time other than setting them up and requesting them, for company records or if required by the customer.

28.1.2 An Assignable Axis System

The first difference one must understand about a CMM compared to any other CNC equipment is the **conformal axis system**, also called a *floating axis set.* Even though the CMM has *X*, *Y*, and *Z* axes, they are machine axes only (Fig. 28-5). The part to be inspected needn't be aligned to them. It can be set up in any convenient orientation—clamped or lightly held or even unrestrained in any way. In some cases, the work is not fastened if it's heavy enough to withstand 3 to 6 grams contact force from an electronic touch trigger probe or when the data will be collected optically.

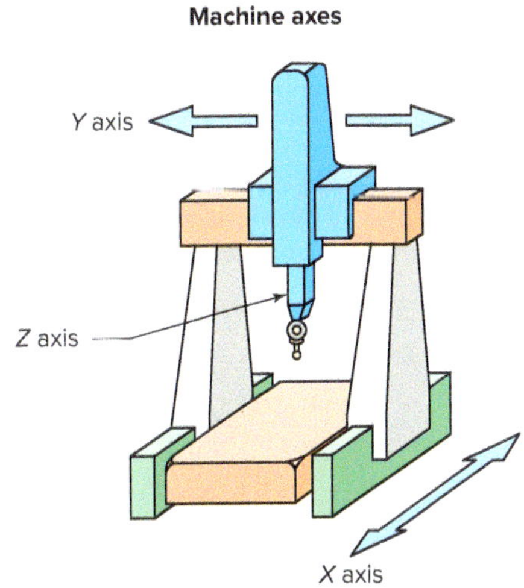

Figure 28-5 CMMs have axes, but the part to be inspected needn't be aligned to them.

Then at the beginning of the inspection, the operator or program determines in what orientation the part lies with respect to the machine axes. Through a routine, it assigns an *X*-*Y* flat plane and an axis direction and origin to the part—very convenient!

> **KEYPOINT**
>
> Objects to be measured do not have to be aligned to the machine's axis system. The machine conforms axes to the part's orientation.

Notice in Fig. 28-6 that the *X* and *Y* axes are nearly reversed relative to the machine axes. This is usually done for convenience. Part axes can be set up in any way useful. That said, it is best when the part is put roughly in line with the CMM axes to avoid accumulated rounding errors.

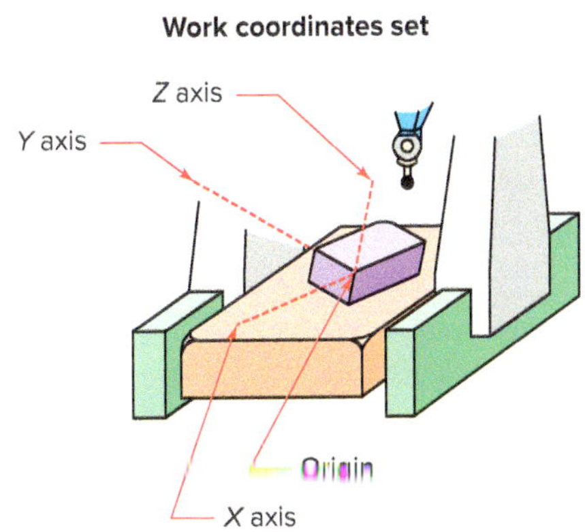

Figure 28-6 The part may be set conveniently on the table.

28.1.3 Data Collection—Probes

Feature data are collected in one of four ways. Understanding the difference is a big part of CMM skills.

- Hard probe–limited usefulness and accuracy
- Touch trigger probes–most common
- 3-D scanning probes–complex geometry or exceptional accuracy
- Nontactile methods–Gaining popularity in many industries where complex shapes are common, data are collected using laser reflection (Fig. 28-7).

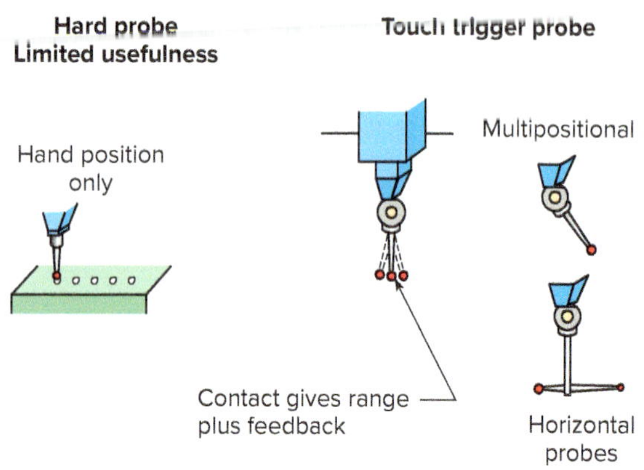

Figure 28-7 Hard probes versus touch trigger probes.

Using a touch trigger probe is by far the more common solution. However, **nontactile** *(not touching)* **measurement** methods consisting of bounced laser or video camera are gaining popularity in certain specialized applications. Touching the probe to the feature surface remains the mainstay of machine shop part evaluation.

> **KEYPOINT**
>
> In most shops the touch trigger probe type (**tactile measurement**—*collecting data by touching*) is by far the more common method.

Hard Probe

A ground sphere of known diameter on a solid steel shank is shown in Fig. 28-7. The precision ground probe sphere is brought into tangency with the object; then the screen position is read, or the data are downloaded to a database. The diameter of the sphere is then accounted within the result.

Hard probes are used on manually operated machines only. Because they do not have any give range (information coming), nor do they provide feedback to CNC controls, they cannot be used when the machine is run automatically. As such, hard probes are not often used on modern CMMs, but they can be useful for quickly locating over holes as shown.

> **KEYPOINT**
>
> Hard probe, approach speed, bounce back, and hand pressure against the work are all controlled by operator skill and experience; thus, they inject human inaccuracy factors into manual CMMs.

Touch Trigger Probes

A touch trigger probe (TTP) features a ruby or other jewel sphere on an omnidirectional joystick that allows a deflection range in every direction including inward along the probe's shank (Figs. 28-8 and 28-9). Electronic touch probes are commonly used for either manual or programmed inspection. They automatically collect data by making single touches against the part surface, called **hits** in shop lingo.

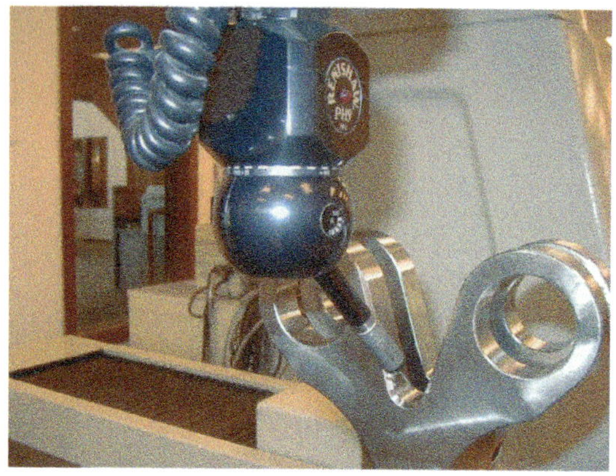

Figure 28-8 Only the CMM can determine if these ears are accurately machined relative to the part's datum axis.

Courtesy J.D. Ott Aerospace Machining/Seattle, WA

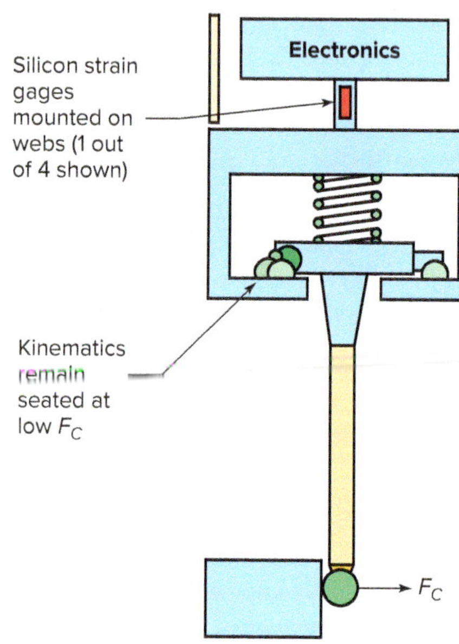

Figure 28-9 A touch trigger probe is not a switch; it's a strain gage that changes values as the prove force F_C changes.

Overshooting or overpressure is not a problem with touch triggers as the probe features a *give range*. After initial contact, the probe deflects.

In CNC mode, the TTP signals the axis drives that the probe sphere is in contact so the control won't overdrive toward the work surface. They can contact the workpiece from any direction—*omnidirectional contact*.

KEYPOINT

Similar to a hard probe, the touch trigger probe takes individual hits. It doesn't sum the entire surface unless an infinite number of hits are taken—an impossibility.

Critical Question?

How can a touch switch be sensitive enough that the instant it contacts, at zero pressure, it changes conditions from off to on? Think this one through; the solution is a brilliant example of computer technology engineering. (See the Shoptalk that follows.)

SHOPTALK

How Does the Probe Signal at Zero Pressure Contact? The secret is that the probe does not signal the computer instantly as it touches the work. It's not a simple switch; it's a strain gage that reports displacement values—something like a test indicator. On contact, it begins to be displaced (moved away from neutral) by a touch in any direction against the tip sphere.

The probe produces a changing value database as it is deflected. Its position is continuously recorded with time at each position, which produces an approach displacement vector. The computer then computes backward in time to the instant that the probe touched the work. The probe controller calculates the exact position where contact was made. The switch needn't be extremely precise or delicate. Computers do the work!

Scanning Probes

Depending on the accuracy required or the functionality being measured, individual touches on the part surface don't always detect the whole answer. Both hard and touch trigger probes do well in many inspection situations, but not all. Irregularities in the surface can be missed. Of course, the inspection routine could include a larger number of hits, but you can see that the data density is incomplete no matter how many hits are taken. Additionally, it's going to take lots of time to collect dense data.

For the best accuracy, an entire element must be collected to determine its absolute shape. Third-generation CMMs use scanning probe technology to solve data density questions. Scanning probes look and work the same as touch triggers. How the machine collects the data makes the difference. The **scanning CMM** automatically traces and records the entire element (Fig. 28-10). It does so by approaching, then staying tangent to a given surface by keeping probe pressure within a range—while still knowing when the probe sphere was at zero pressure! Now that's engineering!

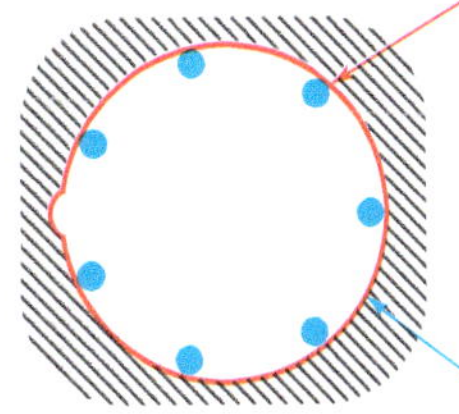

Figure 28-10 Sometimes even a large number of hits can miss an unusual deviation in shape.

KEYPOINT

If features are complex or highly irregular or where absolute shape values are critical, scanning technology should be used to detect the truth.

28.1.4 CMM Differences

The planner, supervisor, and machinist must know which machine is right for a given inspection. Third-generation scanners can do the tasks of all three kinds of CMMs, but they are a large investment if a lower-level machine is right for the job. There are four types of CMMs found in many shops and each has its place in the shop:

Manually operated

Programmed CNC touch probe

Programmed surface scanning (third-generation equipment) including nontactile types.

Radial arms—shown in Fig. 28.11.

Manual CMM

These are the basic CMMs with no axis drives (Fig. 28-11). Their probe must be led to the part by hand, as shown. While less sophisticated than the next two types, they are affordable and reliably robust and have a definite place in machine shops. Often overlooked because of their simplicity, manual CMMs are well suited to one-off and tooling work and where moderate inspection turnaround is acceptable. They are also at home where an affordable, **shop-hardened** CMM is required within the chip-making environment.

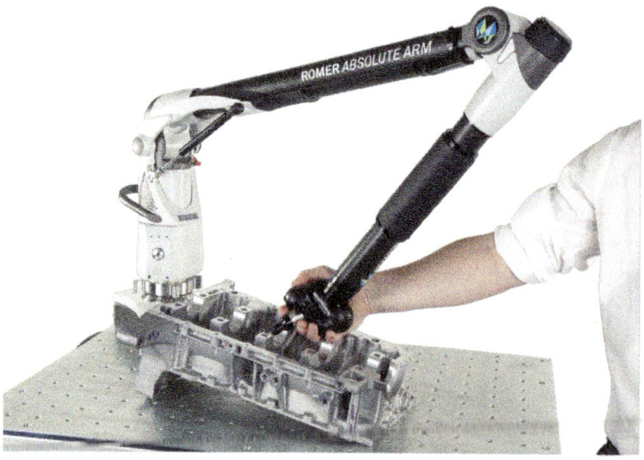

Figure 28-11 This machinist is hand guiding the probe to the machined part.
Courtesy of Hexagon Manufacturing Intelligence

With manual CMMs, parts can be quickly checked (without a program) for basic feature compliance such as angularity, flatness, location, and the size of basic features or roundness. If the shop's general inspection load involves constantly changing parts with simple features or low batch numbers, then the best choice might be a manual machine (Fig. 28-12).

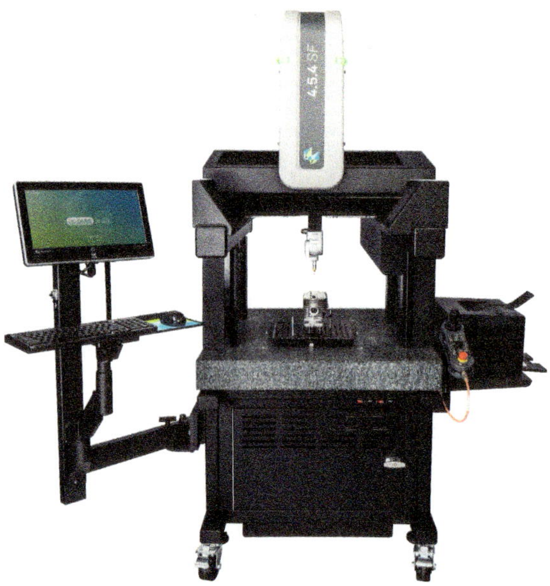

Figure 28-12 The 4.5.4 SF is designed for the shop floor.
Courtesy of Hexagon Manufacturing Intelligence

KEYPOINT

For basic inspection on medium- to low-volume production, or for tooling work, the manual CMM is right for the job. It can also be placed near the mill and lathe for immediate quality control.

Manual Mode on CNC Machines

Often, programmed CMMs can be operated in manual lead mode with the drives disconnected. They can also be operated in jog mode with a joystick control to perform the start of a program, until the control recognizes the *X*-*Y* plane and axis conformality.

Manual mode is also used for simple inspection or for low-volume work where there isn't a need to write a program (Fig. 28-13).

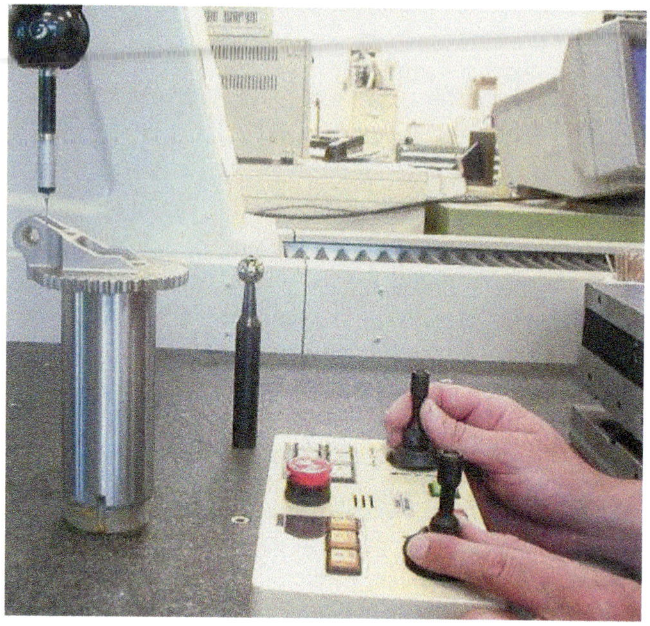

Figure 28-13 Similar to other CNC equipment, the automatic CMM can be driven by manual jog mode.
Courtesy J.D. Ott Aerospace Machining/Seattle, WA

Manually Digitizing a Program

Second- and third-generation CMMs can be put into manual lead mode to compile programs where axis motion or approach vectors are too complex to write with commands. As when you compile a macro routine on your PC, the CMM remembers all axis moves and screen picks, to be played back as a program. This may also be called **teach-learn** mode.

28.1.5 Programmed CMM Inspection

Touch Probe Machines—Second Generation

Besides programmed movements, the touch trigger probe can be moved similar to manual machining equipment with a joystick or another jog device on an operator control pendant (see Fig. 28-13).

The programmed machine (see Fig. 28-14) is best applied to medium- to high-volume production and complex but normal characteristics (not irregular surfaces).

Figure 28-14 This CNC CMM can completely inspect the engine block in minutes.
Courtesy of Hexagon Manufacturing Intelligence

KEYPOINT

CNC CMMs are suited to inspection where the user must be free to concentrate on machining operations or where there are many features to be verified, thus a long routine.

Scanning Third Generation

When geometry is complex, location and orientation tolerances are very tight, or where the entire part envelope must be quantified before testing a feature for close tolerance location and/or orientation, a scanning CMM is right for the job. The scanning CMM shown in Fig. 28-15 is testing the overall shape of a turbine blade; then it can verify its functionally critical position relative to the central axis. Only the scanner can do that.

Although a touch probe CMM could test the surface at many locations and compare them to a database describing the blade, it would never complete a full element analysis. It would still require connecting the dots between hits and could easily miss a bump!

Figure 28-15 To determine shape, size, and location of each blade, a scanning probe is required.
Courtesy Renishaw Inc

UNIT 28-1 REVIEW

Replay the Key Points

- Modern inspection helps pace measuring time to production time.
- A CMM can compare a CAD database or digital shape definition to the actual result.
- The CMM can filter out the value of the required characteristic from other variation–flatness from tilt, for example.
- If the true variation of any process is not known, then the capability of the process (Cp ratio), on which the control is based, is wrong.
- The CMM accepts part axes in whatever position they may be relative to the machine table. Objects to be measured do not have to be true to the machine's axis system.
- Probe approach speed, bounce back, and hand pressure against the work are controlled by operator skill and experience; thus, they are an inaccuracy factor for manual CMMs.
- For basic inspection of medium- to low-volume production, or for tooling work where characteristics are not complex, the manual CMM is just right for the job.
- Touch trigger probes eliminate the inaccuracy of hand pressure, and they eliminate the need for the foot switch.
- CNC touch probe CMMs are suited to inspection where the user needs to be freed from the task to concentrate on machining operations or where there are many features to be verified; thus, a long routine can be taken over by the CPU.
- The scanning CMM tests the entire element, then analyzes it, responding with a near perfect value. It is best suited to complex shapes and where absolute values are required for data.

Respond

1. Identify and describe the three different kinds of CMM.
2. What are the three ways data are collected for CMM inspection?
3. You have three parts to check, with no complex surfaces. Of the three types, which is best suited to this task? Of the three, which could do the job, and how?
4. Identify the two ways CNC CMMs are programmed.
5. Your CNC vertical mill makes one part every 55 seconds. You are using SPC to track progress by checking five key features–a distance, a square corner, and three hole diameters. The sample range is 3 parts in every 50. Which of the three CMMs would be ideal for your use?

Unit 28-2 Set Up a Manual CMM for Inspection

Introduction: In Unit 28-2, we investigate setting up and using a manual CMM equipped with a touch trigger probe. These skills will transfer to higher-level CNC inspection equipment. There are four phases:

1. **Planning the inspection** Not all CMM inspection requires a plan. We'll withhold planning until after setup.
2. **Machine qualification**
3. **Work qualification**
4. **Inspection of characteristics**

TERMS TOOLBOX

Machine axes (CMM) The rigid axes of the CMM but not necessarily of the work.

Machine zero The constant position used to set the axis register to zero.

Qualifying Bringing the machine up to operational level–the "wake-up call."

Qualifying sphere A sphere of known size used to determine the size and position of a probe tip.

Touch probe An artificial ruby sphere that senses contact with the work electronically.

Work axes The floating axis set that can be assigned to the work as it lies within the envelope.

28.2.1 Step 1—Machine Setup

Similar to homing a vertical mill, then setting tool offsets, there are two tasks needed to get a CMM ready to measure: homing its axis system and setting up the probe(s). After a bit of practice, both tasks combined shouldn't take over 5 minutes or so. Unless something unusual happens, this needn't be done more than once per day. If the machine remains powered up, then the start-up is valid until it is shut down.

Setting Machine Axes to Zero

While higher-level CMMs either home themselves or they are absolutely positioned and require no homing routine, the manual CMM must be at **machine zero** based at start-up

(Fig. 28-16). Moving all three **machine axes**, the quill must be positioned at a predetermined location, at the edge of the envelope. This operation might be called **qualifying** the machine. Following a menu routine, or from a keypad, the position registers are set to zero.

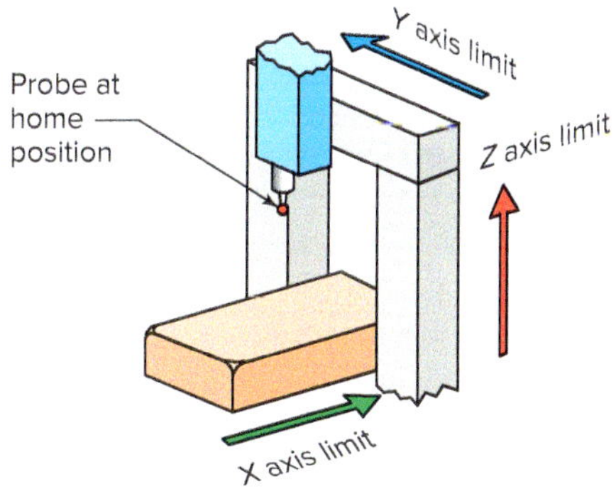

Figure 28-16 Some CMMs require a homing routine when first initialized.

Probe Qualification

The second premeasure activity provides a sense of the size and position of its probe (Fig. 28-17). Each time a probe is changed or when first mounted at initial start-up, it must be registered in the control as to its

Diameter

Position relative to the quill

Shank length—this determines its pivot distance within the probe body

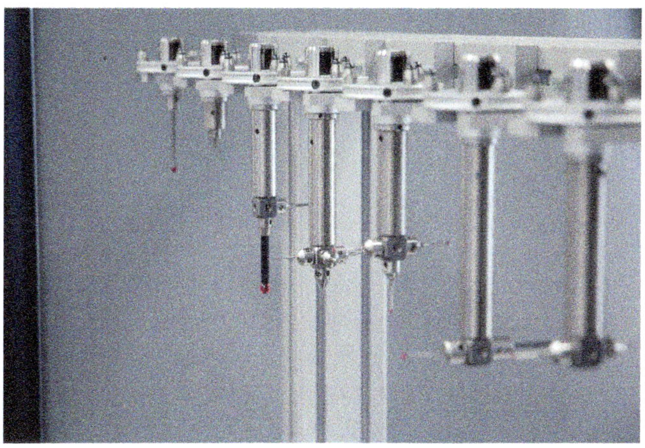

Figure 28-17 Different probes may be qualified at setup time. Note the autochanger storage rack for other probes behind.
Courtesy of Hexagon Manufacturing Intelligence

These three facts can be set ahead of time for each probe, or they can be found one at a time as the new probe is mounted.

> **KEYPOINT**
>
> All three probe factors are found by touching the probe sphere to a precise sphere located within the work envelope (Fig. 28-18).

The Qualifying Sphere

A precision **qualifying sphere**, 19 mm in diameter, is mounted on the machine table (Fig. 28-18). Several touches of the probe around the sphere yield a data set. Since the CMM controller knows the size of the qualifying sphere, and is now in probe qualify mode, from five or more touches around the sphere, it can calculate the exact diameter of the **touch probe** and its position in three-space relative to machine home and the probe body. With these facts known, the CMM is ready to measure.

Figure 28-18 The probe is touched several times to the qualifying sphere.
Courtesy J.D. Ott Aerospace Machining/Seattle, WA

Multiprobe Qualification

According to the inspection plan, all needed probes are usually qualified before the inspection begins. Each is assigned a number in the controller with the three parameters attached.

28.2.2 Step 2—Work Alignment for Inspection

Okay, the machine is ready to go. Inspection begins by conforming the axis system to the part and establishing the reference zero point upon the work. Every part setup is different according to its shape. We'll look at three basic ways in which this can be done.

Work conformation occurs in two steps:

1. Establishing the *X*-*Y* plane.
2. Establishing the *X* and *Y* axes and their value direction; *Z* will be at right angles to the *X*-*Y* plane according to the right-hand rule, thus it's not necessary to establish it.

Since the controller conforms its axes to the **work**, it doesn't matter a great deal how the work is placed upon the table. However, to speed up and simplify three-dimensional calculations within the CPU, it's good practice to place work roughly parallel with the machine axes.

X–*Y* Plane First

As shown in Fig. 28-19, part qualifying mode is selected, then with three (minimum) touches of the probe upon the surface that is bounded by the *X* and *Y* axes, the operator indicates to the computer which work surface represents the *X*-*Y* plane. At this stage, the control now recognizes how the *X*-*Y* plane is oriented within the machine.

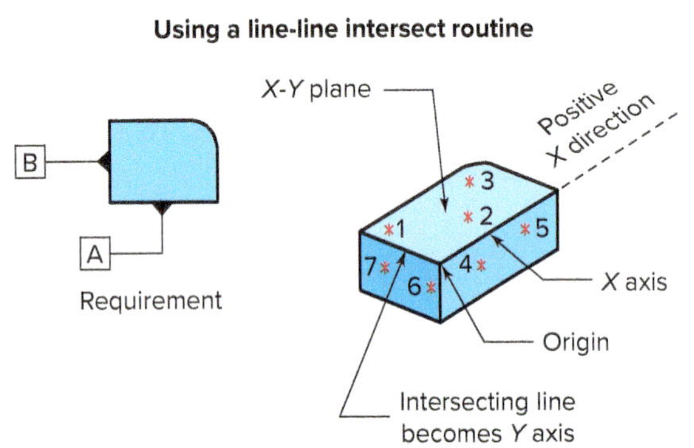

Figure 28-19 Setting up part 1 for inspection.

> **KEYPOINT**
>
> To establish the best flat plane, hits must not be taken close to a straight line and should be spaced as far apart as possible on the surface. More than three touches are recommended if in doubt about flatness.

Note that in the setup sketches, individual hits are indicated by a star. The number shows the sequence in which they would be taken to perform a part alignment.

Minimum Touches and Part Geometry

In many cases, the number of hits required is just the minimum, more would yield better accuracy. For example, 10 hits on each edge of the example would have provided a more accurate picture of how straight or flat the edge was. However, two hits on an edge provide adequate information to establish a line but not the line's geometric characteristics.

> **KEYPOINT**
>
> In some routines there are a required number of hits—no more, no less. In most it's a least number, and more is okay and usually yields better data.

Establishing Axes and Program Reference Zero (PRZ)

X Axis First

The operator has the flexibility to assign any direction on the work, parallel to the *X* axis. This is done by touching the *X* line in two locations–with the first being closer to the origin (*4 in Fig. 28-19); then the next being farther from the origin (*5). That informs the CMM which direction is positive *X*.

> **KEYPOINT**
>
> Establishing the *X* axis is accomplished by first touching the feature closest to the origin, then touching it farther away, which shows both the *X* axis line and which direction is positive.

X–*Y* Origin

The next step depends on the part shape. If it's square as in Fig. 28-19, and it has a *Y* axis, then *Y* can be established in the same way as *X*. Here the *Y* axis is hit twice, going away from zero. At that time, the CMM knows the intersection of each line forms the *X*-*Y* origin.

On part number 2 (Fig. 28-20), the *Y* axis direction is first established by hits 4 and 5, then the origin is established at the intersection of the slanted line using hits 6 and 7. The *Y* axis extends out from the origin according to the right-hand rule, even though it does not exist on the part.

In Fig. 28-21, the center of the hole is to be the origin. That's called a separate origin routine on the Brown & Sharpe Microval®, shown previously. After hits 1, 2, and 3 establish the plane, hits 4 and 5 establish the *X* angle and direction, *but* see the Keypoint.

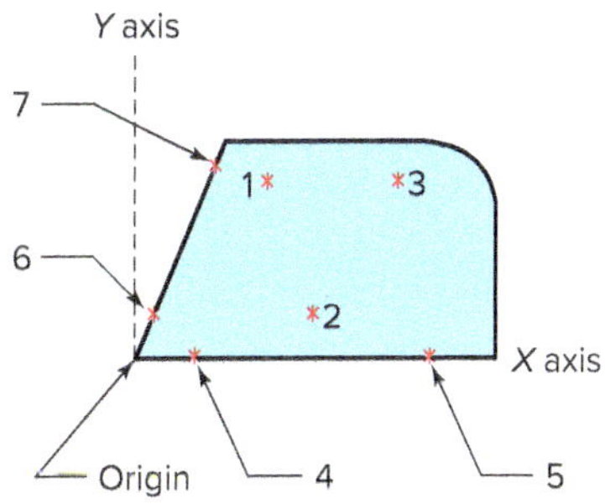

Figure 28-20 On part 2, the intersecting lines determine the origin and the *Y* axis.

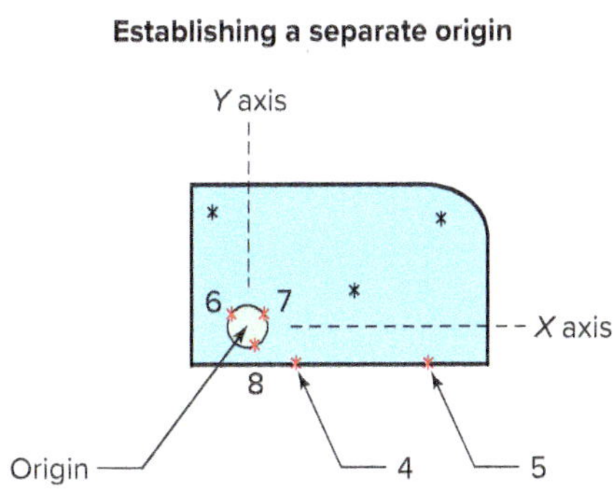

Figure 28-21 On part 3, the hole becomes a separate origin for the *X* and *Y* axes.

KEYPOINT

*4 and *5 do not establish the *X* axis—only its angle and direction relative to the machine.

Using a separate origin menu with touches 6, 7, and 8 on the hole's rim determines its circular element, and thus its center to become the origin. For this part, the *X* and *Y* axes will then be automatically established according to the right-hand rule, at the hole's center.

TRADE TIP

There are times when one or two simple characteristics need to be evaluated without regard to the part's datum relationship, for example, a hole's size or its form but not its position. In these cases, the part need not be put through a setup qualification.

28.2.3 Step 3—Work Inspection and Planning

The work is now qualified and ready to measure. Next a menu is chosen to inspect the characteristics in question, one at a time–for example, the size and position of a hole or the distance between two sides of a part, and so on. Menus will guide you through the needed touches and movements (Fig. 28-22). Using screen icons and help questions, this is easy to do.

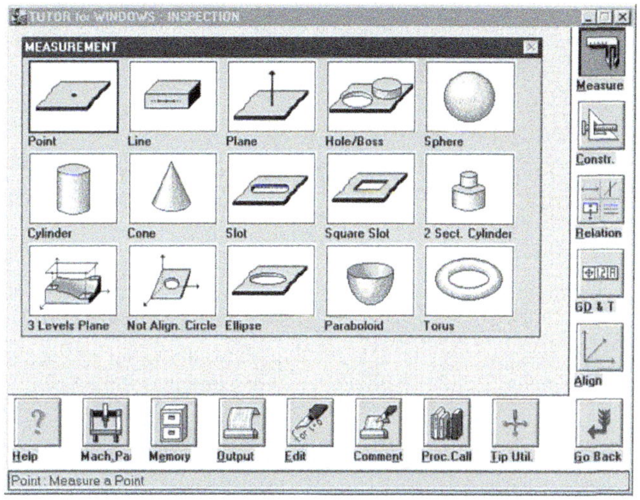

Figure 28-22 A typical screen selection for feature evaluation.

Planning an Efficient Inspection

Unless it's a simple one-off inspection, a little time spent planning can save lots of confusion and mistakes. Look at Fig. 28-23 to see how a CMM inspection for this test bolt project might be planned out. The handwritten notes identify features in their sequence. Notes keep the inspection process on track (and the inspector, too):

Key features to be measured

The sequence in which they will be inspected

The menu picks (to become a program)

KEYPOINT

Without an inspection plan, it's easy to lose track of which dimension represents which feature on the CMM printout report. This is especially true for complex work or where many parts must be checked.

The notes are the routine sequences planned by the operator. They will coincide with the actual report of the job.

TRADE TIP

While writing on a drawing is taboo in many shops unless they use throwaway drawings, an inspection copy is often created that can be used for CMM planning and programming purposes only—but it must be clearly noted as such and never returned to the shop.

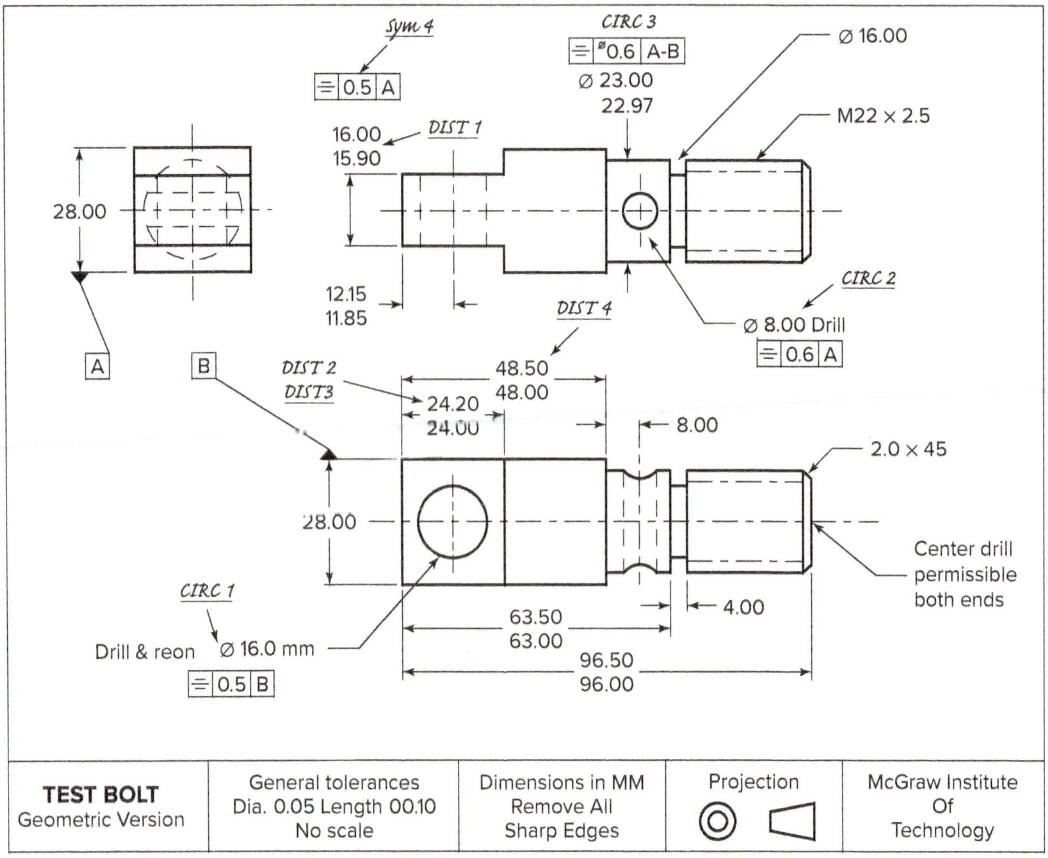

Figure 28-23 The machinist has planned this inspection using notes on the print.

CMM Printout Reports

CMM reports vary widely based on controller manufacturer and the way the operator sets up the program. They can be a simple line-by-line recap of the menu picks and measured results. They can also display tolerances in both coordinate or geometric values. They can be set up with metric or imperial values as well and show them with graphics if needed. They can show compliance or beyond tolerance, and they can be put into SPC Cpk format or even plotted on a control chart!

Figure 28-24 shows an inspection report such as this that would be easy to follow. Notice the SPC bar graph on the bottom. It looks as though a key feature is showing a trend.

Fixture Inspection (Fig. 28-25)

To speed inspection, part holding devices such as precision vises and angle plates can be used, and fixtures can be made similar to those for mill work. They bring three advantages:

1. It stabilizes odd-shaped work that would not easily lie on the table.
2. It speeds repetitive inspection where the same part is checked many times during the day. That way, the initial part qualification can be automatic, too, since it will always lie at the same place and orientation.
3. Two or more fixtures can be preset similar to fixure offsets in machining.

28.2.4 Arm CMMs and Smart Height Gages

Arms

Gaining popularity, these portable measuring instruments work far differently than the *X-Y-Z* machines just explored.

Features

- **Portability** Can be moved quickly to the work and machine.
- **Designed for the Shop** Can measure parts where they are made, too big to fit into the CMM.
- Infinite **Rotation Encoders** in six or seven axes, allowing access to difficult-to-reach areas.
- **Quick Probe Exchange** for tactile measurement (Fig. 28-26).
- **Laser Nontactile Scanning** for surfaces with minimal operator interaction (Fig 28-27).
- **Repeatability** of from 0.003 in. to 0.008 in. depending on arm reach (size). Accuracy diminishes with length.

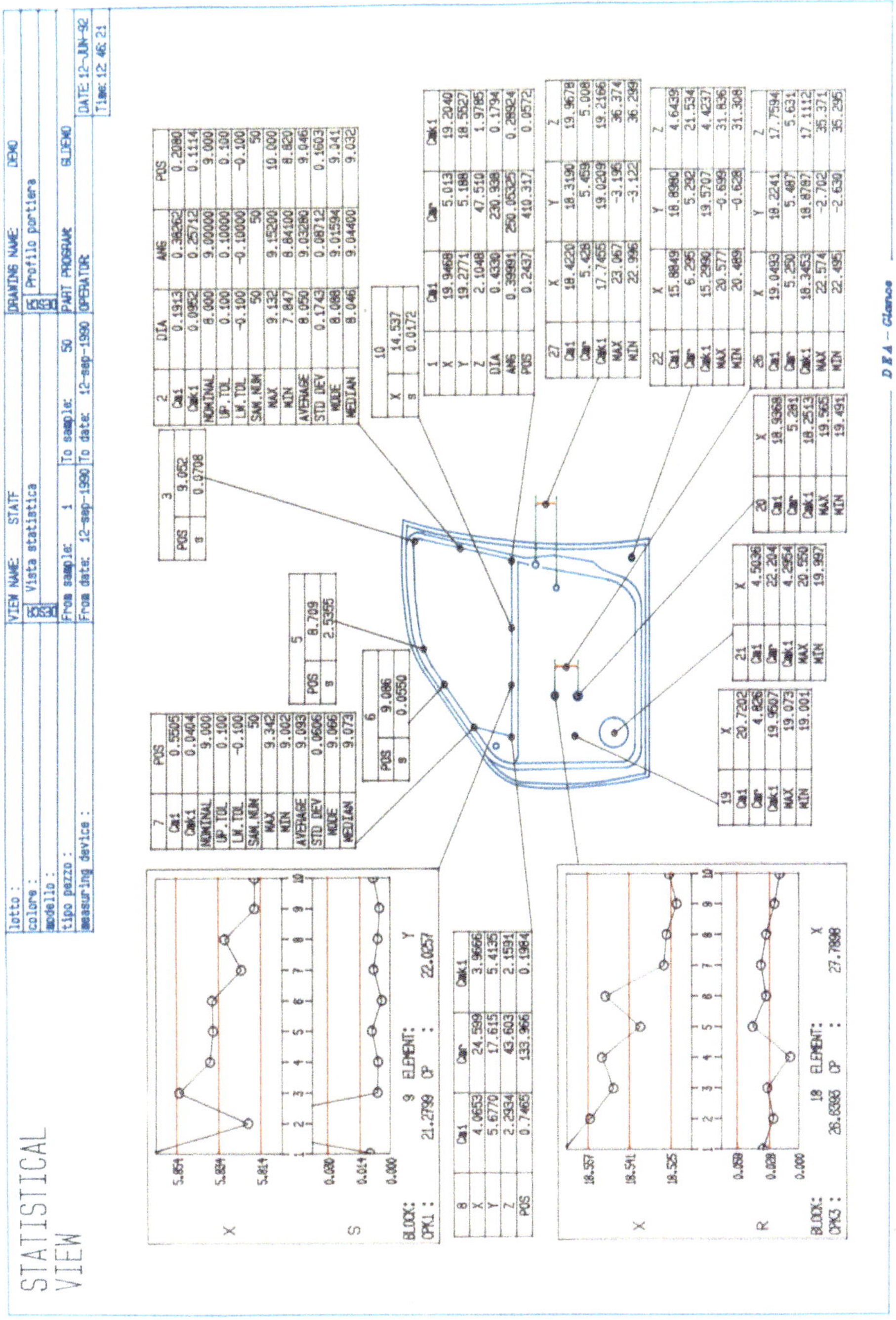

Figure 28-24 A graphic inspection report helps digest the data.

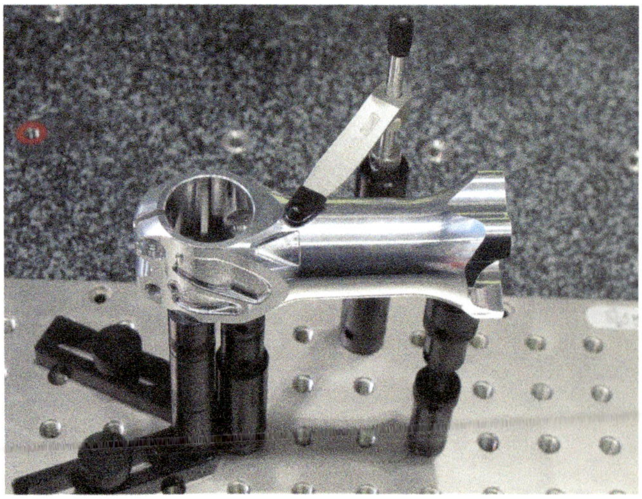

Figure 28-25 Part fixtures are preset to speed the inspection process.
Courtesy of Hexagon Manufacturing Intelligence

Figure 28-26 A six-axis Romer Arm® CMM with touch trigger probe.
Courtesy AJAC Aerospace Joint Apprenticeship Committee, Seattle WA

Smart Height Gages

Featuring onboard microprocessors, these new height gages float on a bed of air across the inspection table or they can be taken to the machine table or set upon their own tripod! Going

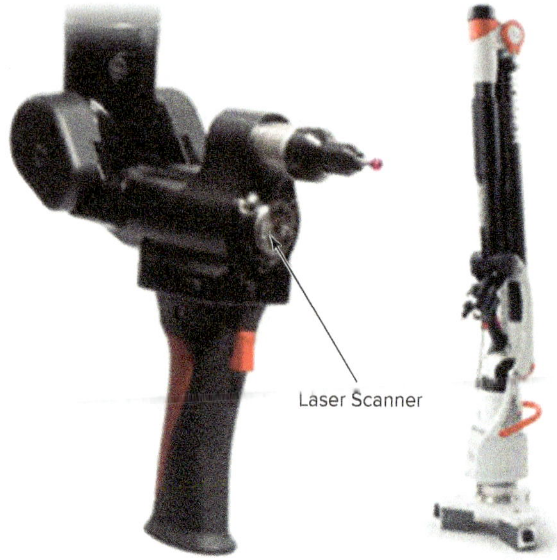

Figure 28-27 A seven-axis Romer Arm® showing the measuring head with probe and laser scanner.
Courtesy of Aerospace Joint Apprenticeship Committee—AJAC Advanced Metrology Training Uni—Seattle, WA

far beyond the normal duties of a height gage, they can perform a range of duties (Fig. 28-28). A basic duty made simple (Fig. 28-29), the user is determining the height of the hole above the datum table with just two touches of the hard probe—top and bottom. The processor calculates height.

Figure 28-28 A Tesa Smart Height Gage performs an array of routines, making inspection quick and easy.
Courtesy of Universal Aerospace, Arlington, WA

Figure 28-29 The genius lies in the onboard processor and preprogrammed routines.
Courtesy Universal Aerospace/ Arlington, WA

Using preprogrammed routines accessed through function buttons (see Fig. 28-29), these instruments simplify complex routines such as a bolt circle inspection. Repeating to 0.0001 in. with a minimum of touches, they report hole size and location!

Features

- To be free of cords, both the processor and air compressor are rechargeable battery powered.
- Fast probe exchange has both horizontal and vertical options.
- Quick initialization or requalification of the probe uses a setup tower and routine.
- Probe pressure control for ultralight touch prevents deflection and makes all touches the same (similar to friction control duty on a micrometer).

KEYPOINT

Making a part alignment each time can be time-consuming for high-speed inspection. Seven to 10 probe touches can be eliminated from the plan (and program) if the part is reliably placed in a precision fixture.

UNIT 28-2 REVIEW

Replay the Key Points

- The needed number of hits necessary to compile a complete geometric image of the object is an operator call based on experience. However, the computer will always require a minimum number for the base function.
- There are four phases to setting up a CMM: planning the inspection (when complex), machine qualification (home and probe), work qualification (X-Y plane and axis set), and inspection of characteristics.
- To speed up and simplify three-dimensional calculations within the CPU, it's good practice to place work roughly parallel with the machine axes.
- To establish a plane, touches must not be in a straight line. More than three touches are recommended if in doubt about flatness.

Respond

1. Describe two tasks that must be done to get a CMM ready to inspect.
2. What do you do when qualifying a probe sphere?
3. Why choose a touch trigger probe over a hard probe?
4. To begin an inspection of a part, what must be done?
5. What advantage does a fixture offer for CMM inspection?

CHAPTER 28 REVIEW

Unit 28-1

Entering into industrial application later than CNC-programmed machine tools, CMM technology is now evolving at a faster pace than most other aspects of modern manufacturing.

Looking closely, one sees that surface finish is actually feature variation at a microinch resolution. As the technology marches on toward absolute accuracy, elements at that level will need to be folded into the overall shape envelope of the feature.

Unit 28-2

With a background in geometrics, CNC machining, shop math, and metrology, you'll have no difficulty learning to operate any CMM. When first learning to use our new manual CMM in our tech school, I was able to set it up and do most of the basic functions all on a Sunday morning! Of course, the higher-level machines require more formal training, but as with all computer equipment, they are becoming ever more user-friendly. Then, if you didn't view the QR tag about portable CMMs back in Chapter 6, type "Romer Arm technology" into your browser. This is a technology to watch.

QUESTIONS AND PROBLEMS

1. A. Of the three CMM types, which would be ideally suited to evaluate the key features in Fig. 28-30, and why?

B. Identify the hits by number and describe their purpose to inspect this drill gage. Follow this plan:

(1) Set up line-to-line intersect origin A at B.

(2) Inspect 1.5000-in. key feature.

(3) Inspect five holes for size, roundness, and position.

2. The CAM software cycle time is 4.5 minutes for the profile, pocket, and hole and you find that it takes you 2 minutes to change parts in the vise. Which CMM is best for the task if the part count is to be 750 total?

3. Identify and describe the three probe types.

4. Which probe or probes take hits?

5. What ability must a probe have to be used with a CNC machine, either a cutting machine or a CMM? Explain.

6. What inaccuracies exist when using a hard probe that do not occur with touch triggers and scanners?

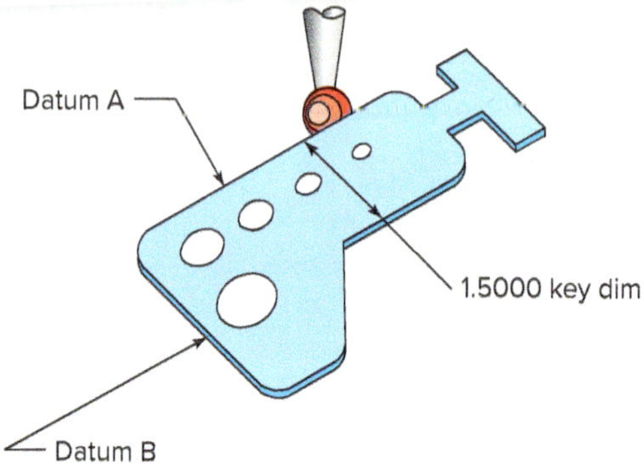

Figure 28-30 Plan to inspect the key features of this drill gage.

CRITICAL THINKING QUESTIONS

7. In your opinion, of the advantages of CMM technology, which is the most important? Explain.

8. Briefly describe how a touch trigger probe computes the zero pressure hit.

9. When during an inspection do you not need to qualify the probe?

10. Can a program be written for a manual CMM? If yes, describe it.

11. You need to know the mean distance and position of a complex contoured surface from a datum. Which CMM can do the job? Explain.

12. When would a hard probe be a good application?

13. What are the two steps required to set up a manual CMM for inspection?

14. Describe qualifying a probe on a CMM.

15. List the three facts that probe qualifying tells the CMM control.

16. True or false? To inspect a hole for size and location, I need not set up (qualify the part). If the statement is false, what would make it true?

17. Describe the number of hits required to determine the corner radii size on the drill gage drawing (see Fig. 28-30).

18. Describe a typical part alignment on a CMM.

19. Define a conformal axis system.

CRITICAL THINKING PROBLEM

20. Add up the total minimum number of hits that would be required to inspect the drill gage in Fig. 28-30.

Appendix I

INCH / METRIC TAP DRILL SIZES & DECIMAL EQUIVALENTS

Small Drills

FRACTION OR DRILL SIZE	DECIMAL EQUIVALENT	TAP SIZE
NUMBER SIZE DRILLS 80	.0135	
79	.0145	
1/64	.0156	
78	.0160	
77	.0180	
76	.0200	
75	.0210	
74	.0225	
73	.0240	
72	.0250	
71	.0260	
70	.0280	
69	.0292	
68	.0310	
1/32	.0312	
67	.0320	
66	.0330	
65	.0350	
64	.0360	
63	.0370	
62	.0380	
61	.0390	
60	.0400	
59	.0410	
58	.0420	
57	.0430	
56	.0465	
3/64	.0469	0-80
55	.0520	
54	.0550	
53	.0595	1-64,72
1/16	.0625	
52	.0635	
51	.0670	
50	.0700	2-56,64
49	.0730	
48	.0760	
5/64	.0781	
47	.0785	3-48
46	.0810	
45	.0820	3-56
44	.0860	
43	.0890	4-40
42	.0935	4-48
3/32	.0938	
41	.0960	
40	.0980	
39	.0995	
38	.1015	5-40
37	.1040	5-44
36	.1065	6-32
7/64	.1094	
35	.1100	
34	.1110	
33	.1130	6-40
32	.1160	
31	.1200	
1/8	.1250	
30	.1285	
29	.1360	8-32,36
28	.1405	

Bigger

FRACTION OR DRILL SIZE	DECIMAL EQUIVALENT	TAP SIZE
9/64	.1406	
NUMBER SIZE DRILLS 27	.1440	
26	.1470	
25	.1495	10-24
24	.1520	
23	.1540	
5/32	.1562	
22	.1570	
21	.1590	10-32
20	.1610	
19	.1660	
18	.1695	
11/64	.1719	
17	.1730	
16	.1770	12-24
15	.1800	
14	.1820	12-28
13	.1850	
3/16	.1875	
12	.1890	
11	.1910	
10	.1935	
9	.1960	
8	.1990	
7	.2010	1/4-20
13/64	.2031	
6	.2040	
5	.2055	
4	.2090	
3	.2130	1/4-28
7/32	.2188	
2	.2210	
1	.2280	
LETTER SIZE DRILLS A	.2340	
15/64	.2344	
B	.2380	
C	.2420	
D	.2460	
1/4 E	.2500	
F	.2570	5/16-18
G	.2610	
17/64	.2656	
H	.2660	
I	.2720	5/16-24
J	.2770	
K	.2810	
9/32	.2812	
L	.2900	
M	.2950	
19/64	.2969	
N	.3020	
5/16	.3125	3/8-16
O	.3160	
P	.3230	
21/64	.3281	
Q	.3320	3/8-24
R	.3390	
11/32	.3438	
S	.3480	
T	.3580	

Largest

FRACTION OR DRILL SIZE	DECIMAL EQUIVALENT	TAP SIZE
23/64	.3594	
LETTER SIZE DRILLS U	.3680	7/16-14
3/8	.3750	
V	.3770	
W	.3860	
25/64	.3906	7/16-20
X	.3970	
Y	.4040	
13/32	.4062	
Z	.4130	
27/64	.4219	1/2-13
7/16	.4375	
29/64	.4531	1/2-20
15/32	.4688	
31/64	.4844	9/16-12
1/2	.5000	
33/64	.5156	9/16-18
17/32	.5312	5/8-11
35/64	.5469	
9/16	.5625	
37/64	.5781	5/8-18
19/32	.5938	
39/64	.6094	
5/8	.6250	
41/64	.6406	
21/32	.6562	3/4-10
43/64	.6719	
11/16	.6875	3/4-16
45/64	.7031	
23/32	.7188	
47/64	.7344	
3/4	.7500	
49/64	.7656	7/8-9
25/32	.7812	
51/64	.7969	
13/16	.8125	7/8-14
53/64	.8281	
27/32	.8438	
55/64	.8594	
7/8	.8750	1-8
57/64	.8906	
29/32	.9062	
59/64	.9219	
15/16	.9375	1-14
61/64	.9531	
31/32	.9688	
63/64	.9844	1 1/8-7
1	1.0000	
1 3/64	1.0469	1 1/8-12
1 7/64	1.1094	1 1/4-7
1 1/8	1.1250	
1 11/64	1.1719	1 1/4-12
1 7/32	1.2188	1 3/8-6
1 1/4	1.2500	
1 19/64	1.2969	1 3/8-12
1 11/32	1.3438	1 1/2-6
1 3/8	1.3750	
1 27/64	1.4219	1 1/2-12
1 1/2	1.5000	

Met. Taps

METRIC TAP DRILL SIZES

METRIC TAP	TAP DRILL mm	DECIMAL EQUIV. Inches
M1.6 x 0.35	1.25	.0492
M1.8 x 0.35	1.45	.0571
M2 x 0.4	1.60	.0630
M2.2 x 0.45	1.75	.0689
M2.5 x 0.45	2.05	.0807
M3 x 0.5	2.50	.0984
M3.5 x 0.6	2.90	.1142
M4 x 0.7	3.30	.1299
M4.5 x 0.75	3.70	.1457
M5 x 0.8	4.20	.1654
M6 x 1	5.00	.1968
M7 x 1	6.00	.2362
M8 x 1.25	6.70	.2638
M8 x 1	7.00	.2756
M10 x 1.5	8.50	.3346
M10 x 1.25	8.70	.3425
M12 x 1.75	10.20	.4016
M12 x 1.25	10.80	.4252
M14 x 2	12.00	.4724
M14 x 1.5	12.50	.4921
M16 x 2	14.00	.5512
M16 x 1.5	14.50	.5709
M18 x 2.5	15.50	.6102
M18 x 1.5	16.50	.6496
M20 x 2.5	17.50	.6890
M20 x 1.5	18.50	.7283
M22 x 2.5	19.50	.7677
M22 x 1.5	20.50	.8071
M24 x 3	21.00	.8268
M24 x 2	22.00	.8661
M27 x 3	24.00	.9449
M27 x 2	25.00	.9843
M30 x 3.5	26.50	1.0433
M30 x 2	28.00	1.1024
M33 x 3.5	29.50	1.1614
M33 x 2	31.00	1.2205
M36 x 4	32.00	1.2598
M36 x 3	33.00	1.2992
M39 x 4	35.00	1.3780
M39 x 3	36.00	1.4173

PIPE THREAD SIZES (NPSC)

THREAD	DRILL
1/8-27	11/32
1/4-18	7/16
3/8-18	37/64
1/2-14	23/32
3/4-14	59/64
1-11 1/2	1 5/32
1 1/4-11 1/2	1 1/2
1 1/2-11 1/2	1 3/4
2-11 1/2	2 7/32
2 1/2-8	2 21/32
3-8	3 1/4
3 1/2-8	3 3/4
4-8	4 1/4

Courtesy L. S. Starrett Co.

Appendix II

Drill Speeds—Common Bit Sizes in Six Materials

Notes

- All speeds are based on high-speed steel (HSS) drill bits.
- If your machine does not have the RPM listed, shift to the next lower speed.
- Data are simplified for learning. For a complete listing of letter, number, fraction, and metric drill RPMs, see *Machinery's Handbook*.
- Or compute your RPM using the following simplified formula:

$$\frac{4 \times \text{Surface speed}}{\text{Diameter of rotating object}}$$

Diam.	Mild Steel	Carbon Steel-Annealed	Aluminum	Soft Brass	Cast Iron	Annealed Stainless
1/8	3,200	2,880	8,000	5,600	3,200	2,880
3/16	2,133	1,920	5,333	3,733	2,133	1,920
1/4	1,600	1,440	4,000	2,800	1,600	1,440
5/16	1,280	1,152	3,200	2,240	1,280	1,152
3/8	1,066	960	2,667	1,867	1,067	960
7/16	914	823	2,285	1,600	914	823
1/2	800	720	2,000	1,400	800	800
9/16	711	640	1,778	1,244	711	711
5/8	640	576	1,600	1,120	640	576
11/16	581	523	1,454	1,018	582	523
3/4	533	480	1,333	933	533	480
13/16	492	443	1,230	862	492	443
7/8	457	411	1,143	800	457	411
15/16	426	384	1,066	747	426	384
1	400	360	1,000	700	400	360
1 1/16	376	339	941	658	376	339
1 1/8	355	320	888	622	355	320
1 3/16	336	303	842	589	336	303
1 1/4	320	288	800	560	320	288

Appendix III

Recommended Cutting Speeds for Six Materials: Surface Speeds in Feet per Minute

Notes

- These numbers are biased toward safe/convenient learning.
- With experience, cutting speeds shown may be exceeded.
- Stronger setups, coolants, and many factors combine to determine the final result.

Cutting Tool	Mild Steel	Carbon Steel-Annealed	Aluminum	Soft Brass	Cast Iron	Annealed Stainless
HSS	100	80	250 to 350	175	100	80 to 100
Carbide	300	200	750 to 1,000	500	250	200 to 250

Appendix IV

Commonly Used Machining-Center Codes

G-Codes

Code	Description
G00	positioning (rapid traverse)
G01	linear interpolation (feed)
G02	circular interpolation CW
G03	circular interpolation CCW
G04	dwell
G07	imaginary axis designation
G09	exact stop check
G10	offset value setting
G17	*X–Y* plane selection
G18	*Z–X* plane selection
G19	*Y–Z* plane selection
G20	input in inch
G21	input in millimeter
G22	stored stroke limit ON
G23	stored stroke limit OFF
G27	reference point return check
G28	return to reference point
G29	return from reference point
G30	return to 2nd, 3rd, and 4th ref. pt.
G31	skip cutting
G33	thread cutting
G40	cutter compensation cancel
G41	cutter compensation left
G42	cutter compensation right
G43	tool length compensation + direction
G44	tool length compensation direction
G45	tool offset increase
G46	tool offset decrease
G47	tool offset double increase
G48	tool offset double decrease
G49	tool length compensation cancel
G50	scaling off
G51	scaling on
G52	local coordinate system setting

Code	Description
G54	work coordinate system 1 select
G55	work coordinate system 2 select
G56	work coordinate system 3 select
G57	work coordinate system 4 select
G58	work coordinate system 5 select
G59	work coordinate system 6 select
G60	single direction positioning
G61	exact stop check mode
G64	cutting mode
G65	custom macro simple call
G66	custom macro modal call
G67	custom macro modal call cancellation
G68	coordinate system rotation ON
G69	coordinate system rotation OFF
G73	peck drilling cycle
G74	counter tapping cycle
G76	fine boring
G80	canned cycle cancel
G81	drilling cycle, spot boring
G82	drilling cycle, counter boring
G83	peck drilling cycle
G84	tapping cycle
G85, G86	boring cycle
G87	back boring cycle
G88, G89	boring cycle
G90	absolute programming
G91	incremental programming
G92	programming of absolute zero point
G94	per minute feed
G95	per revolution feed
G96	constant surface speed control
G97	constant surface speed control cancel
G98	return to initial point in canned cycle
G99	return to R point in canned cycle

Standard M-Codes

M00	program stop
M01	optional stop
M02	end of program (no rewind)
M03	spindle CW
M04	spindle CCW
M05	spindle stop
M06	tool change
M07	mist coolant ON
M08	flood coolant ON
M09	coolant OFF
M19	spindle orientation ON
M20	spindle orientation OFF
M21	tool magazine right
M22	tool magazine left
M23	tool magazine up
M24	tool magazine down
M25	tool clamp
M26	tool unclamp
M27	clutch neutral ON
M28	clutch neutral OFF
M30	end program (rewind stop)
M98	call subprogram
M99	end subprogam

(M19–M28 used for maintenance purposes)

Appendix V

Comparison of Symbols

SYMBOL FOR:	ASME Y14.5	ISO
STRAIGHTNESS	⏤	⏤
FLATNESS	⏥	⏥
CIRCULARITY	○	○
CYLINDRICITY	⌭	⌭
PROFILE OF A LINE	⌒	⌒
PROFILE OF A SURFACE	⌓	⌓
ALL AROUND		
ALL OVER		 (proposed)
ANGULARITY	∠	∠
PERPENDICULARITY	⊥	⊥
PARALLELISM	//	//
POSITION	⌖	⌖
CONCENTRICITY (Concentricity and Coaxiality in ISO)	◎	◎
SYMMETRY	⌯	⌯
CIRCULAR RUNOUT	* ↗	* ↗
TOTAL RUNOUT	* ⌰	* ⌰
AT MAXIMUM MATERIAL CONDITION	Ⓜ	Ⓜ
AT MAXIMUM MATERIAL BOUNDARY	Ⓜ	NONE
AT LEAST MATERIAL CONDITION	Ⓛ	Ⓛ
AT LEAST MATERIAL BOUNDARY	Ⓛ	NONE
PROJECTED TOLERANCE ZONE	Ⓟ	Ⓟ
TANGENT PLANE	Ⓣ	NONE
FREE STATE	Ⓕ	Ⓕ
UNEQUALLY DISPOSED PROFILE	Ⓤ	UZ (proposed)
TRANSLATION	▷	NONE
DIAMETER	Ø	Ø
BASIC DIMENSION (Theoretically Exact Dimension in ISO)	[50]	[50]
REFERENCE DIMENSION (Auxiliary Dimension in ISO)	(50)	(50)
DATUM FEATURE	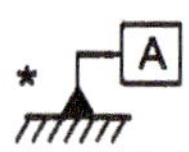	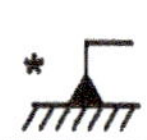 or 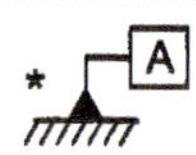
	* May be filled or not filled	

Index

Note: Page numbers followed by f and t indicate figures and tables.

D

N

O

P

Q